Indigenous Fermented Foods

of

South Asia

FERMENTED FOODS AND BEVERAGES SERIES

Series Editors

M.J.R. Nout and Prabir K. Sarkar

Indigenous Fermented Foods of South Asia (2015)
Editor: V.K. Joshi

Fermented Milk and Dairy Products (2015)
Editor: Anil Kumar Puniya

Indigenous Fermented Foods of Southeast Asia (2014)
Editor: J. David Owens

Cocoa and Coffee Fermentations (2014)
Editors: Rosane F. Schwan and Graham H. Fleet

Handbook of Indigenous Foods Involving Alkaline Fermentation (2014)
Editors: Prabir K. Sarkar and M.J.R. Nout

Solid State Fermentation for Foods and Beverages (2013)
Editors: Jian Chen and Yang Zhu

Valorization of Food Processing By-Products (2013)
Editor: M. Chandrasekaran

Indigenous Fermented Foods of South Asia

of

South Asia

Edited by V.K. Joshi

CRC Press
Taylor & Francis Group
Boca Raton London New York

CRC Press is an imprint of the
Taylor & Francis Group, an **informa** business

CRC Press
Taylor & Francis Group
6000 Broken Sound Parkway NW, Suite 300
Boca Raton, FL 33487-2742

First issued in paperback 2019

© 2016 by Taylor & Francis Group, LLC
CRC Press is an imprint of Taylor & Francis Group, an Informa business

No claim to original U.S. Government works

ISBN-13: 978-1-4398-8783-7 (hbk)
ISBN-13: 978-0-367-37707-6 (pbk)

Visit the Taylor & Francis Web site at
http://www.taylorandfrancis.com

and the CRC Press Web site at
http://www.crcpress.com

Contents

V

Series Preface

Natural fermentation precedes human history, and since ancient times, humans have been controlling the fermentation process. Fermentation, the anaerobic way of life, has attained a wider meaning in biotransformation, resulting in a wide variety of fermented foods and beverages.

Fermented food products made with uncontrolled natural fermentations or with defined starter cultures achieve their characteristic flavor, taste, consistency, and nutritional properties through the combined effects of microbial assimilation and metabolite production, as well as from enzyme activities derived from food ingredients.

Fermented foods and beverages span a wide diversity range of starchy root crops, cereals, pulses, vegetables, nuts, and fruits, as well as animal products such as meats, fish, seafood, and dairy.

The science of chemical, microbiological, and technological factors and changes associated with manufacture, quality, and safety is progressing and is aimed at achieving higher levels of control of the quality, safety, and profitability of food manufacture.

Both producer and consumer benefit from scientific, technological, and consumer-oriented research. Small-scale production needs to be better controlled and safeguarded. Traditional products need to be characterized and described to establish, maintain, and protect their authenticity. Medium- and large-scale food fermentations require selected, tailor-made, or improved processes that provide sustainable solutions for the future conservation of energy and water, and responsible utilization of resources and disposal of by-products in the environment.

The scope of the CRC Press book series on "Fermented Foods and Beverages" shall include (i) globally known foods and beverages of plant and animal origin (such as dairy, meat, fish, vegetables, cereals, root crops, soybeans, legumes, pickles, cocoa and coffee, wines, beers, spirits, starter cultures, and probiotic cultures), their manufacture, chemical and microbiological composition, processing, compositional and functional modifications taking place as a result of microbial and enzymic effects, their safety, legislation, development of novel products, and opportunities for industrialization; (ii) indigenous commodities from Africa, Asia (South, East, and Southeast), Europe, Latin America, and Middle East, their traditional and industrialized processes and their contribution to livelihood; and (iii) several aspects of general interest such as valorization of food-processing by-products, biotechnology, engineering of solid-state processes, modern chemical and biological analytical approaches (genomics, transcriptomics, metabolomics, and other -omics), safety, health, and consumer perception.

The seventh book, born in the series, deals with *Indigenous Fermented Foods of South Asia*. The treatise, edited by Professor V.K. Joshi, deals with the indigenous fermented foods of India, Pakistan, Bangladesh, Sri Lanka, Nepal, Bhutan, Maldives, and Afghanistan. The region is known for its large diversity of fermented foods and the editor has made a great effort to represent this diversity at the microbiological and ingredient levels. Thanks to his network of scientists from the South Asian countries, he was able to bring together much of the current knowledge and state of the art. We are convinced that students, researchers, teachers, and the general public will benefit considerably from this compendium.

Rob Nout
Prabhat K. Sarkar

Preface

Humbly dedicated to Professor K.H. Steinkraus, whose research on indigenous fermented foods inspired my scientific journey, and to Professor Stephen Hawkins, whose life instilled in me the confidence and the will power to overcome physical disability in life.

To survive, the most essential requirement for humans or, for that matter, for any living creature, is food. Food symbolizes the culture of a community, reflecting its eating habits, consumption patterns and preferences, health, agricultural systems, marketing strategies, social life, ethnicity, and religious taboos, though it is also a highly personal commodity, deeply embedded in our senses and memories. The history of food is very intimately connected with the history of mankind. The presence of fossil microorganisms found in rocks 3.5 billion years old has revealed their existence then. To locate or grow food has been the major pre-occupation of humans throughout history, but how different people around the world learned to utilize their natural resources, preferred or shunned particular foods, and developed unique food preparation methods reveals a lot about the ability of ancient humans. It is a material basis for rituals through which people celebrate the passage of life stages and their link to divinity. The ancient records reveal that fermentation and drying were the techniques invented by humans to extend the availability of foods. Thus, it can safely be stated that humans might have prepared and consumed indigenous fermented foods long before the dawn of civilization.

The food substrates invaded or overgrown by the edible microorganisms whose enzymes hydrolyze various substrates to non-toxic products with pleasant sensory qualities that are also attractive to the consumers are described as indigenous fermented foods. Diverse microorganisms, such as fungi, yeasts, and bacteria, are involved in the fermentation and production of ethnic fermented foods and alcoholic beverages. These foods have been an integral part of the human diet in many cultures for centuries, and South Asia is no exception. The combination of cultures, ethnic diversity, and biological resources has produced a remarkably diverse food culture, consisting of a wide range of fermented ethnic foods and alcoholic beverages. It is no exaggeration to say that it would have been impossible for humans to survive over the millennia without these foods.

Books are the most important source of documented knowledge and are the strongest link between our past and present. Further, scientific and technical books serve as a vast reservoir of useful technological information. There are a few books that deal with indigenous fermented foods, describing these foods the world-over. Why, then, another book on the same subject? Several reasons can be cited, but most important is to provide the reader ample opportunities for a glimpse into the holistic view of the indigenous fermented foods of South Asia, as there is virtually no text available on the subject. In addition, it aims to consolidate research and identify gaps in that research and missing links in our knowledge so as to stimulate further research in appropriate directions.

In Chapter 1, an overview of indigenous fermented foods of South Asia, India, Pakistan, Sri Lanka, Bangladesh, Afghanistan, Nepal, Bhutan, and the Maldives is presented, including different fermented foods, their producing countries/regions, geography, history, origin, ethnicity, culture, etc. There is, perhaps, no better way to understand a culture, its values and pre-occupations, than by examining its attitudes toward food. Food preferences also serve to separate individuals and groups from each other, and serve as a powerful factor in forming our physical, emotional, and spiritual identity. The role played by food fermentation and indigenous fermented foods in the social fabric and diet of a people is also elaborated. The most important contribution of fermentation is the biological enrichment via fermentation of foods with additional nutrients. How such production has varied from prehistoric time to the present in South Asia is also described.

There is a wide diversity of indigenous fermented foods of South Asia (see Chapter 2) that are made and consumed, including, *idli*, *dhokla*, *dosa*, *nan*, *appam*, *papad* (*papadam*) in India and Sri Lanka. *Nan* (naan) (flat leavened bread) is consumed as a staple food by the people of Afghanistan, Iran, India, and Pakistan, as are *bhaturas* (*pathuras*) and *kulchas*. Similarly, in almost all the tropical places in Asia, palm wine, or *toddy*, is made; *apong* (a local drink made from rice or millet) is part of Arunachali cuisine, while *Kyat* is a local brew made from rice and, prepared in Meghalyan. Sikkimese cuisine has a beverage locally called as *tchang*, a beer served in a bamboo mug. Products such as *gundruk*, *sinki*, dried *kinema*, *churpi*, *suka ko masu*, *sukao maccha*, *gnuchi*, *ngari*, *masauyra*, etc., are all still prepared and consumed by the people of the Himalayan regions.

Knowledge of microorganisms and their biochemical activities is essential to understand the process of fermentation. The microorganisms that conduct the fermentation are edible and present in or on the ingredients and utensils, or in the environment, and are selected either through adaptation to the substrate or by adjusting the fermentation conditions and encouraging growth of the suitable microorganisms and discouraging the growth of the undesirable microorganisms. Which microorganisms are associated with which traditional fermented foods is shown in the text to be due to diverse and wide variations in the agro-climatic conditions of South Asia. Diversity within the species of microorganisms, such as lactic acid bacteria, has created ethnic foods with functions that have imparted a number of important benefits, including health-promoting benefits, bio-preservation of perishables, bio-enrichment of nutritional value, and protective properties, as well as therapeutic values. The description of several compounds formed during fermentation that affect the quality, antimicrobial production and effect on spoilage causing microorganisms is made clear, including chemical/biochemical changes that take place during fermentation (in Chapter 3). Changes during fermentation with respect to nutrients have also been cited.

Chapter 4 discusses the composition and nutritive value of traditional fermented foods. For any food product, quality and safety are indispensable considerations, and this aspect of indigenous fermented foods is elaborated in Chapter 5. Some foods in their raw state have toxins and antinutritional compounds that must be eliminated before consumption, and microbial action during fermentation has proved capable of removing or detoxifying such compounds. Microorganisms during fermentation also induce or produce different types of bioactive compounds, which have extra-nutritional values, such as reducing the effects of aging and degenerative diseases, as well as producing several physiological effects related to signaling, cholesterol lowering, imparting immunity, etc. Research has revealed how indigenous fermented foods serve as therapeutic foods, and how their active ingredients impact human health by producing antibiotic and antioxidant activity, and this is highlighted appropriately in Chapter 6.

The technology or method of preparation of different indigenous fermented foods forms the main theme of Chapters 7 through 12 of the book. Traditionally, different types of cereals and grains are fermented with mixed starter cultures, leading to variability in product quality, nutritive value, and safety. Such foods with their methods of preparation are described in Chapter 7. Similarly, the technology of fermented vegetables and their products with high nutritive value, especially antioxidant activity, and of acidic fermented milk and milk products, which are part of a highly nutritious and healthy diet, besides having therapeutic values, is the subject matter of Chapter 8. Alcoholic beverages have strong ritual importance among the ethnic peoples of South Asia, and their social activities require the consumption of appreciable quantities of alcohol, playing a very important role in their local customs. In South Asia, these beverages are also offered in prayer to the family gods. Methods for their preparation are described in Chapter 9. Chapter 10 describes acetic acid fermented food products, mainly vinegars of different types. Besides the methods of their production, the composition, characteristics, and factors important in this fermentation are also described. Fermented fish and shrimp products are indigenous fermented foods greatly relished by the peoples of South Asia, and are consumed along with their staple food (i.e., rice). These, including peptide sauces, pastes with meat-like flavor, fish sauces and related products, and alkaline fermented foods are described in Chapter 11. In Chapter 12, details of mushroom production and its postharvest technology has been described.

Many developments have taken place in microbiology/biotechnology, biochemistry genetics, engineering, etc., that have made an impact on our ability to scale-up the processes for the production of indigenous fermented foods. Since most of such foods are produced from locally available raw material, using local methods at a home scale, upgrading such traditional processes is essential. Chapter 13 focuses on such issues, with possible improvements in both processes and products. In South Asia, many indigenous fermented foods are yet to be investigated, and the indigenous traditional technologies of such foods are also not easily adapted. At the same time, the people who invented and preserved the age-old traditional food fermentation technology need to be reassured about the worth of their indigenous knowledge. In Chapter 14, the socioeconomic conditions and sustainability of indigenous fermented foods and related issues are discussed, with typical examples.

The book draws upon the expertise of leading researchers and teachers in the field. I am grateful to all the authors who have contributed unique, concise, and excellent articles based on their experience in research and teaching. Since no author was expected to be a specialist in all the areas, different authors were invited to submit manuscripts pertaining to their country or regions, and these have been compiled into discrete chapters. Depending upon the material taken from their contributions, their names have been added in one or more than one chapter and arranged alphabetically by first names. The contributions of some contributors, however, had to be reduced or edited, keeping in view the products of different countries of South Asia and the scope of the book. The entire text of this book is based on the production schedule, illustrated with tables and figures with the latest trends or innovations in the fields. References cited at the end of each chapter enable the reader to search for more information on a particular aspect. The future developments given at the end of each chapter are expected to stimulate further research in the field. Attempts have been made to document and update the indigenous knowledge of production, history, and ethnic value of the fermented foods of South Asia. Most of the information is extracted from the reviews, research papers, books, and the Internet, in addition to indigenous knowledge available locally. While compiling the book, my village background helped me a lot, and I have tried to write down some of my relevant experiences.

At this juncture, I respectfully remember my father, the late Sh. M. L. Joshi, and my mother, the late Mrs. Bimla Joshi, who inculcated in me the sense of creativeness and respect for the principles and dignity of labor that inspire and encourage me a lot. The moral support, stimulatory, healthy and work-conducive environment extended by my wife Mrs. Sushma Joshi, the help rendered by my sons, Dr. Bharat and Er. Sidharath, acting more or less as my assistants, especially to help me with the computer and Internet while editing the manuscript, cannot be expressed in words. With profound respect, I recall our former and founder vice-chancellor, the late Dr. M. R. Thakur, for encouraging and nurturing the "Fermentation Technology" concept into reality at the

University of Horticulture and Forestry, Nauni, Solan (HP), India. I am highly grateful to Professor Rob Nout and Professor Prabir Sarkar, series editors of Fermented Foods and Beverages, for considering me worthy of this task and ignoring my pitfalls. I cannot forget their expert guidance in giving shape to this manuscript, besides constant encouragement, and friendly and stimulatory behavior. I respectfully thank Professor Ruddle, whose pioneering work on fish fermentation has been very inspiring and who gave several constructive suggestions and advice on various issues connected to this manuscript. The pioneering work done by Professor Tamang and several of his students, and co-researchers has been very inspiring and has been quoted in several places in various chapters of the book. I convey my deep sense of appreciation for their way of handling the book production especially conveying the bitter pill with great ease while still making the project go on a defined path to Stephen Zollo and Laurie Oknowsky, acquiring editor and project coordinator, respectively, of Taylor & Francis Group, LLC, Boca Raton, FL and S. M. Syed, deputy manager, project management, Techset Composition, India, for preparing this text so nicely. Their patience as well as advice during the entire period from development of the manuscript to final publishing has been praiseworthy. It is hoped that the reader, besides knowing the academic and practical aspects of indigenous fermented foods of South Asia, will also be encouraged to make and consume these foods and, ultimately, appreciate the indigenous knowledge of the people who developed the technology.

Dr. V.K. Joshi

Professor and Head
Department of Food Science and Technology
Dr. YSP University of Horticulture and Forestry
Nauni, Solan, India

About the Book

Indigenous Fermented Foods of South Asia describes various indigenous fermented foods prepared and consumed in the countries of South Asia, namely, India, Pakistan, Bangladesh, Sri Lanka, Nepal, Bhutan, and the Maldives. Their microbiology, biochemistry, biotechnology, quality, and nutritional value are covered indepth. The general aspects of indigenous foods are discussed in separate chapters, which have been compiled to describe the various fermented foods of South Asia, that are grouped according to the different types of fermentation and the raw material involved in their production. All the chapters have been contributed by well-known experts in the respective fields.

- Chapter 1 is an overview of the fermented foods of South Asia, covering different facets of such foods, including the types of fermented foods, food fermentation, and the role of fermented foods in health, with an brief overview of the countries of South Asia.
- The diversity of South-Asian fermented foods is described in Chapter 2, focussing on the products of different countries and regions.
- Chapter 3 deals with the microbiology and biochemistry of indigenous fermented foods in depth, including the biochemical changes that are brought out by the microorganisms.
- The composition and nutritive value of fermented foods form the subject matter of Chapter 4.
- Chapter 5 is devoted to the aspects related to the quality and safety, including toxicity, of indigenous fermented foods.
- Chapter 6 deals with the health-related issues, especially therapeutic, values of indigenous fermented foods.

- How to prepare indigenous cereal-based fermented foods is described in Chapter 7, while fermented foods involving acid fermentation is the subject of Chapter 8. The indigenous fermentation technology of fruits and vegetables, involving acid fermentation, is also delineated in this chapter.
- Technology of the preparation of indigenous alcoholic fermented products is reviewed in Chapter 9, while methods of preparation of vinegars of various types are covered in Chapter 10.
- Chapter 11 deals with fermented meat and fish products, protein-rich vegetarian meat substitutes, sauces, and pastes, South Asian fish sauces, and other related products.
- Chapter 12 covers the most ancient fermentation-based product—the mushroom—produced by solid-state fermentation, and its postharvest technology.
- Biotechnological aspects of fermented products and molecular techniques employed are described in Chapter 13.
- Chapter 14 focuses on the problems of industrialization, socio-economic conditions, and the sustainability of indigenous fermented foods.

In brief, the book provides in-depth information both about the traditional and state-of-art technology of the production, quality, microorganisms involved, and consumption of the entire spectrum of indigenous fermented foods of South Asia.

Editor

Professor V.K. Joshi, is an eminent scientist and teacher with more than 35 years varied research experience in fruit fermentation technology and fermented foods, food toxicology, biocolor, quality assurance, and waste utilization. He has made a very significant contribution to the development of technology of non-grape fruit wines, apple pomace utilization and lactic acid fermented foods, and indigenous fermented foods. He is a former head of the Department of Postharvest Technology (PHT) and now heads the Department of Food Science and Technology in Dr. YSP University of Horticulture and Forestry, Nauni, Solan (HP), India. He has earned BSc (med), from Guru Nanak Dev, University Amritsar, an MSc (microbiology) from Punjab Agriculture University Ludhiana, and a PhD (microbiology) specializing in food fermentation from Guru Nanak Dev, University, Amritsar (Pb), India. He has received training on several aspects of food such as the sensory evaluation of food, processing and nutrition, and the postharvest technology of fruits and vegetable from prestigious institutes such as the Central Food Technological Research Institute (CFTRI), Mysore, and the Punjab Agricultural University (PAU), Ludhiana. He became a fellow of the Biotechnology Research Society of India in 2005 and a fellow of the Indian Society of Hill Agriculture in 2010. He has authored/edited more than 10 books, 5 practical manuals, 150 research papers in international and national journals, more than 50 book chapters, 35 review/technical articles, and several popular articles, besides presenting more than 50 papers and lead talks at different seminars and conferences. Professor Joshi has edited special issues of the *Journal of Scientific and Industrial Research* and the *Natural Product Radiant* as a guest editor, and is editor-in-chief of the *International Journal of Food Fermentation Technology*, editor of *Indian Food Packer*, the *Mushroom Journal*, and the *Journal of Food and Nutrition*. Two of his books, *Postharvest Technology of Fruits and Vegetables*, and *Biotechnology: Food Fermentation* are prescribed for MSc/Phd courses in postharvest technology, food science and technology, biotechnology, by the Indian Council of Agricultural Research (ICAR) and other universities in India. He has guided several postgraduate students for their dissertations in postharvest technology, and food science and technology. He is a regular reviewer of research projects and research papers, and a paper-setter and thesis-evaluator in varied subjects, including food fermentation, food science, microbiology, and biotechnology.

Professor Joshi has successfully conducted research and handled the research projects of ICAR, the National Horticultural Board (NHB), the Department of Biotechnology (DBT), the Ministry of Food Processing Industry (MOFPL), and the Department of Science and Technology (DST). He remains a member of the Board of Studies of several universities and institutes, a member of the Food Safety and Standard Authority of India, an advisor to the Horticultural Produce Processing and Marketing Corporation (HPMC) and a wine consultant (established wine factory). He has been honored with several awards for his research contributions, such as the N. N. Mohan, Kejeriwal, Pruthi, N. A. Pandit for the best research and review papers, the Himachal Shree for the teaching contribution, among others. Professor Joshi successfully conducted the ICAR sponsored Summer School of 2012 as a course director, and a Khadi and Village Industry Corporation (KVIC) sponsored workshop in March 2013. As a chairman of the Equal Opportunity Cell of the university, he initiated measures for the welfare of physically handicapped persons, including persons with disability (PWD) friendly buildings, grant of conveyance allowance, and employment creation, and helped organize several training programs as well as mock tests for the students of the university for JRF (junior research fellowship), SRF (senior research fellowship), and ARS (agricultural research services). He established the fruit processing unit in the Department of FST for the manufacture of fruits and vegetables products with FPO license that generated income for the university. As principal investigator of the experimental learning programme of ICAR he organized training for the UG students of the university.

Contributors

Ghan Shyam Abrol
Department of Food Science and
    Technology
Dr. Y S Parmar University of
    Horticulture and Forestry
Himachal Pradesh, India

Ome Kalu Achi
Department of Microbiology
Michael Okpara University of
    Agriculture
Abia State, Nigeria

Dorjey Angchok
Defence Institute of High Altitude
    Research
Defence Research and Development
    Organisation
Jammu and Kashmir, India

Kunzes Angmo
Department of Biotechnology
Himachal Pradesh University
Himachal Pradesh, India

Anton Ann
Borneo Marine Research Institute
University Malaysia Sabah
Sabah, Malaysia

B. L. Attri
ICAR-Central Institute of Temperate
    Horticulture
Regional Station
Uttarakhand, India

Laxmikant S. Badwaik
Department of Food Engineering and
    Technology
Tezpur University
Assam, India

Fatimah Abu Bakar
Department of Food Science
University Putra Malaysia
Selangor D.E., Malaysia

K. Lakshmi Bala
Department of Food Process
    Engineering
Sam Higginbottom Institute of
    Agriculture Technology and
    Sciences
Uttar Pradesh, India

Vandana Bali
Biotechnology Research
    Laboratory
Department of Food Engineering
    and Technology
Sant Longowal Institute of
    Engineering and Technology
Punjab, India

Idahun Bareh
Department of Philosophy
North-Eastern Hill University
Meghalaya, India

Tek Chand Bhalla
Department of Biotechnology
Himachal Pradesh University
Himachal Pradesh, India

Keshani Bhushan
Department of Microbiology
Himachal Pradesh Agricultural
    University
Himachal Pradesh, India

Zuberi M. Bira
Agricultural Research Institute
Morogoro, Tanzania

Pallab Kumar Borah
Department of Food Engineering
    and Technology
Tezpur University
Assam, India

María José Grande Burgos
Department of Health Sciences
University of Jaen
Jaén, Spain

Jayasree Chakrabarty
Department of Botany
Cotton College
Assam, India

Reena Chandel
Department of Food Science and
    Technology
Dr. Y S Parmar University of
    Horticulture and Forestry
Himachal Pradesh, India

Vinay Chandel
Department of Food Science and
    Technology
Dr. Y S Parmar University of
    Horticulture and Forestry
Himachal Pradesh, India

Arjun Chauhan
Department of Biotechnology
Dr. Y S Parmar University of
    Horticulture and Forestry
Himachal Pradesh, India

Vandita Chauhan
State Department of Ayurveda
Himachal Pradesh, India

Rupesh S. Chavan
Department of Food Science and
    Technology
National Institute of Food Technology
    Entrepreneurship and Management
Haryana, India

Shraddha R. Chavan
Department of Biotechnology
Junagadh Agricultural University
Gujarat, India

Fook Yee Chye
Department of Food Science and
  Nutrition
University Malaysia Sabah
Sabah, Malaysia

Arup Jyoti Das
Department of Food Engineering and
  Technology
Tezpur University
Assam, India

Sankar Chandra Deka
Department of Food Engineering and
  Technology
Tezpur University
Assam, India

Guru Aribam Shantibala Devi
Department of Life Sciences
Manipur University
Manipur, India

M. Preema Devi
Department of Pomology and Post
  Harvest Technology
Uttar Banga Krishi Viswavidyalaya
West Bengal, India

Deepa H. Diwedi
Department of Applied Plant Science
Baba Saheb Bhim Rao Ambedkar
  University
Uttar Pradesh, India

Antonio Gálvez
Department of Health Sciences
University of Jaen
Jaén, Spain

Neelima Garg
Division of Post Harvest Management
Central Institute for Sub-Tropical
  Horticulture
Uttar Pradesh, India

Neha Gautam
Department of Biotechnology
St Bed's College of Education
Himachal Pradesh, India

Satyendra Gautam
Food and Technology Division
Bhabha Atomic Research Centre
Maharashtra, India

Anupama Gupta
Department of Basic Science
Dr. Y S Parmar University of
  Horticulture and Forestry
Himachal Pradesh, India

Ashok Das Gupta
Department of Anthropology
University of North Bengal
West Bengal, India

Dharmesh Gupta
Department of Plant Pathology
Dr. Y S Parmar University of
  Horticulture and Forestry
Himachal Pradesh, India

Oluwatosin Ademola Ijabadeniyi
Department of Biotechnology and Food
  Technology
Durban University of Technology
Durban, South Africa

Vidhan Jaiswal
Futurebiotics LLC
Hauppauge, New York

Chetna Janveja
Department of Microbiology
Panjab University
Punjab, India

Kumaraswamy Jeyaram
Department of Biotechnology
Institute of Bioresources and Sustainable
    Development
Manipur, India

Deepti Joshi
Department of Food Science and
    Technology
Dr. Y S Parmar University of
    Horticulture and Forestry
Himachal Pradesh, India

S. R. Joshi
Microbiology Laboratory
Department of Biotechnology and
    Bioinformatics
North-Eastern Hill University
Meghalaya, India

V. K. Joshi
Department of Food Science and
    Technology
Dr. Y S Parmar University of
    Horticulture and Forestry
Himachal Pradesh, India

Jahangir Kabir
Department of Post Harvest Technology
    of Hortcultural Crops
Bidhan Chandra Krishi Viswavidyalaya
West Bengal, India

S. S. Kanwar
Department of Microbiology
Himachal Pradesh Agricultural
    University
Himachal Pradesh, India

Shammi Kapoor
Department of Microbiology
Punjab Agricultural University
Punjab, India

Swati Kapoor
Department of Food Science and
    Technology
Punjab Agricultural Punjab
    University
Punjab, India

Manisha Kaushal
Department of Food Science and
    Technology
Dr. Y S Parmar University of
    Horticulture and Forestry
Himachal Pradesh, India

P. K. Khanna
Department of Microbiology
Punjab Agricultural University
Punjab, India

Sushma Khomdram
Department of Life Sciences
Manipur University
Manipur, India

Ashwani Kumar
Department of Food Science and
    Technology
Punjab Agricultural University
Punjab, India

Avanish Kumar
Department of Food Process
    Engineering
Sam Higginbottom Institute of
    Agriculture Technology and
    Sciences
Uttar Pradesh, India

Naveen Kumar
Department of Food Science and
    Technology
Dr. Y S Parmar University of
    Horticulture and Forestry
Himachal Pradesh, India

Suresh Kumar
Horticulture College and Research
    Institute
ICAR-NRC for Banana
Tamil Nadu, India

Vikas Kumar
Department of Food Science and
    Technology
Lovely Professional University
Punjab, India

Anila Kumari
Department of Biotechnology
Himachal Pradesh University
Himachal Pradesh, India

Pooja Lakhanpal
Department of Food Science and
    Technology
Dr. Y S Parmar University of
    Horticulture and Forestry
Himachal Pradesh, India

Ranendra Kumar Majumdar
College of Fisheries (CAU)
Tripura, India

Jaruwan Maneesri
Department of Technology and
    Industries
Prince of Songkla University
Pattani, Thailand

Bijoy Moktan
Department of Botany
University of North Bengal
West Bengal, India

Antonio Cobo Molinos
Department of Health Sciences
University of Jaen
Jaén, Spain

Monika
Department of Biotechnology
Himachal Pradesh University
Himachal Pradesh, India

Ghulam Mueen-ud-Din,
Institute of Food Science and
    Nutrition
University of Sargodha
Punjab, Pakistan

Abu Hassan Siti Nadiah
School of Industrial Technology
Universiti Sains Malaysia
Penang, Malaysia

W. A. Wan Nadiah
School of Industrial Technology
Universiti Sains Malaysia
Penang, Malaysia

Ishige Naomichi
National Museum of Ethnology
Osaka-fu, Japan

A. K. Nath
Department of Biotechnology
Dr. Y S Parmar University of
    Horticulture and Forestry
Himachal Pradesh, India

Prabhat K. Nema
Department of Food Science and
    Technology
National Institute of Food
    Technology Entrepreneurship and
    Management
Haryana, India

Bhanu Neopany
Department of Environment, Science
    and Technology
Himachal Pradesh Government
Himachal Pradesh, India

Samuel Oluwole Ogundele
Department of Archaeology and
    Anthropology
University of Ibadan
Ibadan, Nigeria

Olanrewaju Olaseinde Olotu
Centre for Physical and Nutrition
Auckland University of Technology
Auckland, New Zealand

Lok Man S. Palni
Biotechnological Applications
GB Pant Institute of Himalayan
    Environment and Development
Uttarakhand, India

Anita Pandey
Biotechnological Applications
GB Pant Institute of Himalayan
    Environment and Development
Uttarakhand, India

Parmjit S. Panesar
Biotechnology Research Laboratory
Department of Food Engineering and
    Technology
Sant Longowal Institute of Engineering
    and Technology
Punjab, India

Reeba Panesar
Biotechnology Research Laboratory
Department of Food Engineering and
    Technology
Sant Longowal Institute of Engineering
    and Technology
Punjab, India

Chamgongliu Panmei
Department of Life Sciences
Manipur University
Manipur, India

S. V. Pinto
Dairy Technology Department
Anand Agricultural University
Gujarat, India

J. P. Prajapati
Dairy Technology Department
Anand Agricultural University
Gujarat, India

S. G. Prapulla
Fermentation Technology and
    Bioengineering Department
Central Food Technological Research
    Institute
Karnataka, India

Rubén Pérez Pulido
Department of Health Sciences
University of Jaen
Jaén, Spain

Amit Kumar Rai
Microbial Resource Division
Regional Centre of Institute of
    Bio-Resources and Sustainable
    Development
Sikkim, India

Anup Raj
High Mountain Arid Agriculture
　Research Institute
SKUAST Kashmir
Jammu and Kashmir, India

Dev Raj
Department of Post-Harvest Technology
Navsari Agricultural University
Gujarat, India

George F. Rapsang
Department of Biotechnology and
　Bioinformatics
North-Eastern Hill University
Meghalaya, India

Ramesh C. Ray
ICAR-Central Tuber Crops Research
　Institute (Regional Centre)
Orissa, India

L. V. A. Reddy
Department of Microbiology
Yogi Vemana University
Andhra Pradesh, India

B. Renuka
Fermentation Technology and
　Bioengineering Department
Central Food Technological Research
　Institute, CSIR
Karnataka, India

Md. Shaheed Reza
Department of Fisheries Technology
Bangladesh Agricultural University
Mymensingh, Bangladesh

Ahmad Rosma
Bioprocess Technology Division
School of Industrial Technology
University Sains Malaysia
Penang, Malaysia

Arindam Roy
Department of Microbiology
Ramakrishna Mission Vidyamandira
West Bengal, India

Kenneth Ruddle
Asahigaoka-cho
Hyogo-ken, Japan

K. S. Sandhu
Department of Food Science and
　Technology
Chaudhary Devi Lal University
Haryana, India

Savitri
Department of Biotechnology
Himachal Pradesh University
Himachal Pradesh, India

Ulrich Schillinger
Institute for Microbiology and
　Biotechnology
Max-Rubner Institute
Karlsruhe, Germany

A. K. Senapati
Department of Post-Harvest Technology
Navsari Agricultural University
Gujarat, India

Arun Sharma
Food and Technology Division
Bhabha Atomic Research Centre
Maharashtra, India

Nivedita Sharma
Department of Basic Science
Dr. Y S Parmar University of
　Horticulture and Forestry
Himachal Pradesh, India

Rakesh Sharma
Department of Food Science and
  Technology
Dr. Y S Parmar University of
  Horticulture and Forestry
Himachal Pradesh, India

Sangeeta Sharma
Department of Food Science and
  Technology
Dr. Y S Parmar University of
  Horticulture and Forestry
Himachal Pradesh, India

Somesh Sharma
Department of Food Technology
Shoolini University
Himachal Pradesh, India

Kheng Yuen Sim
Universiti Malaysia Kelantan
Kelantan, Malaysia

Amarjit Singh
Department of Processing and Food
  Engineering
Punjab Agricultural University
Punjab, India

Ningthoujam Sanjoy Singh
Department of Botany
Ghanapriya Women's College
Manipur, India

Raman Soni
Department of Biotechnology
D.A.V. College
Punjab, India

S. K. Soni
Department of Microbiology
Panjab University
Punjab, India

Swati Sood
Department of Microbiology
Himachal Pradesh Agricultural
  University
Himachal Pradesh, India

Aditi Sourabh
Department of Microbiology
Himachal Pradesh Agricultural
  University
Himachal Pradesh, India

Tsering Stobdan
Defence Institute of High Altitude
  Research
Defence Research and Development
  Organisation
Jammu and Kashmir, India

B. C. Suman
Department of Plant Pathology
Dr. Y S Parmar University of
  Horticulture and Forestry
Himachal Pradesh, India

C. K. Sunil
Department of Food Engineering, Post
  Harvest Technology
Indian Institute of Crop Processing
  Technology
Tamil Nadu, India

Manas R. Swain
Department of Biotechnology
College of Engineering and Technology
Orissa, India

Somboon Tanasupawat
Department of Biochemistry and
  Microbiology
Chulalongkorn University
Bangkok, Thailand

Konchok Targais
Defence Institute of High Altitude
    Research (DIHAR)
Defence Research and Development
    Organisation (DRDO)
Jammu and Kashmir, India

Malai Taweechotipatr
Department of Microbiology
Srinakharinwirot University
Bangkok, Thailand

Nisha Thakur
Department of Biotechnology
Dr. Y S Parmar University of
    Horticulture and Forestry
Himachal Pradesh, India

B. M. K. S. Thilakarathne
Institute of Postharvest Technology
Research and Development Centre
Anuradhapura, Sri Lanka

Sharmila Thokchom
Department of Biotechnology and
    Bioinformatics
North-Eastern Hill University
Meghalaya, India

S. S. Thorat
Department of Food Science and
    Technology
Mahatma Phule Krishi Vidyapeeth
Maharashtra, India

Gitanjali Vyas
Department of Basic Science
Dr. Y S Parmar University of
    Horticulture and Forestry
Himachal Pradesh, India

Sohini Walia
Department of Microbiology
Himachal Pradesh Agricultural
    University
Himachal Pradesh, India

Preeti Yadav
Division of Post Harvest Management
Central Institute for Subtropical
    Horticulture
Uttar Pradesh, India

Muhammad Zukhrufuz Zaman
Department of Food Science
Universiti Putra Malaysia
Selangor D.E., Malaysia

INDIGENOUS FERMENTED FOODS OF SOUTH ASIA

An Overview

ANTONIO COBO MOLINOS, ANTONIO GÁLVEZ,
ANUP RAJ, ARJUN CHAUHAN, ASHOK DAS GUPTA,
FOOK YEE CHYE, GEORGE F. RAPSANG,
GHAN SHYAM ABROL, IDAHUN BAREH,
ISHIGE NAOMICHI, KENNETH RUDDLE,
KHENG YUEN SIM, MARÍA JOSÉ GRANDE BURGOS,
RUBÉN PÉREZ PULIDO, SAMUEL OLUWOLE
OGUNDELE, SANGEETA SHARMA,
SHARMILA THOKCHOM, S.R. JOSHI, V.K. JOSHI,
VANDITA CHAUHAN, VIDHAN JAISWAL,
VINAY CHANDEL, AND ZUBERI M. BIRA

Contents

1.1 Introduction

Fermentation is one of the oldest "food biotechnological" processes used to prepare foods and beverages which is recorded in the ancient history of man (Joshi and Pandey, 1999; Ross et al., 2002). Indigenous fermented foods and beverages have been an integral component of the dietary culture of every community in the world ever since the beginning of human civilization (Tibor, 2007). These foods have been prepared and consumed all over the world (Steinkraus, 1979; Sekar and Mariappan, 2007), including South Asia and India (Panesar and Marwaha, 2013). These foods served as dietary staples, adjuncts to staples, pickles, condiments, and beverages (Dahal et al., 2005). These foods are palatable, wholesome, and nutritive confectionery items (Ann Mothershaw and Guizani, 2007), and have unique flavor, aroma, and texture attributes which are much appreciated by consumers (Caplice and Fitzgerald, 1999). The original and primary purpose of fermenting food substrates was to achieve a preservative effect, but with the development of many more effective alternative preservation technologies, the preservation effect is no longer the most pressing requirement.

Fermented foods are made by the action of microorganism(s), either naturally or by adding a starter culture(s), which modify the substrates biochemically and organoleptically into edible product (Pederson, 1960; Holzapfel et al., 1998; Joshi et al., 1999; Tamang and Holzapfel, 1999; Hansen, 2002; Tamang et al., 2007). Out of a long list of indigenous fermented foods, sausages, *dahi*, cheese, beer, wine, distilled liquors, *bhalle*, *papad*, *idli*, and *dosa* can be mentioned. Several types of indigenous fermented foods are made based on raw products, namely, milk, cereals, grains, fruits, and vegetables (Joshi et al., 2012), using local knowledge and locally available raw materials. These foods contribute significantly to the nutritional status of the consumers.

The fermented foods are made through fermentations, such as alcoholic, lactic acid, acetic, and alkaline fermentations, of which, alcoholic and lactic acid fermentation are involved in the production of a large number of products throughout the world (Joshi, 2006a; Ray and Joshi, 2015). Indigenous alkaline fermented foods form an important part of the diet of Asian countries, where the fermentation makes otherwise inedible foods edible (Sarkar and Nout, 2014). Fermentation, in general, enhances the flavor and nutritional value; decreases toxicity; preserves food; decreases cooking time and energy requirements, and generally brings diversity into the kinds of foods and beverages available (Campbell-Platt, 1987; Steinkraus, 1996; Wang and Fung, 1996; Parkouda et al., 2009; Chukeatirote et al., 2010). Indeed, fermentation is a part of ethno-knowledge and constitutes one of the oldest food preparation and preservation technologies (Dirar, 1993; Steinkraus, 1996; Joshi et al., 2012).

The origins of most fermentation technologies have been lost in the mists of history. Some products and practices no doubt fell by the wayside, but those that remain today have survived the onslaught of time. The people in Asia are considered pioneers in the fermentation of vegetable proteins with meat-like flavors (Soni and Sandhu, 1990). The traditional art of fermentation practiced by the common man continues despite the scientific and technological revolution, though remains confined mainly to rural and tribal areas, due to the inaccessibility of commercially made products in the remote areas and their cost, and the sociocultural linkages of indigenous fermented products (Thakur et al., 2004).

Ethnic or traditional foods are those produced by ethnic people using locally available raw materials to make edible products that are culturally and socially acceptable to the consumer (Tamang, 2010a–d). It would be interesting to know when and where these foods originated and what was their historically background. Why foods are fermented and on what principles would be interesting and useful to understand. It is equally pertinent to know how the fermentation originated and contributed to the development of the sensory qualities of the products. The role of fermented foods in the social fabric of the peoples of South Asia is considered in this text. Are these foods safe to consume and to what extent? What constraints these products face, and how science and technology can be applied to them, needs considerable attention. Like other countries of the world, those in South Asia—India, Pakistan, Bangladesh, Srilanka, Nepal, Bhutan, Afghanistan, and the Maldives—have contributed to the production of indigenous fermented foods, and which products are made in which country employing which processes have been documented. All these aspects are reviewed briefly in this chapter.

1.2 Origin, History, and Role of Fermentation

1.2.1 Origin of Microorganisms, Plants, and Foods

The earth is about 4.5 billion years old, and the first forms of life to appear or evolve on earth were microorganisms. Fossil microorganisms have been traced in rocks 3.3–3.5 billion years old (Schopf and Packer, 1987). The universe, the earth, and man originated long before any scientific method or the means to study these phenomena or knowledge of the concept was developed (Steinkraus, 1996). In fact, when we study food, or fermented foods, all aspects, including philosophy, archeology, and anthropology, are involved. There are intimate relationships between man, microbes, and food. There is a never-ending struggle between man and microbes to see which will be first to consume available food supplies, even today. Religion was an attempt by man to explain the unexplainable origin of the universe, the earth, and man long before there was any scientific method or the means to study these difficult problems. At that time, there was also no scientific concept of origin of life.

1.2.2 Food Preservation

It is established that microorganisms were followed by the plants as the next form of life to evolve on earth, and both meet the requirement of food for man (Steinkraus, 1996). Man learned the life of the hunter gatherer, started agriculture, and learned how to process food. It can be speculated that some billions of years ago, there might have been a huge food supply and relatively few humans to consume it so that excess food supplies might have fell on the ground where either the seeds might have germinated or the carbohydrates, proteins, fats, etc. were consumed by microorganisms that might have used enzymes to convert fermentable carbohydrates to alcohol or acids and similar other products and those containing protein into essential amino acids, peptides ammonia, etc. Further, the food producing period dates from 8000 to 10,000 years ago and it can be presumed that the problem of food spoilage and food poisoning might have been encountered during that time (Jay et al., 2004). Thus, when there was excess availability of food, the problem of spoilage might also have occurred. It was Pastuer in 1837 who showed and appreciated the presence and the role of microorganisms in food spoilage and preservation (Stanier et al., 1970; Jay et al., 2004). So a need to protect food from spoilage might have arisen, leading to the development of methods or techniques or simply a set of steps, such as drying, heating, and fermentation, to preserve the foods.

1.2.3 Fermented Food

Unlike drying as a method of preservation, fermentation is known to give food a variety of flavor, taste, texture, and sensory-quality attributes, and nutritional and therapeutic values (Joshi, 2006b; Nout, 2009; Mehta et al., 2012). When any food produced an altered state of awareness or consciousness upon consumption that was noticeable, without any serious toxic side effects (motor impairment), or had a substantial improvement in nutritional value over the unprocessed raw food, it became immediately accepted (Katz and Voight, 1987; http://www.sirc.org/publik/drinking_origins.html). This is exactly what might have happened when barley and wheat were fermented into beer, which might have motivated the people to cultivate wheat and barley. But the consumption of such fermented foods might have led to social disorder, and consequently to the development of regulatory religious and social traditions. All these events might have possibly led to the formation of groups for labor (harvesting, large-scale construction tasks); ritual ceremonies (marriage or a funeral or social gatherings) where such beverages were consumed (Katz and Voight, 1987). Thus, it can fairly be predicted that such beverages were discovered and used at a relatively early time in human evolution, and the most consistently sought-after beverage was alcohol, as in wine and beer.

Wine is perhaps the oldest known fermented product, and traces its antiquity to at least 5000 BC (Amerine et al., 1980; Vine, 1981; Joshi et al., 2011; Reddy and Reddy, 2012). The origin of wine might have been accidental, when fruit juice might have transformed into a beverage having exhilarating or stimulating properties. Practically, every civilization is known to have a characteristic beer, wine or other alcoholic beverage (Vine, 1981), and wine making existed for a long time even before the chronicles found in Egyptian heiroglyphs. In Vedic times, wine was said to have been used extensively in the Durbar of Indira, and is referred to in the ancient scriptures as *Soma* such as *Rigvedas* (Joshi, 1997). Grape growing and use of wine in Biblical times has been referred to in both the Old and New Testaments. Wine is known to have also been prepared by the Assyrians in 3500 BC.

Wang and Hesseltine (1979) noted that probably the first fermentation was discovered accidentally, when salt was incorporated into food material, which selected certain harmless microorganisms that fermented the raw materials to give a nutritious and acceptable food. It might have been followed by the early Chinese, who first inoculated basic foods with molds, which created enzymes in salt-fermented soy foods, such as miso, soy sauce, and fermented tofu. The earliest records of the koji-making process can be traced back to at least 300 BC in China and to the third century AD in Japan. The only traditional East Asian fermented soy food not prepared with molds is Japan's *natto*, and its relatives *thua-nao* in Thailand and *kinema* in Nepal; these are the results of bacterial fermentations (William and Akiko, 2004).

1.2.4 Scientific Journey of Fermentation

To the early societies, the transformation of basic food materials into fermented foods was both a mystery and a miracle. Despite the long history of fermentation, the understanding of the science behind these arts, came quite late (Mehta et al., 2012). The scientific journey began (1632–1723), some 200 years before Pasteur, when Antonie Van Leeuwenhoeck built a new microscope to observe tiny living creatures that he called "animalcules," now called microbes (Stenier et al., 1970). In the late 1700s, Lavoisier showed that in the process of transforming sugar to alcohol and carbon dioxide (as in beer and wine), the weight of the former that was consumed in the process equaled the weight of the latter produced. In 1810, Guy-Lussac summarized the process with the famous equation that $C_6H_{12}O_6$ yielding $C_2H_5OH + 2CO_2$. The entire process at that time was considered to be simply a chemical reaction, and the yeast was thought to play a physical rather than a chemical role, an idea that dated back to the time of George Stahl in 1697. Putrefaction, spoilage, and fermentation were all considered to be the processes of death, not life.

Proof of the living nature of yeast appeared between 1837 and 1838, when three publications by Cagniard *de la tour*, Swann, and Kuetzing, appeared, each of whom independently concluded that yeast was a living organism that reproduced by budding. It was followed soon by the discovery of bacteria. The scientific breakthroughs to unravel the mysteries of fermentation started in the 1830s, primarily by French and German chemists. During the 1800s, the making of wine and beer was refined into the techniques known today. Still one problem remained: 25% of the fermented products turned into vinegar before the end of fermentation, and nobody knew the reason (Stanier et al., 1970; Jay et al., 2004). Later on, in 1857, Louis Pasteur solved the problem of wine spoilage when Napolean III referred the same to him. He found that

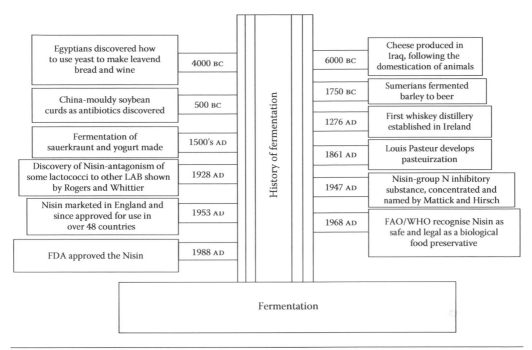

Figure 1.1   A pictorial representation of history of fermentation. (Data from http://en.wikipedia.org/wiki/Fermentation biochemistry).

heating the wine after it was fermented prevented spoilage, thus opening the way for the aseptic conditions used even today in wine making: a process called pasteurization.

Some of the developments in fermentation are depicted in Figure 1.1. In 1880 a Danish botanist, Carlsberg, started looking at the morphology (shape and structure) and the physiology (function) of yeasts, and found differences between various yeasts. In 1883, the Carlsberg brewery of Copenhagen introduced the concept of pure cultures of yeast to the market, and made it possible to brew a beer with only one type of yeast (Stanier et al., 1970; Jay et al., 2004). Several scientists, including Pasteur, had earlier attempted, unsuccessfully, to extract the enzymes involved in fermentation from yeast. However, success came finally in 1897 when the German chemist Eduard Bucchner ground up yeast, extracted a juice from the yeast cells, then found to his astonishment that the so called "dead" liquid fermented a sugar solution, just like any living yeast. In 1907, *Lactobacillus* was isolated from fermented milk by the Russian microbiologist Ellie Metchnikoff (Mehta et al., 2012). That made a beginning to the production of healthy foods with probiotic activity.

It is known that foods invaded by bacteria producing toxins or fungi producing mycotoxins are dangerous to man. In earlier times, if the products of invasion were ill-smelling, off-flavored, or toxic, consumers tried to avoid them, and the foods were described as spoiled (Frazier and Westhoff, 2004; Joshi et al., 2013). On the other hand, when the microbial products were pleasantly flavored, had attractive aromas and textures, and were non-toxic, people accepted them and they were designated as fermented foods (Campbell-Platt, 1987). Certain flavors, such as sweet, sour, alcoholic, and meat-like appealed to large numbers of people. Similarly, milk soured naturally, resulting in curd/*dahi*. Fruit and berry juices rapidly become alcoholic (Steinkraus, 2002; Joshi, et al., 2011) and constituted a very important group of indigenous fermented foods, as alcoholic beverages. Thus, it became apparent that all these products were made by the process of fermentation.

1.2.5 *History of Fermented Foods*

The history of fermented foods is lost in antiquity. May be it was a mere accident when people first experienced the taste of fermented food, that could have started with, for example, the storage of surplus milk, resulting in a fermented product (curd) the next day. Since Neolithic times man is known to have made fermented foods, and the earliest types were beer, wine, and leavened bread (made primarily by yeasts) and cheeses (made by bacteria and molds). These were soon followed by yoghurt and other fermented milk products, pickles, sauerkraut, vinegar (soured wine), butter, and a host of traditional alcoholic beverages. The fermentation of milk started in many places, with evidence of fermented products in use in Babylon over 5000 years ago. Rock drawings discovered in the Libyan desert, believed to have been made about 9000 BC, depict cow worship and cows being milked (Pederson and Albury, 1969). Some of the oldest records suggest the development of dairying in ancient India, Mesopotamia, and Egypt. It is apparent from writings and drawings dating back to 6000 BC from the Sumerians of Mesopotamia, that dairying was highly developed. A sculptured relic dating back to 2900–2460 BC, found at Teil Ubaid in the Middle East in the territory of ancient Babylonia, shows the development of a system for processing milk. A great many of today's fermented milk products were originally developed by nomadic Asian cattle breeders. It is believed that the art of making cheese evolved about 8000 years ago in the "Fertile Crescent" between the rivers Tigris and Euphrates in Iraq (Hill et al., 2002; Fox and McSweeney, 2004). Since the dawn of civilization, methods for the fermentation of milks, meats and vegetables have been described, with earliest records dating back to 6000 BC and the civilizations of the Fertile Crescent in the Middle East (Fox, 1993; Caplice and Fitzerald, 1999). Some of the important milestones in the history of fermented foods are summarized in Table 1.1.

Indians are credited with discovering the method of souring and leavening of cereal legume batters (Padmaja and George, 1999). Thus, it appears that the art of fermentation originated in the Indian sub-continent in the settlements that predate the great Indus Valley civilization. During the Harappan spread or pre-Vedic times, there are indications of a highly developed system of agriculture and animal husbandry. It is also believed that the knowledge written in the four *Vedas* (sacred Hindu writings) came from the experiences, wisdom, and foresightedness of sages, which had been preserved by verbal tradition. As there is no written proof, controversies exist among historians in predicting the probable date of the *Vedas*. Based on astronomy, Lokmanya Tilak estimated it as the period between 6000 and 4000 VP (VP stands for the Hindu

Table 1.1   History of Fermented Foods

DATES	MILESTONE/DEVELOPMENT/LOCATION
ca. 10,000 BC to Middle Ages	Evolution of fermentation from salvaging the surplus, probably by pre-Aryans
ca. 7000 BC	Cheese and bread making practiced
ca. 6000 BC	Wine making in the Near East
ca. 5000 BC	Nutritional and health value of fermented milk and beverages described
ca. 3500 BC	Bread making in Egypt
ca. 1500 BC	Preparation of meat sausages by ancient Babylonians
2000 BC–1200 AD	Different types of fermented milks from different regions
ca. 300 BC	Preservation of vegetables by fermentation by the Chinese
500–1000 AD	Development of cereal-legume based fermented foods
1881	Published literature on *koji* and *sake* brewing
1907	Publication of book *Prolongation of Life* by Eli Metchnikoff describing therapeutic benefits of fermented milks
1900–1930	Application of microbiology to fermentation, use of defined cultures
1970–present	Development of products containing probiotic cultures or friendly intestinal bacteria

Source: Adapted from Pederson, C.S. 1971. Microbiology of Food Fermentations. AVI, Publishing Co., Westport CT, pp. vi, 283; Achaya, K.T. 1994. *Indian Food: A Historical Companion.* Oxford University Press, London; Steinkraus, K.H. 1996. *Handbook of Indigenous Fermented Food.* 2nd edition. Marcel Dekker, Inc., New York, NY; Farnworth, E.R. 2003. *Handbook of Fermented Functional Foods.* CRC Press, New York, NY; Prajapati, J.B. and Nair, B.M. 2008. *Handbook of Fermented Functional Food.* 2nd edition, CRC Press Taylor & Francis, Boca Raton, FL.

calendar of Vikram); using other methods of calculations, it is approximately 2500 VP (Upadhyay, 1967).

As early as 4000 and 3000 BC, fermented bread and beer were known in Pharaonic Egypt and Babylonia (Hutkins, 2006; Prajapati and Nair, 2008; Mehta et al., 2012). Bread-making probably originated in Egypt over 3500 years ago. Several triangular loaves of bread have been found in ancient tombs. There is also evidence of fermented meat products being produced for King Nebuchadnezer of Babylon. China is thought to be the birth-place of fermented vegetables and the use of *Aspergillus* and *Rhizopus* molds to make food. Knowledge about traditional fermentation technologies has been handed down from parent to child, for centuries (Aidoo, 2011).

Alcoholic fermentation involved in the making of wine and brewing is considered to have been developed during the period 2000–4000 BC by the Egyptians and Sumerians. Rigveda (ca. 1500 BC) has mentioned "The Somras"—fermented juice and wine. The fermentation technology started to develop after observations of fermentative changes in fruits and juices (Upadhyay, 1967; Prajapati and Nair, 2008; Mehta et al., 2012). It is documented that fermented drinks were being produced over 7000 years ago in Babylon (now Iraq), 5000 years ago in Egypt, 4000 years ago in Mexico, and 3500 years ago in Sudan (Dirar, 1993). Alcohol has been used since the time of discovery of fermentation of plant products around 6000–4000 BC. The world's earliest known wine jar (dating back to the period 5400–5000 BC) was found in Hajji Firuz Tepe, Iran, by archaeologists from the University of Pennsylvania. Since 3000–2000 BC, various references and descriptions regarding the use of alcoholic beverages are available from almost all the civilizations of world (http://www.searo.who.int). Fermented honey drinks were likely to have been the earliest alcoholic beverages known to mankind. Earliest historical evidence of alcoholic beverages came from an archeological discovery of Stone Age beer jugs from approximately 10,000 years ago, whereas evidence of wine appeared in Egypt about 5000 years ago. Distillation, a process used to produce liquor with high alcohol content, originated in China and India around 800 BC (Dasgupta, 1958).

There are historical records of the consumption of alcoholic beverages in the South Asian region. In Hindu mythology, the use of *soma/somras* as well as *sura*, by various groups in the society around 2000 BC for sedating and calming effects has been mentioned. *Soma*, an alcoholic beverage having a pleasant effect, was the drink of the upper classes of society and the gods, while *sura* was the drink of the lower classes of society, especially the warriors (*kshatriya*), as a relief from their physical hardships (Joshi, 1997; Tamang, 2010a–d). These beverages have their own socio-economic importance, as tribal people use them for religious practices, like offerings to God, and in ceremonies like weddings and local festivals. The use of these beverages by small groups in the Hindu civilization in Asia continued through the ages, but did not dominate society.

1.3 Food Fermentation

Fermentation is one of the important and ancient processes that contribute to the nutritional requirements of millions of individuals (Chojnacka, 2010). A number of microorganisms and enzymes play an important role in the fermentation process, by the effective utilization of available natural food/feed stocks and transformation of waste materials, thereby, contributing to meet the world's food problems (Holzapfel, 2002). Fermentation is the most economical methods of producing and preserving foods (Chavan and Kadam, 1989; Murty and Kumar, 1995; Steinkraus, 1996; Billings, 1998; Gadaga et al., 1999) including the production of alcoholic beverages (Fleet, 1998). The word "fermentation" is derived from the Latin word fermentum meaning "to boil," since the bubbling and foaming of early fermenting beverages seemed closely akin to boiling. It is the chemical transformation of organic substances into simpler compounds by the action of enzymes, the complex organic catalysts which are produced by microorganisms such as molds, yeasts, or bacteria (Petchkongkaew, 2007), and due to the enzymatic activity various by-products are formed (Bisen et al., 2012). It is an energy-yielding process, whereby organic molecules serve both as electron donor and electron acceptor. It covers a wide range of microbial and enzymatic processing of foods and their ingredients to achieve desirable characteristics (Sharma and Kapoor, 1996; Holzapfel, 2002; Giraffa, 2004; Savadogo, 2012). The traditional fermentation of foods serves several functions (Battcock and Azam-Ali, 1998; Joshi and Pandey, 1999).

Food fermentation involves all those processes where either the ultimate product is used directly as a food or as an additive to food, or is a basic ingredient in the food

(Joshi and Pandey, 1999). The growth and activity of microorganisms plays an essential role in the biochemical changes in substrates, such as plant, dairy, meat, and fish products, during fermentation. The microbiota—which may be indigenously present on the substrate, or added as a starter culture (Blandino et al., 2003), or may be present in or on the ingredients and utensils, or in the environment—are selected through adaptation to the substrate and by adjusting the fermentation conditions (Soni and Sandhu, 1990). There are four main fermentation processes, namely, alcoholic, lactic acid, acetic acid, and alkaline fermentation (Soni and Sandhu, 1990; Joshi and Pandey, 1999; McKay et al., 2011; Sarkar and Nout, 2014). Three major types of microorganisms, namely, yeast, bacteria, and fungi, are associated with traditional fermented foods and beverages (Syal and Vohra, 2013). In many of the indigenous fermented foods, yeasts are predominant and functional during fermentation. The diversity of indigenous fermented foods ranges from *nan* to *idli* to alcoholic beverages, such as rice and palm wine (Aidoo et al., 2006). Such aspects have been elaborated upon in Chapter 3 of this text.

Basically, food fermentation can take place if there is a suitable substrate, appropriate microorganism(s) either from nature or by inoculation of specific microorganism, and the necessary environmental conditions for the fermentation to take place. Environmental conditions, like temperature and moisture, need to be optimum for a specific fermentation, as well as the intrinsic factors of fermentation, including the pH, type of sugar, nutrients, availability or otherwise of oxygen, etc. The composition and quality of the raw materials, the microflora involved, the amount of water, type of raw material, and time also influence the fermentation. Preservation of foods by fermentation depends on the principle of oxidation of carbohydrates and related derivatives to generate various products (generally acids, alcohol, and carbon dioxide). These end-products determine the growth of food spoilage microorganisms, and because the oxidation is only partial, the food retains sufficient energy potential to be of nutritional benefit to consumers (Caplice and Fitzerald, 1999).

1.3.1 Classification of Food Fermentations

Food fermentations have been classified in a number of ways (Dirar, 1993): by categories (Yokotsuka, 1982) such as (1) alcoholic beverages fermented by yeasts; (2) vinegars fermented with *Acetobacter*; (3) milks fermented with lactobacilli; (4) pickles fermented with lactobacilli; (5) fish or meat fermented with lactobacilli; and (6) plant proteins fermented with molds with or without lactobacilli and yeasts. These fermentations can also be grouped by the type of product (Campbell-Platt, 1987), such as: (1) alcoholic beverages; (2) cereal products; (3) dairy products; (4) fish products; (5) fruit and vegetable products; (6) legumes; and (7) meat products. Attempts have been made to classify the food fermentations by the type of commodity (Odunfa, 1983), namely: (1) fermented starchy roots; (2) fermented cereals; (3) alcoholic beverages; (4) fermented vegetable proteins; and, (5) fermented animal protein; and by commodity (Kuboye, 1985): (1) cassava based; (2) cereal; (3) legumes; and, (4) beverages.

Steinkraus (1983a, 1996, 2002) classified the food fermentations in detail according to the following categories:

1. Fermentations producing textured vegetable protein meat substitutes in legume/cereal mixtures. Examples are Indonesian *tempe* and *ontjom*.

2. High salt/savory meat-flavored/amino acid/peptide sauce and paste fermentations. Examples are Chinese soy sauce, Japanese *shoyu* and Japanese *miso*, Indonesian *kecap*, Malaysian *ki-cap*, Korean *kanjang*, Taiwanese *inyu*, fish sauces: Vietnamese *nuocmam*, Malaysian *budu*, fish pastes: Philippine *bagoong*, Malaysian *belachan*, Vietnamese *mam*, Cambodian *prahoc* etc. These are predominately oriental fermentations.

3. Lactic acid fermentations. Examples of vegetable lactic acid fermentations, such as: *sauerkraut*, cucumber pickles, olives in the Western world, Indian pickled vegetables and Korean *kimchi*, Chinese *hum-choy*. Malaysian pickled vegetables and Malaysian *tempoyak*. Lactic acid fermented milks include: yogurts in the Western world, Russian *kefir*, Middle-East yogurts, *liban* (Iraq), Indian *dahi*, Egyptian *laban rayab*, *laban zeer*, Malaysian *tairu* (soybean milk) and fermented cheeses in the Western world, yogurt/wheat mixtures: Egyptian *kishk*, Greek *trahana*s, Turkish *tarhana*s. Lactic acid fermented cereals and tubers (cassava): Mexican *pozol*, Ghanian *kenkey*, Nigerian *gari*; boiled rice/raw shrimp/raw fish mixtures: lactic fermented/leavened breads: sour dough breads in the Western world; Indian *idli*, *dhokla*, *khaman*, Sri-lankan *hoppers*, and Ethiopian *enjera*.

4. Alcoholic fermentations. Examples are wines and beers, Mexican *pulque*, honey wines, South American Indian *chichi*, and beers in the Western World; wines and Egyptian *bouza* in the Middle East; palm and jackfruit wines in India, Indian rice beer, Indian *madhu*, Indian *ruhi*; palm wines, Kaffir/*bantu beers*, Nigerian *pito*, Ethiopian *talla*, Kenyan *busaa*, Zambian maize beer; in the Far East, sugar cane wines, palm wines, Japanese sake, and Malaysian tapuy.

5. Acetic acid/vinegar fermentations. Examples are apple cider and wine vinegars in the West; sugarcane and jamun vinegar in india, palm wine vinegars in Africa and the Far East, coconut water vinegar in the Philippines.

6. Alkaline fermentations. Examples are; Nigerian *dawadawa*, Ivory Coast *soumbara*, African *iru*, *ogiri*, Indian *kniema*, Japanese *natto*, etc.

7. Leavened breads. Examples are Western yeast and sour dough breads; Middle East breads.

8. Flat unleavened breads.

The above classes of fermented foods are found around the world. It may be noted that the lines between the various classifications are not always distinct.

1.3.2 Type of Fermentation

The principal fermentations, their products and types of microorganisms are depicted in Figure 1.2.

1.3.2.1 Lactic Acid Fermentation The lactic acid bacteria present in substrates like milk, ferment the lactose in milk to lactic acid, resulting in an indigenous fermented food called *dahi*. If the water in the whey is allowed to escape or evaporate, the residual *curd* becomes a primitive cheese. Vegetable foods and vegetable/fish/shrimp mixtures are also fermented by lactic acid bacteria, and are preserved around the world by lactic acid fermentation (Steinkraus, 1983a,c, 1996, 1997). Pickled vegetables, cucumbers, radishes, carrots, and very nearly all vegetables and even some green fruits, such as olives, papaya, and mango, are all lactic acid fermented in the presence of salt all around the world. Indian *idli*, a sour, steamed bread, and *dosa*, a pancake, are examples of household fermentations. Polished rice and black gram *dahl* are used in it.

The production of cheese is also a typical lactic fermentation of milk, carried out by using a suitable starter culture of lactic acid bacteria (Cogan and Hill, 1993; Cogan and Accolas, 1996) or preservation of staple carbohydrate food by Fijian method of pit preservation (Davuke)(Aabessberg et al., 1988). For more details, see Chapter 7.

1.3.2.2 Alcoholic Fermentations Alcoholic fermentation is one of the most important and the oldest process (Steinkraus, 1979; Amoa-Awua, 2006), involving production of mainly ethanol and carbon dioxide. It results in the production of various

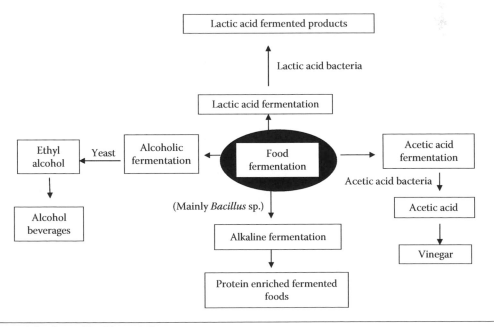

Figure 1.2 Types of food fermentation.

beverages, like wines, beers, and distilled liquors (Fleet, 2006). These are generally yeast fermentations, but can also involve yeast-like molds, such as *Amylomyces rouxii*, and mold-like yeasts such as *Endomycopsis* and sometimes bacteria such as *Zymomonas mobilis*. The substrates include diluted honey, sugar cane juice, palm sap, fruit juices, germinated cereal grains, or hydrolyzed starch, all of which contain fermentable sugars which are rapidly converted to ethanol in natural fermentations by yeasts in the environment. In Asia, there are at least two additional ways of fermenting starchy rice to alcohol: that is, the use of a mold such as *Amylomyces rouxii* that produces amylases, converting starch to sugars, and a yeast such as *Endomycopsis fibuliger*, which converts the glucose/maltose to ethanol. See Chapter 9, for more details.

1.3.2.3 Leavened Bread Fermentation   Breads are also made from yeast by alcoholic fermentation, and ethanol is a minor product in bread due to the short fermentation time. Carbon dioxide produced by the yeasts leavens the bread, producing anaerobic conditions, and baking produces a dry surface resistant to invasion by microorganisms in the environment. Baking also destroys many of the microorganisms in the bread itself. Yeast breads are made by fermentation of wheat and rye flour dough with yeasts, generally *Saccharomyces cerevisiae*. The sour dough breads are fermented both with lactic acid bacteria and yeasts.

1.3.2.4 Acetic Acid/Vinegar Fermentation   It is the fermentation involving the production of acetic acid which yields foods or condiments that are generally considered as safe, as acetic acid is either bacteriostatic or bactericidal, depending upon the concentration employed. When the products of alcoholic fermentation are not kept anaerobic, bacteria belonging to the genus *Acetobacter* present in the environment oxidize portions of the ethanol to acetic acid/vinegar (Conner and Allgeier, 1976; Steinkraus, 1983a, 1996; Steinkraus, 1997; Steinkraus, 2009). Vinegar is a highly acceptable condiment used in pickling and preserving cucumbers and other vegetables.

1.3.2.5 Alkaline Fermentations   Fermented foods involving highly alkaline fermentations are generally considered as safe. Alkaline fermentation is a process in which the pH of the substrate increases to alkaline values as high as 9 (Aniche et al., 1993; Sarkar and Tamang, 1995; Amadi et al., 1999; Omafuvbe et al., 2000; Sarkar and Nout, 2014) due to the enzymatic hydrolysis of proteins from the raw material into peptides, amino acids, and ammonia (Kiers et al., 2000) and/or due to alkali-treatment during production (Wang and Fung, 1996; Parkouda et al., 2009). In several countries, including Asian countries, the traditional diets of the majority of people rely largely on starchy staples such as cereals, cassava, yam, and plantain which are rich in calories but poor in other nutrients (Achi, 2005; Dakwa et al., 2005). The essential microorganisms of alkaline fermentation are *Bacillus subtilis* and related bacilli. Fermented food condiments form a significant part of the diets of many people in developing countries (Parkouda et al., 2010; Tamang, 2010a–d). The addition of salt in the range of 13%

(w/v) or higher to protein-rich substrates results in a controlled protein hydrolysis that prevents putrefaction, prevents development of food poisonings, such as botulism, and yields meaty, savory, amino acid/peptide sauces and pastes that provide very important condiments, particularly for those unable to afford much meat in their diets. Meaty/savory flavored amino acid/peptide sauces and pastes, fish sauces and pastes, made by fermentation of small fish and shrimp, using principally their proteolytic gut enzymes are also included in this type of fermentation (Steinkraus, 1983a,b, 1989; Ebine, 1989; Fukashima, 1979).

1.4 Indigenous Fermented Foods: Their Toxicity and Safety

1.4.1 Indigenous Fermented Foods

Indigenous fermented foods are widely consumed as an important part of the diet of a large population of the world (van Veen 1957; Campbell-Platt, 1994; Ryan, 2003). Foods that were invented centuries ago and even predate written historical records (Hesseltine and Wang, 1980; Tamang, 2009), which can be prepared by household or cottage industry using relatively simple techniques and equipment (Aidoo et al., 2006), are called indigenous fermented foods. These foods and beverages have been defined differently on the basis of various interpretations, as detailed in Table 1.2.

Most of the traditional fermented foods are solid substrate based, where the substrates undergo the process of natural fermentation either naturally or by adding starter cultures. The alcoholic beverages produced generally include beer, wine, and toddy. Indigenous food fermentation depends on the biological activity of microorganisms (Ross et al., 2002) by which the development of fermented foods is achieved (Geisen and Holzapfel, 1996) using locally available ingredient(s) of either plant or animal-origin. These are converted biochemically and organo-leptically into upgraded edible products called fermented foods (Campbell-Platt, 1994; Steinkraus, 1996; Tamang, 2001).

Table 1.2 Some of the Definitions of Indigenous Fermented Foods

- Foods that are fermented till at least one of the constituents has been subjected to the action of microorganism(s) for a period, so that the final products have often undergone considerable changes in chemical composition and other aspects due to microbial and enzymatic changes (van Veen, 1957; Tamang, 2010)
- Traditional fermented foods are those that have been used for centuries, even predating written historical records, and that are essential for the well-being of many people of the world (Hesseltine, 1979; Tamang, 2010)
- These are the foods where microorganisms bring about some biochemical changes in the substrates during fermentation that are nutritional, preservational, sensorial, or detoxificational in nature (Steinkraus, 1996; Tamang, 2010) that may have reduced the cooking time
- These are the foods that have been subjected to the action of microorganisms or enzymes so that the desirable biochemical changes cause a significant modification to the food, and include the direct consumption of fungal fruiting bodies or mushrooms (Campbell-Platt, 1994; Tamang, 2010)
- Fermented foods are palatable and wholesome foods prepared from raw or heated raw materials by microbial fermentation (Holzapfel, 1997; Tamang, 2010)

The processes used for the production of indigenous fermented foods are artisanal in nature. Women using traditional knowledge usually prepared fermented foods (Tamang, 2001). Traditions were however, established by which the handling and storage of certain raw materials in a specific manner resulted in the development of foods with better keeping qualities than the original substrate, with the additional advantage of desirable and pleasing sensory characteristics. The knowledge associated with manufacturing these products used to be transferred from one generation to the next within the local communities, monasteries, and feudal estates (Caplice and Fitzerald, 1999).

Toddy, sometimes referred to as palm wine, is indigenous to those countries where palms are found, including India and Sri Lanka. Different species of palms are used in these countries. The wine is made from fermented palm sap of the inflorescence (Batra and Millner, 1974; Steinkraus, 1983a,b,c; Klanarong and Piyachomkwan, 2013). The other indigenous fermented foods are *idli*, popular in southern India (Reddy et al., 1986), *dhokla* and *khaman*, very popular in northern Indian states; papads or papadams are eaten throughout India (Odunfa, 1985). Damboo based indigenous fermented foods such are Mesu is prepared and consumed (Tamang and Sarkar, 1996) as is the case with many such products of other parts of India (Padmaja and George, 1999; Joshi, 2005; Thapa and Tamang, 2005; Joshi et al., 2012). Some indigenous foods are made from fermented milk, soybean milk, or other legume milk products. Curd or *dahi* is another liquid food popular in the diet of people in the Orient and many other parts of Asia (Rao et al., 1986). Mushrooms are a category by themselves and are consumed as a food. Fermented food condiments form a significant part of the diets of many people in developing countries (Steinkraus, 2002; Parkouda et al., 2009; Parkouda et al., 2010; Tamang, 2010a).

1.4.2 Contamination, Spoilage, and Toxic Microbiological Hazards

The traditional fermentation of cereal products is widely practiced in different countries, including those of South Asia, and usually involves a spontaneous development of different lactic acid bacteria (Muller, 1998). Yeast is also involved in the spoilage of foods (Stratford, 2006). The final bacteriological status of the product, however, is influenced partly by the raw materials and partly by the processing method (Steinkraus, 1983a,b,c; Svanberg et al., 1992), and the fermentation process is considered to be an effective method of preserving these foods (Smith and Palumbo, 1983). However, the extent to which other less acute microbiological problems can occur is difficult to assess, due to the problems of establishing a cause and effect relationship with some types of food-related illnesses, and the general lack of authentic epidemiological surveillance data.

It has long been realized that certain molds and their toxic metabolic products can pose a threat to the fermented food products. Out of various toxins produced, mycotoxins are of considerable significance, as is the case with patulin (Joshi et al., 2013).

These are a structurally diverse group of mainly low molecular weight compounds produced mostly by fungi, as secondary metabolites (Smith and Moss, 1985). The toxins can cause poisoning in humans and animals when low concentrations are consumed or inhaled (Smith et al., 1995; Steyn, 1995; Steyn et al., 2009; Clavel and Brabet, 2013). The most regularly documented mycotoxins produced by the five genera, are Aspergillus toxins: aflatoxins B1, G1 and M1, ochratoxin A, sterigmatocystin, and cyclopiazonic acid; Penicillium toxins: cyclopiazonic acid, citrinin, and patulin; Fusarium toxins: deoxynivalenol, nivalenol, zearalenone, T-2 toxin, diacetoxyscirpenol, moniliform, and fumonisins; Alternaria toxins: tenuazonic acid, alternariol, and alternariol methyl ether; Claviceps toxins: ergot alkaloids. In nature, such toxins are primarily derived from agricultural crops like cereals, oil seeds, and products derived from them, and from animal derived foods, such as milk (Smith et al., 1995).

Mycotoxins can enter the human dietary system by indirect or direct contamination. An indirect source of contamination could be an ingredient of a food or beverage (e.g., cereals or legumes) that has previously been contaminated with toxin-producing molds (Smith et al., 1995). Although the mold may be killed or removed during processing, mycotoxins will still persist in the final product (Miller and Trenholm, 1994; Smith et al., 1995). Further, direct contamination can also occur in two ways by the fermentation process, which may involve a fungus essential for the fermentation that is also capable of producing myctotoxins; second if the process or final product becomes infected with a toxigenic mold, with consequent toxin production. Thus, almost all fermented foods and beverages have the potential to be infected by toxigenic molds at some stage during their production, processing, transport, or storage.

The indigenous fermented foods are also subjected to normal spoilage and contamination problems like any other food product (Joshi et al., 2000). In areas where relative humidity is more than 80%, spoilage of fermented products by unwanted molds is often encountered. In such conditions, proper storage, such as by refrigeration, is essential to prevent spoilage. While the fermented foods which are the mainstay of many developing countries have long been perceived as safe for consumption, as fermentation can reduce the level of toxin (Westby et al., 1997), the presence of many prevalent food-borne bacterial pathogens during the production of such foods has also been described (Farnworth, 2003). Some people, especially children, may suffer bouts of food-related infections or poisonings, depending upon a number of inter-related factors discussed earlier (Farnworth, 2003). The fermented products need to be prepared under good sanitary and hygienic conditions. (Motarjemi et al., 1993) and efforts be must made to avoid these foods being a source of contamination.

1.4.3 Safety of Indigenous Foods

Fermented foods, generally, have a very good safety record even in the developing world, where they are manufactured by people with no formal training in microbiology or chemistry, in unhygienic and contaminated environments. They are consumed by hundreds of millions of people every day, both in the developed and the developing worlds, and there has not been any documentation of serious out-breaks of food poisoning or intoxication. However, the fact that the fermented foods are themselves generally safe, does not solve the problems of contaminated drinking water, environments heavily contaminated with human waste, improper personal hygiene in food handlers, flies carrying disease-causing microorganisms, unfermented foods carrying food poisoning bacteria or other human pathogens, and unfermented foods, even when cooked, if handled or stored improperly (Smith and Fratamico, 1995; Steinkraus, 2002). At the same time, improperly fermented foods can also be unsafe (Steinkraus, 2002).

The safety of food fermentation processes is related to several well-established principles. The first is that the food substrates overgrown with desirable, edible microorganisms become resistant to invasion by spoilage, toxic, or food poisoning microorganisms, so that these other less desirable—possibly disease-producing—organisms find it difficult to compete. From the toxicological angle, however, a valid concern is mycotoxins that are present in many cereal grain and legume substrates before fermentation. They are produced when the cereal grains or legumes are improperly harvested or stored (Steinkraus, 2009). Soaking and cooking of the raw substrates before fermentation, leads to destruction of many potential toxins, such as trypsin inhibitor (TI), phytate, and hemagglutinin, and thus, in general, fermentation tends to detoxify the substrates (see Chapter 3 for more details).

A second principle is that in fermentation several antimicrobial substances are produced, such as in those fermentations that involve production of lactic acid, which are generally safe. The Lactic Acid Bacteria (LAB) fermentation is responsible for processing and preserving vast quantities of human food and ensuring its safety. The excellent safety record of sour milks/yogurts, cheeses, pickles, etc., is well known. Similarly, acetic acid produced during acetic acid fermentation is also an antimicrobial compounds, so in vinegar fermentation it does not allow other microorganisms to grow (Joshi et al., 2000). In case of alcoholic fermentation, for example, the production of ethanol and carbon dioxide, which are antimicrobial substances, inhibits the growth of spoilage causing bacteria. Production of bacteriocins is another example of antimicrobial compounds ensure safety of fermented foods (Abee et al., 1995). Thus, these fermentations lead to the safety of the fermented foods (Joshi et al., 2000).

Socio-economic constraints such as inadequate supplies of safe water, lack of facilities for safe preparation and storage of food, and time constraints for the proper preparation of food prior to each meal, can, however, interfere with the application of the principles discussed earlier. The need for the prevention of contamination of raw materials, proper storage, and fermentation use of proper/pure culture (Tamang and Nikkuni, 1998) fermentation conditions (Joshi and Pandey, 1999; Joshi et al., 2011), discarding any fermented food with atypical characteristics, is, therefore, stressed. These aspects have also been elaborated in Chapters 5 and 13.

1.5 Role of Food Fermentation and Indigenous Fermented Foods

Food fermentation offers several advantages with respect to the improvement in the quality and nutritional value, it makes the foods more digestible, it improves the sensory attributes, it synthesizes vitamins such as B_{12}, it destroys or masks undesirable flavors, and have ability to replenish intestinal microflora. It can improve protein quality and the bioavailability of micronutrients, and can reduce toxic and antinutritional factors (cyanogenic glycosides), (Sanni, 1993; Padmaja, 1995; Iwuoha and Eke, 1996; Tamang, 1998, 2010, 2011; Tamang and Holzapfel, 1999; Sindhu and Kheterpaul, 2001; Jespersen, 2003). Some of the effects of food fermentation on indigenous fermented foods are illustrated in Figure 1.3 and have also been discussed here.

1.5.1 Nutritional Quality

Bioenrichment of food substrates by traditional fermentation enhances the nutritive value of fermented food (Lorri, 1995; Joshi et al., 1999; Joshi, 2006a). Total and

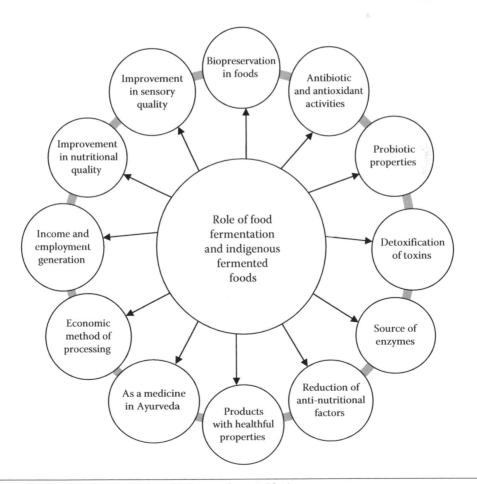

Figure 1.3   Role of food fermentations and indigenous fermented foods.

free amino acids, along with mineral content, increase in fermented soybean foods (Tamang and Nikkuri, 1998, Tamang, 1999, 2010). The microorganisms involved in fermentation synthesize vitamins, proteins, and amino acids. LAB fermentation has been shown to improve both the nutritional value and digestibility of such fermented cereal-based foods (Obiri-Danso et al., 1997; Nout, 2009; Chelule et al., 2010), and the bioavailability of minerals, protein, and simple sugars (Sripriya et al., 1997; Sybesma et al., 2003; Santos et al., 2008; Chelule et al., 2010). Toxicological problems, in addition to spoilage and contamination, are important considerations (Chaval and Brabet, 2013).

Two major food problems exist in the world, namely, starvation in under developed or developing countries and obesity in developed countries. In addition, there are a number of nutritional deficiency diseases in the developing world today, such as Kwashiorkor, xerophthalmia, childhood blindness, *beri-beri pellagra*, rickets, and anemia (Jelliffe, 1968). Fermented foods provide a solution to these problems.

Food fermentations that raise the protein content or improve the balance of essential amino acids and vitamins, or their availability, will have a direct curative effect when the fermented foods are consumed (Steinkraus, 2009). During fermentation, the microorganisms also selectively enrich the substrate, such as rice with lysine, an essential amino acid in rice (Cronk et al., 1977; Steinkraus, 2002), and also improve the protein quality. An increase of 10.6%–60.0% in methionine content during Indian *idli* fermentation is reported (Rajalakshmi and Vanaja, 1967; Steinkraus et al., 1967; Rao, 1986; Steinkraus, 1997). Palm wine has increased thiamine, riboflavin vitamin B-12, and pyridoxine content during fermentation (Van Pee and Swings, 1971; Okafor, 1975; Steinkraus, 1979b).

1.5.2 Sensory Quality Enhancement

Fermentation helps in the development of diversity of flavors, aromas and textures in food substrates and, therefore, adds variety to the diet (Reddy et al., 1982) and makes the food more palatable and, ultimately, more popular than the unfermented food (Caplice and Fitzerald, 1999; Blandino et al., 2003; Chelule et al., 2010). The improvement takes place in diverse ways, resulting in new sensory properties in the fermented product (Leroy and De Vuyst, 2004; Chelule et al., 2010). Alkaline fermentation is known to play an important role in making otherwise inedible foods edible, enhancing flavor and nutritional values and bringing diversity to the kinds of foods and beverages available (Steinkraus, 1996; Wang and Fung, 1996; Parkouda et al., 2009; Chukeatirote et al., 2010).

1.5.3 Biopreservation of Foods/Preservative Properties/Shelf Life

Fermentation is a potent tool to preserve food by producing metabolites inhibitory to pathogenic organisms through the activities of microorganisms involved in the fermentation or their enzymes (Joshi and Pandey, 1999). The preservative activity of LAB has been documented in some fermented products such as cereals, where lowering the pH to below four through acid production inhibits the growth of pathogenic microorganisms responsible for food spoilage, food poisoning, and disease. These bacteria (LAB) also have antifungal activity, thus the shelf-life of fermented food is prolonged (Chelule et al., 2010). The biopreservation aspect of fermentation can be illustrated by the process of common indigenous nonsalted fermented vegetable products (*gundruk* and *sinki*), where species of *Lactobacillus* and *Pediococcus* produce lactic acid and acetic acid which reduce the pH of the substrates, making the products more acidic in nature, that inhibits the growth of pathogenic microorganisms, thus making the foods safe for consumption (Tamang et al., 1988; Tamang, 2010). Due to low pH (3.3–3.8) and high acid content (1.0%–1.3%), these foods (*gundruk* and *sinki*) after sun drying, can be preserved without refrigeration for more than two years without addition of any synthetic preservative (Tamang, 2010). The preservative effect is also due to several antimicrobial compounds synthesized during fermentation. For example, in lactic acid fermentation, the inhibitory compounds produced by lactic acid bacteria against other bacteria include hydrogen peroxide, reutrin, diacetyl, reutericyclin, and bacteriocins. The products of this fermentation also possess antitumor effects (Leroy and De Vuyst, 2004; Chelule et al., 2010). These bacteria also produce fungal inhibitory metabolites including organic acids (propionic, acetic, and lactic acids) that are also a hurdle for nonacid-tolerant bacteria (Mensah, 1997; Chelule et al., 2010). Carbon dioxide, formed from ethanolic or heterolactic fermentation, can directly create an anaerobic environment which is toxic to some aerobic food microorganisms through its action on cell membranes and its ability to reduce both internal and external pH (Eklund, 1984; De Vuyst and Vandamme, 1994). Diacetyl is a product of citrate metabolism (Lindgren and Dobrogosz, 1990; Cogan and Hill, 1993) and several strains of *Leuconostoc, Lactococcus, Pediococcus,* and *Lactobacillus* may produce diacetyl (Cogan, 1986; Jay, 1986, 1996, 2004). Gram-negative bacteria, yeasts, and molds are more sensitive to diacetyl than Gram-positive bacteria (Jay, 1986; Motlagh et al., 1991; De Vuyst and Vandamme, 1994). Reuterin is produced during the stationary phase by the anaerobic growth of *Lactobacillus reuteri* on a mixture of glucose and glycerol or glyceraldehyde. It has a general antimicrobial effect on viruses, fungi, and protozoa, as well as bacteria (Chung et al., 1989; Caplice and Fitzerald, 1999).

LAB are also known to produce proteinaceous antimicrobial agents such as bacteriocins, the peptides that elicit antimicrobial activity against food spoilage organisms and food (Carolissen-Mackay et al., 1997; Abees et al., 1999; Joshi et al., 2006; Chelule et al., 2010). Due to antimicrobial activity, the potential use of lactic acid bacteria, namely, *Lactobacillus delbrueckii* sp. *bulgaricus* CFR 2028 and *Lb. delbrueckii* sp. *lactis* CFR 2023 and a nonantagonistic strain of *Lactococcus lactis* sp. *lactis* CFR 2039 has been attempted in indigenous fermented food, namely, *Kadhi, Dhokla, Punjabi Warri* and have been found to be antagonist to spoilage causing microorganism, namely,

Bacillus laterosporus CFR 1617, *Bacillus licheniformis* CFR 1621, and *Bacillus subtilius* CFR 1604 (Vardaraj et al., 1997). The antimicrobial properties of alcoholic beverages like wine against some pathogenic microorganisms and microbes of public health relevance have also been observed (Joshi and John, 2002). Such antimicrobial properties of fermented foods are useful in traditional food fermentations, making such foods safe to eat (Dewan, 2002; Thapa, 2002; Tamang, 2010). Many of the fermented foods have good keeping qualities, and can be kept without refrigeration for more time than the fresh unfermented foods (Joshi and Pandey, 1999; Jespersen, 2003; Joshi, 2006a). Fermentation increases the shelf-life and decreases the need for refrigeration or other method of food preservation (Cook et al., 1987; Aidoo, 2011) or improve the quality such as in amahewu (Carries and Gqaleni, 2010).

1.5.4 Healthful Effect: Probiotic and Prebiotic Properties

LAB fermentation prevents diarrheal diseases as they modify the composition of intestinal microorganisms, constipation, and abdominal cramps, acting as deterrents for pathogenic enteric bacteria. Such microorganisms are called "Probiotic" (Dewan et al., 2003; Chelule et al., 2010). Mostly species of LAB used are those having "Generally Recognized As Safe" (GRAS) status, thus eliminating any health risk (Nout, 2001; Hansen, 2002). Both the microbes and fermented products have various functionalities in maintaining human health; thus, they have the potential to act as physiologically functional foods (Parigi and Prakash 2012; Joshi et al., 2009). These bacteria are normal residents of the complex ecosystem of the gastrointestinal tract (GIT) (Thapa, 2002; Tamang, 2003). The high degree of hydrophobicity of the isolates found in the indigenous fermented foods of the Himalayas indicates their potential probiotic character (Adams, 1999; Nout, 2001; Hansen, 2002). The probiotic effects and the reduced level of pathogenic bacteria documented in fermented foods and beverages are of special significance when it comes to the developing countries, where fermented foods reduce the severity, duration, and morbidity from diarrhea (Mensah et al., 1990; Kimmons et al., 1999; Jespersen, 2003). Oligosaccharides with prebiotic activity, anticonstipation, peptides with antihypertensive properties, probiotic with improvements of lipid metabolisms, antiobesity, and cholesterol-lowering activity are found in various Asian indigenous fermented foods.

1.5.5 Detoxification of Toxins and Antimicrobial Components

Some fruits and vegetables contain naturally occurring toxins and antinutritional compounds which include unavailable carbohydrates, phytates, trypsin inhibitors, hemagglutinins, goitrogenic factors, cyanogenic glucosides, and saponins present in different legumes and cereals, that may reduce the bioavailability of minerals and inhibit the digestibility of proteins. Fermentation is known to play an important role in detoxifying these compounds (Olukoya et al., 1994; Hill, 2002; Nout, 2009). The raw substrates undergo soaking and hydration during various fermentation processes, and, for example, in the usual cooking many potential toxins (trypsin inhibitor (TI), phytate and hemagglutinin, and cyanogens) in cassava are reduced or destroyed. Foods and feeds are often contaminated with a number of toxins, either naturally or through contamination with microorganisms. (Sweeney and Dobson, 1998; Chelule et al., 2010). Several methods have been reported for degrading toxins from contaminated foods (Gourama and Bullerman, 1995; Mokoena et al., 2005, 2006; Schnürer and Magnusson, 2005; Chelule et al., 2010; Zannini et al., 2012). LAB fermentation has successfully detoxifed cassava toxins (cyanogens) by fermentation (Caplice and Fitzgerald, 1999; Chelule et al., 2010). More of such examples are described in Chapter 13 of this text.

1.5.6 *Source of Enzymes and Antioxidants*

Different enzymes like amylase, glucoamylase, protease, lipase, etc. (Tamang and Nikkuri, 1996; Thapa, 2001; Tamang, 2010), are produced in traditional fermented foods that are beneficial for health, and so the fermented products serve as a source of enzymes. Some of the fermented foods, like wine, have been reported to possess antioxidant properties (Abbas, 2006; Joshi et al., 2009) and, thus, are useful; alkaline fermented foods like Kinema have also been associated with antioxidants property (Sarkar et al., 1998; Sarkar and Nout, 2014).

1.5.7 *Indigenous Fermented Food as Ayurveda Medicine*

Fermentation is practiced all over the world, especially in South Asia, for producing foods with medicinal and therapeutic values for a long time, and India is not an exception (Muralidhar, et al., 2003). *Ayurveda* medicine—an ancient Indian healing system that links, body, temperature, and food—is probably the best known humoral system in Asia (Penny, 2008). It had the vision to exploit the fermentation technique for preparing various products, including alcoholic beverages (Seker, 2007), with a wide variety of therapeutically useful properties. Based on ancient wisdom, a vast database has been created in *Ayurveda* on the useful herbs and the modification in the respective technique of fermentation. One such fermented liquid—*Sirisarista* (Bhaishajya Ratnavali 72/72–74)—is therapeutically useful in treating the disease *Swasa* (bronchial asthma) II C.K.12/48 II. *Smsansia* is one of the drugs widely used in treating the disease *Swasa* in the Southern states of India.

1.5.8 *Fermentation: An Economic Method of Processing*

Fermentation provides an economic means of preserving food and inhibiting the growth of pathogenic bacteria, even under conditions where refrigeration or other means of safe storage are not available. Lactic-acid fermented foods generally require little, if any, heat in their fermentation, and can be consumed without cooking, for example, pickles (Steinkraus, 1983a). The fermentation process is less energy-consuming and requires less costly equipment and, thus, is a cheap and efficient means of preserving perishable raw materials.

1.5.9 Source of Income and Employment Generation

The production of fermented foods provides income and employment to millions of people. In many villages/regions, only women are involved in the production of indigenous fermented foods. In any case, production of such foods is a very important activity and a source of income.

1.6 South Asia: Countries, Their Indigenous Fermented Foods and Origin, Indigenous Technology Knowledge and Cultural Diffusion

1.6.1 South Asian Countries and their Indigenous Fermented Foods

South Asia (Figure 1.4) includes India, Bhutan, Afghanistan, Pakistan, Sri Lanka, the Maldives, Bangladesh, and Nepal (Phadnis and Ganguly, 2001). It is pertinent to say that the size of South Asia is almost the same as that of Europe, but its population is almost twice that of Europe. The countries of South Asia have many religions, ethnic

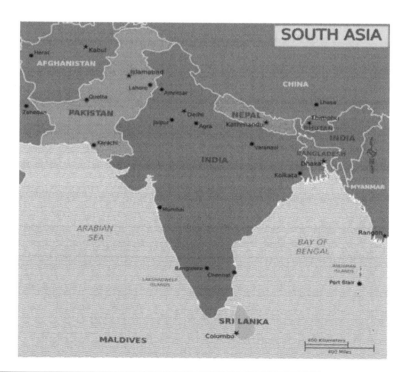

Figure 1.4 Map of South Asia depicting various countries of this region. (http://wikitravel.org/shared/File:Map_of_South_Asia.png.)

groups, cultures, and languages. A wide range of fermented foods and beverages contribute significantly to the diets of many people (Aidoo et al., 2006) including those of South Asia. These foods are generally produced from plant- or animal-based raw materials using microorganisms, which are either present in the natural environment, or added manually, to obtain the desirable end product (Law et al., 2011). Fermented foods are typically unique and vary according to the region, climate, social and cultural pattern, consumption practices, demographic profile, and, most importantly, the availability of raw materials (Nout and Motarjemi, 1997; Law et al., 2011; Gupta and Abu-Ghannam, 2012). Every society and group have their own concepts of food and their own history of food habits, which shape their food culture. A country-wise description of various geographical characteristics and cultures, especially the food culture, is provided in the next section.

1.6.1.1 India The name *India* is derived from *Indus*, which originates from the Old Persian word *Hinduš*. Officially known as the Republic of India, it is the seventh largest country in South Asia by area and the second-most populous democracy in the world. It is bounded by the Indian Ocean on the South, the Arabian Sea on the South-West, and the Bay of Bengal on the South-East. It shares land borders with several countries, namely, Pakistan to the West, China, Nepal, and Bhutan to the North-East, and Burma and Bangladesh to the East. Its coastline measures 7517 km (4700 mi) in length; including the Andaman, Nicobar, and Lakshadweep Island chains (Kumar, 2000). The climate is strongly influenced by the Himalayas and the Thar Desert, which results in summer and winter monsoons (Chang, 1967). Increased agricultural productivity has been brought about by the "Green Revolution" and, consequently, she now has surplus food grain after feeding its population.

India is the ancient Indus Valley Civilization, a region of historic trade routes, vast empires, and commercial and cultural wealth for much of its long history (Stein, 1998). She is a pluralistic, multilingual, and a multi-ethnic society. Indo-Aryan (spoken by about 74% of the population) and Dravidian (24%) are the two major languages, but the largest number of people speak Hindi, which is also the official language (Mallikarjun, 2004; Ottenheimer, 2008), while English is used extensively in technical matters, administration, and business (Government of India, 1960). Indian cultural history is more than 4500 years old (Kuiper, 2010).

Much Indian architecture, including the Taj Mahal, is listed in UNESCO World Heritage, while other works of Mughal and South Indian architecture blend ancient local traditions with imported styles (Kuiper, 2010).

India is the home of several indigenous fermented foods such as *dhokla*, *dosa*, *dahi*, *lassi*, *vada*, *khaman*, *papad*, *kinema*, *idli* and *Jalebis* (Figures 1.5 and 1.6), and of several alcoholic beverages consumed by the tribal people, such as *opo*, *beer*, *sur*, *chhang*, *angoori*, etc. Indians are given credit for inventing the methods of souring and leavening cereal-legume batters (Padmaja and George, 1999; Sarkar et al., 1994; Nema et al., 2003; Pant and Nema, 2003; Rao et al., 2005) and the traditional foods

Figure 1.5 A typical Indian *bajar* with *jalebis* being produced.

represent the heritage of India's multidimensional culture. The production of most of these foods, however, remains confined to households, and only few are prepared on larger scale for sale in restaurants. Nevertheless, traditional foods have evolved into a source of income for women, who cook these foods at home or on small wooden carts. Although the consumption of meat and dairy products as primary sources of protein and calories is common in India, the bulk of the population avoids eating meat and meat products, either due to its high cost or for religious reasons (http://en.wikipedia.org/wiki/India). The frontline staple foods of the Indian population are cereals and legumes, which are consumed in diverse forms, and an important form of consumption is as fermented foods. Several legume-based fermented foods are being prepared by employing traditional methods using natural microflora from the staples and the surroundings. The people of the Indian trans-Himalayan region, such as Ladakh and Lahaul-Spiti, make local alcoholic beverages from barley, called *Chang*, while those in the Kinnaur district of Himachal Pradesh make *Ghanti*, which is made from fermented grapes (Joshi et al., 2003). In Sikkim and Darjeeling, finger millet is the main substrate for their local beverages, called *Kodo ko Jaanr* (Rizvi, 1983; Bajpai, 1987;

Figure 1.6 Indigenous fermented foods of India. (a) *Dhokla*. (b) *Idli*. (c) *Dosa*.

Tamang et al., 1996; Roy et al., 2004). Traditional foods have an important bearing on the dietary habits of the people of Sikkim (Tamang, 2005; Tamang et al., 2012), which is also reflected in the pattern of food production in a mixed farming system (Tamang, 2001). *Bhat-dal-tharkari-achar* (rice-legume soup-curry-pickle) is the basic diet of the Sikkimese meal. The early morning starts with a full mug of tea with sugar or salt with or without milk, with a pinch of hot black pepper. *Odisha* is a small state in the eastern part of India with wide cultural, social, and ethnic, as well as plant and food diversity. In the state, 62 tribal communities, such as *Khanda, Khola, Santala, Juanga, Bhuiyan, Saora, Dharua, Bonda* and others, are found and contribute about 22% of the population of the state. Traditional foods include *bhapa pitha, chakuli, chhuchipatra pitha, chitou, enduri pitha, munha pitha and podo pitha* (Roy et al., 2007). In the Southern state of Tamil Nadu, Madras brings to the minds of people the traditional meals of *Idli* and *Dosa* (Figure 1.6), which have a lot of chillies. The people of *Karnataka* in the Deccan plateau enjoy a wide variety of rice-based preparations, such as *bisebele bath, vanghi bath,* curd rice, lemon rice, etc. West Bengal is in the East of India, and has traditional fermented foods like *toddy.* Other homemade alcoholic brews are prepared and consumed, but remain confined to industrial laborers and the tribal population. About 70% of the inhabitants of the Darjeeling hill district of West Bengal traditionally consume large quantities of fermented foods and beverages (Tamang and Sarkar, 1988; Tamang et al., 1988). In the tribal area of Arunachal Pradesh and Manipur, several traditional fermented foods are prepared and consumed (Singh et al., 2007; Devi and Suresh, 2012). Some ethnic groups are economically dependent upon these products. Although there are some reports of fermented foods from Darjeeling (Batra and Millner, 1976; Hesseltine, 1979; Tamang et al., 1988; Sekar and Mariappan, 2007), such information from other parts of West Bengal, by and large, has not been well documented.

In the north eastern region of India, fish (*Shidal*) is the favorite food item (Figure 1.7). Being highly perishable, this is cured as a means of preservation (Muzaddadi and Mahanta, 2013). The methods of preservation also provide a cultural identity amidst the ethnic groups. Partially cured and fermented fishes permit the constant availability of fish throughout the year. The diversity of indigenous fermented foods in India is detailed in Chapter 2 of this text.

1.6.1.2 Pakistan Pakistan is a sovereign Islamic Federal Parliamentary Republic with four provinces and four federal territories in South Asia. The name *Pakistan* literally means "Land of the Pure" in Urdu and Persian. It is the sixth populous country in the world, with an area covering 796,095 km^2 (307,374 sq mi). The territory of modern Pakistan was a home to several ancient cultures, including the Neolithic Mehrgarh and the Bronze Age Indus Valley civilization. Consequent to the movement led by Muhammad Ali Jinnah and India's struggle for independence, Pakistan became an independent state in 1947. It is an ethnically and linguistically diverse country, with a similar variation in its geography and wildlife. The climate varies from tropical to

Figure 1.7 Shidal producing states of North East India. (From Muzaddadi, A.U. and Mahanta, P. 2013. *African Journal of Microbiology Research* 7(13):1086–1097. With permission.)

temperate, with arid conditions in the coastal south. There is a monsoon season, with frequent flooding due to heavy rainfall, and a dry season, with significantly less rainfall (http://skaphandrus.com/en/scuba-dive-locations/info/country/Pakistan; http://www.headline-news.org/Pakistan).

The site of Alexander's battle on the Jhelum River is in the Punjab province, and the historic city Lahore, has many examples of Mughal architecture, such as Badshahi Masjid, the Shalimar Gardens, the Tomb of Jahangir, and the Lahore Fort. English is the official language of Pakistan. Pakistani cuisine (Figure 1.8) has some similarities with different regions of the Indian sub-continent, originating from the royal kitchens of sixteenth century Mughal emperors. It has a greater variety of meat dishes than the rest of the sub-continent, using large quantities of spices, herbs, and seasoning in the cooking. Garlic, ginger, turmeric, red chilli, and *garam masala* are used in most of the dishes, and home cooking regularly includes curry. *Chapati* is a staple food, served with curry, meat, vegetables, and lentils. Rice is also common, and is served plain or fried with spices, and is also used in sweet dishes (Kathleen, 2001). *Lassi* is a traditional drink in the Punjab region of Pakistan, while fermented cereals products are made and consumed (Parveen and Hafiz, 2003). Pakistan is fourth largest producer of mangoes in the world.

Figure 1.8 Sugarcane juice, Pakistan's national drink. (https://creativecommons.org/licenses/by-sa/2.0/deed.en.)

1.6.1.3 Bangladesh Bangladesh another country of South Asia, located on the fertile Bengal delta. It is bordered by the Republic of India to its North, West, and East, by the Union of Myanmar (Burma) to its South-East, and by the Bay of Bengal to its South. The name is believed to be derived from *Bang/Vanga*, the Dravidian-speaking tribe that settled in the area around the year 1000 BC (Eaton, 1996; Bharadwaj, 2003). The borders of modern Bangladesh however, took shape during the war of Pakistan with India in 1971. Bangladesh is a parliamentary republic with a territory of 56,977 sq mi and is the world's eighth most populous country.

A warm and humid monsoon season lasts from June to October and supplies most of the country's rainfall. It is the world's fifth largest producer of fish, fourth of rice, eleventh of potatoes, ninth of mangoes, sixteenth of pineapples, fifth of tropical fruits, sixteenth of onions, seventeenth of bananas, second of jute, and eleventh of tea. Jute was once the economic engine of the economy (FAOSTAT, 2008). Bangaldesh has the longest natural unbroken sea beach, and five World Heritage sites, such as the famous eighty-one domed Shat Gombuj Mosque in Bagerhat, the world's largest mangrove forest—the Sundarbans, and is renowned for the world-famous Royal Bengal Tiger.

More than 98% of Bangladeshis speak Bengali as their mother tongue, which is also the official language (Central Intelligence Agency, 2012). Bangladesh has a low literacy rate. The culinary tradition of Bangladesh has close relations to the surrounding Bengali and North-East Indian cuisine, as well as having its own unique traits. Rice and fish are the traditional favorites, with vegetables and lentils served with rice as a staple diet. *Biryani* (Figure 1.9) is a favorite dish of Bangladesh, including egg *biryani*, mutton *biryani*, and beef *biryani* (http://en.wikipedia.org/wiki/Bangladesh). Bangladeshi cuisine is known for its subtle flavors, and its huge spread of confectionaries, desserts, and a variety of fermented foods (Hafiz and Majid, 1996; Hassan, 2003). Bangladeshis make distinctive *sweetmeats* from milk products, such as *Rôshogolla, Rasmalai, Rôshomalai, chômchôm, kalojam Chhana, Sandesh*, and *Bakerkhani* (Cadi, 1997).

1.6.1.4 Nepal Nepal is officially called the Federal Democratic Republic of Nepal. It is a landlocked sovereign state located in South Asia, with an area of 147,181 square

Figure 1.9 Bangladeshi biryani.

kilometers (56,827 sq mi) and a population of approximately 27 million (and nearly 2 million absentee workers living abroad). It is located in the Himalayas and bordered in the North by China and in the South, East, and West by India. It is separated from Bangladesh by the narrow Indian Siliguri corridor. Kathmandu is the nation's capital and the largest metropolis. The north of Nepal has eight of the world's ten tallest mountains, including the highest point on Earth, Mount Everest, called *Sagarmatha* in Nepali. Lumbini, the birthplace of Lord Gautam Buddha, is also in Nepal. It is of roughly trapezoidal in shape, 800 km (497 mi) long and 200 km (124 mi) wide, with an area of 147,181 km² (56,827 sq mi). It is commonly divided into three physiographic areas: mountain, hill, and Terai, and has five climatic zones. Nepal experiences five seasons: summer, monsoon, autumn, winter and spring (http://www.spacenepal.com/Nepal.php).

Agricultural produce—mostly grown in the Terai region bordering India—includes tea, rice, corn, wheat, sugarcane, root crops, milk, and water buffalo meat. Industry mainly involves the processing of agricultural produce, including jute, sugarcane, tobacco, and grain. Nepal's diverse linguistic heritage evolved from four major language groups: Indo-Aryan, Tibeto-Burman, Mongolian, and various indigenous languages isolates (http://wikien3.appspot.com/wiki/Nepal). Nepal is the home of the famous Lord Shiva temple, the Pashupatinath Temple. The *Bhutias* and the *Lepchas* are non-vegetarians and prefer beef and pork, though some Nepalis are vegetarians. Non-vegetarians eat chicken, mutton, lamb and pork. Beef is taboo to a majority of Nepalis except Tamangs, Sherpas, and Yolmos. *Newars* prefer to eat buffalo meat (Tamang, 2005).

A typical Nepalese meal is *Dal bhat* (Figure 1.10) which is a spicy lentil soup, served over *bhat* (boiled rice), with *tarkari* (curried vegetables) together with *achar* (pickles) or *chutni* (a spicy condiment made from fresh ingredients). A variety of fermented foods are also made and consumed (Thapa et al., 2003). Mustard oil is the cooking medium, and a host of spices, such as cumin, coriander, black peppers, sesame seeds, turmeric, garlic, ginger, *methi* (fenugreek), bay leaves, cloves, cinnamon, pepper, chillies, mustard seeds, etc., are used in the cooking. The cuisine served on festivals is generally

Figure 1.10 Traditional Nepalese cuisine-(*Dal bhat*).

the best. *Momo* is a type of steamed bun, with or without filling, which has become a traditional delicacy in Nepal (http://associate.com.np/Nepal.php) (Figure 1.11).

1.6.1.5 Bhutan Bhutan is officially called the Kingdom of Bhutan and is also a land-locked country in South Asia, located at the eastern end of the Himalayas. It is bordered in the North by China and in the South, East, and West by India. To the West, it is separated from Nepal by the Indian state of Sikkim, while in the South it is separated from Bangladesh by the Indian states of Assam and West Bengal. Its total area is approximately 38,394 square kilometers (14,824 sq mi). Thimphu is Bhutan's capital and the largest city. In 2008, Bhutan made a transition from absolute monarchy to constitutional monarchy. The national language is Bhutanese (Dzongkha) and the currency is the ngultrum. Agriculture, forestry, livestock, and hydro-electrical power are the mainstay of its economy (Choudhury, 2007). Agricultural produce includes rice, chillies, dairy products (some yak, mostly cow), buckwheat, barley, root crops, apples, and citrus and maize at lower elevations. Industries include cement, wood products, processed fruits, alcoholic beverages (Thapa et al., 2003) and calcium carbide. Rice, buckwheat, and maize are the staples of Bhutanese cuisine. The local diet also includes

Figure 1.11 *Momo*—A famous indigenous fermented food of Nepal.

pork, beef, yak meat, chicken, and mutton. Soups and stews of meat and dried vegetables spiced with chillies and cheese are prepared and consumed. *Emadatshi*, made very spicy with cheese and chillies, can be called the national dish for its ubiquity. Dairy foods, especially butter and cheese from yaks and cows, are also popular, and indeed almost all the milk is turned into butter and cheese. Popular beverages include butter tea, tea, locally brewed *ara* (rice wine) and beer (http://en.wikipedia.org/wiki/ Bhutan; http://saarcradiology.org/Bhutan).

1.6.1.6 The Maldives   The Maldives islands are officially called the Republic of the Maldives. It has a double chain of twenty-six atolls in the Indian Ocean, oriented North–South, that lie between Minicoy Island (the southern most part of Lakshadweep, India) and the Chagos Archipelago. The chains are in the Laccadive Sea, about 700 km (430 mi) South-West of Sri Lanka and 400 km (250 mi) South-West of India. The name *Maldives* may have been derived from Sanskrit *mālā* (garland) and *dvīpa* (island) (Hogendorn and Johnson, 1986), *Maala Divaina* (Necklace Islands) in Sinhala (Deraniyagala et al., 1978). The Maldivian people were called *Dhivehin* and the territory spreads over roughly 90,000 square kilometers (35,000 sq mi), making the country one of the world's most geographically dispersed states. With an average ground level elevation of 1.5 m (4 ft 11 in) above sea level, it is the planet's lowest country (Henley, 2008). Buddhism came to the Maldives at the time of Emperor Ashoka's expansion, and became the dominant religion of the people of the country until the twelfth century AD.

The essential product of the Maldives is *coir*, the fiber of the dried coconut husk. The weather in Maldives is affected by the large landmass of South Asia to the north which causes differential heating of land and water that sets off a rush of moisture-rich air from the Indian Ocean over South Asia, resulting in the southwest monsoon. For many centuries, the Maldivian economy was entirely dependent on fishing and other marine products, which remains the main occupation of the people, and is given priority. A fish canning plant was installed on Felivaru in 1977, as a joint venture with a Japanese firm, while a Fisheries Advisory Board was set up in 1979 for the development of the fisheries. Maldivian culture is heavily influenced by its geographical proximity to Sri Lanka and Southern India.

1.6.1.7 Sri Lanka   Sri Lanka is officially called the Democratic Socialist Republic of Sri Lanka which is an island country in the Northern Indian Ocean off the Southern coast of the Indian sub-continent in South Asia. Known until 1972 as Ceylon, Sri Lanka has maritime borders with India to the Northwest and the Maldives to the southwest. It is also known as "the Pearl of the Indian Ocean" due to its natural beauty, shape and location, and "the nation of smiling people." Known in India as Lanka or Sinhala, *Ceilão* was the name given to Sri Lanka by the Portuguese when they arrived in 1505, which, when translated into English, was *Ceylon* (Deraniyagal, 1992). It achieved independence as the Dominion of Ceylon in 1948. It is the land of the Sinhalese, Sri Lankan Tamils, Moors, Indian Tamils, Burghers, Malays, Kaffirs, and the aboriginal Vedda (Religions—Buddhism: Theravada Buddhism, 2008 usa-harshana.blogsopt.in). The country has a rich Buddhist heritage, and the first known Buddhist writings were composed on the island (*Vedda: Encyclopædia Britannica*, 2008; http://www.globallinktours.com/sri-lanka.html). It is an important producer of tea, coffee, gemstones, coconuts, rubber, and the native cinnamon. One of the first written references to the island is found in the Indian epic *Ramayana*, which provides details of a kingdom named *Lanka* that was created by the divine sculptor *Vishwakarma* for *Kubera*, the Lord of Wealth (Parker, 1992). The climate is tropical and warm, due to the moderating effects of ocean winds. The rainfall pattern is influenced by monsoon winds from the Indian Ocean and the Bay of Bengal. In the nineteenth and twentieth centuries, Sri Lanka became a plantation economy, famous for the production and export of cinnamon, rubber, and Ceylon tea, which remains a trademark national export (Annual Report, 2010a,b). Sinhalese and Tamil are the two official languages of Sri Lanka, which is also a multi-religious country, but with 70% of the population Buddhist.

Most of the traditional fermented foods are authentic, assure health, and provide rich taste (http://www.asian-recipe.com/sri-lanka/lk-information/ancient-food-and-drinks-of-sri-lanka.html). Dishes of Sri Lanka include rice and curry, *pittu*, *Kiribath*, wholemeal *Roti*, *String hoppers*, *wattalapam* (a rich pudding of Malay origin made of coconut milk, jaggery (Sagarika and Pradeepa, 2003), cashew nuts, eggs, and spices, including cinnamon and nutmeg, *kottu*, and *hoppers*. Jackfruit may sometimes replace rice and curries. Traditionally, food is served on a plantain leaf. Middle Eastern influences and practices are found in traditional Moor dishes, while Dutch and Portuguese influences are found with the island's Burgher community, preserving their culture through traditional dishes such as *Lamprais* (rice cooked in stock and baked in a banana leaf), *Breudher* (Dutch Holiday Biscuit), and *Bolo Fiado* (Portuguese-style layer cake). The staple diet of the people of Sri Lanka is rice. Mention is also made of products such as turmeric (*kaha*), ginger (*inguru*), pepper (*gammiris*), and spices (*kulu badu*) which are grown in the hilly regions. Cured and fermented fish are also important items in the diet of the people of Sri Lanka (Jayasinghe and Sagarika, 2003).

1.6.1.8 Afghanistan   The name Afghānistān means "Land of the Afghans" (Banting, 2003) which originates from the ethnonym "Afghan." Historically, the name "Afghan" mainly designated the *Pashtun* people, known to be the largest ethnic group of Afghanistan. It is officially called the Islamic Republic of Afghanistan, and is a land-locked country located in Central/South Asia, and is a part of the Greater Middle East (Afghanistan Country Profile, 2012). Afghanistan borders six countries: that is, Pakistan in the south and the east, Iran in the West, Turkmenistan, Uzbekistan, and Tajikistan in the North, and China in the far Northeast. It has a population of around 30 million inhabiting an area of approximately 652,000 km² (252,000 sq mi). Mostly, the population (90%) is Moslem, with most belonging to the *Sunni* branch, but there are also *Shiite* followers (Haber et al., 2012). Approximately 40% of the country is mountainous, and only about 12% is arable. It has been an ancient focal point of the Silk Road and human migration. She is known for producing some of the finest pomegranates, grapes, apricots, melons, and several other fresh and dry fruits, and nuts (Tobia, 2009). It is mostly a nomadic and tribal society, with different regions of the country having their own traditions (http://en.m.wikipedia.org/wiki/Afghanistan; http://www.flu.ofertyseks.kutno.pl/p-Afghanistan).

1.6.2 Indigenous Technology System and Public Services

Knowledge is a philosophical term and can be conceptualized as a set of various facts and information traits. It is of two types: scientific and indigenous. Both work as systems and, hence, terms like scientific knowledge system (SKS) and indigenous knowledge system (IKS) are used frequently (Gupta, 2013a). These two together constitute a global knowledge system (GKS). Clearly, the proven knowledge is scientific knowledge (SK), whereas the knowledge of the indigenous people is indigenous knowledge (IK), though there is actually no universally accepted definition of IK. Indigenous knowledge traits are oral, undocumented, and simple—dependent on the values, norms, and customs of the folk life, the practice of informal experiments through trial and error, the accumulation of generation-wise intellectual reasoning of day-to-day life experiences, lost, and rediscovered—practical rather than theoretical, and asymmetrically distributed. Indigenous knowledge is also known by several names, such as, folk knowledge, traditional knowledge, local knowledge, indigenous technical knowledge (ITK), and traditional environmental/ecological knowledge (TEK). Indigenous knowledge has certain characteristics, as summarized in Table 1.3.

Table 1.3   Characteristics of Indigenous Knowledge (IK)

Indigenous knowledge is local or specific to a particular geography or microenvironment or ecosystem and the folk people living there, close to nature

Originated through interactions and not at individual level, orally transmitted

Outcomes of informal experiments, intimate understanding of nature, and accumulation of generation-wise intellectual reasoning of day-to-day life experiences, generation-wise intellectual reasoning tested on "religious laboratory of survival"

Empirical rather than theoretical or any abstract scientific knowledge, functional or dynamic and hence, constantly changing, discovered, lost and rediscovered in a new form (open-ended IK)

Culturally embedded (close-ended) where separating the technical from nontechnical, rational from nonrational is problematic, repeating over time (as IK is both cultural and dynamic)

Segmented into social clusters or asymmetrically distributed within a population, by gender and age

Indigenous knowledge shared by many and even by global science

---

Source: Adapted from Ellen, R. and Harris, H. 1996. Concepts of Indigenous Technical Knowledge in scientific and Developmental Studies Literature: A Critical Assessment. www.worldbank.org/afr/ik/basic.htm_68k; Rao, R.E., Varadaraj, M. and Vijayendra, S. 2005. *Food Biotechnology.* 2nd edition. CRC Press, Taylor & Francis, Boca Raton, FL; Joranson, K. 2008. *International Information & Library Review* 40(1):64–72; Gupta, S. and Abu-Ghannam, N. 2012. *Critical Reviews in Food Science and Nutrition* 52(2):183–199; Gupta, A.D. 2013a. *A Three-Level approach with Special Reference to Rajbanshi.* International E-Publication 427, Indore, p. 155.

Folk people have preserved functional IK traits within non-functional symbols of their value-loaded folk life. To gather IK traits, many domains have to be decoded in the folk life, that is, folk music and song, folk proverb, folk etymology and chants, folk tales, folk literature, folk dance, folk painting and sculpture, folk recreation play, folk art and craft, folk cookery, folk settlement and patterns, folk architecture, the notion of time in folk society, weather forecasting, dialectology of folk speech, superstitions, myths, legends, riddles, folk religion, folk lore, norms regarding kinship/relations, and rites of passage. Besides, folk customs regarding household affairs and agricultural operations and the behavior of the folk people, folk dialect and folk technology, various type of organizations (political, economic, religious, and social) and ethno-medicinal practices are also to be taken into account (Gupta and Abu-Ghannam, 2011, 2012).

Indigenous Knowledge System (IKS) is the cognate of Indigenous Knowledge (IK). Folk people are well aware of how to apply IK traits in quite a systematic way so as to gain certain nature-friendly Public Services (PS) form, the so formed Indigenous Knowledge Systems (IKS) (Gupta and Abu-Ghannam, 2012; Gupta, 2013a).

IKS could be divided into various domains, such as agriculture, post-agricultural practices, animal husbandry, poultry, ethno-fishery, hunting and gathering, artisan, disease treatment, handicrafts, tools and techniques, nutrition, natural and biological resource, management of evironmental and biodiversity resources, poverty alleviation, community development, education and communication, and ethno-medicine and folk remedies. Each of these domains is provided with its own respective area and manifestation (Mondal, 2009; Gupta and Abu-Ghannam, 2012; Gupta and Loralie, 2012). Actually, farmers remain no longer passive consumers, but become active problem solvers (Warren, 1991; Gupta and Abu-Ghannam, 2012). So the highest priority is given to the alternative role of IKS rather than a high-cost modern crop production system (Davis and Ebbe, 1993). This, then, only allows a low-level external input into the traditional agriculturists living in remote areas (Haverkort et al., 1992). To document the indigenous technology in the various foods products, like pickles, dry foods, liquor, spices, sun-dried products, and preserved foods, the concept of fresh food and different types of food tastes are very important in the study of IKS regarding folk agriculture and animal husbandry.

1.6.3 Ethnic Values and Cultural Diffusion of Indigenous Fermented Foods

1.6.3.1 Ethnic Values   An ethnic group is defined as either a large or small group of people, in either backward or advance society, who are united by a common inherited culture (including language, music, food, dress, customs, and practices), racial similarity, common religions, and belief in common history and ancestors, who exhibit a strong psychological sentiment of belonging of this group (Phadnis and Gaungly, 2001). Ethnic groups can be of two distinct types; homeland societies and diaspora communities. The former are those with long time occupation of a particular territory, while the later communities are found in foreign countries and are caused by population migration, induced mainly by oppression in their homeland or the attraction of economic prospects. For more details, the readers may consult the literature cited on this aspect (Phadins and Ganguly, 2001). Ethnic identity is a biological given or a natural phenomenon (Geertz, 1963; Issacs, 1974; Bhattacharya, 2003) thus, ethnic groups constitute the network into which human individuals are born and become members of a community coming to acquire with other group member, the group's territory and objective cultural attributes, such as language, race, religion, custom, tradition food, dress and music, ethnic identity, and the nation state (Rickmond, 1978; Rex, 1995; Bhattacharya, 2003).

The traditions of different ethnic groups in South Asia are divergent, influenced by external cultures, especially in the North-Western parts of South Asia (where Turkic and Iranian people have had much influence) and in the border regions and busy ports, where there are greater levels of contact with external cultures. This is particularly true for many ethnic groups in the North-Eastern parts of South Asia, who are ethnically and culturally related to peoples of the Far East. The largest ethnolinguistic group in South Asia is the Indo-Aryans, numbering around 1 billion, and the largest sub-group is the native speakers of Hindi languages, numbering more than 470 million.

For the most part, the cultures of the South Asian countries have left relatively few artefacts as evidence of production of fermented foods, which has led to an overemphasis on the cultural advances of other regions, such as the Middle East, Central America, and even sub-Saharan Africa in comparison to South Asia (Stanton, 1985). The skills of food preservation existed among the native people of many areas of South Asia, but the knowledge was propagated orally. During the middle Ages, the varieties of fermented foods and the drinks developed, depended upon the availability of raw materials, environmental conditions, and the taste preferences of the local people. Indians are given credit for inventing the methods of souring and leavening cereal-legume batters (Padmaja and George, 1999). Traditional foods are a rich heritage of India's multidimensional culture.

Traditional alcoholic beverages constitute an integral part of dietary culture and have strong ritual importance among the ethnic peoples of the world. In the Indian sub-continent, the making and use of fermented food and beverages using local food crops and other biological resources is very common amongst the inhabitants of the high Himalaya. The name of the products and the base raw materials employed, however, vary from region-to-region. More than ten varieties of fermented beverages are consumed in India (Tamang et al., 1996, 2007). In many of the South Asia countries, the alcoholic beverages are culturally and socially accepted products for consumption, entertainment, customary practices, and religious purposes, and, therefore, are of wide interest, enhancing the nutritional significance as well as imparting the pleasure of drinking (Darby, 1979; Tamang, 2010), as well as their use for medicinal, social, religious, or recreational purposes (Tamang et al., 2007).

1.6.3.2 Origins, Diffusion and Cultural Context of Fermented Fish Products   An instance of the origin of indigenous fermented foods can be given by fermented fish. The technology of *Lona ilish* production actually originated in the erstwhile undivided India (now Bangladesh) about 100 years ago, on the banks of the river "Padma" and "Meghna" in the Noakhali district. It is assumed that the technology evolved during a glut period, when there were no preservation techniques known except sun-drying and salting. Sun-drying was not suitable for hilsa, the highly fatty fishes, due to the rapid development of rancidity on being exposed to the sun (Majumdar et al., 2006). In addition, sun-drying was also difficult during the continuous spells of rain in July–August, which corresponds to the main glut period. This way of processing might have been started to quickly preserve large quantities of fish in an inexpensive way. The technology, however, has not changed much since the earlier days, and the practice is still one of the major means of preservation of hilsa by the fishermen community of Bangladesh. Although different technologies, like packaging in an inert atmosphere (Majumdar and Basu, 2009), have come into being, so far, no other preservation techniques except salt drying, and to some extent canning, are in current use. During the partition of India in 1947, the technology gained entry into the Northeast sector of India through migrants. However, large scale production is limited mostly to the Chandpur Sub-division in the Noakhali district of Bangladesh (formerly East Pakistan). Presently, the consumption of *Lona ilishis* is restricted to Bangladesh and Northeast India. How the salting and fermentation of fish was originated and practiced in South Asia is a key issue. Since documentation in the literature is lacking altogether, examples of regions other than those of South Asia which are linked with this region are cited as illustration.

In the South Asian region, especially those of coastal areas eating large quantities of rice, which is a cheap source of energy (Tanasupawat and Visessanguan, 2014), is a routine practice. So a vital individual foodstuff is either a salty side dish or a condiment that facilitates rice consumption. Fermented products are well suited for this purpose, being simple to produce and cook, have a long shelf-life, and imparting *umami* flavor, which is a category recognized by Japanese as the taste of glutamic acid (O'Mahony and Ishii, 1987), and a salty taste to vegetable dishes (Mizutani et al., 1987; Kimizuka et al., 1992; Tamang, 2010). It is no coincidence that the main regions where fermented fish products are consumed overlap with the main regions of irrigated rice cultivation.

1.6.3.3 Generic Fish Products   The term "fermented fish products" is used here to describe the products of freshwater and marine finfish, shell-fish, and crustaceans that are processed with salt to cause fermentation and to prevent putrefaction. Although the same phenomenon occurs with salted fish products also, the state of those products described here is altered intentionally by fermentation (Tamang, 2010, 2012; Tanasupawat and Visessenguan, 2014). A generic classification of fermented fish products in Asia is given in Figure 1.12.

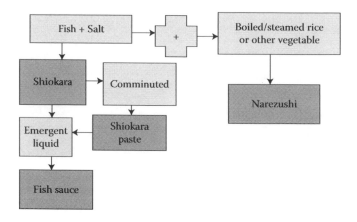

Figure 1.12   Generic classification of fermented fish products. (Adapted from Ruddle and Ishige, 2005. Fermented Fish products in East Asia Hong Kong. International Resource Management Institute).

Thus, the fermented fish products *shidal* (Assam), *nya sode* (Bhutan), and *jadi* (Sri Lanka), among many others, do not fit into the category of intentionally fermented products discussed here (Ruddle and Ishige, 2005; Kose and Hall, 2010). There is no evidence that the fish sauces of Asia originated by diffusion from the Mediterranean Basin or *vice versa*. The origins of these geographically distinct groups appear to be different. The prototypical product is probably the highly salted fish, which in Japan is known as *shiokara*. Since there are no succinct equivalent English terms for these products, simple Japanese terms have been used throughout this text. The product of combining fish and salt (Figure 1.13) that preserves the shape of the original raw fish, is termed as *shiokara* (Tamang, 2010) which can also be comminuted to *shiokara* paste, having a condiment like character.

If no vegetable ingredients are added, the salt fish mixture yields fish sauce, a liquid used as a pure condiment. If cooked vegetable ingredients are added to the fish and salt mixture, it becomes narezushi (Figure 1.14a). Narezushi (Figure 1.14b) results when boiled carbohydrates (normally just rice) are added to the fish and salt mixture, to prepare *shiokara*.

Figure 1.13   Salted fish *Shiokara*. (Adapted from Ruddle and Ishige, 2005. Fermented fish products in East Asia, Hong Kong International Resource Management Institute).

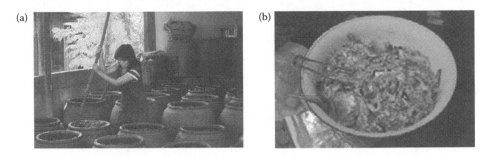

Figure 1.14 (a) Woman stirring fermentation; (b) Narezushi (pla som) in the Kaen Market Thailand.

1.6.3.4 Fermented Shrimp Products Uncomminuted shiokara, shiokara paste, and shiokara sauce are the three main fermented shrimp products produced (Figure 1.15). Superficially, these can be substituted for fish products. Perhaps, during the evolution of fermented fish products, the techniques of making fish *shiokara* were simply applied to the preparation of shrimp paste. However, it is also possible that some shrimp products did not originate as a variant of *shiokara*, as in some areas fermented shrimp paste is produced without salt, as is sometimes with the case of countries including Bangladesh. Some shrimp pastes have a very low salt content compared to the fermented fish products, which may be a result of the different compositions of fish and shrimp. Shrimp has a carapace and a higher watery content than fish (Tanasupawat and Visessanguan, 2014). On the other hand, shrimp shiokara could have originated from the preparation of sun dried shrimp. Only two techniques, fermentation and sun drying, are used to preserve epipelagic shrimp (Figure 1.15). Sun drying is the simplest, but it does not overcome the problem of the coarse texture of the carapace and hence, the need for commenting. Comminuting the sun-dried shrimp without the addition of salt produces an unsalted shrimp paste. However, salting this kind of paste

Figure 1.15 Blocks of sundried fermented shrimp paste (terasi) ready for wrapping at a cottage industry level producer house in Rembang, Central Java, Indonesia. (Adapted from Ruddle, K. and Ishige, N. 2005. Fermented fish products in East Asia. Hong Kong. International Resources Management Institute. http://www.intresmanins.com.)

enhances both the taste and shelf-life. Thus, despite the lack of strong supporting evidence, the origin of shrimp paste from sun dried shrimp is compelling (Tanasupawat and Visessanguan, 2014).

1.6.3.5 Study of Fermented Fish Products   Most of the studies deal with East Asian fermented fish products, so only a brief mention will be made here. The most comprehensive research on this aspect was done in Vietnam during colonial times, and focused mainly on chemical analysis of sauce quality, specifically for taxation purposes, and pertinent literature has been cited for more information (Van Veen, 1953, 1965; Indo-Pacific Fisheries Council, 1967; the Tropical Products Institute, 1982; Steinkraus, 1983a,b, 1985; Lee et al., 1991; Shinoda, 1952, 1957, 1961, 1966, 1978; Ishige and Ruddle, 1990; Segi, 2006). During 1982–1985, Ishige and Ruddle conducted a comprehensive field survey on the fermented fish products industry, from the catching of the raw materials to their culinary use in Bangladesh, Cambodia, China, India, Indonesia, Japan, Korea, Malaysia, Myanmar, the Philippines, Taiwan, Thailand, and Vietnam (Ishige and Ruddle, 1987, 1990; Ruddle and Ishige, 2005) using questionnaires and structured interviews of factory managers, household producers, market vendors, wholesalers, and consumers. Related literature covered the cultural and historical contexts in terms of the origin of products, diffusion, and history, with information being culled from historical cookery books, character dictionaries, general descriptions, and other documents. Due to lack of documentary materials, the cultural history of Southeast Asia was examined. Based on these aspects, it was possible to trace the likely origins and routes of diffusion, together with the development of fermented fish culture in Southeast Asia (Lee and Kim, 2013) and is discussed here for the sake of illustration, as South East Asia is not the subject matter of this text.

1.6.3.6 Culture History and Human Ecology

1.6.3.6.1 Human Migrations in Indo-China   The widest variety of fermented fish products and their principal dietary role occurs in continental Southeast Asia, and so this area should be regarded as one centre of their origin. Early human settlement in this region was in those areas most suited to cultivate irrigated rice, so freshwater fish species naturally occurring in local hydrological systems would have been fermented (Tanasupawat and Visessunguan, 2014). These products continue to be the best developed from the area west of the Annamite Mountains to Lower Burma, where the main populations are Thai-Lao, Burmese, and Khmer (Ishige and Ruddle, 1987, 1990; Ruddle and Ishige, 2005). The Burmese originated from an area in Chinese Central Asia and Tibet where there was no fisheries tradition, so it is unlikely that they prepared fermented fish prior to their southward migration. Further, the Thai-Lao originated in Yunnan, where the only historical reports on fermented fish products concern *narezushi* (Shinoda, 1952; Ishige and Ruddle 1987; Tamang and Samuel, 2010). There are no Chinese historical documents indicating the preparation or use of fermented fish products among the minority ethnic groups that lived south of the

Yangtze River (Shinoda, 1952; Ishige and Ruddle 1987; Tamang and Samuel, 2010). Many of these people were Thai-Lao, who were most likely to have adopted the use of fermented fish from the earlier inhabitants after entering the Indo-Chinese peninsula, so it is safely stated that fermented fish products did not originate in China (Ishige and Ruddle, 1987, 1990; Ruddle and Ishige, 2005; Tamang and Samuel, 2010).

It is probable that fermented fish products could have been made in the Indo-Chinese peninsula before the in-migration of the various ethic groups. However, only philological evidence supports our suggestion that *narezushi* was known in the area, and that the Han Chinese learned of it during the course of their extremely prolonged expansion south of the Yangtze. It appears that *narezushi* was prepared by the rice cultivators of Southeast Asia, and was taken from there to China. *Narezushi* remains common in Laos, Cambodia, and North and Northeast Thailand, that is, in the Mekong Basin. Although the present day inhabitants of this area are Laotians and Khmers, the Mekong Basin was formerly co-extensive with the Khmer civilization (Coedes, 1962; Tamang and Samuel, 2010). In this respect, the hypothesis that rice cultivation originated in Yunnan and spread down the Mekong Valley into Laos, Thailand, and Cambodia, with Myanmar and Vietnam as the branches (Shinoda, 1977), is important, because this coincides with the center of *narezushi* and other fermented fish production.

It is probable that irrigated rice cultivation and the associated rice field fishing originated in Yunnan and diffused southwards down the Mekong Valley. But, given the marked seasonality of fish abundance along the Mekong Valley (Ishige and Ruddle, 1987), it can be assumed that a need arose to preserve fish for the times of scarcity, which eventually gave rise to fish fermentation.

The relationship between seasonal hydrological conditions and inland fisheries in the different agro-ecological zones of rice cultivation is also important. Mountainous areas are inhabited mainly by shifting cultivators, who do not cultivate irrigated rice and who live where the fish fauna is sparse. Early irrigation networks developed in intermountain basins, and alluvial fans within the mountains are flooded in the rainy season and suffer drought in the dry, when the fish populations are limited only to the larger watercourses and to mud in pools and swamps. However, the fish and other aquatic fauna is widely distributed throughout the flooded area in the wet season. Seasonally abundant fish caught at the end of the wet season are therefore, preserved by fermentation for year-round use. In contrast, upper delta areas have large and abundant watercourses, and are widely flooded during the rainy season. Since fresh fish is available throughout the year, there is generally no need for preservation. The same is the case for lower deltas. All these aspects have been illustrated by the example of the Chao Phrya Basin of Thailand, and the rain fed plateau areas as the Korat and Plateau of Northeastern Thailand. This strong seasonal contrast in resource availability makes the preservation of the wet season catch imperative.

Preparation of fermented fish products requires plentiful amounts of salt, which in itself would determine their origin and distribution. In terms of salt supply

(Sinanuwong and Takaya, 1974a,b), it may, therefore, be surmised that the Khmer, Cham, and Mon were the people most intimately concerned with the history of fermented fish products. The center of salt production, the ecological zonation of irrigated rice cultivation, and the seasonal behavior of fish stocks all support this hypothesis.

1.6.3.7 Ethno-linguistic Evidence   In the absence of documentary evidence, a general history of human migration plus ethnolinguistic evidence can be used to reconstruct the probable history of the diffusion of fermented aquatic product production in continental Southeast Asia (Ishige and Sakiyama, 1988). Dictionaries played a vital role in this part of the research (For more details, see Shorto, 1962; Smith, 1967; Moussay, 1971; Sakamoto, 1976a,b; Romah, 1977; Chantrupanth and Phromjakgarin, 1978; Thongkum and Gainey, 1978; Shintani, 1981; Henley, 2008).

Based on the production technique used, it cannot be said if *shiokara* diffused from one or several sources, or originated independently. Apart from continental Southeast Asia, where freshwater fish are common, *shiokara* is made from marine fish. In China, *shiokara* was made along the entire coast; however, it became a "relict" food with scattered distribution. The origin of *shiokara* made of freshwater fish with the addition of *koji* and rice wine is unknown. It is possible that fermentation is not related to the use of *koji* in *shiokara*-making. The deliberate production of fish sauce as a special product is relatively recent. Historically, a liquid natural by-product of *shiokara*-making was used as a condiment prior to the commercial manufacture and wide distribution of fish sauce as a specific product or culinary use. There is no evidence to demonstrate whether fish sauce originated in China or in Vietnam, as is the case with fish sauce made from freshwater fish. *Narezushi* originated in the Mekong Basin and might be of Khmer origin. Shrimp paste originated in continental Southeast Asia, probably among the Cham and Mon people of Indo-china, from where it diffused southwards to insular Southeast Asia.

1.6.4 Cultural Anthropology—Indigenous Food Fermentation and Eating Culture

Anthropology of food is a potentially robust field of research to give the wide varieties of its agricultural and culinary products, and these products are embedded in complex geographical and historical settings that are basic to our understanding and appreciation of the form and content of the grammar of essence. History (with reference to migrations) and internal adjustments combined to shape the formation and culinary identity among other things, Yoruba (Bentley and Ziegler, 2003; Ogundele, 2007: 50–56). A similar situation of practices are seen in some of the rural areas in South Asia, especially India.

Given the fact that the food studies are complex and indeed, tentacular in character, they cannot be confined only to the domains of Human Nutrition and Agriculture without disadvantaging modern humanity a great deal. By this yardstick, food studies among the Yoruba escape disciplinary, cultural, environmental, and temporal boundaries. In the same way, the fermented diets are to be appreciated beyond the sphere of metabolism and/or good health. These foods have social and religious meanings and build bridges between materiality and spirituality. Thus, for example, certain staples from maize also serve as foods for some deities or ancestors. Contemporary scholarship must try to capture this reality. Available evidence from fieldwork has shown in a fascinating manner, that Yorubaland, like most other parts of the global village, was never cut-off from the cross-currents of history in antiquity. This situation underscores the reason why there is a considerable amount of fluidity in human behavior, with particular reference to foods and eating patterns.

The oneness of the universe is incontestable despite its obvious biocultural diversity, which tends to shrink the contemporary range of choices and opportunities. This is hinged on general ignorance arising from gross wrong education about the complex character of cultural change and/or continuity. This adaptation of particular food was not to the detriment of the local initiative and respectability. Methods made use of include literary excavations, mining of spoken words of the locals, and ethnographic observations of, and sometimes direct participation in, cassava and maize fermentation processes.

Maize gained in popularity faster than cassava, understandably, because the former has no problems of hydrocyanic toxicity. Maize can be readily eaten after roasting or cooking. Indeed, by about 1800 AD, maize had become a well established secondary crop in Yorubaland. It can be fermented into a popular staple called *ogi* or *eko* and pito-alcoholic beverages. *Eko* is a form of pap which is a soft and almost semi-liquid food for babies and adults alike. Lime or lemon juice can be added to the pap while sugar or honey is also added sometimes, depending on the taste of the consumer, but most Yoruba take pap with a special bean cake called *akara* or another variety of ground beans delicacy—*moinmoin*. *Akara* is a special form of snacks among the Yoruba, which is a pan-Yoruba culinary delight that has, with the passage of time, become a symbol of the people's cultural identity. This is one good example of how the Yoruba also donated to the emergence of the New World culture.

Newly born babies (from about one or two weeks old) are usually given *ogi* (pap), similar to seera, a fermented food of Himachal pradesh (India), in addition to breast-feeding. The pap (*ogi*) is a favorite breakfast especially on Sunday, when most people do not go to work. So maize has gained in popularity both among the living and the dead or deities, despite its foreignness in terms of origins (Daramola and Jeje, 1975).

Despite the broad acceptance of Western values and value-systems, many parts of Yorubaland, particularly the rural area, still preserve aspects of their serving and eating cultures. Thus, for example, visitors (even without prior notice) are served first, followed by the most senior male member (usually on the basis of age) of the household. Children eat together (three or four) in a common bowl in most cases. Food sharing also features prominently in the scheme of things. This practice promotes group solidarity, harmony and understanding (Ayodele, 2011, personal communication). Further, the texture of group solidarity, harmony and understanding is carefully woven with the thread of this age-old cultural practice. Meals can be taken at any time that is convenient. Such a scenario is one of the variations in culinary behaviors of the Yoruba sub-groups or sub-ethnicities (Ogundele, 2007: 50–56). Thus, food research has the capacity to deepen the all-important discourse on globalization, with a particular reference to the engendering of regional and/or transregional peace and understanding. Here the cultural anthropology occupies a prominent position in pushing back the frontiers of knowledge of how one group contributed and/or received both materially and extra-materially from the other. Food among the Yoruba also reflects symbolic values and is seen as a tool of social stability/group solidarity which is a pre-condition for material and social progress.

1.6.5 Fermented Foods in the Indian Himalayan Region: A lifestyle

The region extending between latitudes 26° 20′ and 35° 40′ North, and longitudes 74° 50′ and 95° 40′ East can be broadly called the Himalayan region. It supports remarkable cultural, ethnical, and biological diversity for every 250–300 km across, stretching over 2500 km from Jammu and Kashmir in the West, to Arunachal Pradesh in the East. It covers partially/fully twelve states of India, namely, Jammu and Kashmir, Himachal Pradesh, Uttarakhand, Sikkim, Arunachal Pradesh, Nagaland, Manipur, Mizoram, Tripura, Meghalaya, Assam, and West Bengal. The people of this region rely largely on surrounding bioresources for sustenance; be it food, fodder, fiber, fuel or medicine, which are intimately associated with the life style of ethnic groups (Samal et al., 2003, 2005).

The Himalayan dietary culture has both rice and wheat/barley/maize as staple foods, along with varieties of fermented foods and beverages made from soybean, vegetables, bamboo, milk, meat/fish, and wild edible plants (Tamang, 2010; Tamang and Samuel, 2010). Ethnic fermented foods and alcoholic beverages have been consumed by the ethnic people of North East India for more than 2500 years (Tamang, 2010). The people of these regions prepare and consumed more than forty varieties of common as well as lesser-known indigenous fermented foods and beverages (Tamang, 2003; Tamang et al., 2012). Several indigenous fermented food like *Kinema* are sold in the local market called *huts*, and there is a good market, but production is at a low level only (Tamang, 1977a). The preparation and consumption of fermented foods, including beverages, based on local resources, have been an integral part of the culture of the people of Indian Himalaya. Their products made from various raw materials (resources), are known by different names.

1.6.5.1 Indigenous Fermented Foods and Culture of Ladakh   Ladakh, truly described as a high altitude cold-arid desert, is one of the most far eastern regions of Jammu and Kashmir State of India (Angchok et al., 2009). It constitutes the easternmost trans-Himalayan part of J&K state of India, bordering Pakistan and China, and comprises of Leh and Kargil district which is sandwiched between the Greater Himalayas in its South and the Karakoram ranges in the North (Angchok and Srivastava, 2012).

Literally Ladakh means "Land of Passes," the epithet truly epitomizing the set of land features, such as rugged terrain, lofty mountains, and numerous high-altitude passes. Zojila pass lies in the west on the Srinagar-Leh highway, and the Rohtang and Baralacha passes lie in the East on the Manali-Leh highway.

With extremely cold winters (−30°C) and summers up to +35°C, and aridity, coupled with large diurnal variations in temperature are the limiting factors affecting agricultural productivity. Peculiarities like remoteness and a low level of market integration limit the choice of foods for locals in Ladakh (Angchok et al., 2009; Dame and Nüsser, 2011). Further, intensive sunlight, a high evaporation rate, and strong winds characterize the general climate. Due to high mountains and heavy snowfall during winter, the area remains inaccessible to the outside world for nearly six months in a year. Over the centuries, the people of Ladakh have developed a farming system uniquely adapted to this environment. The principal crop is barley, the mainstay of traditional *Ladakhi* food. In the valleys, there are orchards, and up on the high pastures, where not even barley grows, people husband yaks, cows, or sheep (Figure 1.16).

The unfavorable and hostile environment prevailing over this region, limits the cultivation to a very low scale (both time and place) (Angchok et al., 2009). Under these drastic conditions, one of the major reasons behind human habitation is the ingenuity of local people, who have devised innovative and sustainable way of living. One major product of this ingenuity is traditional fermented foods and beverages, which over the time have been evolved and established in the food system. Fermented foods enrich the

Figure 1.16 Yak. (From Mahammed Sequib Gheewala.)

food substrate biologically with micronutrients, besides preserving them to be used during the winter. As fuel is scarce, fermented foods have the added advantage of requiring little cooking, so consuming less fuel. Among the fermented food and beverages, *chhang* (a barley based alcoholic beverage) is considered indispensable (Angchok et al., 2009).

The Ladakhi community consists of different ethnic groups, such as the *Bot, Balti, Brokpa, Beda, Gara, Mon, Purik* and *Shina* (Mir and Mir, 2000) and many more inhabiting geographically-distinct locations. William Moorcroft, a veterinary surgeon traveling in the 1820s, was the first Englishman to give a detailed account of Ladakh (Angchock and Srivastava, 2012), describing their food consumption pattern. Typical meals are the preparations based on barley, wheat, peas, potato, turnips, and green leafy vegetables, in addition to milk and meat (Dame and Nüsser, 2011). The availability and affordability of vegetables and fruits decreases in winters, resulting in a pronounced seasonality of dietary patterns (Dame and Nüsser, 2011). So there is increased consumption of cereals and meat, and a reduced intake of fruits and vegetables during winters (Kelly et al., 1996). A variety of preparations, procedural nuances, and a multiplicity of appellation evinces the ethnic diversity. Even religion has its role in the preparation and preferences of these food items. *Chhang*, for example, is neither prepared nor consumed by the Muslims, as it is a food containing blood, which is, otherwise, relished by Buddhists.

1.6.5.2 Fermented Foods of Himachal Pradesh: The Dietary Culture A large part of Himachal Pradesh is scenically beautiful, such as Manali (Figure 1.17), with significant tourist attraction. The people of Himachal Pradesh have developed traditional food processing technologies for preparing fermented foods from locally available substrates largely governed by ethnic preference, agro-climatic conditions, socio-cultural ethos, and religion. A number of traditional fermented products are prepared and consumed, and the types of traditional fermented products are unique and different from other areas (Joshi et al., 2012; Savitri and Bhalla, 2013). *Bhatooru, chilra, seera, siddu, gulgule, marchu, sepubari,* and pickle from various locally available fruits and vegetables, and different alcoholic beverages (Figure 1.18) like *ghanti* or *daru, chhang,*

Figure 1.17 View of Himachal Pradesh (Manali), India.

Figure 1.18 People drinking alcoholic beverages.

sura, behmi, etc., are some of the indigenous fermented products of Himachal Pradesh (Thakur et al., 2004; Joshi, 2005; Savitri and Bhalla, 2013).

These foods have been a part of the staple diet in the rural areas of Himachal, especially the districts of Lahaul and Spiti, Kinnaur, Chamba, Kullu, Mandi and Shimla. Inhabitant of the Lahul and Spiti area of Himachal Pradesh, as a part of their routine diet, extensively consume some of these indigenous fermented foods, with cereals as the main substrate, using simple equipment (Kanwar et al., 2007; Savitri and Bhalla, 2013).

1.6.5.3 Sikkim Various indigenous fermented foods of Sikkim Himayala include *chhurpi, dahi, kinema, gundruk, mohi, serlorti, sinki, sukako ko masu, khalpi, masauyra, mesu, somar, suka machha, bhaate jaanr, kodo jaanr, maki jaanr, raski, marcha* (Tamang, 1977b). *Chhang,* a local beverage (beer) from barley is prepared in Ladakh and Lahaul-Spiti; while *Ghanti,* made from grapes with or without *jaggary* or other fruits, is popular in the Kinnaur district of Himachal Pradesh (Rizvi, 1983; Bajpai, 1987). *Kodo ko jaanr,* made from finger millet, is popular in Sikkim and the Darjeeling hills (Thapa and Tamang, 2004). The cultural adaptation of Himalayan ethnic foods has also been documented (Tamang et al., 2010).

1.6.5.4 Uttarakhand—Indigenous Fermented Foods Culture Starting from the foothills in the South, Uttarakhand extends to snow clad mountains in the North. The entire state forms a part of the Central Himalayas and it is situated geographically in the eastern side of the North-Western Indian Himalayas (Roy et al., 2004). The state is interspersed with rivers, deep valleys, glaciers, flower valleys, and high peaks, and it presents a very pristine, pure and picturesque environment (Figure 1.19). It is no wonder that it is considered to be the abode of Gods and Goddesses.

Major crops are paddy, wheat, maize, oil-seeds, soybean, and pulses. The Berinag area in Pithorgarh district has a great history of tea cultivation, which dates back to 1835.

Figure 1.19   A view Nainital in Uttarakhand (India).

The emergence of the indigenous knowledge system in this part of high altitudes of the Himalaya was due to the cold climatic condition of the *Bhotiya* dominated areas. The way this society carved a niche in making a living on the surrounding natural resources for adaptation to the emerging circumstances in the region is unique. The indigenous knowledge of making fermented foods and beverages developed over a long period of time (Das and Pandey, 2007). *Chakti* in Dharchula, *daru* in Munsyari, and *Chhang* in Chamoli and Uttarkashi are the traditional alcoholic beverages, made mainly from cereals, millets, and fruits, by the local communities in Uttarakhand (Roy et al., 2004).

1.6.5.5 North East States   Ethnic fermented foods of North East are classified into fermented soybean and non-soybean legume foods, fermented vegetables and bamboo shoot foods, fermented cereal foods, fermented and smoked fish products, preserved meat products, and alcoholic beverages (Hayford and Jespersen, 1999; Agrahar-Murugkar and Subbulakshmib, 2006; Jeyaram et al., 2009; Sohliya et al., 2009; Tamang and Tamang, 2009; Tamang, 2010; Tamang et al., 2012). The various ethnic group of the tribes of the Dima Hasao (North Cachar Hills) district of Assam prepare and consume a wide variety of fermented foods and alcoholic beverages, among which are *Judima* (alcoholic beverage), *Humao* (Starter), *Miya-Mikhri* (fermented bamboo shoots), *Ngathu* (fermented fish), and *Honoheingrain* (fermented Pork meat) (Chakrabarty et al., 2013).

Traditional fermentation is a low cost method of fish preservation using artisanal equipment which is readily available and easy to fabricate and repair. Due to their economic viability, drying and other curing techniques are the most popular techniques of food preservation (Baishya and Deka, 2009). The *Khasi* tribes of Meghalaya prepare a fermented food product out of fresh fish species *Puntius sarana* (Ham), and the product is locally known as "Tungtap." Sun-dried *Puntius sarana* is comparatively more stable than other similar varieties of fish. Even when the fish is dried without gutting, it can be preserved for a long duration by sun drying, due to low moisture content. It is sold to the agents at the wholesale market of Jagiroad, a place in the Morigaon district of Assam, from where it is distributed to different places in the North-Eastern region for further processing by indigenous techniques. These partially cured fishes are used as a raw material for the fermentation process by the various ethnic groups of the region.

1.6.6 Culture of Indigenous Fermented Foods among the Agrarian Rajbanshis of North Bengal, India

1.6.6.1 Crop Production, Food Fermentation and Processing by Rajbanshis Rajbanshis in and around the plains of northern West Bengal or North Bengal have their own history of thousands of years (Sanyal, 1965; Gupta and Abu-Ghannam, 2012; Gupta, 2013b,c). They have transformed from a community to a huge complex heterogeneous society, incorporating animism, ancient pre-Vedic versions of Hinduism, Vedic traditions, magico-religious practices, and Buddhism, a Kashyap-Bratya Kshattriya combination, Sufism and Vaishnavism, status mobilization, and folk practices symbolic to agriculture and trade relations (Sanyal, 1965, 2002). To study the indigenousness of Rajbanshis, several terms have been coined, for which reference can be made to the literature cited (Gupta, 2013a).

Rajbanshis preferred rice cultivation in the monsoon season which is unique to South Asia, South East Asia and South China. Since they had very low population and minimum needs they used to leave the cultivation ground for a season or a year or several years, a practice which was considered useful for maintenance and the increasing the soil productivity. Not only this, they did not cultivate at fixed places overtime, and used to change the places of crop-cultivation, a system they called *jhum* cultivation (Gupta, 2013b).

1.6.6.2 Food Fermentation and Preservation *Marua* is cultivated by the tribal people of North Bengal. Toto, the smallest tribal group and also the Primitive Tribal Group (PTG) of West Bengal residing in the Toto para-Ballalguri region of the Madarihat block of the Jalpaiguri district of northern West Bengal, also cultivate the crop. It is actually finger millet (*Elcusine coracocna*) called *ragi*, which is commonly used by Indians both as food and fodder. It is also used to make alcohol: *Eu* for the *Toto*. They ferment this millet in earthen or aluminum pots for 2–3 days. For increased intoxication, they use some herbs and add more water for dilution. *Tari* is another alcoholic beverage prepared from fresh palm or date palm. It is collected during winter in an earthen pot which is tightly bound to the trunk of the tree where the juice is stored while coming out of a cut mark. It is a whole night process and in the early morning, the pot full of juice is collected, and fermented in the same way as *Tari*. Adjacent Santhals, Oraons, and other *Adivasi* brew rice that they treat as *Handia* and distill it to make alcohol called *Chullu*. Rice tablets that also include the dried powder of several plant parts are added to the dried cooked rice in an earthen pot (*handi*) in 1:2 ratio with water, which is closed with a lid and placed in a dark cool room for 5–6 days. In this way, *handia* is produced and from this, *chullu* is made through distillation.

Areca a nut is the main economic resource of every village. *Guwa* or *chegua* or *guai* is the local name of areca. In some cases, these nuts are put into a pitcher of water with a tightly closed neck. In 2–3 days, the fermentation is more or less complete. This type of fermented areca nut, with a strong smell, is very common fermented product among the Rajbanshis.

Curd or *curad* is the most auspicious and popular item for any kind of ceremony. To prepare *curad*, the Rajbanshis used to keep cows milk in an earthen pot for several days in a clean, dust-free, cool dark room (with earthen wall and roof made by jute sticks and straw), hanging from the beams of the roof, and after some days, the *curad* is formed and is then, consumed with salt. No lemon extract or sour fruit substances are added to it, as the Bengalis do, nor is the milk heated. This type of *curad* is called *goleya doi*. Rajbanshis, for festival purposes, again mix more milk with the *curad* and put sugar in it and stir continuously while boiling, producing sweet *curad*. No unsaturated fat is added to this to make it thick and sticky. Rajbanshis usually do not drink milk, but are fond of *curad*.

Buguri or plum is also collected and preserved as a pickle, as is mango. Sun-dried fresh pieces of local varieties of small fishes from ponds and streams, called *shutka*, are dusted in *chham-gyin* from the waxy leaf base of varieties of arum (*Mann* and *kalo kochu* types being used). Mustard oil, garlic, chilli and turmeric are used to prepare fish-balls from this waxy fish dust (*sidal*). The balls are then fermented in tightly closed earthen pots filled up with *chheka* dust. After 5/7 days, the seal is broken and the balls are baked (*autha*) or cooked with curry and water of *chheka*.

The Rajbanshis place an earthenware pot over the new shoots from the bamboo rhizome under an earthen pot. The new shoots that grow under it are known as "ban-skorol" and are a popular food. These shoots are often submerged in water for days before consumption.

The Rajbanshis actually know the best way of preserving biodiversity and utilizing the same without excessive exploitation. Their cultural values are completely directed towards the maintenance of the equilibrium between population size and minimum exploitation of the resources. They also oppose the hegemony of the modern market economy, and their cultural values provide protection for the IKS. Among these, traditional Rajbanshis food fermentation processes are very important.

1.6.6.3 Jaintia Tribal Community of Meghalaya, India About 427 tribal communities live in India, and of these more than 130 major tribal communities inhabit the Northeastern states of India. Tribes residing in Northeast India are generally categorized into two broad ethnic communities, tribes belonging to *Monkhemar* culture of the Austoic dialect, such as the *Khasi* and *Jaintia* tribes of Meghalaya, and tribes that belong to the Tibeto-Burman sub-family of the Tibeto-Chinese group, who are basically Mongol (Jaiswal, 2010).

The Indian state of Meghalaya is divided into three hilly regions—the Garo Hills (Western Meghalaya), the Khasi Hills (Central Meghalaya), and the Jaintia Hills (Eastern Meghalaya) (Jaiswal, 2010). The name Meghalaya, meaning *Adobe of the Clouds*, accurately describes the climatic phenomenon that brings torrents of rain to this region, directly influenced by the Southwestern monsoon that originates in the Bay of Bengal. Meghalaya is one of the wettest places on earth, and the village of Mawsynram in the southern slopes of the Khasi hills district receives the heaviest annual rainfall (1169 cm) in the world (Kumar et al., 2005; Jaiswal, 2010). The *Jaintia*, also known as *Pnar* or *Synteng* are the original inhabitants of Jaintia hills district, which is locally known as *Ka Ri Ki Khadar Doloi* (the land of 12 kingdoms). The *Jaintia* have their own dialect, called *Jainta* or *Pnar.* Traditional *Jaintia* villages usually consist of scattered settlements of houses made of bamboo or timber. The people have traditional *Jaintia* dress. Agriculture is the main occupation of the *Jaintia* and in addition to rice and meat which are the staple foods, a large number of wild plants, including fruits, seeds, tubers, and shoots, contribute significantly to the diet of people in this community (Kayang, 2007; Jaiswal, 2010). *Jaintia* festivals are mainly about agricultural seasons, prosperity, and entertainment, such as the *Shad sukra*, a thanks giving dance festival for happiness before sowing the rice, and the *Beh dien khlam*, celebrated annually in July after sowing the rice; while the *Lahoo* dance festival is devoted to entertainment (Samati and Begum, 2006; Jaiswal, 2010).

1.7 Summary and Future Prospects

Historically speaking, indigenous fermented foods were discovered before developments in science were documented, and have played a major role in every civilization and nation. Most knowledge of indigenous fermented foods is ethnic and was handed over from generation-to-generation. It is a contribution of ethnic people to the ITK.

These foods contribute to about one third of the diet worldwide, and South Asia is no exception. The consumption of many indigenous fermented foods has been related to religious ceremonies, rites, and social occasions in different the countries of south Asia. A survey of various countries of South Asia—India, Pakistan, Bangladesh, Srilanka, Nepal, Bhutan, Afghanistan, and the Maldives reveals a great diversity of indigenous fermented foods, prepared and consumed by the people. Not only the products, but also the methods used, vary from region-to-region, and can inspire producers and researchers in other parts of the World.

Food Fermentation is not only a method of preservation, but is also a tool to improve quality, digestibility and nutrition, and involves mostly natural fermentation with mixed microflora that grow simultaneously or in succession or through inoculated fermentations. To date, there are several indigenous foods all over South Asia which are not yet investigated microbiologically or biochemically, and which offer a potential field for future research. A serious study of traditional fermentation processes may also reveal the intellectual richness of the indigenous people of South Asia, which needs to be documented. Many fermented foods have health-promoting or disease-preventing/curing properties, acting not only as a source of nutrition, but also as functional or probiotic foods, and for this reason, there is an increase in consumption of such foods throughout the world. The efficient utilization of locally available low cost food substrates to prepare functional foods by fermentation has potential for an integrated approach to malnutrition management in several countries of South Asia.

Most of the indigenous fermented foods are produced in homes, villages, and small cottage industries at prices within the means of a majority of the consumers. In line with demand, the future could see the industrialization of indigenous fermented foods in South Asia, which would pave the way for their technological and economic growth. However, poor hygienic conditions and lack of specific education in food safety present potential food safety problems, which are of concern in promoting the technology, particularly in areas where facilities for the safe preparation of food are either scarce or lacking. Future research could be directed to solve these problems. Study of such foods may also provide clues as to how food production and preservation contribute to improved nutrition and, hence, health of people in different countries of South Asia.

References

Aalbersberg, W.G.L., Lovelace, C.E.A., Madheji, K., and Parkinson, S.V. 1988. Davuke, the traditional Fijian method of pit preservation of staple carbohydrate foods. *Ecology of Food and Nutrition* 21:173–180.

Abbas, C.A. 2006. Production of antioxidants, aromas, colours, flavors and vitamins by yeasts. In: Amparo, Q. and Graham, F. (Eds.). *Yeasts in Food and Beverages, The Yeast Handbook.* Vol 2. Springer-Verlag, Berlin Heidelberg, pp. 285–334.

Abee, T., Krockel, L., and Hill, C. 1995. Bacteriocins: Modes of action and potential in food preservation and control of food poisoning. *International Journal of Food Microbiology* 28:169–185.

Achaya, K.T. 1994. *Indian Food: A Historical Companion.* Oxford University Press, London.

Achi, O.K. 2005. Review: Traditional fermented protein condiments in Nigeria. *African Journal of Biotechnology* 4:1612–1621.

Adams, M.R. 1999. Safety of industrial lactic acid bacteria. *Journal of Biotechnology* 68:171–178.

Afghanistan country profile. *BBC News.* 12 January 2012. Archived from the original on November 2010.

Agrahar-Murugkar, D. and Subbulakshmib, G. 2006. Preparation techniques and nutritive value of fermented foods from the Khasi Tribes of Meghalaya. *Ecology of Food and Nutrition* 45(1):27–38.

Aidoo, K.E., Nout, M.J.R., and Sarkar, P.K. 2006. Occurrence and function of yeasts in Asian indigenous fermented foods. *FEMS Yeast Research* 6:30–39.

Aidoo, P.A. 2011. Effect of Brewer's yeast (*Saccharomyces cerevisiae* var. ellipsoideus), Baker's Yeast (*Saccharomyces cerevisiae*) and Dual Culture (*Saccharomyces cerevisiae* var. Ellipsoideus and *Saccharomyces cerevisiae*) on the fermentation of pineapple juice into wine. A thesis submitted to the school of research and graduate studies, Kwame Nkrumah university of

Science and Technology, Kumasi Institute of distance education, in partial fulfillment of the requirements for the award of master of science (postharvest technology) degree.

Amadi, E.N., Barimalaa, I.S., and Omosigho, J. 1999. Influence of temperature on the fermentation of bambara groundnut Vigna subterranea, to produce a dawadawa-type product. *Plant Foods for Human Nutrition* 54:13–20.

Amerine, M.A., Berg, H.W., Kunkee, R.E., Qugh, C.S., Singleton, V.L., and Webb, A.D. 1980. *The Technology of Wine Making.* 4th edition, AVI, Westport, CT.

Amoa-Awua, W.K., Terlabie, N.N., and Sakyi-Dawson, E. 2006. Screening of 42 Bacillus isolates for ability to ferment soybeans into dawadawa. *International Journal of Food Microbiology* 106:343–347.

Angchok, D., Dwivedi, S.K., and Ahmed, Z. 2009. Traditional foods and beverages of Ladakh. *Indian Journal of Traditional Knowledge* 8(4):551–558.

Angchok, D. and Srivastava, R.B. 2012. Technology intervention and repercussion among high altitude community of Ladakh: A case study of Trench Greenhouse. *Indian Research Journal of Extension Education* Special Issue, I, January, 2012, 268–271.

Aniche, G.N., Nwokedi, S.I., and Odeyemi, O. 1993. Effect of storage temperature, time and wrapping materials on the microbiology and biochemistry of Ogiri—A fermented castor seed soup condiment. *World Journal of Microbiology and Biotechnology* 9:653–655.

Ann, M. and Nejib, G. 2007. Fermentation as a Method for Food Preservation In: *Handbook of Food Preservation,* Rahman M.S. (ed), 2nd edition, CRC Press, Boca Raton, FL, pp. 215–236.

Annual Report 2010a. Ministry of Finance—Sri Lanka. 2011. In: Ayres, C.A., Mundt, J.O. and Sandine, W.E. (Eds.). *1980. Microbiology of Foods.* WH Freeman and Co., San Francisco, CA.

Annual Report 2010b. Ministry of Finance—Sri Lanka. 2011. In: *The World Facebook: Rank Order Population.* CIA. Retrieved 14 February 2014.

Arvanitoyannis, I.S. 2009. *HACCP and ISO 22000—Application to Foods of Animal Origin.* Blackwell Publishing Ltd., A John Wiley & Sons, Ltd., Publication, IA.

Baishya, D. and Deka, M. 2009. *Fish Fermentation, Traditional to Modern Approaches.* New India Pub. Agency, New Delhi, p. 89.

Bajpai, S.C. 1987. *Lahaul Spiti: A Forbidden Land in the Himalayas.* Indus Publishing Co., New Delhi.

Banting, E. 2003. *Afghanistan: The Land.* Crabtree Publishing Company, Catharines Canada, pp. 4.

Batra, L.R. and Millner, P.D. 1974. Some Asian fermented foods and beverages and associated fungi. *Mycologia* 66:942–950.

Batra, L.R. and Millner, P.D. 1976. Asian fermented foods and beverages. *Development in Industrial Microbiology* 17:117–128.

Battcock, M. and Azam-Ali, S. 1998. Fermented fruits and vegetables, a global perspective. *FAO Agricultural Services Bulletin,* 134.

Bentley, J.H. and Ziegler, H.F. 2003. *Traditions and Encounters—A Global Perspective on the Past, Vol. 11: From 1500 to the Present.* McGraw Hill Companies, New York, NY.

Bharadwaj, G. 2003. The ancient period. In: Majumdar, R.C. (Ed.). *History of Bengal.* B.R. Publishing Corp. Delhi, India, pp. 730.

Bhattacharya, A. 2003. Conceptualising Uyghur separatism in Chinese nationalism. *Strategic Analysis* 27(3):357–381.

Billings, T. 1998. On fermented foods. http://www.livingfoods.com.

Bisen, P.S., Mousumi, D., and Prasad, G.B.K.S. 2012. Microbes as a tool for industry and research. In: *Microbes-Concepts and Applications.* Wiley Blackwell, London, pp. 681–567.

Blandino, A., Al-Aseeri, M.E., Pandiella, S.S., Cantero, D., and Webb, C. 2003. Cereal-based fermented foods and beverages. *Food Research International* 36(6):527–543.

Cadi, P.B. 1997. Traditional foods of Bangladesh. In: *Proceedings of the International Conference on Traditional Foods at CFTRI,* Mysore, India, 2000, pp. 69–79.

Campbell-Platt, G. 1987. *Fermented Foods of the World—A Dictionary ands Guide.* Butterworths, London, 294: pp. 134–142.

Campbell-Platt, G. 1994. Fermented foods—A world perspective. *Food Research International* 27:253–257.

Caplice, E. and Fitzgerald, G.F. 1999. Food fermentations: Role of micro-organisms in food production and preservation. *International Journal of Food Microbiology* 50:131–149.

Carolissen-Mackay, V., Arendse, G., and Hastings, J.W. 1997. Purification of bacteriocins of lactic acid bacteria: Problems and pointers. *International Journal of Food Microbiology*, 34(1):1–16.

Carries, S. and Gqaleni, N. 2010. Lactic acid fermentation improves the quality of amahewu, a traditional South African maize-based porridge. *Food Chemistry* 122(3):656–661.

Coedes, G. 1962. *Les Peuples de la Peninsule Indochinoises.* Dunod, Paris.

Central Intelligence Agency. 2012. *"Bangladesh". The World Factbook.* Central Intelligence Agency, Langley, VA.

Chakrabarty, J., Sharma, G.D., and Tamang, J.P. 2013. Indigenous technology for food processing by the tribes of of Dima Hasao (North Cachar Hills) District of Assam for Social Seurity. In: Das Gupta, D. (Ed.). *Food and Environmental Foods Security Imperatives of Indigenous Knowledge of Systems.* Agrobios, Jaipur, pp. 32–45.

Chang, J.H. 1967. The Indian summer monsoon. *Geographical Review* 57(3):373–396.

Chantrupanth, D. and Phromjakgarin, C. 1978. *Khmer (Surin) Thai English Dictionary.* Chulalongkorn University Language Institute, Bangkok.

Chavan, J.K. and Kadam, S.S. 1989. *Critical reviews in food science and nutrition.* 28:348–400.

Chelule, P.K., Mokoena, M.P., and Gqaleni, N. 2010. Advantages of traditional lactic acid bacteria fermentation of food in Africa. Current Research, Technology and Education Topics. In: Méndez-Vilas, A. (Ed.). *Applied Microbiology and Microbial Biotechnology* (2nd edition). Formatx, Badajoz, Spain, pp. 1160–1167.

Chojnacka, K. 2010. *Chemical Engineeering and Chemical Process Technology, Vol. V, Fermentation Products.* Encyclopedia of Life Support Systems (EOLSS). 338p.

Choudhury, A.U. 2007. White-winged duck *Cairina (Asarcornis) scutulata* and blue-tailed bee-eater *Merops philippinus*: Two new country records for Bhutan. *Folktail* 23:153–155.

Chukeatirote, E., Dajanta, K., and Apichartsrangkoon, A. 2010. Thua nao, indigenous Thai fermented soybean: A review. *Journal of Biological Science* 10:581–583.

Chung, T.C., Axelsson, L.T., Lindgren, S.E., and Dobrogosz, W.J. 1989. *In vitro* studies on reuterin synthesis by Lactobacillus CIA—The World Factbook. Cia.gov. Retrieved 5 December, 2012.

Clavel, D. and Brabet, C. 2013. Mycotoxin contamination of nuts, Improving the safety and quality of nuts. In: Harris, L.J. (Ed.). *Improving the Safety and Quality of Nuts.* Woodhead Publishing, Cambridge, pp. 88–118.

Cogan, T.M. 1986. The leuconostocs: Milk products. In: Gilliland, S.E. (Ed.). *Bacterial Starter Cultures for Foods.* CRC Press, Boca Raton, FL, pp. 25–40.

Cogan, T.M. and Accolas, J.P. 1996. *Dairy Starter Cultures.* VCH Publishers, Cambridge.

Cogan, T.M. and Hill, C. 1993. Cheese starter cultures. In: Fox, P.F. (Ed.), *Cheese: Chemistry, Physics and Microbiology.* Vol. 1, 2nd edition, Chapman and Hall, London, pp. 193–255.

Conner, H.A. and Allgeier, R.J. 1976. Vinegar: Its history and development. *Advances in Applied Microbiology*, 20:81–133.

Cooke, R.D, Twiddy, R., Reilly, P., and Alan J. 1987. Lactic-acid fermentation as a low-cost means of food preservation in tropical countries. *FEMS Microbiology Letters* 46(3):369–379.

Cronk, T.C., Steinkraus, K.H., Hackler, L.R., and Mattick, L.R. 1977. Indonesia tape ketan fermentation. *Applied and Environmental Microbiology* 33:1067–1073.

Dahal, N., Karki, T.B., Swamylingappa, B. Li, Q., and Gu, G. 2005. Traditional foods and beverages of Nepal—A review. *Food Review International* 21:1–25.

Dakwa, S., Sakyi-Dawson, E., Diako, C., Annan, N.T., and Amoa-Awua, W.K. 2005. Effect of boiling and roasting on the fermentation of soybeans into soy-dawadawa. *International Journal of Food Microbiology* 104(1):69–82.

Dame, J. and Nüsser M. 2011. Food security in high mountain regions: Agricultural production and the impact of food subsidies in Ladakh, Northern India. *Food Security* 3:179–194.

Daramola, O. and Jeje, A. 1975. *Awon Asa ati Orisa Ile Yoruba*. Onibonoje Press Ltd., Ibadan.

Darby, W.J. 1979. The nutrient contributions of fermented beverages. In: Gastineau, C.F., Darby, W.J. and Turner, T.B. (Eds.). *Fermented Food Beverage in Nutrition*. Academic Press, New York, NY, pp. 61–79.

Das, C.P. and Pandey, A. 2007. Fermentation of traditional beverages prepared by Bhotiya community of Uttaranchal Himalaya. *Indian Journal of Traditional Knowledge* 6:136–140.

Dasgupta, A. (Ed.). 1958. Alcohol use and abuse: Past and present. In: *The Science of Drinking*. Rowman & Littlefield Publishers, Lanham, MD, pp. 1–14.

Davis, S.H. and Ebbe, K. (Eds.). 1993. Traditional knowledge and sustainable development. *Environmentally Sustainable Development Proceedings Series No. 4*. The World Bank, Washington D.C.

De Vuyst, L. and Vandamme, J.E. 1994. *Bacteriocins of Lactic Acid Bacteria, Microbiology, Genetics and Applications*. Blackie Academic & Professional, London.

Deraniyagal, S. 1992. *The Prehistory of Sri Lanka*. Department of Archaeological Survey, Colombo, p. 454.

Deraniyagala, P.E.P., Paranavitana, S., Prematilleka, L., Engelberta, J., and Leeuw, L.-D. 1978. In: *Senarat Paranavitana Commemoration*. BRILL Academic Pub Netherlands, p. 52.

Devi, P. and Suresh, P. 2012. Traditional, ethnic and fermented foods of different tribes of Manipur. *Indian Journal of Traditional Knowledge* 11(1):70–77.

Dewan, S. 2002. Microbiological evaluation of indigenous fermented milk products of the Sikkim Himalayas. PhD thesis. Food Microbiology Laboratory, Sikkim Government College, Gangtok.

Dewan, P., Kaur, I., Rusia, U., Faridi, M.M.A., Baweja, R., and Agarwal, K.N. 2003. Fermented Milk (Dahi and Actimel) on diarrhoea and immunological status in protein-energy malnutrition. In: *International Seminar and Workshop on Fermented Foods, Health Status and Social Well-Being* held on Nov 2003, Anand, India, p. 55.

Dirar, H.A. 1993. *The indigenous fermented foods of the Sudan. A study in African food and nutrition*, CAB International, Wallingford, UK.

Eaton, R. 1996. *The Rise of Islam and the Bengal Frontier*. Berkeley: University of California Press. pp. 1204–1704.

Ebine, H. 1989. Industrialization of Japanese miso fermentation. In: Steinkraus, K.H. (Ed.). *Industrialization of Indigenous Fermented Foods*. Marcel Dekker, New York, NY, pp. 89–126.

Eklund, T. 1984. The effect of carbon dioxide on bacterial growth and on uptake processes in the bacterial membrane vesicles. *International Journal of Food Microbiology* 1:179–185.

Ellen, R. and Harris, H. 1996. Concepts of Indigenous Technical Knowledge in Scientific and Developmental Studies Literature: A Critical Assessment. www.worldbank.org/afr/ik/basic.htm_68k

FAOSTAT. 2008. Food Production. faostat.fao.org.

Farnworth, E.R. 2003. *Handbook of Fermented Functional Foods*. CRC Press, New York, NY.

Fleet, G.H. 1998. The microbiology of alcoholic beverages. In: Wood, B.J.B. (Ed.). *Microbiology of Fermented Foods*. 1st edition. Blackie Academic and Professional, London, pp. 217–262.

Fleet, G.H. 2006. The commercial and community significance of yeasts in food and beverage production. In: Amparo, Q. and Graham, F. (Eds.). *Yeasts in Food and Beverages. The Yeast Handbook*. Vol. 2. Springer-Verlag, Berlin Heidelberg, pp. 1–13.

Fox, P.F. 1993. Cheese: An overview. In: Fox, P.F. (Ed.). *Cheese: Chemistry, Physics and Microbiology*. Vol. 1, 2nd edition, Chapman and Hall, London, pp. 1–36.

Fox, P.F. and McSweeney, P.L.H. 2004. Cheese: An overview. Cheese: Chemistry, Physics and Microbiology, Volume 1, General Aspects, 3rd edition. In: Fox P.F., McSweeny P.L.H, Cogan T.M., and Guinee T.P. (Eds.). Elsevier Applied Science, Amsterdam, pp. 1–18.

Frazier, W.C. and Westhoff, D.C. 2004. *Food Microbiology*. TMH, New Delhi.

Fukashima, D. 1979. Fermented vegetable soybean—Protein and related foods of Japan and China. *Journal of American Oil Chemists Society* 56:357–362.

Gadaga, T., Mutukumira, A.N., Narvhus, J.A., and Feresu, S.B. 1999. A review of traditional fermented foods and beverages of Zimbabwe. *International Journal of Food Microbiology* 53:1–11.

Geertz, C. 1963. *Old Societies and New States. The Quest for Modernity in Asia and Africa*. Glencoe & London/UK. pp. 105–157.

Geisen, R. and Holzapfel, W.H. 1996. Genetically modified starters and protective cultures. *International Journal of Food Microbiology* 30:315–324.

Giraffa, G. 2004. Studying the dynamics of microbial populations during food fermentation. *FEMS Microbiology Reviews* 28(2):251–260.

Gourama, H. and Bullerman, L.B. 1995. Antimycotic and antiaflatoxigenic effect of lactic acid bacteria: A Review. *Journal of Food Protection* 58:1275–1280.

Government of India. Ministry of Home Affairs Annual Report 1960.

Gupta, A.D. 2013a. Querying relevance of the term "Indigenous peoples". In: *A Three-Level-Approach with Special Reference to Rajbanshi*. International E-Publication 427, Indore, p. 155.

Gupta, A.D. 2013b. *Indigenous People of Sub-Himalayan North Bengal with Special Reference to the Rajbanshis*. International E-Publication 427, Indore, p. 375.

Gupta, A.D. 2013c. *Migration: An Anthropological Perspective with Special Reference to North Bengal, India*. International E-Publication 427, Indore, p. 166.

Gupta, S. and Abu-Ghannam, N. 2011. Probiotic fermentation of plant based products: Possibilities and opportunities. *Critical Reviews in Food Science and Nutrition*, 52(2): 183–199.

Gupta, S. and Abu-Ghannam, N. 2012. Probiotic fermentation of plant based products: Possibilities and opportunities. *Critical Reviews in Food Science and Nutrition* 52(2):183–199.

Gupta, A. and Loralie, L.J. 2012. Alcohol: Use and abuse. In: *Pharmacogenomics of Alcohol and Drugs of Abuse*. Das Gupta A. and Langman, L.J. (Eds.). CRC Press, Taylor & Francis, Boca Raton, FL, pp. 1–229.

Haber, M., Platt, D.E., AshrafianBonab, M., Youhanna, S.C., and Soria-Hernanz, D.F. 2012. Afghanistan's ethnic Groups share a Y-chromosomal heritage structured by historical events. PLOS One 7(3):e, 34288.

Hafiz, F. and Majid, A. 1996. Preparation fermented food from rice and pulses. *Bangladesh Journal of Scientific and Industrial Research* 31(4):43–61.

Hansen, E.B. 2002. Commercial bacterial starter cultures for fermented foods of the future. *International Journal of Food Microbiology* 78:119–131.

Hassan, M.N. 2003. Fermented Foods of Bangladesh. In: *International Seminar and Workshop on Fermented Foods, Health Status and Social Well-Being* held on Nov 2003, Anand, India, p. 12.

Haverkort, B., Reijntjes, C., and Waters-Bayer, A. 1992. An introduction to low-external input and sustainable agriculture. In: Reijntjes, C., Haverkort, B., and Waters-Bayer, A. (Eds.). *Farming for the Future*. Macmillan, London.

Hayford, A.E. and Jespersen, L. 1999. Characterization of Saccharomyces cerevisiae strains from spontaneously fermented maize dough by profiles of assimilation, chromosome polymorphism, PCR and MAL genotyping. *Journal of Applied Microbiology* 86(2):284–294.

Henley, J. 2008. *The Last Days of Paradise*. The Guardian, London.

Hesseltine, C.W. 1979. Some important fermented foods of Mid-Asia, the Middle East, and Africa. *Journal of the American Oil Chemists' Society* 56:367.

Hesseltine, C.W. and Wang, H.L. 1980. The importance of traditional fermented foods. *BioScience* 30:402–404.

Hill, C., Ross, R.P., and Morgan, S. 2002. Preservation and fermentation: Past, present and future. *International Journal of Food Microbiology* 79:3–16.

Hogendorn, J. and Johnson, M. 1986. *The Shell Money of the Slave Trade.* African Studies Series 49, Cambridge University Press, London, pp. 20–22.

Holzapfel, W.H. 2002. Appropriate starter culture technologies for small-scale fermentation in developing countries. *International Journal of Food Microbiology* 75:197–212.

Holzapfel, W.H., Schillinger, U., Toit, M.D., and Dicks, L. 1997. Systematics of probiotic lactic acid bacteria with reference to modern phenotypic and genomic methods. *Microecology and Therapy* 26:1–10.

Hugas, M., Garriga, M., Aymerich, M.T., and Monfort, J.M. 1993. Biochemical characterization of lactobacilli from dry fermented sausages. *International Journal of Food Microbiology* 18:107–113.

Hutkins, R.W. 2006. *Microbiology and Technology of Fermented Foods.* Blackwell Publishing Professional, Ames, IA, pp. 3–14.

Ishige, N. and Ruddle, K. 1987. Gyosho in Southeast Asia—A study of fermented aquatic products. *Bulletin of the National Museum of Ethnology* 12(2):235–314.

Ishige, N. and Ruddle, K. 1990. *Gyosho to narezushi no kenkyu (Research on fermented fish products and narezushi): Iwanamishoten* (in Japanese), Iwanamishoten, Tokyo.

Ishige, N. and Sakiyama. O. 1988. Gyosho to Narezushi no Kenkyu (An ethnolinguistic study of the nomenclature of Fermented fish products in: Northeast and Southeast Asia). *Bulletin of the National Museum of Ethnology* 13(2):383–406.

Issacs, H. 1974. Basic group identity the idols of the tribes. *Ethinicty* 7:15–42.

Iwuoha, C.I. and Eke, O.S. 1996. Nigerian indigenous fermented foods: Their traditional process operation, inherent problems, improvements and current status. *Food Research International* 29:527–540.

Jaiswal, V. 2010. Culture and ethnobotany of Jaintia tribal community of Meghalaya, Northeast India—A mini review. *Indian Journal of Traditional Knowledge* 9(1):38–44.

Jay, J.M., Loessner, M.J., and Golden, D.A. 2004. *Modern Food Microbiology*, 7th edition. Springer, Corporation, New York, pp. 3–5.

Jay, J.M. 1986. *Modern Food Microbiology.* van Nostrand Reinhold, New York, NY.

Jay, J.M. 1996. *Modern Food Microbiology.* 5th edition. Chapman and Hall, New York, NY.

Jayasinghe, P.S. and Sagarika, E. 2003. *Cured and Fermented Fishery Products of Sri Lanka in International Seminar and Workshop on Fermented Foods, Health Status and Social Well-Being* held on Nov 2003, Anand, India, p. 7.

Jelliffe, D.B. 1968. *Infant Nutrition in the Tropics.* World Health Organization (WHO), Geneva, p. 335.

Jespersen, L. 2003. Occurrence and taxonomic characteristics of strains of Saccharomyces cerevisiae predominant in African indigenous fermented foods and beverages. *FEMS Yeast Research* 3(2):191–200.

Jeyaram, K., Singh, T.A., Romi, W., Devi, A.R., Singh, W.M., Dayanidhi, H., Singh, N.R., and Tamang, J.P. 2009. Traditional fermented foods of Manipur. *Indian Journal of Traditional Knowledge* 8(1):115–121.

Joranson, K. 2008. Indigenous knowledge and the knowledge commons. *International Information and Library Review* 40(1):64–72.

Joshi, V.K. 1997. *Fruit Wines.* 2nd edition. Directorate of Extension Education. Dr. YS Parmar University of Horticulture and Forestry, Nauni, pp. 1–255.

Joshi, V.K. 2005. Indigenous fermented foods of North India. In: *Fermented Foods, Health Status and Social Well Being. First Workshop on Strategic Meeting of the Swedish South Asian Network on Fermented Foods for Policy Makers of the Institutions of Higher Education and Research.* Organized by Institute of Rural Management Anand, Swedish South Asian Network on Fermented Foods, University Sinedesnal Anand University Anand 26–27 May 2005, Anand, India.

Joshi, V.K. 2006a. Food fermentation: Role, significance and emerging trends. In: Trivedi, P.C. (Ed.). *Microbiology: Applications and Current Trends*. Pointer Publisher, Jaipur, pp. 1–35.

Joshi, V.K. 2006b. *Sensory Science: Principles and Applications in Evaluation of Food*. Agro-Tech Publishers, Udaipur, p. 527 + Plates, Figures.

Joshi, V.K. Thakur, N.S., Bhatt, A., and Chayanika, G. 2011. *Handbook of Enology: Principles, Practices and Recent Innovations*. Vol I. Asia-Tech Publisher and Distributors, New Delhi, p. 3–45.

Joshi, V.K., Bhutani, V.P., and Thakur, N.K. 1999. Composition and nutrition of fermented products. In: Joshi, V.K. and Ashok, P. (Eds.). *Biotechnology: Food Fermentation*. Vol. I. Educational Publishers and Distributors, New Delhi, pp. 259–320.

Joshi, V.K., Garg, V., and Abrol, G.S. 2012. Indigenious fermented foods. In: Joshi, V.K. and Singh, R.S. (Eds.). *Food Biotechnology: Principles and Practices*. IK Publishing House, New Delhi, pp. 337–374.

Joshi, V.K., Kaushal, N.K., and Sharma, N. 2000. Spoilage of fruits, vegetables and their products. In: Verma, L.R. and Joshi, V.K. (Eds.). *Postharvest Technology of Fruits and Vegetables*. The Indus Publishing Co., New Delhi, pp. 235–285.

Joshi, V.K., Lakhanpal, P., and Vikas, K. 2013. Occurrence of Patulin, its dietary intake through consumption of apple and apple products and methods of its removal. *International Journal of Food and Fermentation Technology* 3(1):1–14.

Joshi, V.K. and Pandey, A. 1999. Biotechnology: Food fermentation. In: Joshi, V.K. and Pandey, A. (Eds.). *Biotechnology: Food Fermentation*. Vol. I. Educational Publishers and Distributors, New Delhi, pp. 1–24.

Joshi, V.K., Sharma, S., John, S., Kaushal, B., Lal, B., and Rana, N. 2009. Preparation of antioxidant rich apple and strawberry wines. *Proceedings of National Academy Sciences of India, Section B* 79(I):415–420.

Joshi, V.K., Sharma, S., and Rana, N. 2006. Production, purification, stability and efficiency of bacteriocin from the isolate of natural lactic acid fermentation of vegetables. *Food Technology Biotechnology* 4(3):435–439.

Joshi, V.K., Sharma, S., and Thakur, N.S. 2003. Technolgy of fermented fruits and vegetables. In: *International Seminar and Workshop on Fermented Foods, Health Status and Social Well-Being* held on Nov 2003, Anand, India, p. 28.

Joshi, V.K. and Siby, J. 2002. Antimicrobial activity of apple wine against some pathogenic and microbes of public health significance. *Acta Alimentaria* 31(6):67–72.

Kanwar, S.S., Gupta, M.K., Katoch, C., Kumar, R., and Kanwar, P. 2007. Traditional fermented food of Lahaul and Spiti area of Himachal Pradesh. *Indian Journal of Traditional Knowledge* 6(1):42–45.

Kathleen, W.D. 2001. *Countries of the World: Pakistan*. Capstone Publishers, Press, Mankato, US pp. 13–15.

Katz, S.H. and Voight, M.M. 1987. Bread and beer. *Expedition*. 28:23–24.

Kayang, H. 2007. Tribal knowledge on wild edible plants of Meghalaya, Northeast India. *Indian Journal of Traditional Knowledge* 6:177–181.

Kelly, W., Asmundson, R.J.V., and Huang, C.M. 1996. Isolation and characterization of bacteriocin-producing lactic acid bacteria from ready-to-eat food products. *International Journal of Food Microbiology* 33(2–3):209–218.

Kiers, J.L., Van Laeken, A.E.A., Rombouts, F.M., and Nout, M.J.R. 2000. *In vitro* digestibility of Bacillus fermented soya bean. *International Journal of Food Microbiology* 60:163–169.

Kimizuka, A., Tadashi, M., Kenneth, R., and Ishige, N. 1992. Chemical components of fermented fish products. *Journal of Food Composition and Analysis* 5(2):152–159.

Kimmons, J.E., Brown, K.H., Lartey, A., Collison, E., Mensah, P.P., and Dewey, K.G., 1999. The effects of fermentation and/or vacuum flask storage on the presence of coliforms in complementary foods prepared in Ghana. *International Journal of Food Science and Nutrition* 50:195–201.

Klanarong, S. and Piyachomkwan, K. 2013. The outlook of sugar and starchy crops. In: Yang, Hesham, S.-T., El-Enshasy, A. and Tgchul, N. (Eds.). *Biorefinery, Bioprocessing Technologies in Biorefinery for Sustainable Production of Fuels, Chemicals and Polymers*, John Wiley & Sons, Inc., Hoboken, NJ, pp. 27–43.

Kose, S. and Hall, G.M. 2010. Sustainability of fermented fish products In: Hall, G.M. (Ed.). *Fish Processing Sustainability and New Opportunities.* John Wiley & Sons, West Sussex, pp. 138–163.

Kuboye, A.O. 1985. Traditional femented foods and beverages of Nigeria. In: *Development of Indigenous Fermented Foods and Food Technology in Africa.* International Foundation for Science (IFS), Stockholm, pp. 224–237.

Kuiper, K. 2010. *The Culture of India.* Britannica Educational Publishing. p. 346.

Kumar, R., Singh, R.D., and Sharma, K.D. 2005. Water resources of India. *Current Science* 89:794–811.

Kumar, V. 2000. *Vastushastra, All You Wanted to Know About Series,* 2nd edition, Sterling Publishing Pvt. Ltd. Geomancy. 160p.

Law, S.V., Abu Bakar, F., Mat Hashim, D., and Abdul Hamid, A. 2011. Popular fermented foods and beverages in Southeast Asia. *International Food Research Journal*, 18:475–484.

Lee, C.H., Steinkraus, K.H., and Reilly, P.J.A. (Eds.) 1991. *Fish Fermentation Technology.* United Nations University, Tokyo.

Lee, J.O. and Kim, J.Y. 2013. Development of cultural context indicator of fermented food. *International Journal of Bio-Science and Bio-Technology* 5(4):45.

Leroy, F. and De Vuyst, L. 2004. Lactic acid bacteria and functional starter cultures for the food fermentation industry. *Trends in Food Science and Technology* 15:67–78.

Lindgren, S.E. and Dobrogosz, W.J. 1990. Antagonistic activities of lactic acid bacteria in food and feed fermentations. *FEMS. Microbiology Review* 7(1–2):149–163.

Lorri, W. 1995. Review of fermentation in improving nutritional value. Background paper prepared for *WHO/FAO Workshop: "Assessment of Fermentation as a Household Technology for Improving Food Safety,"* December 11–15, 1995, CSIR, Pretoria, South Africa.

Majumdar, R.K. and Basu, S. 2009. Evaluation of the influence of inert atmosphere packaging on the quality of salt-fermented Indian shad. *International Journal of Food Science and Technology* 44(12):2554–2560.

Majumdar, R.K., Basu, S., and Nayak, B.B. 2006. Studies on the biochemical changes during fermentation of salt-fermented Indian Shad (*Tenualosa ilisha*). *Journal of Aquatic Food Product Technology* 15(1):53–69.

Mallikarjun, B. 2004. Fifty years of language planning for modern Hindi—The official language of India. *Language in India* 4(11). http://www.languagemedia.com/feb2004/lucknowpaper.html.

McKay, M., Buglass, A.J., and Lee, C.G. 2011. Fermented beverages: Beers, ciders, wines and related drinks. *Handbook of Alcoholic Beverages: Technical, Analytical and Nutritional Aspects* 1:63–229.

Mehta, M.B., Eldin-Kamal, A., and Iwanski, Z.R. 2012. *Fermentation Effects on Food Properties. Chemical and Functional Properties of Food Components Series.* Taylor & Francis Group, CRC Press, Boca Raton, 399pp.

Mensah, P. 1997. Fermentation-the key to food safety assurances in Africa? *Food Control* 8:271–278.

Mensah, P., Tomkins, A.M., Brasar, B.S., and Harrison, T.J. 1990. Fermentation of cereals for reduction of bacterial contamination of weaning foods in Ghana. *Lancet* 336:140–143.

Miller, J.D. and Trenholm, H.L. 1994. *Mycotoxins in Grains: Compounds Other Than Aflatoxin.* Eagan Press, St.Paul, MN.

Mir, M.S. and Mir, A.A. 2000. Ethnic foods of Ladakh (J&K) India. In: Sharma, J.P. and Mir, A.A. (Eds.). *Dynamics of Cold Arid Agriculture.* Kalyani Publication, New Delhi, pp. 297–306.

Mizutani, T., Akimitsu, K., Kenneth, R., and Naomichi, I. 1987. A chemical analysis of fermented fish products and discussion of fermented flavors in Asian Cuisines, a study of fermented fish products. *Bulletin of the National Museum of Ethnology* 12(3):801–864.

Mokoena, M.P., Chelule, P.K., and Gqaleni, N. 2005. Reduction of Fumonisin B1 and zearalenone by lactic acid bacteria in fermented maize meal. *Journal of Food Protection* 68:2095–2099.

Mokoena, M.P., Chelule, P.K., and Gqaleni, N. 2006. The toxicity and decreased concentration of aflatoxin B1 in natural lactic acid fermented maize meal. *Journal of Applied Microbiology* 100(4):773–777.

Mondal, S.R. 2009. Biodiversity management and sustainable development—The issues of indigenous knowledge system and the rights of indigenous people with particular reference to North Eastern Himalayas of India. In: Das Gupta, D.C (Ed.). Indigenous knowledge systems and common people's Rights Agrobios, Jodhpur, India.

Motarjemi, Y., Kaferstein, F., Moy, G., and Quevedo, F.Q. 1993. Contaminated weaning food: A major risk factor for diarrhoea and associated malnutrition. *Bulletin of the World Health Organization* 71(1):79–92.

Motlagh, A.M., Johnson, M.C., and Ray, B. 1991. Viability loss of foodborne pathogens by starter culture metabolites. *Journal of Food Protection* 54: 873–878.

Moussay, G. 1971. *Dictionaire Cam Vietnamien Francais*. Centre Culturelle Sam, Phan Rang.

Muller, C.P. 1998. *The Microbiology of Low-Salt fermented fish Products* 46:219–229.

Muralidhar, R., Choudhary, A.K., and Prajapati, P.K. 2003. Fermented foods in ayurvedic pharmaceutics W.S.R to Sirisarista. In: *International Seminar and Workshop on Fermented Foods, Health Status and Social Well-Being* held on Nov 2003, Anand, India, p. 128.

Murty, D.S. and Kumar, K.A. 1995. Traditional uses of sorghum and millets. In: Dendy, D.A.V. (Ed.). *Sorghum and Millets: Chemistry and Technology*, American Association of cereal chemists St. Paul, MA, USA. pp. 185–221.

Muzaddadi, A.U. and Mahanta, P. 2013. Effects of salt, sugar and starter culture on fermentation and sensory properties in *Shidal* (a fermented fish product). *African Journal of Microbiology Research* 7(13):1086–1097.

Nema, P.K., Karlo, T., and Singh, A.K. 2003. Opo, the rice beer. In: *International Seminar and Workshop on Fermented Foods, Health Status and Social Well-Being* held on Nov 2003, Anand, India, p. 59.

Nout, M.J. 2009. Rich nutrition from the poorest—Cereal fermentations in Africa and Asia. *Food Microbiology* 26:685–692.

Nout, M.J.R. 2001. Fermented foods and their production In: Adams, M.R. and Nout, M.J.R. (Eds.). *Fermentation and Food Safety*. Aspen Publishers, Inc., Gaithersssburg, MD, pp. 1–30.

Nout, M.J.R. and Motarjemi, Y. 1997. Assessment of fermentation as a household technology for improving food safety. *A Joint FAO/WHO Workshop. Food Control* 8:221–226.

Obiri-Danso, K., Ellis, W.O., Simpson, B.K., and Smith, J.P. 1997. Suitability of high lysine maize, Obatanpa for "kenkey" production. *Food Control* 8(3):125–129.

Odunfa, S.A. 1983. Biochemical changes during production of ogiri, a fermented melon (*Citrullus vulgaris Schrad.*) product. *Plant Foods for Human Nutrition*, 32(1): 11–18.

Odunfa, S.A. 1985. African fermented foods. In: Wood, B.J.B. (Ed.). *Microbiology of Fermented Foods*. Vol. 1. Elsevier, New York, NY, pp. 155–191.

Ogundele, S.O. 2007. Understanding aspects of Yoruba Gastronomic culture. *Indian Journal of Traditional Knowledge* 6(1):50–56.

Okafor, N. 1975. Microbiology of Nigerian palm wine with particular reference to bacteria. *Journal of Applied Bacteriology* 38:81–88.

Olukoya, D.K., Ebigwei, S.I., Olasupo, N.A., and Ogunjimi, A.A. 1994. Production of ogi: An improved ogi (Nigerian fermented weaning food) with potentials for use in diarrhoea control. *Journal of Tropical Pediatrics* 40:108–113.

Omafuvbe, B.O., Shonukan, O.O., and Abiose, S.H. 2000. Microbiological and biochemical changes in the traditional fermentation of soybean for "soy-daddawa"—A Nigerian food condiment. *Food Microbiology* 17:469–474.

O'Mahony, M. and Ishii, R. 1987. The Umami Taste concept: Implications for the Dogma of four basic tastes. In: Kawamura, Y. and Morley, R.K. (Eds.). *Umami: A Basic Taste*. Marcel Dekker, New York and Basel, pp. 75–93.

Ottenheimer, H.J. 2008. *The Anthropology of Language: An Introduction to Linguistic Anthropology*. Cengage.

Padmaja, G. 1995. Cyanide detoxification in cassava for food and feed uses. *Critical Reviews of Food Science and Nutrition* 35:299–339.

Padmaja, G. and George, M. 1999. Oriental fermented foods. In: Joshi, V.K. and Pandey, A. (Eds.). *Biotechnology: Food Fermentation*. Vol. 2, Educational Publishers and Distributors, New Delhi, pp. 523–581.

Panesar, P.S. and Marwaha, S.S. 2013. Biotechnology in food processing. In: *Biotechnology in Agriculture and Food Processing Opportunities and Challenges*, Panesar P.S. and Marwaha, S.S (Eds). CRC Press Taylor & Francis, Boca Raton, FL.

Pant, R.M. and Nema, P.K. 2003. Apong, the rice beer of Arunachal: Some socio-cultural aspects. In: *International Seminar and Workshop on Fermented Foods, Health Status and Social Well-Being* held on Nov 2003, Anand, India, p. 96.

Parigi, R.K. and Prakash, V. 2012. Innovations in functional food industry for health and wellness. In: Ghosh, D., Das, S., Bagchi, D. and Smarta, R.B. (Eds.). *Innovation in Healthy and Functional Foods*. CRC Press Taylor & Francis, Boca Raton, FL, pp. 5–13.

Parker, H. 1992. *Ancient Ceylon*. Asian Educational Services. p. 7.

Parkouda, C., Nielsen, D.S., Azokpota, P., Ouoba, L.I.I., Amoa-Awua, W.K., Thorsen, L., Hounhouigan, J.D. et al. 2009. The microbiology of alkaline-fermentation of indigenous seeds used as food condiments in Africa and Asia. *Critical Review Microbiology* 35:139–156.

Parkouda, C., Thorsen, L., Compaoré, C.S., Nielsen, D.S., Tano-Debrah, K., Jensen, J.S., Diawara, B., and Jakobsen, M. 2010. Micro-organisms associated with Maari, a Baobab seed fermented product. *International Journal of Food Microbiology* 142(3):292–301.

Parveen, S. and Hafiz, F. 2003. Fermented cereal from indigenous raw materials. *Pakistan. Journal of Nutrition* 2(5):289–291.

Pederson, C.S. 1960. Sauerkraut. In: Chichester, C.O., Mark, E.M., and Stewart, G.F. (Eds.). *Advances in Food Research*. Vol. 10. Academic Press, New York, NY, pp. 233–260.

Pederson, C.S. 1971. *Microbiology of Food Fermentations*. AVI, Publishing Co., Westport CT, pp. vi, 283.

Pederson, C.S. and Albury, M.N. 1969. Sauerkraut fermentation. *Food Technology*, 8:1–5.

Penny, E.V. 2008. *Food Culture in South East Asia*. Green wood Press, Westport, CT, p. 112.

Petchkongkaew, A. 2007. Reduction of mycotoxin contaminationlevelduring soybean: A Thesis Submitted in Partial Fulfillment of the Requirements for the Degree of Doctor of Philosophy in Food Technology, Suranaree University of Technology/Academic Year, 2007.

Phadnis, U. and Gaungly, R. 2001. *Ethnicity and Nation-Building in South Asia*. Sage Publications, New Delhi, p. 459.

Prajapati, J.B. and Nair, B.M. 2008. The history of fermented food. In: Farnworth, E.R. (Ed.). *Handbook of Fermented Functional Food* 2nd edition, CRC Press Taylor & Francis, Boca Raton, FL, pp. 1–22.

Rajalakshmi, R. and Vanaja, K. 1967. Chemical and biological evaluation of the effects of fermentation on the nutritive value of foods prepared from rice gram. *British Journal of Nutrition* 21:467–473.

Rao, D.R., Pulusani, S.R., and Chawan, C.B. 1986. Fermented soybean milk and other fermented legume milk products. In: Reddy, N.R., Pierson, M.D., and Salunkhe, D.K. (Eds.). *Legume-Based Fermented Foods*. CRC Press, FL, pp. 119–134.

Rao, R.E., Varadaraj, M. and Vijayendra, S. 2005. Fermentation biotechnology of traditional foods of the Indian Sub-continent. In: Pometto, A., Shetty, K., Paliyath, G., and Levin, R.E. (Eds.). *Food Biotechnology.* 2nd edition. CRC Press, Taylor & Francis, Boca Raton, FL, pp. 1759–1794.

Ray, R.C. and Joshi, V.K. 2015. Fermented foods: Past, present and future. In: Ray, R.C. and Montet, D. (Eds.). *Micro-organisms and Fermentation of Traditional Foods.* CRC Press, Taylor & Francis, New York, NY, pp. 1–36.

Reddy, L.V. and Reddy, O.V.S. 2012. Production and characterization of wine from mango (*Mangifera indica* L.) fruit juice Chapter 14. In: Hui, Y.H. and Özgül, E., Hsui, Y.H. and Özgül Evranuz, E. (Eds.). *Plant Based Fermented Foods.* CRC Press pp. 249–272.

Reddy, N.R., Pierson, M.D., Sathe Shridhar, K., and Salunkhe, D.K. 1982. Legume-based fermented foods: Their preparation and nutritional quality. *Critical Reviews in Food Science and Nutrition* 17(4):335–370.

Reddy, N.R., Pierson, M.D., and Salunkhe D.K. 1986. Idli. In: Reddy, N.R., Pierson, M.D. and Salunkhe, D.K. (Eds.). *Legume-Based Fermented Foods.* CRC Press, FL, pp. 145–160.

Rex, J. 1995. Ethenic identity and Nation State. The poltical sociology of multi-cultural society. *Social Identities* 1(24):25.

Rickmond, A.H. 1978. Migration and race relations. *Ethnic and Racial Studies* 1:60.

Rizvi, J. 1983. *Ladakh: Crossroads of High Asia.* Oxford University Press, New Delhi.

Romah, D. 1977. *Tu Dien Viet Gia Rai (Vietnamese Jarai Dictionary).* Nha Xuat Ban Khoa Hoc Xa Hoi, Hanoi.

Ross, R.P., Morgan, S., and Hill, C. 2002. Preservation and fermentation: Past, present and future. *International Journal of Food Microbiology* 79:3–16.

Roy, A., Moktan, B., and Sarkar, P.K. 2007. Traditional technology in preparing legume-based fermented foods of Orissa. *Indian Journal of Traditional Knowledge* 6:12–16.

Roy, B., Kala, C.P., Farooquee, N.A., and Majila, B.S. 2004. Indigenous fermented food and beverages: A potential for economic development of the high altitude societies in Uttaranchal. *Journal of Human Ecology* 15(1):45–49.

Ruddle, K. and Ishige, N. 2005. Fermented fish products in East Asia. Hong Kong. International Resources Management Institute. http://www.intresmanins.com.

Ryan, J.K. 2003. A thesis submitted in partial fulfilment of the requirements for the degree of Master of Liberal Arts Concentration, Gastronomy, Boston University, 2003.

Sagarika, E. and Pradeepa, J. 2003. Fermented food of Sri Lanka Maldives. In: *International Seminar and Workshop on Fermented Foods, Health Status and Social Well-Being* held on Nov 2003, Anand, India. p. 4.

Sakamoto, Y. 1976a. *Mongo Goishu (A Mon Lexicon).* Institute for the Study of Languages and Culture of Asia and Africa, Tokyo University of Foreign Languages, Tokyo.

Sakamoto, Y. 1976b. Kanbojiago shoo jiten (A Khmer Japanese Dictionary). Tokyo: Institute for the Study of Languages and Culture of Asia and Africa, Tokyo University of Foreign Languages.

Samal, P.K., Palni, L.M.S., and Devendra, K. 2003. Ecology, ecological poverty and sustainable development in Central Himalayan region of India. *International Journal of Sustainable Development and World Ecology* 10(2):157–168.

Samal, P.K., Palni, L.M.S., and Dhyani, P.P. 2005. Status and trends in research and development projects in the mountains: A situational analysis in the Indian Himalaya. *International Journal of Sustainable Development and World Ecology,* 12(4):279–288.

Samati, S. and Begum, S.S. 2006. Plant indicators for agricultural seasons amongst pnar tribe of Meghalaya, India. *Indian Journal of Traditional Knowledge,* 5:57–59.

Sanni, A.I. 1993. The need for optimisation of African fermented foods and beverages. *International Journal of Food Microbiology,* 18:85–95.

Santos, F., Wegkamp, A., de Vos, W.M., Smid, E.J., and Hugenholtz, J. 2008. High-level folate production in fermented foods by the B12 producer Lactobacillus reuteri JCM1112. *Applied and Environmental Microbiology* 74(10):3291–3294.

Sanyal, C.C. 1965. *The Rajbanshi of North Bengal*. Asiatic Society, Kolkata.

Sanyal, C.C. 2002. *The Rajbanshi of North Bengal*. Asiatic Society, Kolkata.

Sarkar, P.K., Morrison, E., Tingii, U., Somerset, S.M., and Craven, G.S. 1998. B-group vitamin and mineral contents of soybeans during kinema production. *Journal of Science Food Agriculture* 78:498–502.

Sarkar, P.K. and Nout, M.J.R. (Eds.). 2014. *Handbook of Indigenous Foods Involving Alkaline Fermentation*. CRC Press, Boca Raton, FL.

Sarkar, P.K. and Tamang, J.P. 1995. Changes in the microbial profile and proximate composition during natural and controlled fermentations of soybeans to produce kinema. *Food Microbiology* 12:317–325.

Sarkar, P.K., Tamang, J.P., Cook, P., and Eand Owens, J.D. 1994. Kinema—A traditional soybean fermented food—Proximate composition and microflora. *Food Microbiology* 11:47–55.

Savadogo, A. 2012. The role of fermentation in the elimination of harmful components present in food raw materials. In: Mehta, B.M., Kamal-Eldin, M.A. and Iwanski, R.Z. (Eds.). *Chemical and Functional Properties of Food Components*. CRC Press, Taylor & Francis, Boca Raton, FL, pp. 169–184.

Savitri, and Bhalla, T.C. 2013. Characterization of *bhatooru*, a traditional fermented food of Himachal Pradesh: microbiological and biochemical aspects. *Biotechnology* 3:247–254.

Schopf, J.W. and Packer, B.M. 1987. Early Archean (3.3-billion to 3.5-billion-year-old) microfossils from Warraweena group, Australia. *Science* 237:70–73.

Schnürer, J. and Magnusson, J. 2005. Antifungal lactic acid bacteria as biopreservatives. *Trends in Food Science & Technology*, 16:70–78.

Segi, S. 2006. A review of Fermented fish products in East Asia: In: *IRMI Research Study 1, Kenneth Ruddle, Naomichi Ishige*. International Resources Management Institute, Hong Kong (2005), *Trends in Food Science and Technology* 17(11):626.

Sekar, S. 2007. Traditional alcoholic beverages from Ayurveda and their role on human health. *Indian Journal of Traditional Knowledge* 6(1):144–149.

Sekar, S. and Mariappan, S. 2007. Usage of traditional fermented products by Indian rural folks and IPR. *Indian Journal of Traditional Knowledge* 6:111–120.

Sharma, A. and Kapoor, A.C. 1996. Levels of antinutritional factors in pearl millet as affected by processing treatment and various types of fermentation. *Plant Food Human Nutrition* 49:241–252.

Shinoda, O. 1952. Chugoku ni Okeru Sushi no Hensen (Sushikoo Sono 1) (Historical Change of Sushi in China [A Study of Sushi 1]). Seikatubunka Kenkyu (*Journal of Domestic Life and Culture*) 1:69–77.

Shinoda, O. 1957. Sushinenpyoo Shina no Bu (Sushikoo Sono 9) (Chronological Table of Sushi in China [A Study of Sushi 9]). Seikatubunka Kenkyuu. *Journal of Domestic Life and Culture* 6:39–54.

Shinoda, O. 1961. Sushinenpyoo sono 2, Nihon no Bu (Sushikoo Sono 10) (Chronological Table of Sushi in Japan [A Study of Sushi 10]). Seikatubunka Kenkyuu. *Journal of Domestic Life and Culture* 10:1–30.

Shinoda, O. 1966. Sushi no Hon *(The Book of Sushi)*. Shibatashoten, Tokyo.

Shinoda, O. 1977. Zotei Komeno Bunkashi (*A Cultural History of Rice*). Shakaisisosha, Tokyo.

Shinoda, O. 1978. Sushi no Hanashi (*The Story of Sushi*). Shinshindoo, Kyoto.

Shintani, T. 1981. Radego Betonamugo Nihongo Goishuu (Rade-Vietnamese Japanese).

Shorto, H.L. 1962. *A Dictionary of Modern Spoken Mon*. Oxford University Press, London.

Sinanuwong, S. and Takaya, Y. 1974a. Saline soils in Northeast Thailand. *Southeast Asia Studies* 12(1):105–119.

Sinanuwong, S. and Takaya, Y. 1974b. Distribution of soils in the Khorat Basin of Thailand. *Southeast Asia Studies* 12(3):365–382.

Sindhu, S.C. and Kheterpaul, N. 2001. Probiotic fermentation of indigenous food mixture: Effect on antinutrients and digestibility of starch and protein. *Journal of Food Composition and Analysis* 14:601–609.

Singh, R.K., Singh, A., and Sureja, A.K. 2007. Traditional foods of monpa tribe of West Kameng, Arunachal Pradesh. *Indian Journal of Traditional Knowledge* 6(1):25–36.

Smith, C.R. and Moss, M.O. 1985. *Mycotoxins: Formation, Analysis, and Significance.* John Wiley & Sons, New York.

Smith, J.E., Solomons, G., Lewis, C., and Anderson, J.G. 1995. Role of mycotoxins in human and animal nutrition and health. *Natural Toxins, Wiley-Liss, Inc., A Wiley Co.* 3(4):187–192.

Smith, J.L. and Fratamico, P.M. 1995. Factors involved in the emergence and persistence of foodborne disease. *Journal of Food Protection* 58:696–708.

Smith, J.L. and Palumbo, S.A. 1983. Use of starter cultures in meats. *Journal of Food Protection* 46:997–1006.

Smith, K.D. 1967. *Sedang Vocabulary.* Bo Giao Duc, Saigon.

Sohliya, I., Joshi, S.R., Bhagobaty, R.K., and Kumar, R. 2009. Tungrymbai—A traditional fermented soybean food of the ethnic tribes of Meghalaya. *Indian Journal of Traditional Knowledge* 8(4):559–561.

Soni, S.K. and Sandhu, D.K. 1990. Indian fermented foods: Microbiological and Biochemical aspects. *Indian Journal of Microbiology* 30(2):135–137.

Sripriya, G., Antony, U., and Chandra, T.S. 1997. Changes in carbohydrate, free amino acids, organic acids, phytate and HCl extractability of minerals during germination and fermentation of finger millet (*Eleusine coracana*). *Food Chemistry* 58(4):345–350.

Stanier, R.Y., Doudoroff, M., and Adelberg, E.A. 1970. *General Microbiology.* Prentice Hall, Inc, Engelwood Cliffs, NJ, Macmillan, New Delhi, pp. 2–10.

Stanton, R.W. 1985. Food fermentation in the tropics. In: Wood, B.J.B. (Ed.). *Microbiology of Fermented Foods.* Elsevier Applied Science Publishers, UK. pp. 696–712.

Stratford, M. 2006. Food and beverage spoilage yeasts. In: Amparo, Q. and Graham, F. (Eds.). *Yeasts in Food and Beverages, The Yeast Handbook.* Vol. 2. Springer-Verlag, Berlin Heidelberg, p. 335.

Stein, B. 1998. *A History of India.* 1st edition. Wiley-Blackwell, Oxford.

Steinkraus, K.H. 1979. Nutritionally significant indigenous foods involving an alcoholic fermentation. In: Gastineau, C.F., Darby, W.J., and Turner, T.B. (Eds.). *Fermented Food Beverages in Nutrition.* Academic Press, London, pp. 35–59.

Steinkraus, K,H. 1983a. *Handbook of Indigenous Fermented Foods.* Marcel Dekker, New York, NY, p. 671.

Steinkraus, K.H. 1983b. Fermented foods, feeds and beverages. *Biotechnology Advance* 1:31–46.

Steinkraus, K.H. 1983c. Lactic acid fermentation in the production of foods from vegetables, cereals and legumes. *Antonie van Leeuwenhoek* 49:337–348.

Steinkraus, K.H. 1985. Bio-enrichment: Production of vitamins in fermented foods. In: Wood, B.J.B. (Ed.). *Microbiology of Fermented Foods.* Vol. 1. Elsevier, New York, NY, pp. 323–343.

Steinkraus, K.H. 1989. Microbial interaction in fermented foods. In: Hattori, T. and others (Eds.). *Recent Advances in Microbial Ecology.* Japan Scientific Society Press, Tokyo, pp. 547–552.

Steinkraus, K.H. 1996. *Handbook of Indigenous Fermented Food.* 2nd edition. Marcel Dekker, Inc., New York, NY.

Steinkraus, K.H. 1997. Classification of fermented foods: Worldwide review of household fermentation techniques. *Food Control,* 8(5–6):311–317.

Steinkraus, K.H. 2002. Fermentations in world food processing. *Comprehensive Reviews in Food Science and Food Safety,* 1(1):23–32.

Steinkraus, K.H. 2009. *Fermented Foods, Encyclopedia of Microbiology*. Academic Press, London, pp. 45–53.

Steinkraus, K.H., Van Veen, A.G., and Thiebeau, D.B. 1967. Studies on Idli-an Indian Fermented Black Gram-rice food. *Food Technology* 21:110–113.

Steyn, P.S. 1995. Mycotoxins, general view, chemistry and structure. *Toxicology Letters* 82/83:843–851.

Steyn, P.S., Gelderblom, W.C., Shephard, G.S., and van Heerden, F.R. 2009. Mycotoxins with a special focus on Aflatoxins, Ochratoxins and Fumonisins. *General and Applied Toxicology* 15th Dec. 2014.

Svanberg, U., Sjögren, E., Lorri, W., Svennerholm, A.M., and Kaijser B. 1992. Inhibited growth of common enteropathogenic bacteria in lactic-fermented cereal gruels. *World Journal of Microbiology and Biotechnology* 8:601–606.

Sweeney, M.J. and Dobson, A.D.W. 1998. Mycotoxin production by Aspergillus, Fusarium and Penicillium species. *International Journal of Food Microbiology* 43(3):141–158.

Syal, P. and Vohra, A. 2013. Probiotic potential of yeasts isolated from traditional Indian fermented foods, International. *Journal of Microbiology Research* 5(2):390–398.

Sybesma, W., Starrenburg, M., Kleerebezem, M., Mierau, I., de Vos, W.M., and Hugenholtz, J. 2003. Increased production of folate by metabolic engineering of Lactococcus lactis. *Applied and Environmental Microbiology* 69(6):3069–3076.

Tamang, B. and Tamang, J.P. 2009. Lactic acid bacteria isolated from indigenous fermented bamboo products of Arunachal Pradesh in India and their functionality. *Food Biotechnology* 23:133–147.

Tamang, J.P. 1977a. Case study on socio-economical prospective of kinema, a traditional fermented soybean food. In: *Proceedings of the International Conference on Traditional Foods at CFTRI*, Mysore, India, 2000, pp. 180–185.

Tamang, J.P. 1977b. Traditional fermented foods and beverages of the Sikkim Himalayas in India: Indigenous process and product characterization. In: *Proceedings of the International Conference on Traditional Foods at CFTRI*, Mysore, India, 2000, pp. 99–118.

Tamang, J.P. 1998. Role of micro-organisms in traditional fermented foods. *Indian Food Industry* 17(3):162–167.

Tamang, J.P. 2001. Food culture in the Eastrern Himalayas. *Journal of Himalayan Research Cultural Foundation* 5(3–4):107–118.

Tamang, J.P. 2003. Indigenous fermented foods of the Himalayas: Microbiology and food safety. *Proceedings of the First International Symposium on Insight into the World of Indigenous Fermented Foods for Technology Development and Food Safety*. Bangkok, Thailand, pp. 1–13.

Tamang, J.P. 2005. *Food Culture of Sikkim*. Sikkim study series. Vol. 4. Information and Public Relation Department, Government of Sikkim, Gangtok.

Tamang, J.P. 2009. Fermented milks. In: Tamang, J.P. (Ed.). *Himalayan Fermented Foods: Microbiology, Nutrition and Ethnic Values*. CRC Press Taylor & Francis, Boca Raton, FL, pp. 216–224.

Tamang, J.P. 2010a. Diversity of fermented foods. In: Tamang, J.P. and Kailasapathy, K. (Eds.). *Fermented Foods and Beverages of the World*. CRC Press, Taylor & Francis, Boca Raton, FL, pp. 41–84.

Tamang, J.P. 2010b. Diversity of fermented beverages and alcoholic drinks. In: Tamang, J.P. and Kailasapathy, K. (Eds.). *Fermented Foods and Beverages of the World*. CRC Press, Taylor & Francis, Boca Raton, FL, pp. 85–125.

Tamang, J.P. 2010c. Front matter. *Fermented Foods and Beverages of the World*. CRC Press, Taylor & Francis, Boca Raton, FL.

Tamang, J.P. 2010d. *Himalayan Fermented Foods: Microbiology, Nutrition and Ethnic Value*. CRC Press, Taylor & Francis Group, USA, New York, NY.

Tamang, J.P. 2011. Prospects of Asian Fermented Foods in Global Markets. In: *The 12th Asean Food Conference*. BITEC Bangna, Bangkok, Thailand.

Tamang, J.P. 2012. Animal-based fermented foods of Asia. In: Hui, Y.H. and Özgül Evranuz, E. (Eds.). *Handbook of Animal-Based Fermented Food and Beverage Technology*. 2nd edition. CRC Press, Taylor & Francis, Boca Raton, FL, p. 814.

Tamang, J.P. and Nikkuni, S. 1996. Selection of starter cultures for the production of kinema, a fermented soybean food of the Himalaya. *World Journal of Microbiology and Biotechnology*, 12(6):629–635. doi: 10.1007/BF00327727.

Tamang, J.P. and Samuel, D. 2010. Dietary cultures and antiquity of fermented foods and beverages. In: Tamang, J.P. and Kailasapathy, K. (Eds.) *Fermented Foods and Beverages of the World*. CRC Press, Taylor & Francis, Boca Raton, FL, pp. 1–40.

Tamang, J.P., Dewan, S., Tamang, B., Rai, A., Schillinger, U., and Holzapfel, W.H. 2007. Lactic acid bacteria in Hamei and Marcha of North East India. *Indian Journal of Microbiology* 47:119–125.

Tamang, J.P., Okumiya, K., and Kosaka, Y. 2010. Cultural adaptation of the Himalayan ethnic foods with special reference to Sikkim, Arunachal Pradesh and Ladakh. *Himalayan Study Monographs* 11:177–185.

Tamang, J.P. and Holzapfel, W.H. 1999. Biochemical identification techniques-modern techniques: Microfloras of fermented foods. In: Robinson, R.K., Batt, C.A. and Patel, P.D. (Eds.). *Encyclopedia of Food Microbiology*. Academic Press, London, pp. 249–252.

Tamang, J.P. and Nikkuni, S. 1996. Selection of starter culture for production of kinema, fermented soybean food of the Himalaya. *World Journal of Microbiology and Biotechnology* 12(6):629–635.

Tamang, J.P. and Nikkuri, S. 1998. Effect of temperatures during pure culture fermentation of Kinema. *World Journal of Microbiology and Biotechnology* 14:847–850.

Tamang, J.P. and Samuel D. 2010. Dietary cultures and antiquity of fermented foods and beverages. In: Tamang, J.P. and Kailasapathy, K. (Eds.). *Fermented Foods and Beverages of the World*. CRC Press, Boca Raton, FL, pp. 1–40.

Tamang, J.P. and Sarkar, P.K. 1988. Traditional fermented foods and beverages of Darjeeling and Sikkim—A review. *Journal of Science Food Agriculture* Vol. 44(4):375–385.

Tamang, J.P. and Sarkar, P.K. 1996. Microbiology of Mesu, a traditional fermented bamboo shoot product. *International Journal Food Microbiology* 29:49–58.

Tamang, J.P., Sarkar, P.K., and Clifford, W. 1988. Traditional fermented foods and beverages of Darjeeling and Sikkim—A review. *Journal of the Science of Food and Agriculture* 44(4):375–385.

Tamang, J.P., Sarkar, P.K., and Hesseltine, C.W. 1988. Traditional fermented foods and beverages of Darjeeling and Sikkim—A review. *Journal of the Science of Food and Agriculture* 44:375–385.

Tamang, J.P., Tamang, N., Thapa, S., Dewan, S., Tamang, B. Yonzan, H., Rai, A.K., Chettri, R., Chakrabarty, J., and Kharel, N. 2012. Micro-organisms and nutritional value of ethnic fermented foods and alcoholic beverages of North East India. *Indian Journal of Traditional Knowledge* 11(1):7–11.

Tamang, J.P., Thapa, N., Rai, B., Thapa, S., Yonzan, H., Dewan, S., Tamang, B., Sharma, R., Rai, A., and Chettri, R. 2007. Food consumption in Sikkim with special reference to traditional fermented foods and beverages: A micro-level survey. *Journal of Hill Research* 20(1):1–37.

Tamang, J.P., Thapa, S., Tamang, N., and Rai, B. 1996. Indigenous fermented food beverages of Darjeeling hills and Sikkim: Process and product characterization. *Journal of Hill Research* 9(2):401–411.

Tanasupawat, S. and Visessanguan, W. 2014. Fish fermentation. Chapter 8. In: Boziaris, I.S., Somboon, T., and Visessanguan, W. (Eds.). *Seafood Processing: Technology, Quality and Safety*. John Wiley & Sons, Chichester, UK.

Thapa, N. 2002. *Studies of Microbial Diversity Associated With Some Fish Products of the Eastern Himalayas*. Ph.D.Thesis, North Bengal University, India.

Thakur, N., Savitri and Bhalla, T.C. 2004. Characterization of some traditional foods and beverages of Himachal Pradesh. *Indian J. Traditional Knowledge* 3(3):325.

Thapa, S. 2001. Microbiological and biochemical studies of indigenous fermented cereal-based beverages of the Sikkim Himalayas. PhD thesis, Food Microbiology Laboratory, Sikkim Government College (under North Bengal University), 190pp.

Thapa, S. and Tamang, J.P. 2004. Product characterization of kodo ko jaanr: Fermented finger millet beverage of the Himalayas. *Food Microbiology* 21: 617–622.

Thapa, S. and Tamang, J.P. 2005. Identification of yeast strains isolated from marcha in Sikkim, a microbial starter for amylolytic fermentation. *International Journal of Food Microbiology* 99:135–146.

Thapa, T.B., Jagat Bahadur, K.C., Gyamtsho, P., Karki, D.B., Rai, B.K., and Limbu, D.K. 2003. Fermented foods of Nepal and Bhutan. In: *International Seminar and Workshop on Fermented Foods, Health Status and Social Well-Being* held on Nov 2003, Anand, India, p. 9.

Thongkum, T.L. and Gainey, J.W. 1978. *Kui (Suai) Thai English Dictionary*. Chulalongkorn University Language Institute, Bangkok.

Tibor. D. 2007. Yeasts in specific types of foods, Chapter 7. In: *Handbook of Food Spoilage Yeasts*. 2nd edition. CRC Press, Boca Raton, FL, pp. 117–201.

Tobia. 2009. Exporting Afghanistan.

Upadhyay, B. 1967. *Vedic Sahitya Aur Sanskruti*. 3rd edition. Sharda Mandir, Varanasi, pp. 436–450.

Van Pee, W. and Swings, J.J. 1971. Chemical and microbiologial studies of Congolese palm wines (*Elaeis guineensis). East African Agriculture Foreign Journal* 36:311–314.

Van Veen, A.G. 1953. Fish preservation in Southeast Asia. *Advances in Food Research* 4:209–232.

Van Veen, A.G. 1957. *Fermented Protein-Rich Foods*. FAO report no. FAO/57/3/1966.

Van Veen, A.G. 1965. Fermented and dried seafood products in Southeast Asia. In: ed. Borgstrom, G. (Ed.). *Fish as Food v.3 Processing*. Pt. 1. Academic Press, New York, NY, pp. 227–250.

Vardaraj, M.C., Balasubramanyam, B.V., and Joseph, R. 1997. Antibacterial effect of *Lactobacillus* spp. on spoilage bacteria in selected Indian fermented foods. In: *Proceedings of the International Conference on Traditional Foods at CFTRI*. Mysore, India, 2000, Mysore India, pp. 210–226.

Vedda. *Encyclopædia Britannica*. 2008. http://www.globallinktours.com/sri-lanka.html.

Vine, R.P. 1981. *Commercial Winemaking*. AVI Publishing Co Inc. Westport, CT, pp. 1–481.

Wang, H.L. and Hesseltine, G.W. 1979. *Prescott and Dunn's Industrial Microbiology Read J*. AVI Publishing Go. Inc. Westport, CT, p. 492.

Wang, J. and Fung, D.Y.C. 1996. Alkaline-fermented foods: A review with emphasis on pidan fermentation. *Critical Review Microbiology* 22(2):101–138.

Warren, D.M. 1991. Using indigenous Knowledge in Agricultural Development. World Bank Discussion Paper No. 127. World Bank, Washington D.C. (www.worldbank.org/afr/ik/basic.htm_68k)

Westby, A. Bainbridge, Z.A., and Reilly, P.J.A. 1997. Review of the effect of fermentation on naturally occurring toxins. *Food Control* 8:329–339.

William, S. and Akiko, A. 2004. *A Brief History of Fermentation, East and West: A Chapter from the Unpublished Manuscript, History of Soybeans and Soyfoods*, 1100 B.C. to the 1980s. Copyright, 2004. Soyfoods Center, Lafayette, CA.

Yokotsuka, T. 1982. Traditional fermented soybean foods. In: Rose, A.H. (Ed.). *Fermented Foods*. Academic Press, London, pp. 180–188.

Zoysa, N.D. 1992. Tapping patterns of the Kitul Palm (*Caryota urens*) in the Sinharaja Area, Sri Lanka. *Principes* 36(1):28–33.

Zannini, E., Pontonio, E., Waters, D.M., and Arendt, E.K. 2012. Applications of microbial fermentations for production of gluten-free products and perspectives. *Applied Microbiology and Biotechnology* 93:473–485.

2

Diversity of Indigenous Fermented Foods and Beverages of South Asia

A.K. SENAPATI, ANTON ANN, ANUP RAJ,
ANUPAMA GUPTA, ARUN SHARMA,
BHANU NEOPANY, CHAMGONGLIU PANMEI,
DEEPA H. DIWEDI, DEV RAJ, DORJEY ANGCHOK,
FATIMAH ABU BAKAR, FOOK YEE CHYE,
GEORGE F. RAPSANG, GITANJALI VYAS,
GURU ARIBAM SHANTIBALA DEVI,
J.P. PRAJAPATI, KHENG YUEN SIM,
KONCHOK TARGAIS, L.V.A. REDDY,
MANAS R. SWAIN, MD. SHAHEED REZA,
MUHAMMAD ZUKHRUFUZ ZAMAN,
NEELIMA GARG, NINGTHOUJAM SANJOY SINGH,
NIVEDITA SHARMA, RAMESH C. RAY,
S.S. THORAT, S.V. PINTO, SATYENDRA GAUTAM,
SHARMILA THOKCHOM, S.R. JOSHI,
SUSHMA KHOMDRAM, AND TSERING STOBDAN

Contents

2.1 Introduction

A large variety of foods (baked products, alcoholic beverages, yoghurt, cheese, soy, fish and meat products, and many others) are derived from food fermentation, not only in households, but also on small-scale as well as large-scale commercial enterprises and food industries. Fermented foods make a major contribution to the human diet all over the world. Fermentation has been a popular method to preserve food since ancient times. Some practices have recently been improved by more modern technology in the developed countries, with the growth of many sophisticated industries. Food fermentation became popular in many civilizations because it not only extended the shelf-life of food, but it also provided a variety of forms, flavors, and other sensory experiences. The primary purpose of fermenting food substrate is to preserve foods that are susceptible to spoilage due to undesirable environment or climate conditions. Traditional fermentation serves as a low cost food processing technique producing desirable products with special characteristics, which develop due to the biochemical reactions of the microorganisms or the enzymes in the food substrates (Fellows,

2000; Campbell-Platt, 1987, 2009). Hence, fermented foods to some extent are more attractive and nutritious than the unfermented food substrates. The traditional way of carrying out fermentation at the household-scale is still followed, using relatively simple processing facilities. These products often contain mixed microbial populations because of lack of sterility and the use of spontaneous fermentations (Nout et al., 2007). The rich variety of fermented foods and beverages are often the pride of many countries and civilizations throughout the history of mankind; for example, the Japanese are proud of their *sake*, the Indonesians of their *tempeh*, the Thais of their fish sauce, the Malaysians of their *belacan*, the Indians of *dosa* and *idli*, the British are proud of their beers, the German of their *sauerkraut*, and Italians of their cheese.

Asians have used fermentative microorganisms to convert agricultural products into foods from ancient time (Nout and Aidoo, 2010). The general characteristics and properties of Asian fermented food products are influenced by ingredients native to their geography, as well as culture, economic, and religion. In many developing countries, fermented foods are important as cheap nutritious food. Protein and vitamin deficiencies are still a major problem in undeveloped and many developing countries. Likewise, approximately three quarters of humanity intermittently or permanently face a shortage of balanced food (Jeyaram et al., 2009) in one or another part of the world. Hence, the production of fermented foods is more important to overcome the shortage of balanced food and endemic malnutrition. Indigenous fermented foods make up a large proportion of the daily food intake of South Asia (Hayford and Jespersen, 1999). Each fermented food is associated with a unique group of microorganisms which increases the concentration of protein, essential amino acids, fatty acids, vitamins, and the availability of minerals (Dahal et al., 2005). Several South Asian fermented foods also have health-promoting benefits and certain curative properties against many diseases and disorders (Singh et al., 2007; Tamang, 2007; Farhad et al., 2010; Devi and Kumar, 2012).

Basically, the technology of fermented foods has developed from the idea of preservation and salvaging surplus produce—for example, surplus milk used to be collected, preferably in earthen pots, and allowed to ferment by natural culture of lactic acid bacteria. The fermented clotted milk (gel) is called *dahi* and consumed as such, or with rice or *chapatti*. Curd when agitated with added salt or sugar, is relished as *lassi*. In the process of country butter making, the *dahi* is churned in big earthen pots and the remaining butter milk (*Chhash*) is consumed with food or is relished as a refreshing beverage. It is also spiced and cooked along with a small portion of gram flour (*besan*) and is called *Kadhi*, which is a carrier of main foods like rice or *chapatti*. To achieve longer shelf-life and year round availability, traditional people developed techniques of sun-drying of fermented cereal–pulse mixtures, resulting in the production of a variety of *warries*, seen in the northern parts of India (Wacher, et al., 2010).

Fermented food products of South Asia are generally produced by ethnic people at household scale, mostly by women. Based on trial and error, they have learnt how to provide favorable conditions for fermentation, so permitting the beneficial microorganisms to flourish to get the desired fermented food products (Tamang and Tamang,

2009). In Asia, there are several fermented food products made by fermentation (Nout and Aidoo, 2010). South Asian countries have a large diversity of fermented food products produced from both animal and plant sources. The diversity of fermented foods in this region is directly related to the food culture of each ethnic community (see Chapter 1) the availability of raw materials and the microorganisms provided by the environment (Tamang, 2010). These foods and beverages of South Asia are mostly prepared through indigenous practices of food processing and preservation (Bulent, 2011) and the knowledge is handed down from generation-to-generation. Nevertheless, the rich diversity in fermented food products is not widely known outside its niche. Most of the indigenous fermented foods are produced in small volumes with inconsistent quality. The extrinsic and intrinsic factors, such as temperature, humidity, hygiene, suitable preserving techniques, and the nature of the raw materials always determine the success of the production batch. In developed countries, many fermented foods and beverages are well investigated and data on production, consumption, microbiology, including starter cultures and nutritional profiles, are available (Agati, 1998; Tamang, 2007). But this is not true of South Asian countries. This chapter reviews the diversity of indigenous fermented foods in South Asia. The production processes and, microbiological and nutritional aspects of these foods will be discussed in detail in the respective chapters of this book, except where, for the sake of continuity of subject matter, these aspects have been briefly described here.

2.2 Diversity and Consumption of Fermented Foods in South Asia

Hundreds of different types of fermented foods are produced in the Southern part of Asia (Bulent, 2011). The rich diversity of these foods is attributed to two main factors: (1) the rich diversity of fermenting microorganisms and their metabolism; (2) the rich diversity of plant and animal products used as substrates in the various fermentation processes (Anonymous, 2008). Millions of people in many Asian countries eat cooked rice as their staple food. Wheat or barley is also consumed as a staple food in many South Asian countries, such as India, Bangladesh, Pakistan, and Sri Lanka (Tamang and Sameul, 2010). The Himalayan dietary culture has both rice and wheat or barley as staple foods, along with varieties of indigenous fermented or non-fermented foods prepared from fish, legumes, cereals, vegetables, and bamboo shoots (Tamang, 2009, 2010). Table 2.1 describes some of the fermented foods and beverages of South Asia (Tamang, 2010).

Fermented foods are widely consumed by people around the world due to unique sensory attributes, nutritional availability, and health-promoting advantages, and they contribute approximately to 20% of the total food consumption (Kwon, 1994). These products have also been introduced to regions others than those where they are produced, and accepted as a part of the diet, although little may be known about a particular food. Fermented foods have been improvised and diversified, resulting from innovations in food production techniques due to advances in science and technology (Tamang, 2010). This transformation has led to the production of certain traditional

Table 2.1 Diversity of Fermented Foods and Beverages of South Asia

PRODUCT	SUBSTRATES	PREDOMINANT MICROORGANISMS	METHOD OF FERMENTATION	USAGE OF PRODUCT	COUNTRY OR REGION OF ORIGIN	REFERENCE
Idli	Rice and dehulled black gram	*Leuconostoc mesenteroides, Streptococcus faecalis*	Natural, ambient temperature for overnight	Pancake for breakfast or snack food	India (especially south region), Sri Lanka	Soni and Sandhu (1989)
Dosa	Rice and dehulled black gram	*Leuconostoc mesenteroides, Saccharomyces cerevisiae*	Natural, ambient temperature for 8–20 h	Snack food	India, Pakistan, Sri lanka	Soni et al. (1986)
Jalebi	Refined wheat flour	*Lactobacillus fermentum, Saccharomyces cerevisiae*	Natural, ambient temperature for few hours	Snack food, served at festive occasion	India, Pakistan, Nepal, Sri Lanka, Bangladesh	Steinkraus (1996)
Hawaijar	Soybean	*Bacillus subtilis, Bacillus licheniformis*	Natural, warm temperature for 3–5 days	Side dish with boiled rice	India, Nepal	Jeyaram et al. (2008)
Kinema	Soybean	*Bacillus subtilis, Enterococcus faecium,* yeasts	Natural, 25–35°C for 2–3 days	Deep fat fried, mixed with other ingredient to prepare curry	West Bengal and Sikkim in India, Nepal	Sarkar and Tamang (1995); Singh et al. (2007)
Fish sauce	Anchovy, small pelagic fish	Halophilic and halotolerant *Micrococci, Staphylococci* and lactic acid bacteria	Natural, ambient temperature (30–35°C) for 4–12 months	Condiment in cooking and dipping sauce	India	Wongkhalaung (2004)
Ngari	Small fish *Puntius sophore*	*Lactobacillus lactis, Lactobacillus plantarum*	Natural, ambient temperature for 4–6 months	Side dish locally called *ronba*	Manipur in Northeast India	Jeyaram et al. (2009)
Hentak	Small fish *Esomus danricus* with petiole of aroid plant	*Lactobacillus lactis, Lactobacillus plantarum*	Natural, ambient temperature for 7–9 days	Condiment, eaten as curry prepared with other ingredient	Manipur in Northeast India	Thapa et al. (2004); Jeyaram et al. (2009); Soni et al. (2013)
Sinki	Radish tap root	*Lactobacillus fermentum, Lactobacillus brevis* and *Lactobacillus plantarum*	Natural, warm dry place for 15–30 days	Appetizer as a base of soup or pickle	Northeast India, Nepal, Bhutan	Tamang and Sarkar (1993)
Gundruk	Rayo-sag (*Brasicca rapa*) or other vegetables	*Lactobacillus brevis, Pediococcus pentosaceus, Lactobacillus plantarum*	Natural, ambient temperature for 15–22 days	Appetizer as a base of soup or pickle	Himalayan region of India, Nepal and Bhutan	Tamang and Tamang (2010)
Goyang	Leaves of magane-saag (*Cardamine macrophyla*)	*Lactobacillus plantarum, Lactobacillus brevis, Lactococcus lactis*	Natural, ambient temperature (15–25°C) for a month	Side dish	Darjeeling hills and Sikkim, India	Tamang and Tamang (2010)

(Continued)

Table 2.1 (*Continued*) Diversity of Fermented Foods and Beverages of South Asia

PRODUCT	SUBSTRATES	PREDOMINANT MICROORGANISMS	METHOD OF FERMENTATION	USAGE OF PRODUCT	COUNTRY OR REGION OF ORIGIN	REFERENCE
Khalpi	Cucumber (*Cucumber sativus* L.)	*Leuconostoc fallax, Lactobacillus brevis, Lactobacillus plantarum*	Natural, ambient temperature for 4–7 days	Eaten as pickle along with boiled rice	Sikkim State, India	Tamang and Tamang (2010)
Mesu	Young bamboo shoot	*Pediococcus pentosaceus, Lactobacillus brevis, Lactobacillus plantarum*	Natural, ambient temperature of (20–25°C for 7–15 days	Side dish or pickle	West Bengal and Sikkim, India	Tamang et al. (2008)
Soibum	Succulent bamboo shoot	*Pediococcus pentosaceus, Lactobacillus brevis, Lactobacillus plantarum*	Natural, ambient temperature for 20 days	Pickle, curry mixed with fermented fish	Manipur State, India	Sarangthem and Singh (2003); Choudhury et al. (2012)
Soidon	Tip of matured bamboo shoot	*Lactobacillus brevis, Lactobacillus plantarum, Leuconostoc fallax*	Starter from previous batch is used, ambient temperature for 5–7 days	Eaten as curry	Manipur State, India	Jeyaram et al. (2010)
Palm wine	Sap of coconut pal tree	*Saccharomyces cerevisiae, Saccharomyces chevalieri, Acetobacter aceti*	Natural, ambient temperature for 2–5 days	Alcoholic beverage	India, Sri Lanka	Shamala and Sreekantiah (1988)
Atingba	Glutinous rice	*Saccharomyces cerevisiae, Aspergillus niger, Bacillus* sp.	Starter *hamei* is used, ambient temperature for 3–7 days	Alcoholic beverage	Manipur State, India	Jeyaram et al. (2009)
Yu	Glutinous rice	*Saccharomyces cerevisiae, Aspergillus niger, Bacillus* sp	Distillation product of Atingba	Alcoholic beverage	Manipur State, India	Jeyaram et al. (2009)
Dahi	Cow or goat milk	*Lactococcus lactis, Lactobacillus acidophilus, Lactobacillus casei*	Starter probiotic is used, ambient temperature for overnight	Yoghurt	India, Pakistan	Raju and Pal (2009)
Lassi	Cow or goat milk	*Streptococcus thermophilus, Lactobacillus acidophilus, Lactobacillus casei*	Starter probiotic is used, ambient temperature for overnight	Refreshig yoghurt beverage	India, Pakistan	George et al. (2012)

fermented foods at a larger scale and with better formulation, and this can enhance biofunctionality of these foods against life-threatening ailments (Rao et al., 2006).

Fermented cereals such as *idli*, *dosa*, and *dhokla* are eaten as breakfast or snack food along with vegetable stew (Sekar and Mariappan, 2007). Fermented vegetables such as *sinki*, *gundruk*, and *khalpi* are consumed as a base soup or pickle by adding mustard oil, salt, and powdered chillies in a meal with boiled rice (Tamang, 2009; Tamang and Tamang, 2009). Fermented fish products, such as fish sauce and shrimp paste, are used as a condiment in the cuisine of many South Asian peoples (Gildberg and Thongthai 2005; Panda et al., 2011). Fermented fish, like *ngari* and *hentak*, is eaten daily as a side dish with cooked rice by the people in northern India and Nepal (Jeyaram et al., 2009). Fermented legume products, such as *kinema* and *hawaijar*, are also consumed as a side dish, along with rice and vegetables. Other fermented food products from different raw materials i.e. black beans (Granito and Álvarez, 2006) are eaten either as main dishes, side dishes, snack foods, or used as condiment for cooking, in certain seasons.

Alcoholic drinks have been widely consumed since pre-Vedic times in India, and specific reference to their consumption among the tribal peoples has been made in the Ramayana (300–75 BC). *Kodo ko jaanr* is the most common fermented alcoholic beverage, prepared from the dry seeds of finger millet, rich in crude fiber, locally called "*kodo*" in the Eastern Himalayan regions of the Darjeeling hills and Sikkim in India, Nepal, and Bhutan (Thapa and Tamang, 2004). Similarly, *Bhaati jaanr* is an inexpensive high calorie mildly alcoholic beverage prepared from steamed glutinous rice by fermentation with molds, yeasts, and lactic acid bacteria (LAB) consumed as a staple food beverage in the Eastern Himalayan regions of Nepal, India, and Bhutan (Thapa and Tamang, 2004). *Chhang, lugari, aara, chiang, daru*, etc., are other alcoholic beverages consumed by ethnic groups in the Lahaul and Spiti region of Himachal Pradesh, India (Kanwar et al., 2007).

2.3 Type of Indigenous Fermented Foods of South Asia

Different indigenous fermented foods can be grouped together, based on the substrates, as described here, with a few typical examples.

2.3.1 Fish-Based Fermented Foods

2.3.1.1 Fish Sauce Fish sauce is clear amber to reddish brown liquid with a salty taste, mild fishy flavor, and characteristic aroma. It is commonly used as a condiment either for cooking or dipping (Wongkhalaung, 2004). Fish sauce is locally known as *colombo-cure* in India and Pakistan, where fish sauces are less extensively consumed (Lopetcharat et al., 2001; Wongkhalaung, 2004; Thongthai and Asbjørn, 2005; Jetsada et al., 2013).

Anchovy (*Stoleporus* spp.) is the most frequently used raw material in fish sauce production. Other marine fish, like herring (*Clupea* spp.), mackerel (*Rastrelliger* spp.), sardine (*Sardinella gibbosa*), and capelin (*Mallotus villosus*) are also used (Lopetcharat et al., 2001; Wongkhalaung, 2004; Joshi and Zulema, 2012). Most of the fishes used as raw materials are rich in protein, and this is reflected in the protein content of fish sauce. Instead of marine fish, fresh water fish such as mud carp (*Cirrhinus* spp.) and silver carp (*Hypophthalmichthys molitrix*) are occasionally used (Wongkhalaung, 2004; Uchida et al., 2005).

Salt is the second most common raw material for fish sauce fermentation. Sea salt is preferred by the fish sauce industry because of its abundant availability (Choi, 2001; Lopetcharat et al., 2001). Salt controls the type of microbial flora and retards some pathogenic microbes during fermentation (Lopetcharat et al., 2001; Joshi and Zulema, 2012). Fish sauce fermentation takes a fairly long time, since the hydrolysis depends on just the enzyme acivities in fish or microbial flora. Nowadays, enzymes such as bromelin, ficin, and papain, as well as other natural enzymes, are often added to accelerate the fermentation process and reduce the time (Reddy et al., 1983; Wongkhalaung, 2004; Yongsawatdigul et al., 2007).

2.3.1.2 Shrimp Paste and Crab   Shrimp paste is a fermented pasty product which is prepared from the planktonic shrimp *Acetes intermedius*. Shrimp paste is locally known as *terasi* in Indonesia and *belacan* in Malaysia, and for more information on its preparation see the literature cited (Putro, 1993; Surono and Hosono, 1994a,b; Salampessy et al., 2010).

Another favorite food item of several Naga tribes, such as the "Lotha," the "Mao," and the "Angami," is made from crab (Mao and Odyuo, 2007). The black species of crab with a hard shell is preferred, as it produces a good aroma and flavor.

2.3.1.3 Ngari and Hentak   "*Ngari*" is an indigenous fermented fish product of Manipur, North East India (Suchitra and Sarojnalini, 2009). It is consumed as a side dish called *ironba* (mixed with potato, chillies, etc.) with cooked rice (Tamang, 2009). It is prepared from small, low-priced sun dried fishes such as *Puntius sophore* (Ham) and *Puntius ticto* (Ham) locally called *phabou*, subjected to fermentation in the absence of salt for 5–6 months or more at room temperature. The process of *"Ngari"* preparation involves a brief washing of the sun-dried fishes, followed by draining and drying for 24–48 h (Jeyaram et al., 2009; Suchitra and Sarojnalini, 2009; Devi and Suresh Kumar, 2012). Because of the special flavor of fermented and dried fish, it is used as a compulsory item in daily curry preparation (Sarojnalini and Vishwanath, 1988; Wang, 2011) and considered as an important commodity amongst the people of North-East India (Suchitra and Sarojnalini, 2009; Tanasupawat and Visessangua, 2014) (For more details, see Chapter 11 of this text).

Hentak is a traditional fermented fish paste of Manipur (India). During its preparation, sun dried fish (*Esomus danricus*) is crushed into powdered form. The petioles of aroid plants (*Alocasia macrorhiza*) are crushed after washing and cutting into pieces (Tamang, 2009). The powdered fish and crushed petioles are mixed in equal proportion and a ball of thick paste is made (Salampessy et al., 2010) and the mixture is fermented naturally (Jeyaram et al., 2009; Tamang, 2009; Devi and Suresh Kumar, 2012). It is consumed as a curry as well as a condiment with boiled rice, and even given to mothers in confinement and patients in convalescence (Tamang, 2009; Cagno et al., 2013).

2.3.2 Cereal-Based Fermented Foods

Some of the cereal-based indigenous non-alcoholic fermented foods and beverages prepared along with the substrate used and consumed in various parts of South Asian countries are listed in Table 2.2.

Table 2.2 Some of the Cereal-Based Non-Alcoholic Fermented Foods and Beverages of South Asia

FERMENTED PRODUCT	SUBSTRATE	NATURE OF PRODUCT/USE	COUNTRY/AREA
Ambali	Millet flour	Semi-solid/all time food	India
Appa (hopper)	Rice or white wheat flour	Semi-solid breakfast food	Sri Lanka
Bhatura/kachauri	Wheat flour	Deep-fried *roties* used as breakfast food	India
Bread	Wheat, rye, other grains	Solid bread used as staple food	All over the world
Dhokla	Bengal gram and wheat	Spongy cake used as staple food	India
Dosa (dosai)	Black gram and rice	Spongy cake used as breakfast food	India
Fermented rice	Rice	Semi-solid breakfast food	India
Handwa	Rice, redgram, and Bengal gram flour	Baked food	Western India
Hopper (appa)	Rice or wheat flour and coconut water	Semi-solid breakfast food	Sri Lanka
Idli	Rice and black gram	Spongy steamed bread	Southern India
Imrati	Wheat flour	Sweet snack food	India, Pakistan, and Nepal
Jalebi	Wheat flour	Deep fried pretzels, confectionery	India, Pakistan, and Nepal
Kanji	Rice and carrots	Sour liquid added to vegetables	India
Nan	Wheat flour	Flat bread used as staple food	India, Pakistan, Iran, and Afghanistan
Pitha	Legumes and cereals	Fermented cakes	India
Rabadi	Maize and buttermilk	Semi-solid product eaten with vegetables	India
Selroti	Rice flour	Deep-fried, spongy, pretzel-like product commonly consumed as confectionery bread	Nepal
Torani	Rice	Liquid used for seasoning for vegetables	India

Source: Adapted from Beuchat, L.R. 1983. *Biotechnology: Food and Feed Production with Microorganisms*, Vol. 5. Verlag Chemie, Weinheim, pp. 477–528; Padmaja and George, M. 1999. *Food Processing: Biotechnological Applications.* Asia Tech Publishers Inc, New Delhi, pp. 143–189; Beuchat, L.R. 2001. *Food Microbiology: Fundamentals and Frontiers.* ASM Press, Washington, DC, pp. 701–719; Adams, M.R. and Moss, M.O. 1996. *Food Microbiology*, 1st edition. New Age International Publishers, New Delhi, pp. 252–302; Dahal, N.R. 2005. *Food Reviews International*, 1(21), 1–25.

2.3.2.1 Idli   *Idli* is a popular cereal- or legume-based fermented product of South India and Sri Lanka, which possesses a sour and salty taste. It is consumed typically as a breakfast or snack food, and is served with *sambar* (stew of tamarind and pigeon pea) and coconut chutney (Tamang, 2009). It is made through natural fermentation of rice (*Oriza sativa*) and dehulled black gram (*Phaseolus mungo*). To prepare *idli*, rice and black gram are soaked separately in water for around 4 h at room temperature; the water is then, drained (Blandino et al., 2003; Sridevi et al., 2010; Ghosh and Chattopadhyay, 2011) and the rice is coarsely ground to get a coarse flour, while the black gram is finely ground to get smooth, mucilaginous paste. Rice and black gram batter, in a ratio of 2:1 (may vary between 4:1 and 1:4), are then mixed together with the addition of a little salt and allowed to ferment overnight (20–22 h) at room temperature (Hesseltine, 1979; Blandino et al., 2003; Nout, 2009). During fermentation, the batter undergoes several changes such as leavening, acidification, and pleasant flavor enhancement, as a result of microbial activities (Steinkraus, 1996; Ghosh and Chattopadhyay, 2011). The fermentation of *idli* is natural, and the microorganisms involved originate from raw materials, although sometimes sour buttermilk or yeast is added to enhance the fermentation process. For details, see the separate chapter on cereal-based fermented foods in this text, and the literature cited (Soni and Sandhu, 1989; Ramakrishnan, 1993; Steinkraus, 1996; Nout et al., 2007; Sekarand Mariappan, 2007; Nout, 2009).

2.3.2.2 Dosa   *Dosa* is a thin and fairly crisp pancake-like food from India (Soni et al., 2013). It is a fermented product similar to *idli*, but the rice and black gram are finely ground and the leavened batter, instead of being steamed, is heated on a hot flat plate with a little oil (Salampessy et al., 2010). To prepare a *dosa* batter, wet rice and black gram are finely ground separately (Blandino et al., 2003). Both materials are then, mixed together and allowed to undergo natural fermentation at room temperature for 8–20 h, after which the mixture is spread in a thin layer (1–5 mm thickness) on a flat heated plate smeared with oil or fat (Bhattacharya and Bhat, 1997; Das et al., 2012). The nutritive value of *dosa* is quite similar to *idli* (Ramakrishnan, 1993; Sekar and Mariappan, 2007).

2.3.2.3 Kancheepuram Idli   *Kancheepuram idli* is a popular type of *idli* among indigenous Tamil people. During its cooking, cashew nuts, ghee, salt-pepper, ginger, and cumin are added to enhance the taste. In *idli* preparation, instead of rice, *kodri* (*Paspalum scorbiculatum* Steud.) and soybean or green gram (*Phaseolus aureus* Roxb.) instead of black gram are also used (Sekar and Mariappan, 2007).

2.3.2.4 Dhokla   *Dhokla* is also similar to *idli*, except that black gram is replaced by Bengal gram (*Cicer arietinum* Linn.) in its preparation (Bhattacharya and Bhat, 1997; Das et al., 2012). A batter of rice (*Oryza sativa* Linn.) and chickpea flour is also used as a substrate for the fermentation (Soni et al., 2013). To make a *dhokla*, the fermented batter is poured into a greased pie tin and steamed in an open steamer (Ramakrishnan, 1993). A significant improvement in the nutritive value and protein utilization of *dhokla* due to fermentation has been reported (Sands and Hankin, 1974; Aliya and Gervani, 1981; Chavan et al., 1989; Das et al., 2012). In *dhokla* preparation, coarsely ground meals of wheat, maize, or *kodri* instead of rice, and soybean, peas, red gram, moth beans instead of Bengal gram are also used (Sekar and Mariappan, 2007). *Dhokla* is a lactic acid fermented cake, having its origin in Gujarat, India (Joshi et al., 1989; Roy et al., 2009).

2.3.2.5 Jalebi   *Jalebi* is a sweetened fermented product made from *maida* (refined wheat flour), *dahi* and water (Sekar and Mariappan, 2007). It is a popular Indian sweet and every Indian, rich or poor is fond of it. It has evolved into a ubiquitous sweet over several centuries. The *jalebi* batter is fermented for a few hours at room temperature resulting in a decrease of pH from 4.4 to 3.3 and an increase in the batter volume of up to 9%, with decrease in amino nitrogen and free sugars during fermentation (Steinkraus, 1996; Das et al., 2012). To prepare *jalebi*, the fermented batter is deep fried in oil in spiral shapes and immersed in sugar syrup for few minutes (Das et al., 2012). It is consumed in the countries of the Indian sub-continent, including India, Pakistan, Nepal, Sri Lanka, and Bangladesh. This traditional food is commonly served at various celebrations like weddings and national holidays in India (Das et al., 2012).

Traditionally, *maida jalebi* is also prepared from a batter made with *maida* (refined wheat flour), water and *dahi* (curds), with *khoa* being used; the rest of the procedure is the same as for *Jalebi* (See Chapter 7 of this text).

2.3.3 Legume-Based Fermented Foods

2.3.3.1 Hawaijar   *Hawaijar* is a non-salted fermented soybean product from the northeastern part of India, especially Manipur. It is consumed as a side dish or is added while cooking other vegetables, such as mustard leaves and cauliflower, to make the curry softer and give a unique taste (Jeyaram et al., 2008; Premarani and Chhetry, 2010; Wacher et al., 2010). It is prepared by boiling medium or small sized soybean seeds for 2 h in the traditional method, or for 1 h in a pressure cooker. The beans are then, washed with hot water and packed tightly in a small bamboo basket layered with leaves of *Ficus hispida* (local name: *Asssee heibong*) or banana (*Musa* spp.) (Anonymous, 2008). The fermentation is held at above ambient temperature by keeping the baskets near to an earthen oven, or covered with gunny bags (Jeyaram, 2008; Premarani and Chhetry, 2010). The palatable stage of fermented soybean will be achieved within 3–5 days (Jeyaram et al., 2008). Fermentation of *hawaijar* is natural; that is, without the addition of a starter cultures. *Bacillus subtillis*, *Bacillus licheniformis*, and *Bacillus cereus* are the predominant fermenting bacteria in *hawaijar* (Jeyaram, 2008; Premarani and Chhetry, 2008).

The quality of *hawaijar* is characterized by its alkalinity (pH 8.0–8.2), ammonia odor, and mucilaginous fiber production (Jeyaram, 2008). *Hawaijar* prepared with *Asssee heibong* as lining material is preferred as it has a better taste, is dark brown in color, has a very strong aroma, and a very soft and slimy texture (Premarani and Chhetry, 2010). The shelf-life of *hawaijar* is generally 3–4 days at ambient temperature.

2.3.3.2 Kinema   *Kinema* is an alkaline soybean fermented product, which is commonly consumed by indigenous people of Nepal, the Darjeeling hills of West Bengal, and Sikkim, in India. It is a product similar to Japanese *natto*, Korean *chungkukjang*, and Chinese *schuidouchi* (Sarkar et al., 1993; Moktan et al., 2008). To prepare *kinema* traditionally, yellow seeded soybeans are washed, soaked overnight (12–20 h) in water at ambient temperature, cooked by boiling until softened, crushed lightly to grits, wrapped in fresh fern leaves and sackcloth and left to ferment (25–35°C) for 1–3 days (Sarkar and Tamang, 1995; Tamang and Nikkuri, 1998; Dahal et al., 2005; Moktan et al., 2008). Instead of fern leaves, *Ficus* or banana leaves are also used as a wrapping material. Fresh *kinema* is deep-fried and mixed with vegetables, spices, salt, and water to prepare a thick curry, consumed as a side dish with boiled rice (Sarkar and Tamang, 1995; Sarkar et al., 1997, 1998; Singh et al., 2007; Moktan et al., 2008). The liquid has the property of forming long, string threads when touched with the finger—the longer the strings, the better the quality of the *kinema*. Fresh *kinema* of good quality has a nutty flavor accompanied by a mild smell of ammonia, a greenish brown color, and is semi-hard, like raisins. *Bacillus subtilis* is the predominant fermenting bacteria in *kinema*. *Enterococcus faecium*, *Candida parapsilosi*, and *Geotrichum candidum* are the accompanying flora, occurring, respectively, in 100%, 50%–80%, and 40%–50% of *kinema* samples from the market (Sarkar et al., 1994; Sarkar and Tamang, 1995; Tamang, 2003; Sekar and Mariappan, 2007). *Kinema* is a low cost nutritious product (Sarkar et al., 1996, 1997, 1998; Moktan et al., 2008). Fermentation of soybean with *Bacillus subtilis* to *kinema* contributes to enhance the free radical scavenging activity, and thus, is employed to reduce oxidative stress, metal chelating ability, and lipid peroxidation inhibitory activity (Moktan et al., 2008).

2.3.4 Vegetable-Based Fermented Foods

2.3.4.1 Sinki   *Sinki* is a non-salted fermented radish tap root consumed as a base of soup or as a pickle by the indigenous peoples of northeastern states of India, Nepal, and a few places in Bhutan. The soup is made by soaking *sinki* in water for about 2 min. before it is fried with salt, onion, and green chili (Sarkar and Mariappan, 2007). The fried mixture is then, boiled in rice water and consumed as soup, which may be used as a remedy for indigestion (Sarkar and Mariappan, 2007). The pickle is prepared by soaking *sinki* in water, squeezing out the water and then, mixing it with salt, mustard oil, onion, and green chilli (Tamang and Sarkar, 1993). During the traditional preparation of *sinki*, fresh radish tap roots of (*Raphanus sativus* L.) are washed, wilted by sun drying for 2–3 days, shredded, washed again, and placed tightly into an earthen jar covered by radish leaves with an earthen lid at the top (Tamang and Sarkar, 1993; Das et al., 2005). Fermentation is carried out in a warm, dry place for 15–30 days (Tamang and Sarkar, 1996; Tamang and Tamang, 2009). In some places, *sinki* is made by unique pit fermentation. For more information, see a separate Chapter 8 of this text and the literature cited (Tamang and Sarkar, 1993; Tamang et al., 2005; Tamang, 2009).

2.3.4.2 Gundruk   *Gundruk* is a fermented and acidic vegetable product of the Himalayan regions of India, Nepal, and Bhutan. All Nepalese, irrespective of wealth or status, relish *gundruk*. Thus, it has an important bearing in the Nepalese diet. Unlike sauerkraut and pickles, *gundruk* is used as a condiment to enhance the overall flavor of the meal. It is, generally, produced and consumed during the winter, when perishable leafy vegetables are abundant, and is similar to other fermented vegetable products, such as the *kimchi* of Korea, the *sauerkraut* of Germany, the *sunki* of Japan, and the *suan-cai* of China (Tamang and Tamang, 2009; Tamang, 2012). But unlike *kimchi* and *sauerkraut*, freshly prepared wet *gundruk* is not normally eaten alone, but is consumed as a soup and pickle with boiled rice. It is a good appetizer in a bland, starchy diet. During the preparation of *gundruk*, the leaves of a local vegetable known as *rayo-sag* (*Brasicca rapa* L. ssp. *Campestris* (L), *Clapam* var. *crucifolia* Roxb), mustard (*Brasicca juncea* (L.) Czern.), radish (*Raphanus sativus*), or cauliflower (*Brasicca oleracea* L. var. *botrytis*) are wilted for 1–2 days and then, crushed lightly and pressed into an earthen jars or containers, made air tight and then, fermented naturally for about 15–22 days (Wacher et al., 2010) and sun dried for 3–4 days (Tamang et al., 2005; Tamang and Tamang, 2009, 2010). Gundruk fermentation is initiated with *Lactobacillus brevis*, *Pediococcus pentosaceusus*, and, finally, dominated by *Lactobacillus plantarum* (Tamang and Tamang, 2010). Karki et al. (1983) have observed that lactic acid bacteria such as *Lactobacillus plantarum*, *Lactobacillus casei* subsp. *casei*, *Lactobacillus casei* subsp. *pseudo-plantarum*, *Lactobacillus fermentums*, and *Pediococcus pentosaceus* dominate the micro-flora in *gundruk* from Nepal (Tamang et al., 2005). After complete drying, it is served as a chutney or used as a curry.

2.3.4.3 Khalpi   *Khalpi* is a fermented cucumber (*Cucumber sativus* L.) product consumed by the *Brahmin* Nepalis in Sikkim, usually as pickle by adding mustard oil, salt, and powdered chillies, in a meal with boiled rice (Tamang et al., 2005; Tamang and Tamang, 2010). To prepare *khalpi*, matured and ripened cucumber is cut into suitable pieces, sun dried for 2 days, and then, put into a bamboo vessel, locally called a *dhungroo*, made air tight, and fermented at room temperature for 4–7 days. Hetero-fermentative lactic acid bacteria such as *Leuconostoc fallax*, *Lactobacillus brevis*, and *Pediococcus pentosaceus* initiated the fermentation of *khalpi*, which is finally completed by *Lactobacillus plantarum*.

2.3.5 Fermented Bamboo Shoot Products

Fermented bamboo shoot products are popular traditional foods of the indigenous people living in the Himalayan regions of northeast India, Nepal, and Bhutan, where bamboo trees are plentiful (Bhatt et al., 2003). The products of fermented bamboo shoot in the regions including *mesu, soibum, soidon,* and *soijim* (Tamang and Sarkar, 1996; Tamang et al., 2008; Jeyaram et al., 2010).

2.3.5.1 Mesu   *Mesu* is a non-salted fermented bamboo shoot product of the Darjeeling hills of West Bengal and Sikkim in India (Tamang and Sarkar, 1996; Rao et al., 2006). It is commonly eaten as a curry, pickle, or soup. It is produced and sold in the local market during the months of July and September, when young bamboo shoots sprout. During its traditional preparation, young edible bamboo shoots (*Dendrocalamus sikkimensis, Bambusa tulda,* and *Dendrocalamus hamiltonii*) are collected, defoliated, chopped, and washed thoroughly with clean water (Tamang, 2012). After draining, the chopped shoots are pressed tightly into a cylindrical bamboo container and left to ferment at ambient temperature (20–25°C) for 7–15 days (Tamang and Sarkar, 1993, 1996; Rao et al., 2006; Tamang et al., 2008). The end of fermentation is indicated by the typical *mesu* flavor. The shelf-life of *mesu* is only about a week, but when it is pickled by mixing it with salt, mustard oil, and green chillies, the shelf-life is extended for a year or more (Tamang and Sarkar, 1996; Rao et al., 2006). Many species of bacteria, especially lactic acid bacteria, are involved in the fermentation of bamboo shoots for *Mesu,* include *Pediococcus pentosaceus* initially, followed by *Lactobacillus brevis,* and, finally, succeeded by *Lactobacillus plantarum* (Tamang and Sarkar, 1996; Sekar and Mariappan, 2007; Wacher et al., 2010) and for more details see the literature cited (Sarangthem and Singh, 2003; Jeyaram et al., 2008; Tamang et al., 2008). A decline in pH value from 6.4 to 3.8 due to the increase of titratable acidity from 0.04% to 0.95% has been documented during *mesu* fermentation (Tamang and Sarkar, 1996; Klanbuta et al., 2002).

2.3.5.2 Soibum   *Soibum* is a fermented food product of Manipur state in India, which is prepared from succulent bamboo shoots of *Bambussa balcooa, Dendrocalamus strictus,* and *Melocana baccifera* (Sarangthem and Singh, 2003; Tamang et al., 2005, 2008; Tamang, 2010, 2012). *Soibum* has a whitish color, faint aroma, and sour taste. To prepare *soibum* traditionally, the inner part of young bamboo shoots are chopped into pieces, washed, and transfered to a covered earthen pot to ferment at ambient temperature for 20 days. It is sold in local vegetable market, and is consumed as pickle and curry, mixed with fermented fish (Tamang et al., 2008; Tamang, 2009).

2.3.5.3 Soidon   *Soidon* is a non-salted vegetable fermented food product of Manipur, India prepared from the tips of matured bamboo shoots. It is commonly eaten as a curry (local name: *Eronba*) by mixing with potato, green chillies, salt, and fermented fish (Agati, 1998; Tamang et al., 2008; Jeyaram et al., 2009, 2010). During its preparation, the outer casing of the tips is removed and the inner part is chopped into pieces, which are then, transferred into an earthen pot containing water to undergo submerged fermentation at room temperature for 5–7 days in winter or 2–3 days in summer (Jeyaram et al., 2009). A milky fermented juice from a previous batch (1:1 dilution) is used as a starter for this fermentation. This starter is an acidic juice extract from 1 to 1.5 kg of *Garcinia pedunculata* (local name: *Heibung*) fruit with 10–15 L of washed rice water (Jeyaram et al., 2010). After fermentation, the chopped bamboo shoots called *soidon* are taken out, while the acidic liquid portion called *soijim* is kept in bottles and used as a condiment to supplement the sour taste in curry (Jeyaram et al., 2008; Tamang et al., 2008). *Lactobacillus brevis, Leuconostoc fallax*, and *L. lactis* are the microorganisms involved in the preparation of it (Tamang and Tamang, 2009).

2.3.5.4 Lungsiej   Tender bamboo shoots (*Dendrocalamus hamiltonii*) are selected and cut from the bamboo groves. Only tender shoots of about 0.5 m in length are selected for this purpose and cut-off from the main stem. The skin and hairs are removed carefully by cutting till the fleshy white portion is obtained (Murugkar, 2006). The properly cleaned and washed bamboo shoots are sliced or cut into small pieces and fermented for 1–2 months. Care is taken that the water of the stream where bamboo shoots are kept for fermentation is cold, otherwise it is believed that the shoots will not ferment properly and will spoil (Murugkar, 2006).

2.3.5.5 Eup   *Eup* is consumed as a curry along with meat, fish, or vegetables, and with meat is considered highly delicious by the people of Arunachal Pradesh (Tamang, 2009). Edible bamboo shoots (*Dendrocalamus hamiltonii* Nees. Et Arn. Ex Munro, *Bambusa balcooa* Roxb., *Dendrocalamus giganteus* Munro, *Phyllostachys assamica* Gamble ex Brandis, and *Bambusa tulda* Roxb.) are collected, the outer casings are peeled-off, and they are washed and chopped into small pieces and fermented, similarly to *ekung*, taking a bout 1–3 months. Unlike *ekung, eup* is a dry product, and is cut into smaller pieces and then dried in the sun for 5–10 days until its color changes from whitish to chocolate brown. It can be kept up to two years at ambient temperature. *Lactobacillus plantarum* and *L. fermentum* are the main microorganisms involved in its fermentation (Tamang, 2009, 2012).

2.3.5.6 Ekung   *Ekung* is used on many cultural occasions and festivals in Manipur (Singh et al., 2007). *Ekung* is consumed raw or cooked with meat, fish, and vegetables by the Nyishing (Tamang, 2009). The women of Adi and Meitei communities prepare *ekung* using bamboo shoots. Due to the problem of toxicity (presence of cyanide in the shoots) that causes several diseases/disorders related to the nervous system, miscarriages, abnormal child birth, and goiter problems, the old-aged women advise pregnant women not to eat any bamboo-based food product unless it is vigorously processed (Singh et al., 2007). Fermentation is performed, coupled with the removal of toxicity by washing with running water. The small packets of already semi-fermented shoots are wrapped in *ekkam* leaves and made airtight using cane or bamboo rope. These packets are then, pressed under a stone near a stream coming from the top of the hills, for several months (3–4 months) to reduce the bitterness as well as to reduce the cyanogenic glycoside by the activity of yeast and bacteria (*Lactobacillus lactis*). Lactic acid bacteria (*Lactobacillus plantarum, L. brevis, L. casei, Tetragenococcus halophilus*) and yeast are involved in the fermentation (Singh et al., 2007; Tamang, 2009). During cooking, small fishes break easily into pieces. To minimize this, and to improve the taste, *Adi* women add *ekung*, which is part and parcel of their diet (Singh et al., 2007).

2.4 Fermented Alcoholic Beverages/Products

For centuries, alcoholic beverages have been produced in various societies and have been part and parcel of many personal and social ceremonies, both in modern as well as less literate societies. In traditional ceremonies, such as child naming, marriage, and feasts, alcoholic beverages are used. The alcoholic beverages include *Apong* or *Opo* rice beer, *Chhang* palm wine etc., which are well-known product prepared and consumed (Nema et al., 2003; Pant and Nema, 2003; Ciani et al. 2012; Joshi et al. 2012).

2.4.1 Palm Wine

The abundance of tropical plants from the *Palmae* family, such as the coconut palm (Ciani et al., 2012) (*Cocos nucifera*), the date palm (*Phoenix sylvestris*), the nipa palm (*Nypa fruticans*), the palmyra palm (*Borassus flabellifer*) and the kithul palm (*Caryota urens*) in South Asian countries contribute to the variety of palm wines produced in the region (Atputharajah et al., 1986; Shamala and Sreekantiah, 1988; Sekar and Mariappan, 2007; Gupta and Kushwaha, 2011; Law et al., 2011). Palm wine is a fermented sap which is obtained from the unopened inflorescence of palm trees through a method called tapping. Different methods of tapping had been applied to achieve the optimum quantity and quality of sap from different palm trees (Grimwood and Ashman, 1975; Zoysa, 1992). Most tapped palm trees give a sap very rich in sugar (10%–20%), which can either be consumed as fresh juice, sugar syrup, or fermented into wine (Sekar and Mariappan, 2007).

The fermented palm sap is known as *toddy* in India, Sri Lanka, Bangladesh, and Malaysia (Shrestha, 2002; Law et al., 2011). Fresh palm sap has a dirty brown color, but it turns to milky white after the fermentation due to the growth of yeasts and other microorganisms (Steinkraus, 1983; González and De Vuyst, 2009). Palm wine is thus a vigorously effervescent alcoholic beverage (alcohol content 1.5%–2.1%) a typically sweet taste, a milky white color, and a faint sulfur like odor (Sekar and Mariappan, 2007), with suspended microorganisms such as yeasts (Rokosu and Nwisienyi, 1979).

2.4.2 Sur

Sur is a finger millet (ragi; *Eleucine coracana*) based fermented beverage mostly prepared in the Lug valley of the Kullu district and the Chhota Bhangal area of Kangra District (Hulse et al., 1980; Bhalla et al., 2004; Joshi et al., 2012). No specific inocula are used for its preparation. The natural microflora carry out starch hydrolysis and ethanol fermentation. A herbal mix in *sattu* (flour of roasted barley) base (called as *dhaeli*) by the folk people is added during fermentation (Bhalla et al., 2004; Thakur et al., 2004). After the completion of fermentation, *sur* is obtained (Thakur et al., 2004; see another chapter of this text for more information on the preparation of sur).

2.4.3 Atingba *and* Yu

Atingba and *Yu* are popular fermented rice wines of India, especially in Manipur state. *Hamei*, a natural starter culture with a flat rice-cake form, is used, as in the fermentation of *Atingba* and *Yu* (Jeyaram, 2008). *Hamei* is similar to the *Ragi* of Indonesia, the *Budop* of the Philippines, the *Chu* of China, and the *Naruk* of Korea, which have been used as starters in the fermentation of rice wine. To prepare *atingba* and *yu*, *hamei* is crushed into powder and mixed with cooked and cooled rice in the proportion of 40–50 g *hamei* per kg rice and wrapped with *Ficus hispida*, teak, or banana leaves (Singh and Singh, 2006; Jeyaram et al., 2009). The mixture is then, transferred into *Ficus hispida*/teak/banana leaves wrapped earthern/aluminum pot in summer and in bamboo basket in the winter. After covering the open mouth of the basket or pot with a coarse clean cloth, fermentation is carried out in direct sunlight for 3–4 days during the summer or 5–7 days during the winter (Singh and Singh, 2006; Jeyaram et al., 2009; Tamang, 2009).

The taste of rice undergoes three successive stages during fermentation. The total fermentation period, including the stages of alkaline taste, sweet taste, and bitter taste, takes about two days (Singh and Singh, 2006). The sweet and bitter products are filtered, and the filtrate is called *Atingba-Yu* or *Atingba*. The bitter product is subjected to distillation, and the distillate is called *Yu* (Singh and Singh, 2006; Devi and Suresh, 2012; Tamang, 2012). Under prolonged fermentation, the product becomes poor in quantity and quality, and a sour taste is produced, which makes it unacceptable (Singh and Singh, 2006). *Atingba* can be consumed for only 1–2 days after fermentation, but can be kept for 1–2 months to be used to prepare *Yu* (Devi and Suresh Kumar, 2012).

2.5 Fermented Milk Products

2.5.1 Dahi

Dahi is an Indian traditional fermented dairy product, similar to the *yoghurt* of western world. *Dahi* (Sanskrit: *dadhi*) is quite analogous to plain yogurt in appearance and

Figure 2.1   Lord Krishna eating butter (a fermented food) as per ancient epics of India. (Courtesy of Yogendra Rastogi.)

consistency (Yadav et al., 2007). It is popular with consumers due to its distinctive flavor and a belief in its good nutritional and therapeutic value; is utilized in various forms in many Indian culinary preparations (Nair and Prajapati, 2003). The use of *dahi* in India has been prevalent since Vedic times, and it is mentioned in ancient scriptures like the Vedas, Upanishads, and various hymns. During Lord Krishna's time (ca. 3000 BC), *dahi*, butter milk, and country butter were highly regarded (see Figure 2.1).

Dahi is also traditionally used in rituals and as an ingredient of *panchamrut* (five nectors) (Farnworth, 2008). It is prepared traditionally by boiling milk with a preferred quantity of sugar, and partially concentrated by simmering over a low fire, during which the milk develops light caramel color and flavor (Raju and Pal, 2009, 2011). It is then, cooled and inoculated at ambient temperature with lactic culture, usually the *dahi* from previous day's fermentation, and is poured into earthern cups and left undisturbed overnight for fermentation (Raju and Pal, 2009, 2011). When a firm-bodied curd has set, it is stored at a chilled temperature. Such a *Dahi* after sweatening is called *misti dahi*. These days fermentation is usually carried out by inoculation with selected cultures of lactic acid bacteria.

Lactococcus lactis, *Lactobacillus acidophilus*, and *Lactobacillus casei* are used as starters in *dahi* fermentation (Yadav et al., 2007). Mesophilic *Lactococci* is usually used as a starter, to which, sometimes, *Leuconostoc* or *Lactobacillus* is added as adjunct microorganisms (Raju and Pal, 2011; Routray and Mishra, 2012), Beside its nutritive value, *dahi* has several health benefits, and is considered to be effective in both the prevention and treatment of gastro-intestinal disorders (Deeth and Tamime, 1981; Yadav et al., 2008; Arvind et al., 2009). Ayurveda, the traditional system of Indian medicine, in its treatises *Charaka Samhita* and *Sushruta Samhita*, discusses various properties of cow and buffalo milk *dahi*, and emphasizes the therapeutic characteristics (Nair and Prajapati,

2003). Besides *dahi*, various types of *chhash* (stirred diluted *dahi*), and their role in the control of intestinal disorders, have been described (Farnworth, 2008). *Dahi*, which came into use as a means of preserving milk nutrients, was probably used by Aryans in their daily diet, as it reduced putrefactive changes and provided an acidic, refreshing taste. It is consumed with rice in South India, and with wheat preparations in the north. It is also used as a beverage or dessert. *Dahi* from the milk of the yak and/or the zomo is also made in the Himalayas (Nair and Prajapati, 2003). *Dahi* is still made by local *halwais* shops and restaurants, and in homes by traditional methods. However, some dairies have also started their commercial manufacture in India (FAO, 2000; Farnworth, 2008).

2.5.2 Shrikhand

Shrikhand is sweetened dewatered *dahi*, and is very popular in western and some parts of southern India (Steinkraus, 1996). It has a distinctive, rich flavor and a fairly long shelf-life (Anonymous, 2010). *Chakka* is a concentrated product obtained after draining the whey from *dahi*. When it is blended with sugar and other condiments, it becomes *shrikhand*, referred to as *shikhrini* in the old Sanskrit literature (Farnworth, 2008). To prepare *shrikhand*, *dahi* is suspended in a muslin cloth until all the free water has drained-off. The semi-solid mass is then, whisked with sugar through a fine cloth, colored and scented with saffron or rose water, and flavored with cardamom, if desired. To further extend the shelf life of *shrikhand*, a preparation known as *shrikhandvadi*, which is essentially a desiccated *shrikhand*, is also made (Anonymous, 2010).

2.5.3 Lassi

Lassi is a popular and traditional fermented dairy product of the Indian subcontinent, usually consumed in summer as a cold, refreshing therapeutic beverage. *Lassi* is a white to creamy-white viscous liquid with a sweetish rich aroma, and a mild to acidic flavor, which make the product palatable (Behare et al., 2010; George, 2010; George et al., 2012). It is prepared from milk with 1.5%–4.5% fat after making a set curd product, such as *dahi*, followed by vigorous stirring to break the curd into fine particles and, then, the addition of sugar syrup and flavoring. *Streptococcus thermophilus* is the predominant bacteria in *lassi*. Several species of *Streptococcus*, *Lactobacillus*, and *Enterococcus* have been found in this drinkable yoghurt (Behare et al., 2010).

Lassi is a by-product of the preparation of country butter (*ghee*) from *dahi* by indigenous methods. *Dahi* is churned with frequent additions of water until butter granules are formed (Figure 2.2). The product obtained after the manual removal of the butter granules is called *lassi*. The term *lassi* is also used in some parts of Northern India to denote a cold refreshing beverage obtained by blending *dahi* with water and sugar (Anonymous, 2010).

Figure 2.2   Production of *lassi* in typical Indian village. (Courtesy of P.K. Rawat.)

2.6 Indigenous Fermented Foods of Different Countries of South-Asia

Most of the indigenous fermented foods are specific to a particular area or country. A brief description to signify this aspect is made here.

2.6.1 Indigenous Fermented Foods of India

India has a rich knowledge of fermented foods from milk, cereals, pulses, fruits, fish, etc. (Rao et al., 2005). Fermented milks, butter milk (*chhash*), and *lassi* are produced and consumed all over the country. Fermented products like *dhokla* are very popular in western and southern India. However, most of Indians enjoy the tradition of *dahi* and fermented rice-based food and beverages.

There are hundreds of varieties of fruits and vegetables which are used as pickles all over India as taste enhancers. Pickles find an important place in Indian dishes as spicy taste enhancers. Mangoes are the main fruits used for the purpose, while the main vegetables include carrots, chillies, etc. Additionally, a number of refreshing beverages obtained from fermented milk-based or cereal-based products have originated from India. Consumption of milk–cereal-based products, like curd-rice, are very common in the meals of southern India. Himachal Pradesh has several products in common (Figure 2.3) with Northern India, may be with some modifications, depending upon the availability of raw material and tastes of consumers. At the same time, there are several fermented products which are unique to the Himalayan region, while others are specific to Himachal Pradesh, especially to the tribal areas (Gupta,

Figure 2.3 Some of the fermented foods of Himachal Pradesh. (Courtesy of Bhalla, T.C. et al. 2004. *Indian Journal of Traditional Journal*, 3(3), 325–335.)

2011). In Table 2.3, some of the products are listed. Some of the regions of Himachal Pradesh are known for typical products, like the *Angoori* of Kinnour district, while *Chhang* is known throughout the Himalayan region.

In the eastern part of India, there is a tradition of making *dahi* from concentrated milk with a large amount of sugar (upto 15%–25%), which looks like strong brown junket. Traditionally, this was made in small earthen pots, but it is now available in plastic containers. In the western part of India, there is a tradition of removing excess whey from the curd by hanging it in a muslin cloth. The concentrated curd mass is called *Chhakka*. It is kneaded, usually with an equal amount of sugar, and different flavors, colors, fruits, nuts, etc. are added; it is relished as a sweetmeat called *Shrikhand*, as described earlier.

India grows a variety of legumes like red gram (*tut*), black gram (*urd*), Bengal gram (*gram*), green gram (*moong*), cow pea (*chora*), moth beans (*muth*), *massor*, peas, etc. (Reddy et al., 1983). Indians have developed technologies to make them more nutritious and easily digestible by fermenting them with rice and cooking into various delicacies. The fermentation also help in reducing the antinutritional factors from pulses and improves the taste (Gupta and Kushwaha, 2011). The most common foods, such as *idli* and *dosa*, are prepared from rice and *urd dal* mixture. The mixture of rice and various types of pulses, depending upon the availability and taste, are mixed, wet ground, and fermented with the addition of butter milk or *dahi*, and the leavened paste is spiced and steamed to make *khaman* and *dhokla*, which are very popular in Gujarat State. *Handwa* is a spiced cake made from such a mixture. *Bhalle* or *Vada* are deep fried patties, usually prepared from black gram and served after soaking in *dahi*. All these products involve the fermentation of carbohydrates by lactic acid bacteria and yeasts (Tamang and Fleet, 2009).

Roti or flattened bread is the most common staple food in India. The baked staple food made from raised dough of wheat flour with salt and the previous day's dough or *dahi* as inoculum is called a *nan*, which is very popular throughout the country.

Table 2.3   Common Traditional Fermentable Foods/Beverages of India

FOOD	SUBSTRATE	MICROORGANISM	NATURE AND USE
TRADITIONAL FERMENTED PRODUCTS OF SOUTH INDIA			
Idlis	Rice, black gram	LAB, yeasts	Steamed, spongy cake, breakfast food
Dosa	Rice, black gram	LAB, yeasts, *B. amyloliquefaciens*	Spongy pan cake, shallow-fried staple food
Dahi	Milk	LAB	
Butter milk		LAB, *Bacillus* ssp.	Sweet/sour beverage used to be taken along with rice
Lassi		LAB, *Bacillus* ssp.	Sweet and flavored beverage
Ambali	Millet, rice	LAB, yeast and *Bacillus* ssp.	Steamed sour cake, staple food
Kanji	Carrot or beetroots	Starter culture used is TORANI which contains LAB, yeasts	Strong flavored alcoholic beverages
Kali			
Papadam	Black gram	LAB	
Pickles	Matured fruits and vegetables		
Toddy/Palm wine	Sap from inflorescence/trunk of palm plants		
Sara	Black jaggery and plant bark	Yeast, LAB and *Acetobacter*	
TRADITIONAL FERMENTED PRODUCTS OF NORTH INDIA			
Ballae	Black gram	LAB, yeasts, *B. subtilis*	Deep-fried patties, snack
Vada	Black gram	LAB, yeasts, *B. subtilis*	Deep-fried patties, snack
Papad	Black gram	LAB, yeasts	Circular wafers, snack
Wari	Black gram	LAB, yeasts	Ball-like hollow, brittle, condiment
Bhature	Wheat	LAB	Flat deep-fried, leavened bread, snack
Nan	Wheat	Yeasts, LAB	Leavened flat baked bread, staple food
Jalebi	Wheat	LAB, Yeasts	Crispy, deep-fried, pretzel sweet confectionery
Paneer	Milk	LAB	Soft mild flavored cheese, fried, curry
TRADITIONAL FERMENTED PRODUCTS OF WESTERN REGIONS OF A INDIA			
Dhokla	Bengal gram	LAB, yeasts	Steamed, spongy cake, snack
Khamam	Bengal gram	LAB	Spongy cake, breakfast food
Rabadi	Wheat/pear-millet/maize/barley-buttermilk mixture	LAB, *Bacillus* spp.	Cooked paste, staple food
Sri Khand	Milk	LAB	Concentrated sweetened, savory
TRADITIONAL FERMENTED PRODUCTS OF EASTERN REGIONS OF INDIA			
MistidaJJi	Milk	LAB	Thick gel, sweet savory
Tari	Date palm	Yeasts, LAB	Sweet cloudy white alcoholic beverage

(*Continued*)

Table 2.3 (*Continued*) Common Traditional Fermentable Foods/Beverages of India

FOOD	SUBSTRATE	MICROORGANISM	NATURE AND USE
TRADITIONAL FERMENTED PRODUCTS OF THE HIMALAYA			
Kinema	Soybeans	*Bacillus subtilis, Enterococcus faecium,* Yeasts	Sticky with typical flavor, side-dish curry
Hawaijar	Soybeans	*Bacillus* spp.	Sticky with typical flavor, side-dish curry. Fish substitute
Gundruk	Leafy vegetables	LAB	Sun-dried, sour-acidic taste, soup/pickle
Sinki	Radish tap root	LAB	Sun-dried, sour-acidic taste, soup/pickle
Mesu	Bamboo shoot	LAB	Sour-acidic pickle
Laanr	Finger-millet/rice maize/barley	Starter culture used is *Mrrcha* which contains filamentous molds, Yeasts, LAB	Mild alcoholic, slightly sweet-acidic beverage
TRADITIONAL FERMENTED PRODUCTS OF EASTERN REGIONS OF INDIA			
Mistidall	Milk	LAB	Thick gel, sweet savory
Tari	Date palm	Yeasts, LAB	Sweet cloudy white alcoholic beverage

Source: Adapted from Soni, S.K. and Sandhu, D.K. 1989. *J Cereal Sci*, 10(3), 227–238; Tamang, J.P. 1998. *Indian Food Industry*, 17(3), 162–167; Tamang, J.P. 2009a. *Himalayan Fermented Foods Microbiology Nutrition and Ethnic Values*, CRC Press, Taylor & Francis Group, New York, pp. 139–158; Tamang, J.P. 2009b. *Himalayan Fermented Foods Microbiology Nutrition and Ethnic Values*. CRC Press, Taylor & Francis Group, New York, pp. 25–64; Tamang, J.P. 2010. *Himalayan Fermented Foods: Microbiology, Nutrition and Ethnic Values*. CRC Press, New York; Tamang, J.P. 2012. *Handbook of Plant-based Fermented Food and Beverage Technology*, 2nd edition. CRC Press, Taylor & Francis Group, New York, NY, pp. 49–90; Farnworth, E.R. 2008. *Handbook of Fermented Functional Foods*. 2nd edition. *Functional foods and Nutraceuticals Series*. Taylor & Francis, CRC Press, Boca Raton, FL; Rao, R.E. et al. 2006. *Food Biotechnology*. CRC press, Taylor & Francis, Boca Raton, FL, pp. 1759–1794.

These are baked in a special oven called a *tandury*. Another product is *bhatura*, which is more digestible and palatable (Figure 2.4). Jalebis are confectioneries prepared from fermented wheat maida, which are deep fried and dipped in sugar syrup (Figure 2.4). Some of the population of India also enjoys fermented fish and sausages, especially Goan sausages. In the eastern and hilly regions of India, rice is the main crop. Apart

Figure 2.4 Some of the cereal-based fermented foods.

from making fermented rice, it is also used to make traditional beer called *opo*. Products like *Chhurpi, Chhang, Shosim, Kinema*, etc., are also popular in the region. Most of the traditional fermented foods of Andhra Pradesh include a few common fermented foods such as *idli, dosa, dahi*, etc.

Indigenous fermented foods also play a major role in the daily food intake of the north eastern regions. Ethnic or indigenous fermented foods are produced based on the indigenous knowledge (IK) of the ethnic peoples, from locally available raw materials of plant or animal sources, either naturally or by adding starter culture(s) containing functional microorganisms that modify the substrates biochemically and organoleptically into edible products that are culturally and socially acceptable to consumers (Tamang, 2009, 2010). A traditional fermented food made from soybean fermented with *Bacillus subtilis*, namely *Hawaijar*, is indigenous to the Manipur valley (Holzapfel, 2002). *Hawaijar* uses naturally occurring microorganism for fermentation (Feng, 2007). Fermented soybean is most popular in north eastern regions of India, particularly Manipur, Sikkim, Nagaland, Arunachal Pradesh, and Mizoram (Rao et al., 2006). The different ethnic groups of the district of Assam prepare and consume different types of fermented foods and alcoholic beverages using dry mixed starter cultures (*Dimasa* or *Humao*), in accordance with their indigenous traditional technology (Tamang and Sarkar, 1996).

The fermented foods are an integral part of the diet of man in northern India also (Blandino, 2003). Some of the common indigenous fermented foods prepared and consumed in India are listed in Table 2.3 while those made in the North East Himalayas and Himachal Pradesh are described in Tables 2.4 and 2.5. There are several fermented foods which are prepared and consumed in Himachal Pradesh (India) (Figure 2.3) these include *Bhalle, Warries, Kulche*, and *Jalebis* (Soni and Sandhu, 1999; Joshi et al., 2012; Soni et al., 2013). The products like *Bhaturu, Siddu, Chilra, Marchu, Manna, Dosha, Pinni/Bagpinni, Seera*, etc., are unique to Himachal Pradesh. A large variety of fermented foods are prepared either daily, during special occasions, or for consumption during journeys. Traditional starter cultures like "Malera" and "Treh" are used as inocula in making these fermented foods. However, the natural fermentation (without the addition of inoculum, as microorganisms are present in the raw materials) is used in the production of *Seera, Sepubari, Bohre*, etc. (Bhalla et al., 2004).

Of course, a variety of alcoholic beverages are also produced and consumed in Punjab and in the adjoining area of Northern India. Fermented milk *dahi* is a universal fermented product prepared and consumed in Northern India, as elsewhere in the world. Another fermented product, *Lassi*, is common in rural India, and Northern India is not an exception.

Many of the fermented products are well known as house-hold items, while a few are prepared at cottage scale. Industrialization of these products has not taken place to any great extent, although some of these products have already found a place in the menu of several restaurants and hotels (Nout and Motarjemi, 1997). Mostly these foods are cereal-based (wheat/barley/buckwheat ragi) but some are legume-based (black gram), and milk-based fermented foods are also common.

Table 2.4 Common Indigenous Fermented Foods of the North East Himalayas (Sikkim)

FOOD	SUBSTRATE	FUNCTIONAL MICROORGANISMS
Kinema	Soybean	*Bacillus subtilis* + LAB
Turangbai	Soybean	*Bacillus subtilis* + LAB
Hawaijar	Soybean	*Bacillus subtilis* + LAB
Aakhuni	Soybean	*Bacillus subtilis* + LAB
Bekanthu	Soybean	*Bacillus subtilis* + LAB
Masaurya	Black gram	Yeasts + LAB
Gundruk	Leafy vegetable	LAB
Sinki	Radish tap-root	LAB
Mesu	Bamboo shoot	LAB
Lung-siej	Bamboo shoot	LAB
Soibum	Bamboo shoot	LAB
Soidon	Bamboo shoot	LAB
Khalpi	Cucumber	LAB
Selroti	Rice/wheatmilk	LAB + yeasts
Dahi	Cow milk	LAB + yeasts
Gheu	Cow milk	LAB + yeasts
Mohi	Cow milk	LAB
Soft Chhurpi	Cow milk	LAB + yeasts
Chhu	Cow milk	LAB + yeasts
Somar	Cow milk	LAB
Dudh chhurpi	Cow milk	LAB + yeasts
Hard chhurpi	Yak milk	LAB + yeasts
Philu	Cow/yak milk	LAB + yeasts
Sukako masu	Mutton/pork	Unknown
Sha-kampo	Beef/Yak	Unknown
Marcha/Hamai/Phab	Rice, wild herbs, spices	Molds + yeasts + LAB
Kodo ko jaanr or Chhyang	Fingermillet	Molds + yeasts + LAB
Bhaati jaanr	Rice	Molds + yeasts + LAB
Makai ko jaanr	Maize	Molds + yeasts + LAB
Sidra	Fish	LAB + *Bacillus* spp. + *Micrococcus* spp. + yeasts
Sukuti	Fish	LAB + *Bacillus* spp. + *Micrococcus* spp. + yeasts
Tungtap	Fish	LAB + *Bacillus* spp. + *Micrococcus* spp. + yeasts
Hentak	Fish	LAB + *Bacillus* spp. + *Micrococcus* spp. + yeasts
Ngari	Fish	LAB + *Bacillus* spp. + *Micrococcus* spp. + Yeasts
Karoti	Fish	LAB + *Bacillus* spp. + *Micrococcus* spp. + Yeasts
Bardia	Fish	LAB + *Bacillus* spp. + *Micrococcus* spp. + yeasts

Source: Adapted from Tamang, J.P. 2009a. *Himalayan Fermented Foods Microbiology Nutrition and Ethnic Values.* CRC Press, Taylor & Francis Group, New York, NY, pp. 139–158; Tamang, J.P. 2009b. *Himalayan Fermented Foods Microbiology Nutrition and Ethnic Values.* CRC Press, Taylor & Francis Group, New York, NY, pp. 25–64; Tamang, J.P. 2009c. *Himalayan Fermented Foods Microbiology Nutrition and Ethnic Values.* CRC Press, Taylor & Francis Group, New York, NY, pp. 1–24. *Himalayan Fermented Foods: Microbiology, Nutrition and Ethnic Values.* CRC Press, New York.

Table 2.5   Some of the Traditional Fermented Foods of Himachal Pradesh

FERMENTED FOOD	SUBSTRATE	AREA
Aska	Rice flour	La haul
Bah	Black gram	Kullu, Kangra, Mandi, Bilaspur
Bedvin roti	Wheat floor, opium, seeds, linseeds	Kullu, Kangra, Mandi, Chamba, Shimla
Bhaturu	Wheat flour	Kullu, Kangra, Mandi, Chamba, Shimla
Borhe	Black gram	Spiti
Babru/Suhala	Wheat flour	Kullu, Kangra, Mandi
Chilra (Lwar)	Rice/Wheat/Buck-wheat/Barley	Lahaul, Kangra
Dosa	Wheat flour	Lahaul
Gulguie	Wheat flour	Kullu, kangra, Mandi, Chamba, Spiti
Jute	Buck wheat flour	Kinnour
Khoblu	Wheet flour, buttermilk, Lassi	Mandi, Suket
Manna	Wheat flour	Lahaul
Marchu	Wheat flour	Lahaul
Pakk	Barley, butter, and lassi	Kullu, Kangra, Mandi, Bilaspur
Pinni/Bagpinni	Barley flour	Kullu, Kangra, Mandi, Chamba, and Shimla
Seera/Kheera	Wheat grains	Kullu, Kangra, Mandi, Chamba, Spiti
Sepubari	Black gram	Kullu, Kangra, Mandi, Bilaspur
Siddu	Wheat flour	Kullu, Kangra, Shimla
Thuktal	Barley	Bilaspur

Source: Adapted from Thakur, N. et al. 2004. *Indian J Trad Know*, 3(3), 325–335; Thakur, N.S. 2005. *Personal communication*; Savitri and Bhalla, T.C. 2013. *3 Biotech*, 3(3), 247–254; Kanwar, S.S. et al. 2007. *Indian J Trad Know*, 02, 6, 42–45; Bhalla et al. 2004. *Indian J Trad Know*, 3(3), 325–335.

Manipur, one of the states of North Eastern India, is inhabited by different communities such as the Meitei, Nagas, Kukis, and Meitei Pangals, belonging to Mongoloid and Indo-Aryan stocks. Though most of the Meitei community follow Hinduism, which forbids alcohol consumption, certain sections of the community still follow pre-Hindu religious traditions, and practice alcoholic fermentation. Although most

Table 2.6   Some of the Traditional Alcoholic Beverages Used in Manipur, India

TRADITIONAL ALCOHOLIC BEVERAGES	COMMUNITY	RAW MATERIAL
Yu, Kalei, Wanglei	Meitei	Rice
Waiyu	Meitei	Rice and husk
Zupar, zuhrin	Anal	Rice
Yuh	Chothe	Rice
Zu	Hmar	Rice
Zou	Kabui	Rice
Zoungou	Zemei, Liangmai	Rice
Vai Zu	Kom	Rice
Armoon Yu	Lamkang	Job's tears
Azual	Maram	Rice
Noom Yu	Monsang	Job's tears
Zam	Tangkhul	Rice

of them use similar substrates for fermentation, the methods for wine and beverage production differ among the communities, as they follow their own indigenous protocols (Tamang, 2010). Beverages used by the Meitei, Naga, and Kuki communities are mainly derived from rice as the raw materials (Table 2.6). Some tribal groups also use Job's tears (*Coix lacryma-jobi*) as the substrate for fermentation.

Ladakh, truly described as a high altitude cold-arid desert, is one of the most eastern regions of Jammu and Kashmir State of India. The people have developed traditional foods and beverages which over time have been evolved (through outside influence and local resources availability) and established in the food system of Ladakh (Angchok et al., 2009). The traditional fermented foods of Ladakh include fermented cereal products, vegetable products, dairy products, meat products, and beverages (Guizani and Mathershaw, 2006).

2.6.2 Fermented Foods of Sri Lanka and the Maldives

In Sri Lanka, fermented foods have been used as meals or parts of a meal for a long time. A well-known indigenous fermented food is curd, made from buffalo milk and considered a delicacy from ancient times (Wikramanayake, 1996; Rao et al., 2006). These foods are an integral part of the meal pattern in a variety of ways. With a major portion of rice, one or two vegetables (green leafy vegetable and legume or potatoes), fish, or meat curry, with condiments, are common in all parts of Sri Lanka (Sagarika and Pradeepa, 2003). However, most Buddhists avoid eating meat, while beef is avoided by most of the Hindus. String hoppers, hoppers, *roti*, *Dosa*, and *idli*, together with one or two side dishes, including fermented food, are some dishes that make breakfast and dinner. In particular, *Dosa* and *idli*, together with *Vada*, are common foods in the Tamil community of Sri Lanka. Others common foods—for Sinhalase, Muslims, and others—are *Dosa* and *idli*. However, the foods that were earlier restricted to different ethnic groups are now popular in all the communities (Figure 2.4).

Fermentation due to lactic acid bacteria in various foods, including curd, is a natural process brought about by the lactic acid bacteria present in the raw food, or those derived from a starter culture, is a common practice used in the households of Sri Lanka for a long time (Nout, 1997). Hopper is a popular product of Sri Lanka. In hopper making, rice grains are soaked overnight, during which the naturally occurring microorganism population is dominated by lactic acid bacteria (Agati et al., 1998). Breads and other products (different types of buns) made from cereal meal or flour, are leavened by the action of yeasts or mixed yeast–lactic acid bacteria populations. The main function of fermentation is to produce carbon dioxide to leaven and condition the dough. Yeast and other *Lactobacillus* species in dough contribute to the flavor of the bread (Singh et al., 2007).

Toddy is a common fermented alcoholic beverage popular in the villages, made by the fermentation of the sap from coconut (commonest), palmyrah (northern parts), and *kithul* (wet zone) palms (Gupta and Kushwaha, 2011). The sap is collected by slicing-off the tip of an unopened flower. The sap oozes out and is collected twice daily in a small pot tied underneath the flower. The fermentation starts as soon as the sap collects in the pots on the palms, and after straining it is sold on the same day. It is white and sweet, with a characteristic flavor (4%–6% alcohol), and has a shelf-life of about 24 h. Coconut toddy vinegar is another fermented product produced throughout South Asia, particularly in Sri Lanka. It is a clear liquid with a strong acetic acid flavor with a hint of coconut flavor. The fresh toddy is strained and natural yeast fermentation is allowed to occur for 48–72 h. After 2–4 weeks of settling, the fermented toddy is placed in barrels. The alcohol is then oxidized into acetic acid by acetic acid bacteria, which are naturally present, in about 2 months. Ageing for 6 months results in a pleasantly flavored final product (Anonymous, 2010).

Lime pickle is prepared and is consumed usually as a condiment and is favored by many Sri Lankans (Wacher et al., 2010). Limes are cut into quarters and placed in a layer, approximately 2.5 cm deep, into the fermenting container and salt added (4:1 ratio). Lime and salt are layered until the container is three quarters full. A cloth is placed above the limes and a weight is placed to compress them and assist in the formation of brine which takes about 24 h. As soon as the brine is formed, fermentation (for 1–4 weeks) starts (Anonymous, 2010) and the final product is a sour lime pickle. Spices are added depending on local preference.

Salami and *pepperoni*, produced by using starter cultures on meat, are also gaining popularity, although they are not traditional products of Sri Lanka. All fermented fishery products in Sri Lanka are salt-based. Of these, nearly 75% is Maldive fish. Earlier, this product used to be imported from the Maldives islands, but now most of it comes from India and local production. It is a slightly salted, smoked, dried form of tuna and is widely used as a flavoring agent in most of the local products. Dry fish marketed in Sri Lanka generally cannot be categorized as a fermented fishery product due to the process of salting involved, and the product is in various degrees of fermentation. *Jaadi*—pickled fish—is another important fermented fishery product and is mainly produced locally on a cottage scale using seasonal gluts of fish. It is a high salted fermented fishery product consisting of partially hydrolysed fish immersed in their own liquid exudates (Tanasupawat and Visessangua, 2014). A low pH is maintained by addition of ripe goraka pod (*Garcinia gamboges*). In the Maldives, hoppers, *Dosa*, *ktiy*, *Vada*, curd, *toddy*, *jaadi* (cured fish), lime pickle, dry fish, and bread are the common fermented foods that have been utilized for a long time. Today fermented fish sausages, fish sauces, and fish pastes are coming up as new arrivals in the commercial market.

2.6.3 Fermented Foods of Nepal and Bhutan

Fermented foods have also been a part of staple foods in the mountainous region of Nepal and Bhutan, as well as for the community of Nepalese origin ethnic living in different parts of South Asia, including Sikkim, West Bengal Hills, Dehradun, and so on (Thapa et al., 2004). The most popular fermented foods and beverages of these areas are *kinema, gundruk, sinki, tama, dahi, mohi, sher, shergum, chhurpi, selroti, rakshi, tumba*, etc. Out of these foods, *gundruk* is indigenous to the mountains and is recognized as a symbol of identity for the Nepalese people (Thapa et al., 2004; Dahal et al., 2005; Savitri and Bhalla, 2013).

These foods are important components of the staple diet of these people, and are used as adjuncts to staples, condiments, and beverages. The most popular fermented foods are *Kinema* (fermented soyabean), *Gundruk* (*Brassica compastris* leaves), *Sinki* (*Raphanus sativus*), (succulent bamboo shoot), *Dahi* (similar to yoghurt), *Mohi* (buttermilk), *Sher, Shergum* (soft cheese from buttermilk), *Chhurpi* (dried cheese from buttermilk), *Selroti* (deep-fried preparation from rice flour), *Jand* (local beer from rice/maize), *Rakshi* (alcohol distilled from fermented rice, maize, or millet), *Tumba* (fermented millet drink), and varieties of fermented acidic pickles (*Achar*). Fermented milk products are other popular food products in this mountainous region, including *Dahi, Mohi, Sher, Sherghum*, and *Chhurpi*, and *yak*, and other cheeses. Nepal produces around 200 metric tones of *yak* cheese, annually, which is very typical to Nepal, and the production is some five decades old. In the same way, *Sher, Shergum, Shosim, Churtsi, Chhuga*, and *Chhurpi* are produced from *Nak* (yak) milk in the high altitude regions of Bhutan (FAO, 1990). In the mostly household processing, milk is boiled, fermented, and churned, and the butter is taken out, and the buttermilk is a by-product which is used as a raw material for producing *Sher, Sherghum*, and *Chhurpi*, as well as being used as a staple food and consumed with rice, maize meal, and bread. Most milk is informally processed and converted into indigenous fermented milk products in Bhutan (IDA, 2003).

2.6.4 Fermented Foods of Bangladesh

Fish and rice are the staple foods in Bangladeshi cuisine, and it is said that these two food items make a Bengali (*Machh-e-Bhat-e-Bangali*). The majority of these foods are consumed fresh after cooking, although a small proportion is used to prepare various indigenous fermented foods, including fermented beverages from rice and various fermented fishery products. The former is largely consumed by the tribal villagers in the Chittagong hill tracts, while the latter is consumed in different parts of the country, including Mymensingh, Comilla, Sylhet, Rangpur, Gazipur, and Chittagong. In addition to fermented rice and fishery products, a wide variety of traditional fermented products are available in the country, including *Dahi* (Yoghurt), *Modhubhat* (fermented food prepared from germinated rice powder and boiled rice), *Kanjibhat* (fermented rice), *Pantabhat* (fermented cooked rice), *Zilapi* (savoury from fermented Bengal gram), *Vapa pitha* (fermented snacks), and *Bundia* (fermented Bengal gram). See Hafiz and Majid (1996) and Parveen and Hafiz (2003) for detailed information on preparation of fermented food from rice and pulses in Bangladesh.

Fermented foods, especially prepared from milk or its by-products, are very popular in Bangladesh, including *dahi, yoghurt*, acidophilus milk, *matha*, and fermented whey drinks, and they are widely manufactured throughout Bangladesh. *Dahi* (usually sweet curd) is prepared by using a mixed culture of *Streptococcus lactis, Streptococcus thermophilus, Streptococcus citrophilus, Lactobacillus bulgaricus, Lactobacillus plantarum*, etc. The same is true for *matha*, which is prepared from butter milk and sour in taste; a small amount (0.5%) of salt may also be added to increase the palatability. Acidophilus milks and fermented whey drinks are new additions to the fermented food industry of Bangladesh. *Lassi*, a refreshing beverage made from sweet curd, sugar syrup, and rose flavoring, is liked by all, especially during the summer when served with ice (Qureshi et al., 2011).

In Bangladesh, fermented food products include some food items manufactured by the sweetmeat industry which are widely accepted and have been consumed since time immemorable. Among all these foods, *dahi*, especially the sweet variety, is the most popular item. It is a custom in Bangladesh to offer a handsome amount of sweet *dahi* after any feast, and it is a part and parcel of marriage ceremonies. It is also offered after the banquet and sour *dahi* is used as a raw material to cook several of the key dishes. Another very popular beverage called "*borhani*," hot and sour in taste, is served along with the courses. Sweet *dahi* is mostly prepared from whole cow's milk, whereas majority of sour *dahi* is prepared from skimmed cow or buffalo milk. In Bangladesh, the *dahi* of Bogra district is famous for its overall acceptability. A very refreshing beverage, commonly called *lassi* is prepared from sweet *dahi*. Items like *matha* are sold by street vendors, mostly in the early morning. Yet another beverage is made by using sour *dahi* almost in the same way, which is called "*ghole*." *Lassi* and *ghole* are normally served during summer. However, *lassi* is usually marketed by confectioners on a quick serve basis or by large *dahi* production plants, who market it in polypacks.

2.7 Summary and Future Prospectives

There is great diversity in the fermented foods of South Asia, both in the quality as well as in the methods of preparation. However, these foods are not confined to a particular purpose, they form a large and important sector of the food industry, and they are acknowledged as more appetizing and digestible. Some of the indigenous fermented foods have been documented. Still there is a lack of complete documentation of the production, processing, storage, and quality aspects of several important indigenous fermented food products. Consequently, some of these popular traditional fermented foods are gradually diminishing due to the increased rate of urbanization, and indigenous technology not being transferred to the new generation, either due to the lack of interest on the part of the younger generation, or lack of the necessary time. It is a challenge to the food scientists and technologists not to miss out such an important commodity, of historical significance as well as nutritional and diversity value. It is also a way to help the ethnic communities and regions to improve the quality and quantity, so that a larger segment of consumers can be benefited. Important issues related to indigenous fermented foods produced and consumed in the countries of South Asia, also need to be addressed to improve upon both their quality and marketability. Many of these foods are now ceasing to be indigenous, as they are gradually introduced into the countries other than the original producer and are produced at large scale. There is a great prospect for large-scale commercialization of fermented foods in South Asia as a large ready market, wide popularity of the products, and the traditional approach to manufacturing at small scale is being upgraded to large scale commercial production is awaiting such products. There are many new inventions to ensure the best quality of fermented foods, and several new developments, including mechanization of the entire production operation, which can be adopted for large scale commercial production. The future needs to preserve both the indigenous knowledge of production of indigenous fermented foods and their diversity with respect to the use of raw materials and the methods of production of such foods.

References

Adams, M.R. and Moss, M.O. 1996. Fermented and microbial foods. In: Adams, M.R. and Moss, M.O. (eds) *Food Microbiology*, 1st edition. New Age International Publishers, New Delhi, pp. 252–302.

Agati, V., Guyot, J.P., Guyot-Morlon, J., Talamond, P., and Hounhouigan, D.J. 1998. Isolation and characterization of new amylolytic strains of *Lactobacillus* fermentum from fermented maize doughs (mawe and ogi) from Benin. *J Appl Microbiol*, 85(5), 512–520.

Aliya, S. and Gervani, P. 1981. An assessment of protein quality and vitamin B content of commonly used fermented products legumes and millet. *J Sci Food Agric*, 32, 837–842.

Angchok, D., Dwivedi, S.K., and Ahmed Z. 2009. Traditional foods and beverages of Ladakh: Ladakh traditional foods beverages Ladakhi Kholak Paba. *IJTK*, 08(4), 551–558.

Anonymous. http://fermentationtechnology.blogspot.com/2008_04_01_archive.html.

Anonymous. http://microbewiki.kenyon.edu/index.php?title=Soybean&redirect=no.

Anonymous. http://www.bdu.ac.in/schools/biotechnology/industrial_biotechnology/sekardb/pdf/food/4.pdf 2010.

Arvind, S.P.R., Singh, N.K., and Kumar, R. 2009. Effect of *Acidophilus casei* dahi (Probiotic curd) on lipids in 1,2-dimethylhydrazine induced intestinal cancer in rats. *Int J Probiot Prebiot*, 4, 195–200.

Atputharajah, J.D., Widanapathirana, S., and Samarajeewa, U. 1986. Microbiology and biochemistry of natural fermentation of coconut palm sap. *Food Microbiol*, 3(4), 273–280.

Behare, P.V., Singh, R., Tomar, S.K., Nagpal, R., Kumar, M., and Mohania, D. 2010. Effect of exoplysaccharide-producing strains of *Streptococcus thermophilus* on technological attributes of fat-free lassi. *J Dairy Sci*, 93, 2874–2879.

Beuchat, L.R. 1983. Indigenous fermented foods. In: Reed, G. (ed) *Biotechnology: Food and Feed Production with Micro-organisms*, Vol. 5. Verlag Chemie, Weinheim, pp. 477–528.

Beuchat, L.R. 2001. Traditional fermented foods. In: Doyle, M.P., Beuchat, L.R. and Montville, T.J. (eds) *Food Microbiology: Fundamentals and Frontiers*. ASM Press, Washington, DC, pp. 701–719.

Bhalla, T.C., Thakur, N., and Savitri. 2004. Characterization of some traditional fermented foods and beverages of Himachal Pradesh. *Indian J Trad Know*, 3(3), 325–335.

Bhatt, B.P., Singha, L.B., Singh, K., and Sachan, M.S. 2003. Some commercial edible bamboo species of Norteast India: Production, indigenous uses, cost-benefit and management strategies. *Bamboo Sci Culture*, 17(1), 4–20.

Bhattacharya, S. and Bhat, K.K. 1997. Steady shear rheology of rice-blackgram suspensions and suitability of rheological models. *J Food Eng*, 32(3), 241–250.

Blandino, A., Al-Aseeri, M.E., Pandiella, S.S., Cantero, D., and Webb, C. 2003. Cereal-based fermented foods and beverages. *Food Res Int*, 36(6), 527–543.

Cagno, D.R., Coda, R., Angelis, D.M., and Gobbetti, M. 2013. Exploitation of vegetables and fruits through lactic acid fermentation. *Food Microbiol*, 33(1), 1–10.

Campbell-Platt, G. 1987. *Fermented Foods of the World—A Dictionary and Guide*. Butterworths, London, 294, 134–142.

Campbell-Platt, G. (Ed.). 2009. *Food Science and Technology*. Wiley–Blackwell, John Wiley & Sons, Hoboken, NJ, p. 520.

Chavan, J.K., Kadam, S.S., and Beuchat, L.R. 1989. Nutritional improvement of cereals by fermentation. *Crit Rev Food Sci Nutr*, 28(5), 349–400.

Choi, Y. 2001. Fish sauce products and manufacturing: A review. *Food Rev Int*, 1/1, 65–88.

Choudhury, D., Sahu, J.K., and Sharma, G.D. 2012. Value addition to bamboo shoots: A review. *J Food Sci Technol*, 49(4), 407–414.

Ciani, M., Stringini, M., and Comitini, F. 2012. Palm wine. In: Hui, Y.H. and Özgül, E. (eds) 2nd edition, *Handbook of Plant-based Fermented Food and Beverage Technology*. CRC Press Taylor and Francis Group, Boca Raton, FL, pp. 631–638.

Dahal, N.R., Karki, T.B., Swamylingappa, B., Li, Q., and Gu, G. 2005. Traditional foods and beverages of Nepal. A review. *Food Rev Int*, 1(21), 1–25.

Das, A., Raychaudhuri, U., and Chakraborty, R. 2012. Cereal based functional food of Indian subcontinent: A review *J Food Sci Technol*, 49(6), 665–672.

Das S., Holland R., Crow V.L., Bennett R.J., and Manderson G.J. 2005. Effect of yeast and bacterial adjuncts on the CLA content and flavour of a washed-curd, dry-salted cheese. *Int Dairy J*, 15, 807–815.

Deeth, H.C. and Tamime, A.Y. 1981. Yoghurt: Nutritive and therapeutic aspects. *J Food Prot*, 44, 78–86.

Devi, P. and Suresh Kumar, P. 2012. Traditional, ethnic and fermented foods of different tribes of Manipur. *Indian J Trad Know*, 11(1), 70–77.

FAO. 1990. The Technology of Traditional Milk Products in Developing Countries. FAO Animal Production and Health Paper Series, 85.

FAO. 2002. Report on the FAO E-mail conference on Small-scale Milk Collection and Processing in Developing Countries; Conference moderated by Anthony Bennett, Jurjen Draayer, Brian Dugdill, Jean-claude Lambert and Tek Thapa; AGAP, FAO, Rome, Italy (May–July 2000).

Farhad, M., Kailasapathy, K., and Tamang, J.P. 2010. Health aspects of fermented foods. In: Tamang, J.P. and Kailasapathy, K. (eds) *Fermented Foods and Beverages of the World*. CRC Press, New York, NY, pp. 391–414.

Farnworth, E.R. 2008. Handbook of fermented functional foods. In: Mazza, G. (ed), 2nd edition. *Functional Foods and Nutraceuticals Series*. Taylor and Francis, CRC Press, Boca Raton, FL, p. 533.

Fellows, P.J. 2000. Fermentation and enzyme technology. In: *Food Processing Technology*. Wood Head Publishing Limited, Cambridge, England.

Feng, X.M. 2007. Image analysis for monitoring the barley tempeh fermentation process. *J Appl Microbiol*, 6, 21.

George, V. 2010. Stability, physico-chemical, microbial and sensory properties of sweetener/ sweetener blends in lassi during storage. *Food Bioproc Technol*, 1, 27.

George, V., Arora, S., Sharma, V., Wadhwa, B.K., and Singh, A.K. 2012. Stability, physico-chemical, microbial and sensory properties of sweetener/sweetener blends in *lassi* during storage. *Food Bioproc Technol*, 5, 323–330.

Ghosh, D. and Chattopadhyay, P. 2011. Preparation of idli batter, its properties and nutritional improvement during fermentation. *J Food Sci Technol*, 48(5), 610–615.

Gildberg, A. and Thongthai, C. 2005. Asian fish sauce as a source of nutrition. In: Shi, J., Ho, C.T. and Shahidi, F. (eds) *Asian Functional Foods*, CRC Press, Boca Raton, Florida, pp. 215–266.

González, Á. and De Vuyst, L. 2009. Vinegars from tropical Africa. In: Solieri, L. and Giudici, P. (eds) *Vinegars of the World*. Springer-Verlag, Italia, pp. 209–221.

Granito, M. and Álvarez G. 2006. Lactic acid fermentation of black beans (*Phaseolus vulgaris*): Microbiological and chemical characterization. *J Sci Food Agric*, 86(8), 1164.

Grimwood, B.E. and Ashman, F. 1975. Coconut palm production: Their process in developing countries. *Food Agric Organ*, 189–190.

Guizani, N. and Mothershaw, A. 2006. Fermentation. In: Hui, Y.H. (ed.) *Handbook of Food Science, Technology and Engineering*, Vol. 2. CRC Press, Taylor & Francis, London, New York. pp. 63.1–63.30.

Gupta, N. and Kushwaha, H. 2011. Date palm as a source of bioethanol: Producing micro-organisms. In: Jain, S.M., Al-Khayri, J.M. and Dennis V. J. (eds) *Date Palm Biotechnology*. Springer Publishers, London, New York. pp. 711–721.

Hafiz, F. and Majid, A. 1996. Preparation of fermented food from rice and pulses. Bangladesh. *J Sci Ind Res*, 31(4), 43–61.

Hayford, A.E. and Jespersen, L. 1999. Characterization of *Saccharomyces cerevisiae* strains from spontaneously fermented maize dough by profiles of assimilation, chromosome polymorphism, PCR and MAL genotyping. *J Appl Microbiol*, 86, 284–294.

Hesseltine, C.W. 1979. Some important fermented foods of Mid-Asia, the Middle East, and Africa. *J Am Oil Chem Soc*, 56(3), 367–374.

Holzapfel, W.H. 2002. Appropriate starter culture technologies for small-scale fermentation in developing countries. *Int J Food Microbiol*, 75(3), 197–212.

Hulse, J.H., Laing. E.M., and Pearson, O.E. 1980. *Sorghum and the Millets: Their Composition and Nutritive Value*. Academic Press, London, p. 997.

IDA. 2003. *Indian Dairyman*, March 2003 Issue (pp. 139–142).

Jeyaram, K., Anand Singh, T., and Romi, W. 2009. Traditional fermented foods of Manipur. *Indian J Trad Know*, 8(1), 115–121.

Jeyaram, K., Romi, W., Anand Singh, T., Ranjita Devi, A., and Soni Devi, S. 2010. Bacterial species associated with traditional starter cultures used for fermented bamboo shoot production in Manipur state of India. *Int J Food Microbiol*, 143(1–2), 1–8.

Jeyaram, K., Singh, W.M., and Premarani, T. 2008. Molecular identification of dominant microflora associated with "Hawaijar"—A traditional fermented soybean (*Glycine max* L.) food of Manipur India. *Int J Food Microbiol*, 122(3), 259–268.

Jetsada, W., Posri, W., Assavanig, A., Thongthai, C., and Lertsiri, S. 2013. *Categorization of Thai Fish Sauce Based on Aroma Characteristics: J. Food Quality* 36(2), 91–97.

Joshi, N., Godbole, S.H., and Kanekar, P. 1989. Microbial and biochemical changes during *dhokla* fermentation with special reference to flavor compounds. *J Food Sci Technol*, 26, 113–115.

Joshi, N.H. and Zulema, P. 2012. Fermented seafood products. In: Mehta, B.M., Kamal-Eldin A., Iwanski R.Z. (eds) *Fermentation Effects on Food Properties*. CRC Press, Taylor and Francis Group, New York, NY, pp. 285–308.

Joshi, V.K., Garg, V., and Abrol, G.S. 2012. Indigenous fermented foods. In: Joshi, V.K. and Singh, R.S. (eds) *Food Biotechnology: Principles and Practices*. I K International Publishing House, New Delhi, pp. 337–374.

Kanwar, S.S., Gupta, M.K., Katoch, C. Kumar, R., and Kanwar, P. 2007. Traditional fermented foods of Lahaul and Spiti area of Himachal Pradesh. *Indian J Trad Know*, 02(6), 42–45.

Karki, T., Okada, S., Baba, T., Itoh, H., and Kozaki, M. 1983. Studies on the microflora of Nepalese pickles gundruk. *Nippon Shokuhin Kogyo Gakkaishi*, 30, 357–367.

Klanbuta, K., Chantawannakula, P., Oncharoena, A., Chukeatiroteb, E., and Lumyong, S. 2002. Characterization of proteases of *Bacillus subtilis* strain 38 isolated from traditionally fermented soybean in Northern Thailand. *Science Asia*, 28, 241–245.

Kwon, T.W. 1994. The role of fermentation technology for the world food supply. In: Lee, C.H., Adler-Nissen, J., and Barwald, G. (eds) *Lactic Acid Fermentation of Non-Dairy Food and Beverages*, Seoul, South Korea: Ham Lini Won. pp. 1–7.

Law, S.V., Abu Bakar, F., Mat Hashim, D., and Abdul Hamid, A. 2001. Popular fermented foods and beverages in Southeast Asia. *Int Food Res J*, 18, 475–484.

Lopetcharat, K., Choi, Y.J., Park, J.W., and Daeschel, M.A. 2001. Fish sauce products and manufacturing: A review. *Food Rev Int*, 17(1), 65–88.

Mao, A.A. and Odyuo, N. 2007. Traditional fermented foods of the Naga tribes of Northeastern India. *Indian Journal of Traditional Knowledge*. 6 (1), 37–41.

Moktan, B., Saha, J., and Sarkar, P.K. 2008. Antioxidant activities of soybean as affected by *Bacillus*-fermentation to kinema. *Food Res Int*, 41(6), 586–593.

Murugkar, A.D. and Subbulakshmi, G. 2006. Preparation techniques and nutritive value of fermented foods from the *Khasi* tribes of meghalaya. *Ecol Food Nutr*, 45(1), 27–38.

Nair, B.M. and Prajapati, J.B. 2003. The history of fermented foods. In: Farnwoth, E. and Boston, R. (eds.) *Handbook of Fermented Foods*. CRC Press, Taylor and Francis Group, Boca Raton, FL, pp. 1–25.

Nand, K., Rati, E.R., and Shrestha, H. 2002. Microbiological profile of murcha starters and physico-chemical characteristics of *poko*, a rice based traditional fermented food product of Nepal. *Food Biotechnol*, 16(1), 1–15.

Nema, P.K., Karlo, T., and Singh, A.K. 2003. Opo, The Rice Beer in *International Seminar and Workshop on Fermented Foods, Health Status and Social Well-being* held on November-2003 at Anand, India. p. 59.

Nout, M.J.R. 2009. Rich nutrition from the poorest—Cereal fermentations in Africa and Asia. *Food Microbiol*, 26(7), 685–692.

Nout, M.J.R. and Aidoo, K.E. 2010. Asian fungal fermented food, industrial applications. *Mycota*, 10, 29–58.

Nout, M.J.R. and Motarjemi Y. 1997. Assessment of fermentation as a household technology for improving food safety: A joint FAO/WHO workshop. *Food Control*, 8, 221–226.

Nout, M.J.R., Sarkar, P.K., and Beuchat, L.R. 2007. Indigenous fermented foods. In: Doyle, M.P. and Beuchat, L.R. (eds) *Food Microbiology: Fundamentals and Frontiers*. ASM Press, Washington, DC, pp. 817–835.

Padmaja, G. and George, M. 1999. Oriental fermented foods; Biotechnological approaches. In: Marwaha, S.S. and Arora, J.K. (eds) *Food Processing: Biotechnological Applications*. Asia Tech Publishers Inc, New Delhi, pp. 143–189.

Panda, S., Aly, S., Didier, M., and Wanchai, W. 2011. Fermented fish and fish products: An Overview. *Aquac Microbiol Biotechnol*, 2, p. 306.

Panda, S.H., Ray, R.C., EL-Sheikha, A.F., Montet, D., and Worawattanamateekul, W. 2011. Fermented fish and fish products: An overview. *Aquac Microbiol Biotechnol*, 2, 132–172.

Pant, R.M. and Nema, P.K. 2003. Apong, the rice beer of arunachal: Some socio-cultural aspects in *International Seminar and Workshop on Fermented Foods, Health Status and Social Well-being* held on November-2003 at Anand, India, p. 96.

Parveen, S. and Hafiz, F. 2003. Fermented cereal from indigenous raw materials. *Pak J Nutr*, 2(5), 289–291.

Premarani, T. and Chhetry, G.K.N. 2008. Microbiota associated with natural fermentation of *Hawaijar* (an indigenous fermented soyabean product) of Manipur and their enzymatic activity. *J Food Sci Technol*, 45(6), 516–519.

Premarani, T. and Chhetry, G.K.N. 2010. Evaluation of traditional fermentation technology for the preparation of *Hawaijar* in Manipur. *Assam Univ J Sci Technol*, 6(1), 82–88.

Putro, S. 1993. Fish fermentation technology in Indonesia. In: Lee, C.H., Steinkhaus, K.H., and Reilly, P.J.A. (eds) *Fish Fermentation Technology*. United Nations University Press, Tokyo, Japan, pp. 107–128.

Qureshi, M.S., Hassan, Z., Sadiq, U., Omer, M.V., Afridi, S.S., Rahman, A., and Iqbal, A. 2011. Proceedings of the International Workshop on Dairy Science Park. In: *International Workshop on Dairy Science Park*, Nov 21–23, 2011, Agricultural University Peshawar, Pakistan. Faculty of Animal Husbandry and Veterinary Sciences, Khyber Pakhtunkhwa Agricultural University, Peshawar-25120, Pakistan.

Raju, P.N. and Pal, D. 2009. The physic-chemical, sensory, and textural properties of *misti dahi* prepared from reduced fat buffalo milk. *Food Bioproc Technol*, 2, 101–108.

Raju, P.N. and Pal, D. 2011. Effect of bulking agents on the quality of artificially sweetened *misti dahi* (caramel colored sweetened yoghurt) prepared from reduced fat buffalo milk. *LWT—Food Sci Technol*, 44(9), 1835–1843.

Ramakrishnan, C.V. 1993. Indian idli, dosa, dhokla, khaman, and related fermentations, In: Steinkraus, K.H. (ed) *Handbook of Indigenous Fermented Foods*. Marcel Dekker, New York, NY, pp. 149–165.

Rao R.E., Vijayendra, S.V.N., and Varadaraj M.C. 2006. Fermentation biotechnology of traditional foods of the Indian subcontinent. In: shetty K., Paliyath G., Pometto A., and Levin R.E. (eds) 2nd edition, *Food Biotechnology*. CRC Press Taylor and Francis, Boca Raton, FL, pp. 1759–1794.

Reddy, N.R. et al. 1982 Legume-based fermented foods: Their preparation and nutritional quality. *Crit Rev Food Sci Nutr*. 17(4), 335–370.

Rokosu, A.A. and Nwisienyi, J.J. 1979. Variation in the components of palm wine during fermentation. *Enzyme Microb Technol*, 2(1), 63–65.

Routray, W. and Mishra, H.N. 2012. Sensory evaluation of different drinks formulated from *dahi* (Indian yogurt) powder using fuzzy logic. *J Food Process Preserv*, 36, 1–10.

Roy, A., Bijoy, M., and Prabir, S.K. 2009. Survival and growth of foodborne bacterial pathogens in fermenting batter of dhokla. *J Food Sci Technol*, 46, 132–135.

Sagarika, E. and Pradeepa, J. 2003. Fermented food of Sri Lanka Maldives in *International Seminar and Workshop on Fermented Foods, Health Status and Social Well-being* held on November-2003 at Anand, India. p. 4.

Salampessy, J., Kailasapathy, K., and Thapa, N. 2010. Fermented fish products. In: Tamang, J.P. and Kailasapathy, K. (eds.) *Fermented Foods and Beverages of the World*. CRC Press Taylor and Francis Group, New York, NY, pp. 289–307.

Sands, D.C. and Hankin, L. 1974. Selecting lysine excreting mutans of lactobacilli for use infood and feed enrichment. *J Appl Microbiol*, 28, 523–534.

Sarangthem, K. and Singh, T.N. 2003. Microbial bioconversion of metabolites from fermented succulent bamboo shoots during fermentation. *Curr Sci*, 84(12), 1544–1547.

Sarkar, P.K., Cook, P.E., and Owens, J.D. 1993. Bacillus fermentation of soybeans. *World J Microbiol Biotechnol*, 9(3), 295–299.

Sarkar, P.K., Jones, L.J., Craven, C.S. Somerset, S.M., and Palmer, C. 1997. Amino acid profiles of kinema, a soybean-fermented food. *Food Chem*, 59(1), 69–75.

Sarkar, P.K., Jones, L.J., Gore, W., Craven, G.S., and Somerset, S.M. 1996. Changes in soya bean lipid profiles during kinema production. *J Sci Food Agric*, 71, 321–328.

Sarkar, P.K., Morrison, E., Tinggi, U., Somerset, S.M., and Craven, G.S. 1998. B-Group vitamin and mineral contents of soybeans during kinema production. *J Sci Food Agric*, 78, 498–502.

Sarkar, P.K. and Tamang, J.P. 1995. Changes in the microbial profile and proximate composition during natural and controlled fermentations of soybeans to produce kinema. *Food Microbiol*, 12, 317–325.

Sarkar, P.K., Tamang, J.P., Cook, P.E., and Owens, J.D. 1994. Kinema—a traditional soybean fermented food: Proximate composition and microfora. *Food Microbiol*, 11, 47–55.

Sarojnalini, C. and Vishwanath, W. 1988. Studies on the biochemical composition of some freshwater fishes of Manipur. *J Adv Zool*, 9, 1–5.

Savitri, and Bhalla, T.C. 2013. Characterization of bhatooru, a traditional fermented food of Himachal Pradesh: Microbiological and biochemical aspects. *Biotechnology*, 3(3), 247–254.

Sekar, S. and Mariappan, S. 2007. Usage of traditional fermented products by Indian rural folks and IPR. *Indian J Trad Know*, 6(1), 111–120.

Shamala, T.R. and Sreekantiah, K.R. 1988. Microbiological and biochemical studies on traditional Indian palm wine fermentation. *Food Microbiol*, 5(3), 157–162.

Shrestha, H. 2002. Microbiological profile of murcha starters and physico-chemical characteristics of poko, a rice based traditional fermented food product of Nepal. *Food Biotechnol*, 16(1): 1–15.

Singh, A., Singh, R.K., and Sureja, A.K. 2007. Cultural significance and diversities of ethnic foods of Northeast India. *Indian J Trad Know*, 6(1), 79–94.

Singh, P K. and Singh, K.I. 2006. Traditional alcoholic beverage, *Yu* of *Meitei* communities of Manipur. *Indian J Trad Know*, 5(2), 184–190.

Singh, R.K., Singh, A., and Sureja, A.K. 2007. Traditional Foods of *Monpa* tribe of West Kameng, Arunachal Pradesh. *Indian J Trad Know*, 6(1), 25–36.

Soni, S.K. and Sandhu, D.K. 1989. Fermentation of Idli: Effects of changes in raw material and physico-chemical conditions. *J Cereal Sci*, 10(3), 227–238.

Soni, S.K., Sandhu, D.K., Vilkhu, K.S., and Kamra, N. 1986. Microbiological studies on *dosa* fermentation. *Food Microbiol*, 3, 45–53.

Soni, S.K., Soni, R., and Janveja, C. 2013. Production of fermented foods. In: Panesar, P.S. and Marwaha, S.S. (eds) *Biotechnology in Agriculture and Food Processing Opportunities and Challenges*. CRC Press, Taylor & Francis Group, Boca Raton, FL, pp. 219–278.

Sridevi, J., Halami, P.M., and Vijayendra, S.V.N. 2010. Selection of starter cultures for idli batter fermentation and their effect on quality of *idlis*. *J Food Sci Technol*, 47(5), 557–563.

Steinkraus, K.H. 1983. Handbook of indigenous fermented foods. *Microbiol Ser*, 19, 315–328.

Steinkraus, K.H. 1996. *Handbook of Indigenous Fermented Foods*, 2nd edition. Marcel Dekker, New York, NY.

Suchitra, T. and Sarojnalini, C.H. 2009. Microbial profile of starter culture fermented fish product "Ngari" of Manipur. *Indian J Fish*, 56(2), 123–127.

Surono, I.S. and Hosono, A. 1994a. Chemical and aerobic bacterial composition of "Terasi", a traditional fermented product from Indonesia. *J Food Hyg Soc Japan*, 35(3), 299–304.

Surono, I.S. and Hosono, A. 1994b. Microflora and their enzyme profile in "Terasi" starter. *Biosci Biotech Biochem*, 58(6), 1167–1169.

Tamang, B. and Tamang, J.P. 2009. Traditional knowledge of biopreservation of perishable vegetable and bamboo shoots in Northeast India as food resources. *Indian J Trad Know*, 8(1), 89–95.

Tamang, B. and Tamang, J.P. 2010. *In situ* fermentation dynamics during production of gundruk and khalpi, ethnic fermented vegetable products of the Himalayas. *Indian J Microbiol*, 50, 93–98.

Tamang, B., Tamang, J.P. Schillinger, U., Franz, C.M.A.P., Gores, M., and Holzapfel, W.H. 2008. Phenotypic and genotypic identification of lactic acid bacteria isolated from ethnic fermented bamboo tender shoots of North East India. *Int J Food Microbiol*, 121(1), 35–40.

Tamang, J.P. 2003. Native micro-organisms in the fermentation of Kinema. *Int J Food Microbiol*, 43, 127–130.

Tamang, J.P. 2007. Fermented foods for human life, In: Chauhan, A.K., Verma, A. and Kharakwal, H. (eds) *Microbes for Human Life*. I.K. International Publishing House, New Delhi, pp. 73–87.

Tamang, J.P. 2009a. Ethnic fish products. In: *Himalayan Fermented Foods Microbiology Nutrition and Ethnic Values*, CRC Press, Taylor & Francis Group, New York, pp. 139–158.

Tamang, J.P. 2009b. Fermented vegetables. In: Tamang J.P. (ed) *Himalayan Fermented Foods Microbiology Nutrition and Ethnic Values*. CRC Press, Taylor & Francis Group, New York, pp. 25–64.

Tamang, J.P. 2009c. The Himalayas and food culture. In: Tamang, J.P. (ed) *Himalayan Fermented Foods Microbiology Nutrition and Ethnic Values*. CRC Press, Taylor & Francis Group, New York, pp. 1–24.

Tamang, J.P. 2010. *Himalayan Fermented Foods: Microbiology, Nutrition and Ethnic Values*. CRC Press, New York, pp. 1–295.

Tamang, J.P. 2012. Plant-based fermented foods and beverages of Asia. In: Hui, Y.H. and Özgül, E. (ed) *Handbook of Plant-based Fermented Food and Beverage Technology*, 2nd edition. CRC Press, Taylor & Francis Group, New York, pp. 49–90.

Tamang, J.P. and Delwen, S. 2010. Dietary cultures and antiquity of fermented foods and beverages. In: Tamang, J.P. and Kailashpathy, K. (eds) *Fermed Foods and Bever World*, CRC Press, Taylor & Francis, Boca Raton, FL. p. 1–40.

Tamang, J.P. and Fleet, G.H. 2009. Yeasts diversity in fermented foods and beverages. In: Satyanarayana, T. and Kunze, G. (eds) *Yeast Biotechnology Diversity and Applications*. Springer, Netherland, pp. 170–193.

Tamang, J.P and Nagai, T. 2010. Fermented legumes: Soybean and non-soybean products. *Ferm Foods Bever World*.

Tamang, J.P. and Nikkuri, S. 1998. Effect of temperatures during pure culture fermentation of Kinema. *World J Microbiol Biotechnol*, 14, 847–850.

Tamang, J.P. and Samuel, D. 2010. Dietary cultures and antiquity of fermented foods and beverages. In: Tamang, J.P. and Kailashpathy, K. (eds) *Fermented Foods and Beverages of the World*. CRC Press, Taylor and Francis Group New York, NY, pp. 1–40.

Tamang, J.P. and Sarkar, P.K. 1993. Sinki: A traditional lactic acid fermented radish tap root product. *J Gen Appl Microbiol*, 39, 395–408.

Tamang, J.P. and Sarkar, P.K. 1996. Microbiology of mesu, a traditional fermented bamboo shoot product. *Int J Food Microbiol*, 29(1), 49–58.

Tamang, J.P. and Sarkar P.K. 2010. Diversity of fermented foods. In: Tamang, J.P. and Kailashpathy, K. (eds) *Fermented Foods and Beverages of the World*. CRC Press, Taylor and Francis Group, New York, NY, pp. 41–84.

Tamang, J.P., Tamang B., Schillinger U., Franz, C.M.A.P., Gores, M., and Holzapfel, W.H. 2005. Identification of predominant lactic acid bacteria isolated from traditionally fermented vegetable products of the Eastern Himalayas. *Int J Food Microbiol*, 105, 347–356.

Tanasupawat, S. and Visessangua, W. 2014. Fish fermentation. In: Boziaris, I.S. (ed) *Seafood Processing; Technology, Quality and Safety*. John Wiley & Sons, Chichester, pp. 177–207.

Thakur, N., Savitri and Bhalla, T.C. 2004. Characterization of some traditional fermented foods and beverages of Himachal Pradesh. *Indian J Trad Know*, 3(3), 325–335.

Thapa, N., Pal, J., and Tamang, J.P. 2004. Microbial diversity in ngari, hentak and tungtap, fermented fish products of North-East India. *World J Microbiol Biotechnol*, 20(6), 599–607.

Thapa, S. and Tamang, J.P. 2004. Product characterization of *kodo ko jaanr*: Fermented finger millet beverage of the Himalayas. *Food Microbiol*, 21, 617–622.

Thongthai, C. and Asbjørn. G. 2005. Asian fish sauce as a source of nutrition. In: Shi, J., Ho, C.T. and Shahidi, F. (eds) *Asian Functional Foods*. CRC Press, Taylor & Francis, London: Boca Raton, FL, pp. 215–265.

Uchida, M., Ou, J., and Chen, B.W. 2005. Effect of soy sauce koji and lactic acid bacteria on the fermentation of fish sauce from freshwater silver carp *Hypophtalmichtys molitrix*. *Fisheries Sci*, 71, 422–430.

Wacher, C., Díaz-Ruiz, G., and Tamang, J.P., 2010. Fermented vegetable products. In: Tamang, J.P and Kailashpathy, K. (eds) *Fermented Foods and Beverages of the World*. CRC Press, Taylor & Francis Group, New York, pp. 149–190.

Wang, N., Yan, Z., Li, C., Jiang, N., and Liu, H.J. 2011. Antioxidant activity of peanut flour fermented with lactic acid bacteria: Antioxidant activity of fermented peanut flour. *J Food Biochem*, 35(5), 1514–1521.

Wikramanayake, T.W. 1996. *Food and Nutrition*. Hector Kobbekaduwa Agrarian Research and training Institute, Colombo 07, Sri Lanka.

Wongkhalaung, C. 2004. Industrialization of Thai fish sauce (Nam Pla). In: Steinkraus, K.H. (ed) *Industrialization of Indigenous Fermented Foods*. Marcel Dekker, New York, NY, pp. 647–705.

Yadav, H., Jain, S., and Sinha, P.R. 2007. Evaluation of changes during storage of probiotic dahi at 7°C. *Int J Dairy Technol*, 60(3), 205–210.

Yadav, H., Jain, S., and Sinha, P.R. 2008. Oral administration of dahi containing probiotic *Lactobacillus acidophilus* and *Lactobacillus casei* delayed the progression of streptozotocin-induced diabetes in rats. *J Dairy Res*, 75, 189–195.

Yongsawatdigul, J., Rodtong, S., and Raksakulthai, N. 2007. Acceleration of thai fish sauce fermentation using proteinases and bacterial starter cultures. *J Food Sci*, 72(9), 382–390.

Zyosa, N.D. 1992. Tapping patterns of the kitul palm (*Caryota urens*) in Sinharaja area, Sri Lanka. *Principes*, 36(1), 28–33.

3

MICROBIOLOGY AND BIOCHEMISTRY OF INDIGENOUS FERMENTED FOODS

ANILA KUMARI, ANITA PANDEY, ANTON ANN,
ANTONIO COBO MOLINOS, ANTONIO GÁLVEZ,
ARUP JYOTI DAS, CHETNA JANVEJA, DEEPTI JOSHI,
FOOK YEE CHYE, JAYASREE CHAKRABARTY,
KESHANI BHUSHAN, KHENG YUEN SIM,
KUNZES ANGMO, LOK MAN S. PALNI,
MARÍA JOSÉ GRANDE BURGOS,
MONIKA, NEHA GAUTAM, RAMAN SONI,
RUBÉN PÉREZ PULIDO, S.K. SONI, S.S. KANWAR,
SANKAR CHANDRA DEKA, SAVITRI,
SOMBOON TANASUPAWAT, TEK CHAND BHALLA,
V.K. JOSHI, AND VIKAS KUMAR

Contents

3.1 Introduction

Fermentation is a part of ethno-knowledge and constitutes one of the oldest forms of food preparation and preservation (Dirar, 1993). It is a critical component to food safety beyond preservation in many cases. It is generally carried out to bring diversity to the kinds of foods and beverages available to preserve food, decrease cooking times and energy requirements, bioenrich the foods, as well as several other advantages (Steinkraus, 1995, 1988, 1996, 1998). A number of fermented foods are prepared including bread, alcoholic beverages, curd, pickles, fermented fish products, vegetable products, cereal products, indigenous alkaline-fermented food condiments, etc. (Steinkraus 1988; Chaven and Kadam, 1989; Aniche et al., 1993; Sarkar and Tamang, 1995; Hafiz and Majid, 1996; Rose, 1982; Anonymous, FAO, 1998, 1999a,b; Amadi et al., 1999; Omafuvbe et al., 2000; Sekar and Mariappan, 2005; Steinkraus 2002; Sarkar and Nout, 2014). The increase in pH of alkaline fermented foods is due to the degradation of proteins from the raw material into peptides, amino acids, and ammonia (Kiers et al., 2000) or due to alkali-treatment during production (Wang and Fung, 1996; Sarkar and Nout, 2014). Thus, animal or plant tissues subjected to the action of microorganisms and/or enzymes to give desirable biochemical changes and significant modification of food quality are essential for the fermentation and, consequently, to the production to the fermented foods (Campbell-Platt, 1994; Anonymous, FAO, 1998, 1999a,b).

Another facet of indigenous fermented foods includes the microbiological risk factors associated with fermented foods, and a concern for the safety of such foods (Nout, 1995). Cases of food-borne infection, and intoxications due to microbial metabolites such as mycotoxins, ethyl carbamate, and biogenic amines are other aspects of food fermentation technology having ramification in indigenous fermented foods (Anonymous, FAO, 1998, 1999a,b).

Traditionally fermented foods are processed by naturally occurring microorganisms. Which microorganisms are associated with which traditional fermented foods, and what is their role and population sequence in the fermented foods, are interesting questions. What are the metabolites they produce in the food, and what is their significance? What is their physiology, which can influence both the food properties, quality, and safety? These are important topics for research. From the biochemical point of view, it is essential to document various metabolites produced by the microorganisms and their role in fermented foods. It is well known that some microorganisms produce metabolites that impact flavor, and are thus, very important for the sensory quality of the product (Beaumont, 2002; Joshi, 2006). At the same time, it is pertinent to understand that these are some of the metabolites that are toxic or undesirable. What are the microorganisms that produce such metabolites, and under what conditions, are equally relevant to the safety of indigenous fermented foods? Some of the microorganisms may not produce toxins, but can certainly spoil the food (Joshi et al., 2000). This information is of paramount significance in achieving food without spoilage. Though indigenous fermented foods are produced mostly by natural fermentation i.e. without inoculation either by back or standard culture, the process is quite complex and has remained unexplained. Modern methods of production for fermented foods generally make use of defined starter cultures to ensure the consistency and quality of the final product (Ross et al., 2002).

Food fermentation is affected by a number of factors, including the types of microorganisms, the substrates, conditions of fermentation, and further processing and storage (Joshi and Pandey, 1999; Frazier and Westhoff, 2004). Microorganisms associated with traditional fermented foods are selected through adaption to the substrate and adjustment to the fermentation conditions. The fermented products are the result of the type of microorganisms involved, and their interactions in the process. Thus, the microbial ecology plays an equally important role in the preparation of indigenous fermented foods (Scott and Sullivan, 2008).

The prospects for applying advanced fermentation technologies to indigenous fermented foods (Wood, 1994) and to the production of value-added products, additives using microorganisms (colors, flavors, enzymes, antimicrobials), and health products (Sarkar et al., 1994; Anonymous, FAO, 1998, 1999a,b; Joshi and Pandey, 1999) during food fermentations have also been explored. All these aspects are described in this chapter from the microbiology and biochemistry angles, especially the changes brought about by the microorganisms during fermentation.

3.2 Microorganisms, their Ecology, Role and Significance in Indigenous Fermented Foods

3.2.1 Taxonomical Considerations of Microorganisms in Indigenous Fermented Foods

The indigenous fermented foods are the result of natural fermentation, so the type of microflora and the growth kinetics of by-products formed vary from food-to-food. Classification of the microorganisms forms the back-drop of such knowledge of indigenous fermented foods. The protocol developed by Linneaus to name eukaryotes was quickly adapted by microbiologists for prokaryotes that gave a uniformity and logic to nomenclature within the entire biological sciences. The definition of a species as an organism which is, and remains, distinct because it does not normally interbreed with other organisms is not a particularly good description for a bacterial "species" (Gray, 1967). As a consequence, classification was largely based upon phenotypic characteristics, as measured under laboratory conditions, coupled with differentiation methods such as serotyping, phage typing, biochemical characteristics, and, eventually, aspects of the genetic material itself were included (Kilsby, 1999).

The basis of all such taxonomic efforts revolves around the notion that the "cell" in the test tube has some taxonomic significance and a "type," and would be viable if provided with suitable life support systems (Kilsby, 1999). In such a system, exchange of genetic material in their normal habitat as freely as in the test tube is considered normal for them. As such, they have no species status when in our test tubes, but the story with emerging pathogens is reported to be different, where a change in the environment produces selective pressure with the probability that a new pathogen will arise (Armstrong et al., 1996). Thus, the question that emerges is whether *E.coli* O157:H7 is a new pathogen, or is the "newness" merely due to its recognition in humans (Armstrong et al., 1996; Kilsby, 1999). The most probable source of such organisms is recombinants from the wild genetic pool (Dietsch et al., 1997). This concept, when applied to indigenous fermented food in terms of known food borne pathogens (*Campylobacter jejuni, Clostridium botulinum, Escherichia coli* O157:H7, *Listeria monocytogenes, Salmonella enteritidis, Vibrio cholera, Vibrio vulnificus, Yersinia enterocolitica,* and Norwalk-like viruses) that have emerged out during the past 20 years (Fratamico, 1995; Smith and Fratamico, 1995) becomes a big concern for food safety. This aspect is illustrated by the *Citrobacter freundii* implicated in gastro-enteritis cases in children (Tschape et al., 1995) where genes shared between *C. freundii* and *Escherichia coli* was the most likely origin, and becomes even more of a threat when we accept the idea of "pathogenicity islands" as particular regions of the genome which are easily transferred (Hacker et al., 1997; Kilsby, 1999). So the biotype that we may have been discussing might be a recombinant obtained by the selective pressure which is usually applied in our recovery techniques. So there is a possibility of some uncertainty in our results with laboratory strains, for example, when developing preservation systems or modelling kinetics (Kilsby, 1999). Such aspects call for serious consideration.

3.2.2 Microflora Associated with Traditional Fermented Foods

3.2.2.1 Types of Microflora and their Characteristics   Many desirable species and strains of bacteria, yeasts, and molds are associated with the fermentation of foods (Joshi et al., 2000). Despite their presence, the role of yeasts in the fermented food products is often poorly understood. The most dominant yeast species associated with indigenous fermented foods and beverages is *Saccharomyces cerevisiae*. Depending on the product, fermentation may be achieved by a single predominating species and strain, but in most fermentations, a mixed population of several bacterial species and strains, or even bacteria and yeasts, or bacteria and molds, is involved. In such cases, the members should not be antagonistic toward one another; rather, they should be synergistic. Optimum growth of a desirable microorganism and optimum fermentation rate are dependent on environmental parameters such as nutrients, temperature of incubation, oxidation-reduction potential, and pH (Frazier and Westhoff, 2004). During the fermentation, if the different species in a mixed population require different environmental conditions (e.g., temperature of growth), a compromise is made to facilitate growth of all the species at a moderate rate. Depending on raw or starting material and specific need, carbohydrates (dextrose in meat fermentation), salts, citrate, and other nutrients are added. In some natural fermentations, several species may be involved to produce the final desirable characteristics of the product. But, instead of growing at the same time, they appear in sequence, with the consequence that a particular species predominates at a certain stage during fermentation. But analysing the final product to isolate the species involved in fermentation of such a food does not always give the correct picture. Instead, samples need to be analysed at intervals to determine predominant types at different times, and to understand the sequence of their appearance. Finally, some minor flora (secondary flora) can be present at a very low level in a raw material and the finished product, and might not be detected during regular analysis. Nevertheless, they may contribute to the desirable characteristics, particularly some unique aroma, of the fermented product.

The most important bacteria in desirable food fermentations are the lactic acid bacteria with the ability to produce lactic acid from carbohydrates (Pederson, 1973; Nout, 2005). Other important bacteria, especially in the fermentation of fruits and vegetables, are the acetic acid bacteria that produce acetic acid (Frazier and Westhoff, 2004; Joshi and Sharma, 2009; Soni et al., 2012). Due to the wide variation in agro-climatic conditions in different countries of South Asia, and the diverse forms of dietary culture of the various ethnic groups, all the three major types of microorganisms are encountered in traditional fermented foods and beverages (Tamang, 1998).

Microorganisms associated with traditional fermented foods are present in or on the ingredients and utensils, and in the environment. They are selected through adaptation to the substrate and by adjusting the fermentation conditions (Tamang, 1998). Indigenous natural fermentation takes place in a mixed colony of microorganisms, such as molds, bacteria, and yeasts. Thus, the fermentation products are based on the type of microorganisms involved in the process. Such compounds formed during fermentation include organic acids (e.g., pyruvic, lactic, acetic, propionic, and butyric acids), alcohols (mainly ethanol but include others like methanol, Propanol, hexamol etc), aldehydes, and ketones (acetaldehyde, acetoin, 2-methylbutanol) (Chelule, 2010). A brief account of the microorganisms associated with indigenous food is given here. For more details, the readers can refer to a number of standard texts on these aspects.

3.2.2.2 Bacteria   Lactic Acid Bacteria: The lactic acid bacteria (LAB) represent a group of Gram positive, non-respiring, non-spore forming, cocci, or rods. The LAB are very significant as they are involved in production of many fermented foods (Table 3.1).

They have the ability to produce lactic acid during homo- or hetero-fermentative metabolism. See for more details on morphology and physcology (Carr et al., 2002;

Table 3.1   Main Lactic Acid Bacteria Involved in Industrial-Scale Fermented Food Products

FERMENTED PRODUCT	MAIN LAB INVOLVED
Cheese	*Lactococcus lactis*, together with other LAB (*Streptococcus thermophilus, Lactobacillus bulgaricus, Lactobacillus helveticus, Leuconostoc cremoris*), depending on type of cheese
Yogurt	*Streptococcus thermophilus, Lactobacillus bulgaricus*
Butter and butter milks	*Lactococcus lactis* (ssp. *lactis*, ssp. *cremoris*, ssp. *diacetylactis*), *Leuconostoc mesenteroides*
Acidophilus milk	*Lactobacillus acidophilus*
Kefir	*Lactobacillus brevis, Lactobacillus plantarum, Lactobacillus caucasicus, Lactobacillus kefiri, Lactobacillus kefiranofaciens, Lactobacillus bulgaricus, Leuconostoc*
Sausages and other fermented meats	*Lactobacillus sakei, Lactobacillus curvatus, Pediococcus acidilactici, Pediococcus pentosaceus*
Fermented fish	*Tetragenococcus halophilus, Pediococcus halophilus*
Cucumbers	*Lactobacillus plantarum, Pediococcus acidilactici*
Sourdough	*Lactobacillus sanfranciscensis*

Lechardeur et al., 2011; Salminen et al., 2000). The acidification and enzymatic processes accompanying their growth impart the key flavor, texture, and preservative qualities to a variety of fermented foods (Pederson, 1973; Carr et al., 2002; Joshi and Sharma, 2012). At present, bacterial species from 12 genera are included in the group designated LAB. The genera include: *Lactococcus, Leuconostoc, Pediococcus, Streptococcus, Lactobacillus, Enterococcus, Aerococcus, Vagococcus, Tetragenococcus, Carnobacterium, Weissella*, and *Oenococcus*. Many of the genera have been identified recently from previously existing genera, and include one or a few species. For example, *Lactococcus* and *Enterococcus* were previously classified as *Streptococcus* Group N and Group D, respectively. *Vagococcus* is indistinguishable from *Lactococcus*, except that they are motile. *Weissella* and *Oenococcus* have been separated from *Leoconostoc*. *Tetragenococcus* includes a single species that was previously included with *Pediococcus* (*P. halophilus*). *Carnyobacterium* was created to include a few species that were previously in the genus *Lactobacillus* and are obligatorily heterofermentative. The LAB have six key beneficial and nonpathogenic species; *Lactococcus, Lactobacillus, Leuconostoc, Pediococcus, Oenococcus oeni*, and *Streptococcus thermophilus* (Soni et al., 2013). Lactic acid bacteria employed as functional starter cultures in the food fermentation industry (De Vuyst, 2004). These lactobacilli occupy important niches in the gastrointestinal tracts of humans and animals, exert a positive influence on the normal microflora, competitively exclude pathogens, and stimulate mucosal immunity, and are considered to offer a number of probiotic benefits to the general health and well being of man (Hofvendahl and Hahn-Hägerdal, 2000; Anonymous, 2002) and the concept has been employed for functional (Shah, 2000).

Lactococcus lactic, used widely in dairy, fermentation has three subspecies (ssp.), i.e. ssp. *lactis*, ssp. *cremoris*, and ssp. *hordniae*, but only the first two are used in dairy fermentation. The cells are ovoid, ca. 0.5–1.0 mm in diameter, present in pairs or short chains, non-motile, non-sporulating, and facultative anaerobic to microaerophilic, have optimum temperature between 20°C and 30°C, but do not grow in 6.5% NaCl or at pH 9.6. In a suitable broth, they can produce about 1% L(+)-lactic acid and reduce the pH to about 4.5. The ssp. *cremoris* can be differentiated from ssp. *lactis* by its inability to grow at 40°C, in 4% NaCl, ferment ribose, and hydrolyze arginine to produce NH_3. Biovar diacetylactis, as compared to others, produces large amounts of CO_2 and diacetyl from citrate and generally, hydrolyze lactose and casein, besides fermenting galactose, sucrose, and maltose. Natural habitats are green vegetation, silage, the dairy environment, and raw milk. For phylogenetic analysis of genus *Lacto bacillus* and related lactic acid bacteria, see Collins et al., 1991.

> *Streptococcus*: These are used in dairy fermentation and are Gram-positive cells, spherical-to-ovoid, 0.7–0.9 mm in diameter, and exist in pairs or long chains, grow well at 37–40°C (Harrigan and McCance, 1966), but can also grow at 52°C. They are facultative anaerobes and in glucose broth can reduce the pH to 4.0 and produce L(+)-lactic acid by fermenting fructose, mannose, and lactose,

but generally not galactose and sucrose. Cells survive at 60°C for 30 minutes. Their natural habitat is unknown, although they are found in milk.

Lactobacillus: The genus *Lactobacillus* includes a heterogenous group of Gram-positive, rod-shaped, usually nonmotile, nonsporulating, facultative anaerobic species that vary widely morphologically in growth and metabolic characteristics (Salminen et al., 2004). Cells vary from very short (almost coccoid) to very long rods, slender or moderately thick, often bent, and can be present as single cells or in short to long chains (Harrigan and McCance, 1966). While growing on glucose, they produce either only lactic acid D(–), or DL, or a mixture of lactic acid, ethanol, acetic acid, and CO_2, while others produce diacetyl. Many species utilize lactose, sucrose, fructose, or galactose, while a few species can ferment pentoses also. Growth temperature can vary from 1°C to 50°C, but most of those used as starter cultures in controlled fermentation of foods grow well from 25°C to 40°C. Several species involved in natural fermentation of some foods at low temperature can grow well from 10°C to 25°C. *Lactobacillus delbrueckii* ssp. is used in the fermentation of dairy products, such as some cheeses and yogurt. *L. acidophilus* and *L. reuteri* are considered beneficial intestinal microbes (probiotic), and are present in the small intestine. *L. acidophilus* is used to produce fermented dairy products, while *L. plantarum* is used for vegetables and meat fermentation and consumed as probiotics. *Lactobacilli* are particularly beneficial microorganisms. They cause spoilage of low-alcoholic beverages, but are involved in the reduction of the acidity of high acid wines to an acceptable level (Frazier and Westhoff, 2004).

Pediococcus: The cells are spherical and form tetrads or are in pairs; single cells or chains are absent. They are Gram-positive, nonmotile, nonsporulating, facultative anaerobes, and grow well between 25°C and 40°C, while some species grow at 50°C. They ferment glucose to or DL-lactic acid, and some species can reduce the pH to 3.6. Depending on the species, they can ferment sucrose, arabinose, ribose, and xylose. Lactose is not generally fermented, especially in milk, and milk is not curdled. Depending on the species, they are found in plants, vegetables, silage, beer, milk, and fermented vegetables, meats, and fish (Frazier and Westhoff, 2004). Some are found in the digestive tract of humans, animals, and birds, while others have been associated with the spoilage of foods. *P. pentosaceus* and *P. acidilactici* are used in vegetables, meat, and cereal, and are also implicated in the ripening and flavor production of some cheeses, as secondary cultures (Reps, 1987).

Oenococcus: *O. oeni*, previously designated as *L. oeni*, has the general characteristics of *Leuconostoc* spp. and is found in the winery environment. It is sometimes used to accelerate malo-lactic fermentation in wine. The cells transport malate in wine and metabolize it into lactic acid and CO_2 thus, reducing the acidity of wine.

Leuconostoc: The genus *Leuconostoc* encompasses a phylogenetically coherent group of LAB and, currently, consists of eight species, namely *Leuconostoc mesenteroides, L. lactis, L. gelidum, L. carnosum, L. citreum, L. pseudomesenteroides, L. fallax* and *L. argentinum* (Holzapfel and Schillinger, 1992; Dicks et al., 1995). *L. mesenteroides* contains three sub-species, *L. mesenteroides* ssp. *mesenteroides*, *L. mesenteroides* ssp. *dextranicum*, and *L. mesenteroides* ssp. *cremoris* (Garvie, 1983). These Gram-positive cells are spherical to lenticular, arranged in pairs or in chains, non-motile, non-sporulating, catalase negative, and facultative anaerobes (Frazier and Westhoff, 2004). The species grow well between 20°C and 30°C, with a range of 1–37°C. Glucose is fermented to D(–)-lactic acid, CO_2, ethanol, or acetic acid, with the pH reduced to 4.5–5.0. The species grow in milk but may not curdle it, and hydrolyze arginine. Many form dextran while growing on sucrose and utilize citrate to produce diacetyl and CO_2. Some species can survive at 60°C for 30 minutes. *Leuconostoc* species are found in plants, vegetables, silage, milk and some milk products, and raw and processed meats. Facultatively anaerobic, asporogenous, produce lactic acid as a main end-product of fermentation, containing DNA with relatively low G + C content (37–45 mol%), which in many cases produce dextran also. They thrive in a variety of environments including fermented foods, like dairy and meat products (Garvie, 1986; Holzapfel and Schillinger, 1992), while others play an important role in the fermentation of plant materials, such as *L. mesenteroides* ssp. *mesenteroides*, which initiates the fermentation of sauerkrout and a number of traditional fermented foods in tropical regions (Stamer, 1975; Puspito and Fleet, 1985; Gashe, 1987).

Bifidobacterium: Morphologically similar to some *Lactobacillus* spp., these were previously included in the genus *Lactobacillus*. The cells are Gram-positive, are rods of various shapes and sizes, and are present as single cells or in chains of different sizes (Harrigan and McCance, 1966). They are non-spore forming, nonmotile, and anaerobic, although some can tolerate O_2 in the presence of CO_2. The species will grow optimally at 37–41°C, with a growth temperature range of 25–45°C, but usually will not grow at a pH above 8.0 or below 4.5. They ferment glucose to produce lactic and acetic acid in a 2:3 molar ratio, without producing CO_2, as well as lactose, galactose, and some pentoses. They have been isolated from the feces of humans, animals, and birds, and are considered beneficial for the normal health of the digestive tract. They are also present in large numbers in the feces of infants within 2–3 days of birth, and usually present in high numbers in breast-fed babies.

Propionibacterium: The genus includes species in the classical or dairy propionibacterium group and the cutaneous or acne propionibacterium group (Frazier and Westhoff, 2004). The cells are Gram-positive, pleomorphic thick rods, 1–1.5 mm in length, and occur in single cells, pairs, or short chains with different configurations. They are non-motile, non-sporulating, anaerobic (can also tolerate air), catalase positive, and ferment glucose to produce large amounts of propionic acid and acetic acid. Depending on the species, they can ferment lactose, sucrose, fructose, galactose, and some pentoses, and grow optimally at 30–37°C. Some species also form pigments. They have been isolated from raw milk, some types of cheese, dairy products, and silage. At present, four species of dairy propionibacterium are included in the genus: *Propionibacterium freudenreichii*, *P. jensenii*, *P. thoenii*, and *P. acidipropionici*.

Brevibacterium: The genus contains a mixture of coryniform bacterial species, some of which have important applications in cheese production and other industrial fermentations. The cells are non-motile, Gram-positive, and capable of growing in high salt and wide pH ranges.

3.2.2.2.1 Acetic Acid Bacteria A second group of bacteria of importance in food fermentations are the acetic acid producing *Acetobacter* species. These bacteria belong to genus *Acetobacter* and *Gluconobacter*. *Acetobacter* is important in the production of vinegar from alcoholic fermented juices (Anonymous, FAO, 1998, 1999a,b; Joshi and Thakur, 2000; Joshi and Sharma, 2009; Soni et al., 2012). The most desirable action of acetic acid bacteria is in the production of vinegar. *Acetobacter* genus includes three distinct species: *A. aceti*, *A. pasteurianus*, and *A. peroxydans*. These bacteria are 0.5×10 μm, non-motile, occurring in pairs or chains, and are characterized by their ability to oxidize ethyl alcohol into acetic acid and subsequently, to carbon dioxide and water. These bacteria are of great concern to wine makers, as production of acetic acid increases the volatile acidity of wine which affects its sensory quality. The maximum ethanol tolerance of acetic acid bacteria lies between 14% and 15%.

3.2.2.2.2 Bacteria of Alkaline Fermentations A third group of bacteria, which includes the *Bacillus* species, are responsible for alkaline fermentations. *Bacillus subtilis*, *B. licheniformis*, and *B. pumilius* are the dominant species, causing the hydrolysis of protein to amino acids and peptides, and releasing ammonia, which increases the alkalinity and makes the substrate unsuitable for the growth of spoilage microorganisms (Soni et al., 2012, 2013; Sarkar and Nout, 2014). These are more common with protein rich foods such as soybeans and other legumes.

3.2.2.3 Yeasts and Molds Many yeasts and molds are important in food being involved in fermentation to make fermentation foods, but many are involved in spoilage of food and mycotoxin production (by molds). The molds are involved in food fermentation especially where starchy material is involved, such as rice, cereals. The most important microorganism associated with fermentation are yeasts, which are a phylogenetically diverse group of unicellular fungi (Frazier and Westhoff, 2004). These are widely distributed in nature and have been isolated from a number of substrates, such as foods, beverages, flowers, insects, soil, and seawater (Phaff et al., 1978). As a group of microorganisms, they have been commercially exploited as a fermentative species necessary to carry out alcoholic fermentation, and the factors governing their growth, survival, and biological activities in different ecosystems, and foods and beverages have been studied (Phaff et al., 1978; Fleet, 1985; Querol and Fleet, 2006; Joshi, 2006).

3.2.2.3.1 Yeasts   Among the many types of yeasts, only a few have been associated with the fermentation of foods and alcohol production, enzyme production or for use in food, or production of SCPs, or as additives to impart desirable flavors in some foods (Aidoo et al., 2006; Querol and Fleet, 2006). Even these have been found as probiotic (Czerucka et al., 2007). These include *Debaromyces*, *Candida*, *Geotrichum*, *Hansenula*, *Kluyveromyces*, *Pichia*, *Saccharomyces*, *Saccharomycopsis*, *Torulopsis*, and *Zygosaccharomyces*. The most important genus and species used is *Saccharomyces cerevisiae*. It has been used to leaven bread and produce beer, wine, distilled liquors, and industrial alcohol; to produce invertase (enzyme); and to flavor some foods (soups). The different yeast strains may be distinguished by various systems and methods, including the giant colony morphology method (Gilliland, 1980; Barnett et al., 1983; Priest and Campbell, 1987; Stewart, 1987). Detailed systematics and taxonomy of yeasts and related species, can be found in a number of references (Lodder, 1970; Von Arx, 1977; Miller, 1982; Barnett et al., 1983). These yeasts, thus, play a very important role in food fermentation (Rose and Harrison, 1970; Kreger-van Rij, 1984). Description of some of the yeasts, associated with food fermentation are described here.

> *Saccharomyces*: Morphologically, this group of yeasts appears spherical to ellipsoidal in shape with approximate dimension of 8×7 μm, depending on the organism and the growth medium (Lodder, 1970). Asexual reproduction is by multilateral budding (Figure 3.1). The cells are round, oval, or elongated and multiply by multipolar budding or by conjugation and formation of ascospores. Most strains of *S. cerevisiae* are capable of producing alcohol levels up 16% (v/v). Although these yeasts are usually considered as fermentative, they may also grow oxidatively as a part of the surface film community; for example, *S. beticus*, in sherry preparation. The conversion of sugar into ethanol efficiently is the most important characteristics of species *Saccharomyces cerevisiae* (Lodder, 1970). Thus, it is used in wine, sparkling wine (Amerine et al., 1967; 1980), as the baker's yeast, and as beer yeast. Beer yeast is further divided into top and bottom fermentors.
>
> *Brettanomyces*: Microscopically, these yeasts resemble *S. cerevisiae*, but are somewhat smaller in size with vegetative cells which are ogival in shape (Figure 3.1). *Brettanomyces* and its sporulating counterpart, *Dekkera*, are capable of producing 10%–11% alcohol. Insects, particularly fruit flies, are important vectors of this yeasts, besides the insanitary conditions of crushing equipment and the resulting contamination.

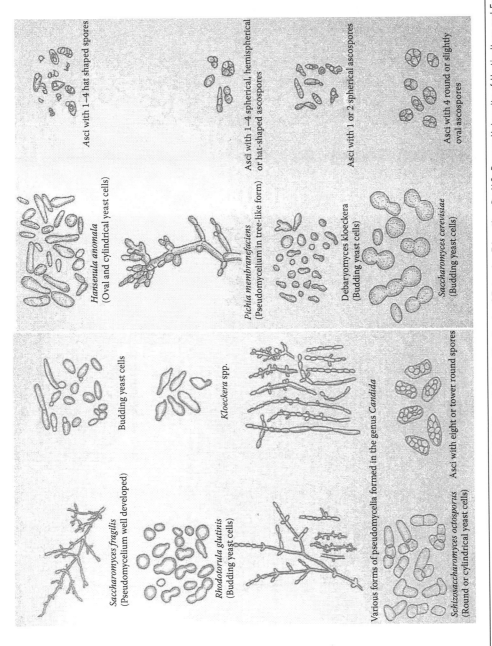

Figure 3.1   Different types of Yeasts. (From Joshi, V.K. 1997. *Fruit Wines*. 2nd edn. Directorate of Extension Education, Dr. Y.S. Parmar University of Horticulture and Forestry, Nauni-Solan, India; Lodder, J. (Ed.) 1970. *The Yeasts*. North-Holland Publ., Amsterdam, pp. 1385.)

Kloeckera/Hanseniaspora: The apiculate (lemon) shape cells (Figure 3.1) characteristic of this group arise from repeated budding at both the poles (Frazier and Westhoff, 2004). Both *Kloeckera* and *Hanseniaspora* are capable of producing substantial and inhibitory levels of acetic acid and ethyl acetate.

Zygosaccharomyces: In the case of wines sweetened with concentrate, the osmophilic yeast *Zygosaccharomyces bailli* may be troublesome, due to its survival at pH of below 2.0 and in alcohol levels greater than 15% (Frazier and Westhoff, 2004).

Candida: These yeasts appear as long cylindrical cells, globose ovoid to elongated, of approximate dimensions 3–10 × 2–4 μm under the microscope. Asexual reproduction takes place by multipolar budding (Lodder, 1970). The asporogeneous (imperfect) genus *Candida* includes numerous species. *Candida* are considered weak fermentors. In some species mycelia may be present, and under certain conditions chlamydospores may also develop. Fermentative ability is not found in every species (Lodder, 1970). *Candida utilis* has been used to produce SCPs. It reproduces by budding (not by conjugation). The cells are oval to elongated and form hyphae with large numbers of budding cells (Harrigan and McCance, 1966). They are also involved in food spoilage. This yeast is involved in the fermentation of palm to make toddy (Batra and Millner, 1976; Steinkraus, 1983a,b,c).

Pichia: Important species of this ascospore-producing yeast include *Pichia membranaefaciens*, *P. vini*, and *P. farinosa*. Most species are inhibited by alcohol levels of about 10% (Lodder, 1970). Growth on the surface of alcoholic beverages such as wine appears as a very heavy "balbon-like" chalky film, imparting an aldehydic character to the wine. Actively growing cells appear as short ellipsoidal to cylindrical-shaped rods. Reproduction is usually by multilateral budding leading to an extensive pseudomycelium development.

Hansenula: Microscopically, *Hansenula* appears as an oval to oblong cell of approximate dimensions 2.5 × 5 μm. Members of the genus reproduce by asexual budding (Figure 3.1). *Hansenula anomala* has both oxidative and fermentative metabolism. Most species of *Hansenula* form a film on liquid culture media and on wines of low alcohol content. *Hansenula* sp. form large amounts of acetic acid, ranging from 1 to 2 g/L, as well as volatile esters, particularly ethyl acetate up to 2.150 mg/L, but isoamyl acetate also. Thus, this yeast might be useful in increasing the ester content (known for a fruity aroma) of wine in mixed culture.

Fission yeast: *Schizosaccharomyces*, like *Saccharomyces*, is an ascomycetous yeast in the same family (Figure 3.2). The former reproduces asexually by fission and the latter by budding, but sexual reproduction is similar in both the yeasts; that is, vegetative cells conjugate to produce asci, each of which contains four to eight ascospores (Kreger-van Rij, 1984). All species of *Schizosaccharomyces* are fermentative. This genus comprises four species out of which *Schizosaccharomyces*

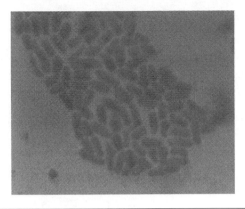

Figure 3.2 Cells of *Schizosaccharomyces pombe* 1 × 1600. (Courtesy of Dr. V.K. Joshi, Fermentation Lab.)

pombe is well studied for its ability to utilize malic acid, including that of grape juice, fermentatively, with the production of ethanol, though at a slow rate. The possibility of reducing acidity in plum wine by the use of this yeast has also been explored, wherein the overall acidity in the wines made from fruits with high malic acid content such as plum was reduced (Vyas and Joshi, 1982; Joshi et al., 1991). The de-acidification by *S. pombe* in plum fermentation is affected by the pH, SO_2, nitrogen source, and concentration of ethyl alcohol in the must (Joshi et al., 1991). It has potential for deacidification to improve the quality of wine from high malic content fruits in future. As a wild yeast, it is involved in the fermentation of toddy.

Kluyveromyces: *Kluyveromyces marxianus* and *K. marxianus* var *lactis* can hydrolyze lactose and have been associated with natural fermentation (Lodder, 1970), along with other yeasts and LAB, of alcoholic dairy products, such as *kefir*. They have also been associated with the spoilage of some dairy products. At present, they are used to produce β-galactosidase (lactase) for use commercially to hydrolyze lactose to produce low-lactose milk.

Geotrichum: *Geotrichum* is also a member of the hyphomycetes and is characterized by having creeping, submerged hyphae, as well as aerial ones, which break up into arthrospores (Harrigan and McCance, 1966). The genus has an ascomycete relationship as the teleomorph of the very common species *G. candidutn* is Dipodascus (Galaciomyces) geotrichum (Samson and Van Reenen-Hoekstra, 1988). It is involved in the production of toddy (Batra and Millner, 1976; Steinkraus, 1983a,b,c; Stewart, 1987).

3.2.2.3.1.1 Fungi/Mold   In general, molds are multicellular, filamentous fungi. The filaments (hyphae) can be septate or nonseptate and have nuclei (Frazier and Westhoff, 2004). Division is by elongation at the tip of a hypha (vegetative reproduction) or by forming sexual or asexual spores on a spore-bearing body. Most molds are associated with food spoilage, and many form mycotoxins while growing on foods, but other species and strains are used in the processing of foods, and to produce additives and enzymes for use in foods (Joshi et al., 2000). Species of *Aspergillus*, *Amylomyces*, *Actinomucor*, *Monascus*, *Mucor*, *Neurospora*, *Penicillium*, and *Rhizopus* are generally associated with food, either for production of indigenous fermented foods or spoilage of foods.

Several species from the genera *Aspergillus* and *Penicillium*, and a few from *Rhizopus* and *Mucor*, have been used for beneficial purposes in food, but these strains should not produce mycotoxins, and this is an important consideration in the selection of strains for use in controlled fermentation.

Aspergillus: *Aspergillus oryzae* is used in fermentation of several local foods, besides being a source of several food enzymes, such as pectinase and amylase. *A. niger* is also used to produce citric acid and gluconic acid from sucrose. *A. Oryzae* is used in making of Japanese sake, and *Rhizopus arrhizus* or *R. oryzae* or *Amylomyces rouxii* in the making of rice wine. These fungi produce the amylases which break down the rice starch into dextrins and sugars which can be converted into alcohol by fermentation with the yeast such as *S. cerevisiae*. The fungus *A. oryzae* is a member of the hyphomycetes of the class Deuteromycotina (fungi imperfecti). In its conidial (asexual) state (anamorph) it produces enormous masses of conidia which form the inocula used in the fermentation process, and are recognized microscopically by a swollen vesicle at the end of a long conidiophore, and on this vesicle are one or two rows of sterigmata on which long chains of conidia are borne. Its gross colony color on Czapek Dox agar is a pale greenish-yellow color that may shift to brown in older colonies. There are many strains of this species, used for centuries, especially in the Orient in various food fermentations. Consequently, its synonyms form an extremely long list (Raper and Fennell, 1965). This mold is said to form a part of the *A. flavus–oryzae* complex (Thorn and Rapor, 1945; Klich and Pitt, 1985) and the group is separated into two major series centered upon *A. flavus* and *A. oryzae*, based on certain morphological and colony characters. The group of yellow-green *Aspergillus* strains used in Asian soybean fermentations has been reviewed earlier (Flegel, 1988). The strains of *A. oryzae* used in food fermentation have not been found to produce aflatoxins (Matsuura, 1970; Matsuura et al., 1970; Murakami and Suzuki, 1970; Maing et al., 1973; Moss, 1977; Manabe et al., 1984), though some strains of the species do produce other kinds of mycotoxins, such as cyclopiazonic acid (Orth, 1977). The *Aspergillus* species are often responsible for undesirable changes in foods, and are frequently found in foods, since they can tolerate high concentrations of salt and sugar. It is aerobic and, thus, needs oxygen for growth. They also have the greatest array of enzymes, and can colonize and grow on most types of food (Anonymous, FAO, 1998, 1999a,b). The mold used in soy sauce manufacturing is *A. oryzae* or *A. soyae*. The latter species is mentioned as a synonym of the former (Raper and Fennell, 1965), as has been described earlier.

Rhizopus: This is a mold used in rice wine fermentation. These fungi are members of the Zygomycotina, order Mucorales, family Mucoraceae. Morphologically, it possesses filamentous threads or hyphae without cross-walls, except in very old hyphae or when reproductive structures develop. More details of its characteristic features and taxonomy have been described elsewhere (Inui et al., 1965; Schipper, 1984). In the asexual state (anamorph), sporangia are produced, containing enormous numbers of spores, which often serve as the inocula for the fermentation process. At the opposite end to the sporangia, which is borne on a long sporangiophore, rhizoids are present, which serve to anchor the mold to the substrate. Many *Rhizopus* spp. produce several enzymes (Wang and Hesseltine, 1970; Lim et al., 1987), including amylases which break down starch and are used in various indigenous fermentations where carbohydrates are broken down to sugars which are fermented by yeasts to produce alcoholic beverages. However, in many indigenous fermentations to produce rice wine, pure cultures of *Rhizopus* are not used; rather a starter culture called *ragi*—generally a mixture of *Rhizopus* species and yeasts—is used (Ko, 1972; Saono et al., 1974). The fungus or mold is also called tempeh fungus, as it used in the preparation of soybean tempeh (Stahel, 1946; Van Veen and Schaefer, 1950). Later on, others identified the principal mold to be *R. oligosporus* (Hesseltine and Wang, 1967). Other species of *Rhizopus* which can also produce tempeh are *R. arrhizus*, *R. oryzae*, and *R. stolonifer* (Hesseltine, 1965).

Actinomucor, Mucor, Rhizopus: *Actinomucor*, along with *Mucor* or *Rhizopus*, as members of the order Mucorales of the class zygomycetes, are used in the production of *Sufu*. The species reported are *Mucor hiernalis*, *M. praini*, *M. siivaticust*, and *Rhizopus chinensis* var. *chungyuen* (Su, 1977, 1986). As all three genera belong to the same family, Mucoraceae, they share some common features, especially in their mycelium and asexual reproductive apparatus.

Monascus: *Monascus purpureus*, responsible for making red rice, is a member of the ascomycetes, and belongs to the monogeneric family Monascaceae (Hesseltine, 1983). It produces a reddish-purple pigment, and thus, *ang kak* (fermented rice) is used both as a flavoring and coloring agent in food (Wang and Hesseltine, 1979; Hesseltine, 1983; Samsonand Van Reenen-Hoekstra, 1988) (Figure 3.3).

Penicillium, Trichoderma, Curvularia: *Penicillium roquefortii* is used in the ripening of Roquefort, Gorgonzola, and other blue cheeses, but some strains can produce the neurotoxin roquefortin. In the selection and development of strains for use in cheese, this aspect needs careful consideration. *P. camembertii* is used in Camembert cheese and *P. caseicolum*is is used in Bric cheese, as well as the production of glucose oxidase enzyme. *Aspergillus* in

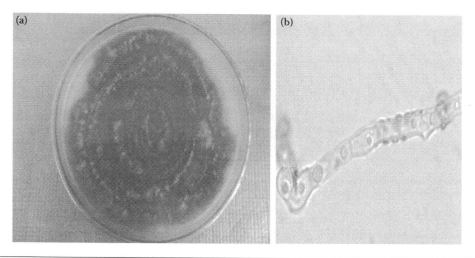

Figure 3.3 *Monascus purpurea.* (a) Growth on petriplate (b) hyphae with a pigment under a microscope (80 × 10). (Courtesy of Ms Sangeeta Sharma.)

rice wine (samsu, sake) production causes saccharification of the rice starch. The fungi that produce such aromatic volatiles range from members of the ascomycetes to basidiomycetes and fungi imperfecti. Among the fungi imperfecti, the species reported are *P. decumbens*, *Trichoderma viride* (Schindler and Schmid, 1982; Yong and Lim, 1985, 1986; Young et al., 1989), *P. italicum* (Kok, 1987), and *Curvularia lunata*. These fungi are useful sources of aromatic volatiles, as shown by *P. italicum*, from which linalool, an orange-fruit aroma can be obtained (Kok, 1987).

Mushroom Fungi: Among the basidiomycetes, quite a large number of the polypores and mushrooms have been reported. Some of these polypores are species of *Polyporus*, *Stereum*, and *Trametes*, and some of the mushrooms are *Coprinus* and *Pleurotus* (Schindler and Schmid, 1982; Drawert et al., 1983). The common mushrooms cultivated include the button mushroom *Agricus bisporus* straw mushroom, *Volvariella volvacea* (Chang, 1974, 1977, 1982), the oyster mushroom, *Pleurotus florida*, *P. flabellatus*, and *P. sajor-caju* (Chang and Quimio, 1982), the abalone mushroom, *P. cystidiosus*, the wood ear or black jelly fungus, *Auricularia* spp. (Yong and Leong, 1983), and the Shiitake mushroom, *Lentinus edodes*.

3.2.3 Microbial Ecology and Indigenous Fermented Foods

The microbiology of many of Indigenous fermented products is quite complex and remains unexploited. Mostly the fermentations are natural and involve mixed cultures of yeasts, bacteria, and fungi. In such fermentations, microorganisms may participate in parallel with one another, while in others they act in a sequential manner with a changing dominant biota during the course of fermentation. The common bacteria involved in fermentation include species of *Leuconostoc, Lactobacillus, Streptococcus, Pediococcus, Micrococcus,* and *Bacillus.* The fungal genera are mainly representatives of *Aspergillus, Paecilomyces, Cladosporium, Fusarium, Penicillium,* and *Trichothecium,* where the most common fermenting yeast species is *Saccharomyces,* which contributes to alcoholic fermentation (Steinkraus, 1998; Blandino et al., 2003). Yeasts have been reported to be involved in several different types of indigenous fermented foods and beverages (Amoa-Awua and Jakobsen, 1996; Halm and Olsen, 1996; Holzapfel, 1997; Blanco et al., 1999; Gadaga et al., 2001; Jesper, 2003).

In indigenous fermented food, the microbial ecology is of considerable significance in developing a strategy to overcome the risk of toxicity from indigenous fermented foods (Scott and Sullivan, 2008). These ecological principles are the foundations upon which risk analysis (Lund, 1992; Notermans and Teunis, 1996; Fleet, 1997; van Gerwen et al., 1997) strategies have been developed to prevent the outbreaks of food spoilage and food-borne disease. Not only this, it can promote the functional uses of microorganisms in the production of fermented foods and beverages for use as probiotic and biocontrol agents (Giraffa, 2004). The growth, survival, and activity of any one species or strain, involved in spoilage or pathogenicity, or as desirable biocontrol or probiotic organisms, will, in most cases, be determined by the presence of other species in the environment, thus will involve the total ecology. This concept has been applied to commodities such as cheese and wine (Fleet, 1999). To apply it to other foods, it is essential to examine the various methods employed to analyse the various microorganisms involved.

3.2.3.1 Methods of Microbiological Analysis The basic purpose of microbiological analysis of any raw material and product is either to profile the diversity of species occurring in a product or to find out the presence or absence of specific pathogens in a food. The examination of foods for total or specific microflora involves basic operations such as, maceration/blending of the sample, dilution of the homogenate, plating of dilutions onto appropriate agar media, and isolation and identification of colonies (Harrigan and McCance, 1966; Fleet, 1999). Pre-enrichment and selective enrichment of the culture before plating will be needed to recover species present at low populations. While this basic approach has been accepted for a long time, and is generally successful, there are same inherent problems that are worthy of re-consideration, including maceration, dilution, enrichment cultures, anaerobes, unknown and nonculturable microorganisms, and quantitative data (Fleet, 1999). During microbiological analysis, while macerating, some of the microbial cells present on the surface of the products suddenly become exposed to a greatly different chemical environment which may be toxic or inhibitory to some species. Thus, the results obtained might not be reliable. During this operation, more rigorous scrutiny is necessary, as it is already known that extracts of vegetables, herbs, and other species are toxic to some microorganisms (Kyung and Fleming, 1997; Fleet, 1999; Joshi et al., 2011a). So these types of substances are likely to change the results of microbiological examination.

The dilution is an essential process which is considered harmless with respect to an ecological outcome. It is carried out with the purpose of the dispersion of cell clumps and should not normally affect the cell viability (Harrigan and McCance, 1966). But the composition of diluents and the time lapse between dilution and plating have a significant influence on viable plate counts. Usually, 0.1% peptone has been more or less standardized using as the usual dilutant (Straka and Stokes, 1957). For the isolation of yeasts and molds, the dilutants commonly used include distilled water, saline, phosphate buffer, and 0.1% peptone, with variable outcomes (Mian et al., 1997; Fleet, 1999). Besides the stage of cell life cycle, cell stress prior to dilution, degree of cell clumping and aggregation, shear forces during shaking, presence of contaminating metal ions, pH, and temperature could all effect the survival of the yeast cells or bacteria and filamentous fungi during the dilution operation (Fleet, 1999).

The culture enrichment technique is known to enhance the cell population and detectability of minority species, especially pathogens such as *Salmonella*, *Listeria monocytogenes*, *Escherichia coli*, and *Campylobacter jejuni* in various food products. It is especially true with respect to ecological survey of food for the presence of foodborne pathogens. Many factors like medium composition, time and temperature of incubation, degree of aeration, food components, competition from non-target flora in the food, the possible influences of bacteriophages, and extent of any sub-lethal injury can effect the results, and the reliability and limitations of enrichment technique becomes questionable (Hawa et al., 1984; Fleet et al., 1991). The obligate anaerobes can also contribute to the anaerobic microflora of many foods and beverages and, therefore, a good methodology for microorganisms needs to be adopted to successfully isolate these organisms (Anderson and Fung, 1983).

The plate culturing techniques normally reveal only 1%–10% (or less) of the true microbial population in many environments (Amman, 1995; Fleet, 1999), The presence of unknown, novel species that are not culturable by existing methods, and the presence of known species that are metabolically active and viable, but have entered a non-culturable state, are not accounted for. The application of molecular methods, however, can detect the non-culturable species as the method is based on analysis of the total DNA extracted from the ecosystem (See Chapter 13 for details). The microbial ribosomal (r)-DNA in the extract is specifically amplified, cloned, and then, sequenced using PCR technology (Fleet, 1999) or the use of 16 sr RNA technique (Collins et al., 1999), and thus, it can be applied to the investigation of the microbiology of food ecosystems. Consequently, the microbial composition of some foods, especially fresh products such as vegetables and meats, or complex fermented products has not so far been documented, where a diversity of species may be present but remain unknown (See, for more detail, Chapter 13 of this book). The viable but non-culturable (VBNC) phenomenon is well known, where adverse conditions can cause healthy, culturable cells to enter a phase which does not produce colonies on media that normally support their growth. Such cells remain metabolically active so can cause infection, when given optimum conditions (McKay, 1992; Kell et al., 1998; McDougald et al., 1998; Fleet, 1999).

Risk of safety from microorganisms in foods depends not only on the species present, but also on their populations. The number of microbial cells ultimately determines whether or not the product will cause an outbreak of disease or develop an off-flavor or spoil. Similar aspects can be considered with respect to the sequential development of species and strains which also occur in many indigenous fermented products. To achieve dependable assessments of public health and spoilage risks, quantitative, reliable ecological data are essential (Whiting and Buchanan, 1997a,b; ICMSF, 1998b; Roever, 1998).

3.2.4 Eco-Biochemistry and Physiology and Interactions

3.2.4.1 Eco-Biochemistry and Physiology To determine microbial growth in a particular food ecosystem, requires biochemical and physiological knowledge to understand how specific microorganisms impact on food quality, so as to be able to develop processes for managing this growth. Based on pure culture studies of microbial isolates in a laboratory media, biochemical reactions taking place as regulated by the food properties have been observed in some foods. But how the unique reactions of the *in-situ* environment take place, which will simultaneously harbor microbial cells in a diversity of physiological states-growing, non-growing, dead, and autolyzed, is not known (Fleet, 1999).

It has been clearly demonstrated the microbial cells specifically immobilized in alginate, polyacrylamide, etc., have their properties of growth, survival, tolerance of extremes, biochemical activity, and even cell composition significantly changed from cells growing freely in liquid culture (Rehm and Omar, 1993; Norton and D'Amore, 1994; Fleet, 1999). The physical closeness of cells in the environments of solid phase association, cell–cell signaling, and other communication mechanisms could also influence these responses (Fleet, 1999).

Depending upon the nutrient availability, its uptake and transport in the microbial cells, and environmental factors, the growing or exponential state probably presents only a short time span throughout the total association with the ecosystem, and a cyclical movement between exponential and non-proliferating stages is likely to occur (Boddy and Wimpenny, 1992; Fleet, 1999). It is well known in the fermentation of alcoholic beverages by yeasts that sugar transport into the yeast cells can be a significant rate-limiting reaction (Bisson, 1993; Fleet, 1998) that also explains the growth of microorganisms like LAB (Stiles and Holzapfel, 1997), pseudomonads (Palleroni, 1993), and yeasts (Praphailong and Fleet, 1998). At the same time, it is pertinent to know that microbial cells may be simultaneously exposed to a combination of stresses (e.g., low pH, low temperature; high NaCl concentration, low temperature, low pH; ethanol, acetic acid), the net result of which on growth and survival of particular microorganisms may be additive or synergistic (interactive) (Gould and Jones, 1989; Gould, 1992; Leistner and Gorris, 1995; Fleet, 1999). However, the severity of the stress can change with time, as will the physiological state of the microbial cells (Watson, 1990;

Kolter et al., 1993). Weakly or nonproliferating cells, in a resting or stationary phase of growth, probably dominate in many foods, but are still metabolically active and can carry out biochemical reactions, like development of flavors, that are distinctively different from the cells growing exponentially (Kolter et al., 1993; Rees et al., 1995; Fleet, 1999). The non-proliferating phase is eventually followed by cell death and cell autolysis. Consequently, most "long-term" food ecosystems will harbor a substantial population of dead, autolyzed cells. Autolysis is characterized by extensive loss of cellular structure and function, and enzymatic breakdown of cell proteins, nucleic acids, lipids, and polysaccharides (Charpentier and Feuillat, 1993; Hernawan and Fleet, 1995; Kang et al., 1998; Fleet, 1999). These degradation products become part of the ecosystem, serving as nutrients or antagonists for other microorganisms, and they also impact on the sensory properties of food. The production sparkling wine illustrates this aspect (Joshi, 1977; Amerine et al., 1980; Jendeat et al., 2011).

3.2.4.2 Microbial Interactions   Most foods, except processed products, harbor a mixture of microorganisms. To achieve survival, growth, and dominance, interactions will occur between these different strains and species, the net result of which will determine the population levels of any particular organism at any given time. There are a number of interactive associations, like competitiuo, amensalism, antagonism, commensalism, mutualism, and parasitism or predation (Fredrickson, 1977; Boddy and Wimpenny, 1992; Fleet, 1999; Giraffa, 2004). These could occur both within and between different microbial groups in a food. Some of the applications of microbial interactions are summarized in Table 3.2.

Besides antagonism, commensalistic microbial interactions frequently occur in food environments, such as in the degradation of complex proteins and carbohydrates by some species to produce simple substrates for the growth of other species; the

Table 3.2   Microbial Interaction and Their Applications in Food

- Antagonism is probably the best known example of microbial interaction in food ecosystems. It can be applied to enhance food quality and safety such as production of bacteriocins that can be used to control spoilage and pathogenic bacteria (Barnby-Smith, 1992; Muriana, 1996; Montville and Winkowski, 1997; Joshi et al., 2006).
- Besides "bacteria–bacteria" interaction, there are instances where bacteriocins will inhibit yeasts (Dielbandhoesing et al., 1998). The production of killer toxins by yeasts (Young, 1987; Shimizu, 1993; Wickner, 1993) is somewhat analogous to bacteriocin production by bacteria (See chapter 13 of this text). These toxins are extracellular proteins or glycoproteins that disrupt cell membrane function in susceptible yeasts.
- A less recognized form of antagonism is the production of cell wall lytic enzymes such as the production of $3\text{-}(1 \to 3)$-glucanases by bacterial and yeast species that destroy the $3(1 \to 3)$-glucans in the cell walls of fruit spoilage fungi such as *Penicillium expansum* and *Botrytis cinerea* (Wisniewski et al., 1991).
- Another less familiar form of microbial interaction that could be significant in food systems is the ability of yeasts and bacterial cells to agglutinate and aggregate.
- Most species of Enterobacteriaceae and some LAB will agglutinate *Saccharomyces cerevisiae* by reaction with the surface mannoproteins of the yeast (Mirelman et al., 1980).

Source: Adapted from Fleet, G.H. 1999. *Int J Food Microbiol* 50(1–2): 101–117; Giraffa, G. 2004. *FEMS Microbiol Rev* 28(2): 251–260.

utilization of organic acids by yeasts and molds favor the growth of bacteria, the auto-lytic release of nutrients by dead cells, and the production of vitamins, specific amino acids, carbon dioxide, and other micro-nutrients by some species that will assist the growth of other species (Boddy and Wimpenny, 1992; Fleet, 1999).

The role played by bacteriophages in food environments, however, needs greater consideration than it is receiving at present (Arman and Ketty, 1996). Their ability to destroy starter cultures of LAB used in milk fermentations has been considered (Frank, 2001), but there are only a few reports on the isolation of bacteriophages from foods, including phages for *Pseudomonas* spp. from refrigerated meat (Greer and Dilts, 1990), *Leuconostoc oenos* from wines (Davis et al., 1985a; Fleet, 1999), *Propionibacterium* spp. from cheese (Gautier et al., 1995), and *V. vulnificus* from oysters (De Paola et al., 1998). Phage activity would certainly cause bacterial cell lysis and nutrient release in ecosystems, and possibly account for the variable population data often obtained from food samples. For more information on bacteria-phages see Sand meier and Meyer, 1995.

So to understand the role of microorganisms in food safety and preservation, techniques that guarantee that the biomass being studied truly represent the ecosystem of interest are needed (Kilsby, 1999). To achieve this, microbial ecology assumes a great significance. Therefore, a precise understanding of the impact of an environmental selective pressure anywhere in the food chain on the microbial flora needs to be made for proper assessment of the risk of food safety. It is particularly true of indigenous fermented foods.

3.2.5 Microorganisms and their Role in Indigenous Fermented Foods

Out of various causes of spoilage of foods, microorganisms cause spoilage to the greatest extent (Brackett, 1997; Joshi et al., 2000). Not only this, they are responsible for causing food poisoning and food infection. From safety point of view, the latter aspects are of great importance. Nevertheless, spoilage results in huge economic loss, so cannot be ignored. Thus, the microbiology of indigenous fermented foods is of immense significance with respect to the understanding and prevention of spoilage.

3.2.5.1 Pathogenic Microorganisms

Food products like milk can serve not only as a potential vehicle for transmission of some pathogens, but also can allow these organisms to grow, multiply, and produce toxins. A variety of pathogenic organisms may gain access into milk and milk products from different sources, and cause different types of food borne illnesses. These products may also carry toxic metabolites from different organisms growing in it. Ingestion of products, contaminated with these metabolites causes food poisoning and intoxication (Frazier and Westhoff, 2004). On the other hand, the ingestion of viable pathogenic bacteria along with the food product could lead to food borne infection. Sometime these organisms undergo lysis in the gastrointestinal tract and liberate toxic substances which are detrimental to the health of the consumers (Aneja et al., 2002; Kumbhar et al., 2009). Recent development of Quality and Safety Management Systems such as ISO and Hazard Analysis Critical Control Point (HACCP) and their application has reduced the occurrence of such incidences. Similar examples from other foods can also be cited. Another problem of concern from food safety is the microbial bio-film (Zottola and Sasahara, 1994).

3.2.5.2 Spoilage Causing Microbes Most of the indigenous fermented products are prepared by traditional methods in the unorganized industrial sector. Such methods often bring about contamination of various microorganisms, including pathogens which can be prevented by automating the traditional methods in different organized sectors, that can improve the microbiological quality of the products. However, due to the prohibitive cost of capital investment, a sizeable portion of these products are still produced in the unorganized sectors, without the necessary licenses to do so, and, hence, are not bound by any law of product safety (Kumbhar et al., 2009). The consequence is poor microbiological quality. Examples can be cited of spoiled pickles (Sauerkrant) and jam (Figure 3.4).

Out of various indigenous fermented foods prepared and consumed by man, alcoholic beverages constitute a big part. Unlike wine and wine-like beverages, distilled alcoholic beverages have high alcoholic content that inhibits the growth of microorganisms, and so are not spoiled microbiologically. But low alcoholic beverages without any preservatives are prone to microbiological spoilage, while those with ethanol content below 15% (Frazier and Westhaff, 1997; Joshi, 1997) are not ordinarily spoiled.

Low alcoholic beverages like wines, even after fermentation, may contain sufficient nutrients to support the growth of a range of spoilage causing yeasts and bacteria. Some species will be tolerant of the combined effects of low pH (3.0–4.0) and high

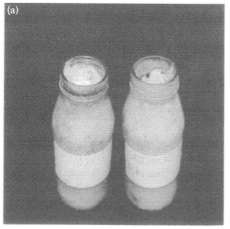 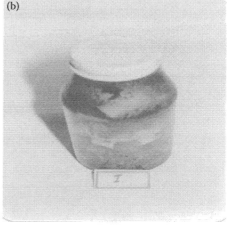

Figure 3.4 Spoilage of food products by microbes. (a) Spoiled Sauerkrout (b) Spoiled of jam.

ethanol concentrations (10%–15% w/v). Yeasts include oxidative species of *Pichia* and *Candida*, and fermentative species of *Zygosaccharomyces*, *Saccharomycodes*, and *Brettanomyces/Dekkera*, and LAB, including species of *Lactobacillus* and *Pediococcus*, are the normal spoilage causing microflora. The acetic acid bacteria, *Acetobacter pasteurianus* and *A. aceti*, are well known for their ability to oxidize ethanol at the wine–air interface to give vinegary (acetic acid) spoilage (Joshi, 1997). But their involvement is probably more complex, since species are frequently isolated from the middle of barrelled wines where little or no oxygen is present (Drysdale and Fleet, 1989). However, acid-tolerant, ethanol-tolerant species of *Bacillus* and *Clostridium* can grow in wines (Fleet, 1997, 1998), and thus, should be considered from a safety point of view.

3.2.5.3 Physiology of Food Spoilage Causing Microorganisms Microbial food spoilage costs the food industry (and indirectly, the consumer) an enormous amount annually, and ultimately represents a colossal waste of a valuable resource. Food losses begin on the farm itself and continue throughout post-harvest storage, distribution, processing, wholesaling, retailing, and use in the home and in catering (Brackett, 1997; Roller, 1999; Verma and Joshi, 2000). An alternate system of curing of meat has been developed, as a step for better preservation (Shahidi, 1991). Many indigenous fermented foods have been and still prepared as a house technology. Assessment of fermentation technology as a house technology for improving food safety has been made (Nout and Motorjemi, 1998). It is technically possible to produce food entirely free of microbial contamination by preservation methods such as gamma irradiation, but such a severe approach would result in the loss of several nutrients and is contrary to the current consumer demand for minimally processed foods, perceived as "fresh". It may be added that most of the traditional food preservation processes were developed empirically without a full understanding of the mechanisms of action of the antimicrobial agents used. It could be argued that truly novel combination systems of food preservatives can only be developed logically if they are based on a thorough understanding of the physiology of microorganisms in foods. Such an approach can reduce the cost of safety testing now required for all new food additives. A thorough understanding of the physiological responses of microorganisms to the stresses imposed during food preservation is essential if novel combination systems based on mild food processing procedures are to be developed effectively. The intrinsic characteristics of microbes, and external factors like water activity, temperature, preservatives, composition of the gaseous atmosphere, etc., influence the stress response of microorganisms. How and what are the interaction of spoilage organisms with each other as well as with food pathogens that take place, are the ultimate consequences for food safety and quality which need to be explored (Roller, 1999). Thus, by systematic study of the stress responses of microorganisms, together with detailed assessment of technological performance of available preservatives and preservation technologies in real food formulations, much more sophisticated usage will emerge, resulting in development of new processes and products (Roller, 1999).

3.2.6 Control of Pathogenic Microorganisms through Interference and Production of Antimicrobial Agents during Fermentation

3.2.6.1 Microbial Interference   General microbial interference is an effective nonspecific control mechanism which is common to all the populations and environments, including foods, where the inhibition of growth of certain microorganisms by other members of a habitat take place. It was first used to describe the suppression of virulent staphylococci by avirulent strains (Shinefield et al., 1971; Caplice and Fitzgerald, 1999). To meet the objective, generally the microflora of the habitat needs to outnumber the target host many times, which can be achieved by mechanisms involving nutrient competition, generation of an unfavorable environment, and competition for attachment/adhesion sites, which are common to all the microorganisms (Caplice and Fitzgerald, 1999).

3.2.6.2 Mechanisms of Antibiosis   This type of antibiosis is well illustrated by the LAB. The specific antimicrobial mechanisms of LAB exploited in the biopreservation of foods include the production of organic acids, hydrogen peroxide, carbon dioxide, diacetyl, broad-spectrum antimicrobials such as reuterin, and the production of bacteriocins (De Vuyst and Vandamme, 1994; Joshi and Sharma, 2012).

3.2.6.3 Production of Antimicrobial Compounds

3.2.6.3.1 Organic Acids, Acetaldehyde, and Ethanol   These compounds are produced in the normal lactic acid fermentation and contribute to extend the shelf-life of food. The direct antimicrobial effects of organic acids, including lactic, acetic, and propionic acids, which may be produced by LAB fermentation of foods, are well known (Davidson et al., 1995; Caplice and Fitzgerald, 1999). It is believed to result from the action of the acids on the bacterial cytoplasmic membrane that interferes with the maintenance of membrane potential and inhibits active transport (Sheu and Freese, 1973; Eklund, 1989; De Vuyst and Vandamme, 1994; Caplice and Fitzgerald, 1999), and may be mediated both by dissociated and undissociated acid (Cherrington et al., 1991). It may be noted that the antimicrobial activity of each of the acids at a given molar concentration is not equal. Acetic acid is more inhibitory than lactic acid, and can inhibit yeasts, molds, and bacteria (Blom and Mortvedt, 1991), while propionic acid inhibits fungi and bacteria (Mayra-Makinen and Suomalainen, 1995). The contribution of acetaldehyde to biopreservation is not very significant, as the flavor threshold is much lower than the levels that are considered necessary to achieve inhibition of microorganisms (Kulshrestha and Marth, 1974). Same is true of ethanol (Caplice and Fitzgerald, 1999).

3.2.6.3.2 Hydrogen Peroxide   LAB do not have true catalase to break down the hydrogen peroxide generated in the presence of oxygen, and thus, H_2O_2 is accumulated, which is inhibitory to some microorganisms (Condon, 1987). The inhibition is mediated through the strong oxidizing effect on membrane lipids and cell proteins (Morris, 1976; Lindgren and Dobrogosz, 1990). In addition to this, it may activate the lactoperoxidase system of fresh milk with the formation of hypothiocyanate and other antimicrobials (Reiter and Harnulv, 1984; Pruitt et al., 1986; Condon, 1987; De Vuyst and Vandamme, 1994; Caplice and Fitzgerald, 1999). However, due to the ability of other enzyme systems, such as flavoproteins and peroxidases, to breakdown H_2O_2, it is not clear what, if any, the *in vivo* contribution of H_2O_2 is to antibacterial activity (Nagy et al., 1991; Fontaine et al., 1996).

3.2.6.3.3 Bacteriocins Bacteriocins of LAB, according to the classification procedure proposed by Klaenhammer (1993) and modified by Nes et al. (1996), are divided into four major subclasses, but the majority of those produced by bacteria associated with food belong to classes I and II. The most extensively characterized class I bacteriocin is nisin, with GRAS status for use as a direct human food ingredient (Federal Register, 1988; Hugenholtz and De Veer, 1991). It is produced by strains of *L. lactis* ssp. *lactis*, and has a broad inhibitory spectrum against Gram-positive bacteria, including many pathogens, and can prevent the outgrowth of *Bacillus* and *Clostridium* spores (Daeschel, 1989). It is able to sensitize spores of *Clostridium* to heat, allowing a reduction in thermal processing time (Hurst, 1981; Caplice and Fitzgerald, 1999). Nisin is approved as a component of the preservation for processed and fresh cheese, canned foods, processed vegetables, and baby foods, in upto 50 countries (Hurst, 1981; Delves-Broughton, 1990; De Vuyst and Vandamme, 1994; Caplice and Fitzgerald, 1999) at levels ranging between 2.5 and 100 ppm. It is most stable in high-acid foods. However, nisin addition could not inhibit the growth of *Clostridium botulinum* in a model cured meat system (Rayman et al., 1983). Another bacteriocin propionicin PLG-1 has been documented to be produced by Propioni bactorium with ability to inhibit the growth of psychrotrophic organisms (Lyon et al., 1993).

3.2.6.4 Fermentation in Biopreservation Foods preservation by fermentation is a widely practiced but as an ancient technology (Borgstrom, 1968). Fermentation ensures not only increased shelf-life and microbiological safety of a food but may also make some foods more digestible and, in the case of cassava fermentation, reduce toxicity of the substrate. In alcoholic fermentation, the production of ethanol and CO_2 contribute towards preservative effects on food. In alkaline fermentation, increase in pH of the food takes place. Since it is inhibitory to spoilage-causing bacteria, it helps preserve food. LAB, because of their unique metabolic characteristics, are involved in different fermentation processes of milk, meats, cereals, and vegetables. Lactic acid fermentation contributes towards the safety, nutritional value, shelf-life, and acceptability of a wide range of cereal-based foods (Oyewole, 1997). For more details, see the literature cited (Oye wole, 1997; Joshi and Pandey, 1999; Soni and Sandhu, 1999; Verma and Joshi, 2000; Ross et al., 2002; Nout, 2005; Joshi and Sharma, 2012).

3.3 Microbiology of Different Fermentations and Associated Microorganisms

3.3.1 Microbiology of Fermentation

Many fermentations are traditionally dependent on inocula from a previous batch, but starter cultures are also available for many commercial processes, such as cheese manufacture, thus ensuring consistency of process and product quality (Caplice and Fitzgerald, 1999). Fermented foods and beverages harbor diverse microorganisms from the environment. The common fermenting bacteria are species of *Leuconostoc*, *Lactobacillus*, *Streptococcus*, *Pediococcus*, *Micrococcus*, and *Bacillus*. The fungi of genera *Aspergillus*, *Paecilomyces*, *Cladosporium*, *Fusarium*, *Penicillium*, and *Trichothecium* are the most frequently found in certain products (Soni and Sandhu, 1990a; Soni and Arora, 2000). The common fermenting yeasts are species of *Saccharomyces* in alcoholic fermentation (Steinkraus, 1998; Blandino et al., 2003). The type of bacterial flora developed in each fermented food depends on the water activity, pH, salt concentration, temperature, and the composition of the food matrix. Most fermented foods are dependent on LAB to mediate the fermentation process (Conway, 1996; Blandino et al., 2003). In the case of cereal grains, after cleaning, these are soaked in water for a few days, during which time a succession of naturally occurring microorganisms will result in a population dominated by LAB. In such fermentations, endogenous grain amylases generate fermentable sugars that serve as a source of energy for the LAB (Blandino et al., 2003).

During the fermentation process, microorganism transforms the chemical constituents of raw materials including the enhancement of the nutritive value of the product, improve flavor and texture, preserve perishable foods, improve digestibility, and stimulate probiotic functions (Sidhu and Al-Hooti, 2005; Tamang and Kailasapathy, 2010).

3.3.2 Fermentation and Its Types

Fermentation is a biological process brought about by the activity of a whole range of microorganisms and their enzymes, wherein the chemical transformation of complex organic substances into simpler compounds takes place. During fermentation, important by-products, many of commercial value, such as alcohol, acids, flavoring compounds, and various gases are produced (Bibek, 2004). The well known microorganisms responsible for fermentation are molds (e.g., *Aspergillus*, *Mucor*, *Neurospora*, and *Rhizopus*), yeasts (e.g., *Saccharomyces*), and bacteria (e.g., *Bacillus*, *Lactobacillus*, and *Pediococcus*). In fermentation, the microbial activity results in production of a range of metabolites that can suppress the growth and survival of undesirable microflora, including pathogenic ones, in foodstuffs (Ross et al., 2002). In natural fermented foods, the fermentation is directed by indigenous microflora. While carbohydrate rich foods are favored by the growth of yeasts resulting in alcoholic fermentation, foods rich in proteins, vitamins, nitrogenous constituents, and carbohydrates are favored for the growth of LAB, and fermentation is referred to as the lactic acid fermentation.

Interest and research into indigenous fermented foods have seen a marked growth in the last few decades (Dirar, 1993; Wang and Fung, 1996; Steinkraus, 2004). Accordingly, a number of studies on microbial cultures involved in these traditional fermentations, processing equipment, nutritional aspects, and methods for optimizing fermentation conditions have been carried out, along with the use of molecular biology-based methods for identification of microbial species (Ouoba et al., 2004; Dakwa et al., 2005; Azokpota et al., 2006; Terlabie et al., 2006; Moktan and Sarkar, 2007; Roy, 2007).

Different types of food fermentations have been classified based on the major end products of the fermentation process, such as:

- Lactic acid fermentation
- Alcoholic fermentation
- Alkaline fermentation
- Acetic acid fermentation
- Pit fermentation

3.3.2.1 Lactic Acid Fermentation Lactic acid fermentation is employed in the homemade and commercial production of many fermented foods. LAB are found in low numbers on the surface of raw materials like vegetables and when stored under suitable conditions (microaerophilic or anaerobic), the population of LAB increases rapidly. The presence of brine, an acidic pH, or spices may further favor the growth of these bacteria. At early stages of fermentation, oxygen is rapidly consumed and acid production begins, inhibiting the growth of aerobic microflora, followed by a decrease in pH. Further, nutrient competition exerted by LAB starts the production of lactic acid in the medium which increases the acidity and decreases the pH, thus creating selective conditions for the growth of LAB and the inhibition of other competitors (Wood, 1997; Hutkins, 2006).

During the lactic acid fermentation, there is often a succession of LAB populations, depending upon on the food substrate and the fermentation conditions. Heterofermentative and less acid tolerant LAB are often found at early stages of fermentation, but are displaced by homo-fermentative species (more acid tolerant) in the later stages. At the late stages of fermentation, however, a high concentration of lactic acid and a low concentration of sugars in the medium, causes the growth of LAB to stop. Other microorganisms, like yeasts, may also be found in the fermentation with LAB, releasing vitamins and thus, further stimulating the growth of LAB in fermented foods (Hurtado, 2010). In other cases, yeasts (and other acid-tolerant bacteria) can grow at late stages of fermentation at the expenses of the organic acids present, thus lowering the acidity, which may in turn spoil the fermented products.

The lactic acid fermentation of foods can be achieved by the natural microbiota found in the raw materials and fermentation settings (spontaneous fermentation), or it can be accelerated by adding specific inocula or starter cultures (Hutkins, 2006). A common traditional practice, known as backslopping, used for centuries in the preparation of traditional fermented foods, can also be practiced. Food products prepared by lactic acid fermentation are naturally preserved and are safer than the raw materials. Low pH due to the production of lactic acid hinders the growth of various food spoilage and pathogenic organisms. Besides low pH, organic acids produced during the fermentation also elicit antimicrobial activity. As described earlier, the LAB may also produce a variety of compounds with antimicrobial activity, such as diacetyl, acetoin, hydrogen peroxide, ethanol, CO_2, reuterin, or reutericyclin (Caplice and Fitzgerald, 1999; Gálvez et al., 2007). Besides, antimicrobial peptides, also known as bacteriocins, having antifungal or antibacterial activity, are also produced. These have been studied extensively for industrial application as natural preservatives, either by using bacteriocin-producing strains as starter cultures, or as partially purified bacteriocin preparations (Joshi et al., 2006; Gálvez et al., 2007, 2011). Lactic acid, together with other antimicrobials produced by LAB, contributes to preservation and, therefore, sound knowledge of the metabolism of LAB is essential to understand their role in lactic acid fermentation and, hence, in the production of fermented foods with unique flavors and textures.

Traditionally, LAB have been widely used as starter bacteria in the fermentation of milk, meat, fish, and vegetable products (Davidson et al., 1995). LAB are also essential to the fermentation of a number of vegetables, like cabbage, cucumbers, table olives, sourdoughs, and many traditional fermented foods (Daeschel et al., 1987). Due to the low buffering capacity of vegetable substrates, lactic acid fermentation results in a rapid acidification and consequent inhibition of the competing microbiota. More than 40 different species of LAB have additionally been isolated from various sourdoughs, and are generally from the genera *Lactobacillus*, *Leuconostoc*, *Pediococcus*, or *Weissella*, and the majority of strains belongs to the genus *Lactobacillus* (Gänzle et al., 2007), although *Lactobacillus sanfranciscensis* is considered the key organism for the fermentation of sourdoughs. The acidic and sensory properties of fermented foods result from the metabolic activities of these microorganisms. Foods such as ripened cheeses, fermented sausages, sauerkrout, and pickles have not only a greatly extended shelf-life over the raw materials used in their preparation, but also have improved aroma and flavor characteristics.

Acidification has several consequences on protein structure and stability, including destabilization of protein structure, protein denaturation, precipitation, and loss of bound water molecules. These modifications are important in the development of the texture of the final product, such as the coagulation of milk proteins in fermented milk products, the fermentation of meat products, where protein precipitates and dehydration facilitates drying and texture development in the final product. Acidification also increases the activity of enzymes during the ripening of fermented foods, and hence, in the development of flavor.

Besides lactic acid, other end products from lactic fermentation, such as acetoin, butanediol, diacetyl, and acetic or formic acid impart distinctive flavors to the final products (Neves et al., 2005). The specific contribution of each of these components, however, depends on the type of fermented food (See later section for more details). Some LAB have the capacity to metabolize citrate by converting it to pyruvate, which is important for the production of diacetyl, the compound responsible for a buttery flavor in dairy products. Similarly, some lactic acid bactic influence the phenolic compounds (Rodriquez et al., 2009). The phenolic compounds are known to be have antioxidant activity. Another aspect of lactic acid fermentation is malo-lactic fermentation for decarboxilation of malic acid to lactate (Liu, 2002), very important in highly acidic wines. The peptidolytic system of LAB is involved in the generation of flavors, especially during the lactic fermentation of protein-rich substrates such as milk. Due to proteolysis, the release of peptides and/or amino acids, conferring, for example, bitter or umami flavors, are produced. Branched-chain amino acids during metabolism of amino acids are converted into compounds contributing to malty, fruity, and sweaty flavors; catabolism of aromatic amino acids (Phe, Tyr, Trp) produces floral, chemical, and fecal flavors, while aspartic acid (Asp) is converted into buttery flavors and sulfur-containing amino acids (Met, Cys) which are transferred into compounds contributing to boiled cabbage, meaty, and garlic flavors (Marilley and Casey, 2004). Besides, cell lysis of LAB during the ripening of certain fermented foods (cheese) accelerates the reactions responsible for flavor generation. These bacteria also contribute significantly to color stabilization, texture improvement, prevention of oxidation of unsaturated free fatty acids (as in fermented meats), and strongly influence the composition of nonvolatile and volatile compounds, peptides, and amino acids through release/degradation of free amino acids (Talon et al., 2002; Fadda et al., 2010). Similarly, the typical flavor of dry cured fermented sausages occurs during fermentation and ripening (Fadda et al., 2010).

The biosynthesis of extracellular long-chain polysaccharides, important in the texture of fermented foods such as *yogurt* or *kefir*, is accomplished by LAB, which can be loosely attached to the cell surface or secreted in the environment (Mayo et al., 2010). The metabolic role of LAB in sourdough fermentation has also been reviewed earlier (Gobbeti et al., 2005; Gänzle et al., 2007). Proteolysis, followed by peptide or amino acid metabolism by LAB, is one of the key routes of flavor generation in bread, beside producing homo-polysaccharides, relevant to the texture of doughs.

3.3.2.2 Alcoholic Fermentation This is the fermentation in which ethyl alcohol is the principal metabolite resulting in production of a variety of alcoholic beverages whoes preparation and consumption has been carried out from the time immemorial (Amerine et al., 1967; Joshi, 1997, 1980; Joshi et al., 1999). Most of the indigenous alcoholic beverages are produced by natural fermentation. The natural fermentation of traditional alcoholic beverage involves a large number of microorganisms including yeast and bacteria. Among the prominent microflora is the yeast *Saccharomyces cerevisiae*, well known to produce ethyl alcohol. The other yeasts like *Candida, Pichia, Klueveromyces*, etc. have been associated with alcoholic fermentation to varying degrees. *Zymomonas* has also been involved in such fermentation. The association of

LAB and acetic acid bacteria has also been found very frequently with alcoholic fermentation of traditional fermented foods. These have been described under a diversity of microflora of indigenous alcoholic fermented foods.

Most of the ethnic fermented foods and beverages are produced by natural fermentation, except the alcoholic beverages made in Asian countries, which are made using a consortium of microorganisms in the form of dry, cereal-based, starter cakes. The fermentation of starchy rice to alcoholic end products is done in two ways, firstly by the use of molds such as *Amylomyces rouxii*, which produces amylase for saccharification of starch, and the yeast *Endomycopsis fibuligera*, which converts sugars (glucose/maltose) to ethanol (Steinkraus, 1997; Bhalla and Thakur, 2011), and secondly, by simultaneous saccharification and fermentation (yeast, molds, and amylase) of rice to rice wine, as is done in the preparation of Japanese *koji* (Yoshizawa and Ishikawa, 1989), yielding a product with ethanol content up to as high as 23% v/v. The method through which the degradation of carbohydrate into simple sugar followed by alcoholic fermentation *via* yeast is more commonly practiced in rice wine production in the South Asia region (Lee and Lee, 2002; Law et al., 2011). These starters or inocula are commercially produced and given different names in the different South Asian regions. Studies on the microflora of the starter cultures have been widely reported (Ko, 1972; Batra and Millner, 1974; Cronk et al., 1977; Ardhana and Fleet, 1989; Lee, 1990; Limtong et al., 2002; Chiang et al., 2006; Law et al., 2011).

Diversity within the species or strains of several genera of dominant microorganisms has created ethnic foods with different sensory characteristics (Savadogo et al., 2004). Acetic acid fermentation is also found in a very small quantity, and is considered safe. Acetic acid has bacteriostatic to bactericidal properties, depending upon the acetic acid found in fermented beverages. Some of the alcoholic fermentations of South Asian products, along with microflora involved, are shown in Table 3.3.

The alcoholic fermentation is dominated by the growth of yeasts because of their ability to rapidly develop at the low pH (3.0–3.5) of the juice, and produce ethanol that inhibits the growth of filamentous fungi and bacteria. It has been shown that *S. cerevisiae* predominates in almost every alcoholic fermentation such as that of wine, but other yeast species also contribute to the overall fermentation, which has been realized only in recent years (Fleet et al., 1984; Heard and Fleet, 1985, 1988; Fleet and Heard, 1993). The first 2–4 days of alcoholic fermentation are characterized by the growth of various species of *Kloeckera/Hanseniaspora*, *Candida*, *Metschnikowia*, *Pichia*, and *Kluyveromyces*, which achieve populations of about 10^7 cfu/mL before progressively dying-off, depending upon their tolerance of accumulating concentrations of ethanol, and generating sufficient amounts of end-products to have an impact on the wine character. The application of molecular techniques to such fermentation has also been made (Querol and Ramon, 1996). Strains of *Sacchromyces* with killer activities are commonly isolated from wine fermentations and also contribute to the changing profile (Shimizu, 1993). The temperature of alcoholic fermentation, in particular, can have a profound impact (Heard and Fleet, 1988).

Table 3.3 Fermented Beverages of South Asia along with their Microflora

BEVERAGES	COUNTRY	INGREDIENTS	FUNCTIONAL MICROFLORA	REFERENCES
Jnard/jaanr	Nepal, India, Bhutan	Finger millet/rice/maize/wheat	*Mucor* spp., *Rhizopu* spp., *Saccharomycopsis fibuligera*, *Pichia anomala*, *Saccharomyces cerevisiae*, LAB	Tamang et al. (1988)
Palm wines (Toddy/tari, Tuack, Tuba)	India, Bangladesh, Sri Lanka, Thailand, Malaysia, Indonesia, Philippines	Sap of coconut, date or palmyra palm	LAB, AAB, *Saccharomyces cerevisiae*, *Schizosaccharomyces pombe*, *Kodamaea ohmeri* and other yeasts	Batra and Millner (1974); Joshi et al. (1999)
Murcha/marcha	India, Nepal	Rice and wild herbs	*Mucor* spp., *Rhizopus* spp., *Pichia burtonii*, *Saccharomyces cerevisiae*	Shrestha et al. (2002); Tsuyoshi et al. (2005)
Raksi/arak	India, Nepal, Bhutan, Srilanka, Bangladesh	Cereals	—	Tsuyoshi et al. (2005)
Sura	India	Millet flour	*S. cerevisiae*, *Zygosaccharomyces bisporous*, *Leuconostoc sp.*	Thakur et al. (2004)
Channg	India and Nepal	Rice/barley	*Saccharomyces cerevisiae*, *Leuconostoc* sp., *Lactobacillus* sp, *Candida cacoi*	Thakur et al. (2004)
Daru/chakti	India	Cereal and jaggery	*Saccharomyces cerevisiae*, *Candida famata*, *Candida valida*, *Kluyveromyces thermotolerance*	Thakur et al. (2004)
Angoori/chulli	India	Grapes/Wild apricot	*S. cerevisiae*	Thakur et al. (2004)
Behmi	India	Behmi	*S. cerevisiae*	Thakur et al. (2004)
Jann	India	Rice/wheat/jau	—	Roy et al. (2004)
Soor	India	Rice and fruits	—	Rana et al. (2004)
Fenny	India	Cashew apple	—	—
Jackfruit wine	Sri Lanka, India	Jackfruit pulp	—	Dahiya and Prabhu (1977)

Source: Adapted from Bhalla, T.C. and Thakur, N. 2011. *Bio-Processing of Foods.* Asiatech Publishers, Inc., New Delhi, pp. 29–38.
LAB: lactic acid bacteria, AAB: acetic acid bacteria.

In alcoholic lactic acid fermentation, both lactic and ethyl alcohol are produced. This is illustrated by palm sap fermentation involving alcoholic–lactic–acetic acid fermentation, caused by the presence of mainly yeasts and LAB. Aidoo et al. (2006) concluded that the *Saccharomyces* spp. is present in the natural fermented palm sap and is important for the formation of the characteristic aroma of the product. Both *S. cerevisiae* and *S. pombe* have been reported to be the dominant yeast species (Odunfa and Oyewole, 1998; Law et al., 2011) along with other yeast species such as, *Candida* spp. and *Pichia* spp. (Atacador-Ramos, 1996; Law et al., 2011).

Lactic acid and acetic acid bacteria such as *Lactobacillus plantarum*, *L. mesenteroides*, *Acetobacter* spp., and *Zymomonas mobilis* have also been documented. The microorganism originates from the palm tree and the gourd used for sap collection and fermentation, or the tapping equipment. Seventeen species of yeasts and seven genera of bacteria occur in natural fermented coconut palm sap. The yeast species include the *Candida paropsilosis*, *C. tropicalis*, *C. valida*, *K. javanica*, *Pichia etchellsii*, *P. farinose*, *P. guilliermondi*, *P. membranaefciens*, *P. ohmeri*, *Rhodotorula glutinis*, *Saccharomyces chevalieri*, *S. ludwigii*, *S. bailii*, *Schizosaccharomyces pombe*, *Sporobolomyces salmonicolor*, and *Torulopsis* spp. *Bacillus* is the predominant bacteria genus, while others included *Enterobacter*, *Leuconostoc*, *Micrococcus*, and *Lactobacillus* (Atputharajah et al., 1986) Shamala and Sreekantiah (1988) have reported that the fermentation of coconut sap produces mainly ethanol, acetic acid, and lactic acid. The pH of the sap drops rapidly from around 7.2 to 5.5 due to formation of acetic acid, and the ethanol content increases drastically, to 5% (v/v) within 8 h (Law et al., 2011).

Many wines undergo a natural secondary fermentation about 2–4 weeks after completion of the alcoholic fermentation, called the malolactic fermentation (MLF), which is conducted by LAB (Davis et al., 1985b; Wibowo et al., 1985; Martineau and Henick-Kling, 1995). In this fermentation, the stoichiometric decarboxylation of L-malic acid to L-lactic acid and carbon dioxide takes place, causing de-acidification of the wine and an increase in pH by about 0.3–0.5 unit. *Leuconostoc oenos* is the main species that conducts MLF (Dicks et al., 1995), and it is uniquely found in the winery environment. Other species of LAB, namely *Pediococcus parvulus*, *P. pen-tosaceus*, *P. damnosus*, and various *Lactobacillus* spp., can also conduct this fermentation.

3.3.2.3 Alkaline Fermentation   Alkaline fermentation is a process during which the pH of the substrate increases to alkaline values, which may be as high as 9 (Aniche et al., 1993; Sarkar and Tamang, 1995; Amadi et al., 1999; Omafuvbe et al., 2000) due to enzymatic hydrolysis of proteins from the raw material into peptides, amino acids, and ammonia (Allagheny et al., 1996; Kiers et al., 2000), and/or due to alkali-treatment during production (Wang and Fung, 1996; Parkouda et al., 2009). Alkaline-fermented foods constitute a group of less-known food products made from various ingredients, such as cooked soyabeans, that are widely consumed in several countries, including those in South Asia. Despite the nutritional, cultural, and socioeconomic importance of the alkaline fermented condiments (Wang and Fung, 1996; Diawara and Jakobsen, 2004), these foods constitute a group of poorly characterized food products (Wang and Fung, 1996; Steinkraus, 1997). Most alkaline fermentations are achieved spontaneously by mixed bacterial cultures, principally dominated by *Bacillus subtilis*. It is believed that the microorganisms associated with alkaline fermented foods originate from a range of sources, such as handling, raw seeds, utensils, containers, etc. (Odunfa, 1981; Parkouda et al., 2009). A wide range of microorganisms have been found to be involved in the production of alkaline fermented foods. Various aerobic endospore forming bacteria of different genera are generally identified as predominant during fermentation, such as species of the genus *Bacillus*, including *B. subtilis*, *B. licheniformis*, *B. pumilus*, *B. cereus*, *B. coagulans*, *B. megaterium*, and *B. circulans*, and the genus *Lysinibacillus*, including *L. fusiformis*, and *L. sphaericus* (Odunfa and Oyewole, 1986; Sanni et al., 2002; Sarkar et al., 2002; Chukeatirote et al., 2006; Parkouda et al., 2009). Besides, nonspore-forming bacteria, such as *Staphylococcus sciuri*, *S. gallinarium*, *Acinetobacter calcoaceticus*, *Enterococcus faecium*, *Enterococcus avium*, *Enterococcus casseliflavus*, *Pediococcus acidilactici*, *Pediococcus pentosaceus*, and occasionally, *Weisella* and *Lactobacillus* spp., are frequently isolated during processing and from the final products (Sarkar et al., 1994; Parkouda et al., 2009, 2010; Ouoba et al., 2010). The presence of *B. cereus* is of potential concern, given its ability to produce emetic and diarrhoeal toxins (Ouoba et al., 2008; Thorsen et al., 2010, 2011).

3.3.2.4 Acetic Acid Fermentation   Acetic acid fermentation involves the production of acetic acid by acetic acid bacteria and, in terms of indigenous fermented foods, leads to the production of vinegar. Vinegar is regarded as a safe product, besides being bacteriostatic or bactericidal, depending upon its concentration. When the products of alcoholic fermentation are not kept anaerobic, there is fermentation to acetic acid, commonly called acetification. *Acetobacter* present in the environment oxidize ethanol in alcoholic products to acetic acid if anaerobic conditions are not maintained (Conner and Allgier, 1976; Steinkraus, 1996). Bacteria belonging to the genus *Acetobacter* present in the environment oxidize portions of the ethanol to acetic acid/vinegar (Conner and Allgier, 1976; Steinkraus, 1983b, 1996). Fermentations involving the production of acetic acid yield foods or condiments. In many indigenous fermented alcoholic beverages, like palm wines and kaffir beers, not only there is ethanol, but also acetic acid. Vinegar is a highly acceptable condiment used in pickling and preserving cucumbers and other vegetables. Acetic acid fermentations can be used to ensure safety in other foods.

Acetobacter strains are widely distributed in nature and have been isolated from many sources, including vinegar generators, beer, wine, yeast, wort, apple juice, cane sugar juice, flowers, vegetables, and many fruits. Acetic acid bacteria responsible for acetic fermentation are classified into two genera, *Acetobacter* and *Gluconobacter*, belonging to the *Acetobacteriaceae* family (Gillis and De Ley, 1980). These bacteria are very resistant to acetic acid concentrations; rather, they are activated by the latter up to 2.0 acidity (Anonymous, 1971). *Acetobacter*, capable of oxidizing ethanol not only to acetic acid but also totally to carbon dioxide, are used for vinegar production in modern submerged culture fermentors. In actual practice, most of the vinegar manufacturers probably do not select particular bacterial strains but use a consortia.

3.3.2.5 Pit Fermentation   The "pit" fermentations are practised in the South Pacific Islands but are uncommon in South Asian countries. These have been used for centuries by the Polynesians to store and preserve breadfruit, taro, banana, and cassava tubers (Steinkraus, 1986; Aalbersberg et al., 1988). The materials are placed in leaf-lined pits which are covered with leaves and are sealed. It has been found that pit fermentations are lactic acid fermentations (Aalbersberg et al., 1988).

3.4 Microbiology of Typical Indigenous Fermented Foods of South Asia

3.4.1 The Type of Microflora and Its Determination

Food fermentation is a process in which raw materials such as milk, meat, fish, vegetables, fruits, cereal grains, seeds, and beans, fermented individually or in combination are converted to fermented foods by the growth and metabolic activities of desirable microorganisms (Joshi and Pandey, 1999). The microorganisms utilize some components present in the raw materials as substrates to generate energy and synthesize cellular components, to increase population, and to produce many usable by-products (also called end products) that are excreted into the environment. The unused components of the raw materials and the microbial by-products (and sometimes microbial cells) together constitute fermented foods. Fermented foods and beverages harbor diverse microorganisms—molds, yeasts, and bacteria—from the environment. During the fermentation process, microorganisms transform the chemical constituents of raw materials into fermented foods, which have several advantages as discussed earlier (Sidhu and Al-Hooti, 2005; Tamang and Kailasapathy, 2010).

The ancient civilizations employed natural fermentation in the production of fermented foods which is used even today to produce such foods. Either the desirable microbial population is naturally present in the raw materials, or some products containing the desirable microbes from a previous fermentation (called back slopping), are added to the raw materials in this technique. Then, the fermentation conditions are set so as to favor the growth of the desirable types, but prevent or retard the growth of undesirable types that could be present in the raw materials. However, in controlled or pure culture fermentation, the microorganisms associated with fermentation of a food are first purified from the food, identified, and maintained in the laboratory. When needed for fermentation, they are grown in large volume to produce a "starter culture," and are then, added to the raw materials. The fermentation conditions are set in this case also in such way that these microorganisms grow preferentially to produce the desired product. To cite one example, *idli* fermentation in India is caused at least in part by *Leuconostoc mesenteroides* (Tsenk.) van Tieghem, which is one of several bacterial species present in *idli* (Mukherjee et al., 1965), besides *Streptococcus faecalis* and *Pediococcus cerevisiae*. A number of steps, however, are required to determine the type of microorganism(s) responsible for traditional fermentations (Hesseltine, 1983) as listed in Table 3.4.

Yeasts occur in a wide range of fermented foods, made from ingredients of plant as well as animal origin. When yeasts are abundant, alone or in stable mixed populations with mycelial fungi or with (usually lactic acid) bacteria, they have a significant impact on food quality parameters such as taste, texture, odor, and nutritive value. Among the different Asian indigenous fermented food products, pancakes and bread, amylolytic

Table 3.4 Steps to Determine the Type of Microflora

- Information must be obtained on the method used by the local people to prepare the food, such as moisture content of substrate, temperature of incubation, and time of fermentation needed.
- Determine dominant microorganism(s) in food by isolation from at least 10 fresh fermented food samples collected from different locations or factories.
- Test each of the isolate by producing the native product in the traditional way.
- Cultures that produce the proper product should be preserved.
- Proven product strains need to be studied comparatively, given a description covering the variations observations obtained, a comparison made with type cultures or descriptions, and a scientific name assigned.

Source: Adapted from Hesseltine, C.W. 1983. *Annu Rev Microbiol* 37: 575–601.

fermentation starters, alcoholic snacks and beverages, and condiments illustrate the diversity of ingredients used. Yeast fermentation plays an important role in the home kitchen, and several products are presently manufactured at medium or large indus-trial scale (Steinkraus, 1989; Aidoo et al., 2006). Yeasts have been found in different soybean-based fermented food products, including alkaline fermented foods, also as a part of a highly complex microbial consortium which changes over time according to the changing environmental conditions. Depending on the product, the yeast meta-bolic activities during specific stages of fermentation and maturation in production of fermented foods may improve the nutritional and sensory quality or, on the contrary, may contribute to spoilage of the food. Some of the yeast based fermentations are summarized in Table 3.5.

3.4.2 Microbiology of Some Typical Fermented Foods

3.4.2.1 Idli *Fermentation* Idli is a naturally fermented food where both bacteria and yeasts are introduced, generally by the two main ingredients, and participate in the fermentation. Ordinarily, the microorganisms developing during the initial soaking of ingredients and later are sufficient to bring about the fermentation. The acid and gas production has been attributed mostly to the bacterial growth. Black gram, the leguminous component of *idli* batter, serves not only as an effective substrate, but also provides the greatest number of microorganisms for fermentation. Black gram *dhal* soaked in water has a high concentration of soluble nutrients to support the growth of LAB (Soni et al., 2013). Yeasts and buttermilk as the inocula are also sometimes added, but *idli* prepared with added yeast and/or sour butter milk inoculum is generally similar to that prepared without inocula. A bit of freshly fermented batter ("backslop") is often added to the newly ground batter. With the progress of fermentation, both bacterial and yeast cell numbers increase significantly, with a concomitant decrease in pH, and an increased volume of batter, amylase, and protease activity. *Leuconostoc mesenteroides* is the most commonly encountered bacterium (Nout and Sarkar, 1999; Aidoo et al., 2006). Bacteria identified as a part of the microflora responsible for the production of good *idli* include *Leuconostoc mesenteroides, Lactobacillus coryne-formis, L. delbrueckii, L. fermentum, L. lactis, Streptococcus faecalis,* and *Pediococcus*

Table 3.5   Yeast-Based Traditional Fermented Foods and Beverages of South Asia and the Microorganisms Involved

NAME OF THE FOOD	COUNTRY	MAJOR INGREDIENTS	FUNCTIONAL MICROFLORA	FERMENTATION CONTRIBUTES TO	REFERENCES
PANCAKES AND LEAVENED LOW-SALT BREAD					
Idli	India, Sri Lanka	Rice and black gram dal	LAB, *Saccharomyces cerevisiae*	Flavor, texture, nutritional value	Soni and Sandhu (1991)
Dhokla	India	Rice and Bengal gram	LAB, *Pichia silvicola*	Flavor, texture, nutritional value	Kanekar and Joshi (1993)
Nan, kulcha and *bhatura*	India, Pakistan, Afghanistan, Iran	Wheat flour	LAB, *Saccharomyces cerevisiae* and other yeasts	Texture, flavor	Sandhu et al. (1986)
AMYLOLYTIC FERMENTATION STARTERS					
Murcha/ marcha	India, Nepal	Rice	*Mucor* spp., *Rhizopus* spp., *Pichia burtonii*, *Saccharomyces cerevisiae*	Starch degradation, alcoholic fermentation	Shrestha et al. (2002); Tsuyoshi et al. (2005)
SWEET: LOW ALCOHOLIC SNACKS FERMENTED WITH AMYLOLYTIC STARTERS					
Jnard/jaanr/ thumba	Nepal, India, Bhutan	Finger millet/ rice/maize/ wheat	*Mucor* spp., *Rhizopus* spp., *Saccharomycopsis fibuligera*, *Pichia anomala*, *Saccharomyces cerevisiae*, LAB	Saccharification, alcohol, flavor	Tamang et al. (1988)
BEVERAGES FERMENTED FROM SUGARY JUICES					
Palm wines (Toddy/tari, Tuack, Tuba)	India, Bangladesh, Sri Lanka, Thailand, Malaysia, Indonesia, the Philippines	Sap of coconut, date or palmyra palm	LAB, AAB, *Saccharomyces cerevisiae*, *Schizosaccharomyces pombe*, *Kodamaea ohmeri* and other yeasts	Alcohol and flavor production	Batra and Millner (1974); Joshi et al. (1999)
CONDIMENTS					
Wadi	India, Pakistan	Blackgram dal	*Candida krusei*, LAB	Acidification, leavening, nutritional value	Sandhu and Soni (1989)
Papad/ papadam	India	Blackgram dal	*Candida krusei*, *Saccharomyces cerevisiae*	Texture, flavor	Shurpalekar (1986)

AAB: acetic acid bacteria, LAB: lactic acid bacteria.

cerevisiae, while the yeast flora generally involved in fermentation include *Torulopsis, Candida, Trichosporon pullulans, Candida cacaoi, Debaryomyces tamarii, Rhodotorula graminis, Candida fragicola, C. kefyr, Wingea robertsii, Issatchenkia terricola, Hansenula anomala, Candida glabarata, C. tropicalis, C. sake, C. krusei,* and *Torulopsis holmii* (Batra and Miller, 1976; Sandhu and Waraich, 1984; Venkatasubbaiah et al., 1984, 1985;

Soni and Sandhu, 1989a,b, 1991; Soni et al., 2013). During fermentation, along with *L. mesenteroides*, *Saccharomyces cerevisiae*, *Debaryomyces hansenii*, *Pichia anomala*, and *Geotrich candidum* are predominant among the yeasts appearing first, and *Trichosporon cutaneum* and *Trichosporon pullulans* develop, subsequently. These organisms have been found to be naturally present in black gram cotyledons and rice (Chavan and Kadam, 1989). Eventually, only *S. cerevisiae* persists (Soni and Sandhu, 1991), though its role in *idli* batter fermentation is controversial. The fermentation has been reported (Ramakrishnan, 1979; Aidoo et al., 2006) to be entirely due to hetero-fermentative *L. mesenteroides*, but later work has shown the involvement of yeast in fermentation (Venkatasubbaiah et al., 1984; Aidoo et al., 2006). The major functions of the fermentation include the leavening of the batter and the improvement of taste and nutritional value of *idli*. The role of LAB is to reduce the pH of the batter to an optimum level (pH 4.1–4.5) for yeast activity. Yeast helps in the degradation of starch (not possible by *L. mesenteroides*) into maltose and glucose by producing extracellular amylolytic enzymes. Besides, production of carbon dioxide, the yeast plays a significant role in the leavening process, texture development, and synthesis of B vitamins and free amino acids (Venkatasubbaiah et al., 1984, 1985). The sources of the yeast strains are the surface of the stone grinders used for preparation of batter, and the rice used in the batter. Not only the load, but also the diversity, of yeasts in the fermenting batter is greater in winter than in summer. Fermentation of batter by inoculating the ingredients with individual yeasts and in combination with *L. mesenteroides* have revealed that the yeasts contribute towards desirable sensory qualities besides other characteristics outlined earlier.

The initial soaking of the ingredients generally brings about a rapid disappearance of common aerobic contaminants. *Leuconostoc mesenteroides* and *Streptococcus faecalis* develop concomitantly during soaking and then, continue to multiply following grinding, and play a significant role in *idli* fermentation (Mukherjee et al., 1965; Rao et al., 2011). The essential roles played by *Leuconostoc mesenteroides* and *Lactobacilli* in the leavening of *idli* batter and production of acidity have been substantiated by several workers (Ramakrishnana, 1979; Batra, 1981; Venkatasubbiah et al., 1984; Sarasa et al., 1985; Soni and Sandhu, 1990a; Soni and Sandhu, 1991). In contrast, Batra and Miller (1974, 1976) reported that yeasts *Torulopsis*, *Candida*, and *Trichosporon pullulans* are sufficient for proper fermentation. Venkatasubbaiah et al. (1984) observed the involvement of both bacteria and yeasts in gas production during *idli* batter fermentation, but found that the total LAB are always higher than the yeasts. Further, parboiled rice was found to support the growth of yeasts which were absent in the batter when raw rice alone was used.

The microbiological evaluation of traditional commercially fermented *idli* batter samples and those experimentally fermented in the laboratory (Table 3.6) revealed the involvement of several bacteria along with yeasts (Soni and Sandhu, 1989a,b, 1990a). It was found that the market batter samples contained bacteria in the range of 10^6–10^8/g belonging to four genera. *Leuconostoc mesenteroides* was the most commonly

Table 3.6 Occurrence of Bacteria and Yeasts in Fermented *Idli* Batters

SOURCE	SEASON	ATMOSPHERIC TEMPERATURE (°C)	NO OF SAMPLES	RANGE OF CFU/G	NO OF ISOLATES		POSTIVE SAMPLES		RANGE OF CFU/g	NO OF ISOLATES	
					Genera	Species	No.	(%)		Genera	Species
Local market											
	Summer	35–42	10	10^7–10^8	4	4	6	60	0–10^3	2	2
	Winter	15–25	10	10^6–10^7	2	2	9	90	0–10^6	4	4
Total and range		15–42	20	10^6–10^8	4	4	15	75	0–10^6	4	4
Laboratory											
	Summer	35–42	8	10^8–10^9	5	6	3	37	0–10^3	2	3
	Winter	15–25	7	10^7–10^8	4	4	6	83	0–10^6	5	6
Total and range		15–42	15	10^7–10^9	5	6	9	60	0–10^6	5	6
Overall total and range		15–42	35	10^6–10^9	5	6	24	68	0–10^6	5	6

Source: Adapted from Soni, S.K. and Sandhu, D.K. 1989a. *J. Cereal Sci.* 10: 227–238.
Note: CFU = Colony forming units.

encountered bacterium, followed by *Lactobacillus fermentum*, *Streptococcus faecalis*, and *Pediococcus cerevisiae*. Yeasts varied from 0 to 10^6/g and, of these, *Saccharomyces cerevisiae* and *Debaryomyces hansenii* were the predominant species isolated from 65% and 60% of samples, respectively, followed by *Hansenula anomala* (35%) and *Trichosporon beigelii* (10%). The laboratory fermented samples had comparatively higher microbial loads, including bacteria (10^7–10^9/g; belonging to five genera and six species) and yeasts (0–10^7/g; present in 60% samples). In addition to the common microbial types found in commercial samples, laboratory fermentations also exhibited the predominance of *Bacillus amyloliquefaciens*, *Lactobacillus delbrueckii*, *Torulopsis candida*, and *Trichosporon pullulans*.

The prevalence of higher bacterial load (10^7–10^8/g) during summer as compared to that observed during winter (10^6–10^7/g) was also noted (Table 3.6). *Leuconostoc mesenteroides* and *Lactobacillus fermentum* were found to be involved during both the seasons, while *Streptococcus faecalis* and *Pediococcus cerevisiae* were observed during summer only. Yeasts were isolated from 60% of the summer samples (0–10^3/g; two isolates) and 90% winter samples (0–10^6/g; four isolates). *Saccharomyces cerevisiae* and *Hansenula anomala* were the principal yeasts occurring during both the seasons, while *Debaryomyces hansenii* and *Trichosporon beigelii* were encountered during winter only. Both the bacteria and the yeasts, whenever present, tended to increase significantly with the progress in fermentation (Soni and Sandhu, 1989a, 1990a). The bacteria appearing initially were *Leuconostoc mesenteroides*, *Bacillus subtilis*, *Lactobacillus delbrueckii*, and *L. fermentum*. *Streptococcus faecalis* and *Pediococcus cerevisiae* developed subsequently, and persisted till the end of fermentation, along with *Leuconostoc mesenteroides*. *Debaryomyces hansenii*, *Saccharomyces cerevisiae*, *Trichosporon beigelii*, and *Hansenula anomala* were predominant among the yeasts appearing first, while *Trichosporon beigelii* developed, subsequently, but eventually only *Saccharomyces cerevisiae* was recorded (Aidoo et al., 2006). Successive changes in microflora and

Table 3.7   Successive Changes in Microflora during *Idli* Fermentation

INCUBATION PERIOD (H)	pH	VOLUME (ML)	BACTERIA		YEASTS	
			CFU/g	PREDOMINANT TYPES	CFU/G	PREDOMINANT TYPES
0	6.58	200	1.7×10^6	*Leuconostoc mesenteroides* *Bacillus subtilis* *Lactobacillus delbrueckii* *L. fermentum*	$0-6.6 \times 10^2$	*Debaryomyces hansenii* *Saccharomyces cerevisiae* *Trichosporon beigelii* *Hansenula anomala*
6	6.05	240	7.3×10^6	*Leuconostoc mesenteroides* *Bacillus subtilis*	$0-9.3 \times 10^2$	*Debaryomyces hansenii* *Saccharomyces cerevisiae* *Trichosporon pullulans* *Trichosporon pullulans*
12	5.60	300	3.6×10^7	*Leuconostoc mesenteroides*	$0-4.3 \times 10^3$	*Debaryomyces hasenii* *Saccharomyces cerevisiae* *Trichosporon pullulans*
18	5.10	340	8.9×10^7	*Leuconostoc mesenteroides* *Streptococcus faecalis* *Lactobacillus fermentum* *Pediococcus cerevisiae*	$0-8.8 \times 10^3$	*Saccharomyces cerevisiae* *Debaryomyces hansenii* *Trichosporon pullulans*
24	4.95	390	6.3×10^8	*Leuconostoc mesenteroides* *Streptococcus faecalis* *Pediococcus cerevisiae*	$0-7.1 \times 10^4$	*Saccharomyces cerevisiae*
30	4.92	410	4.7×10^9	*Streptococcus faecalis* *Pediococcus cerevisiae*	$0-7.1 \times 10$	*Saccharomyces cerevisiae*

Source: Adapted from Soni, S.K. and Sandhu, D.K. 1989a. *J. Cereal Sci.* 10: 227–238.

the prevalence of different bacteria and yeasts during *idli* fermentation are shown in Tables 3.7 and 3.8, respectively.

3.4.2.2 Dhokla *and* Khaman *Fermentation*   *Dhokla*, popular all over India especially Gujarat, is similar to *idli*, except that the *dhal* used is of Bengal gram (*Cicer arietinum*). The microorganisms involved in the fermentation are *Lactobacillus fermentum*, *L. mesenteroides*, and *Pichia silvicola* (up to 10^7 g^{-1}) (Joshi et al., 1989; Aidoo et al., 2006). The LAB contribute lactic acid and acetoin, imparting a pleasant sour flavor. The yeast produces folic acid and raises the volume of the batter, imparting sponginess to the product (Kanekar and Joshi, 1993; Aidoo et al., 2006).

Rajalakshmi and Vanaja (1967) reported that *Lactobacillus* spp. were the predominant organism in *khaman*. The bacterial species which were present in the raw ingredients in appreciable amounts could not be detected in the fermented batters of *dhokla* and *khaman*. On the other hand, species like *L. mesenteroides* and *L. fermenti* increased during fermentation.

3.4.2.3 Bhaturas/Kulcha   *Bhaturas* (*pathuras*) and *kulchas* of northern India and Pakistan are prepared from leavened doughs; they are, respectively, deep-fried in

Table 3.8   Prevalence of Different Bacteria and Yeasts in Fermented *Idli* Batters

		MICOORGANISMS				
		POSITIVE SAMPLES			POSITIVE SAMPLES	
SOURCE	BACTERIA	NO	(%)	YEASTS	NO	(%)
Market	*Leuconostoc mesenteroides* (S, W)	20	100	*Saccharomyces cerevisiae* (S, W)	13	65
	Lactobacillus fermentum (S, W)	14	70	*Debaryomyces hansenii* (W)	12	60
	Streptococcus faecalis (S)	13	65	*Hansenula anomala* (S, W)	7	35
	Pediococcus cerevisiae (S)	5	25	*Trichosporon beigelii* (W)	2	10
Laboratory	*Lactobacillus delbrueckii* (S)	15	100	*Saccharomyces cerevisiae* (S, W)	8	53
	Streptococcus faecalis (S, W)	15	100	*Torulopsis candida* (W)	7	46
	Bacillus amyloliquefaciens (W)	14	93	*Debaryomyces hansenii* (W)	6	40
	Leuconostoc mesenteroides (S, W)	14	93	*Trichosporon beigelii* (S, W)	5	33
	Lactobacillus fermentum (S, W)	11	73	*Hansenula anomala* (W)	4	26
	Pediococcus cerevisiae (S, W)	11	73	*Trichosporon pullulans* (S, W)	4	26

Source: Adapted from Soni, S.K. and Sandhu, D.K. 1989a. *J. Cereal Sci.* 10: 227–238.
S: present in summer samples, W: present in winter samples, S, W: present during both the seasons.

oil or prepared on a griddle. *Kulcha* dough contains yeasts and LAB. The yeasts belong mainly to the genera *Saccharomyces, Candida, Hansenula, Saccharomycopsis, Kluyveromyces, Rhodotorula, Pichia, Torulopsis, Trichosporon,* and *Debaryomyces* (Sandhu et al., 1986; Soni and Sandhu, 1999; Aidoo et al., 2006).

3.4.2.4 Microbiology of Dosa *Fermentation Leuconostoc mesenteroides, Streptococcus faecalis,* and *Lactobacillus fermentum* are the predominant microflora among the bacteria, while *Saccharomyces cerevisiae, Debarvomvces hansenii, Trichosporon, Candida glabrata, C. tropicalis, C. sake, C. kruzsei, Hansenula,* and *Trichosporon beigelii* are the common yeasts involved in *dosa* fermentation (Soni and Marwaha, 2003; Soni et al., 2013). Microbes present in *dosa* batter and the microbes that contributed from two substrates are as listed in Table 3.9. *Dosa* fermentation is accompanied by an increase in total acids, total volume, total solids, soluble solids, non-protein nitrogen, free amino acids, amylases, proteinases, and vitamins B1 and B2 (Soni et al., 1985, 1986; Sandhu et al., 1986). Total nitrogen and total proteins do not alter significantly, whereas reducing sugars and soluble proteins, after declining initially, tend to increase afterwards with the progress of fermentation.

The production of acid and gas in *dosa* is dependent upon bacteria, while the yeasts contribute enzymes and esters, which impart desirable flavors (Radhakrishnamurthy et al., 1961; Steinkraus et al., 1967; Soni et al., 1986) (Figure 3.5). The biochemical and physiological characterization of the predominant microorganisms showed that the yeasts, like *Saccharomyces, Debaryomyces, Trichosporon,* and *Hansenula,* help in the degradation of starch into maltose and glucose by producing extracellular amylolytic enzymes which are utilized by the developing microbial load. Some strains of

Table 3.9 Prevalence of Different Bacteria and Yeasts in Fermented *Dosa* Batter

MICROORGANISM	NO.	(%)
BACTERIA		
Leuconostoc mesenteroides	33	94
Streptococcus faecalis	30	85
Lactobacillus fermentum	25	71
Bacillus arnyloliquefaciens	20	57
Lactobacillus delbrueckii	13	37
Bacillus polymyxa	5	14
Pediococcus cerevisiae	5	14
Bacillus subtilis	5	14
Micrococcus varians	3	8
Enterobacter sp.	3	8
YEASTS		
Saccharomyces cerevisiae	18	51
Debaryomyces hansenii	15	43
Trichosporon beigeUi	9	26
Hansenula anomala	6	17
Torulopsis candida	5	14
Oosporidium margaritiferium	4	11
Candida robusta	3	9
Trichosporon pullulans	2	6
Kluyveromyces marxianus	2	6
Candida kefyr	2	6
Candida krusei	2	6
Rhodotorula mucilaginosa	1	3
Rhodotorula glutinis	1	3

Source: Adapted from Soni, S.K. et al. 1986. *Food Microbiol.* 3:45–53.

Saccharomy cescerevisiae also produce acid and gas from starch itself, and may thus, play a major role in *dosa* fermentation. Due to the ability of some strains to grow at high temperature (42°C), *Saccharomyces* and *Trichosporon*, along with bacteria, also play a role in fermentation during summer, when the ambient temperature often exceeds 40°C (Soni et al., 1986). The prevalence of different bacteria and yeasts in fermented *dosa* batter is shown in Table 3.10, while the contribution of micro-organisms by rice and black gram in *dosa* fermentation are described in Table 3.11.

3.4.2.5 Wadi/Warri *Wadi* or warri are prepared by fermentation of *dals*, generally from black gram. The surface of the cones or balls become covered with a mucilaginous coating that helps to retain the gas formed during their fermentation (Soni and Marwaha, 2003). The escape of gas creates the holes. The *wadi/warri* look hollow, with many air pockets and yeast spherules in the interior and a characteristic surface crust. Initially, the microflora of *wadi* is diverse and contains LAB, bacilli, flavobacteria, and yeasts (Aidoo et al., 2006). Several bacteria and yeasts constituting

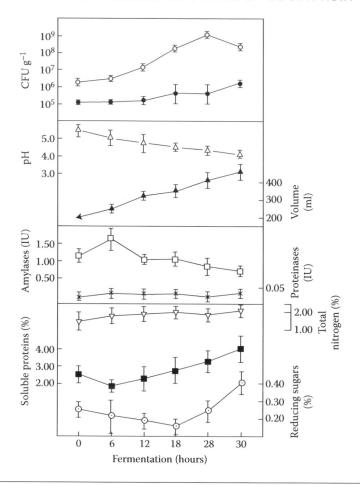

Figure 3.5 Various microbial, physico-chemical and biochemical changes accompanying dosa batter fermentation. (Adapted from Soni, S.K. et al. 1986. *Food Microbiol.* 3:45–53.)

the natural flora of black grams, spices added to the dough, and the surroundings are associated with *warri* fermentation. The microbial population dynamics, the predominant types, their prevalence, and succession suggest the involvement of bacteria alone or in combination with lesser proportions of yeasts during the *warri* fermentation (Sandhu and Soni, 1989; Soni and Sandhu, 1990a; Soni and Arora, 2000; Soni et al., 2013). Gradually, *L. mesenteroides*, *L. fermentum*, *S. cerevisiae*, and *Tr. cutaneum* become dominant. *Candida vartiovaarae* and *K. marxianus* are also often found. The development and prevalence of microflora are affected by the seasons, summer being more favorable for bacteria and winter for yeasts. The LAB are mainly responsible for the acidification of the dough, favorable conditions for the yeasts to grow and become active for leavening. Seasonal variations generally affect the development and prevalence of microorganisms during fermentations (Table 3.11); summer was found to favor a higher bacterial load (10^{10}–10^{12}/g) than winter (10^9–10^{10}/g). It was also found that *Bacillus subtilis* occurs mainly during summer, whereas *Leuconostoc mesenteroides*, *Streptococcus faecalis*, and *Lactobacillus fermentum* are the principal bacteria during both

Table 3.10 Contribution of Microorganism by Rice and Black Gram in *Dosa* Fermentation

| | BACTERIA | | | | | YEAST | | | | |
| | Range of total count at the start of fermentation A (g⁻¹ dry matter) | Range of total count at the end of fermentation (g⁻¹ dry matter) | No of isolates | | Predominant species | Range of total count at the start of fermentation A (g⁻¹ dry matter) | Range of total count at the end of fermentation (g⁻¹ dry matter) | No of isolates | | Predominant species |
Ingredient			Genera	Specie				Genera	Species	
Rice	1.0×10^{5}–	1.0×10^{8}–	5	5	*Leuconostoc mesenteroides*	$0{-}5.6 \times 10^{3}$	$0{-}4.1 \times 10^{4}$	5	6	*S. cerevisiae*
	1.0×10^{6}	1.0×10^{8}			*Streptococcus faecalis*					*Deb. hansenii*
					Lactobacillus delbrueckii					*Trichosporon beigelli*
					Micrococcus spp.					
					Bacillus spp.					
Black gram	1.0×10^{6}–	5.0×10^{8}–	7	7	*Leuconostoc mesenteroides*	$0{-}5.0 \times 10^{3}$	$0{-}1.6 \times 104$	6	6	*Deb. hansenii*
	8.0×10^{6}	1.0×10^{9}			*Enterobacter* spp.					*Torulopsis candida*
					Flavobacter spp.					*Klutueromyces marxianus*
					Bacillus spp.					*Candida krusei*
					Lactobacillus spp.					*C. kefyr*

Source: Adapted from Soni, S.K. et al. 1986. *Food Microbiol.* 3:45–53.

Table 3.11   Prevalence of Bacteria and Yeasts in Indian Fermented Foods in Different Seasons

		MICROORGANISMS					
		BACTERIA			YEASTS		
FOOD	SEASON	RANGE OF TOTAL (CFU/G DRY MATTER)	NO OF ISOLATES		RANGE OF TOTAL (CFU/G DRY MATTER)	NO OF ISOLATES	
			GENERA	SPECIES		GENERA	SPECIES
Warri	Summer	10^{10}–10^{12}	6	6	0–10^4	6	6
	Winter	10^9–10^{10}	5	5	0–10^7	8	8
Papadam	Summer	10^9–10^{11}	3	3	0–10^2	3	3
	Winter	10^8–10^{10}	3	3	0–10^3	6	6
Bhallae	Summer	10^7–10^9	6	6	0–10^4	3	3
	Winter	10^6–10^8	4	4	0–10^5	7	8
Vada	Summer	10^9–10^{10}	5	5	0–10^4	4	4
	Winter	10^7–10^9	4	4	0–10^6	7	8
Idli	Summer	10^7–10^8	5	6	0–10^3	2	3
	Winter	10^6–10^8	4	4	0–10^6	5	6
Dosa	Summer	10^7–10^9	6	8	0–10^4	4	5
	Winter	10^4–10^8	4	4	0–10^7	7	9

Source: Adapted from Soni, S.K. and Sandhu, D.K. 1990. *Indian J Microbiol* 30: 135–157; Soni, S.K. and Arora, J.K. 2000. *Food Processing: Biotechnological Applications.* Asiatech Publishers Pvt Ltd., New Delhi, India, pp. 143–190.

the seasons. Yeasts including *Candida vartiovaarai*, *Kluyeromyces marxianus*, and *C. krusei* are largely encountered during the winter, while *Saccharomyces cerevisiae*, *Pichia membranaefaciens*, *Trichosporon beigelii*, and *Hansenula anomala* are commonly involved during both the seasons (Soni et al., 2013). *Wadis* or *warrie* made by inoculating sterilized ingredients with a mixed culture of *C. krusei* (anamorph of *Issatchenkia orientalis*) and *L. mesenteroides* were similar to the market ones. In contrast, the uninoculated controls were hard and compact and, when broken, had a glistening surface (Batra and Millner, 1974; Aidoo et al., 2006). Studies on microbial dynamics and their successional changes accompanying *warri* fermentation carried out by Sandhu and Soni (1989) and Soni and Sandhu (1990a) indicated that both the bacteria and yeasts, whenever present, increased significantly with the progress in fermentation, followed by a decrease in pH and a rise in volume of the *warri* dough (Figure 3.6). The succession of various bacteria and yeasts in *warri* fermentation are listed in Table 3.12.

3.4.2.6 Seera   *Seera* is a nutritious, easily digestible, traditional fermented food made from wheat grains in the Bilaspur, Kangra, Hamirpur, Mandi, and Kullu districts of Himachal Pradesh, India. Samples taken during *seera* fermentation were analysed for various microbiological and biochemical parameters (Savitri et al., 2012). *Seera* is also called *Nishasta*. The microbial counts increased substantially during fermentation (Figure 3.7). The microflora isolated from *seera* mainly comprised of *Saccharomyces cerevisiae*, *Cryptococcus laurentii*, and *Torulospora delbrueckii* among the yeasts, and *Lactobacillus amylovorus*, *Cellulomonas* spp., *Staphylococcus sciuri*, *Weisella cibaria*, *Bacillus* spp., *Leuconostoc* spp., and *Enterobacter sakazakii* among bacteria (Table 3.13).

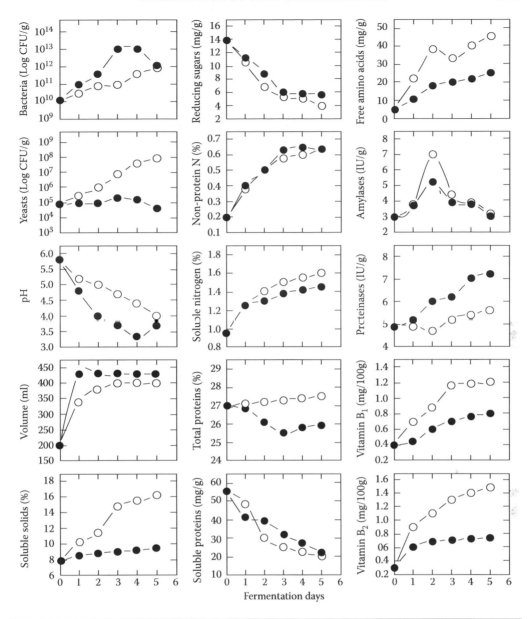

Figure 3.6 Effect of temperature on *warri* dough fermentations. (From Soni, S.K., Sandhu, D.K., and Soni, R. 2001. *J. Food Sci. Technol.* 47: 510–514.) ○ = Incubation at 28°C, ● = Incubation at 40°C.

Out of these, *Saccharomyces cerevisiae* and *Lactobacillus* spp. persisted in the final product. LAB have been reported as the most important group of microorganisms involved in spontaneous or natural fermentation of foods (Kalui et al., 2010). *Bacillus* spp., *Enterobacter sakazaki*, and *Staphylococcus sciuri* were also present in early stages of fermentation, probably originating from the air or the water used for steeping of the wheat. *Lactobacillus* spp. persisted in the final product. Increase in total bacterial count of *seera* was higher than the yeast count. Bacterial fermentation during soaking was important for the acidification. *Lactobacillus* and *Leuconostoc* spp. predominate the

Table 3.12   Successive Changes in Microflora during *Warri* Fermentation

INCUBATION PERIOD (DAYS)	pH	VOLUME (ML)	BACTERIA		YEASTS	
			COUNT/g	PREDOMINANT TYPES	COUNT/g	PREDOMINANT TYPES
0	5.65	200	1.3×10^{10}	*Leuconostoc mesenteroides* *Lactobacillus delbrueckii* *Lactobacillus fermentum* *Bacillus subtilis* *Flavobacter* sp.	$0–8.0 \times 10^4$	*Trichosporon beigelii* *Saccharomyces cerevisiae* *Candida krusei* *Pichia membranaefaciens* *Hansenula anomala*
1	4.70	420	2.1×10^{12}	*Leuconostoc mesenteroides* *Lactobacillus fermentum* *Streptococcus faecalis* *Bacillus subtilis*	$0–1.7 \times 10^6$	*Trichosporon beigelii* *Saccharomyces cerevisiae* *Candida krusei* *Pichia membranaefaciens*
2	3.90	420	3.0×10^{12}	*Leuconostoc mesenteroides* *Lactobacillus fermentum* *Streptococcus faecalis*	$0–2.1 \times 10^6$	*Trichosporon beigelii* *Saccharomyces cerevisiae* *Candida kruseil*
3	3.50	420	4.1×10^{12}	*Leuconostoc mesenteroides* *Lactobacillus fementum* *Streptococcus faecalis*	$0–9.1 \times 10^6$	*Trichosporon beigelii* *Saccharomyces cerevisiae*
4	3.25	420	4.7×10^{12}	*Leuconostoc mesenteroides* *Lactobacillus fermentum*	$0–6.9 \times 10^7$	*Saccharomyces cerevisiae* *Trichosporon beigelii*
5	3.20	420	6.5×10^{12}	*Leuconostoc mesenteroides* *Lactobacillus fermentum*	$0–9.6 \times 10^6$	*Saccharomyces cerevisiae* *Trichosporon beigelii*

Source: Adapted from Sandhu, D.K. and Soni, S.K. 1989. *J. Food Sci. Technol.* 26: 21–25; Soni, S.K. and Arora, J.K. 2000. *Food Processing: Biotechnological Applications.* Asiatech Publishers Pvt. Ltd., New Delhi, India, pp. 143–190.

microflora as the fermentation progresses. Similar profile of LAB, aerobic mesophiles, and enterobacteriaceae, which constituted the primary microflora of *pozol* (prepared by soaking of maize) have been reported by Wacher et al. (1993). The LAB present in fermented foods have been found to produce antimicrobial products that lead to safe and long storage of foods, as observed earlier also (Steinkraus, 2002; Parvez et al., 2006;

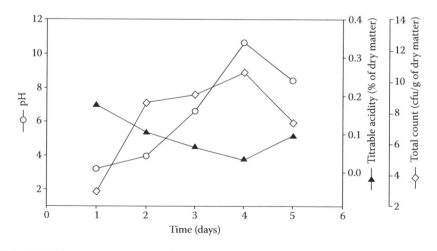

Figure 3.7   Changes in pH, titrable acidity, and total count during seera fermentation. (From Savitri et al., 2012. *Intl. J. Food Ferment.* 2(1), 49–56. With permission.)

Table 3.13 Microbial Population during *Seera* Fermentation

TIME (DAYS)	PREDOMINANT MICROORGANISMS
1	*Bacillus* sp., *Enterobactersakazakii*, *Cellulomonas* sp., *Saccharomyces cerevisiae*, *Staphylococcus sciuri*
2	*Leuconostoc* sp., *Lactobacillus amylovorus*, *S. cerevisiae*, *Cryptococcus laurentii*, *Torulspora delbrueckii*, *Staphylococcus sciuri*
3	*Leuconostoc* sp., *L. amylovorus*, *S. cerevisiae*, *C. laurentii*, *T. delbrueckii*, *Weisella cibaria*
4	*Leuconostoc* sp., *L. amylovorus*, *S. cerevisiae*
5	*S. cerevisiae*, *Leuconostoc* sp., *Lactobacillus* sp. (*Seera*) *Lactobacillus* sp., *S. cerevisiae*

Source: From Savitri et al., 2012. *Intl. J. Food Ferment.* 2(1), 49–56. With permission.

Corgan et al., 2007; Kalui et al., 2009). The presence of *Enterobacter sakazakiiin* in the case of *seera* fermentation is not a matter of concern, as its population starts declining with the fall in pH and, finally, it disappears (Savitri, 2007).

3.4.2.7 Papad   It is another fermented product and consumed widely in India. *Papad* was found to be alkaline (pH 8.7) (Roy et al., 2007). Every sample was found contaminated with total aerobic mesophilic bacteria (detection limit, 10 cfu g^{-1}), and 38% (40/105) of the samples contained more than 106 cfu g^{-1}. Aerobic mesophilic bacterial spores were found in 88% (92/105) of the samples (detection limit, 100 cfu g^{-1}), whereas their anaerobic counterparts were present in 39% (41/105) of the samples (detection limit, 10 cfu g^{-1}). Although all the samples, except one, were free from *Staphylococcus aureus* (detection limit, 100 cfu g^{-1}), 20% (21/105) of the samples were found contaminated with *Bacillus cereus* (detection limit, 100 cfu g^{-1}). Enterobacteriaceae were found in 46% (48/105) of the samples (detection limit, 10 cfu g^{-1}). Of the Enterobacteriaceae isolates, 92% were coliforms and 57% were fecal coliforms. *Escherichia coli* (detection limit, 10 cfu g^{-1}) was found in only one sample each of *wadi* and *idli*, at a load of 103–104 g^{-1}. *Candida krusei* and *S. cerevisiae* have been found involved in the preparation of *papad* (Shurpalekar, 1986).

3.4.2.8 Palm Wine   The palm wine fermentation is always an alcoholic–lactic–acetic acid fermentation, involving mainly yeasts and LAB. In the fermenting sap, *S. cerevisiae* is invariably present, but LAB such as *Lactobacillus plantarum*, *L. mesenteroides*, or other species of bacteria like *Zymomonas mobilis* and *Acetobacter* spp., vary. The other yeast types include *Schizosaccharomyces pombe*, *Saccharomyces chevalieri*, *Saccharomyces exiguus*, *Candida* spp., and *Saccharomycodes ludwigii* in samples of coconut palm wine (toddy). *Saccharomyces cerevisiae*, *Schizo pombe*, *Kodamaea ohmeri*, and *Hanseniaspora occidentalis* are characterized as maximum ethanol producers in toddy, as reviewed earlier (Joshi et al., 1999; Aidoo et al., 2006). The yeasts, especially *Saccharomyces* spp., are largely responsible for the characteristic aroma of palm wine (Uzochukwu et al., 1999; Aidoo et al., 2006).

3.4.2.9 Rice Wine   Molds belonging to the genera *Mucor* and *Rhizopus* are usually the main enzyme producers for the production of rice wines in India and Nepal (Shrestha et al., 2002). The main yeasts which ferment saccharified rice starch to alcohol are *Pichia burtonii*, *Sachharomycopsis fibuligera*, *S. cerevisiae*, *Candida glabrata*, and *Candida lactosa*, while *S. fibuligera* produces amylolytic enzymes as well (Tsuyoshi et al., 2005). In *murcha* from India, *S. capsularis* and *P. burtonii* also contribute to the degradation of starch (Tsuyoshi et al., 2005; Aidoo et al., 2006). Other yeast species, namely *Hansenula* spp., *Pichia* spp., and *Torulopsis* spp., have also been isolated from the rice wine. The role of *Sachharomyces fibuligera* is two-fold: it is an important amylolytic species (Dansakul et al., 2004) and it also produces alcohol, *albeit* at relatively low levels (Limtong et al., 2002; Aidoo et al., 2006).

The mixed population of molds, yeasts, and LAB in *murcha* starter (Tamang et al., 1988; Aidoo et al., 2006) soon becomes active, bringing about changes in the substrate and increasing the temperature of the fermenting mass by 4°C over the ambient (20–25°C) within 2 days of fermentation. Amylase activity reaches its peak on the second day of fermentation; mucoraceous fungi have an active role in saccharification and liquefaction of starch. The molds *Mucor circinelloides* and *Rhizopus chinensis*, and the yeast *Saccharomycopsis fibuligera*, which are dominant at the start, disappear within 12 h of fermentation. Mature *jnard* contains *P. anomala*, *S. cerevisiae*, and *C. glabrata* yeasts and LAB (10^5–10^6 cfu/g^{-1}), which include *Pediococcus pentosaceus* and *Lactobacillus bifermentans*. These three yeast species increase from 10^5 to 10^7 cfu g^{-1} within 2 days, and the population then, remains the same until the end of fermentation. The titratable acidity and alcohol content of the fermenting mass increase significantly during fermentation (Aidoo et al., 2006).

3.4.2.10 Kinema   *Kinema* is a fermented soya bean food of Nepal and the hilly regions of Northeastern States of India. Generally, the fermentation is dominated by *Bacillus* spp. that often cause alkalinity and a desirable stickiness in the product (Anonymous, *kinema*). It is used as a meat substitute in the eastern Himalayan regions (Nepal, Darjeeling hills, Sikkim, and South Butan) and has a final pH of 7.71–7.94 (Sarkar and Tamang, 1994, 1995; Sarkar et al., 1994, 2002; Tamang et al., 2012). Although *B. subtilis* is considered as the main fermenting microorganism (Sarkar and Tamang, 1994), *C. parapsilosis* and *G. candidum* have been isolated from different *kinema* productions (Sarkar et al., 1994; Sarkar and Tamang, 1995; Tamang, 2003). Levels of 4.1–8.7 × 10^4 cfu/g of *C. parapsilosis* and 0.8–4 × 10^4 cfu/g of *G. candidum* were detected in 60%–80% and 40%–50%, respectively, of market samples obtained from different locations (Darjeeling, Kalimpong, Gangtok, and Rongli) (Sarkar et al., 1994). The origin of yeast species has been traced to the mortar and pestle used for cracking the cooked soybeans before fermentation (Tamang, 2003). It has been concluded the yeast has a negative effect on the nutritional and organoleptic quality of *kinema*. For example, the presence of *C. parapsilosis* resulted in significantly lower levels of free fatty acids, namely C18:1 (oleic), C18:2 (linoleic), and C18:3 (linolenic) (Sarkar et al., 1996), as well as significantly lower levels of free amino acids (e.g., lysine) than with the fermentations performed with *B. subtilis* as a monoculture (Sarkar et al., 1997). *Kinema* production with *C. parapsilosis* and *G. candidum* was further associated with a rancid and objectionable smell (Sarkar and Tamang, 1994). Nout et al. (1998) undertook the study in a limited number of commercial (market) *kinema* samples to test for the presence of food-borne pathogens and their properties. *Bacillus cereus* was present in numbers exceeding 10^4 cfu/g in five of the tested 15 market samples. Enterobacteriaceae and coliform bactetia exceeded 10^5 cfu/g in 10 of the 15 samples. *Escherichia coli* exceeding 10^5 cfu/g was found in two samples. *Staphylococcus aureus* was not detected in any of the tested samples. Of 31 isolated typical and atypical strains of *B. cereus*, 18 representative strains were tested qualitatively for the ability to produce diarrhoeal type *Baccilus cereus* enterotoxin (BCET), using an Oxoid BCET-RPLA test kit. Overall, BCET was formed by 12 strains in BHIG (brain heart infusion broth + 1% glucose), by seven strains on sterilized cooked rice, and by five strains on sterilized cooked soya beans. Semi-quantitative tests on BCET revealed that levels exceeding 256 ng/g soya beans, produced by single pure culture inoculation with the isolated *B. cereus* strains, were reduced to 18 ng/g by frying *kinema* in oil, a common procedure when making *kinema* curry. It was also shown in a mixed pure culture experiment that a *kinema* strain, *B. subtilis* DK-Wl, was able to suppress the growth and BCET formation by a selected toxin producing strain (BC7-5) of *B. cereus*. It is further concluded that the traditional way of making *kinema* and its culinary use in curries is safe. However, for novel applications of *kinema*, safety precautions are advisable. One such approach in production of Kinema is the use of pulverized starter (Tamang, 1999).

3.4.2.11 Tungrymbai   This is produced by the Khasi tribe in Meghalaya, India, from dried soybeans employing fermentation in a basket lined with fresh leaves of *Phrynium pubinerve* Blume (Marantaceae). The use of *tungrymbai* differs from *natto* and *kinema* (described below) in that the fermented soybeans are ground in a mortar, before being fried together with ginger, garlic, and black beefsteak seeds (*Perilla frutescens* (L.) Britton) in mustard oil to produce the ready-to-eat *tungrymbai* (Thokchom and Joshi, 2012). The pH of *tungrymbai* has been reported to be slightly alkaline (6.95–7.05) (Thokchom and Joshi, 2012) to alkaline (7.6) (Tamang et al., 2012). Yeasts, *Bacillus subtilis* group spp., LAB, and enterobacteria have been isolated from *tungrymbai* (Thokchom and Joshi, 2012). Yeasts have been isolated from the final product (5.6×10^6 cfu/g) and also from the soaked soybeans (6.7×10^5 cfu/g) and from the mortar and pestle used for grinding the fermented soybeans (Sohliya et al., 2009). The detected yeast species include *S. cerevisiae*, *D. hansenii*, *Candida variabilis* (formerly *Pichia burtonii*) (Tamang et al., 2012), *Geotrichum candidum*, *Candida parapsilosis*, *Saccharomyces bayanus*, and *Saccharomycopsisfibu-ligera* (Sohliya et al., 2009). The role of the yeasts in *tungrymbai* has not been established, but the yeast species occurring in *tungrymbai* are known from other foods, where they are known to influence the quality of these products. As an example, *C. variabilis* and *S. fibuligera* are considered important amylolytic yeasts in different indigenous fermentations, for example, in Himalayan *murcha* (an amylolytic starter for alcoholic beverages) (Tsuyoshi et al., 2005; Takeuchi et al., 2006), and *S. bayanus* is associated with the ethanolic fermentation for production of different wines (Rojas et al., 2012).

3.4.2.12 Pickled Brine   Maneewatthana et al. (2000) demonstrated that yeast occurred in 105 pickled brine samples from retail stores in the markets and fermented vegetable factories. Many species of yeasts were identified, such as *Saccharomyces cerevisiae, Pichia etchell/carsonii, Candida krusei, C. tropicalis, C. lipolytica, C. rugosa, C. holmii, C. utilis, Rhodotorula glutinis, Trichosporon cutaaneum,* and *Zygosaccharomyces* spp. The source of yeast contamination might be the vegetables, humans, soil, or insects.

3.5 Diversity of Microflora in Indigenous Fermented Food of South Asia

The common fermenting bacteria associated with indigenous fermented foods are species of *Leuconostoc, Lactobacillus, Streptococcus, Pediococcus, Micrococcus,* and *Bacillus.* The fungi of genera *Aspergillus, Paecilomyces, Cladosporium, Fusarium, Penicillium,* and *Trichothecium* are the most frequently found in certain products (Soni and Sandhu, 1989a,b; Soni and Arora, 2000; Gupta and Abu-Ghannam, 2012). The common fermenting yeasts are species of *Saccharomyces,* which result in alcoholic fermentation (Steinkraus, 1998; Blandino et al., 2003). The type of bacterial flora developed in each fermented food depends on the water activity, pH, salt concentration, temperature, and the composition of the food matrix. Most fermented foods, are dependent on LAB to mediate the fermentation process (Conway, 1996; Blandino et al., 2003). In many of those processes, cereal grains, after cleaning, are soaked in water for a few days, during which a succession of naturally occurring microorganisms will result in a population dominated by LAB. In such fermentations, endogenous grain amylases generate fermentable sugars which serve as a source of energy for the LAB (Blandino et al., 2003). A variety of microorganisms, including yeast, bacteria, and fungi, have been isolated by several workers from various fermented foods (Blandino et al., 2003; Mugula et al., 2003; Muyanja et al., 2003) and has been described here.

3.5.1 Microorganisms Isolated and their Characterization

In different parts of South Asian countries, various types of microorganisms have been isolated and documented, depending upon the products. Wide varieties of traditional fermented foods have been reported throughout South Asia involving various substrates and microorganisms (Roy et al., 2004). In various countries of South Asia, concerted efforts have also been made to conduct microbiological studies, isolate typical microorganisms, study their characteristics and, where possible, conduct experimental production of indigenious fermented foods, as discussed in earlier section (Soni and Sandhu, 1989a,b; Joshi and Sandhu, 2000; Joshi et al., 2000;

Soni and Arora, 2000; Tamang et al., 2000; Tamang, 2001; Dewan, 2002; Kanwar et al., 2007), but lack of published information in some cases has been a big hurdle.

3.5.1.1 Microbiological Analysis of Inocula

3.5.1.1.1 Malera The microbiological analysis of *malera*, a traditional inoculum for *bhatooru*, revealed a consortium of microorganisms which mainly consisted of LAB—*Lb. plantarum* (MTCC 8296), *Leu. mesenteroides*, and *Saccharomyces cerevisiae* (MTCC 7840) as yeast. The fermented dough microflora was mainly dominated by yeast and *L. acidophilus*, *L. lactis*, *Lb. plantarum*, and *Bacillus* spp. Some species of *Bacillus* and other bacteria, such as *Kocuria rhizophila*, *Pseudomonas synxantha*, and *Microbacterium saperdae*, were also found during the initial stages of *bhatooru* fermentation, which gradually disappeared. Probiotic bacterial strains associated with *chilra* were *L. acidophilus*, *Lactococcus lactis*, and *Leuc. mesenteroides*, while the latter two were not observed in *khameer*. In *Babroo*, the predominant bacteria were *Lb. plantarum* and *L. Lactis* (Kanwar et al., 2007). In another study, bacteria were characterized from various fermented food products (*bhaturu*, *chilra*, *lugri*, fermented milk, *seera*, cheese, *janchang*) of tribal areas of Western Himalayas on the basis of 16S rRNA sequence analysis. Isolates were identified as *Enterococcus faecium*, *B. coagulans*, *Lb. plantarum*, and *Lb. fermentum*. These isolates were found to be intrinsically tolerant to upper gastrointestinal transit. Viability reduction was greater in simulated gastric juice of pH 2 than in pH 3 and 4. Most of the isolates showed a good degree of adhesion, as observed with n-hexadecane, and auto-aggregation capability (Sourabh et al., 2010). In another study, the hydrophobicity of various isolates from ethnic source of inocula was determined, as shown in Figure 3.8.

3.5.1.1.2 Humao In a study conducted by Chakrabarty et al. (2014), one dry mixed starter (*humao*), and fermented beverage (*judima*) were studied, concerning

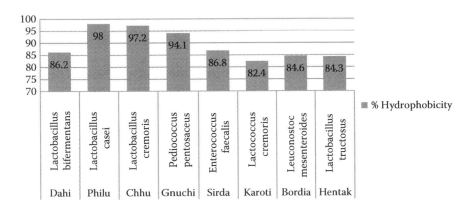

Figure 3.8 Percentage hydrophobicity of LAB strains isolated from fermented foods. (Adapted from Tamang, J.P. et al. 2000. *Food Biotechnol,* 14(1–2): 99–112; Dewan, S. 2002. *Microbiological evaluation of indigenous fermented milk products of the Sikkim Himalayas.* PhD thesis, Food Microbiology Laboratory, Sikkim Government College, North Bengal University, India.)

traditional methods of preparation, their mode of consumption, socioeconomy, microbiological profiles, and nutritional value. *Humao* is the traditionally prepared mixed amylolytic dough inocula used as a starter for the production of various indigenous alcoholic beverages and drinks. *Judima* or *zu* is a mildly alcoholic beverage, with a distinct sweet aroma, prepared from steamed glutinous rice. The microbiological analysis of the dry mixed starter revealed the presence of bacteria as *B. stearothermophilus*, *P. pentosaceous*, *B. circulans*, *B. laterosporus*, yeasts *D. hansenii* and *S. cerevisiae*, and the filamentous fungi *Mucor* and *Rhizopus*. Characteristics of the spore-forming rods and the yeast isolated from *Humao* and *Judima* are listed in Tables 3.14 through 3.18.

In *Judima* and *Juharo*, no mold was found (Table 3.18).

3.5.1.1.3 Marcha   The microflora of *marcha* (the inoculum) consists of filamentous molds, such as *Mucor circinelloides* and *Rhizopus chinensis*, yeasts such as *Saccharomycopsis fibuligera*, and *Pichia anomala*, and bacteria such as *Pediococcus pentosaceus* (Tamang and Sarkar, 1996; Tsuyoshi et al., 2005). Sharma and Gautam (2008) observed the antagonism of different isolates from *marcha* against *L. monocytogenes*, *L. mesenteroides*, *S. aureus*, *E. faecalis* and *Lactobacillus* spp.

3.6 Microorganisms in Indigenous Fermented Foods of the Himalayas

This region in India is known for several indigenous fermented foods being produced and consumed. Many of the products of this regions have been thoroughly documented and investigated in depth and some of these studies are being discussed separately in this section.

3.6.1 North-Eastern Himalayan Region

Indigenous knowledge of ethnic people living in the North-Eastern Himalayan regions of India about traditional fermented foods adapted to harsh conditions and environment is worth documentation (Tamang, 2005; Tamang et al., 2010). More than 250 different types of familiar and less-familiar ethnic fermented foods and alcoholic beverages are prepared and consumed by the different ethnic people of North East India, which include milk, vegetables, bamboo, soybean, meat, fish, cereals, and alcoholic beverages (see Chapter 2). Diverse microorganisms ranging from filamentous fungi to enzyme and alcohol producing yeasts, LAB, bacilli and micrococci are associated with fermentation and the production of ethnic foods and alcoholic drinks (Tamang et al., 2012). Filamentous molds, yeasts, and bacteria constitute the microbiota in indigenous fermented foods and beverages, which are present in or on the ingredients, utensils, or in the environment, and are selected through adaptation to the substrate (Tamang, 1998). Major functional microorganisms isolated from a variety of indigenous fermented foods and beverages of the Himalayas (Tamang and

Table 3.14 Characteristics of Spore-Forming Rods Isolated from *Humao* and *Judimaa* (Based on the Selected Taxonomical Keys of Slepecky and Hemphill (1992))

STRAIN CODE	CELL SIZE (MM)	VOGES-PROSKAUER REACTION	ANAEROBIC GROWTH	GROWTH AT 65°C	GROWTH AT 50°C	STARCH HYDROLYSIS	ANAEROBIC GROWTH	ACID AND GAS FROM GLUCOSE	GROWTH AT NACL 7%	pH IN VP BROTH	IDENTITY
HUMAO											
NCS:B1	l = 1.8 (1.4–2.2) b = 0.7 (0.6–0.9)	+	+		–			+		7.1	*B. polymyxa*
NCS:B2	l = 2.2 (1.9–2.3) b = 0.7 (0.6–0.8)	–		+						4.9	*B. stearothermophilus*
NCS:B5	l = 1.9 (1.4–2.4) b = 0.6 (0.5–0.9)	+	+		+				+	7.1	*B. licheniformis*
NCS:B6	l = 1.9 (1.5–2.1) b = 0.6 (0.4–0.7)	+	+		+				+	7.2	*B. licheniformis*
NCS:B7	l = 2.0 (1.6–2.4) b = 0.7 (0.5–1.1)	+	+		–			+		7.8	*B. polymyxa*
JUDIMA											
NCJ:B1	l = 2.2 (2.0–2.3) b = 0.7 (0.6–0.8)	–		–		+		–		7.6	*B. circulans*
NCJ:B4	l = 1.8 (1.1–2.5) b = 0.8 (0.7–0.9)	–		–		–	+			5.8	*B. laterosporus*
NCJ:B5	l = 1.7 (1.0–2.4) b = 0.6 (0.4–0.8)	+	–			–				5.5	*B. pumilus*
NCJ:B7	l = 2.0 (1.7–2.2) b = 0.8 (0.7–0.9)	–		–		–	+			5.4	*B. laterosporus*
NCJ:B8	l = 1.9 (1.4–2.4) b = 0.7 (0.5–0.9)	–		–		+		–		5.1	*B. firmus*
NCJ:B9	l = 1.9 (1.3–2.5) b = 0.8 (0.6–0.9)	–		–		+				5.1	*B. firmus*
NCJ:B10	l = 2.1 (1.9–2.2) b = 0.7 (0.6–0.8)	–		–		+		–		5.3	*B. firmus*
NCJ:B11	l = 2.2 (1.9–2.5) b = 0.7 (0.5–0.9)	+	–			–				5.2	*B. pumilus*

Source: Adapted from Chakrabarty, J., Sharma, G.D., and Tamang, J.P. 2009. *Assam University Journal of Science and Technology: Biological Science* 4(1): 66–72.

Note: l: length, b: breadth. All isolates were rods, Gram-positive, catalase-positive, motile and sporeformers. NC = North Cachar Judima.

[a] Only main characteristic tests of specific species of *Bacillus* were performed, other tests were ignored, hence, blank in each row.

Table 3.15 Identification of Yeasts Isolated from *Humao*

ISOLATE CODE	CELL SIZE (MM)	ASCOSPORE	GROWTH AT 37°C	SUGARS FERMENTED GLUCOSE	GALACTOSE	LACTOSE	MALTOSE	RAFFINOSE	SUCROSE	STARCH	TREHALOSE	SUGARS ASSIMILATED ARABINOSE	CELLOBIOSE	GALACTOSE	GLYCEROL	INOSITOL	LACTOSE	MALTOSE	MELIBIOSE	MANNITOL	RAFFINOSE	RHAMNOSE	SUCROSE	STARCH	TREHALOSE	XYLOSE	IDENTITY
NCS:Y1	l = 1.6 (1.2–2.3) b = 1.4 (1.2–2.2)	Spheroidal	–	–	–	–	–	–	–	–	–	–	+	+	+	–	–	+	–	w	+	–	+	+	+	+	*Debaryomyces hansenii*
NCS:Y4	l = 3.3 (1.6–4.1) b = 2.9 (1.6–3.3)	Globose	+	+	+	–	+	+	+	+	–	–	–	+	+	–	–	+	+	–	+	–	+	+	+	–	*Saccharomyces cerevisiae*
NCS:Y5	l = 4.8 (3.2–5.5) b = 2.9 (1.6–4.7)	Globose	+	+	+	–	+	+	+	+	–	–	–	+	+	–	–	w	w	–	+	–	+	w	+	–	*Saccharomyces cerevisiae*
NCS:Y13	l = 3.3 (1.6–5.6) b = 2.0 (1.6–3.4)	Spheroidal	–	–	–	–	–	–	–	–	–	–	+	+	+	–	–	+	w	+	+	+	+	+	+	+	*Debaryomyces hansenii*
NCS:Y14	l = 4.8 (3.2–5.6) b = 2.4 (1.6–4.8)	Globose	+	+	+	–	+	+	+	+	–	–	–	+	+	–	–	+	+	–	+	–	+	+	+	–	*Saccharomyces cerevisiae*
NCS:Y16	l = 3.5 (1.6–4.1) b = 2.7 (1.6–3.2)	Globose	+	+	+	–	+	+	+	+	–	–	–	+	+	–	–	+	w	–	+	–	+	+	+	–	*Saccharomyces cerevisiae*
NCS:Y17	l = 3.3 (1.6–4.0) b = 2.9 (1.5–3.2)	Globose	+	+	+	–	–	+	+	+	–	–	–	+	+	–	–	+	+	–	+	–	+	+	+	–	*Saccharomyces cerevisiae*
NCS:Y19	l = 3.8 (1.5–4.2) b = 2.6 (1.2–3.3)	Globose	+	+	+	–	–	+	+	+	–	–	–	+	+	–	–	+	w	–	+	–	+	+	+	–	*Saccharomyces cerevisiae*

Source: Adapted from Chakrabarty, J., Sharma, G.D., and Tamang, J.P. 2009. *Assam University Journal of Science and Technology: Biological Science* 4(1): 66–72.
All yeast isolates were oval to ellipsoidal in shape, produced pseudo-mycelium and did not reduce nitrate. l: length, b: breadth, w: weak, +: positive, –: negative, NCS stands for North Cachar starter.

Table 3.16 Identification of Yeasts Isolated from *Judima*

ISOLATE CODE	CELL SIZE (μM)	ASCOSPORE	GROWTH AT 37°C	SUGARS FERMENTED								SUGARS ASSIMILATED															IDENTITY
				GLUCOSE	GALACTOSE	LACTOSE	MALTOSE	RAFFINOSE	SUCROSE	STARCH	TREHALOSE	ARABINOSE	CELLOBIOSE	GALACTOSE	GLYCEROL	INOSITOL	LACTOSE	MALTOSE	MELIBIOSE	MANNITOL	RAFFINOSE	RHAMNOSE	SUCROSE	STARCH	TREHALOSE	XYLOSE	
NCJ:Y1	l = 3.5(1.7–5.6) b = 2.4 (1.5–3.8)	Spheroidal	–	–	–	–	–	–	–	–	–	–	+	+	+	–	w	+	+	+	+	w	+	+	+	w	*Debaryomyces hansenii*
NCJ:Y3	l = 3.8 (1.6–5.2) b = 2.7 (1.6–4.2)	Globose	+	+	+	–	+	+	+	+	–	–	–	+	+	–	–	+	+	–	w	–	+	w	+	–	*Saccharomyces cerevisiae*
NCJ:Y10	l = 3.3 (1.6–4.8) b = 2.9 (1.6–3.6)	Globose	+	+	+	–	+	+	+	+	–	w	–	+	+	–	–	+	+	–	+	–	+	+	+	–	*Saccharomyces cerevisiae*
NCJ:Y11	l = 4.9 (3.1–5.9) b = 3.2 (2.1–4.4)	Globose	+	+	+	–	–	+	+	+	–	w	–	+	+	–	–	+	+	–	+	–	+	+	+	–	*Saccharomyces cerevisiae*
NCJ:Y14	l = 3.0 (1.6–4.5) b = 2.8(1.6–3.1)	Globose	+	+	+	–	+	+	+	+	–	–	–	+	+	–	–	+	+	–	+	–	w	+	+	–	*Saccharomyces cerevisiae*
NCJ:Y15	l = 3.2 (1.6–4.8) b = 2.1 (1.2–3.2)	Globose	+	+	+	–	+	+	+	+	–	–	–	+	+	–	–	+	+	–	+	–	+	+	+	–	*Saccharomyces cerevisiae*
NCJ:Y16	l = 4.1 (1.5–5.4) b = 2.5(1.6–4.1)	Globose	+	+	+	–	+	+	+	+	–	w	–	+	+	–	–	+	+	–	+	–	+	w	+	–	*Saccharomyces cerevisiae*

Source: Adapted from Chakrabarty et al. (2009)

All yeast isolates were oval to ellipsoidal in shape, produced pseudo-mycelium and did not reduce nitrate. l: length, b: breadth, w: weak, +: positive, –: negative, NCJ = North Cachar Judima.

Table 3.17   Characteristics of Filamentous Molds Isolated from *Humao*

| STRAIN | SPORANGIUM | | SPORANGIOSPORE | | IDENTIFICATION |
	SHAPE	DIAMETER (μm)	SHAPE	DIAMETER (μm)	
HUMAO					*Mucor*
NCS:M1	Globose, borne circinately	28–78	Ellipsoidal to oval	4.1–5.6	
NCS:M3	Globose, borne circinately	30–80	Ellipsoidal to oval	4.2–5.5	
NCS:M9	Globose, borne circinately	32–64	Ellipsoidal to oval	4.4–5.6	
HUMAO					*Rhizopus*
NCS:M7	Globose	56–120	Round	3.0–6.1	
NCS:M10	Globose	70–156	Round	2.0–5.2	

Source: Adapted from Chakrabarty, J., Sharma, G.D., and Tamang, J.P. 2009. *Assam University Journal of Science and Technology: Biological Science* 4(1): 66–72.

Table 3.18   Profile of Microorganisms Isolated from *Humao, Judima,* and *Juharo* of the Dima Hasao District of Assam

| PRODUCT | MICROORGANISMS | | |
	BACTERIA	YEAST	MOLDS
Humao	*P. pentosaceous* *B. polymyxa, B. licheniformis* *B. subtilis*	*D. hansenii,* *S. cerevisiae*	*Mucor, Rhizopus*
Judima	*P. pentosaceous* *B. licheniformis* *B. pumilus, B. firmus*	*D. hansenii* *S. cerevisiae*	–
Juharo	*P. pentosaceus* *B. licheniformis* *B. firmus*	–	–

Source: Adapted from Chakrabarty, J., Sharma, G.D., and Tamang, J.P. 2009. *Assam University Journal of Science and Technology: Biological Science* 4(1): 66–72.

Sarkar, 1993, 1996; Tamang and Nikkuni, 1996; Tamang et al., 2000; Thapa, 2001, 2002; Dewan, 2002) Chakrabarty et al. (2009) are listed below:

Lactic acid bacteria: *Lactobacillus (L.) plantarum, L. brevis, L. fermentum, L. bifermentans, L. curvatus, L. lactis* ssp. *cremoris, L. casei* ssp. *pseudoplantarum, L. casei* ssp. *casei, L. coryniformis* ssp. *torquens, L. alimentarius, L. farciminis, L. salivarius, L. hilgardii, L. kefir, L. confuses, L. fructosus, L. amylophilus, Leuconostoc mesenteroides, Pediococcus pentosaceus, Lactococcus lactis* ssp. *lactis, Lactococcus plantarum, Enterococcus faecium, Enterococcus faecalis*

Endospore-forming rods: *Bacillus subtilis, Bacillus pumilus*; *Aerobic coccus*; *Micrococcus* spp.

Yeasts: *Saccharomycopsis fibuligera, Saccharomycopsis crataegensis, Saccharomycopsis capsularis, Pichia anomala, Pichia burtonii, Saccharomyces cerevisiae, Saccharomyces bayanus, Geotrichum candidum, Candida glabrata, Candida parapsilopsis, Candida bombicola, Candida chiropterorum, Candida castellii.* Filamentous molds: *Mucor*

Table 3.19 Microbial Diversity Associated with the Ethnic Fermented Soybean and Non-Soybean Legume Foods of North-East India

PRODUCT	RAW MATERIAL	MICROORGANISMS	REFERENCES
Kinema	Soybean	*Bacillus subtilis, Enterococcus faecium, Candida parapsilosis, Geotrichum candidum*	Sarkar et al. (1994)
Hawaijar	Soybean	*Bacillus subtilis, B. Licheniformis, B. Cereus, Staphylococcus aureus, S. Sciuri, Alkaligenes* sp. *Providencia rettger*	Jeyaram et al. (2008a)
Tungrymbai	Soybean	*B. subtilis, B. pumilus, B. licheniformis, E. faecium, E. hirae, E. raffinossus, E. durans, E. cecorum, Lb. brevis* andyeasts	Sohliya et al. (2009); Sarkar et al. (1994); Tamang et al. (2012)
Bekang	Soybean	*Bacillus subtilis, B. pumilus, B. licheniformis, B. sphaericus, B. brevis, B. coagulans, B. circulans, Enterococcus faecium, E. hirae, E. raffinossus, E. durans, E. cecorum,* yeasts	Tamang et al. (2009); Chettri and Tamang, 2008
Peruyyan	Soybean	LAB, *Bacillus subtilis,* other *Bacillus* sp.	Tamang et al. (2009)
Tungtoh	Soybean	*Bacillus* spp.	Tamang et al. (2012)
Bemerthu	Soybean	*Bacillus* spp.	Tamang et al. (2012)
Maseura	Black gram	*B. subtilis, B. mycoides, B. pumilus, B. laterosporus, Pediococcus acidilactici, P. pentosaceous, Enterococcus durans, Lb. fermentum, Lb. salivarius,* yeasts	Tamang et al. (2012); Chettri and Tamang (2008)

circinelloides, Mucor hiemalis, Rhizopus chinensis, Rhizopus oryzae, Rhizopus stolonifer var. *lyococcus.*

Detailed diversity of microorganisms product-wise is described in Tables 3.19 through 3.25.

Vegetables: *L. mesenteroides* has been found to be important in the initiation of fermentation of many vegetables, that is, cabbages, beets, turnips, cauliflower, green beans, sliced green tomatoes, cucumber, olives, and sugar beet silages (Rhee et al., 2011). Fermentation dynamics during the production of *gundruk* and *khalpi* (ethnic fermented vegetable products of the Himalayas) revealed *L. brevis, P. pentosaceus* during the initial steps and then *L. plantarum.* Similarly in *khalpi* fermentation, hetero-fermentative LAB, such as *Leu. fallax, L. brevis,* and *P. pentosaceus,* initiated the fermentation which was finally completed by *L. plantarum* (Tamang and Kailasapathy, 2010). Other fermented vegetable products of the Himalayas are *sinki* and *zing-sang.*

Fish and Meat: Fresh fish and meat are extremely perishable proteinaceous foods that spoil due to the metabolism of spoilage microbiota. Their storage life can be extended by acid-fermentation with added carbohydrates and salts. *Ngari, hentak,* and *tungtap* are traditional fermented fish products of North-East India. LAB were identified as *L. lactis* ssp. *cremoris, L. plantarum, E. faecium, L. fructosus, L. amylophilus, L. coryniformis* ssp. *torquens,* and *L. plantarum.* Endospore-forming rods were identified as *B. subtilis* and *B. pumilus,* aerobic coccal strains were identified as *Micrococcus.* Most strains of LAB had a high degree of hydrophobicity, indicating their "probiotic" characters (Thapa et al., 2004).

Suka ko maacha, gnuchi, sidra, and *sukuti* are traditional smoked and sun-dried fish products of the Eastern Himalayan regions of Nepal and India. LAB were identified

Table 3.20   Microbial Diversity Associated with the Ethnic Fermented Milk Products of North-East India

PRODUCT	RAW MATERIAL	MICROORGANISMS	REFERENCES
Dahi	Cow milk	*L. bifermentans, L. alimentarius, L. paracasei* ssp. *pseudoplantarum, L. lactis* ssp. *lactis, L. lactis* ssp. *cremoris,* yeasts	Tamang et al. (2012)
Misti dohi	Cow/buffalo milk	LAB, yeasts	Tamang et al. (2012)
Shyow	Yak milk	LAB, yeasts	Tamang et al. (2012)
Gheu	Cow milk	LAB, yeasts	Tamang et al. (2012)
Maa	Yak milk	LAB, yeasts	Tamang et al. (2012)
Mohi	Cow milk	LAB, yeasts	Tamang et al. (2012)
Lassi	Cow milk	LAB, yeasts	Tamang et al. (2012)
Chhurpi (soft)	Cow milk	*L. farciminis, L. paracasei* ssp. *paracasei, L. confuses, L. bifermentans, L. plantarum, L. curvatus, L. fermentum, L. paracasei* ssp. *pseudoplantarum, L. alimentarius, L. kefir, L. hilgardii, E. faecium* and *Leu. mesenteroides*	Tamangand Tamang (2009a); Tamang et al. (2000)
Chhurpi (hard)	Yak milk	*L. farciminis, L. casei* ssp. *casei, L. confuses* and *L. bifermentans*	Tamang et al. (2012); Dewan (2002)
Dudh chhurpi	Cow milk	LAB, yeasts	Tamang et al. (2012)
Phrung	Yak milk	Unknown	Tamang et al. (2012)
Chhu or *sheden*	Cow/Yak milk	*L. farciminis, L. brevis, L. alimentarius, L. salivarius, Lc. Lactis* ssp. *cremoris,* yeasts	Tamang et al. (2012); Dewan and Tamang (2006)
Chur yuupa	Yak milk	Unknown	Tamang et al. (2012)
Somar	Cow/Yak milk	*L. paracasei* ssp. *pseudoplantarum* and *L. lactis* ssp. *cremoris*	Tamang et al. (2012); Dewan and Tamang (2007)
Dachi	Cow/Yak milk	LAB	Tamang et al. (2012)
Philu	Cow/Yak milk	*L. paracasei* ssp. *paracasei, L. bifermentans* and *E. faecium*	Tamang et al. (2012); Dewan and Tamang (2007)
Pheuja or *suja*	Tea-yak butter	Unknown	Tamang et al. (2012)
Rasogolla	Cow/buffalo milk	LAB, yeasts	Tamang et al. (2012) Chakrabarty et al. (2014)

on the basis of phenotypic characters, including API system, as *L. lactis* ssp. *cremoris, L. lactis* ssp. *lactis, L. plantarum, Leu. mesenteroides, E. faecium, E. faecalis, P. pentosaceus,* and *W. confusa.* Some strains of LAB showed antagonistic properties against pathogenic strains (Thapa et al., 2006).

Drying, smoking, or fermentation of meat are critical steps in its traditional processing in the Himalayas (Tamang and Kailasapathy, 2010). Ethnic peoples of the Western Himalayan regions of India and Nepal prepare and consume meat products such as *chartayshya, arjia,* and *jamma. Chartayshya* is a chevon (goat) meat product consumed by the ethnic people of Uttarakhand state of India and West Nepal. *Jamma* or *geema* and *arija* are ethnic fermented sausages of the Western Himalayas prepared from chevon meat. 16S rRNA and phenylalanyl-tRNA synthase (pheS) genes sequencing confirmed the involvement of *E. durans, E. faecalis, E. faecium, E. hirae, Leu. citreum, Leu. mesenteroides, P. pentosaceus,* and *W. cibaria.* Enterobacteriaceae, *Listeria* sp., *Salmonella* sp., and *Shigella* sp. were however, not detected in samples, indicating their safe consumption (Oki et al., 2011).

Table 3.21 Microbial Diversity Associated with the Ethnic Fermented Vegetable and Bamboo Shoot Foods of North-East India

PRODUCT	RAW MATERIAL	MICROORGANISMS	REFERENCES
Soibum	Bamboo shoot	L. plantarum, L. brevis, L. coryniformis, L. delbrueckii, L lactis, Leu. fallax, Leu. mesentroides, E. durans, Streptococcus lactis, B. subtilis, B. licheniformis, B. coagulans, yeasts	Tamang et al. (2008); Jeyaram et al. (2010)
Soidon	Bamboo shoot tips	L. brevis, Leu. fallax, Leu. Lactis	Tamang et al. (2008)
Mesu	Bamboo shoot	L. plantarum, L. brevis, L. pentosaceus, L. pentosaceus L. brevis, L. plantarum	Tamang and Sarkar (1996)
Ekung	Bamboo shoot	L. plantarum, L. brevis, L. casei, Tetragenococcus halophilus	Tamang and Tamang (2009a)
Khalpi	Cucumber	L. plantarum, L. brevis, Leu. Fallax	Tamang et al. (2012); Tamang and Tamang (2009a)
Goyang	Green vegetable	L. plantarum, L. brevis, L. lactis, E. faecium, P. Pentosaceus, yeasts	Tamang et al. (2012); Tamang and Tamang (2007)
Inziang-sang	Mustard leaves	L. plantarum, L. brevis, P. acidilactici	Tamang et al. (2012); Tamang et al. (2005)
Inziang-dui	Mustard leaves	LAB	Tamang et al. (2012)
Mesu	Bamboo shoot	L. plantarum, L. brevis, L. curvatus, Leu. citreum, P. pentosaceus	Tamang et al. (2012); Tamang and Sarkar (1996); Tamang et al. (2008)
Soijim	Bamboo shoot	LAB	Tamang et al. (2012)
Hitch/hitak	Only tips of bamboo shoot	LAB	Tamang et al. (2012)
Eup	Bamboo shoot	Lb. plantarum, Lb. fermentum	Tamang et al. (2012); Tamangand Tamang (2009)
Hiring	Bamboo shoot	Lb. plantarum, Lactococcus lactis	Tamang et al. (2012); Tamangand Tamang (2009)
Nogom	Bamboo shoot	LAB	Tamang et al. (2012)
Tuaithur	Bamboo shoot	L. plantarum, L. brevis, P. pentosaceou, L. lactis, B. circulans, B. firmus, B. sphaericus, B. subtilis	Tamang et al. (2012) Chakrabarty et al. (2014)

Table 3.22 Microbial Diversity Associated with the Ethnic Fermented Cereal Foods of North-East India

PRODUCT	RAW MATERIAL	MICROORGANISMS	REFERENCES
Selroti	Rice-wheat flour-milk Pretzel-like, deep-fried; bread	Leu. mesenteroides, E. faecium, P. pentosaceus, L. curvatus, yeasts	Tamang et al. (2012); Yonzan and Tamang, (2010)
Jalebi	Wheat flour	Yeasts, LAB	Tamang et al. (2012)
Hakua	Rice	Unknown	Tamang et al. (2012)
Ipoh	Rice, wild herbs	Saccharomyces cerevisiae, Hanseniaspora sp., Kloeckera sp, Pischia sp., Candida sp.	Tanti et al. (2010)
Atingba	Rice	L. plantarum, P. pentosaceus, yeasts	Tamang et al. (2007); Jeyaram et al. (2009)

Table 3.23   Microbial Diversity Associated with the Ethnic Preserved Fish Products of North-East India

PRODUCT	RAW MATERIAL	MICROORGANISMS	REFERENCES
Ngari	Fish	*L. lactis* ssp. *cremoris, L. plantarum, E. faecium, Lb. fructosus, Lb. amylophilus, Lb. corynifomis* ssp. *torquens, Lb. plantarum; B. subtilis, B. pumilus Micrococcus,* yeasts	Thapa et al. (2004)
Hentak	Fish and petioles of aroid plants	*L. lactis* ssp. *cremoris, L. plantarum, E. faecium, Lb. fructosus, Lb. amylophilus, Lb. corynifomis* ssp. *torquens, Lb. plantarum; B. subtilis, B. pumilus Micrococcus,* yeasts	Thapa et al. (2004)
Tungtap	Fish	*L. coriniformis, L. lactis, L. fructosus, B. cereus, B. subtilis,* yeasts	Thapa et al. (2004)
Gnuchi	River fish	*Enterococcus faecium, Pediococcus pentosaceus, Bacillus subtilis, Micrococcus* sp., *L. plantarum, L. lactis* ssp. *cremoris, L. lactis* ssp. *lactis, Leu. mesenteroides, E. faecalis, P. pentosaceus,* yeasts	Thapa et al. (2006)
Suka ko maacha	River fish	*L. lactis* ssp. *cremoris, L. lactis* ssp. *lactis, L. plantarum, Leu. mesenteroides, E. faecium, E. faecalis, P. pentosaceus,* yeasts	Thapa et al. (2006)
Sidra	Fish	*L. lactis* ssp. *cremoris, L. Lactis* ssp. *lactis, L. plantarum, Leu. mesenteroides, E. faecium, E. faecalis, P. pentosaceus, Weissella confuse,* yeasts	Thapa et al. (2006)
Sukuti	Fish	*L. lactis* ssp. *cremoris, L. Lactis* ssp. *lactis, L. plantarum, Leu. mesenteroides, E. faecium, E. faecalis, P. pentosaceus,* yeasts	Thapa et al. (2006)
Karati	Fish	*L. lactis* ssp. *cremoris, Leu. mesenteroides, Lb. plantarum, B. subtilis, B. pumilus,* yeasts	Thapa et al. (2007)
Bordia	Fish	*L. lactis* ssp. *cremoris, Leu. mesenteroides, L. plantarum, B. subtilis, B. pumilus,* yeasts	Thapa et al. (2007)
Lashim	Fish	*L. lactis* ssp. *cremoris, Leu. mesenteroides, Lb. plantarum, B. subtilis, B. pumilus,* yeasts	Thapa et al. (2007)

Table 3.24   Microbial Diversity Associated with the Ethnic Preserved Meat Products of North-East India

PRODUCT	RAW MATERIAL	MICROORGANISMS	REFERENCES
Kargyong: lang, yak, faak	Beef, yak, pork, respectively	*L. sake, L. divergens, L. carnis, L. sanfransisco, L. curvatus, Leu. mesenterioides, E. faecium, B. subtilis, B. mycoides, B. thuringiensis, Staphylococcus aureus, Micrococcus,* yeasts	Rai et al. (2010)
Satchu: lang, yak	Beef, yak, respectively	*P. pentosaceous, L. casei, L. carnis, E. faecium, B. subtilis, B. mycoides, B. lentus, S. aureus, Micrococcus,* Yeasts	Rai et al. (2010)
Suka ko masu	Buffalo meat	*L. carnis, E. faecium, L. plantarum, B. subtilis, B. mycoides, B. thuringiensis, S. aureus, Micrococcus,* yeasts	Rai et al. (2010)
Lang chilu	Beef fat	LAB	Tamang et al. (2012)
Luk chilu	Sheep fat	LAB	Tamang et al. (2012)
Yak kheuri	Yak	LAB	Tamang et al. (2012)
Lang kheuri	Beef	LAB	Tamang et al. (2012)
Honohein grain	Pig/boar meat	LAB, bacilli, micrococci	Tamang et al. (2012)

Table 3.25 Morphological and Biochemical Characterization of Bacteria Isolated from Least Explored Sources

SR. NO.	NAME OF ISOLATE	FOOD SOURCE	SHAPE	GRAM'S STAINING	CATALASE TEST	TENTATIVE IDENTIFICATION
1.	GK-I	Gundruck	Rods	+ve	−ve	*Lactobacillus*
2.	GK-II	Gundruck	Rods	+ve	−ve	*Lactobacillus*
3.	GK-V	Gundruck	Rods	+ve	+ve	*Bacillus*
4.	MA-I	Marcha	Rods	+ve	−ve	*Lactobacillus*
5.	MA-II	Marcha	Rods	+ve	+ve	*Bacillus*
6.	MA-V	Marcha	Rods	+ve	+ve	*Bacillus*
7.	SW-I	Sinki	Rods	+ve	−ve	*Lactobacillus*
8.	SW-II	Sinki	Rods	+ve	−ve	*Lactobacillus*
9.	SW-IV	Sinki	Circular	+ve	−ve	*Lactococcus*
10.	SW-VI	Sinki	Rods	+ve	−ve	*Lactobacillus*
11.	SW-VII	Sinki	Circular	+ve	−ve	*Lactococcus*

The dominant endospore-forming and rod-shaped bacteria studied phenotypically and phylogenetically in *Kinema* were identified as *B. subtilis*. Another non-bacilli bacterium isolated from *kinema* is *E. faecium*, along with yeasts (Tamang et al., 2002). Ethnic people living in the sub-Himalayan regions of North East India, Nepal and Bhutan consume a variety of domesticated and wild tender bamboo shoots and their fermented products for centuries (Sharma, 1989; Tamang, 2001, 2007). The predominant functional LAB strains associated with the fermented bamboo shoot products, as shown by genotypic characterization based on RAPD-PCR, rep PCR, species-specific PCR techniques, 16S rRNA gene sequencing, and DNA–DNA hybridization are *L. brevis*, *L. plantarum*, *L. curvatus*, *P. pentosaceus*, *Leu. mesenteroides* ssp. *mesenteroides*, *Leu. fallax*, *Leu. lactis*, *Leu. Citreum*, and *E. durans* (Tamang, 2007; Tamang et al., 2008).

3.6.1.1 North Cachar Hills (NC) (Dima Hasao) of Assam In the North East States of India, including Assam, a large number of indigenous fermented foods are prepared and consumed, especially by the tribal people (Chakrabarty et al. 2013). These include *Honoheingrain, Humao,* and *Judima*. Their microbiological analyses have revealed that LAB play an important and complex role in these tradition fermentations. The different microorganisms reported were: LAB, namely *Pediococcus pentosaceous, Lactobacillus plantarum, Enterococcus faecium, Leuconostoc mesenteroides, Lactobacillus breivis, Bacillus subtilis, Bacillus licheniformis, B. firmus,* and *Bacillus cereus,* and the yeasts *Debaryomyce shanseni* and *Saccharmyces cerevisiae*. In another study, Chakrabarty et al. (2014) selected two popular ethnic fermented bamboo shoot products (wet-*Tuaithur* and dry-*Tuairoi*), an ethnic fermented pork/boar meat product (*Honoheingrain*), a non-food dry mixed amylolytic starter (*Humao*), and an ethnic fermented beverage (*Judima*) of NC (North Cachor) Hills of Assam for study. The microorganisms isolated and identified from *Tuaithur, Tuairoi, Honoheingrain, Humao,* and *Judima* were: bacteria-*Lactobacillus brevis,*

Lb. plantarum, Enterococcus faecium, Leuconostoc mesenteroides, Pediococcus pentosaceous, Lactococcus lactis, Bacillus subtilis, B. cereus, B. circulans, B. firmus, B. pumilus, B. licheniformis, B. stearothermophilus, B. sphaericus, B. laterosporus, B. polymyxa, Staphylococcus aureus, and *Micrococcus* spp., the yeasts *Debaryomyces hansenii* and *Saccharomyces cerevisiae,* and *Mucor* and *Rhizopus* molds. Characteristics of spore-forming bacteria and the yeast isolated from *Humao* and *Judima* are listed in Tables 3.15 through 3.18.

3.6.2 Himachal Pradesh

Thakur et al. (2004) characterized some traditional fermented foods and beverages of Himachal Pradesh (India) and found some of the predominant yeast species, namely *Saccharomyces cerevisiae, Candida* sp., *Zygosaccharomyces bisporus,* and *Kluveromyces thermotolerance.* The predominant microflora (yeasts) was characterized on the basis of microscopic, physiological, and biochemical characteristics (Table 3.26). Savitri (2007) characterised microbiologically and biochemically some of the fermented foods of Himachal Pradesh. Basappa (2002) studied the microbial diversity of *Chhang,* a common beverage of sub-Himalayan region, along with its traditional inoculum *"Phab."* The starter culture (*Saccharomyces cerevisiae, Endomycopsis fibuligera,* and *Lactobacillus*) and *phab* were equally effective in the fermentation of malted *ragi.* The fermented product was nutritionally superior and organoleptically comparable to the traditional one. The yeasts *Saccharomyces* spp., *Candida,* and *Endomycopsis fibuligera,* a predominance of LAB, and two molds, namely *Mucor* and *Rhizopus,* were identified. The traditional foods and beverages of Himachal Pradesh (India) have also been studied and documented (Savitri and Bhalla, 2007).

Kanwar et al. (2007) documented some of the traditional fermented foods of the Lahaul and Spiti areas of Himachal Pradesh (India) and explored them microbiologically. They documented *Chilra, Jhan Chhang, Babru, Bhaturu,* and *Seera,* which are prepared using a traditional inocula, that is, *Khameera/Malera* or *Phab,* as a starter culture. Microbiological examination of these foods and their source of inoculum revealed the dominance of yeasts, mainly of genera *Saccharomyces, Debaromyces,* and *Schizosaccharomyces* (Table 3.26), while the bacteria were mainly from the genera *Lactobacillus, Lactococcus,* and *Leuconostoc.*

Table 3.26 Microorganism Isolated from Fermented Beverage of Himachal

FERMENTED BEVERAGE	MICROORGANISMS ISOLATED
Sur	*Saccharomyces cerevisiae, Zygosaccharomyces bisporus*
Lugri	*S. cerevisiae, Leuconostoc*
Angoori	*S. cerevisiae*
Daru	*S. cerevisiae, Candida famata, Candida valida, Kluyveromyces*
Chhang	*Candida cocoi, S. cerevisiae*

Source: Adapted from Thakur, N., Savitri, and Bhalla, T.C. 2004. *Indian J. Tradit. Knowl.* 3(3): 325; Kanwar, S.S. et al. 2007. *Indian J. Tradit. Knowl.* 6(1): 42–45.

In one of the studies, the microbiological exploration of various fermented food products (*babru, bhaturu,* and c*hilra*), alcoholic beverages (c*hhang, lugari, aara, chiang, apple wine,* and *chulli*) and traditional inocula (*khameer, phab,* and *dhaeli*) of tribal areas of Himachal Pradesh, representing the regions of Lahaul and Spiti, Sangla, Bharmour, Pangi, Chauntra, and Kinnaur, showed the predominance of yeast (43) microflora (Pathania et al., 2010). As these isolates were of different geographical origin and from different sources, there is a strong possibility of the existence of strain-level differences between them, and the results showed that there is a great diversity existing in yeast strains isolated from various traditional fermented foods in different regions.

Using molecular characterization, out of 43 yeast isolates from alcoholic beverages, fermented foods, and traditional inocula, 23 isolates were identified as *Saccharomyces cerevisiae;* 4 isolates from *chulli, khameer, bhaturu,* and *phab* were identified as *Saccharomyces fermentati,* 1 isolate from *dhaeli* was identified as *Endomyces fibuliger,* 6 isolates from beverages and fermented foods were identified as *Debaromyces hansenii,* 2 isolates from fermented foods were identified as *Schizosaccharomyces pombe,* 5 isolates from apple wine were identified as *Issatchenkia orientalis,* and 2 isolates from fermented foods were identified as *Brettanomyces bruxellenis* and *Candida tropicalis* (Table 3.27).

3.7 Diversity in the Microbial Ecology of typical Fermented Foods

The microbial ecology of most of the fermented foods, when examined critically, shows that it is more diverse and complex than generally thought, and there are still many gaps in microbiological knowledge and understanding (Fleet, 1999). In production of indigenous fermented foods, the diversity of microbial ecology might be playing a great role, being microbiologically, biochemically and sensorially complex products. But these is no documentation for these products. These aspects are illustrated by citing the examples of two typical products: cheese and wine. Wine is a typical product, and the behavior of other alcoholic beverages, except distilled alcoholic products, should be similar. The microflora affects the fermentability, alcohol level, and other products processed during fermentation which affect the sensory quality of indigenous alcoholic beverages (Fleet, 1999) as discussed in the subsequent section. Similar is the case with the cheese another fermented product.

3.7.1 Cheese

Cheese represents an excellent example as it includes many aspects of interest, such as fermentation, spoilage, safety, biocontrol, probiotics, etc. Its production involves several steps as shown in Figure 3.9.

Cheese production is principally the fermentation of the milk by LAB, and it also involves both microbiological and biochemical changes that occur during maturation. The fermentation of the milk to cheese is conducted by LAB by

Table 3.27   Yeasts Associated with Some Traditional Fermented Foods and Beverages of Himachal Pradesh

SOURCE/PRODUCT	PLACE OF COLLECTION	YEAST CLASSIFICATION
Chhang	Lahaul and Spiti	*Saccharomyces cerevisiae*
Lugari	Lahaul and Spiti	*Saccharomyces cerevisiae*
Dhaeli	Lahaul and Spiti	*Saccharomyces cerevisiae*
Aara	Lahaul and Spiti	*Saccharomyces cerevisiae*
Chiang	Lahaul and Spiti	*Saccharomyces cerevisiae*
Chilra	Lahaul and Spiti	*Saccharomyces cerevisiae*
Bhaturu	Lahaul and Spiti	*Saccharomyces cerevisiae*
Babru	Lahaul and Spiti	*Saccharomyces cerevisiae*
Khameer	Lahaul and Spiti	*Saccharomyces cerevisiae*
Faasur	Sangla	*Saccharomyces cerevisiae*
Chulli	Sangla	*Saccharomyces cerevisiae*
Apple wine	Sangla	*Saccharomyces cerevisiae*
Beverage	Bharmour	*Saccharomyces cerevisiae*
Beverage	Pangi	*Saccharomyces cerevisiae*
Beverage	Chauntra	*Saccharomyces cerevisiae*
Beverage	Chauntra	*Saccharomyces cerevisiae*
Wine	Sangla	*Saccharomyces cerevisiae*
Wine	Sangla	*Saccharomyces cerevisiae*
Wine	Sangla	*Saccharomyces cerevisiae*
Fermented food	Kinnaur	*Saccharomyces cerevisiae*
Fermented food	Karnal	*Saccharomyces cerevisiae*
Phab	Lahaul and Spiti	*Saccharomyces cerevisiae*
Bhaturu	Lahaul and Spiti	*Saccharomyces cerevisiae*
Khameer	Lahaul and Spiti	*Saccharomyces cerevisiae*
Chulli	Sangla	*Saccharomyces cerevisiae*
Dhaeli	Lahaul and Spiti	*Endomyces fibuliger*
Chilra	Lahaul and Spiti	*Debaromyces hansenii*
Bhaturu	Lahaul and Spiti	*Debaromyces hansenii*
Babru	Lahaul and Spiti	*Debaromyces hansenii*
Wine	Sangla	*Debaromyces hansenii*
Wine	Sangla	*Debaromyces hansenii*
Fermented food	Kinnaur	*Debaromyces hansenii*
Chilra	Lahaul and Spiti	*Schizosaccharomyces pombe*
Fermented food	Kinnaur	*Schizosaccharomyces pombe*
Faasur	Sangla	*Issatchenkia orientalis*
Chulli	Sangla	*Issatchenkia orientalis*
Apple wine	Sangla	*Issatchenkia orientalis*
Apple wine	Sangla	*Issatchenkia orientalis*
Apple wine	Sangla	*Issatchenkia orientalis*
Fermented food	Kinnaur	*Brettanomyces bruxellensis*
Fermented food	Kinnaur	*Candida tropicalis*

Source: Adapted from Kanwar, S.S. et al. 2007. *Indian J. Tradit. Knowl.* 6(1): 42–45.; Pathania, N. et al. 2010. *World J. Microbiol. Biotechnol.* 26: 1539–1547.

Cheese processing

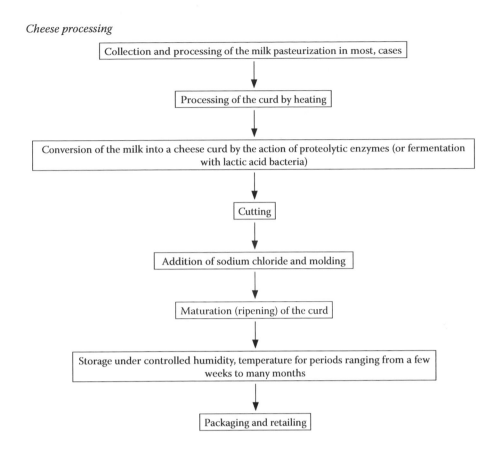

Figure 3.9 The basic steps involved in cheese manufacture. (Adapted from Fleet, G.H. 1999. *Int J Food Microbiol.* 50(1–2): 101–117; Chapman, H.R. and Sharpe, M.E. 1990. *Dairy Microbiology: The Microbiology of Milk Products.* Vol. 2, Elsevier Applied Science, London, pp. 203–289; Olson, N.F. 1995. *Biotechnology.* Vol. 9, 2nd edn, Enzymes, Biomass, Food and Feed, VCH, Weinheim, pp. 353–384; Banks, J.M., 1998. *The Technology of Dairy Products.* 2nd edn, Blackie Academic and Professional, London, pp. 81–122; Stanley, G. 1998a. *The Taxonomy of Dairy Products.* 2nd edn, Blackie Academic and Professional, London, pp. 50–80; Stanley, G. 1998b. *Microbiology of Fermented Foods.* Vol. 1, 2nd edn, Blackie Academic and Professional, London, pp. 263–307.)

inoculation with commercially-produced starter cultures of *Lactococcus lactis* and, in some cases, *Leuconostoc mesenteroides, Streptococcus thermophilus, Lactobacillus helveticus,* and *Lactobacillus delbrueckii* spp. *bulgaricus* (Law, 1982; Choisy et al., 1987; Stanley, 1998a,b). These species are added to milk at initial populations of 10^6 cfu/ mL or more and, the milk is pasteurized in most instances. Since the time of action is relatively short (1–4 h), the ecology of this fermentation is not distinct (Fleet, 1999). Futher, the curd harbors very high populations (10^9–10^{10} cfu/g) of entrapped, essentially immobilized LAB in a non-proliferating phase of growth. The occurrence of bacteriophages at this stage destroys the LAB and disrupts the fermentation (Johnson and Steele, 1997; Fleet, 1999). The commencement of essential lactic acid fermentation rapidly produces sufficient lactic acid, bacteriocins, and other metabolites to prevent the growth of spoilage and pathogenic bacteria (Marshall and Tamine, 1997; Fleet, 1999).

Many cheeses have high populations (10^6–10^9 cfu/g) of microorganisms that are not the LAB added as the starter to ferment the milk, which are referred to as secondary or adventitious microflora. These microorganisms have been found generally to contribute positively to the maturation process (Law, 1982; Chapman and Sharpe, 1990; Stanley, 1998a,b). However, in some cases, microorganisms are deliberately added as a part of the maturation process, such as the use of the filamentous molds *Penicillium camamberti* and *P. roqueforti* and other molds, namely, *Geotrichum candidum*, in the maturation of cheeses (Choisy et al., 1987; Fleet, 1999). However, even in these cases, the maturation microflora comprises a complex mixture of wild bacteria, yeasts, and bacteriophages, apart from the added organisms. Such complexity develops whether the cheese is produced from pasteurized or non-pasteurized milk, or whether or not starter cultures are used to ferment the milk (Gripon, 1987; Reps, 1987; Kaminarides et al., 1992; Macedo et al., 1993; Cuesta et al., 1996, 1997; Valdes-Stauber et al., 1997; Giraffa et al., 1998). The sources of these microorganisms originating as natural contaminants of the process include the milk, added proteolytic enzymes, brine (NaCl) solutions, the surrounding air, and contact with the equipment. When the conditions within the curd become favorable, they initiate growth (Fleet, 1999). The bacteria associated with maturation have also been documented earlier (Choisy et al., 1987; de Boer and Kuik, 1987; Nooitgedagt and Hartog, 1988; Bhowmik and Marth, 1990; Chapman and Sharpe, 1990; Vivier et al., 1994; Giraffa et al., 1997; Valdes-Stauber et al., 1997; Altekruse et al., 1998; ICMSF, 1998a).

Occasionally, spoilage-causing species of clostridia (*Clostridium tyrobutyricum*), coliforms, and Bacillus, and pathogenic species of *Salmonella*, *L. monocytogenes*, and *E. coli* can develop during the maturation process. This happens when there is failure, but natural "biocontrol" of these adverse species by bacteriocins produced by other microflora (e.g., *Enterococcus* spp., Giraffa et al., 1995; *B. linens*, Eppert et al., 1997; Fleet, 1999) is an important but underestimated function of the maturation process.

Yeasts have emerged as significant organisms in the maturation process, although their precise role is still not clear. Predominant species include *D. hansenii*, *Yarrowia lipolytica*, and *Kluyveromyces marxianus*, which are often present in cheeses at populations of 10^6–10^9 cfu/g (Fleet, 1990; Jakobsen and Narvhus, 1996; Roostita and Fleet, 1996). Isolates of *D. hansenii* from cheese exhibited killer activity, the action of which is enhanced by the presence of NaCl (Llorente et al., 1997). It has been suggested that the yeast *Debaryomyces hansenii* can inhibit the growth of spoilage-causing clostridia (Fleet, 1990).

Depending upon intrinsic factors, such as moisture content, salt (NaCl) concentration, and pH of the curd, as well as availability of oxygen, the growth and survival and biochemical activities of the maturation flora take place. These properties, however, vary throughout the curd and change with time. Mostly the microorganisms are distributed throughout the curd (Marcellino and Benson, 1992; Parker et al., 1998), but substantially higher populations are located on the outer surface, due to the availability of oxygen. Thus, oxidative microorganisms (brevibacteria, micrococci, *D. hansenii*, *Y. lipolytica*, molds) are more prevalent on the curd surface, while fermentative species (LAB, *K. marxianus*) are more predominant within the curd. A number of cheeses are produced indigenously, and the microflora of their associated biochemical characteristics need in-depth investigation.

3.7.2 Wine/Low Alcoholic Beverage

Louis Pasteur, about 150 years ago, showed that wine was the product of an alcoholic fermentation of grape juice by yeasts, which seems a very simple process (Amerine et al., 1980). However, it has been found that it extends far beyond a simple alcoholic fermentation and involves complex interactive contributions from yeasts, filamentous fungi, LAB, acetic acid bacteria, other bacterial groups, and even bacteriophages (Fleet and Heard, 1993; Joshi, 1997; Joshi et al., 2011b; Joshi and Suigh, 2012). The various unit operations involved in the preparation of white and red wine are shown in Figure 3.10. The production of other types of wines involves additional specialized operations, but basically remains the same. Nevertheless, microbial growth and activity can be significant at all stages of wine production (Boulton et al., 1994; Fleet, 1997, 1998).

The microflora of wine is significant from several angles (Joshi, 1997). To understand the microbiology of wine making, or for that matter of any alcoholic beverage, it is essential to understand the process of quality wine production. In wine making, different microorganisms—molds, yeasts, and bacteria—are involved, as summarized in Table 3.28. Besides damaging fruits like grapes (Martini, 1996), molds can also effect the flavor of the wine (Doneche, 1993) by their growth on the cooperage. Therefore, the control of molds in wine making is the most important aspect. Yeasts are important microorganisms associated not only with the production of alchololic beverages including wine but also in its spoilage, too. While the yeast strains of *Saccharomyces* predominantly carry out the alcoholic fermentation of fruit juices, yeasts other than *Saccharomyces* (wild yeasts) are responsible for the spoilage of wine (Khalid, 2011). Yeasts such as *Schizosaccharomyces* found in indigenous fermented foods where alcohol is the main ingredient can degrade the malic acid and reduce the acidity of wines such as plum wine (Vyas and Joshi, 1982). The growth of wild yeast during fermentation, especially of apiculate yeast, is considered deterimental to the wine quality. Besides, the yeast might cause cloudiness and microbiological instability (Joshi, 1997; Joshi et al. 2011c). Acetic acid bacteria and LAB are the important groups of bacteria influencing the quality of wine, either through spoilage or by improving the quality of wine by malo-lactic acid fermentation (Fleet and Heard, 1993; Joshi, 1997). The bacteria in wine making are more associated with spoilage than with production.

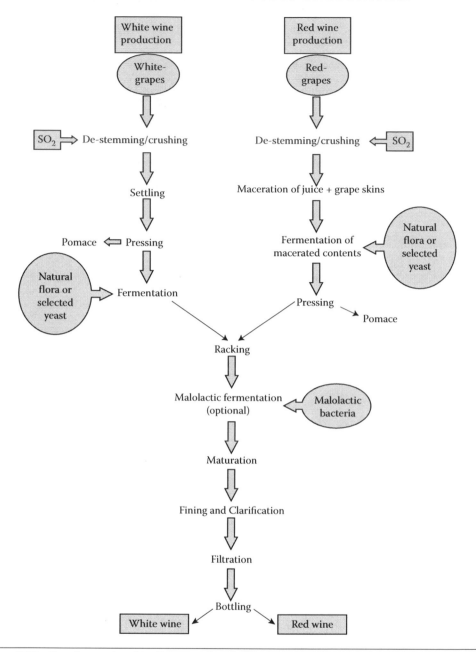

Figure 3.10 Flow diagram describing the preparation of white and red wine from grapes. (Adapted from Nigam, P.S. 2011. *Handbook of Enology: Principles, Practices and Recent Innovations.* Vol. II. Asia Tech. Publ., New Delhi. pp. 383–405.)

3.8 Fermentation Processes Associated with Indigenous Fermented Foods

3.8.1 Types of Food Fermentation Processes

On the basis of the fermentation pattern, traditional fermentation processes can be broadly categorized into solid state (without free water) and submerged (with free water) fermentation (Ray and Sivakumar, 2009). Solid state fermentation (SSF) is

Table 3.28   The Diversity and Significance of Microorganisms Associated with Wine Making

MICROBIAL GROUP	SOURCE	SIGNIFICANCE
FUNGI		
Botrytis cineria	Fruits, winery, plant equipments	Spoilage of fruits.
Fruit		Botrytis useful in botrytized wine
Aspergillus spp., *Penicillium* spp.		Spoilage corky taints.
YEASTS		
Saccharomyces cerevisiae	Fruits, winery, equipments	Alcoholic fermentation, inoculum
Schizosaccharomyces pombe		Deacidification.
Candida spp. *Torulopsis* spp. *Brettanomyces* spp.		Spoilage, autolysis
Hansenula spp. *Kloeckera* spp. *Pichia* spp.		
Killer yeasts		
BACTERIA		
Lactic acid bacteria	Fruits, winery equipments	Inoculums malolactic fermentation
Culture		
Leuconostoc oenos		
Pediococcus pentosaces		
P. parvulus, Lactobacillus plantarum		Spoilate; autolysis
L. fermentum, L. trichodes		
Acetic acid bacteria	Fruits, winery, equipments, barrels	Vinegary, spoilage struck
Acetobacter		fermentation
Gluconobacter		
Clostridium spp., *Bacillus*		Spoilage
ACTINOMYCETES		
Streptomyces spp.		Earthy, corky taints

Source: Adapted from Joshi, V.K. 1997. *Fruit Wines*. 2nd edn. Directorate of extension education, Dr. Y.S. Parmar University of Horticulture and Forestry, Nauni-Solan, India.

the process where microbial growth and product formation occurs on the surface of solid materials, in the absence of "free" water, where the moisture is absorbed by the solid matrix. It has a number of advantages over submerged fermentation—lower cost, improved product characteristics, higher product yield, and reduced energy requirement. Examples of natural solid-state fermented foods include *idli, dosa* and *mushrooms*. In contrast to SSF, submerged fermentation (SmF) is the process where the growth and anaerobic/partially anaerobic decomposition of the carbohydrates or other substractes by microorganisms occur with plenty of free water. Traditional fermented foods like and yoghurt, and beverages like fermented alcoholic beverages and *pito* are the products of SmF (Ray and Sivakumar, 2009).

3.8.2 Types of Fermentation Based on the Use of Microorganisms

On the basis of the use of microorganism(s), there are two types of food fermentation, namely, spontaneous fermentations and directed or controlled fermentations (Tamang, 2012).

3.8.2.1 Spontaneous Fermentations   Use of microorganisms in preparing foods from locally available plant and animal materials has been a traditional practice since pre-historic times. Spontaneous or natural fermentations are carried out by the microorganisms occurring on the raw material and in the environment of the production site (Oyewole, 1997; Achi, 2005). The development of spontaneous food fermentation was primarily governed by climatic conditions, the availability of typical raw materials, the sociocultural ethos, and ethnic preferences. The growth and activity of microorganisms play an essential role in biochemical changes in the substrates during fermentation (Tamang, 1998). A large number of fermented foods and beverages are made by this process, such as *curd*, alcoholic beverages, *idli*, *dosa*, etc. The quality of spontaneously fermented products is dependent on the microbial load and spectrum of the raw material, but it is always unpredictable because of the diverse microbiota initially present (De Vuyst et al., 2009).

Spontaneous or natural fermentation results from the competitive activities of a variety of contaminating microorganisms. Those best adapted to the conditions during the fermentation process will eventually dominate. Initiation of a spontaneous process takes a relatively long time, with a high risk of failure, low yield, and poor product quality (Rodriguez et al., 2009). The failure of fermentation processes can cause spoilage and/or the survival of pathogens, thus creating unexpected health risks and poor food product quality (Sanni et al., 1998; Holzapfel, 2002).

Many traditional fermented foods are still prepared today by an artisan mode of production based on experience gained through many years of trial and error which has been handed down from generation to generation within local communities (Oyewole, 1997). However, it is often empirically applied without recourse to a comprehensive understanding of the underlying principles of the fermentation process (Blandino et al., 2003). It involves either soaking of the raw materials, submerged in water contained in a fermenting vat, for example, clay pots, for a length of time, or an initial size reduction of the raw material by grating or milling in the wet form, before fermentation (Odunfa, 1985b; Oyewole, 1997; Odunfa and Oyewole, 1998). It is an inexpensive and low-cost technique, which can be applied in simple environments, but is often inappropriate for the development of any large-scale industrial process. The industrial design of fermentation processes requires knowledge of the physiology, metabolism, and the genetic properties of the incorporated starter inocula (Navarrete-Bolaños, 2012), so as to satisfy the requirements for their growth and efficient metabolite production (Giraffa, 2003).

The spontaneous traditional fermentation is a complex microbial process operated by both cultivable and non-cultivable microbial species, growing in succession or in combination throughout the fermentation process, exhibiting different metabolic patterns that can be either beneficial or detrimental to the product quality. In such conditions, the substrates and coexisting strains often interact through trophic or nutritional relations *via* multiple mechanisms (Navarrete-Bolaños, 2012). But optimization of spontaneous fermentation processes has been made possible through rebuilding systems that were based on a previous batch, during which a small part of a previously successful spontaneous fermentation is used as an inoculum for the next fermentation.

After a few refreshments, repeated rebuilding or "backslopping" leads to a stable microbiota, giving rise to a selective enrichment of the best adapted strains (Salovaara, 2004; Ravyts et al., 2012). Based on spontaneous methods, the back-slopping technique is used in the preparation of several traditional fermented products, and entails inoculation of raw material with residue from the previous batch (Holzapfel, 1997; Soni et al., 2013). It is known to accelerate the initial fermentation phase and control desirable changes in the process, and is applied in many traditional African fermented foods in addition to other types of "inoculation" methods (Nout, 1995; Holzapfel, 1997).

3.8.2.2 Inoculated and Inoculants of Fermentation Processes, Controlled Preparation With the advent of microbiology as a science, the development of pure cultures allowed the application of starter cultures for food fermentation. A starter culture is a microbiological culture that actually performs fermentation or assists the beginning of the fermentation process in preparation of various foods and fermented beverages (Anonymous, Rice vinegar). Microorganisms used for starter cultures include bacteria, yeasts, and molds. These could be single strains or combinations of multiple strains (Holzapfel, 2002) and serve to accelerate the fermentation process (Halm and Olsen, 1996; Hounhouigan et al., 1999; Mugula et al., 2003), leading to improved and more predictable fermentation products, improved safety, and the reduction of hygienic risks (Kimaryo et al., 2000), and contribute to desirable sensory quality attributes (Soni et al., 2013). Many products are fermented with LAB and, in some, yeasts and molds are utilized as starter cultures and are known to make substantial contributions to the sensory quality, nutrition, digestibility, and safety of the end products. Starter cultures have high numbers of live microorganisms, which may be inoculated into a food raw material to produce desirable changes (Holzapfel, 1997).

Industries specialized in the preparation of starter cultures often supply such cultures to the food fermentation industries. Commercial starter culture preparations may come in the form of live cell suspensions, frozen stocks, or lyophilized preparations, which need to be propagated before inoculation. Commercial starter cultures are now available for a wide variety of fermented milk products and meats, but to a much lesser extent for some fermented vegetable foods. There is also a growing interest in the development of starter cultures consisting of LAB strains with functional properties such as vitamin production or probiotic activity, resulting in new functional fermented food products (Heller, 2001; Leroy and De Vuyst, 2004). In the Western World, commercial starter cultures are available in freeze-dried form (Leroy and De Vuyst, 2004).

3.8.3 Benefits of Using Starter Cultures

The primary aim of using starter cultures is to improve the fermentation process. The starter cultures help to (i) drive the fermentation process, (ii) rapidly reduce the pH of the product, (iii) provide better aromatic compound accumulation during product ripening, (iv) provide beneficial health effects, (v) improve the commercial and hygienic quality of the product, and (vi) promote overall process standardization. The use of starter cultures has solved many of the problems of spontaneous fermentations, including avoiding stuck fermentations and batch-to-batch variations, resulting in more homogeneous fermented food products (Hansen, 2002; Cogan et al., 2007). It also allows a shorter fermentation time and an increase in industrial productivity. The microorganisms used mostly originate from the foods to which they are applied, and they are selected based on viability, competitiveness, adaptability to the substrate, and desired properties (Holzapfel, 1997, 2002; Soni et al., 2013). Fermented products produced using starters are usually of consistent quality (Soni et al., 2013).

3.8.4 Controlled Fermentations

Modern fermentation is characterized by controlled or directed fermentation, where the substrate to be fermented is inoculated with defined pure culture(s) in order to obtain desirable changes (Holzapfel, 1997, 2002; Nout, 2005; Ravyts et al., 2012). It has enabled the improvement of the production of fermented products such as beer, vinegar, and baker's yeast (Hansen, 2002). The addition of a large number of microbial cells to the unfermented material accelerates and guides the traditional fermentation more than the fewer cells that occur in spontaneously fermented materials (Leroy et al., 2012). It has ensured a high degree of control over the fermentation process, and consistency of the end product (Caplice and Fitzgerald, 1999). The LAB are the most important organisms in traditional food fermentations, with a long and safe history of application and consumption in the production of fermented foods and beverages (Ray, 1992; Wood and Holzapfel, 1995). In addition, starter cultures are being developed to further optimize the process and to yield additional nutritional, safety, and quality benefits, besides having functional properties (Leroy and De Vuyst, 2004).

3.8.5 Mixed Culture Fermentations

Research on starter cultures has normally been aimed at the performance of a single strain and its interactions with the food matrix. However, traditional food fermentations are typically carried out by mixed cultures, where the inoculum always consists of two or more organisms (Sieuwerts et al., 2008), which could be all of one microbial group—all bacteria—or might consist of a mixture of organisms, such as fungi and bacteria, or fungi and yeasts, or other combinations in which the components are quite unrelated (Hesseltine, 1983). The mixed culture system has been recognized to be effective for certain fermentations. The use of mixed cultures is a common practice in the production of commercially available dairy-based probiotic formulations. Mechanisms for growth and substrate production of single culture fermentations have been well established, but the potential effects of the use of combined cultures are not well understood. The mixture of microorganisms does not seem to greatly affect cell populations, but it does affect the production of flavor attributes, suggesting that the functional and sensory properties of fermented food could also be considerably modified through changes in the substrate or inocula composition. This could have a considerable effect on consumer acceptance of these products and on the future developments of novel probiotic beverages (Sieuwerts et al., 2008).

Population dynamics plays a crucial role in the performance of mixed-culture fermentations (Sieuwerts et al., 2008). The existence of symbiotic relationship among various bacteria has been clearly demonstrated (Moon and Reinbold, 1976; Kostinek et al., 2005; Lee, 2005). There is potential interest in mixed culture in the biological processing of traditional fermentations due to the sophisticated compositional and structural expectations of food products.

3.9 Biochemical Transformational Aspects of Indigenous Fermented Foods

Indigenous fermented foods are the consequence of fermentation transforming the substrates, depending upon the type of fermentation, microorganisms, and fermentation and processing conditions. Various biochemical transformations taking place during different fermentations are summarized here. Aspects connected with maturation in terms of biochemistry are also illustrated, along with specific changes taking place during fermentation with respect to composition and nutrition. No attempts, however, are made to discuss the detailed basics of biochemical aspects here, and for this the reader is referred to the literature cited (Goyal, 1999; Rana and Rana, 2011).

The yeast, predominantly *Saccharomyces cerevisiae*, is the microorganism of primary importance in producing various products, including alcoholic beverages and bakery products. Similarly, LAB are responsible for lactic acid fermentation, while acetic acid bacteria cause the other fermentation responsible for the production of vinegar and acetic acid. The biochemical changes would focus illustrating these fermentation processes.

3.9.1 Metabolism of Carbon

3.9.1.1 Glycolysis in Alcoholic and Lactic Acid Fermentation In this pathway, operative in many microorganisms including yeast and bacteria, monosaccharides like glucose or fructose are converted into pyruvate. The sequence of reactions and the enzymes involved are shown in Figure 3.11.

3.9.1.2 Metabolisms of Pyruvate After glycolysis, pyruvate is a central metabolite from which a number of products are formed by various enzymes found in different microorganisms (Goyal, 1999). For example, in *Saccharomyces cerevisiae* pyruvate is converted into ethanol (a main component of alcoholic beverages) and carbon dioxide. The ethanolic pathway in the yeast gives rise to the production of ethanolic beverages, as discussed in other parts of the text. Pyruvate is converted into ethanol *via* acetaldehyde from pyruvate (Figure 3.12).

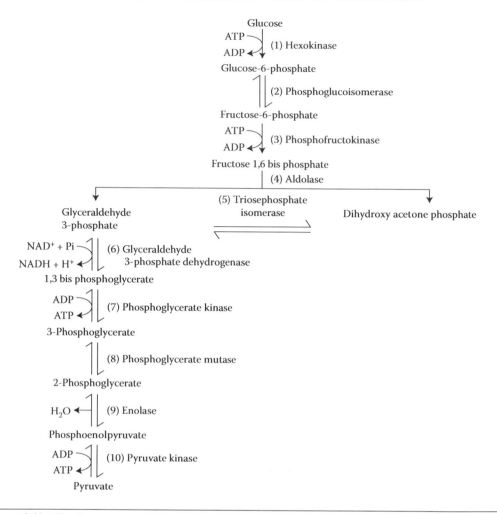

Figure 3.11   Flow diagram representation of the glycolytic pathway. (From Rana, N. and Rana, V. 2011. *Handbook of Enology.* Vol. 2, Asiatech Publishers, Inc., New Delhi, pp. 618–666. With permission.)

Fermentation products of the different metabolic pathways of pyruvate are also depicted in Figure 3.13. In a typical hetero-lactic acid fermentation, products are formed from pyruvate during lactic acid fermentation and, hence, play an important role in flavor development, such as diacetyl acetoin in fermented milk products. Separate products can be formed from pyruvate by *Acetobacter* or *Gluconobacter,* as shown Figure 3.13. As can be seen, the formation of entirely different products from pyruvate by *E. coli* takes place.

3.9.2 Malate Synthesis

Some of the yeasts have the capacity to synthesize malate during fermentation (Goyal, 1999), so will increase the acidity of the fermented products; consequently, the pH is decreased which gives additional advantages of preservation of the fermented foods.

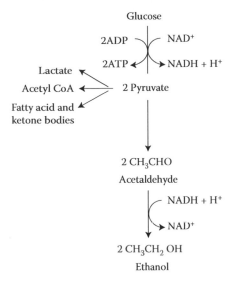

Figure 3.12 NAD^+ molecule regeneration and conversion to pyruvate to ethanol during alcoholic fermentation. (From Goyal, R.K. 1999. *Biotechnology: Food Fermentation.* Vol. 1, 87–137. With permission.)

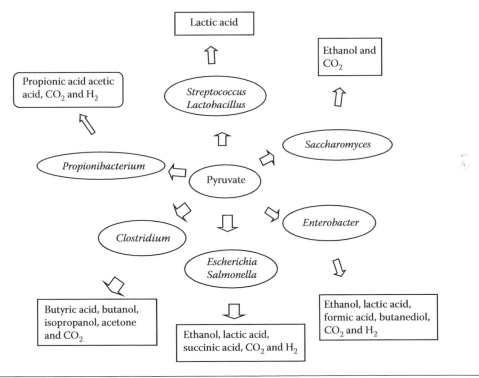

Figure 3.13 Various products of fermentation by various organisms by utilizing pyruvate as starting material. (From Bouthyette, P.Y. 1980. *The Chloroplast of Pisum Sativum: A Three Part Study—The Synthesis of CF?? The Effect of Ca++ on Protein Synthesis—Oligomycin Sensitivity of the Thylakoid ATPase.* Cornell University, pp. 330. With permission.)

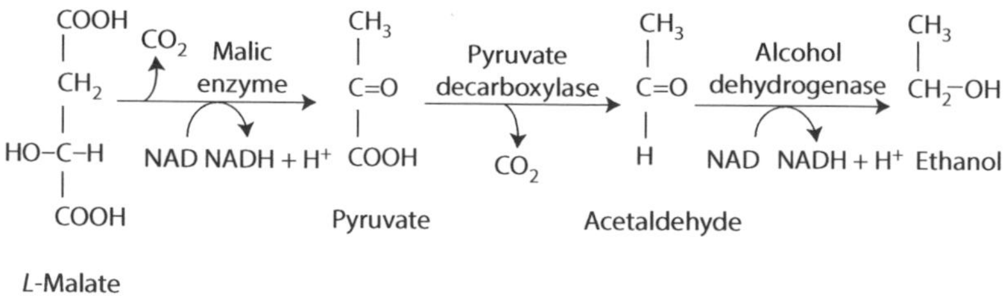

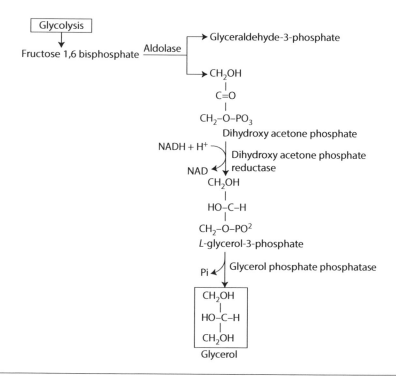

Figure 3.14 Pathway showing metabolism of malate to ethanol in *S. cerevisiae* and *Schizosaccharomyces pombe*.

3.9.3 De-Acidification of Juices

Some fruits like plums, apples, and grapes have malic acid as the dominant acid, and in the alcoholic fermented beverages the high acidity makes such beverages unpalatable (Joshi et al., 1991). Fermentation of juices from such fruits can be conducted by making use of another yeast, *Schizosaccharomyces pombe*, which has the capacity to degrade the malic acid into ethanol, thus reducing the acidity of the final product (Figure 3.14).

3.9.4 Production of Glycerol during Fermentation by Yeast

In alcoholic fermentation by the yeast like *Saccharomyces cerevisiae*, a very important by-product, glycerol, is produced (Figure 3.15). Glycerol has a significant role in

Figure 3.15 Pathway showing glycerol production. (From Goyal, R.K. 1999. *Biotechnology: Food Fermentation*. Vol. 1, pp. 87–137. With permission.)

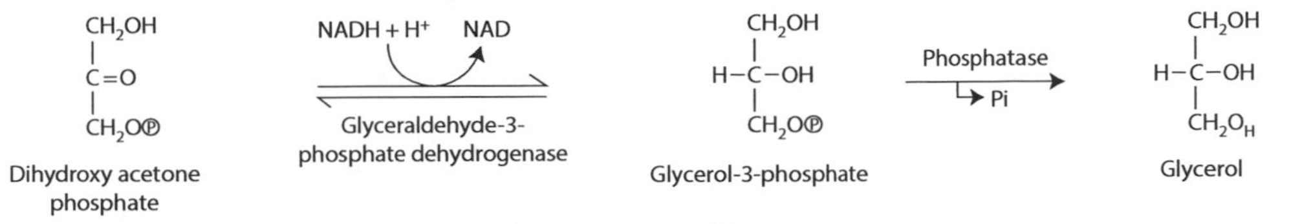

Figure 3.16 Neubergs third form of fermentation.

improving the taste of alcoholic fermented products, though produced only in trace amounts (Goyal, 1999).

This fermentation which results in the production of glycerol is also called Neubergs fermentation (Goyal, 1999). The Neubergs third fermentation is accomplished by addition of sulfite to the fermentation medium, when glyceraldehydes 3-phosphate is converted into glycerol rather than dihydroxy acetone, as is the normal process in the fermentation to produce pyruvate (Figure 3.16).

3.9.5 Production of Organic Acids

During alcoholic fermentation, the yeast is able to synthesize organic acid by degradation of the glutamate (Rana and Rana, 2011) as shown in Figure 3.17.

3.9.6 Formation of Ethyl Carbamate

Ethyl carbamate is a carcinogenic compound found in alcoholic beverages. The formation of ethyl carbonate is shown in Figure 3.18. The formation takes place from the usage of urea which, by reaction with ethyl alcohol, form carbamate.

3.9.7 Metabolic Pathways of LAB

Simple carbohydrates (hexoses, pentoses) are the main carbon sources used by LAB, but disaccharides such as lactose, maltose, or sucrose can also be metabolized.

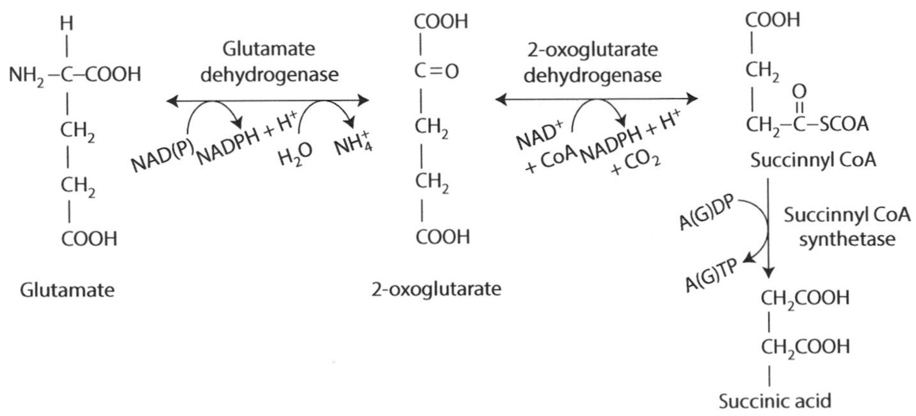

Figure 3.17 Oxidative pathway of synthesis for succinic acid by yeast.

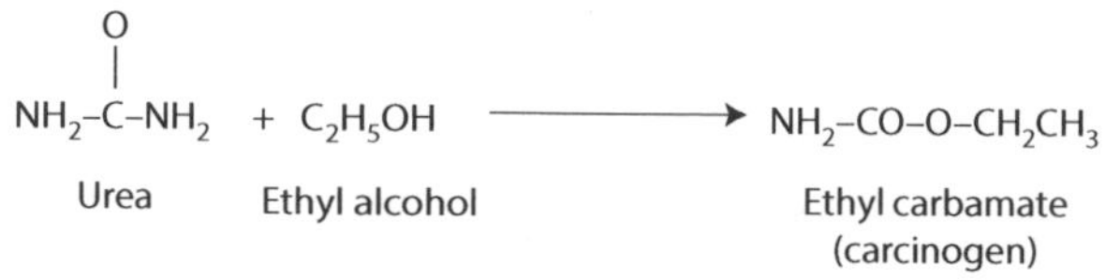

$$NH_2-C-NH_2 \; + \; C_2H_5OH \longrightarrow NH_2-CO-O-CH_2CH_3$$

Urea   Ethyl alcohol   Ethyl carbamate
                                              (carcinogen)

Figure 3.18 Formation of ethyl carbamate from urea.

A limited number of LAB species (known as amylolytic LAB) can degrade starch and are relevant in the fermentation of substrates rich in starch (cassava, cereals, maize, and sorghum). The different LAB groups however, differ with respect to carbohydrate metabolism, not only concerning the substrates that can be metabolized, but also in the final fermentation products. Consequently, each food substrate is usually fermented by a specific group of LAB and, hence, the profile of fermentation products is also distinctive and can only obtain ATP (Adenosin Triphosphate) by fermentation, usually of sugars (Axelsson, 2004) (Figure 3.19).

Two main sugar fermentation pathways (homo-fermentative and hetero-fermentative) can be distinguished among LAB (Axelsson, 2004; Mayo et al., 2010).

The carbohydrate metabolism of different species of LAB in various types, and the biochemismtry of lactic acid producing fermentation products vary with the

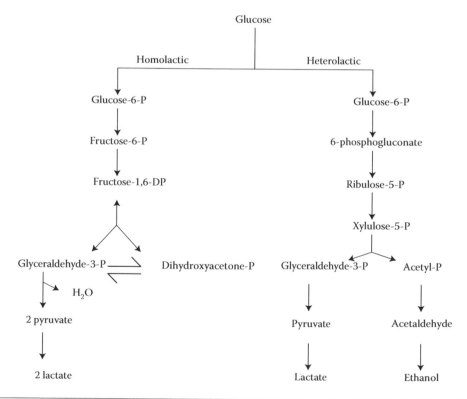

Figure 3.19 Formation of different products in homolactic or heterolactic pathways of lactic acid bacteria (Adapted from Caplice, E. and Fitzgerald, G.F. 1999. *Int. J. Food Microbio* 50: 131–14.)

ability to produce different enzymes by LAB and the available carbon sources. Homo-fermentative LAB conduct the anaerobic fermentation of hexoses *via* the Embden–Meyerhof–Parnas (EMP) pathway until pyruvic acid formation, and then, being divided by pyruvate decarboxilase, the pyruvic acid becomes the acceptor of hydrogen (Kandler, 1983; Buruleanu et al., 2010).

The reduction of the pyruvic acid happens by the action of lactatedehydrogenase, which needs NAD + /NADP+ as coenzymes. In the homolactic fermentation the predominant form of lactic acid is dependant not only on the lactatedehydrogenase specificity (*Lactococcus* form L(+)-lactat and *Lactobacillus* D(–)-lactat), but also on the presence of lactate racemase which, when active in the microorganisms' cells, results in racemic (D, L-lactat) by fermentation (Buruleanu et al., 2010). Hetero-fermentative LAB produce the anaerobic fermentation of glucides via the pentoso-phosphate (6P-gluconat) pathway. The kind of fermentation products depends on the involved species (Figure 3.20). Thus, *Lactobacillus brevis* produces, through heterolactic fermentation, lactic acid, acetic acid, and carbon dioxide, while *Leuconostoc mesenteroides* produces lactic acid, ethanol, and carbon dioxide. Glycolysis (the EMP pathway) results in almost exclusively lactic acid as the end product under normal conditions, and this type of metabolism is referred to as homolactic fermentation (Neti et al., 2011).

Obligate homo-fermentative LAB (*Lactococcus, Streptococcus, Pediococcus, Enterococcus,* and some species of *Lactobacillus*) ferment sugars by the EMP pathway to pyruvate, which is converted into lactic acid (Mayo et al., 2010). Pyruvate can also be converted into diacetyl and acetoin/2,3-butanediol (through the diacetyl/acetoin branch) (Figure 3.21).

Wheat flour contains carbohydrates like maltose, sucrose, glucose, and fructose, along with some trisaccharides like maltotriose and reffinose. The yeasts present in sourdough are not able to ferment maltose. Homo-fermentative LAB mainly produce lactic acid through glycolysis (homolactic fermentation), while hetero-fermentative LAB produce, besides lactic acid, CO_2, acetic acid, and/or ethanol through the 6-phosphogluconate/phosphoketolase (6-PG/PK) pathway (heterolactic fermentation). The utilization of co-substrates such as oxygen or fructose as electron acceptors by obligate hetero-fermentative *Lactobacilli* is coupled to an increased production of acetate in dough. In hexose fermentation, facultative and obligately hetero-fermentative LAB use the EMP and phoshogluconate pathways, respectively. Lactate degradation

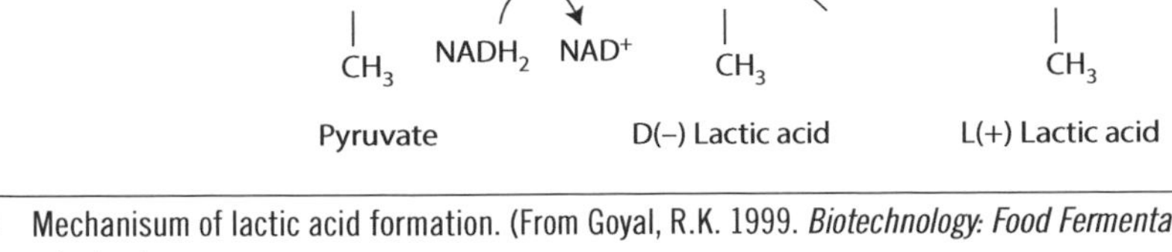

Figure 3.20 Mechanisum of lactic acid formation. (From Goyal, R.K. 1999. *Biotechnology: Food Fermentation.* Vol. 1, pp. 87–137. With permission.)

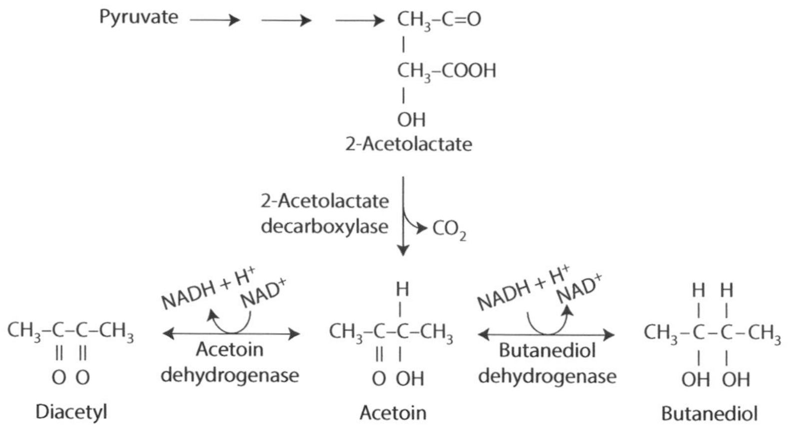

Figure 3.21 Metabolism of acetolactate to diacetyl and butanediol by LAB.

during sourdough fermentation does have an impact on sourdough flavor and texture due to the formation of acetate and CO_2 (Liu, 2003).

The metabolism of the disaccharide lactose is of primary importance in those LAB used in dairy fermentations, as was reviewed earlier (Fox et al., 1990; Axelsson, 1998). Lactose may enter the cell using either a lactose carrier, lactose permease, followed by cleavage to glucose and galactose, or *via* a phosphoenolpyruvate-dependent phosphotransferase (PTS) followed by a cleavage to glucose and galactose-6-phosphate. Glucose is metabolized *via* the glycolytic pathway, galactose *via* the Leloir pathway, and galactose-6-phosphate *via* the tagatose-6-phosphate pathway. Most *L. lactis* strains used as starters for dairy fermentations use the lactose PTS (Thomas and Crow, 1984; Hutkins and Morris, 1987; Cogan and Hill, 1993).

Citrate metabolism is important among *L. lactis* ssp. *lactis* (biovar diacetylactis) and *Lc. mesen-teroides* ssp. *cremoris* strains used in the dairy industry, as it results in excess pyruvate in the cell which is converted *via* α-acetolactate to diacetyl (Cogan and Hill, 1993).

The second fermentation route or 6-phsphogluconate/phosphoketolase (6-PG/PK) pathway is known as heterolactic fermentation. Obligate hetero-fermentative LAB (such as *Leuconostoc*, *Oenococcus*, and also certain *Lactobacillus* species) ferment sugars by the 6-PG/PK pathway (Axelsson, 2004; Mayo et al., 2010). Glucose is first phosphorylated, as in glycolysis. Hetero-fermentative LAB can also grow on pentoses, which are transported into cells by specific permeases, and, subsequently, phosporylated inside the cells to ribulose-5-phosphate or xylulose-5-phosphate by isomerases or epimerases, which are metabolized through the lower part of the 6-PG/PK pathway (Axelsson, 2004; Mayo et al., 2010). This heterolactic fermentation of pentoses results in the production of lactate and acetate as main end-products. There is no formation of CO_2, and no reduction of acetyl phosphate to ethanol. Instead, ATP and acetate are formed from acetyl phosphate by acetate kinase (Axelsson, 2004). Hetero-fermentative LAB are responsible for the production of

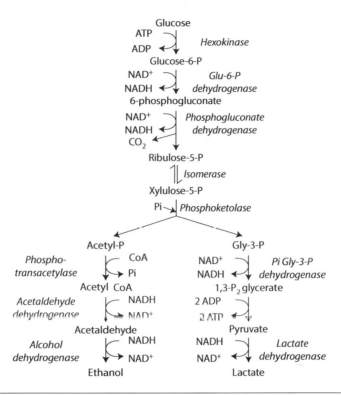

Figure 3.22 Formation of ethanol and lactic acid by heterofermentative LAB.

ethanol and lactic acid, both of which have considerable significance in fermented food products. The pathway of their fermentation from glucose is shown in Figures 3.22 and 3.23.

3.9.8 Metabolism of Nitrogenous Compounds/Proteins

3.9.8.1 Formation of Higher Alcohols   In the normal alcoholic fermentation of carbohydrates by *Saccharomyces cerevisiae,* the production of alcohol other than ethanol also takes place (Amerine et al., 1980). These alcohols with carbon numbers greater than two are called higher alcohols, such as propyl alcohol, butyl alcohol, amyl alcohol, etc. These alcohols are produced as by-products of ethanolic fermentation. Mechanism of formation of these high alcohols from the metabolism of amino acid is shown in Figure 3.24 as an illustration. Higher alcohols are significant from the development of flavor when at low concentration (Joshi, 1997), but higher concentration, usually of more than 400 mg/L, is considered undesirable, being cause of hangovers or headaches.

Efforts are made to reduce their content during fermentation. The higher alcohols are made from deamination of amino acids of the must as described earlier, if the exogenous supply of nitrogen compounds is not available in any fermentation that is, wine fermentation by *Saccharomyces cerevisiae* var *ellipsoids.* When diammonium

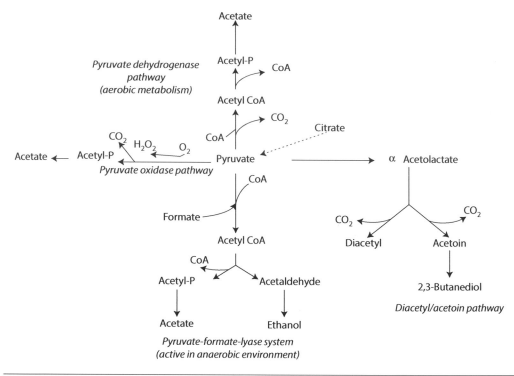

Figure 3.23 Formation of different products in lactic acid fermentation. (From Caplice, E. and Fitzgerald, G.F. 1999. *Int. J. Food Microbiol.* 50: 131–149. With permission.)

hydrogen phosphate as a nitrogen source is added and the nitrogenous compounds of the must are not utilized, higher alcohols are not produced and the quality of alcoholic beverage does not deteriorate (Joshi et al., 1999, 2011).

3.9.9 Amines

During lactic acid fermentation, various amines are synthesized by LAB. Some of these amines are listed in Table 3.29. These amines are known to have allergic reactions.

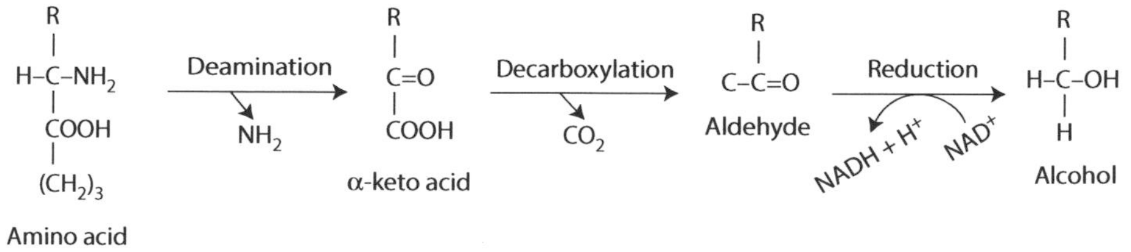

Figure 3.24 Pathway of higher alcohol formation from amino acid. (From Rana, N. and Rana, V. 2011. *Handbook of Enology.* Vol. 2, Asiatech Publishers, Inc., New Delhi, pp. 618–666. With permission.)

Table 3.29 Some of the Amines Found in Wines

Ammonia	Methylamine
Butylamine	2-Methylbutylamine
Cadaverine	Morpholine
1,5-Diaminopentane	Phentylamine
Diethylamine	Phenethylamine
Dimethylamine	Piperidine
Ethanolamine	Propylamine
Ethylamine	Putrescine
Hexylamine	Pyrrolidine
Histamine	2-Pyrrolidine
Indole	Serotonin
Isopentylamine	Tyramine
Isopropylamine	

Source: Adapted from Lang, J.M. and Cirillo, V.P. 1987. *J. Bacteriol*, 169, 2932–2937.

3.9.10 Aroma/Flavor Compound Formation

Indigenous alcoholic beverages are known to possess flavor compounds/constituents. These are common to various alcoholic beverages, including wine and beer. Some of these compounds are formed during alcoholic fermentation, while others, like esters, are formed during maturation. These compounds include esters of various alcohols and acids, alcohols including higher alcohols, carbonyl compounds, sulfur derivatives, and phenolic compounds. The subject is so vast that it is just not possible to describe each and every flavor component of alcoholic beverages (Rana and Rana, 2011). Though not investigated separately, indigenous fermented beverages are likely to have similar compounds. It is an exciting field for the researchers.

The degradation of wheat and rye proteins is of crucial importance for bread flavor, volume, and texture. Gluten proteins, glutenins, and gliadins, are the major storage proteins of the wheat grain, whereas in the case of rye they are secalins (Gänzle et al., 2008). These proteins make them more susceptible for proteolytic degradation. Amino acids serve as substrates for microbial conversions into flavor compounds during baking; accordingly, a limited extent of proteolysis during fermentation improved bread flavor (Gänzle et al., 2008).

The proteolytic system of *Lactococcus* has been investigated in detail due to its pivotal role in allowing growth in milk and the development of flavor and texture in cheese. For more details, see the literature cited (Kunji et al., 1996; Caplice and Fizgerald, 1999). The lysis of starter cultures of lactic acid during cheese ripening leads to increased proteolytic activity in cheese and, thus, theoretically accelerates ripening and/or a better cheese flavor, such as in Swiss cheese, where autolytic propionicbacteria increase the amount of free proline in the cheese, aiding flavor development. Bacteriocin-induced lysis of a starter to produce cheese with a higher concentration of free amino acids has

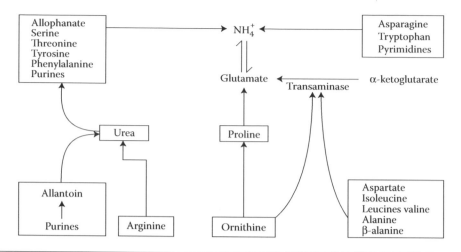

Figure 3.25 Major products of nitrogenous compound degradation. (Adapted from Rana, N. and Rana, V. 2011. *Handbook of Enology*. Vol. 2, Asiatech Publishers, Inc., New Delhi, pp. 618–666..)

been made, which decreases the bitterness levels (Lemee et al., 1994; Dupuis et al., 1995; Ostlie et al., 1995; Morgan et al., 1997).

In the fermentation of foods rich in protein, break down of the several nitrogen compounds takes place (Figure 3.25). The degradation products have a great significance in flavor development in the fermented foods, amino acids, peptides, etc.

In alkaline fermentation, the break down of proteins mainly by *Bacillus* sp., leads to the fermentation of amino acids, which further breaks down into the typical ammonia flavor compounds of alkaline fermented food (Odunfa, 1985a; Parkouda et al., 2009; Sarkar and Nout, 2014).

3.9.11 Biological Conversion of Ethanol to Acetic Acid

Formation of acetic acid during alcoholic fermentation, may be in small amounts, takes place by the pathway shown in Figure 3.26. But in *Acetobactor* and *Gluconobactor*, acetic acid is the major product, and fermentation is called acetic acid fermentation, and is discussed in detail here.

The first step in the conversion of ethanol to acetic acid is the formation of acetaldehyde according to the equation:

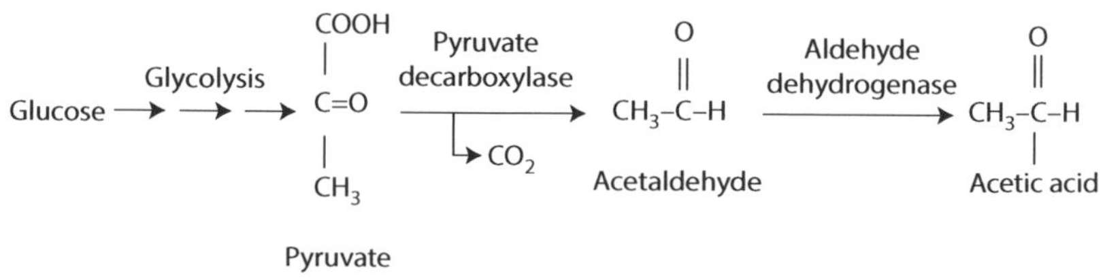

$$\text{Glucose} \xrightarrow{\text{Glycolysis}} \underset{\substack{|\\CH_3\\\text{Pyruvate}}}{\overset{\substack{COOH\\|\\C=O\\|}}{}} \xrightarrow[\substack{\searrow CO_2}]{\substack{\text{Pyruvate}\\\text{decarboxylase}}} \underset{\text{Acetaldehyde}}{CH_3\text{–}\overset{\overset{O}{\|}}{C}\text{–H}} \xrightarrow{\substack{\text{Aldehyde}\\\text{dehydrogenase}}} \underset{\text{Acetic acid}}{CH_3\text{–}\overset{\overset{O}{\|}}{C}\text{–H}}$$

Figure 3.26 Mechanism of acetic acid synthesis in acetic acid bacteria.

$$CH_3CH_2OH + O_2 \xrightarrow{\text{Alcohol dehydrogenase}} CH_3CHO + H_2O \text{ (I)}$$
$$\underset{\text{Ethanol}}{} \qquad \underset{\text{Acetaldehyde}}{}$$

The second step is the formation of acetic acid from acetaldehyde. The latter first reacts with water to yield hydrated acetaldehyde, which in turn is oxidized or dehydrogenated to yield acetic acid, as shown below:

$$CH_3CHO + H_2O \longrightarrow CH_3 - \overset{\overset{\displaystyle H}{\displaystyle |}}{\underset{\underset{\displaystyle OH}{\displaystyle |}}{C}} - OH + [0] \xrightarrow{\text{Alcohol dehydrogenase}} CH_3COOH + H_2O \text{ (II)}$$

See the literature cited for more information (King and Chedelin, 1952; Ebner, 1982; King, 1982).

3.10 Changes in Chemical Composition and Nutritive Values by Fermentation

3.10.1 General Effect of Fermentation on Food Components

Microbial growth causes complex changes in the nutritive value of fermented foods by changing the composition of proteins, fats, and carbohydrates, and by the utilization or secretion of vitamins. Microorganisms absorb fatty acids, amino acids, sugars, and vitamins from the food, but in many fermentations, microorganisms also secrete vitamins into the food and improve nutritive value (http//:www.Studynation.com-2014). The polymeric compounds are also hydrolyzed to produce substrates for cell growth, thus, increasing the digestibility of proteins and polysaccharides (Dworschak, 1982; Ray and Sivakumar, 2009). In general, the mild conditions used in food fermentations produce few deleterious changes to nutritional quality and sensory characteristics. An increase in saltiness in some foods (pickles, soy sauce, fish and meat products) due to salt addition, and a reduction in bitterness of some foods due to the action of debittering enzymes, take place. For example, in bread and cocoa, the subsequent operations of baking and roasting produce the characteristic aromas (http//:www.Studyblue. com-2014).

Several biochemical changes that reportedly occur during food fermentation include an increase in volume, total acids, soluble solids, reducing sugars, non-protein, soluble nitrogen, amylases, proteinases, free amino acids, and water-soluble vitamins (Wang and Hesseltine, 2004; Aidoo et al., 2006). Food fermentations raise the protein content or improve the balance of essential amino acids or their availability, and thus, have a direct curative effect on these parameters (Steinkraus, 1997). Not only this, vitamins such as thiamin, riboflavin, niacin, and folic acid, that can have a profound direct effect on health of the consumers of such foods, are also increased. Most of the

Table 3.30   Compounds Formed during Cereal Fermentation

ORGANIC ACIDS		ALCOHOLS	ALDEHYDES AND KETONES	CARBONYL COMPOUNDS
Butyric	Hcplanoic	Ethanol	Acetaldehyde	Furfural
Succinic	Isovaleric	n-Propanol	Formaldehyde	Methional
Formic	Propionic	Isobutanol	Isovaleraldehyde	Glyoxal
Valeric	n-Butyric	Amyl alcohol	n-Valderaldehyde	3-Methyl butanal
Caproic	Iso-butyric	Isoamyl alcohol	2-Methyl butanol	2-Methyl butanal
Lactic	Caprylic	2,3-Butanediol	n-Hexaldehyde	Hydroxymethyl furfural
Acetic	Isocaproie	5-Phenylethyl alcohol	Acetone	
Capric	Pleargonic		Propionaldehyde	
Pyruvic	Levulinic		Isobutyl aldehyde	
Palmitic	Myristic		Methyl ethyl ketone	
Crotonic	Hydrocinnamic		2-Butanone	
L-Taconic	Benzylic		Diacetyl	
Lauric			Acetoin	

Source: Adapted from Campbell-Platt, G. 1994. *Food Res. Int.* 27: 253–257.

food fermentations of roots and tubers are associated with LAB such as *Lactobacillus, Leuconostoc, Streptococcus,* etc., and yeasts (*S. cerevisiae*), which are commonly considered as "probiotics" (Agrawal and Whorwell, 2006). During cereal fermentations, several volatile compounds are formed, which contribute to a complex blend of flavors in the products (Chavan and Kadam, 1989). The presence of aromatic compounds like diacetyl, acetic acid and butyric acid make fermented cereal-based products more appetizing (Table 3.30) (Tamime, 2002).

Complex changes in proteins and carbohydrates soften the texture of fermented products, but changes in flavor and aroma are complex and, in general, are poorly documented. The color of many fermented foods is retained owing to the minimal heat treatment and/or a suitable pH range for pigment stability. However, color may also change owing to the formation of brown pigments by proteolytic activity, degradation of chlorophyll, and enzymic browning.

The content and quality of cereal proteins may be improved by fermentation (Wang and Fields, 1978; Chavan et al., 1988). Natural fermentation of cereals increases their relative nutritive value and available lysine (Hamad and Fields, 1979). Bacterial fermentations involving proteolytic activity are expected to increase the biological availability of essential amino acids more so than yeast fermentations, which mainly degrade carbohydrates (Chaven and Kadam, 1989; Anon-FAO, 1994). Starch and fiber tend to decrease during fermentation of cereals (Idris, 2005). Fermentation cannot alter the mineral content of the product, but the hydrolysis of chelating agents, such as phytic acid, during fermentation, improves the bioavailability of minerals. Changes in the vitamin content of cereals with fermentation vary according to the fermentation process, and the raw material used in the fermentation. B group vitamins generally show an increase on fermentation (Chavan et al., 1989).

3.10.2 Biochemical and Sensory Changes during Various Indigenous Food Fermentations

Very few traditionally fermented foods have been studied for changes in composition and nutritive value during their production. Recently, effect of germination and probiotic fermentation on nutritional and organoleptic acceptability of cereal based food mixtures has been reported (Khauna and Dhaliwal, 2013). Wherever available, the information from indigenous fermented foods has been included.

3.10.2.1 Cereal and Legume Fermented Foods

Idli *fermentation*: Total nitrogen and total proteins of *idli* do not alter significantly, whereas reducing sugars and soluble proteins, after declining initially, tend to increase afterwards with the progress of fermentation (Soni and Sandhu 1989a; Soni and Sandhu, 1999).

Dosa *fermentation*: *Dosa* fermentation is accompanied by an increase in total acids, total volume, total solids, soluble solids, nonprotein nitrogen, free amino acids, amylases, proteinases, vitamins B1 and B2 (Soni et al., 1985, 1986; Soni and Sandhu, 1999; Sandhu et al., 1986). Total nitrogen and total proteins do not alter significantly, whereas reducing sugars and soluble proteins, after declining initially, tend to increase afterwards with the progress of fermentation.

Dhokla *and* Khanam: Several researchers (Rajalakshmi and Vanaja, 1967; Ramakrishnan, 1977; Steinkraus, 1983a,b,c) observed an increase in batter volume (leavening), amino nitrogen, soluble acids, and increase in total sugars and pH as a result of fermentation of *dhokla* and *khanam* batters. Acidification and leavening (change in batter volume due to gas production) are the two important changes that occur during fermentation of *dhokla* and *khanam* batters, and these are reported to be responsible for the sour taste and porous texture of the products. The other desirable changes that occur during fermentation of *dhokla* and *khaman* include a partial breakdown of protein, starch, and some of the antinutritional factors.

Unfermented legumes, such as Bengal gram cotyledons, are generally considered hard to digest, but the *dhokla* and *khanam* prepared from them are sutitable even for young children and patients with digestive disorders (Ramakrishana, 1979). The changes brought about during the fermentation of *dhokla* and *khanam* result in increased digestibility and greater protein value. A similar increase in carbohydrate digestibility can also be expected in these foods as a result of fermentation.

Warri: The fermentation of *warri* brings about a significant increase in soluble solids, nonprotein nitrogen, soluble nitrogen, free amino acids, proteolytic activity, and B vitamins, including thiamine, riboflavin, and cyanocobalamine. On the other hand, the levels of reducing sugars and soluble protein decrease. Amylase activity increases initially, but declines thereafter (Batra and Millner, 1974; Sandhu et al., 1986; Soni and Sandhu, 1990b; Aidoo et al., 2006).

3.10.2.2 Bhatooru, Siddu *and* Seera   Fermentation causes a significant increase in vitamins (thiamine, riboflavin, nicotinic acid, cyanocobalamin) and amino acids (methionine, phenylalanine, threonine, lysine, and leucine) during *bhatooru, siddu,* and *seera* preparation (Savitri, 2007; Savitri and Bhalla, 2013). In the case of *seera,* a traditional fermented food of Himachal Pradesh (India), there is a sharp decrease in initial pH from the first to the second day (Figure 3.27). The low pH supports the growth of various yeasts and LAB important in food fermentation. With the decrease in pH, total acidity increased from 0.009% initially to 0.45% in the final product (Table 3.31). Acidification of wheat-based substrates during food fermentation is an essential requirement for the prevention of harmful microorganisms and the associated risks of poisoning and spoilage (Katina, 2005). The change in total titrable acidity has been reported earlier in the fermentation of *fufu* (Oyewole, 1990). Similar changes have also been reported during the preparation of Indian *warri, oncom, tempeh, gari, pozol, miso, mahewas, kenkey,* and *idli* (Beuchat, 1983; Soni and Sandhu, 1990b, 1999; Soni et al., 2001). The protein content decreased from 14.9% on day one 1% to 8.2% on 5th day of fermentation (Savitri et al., 2012), possibly due to the increase in the activity of proteolytic enzymes with the progress of fermentation. However, the higher protein content in *seera* in the dried product after fermentation, as compared to the 5th day of fermentation, was due to the concentration of various components during its drying (Table 3.31). It has been reported that the controlled fermentation can be applied for the production of physiologically beneficial components, that is, for increasing the amount of proteins (Niba, 2003).

With the progress of fermentation, the level of total sugars decreased from 74.8% to 67.5%. The sugars are rapidly metabolized to acids, ethanol, biomass, carbon dioxide, and other metabolites required for the growth of microorganisms, with a

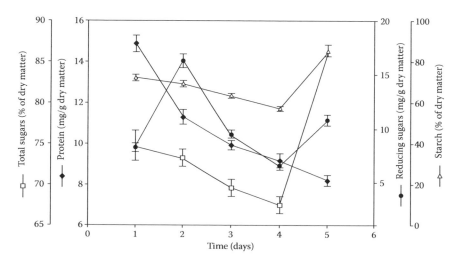

Figure 3.27   Change in total protein, total sugars, reducing sugars, and starch during *seera* fermentation. (From Savitri et al. 2012. *Intl. J. Food Ferment.*, 2(1), 49–56. With permission.)

Table 3.31   Biochemical Analysis of *Seera*

PARAMETERS	VALUES[a]
pH	3.45 ± 0.05
Titrable acidity (% of dry matter)	0.44 ± 0.006
Total count (log cfu/g dry matter)	5.39
Protein (mg/g of dry matter)	10.4 ± 0.20
Total sugars (% of dry matter)	89.0 ± 0.43
Reducing sugars (mg/g of dry matter)	11.9 ± 0.53
Starch (% of dry matter)	87.4 ± 1.51
Amylase[b] (U/g of dry matter)	3.6 ± 0.36
Protease[c] (U/g of dry matter)	1.02 ± 0.05

Source: Adapted from Savitri et al. 2012. *Intl. J. Food Ferment.*, 2(1), 49–56. With permission.

[a] Values are mean ± SD of three observations.

[b] One unit of amylase activity was defined as the amount of enzyme required to release one μg of maltose/mg of substrate/min under the assay conditions.

[c] One unit of protease activity was defined as the amount of enzyme required to release one μg of tyrosine/mg of substrate /min under the assay conditions.

concomitant decrease in total sugars during fermentation (Mensah, 1997). However, higher total sugar content (Table 3.31) in the final product is due to its concentration during drying. Mosha and Svanberg (1983) have reported the hydrolysis of starch and oligosaccharides present in the substrates of fermentation, resulting in an increase in reducing sugars due to the activity of amylases present in cereal grains or secreted by microorganisms. The decrease in reducing sugars with prolonged fermentation was attributed to their utilization by fermenting microflora, as observed by Daeschel et al. (1987). The starch content during fermentation of decreased initially from 72.1% to 57.0% (w/w) from day 1 to day 4. After steeping and drying of the *seera*, the starch content increased to 87.4% (w/w) on a dry weight basis. Giraud et al. (1994) reported the degradation of cassava starch by the amylolytic activities of *Lactobacillus plantarum* during fermentation of raw cassava.

3.10.2.3 Alcoholic Fermented Beverages   Tamang and Thappa (2006) studied the fermentation dynamics during the production of *bhatti jannr*, a rice-based alcoholic beverage, and found that the pH decreased from 6.10 to 3.21 in 2 days and showed an increment to 3.96 on the 10th day, titrable acidity (as % lactic acid) increased from 0.01 to 0.20 in 4 days, followed by a decrease to 0.18%, but remained constant at the same value until 9th day of fermentation. However, it again declined to 0.17% after 10 days of fermentation. The reducing sugar content increased during the first 3 days from 0.01% to 12.60%, and then, it decreased to 0.2% after 10 days of fermentation. The ethanol content increased gradually and reached 10.10% (v/v) after 10 days of fermentation.

Table 3.32   Nutritional Composition of Unfermented and Fermented *Ragi (Chaang)*

COMPONENTS	UNFERMENTED RAGI (g/100 g)	FERMENTED RAGI STARTER CULTURE (g/100 g)
Carbohydrates (starch and sugar)	78.0 + 3.4	35.0 + 2.5
Proteins	7.6 + 0.34	10.5 + 0.50
Fat	1.1 + 0.05	1.0 + 0.05
Phosphorous	0.3 + 0.05	0.3 + 0.05
Calcium	0.4 + 0.06	0.4 + 0.08
Iron	0.004 + 0.009	0.004 + 0.0004
Ethanol	–	16.0 + 0.68
Total acidity (lactic acid)	–	3.0 + 0.27
Microbial biomass	–	2.7 + 0.36
Volatile acidity	–	0.02 + 0.004
Thiamin	0.48 + 0.08	0.50 + 0.09
Riboflavin	0.20 + 0.01	0.60 + 0.10
Niacin	1.00 + 0.05	4.20 + 0.45
Pantothenic acid	0.40 + 0.07	1.60 + 0.08
Folic acid	0.006 + 0.001	0.010 + 0.001
Cyanocobalamin	–	0.04 + 0.004

Source: Adapted from Basappa, S.C. et al. 1997. *Int. J. Food Sci. Nutr.* 48(5): 313–319.
Note: On dry weight basis, all values means + standard deviation (SD).

Thakur et al. (2004) after conducting a survey of various alcohol-producing areas of Himachal, and collecting and analysing the samples of some of the popular indigenous alcoholic beverages, found that the *sura/sur* sample had the highest (15.28%) ethanol content. A comparision of nutritional composition of fermented (*chaang*) with that of unfermented *ragi* is given in Table 3.32.

3.10.2.4 Fermented Pickles   Pickling combines salting to selectively control micro-organisms and fermentation to stabilize the treated materials, such as fruit and vegetables (Frazier and Westhoff, 1978). One of the important changes that occurs in the pickling process is that the fermentable carbohydrate reserve is changed to acid, ranging from 0.6% to 2.5% lactic acid. However, protein content in the fermented cucumber is decreased, while ash and fat content increases when compared to fresh cucumbers.

3.10.2.5 Fermented Milk-Based Products   The progressive fermentation of milk into *dahi* brings about little or no change in the total solids, fat, proteins and milk salts, but causes a significant decrease in the lactose content and pH value (www.bdu. ac-2014). *Shrikhand*, prepared either from buffalo or cow's milk, retains quantitatively all the constituents in proportion to the addition of sugar. Little or no change in pH is seen during the conversion of *chakka* into *shrikhand*. The mineral make-up remains more or less unchanged, except citrate which disappears completely at the *dahi* stage of both buffalo and cow's milk. Despite the greater part of the solubilized minerals being drained along with the whey, a considerable amount of all the minerals remains aggregated into *chakka* from both milk systems. All the minerals are quantitatively transferred from *chakka* into *shrikhand* for both buffalo and cow's milk, though diluted in proportion to the addition of sugar. The fermentation of milk brings about highly significant changes in the mineral content during transition from the colloidal to the soluble phase, which is faster and greater for buffalo milk (Boghra and Mathur, 2000).

In all these milk-based products, the main biochemical change is the production of lactic acid from lactose by varieties of LAB belonging to *Lactococci*, *Streptococci*, *Leuconostocs*, *Lactobacilli*, and a few others. The metabolism of citrate by some of these bacteria leads to production of diacetyl, which gives a typical pleasant flavor. The growth of LAB also makes the products safe by controlling the growth of unwanted spoilage microorganisms and pathogens.

3.10.2.6 Fermented Fish Products Different changes occurring during fermentation of fish are summarized in Tables 3.33.

3.10.2.7 Changes during Alkaline Fermentation Often during fermentation dominated by the genus *Bacillus*, the organoleptic and nutritional properties of the raw materials undergo massive changes, leading to products with desirable nutritional and sensory properties (Leejeerajum et al., 2001; Beaumount, 2002; Parkouda et al., 2009; Sarkar and Nout, 2014).

3.11 Effect of Fermentation on Antinutrient Factors

3.11.1 Antinutrients in Foods

Cereals and other plant foods may contain significant amounts of toxic or antinutritional substances. Legumes, in particular, are a rich source of natural toxic agents, including protease inhibitors, amylase inhibitors, metal chelaters, flatus factors, hemaglutinins, saponins, cyanogens, lathyrogens, tannins, allergens, acetylenic furan, and isoflavonoid phytoalexins (Anon-FAO, 1994; Motarjemi and Nout, 1996; Pariza, 1996; Svanberg and Lorri, 1997). Crops like sorghum and millet contain large amounts of phytates, enzyme inhibitors, and of polyphenols and tannins (Salunkhe et al., 1990)

Table 3.33 Microbial, Chemical and Sensory Changes during Fermentation of Fish

PARAMETER	MALDIVE FISH	AMBULTHIYA	DRIED FISH	JADDI	FISH SAUCE
TVN (total volatile nitrogen) mg/100g	29.58	24.56	62.3	52.3	240
Peroxide value (%)	246.51	28.5	109.5	0	
Free fatty acids (%)	52.59	56.7	54.7	18.81	
Total bacterial count/g	6.30×105	6.72×103	1.97×105	1.986×103	2.5×103
Mold count/g	2.99×102	1.52×102	2.4×105	1.13×103	1.6×103

which decrease the nutritional value of foods by interfering with mineral bioavailability and the digestibility of proteins and carbohydrates. Since legumes are often consumed together with cereals, proper processing of cereal–legume mixtures could eliminate the antinutrients before consumption (Chaven and Kadam, 1989; Reddy and Pierson, 1994; Anonymous, FAO, 1998, 1999a,b). The phytates present in cereals forms complexes with protein or polyvalent cations such as iron, zinc, calcium, and magnesium, which are not digestible in this form. Seeds have a natural content of phytases, which can make the minerals bioavailable, so sprouting and fermentation can be used to decrease phytase activity (Svanberg and Lorri, 1997). Millet is a rich source of dietary fibres and primary nutrients, in addition to minerals, but the bioavailability is low, due to the presence of antinutritional factors, such as phytate, phenols, tannins, and trypsin inhibitors (Idris et al., 2005). How the fermentation effct the anti-nutrient is being discussed in the subsequent section.

3.11.2 Effects of Fermentation

Reddy and Pierson (1994) reviewed the effect of fermentation on antinutritional and toxic components in plant foods (Anon FAO, 1994). Fermentation of corn meal and soybean-corn meal blends lowers flatus-producing carbohydrates, trypsin inhibitors, and phytates (Chompreeda and Fields, 1981; 1984). However, fermentation of cereals with fungi, such as *Rhizopus oligosporus*, released bound trypsin inhibitors, thus increasing their activity (Wang et al., 1972). Fungal and lactic acid fermentations reduce aflatoxin B_1, sometimes by opening of the lactone ring, which results in complete detoxification (Nout, 1994). Studies have shown that both spontaneous fermentations as well as fermentations with starter cultures significantly reduce the content of phytic acid in millet (Sharma and Kapoor, 1996, Elyas et al., 2002; Murali and Kapoor, 2003). It has been found that the starter culture fermentations were more effective than spontaneous fermentations (Murali and Kapoor, 2003). As a result of lactic acid fermentation, the protein digestibility can also be improved (Motaryemi and Nout., 1996; Antony and Chandra, 1998; Taylor and Taylor, 2002; Ali et al., 2003; Onyango et al., 2004) and the tannin content may be reduced in some cereals, leading to the increased absorption of iron (Khetarpaul and Chauhan, 1989, 1990; Motarjemi and Nout, 1996; Antony and Chandra, 1998; Sanni et al., 1999b; Elyas et al., 2002; Onyango et al., 2005). Studies on nutritional changes in fermented millet have found improvement of the *in vitro* protein digestibility (Antony and Chandra, 1998; Ali et al., 2003) and a significant reduction in total polyphenols and phytic acid content (Obizoba and Atii, 1994; Sharma and Kapoor, 1996; Antony and Chandra, 1998; Elyas et al., 2002; Tou et al., 2006). The effects of fermentation on tannins are, however, variable (Antony and Chandra, 1998; Elyas et al., 2002). An increase in starch digestibility, increase in total free amino acids, and a reduction in trypsin inhibitor activity have been found after fermention in millet (Antony and Chandra, 1998).

The fate of some the antinutrients and toxicants in traditional fermented foods is summarized here.

3.11.2.1 Phytates Phytic acid is the 1,2,3,4,5,6-hexaphosphate of myoinositol that occurs in discrete regions of cereal grains and accounts for as much as 85% of the total phosphorous content of these grains (Anonymous, FAO, 1998, 1999a,b). It reduces the bioavailability of minerals, and the solubility, functionality, and digestibility of proteins and carbohydrates (Reddy et al., 1989). Fermentation of cereals reduces phytate content by the action of phytases that catalyze conversion of phytate to inorganic orthophosphate and a series of myoinositols, lower phosphoric esters of phytate. A 3-phytase is possibly characteristic of microorganisms, while a 6-phytase is found in cereal grains and other plant seeds (Reddy and Pierson, 1994), so the reduction takes place in fermentation.

3.11.2.2 Tannins Oligomers of flavan-3-ols and flavan-3,4-diols, called condensed tannins, occur widely in cereals and legumes, but are concentrated in the bran fraction of cereals (Salunkhe et al., 1990). Tannin–protein complexes can cause inactivation of digestive enzymes and reduce protein digestibility by interaction of the protein substrate with ionizable iron (Salunkhe et al., 1990; Anonymous, FAO, 1998, 1999a,b), and can, therefore, lower feeding efficiency, depress growth, decrease iron absorption, damage the mucosal lining, of the gastrointestinal tract, alter excretion of cations, and increase excretion of proteins and essential amino acids (Reddy and Pierson, 1994). Fermentation, among other methods, reduces the tannin content of cereals and other foods.

3.11.3 Enzyme Inhibitors

Protease and amylase inhibitors widely occur in cereal grains (Anonymous, FAO, 1998, 1999a,b). Trypsin, chymotrypsin, subtilisin-inhibitor, and cysteine-protease inhibitors were found to be present in all the major rice cultivars grown in California, although the individual amounts of inhibitors was quite varaiable, but were concentrated in the bran fraction (Izquierdo-Pulido et al., 1994), where they are believed to cause growth inhibition by interfering with digestion, causing pancreatic hypertrophy and metabolic disturbance of sulfur amino acid utilization (Reddy and Pierson, 1994). Although these inhibitors tend to be heat stable, the trypsin inhibitor (TI), chymotrypsin inhibitor (CI) and amylase inhibitor (AI) levels are reduced considerably during fermentation (Chaven and Kadam, 1989; Reddy and Pierson, 1994).

3.12 Summary and Future Prospectives

Food fermentations have been and are indispensable parts of ancient human civilization technology and are used by people all over the world, and South Asia is no exception. It is apparent that there is a large diversity of products made with these fermentations, and the same is the case with the microflora associated with these fermentations and the metabolites produced by them. In the traditional methods of production of fermented foods, indigenous knowledge utilized the ecology of the food and obtained s product with acceptable qualities, as there was no information available with respect to either microbiology or biochemistry. Thus, the ecology of the fermented foods held a special significance and this is especially true in the development of desirable physico-chemical and sensory qualities in fermented foods. However, the ecology of any indigenous fermented products is not fully understood, and so obtaining this information assumes a great significance, so that the processes can be managed to encourage growth of the desired species and prevent the growth of undesirable species. However, this task is further complicated by the very complexity of the microflora of indigenious fermented foods, and the limitations of cultural methodologies in analysing complex ecosystems, molecular ecological techniques, the microbiological mysteries of fermented products, etc.

There are several fermentations which have not been investigated for the type and number of microorganisms, the biochemical pathways involved, or the changes involved in their production. Further research therefore should be directed towards identifying the benefits and risks associated with specific indigenous fermented cereals, and elucidating the contributions of microorganisms, enzymes, and others factors in the fermentation process. Many fermentation techniques involve the use of starter cultures, enzyme additives, and controlled environmental conditions, but others may benefit from genetic modifications of the cereal or starter bacteria. Developing starter cultures, unique microbial strains for nutritive improvement and detoxification, and the testing of new cereal varieties for their suitability as fermentation substrates also needs consideration. Such types of approaches can make indigenous fermented foods into a useful vehicle to fight malnutrition, ensure better health, and provide more food for hungry mouths. Since food is a basic necessity of man, more efforts are needed to develop techniques and methods so that the basic product is not altered but is made safe and more nutritious. From a microbiological point of view, the diversity of microorganisms, their enzyme compliment, substrate utilization, and product quality assume a great significance, and should be focused on in research to achieve these objectives. Genetic engineering may also be very useful.

The issue of fungal contamination and control needs systematic study, since it has a profound impact on the fermentation process. What are its toxic metabolites, and their significance, needs to be determined. It is now clear from the microbiology of indigenous fermented foods that yeast species other than *S. cerevisiae* are also significant in alcoholic fermentation. More data are needed about the ecological and biochemical contributions made by different microflora, especially different yeasts and bacteria, especially at the strain level. These studies are likely to reveal the presence of other novel species in indigenous fermented foods.

The alcoholic beverages in the modern context are known to play a significant role in reducing cardiovascular diseases. Research is needed to employ the microorganisms that could retain the antioxidant activity of these beverages while the production process is developed to suit the requirement of healthful beverage production. Nevertheless, the significance of fungi in the production of enzymes, flavor, other metabolites should not be undermined.

With respect to food fermentation, the dynamics of growth, survival, and biochemical activity of the microflora reflect a number of of stress reactions in response to the changing conditions of salt and pH or other ingredients added to the food. In the context of food fermentation, beyond the concept of bacteriocins, the influences of bacteriophages, killer yeasts, and cross-group (e.g., yeast–bacteria, bacteria–fungi, etc.) responses should be included for future study. The most important aspect of food fermentation is to produce novel foods. It is known that indigenous fermented food contain many such compounds. To be precise, many bioactive compounds are produced in the fermentation of indigenous fermented foods that have positive influences on the health of consumers. Investigations need to be focused on such aspects in future. Another facet of food fermentation is the development of natural flavors, colors, antiomicrobial substrates, and preservatives for use in the foods.

The future foresees the concentrated efforts of researchers, with modern tools in their hands, including molecular techniques for identification and bioreactor technology for commercial production of these foods, which could prove to be a boon for the future generations.

References

Aalbersberg, W.G.L., Lovelace, C.E.A., Madheji, K., and Parkinson, S.V. 1988. Davuke, the traditional Fijian method of pit preservation of staple carbohydrate foods. *Ecol. Food Nutr.* 21: 173–180.

Achi, O.K. 2005. Review: Traditional fermented protein condiments in Nigeria. *Afr. J. Biotechnol.* 4: 1612–1621.

Agrawal, A. and Whorwell., P.J. 2006. Irritable bowel syndrome: diagnosis and management. *Br. Med. J.* 332(7536): 280–283.

Aidoo, K.E., Nout, M.J.R., and Sarkar, P.K. 2006. Occurrence and function of yeasts in Asian indigenous fermented foods. *FEMS Yeast Res.* 6(1): 30–39.

Ali, M.A.M., El Tinay, A.H., and Abdalla, A.H. 2003. Effect of fermentation on the *in vitro* protein digestibility of pearl millet. *Food Chem.* 80: 51–54.

Allagheny, N., Obanu, Z.A., Campbell Platt, G., and Owens, J.D. 1996. Control of ammonia formation during *Bacillus subtilis* fermentation of legumes. *Int. J. Food Microbiol.* 29: 321–333.

Altekruse, S.F., Timbo, B.B., Mowbray, J.C., Bean, N.H., and Potter, M.E. 1998. Cheese associated outbreaks of human illness in the United States, 1973–1992: Sanitary manufacturing practices protect consumers. *J. Food Prot.* 61: 1405–1407.

Amadi, E.N., Barimalaa, I.S., and Omosigho, J. 1999. Influence of temperature on the fermentation of bambara groundnut *Vigna subterranea*, to produce a dawadawa-type product. *Plant Foods Hum. Nutr.* 54: 13–20.

Amerine, M.A., Berg, H.W., and Cruess, W.V. 1967. *The Technology of Wine Making.* AVI Publishing Co., Westport, CT.

Amerine, M.A., Berg, M.W., Kunkee, R.E., Ough, C.S. Suigleton, V.L. and Webb, A.D. 1980. The technology of wine making. 4th edn. AVI, Westport, CT.

Amman, R.I. 1995. Fluorescently labelled r-RNA-targeted oligo-nucleotide probes in the study of microbial ecology. *Mol. Ecol.* 4: 543–554.

Amoa-Awua, W.K.A. and Jakobsen, M. 1996. Activities of molds during the fermentation of Cassava into Agbelima. Paper presented at *Seminar on Fermented Food Processing in Africa, 1996, Accra.*

Anderson, K.L. and Fung, D.Y.C. 1983. Anaerobic methods, techniques and principles for food bacteriology: A review. *J. Food Prot.* 46: 811–822.

Aneja, R.P., Mathur, B.N., Chandan, R.C., and Banerjee, A.K. 2002. *Technology of Indian Milk Products.* A Dairy India Publication, Delhi.

Aniche, G.N., Nwokedi, S.I., and Odeyemi, O. 1993. Effect of storage temperature, time and wrapping materials on the microbiology and biochemistry of *Ogiri*—A fermented castor seed soup condiment. *World J. Microbiol. Biotechnol.* 9: 653–655.

Anon. 1994. *Vanuatu National Agricultural Census.* Statistics Office, Port Vila Vanuatu, 190pp.

Anonymous. 1971. Continuous direct acidification system for producing Mozzarella cheese. *Food Trade Rev.* 41: 28–31.

Anonymous FAO. 1998. *Fermented Frutis and Vegetables. A Global Perspective. Bacterial Fermentations.* FAO, Rome.

Anonymous FAO. 1999a. *Fermented Cereals. A Global Perspective.* FAO, Rome.

Anonymous FAO. 1999b. *Fermented Cereals: A Global Perspective. Cereal Fermentations in Countries of the Asia-Pacific Region.* FAO, Rome.

Anonymous. 2002. Report of a Joint FAO/WHO Working Group on Drafting Guidelines for the Evaluation of Probiotics in Food London, Ontario, Canada, April 30–May 1, 2002.

Anonymous *kinema.* http://localnepalifood.wordpress.com/local-nepali-food/*kinema*/.

Anonymous rice vinegar. http://www.absoluteastronomy.com/topics/Rice_vinegar.

Antony, U. and Chandra, T.S. 1998. Antinutrient reduction and enhancement in protein, starch, and mineral availability in fermented flour of finger millet (*Eleusine coracana*). *J. Agric. Food Chem.* 46: 2578–2582.

Ardhana, M.M. and Fleet, G.H. 1989. The microbial ecology of tape ketan fermentation. *Int. J. Food Microbiol.* 9: 157–165.

Arman, R. and Kott, Y. 1996. Bacteriophages as indicators of pollution. *Crit. Rev. Environ. Sci. Technol.* 26: 299–335.

Armstrong, G.L., Hollingsworth, J., and Morris, J. 1996. Emerging food-borne pathogens: *Escherichia coli* O157:H7 as a model of a new pathogen into the food supply of the developed world. *Epidemiol. Rev.* 18: 29–51.

Atacador-Ramos, M. 1996. Indigenous fermented foods in which ethanol is a major product. In: Steinkraus, K.H. (Ed.), *Handbook of Indigenous Fermented Foods*, 2nd edn. Marcel Dekker, New York, pp. 363–508.

Atputharajah, J.D., Widanapathirana, S., and Samarajeewa, U. 1986. Microbiology and biochemistry of natural fermentation of coconut palm sap. *Food Microbiol.* 3: 273–280.

Axelsson, L. 1998. Lactic acid bacteria: Classification and physiology. In Salminen, S. and von Wright, A. (Eds.), *Lactic Acid Bacteria: Microbiology and Functional Aspects*, 2nd edn., revised and expanded. Marcel Dekker, Inc., New York, pp. 1–72.

Axelsson, L.T. 2004. Lactic acid bacteria: Classification and physiology. In: Salminen, S., von Wright, A., and Ouwehand, A. (Eds.), *Lactic Acid Bacteria. Microbiological and Functional Aspects.* 3rd edn, Revised and Expanded, Marcel Dekker, Inc., New York, NY, pp. 1–66.

Azokpota, P., Hounhouigan, D.J., and Nago, M.C. 2006. Microbiological and chemical changes during the fermentation of African locust bean (*Parkia biglobosa*) to produce afitin, iru and sonru, three traditional condiments produced in Benin. *Int. J. Food Microbiol.* 107: 304–309.

Banks, J.M. 1998. Cheese. In: Early, R. (Ed.), *The Technology of Dairy Products.* 2nd edn, Blackie Academic and Professional, London, pp. 81–122.

Barnby-Smith, F.M. 1992. Bacteriocins: Applications in food preservation. *Trends Food Sci. Technol.* 3: 133–137.

Barnett, J.A., Payne, R.W., and Yarrow, D. 1983. *Yeasts: Characteristics and Identification.* Cambridge University Press, Cambridge.

Basappa, S.C., Somashekar, D., Agarwal, R., Suma, K., and Bharthi, K. 1997. Nutritional composition of fermented ragi (chhang) by phab and defined starter cultures ascompared to unfermented ragi (*Eleusine coracana* G.). *Int. J. Food Sci. Nutr.* 48(5): 313–319.

Batra, L.R. 1981. Fermented cereals and grain legumes of India and vicinity. In Moo-Young, M., and Rovinson, C.W. (Eds.), *Advances in Biotechnology*, Vol. 3. Pergamon Press, Toronto, 547pp.

Batra, L.R. and Millner, P.D. 1974. Some Asian fermented foods and beverages and associated fungi. *Mycologia* 66: 942–950.

Batra, L.R. and Millner, P.D. 1976. Asian fermented foods and beverages. *Dev. Indus. Microbiol.*, 17: 117–128.

Beaumont, M. 2002. Flavoring composition prepared by fermentation with *Bacillus* spp. *Int. J. Food Microbiol.* 75: 189–196.

Beuchat, L.R. 1983. Indigenous fermented foods. In: Reed G. (Ed.), *Biotechnology: Food and Feed Production with Microorganisms*. Vol. 5, Verlag Chemie, Weinheim, pp. 477–528.

Bhalla, T.C. and Thakur, N. 2011. Traditional fermented cereal based alcoholic beverages. In: Panesar, P.S., Sharma, H.K., and Sarkar, B.C. (Eds.), *Bio-Processing of Foods*. Asiatech Publishers, Inc., New Delhi, pp. 29–38.

Bhowmik, T. and Marth, E.H. 1990. Role of *Micrococcus* and *Pediococcus* species in cheese ripening—A review. *J. Dairy Sci.* 73: 859–866.

Bibek, R. 2004. Fundamental food microbiology. In: *Microbiology in Food Fermentation*, CRC Press, London, pp. 125–135.

Bisson, L. 1993. Metabolism of sugars. In: Bibek Ray, Arun Bhunia, and Fleet, G.H. (Eds.), *Wine Microbiology and Biotechnology*. Harwood Academic, Chur, Switzerland, pp. 55–76.

Blanco, J.L., Carrion, B.A., Liria, N., Diaz, S., Garcia, M.E., Domineuez, L., and Suarez. C.L. 1999. Behavior of aflatoxins during manufacture and storage of yoghurt. *Milchtvissenschaft* 48: 385–387.

Blandino, A., Al-Aseeri M.E., Pandiella, S.S., Cantero, D., and Webb, C. 2003. Review: Cereal-based fermented foods and beverages. *Food Res. Int.* 36: 527–543.

Blom, H. and Mortvedt, C. 1991. Anti-microbial substances produced by food-associated microorganisms. *Biochem. Soc. Trans.* 19: 694–698.

Boddy, L. and Wimpenny, J.W.T. 1992. Ecological concepts in food microbiology. *J. Appl. Bacteriol.* 73(Suppl.): 23S–38S.

Boghra, V.R. and Mathur, O.N. 2000. Physico-chemical status of major milk constituents and minerals at various stages of shrikhand preparation. *J. Food Sci. Technol.* 37: 111–115.

Borgstrom, G. 1968. *Principles of Food Science, Food Microbiology and Biochemistry*. Macmillan, New York, NY.

Boulton, C.L., Irving, A.J., Southam, E., Potier, B., Garthwaite, J., and Collingridge, G.L. 1994. The nitric oxide-cyclic GMP pathway and synaptic depression in rat hippocampal slices. *Eur. J. Neurosci.* 6: 1528–1535.

Bouthyette, P.-Y. 1980. *The Chloroplast of Pisum Sativum: A Three Part Study—The Synthesis of CF?? The Effect of Ca++ on Protein Synthesis—Oligomycin Sensitivity of the Thylakoid ATPase*. Cornell University, p. 330.

Burulearu, L., Nicolescu, C.L., Bratu, M.G., Manea, J. and Avram, D. 2010. Study regarding some metabolic features during lactic and fermentation at vegetable juices. *Rom. Biotechnol. Lett.* 15, 5177–5188.

Brackett, R.E. 1997. Fruits, vegetables and grains. In: Doyle, M.P., Beuchat, L.R., Montville, T.J. (Eds.), *Food Microbiology Fundamentals and Frontiers*. ASM, Washington D.C., pp. 117–126.

Campbell-Platt, G. 1994. Fermented foods—A world perspective. *Food Res. Int.* 27: 253–257.

Caplice, E. and Fitzgerald, G.F. 1999. Food fermentations: Role of microorganisms in food production and preservation. *Int. J. Food Microbiol.* 50: 131–149.

Carr, F.J., Chill, D., and Maida, N. 2002. The lactic acid bacteria: A literature survey. *Crit. Rev. Microbiol.* 28: 281–370.

Chakrabarty, J., Sharma, G.D. and Tamang, J.P. 2009. Substrate utilization in traditional fermentation technology practiced by tribes of North *Cachar Hills district of Assam. Assam University Journal of Science and Technology: Biological Science* 4(1): 66–72.

Chakrabarty, J., Sharma, G.D., and Tamang, J.P. 2013. Indigenous technology for food processing by the tribes of of Dima Hasao (North Cachar Hills) District of Assam for social security. In: Das Gupta, D. (Ed.), *Food and Environmental Foods Security Imperatives of Indigenous Knowledge of Systems*. Agrobios, Jaipur, India, pp. 32–45.

Chakrabarty, J., Sharma, G.D., and Tamang, J.P. 2014. Traditional technology and product characterization of some lesser-known ethnic fermented foods and beverages of North Cachar Hills District of Assam. *Indian J. Tradit. Knowl.* 13(4): 706–715.

Chang, S.T. 1974. Production of the straw mushroom (*Volvariella volvacea*) from cotton wastes. *Mushroom Journal* 21: 348–353.

Chang, S.T. 1977. The origin and development of straw mushroom cultivation. *Econ. Bot.* 31: 374–376.

Chang, S.T. 1982. Mushroom spawn. In: Chang, S.T., Quimio, T.H. (Eds.), *Tropical Mushrooms—Biological Nature and Cultivation Methods*. Chinese Univ. Press, Hong Kong.

Chang, S.T. and Quimio, T.H. (Eds.) 1982. Cultivation of Volvariella mushrooms in Southeast Asia. *Tropical Mushrooms*. The Chinese Univ. Press, Hong Kong, pp. 349–361.

Chapman, H.R. and Sharpe, M.E. 1990. Microbiology of cheese. In: Robinson, R.K. (Ed.), *Dairy Microbiology: The Microbiology of Milk Products*. Vol. 2, Elsevier Applied Science, London, pp. 203–289.

Charpentier, C. and Feuillat, M. 1993. Yeast autolysis. In: Fleet, G.H. (Ed.), *Wine Microbiology and Biotechnology*. Harwood Academic, Chur, Switzerland, pp. 25–242.

Chavan, U.D., Chavan, J.K., and Kadam, S.S. 1988. Effect of fermentation on soluble proteins and in vitro protein digestibility of sorghum, green gram and sorghum-green gram blends. *J. Food Sci.* 53: 1574.

Chaven, J.K. and Kadam, S.S. 1989. Nutritional improvement of cereals by fermentation. *CRC Crit. Rev. Food Sci. Technol.* 28(5): 349.

Chelule, P.K., Mokoena, M.P., and Gqaleni, N. 2010. Advantages of traditional lactic acid bacteria fermentation of food in Africa. In: Méndez-Vilas, A. (Ed.), *Current Research, Technology and Education Topics in Applied Microbiology and Microbial Biotechnology*. FORMATEX, Spain, pp. 1160–1167.

Cherrington, C.A., Hinton, M., Mead, G.C., and Chopra, I. 1991. Organic acids: Chemistry, antibacterial activity and practicalapplications. *Adv. Microbiol. Phys.* 32: 87–108.

Chettri, R., and Tamang, J.P. 2008. Microbiological evaluation of maseura, an ethnic fermented Icgumcbuxcil condiment of Sikkim. *J. Hill Res.* 21: 1–7.

Chiang, Y., Chang, L.T., Lin, C.W., Yang, C.Y., and Tsen, H.Y. 2006. PCR primers for the detection of *Staphylococcal enterotoxins* (SEs) K, L, M and survey of SEs types in Staphylococcus aureus isolates from food-poisoning cases in Taiwan. *J. Food Prot.* 69: 1072–1079.

Choisy, C., Gueguen, M., Lenoir, J., Schmidt, J.L., and Tournier, C. 1987. Microbiological aspects. In: Eck, A. (Ed.), *Cheesemaking, Science and Technology*. Lavoisier Publishing, New York, NY, pp. 259–292.

Chompreeda, P.T. and Fields, M.L. 1981. Effects of heat and fermentation on the extractability of minerals from soybean meal and corn meal blends. *J. Food Sci.* 49: 566.

Chompreeda, P.T. and Fields, M.L. 1984. Effect of heat and fermentation on amino acids, flatus producing compounds, lipid oxidation and trypsin inhibitor in blends of soybean and corn meal. *J. Food Sci.* 49: 563.

Chukeatirote, E., Chainun, C., Siengsubchart, Moukamnerd, C., and Chantawannakul, P. 2006. Microbiological and biochemical changes in Thua nao fermentation. *Res. J. Microbiol.* 1: 38–44.

Cogan, T.M., Beresford, T.P., Steele, J., Broadbent, J., Shah, N.P., and Ustunol, Z. 2007. Invited review: Advances in starter cultures and cultured foods. *J. Dairy Sci.* 90: 4005–4021.

Cogan, T.M. and Hill, C. 1993. Cheese starter cultures. In: Fox, P.F. (Ed.), *Cheese: Chemistry, Physics and Microbiology.* Vol. 1, 2nd edn, Chapman and Hall, London, pp. 193–255.

Collins, M.D., Rodrigues, U., Ash, C. et al. 1991. Phylogenetic analysis of the genus *Lactobacillus* and related lactic acid bacteria as determined by reverse transcriptase sequencing of 16S rRNA. *FEMS Microbiol. Lett.* 77: 5–12.

Condon, S. 1987. Responses of lactic acid bacteria to oxygen. *FEMS Microbiol. Rev.* 46: 269–280.

Conner, H.A. and Allgier, F.J. 1976. Vinegar: Its history and development. *Adv. Appl. Microbiol.* 20: 81–133.

Conway, P.L. 1996. Selection criteria for probiotics microorganisms. *Asia Pac. J. Clin. Nutr.* 5: 10–14.

Coppola, R., Nanni, M., Iorizzo, M., Sorrentino, A., Sorrentino, E., and Grazia, L. 1997. Survey of lactic acid bacteria isolated during the advanced stages of the ripening of Parmigiana Reggiano cheese. *J. Dairy Res.* 64: 305–310.

Corgan, T.M., Bresford, T.P., Steele, J., Broadbent, J., Shah, N.P., and Ustunol, Z. 2007. Advances in starter cultures. *J. Dairy Sci.* 90: 4005–4021.

Cronk, T.C., Steinkraus, K.H., Hackler, L.R., and Manick, L.R. 1977. Indonesian tape ketan fermentation. *Appl. Microbiol.* 33: 1067–1073.

Cuesta, P., Fernandez-Garcia, E., Llano, D.G.D., Montilla, A., and Rodriguez, A. 1996. Evolution of the microbiological and biochemical characteristic of Afuega's Pitu cheese during ripening. *J. Dairy Sci.* 79: 1693–1698.

Czerucka, D., Piche, T., and Rampal, P. 2007. Review article: Yeast as probiotics—*Saccharomyces boulardii*. *Aliment Pharmacol. Ther.* 26: 767–778.

Daeschel, M.A. 1989. Antibacterial substances from lactic acid bacteria for use as food preservatives. *Food Technol.* 43: 164–167.

Daeschel, M.A., Andersson, R.E. and Fleming, H.P. 1987. Microbial ecology of fermenting plant materials. *FEMS Microbiol. Rev.* 46: 357–67.

Dahiya, D.S. and Prabhu, K.A. 1977. Indian jackfruit wine. In: *Symposium on Indigenous Fermented Food*, Bangkok, Thailand.

Dakwa, S., Sakyi-Dawson, E., Diako, C., Annan, N.T., and Amoa-Awua, W.K. 2005. Effect of boiling and roasting on the fermentation of soybeans into dawadawa soy-dawadawa. *Int. J. Food Microbiol.* 104: 69–82.

Dansakul, S., Charoenchai, C., Urairong, H., and Leelawatcharamas, V. 2004. Identification of yeasts isolated from Thai fermented foods by sequence analysis of rDNA. *Poster* 11–12: 263.

Davidson, B.E., Llanos, R.M., Cancilla, M.R., Redman, N.C., and Hillier, A.J. 1995. Current research on the genetics of lactic acid production in lactic acid bacteria. *Int. Dairy J.* 5: 763–784.

Davis, C.R., Silveira, N.F.A., and Fleet, G.H. 1985a. Occurrence and properties of bacteriophages of *Leuconostoc oenos* in Australian wines. *Appl. Environ. Microbol.* 50: 872–876.

Davis, C.R., Wibowo, D., Eschenbruch, R., Lee, T.H., and Fleet, G.H. 1985b. Practical implications of malolactic fermentation: A review. *Am. J. Enol. Vitic.* 36: 290–301.

de Boer, E. and Kuik, D. 1987. A survey of the microbiological quality of blue-veined cheese. *Neth. Milk Dairy J.* 41: 227–237.

Delves-Broughton, J. 1990. Nisin and its uses as a food preservative. *Food Technol.* 44: 100–117.

De Paola, A., Motes, M.L., Chan, A.M., and Suttle, A. 1998. Phages infecting *Vibric vulnificus* are abundant and diverse in oysters. (*Crassostrea virginica*) collected from the Gulf of Mexico. *Appl. Environ. Microbiol.* 64: 346–351.

De Vuyst, L. 2004. Lactic acid bacteria as functional starter cultures for the food fermentation industry. *Trends Food Sci. Technol.* 15: 67–78.

De Vuyst, L. and Vandamme, E.J. 1994. Antimicrobial potential of lactic acid bacteria. In: De Vuyst, L. and Vandamme, E.J. (Eds.), *Bacteriocins of Lactic Acid Bacteria*. Blackie Academic and Professional, London, pp. 91–149.

De Vuyst, L., Vranckcn, G., Ravyts, R., Rimaux, T., and Weckx, S. 2009. Biodiversity, ecological determinants, and metabolic exploitations of sourdough microbiota. *Food Microbiol.* 26: 666–675.

Dewan, S. 2002. *Microbiological evaluation of indigenous fermented milk products of the Sikkim Himalayas*. PhD thesis, Food Microbiology Laboratory, Sikkim Government College, North Bengal University, India.

Dewan, S. and Tamang, J.P. 2006. Microbial and analytical characterization of Chhu, a traditional fermented milk product of the Sikkim Himalayas. *J. Sci. Ind. Res.* 65: 747–752.

Dewan, S., and Tamang, J.P. 2007. Dominant lactic acid bacteria and their technological properties isolated from the Himalayan ethnic fermented milk products. *Anton van Leeuwenhoek Int. J. General Mol. Microbiol.* 92(3): 343–352.

Diawara, B. and Jakobsen, M. 2004. Valorisation technologique et nutritionnelledu néré ou *Parkia biglobosa* Jacq, Benth: une espèceagroforestière Ouagadougou.

Dicks, L.M.T., Dellaglio, F., and Collins, M.D. 1995. Proposal to reclassify *Leuconostoc oenos* as *Oenococcus oeni* [corrig.] gen. nov., comb. nov. *Int. J. Syst. Bacteriol.* 45: 395–397.

Dielbandhoesing, S.K., Zhang, H., Caro, L.H.P. et al. 1998. Specific cell wall proteins confer resistance to nisin upon yeast cells. *Appl Environ Microbiol.* 64: 4047–4052.

Dietsch, K.W., Moxon, E.R., and Wellems, T.E. 1997. Shared themes of antigenic variation and virulence in bacterial, protozoal and fungal infections. *Microbiol. Mol. Biol. Rev.* 61: 281–293.

Dirar, H.A. 1993. The indigenous fermented foods of the Sudan. *A Study in African Food and Nutrition*. CAB International, Wallingford, Oxon, U.K.

Doneche, B. 1993. Botrytised wines. In: Fleet, G.H. (Ed.), *Wine Microbiology and Biotechnology*. Harwood Academic, Chur, Switzerland, pp. 327–352.

Drawert, F., Berger, R.G., and Neuhauser, K. 1983. Biosynthesis of flavor compounds by microorganisms. 4. Characterization of the major principles of the odour of *Pleurotus euosmus*. *Eur. J. Appl. Microbiol. Biotechnol.* 18: 124–127.

Drysdale, G.S. and Fleet, G.H. 1989. The effect of acetic acid bacteria upon the growth and metabolism of yeasts during the fermentation of grape juice. *J. Appl. Bacteriol.* 67: 471–481.

Dupuis, C., Corr, C., and Boyaval, P. 1995. Proteinase activity of dairy propionibacteria. *Appl. Microbiol. Biotechnol.* 42: 750–755.

Dworschak, E. 1982. Effect of processing on nutritive value of Fermentation. In: Rechciql, M. Jr. (Ed.), *Handbook of Nutritive Value of Processed*. Vol. 1, *Food for Human Use*, CRC Press, Cleveland, OH, pp. 63–76.

Ebner, H. 1982. Vinegar. In: Reed, G. (Ed.), *Prescott and Dunn's Industrial Microbiology*. 4th edn, Avi Pub. Co. Inc., Westport, CT, pp. 802–834.

Eklund, T. 1989. In: Gould, G.W. (ed.), *Mechanisms of Action of Food Preservation Procedures*. Elsevier, London.

Elyas, S.H.A., El Tinay, A.H., Yousif, N.E., and Elshelkh, E.A.E. 2002. Effect of natural fermentation on nutritive value and *in vitro* protein digestibility of pearl millet. *Food Chem.* 78: 75–79.

Eppert, I., Valdes-Stauber, N., Gotz, H., Busse, M., and Scherer, S. 1997. Growth reduction of *Listeria* spp. caused by undefined industrial red smear cheese cultures and bacterocin-producing *Brevibacterium linens* as evaluated in situ on soft cheese. *Appl. Environ. Microbiol.* 63: 4812–4817.

Fadda, S., López, C., and Vignolo, G. 2010. Role of lactic acid bacteria during meat conditioning and fermentation: Peptides generated as sensorial and hygienic biomarkers. *Meat Sci.* 86(1): 66–79.

Federal Register. 1988. Nisin preparation: Affirmation of GRAS status as a direct human food ingredient. *Fed. Reg.* 53: 11247–11251.

Fleet, G.H. 1990. Yeast in dairy products. *J. Appl. Bacteriol.* 68: 199–211.

Fleet, G.H. 1997. Wine. In: Doyle, M.P., Beuchat, L., and Montville, T. (Eds.), *Food Microbiology Fundamentals and Frontiers.* American Society for Microbiology, Washington, D.C., pp. 671–694.

Fleet, G.H. 1998. The microbiology of alcoholic beverages. In: *Microbiology of Fermented Foods.* 2nd edn, Blackie Academic and Professional, New York, pp. 217–262.

Fleet, G.H. 1999. Microorganisms in food ecosystems. *Int. J. Food Microbiol.* 50(1–2): 101–117.

Fleet, G.H. and Heard, G.M. 1993. Yeasts-growth during fermentation. In: Fleet, G.H. (Ed.), *Wine Microbiology and Biotechnology.* Harwood, Chur, Switzerland, pp. 27–55.

Fleet, G.H., Karalis, T., Hawa, A., and Lukondeh, T. 1991. A rapid method for enumerating Salmonella in milk powders. *Lett. Appl. Microbiol.* 13: 255–259.

Fleet, G.H., Lafon-Lafourcade, S., and Ribereau-Gayon, P. 1984. Evolution of yeasts and lactic acid bacteria during fermentation and storage of Bordeaux wines. *Appl. Environ. Microbiol.* 48: 1034–1038.

Flegel, T.W. 1988. Yellow-green *Aspergillus* strains used in Asian soybean fermentations. *ASEAN Food J.* 4: 14–30.

Fontaine, T., Hanlami, R.P., Beauvais, A., Diaquin, M., and Latge, J.P. 1996. Purificaiion and characterization of on endo-lS-1,3-glucanase from *Aspergillus fumigatus. Eur. J. Biochem.* 243: 315–321.

Fox, P.F., Lucey, J.A., and Cogan, T.M. 1990. Glycolysis and related reactions during cheese manufacture and ripening. *CRC Crit. Rev. Food Sci. Nutr.* 29: 237–253.

Frank, J.F. 2001. Milk and dairy products. In: Doyle, M.P., Beuchat, L.R., and Montville, T.J. (Eds.), *Food Microbiology: Fundamentals and Frontiers.* 2nd edn. ASM Press, Washington, D.C., pp. 111–126.

Fratamico, P.M. 1995. Factors involved in the emergence and persistence of foodborne disease. *J. Food Prot.* 58: 696–708.

Frazier, W.C. and Westhoff, C.D. 1997. *Food Microbiology,* 4th edn. Tata McGraw-Hill Book Company, New Delhi, pp. 5–10.

Frazier, W.C. and Westhoff, D.C. 1978. *Food Microbiology,* 3rd edn. Tata McGraw Hill Publ. Co. Ltd., New Delhi, pp. 17–34.

Frazier, W.C. and Westhoff, D.C. 2004. *Food Microbiol.* TMH, New Delhi.

Fredrickson, A.G. 1977. Behavior of mixed cultures of microorganisms. *Annu. Rev. Microbiol.* 31: 63–87.

Gadaga, T.H., Mutukumira, A.N., and Narvhus, J.A. 2001. Growth characteristics of *Candida kefyr* and two strains of *Lactococcus lactis* subsp *lactis* isolated from Zimbabwean naturally fermented milk. *Int. J. Food Microbiol.* 70: 11–19.

Gálvez, A, Abriouel, H., Ben Omar, N., and Lucas, R. 2011. Bacterioans: Food applications and regulations. In: Drider, D. and Rebuffat, S. (Eds.), *Prokaryotic Antimicrobiol Peptides: From genus to Applications.* Springer, New York, pp. 353–390.

Gálvez, A., Abriouel, H., Lucas, R., and Bin Omar, N. 2007. Bacteriocin-based strategies for food biopreservation. *Int. J. Food Microbiol.* 120: 51–70.

Gänzle, M.G., Loponen, J., and Gobbeti, M. 2008. *Trends Food Sci. Technol.* 19(10): 513–521.

Gänzle, M.G., Vermeulen, N., and Vogel, R.F. 2007. Carbohydrate, peptide and lipid metabolism of lactobacilli in sourdough. *Food Microbiol.* 24(2): 128–138.

Garvie, E.I. 1983. *Leuconostoc mesenteroides* ssp. *cremoris* (Knudsen and Sørensen) comb. nov. and *Leuconostoc mesenteroides* ssp. *dextranicum* (Beijerinck) comb. nov. *Int. J. Syst. Bacteriol.* 33: 118–119.

Garvie, E.I. 1986. Genus Leuconostoc van Tieghem 1878, 198 AL emend mut. char. Hucker and Pederson 1930, 66AL. In: Sneath, P.H.A., Mair, N.S., Sharpe, M.E., and Holt, J.G. (Eds.), *Bergey's Manual of Systematic Bacteriology*. Vol. 2, Williams & Wilkins, Baltimore, pp. 1071–1075.

Gashe, B.A. 1987. Kocho fermentation. *J. Appl. Bacteriol.* 62: 473–478.

Gilliland, R.B. 1980. Strain variation in brewing yeasts. In: Kirsop, B.E. (Ed.), *The Stability of Industrial Organisms*. CMI (CAB), Kew, pp. 15–19.

Gillis, M. and DeLey, J. 1980. Intra-and intergeneric similarities of the ribosomal ribonucleic acid cistrons of *Acetobacter* and *Gluconobacter*. *Int. J. Syst. Bacteriol.* 30: 7–27.

Giraffa, G. 2003. Functionality of enterococci in dairy products. *Int. J. Food Microbiol.* 88: 215–222.

Giraffa, G. 2004. Studying the dynamics of microbial populations during food fermentation. *FEMS Microbiol. Rev.* 28(2): 251–260.

Giraffa, G., Carminati, D., and Neviani, E. 1997. Enterococci isolated from dairy products: A review of risks and potential technological use. *J. Food Prot.* 60: 732–738.

Giraffa, G., Picchioni, N., Ncviani, E., and Canninati, D. 1995. Production and stability of an *Enterococcus faecium* bacteriocin during Taleggio cheese making and ripening. *Food Microbiol.* 12: 301–307.

Giraffa, G., Vecchi, P., Rossi, R., Nicastro, G., and Fortina, M.G. 1998. Genotypic heterogeneity among *Lactobacillus helveticus* strains isolated from natural cheese starters. *J. Appl. Microbiol.* 85: 411–416.

Giraud, E.L., Champailler, A., and Raimbault, M. 1994. Degradation of raw starch by a wild amylolytic strain of Lactobacillus plantarum. *Appl. Environ. Microbiol.* 60: 4319–4323.

Gobbeti, M., De Angelis, M., Corsetti, A., and Di Cagno, R. 2005. Biochemistry and physiology of sourdough lactic acid bacteria. *Trends Food Sci. Technol.* 16(1–3): 57–69.

Gould, G.W. 1992. Ecosystem approaches to food preservation. *J. Appl. Bacteriol.* 73(Suppl.): 58S–68S.

Gould, G.W. and Jones, M.V. 1989. Combination and synergistic effects. In: Gould, G.W. (Ed.), *The Mechanisms of Action of Food Preservation Procedures*. Elsevier Applied Science, London, U.K., pp. 401–421.

Goyal, R.K. 1999. Biochemistry of fermentation. In: *Biotechnology: Food Fermentation*, V.K. Joshi and Ashok Pandey, (Eds.), Educational Publisher and Distributor, Ernakulam and New Delhi vol. 1, pp. 87–137.

Gray, P. 1967. The dictionary of biological sciences. *Mol. Microbiol.* 23: 1089–1097.

Greer, G.G. and Dilts, B.D. 1990. Inability of a bacteriophage pool to control beef spoilage. *Int. J. Food Microbiol.* 10: 331–342.

Gripon, J.C. 1987. Mold-ripened cheeses. In: Fox, P.F. (Ed.), *Cheese: Chemistry, Physics and Microbiology*. Vol. 2, Major Cheese Groups, Elsevier Applied Science, London, pp. 121–149.

Gupta, S. and Abu-Ghannam, N. 2012. Probiotic fermentation of plant based products: Possibilities and opportunities. *Crit. Rev. Food Sci. Nutr.* 52(2): 183–199.

Hacker, J., Blum-Oehler, G., Muhidorfer, I., and Tschape, H. 1997. Pathogenicity islands of virulent bacteria: Structure function and impact on microbial evolution. *Mol. Microbiol.* 23: 1089–1097.

Hafiz, F. and Majid, A. 1996. Preparation of fermented foods from rice and pulses. *Bangladesh J. Sci. Ind. Res.* 31(4): 43–61.

Halm, M. and Olsen, A. 1996. The inhibitory potential of dominating yeasts and molds in maize fermentation. In: Halm, M. and Jakobsen, M. (Eds.), *3rd Biennial Seminar on African Fermented Food: Traditional Food Processing in Africa*. KVL, Copenhagen, Denmark, pp. 33–39.

Hamad, A.M. and Fields, M.L. 1979. Evaluation of the protein quality and available lysine of germinated and fermented cereals. *J. Food Sci.* 44: 456–459.

Hansen, E.B. 2002. Commercial: Bacterial starter cultures for fermented foods of the future. *Int. J. Food Microbiol.* 78: 119–131.

Harrigan, W.F. and McCance, M.E. (1966). *Laboratory Methods in Microbiology*, Academic Press, London and New York, pp. 199–229.

Hawa, S.G., Morisson, G.J., and Fleet, G.H. 1984. Method to rapidly enumerate *Salmonella* on chicken carcasses. *J. Food Prot.* 47: 932–936.

Heard, G.M. and Fleet, G.H. 1985. Growth of natural yeast flora during the fermentation of inoculated wines. *Appl. Environ. Microbiol.* 50: 727–728.

Heard, G.M. and Fleet, G.H. 1988. The effects of temperature and pH on the growth of yeasts species during the fermentation of grape juice. *J. Appl. Bacteriol.* 65: 23–28.

Heller, K.J. 2001. Probiotic bacteria in fermented foods: Product characteristics and starter organisms. *Am. J. Clin. Nutr.* 73(2 Suppl.): 374S–379S.

Hernawan, T. and Fleet, G.H. 1995. Chemical and cytological changes during the autolysis of yeasts. *J. Ind. Microbiol.* 14: 440–450.

Hesseltine, C.W. 1965. A millennium of fungi, food and fermentation. *Mycologia* 57(2): 149–197.

Hesseltine, C.W. 1983. Microbiology of oriental fermented foods. *Annu. Rev. Microbiol.* 37: 575–601.

Hesseltine, C.W. and Wang, H.L. 1967. Traditional fermented foods. *Biotechnol. Bioeng.* 9: 275–288. doi: 10.1002/bit.260090302.

Hofvendahl, K. and Hahn-Hägerdal, B. 2000. Factors affecting the fermentative lactic acid production from renewable resources (1). *Enzyme Microb. Technol.* 26(2–4): 87–107.

Holzapfel, W.H. 1997. Use of starter cultures in fermentation on a household scale. *Food Control* 8: 241–258.

Holzapfel, W.H. 2002. Appropriate starter culture technologies for small-scale fermentation on developing countries. *Int. J. Food Microbiol.* 75: 197–212.

Holzapfel, W.H. and Schillinger, U. 1992. The genus *Leuconostoc*. In: Balows, A., Truuper, H. G., Dworkin, M., Harder, W., and Schleifer, K.H. (Eds.), *The Prokaryotes*. Springer, New York, pp. 1508–1534.

Hounhouigan, D.J., Kayode, AP., Mestres, C., and Nago, C.M. 1999. Etude dc la mecanlsation du dccortlcagc du nuns pour la production du mawc. *Ann. Sci. Agronom. Benin.* 2: 99–113.

Hugenholtz, J. and De Veer, G.J.C.M. 1991. Application of nisin A and nisin Z in dairy technology. In: Jung, G. and Sahl, H.-G. (Eds.), *Nisin and Novel Lantibiotics*. ESCOM, Leiden, pp. 440–448.

Hurst, A. 1981. Nisin. *Adv. Appl. Microbiol.* 27: 85–123.

Hurtado, A., Reguant, C., Bordons, A., and Rozès, N. 2010. Evaluation of a single and combined inoculation of a *Lactobacillus pentosus* starter for processing cv. *Arbequina* natural green olives. *Food Microbiol.* 27: 731–40.

Hutkins, R.W. 2006. *Microbiology and Technology of Fermented Foods*. John Wiley and Sons, New York, NY.

Hutkins, R.W. and Morris, H.A. 1987. Carbohydrate metabolism in *Streptococcus thermophilus*: A review. *J. Food Prot.* 50: 876–884.

ICMSF. 1998a. *Microorganisms in Foods. 6. Microbial Ecology of Food Commodities*. Aspen Publishers Inc., Gaithersburg, MD. ISBN 0-8342-1825-9.

ICMSF. 1998b. Potential application of risk assessment techniques to microbiological issues related to international trade I: Food and food products. *J. Food Prot.* 61: 1075–1086.

Idris, W.H., Hassan, A.B., Babiker, E.E., and Tinay, A.E. 2005. Effect of malt pretreatment on antinutritional factors and HCl extractability of minerals of sorghum cultivars. *Pak. J. Nutr.* 4(6): 396–401.

Inui, T., Takeda, Y., and Iizuka, H. 1965. Taxonomical studies on genus *Rhizopus*. *J. Gen. Appl. Microbiol.* 11: 1–121.

Izquierdo-Pulido, M.L., Marine-Font, A., and Vjdal-Carou, M.C. 1994. *Food Sci.* 59: 1104–1107.

Jakobsen, M. and Narvhus, J. 1996. Yeasts and their possible beneficial and negative effects on the quality of dairy products. *Int. Dairy J.* 6: 755–768.

Jeandet, P., Vasserot, Y., Liger-Belair, G., and Marchal, R. (2011). Sparkling wine production. In: V.K. Joshi, (Ed.), *Handbook of Enology*. Vol II, Asia Tech Publishing Co., New Delhi, pp. 1064–1115.

Jespersen, L. 2003. Occurrence and taxonomic characteristics of strains *Saccharomyces cerevisiae* predominant in African indigenous fermented foods and beverages. *FEMS Yeast Res.* 3: 191–200.

Jeyaram, K., Komi, W., Singh, T.A., Devi, A.R., and Devi, S.S. 2010. Bacterial species associated with traditional starter cultures used for fermented bamboo shoot production in Manipur state of India. *Int. J. Food Microbiol.* 143: 1–8.

Jeyaram, K., Singh, W.M., Premarani, T., Devi, A.R., Chanu, K.S., Talukdar, N.C. and Singh, M.R. 2008a. Molecular identification of dominant microflora associated with "Hawaijar"—A traditional fermented soybean (*Glycine max* (L.)) food of Manipur, India. *Int. J. Food Microbiol.* 122: 259–268.

Jeyaram, K., Singh, A., Romi, W., Devi, A.R., Singh, W.M., Dayanithi, H., Singh, N.R., Tamang, J.P. et al. 2009. Traditional fermented foods of Manipur. *Indian J. Trad. Knowledge* 8(1): 115–121.

Johnson, M. and Steele, J. 1997. Fermented dairy products. In: Doyle, M., Beuchat, L., and Montville, T. (Eds.), *Food Microbiology—Fundamentals and Frontiers*. ASM Press, Washington, DC, pp. 581–594.

Joshi, N., Godbole, S.H., and Kanekar, P. 1989. Microbial and biochemical changes during dhokla fermentation with special reference to flavor compounds. *J. Food Sci. Technol.* 26: 113–115.

Joshi, V.K. 1997. *Fruit Wines*. 2nd edn, Directorate of extension education, Dr. Y.S. Parmar University of Horticulture and Forestry, Nauni-Solan, India.

Joshi, V.K. 2006. Food fermentation: Role, significance and emerging trends. In: Trivedi, P.C. (Ed.), *Microbiology: Applications and Current Trends*. Pointer Publisher, Jaipur, pp. 1–35.

Joshi, V.K., Attri, D. Singh, T.K., and Abrol, G.S. 2011c. Fruit wines: Production Technology. In V.K. Joshi (Ed.), *Handbook of Enology: Principles, Practices and Recent Innovations*. Vol II. AsiaTech Publ., New Delhi, pp. 1177–1221.

Joshi, V.K., Kaushal, N.K and Sharma, N. 2000. Spoilage of fruits, Vegetables and their processed products. In: L.R. Verma and V.K. Joshi (Eds). Vol. II Indus Publi., New Delhi, pp. 235–284.

Joshi, V.K. and Sandhu, D.K. 2000. Quality evaluation of naturally fermented alcoholic beverages, microbiological examination of source of fermentation and ethanolic productivity of the Isolates. *Acta Aliment.* 29(4): 323–334.

Joshi, V.K., Sandhu, D.K., and Thakur, N.S. 1999. Fruit based alcoholic beverages. In: Joshi, V.K., Pandey, A. (Eds.), *Biotechnology: Food Fermentation. Microbiology, Biochemistry and Technology*. Vol. II, Educational Publishers and Distributors, New Delhi, Ernakulam and, Calcutta, India, p. 647.

Joshi, V.K. and Pandey, A. 1999. Biotechnology: Food fermentation. In: V.K. Joshi and A. Pandey (Eds.), *Biotechnology: Food fermentation: Microbiology, Biochemistry and Technology*. Vol. 1. Educational Publishers and Distributors, New Delhi, Ernakulam, India. pp. 1–37.

Joshi, V.K., Sharma, P.C., and Attri, B.L. 1991. A note on the deacidification activity of *Schizo saccharomyces pombe* in plum must of variable composition of composition. *J. Appl. Bacteriol.*, 70: 386–390.

Joshi, V.K. and Sharma, S. 2009. Cider vinegar: Microbiology, technology and quality. In: Solieri, L. and Gludici, P. (Eds.), *Vinegars of the World*. Springer-Verlag, Italy, pp. 197–207.

Joshi V.K. and Sharma, S. 2012. A panorama of lactic acid bacterial fermentation of vegetables. *Int. J. Food Ferment. Technol.* 2(1): 1–12.

Joshi, V.K., Sharma, R., and Kumar, V. 2011a. Antimicrobiol activity of essential oil: A review. *Intl. J. Food Ferm. Technol.*, 1(2): 161–172.

Joshi, V.K., Sharma, S., and Neerja, R. 2006. Production, purification, stability and efficiency of bacteriocin from the isolate of natural lactic acid fermentation of vegetables. *Food Technol. Biotechnol.* 4(3): 435–439.

Joshi, V.K. and Singh, R.S. (Eds.) 2012. *Food Biotechnology: Principles and Practices.* IK International Publishing House, New Delhi, p. 920.

Joshi, V.K. and Thakur, N.S. 2000. Vinegar: Composition and production. In: L.R. Verma, and V.K. Joshi, (Eds.), *Postharvest Technology of Fruits and Vegetables.* Indus Publishing Co., New Delhi. pp. 1128–1170.

Joshi, V.K., Thakur, N.S., Bhatt, A., and Garg, C. 2011b. Wine and brandy. In: Joshi, V.K. (Ed.), *Handbook of Enology: Principles, Practices and Recent Innovations.* Vol 2, Asia-Tech Publisher and Distributors, New Delhi, pp. 1–45.

Kalui, C.M., Julius, M.M., and Kutima, P.M. 2010. Probiotic potential of spontaneously fermented cereal based foods—A review. *Afr. J. Biotechnol.* 9(17): 2490–2498.

Kalui, C.M., Mathara, J.M., Kutima, P.M., Kiiyukia, C., and Wongo, L.E. 2009. Functional characteristics of *Lactobacillus plantarum* and *Lactobacillus rhamnosus* from ikii, a Kenyan traditional fermented maize porridge. *Afr. J. Biotechnol.* 8(17): 4363–4373.

Kaminarides, S.E., Amfantakis, E.M., and Balis, C. 1992. Changes in Kopanisti cheese during ripening using selected pure microbial cultures. *J. Sci. Dairy Technol.* 45: 56–60.

Kanekar, P. and Joshi, N. 1993. *Lactobacillus fermentum, Leuconostoc mesenteroides* and *Hansenula silvicola* contributing to acetoin and folic acid during "dhokla" fermentation. *Indian J. Microbiol.* 33: 111–117.

Kang, O.J., Vezinz, L.P., Laberge, S., and Simard, R.E. 1998. Some factors influencing the autolysis of *Lactobacillus bulgaricus* and *Lactobacillus casei. J. Dairy Sci.* 81: 639–646.

Kanwar, S.S., Gupta, M.K., Katoch, C., Kumar, R., and Kanwar, P. 2007. Traditional fermented foods of Lahaul and Spiti area of Himachal Pradesh. *Indian J. Tradit. Knowl.* 6(1): 42–45.

Karr-Lilienthal, L.K., Kadzere, C.T., Grieshop, C.M., and Fahey, G.C. 2005. Chemical and nutritional properties of soybean carbohydrates as related to nonruminants: A review. *Livestock Prod. Sci.* 97: 1–12.

Katina, K. 2005. Sourdough: A tool for the improved flavor, texture and shelf-life of wheat bread. *VTT Publ.* 569: 1–92.

Kandler, O. 1983. Carbohydrate metabolism in Lactic acid bacteria. *Antonie* van Leeuwenhoek, 49: 209–224.

Kell, D.B., Kaprelyants, A.S., Weichart, D.H., Harwood, C.R., and Baier, M.R. 1998. Viability and activity in readily culturable bacteria: A review and discussion of practical issues. *Antoin. Leeuw.* 73: 169–187.

Khalid, K. 2011. An overview of lactic acid bacteria. *Int. J. Biosci.* 1(3): 1–13.

Khanna, P. and Dhaliwal, Y.S. 2013. Effect of germination and probiotic fermentation on the nutritional and organoleptic acceptability value of cereal based food mixtures. *Scholarly J. Agric. Sci.* 3(9): 367–373.

Khetarpaul, N. and Chauhan, B.M. 1989. Effect of fermentation by pure cultures of yeasts and lactobacilli on phytic acid and polyphenol content of pearl millet. *J. Food Sci.* 54: 780–781.

Khetarpaul, N. and Chauhan, B.M. 1990. Effect of germination and fermentation on in vitro starch and protein digestibility of pearl millet. *J. Food Sci.* 55: 883–884.

Kiers, J.L., Van Laeken, A.E.A., Rombouts, F.M., and Nout, M.J.R. 2000. *In vitro* digestibility of *Bacillus* fermented soya bean. *Int. J. Food Microbiol.* 60: 163–169.

Kimaryo, V.M., Massawe, G.A., Olasupo, N.A., and Holzapfel, W.H. 2000. The use of a starter culture in the fermentation of cassava for the production of kivunde, a traditional Tanzanian food product. *Int. J. Food Microbiol.* 56: 179–190.

King, A.D. 1982. Identification of some volatile constituents of *Aspergillus clavatns. J. Agric. Food Chem.* 30: 786–790.

King, T.E. and Chedelin, V.H. 1952. Oxidative dissimilation in *Acetobacter suboxydans. J. Biol. Chem.* 198: 127–133.

Kilsby, D.C. 1999. Food microbiology: The challenges for the future. *Int. J. Food Microbiol.* 15, 50(1–2): 59–63.

Klaenhammer T.R. 1993. Genetics of bacteriocins produced by lactic acid bacteria. *FEMS Microbiol. Rev.* 12: 39–85.

Klich, M.A. and Pitt, J.I. 1985. The theory and practice of distinguishing species of the *Aspergillus flavus* group. In: Samson, R.A. and Pitt, J.I. (Eds.), *Advances in Penicillium and Aspergillus Systematics.* Plenum Press, New York, NY, pp. 211–220.

Ko, S.D. 1972. Tape fermentation. *Appl. Microbiol.* 23: 976–978.

Kok, M.F. 1987. *The growth and linalool production of Penicillium italicum.* MSc thesis, Nat. Univ. Singapore, Singapore.

Kolter, R., Siegala, D.A., and Tomo, A. 1993. The stationary phase of the bacterial life cycle. *Annu. Rev. Microbiol.* 47: 855–874.

Kostinek, M., Specht, I., Edward, V.A., Schillinger, U., Hertel, C., Holzapfel, W.H., and Franz, C.M.A.P. 2005. Diversity and technological properties of predominant lactic acid bacteria from fermented cassava used for the preparation of Gari, a traditional African food. *Syst. Appl. Microbiol.* 28: 527–540.

Kreger-van Rij, N.J.W. (Ed.) 1984. *The Yeasts, A Taxonomic Study.* Elsevier, Amsterdam.

Kulshrestha, D.C. and Marth, E.H. 1974. *J. Milk Food Technol.* 37: 606–611.

Kumbhar, S.B., Ghosh, J.S., and Samudre, S.P. 2009. Microbiological analysis of pathogenic organisms in indigenous fermented milk products. *Adv. J. Food Sci. Technol.* 1(1): 35–38.

Kunji, E.R.S., Mierau, I., Hagting, A., Poolman, B., and Konings, W.N. 1996. The proteolytic system of lactic acid bacteria. *Anton. Leeuw.k* 70: 187–221.

Kyung, K.H. and Fleming, H.P. 1997. Antimicrobial activity of sulfur compounds derived from cabbage. *J. Food Prot.* 60: 67–71.

Lang, J.M. and Cirillo, V.P. 1987. Glucose transport in Kinase less *Saccharomyces cerevisiar. J. Bacteriol.*, 169, 2932–2937.

Law, B.A. 1982. Cheeses. In: Rose, A.H. (Ed.), *Fermented Foods.* Vol. 7, *Economic Microbiology* Academic Press, London, pp. 148–198.

Law, S.V., Abu Bakar, F., Mat Hashim, D., and Abdul Hamid, A. 2011. Mini review: Popular fermented foods and beverages in Southeast Asia. *Int. Food Res. J.* 18: 475–484.

Lechardeur, D., Cesselin, B., Fernandez, A. et al. 2011. Using heme as an energy boost for lactic acid bacteria. *Curr. Opin. Biotechnol.* 22: 143–149.

Lee, C.H. 1990. Fish fermentation technology—A review. *Proceedings of the Workshop on Post-Harvest Technology, Preservation and Quality of Fish in Southeast Asia.* Echanis Press, Manila, Philippines, pp. 1–13.

Lee, C.H. 1990. Fish fermentation technology—A review. In: Reilly, P.J.A., Parry, R.W.H., and Berley, E. (Eds.), *Proceedings of the Workshop on Post-Harvest Technology, Preservation and Quality of Fish in Southeast Asia.* Echanis Press, Manila, Philippines, pp. 1–13.

Lee, H.J. and Lee, S.J. 2002. Characterization of cytoplasmic alpha-synuclein aggregates. Fibril formation is tightly linked to the inclusion-forming process in cells. *J. Biol. Chem.* 277: 48976–48983.

Lee, K.B. 2005. Comparison of fermentative capacities of lactobacilli in single and mixed culture in industrial media. *Orig. Res. Art. Process Biochem.* 40(5): 1559–1564.

Leejeerajumnean, A., Duckham, S.C., Owens, J.D., and Ames, J.M. 2001. Volatile compounds in *Bacillus* fermented soybeans. *J. Sci. Food Agric.* 81: 525–529.

Leistner, L. and Gorris, L.G.M. 1995. Food preservation by hurdle technology. *Trends Food Sci. Technol.* 6: 41–46.

Lemee, R., Lortal, S., Cesselin, B., and Van Hrijenoort, J. 1994. Involvment of an N-acetylglucosamine in autolysis of *Propionibacterium freudenreichii* CNRZ 725. *Appl. Environ. Microbiol.* 60: 4351–4356.

Leroy, F. and De Vuyst, L. 2004. Functional lactic acid bacteria starter cultures for the food fermentation industry. *Trends Food Sci. Technol.* 15: 67–78.

Levroy, S., Autin, J., Razin–Julia, P., Bache, F., d'Acremont, E., Watremez, L., Robinet, J. et al. 2012. From rifting to oceanic spreading in the Gulf of Aden: A synthesis. *Arabian J. Geosci.*, 5: 859–901. http://dx.doi.org/10.1007/s12517-011-0475-4

Lim, G., Tan, T.K., and Rahim, N.A. 1987. Variations in amylase and protease activity among *Rhizopus* isolates. *MIRCEN J. Appl. Microbiol. Biotechnol.* 3: 319–322.

Limtong, S., Sintara, S., Suwanarit, P., and Lotong, N. 2002. Yeast diversity in Thai traditional alcoholic starter. *Kasetsart J. (Nat. Sci.).* 36: 149–158.

Lindgren, S.E. and Dobrogosz, W.J. 1990. Antagonistic activities of lactic acid bacteria in food and feed fermentations. *FEMS Microbiol. Rev.* 87: 149–163.

Liu, S.Q. 2002. A review: Malolactic fermentation in wine—Beyond deacidification. *J. Appl. Microbiol.* 92: 589–601.

Liu, S.Q. 2003. Practical implications of lactate and pyruvate metabolism by lactic acid bacteria in food and beverage fermentations. *Int. J. Food Microbiol.* 83: 115–131.

Llorente, P., Marquina, D., Santos, A., Peinado, J.M., and Spencer-Martins, I. 1997. Effect of salt on the killer phenotype of yeasts from olive brines. *Appl. Environ. Microbiol.* 63: 1165–1167.

Lodder, J. (Ed.) 1970. *The Yeasts.* North-Holland Publ., Amsterdam, p. 1385.

Lund, B.M. 1992. Ecosystems in vegetable foods. *J. Appl. Bacteriol.* 73(Suppl.): 115S–126S.

Lyon, W.J., Sethi, J.E., and Glatz, B.A. 1993. Inhibition of psychrotrophic organisms by propionicin PLG-1, a bacteriocin produced by *Propionibacterium thoenii*. *J. Dairy Sci.* 76: 1506–1513.

Macedo, A.C., Malcata, F.Y., and Oliveira, J.C. 1993. The technology chemistry and microbiology of Serra cheese: A review. *J. Dairy Sci.* 76: 1725–1739.

Maing, H.Y., Ayers, J.C., and Koethler, P.E. 1973. *Appl. Microbiol.* 25: 1015.

Manabe, M., Tanaka, K., Goto, T., and Matsuura, S. 1984. Producing capability of kojic acid and aflatoxin by koji mold. In: Kurata, H. and Ueno, Y. (Eds.), *Toxi-Genic Fungi—Their Toxins and Health Hazard.* Elsevier, New York, NY, pp. 4–14.

Maneewatthana, D., Rapeesak, T., and Suntornsuk, W. 2000. Isolation and identification of yeasts from fermented vegetable brine. *KMUTT Res. Dev. J.* 23(3): 47–62.

Marcellino, S.N. and Benson, D.R. 1992. Scanning and light microscopy of microbial succession on Bethlehem St. Nectaire cheese. *Appl. Environ. Microbiol.* 58: 3448–3454.

Marilley, L. and Casey, M.G. 2004. Flavors of cheese products: Metabolic pathways, analytical tools and identification of producing strains. *Int. J. Food Microbiol.* 90: 139–59.

Martini, A., Ciani, M., and Scorzetti, G. 1996. Direct enumeration and isolation of wine yeasts from grape surfaces. *Am. J. Enol. Viticult.* 47: 435–440.

Martineau, B. and Henick-Kling, T. 1995. Performance and diacetyl production of commercial strains of malolactic bacteria in wine. *J. Appl. Bacteriol.* 78: 526–536.

Matsuura, S. 1970. Aflatoxins and fermented foods in Japan. *JARO* 5: 46–51.

Matsuura, S. Manabo, M., and Saro, T. 1970. *Proc. U.S.-Japan Conf. Toxic Mtcro-Org 1968*, pp. 48–55.

Mayo, B., Piekarczyk, T.A., Fernández, M. et al. 2010. Updates in the metabolism of lactic acid bacteria. Chapter 1. In: Mozzi, F., Raya, R.R., and Vignolo, G.M. (Eds.), *Biotechnology of Lactic Acid Bacteria. Novel Applications.* Wiley-Blackwell, Singapore, pp. 3–33.

Mayra-Makinen, A. and Suomalainen, T. 1995. *Lactobacillus casei* spp. *rhamnosus,* bacterial preparations comprising said strain and use of said strain and preparations for the controlling of yeast and moulds. U.S. Patent US 5: 378–458.

McDougald, D., Rice, S.A., Weichart, D., and Kjelleberg, S. 1998. Nonculturability: Adaptation or debilitation. *FEMS. Microbiol. Ecol.* 25: 1–9.

McKay, A.M. 1992. Viable but non-culturable forms of potentially pathogenic bacteria in water. *Lett. Appl. Microbiol.* 14: 129–135.

Mensah, P. 1997. Fermentation-the key food safety assurance in Africa? *Food Control* 8: 271–278.

Mian, M., Fleet, G.H., and Hocking, M.D. 1997. Effect of diluent type on viability of yeasts enumerated from foods or pure culture. *Int. J. Food Microbiol.* 35: 103–107.

Miller, M.W. 1982. Yeasts. In Reid, G. (Ed.), *Prescott and Dunn's Industrial Microbiology*, 4th edn. AVI Publishing Co., Westport, Connecticut.

Mirelman, D., Altman, G., and Eshdat, Y. 1980. Screening of bacterial isolates for mannose-specific lectin activity by agglutination of yeasts. *J. Clin. Microbiol.* 11: 328–331.

Moktan, B. and Sarkar P.K. 2007. Characteristics of *Bacillus cereus* isolates from legume-based Indian fermented foods. *Food Control.* 18: 1555–1564.

Montville, T.J. and Winkowski, K. 1997. Biologically based preservation systems and probiotic bacteria. In: Doyle, M.P., Beuchat, L.R., and Montville, T.J. (Eds.), *Food Microbiology: Fundamentals and Frontiers.* ASM Press, Washington, D.C., pp. 557–578.

Moon, N.J. and Reinbold, G.W. 1976. Commensalism and competition in mixed cultures of *Lactobacillus bulgaricus* and *Streptococcus thermophilus. J. Milk Food Technol.* 39: 337–341.

Morgan, R.J.M., Williams, F., and Wright, M.M. 1997. An early warning scoring system for detecting developing critical illness. *Clin. Intensive Care* 8: 100.

Mosha, A.C. and Svanberg, U. 1983. Preparation of weaning foods with high nutrient density using flour of germinated cereals. *Food Nutr. Bull.* 5: 10–14.

Moss, M.O. 1977. Aspergillus mycotoxins. In: Smith, J.E., and Paterman, J.A. (Eds.), *Genetics and Physiology of Aspergillus.* Acad. Press, New York, NY, pp. 499–538.

Motarjemi, Y. and Nout, M.J.R. 1996. *Food Fermentation: A Safety and Nutritional Assessment.* Bulletin of the World Health Organization.

Mugula, J.K., Narvhus, J.A., and Sørhaug, T. 2003. Use of starter cultures of lactic acid bacteria and yeasts in the preparation of togwa, a Tanzanian fermented food. *Int. J. Food Microbiol.* 83(3): 307–318.

Mukherjee, S.K., Albury, M.N., Pederson, C.S., Van Veen, A.G., and Steinkraus, K.H. 1965. Role of *Leuconostoc mesenteroides* in leavening the batter of *idli*—a fermented food of India. *Appl. Microbiol.* 13: 227–231.

Murakami, H. and Suzuki, M. 1970. Mycological differences between the producer and non-producer of afla-toxin of Aspergillus. In: Herzberg, M. (Ed.), *Proc. 1st U.S.-Japan Conf. on Toxic Microorganisms.* U.S. Dept. of the Int., USA, pp. 198–201.

Murali, A. and Kapoor, R. 2003. Effect of natural and pure culture fermentation of finger milleton zinc availability as predicted from HCl extractability and molar ratios. *J. Food Sci. Technol.-Mysore* 40: 112–114.

Muriana, P.M. 1996. Bacteriocins for control of *Listeria* spp. in food. *J. Food Protect. Suppl.* 54–63.

Muyanja, B.K., Naruhus, J.A., and Langsrud, T. 2003. Isolation, characterization and identification of lactic acid bacteria from Bushera: A Ugandan traditional fermented beverage. *Int. J. Food Microbiol.* 80(3): 201–210.

Nagy, E., Peterson, M., and Maardh, P.-A. 1991. Antibiosis between bacteria isolated from the vagina of women with and without signs of bacterial vaginosis. *Acta Pathol. Microbiol. Immunol. Scand.* 99: 739–744.

Navarrete-Bolaños, J.L. 2012. Improving traditional fermented beverages: How to evolve from spontaneous to directed fermentation. *Eng. Life Sci.* 12: 410–418. doi: 10.1002/elsc.201100128.

Nes, I.F., Diep, D.B., Havarslien, L.S., Brurberj, M.B., Eijsink, V., and Holo, H. 1996. Biosynsthesis of acteriocins in lactic acid bacteria. *Antoin. Leeuw.* 70: 113–128.

Neti, Y., Erlinda, I.D., and Virgilio, V.G. 2011. The effect of spontaneous fermentation on the volatile flavour constituents of durian. *Int. Food Res. J.* 18: 625–631.

Neves, A.R., Pool, W.A., Kok, J., Kuipers, O.P., and Santos, H. 2005. Overview on sugar metabolism and its control in *Lactococcus lactis*—The input from *in vivo* NMR. *FEMS Microbio. Rev.* 29: 531–554.

Niba, L. 2003. The relevance of biotechnology in the development of functional foods for improved nutritional and health quality in developing countries. *Afr. J. Biotechnol.* 2: 631–635.

Nigam, P.S. 2011. Microbiology of Wine Making. In: *Handbook of Enology: Principles, Practices and Recent Innovations*. V.K. Joshi, ed. Vol. II. Asia Tech. Publ. New Delhi. pp. 383–405.

Nooitgedagt, A.J. and Hartog, B.J. 1988. A survey of the microbiological quality of Brie and Camembert cheese. *Neth. Milk Dairy J.* 42: 57–72.

Norton, S. and D'Amore, T. 1994. Physiological effects of yeast cell immobilisation-applications for brewing. *Enzyme Microbial. Tech-nol.* 16: 365–375.

Notermans, S. and Teunis, P.F.M. 1996. Quantitative risk analysis and the production of microbiologically safe food: An introduction. *Int. J. Food Microbiol.* 30: 3–7.

Nout, M.J.R. 1995. Useful role of fungi in food processing. In: Samson, R.A., Hoekstra, E., Frisvad, J. C., and Filtenborg, O. (Eds.), *Introduction to Food-Borne Fungi*. pp. 295–303.

Nout, M.J.R., Bakshi, D., and Sarkar, P.K. 1998. Microbiological safety of *kinema*, a fermented soya bean food. *Food Control* 9: 357–362.

Nout, M.J.R. and Motarjemi, Y. 1998. Assessment of fermentation as a household technology for improving food safety: A joint FAO/WHO workshop. *Food Control* 8(5–6): 221–226.

Nout, M.J.R., Nche, P.F., and Hollman, P.C.H. 1994. Investigation of the presence of biogenis amines and ethyl carbamate in kenkey made with maize and maize cowpea mixtures as influenced by process conditions. *Food Addit. Contam.* 11: 397–402.

Nout, M.J.R. and Sarkar, P.K. 1999. Lactic acid food fermentation in tropical climates. *Anton van Leeuwenhoek*, 76: 395–401.

Nout, R. 2005. Food fermentation: An introduction. In: Nout, R.M.J., de Vos, W.M., and Zwietering, M.H. (Eds.), *Food Fermentation*. Academic Publishers, Wageningen.

Obizoba, C.I. and Atii, V.J. 1994. Evaluation of the effect of processing techniques on the nutrient and antinutrient of pearl millet (*Pennisetum glaucum*) seeds. *Plant Food Human Nutr.* 45: 23–34.

Odunfa, S.A. 1981. Microorganisms associated with the fermentationof African locust bean *Parkia filicoidea*, during "iru" fermentation. *J. Plant Foods* 3: 245–250.

Odunfa, S.A. 1985a. Biochemical changes in fermenting African locust bean (*Parkia biglobosa*) during iru fermentation. *J. Food Technol.* 20: 295–303.

Odunfa, S.A. 1985b. Microbiological and toxicological aspect of fermentation of castor oil seeds for ogiri production. *J. Food Sci.* 50: 1758–1759.

Odunfa, S.A. and Oyewole, O.B. 1986. Identification of Bacillus species from iru, a fermented African locust bean product. *J. Basic Microbiol.* 26: 101–108.

Odunfa, S.A. and Oyewole, O.B. 1998. African fermented foods. In: Wood, B.J.B. (Ed.), *Microbiology of Fermented Foods*. Vol. 2, 2nd edn, Blackie Academic and Professional, London, pp. 713–752.

Oki, K., Rai, A.K., Sato, S., Watanabe, K., and Tamang, J.P. 2011. lactic acid bacteria isolated from ethnic preserved meat products of the Western Himalayas. *Food Microbiol.* 28: 1308–1315.

Olson, N.F. 1995. Cheese. In: Reed, G., Nagodawithana, T.W. (Eds.), *Biotechnology*. Vol. 9, 2nd edn, *Enzymes, Biomass, Food and Feed*, VCH, Weinheim, pp. 353–384.

Omafuvbe, B.O., Shonukan, O.O., and Abiose, S.H. 2000. Microbiologicaland biochemical changes in the traditional fermentation ofsoybean for "*soy-daddawa*"—Nigerian food condiment. *Food Microbiol.* 17: 469–474.

Onyango, C., Noetzold, H., Bley, T., and Henle, T. 2004. Proximate composition and digestibility of fermented and extruded uji from maize-finger millet blend. *Lebensmittel Wiss.-Technol. (Food Sci. Technol.)* 37: 827–832.

Onyango, E.M., Bedford, M.R., and Adeola, O. 2005. Efficacy of an evolved *Esherichia coli* phytase in diets of broiler chicks. *Poult Sci.* 84: 248–255.

Orth, R. 1977. Mycotoxins of aspergillus oryzae. Strains for use in the food industry as starters and enzyme producing molds. *Ann. Nutr. Aliment.* 31: 617–624.

Ostlie, H., Vegarud, G., and Langsrud, T. 1995. Autolysis of lactococci: Detection of lytic enzymes by polyacrylamide gel electrophoresis and characterization in buffer systems. *Appl. Environ. Microbiol.* 61: 3598–3603.

Ouoba, L.I.I., Diawara, B., Moa-Awua, W.K., Traore, A.S., and Moller, P.L. 2004. Genotyping of starter cultures of *Bacillus subtilis* and *Bacillus pumilus* for fermentation of African locust bean *Parkia biglobosa*, to produce soumbala. *Int. J. Food Microbiol.* 90: 197–205.

Ouoba, L.I.I., Lei, V., Jensen, L.B. 2008. Resistance of potential probiotic lactic acid bacteria and bifidobacteria of African and European origin to antimicrobials: Determination and transferability of the resistance genes to other bacteria. *Int. J. Food Microbiol.* 121: 217–224.

Ouoba, L.I.I., Nyanga-Koumou, C.A.G., Parkouda, C., Sawadogo, H., Kobawila, S.C., Keleke, S. et al. 2010. Genotypic diversity of lactic acid bacteria isolated from African traditional alkaline-fermented foods. *J. Appl. Microbiol.* 108: 2019–2029. 10.1111/j.1365-2672.2009.04603.x.

Oyewole, O.B. 1990. Optimization of cassava fermentation for fufu production: Effect of single starter cultures. *J. Appl. Bacteriol.* 68: 49–54.

Oyewole, O.B. 1997. Lactic fermented foods in Africa and their benefits. *Food Control* 8: 289–297.

Palleroni, N.J. 1993. Pseudomonas classication. A new case history in the taxonomy of Gram-negative bacteria. *Anton. Leeuw. Int. J. Microbiol.* 64: 231–251.

Pariza, M. 1996. Processed food and their toxic constituents of animal food stuffs. *Biochem. J.* 219: 1–14.

Parker, M.L., Gunning, P.A., Macedo, A.C., Malcata, F.X., and Brocklehurst, T.F. 1998. The microstructure and distribution of microorganisms within mature serra cheese. *J. Appl. Bacteriol.* 84: 523–530.

Parkouda, C., Nielsen, D.S., Aokpota, P., Ouoba, L.I.I., Amoa-Awua, W.K., and Thorsen, L. 2009. The microbiology of alkaline-fermentation of indigenous seeds used as food condiments in Africa and Asia. *Crit. Rev. Microbiol.* 35: 139–156.

Parkouda, C., Thorsen, L., Compaorc, C.S., Nielsen, D.S., Tano-Dcbrah, K., and Jensen, J. 2010. Microorganisms associated with Maari, a baobab seed fermented product. *Int. J. Food Microbiol.* 142: 292–301.

Parvez, S., Malik, K.A., Ah Kang, S., and Kim, H.Y. 2006. Probiotic and their fermented food products are beneficial for health. *J. Appl. Microbiol.* 100: 1171–1185.

Pathania, N., Kanwar, S.S., Jhang, T., Koundal, K.R., and Sharma, T.R. 2010. Application of different molecular techniques for *deciphering* genetic diversity among yeast isolates of traditional fermented food products of Western Himalayas. *World J. Microbiol. Biotechnol.* 26: 1539–1547.

Pederson, C.S. 1973. Microbiology of Food Fermentation. AVI Publishing Co., Westport, CT.

Phaff, H.J., Starmer, W.T., Miranda, M., and Miller, M.W. 1978. *Pichia heedii*, a new species of yeast indigenous to necrotic cacti in the North American Sonoran desert. *Int. J. Syst. Bacteriol.* 28: 326–331.

Praphailong, W. and Fleet, G.H. 1998. The effect of pH, sodium chloride, sorbate and benzoate on the growth of food spoilage yeasts. *Food Microbiol.* 14: 459–468.

Priest, F.G. and Campbell, I. (Eds.) 1987. *Brewing Microbiology*, Elsevier Applied Science Ltd,, England.

Pruitt, K.M., Tenovuo, J., Mansson-Rahemtulla, B., Harrington, P., and Baldone, D.C. 1986. Is thiocyanate peroxidation at equilibrium in vivo? *Biochim. Biophys. Acta* 870: 385–391.

Puspito, H. and Fleet, G.H. 1985. Microbiology of Sayur-Asin fermentation. *Appl. Microbiol. Biotechnol.* 22: 442–445.

Querol, A. and Fleet, G. 2006. *Yeasts in Food and Beverages*. Springer-Verlag, Berlin, Germany.

Querol, A. and Ramon, D. 1996. The application of molecular techniques in wine microbiology. *Trends Food Sci. Technol.* 7: 73–78.

Radhakrishnamurthy, R., Desikachar, H.S.R., Srinivasan, M., and Subrahmanyan, V. 1961. Studies on idli fermentation II. Relative participation of blackgram flour and rice semolina in the fermentation. *J. Sci. Ind. Res.* 20: 342–344.

Rai, A.K., Tamang, J.P., and Palni, U. 2010. Microbiological studies of ethnic meat products of the Eastern Himalayas. *Meat Sci.* 85: 560–567.

Rajalakshmi, R. and Vanaja, K. 1967. Chemical and biological evaluation of the effects of fermentation on the nutritive value of foods prepared from rice and grams. *Br. J. Nutr.* 22: 467–473.

Ramakrishnan, 1977. The use of fermented foods in India. *Symposium on Indigenous Fermented Foods*, Bangkok, Thailand.

Ramakrishnan, C.V. 1979. The studies on Indian fermented food. *Baroda J. Nat.* 6: 1–57.

Rana, N. and Rana, V. 2011. Biochemistry of wine preparation. In: Joshi, V.K. (Ed.), *Handbook of Enology*. Vol. 2, Asiatech Publishers, Inc., New Delhi, pp. 618–666.

Rana, T.S., Datta, B., and Rao, R.R. 2004. Soor. A traditional alcoholic beverage in Tons Valley, Garhwal, Himalaya. *Indian J. Tradit. Know.,* 3: 59–65.

Rao, B.R.P., Babu, M.V.S., Reddy, M.S., Reddy, A.M., Rao, V.S., and Sunitha, S. 2011. Sacred groves in southern Eastern Ghats, India: Are they better managed than forest reserves. *Trop. Ecol.* 52(1): 79–90.

Raper, K.B. and Fennell, D.L. 1965. *The Genus Aspergillus*. The Williams & Wilkins Co., Baltimore, MD.

Ravyts, F., De Vuyst, L., and Leroy, F. 2012. Bacterial diversity and functionalities in food fermentations. *Eng. Life Sci.* 12: 1–12.

Ray, B. 1992. The need for food bio preservation. In: Ray, B. and Daeschel, M. (Eds.), *Food Bio Preservatives of Microbial Origin*, CRC Press, Boca Raton, Florida, pp. 1–23.

Ray, R.C. and Sivakumar, P.S. 2009. Traditional and novel fermented foods and beverages from tropical root and tuber crops: Review. *Int. J. Food Sci. Technol.* 44: 1073.

Rayman, K., Malik, N., and Hurst, A. 1983. Failure of nisin to inhibit outgrowth of *Clostridium botulinum* in a model cured meat system. *Appl. Environ. Microbiol.* 46: 1450–1452.

Reddy, N.R. and Pierson, M.D. 1994. *Food Res. Int.* 27: 281.

Reddy, N.R., Pierson, M.D., Sathe, S.K., and Salunkhe, D.K. 1989. *Phytates in Cereals and Legumes*. CRC Press, Inc., Boca Raton, Florida, 152pp.

Rees, C.E.D., Dodd, C.E.R., Gibson, P.T., Booth, I.R., and Stewart, G.S.A.B. 1995. The significance of bacteria in stationary phase to food microbiology. *Int. J. Food Microbiol.* 28: 263–275.

Reiter, B. and Harnulv, G. 1984. Lactoperoxidase antibacterial system: Natural occurrence, biological functions and practical applications. *J. Food Prot.* 47: 724–732.

Rehm, H.J. and Omar, S.A. 1993. Special morphological and metabolic behavior of immobilised microorganisms. In: Sahm, H. (Ed.), *Biotechnology*. Vol. 1, 2nd edn, *Biological Fundamentals*, VCH, Weinheim, pp. 223–248.

Reps, A. 1987. Bacterial surface-ripened cheeses. In: Fox, P.F. (Ed.), *Cheese: Chemistry, Physics and Microbiology*. Vol. 2, *Major Cheese Groups*, Elsevier Applied Science, London, pp. 151–184.

Rhee, J., Kim, R., Choi, H.G., Lee, J., and Lee, Y. 2011. Molecular and biochemical modulation of heat shock protein 20 (Hsp20) gene by temperature stress and hydrogen peroxide (11,0,) in the monogonont rotifer. *Brach'umus sp. Comp. Biochem. Physiol. C-Toxicol. Pharmacol.* 154: 19–27.

Rodríquez, H., Curiel, J.A., Landete, J.M. et al. 2009. Food phenolics and lactic acid bacteria. *Int. J. Food Microbiol.* 132(2–3): 79–90.

Roever, C. 1998. Microbiological safety evaluations and recommendations on fresh produce. *Food Control* 9: 321–348.

Rojas, L.B., Smith, P.A., and Bartowsky, E.J. 2012. Influence nee of choice of yeasts on volatile fermentation-derived compounds, color and phenolics composition in Cabernet Sauvignon wine. *World J. Microbiol. Biotechnol.* 28: 3311–3321.

Roller, S. 1999. Physiology of food spoilage organisms. *Int. J. Food Microbiol.* 50(1–2): 151–153.

Roostita, R. and Fleet, G.H. 1996. Growth of yeasts in milk and associated changes to milk composition. *Int. J. Food Microbiol.* 31: 205–219.

Rose, A.H. 1982. *Fermented foods—Economic Microbiology.* Academic Press, London, pp. 250.

Rose, A.H. and Harrison, J.S. (Eds.) 1970. *The Yeasts.* Vol. 3, Yeast Technology. S., 25 Abb., 52 Tab. London/New York, 11 + 590p.

Ross, R.P., Morgan, S., and Hill, C. 2002. Preservation and fermentation: Past, present and future. *Int. J. Food Microbiol.* 79: 3–16.

Roy, A., Moktan, B., and Sarkar, P.K. 2007. Microbiological quality of legume-based traditional fermented foodsmarketed in West Bengal, India. *Food Control* 18: 1405–1411.

Roy, B., Prakash, K.C., Nehal, A.F., and Majila, B.S. 2004. Indigenous fermented food and beverages: A potential for economic development of the high altitude societies in Uttaranchal. *J. Hum. Ecol.* 15: 45–49.

Salminen, S., von Wright, A., and Ouwehand, A. 2004. *Lactic Acid Bacteria: Microbiological and Functional Aspects.* 3rd edn, Revised and Expanded Edition, Marcel Dekker, Inc., New York, NY.

Salovaara, H. 2004. Lactic acid bacteria in cereal-based products. In: Salminen, S., von Wright, A., Ouwehand, A. (Eds.), *Lactic Acid Bacteria, Microbiological and Functional Aspects*, 3rd edn. Marcel Dekker, New York, pp. 431–452.

Salunkhe, D.K., Chavan, J.K., and Kadam, S.S. 1990. *Dietary Tannins: Consequences and Remedies.* CRC Press, Boca Raton, Florida.

Samson, R.A. and Van Reenen-Hoekstra, E.S. 1988. *Introduction to Food-borne Fungi.* 3rd edn, Centraalbureau voor Schimmclcullures, Baarn.

Sandhu, D.K. and Soni, S.K. 1989. Microflora associated with Indian Punjabi warri fermentation. *J. Food Sci. Technol.* 26: 21–25.

Sandhu, D.K., Soni, S.K., and Vikhu, K.S. 1986. Distribution and role of yeast in Indian fermented foods. *Proceedings of the National Symposium on Yeast Biotechnology*, Haryana Agriculture University, Hissar, December, pp. 142–148.

Sandhu, D.K. and Waraich, M.K. 1984. Distribution of yeasts in indigenous fermented foods with a brief review of literature. *Kawaka* 12: 73–85.

Sandmeier, H. and Meyer, J. 1995. Bacteriophages. In: Sahm, H. (Ed.), *Biotechnology*. Vol. 1, 2nd edn, Biological Fundamentals, VCH, Weinheim, pp. 545–575.

Sanni, A.I., Asieduw, M., and Ayernorw, G.S. 2002. Microflora and chemical composition of momoni, a Ghanaian fermented fish condiment. *J. Food Compos. Anal.* 15: 577–583.

Sanni, A.I., Onilude, A.A., Fadahunsi, I.F., and Afolabi, R.O. 1998. Microbial deterioration of traditional alcoholic beverages in Nigeria. *Food Res. Int.* 32: 163–167.

Sanni, A.I., Onilude, A.A., and Ibidapo, O.T. 1999b. Biochemical composition of infant weaning food fabricated from fermented blends of cereal and soybean. *Food Chem.* 65: 35–39.

Saono, S., Gandjar, I., Basuki, T., and Karsono, H. 1974. Myco-flora of "ragi" and some other traditional fermented foods of Indonesia. *Ann. Bogor.* 5: 187–204.

Sarasa, M.S., Ingle, A.O., and Nath, N. 1985. Microbiological examination of certain new idli batters during fermentation. *Indian J. Microbiol.* 25: 212–220.

Sarkar, P.K., Hasenack, B., and Nout, M.J.R. 2002. Diversity and functional of *Bacillus* and related genera isolated from spontaneously fermented beans (Indian *kinema*) and locust beans (African soumbala). *Int. J. Food Microbiol.* 77: 175–186.

Sarkar, P.K., Jones, L.J., Craven, G.S., Somerset, S.M., and Palmer, C. 1997. Amino acid profiles of *kinema*, a soybean fermented food. *Food Chem.* 59(1): 69–75.

Sarkar, P.K., Jones, L.J., Gore, W., Craven, G.S., and Somerset, S.M. 1996. Changes in soya bean lipid profiles during *kinema* production. *J. Sci. Food Agric.* 71: 321–328.

Sarkar, P.K. and Nout, M.J.R. (Eds.) 2014. *Handbook of Indigenous Foods Involving Alkaline Fermentation.* CRC Press, Boca Raton, FL, USA.

Sarkar, P.K. and Tamang, J.P. 1994. The influence of process variables and inoculum composition on the sensory quality of *kinema*. *Food Microbiol.* 11: 317–325.

Sarkar, P.K. and Tamang, J.P. 1995. Changes in the microbial profile and proximate composition during natural and controlled fermentations of soybeans to produce *kinema*. *Food Microbiol.* 12: 317–325.

Sarkar, P.K., Tamang, J.P., Cook, P.E. and Owens, J.D. 1994. *Kinema*—A traditional soybean fermented food: Proximate composition and microflora. *Food Microbiol.* 11: 47–55.

Savadogo, A., Ouattara Cheik, A.T., Savadogo Paul, W., Baro Nicolas, O., Aboubacar, S., and Traore Alfred, S. 2004. Microorganism involved in Fulanifermented milk in Burkina Faso. *Pak. J. Nutr.* 3: 134–139.

Savitri, 2007. *Microbiological and biochemical characterization of some fermented foods (bhatooru, seera and siddu) of Himachal Pradesh*. PhD thesis, Faculty of Life Sciences, Himachal Pradesh University, Shimla.

Savitri, and Bhalla, T.C. 2007. Traditional foods and beverages of Himachal Pradesh. *Indian J. Tradit. Knowl.* 6(1): 17–24.

Sekar, S. and Mariappan, S. 2005. Usage of traditional fermented products by Indian rural folks and IPR. *Indian J. Tradit. Knowl.* 6(1): 111–120.

Schindler, J. and Schmid, R.D. 1982. Fragrance or aroma chemicals—Microbial synthesis and enzymatic transformation—A review. *Process Biochem.* 17: 2–8.

Schipper, M.A.A. 1984. A revision of the genus *Rhizopus*. 1. The *Rh. stolonifer* group and *Rh. oryzae. Stud. Mycol.* 25: 1–19.

Scott, R. and Sullivan, C. 2008. Ecology of Fermented foods. *Hum. Ecol. Rev.* 15: 25.

Shah, N.P. 2001. Functional foods from probiotics and prebiotics. *Food Technol.* 55(11): 46–53.

Shahidi, F. 1991. Developing alternative meat-curing systems. *Trends Food Sci. Technol.* 2: 219–222.

Shamala, T.R. and Srikantiah, K.R. 1988. Microbial and biochemical studies on traditional Indian palm wine fermentation. *Food Microbiol.* 5: 157–162.

Sharma, A. and Kapoor, A.C. 1996. Levels of antinutritional factors in pearl millet as affected by processing treatments and various types of fermentation. *Plant Foods Hum. Nutr.* 49: 241–252.

Sharma, N. and Gautam, N. 2008. Antibacterial activity and characterization of bacteriocin of *Bacillus mycoides* isolated from whey. *Ind. J. Biotechnol.* 7: 117–121.

Sharma, O.P. 1989. *Dendrocalamus hamiltonii* Munro, the Himalayan miracle bamboo. In: Trivedi, M.L., Gill, B.S., Saini, S.S. (Eds.), *Plant Science Research in India*. Today & Tomorrow's Printers & Publishers, New Delhi, pp. 189–195.

Sheu, C.W. and Freese, E. 1973. Lipopolysaccharide layer protection of gram-negative bacteria against inhibition by long-chain fatty acids. *J. Bacteriol.* 115: 869–875.

Shimizu, K. 1993. Killer yeasts. In: Fleet, G.H. (Ed.), *Wine Microbiology and Biotechnology*. Harwood Academic, Chur, Switzerland, pp. 234–264.

Shinefield, H.R., Ribble, J.C., and Boris, M. 1971. Bacterial interference between strains of *Staphylococcus aureus*, 1960 to 1970. In: Cohen, J.O. (Ed.), *The Staphylococci*. Wiley-Interscience, New York, pp. 503–515.

Shrestha, P., Gautam, R., Rana, R.B., and Sthapit, B. 2002. Homo gardens in Nepal: Status and scope for research and development. In: Watson, J., and Eyzaguirre, P.B. (Eds.), *Home Gardens and in situ Conservation of Plant Genetic Resources in Farming Systems. Proceedings of the Second International Home Gardens Workshop*, July 17–19, 2001, Witzenhausen, Germany. IPGRI, Rome, pp. 105–124.

Shurpalekar, S.R. 1986. Papads. In: Reddy N.R., Pierson, M.D., and Salunkhe, D.K., (Eds.), *Legume-Based Fermented Foods*. CRC Press, Boca Raton, FL, pp. 191–217.

Sidhu, J.S. and Al-Hooti, S.N. 2005. Functional foods from Dale fruits. In: Shi, J., Ho, C.T., and Shahidi, F. (Eds.), *Asian Functional Foods*. Marcel and Dekker Inc., New York, pp. 491–524.

Sieuwerts, S., de Bok, F.A., Mols, E., de Vos, W.M., and van Hylckama Vlieg, J.E. 2008. A simple and fast method for determining colony forming units. *Lett. Appl. Microbiol.* 47(4): 275–278.

Slepecky, R.A. and Hemphill, H.E. 1992. The genus Bacillus—nonmedical. In: Balows, A., Trüper, H.G., Dworkin, M., Harder, W., and Schleifer, K.H. (Eds.), *The Prokaryotes*. 2nd edn, Springer, New York, NY, pp. 1663–1696.

Smith J.L. and Fratamico P.M. 1995. Factors involved in the emergence and persistence of food-borne diseases. *J. Food Prot.* 58(6): 696–708.

Sohliya, I., Joshi, S.R., Bhagobaty, R.K. and Kumar, R. 2009. *Tungrymbai*—A traditional fermented soybean food of the ethnic tribes of Meghalaya. *Indian J. Tradit. Knowl.* 8(4): 559–561.

Soni, S.K. and Arora, J.K. 2000. Indian fermented foods: Biotechnological approaches. In: Marwaha, S.S. and Arora, J.K. (Eds.), *Food Processing: Biotechnological Applications.* Asiatech Publishers Pvt. Ltd., New Delhi, India, pp. 143–190.

Soni, S.K. and Marwaha, S.S. 2003. Cereal products: Biotechnological approaches for their production. In: Marwaha, S.S. and Arora, J.K. (Eds.), *Biotechnology: Strategies in Agro-Processing.* Asiatech Publishers Pvt. Ltd., New Delhi, India, 236–266.

Soni, S.K. and Sandhu, D.K. 1989a. Fermentation of idli: Effects of changes in the raw materials and physio-chemical conditions. *J. Cereal Sci.* 10: 227–238.

Soni, S.K. and Sandhu, D.K. 1989b. Nutritive improvement of Indian dosa batters by yeast enrichment and black gram replacement. *J. Ferment. Bioeng.* 68: 52–5.

Soni, S.K. and Sandhu, D.K. 1990a. Indian fermented foods: Microbiological and biochemical aspects. *Indian J. Microbiol.* 30: 135–157.

Soni, S.K. and Sandhu, D.K. 1990b. Biochemical and nutritional changes associated with Indian Punjabi warri fermentation. *J. Food Sci. Technol.* 27: 82–85.

Soni, S.K. and Sandhu, D.K. 1991. Role of yeast domination in Indian *idli* batter fermentation. *World J. Microbiol. Biotechnol.* 7: 505–507.

Soni, S.K. and Sandhu, D.K. 1999. Fermented cereal products. In: Joshi, V.K. and Pandey, A. (Eds.), *Biotechnology: Food Fermentation Microbiology, Biochemistry and Technology.* vol. 2, Applied, Educational Publishers and Distributors, New Delhi, India, pp. 895–950.

Soni, S.K., Sandhu, D.K., and Soni, R. 2001. Effect of physio-chemical conditions on yeast enrichment in Punjabi warri fermentation. *J. Food Sci. Technol.* 47: 510–514.

Soni, S.K., Sandhu, D.K., Vikhu, K.S., and Karma, N. 1985. Studies on dosa—An indigenous Indian fermented food: Some biochemistry and changes occurring during natural fermentation. *Food Microbiol.* 3(1): 45. doi: 10.1016/S0740-0020(86)80025-9.

Soni, S.K., Sandhu, D.K., Vikhu, K.S., and Karma, N. 1986. Microbiological studies on dosa natural fermentation. *Food Microbiol.* 3: 45–53. doi: 10.1016/S0740-0020(86)80025-9.

Soni, S.K., Soni, R., and Bansal, N. 2012. Genetic manipulation of food microorganisms. In: Joshi, V.K. and Singh, R.S. (Eds.), *Food Biotechnology: Principles and Practice.* I.K. International Publishing House, India, pp. 157–232.

Soni, S.K., Soni, R. and Janveja, C. 2013. Production of fermented foods. In: Panesar P.S. and Marwaha, S.S. (Eds.), *Biotechnology in Agriculture and Food Processing Opportunities and Challenges.* CRC Press, Taylor & Francis, Boca Raton, FL, pp. 219–278.

Sourabh, A., Kanwar, S.S., and Sharma, O.P. 2010. Screening of indigenous yeast isolates obtained from traditional fermented foods of Western Himalayas for probiotic attributes. *J. Yeast Fungal Res.* 2(8): 117–126.

Stahel, G. 1946. Foods from fermented soybeans II. Tempe, a tropical staple. *J.N.Y. Bot. Card.* 47: 285–296.

Stamer, J.R. 1975. Recent developments in the fermentation of sauerkraut. In: Carr, J.G., Cutting, C.V., and Whitting, G.S. (Eds.), *Lactic Acid Bacteria in Beverages and Food.* Academic Press, , New York, NY, pp. 267–280.

Stanley, G. 1998a. Microbiology of fermented milk products. In: Early, R. (Ed.), *The Taxonomy of Dairy Products.* 2nd edn, Blackie Academic and Professional, London, pp. 50–80.

Stanley, G. 1998b. Cheeses. In: Wood, B.J. (Ed.), *Microbiology of Fermented Foods.* Vol. 1, 2nd edn, Blackie Academic and Professional, London, pp. 263–307.

Steinkraus, K.H. 1983a. Fermented foods, feeds and beverages. *Biotechol. Adv.* 1: 31–46.

Steinkraus, K.H. 1983b. *Handbook of Indigenous Fermented Foods.* Marcel Dekker, Inc., New York, NY, p. 671.

Steinkraus, K.H. 1983c. Lactic acid fermentation in the production of foods from vegetables, cereals and legumes. *Anton. Leeuw.* 49: 337–348.

Steinkraus, K.H. 1986. Production of vitamin B12 in tempe. *Proceedings of the Asian Symposium on Non-salted Soybean Fermentation*, July 15–17, 1985. Tsukuba Science City.

Steinkraus, K.H. 1988. *Handbook of Indigenous Fermented Foods.* Marcel Dekker, Inc., New York, NY.

Steinkraus, K.H. 1989. *Industrialization of Indigenous Fermented Foods.* Marcel Dekker, New York, 439pp.

Steinkraus, K.H. 1995. *Handbook of Indigenous Fermented Foods.* Marcel Dekker, Inc., New York, NY.

Steinkraus, K.H. 1996. *Handbook of Indigenous Fermented Food.* 2nd edn, Marcel Dekker, Inc., New York, NY.

Steinkraus, K.H. 1997. Classification of fermented foods: Worldwide review of house-hold fermentation techniques. *Food Control* 8: 311–317.

Steinkraus, K.H. 1998. Bio-enrichment: Production of vitamins in fermented foods. In: Wood, B.J.B. (Ed.), *Microbiology of Fermented Foods.* 2nd edn, Blackie Academic and Professional, London, pp. 603–619.

Steinkraus, K.H. 2002. Fermentations in world food processing. *Compr. Rev. Food Sci. Food Technol.* 2: 23–32.

Steinkraus, K.H. 2004. *Industrialization of Indigenous Fermented Foods*, 2nd edn. Marcel Dekker, New York.

Steinkraus, K.H., Van Veen, A.G., and Thiebeau, D.B. 1967. Studies on 'idli'—An Indian fermented Black gram-rice food. *Food Technol.* 21: 916–991.

Stewart, G.G. 1987. Alcoholic beverages. In: Beuchat, L.R. (Ed.), *Food Beverage Mycology.* 2nd edn, AVI, New York, NY, pp. 307–354.

Stiles, M.E. and Holzapfel, W.H. 1997. Lactic acid bacteria of foods and their current taxonomy. *Int. J. Food Microbiol.* 36: 1–29.

Straka, R.P. and Stokes, J.L. 1957. A rapid method for the estimation of the bacterial content of precooked frozen foods. *J. Food Sci.* 22: 412–419.

Su, Y.C. 1977. *Sufu (tao-hu-yi) production in Taiwan. Symp. Indig. Ferm. Foods.* Bangkok, Thailand.

Su, Y.C., Sufu Reddy, N.R., Pierson, M.D., and Salunkhe, D.K. (Eds.) 1986. Sufu. *Legume-Based Fermented Foods.* CRC Press, Boca Raton, FL, pp. 69–83.

Svanberg, U. and Lorri, W. 1997. Fermentation and nutrient availability. *Food Control* 8(5/6): 319–327.

Takeuchi, A., Shimizu-Ibuka, A., Nishiyama, Y., Mura, K., Okada, S., Tokue, C., and Arai, S. 2006. Purification and characterization of an α-amylase of *Pichia burtonii* isolated from traditional starter, "Murcha" in Nepal. *Biosci. Biotechnol. Biochem.* 70: 3019–3024.

Talon, R., Leroy-Sétrin, S., and Fadda, S. 2002. Bacterial starters involved in the quality of fermented meat products. Chapter 10. In: Toldrá, F. (Ed.), *Research Advances in the Quality of Meat and Meat Products.* Research Singpost, Trivandrum, India, pp. 175–191.

Taylor, J. and Taylor, J.R.N. 2002. Alleviation of the adverse effect of cooking on sorghum protein digestibility through fermentation in traditional African porridges. *Int. J. Food Sci. Technol.* 37: 129–137.

Tamang, B. and Tamang, J.P. 2009a. Traditional knowledge of bio-preservation of perishable vegetables and bamboo shoots in Northeast India as food resources. *Indian J. Trad. Know.* 8(1): 89-95.

Tamang, B., Tamang, J.P., Schillinger, U., Franz, C.M.A.P., Gores, M., and Holzaphel, W.H. 2008. Phenotypic and genotypic identification of lactic acid bacteria isolated from ethnic fermented tender bamboo shoots of North-East India. *Int. J. Food Microbiol.* 121: 35–40.

Tamang, J.P. 1999. Development of pulverized starter for *kinema* production. *J. Food Sci. Technol.* 36(5): 475–478.

Tamang, J.P. 2001. Kinema. *Food Cult.* 3: 11–14.

Tamang, J.P. 2003. Native microorganisms in the fermentation of kinema. *Indian J. Microbiol.* 43(2): 1–4.

Tamang, J.P. 2005a. Ethnic fermented foods of the Eastern Himalayas. In: *The Proceeding of the Second International Conference on "Fermented Foods, Health Status and Social Well-being,"* December 17–18, 2005, Swedish South Asian Network on Fermented Foods and Anand Agricultural University, Gujarat, pp. 19–26.

Tamang, J.P. 2007. Fermented foods for human life. In Chauhan, A.K., Verma, A., and Kharakwal, H., (Eds.), *Microbes for Human Life*. International Publishing House Pvt. Limited, New Delhi, pp. 73–87.

Tamang, J.P., Chettri, R., and Sharma, R.M. 2009. Indigenous knowledge on North-East women on production of ethnic fermented soybean foods. *Indian J. Trad. Know.* 8(1): 122–126.

Tamang, J.P., Dewan, S., Thapa, S., Olasupo, N.A., Schillinger, U., and Holzapfel, W.H. 2000. Identification and enzymatic profiles of predominant lactic acid bacteria isolated from soft-variety chhurpi, a traditional cheese typical of the Sikkim Himalayas. *Food Biotechnol.* 14(1–2): 99–112.

Tamang, J.P. and Kailasapathy, K. (Eds.) 2010. Fermented foods. *Fermented Foods and Beverages of the World*. CRC Press, Taylor & Francis Group, New York, NY p. 448.

Tamang, J.P. and Nikkuni, S. 1996. Selection of starter culture for production of *kinema*, fermented soybean food of the Himalaya. *World J. Microbiol. Biotechnol.* 12(6): 629–635.

Tamang, J.P. and Sarkar, P.K. 1993. Sinki: A traditional lactic acid fermented radish tap root product. *J. General Appl. Microbiol.* 39: 395–408.

Tamang, J.P. and Sarkar, P.K. 1996. Microbiology of mesu, a traditionally fermented bamboo shoot product. *Int. J. Food Microbiol.* 29(1): 49–58.

Tamang, J.P., Sarkar, P.K., and Hesseltine, C.W. 1988. Traditional fermented foods and beverages of Darjeeling and Sikkim—A review. *J. Sci. Food Agric.* 44: 375–385.

Tamang, J.P., Tamang, N., Thapa, S., Dewan, S., Tamang, B., Yonzan, H., Rai, A.K., Chettri, R., Chakrabarty, J., and Kharel, N. 2012. Microorganisms and nutritional value of ethnic fermented foods and alcoholic beverages of North East India. *Indian J. Tradit. Knowl.* 11: 7–25.

Tamang, J.P. and Thapa, S. 2006. Fermentation dynamics during production of bhaati jaanr, a traditional fermented rice beverage of the Eastern Himalayas. *Food Biotechnol.* 20(3): 251–261.

Tamang, J.P., Thapa, S., Dewan, S., Yasuka, J., Fudou, R., and Yamanaka, S. 2002. Phylogenetic analysis of Bacillus strains isolated from fermented soybean foods of Asia: Kinema, Chungkokjang and Natto. *J. Hill Res.* 15(2): 56–62.

Tamime A.Y. 2002. Fermented milks: A historical food with modern applications—A review. *Eur. J. Clin. Nutr.* 56(4 Suppl.): S2–S15.

Terlabie, N.N., Sakyi-Dawson, E., and Amoa-Awua, W.K. 2006. The comparative ability of four isolates of *Bacillus subtilis* to ferment soybeans into dawadawa. *Int. J. Food Microbiol.* 106: 145–152.

Thakur, N., Savitri, and Bhalla, T.C. 2004. Characterization of some traditional foods and beverages of Himachal Pradesh. *Indian J. Tradit. Knowl.* 3(3): 325.

Thapa, N. 2002. Studies of microbial diversity associated with some fish products of the Eastern Himalayas. PhD thesis, North Bengal University, India.

Thapa, N. and Pal, J. 2007. Proximate composition of traditionally processed fish products of the Eastern Himalayas. *J. Hill Res.* 20: 75–77.

Thapa, N., Pal, J., and Tamang, J.P. 2006. Phenotypic identification and technological properties of lactic acid bacteria isolated from traditionally processed fish products of the Eastern Himalayas. Int. *J. Food Microbiol.* 107: 33–38.

Thapa, S. 2001. Microbiological and biochemical studies of indigenous fermented cereal-based beverages of the Sikkim Himalayas. PhD thesis. Food Microbiology Laboratory, Sikkim Government College (North Bengal University), Gangtok, India, p. 100.

Thapa, N., Pal, J., and Tamang, J.P. 2004. Microbial diversity in ngari, hentak and tung-tap, fermented fish products of North East India. *World J. Microbiol. Biotechnol.* 20: 599–607.

Thokchom, S. and Joshi, S.R. 2012. Microbial and chemical changes during preparation in the traditionally fermented soybean product *Tungrymbai* of ethnic tribes of Meghalaya. *Indian J. Tradit. Knowl.* 11: 139–142.

Thomas, T.D. and Crow, V.F. 1984. Selection of galactose fermenting *Streptococcus thermophilus* in lactose-limited chemostat cultures. *Appl. Environ. Microbiol.* 48: 186–191.

Thorn, C. and Rapor, K.B. 1945. *Manual of Aspergilli*. William & Wilkins Co., Baltimore, MD.

Thorsen, L., Azokpota, P., Hansen, B.M., Hounhouigan, D.J. and Jakobsen, M. 2010. Identification, genetic diversity and cereulide producing ability of *Bacillus cereus* group strains isolated from Beninese traditional fermented food condiments. *Int. J. Food Microbiol.* 142: 247–250.

Thorsen, L., Azokpota, P., Hansen, B.M., Rønsbo M.H., Nielsen K.F., Hounhouigan D.J., and Jakobsen M. 2011. Formation of cereulide and enterotoxins by *Bacillus cereus* in fermented African locust beans. *Food Microbiol.* 28: 1441–1447.

Tou, E.H., Guyot, J.P., Mouquet-Rivier, C., Rochette, I., Counil, E., Traore, A.S., and Treche, S. 2006. Study through surveys and fermentation kinetics of the traditional processing of pearl millet (*Pennisetum glaucum*) into ben-saalga, a fermented gruel from Burkina Faso. *Int. J. Food Microbiol.* 106: 52–60.

Tschape, H., Prager, R., Strekel, A., Fruth, A., Tietze, E., and Bohme, G. 1995. Verotoxinogenic Citrobacter freundii associated with severe gastro-enteritis and cases of haemolytic uraemic syndrome in a nursery school: Green butter as the infection route. *Epidemiol. Infect.* 114: 441–450.

Tsuyoshi, N., Fudou, R., Yamanaka, S., Kozaki, M., Tamang, N., Thapa, S., and Tamang, J.P. 2005. Identification of yeast strains isolated from marcha in Sikkim, a microbial starter for amylolytic fermentation. *Int. J. Food Microbiol.* 99: 135–146.

Uzochukwu, S., Balogh, E., Tucknot, O.G., Lewis, M.J., and Ngoddy, P.O. 1999. Role of palm wine yeasts and bacteria in palm wine aroma. *J. Food Sci. Technol.* 36: 301–304.

Valdes-Stauber, N., Gotz, H., Busse, M., and Scherer, S. 1997. Growth reduction of Listeria spp. caused by undefined industrial red smear cheese cultures and bacterocin-producing. *Appl. Environ. Microbiol.* 63(12): 4812–4817.

van Gerwen, S.J.C., de Wit, J.C., Notermans, S., and Zwietering, M.H. 1997. An identification procedure for foodborne microbial hazards. *Int. J. Food Microbiol.* 38: 1–15.

Van Veen, A.G. and Schaefer, G. 1950. The influence of the 11 tempeh fungus on the soya bean. *Documenta Neerlandica Indonesia de Morbis Tropics.* 2: 270–281.

Venkatasubbaiah, P., Dwarkanath, C.T., and Sreenivasamurthy, V. 1984. Microbiological and physico chemical changes in 'idli' batter during natural fermentation. *J. Food Sci. Technol.* 21: 59–56.

Venkatasubbaiah, P., Dwarkanalh, C.M., and Sreenivasamurthy, V. 1985. *Food Sci. Technol.* 22: 88–90.

Verma, L.R. and Joshi, V.K. 2000. In: Verma, L.R. and Joshi, V.K. (Eds.), *Postharvest Technology of Fruits and Vegetables*. Indus Publishing Co., New Delhi. Vol. 2, pp. 1–1144.

Vivier, D., Ratomahenina, R., and Galzy, P. 1994. Characteristics of micrococci from the surface of Roquefort cheese. *J. Appl. Bacteriol.* 76: 546–552.

Von Arx, J.A. 1977. Notes on Dipodoscus, Endomyces and Geotrichum with the description of two new species. *Antonie van Leeuwenhoek*, 43: 333–340.

Vyas, K.K. and Joshi, V.K. 1982. Plum wine making: Standardization of a methodology. *Indian Food Packer* 36(6): 80–86.

Wacher, C., Cartas, A., Cook, P.E., Barzana, E., and Owens, J.D. 1993. Sources of micro-organisms in pozol, a traditional Mexican fermented maize dough. *World J. Microbiol. Biotechnol.* 9: 269–274.

Wang, H.L. and Hesseltine, C.W. 1970. Sufu and lao-chao. *J. Agric. Food Chem.* 18: 572–575.

Wang, H.L. and Hesseltine, C.W. 1979. Mold-modified foods. In: Peppier, H.J. and Perlman, D. (Eds.), *Microbial Technology*, 2nd edn., Vol. II, Academic Press, London, pp. 95–129.

Wang, H.L. and Hesseltine, C.W. 2004. Traditional fermented foods. *Biotechnol. Bioeng.* 9:275–288. Article first published online: Feb. 18, 2004, doi: 10.1002/bit.260090302.

Wang, H.L., Vespa, J.B., and Hesseltine, C.W. 1972. Release of bound trypsin inhibitors in soybeans by *Rhizopus oigosporus*. *J. Nutr.* 102: 1495.

Wang, J. and Fung, D.Y.C. 1996. Alkaline-fermented foods: A review with emphasis on pidan fermentation. *Crit. Rev. Microbiol.* 22: 101–138.

Wang, Y.-Y.D. and Fields, M.L. 1978. Germination of corn and sorghum in the home to improve nutritive value. *J. Food Sci.* 43: 1113–1115.

Watson, K. 1990. Microbial stress proteins. *Adv. Microbial. Physiol.* 31: 183–223.

Whiting, R.C. and Buchanan, R.L. 1997a. Predictive modelling. In: Doyle, M.P., Beuchat, L.R., and Montville, T.J. (Eds.), *Food Microbiology Fundamentals and Frontiers*. ASM, Washington, D.C., pp. 728–739.

Whiting, R.C. and Buchanan, R.L. 1997b. Development of a quantitative risk assessment model for *Salmonella enteritidis* in pasteurised liquid eggs. *Int. J. Food Microbiol.* 36: 111–125.

Wibowo, D., Eschenbruch, R., Davis, C.R., Fleet, G.H., and Lee, T.H. 1985. Occurrence and growth of lactic acid bacteria in wine: A review. *Am J Enol Vitic.* 36: 302–313.

Wickner, R.B. 1993. Double-stranded and single-stranded RNA viruses of *Saccharomyces cerevisiae*. *Annu. Rev. Microbiol.* 46: 347–375.

Wisniewski, M., Biles, C., Droby, S., McLaughlin, R., Wilson, C., and Chalutz, E. 1991. Mode of action of the postharvest biocontrol yeast, *Pichia guilliermondii*. I. Characterisation of attachment to *Botrytis cinerea*. *Physiol. Mol. Plant Pathol.* 39: 245–258.

Wood, B.J. 1997. *Microbiology of Fermented Foods*. Vols. 1–2, Springer.

Wood, B.J.B. 1994. *Microbiology of Fermented Foods*. 2nd edn. Blackie Academic and Professional, London, UK.

Wood, B.J.B. and Holzapfel, W.H. 1995. *The Genera of Lactic Acid Bacteria*, Vol. 2. Blackie Academic and Professional, Glasgow, London.

Yong, F.M. and Lim, G. 1985. Production of aroma chemicals by fungi. In: Lim, G. and Nga, B.H. (Eds.), *Proc. Internat. Conf. on Microbiology in the Eighties*. Singapore Soc. Microbiol., Singapore, pp. 365–375.

Yong, F.M. and Lim, G. 1986. Effect of carbon source on aroma production by *Trichoderma viride*. *MIRCEN J. Appl. Microbiol. Biotechnol.* 2: 483–488.

Yong, T.A. and Leong, P.C. 1983. *A Guide to Cultivation of Edible Mushrooms in Singapore*. Agriculture Handbook.

Yonzan, H. and Tamang, J.P. 2010. Microbiology and nutritive value of selroti, an ethnic fermented food of the Nepalis. *Indian J. Tradit. Knowl.* 8(1): 110–114.

Yoshizawa, K. and Ishikawa, T. 1989. Industrialization of *Sake* manufacture. In: Steinkraus, K.H. (Ed.), *Industrialization of Indigenous Fermented Foods*. Marcel Dekker, Inc., New York, NY, pp. 127–168.

Young, N.M., Johnston, R.A.Z., and Richards, J.C. 1989. Purification of the α-L-rhamnosidase of *Penicillium decumbens* and characteristics of two glycopeptide components. *Carbohydrate Res.* 191: 53–62.

Young, T.W. 1987. Killer yeasts. In: Rose, A.H., and Harrison, J.S. (Eds.), *The Yeasts*. Vol. 2, 2nd edn, *Yeasts and the Environment*, Academic, London, pp. 131–164.

Zottola, E.A. and Sasahara, K.C. 1994. Microbial biofilms in the food processing industry—Should they be of concern. *Int. J. Food Microbiol.* 23: 125–148.

4

TRADITIONAL FERMENTED FOODS

Composition and Nutritive Value

A.K. SENAPATHI, ANILA KUMARI, DEV RAJ,
J.P. PRAJAPATI, K.S. SANDHU, KUNZES ANGMO,
MALAI TAWEECHOTIPATR, MONIKA,
REENA CHANDEL, S.V. PINTO, SAVITRI,
SOMBOON TANASUPAWAT, SOMESH SHARMA,
S.S. THORAT, AND TEK CHAND BHALLA

Contents

4.1 Introduction

Fermented foods including alcoholic beverages are an indispensable part of diet of man all-over the world (Arrmoto, 1961; Pederson, 1971; Joshi et al., 1999), especially in Oriental countries. In the present strongly inter-connected world, no country or people have remained in isolation, and fermented foods have significantly contributed to this relationship. Soy-sauce is not confined to China or Japan, but is relished in other, Western countries, as are *idli* and *dosa*, the delicious fermented foods of Indian origin. The importance of traditional fermented foods has been reviewed extensively (Pederson, 1971; Steinkraus, 1994; Gadaga et al., 1999; Haard et al., 1999; Blandino et al., 2003; Ramaite and Cloete, 2006; Anukam and Reid, 2009; Chelule et al., 2010). Since fermented foods constitute an integral part of the human diet, knowledge of their nutritional composition is essential, especially the presence of nutrients in the original raw materials and those in the fermented products. The chemical composition and nutritive value of fermented cereals, fruit and vegetables, milk products, and fish and various meat products are presented in detail in this chapter. Various components of indigenous fermented foods, including protein, carbohydrate, lipids, etc., their significance, and the factors influencing their contents, such as pH, water content, fermentation, etc., are also discussed.

4.2 Composition and Nutritive Value of Indigenous Fermented Foods

4.2.1 Cereal-Based Fermented Foods

Cereal grains were one of man's earliest sources of food (Matz, 1971) and are an important source of dietary proteins, carbohydrates, vitamins, minerals, and fibers. One way of processing the grains into food is through fermentation (Akinrele et al., 1970; Akingbala, et al., 1981; Fields et al., 1981; Adeyemi and Beckley, 1986; Blandino et al., 2003). In terms of texture, fermented cereal foods are either liquid (porridge) or stiff gels (solid). Fermentation improves the nutritional properties of cereals (Blandino et al., 2003) but decreases carbohydrates, non-digestible poly- and oligosaccharides, certain amino acids, and the availability of B group vitamins (Chavan et al., 1988; Blandino et al., 2003). Degradation of phytate during fermentation may increase the amount of soluble iron, zinc, and calcium several folds (Khetarpaul and Chauhan, 1990; Nout and Motarjemi, 1997a,b; Haard et al., 1999; Blandino et al., 2003). Corn meal fermentation leads to increased availability of lysine, methionine, and tryptophan (Nanson and Field, 1984; Blandino et al., 2003). Barley (*Hordeum vulgare*) is an excellent source of B-complex vitamins, minerals, and complex carbohydrates (Kalra and Jood, 2000), but needs more cooking time and has relatively poor digestibility and availability of minerals

(Arora et al., 2010). A combination of germination and fermentation significantly decreases the crude protein, crude fiber, starch, and total and insoluble dietary fiber content, while improvements in reducing sugar, thiamine, niacin, lysine, and soluble dietary fiber content have been documented (Arora et al., 2010). Increased level of vitamins has also been observed during the fermentation process, along with large quantities of thiamin, nicotinic acid, and biotin, by the action of *Saccharomyces cerevisiae*.

4.2.1.1 Idli   *Idli*, a low calorie, starchy and nutritious food of South India and Sri Lanka, consumed as a breakfast, is a fermented product of rice and dehulled black gram. It is rich in protein (3.4%), carbohydrate (20.3%) and water (70%) (Teniola and Odunfa, 2001; Blandino et al., 2003). Nutritive value is almost equal to that of the unfermented mixture of rice and blackgram originally used for preparing it (Table 4.1).

4.2.1.2 Dosa   *Dosa* contains 35%–40% water (on dry matter basis), 15%–20%, fat 15%–25%, carbohydrates 40%–50% fibers 5%–8%, ash 4%–6%, and energy 400–450 kcal; vitamin B, niacin, thiamine concentrations have been found to increase during fermentation (Soni et al., 1985).

4.2.1.3 Selroti   *Selroti* is a popular fermented rice product of the *Gorkha*/ethnic Nepalese which is a ring shaped, spongy, pretzel-like, deep-fried food. The composition of it can be summarized as: moisture 42.5%, pH 5.8 acidity, 0.08% ash 0.8% Dry Matter (DM), fat 2.7% DM, protein 5.7% DM, carbohydrate 91.3% DM, food value 410.3 kcal/100 gm DM, Na 8.9 mg/100 g, P 29.7 mg/100 g, and Ca 23.8 mg/100 g (Yonzan and Tamang, 2010).

4.2.1.4 Kinema   *Kinema* is a sticky fermented soybean food of the *Gorkha*/ethnic Nepalis of the North East, produced by natural fermentation. It is similar to other Asians bacillus-fermented sticky soybean foods, such as *natto* of Japan, *chungkukjang* of Korea, *thuanao* of northern Thailand, *pe-poke* of Myanmar, and *seing* of Cambodia (Nagai and Tamang, 2010; Tamang et al., 2012). Compostion and nutritional value of *kinema* has been given as: pH 7.9, food value 454 kcal/100 g Dry Matter (DM), total

Table 4.1 Approximate Composition of Optimized *Idli*

PARAMETERS (G% ± SD)	RICE (VARIETY *IR*20)	BLACK GRAM (VARIETY ADT3)	*IDLI* (3:1.18)
Starch	79.50 ± 2.90	52.00 ± 1.21	75.00 ± 2.84
Amylose	32.00 ± 1.60	17.00 ± 0.74	31.00 ± 1.42
Amylopectin	47.50 ± 2.37	35.00 ± 1.62	44.00 ± 2.08
Total carbohydrates	84.00 ± 3.52	65.80 ± 3.02	81.60 ± 3.42
Protein	06.46 ± 0.32	24.16 ± 1.20	10.21 ± 0.50
Fat	00.27 ± 0.13	00.87 ± 0.02	00.10 ± 0.01
Crude fiber	00.20 ± 0.01	00.70 ± 0.03	00.28 ± 0.01

Source: From Steinkraus K.H., Van Veen A.G., and Thiebean D.B. 1967. *Food Technol* 21:916–919. With permission.

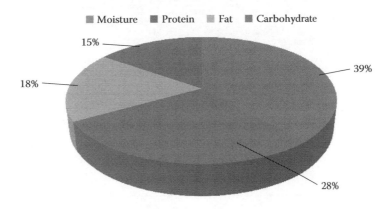

Figure 4.1 Nutritional composition of Kinema. (Adapted from Sarkar, P.K. et al., 1994. *Food Microbiol* 11(1):47–55.)

amino acids 42618 mg/100 g, free amino acids 5129 mg/100 g, Ca 432 mg/100 g, Na 27.7 mg/100 g, Fe 17.7 mg/100 g, Mn 5.4 mg/100 g, and Zn 4.5 mg/100 g (Sarkar et al., 1994), and the rest of the composition is given in Figure 4.1.

It also has B group vitamins and minerals (Sarkar et al., 1996, 1998; Sarkar and Mishra, 1997; Tamang et al., 2012). It is rich in all essential amino acids which constitute 52.6% of its protein, comparable to that of hen's egg and milk proteins (Sarkar and Mishra, 1997). Traditionally prepared *kinema*, contains 8 mg B1, 12 mg B2, 45 mg B3, 683 mg Ca, 4 mg Cu, 18 mg Fe, 494 mg Mg, 10 mg Mn, 1257 mg P, 2077 mg K, 13 mg Zn, and <0.5 mg of Cd, Cr, Pb, Ni, and Na per kg dry matter, while the vitamin B1 content was significantly ($P < 0.05$) higher, the contents of vitamins B2 and B3 were significantly lower ($P < 0.05$) in raw soybeans than in *kinema* (Sarkar et al., 1998). Mineral concentrations were 3.1–8.3 times higher in raw soybeans than in *kinema*. The fresh kinema contains 39.1%w/w total dry matter, water soluble dry matter 25.6% of total dry matter, pH 6.25, free ammonia nitrogen 1.19 mM/g of dry matter (Nikkuni, 1997; Nout et al., 1998).

4.2.1.5 Hawaijar *Hawaijar* is a sticky fermented soybean food commonly eaten in Manipur (India). The composition of *hawaijar* has been described as: pH 7.4, food value 521.2 kcal/100 gm DM, Ca 357.8 mg/100 gm, Na 88.7 mg/100 gm, Fe 92.3 mg/100 gm, K 835.1 mg/100 gm, and Zn 63.0 mg/100 gm, and the major components of its composition are given in Figure 4.2 (Tamang et al., 2012).

In general, 3-day fermented *hawaijar* contains 62% water, 1.42% ash, 8.2% crude fiber, 26.02% soluble protein, 3.8% free amino acid, 24.36% fat, 0.9% total soluble sugar, and 0.23% reducing sugar (Premarani and Chhetry, 2011).

4.2.1.6 Tungrymbai *Tungrymbai* is an ethnic fermented soybean food of the *Khasi* in Meghalaya. The composition of *tungrymbai* has been determined as: Water 60.0%, pH 7.6, protein 45.9 g/100 g, fat 30.2 g/100 g, fiber 12.8 g/100 g, carotene 212.7 µg/100 g and folic acid 200 µg/100 g antioxidant activities measured as DPPH scavenging activity as 670.9 µg/mL, ABTS radical scavenging activity of 190.9 µg/mL, total phenols as 2.6 mg Gallic Acid Equivalent (GAE)/g fresh weight. The saponin content

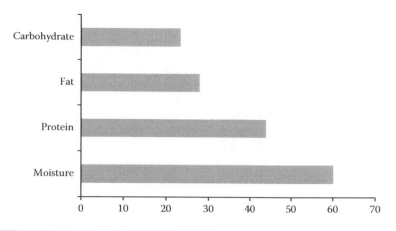

Figure 4.2 Nutritional composition of *Hawaijar*. (From Anonymous, Annual Report 2009–2010. Institute of Bioresources and Sustainable Development (IBSD), Imphal, 2010. Composition of protein, carbohydrate and fats is on dry matter basis. With permission.)

(Group B) of this product was determined as 447.9 mg/100 g (Omizu et al., 2011; Tamang et al., 2012).

4.2.1.7 Bekang *Bekang* is an ethnic fermented soybean food of Mizoram (India) (Nagai and Tamang, 2010). The composition of *Bekang* has been summarized as: water 63.5%, pH 7.1, antioxidants activities such as DPPH scavenging activity 477.2 µg/mL, ABTS radical scavenging activity 158.9 µg/mL, total phenols 3.8 mg GAE/g fresh weight. Saponin content (Group B) was found as 331.3 mg/100 g, which is an anti-nutritional factor (Omizu et al., 2011; Tamang et al., 2012).

4.2.2 Fermented Non-Soybean Legume Foods

4.2.2.1 Maseura *Maseura* or *masyaura* is an ethnic fermented non-soybean product prepared from black gram by the Gorkhas of the North East parts of Himalaya. It is a cone-shaped hollow, brittle, and friable product. *Maseura* is similar to North Indian *wari* or *dal badi* and South Indian *sandige* (Tamang et al., 2012). The nutritional value of *Maseura* has been found to be water 8%–10%, pH 5.6–6.3, protein 8%–10%, and carbohydrate 67%–70% (Dahal et al., 2003; Chettri and Tamang, 2008). An increase in soluble protein, amino nitrogen, nonprotein nitrogen, thiamin and riboflavin has been observed in *maseura* during fermentation (Dahal et al., 2003; Nagai and Tamang, 2010).

4.2.2.2 Warries and Papadam *Warries* are spicy, hollow, brittle, friable balls 5–8 cm in diameter, very popular in Northern India, especially in Amritsar city (Punjab India) and are used as condiment in cooking with vegetables, legumes or rice (Yadav and Khetarpaul, 1994; Soni et al., 2013).

4.2.3 Traditional Fermented Fruit and Vegetable Foods

4.2.3.1 Gundruk Gundruk is a common non-salted dried fermented leafy vegetable food of the *Gorkha* of the North East. Though dry in nature, *gundruk* is similar to other wet fermented vegetable foods such as *kimchi* of Korea and sauerkraut of Germany (Tamang et al., 2012). The nutritional value of *Gundruk* can be summarized as: water 15%, pH 5.0, acidity 0.49%, ash 22.2% Dry Matter (DM), protein 38.7% DM, fat 2.1% DM, carbohydrate 38.3% DM, food value 321.9 kcal/100 g DM, Ca 234.6 mg/100 g, Na 142.2 mg/100 g, and K 677.6 mg/100 g (Tamang, 2006; Tamang et al., 2012).

The pH drops slowly to a final value of 4.0 during fermentation of *Gundruk*, and the amount of acid (as lactate) increases to about 1% on the 6th day (Anonymous FAO, 1998). Glutamic acid, aspartic acid, glycine, alanine, isoleucine, phenylalanine and lysine content increase considerably, but with substantial decreases in asparagine, serine, glutamine, proline, tyrosine, and histidine content. Lactic acid and acetic acid are found to be the major organic acids in this product. *Gundruk* prepared from mustard, rape, and radish leaves showed pH values of 4.0, 4.3, and 4.1, respectively. The lactic acid contents were 1.0%, 0.8%, and 0.9%, respectively. One of the disadvantages of the traditional process of *Gundruk* fermentation is the loss of 90% of the carotenoids, probably during sun drying. Improved methods of drying might reduce the vitamin loss and so improve the vitamin content (Dahal et al., 2005; Anonymous 1998).

4.2.3.2 Sinki Sinki, a non-salted fermented radish tap root of the Gorkha, is prepared by pit fermentation. When the leaves of the radish are fermented it is *gundruk*, but when the tap root is fermented, it is called *sinki* (Tamang et al., 2012). The nutritional value of *sinki* has been described as: water 22.8%, pH 4.1, acidity 0.65%, ash 15.6% DM, protein 14.9% DM, fat 1.4% DM, carbohydrate 68% DM, food value 344.2 kcal/100 gm DM, Ca 223.9 mg/100 gm, Na 737.3 mg/100 gm, and K 2320.4 mg/100 gm (Tamang, 2006; Tamang et al., 2012). *Sinki* has 14.5% protein, 2.5% fat and 11.3% minerals on a dry weight basis (DWB). During fermentation, the pH of radish tap root drops from 6.7 to 3.3 due to the increase of titratable acidity from 0.04% to 1.28%. The increase in acidity or drop in the pH is the major contribution of fermentation to the preservation of this food (Tamang, 1993; Tamang et al., 2012).

4.2.3.3 Goyang Goyang is a fermented vegetable food of the Sherpa, living in Sikkim. The nutritive value of *goyang* is: water 92.5%, pH 6.5, acidity 0.13%, ash 12.9% DM, protein 35.9% DM, fat 2.1% DM, carbohydrate 48.9% DM, food value 357.2 kcal;/100 g, Ca 92.2 mg/100 g, Na 6.7 mg/100 g, and K 268.4 mg/100 g (Tamang and Tamang, 2007; Tamang et al., 2012).

4.2.3.4 Inziangsang   *Inziangsang* or *ziansang* is an ethnic fermented leafy vegetable product of the *Nagas* living in Nagaland and Manipur. It is very similar to *gundruk*. The composition of this product has also been reported, and can be summarized as: water 17.6%, pH 4.8, acidity 0.50%, ash 16.9% DM, protein 38.7% DM, fat 3.2% DM, carbohydrate 41.2% DM, food value 348.4 kcal/100 g, Ca 240.4 mg/100 g, Na 133.7 mg/100 g, and K 658.4 mg/100 g (Tamang, 2006; Tamang et al., 2012).

4.2.3.5 Khalpi   *Khalpi* is a fermented cucumber (*Cucumis sativus L.*) product, consumed by the *Gorkha* of the North East. The nutritional value of *Khalpi* is: water 91.4%, pH 3.9, acidity 0.95%, ash 14.2% DM, fat 2.6% DM, protein 12.3% DM, carbohydrate 70.9% DM, food value 356.2 kcal/100 gm, and K 125.1 mg/100 g (Tamang, 2006; Tamang et al., 2012).

4.2.4 Fermented Bamboo Shoot Foods

4.2.4.1 Mesu   *Mesu* is a fermented bamboo shoot consumed by the *Gorkha* of Sikkim. Its composition has been summarized as: water 89.9%, pH 3.9, acidity 0.88%, ash 15.0% DM, fat 2.6% DM, protein 27.0% DM, carbohydrate 55.6% DM, calorific value 352.4 kcal/100 g DM, Ca 7.9 mg/100 g, Na 2.8 mg/100 g, and K 282.6 mg/100 g (Tamang, 2006; Tamang et al., 2012).

4.2.4.2 Soibum   *Soibum* is a fermented tender bamboo shoot food produced and eaten by the *Meitei* tribe of Manipur. It is produced from bamboo shoots *Dendrocalamus hamiltonii* (Wanap, Unap, Pecha), *D. sikkimensis*, and *D. giganetus* (Maribop), *Melacona bambusoide* (Moubi/Muli), *Bambusa tulda* (Utang), *B. balcona* (Ching saniebi), etc., by natural fermentation. The nutritive value of *soibum* has been determined as: water 92.0%, pH 3.9, acidity 0.98%, ash 13.3% DM, fat 3.2% DM, protein 36.3% DM, carbohydrate 47.2% DM, calorific value 362.8 kcal/100 g DM, Ca 16.0 mg/100 g, Na 2.9 mg/100 g, and K 212.1 mg/100 g (Tamang, 2006; Tamang et al., 2012).

4.2.4.3 Soidon   *This* is the popular fermented tip of matured bamboo shoots, and is a product of Manipur made by the *Meitei*. The nutritional value of *soidon* is: water 92.2%, pH 4.2, acidity 0.96%, ash 13.1% DM, fat 3.1% DM, protein 37.2% DM, carbohydrate 46.6% DM, calorific value 363.1 kcal/100 g DM, Ca 18.5 mg/100 g, Na 3.7 mg/100 g, and K 245.5 mg/100 g (Tamang, 2006; Tamang et al., 2012).

4.2.4.4 Ekung   *Ekung* is an ethnic fermented tender bamboo shoot product of Arunachal Pradesh, produced by the *Nishi*. The word "*ekung*" is derived from the *Nishi dialect*, the *Adi* call it *iku* and the *Apatani* call it *hikku* (Tamang, 2010). The nutritional value of *ekung* has been determined as: water 94.7%, pH 3.9, acidity 0.94%, ash 14.0% DM, protein 30.1% DM, fat 3.8% DM, carbohydrate 52.1% DM, calorific value 363.0 kcal/100 gm DM, Ca 35.4 mg/100 gm, Na 10.9 mg/100 gm, and K 168.6 mg/100 gm (Tamang, 2006; Tamang et al., 2010).

4.2.4.5 Eup *Eup*, a word derived from the *Nishi dialect*, is a dry fermented bamboo tender shoot product of Arunachal Pradesh. *Eup* has synonyms such as *hi* by the *Apatani*, *nogom* by the *Khampti*, and *ipe* by the *Adi* (Tamang, 2010). The nutritional value of *eup* has been summarized as: water 36.8%, pH 4.1, acidity 0.80%, ash 18.2% DM, protein 33.6% DM, fat 3.1% DM, carbohydrate 45.1% DM, calorific value 342.7 kcal/100 gm DM, Ca 76.9 mg/100 gm, Na 3.4 mg/100 gm, and a K of 181.5 mg/100 gm (Tamang, 2006; Tamang et al., 2012).

4.2.4.6 Hirring *Hirring* is also a fermented bamboo shoot prepared by the *apatani* of Arunachal Pradesh (India). The *Nishi* call it *hitch* or *hitak*. The nutritional value of *hirring* is: water 88.8%, pH 4.0, acidity 0.81%, ash 15.0% DM, protein 33.0% DM, fat 2.7% DM, carbohydrate 49.3% DM, calorific value 353.5 kcal/100 g DM, Ca 19.3 mg/100 g DM, Na 3.4 g/100 g, and K 272.4 mg/100 g (Tamang, 2006; Tamang et al., 2012).

4.2.4.7 Tuaithur *Tuaithur* is an ethnic fermented bamboo shoot product prepared and consumed by the *Hrangkhol*, the *Baite*, and the *Hmar* of Assam. The nutritional value of *tuaithur* is: water 92.3%, pH 4.0, acidity 0.83%, ash 4.6% DM, fat 3.4% DM, protein 4.6% DM, carbohydrate 87.4% DM, and calorific value 398.6 kcal/100 gm DM (Chakrabarty, 2011).

4.2.5 Fermented Milk and Milk Products

Milk products are nutrient dense and provide abundant amount of protein, vitamins, and minerals. Fermented milks and milk products are made more digestible by the breakdown of the protein casein by microbes during fermentation. The major constituents of milk are water, fat, protein, sugar or lactose, minerals, and ash, as given in Table 4.2.

The concentration of lactose in milk is relatively constant and averages to about 5% (4.8%–5.2%), which is a major source of calories. The fermentation of lactose during processing lowers its concentration in many dairy products, especially in yoghurts and cheeses (Tamime and Robinson, 1999; Watteanx, 2012). Some people are lactose intolerant. During the fermentation of milk lactose is converted into lactic acid, and such milk, products can be consumed by lactose-intolerant people. Whey proteins, including α-lactalbumins β-lactoglobulins, are present in colloidal form and are easily coagulated by heat. The fat portion of the milk contains fat-soluble vitamins. Milk is also a rich source of minerals and vitamins. It is a rich source of calcium, phosphorus, sodium, magnesium, chloride, etc. Iron is also present in milk, though in a very small amount. The minerals are found in association with the casein. Vitamins present in

Table 4.2   Chemical Composition of Fresh Milks

NUTRIENTS	SOURCE				
	COW	BUFFALO	CAMEL	GOAT	HUMAN
Water (%)	86–88	82–84	87.05	87.38	87.5
Energy (kcal)	64–71	97–105	70.09	67.36	70–74
Proteins (g)	3.2–3.48	3.7–4.02	3.27	3.32	1.0
Fat (g)	3.4–4.14	6.9–7.5	4.2	4.04	4.1–4.4
Lactose (g)	4.7	5.2	4.31	4.27	6.9–7.12
Minerals (g)	0.72	0.79	–	–	0.20

Source: Adapted from Abou-Soliman N.H. 2005. Studies on goat milk proteins: Molecular and immunological characterization with respect to human health and nutrition. PhD thesis, Alexandria University, Egypt.

milk are fat-soluble vitamins (A, D, E, and K) while water soluble vitamins include vitamin B complex (Wattiaux, 2012).

4.2.5.1 Dahi   *Dahi* (curd) is a popular fermented milk product from several countries of South Asia, for direct consumption as well as for the preparation of various ethnic milk products such as *gheu, mohi, chhurpi,* etc. (Tamang et al., 2012). The color of *dahi* should be pleasing, attractive, and uniform, without showing any signs of visible foreign matter. The color ranges from creamish yellow for cow milk to creamish white for buffalo milk, but should not be brown, with a smooth and glossy surface, without the appearance of any free whey on top. It should have pleasant sweetish aroma and a mild, clean, acidic taste. A good pleasant diacetyl flavor is desired in *dahi*. It should however, not show any signs of bitterness, saltiness, or other off-flavors.

The nutritional value of *dahi*: It has water 84.8%, pH 4.2, acidity 0.73%, ash 4.7% DM, fat 24.5% DM, protein 5.7% DM, carbohydrate 22.5% DM, and food value 503.6 kcal/100 g DM (Dewan, 2002; Tamang et al., 2012). The chemical composition of dahi has been reported as fat ranging from 3%–7%, protein 3.3%–3.4%, ash 0.75%–0.79%, lactic acid 0.5%–1.1%, calcium 100 mg, minerals 0.4%, phosphorus 50 mg, vitamin A 100 (IU), biotin 15 µg, and ascorbic acid 1 mg/100 mL. The factors affecting composition of *dahi* are; the type of milk used, the extent of dilution during churning, and the efficiency of fat removal (Laxminarayan et al., 1952) (Table 4.3).

4.2.5.1.1 Mishti Dahi   *Mishti dahi* is a fermented milk product, having creamish to light brown color, firm body, smooth texture, sweet–acidic flavor, and a pleasant aroma. In the absence of PFA/BIS standards, there could be different grades of *mishti dahi* depending on the percentage of fat, SNF, and sugar in the product to meet the wide variety of consumer tastes. *Mishti dahi* preparation is mainly confined to domestic or cottage-scale operations. Due to a lack of standardization in the composition and quality of raw materials and manufacturing techniques, the market quality of the

Table 4.3 List of Constituents and Concentration of *Dahi*

CONSTITUENTS	CONCENTRATION (%)
Water (%)	85–88
Fat (%)	5–8
Protein (%)	3.2–3.4
Lactose (%)	4.6–5.2
Ash (%)	0.7–0.75
Lactic acid (%)	0.5–1.0
Ca (%)	0.12–0.14
P (%)	0.09–0.11

Source: From Batish V.K. et al. 1999. *Biotechnology: Food Fermentation Microbiology, Biochemistry and Technology.* Educational Publishers and Distributors, Kerala, India, pp. 781–864. With permission.

product varies considerably. Wide variations in total solids content (27%–43%), non-fat milk solids (11%–16%) and sucrose (13%–19%) in market samples of *mishti dahi* sold in Calcutta have been reported (Ghosh and Rajorhia, 1987). Many flavor defects, such as fruity, alcoholic, highly acidic, flat, etc, and body-texture defects like gassiness, thin, whey, or a thick crust on the top surface were observed in most of the market samples. However, standardized *mishti dahi* from modern plants is now being marketed in metropolitan cities. *Typical* composition of *mishti dahi* is given in Table 4.4.

4.2.5.2 Lassi This is a popular fermented milk products and is consumed widely in the Indian sub-continent. The composition of *lassi* is: water 96.2%, fat 0.8%, protein 1.29%, lactose 1.2%, lactic acid 0.44%, ash 0.4%, calcium 0.6%, and phosphorus 0.04% (Rangappa and Achaya, 1974). The factors affecting its composition are; the type of milk used, extent of dilution during churning, and the efficiency of fat removal. The composition of *lassi or chhas* varies widely from one locality to another, as it is made under different market and domestic conditions, as well as with different types of milk, with variable chemical and bacteriological quality (Table 4.5).

Table 4.4 Composition of Different Grades of *Mishti Dahi*

CONSTITUENT	LOW FAT	MEDIUM FAT	HIGH FAT
Milk fat	2–3	4–5	8–9
Milk SNF	13–14	11–13	1041
Sugar	17–19	17–18	17–18
Total solids	32–35	32–36	35–38

Source: From Aneja R.P. et al. 2002. *Cultured/Fermented Products, Cited in Technology of Indian Milk Products.* A dairy India publication, Delhi, India, p. 165. With permission.

Table 4.5   Average Composition of *Lassi*

CONSTITUENTS	CONCENTRATION (%)
Water	96.2
Fat	0.80
Protein	1.29
Lactose	1.2
Lactic acid	0.44
Ash	0.40
Ca	0.60
P	0.04

Source: From Rangappa K.S. and Achaya K.T. 1974. *Indian Dairy Products*. Asia Publishing House, Bombay, pp. 119–124; Rangappa K.S. and Achaya K.T. 2014. *Dairy Technology. Dairy Products & Quality Assurance*. Vol. 2. New India Publishing Agency, New Delhi, India, p. 133. With permission.

4.2.5.3 Chhurpi   Two types of *chhurpi*—hard and soft—are popular among the ethnic people of Sikkim and Arunachal Pradesh in India. Hard-variety *chhurpi* is prepared from yak milk at high altitudes (2100–4500 m) and has a characteristic gumminess and chewiness. Soft *chhurpi* is a cheese-like fermented milk product (Tamang et al., 2000, 2012). The nutritional value of hard *chhurpi* can be summarized as: water 3.9%–13%, pH 5.3, acidity 0.3%, ash 6.6%–7.7%, fat 7.7%–12.3%, protein 53.4%–68.5%, and carbohydrate 20.4%–23.2% (Katiyar et al., 1991; Pal et al., 1996).

The composition of soft-*chhurpi* has been determined as: water 73.8%, pH 4.2, acidity 0.61%, ash 6.6% DM, fat 11.8% DM, protein 65.3% DM, carbohydrate 16.3% DM, Ca 44.1 mg/100 gm, Fe 1.2 mg/100 gm, Mg 16.7 mg/100 gm, Mn 0.6 mg/100 gm, and Zn 25.1 mg/100 gm (Dewan, 2002).

4.2.5.4 Chhu   *Chhu* or *sheden*, an ethnic fermented milk product of the *Bhutia*, the *Lepcha*, the *Monpa*, the *Sherdukpen*, the *Khamba*, the *Memba*, and Tibetans living in the North East, is a strong-flavored traditional cheese-like product prepared from yak milk. It has a rubbery texture with a slightly sour taste and a strong flavor. A traditional fermented dairy product *chhu* prepared from cow milk or yak milk reportedly contains lactic acid bacteria (*L. alimentarius, L. farciminis, L. salivarius, L. bifermentans, L. brevis,* and *Lactococcus lactis*) and yeast (*Saccharomycopsis crataegensis* and *Candida castellii*) with improved functional properties like hydrophobicity, a large number of enzymes, and antagonistic and probiotic properties (Dewan and Tamang, 2006, 2007; Tamang et al., 2012).

Physico-chemical analysis of *chhu* has also been carried out, showing: water 38.2%, pH 4.3, acidity 0.61%, ash 3.6% DM, fat 32% DM, protein 52% DM, carbohydrate 12.5% DM, Ca 34.9 mg/100 g, Fe 0.8 mg/100 g, Mg 16.9 mg/100 g, Mn 0.9 mg/100 g, and Zn 27.1 mg/100 g (Dewan, 2002; Tamang et al., 2012).

4.2.5.5 Cultured Butter Milk Cultured butter milk is rich in protein, vitamins, and iron content. Acidophilus milk, prepared from *L. acidophilus*, generates acidic conditions in the intestine, thereby suppressing the growth of gas forming organisms (Soni and Arora, 2000).

4.2.5.6 Shrikhand *Shrikhand* is a delicious product very popular with many Indians, who consume it regularly during various festivals. *Shrikhand* is a semi-soft sweetish–sour milk product made from lactic acid fermented curd (see Chapter 8 for details). The nutritive value of *shrikhand* and its acceptability increases with the fat content, and *shrikhand* with 3% fat was found to be palatable and acceptable (Ingle and Joglekar, 1974). The titrable acidity of the *shrikhand* is the measure of lactic acid formed, and it should be lower than 1.4%.

4.2.5.6.1 Composition of Shrikhand The average composition of *shrikhand* shows: fat 5.0%, sugar 42.0%, and total solids 60.0%, while s*hrikhand* made in the laboratory contained on an average 5.16% fat, 6.59% protein, 1.68% reducing sugar, 39.37% non-reducing sugar, and 42.24% water (Upadhyay, 1981). *Shrikhand* containing about 6% fat, 35%–40% water, and 40% sugar was highly preferred with respect to sensory profile and consistency of the product (Miyani, 1982; Patel and Chakraborty, 1985b). Reddy et al. (1984) reported that *shrikhand* of superior quality can be prepared from buffalo milk (6% fat and 9% SNF) treated with sodium citrate (0.25%), while Desai et al. (1985) reported that the homogenization of milk can be used to improve the sensory qualities, (flavor, body, and texture) of *shrikhand*. The most desirable *shrikhand* has 6% fat, 41% sugar, and 40% water (Patel and Abd-El-Salam, 1986; Patel and Chakraborty, 1988). An acceptable quality *shrikhand* has been made using *Streptococcus cremoris* as the starter culture, with the addition of 40% sugar and 20% cream (40% fat) reported by Rao et al. (1986).

Upadhyay et al. (1975a) analysed market samples of *shrikhand* and found average composition of the market *shrikhand* to be: fat 1.93%–5.56%, water 66%, total solids 64.34%–64.52%, protein 5.33%–6.13%, reducing sugar 1.56%–2.18%, and non-reducing sugar 52.55%–53.76%. The market samples of *shrikhand* were also examined for microbiological quality by Upadhyay et al. (1975b), who found a wide variation in the count of various groups of microorganisms. Counts of *shrikhand* per gram were shown to be 2.4×10^4 to 1.6×10^6, 0–930 coliforms, $0–1 \times 10^4$ psychrotrophs, and $0–1.8 \times 10^5$ yeast and molds. All the samples were, however, free of enteric pathogens. A large variations in the composition of market samples of *shrikhand* were found, such as fat 4.5%–11.4%, protein 3.4%–15.7%, lactose 0.66%–2.79%, sucrose 38.8%–57.10%, and water 25.4%–40.8% (Sharma and Zariwala, 1980). The sensory quality of market *shrikhand* sold in Gujarat State, India was also evaluated and found to vary significantly among the manufacturers, both within the city and between the cities. Flavor and total scores of traditionally made *shrikhand* were significantly higher than those for *shrikhand* made by mechanized methods (Jain et al., 2003).

The composition of *shrikhand* is summarized as: water 34.48%–35.66%, fat 1.93%–5.56%, protein 5.33%–6.13%, reducing sugar 1.56%–2.18%, and nonreducing sugar (sucrose) 55.55%–53.76% (Mital, 1977). The composition of *shrikhand vadi* has been reported as: water 6.5%, fat 7.4%, protein 7.7%, ash 0.8%, reducing sugar 15.9%, and non-reducing sugar 68.9% (Date and Bhatia, 1955). The progressive fermentation of milk into *dahi* brought about little or no change in the total solids, fat, proteins, and milk salts, but significant decreases in the lactose content and pH value occured (Table 4.6). The chemical composition of *shrikhand* sold in Gujarat is shown in Table 4.7.

4.2.5.7 Yoghurt Common fermented milk products including yoghurt, a low fat food, have a wide range of nutrients, including increased concentrations of lactic acid, galactose, free amino acids, and fatty acids due to fermentation (Tamine and Robinson, 1999; Gurr, 1987; Hattingh and Viljoen, 2001). Fermented milk product in general showed an increase in folic acid content but a slight decrease in vitamin B_{12}, while other B vitamins were affected only slightly (Alm, 1982) in comparison to the raw milk (Sahlin, 1999).

Cultured milk, yoghurt, and curd contain microbes producing a certain amount of enzyme β-galactosidase, which help in tackling the problem of lactose intolerance (Sarkar, 2006b). Nutrient composition of commercial yoghurt and probiotics is shown in Table 4.8.

4.2.5.8 Cheese Cheese is a rich source of minerals (particularly calcium and phosphorus), has a high protein content, and a wide range of vitamins produced by the microflora during cheese production (Soni et al., 2000).

4.2.5.9 Whey Whey is the major by-product of the preparation of dairy products such as cheese, *channa*, *paneer*, and *shrikhand*. Whey contains a pool of nutrients, such as lactose and growth factors, that have the potential to stimulate the growth of

Table 4.6 Average Composition of *Dahi*, *Chakka*, and *Shrikhand*

CONSTITUENTS (%)	*DAHI* (SKIMMED MILK)	*CHAKKA*	*SHRIKHAND*
Total solids	9–10	22–23	57–60
Fat	Trace	Trace	5–6
Protein	3.7–3.9	13.5-/4.0	6.5–7.0
Sucrose	–	–	40–43
Ash	0.72	0.95–1.08	0.49–0.55
Titratable acidity	0.9–0.95	2.1–2.2	1.05–1.10
pH	4.4–4.6	4.4–4.6	4.4–4.6

Source: Adapted from Singh S. 2014a. *Dairy Technology. Dairy Products & Quality Assurance.* Vol. 2. New India Publishing Agency, New Delhi, India, p. 117; Sharma U.P. and Zariwala I.T. 1980. *Indian J Dairy Sci* 33(2):223–231; Upadhyay S.M., Dave J.M., and Sannabhadti S.S. 1984. *J Food Sci Technol.* 21(4):208; Salunkhe P., Patel H.A., and Thakur, P.N. 2006. *J Food Sci and Technol.* 43(3):276.

Table 4.7 Physico-Chemical Characteristics and Chemical Compostion of *Shrikhand* Sold in Gujrat (India)

CONSTITUENTS (%)	TRADITIONAL *SHRIKHAND*	COMMERCIAL *SHRIKHAND*
Water	40.52	35.11
Fat (DM*)	8.27	9.25
Protein (DM)	9.53	12.53
Lactose (DM)	3.74	2.59
Sucrose (DM)	76.25	70.00
Ash	0.530	0.683
Titratable acidity	1.075	1.310
pH	4.10	4.33
Soluble nitrogen	0.290	0.286
FFA (% oleic acid)	0.395	0.265

Source: Adapted from Singh S. 2014a. *Dairy Technology. Dairy Products and Quality Assurance.* Vol. 2. New India Publishing Agency, New Delhi, India, p. 117; Sharma U.P. and Zariwala I.T. 1980. *Indian J Dairy Sci* 33(2):223–231; Upadhyay S.M., Dave J.M. and Sannabhadti S.S. 1984. *J Food Sci Technol.* 21(4):208; Salunkhe P., Patel H.A. and Thakur, P.N. 2006. *J Food Sci and Technol.* 43(3):276.

DM = Dry matter.

desirable microorganisms in the intestine, producing acid that may inhibit the growth of putrefying microorganisms and promote a healthy intestinal flora (Lifran et al., 2000). Sialic acid is another whey-derived carbohydrate component with prebiotic activity (Naidu et al., 1999; Goyal and Gandhi, 2008).

4.2.5.10 Khoajalebi Khoajalebi is a very popular and delicious product liked by many Indians, who consume it regularly during various festival occasions. The chemical compostion of *khoajalebi* is given in Table 4.9.

Table 4.8 Nutrient Composition of Commercial Yoghurt and Probiotic Yoghurt

NUTRIENTS (ESTIMATED)	COMMERCIAL YOGHURT	PROBIOTIC YOGHURT	INCREASE/ DECREASE
Macronutrients			
Protein (g)	3.9 ± 0.26	5.25 ± 0.05	35.38% increase
Fat (g)	3.375 ± 0.02	2.115 ± 0.005	− 37.65% decrease
Carbohydrate	4.96 ± 0.2	3.75 ± 0.05	− 24.2% decrease
Minerals			
Calcium (mg)	120.75 ± 0.25	152.25 ± 0.05	26.09% increase
Phosphorous (mg)	101.95 ± 0.05	114 ± 0.25	11.89% increase
Iron (mg)	0.8 ± 0.05	0.91 ± 0.02	13.75% increase
Vitamins			
Vit. A (IU)	140 ± 0.5	204.5 ± 0.2	
Vit. C (mg)	0.7 ± 0.1	0.9 ± 0.53	
Riboflavin (mg)	0.275 ± 0.015	0.32 ± 0.01	

Source: From Steinkraus K.H. 1994. *Food Research International* 27(3), 259–267. With permission.

Table 4.9 Chemical Composition of *Khoajalebi* Collected from Different Markets

S. NO.	CHEMICAL COMPOSITION (%)	RANGE	AVERAGE[a] ± SD
1	Fat	12.33–16.77	14.932 ± 3.4
2	Protein	2.78–6.98	4.892 ± 2.0
3	Lactose	7.39–13.13	11.274 ± 2.2
4	Sucrose	15.20–48.07	27.7 ± 12.5
5	Water	18.21–28.69	22.788 ± 4.7
6	Minerals	0.60–1.29	0.91 ± 0.3
7	Total solids	71.31–81.79	77.21 ± 4.7

Source: From Pagote C.N. and Rao K.J. 2012. *Indian J. Traditional Knowledge* 11:96–102. With permission.
[a] Average of the market samples.

4.2.6 *Ethnic Fermented Fish Products*

A comparison of various fermented meat and fish product for nutritive value is being made in Table 4.10.

4.2.6.1 Ngari   *Ngari* is a fermented fish product of Manipur (India) where it is traditionally consumed by the *Meitei*. Physico-chemical characteristics and composition of *ngari* can be summarized as: water 33.5%, pH 6.2, acidity 0.61%, ash 21.1%, fat 13.2% DM, protein 34.1%, carbohydrate 31.6%, food value 381.6 kcal/100 g, Ca 41.7 mg/100 g, Fe 0.9 mg/100 g, Mg 0.8 mg/100 g, Mn 0.6 mg/100 g, and Zn 1.7 mg/100 g (Tamang et al., 2002; Thapa and Pal, 2007).

4.2.6.2 Hentak   *Hentak* is a fermented fish paste prepared from a mixture of sundried fish powder and the petioles of aroid plants in Manipur (India). The composition of *hentak* can be summarized as: water 40.0%, pH 6.5, ash 15.0%, fat 13.6%,

Table 4.10   Fermented Meat/Fish Products and Their Composition and Food Values

FOOD PRODUCT	AREA OF USAGE	PROTEIN (%)	FAT (%)	CARBOHYDRATE (%)	ENERGY VALUE (KCAL/100 G)
Ngari	Manipur	34.1	13.2	31.6	382.6
Hentak	Manipur	32.7	13.6	38.7	408
Tungtap	Meghalya	32.0	12.0	37.1	384.4
Gnuchi	Sikkim	21.3	14.5	47.3	404.9
SukakoMoacha	Eastern India	35.0	12	36.8	395.2
Sidra	Eastern India	25.5	12.2	45.7	394.6
KaratiBordiaLashim	Assam	24.–35.0	11.8–12.4	38.1–47.9	400–407.8
Kargyong	Arunachal Pradesh, Sikkim	16.0	49.1	32.0	634.5
Sukokamasu	NA	44.8	2.0	51.4	403.1

Source: Adapted from Tamang J.P. et al. 2012. *Indian Journal of Traditional Knowledge* 11(I):7–25.

protein 32.7%, carbohydrate 38.7%, food value 408.0 kcal/100 g, Ca 38.2 mg/100 g, Fe 1.0 mg/100 g, Mg 1.1 mg/100 g, Mn 1.4 mg/100 g, and Zn 3.1 mg/100 g (Thapa and Pal, 2007).

4.2.6.3 Tungtap   *Tungtap* is a fermented fish paste of the *Khasi* tribe in Meghalaya (India). The composition of *tungtap* is: water 35.4%, pH 6.2, ash 18.9%, fat 12%, protein 32.0%, carbohydrate 37.1%, food value 384.4 kcal/100 gm, Ca 25.8 mg/100 gm, Fe 0.9 mg/100 gm, Mg 1.6 mg/100 gm, Mn 0.8 mg/100 gm, and Zn 2.4 mg/100 gm (Thapa and Pal, 2007; Tamang et al., 2012).

4.2.6.4 Fish Sauce   This is considered as an important source of dietary protein and amino acids. It contains about 20 g/L of nitrogen, of which 80% is in the form of amino acids (Zaman et al., 2009). Amino acid profiles of fish sauce are often dominated by glutamic acid, lysine, aspartic acid, alanin, valine, and histidine (Ijong and Ohta, 1995; Park et al., 2001; Osako et al., 2005). In addition, it has a considerable amount of water soluble vitamins, including thiamin, riboflavin, niacin, and vitamin B_6 and B_{12} (Areekul et al., 1972, 1974).

    Fish sauce—nam-pla: This product contains: 70.6%–76.7% water, 1.8%–2.2% protein, 0.7%–4.7% fat, 25.6%–29.4% ash, 22.8%–26.2% NaCl, and pH 5.7–6.0. The standard chemical composition of grade 1 fish sauce contained 28.15% NaCl, 1.92% total nitrogen, 1.13% formaldehyde nitrogen, 1.64% organic nitrogen, 0.28% ammonia nitrogen, 0.85% amino nitrogen, pH 5.3–6.6, and 1.19–1.24 specific gravity (Phithakpol et al., 1995).

4.2.6.5 Shrimp Paste   Shrimp (*terasi*) paste contains (per 100 g) 25.42 g protein, 16.75 g sodium chloride, 1.94 g carbohydrate, 6.11 g lipid, 29.12 g ash (including salt), 37.41 g water, with pH 7.53 (Surono and Hosono, 1994a).

4.2.7 Ethnic Smoked and Dried Fish Products

4.2.7.1 Gnuchi   *Gnuchi* is a traditional smoked fish product of the Lepcha of Sikkim, and its composition can be summarized as: water 14.3%, pH 6.3, ash 16.9%, fat 14.5%, protein 21.3%, carbohydrate 47.3%, food value 404.9 kcal/100 g, Ca 37.0 mg/100 g, Fe 1.1 mg/100 g, Mg 8.8 mg/100 g, Mn 1.1 mg/100 g, and Zn 7.5 mg/100 g (Thapa and Pal, 2007; Tamang et al., 2012).

4.2.7.2 Suka ko maacha   This traditional smoked fish product is called *suka ko maacha* by the *Gorkha*, and its composition is summarized as: water 10.4%, pH 6.4, ash 16.2%, fat 12.0%, protein 35.0%, carbohydrate 36.8%, food value 395.2 kcal/100 g, Ca 38.7 mg/100 g, Fe 0.8 mg/100 g, Mg 5.0 mg/100 g, Mn 1.0 mg/100 g, and Zn 5.2 mg/100 g (Thapa and Pal, 2007; Tamang et al., 2012).

4.2.7.3 Sidra   *Sidra* is a sun-dried fish product commonly consumed by the *Gorkha*. The product is nutrititive, and its composition has been determined as: water 15.3%, pH 6.5, ash 16.6%, fat 12.2%, protein 25.5%, carbohydrate 45.7%, food value 394.6 kcal/100 g, Ca 25.8 mg/100 g, Fe 0.9 mg/100 g, Mg 1.6 mg/100 g, Mn 0.8 mg/100 g, and Zn 2.4 mg/100 g (Thapa and Pal, 2007; Tamang et al., 2012).

4.2.7.4 Sukuti   It is a popular sun-dried fish cuisine of the *Gorkha*. *Sukuti* is nutritive and its composition is summarized as: water 12.7%, pH 6.4, ash 13.6%, fat 11.4%, protein 36.8%, carbohydrate 38.2%, food value 402.6 kcal/100 g, Ca 17.7 mg/100 g, Fe 0.3 mg/100 g, Mg 1.4 mg/100 g, Mn 0.2 mg/100 g, and Zn 1.3 mg/100 g (Thapa and Pal, 2007).

Chepa shutki: It is a fermented fish product. Water content in traditional *chepa shutki* varies from 39.62% to 46.89% (Nayeem et al., 2010). The lowest water content was recorded for the product at the producer level, while the highest value was found in the product at retailer level. Thus, *chepa shutki* absorbs water from the environment during storage, since, in most of the places, they are stored in a bamboo basket or clay vat in the open air where there is always chance of absorbing water. The composition of this product has also been evaluated. Protein, lipid, and ash content ranged from 32.46% to 33.83%, 19.25% to 24.97%, and 0.81% to 1.01%, respectively, while the amount of sodium, potassium, calcium, phosphorus, magnesium, zinc, and iron content in 100 g product were recorded to be 247, 163, 621, 440, 50, 3.4, and 3 mg, respectively. Clearly *chepa shutki* is a good source of nutrition (Mansur et al., 2000; Nayeem et al., 2010).

4.2.7.5 Karati, Bordia and Lashim   These are sun dried and salted fish products of Assam, and their nutritive value has been summarized as: water 9.6%–12.0%, pH 6.3–6.4, ash 12.8%–15.3%, fat 11.8%–12.4%, protein 24.5%–35.0%, carbohydrate 38.1%–47.9%, food value 400.0–407.8 kcal/100 g (Thapa and Pal, 2007; Tamang et al., 2012).

4.2.7.6 Fermented Fish   Fermentation of fish is carried out in Sri lanka and the Maldives. A comparison of fermented fish for composition is made in Table 4.11.

4.2.7.7 Shidal   The nutritional value of *shidal* was reported by Majumdar et al. (2006). The water, ash, protein and lipid content of *shidal* collected from the markets of Tripura were found to be: 18.84%, 16.3%, 38.93%, and 16.73%, respectively. The pH and total titratable acidity (TTA) were observed as 6.9% and 1.66%, respectively. Out of total nitrogen, the protein nitrogen and non-protein nitrogen (NPN) were found to be 46% and 54%, respectively. The Total Volatile Base Nitrogen (TVBN) content of the product was also high, and found to be 509 mg/100 g, but it usually does not manifest

Table 4.11 Chemical Composition of Cured and Fermented Fish Products in Sri Lanka

PARAMETERS	RAW FISH	AMBULTHIYAL	DRIED FISH	MALDIVE FISH	*JAADI* 4.0-PH	FISH SAUCE (5.9 > pH)
Water (%)	70–80	58.52	28.0	15.73	48.5	59.12
Ash (%)	2	5.9	3.16	6.40	22	25
Crude protein (%)	15–24	28.93	64.51	68.11	2	*5
Fat (%)	0.1–2	1.78	3.5	2.46	1 2-	′5
Salt (%)	0.2	4.77	17.13	5.98		29.5
Shelf-life (days)	1	7–180	80	120	35	
Water activity	0.96	0.95	0.56	0.76		1—

Source: From Jayasinghe P.S. and Sagarika E. 2003. Cured and fermented fishery Products of Sri Lanka. *International Seminar and Workshop on Fermented Foods, Health Status and Social Well-Being*, Nov 2003, Anand, India, p. 7. With permission.

any ammonia-like odor in the product, probably due to masking of ammonia odor by the characteristic strong odor of the s*hidal* itself. The free alpha amino nitrogen content (79.54 mg/100 g) of the product was in the moderate range. The contents of non-protein nitrogen (NPN) and Total Volatile Base Nitrogen (TVBN) of the product were an indication of the high degree of fermentation. Since fermentation is carried out in an anaerobic environment in salt-free conditions, a moderate peroxide value (18.1 milliequivalent O_2/kg fat) and thiobarbituric acid (TBA) value (0.41) has been found. The absence of salt (a potential pro-oxidant) and metals (as the fermentation is carried out in earthen container) in the system may account for such low values of Peroxide Value (PV) and Thisbar-Bituric Acid (TBA). Amino acids like glutamic acid, aspartic acid, leucine, alanine, and lysine were in the higher proportion, while others, such as tyrosine, histidine, arginine, and tryptophan were in very low quantities and proline was not detected. The free fatty acid value was found to be moderate (16.21%, as oleic acid). Among the saturated fatty acids, palmitic acid (25.83%) was found to be the highest, followed by stearic acid (10.81%). Oleic acid (25.33%) and palmitoleic acid (4.59%) were the most prominent among the mono-enoic acids. The *n*-6 PUFAs in *shidal* were composed mainly of linoleic acid (10.77%) and a very low amount of arachidonic acid (0.71%). Among the *n*-3 PUFAs, linolenic acid (4.18%), EPA (0.41%) and DHA (0.57%) along with two others (C 18:4 and C 22:5) were detected (Muzaddadi and Prasanta, 2013).

4.2.8 Ethnic Preserved Fermented Meat Products

The nutritive value of sausage is almost equal to the raw meat and the ingredients from which these are prepared. Little nutritional changes are introduced by fermentation (Pederson, 1971). In addition, fermented yak (*yak kargyong, yak satchu, yak kheuri*), beef (*lang kargyong, lang satchu, lang kheuri*), buffalo (*mogong-grain*) and sheep fat (*luk chilu*) are used in North Eastern India (Tamang et al., 2012).

4.2.8.1 Kargyong *Kargyong* is a sausage-like meat product of Sikkim and Arunachal Pradesh prepared from meat. *Kargyong* is nutritive and its composition is: water 21.9%, pH 6.9, ash 2.8% DM, fat 49.1% DM, protein 16.0% DM, carbohydrate 32.0% DM, and calorific value 634.5 kcal/100 g (Rai et al., 2010).

4.2.8.2 Satchu *Satchu* is an ethnic dried meat (beef/yak/pork) and is consumed by the Tibetan, the *Bhutia*, the *Lepcha*, the *Sherdukpen*, and the *Khamba*. The nutritive value of *satchu* has been found to be: water 23.7%, pH 5.7, ash 7.3% DM, fat 4.7% DM, protein 51.0% DM, carbohydrate 37.0% DM, and food value 405.8 kcal/100 gm (Rai et al., 2010; Tamang et al., 2012).

4.2.8.3 Suka ko masu *Suka ko masu* is a dried or smoked meat product from buffalo meat or chevon (goat meat). It is consumed by the non-vegetarian *Gorkha*. The nutritional value of *suka ko masu* can be summarized as: water 23.2%, pH 5.2, ash 1.8% DM, fat 2.0% DM, protein 44.8% DM, carbohydrate 51.4% DM, and food value 403.1 kcal/100 gm (Rai et al., 2010; Tamang et al., 2012).

4.2.9 Alcoholic Beverages

A variety of alcoholic beverages is produced by many tribes and has remained an integral part of many cultures all around the world from ancient times.

4.2.9.1 Sur It is a finger millet-based (*Eleucine coracana*) fermented beverage mostly prepared in the Lug valley of Kullu district and the *Chhota* Bhangal area of Kangra District of Himachal Pradesh, India (Thakur et al., 2004; Joshi et al., 2015). Finger millet is also known as *ragi* in India (Hulse et al., 1980). A herbal mix in *sattu* (flour of roasted barley) base (called as *dhaeli*) by the folk people is reported to be added during fermentation (Thakur et al., 2004). The physico-chemical characteristics of *sur* are summarized in Table 4.12.

4.2.9.2 Kodo ko jaanr The most popular fermented finger millets-based mild alcoholic beverage with a sweet–sour and acidic taste is *kodo ko jaanr* or *chyang* or *chee*, prepared and consumed by the *Gorkha*, the *Bhutia*, the *Lepcha*, the *Monpa*, and many other ethnic groups of North East India. The nutritional value of *Kodo ko jaanr* can be summarized as: water 69.7%, pH 4.1, alcohol 4.8%, ash 5.1% DM, fat 2.0% DM, protein 9.3% DM, carbohydrate 83.7% DM, calorific value 398.0 kcal/100 g, Ca 281.0 mg/100 g, K 398 mg/100 g, P 326.0 mg/100 g, Mg 118 mg/100 g, Mn 9.0 mg/100 g, and Zn 1.2 mg/100 g (Thapa and Tamang, 2004; Tamang et al., 2012).

4.2.9.3 Bhaati jaanr *Bhaati jaanr* is an ethnic fermented rice beverage, consumed as a staple food beverage by the *Gorkha* in North Eastern India. Composition of *bhaati jaanr* is as follows: water 83.4%, pH 3.5, alcohol 5.9%, ash 1.7% DM, fat 2.0%

Table 4.12 Physico-Chemical Characteristics of the *Sur* Samples Collected from Different Regions of Himachal Pradesh (India)

DISTRICT	TSS (°B)	REDUCING SUGARS (%)	TOTAL SUGARS (%)	ETHANOL (% V/V)	COLOUR (O.D. AT 440 NM)	pH	TITRABLE ACIDITY (% AS L.A.)	VOLATILE ACIDITY (% AS A.A.)	TOTAL PROTEINS (%)	FREE AMINO ACIDS (%)	TOTAL PHENOLS (mg/L)	TOTAL ESTERS (mg/L)	HIGHER ALCOHOLS (mg/L)	ACETALDEHYDES (mg/L)	METHANOL (mg/L)
Kangra	8.74 ± 1.8	0.25 ± 0.10	0.44 ± 0.129	9.58 ± 3.12	3.66 ± 3.1	3.65 ± 0.22	2.02 ± 0.62	0.022 ± 0.01	0.95 ± 0.12	0.09 ± 0.01	330.48 ± 56.67	144.82 ± 20.42	126.86 ± 40.76	13.78 ± 2.71	428.61 ± 138.58
Kullu	9.8 ± 3.9	0.47 ± 0.27	0.49 ± 0.410	8.97 ± 3.28	3.12 ± 2.1	3.90 ± 0.29	2.47 ± 1.54	0.056 ± 0.06	0.85 ± 0.13	0.10 ± 0.02	284.85 ± 26.67	184.386 ± 86.96	141.18 ± 66.72	14.59 ± 3.09	391.85 ± 107.56
Mandi	9.1 ± 4.00	0.36 ± 0.33	0.56 ± 0.685	10.53 ± 3.29	2.57 ± 1.97	3.83 ± 0.27	1.54 ± 0.86	0.027 ± 0.02	0.79 ± 0.15	0.10 ± 0.02	272.57 ± 59.88	148.84 ± 76.34	137.62 ± 71.76	14.19 ± 5.57	608.95 ± 112.15
Sirmaur	6.76 ± 0.06	0.11 ± 0.01	0.10 ± 0.008	5.42 ± 0.188	0.883 ± 0.02	3.46 ± 0.06	2.15 ± 0.07	0.027 ± 0.0009	0.84 ± 0.21	0.08 ± 0.02	216.451 ± 1.93	177.75 ± 0.65	60.905 ± 0.58	28.16 ± 0.88	588.7 ± 13.01

Source: Adapted from Joshi V.K. et al., 2015. *Indian J Tradit Know* (in press).
SD = Standard deviation, AA = Acetic acid, OD = Optical Density, LA = Lactic acid.

Table 4.13   Composition of Wild Apricot Wine

SUBSTANCE	CONCENTRATION
Alcohol (% v/v)	10.6
Titratable acidity[a] (% MA)[1]	0.75
Total phenols (mg/L)	240
Volatile acidity[b] (%)	0.08
TSS (°B)	6.8

Source: Adapted from Joshi V.K., Bhutani V.P., and Sharma R.C. 1990. *Am. J. Enol. Vitic.* 41(3):239–241.
[a] Titratable acidity expressed as malic acid equivalent g/100 mL.
[b] Volatile acidity expressed as acetic acid g/100 mL.

DM, protein 9.5% DM, crude fiber 1.5% DM, carbohydrate 86.9% DM, calorific value 404.1 kcal/100 g, Ca 12.8 mg/100 g, K 146.0 mg/100 g, P 595.0 mg/100 g, Fe 7.7 mg/100 g, Mg 50.0 mg/100 g, Mn 1.4 mg/100 g, and Zn 2.7 mg/100 g (Tamang and Thapa, 2006; Tamang et al., 2012).

4.2.9.4 Wild Apricot Wine   Wild apricots are utilized by the tribes of the dry and temperate region of Himachal Pradesh, India for the production of alcoholic beverages including cwine. The fermentation is carried out by the natural microflora and in most of the cases the wine is distilled to make liquor. The composition of wild apricot wine made experimentally is given in Table 4.13.

The mineral content present in wild apricot wine is shown in the Table 4.14.

4.2.9.5 Chhang/Lugri Chakhti/Jhol   The *Chhang/Lugri* is a very popular fermented beverage which is served during *Phagli* (traditional New Year of Lahulis) and marriage ceremonies to guests. It is an indigenous rice beer made in the tribal belt of Lahaul and Spiti, Kullu and Kangra. The huskless Sherokh variety of barley, locally called *grim*, is used in its preparation (Bhatia et al., 1977). *Chhang* is also called *jhol or*

Table 4.14   Mineral Contents (mg/L) of Wild Apricot Wine

MINERALS	CONCENTRATION
Potassium (K)	2775
Sodium (Na)	44
Calcium (Ca)	86
Magnesium (Mg)	24
Zinc (Zn)	2.53
Iron (Fe)	4.10
Manganese (Mn)	0.98
Copper (Cu)	0.29

Source: Adapted from Joshi V.K., Bhutani V.P., and Sharma R.C. 1990. *Am. J. Enol. Vitic.* 41(3):239–241.

chakti in Kullu. A distilled form is known as *sura* in the Lahaul valley. A comparison of unfermented and fermented ragi is made in Table 4.15.

4.2.10 Vinegar

The major component of vinegar, apart from water, is self-evidently acetic acid (Lea, 1988). Data on minor components are relatively sparse. Cider vinegar is popular in fold medicine and is suggested as a remedy to various diseases (Joshi and Sharma, 2009). Evidence linking use of vinegar to reduced risk for hypertension and cancer is equivocal. It has been documented that vinegar ingestion reduces the glucose response to a carbohydrate load in healthy adults and in individuals with diabetes. There is also some evidence that vinegar ingestion increases short-term satiety and also influences glycemic control. The composition of cider vinegar is given in Table 4.16.

4.2.10.1 Cider Vinegar is made from apple by natural fermentation at home scale level. The fermentation is usually very slow and vinegar is ready in 3-4 months period (Joshi and Sharma, 2009).

4.2.10.2 Coconut Vinegar Coconut vinegar, made from the sap or "toddy" of the coconut palm, is produced and used extensively in South Asian countries, including India (Murooka et al., 2009). A cloudy, white liquid, coconut vinegar has a particularly sharp, acidic taste (4% acetic acid) with a slightly yeasty flavor. Coconut water

Table 4.15 Nutritional Composition of Unfermented and Fermented *Ragi (chhang)*

COMPONENTS	UNFERMENTED *RAGI* (g/100 g)	FERMENTED *RAGI* BY STARTER CULTURE (g/100 g)
Carbohydrates (Starch and sugar)	78.0 ± 3.4	35.0 ± 125
Proteins	7.6 ± 0.34	10.5 ± 0.50
Fat	1.1 ± 0.06	1.0 ± 0 05
Phosphorus	0.3 ± 0.05	0.3 ± 0 05
Calcium	0.4 ± 0.06	0.4 ± 0.08
Iron	0.004 ± 0.009	0.004 ± 0004
Ethanol	–	16.0 ± 0.68
Total acidity (lactic acid)	–	3.0 ± 0 27
Microbial biomass	–	2.7 ± 0.36
Volatile acidity	–	002 ± 0.004
Thiamin	0.48 ± 0.08	0.50 ± 0.09
Riboflavin	0.20 ± 0.01	0.60 ± 0.10
Niacin	1.00 ± 0.05	4.20 ± 0.45
Pantothenic acid	0.006 ± 0.07	1.60 ± 0.06
Folic acid	0006 ± 0.001	0.010 ± 0 001
Cyanocobalamin	0.00	0.04 ± 0.004

Source: From Basappa S.C. et al. 1997. *Int J Food Sci Nutri* 48:313. With permission.
Note: On dry weight basis All values mean ± standard deviation (SD).

Table 4.16 Composition of Cider Vinegar

COMPOSITION	VALUE
Total acid as acetic (% w/v)	3.3–9
Nonvolatile acids as malic (% w/v)	0.03–0.4
Total solids (% w/v)	1.3–5.5
Total ash (% w/v)	0.2–0.5
Non-sugar solids (% w/v)	1.2–2.9
Total sugar (% w/v)	0.15–0.7
Alcohol (% w/v)	0.03–2.0
Protein (%)	0.03
Glycerol (% w/v)	0.23–0.46
Sorbitol (% w/v)	0.11–0.64

Source: Adapted from Lea A.G.H. 1988. *Processed Apple Products*. Van Nostand Reinhold, New York, pp. 279–301.

vinegar is made from coconut water and contains between 3% and 4% (v/v) acetic acid and is an indispensable commodity in many households (Sanchez, 1990).

4.2.10.3 Sugar Vinegar   Cane sugar vinegar (sugarcane juice) and palm vinegars are popular in India and other countires of South Asia. They range in color from dark yellow to golden brown and have a mellow flavor.

4.2.10.4 Mango Vinegar   Vinegar is produced from mango pulp by alcoholic fermentation and then, acetification. The fermentation results in production of around 8% (v/v) of ethanol (Akubor, 1996) and the final mango vinegar has an acidity of around 5.3% (v/v) as acetic acid.

4.2.10.5 Palm Vinegar   The palms most frequently tapped for sap are raphia palms (*Raphia hookeri* and *Raphia vinifera*) and the oil palm (*Elaeis guineese*). The acetic acid concentrations of palm vinegar are around 4% (v/v) (Battcock and Azam-Ali, 1998; González and VuystLuc, 2009).

4.2.10.6 Date Vinegar   For the production of date vinegar, water is added to the dates and the mixture is boiled and alcoholic fermentation is carried out (Ali and Dirar, 1984). Following acetification, the final vinegar has a concentration of acetic acid of 4%–5% (v/v) (González and VuystLuc, 2009).

4.3 Summary and Future Prospectus

Fermentation technology is of great significance in ensuring nutritive, safe, better preserved, and flavored food. The chemical composition and nutritive value of indigenous fermented food products, including fermented soybean products, show the availability of protein, carbohydrates, fats, and fibers, as well as the mineral constituents, in significant quantities. But a large number of indigenous fermented foods are still to be documented for the composition and nutritive values. A review of the literature revealed that there is no information available regarding the detailed composition of alcoholic products prepared indigenously. Similarly in other cases information regarding the product is available, but it is not complete and much has to be done to improve the quality of such products. At the same time, many fermented foods have never been investigated at all for their nutritional content. With the development of highly sophisticated equipment and accurate techniques of analysis, extensive evaluation of fermented products should not be a problem. Further, more intensive evaluation of the fermented foods for toxic compounds is also needed. Together with food safety, the nutritional and flavor profile of the indigenous fermented products needs to be made so as to meet the expectations of modern consumers. This technology needs to be developed further to enhance the safety and ease of application in a rural resource-poor setting.

Needless to say, improved products and practices are readily taken up by people and the industry, as there is no problem of taste development, and so the introduction of such products into the market is quite easy. The need today is not only for the production of more food, but also of nutritionally rich, better preserved food, safe for consumption, as it is useless to produce more food unless it can be preserved, and made available to the population. There is a need for research to be directed towards identifying the benefits and risks associated with specific indigenous fermented foods.

Many countries of South Asia are rich in natural resources, with excellent soil and abundant rainfall, and should not have any problems in feeding their people, but for the difficulties of proper preservation and transportation, which are a deterrent to nutritious food provision, and which need to be solved. Indigenous fermented foods are the big tool for nutritional enrichment and taste improvement.

References

Abou-Soliman N.H. 2005. *Studies on goat milk proteins: Molecular and immunological characterization with respect to human health and nutrition.* PhD thesis, Alexandria University, Egypt.

Adeyemi I.A. and Beckley O. 1986. Effect of period of maize fermentation and souring on chemical properties and amylograph pasting viscosities of ogi. *Cereal Sci.* 4:353–360.

Akingbala J.O., Rooney K.W., and Faubion J.M. 1981. A laboratory procedure for the preparation of ogi; a Nigerian fermented food. *Food Sci.* 45(5):1523–1526.

Akinrele I.A., Adeyinka O., Edwards C.C.A., Olatunji F.O., Dina J.A., and Koleoso A.O. 1970. The development and production of soy-ogi—A corn based complete protein. FIRO, Research Report No. 42.

Akubor P.I. 1996. The suitability of African bush mango juice for wine production. In: *Plant Foods.* Alexandria University, Egypt. pp. 213–219.

Ali M.Z. and Dirar H.A. 1984. A microbiological study of Sudanese date wines. *J Food Sci* 49:459–460.

Alm L. 1982. Effect of fermentation on B-vitamin content of milk in Sweden. *J Dairy Sci* 65:353–359.

Aneja R.P., Mathur B.N., Chandan R.C., and Banerjee A.K. 2002. *Cultured/Fermented Products, Cited in Technology of Indian Milk Products.* A dairy India publication, Delhi, India, p. 165.

Anonymous IV. Milk based fermented foods www.bdu.ac.in/schools/biotechnology/industrial.../pdf/food/4.pdf

Anonymous 1998. FAO: Products of bacterial fermentations Fermented fruits and vegetables a global perspective Mike Battcock and Sue Azam-Ali. FAO agricultural services bulletin no. 134, Food and Agriculture Organization of the United Nations Rome, 1998.

Anonymous, Annual Report 2009–2010. Institute of Bioresources and Sustainable Development (IBSD), Imphal, 2010.

Anonymous 1998. Intermediate Technology Food Chain 23 Improving *Madila*—A traditional fermented milk from Botswana https://practicalaction.org/docs/agroprocessing/food_chain_23.pdf

Anukam K.C. and Reid G. 2009. African Traditional Fermented Foods and Probiotics. *J Med Food* 12(6):177–1184.

Areekul S., Boonyananta C., Matrakul D., and Chantachum, Y. 1972. Determination of vitamin B12 in fish sauce in Thailand. *J Med Assoc Thai*, 55:243–247.

Areekul S., Thearawibul R., and Matrakul, D. 1974. Vitamin B12 contents in fermented fish, fish sauce and soya-bean sauce. *Southeast Asian J Trop Med Public Health* 5:461.

Arora S., Jood S., and Khetarpaul N. 2010. Effect of germination and probiotic fermentation on nutrient composition of barley based food mixtures. *Food Chem.* 119(2):779–784.

Arrmoto K. 1961. Nutritional research on fermented soyabean products. In: *Meeting the Protein Needs of Infants and Preschool Children.* Publ. 843. Natl Acad Sci—Natl Res Council, Washington, DC.

Basappa S.C., Someshekar D., Agrawal R., Suma D., and Bharthi K. 1997. Nutritional composition of fermented Ragi (chhang) by phab and defined starter culture as compared to unfermented ragi (*eleucinecoracana G*). *Int J Food Sci Nutria* 48:313.

Batish V.K., Grovers S., Patnaik P., and Ahmed N. 1999. Fermented milk products. In: *Biotechnology: Food Fermentation Microbiology, Biochemistry and Technology.* Joshi V.K., Pandey A. (eds). Educational Publishers and Distributors, Kerala, India, pp. 781–864.

Battcock M. and Azam-AIi S. 1998. Fermented fruits and vegetables: A global perspective. *FAO* beverage of the Himalayas. *Food Microbiol* 21:617–622.

Bhatia A.K., Singh R.P., and Atal C.K. 1977. *Chhang*—the fermented beverage of Himalayan folk. *Indian Food Packer* 31:32–39.

Blandino A., Al-Aseeri M.E., Pandiella S.S., Cantero D., and Webb C. 2003. Cereal-based fermented foods and beverages. *Food Res Int.* 36(6):527–543.

Chakrabarty J. 2011. *Microbiological and nutritional analysis of some fermented foods consumed by different tribes of North Cachar Hills District of Assam.* PhD thesis, Food Microbiology Laboratory, Sikkim Government College, Gangtok (Assam University),

Chavan, U.D., Chavan J.K., and Kadam, S.S. 1988. Effect of fermentation on soluble proteins and *in vitro* protein digestibility of sorghum, green gram and sorghum-green gram blends. *J. Food Sci.* 53:1574.

Chelule P.K., Mokoena M.P., and Gqaleni N. 2010. Advantages of traditional lactic acid bacteria fermentation of food in Africa. In: *Current Research, Technology and Education Topics in Applied Microbiology and Microbial Biotechnology.* Méndez-Vilas A. (ed.), www.formatex.info/microbiology,2/1160-1167.

Chettri R. and Tamang J.P. 2008. Microbiological evaluation of *inaseura*, an ethnic fermented legume-based condiment of Sikkim. *J Hill Res* 21(1):1–7.

Dahal N., Rao E.R., and Swamylingappa. 2003. Biochemical and nutritional evaluation of masyaura—A legume based traditional savoury of Nepal. *Food Sci Technol* 40:17–22.

Dahal N.R., Karki T.B., Smylingappa B., Qi Li, and Gu Guoxian. 2005. Traditional foods and beverages of Nepal—A review. *Food Rev Int* 21(1):1–25.

Date W.B. and Bhatia D.S. 1955. Preservation of Indian milk sweets: Some preliminary studies on shrikhand vadi and milk burfee. *Indian J Dairy Science* 8:61–66.

Desai H.K., Vyas S.H., and Upadhyay K.G. 1985. Influence of homogenization of milk on the quality of *chakka* and *Shrikhand*. *Indian J Dairy Sci* 38 (2): 102–106.

Dewan S. 2002. Microbiological evaluation of indigenous fermented milk products of the Sikkim Himalayas. PhD thesis, Food Microbiology Laboratory, Sikkim Government College, Gangtok (North Bengal University).

Dewan S. and Tamang J.P. 2006. Microbial and analytical characterization of *Chhu*, a traditional fermented milk product of the Sikkim Himalayas. *J Sci Ind Res* 65:747–752.

Dewan S., Tamang, J.P. 2007. Dominant lactic acid bacteria and their technological properties isolated from the Himalayan ethnic fermented milk products. *Antonie van Leeuwenhoek Int J Gen Mol Microbiol* 92(3):343–352.

Fields M.L., Hamad A.M., and Smith, D.K. 1981. Natural lactic acid fermentation of corn meal. *J Food Sci* 46:900–902.

Gadaga T.H., Mutukumira A.N., Narvhus J.A., and Feresu S.B. 1999. A review of traditional fermented foods and beverages of Zimbabwe. *Int J Food Microbiol* 53(1):1–11

Ghosh J. and Rajorhia G.S. 1987. Chemical microbiological and sensory properties of *misti dahi* sold in Calcutta. *ASIN J Dairy Res.* 6(1):11.

González Á. and VuystLuc De. 2009. Vinegars from Tropical Africa. In: *Vinegars of the World.* Solieri L., Giudici P. (eds). Springer-Verlag, Italia, pp. 209–221.

Goyal N. and Gandhi D.N. 2008. Whey, a carrier of probiotics against diarrhoea. [On-line]. Available from: http://www.dairyscience.info/probiotics/110-whey-probiotics.html? show-all=1. Accessed: October , 2014.

Gurr M.I. 1987. Nutritional aspects of fermented milk products. *FEMS Microbiol Letts* 46(3):337–342.

Haard N.F. and Odunfa S.A., Lee1 C.H. 999. *Fermented Cereals. A Global Perspective.* FAO, Rome, Italy.

Hattingh A.L. and Viljoen B.C. 2001. Yogurt as probiotic carrier food. *Int Dairy J* 11:1–17.

Hulse J.H., Laing E.M., and Pearson O.E. 1980. *Sorghum and the Millets: Their Composition and Nutritive Value.* Academic Press, London, p. 997.

Ijong F.G. and Ohta Y. 1995. Amino acid compositions of bakasang, a traditional fermented fish sauce from Indonesia. *Food Sci Technol* 28:236–237.

Ingle U.M. and Joglekar N.V. 1974. *J Food Sci Technol.* 2:54–58.

Jain A., Desai H.K., and Upadhyay K.G. 2003. Sensoric profile of market *shrikhand* sold in Gujarat State. *Indian J Dairy Sci* 56(5):292–294.

Jayasinghe P.S. and Sagarika E. 2003. Cured and fermented fishery Products of Sri Lanka. *International Seminar and Workshop on Fermented Foods, Health Status and Social Well-Being,* Nov 2003, Anand, India, p. 7.

Joshi V.K., Bhutani V.P., and Sharma R.C. 1990. The effect of dilution and addition of nitrogen source on chemical, mineral and sensory qualities of wild apricot wine. *Am. J. Enol. Vitic.* 41(3):239–241.

Joshi V.K., Ashwani K., and Thakur N.S. 2015. Technology and production of *sur* in Himachal Pradesh. *Indian J Tradit Know* (in press).

Joshi V.K., Bhutani, V.P., and Thakur, N.K. 1999. In: *Biotechnology: Food Fermentation Microbiology, Biochemistry and Technology.* Joshi V.K. and Pandey A. (eds). Educational Publishers and distributors, Kerala, India, pp. 1291–1348.

Joshi, V.K. and Sharma, S. 2009. Cider Vinegar: Microbiology, Technology and Quality. In: Solieri, L. and Springer-Verlag, Italy, pp 197–207.

Kalra S. and Jood S. 2000. Cholesterol-lowering effect of dietary barley β-glucan in rats. *J Cereal Sci* 31:141–145.

Katiyar S.K, Bhasin A.K., and Bhatia A.K. 1991. Traditionally processed and preserved milk products of Sikkimese Tribes. *Sci Cul* 57(10,11):256–258.

Khetarpaul N. and Chauhan B.M. 1990. Effect of fermentation by pure cultures of yeasts and lactobacilli on the available carbohydrate content of pearl millet. *Food Chem* 36: 287–293.

Laxminarayana H., Nambudripad V.K.N., Laxmi N.V., Anantaramiah S.N., and Srinvasmurty V. 1952. Studies on dahi.II. General survey of market dahi. *Indian J Vet Sci Ani Husband* 22:13.

Lea A.G.H. 1988. Cider vinegar. In: Downing D.L. (ed.). *Processed Apple Products*. Van Nosand Reinhold, New York, pp. 279–301.

Lifran E.V. Hourigan J.A., Sleigh R.W., and Johnson, 2000. New whey for lactose. *Food Aust* 52(4):120–125.

Majumdar R.K., Basu S., and Nayak B.B. 2006. Studies on thebiochemical changes during fermentation of salt fermented Indian *shad*. *J. Aquatic Food Product Technol*. 15(1):53–69.

Mansur M.A., Islam M.N., Bhuyan A.K.M.A., and Haq M.E. 2000. Nutritional composition, yield and consumer response to semi-fermented fish product prepared from underutilized fish species of Bangladesh coastline. *Indian J. Mar. Sci*. 29:73–76.

Matz S.A. 1971. *Cereal Science*. AVI Publishing Co., Westport, CT, p. 241.

Mital B.K. 1977. Indian fermented milks. *Symposium on Indigenous Fermented Foods*, Bangkok, Thailand.

Miyani R.V. 1982. Effect of different levels of moisture, sugar and fat on consistency and acceptability of Shrikhand. MSc thesis, Guj. Agril. Uni. S.K. Nagar.

Murooka Y., Nanda K., and Yamashita M. 2009. Rice vinegars. In: *Vinegars of the World*. Solieri L, Giudici, P. (eds). Springer-Verlag, Italia, pp. 121–133.

Muzaddadi A.U. and Prasanta M. 2013. Effects of salt, sugar and starter culture on fermentation and sensory properties in *Shidal* (a fermented fish product). *Afr. J. Microbiol. Res*. 7(13):1086–1097.

Nagai T. and Tamang J. 2010. Fermented legumes: Soybean and non-soybean products. In: *Fermented Foods and Beverages of the World*. J.P. Tamang and K. Kailasapathy (eds). Taylor & Francis, CRC Press, Boca Raton, FL, pp. 191–224.

Naidu A.S., Bidlack W.R., and Clemens R.A. 1999. Probiotic spectra of lactic acid bacteria (LAB). *Crit Rev Food Sci Nutr* 38(1):13–126.

Nanson N.J. and Field M.L. 1984. Influence of temperature on the nutritive value of lactic acid fermented cornmeal. *J Food Sci* 49:958–959.

Nayeem M.A., Pervin K., Reza M.S., Khan M.N.A., Islam M.N., and Kamal M. 2010. Quality assessment of traditional semi-fermented fishery product (*Chepa Shutki*) of Bangladesh collected from the value chain. *Bangladesh Res. Pub. J*. 4(1):41–46.

Nikkuni S. 1997. *Natto, kinema* and *thua-nao*: Traditional non-salted fermented soybean foods in Asia. *Farming Japan* 31(4): 27–36.

Nout M.J.R., Bakshi D., and Sarkar P.K. 1998. Microbiological safety of kinema, a fermented soya bean food. *Food Control* 9(6):357–362.

Nout M.J.R. and Motarjemi Y. 1997a. Assessment of fermentation as a household technology for ogi; a Nigerian fermented food. *Food Sci* 45(5):1523–1526.

Nout M.J.R., and Motarjemi Y. 1997b. Assessment of fermentation as a household technology for improving food safety: A joint FAO/WHO workshop. *Food Control* 8(5–6): 221–226.

Omizu Y., Tsukamoto C., Chettri R., and Tamang J.P. 2011. Determination of saponin contents in raw soybean and fermented soybean foods of India. *J Sci Indus Res* 70:533–538.

Osako K., Hossain M.A., Kuwahara K., Okamoto A., Yamaguchi A., and Nozaki K. 2005. Quality aspectof fish sauce prepared from underutilized fatty Japanese anchovy and rabbit fish. *Fish Sci* 71:1347–1355.

Pagote, C.N. and Rao, K.J. 2012. *Khoa-Jalebi*-a unique traditional product of Central India. *Indian J Traditional Knowledge* 11, 96–102.

Pal P.K., Hossain S.A., and Sarkar P.K. 1996. Optimisation of process parameters in the manufacture of churpi. *J Food Sci Technol* 33:219–223.

Park J.N., Fukumoto Y., and Fujita E. 2001. Chemical composition of fish sauce produced in southeast and East Asian countries. *J Food Compos Anal* 14:113–125.

Patel R.S. and Abd-El-salam. 1986. *Shrikhand*—An Indian analogue of Western quarg. *Cultured Dairy Products J* 21(1):6–7.

Patel R.S. and Chakraborty B.K. 1985b. Factors affecting the consistency and sensory properties of *Shrikhand*. *Egyptian J Dairy Sci* 12:73–78.

Patel R.S. and Chakraborty B.K. 1988. *Shrikhand*: A review. *Ind J Dairy Science* 41:126.

Pederson C.S. 1971. *Microbiology of Food Fermentation*. AVI Pub Co Inc. Westport, CT.

Phithakpol B., Varanyanond W., Reungmaneepaitoon S., and Wood H. 1995. The Traditional Fermented Foods of Thailand, Institute of Food Research and Product Development, Kasetsart University, Bangkok, 157 p.

Premarani T. and Chhetry G.K.N. 2011. Nutritional analysis of fermented soybean *(Hawaijar)*. *Assam Univ J Sci Technol* 7(1):96–100.

Rai A.K, Tamang J.P., and Palni U. 2010. Microbiological studies of ethnic meat products of the Eastern Himalayas. *Meat Sci* 85:560–567.

Ramaite R.A.A. and Cloete T.E. 2006. *Traditional African fermentations*. Van Schaik, Pretoria.

Rangappa K.S. and Achaya K.T. 2014. Composition of lassi. In: *Dairy Technology. Dairy Products and Quality Assurance*. Vol-2. New India Publishing Agency, New Delhi, India, p. 133.

Rangappa K.S. and Achaya K.T. 1974. *Indian Dairy Products*. AsiaPublishing House, Bombay, pp. 119–124.

Rao H.G.R., Thygraj N., and Puranik D.B. 1986. Standardised methods for preparation of *Shrikhand* A popular fermented milk product. *Dairy Guide* 8(11):35.

Reddy K.K., Ali M.P., Rao B.V., and Rao T.J. 1984. Studies on the production and quality of *shrikhand* from buffalo milk. *Indian J Dairy Sci* 37(4):293–296.

Sahlin P. 1999. *Fermentation as a method of food processing production of organic acids, pH-development and microbial growth in fermenting cereals*, Licentiate thesis. Division of Applied Nutrition and Food Chemistry, Center for Chemistry and Chemical Engineering, Lund Institute of Technology, Lund University.

Salunkhe P., Patel H.A., and Thakur, P.N. 2006. Physico-chemical properties of shrikhand sold in Maharashtra state. *J Food Sci and Technol*. 43(3):276.

Sanchez P.C. 1990. Vinegar. In: *Coconut as Food*. Coconut Research and Development Foundation (ed). Philippine Inc Publication, Manila, pp. 151–161.

Sarkar P.K., Jones L.J., Gore W., and Craven G.S. 1996. Changes in soya bean lipid profiles during kinema production. *J Sci Food Agric* 71:321–328.

Sarkar P.K., Morrison E, Tingii U., Somerset S.M., and Craven G.S. 1998. B-group vitamin and mineral contents of soybeans during kinema production. *J Sci Food Agri* 78:498–502.

Sarkar, P.K., Tamang, J.P., Cook, P.E., and Owens, J.D. 1994. Kinema—A traditional soybean fermented food: Proximate composition and microflora. *Food Microbiol*. 11(1):47–55.

Sarkar S. 2006a. Potential of soyoghurt as a dietetic food. *Nutr Food Sci* 36(1):43–49.

Sarkar S. 2006b. Shelf-life extension of cultured milk products. *Nutr Food Sci* 36(1):24–31.

Sarkar S. and Mishra A.K. 1997. *Indian Food Industry* 16:110–117.

Sharma U.P. and Zariwala I.T. 1980. Deterioration of *Shrikhand* during storage. *Indian J Dairy Sci* 33(2):223–231.

Singh S. 2014a. Shrikhand. In: *Dairy Technology. Dairy Products and Quality Assurance*. Vol. 2. New India Publishing Agency, New Delhi, India, p. 117.

Singh S. 2014b. Classification & Composition of cheese. In: *Dairy technology. Dairy products and Quality Assurance*. Vol. 2. New India Publishing Agency, New Delhi, India, p. 174.

Soni S.K. and Arora J.K. 2000. Indian fermented foods: Biotechnological approaches. In: *Food Processing: Biotechnological Applications.* Marwaha S.S., Arora J.K. (eds). Asiatech Publishers Inc, New Delhi, India, pp. 143–190.

Soni S.K., Gupta L.K., Marwaha S.S., and Arora J.K. 2000. Cheese production technologies. In: *Food Processing: Biotechnological Applications.* Marhawa S.S., Arora J.K (eds). Asiatech Publishers Inc., New Delhi, pp. 221–240.

Soni S.K., Sandhu D.K., and Vilkhu K.S. 1985. Studies on dosa—An indigenous Indian fermented food: some biochemical changes accompanying fermentation. *Food Microbiol* 2(3):175–181.

Soni S.K., Soni R., and Janveja C. 2013. *Production of Fermented Foods.* In: *Biotechnology in Agriculture and Food Processing Opportunities and Challenges.* Panesar P S. and Marwaha S S. (eds). Taylor & Francis, CRC Press, BocaRaton, FL. pp, 219–278.

Steinkraus K.H., 1994. Nutritional significance of fermented foods. *Food Res Int* 27(3):259–267.

Steinkraus K.H., Van Veen A.G. and Thiebean D.B. 1967. Studies on idli—An indian fermented black grain-rice food. *Food Technol* 21:916–919.

Surono I.S. and Hosono A. 1994a. Chemical and aerobic bacterial composition of "Terasi," a traditional fermented product from Indonesia. *J Food Hyg Soc Japan* 35(3):299–304.

Tamang B. 2006. Role of lactic acid bacteria in fermentation and biopreservation of traditional vegetable products. PhD thesis, Food Microbiology Laboratory, Sikkim Government College, Gangtok (North Bengal University).

Tamang B. and Tamang J.P. 2007. Role of lactic acid bacteria and their functional properties in *Goyang,* a fermented leafy vegetable product of the Sherpas. *J Hill Res* 20(20):53–61.

Tamang J.P. 1993. Sinki: A traditional lactic acid fermented radish taproot product. *J Gen Appl Microbiol* 39:395–408.

Tamang, J.P. 2010. *Himalayan Fermented Foods: Microbiology, Nutrition, and Ethnic Values.* CRC Press, Taylor & Francis, New York.

Tamang, J.P., Dewan, S., Thapa, S., Olasupo. A.N., Schillinger, U., and Holzapfel, W.H. 2000. Identification and enzymatic profiles of predominant lactic acid bacteria isolated from soft-variety *chhurpi,* a traditional cheese typical of the Sikkim Himalayas. *Food Biotechnol.* 14(1–2): 99–112.

Tamang J.P, Okumiya K., and Kosaka Y. 2010. Cultural adaptation of the Himalayan ethnic foods with special reference to Sikkim, Arunachal Pradesh and Ladakh. *Himalayan Study Mon RIHN, Kyoto, Japan* 11:177–185.

Tamang J.P. and Thapa S. 2006. Fermentation dynamics during production of *bhaati jaanr,* a traditional fermented rice beverage of the Eastern Himalayas. *Food Biotechnol* 20(3):251–261.

Tamang J.P., Tamang N., Thapa S., Dewan S., Tamang B., Yonzan H., Rai A.K., Cheltri R., Chakrabarty J., and Kharel N. 2012. Microorganisms and nutritional value of ethnic fermented foods and alcoholic beverages of North East India. *Indian J Trad Know* 11(I): 7–25.

Teniola O.D. and Odunfa S.A. 2001. The effects of processing methods on the level of lysine and methionine and the general acceptability of ogi processed using starter cultures. *Int J Food Microbiol* 63:1–9.

Thakur N., Savitri, and Bhalla T.C. 2004. Characterizaiton of some traditional fermented foods and beverages of Himachal Pradesh. *Indian J Trad Know* 3(3):325–335.

Thapa N. and Pal J. 2007. Proximate composition of traditionally processed fish products of the Eastern Himalayas. *Hill Res* 20(2):75–77.

Thapa S., Tamang J.P. 2004. Product characterization of *kodo ko jaanr*: Fermented finger millet beverage of the Himalayas. *Food Microbiol* 21:617–622.

Upadhyay S.M., Dave J.M., and Sannabhadti S.S. 1984. Microbiological changes in stored shrikhand and their application in predicting the sensory quality of product. *J Food Sci Technol* 21(4):208.

Upadhyay K.G., Vyas S.H., Dave J.M., and Thakar P.N. 1975a. Studies on chemical composition of market samples of *shrikhand. J Food Sci Technol* 12:190–194.

Upadhyay K.G., Vyas S.H., Dave J.M., and Thakar P.N. 1975b. Studies on microbiological quality of market *shrikhand*. *Indian J Dairy Sci* 28:147–149.

Upadhyay S.M. 1981. Assessing the suitability of different microbiological and chemical tests as keeping quality tests for *Shrikhand*. MSc thesis, Gujarat Agricultural University, S.K. Nagar.

Wattiaux M.A. 2012. Milk composition and nutritional value, Babcock Institute for International Dairy Research and Development Dairy Essentials University of Wisconsin-Madison.

Yadav S. and Khetarpaul N. 1994. Indigenous legume fermentation: Effect on some antinutrients and in-vitro digestibility of starch and protein. *Food Chem* 50(4):403–406.

Yonzan H. and Tamang J.P. 2010. Microbiology and nutritional value of *selroti*, an ethnic fermented cereal food of the Himalayas. *Food Biotechnol* 24(3):227–247.

Zaman, M.Z., Abdulamir, A.S., Bakar, F.A., Selamat, J., and Bakar, J. 2009. Microbiological, physicochemical and health impact of high level of biogenic amines in fish sauce. *Am J App Sci* 66: 1199–1211.

5

QUALITY AND SAFETY OF INDIGENOUS FERMENTED FOODS

A.K. SENAPATI, AHMAD ROSMA,
ABU HASSAN SITI NADIAH, DEV RAJ,
J.P. PRAJAPATI, MANISHA KAUSHAL,
OLANREWAJU OLASEINDE OLOTU,
OLUWATOSIN ADEMOLA IJABADENIYI,
POOJA LAKHANPAL, S.V. PINTO,
SOMBOON TANASUPAWAT, V.K. JOSHI,
AND W.A. WAN NADIAH

Contents

5.1 Introduction

Indigenous fermented foods are typically unique and their quality characteristics vary according to the regions they are produced. These are important components of the diets in South Asian countries (Lazos et al., 1993) and provide diversity of flavors, aromas, and textures which help to enrich the diet (Ijabadeniyi, 2007; Joshi, 2006a; Joshi and Singh, 2012; Steinkraus, 1997), as discussed earlier also. These foods are produced by local people who have acquired these skills and continue to produce these products at a small scale and household levels and have remained un-commercialized for several reasons (Arogba et al., 1995), including very short shelf-life, objectionable packaging materials, stickiness, and, sometimes, characteristic putrid or offensive odors. Besides, many small-scale processors of locally fermented foods are also reluctant to accept change and modification of the fermentation process (Valyasevi and Rolle, 2002).

Quality and safety of the foods that are consumed in large quantities is of great concern from the safety point of view. In South Asia the quality of these products is affected by climate, social patterns, raw materials, processing technology and availability of reference quality standard for each product, consequently many of the fermented foods made in the South Asia do not have any product specification or quality standard. The manufacturers of these food products rely on the environmental microorganism and climatic conditions (see Chapter 3). Furthermore, the majority of the people involved in the production of these foods are poor, illiterate and may not even be aware of sanitation and hygiene. In addition, there is lack of water of good quality and also a prevalence of failing sewage disposal systems, if present at all, in most of the village settlements, poor communities, and informal settlements. All these factors contribute to the uncertain, if not actually poor quality. In this chapter, a focus is made on the quality and safety of indigenous fermented foods and related issues.

5.2 Quality of Indigenous Fermented Foods

The quality of any food or food product has been defined as a combination of attributes or characteristics of the product that have significance in determining the degree of acceptability of the product to a user (Gould, 1977). In general, factors that affect quality are cultivar, stage of crop maturity used for processing, cultural practices employed, processing procedure, shelf-life, etc. Out of these factors, processing methods and conditions employed affect the quality of the resulting product. Though the processor/manufacturer cannot change greatly the status of raw material, he can certainly enhance the product quality by using appropriate processing techniques and hygienic conditions. Proper control of all in-plant variables is thus, necessary for quality retention (Anon. http://www.crcpress.com/). Not only this, the storage temperature of the food also influences retention of quality of the finished products. A large variety of food products all over the world are prepared by fermentation of various raw materials (Joshi, 2006b; Joshi and Pandey 1999). Fermentation impacts the food properties by influencing the fermentation reactions in the chemical, functional, and sensory quality of food components and biological activity (Joshi, 2006b; Zambonelli et al., 2002). To be precise, quality is further identified based on physical, chemical, microbiological, and sensory characteristics of the food or food products (Figure 5.1).

Physical methods involve the appearance and color, texture, shape, etc. Chemical methods of measurement include moisture, enzyme, fiber, total soluble solids (TSS), sugars, alcohol, esters, and pH or acidity (Ranganna, 1986). The test for nutrition includes determination of energy, protein, carbohydrates, fats, minerals, and vitamins. The microbiological methods encompass direct microscopic examination of food, estimation of microbial population, determination of their characteristics, especially their pathogenicity or ability to cause spoilage, and this require considerable training

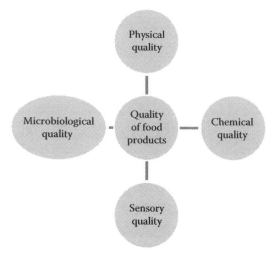

Figure 5.1 Components of quality of food products.

to properly interpret the results. The sensory qualities of the food products are of great significance from consumer acceptability considerations (Joshi 2006a; Joshi and Bhushan, 2000) and include measurement of color, body, taste, flavor, and texture. All these aspects are discussed here.

5.2.1 Physical Characteristics

Physical attributes are the quickest to be assessed and require least training for evaluation. These are concerned with attributes of size, texture, color, consistency, or process variables like head-space (in case of bottled or canned products), drained weights, vacuum (canned products), etc. In case of fermented products, body, consistency, color, and appearance are taken into account to determine the quality.

5.2.2 Chemical Characteristics

A number of chemical parameters are determined using the standard procedures (Amerine et al., 1980; Ranganna, 1986). The compositional characteristics, namely, moisture, carbohydrates, proteins, fats, vitamins, and minerals are generally determined. In the case of alcoholic beverages, alcohol content, volatile acidity, pH, acidity, methanol, total esters, and sugar are the usual parameters needed to define quality (Joshi and Sharma, 2009; Joshi et al., 1990, 1999, 2004, 2011). As an example, the quality characteristics of some typical products are given in Tables 5.1 and 5.2.

The chemical analysis with respect to safety is also undertaken to ascertain the levels of toxins like aflatoxin, antinutritional factors, heavy metals, etc. In some instances, such as alcoholic beverages, it is necessary to subject the products to long aging periods to achieve the quality and typical characteristics essential for the quality of these products (Joshi, 1997). Many post-fermentative composition variations can be attributed to the enzymes already present in the product before the beginning of fermentation, such

Table 5.1   Physico-Chemical Composition of Apricot Wine

CHARACTERISTICS	MEAN ± SD* APRICOT (NEW CASTLE)
TSS (°B)	8.20 ± 0.07
Titratable acidity (%MA)	0.76 ± 0.02
pH	3.15 ± 0.02
Ethanol (% v/v)	10.64 ± 0.09
Reducing sugars (%)	0.34 ± 0.01
Total sugars (%)	1.11 ± 0.02
Volatile acidity (%AA)	0.025 ± 0.002
Total phenols (mg/L)	253.60 ± 0.8
Total esters	120.6 ± 0.6
Color (units)	
Red	0.70 ± 0.05
Yellow	4.30 ± 0.08
Blue	0.60 ± 0.05

Source: Adapted from Joshi, V.K., Sandhu, D.K., and Thakur, N.S. 1999. *Biotechnology: Food Fermentation.* Educational Publishers and Distributors, New Delhi, pp. 647–744.
SD: standard deviation, MA = Malic acid, v/v = volume/volume, AA = Acetic acid.

Table 5.2   Specifications for Cider Vinegar

CHARACTERISTICS	LIMIT	TEST METHOD
Acid content	4.0 g/100 mL minimum expressed as acetic acid	AOAC 30.071
Color	Light to medium amber as per reference sample	AOAC 30.062
Trace metals		
Copper	5.0 ppm max	AOAC 30.035
Iron	10.0 ppm max	AOAC 30.079
Heavy metal	1.0 ppm max	AOAC 30.058
Alcohol content	0.5% by volume, max	AOAC 30.078

Source: Adapted from Springer Science+Business Media: *Vinegars of the World*, Cider vinegar: microbiology, technology and quality, 2009, pp. 197–207, Joshi, V.K. and Sharma, S.

as enzymes brought by the rennet for cheese, or grape enzymes in wine preparation. A predominant role is played by the enzymes released by bacteria, especially lactic acid bacteria (LAB), when their growth is completed. Milk and milk products are subjected to several tests, including fats, solid not fat (SNF), lactose, minerals, vitamins, acids like lactic acid, pH, and total and reducing sugars. In addition, testing for safety or toxicity is also carried out, that is, for carbamide, methanol, etc. in alcoholic beverages.

5.2.3 Microbiological Characteristics

Microbiological quality of the food includes the study (Anon. http://www. bestmedicalhealth.com/c/yersiniosis) of the microorganisms that inhabit, create, or contaminate the food during production and processing. It also includes the study of microorganisms causing food spoilage and focuses on the general biology of the microorganisms that are found in foods, including: their growth characteristics, identification, and pathogenesis (Frazier and Westhoff, 1978). Microbiological tests are conducted to determine the quality of the food under consideration. Tests for the presence or absence of pathogenic microbes, including *Clostridium*, *Staphylococcus*, *Salmonella*, etc., are carried out.

The quality of water used is determined by coliform count of the food samples, which also determines the quality of raw materials and the efficacy of thermal processing or other techniques employed for food processing. The fermentation should have a proper sequence of microbial accession so that fermentation is completed properly and the product is safe (Fleet, 1990). Earlier studies have shown that the yeasts are responsible for the alcoholic fermentation of grape juice into wine, although certain species of bacteria can also grow and cause its spoilage (Anon. http://thedailyomnivore.net/category/food/). Pathogenic bacteria, viruses, and toxins produced by microorganisms are all possible contaminants of food. However, microorganisms and their products such as bacteriocim (Joshi et al., 2006; Kabore et al., 2012) can also be used to combat these pathogenic microbes. Microbiological analysis also reflects the storability of the product and the impact of storage conditions on the quality of products (Yildirim, 2009). Consequently, microbial contamination leads to changes in the chemical composition and ultimately, the sensory quality. Table 5.3 gives an idea of the diversity of microorganisms significant in alcoholic beverage production.

5.2.4 Sensory Quality

The sensory quality is ultimately the parameter for acceptance of any product by the consumer. Without sensory quality a product may contain any amount of nutritive component, but is of no significance as it will not be consumed. Sensory characteristics include color, taste, appearance, texture/consistency, aroma, and overall acceptability (Joshi, 2006a). Sensorial qualities of fermented food can be divided into three types:

Table 5.3 Different Microorganisms Associated with Wine Making and Their Significance

MICROBIAL GROUP	SIGNIFICANCE
Yeasts	Alcoholic fermentation, spoilage, autolysis, deacidification
Killer yeasts	Alcoholic fermentation, spoilage control
Lactic acid bacteria	Malolactic fermentation, spoilage, autolysis
Acetic acid bacteria	Struck fermentations, spoilage
Fungi	Spoilage
Bacillus clostridium	Spoilage

Source: Adapted from Joshi, V.K. 1997. *Fruit Wines.* 2nd edn. Directorate of Extension Education, Dr. Y.S. Parmar University of Horticulture and Forestry, Nauni, Solan, HP, 255 p.; Joshi, V.K., Sandhu, D.K., and Thakur, N.S. 1999. *Biotechnology: Food Fermentation.* Vol. II. Educational Publishers and Distributors, Kerala, India, pp. 647–744.

pre-fermentative, fermentative, and post-fermentative. Pre-fermentative qualities are those typical of and proper to the raw material, and can be quite homogenous or vary with the source. Fermentative qualities are those contributed by the fermenting agent, but can also be due to primary or secondary fermentation products or the transformation of compounds in the raw material used.

Post-fermentative qualities are those characteristics that appear during aging or ripening, when microbial growth has ceased. In many cases, post-fermentative qualities do not have enough time to appear because the product is consumed shortly after the end of fermentation (e.g., any type of cheese and wine). Generally, in case of alcoholic beverages and cheeses, the post fermentative changes improve the quality characteristics considerably (Tesfaye et al., 2002). Nevertheless, sensory analysis requires a well trained testing panel, and concrete and adequate attributes which are easily perceived and able to be differentiated (Amerine et al., 1980; Joshi, 2006a; Joshi and Bushan, 2000).

5.3 Quality Characteristics of Some Indigenous Fermented Food

5.3.1 Indigenous Alcoholic Beverages

5.3.1.1 Naturally Fermented Alcoholic Beverages-Ghanti Preparation of customary alcoholic beverages (naturally fermented beverages) using natural sources of microflora as the fermenting agent, and locally available raw materials (ragi, apricot, apple, etc.) in tribal areas of Himachal Pradesh (India) is commonly practised. These beverages are known mostly as ghanty. However, due to lack of proper scientific technology, many times, the quality of such beverages is uncertain and their consumption leads to a high incidence of fatal illnesses among the consumers (Sharma, 1986). A lack of quality control results in many impurities and adulterants in these beverages, such as heavy metals like lead and arsenic, organic solvents, and sometimes sedative drugs like benzodiazepines and barbiturates (Saxena, 2000). The presence of some compounds, besides affecting the quality, can make the natural fermented beverages toxic. Among the toxic compounds in alcoholic beverages is methanol which occurs ranging from 32 to 238 mg/L in grape wine, 410–700 mg/L in French Cognac, 180–500 mg/L in Spanish distillates, 390–920 mg/L in German distillates, 1200 g/hL in plum, and 700 g/hL in pear (Lisle et al., 1978; Postel and Adam, 1979, 1980). A study conducted (Joshi and Sandhu, 2000) on samples (22) of naturally fermented alcoholic (NFA) beverages produced in the tribal district of Kinnaur of Himachal Pradesh (India) showed appreciable differences among various types of the beverages for methanol content (Table 5.4).

In comparison to the experimental wines, NFA beverages had 14 times more methanol, while the distilled liquors prepared from the NFA beverages in the tribal areas contained about 40 times more methanol than their experimental counterparts. However, the cereal based NFA beverages, irrespective of being distilled or not had considerably lower methyl alcohol content. The higher methanol content recorded in

Table 5.4   Methanol Content (μl/L) in Different Types of Alcoholic Beverages

TYPE OF ALCOHOLIC BEVERAGES	MEAN ± SD
Naturally fermented wines	772 ± 156.4
Naturally fermented distillates	4414 ± 2463
Naturally fermented cereal products	206 ± 223.6
Naturally fermented cereal distillates	448 ± 559.7

Source: Adapted from Joshi, V.K. and Sandhu, D.K. 2000. *Acta Aliment* 29(4): 323–334.

NFA beverages has been attributed to the activity of mixed natural microflora including fungi and bacteria involved in such fermentations, pectolytic enzyme activity of the source of fermentation, or the raw material and the method of distillation (Ali and Dirar, 1984; Amerine et al., 1980; Frazier and Westhoff, 1978; Lee et al., 1975). The higher methanol content in natural sources of fermented beverages or home brewed beverages is of great concern due to its toxicity, causing suspected bowel ulceration, blindness, or death, although the lethal properties of some of these country made wines are legendary (Fowles, 1989). The human body can metabolize 340 mg/kg of body weight of methanol compared to 1400 mg/kg body of ethanol (Christensen, 1973). The most serious effects are seen when high levels of methyl alcohol are found in such beverages and its metabolism by the body results in the production of formaldehyde, which is poisonous (WHO, 1993). The alcoholic beverages, thus should have methanol content as low as possible for safety considerations.

A large variation among the samples of naturally fermented alcoholic beverages tested for ethanol, volatile acidity and pH were noted. The very low ethanol content in non-distilled NFA beverages is expected, but not in the distilled NFA beverages, where it indicates faulty method of fermentation and distillation. About 80% of the total samples of NFA had volatile acidity of more than 0.04%, indicating the occurrence of acetification after the fermentation (Joshi and Sandhu, 2000). The large variations in ethanol content of naturally fermented beverage might be due to a lack of standardized technology (Thakur et al., 2004), low content of fermentable sugars in the raw material, and the inefficacy of the fermentative microorganisms to carry out fermentation (Joshi and Sandhu, 2000). In another study, the fermented beverage samples were analysed for ethanol content, which also exhibited a large variation (Thakur et al., 2004). The highest ethanol content was recorded in *Sura* (a millet-based naturally fermented beverage), followed by *Chhang* and *Lugri*. *Behmi*, *Chulli* and *Angoori* had a low content of alcohol, while jaggery and red grape based beverage contained 5% ethanol (Thakur et al., 2004).

Analysis of the material commonly used by local tribals as a source of fermentation/inoculum (SN1 and SN2) in the preparation of such beverages showed the presence of yeast, bacteria, and fungi in both the samples of natural source of fermentation (NSF). Table 5.5 shows the microbial profile of naturally fermented alcoholic beverages by the tribal people.

Table 5.5 Microbiological Profile of Natural Sources of Fermentation Used by the Tribals

	LOG COLONY FORMING UNIT (CFU/g)			
SAMPLE TYPE	BACTERIA	YEAST	FUNGI	TOTAL
Natural source 1 (SN1)	2.47	6.30	3.02	11.52
Natural source 2 (SN2)	2.69	3.69	3.60	9.04

Source: Adapted from Joshi, V.K. and Sandhu, D.K. 2000. *Acta Aliment* 29(4): 323–334.

Out of various samples of NFA beverages collected, four (18.2%) were found to be positive for microbial contamination, showing standard plate counts (SPC) ranging from 10 to 3×10^4 CFU/mL showing the presence of bacteria in all the samples except one, which had yeast almost in a pure culture. The bacteria belonged to the coliform group, while the yeast was identified as belonging to *Brettanomyces* spp. (Joshi and Sandhu, 2000). The presence of coliform bacteria in these beverages also points out to the unhygienic conditions used in the preparation of such beverages in tribal areas (Joshi and Sandhu, 2000).

The source of fermentation normally employed in preparation of such beverages was also tested. Two samples SN1 and SN2 were analysed. SN1 had the higher population of microorganism, with a higher count of yeast than SN2. Among the fungi, *Aspergillus flavus*, *Aspergillus oryzae*, *Mucor* spp., and *Rhizopus* spp., in order of decreasing occurrence, were recorded. Among the yeasts, *Sacccharomyces* spp. were predominant followed by (*Torula*) and *Rhodotorula*. Among the bacteria, *Pediococcus*, *Leuconostoc*, and *Pseudomonas fluorescens* were isolated (Joshi and Sandhu, 2000). The occurrence of fungi, yeast, and bacteria in the NSF is due to the absence of any preservative; microorganisms present on the surface of the raw materials used very frequently initiate such fermentations.

Microbiological studies have revealed *Saccharamyces cerevisiae* to be a dominant microorganism in such fermentations, along with species of *Candida*, *Zygosacchromyces bisporus*, *Kluveromyces thermotolerance*, *Leuconostoc*, and *Lactobacillus* (Thakur et al., 2004). The bacterial species of *Lactobacillus* and *Leuconostoc* are prominently involved in fermentation, causing acidification and leavening (Thakur et al., 2004).

Traditional fermented foods and beverages are also very popular in the tribal and rural areas of Western Himalayas (www.niscair.res.in). Traditional starter cultures like "*Phab*" (dehydrated yeast formulations), "*Treh*" (previously fermented wheat flour slurry), and "*Malera*" (previously fermented wheat flour dough) are inocula used in preparing these fermented products. "*Phab*" is known to contain yeast, bacteria, and actinomycetes, the proportion of which, however, varies from one batch to another, as no stringent microbiological precautions are observed during their preparation and formulation (Thakur et al., 2004). It not only contains fermentative microorganisms but also saccharolytic and acidogenic microorganisms which are responsible for the release of fermentable sugars from starch-based raw materials like rice in *Chhang* preparation.

5.3.1.2 Sur The inoculum used in *sur* production is called as *dhaeli* (Figure 5.2) and it is also known as *roat* in the Sirmaur district in Himachal Pradesh (India). The additive used in its preparation is mostly jaggery (gur). The *dhaeli* procured from the different regions were analysed for the total plate count and the results are depicted in Table 5.6.

The physico-chemical characteristics of *sur* samples collected from different locations (Table 5.7) showed that the maximum TSS 9.8 ± 3.9°B was recorded in the samples collected from the Kullu district of Himachal Pradesh, while the minimum was recorded in the samples from the Sirmour district. The percent ethanol content (% v/v) was found to be maximum—10.53 ± 3.29—in *sur* samples collected from Mandi district while the minimum ethanol content—5.42 ± 0.188—was found in samples collected from Sirmour district (Joshi et al., 2015).

Figure 5.2   *Dhaeli* procured from different regions: (a) Kullu, (b) Mandi, (c, d) Kangra. (Courtesy of A. Sharma. Preparation and evaluation of *sur* in Himachal Pradesh. Msc thesis, UHF, NAUNI, Solan, India. With permission.)

Table 5.6   Total Plate Count of the Different *Dhaeli* Samples Using Different Growth Media

PLACE	TOTAL NA COUNT (CFU/mL)	TOTAL YEMA COUNT (CFU/mL)	TOTAL PDA COUNT (CFU/mL)	TOTAL EMB COUNT (CFU/mL)
Luharati	3.47	5.55	5.70	3.34
Multhan	3.81	6.21	6.38	–
Kullu	6.15	–	Fungi	5.89
Baijnath	6.09	5.75	5.53	–

Source: Adapted from Joshi, V.K., Kumar, A., and Thakur, N.S. 2015. *Indian J Trad Know* (in press).

Table 5.7 Physico-Chemical Characteristics of the *Sur* Samples Collected from Different Regions

DISTRICT	TSS (°B)	REDUCING SUGARS (%)	TOTAL SUGARS (%)	ETHANOL (% V/V)	COLOR (OD AT 440 NM)	pH	TITRATABLE ACIDITY (% AS LA)	VOLATILE ACIDITY (% AS AA)	TOTAL PHENOLS (MG/L)	TOTAL ESTERS (MG/L)	HIGHER ALCOHOLS (MG/L)	ACETALDEHYDES (MG/L)	METHANOL (µL/L)
Kangra	8.74 ± 1.8	0.25 ± 0.10	0.44 ± 0.129	9.58 ± 3.12	3.66 ± 3.1	3.65 ± 0.22	2.02 ± 0.62	0.022 ± 0.01	330.48 ± 56.67	144.82 ± 20.42	126.86 ± 40.76	13.78 ± 2.71	428.61 ± 138.58
Kullu	9.8 ± 3.9	0.47 ± 0.27	0.49 ± 0.410	8.97 ± 3.28	3.12 ± 2.1	3.90 ± 0.29	2.47 ± 1.54	0.056 ± 0.06	284.85 ± 26.67	184.386 ± 86.96	141.18 ± 66.72	14.59 ± 3.09	391.85 ± 107.56
Mandi	9.1 ± 4.00	0.36 ± 0.33	0.56 ± 0.685	10.53 ± 3.29	2.57 ± 1.97	3.83 ± 0.27	1.54 ± 0.86	0.027 ± 0.02	272.57 ± 59.88	148.84 ± 76.34	137.62 ± 71.76	14.19 ± 5.57	608.95 ± 112.15
Sirmaur	6.76 ± 0.06	0.11 ± 0.01	0.10 ± 0.008	5.42 ± 0.188	0.883 ± 0.02	3.46 ± 0.06	2.15 ± 0.07	0.027 ± 0.0009	216.451 ± 1.93	177.75 ± 0.65	60.905 ± 0.58	28.16 ± 0.88	588.7 ± 13.01

Source: Adapted from Sharma, A. 2013. *Preparation and evaluation of Sur Production in Himachal Pradesh.* MSc thesis, Dr. Y.S. Parmar University of Horticulture and Forestry, Nauni, Solan, HP; Joshi, V.K., Kumar, A., and Thakur, N.S. 2015. *Indian J Trad Know* (in press).

SD = standard deviation.

The quality of such traditionally prepared products can be enhanced by disseminating knowledge of isolation, selection, preservation, or collection of strains and formulating starters for use as inocula. Over the years, there has been a systematic and deliberate effort by researchers to introduce production technology through hygienic utilization of pure microbial starter cultures and optimum process control. With respect to alcoholic beverages, the use of pure cultures and vinification practices have been optimized to prepare alcoholic beverages with improved quality (Joshi, 1997; Joshi et al., 1990, 1999, 2006, 2011). Using the optimized conditions, low alcoholic beverages (wine) have been produced with improved chemical and microbiological quality characteristics from wild apricot (chulli) (Joshi et al., 1990), which is being produced traditionally in the Kinnaur district of Himachal Pradesh (India).

5.3.1.3 Toddy Toddy is a popular indigenous fermented beverage made and consumed in South India, Sri Lanka, and other tropical countries. It is prepared by the fermentation of palm juice and represents an excellent nutritive beverage. Fresh coconut sap contains 12%–15% of sucrose (by weight) with trace amounts of reducing sugar including glucose, fructose, maltose, and raffinose. The sap contains approximately 0.23% protein and 0.02% fat. Half of the total sugars are fermented during the first 24 h, and the ethanol content of the fermented palm sap reaches a maximum value of 5.0%–5.28% (v/v) after 48 h (Law et al., 2011; Sekar and Mariappan, 2005). During fermentation, the thiamin content increases from 25 to 150 µg/L, riboflavin content ranges between 0.35 and 0.50 µg/L and pyridoxine between 4 and 18 µg/L. It is generally stored at ambient temperature, which affects its physicochemical characteristics, like pH, alcohol, and acidity, along with microbial count (Bazirake, 2008). Freshly prepared toddy has an average alcohol content of 7%–7.9% (v/v) (Bazirake, 2008; Joshi et al., 1999; Law et al., 2011). The ethanol content of a naturally fermented coconut palm sap reaches maximum concentration (approximately 9% v/v) after five days fermentation (Atputharajah et al., 1986; Law et al., 2011). For commercial toddy or palm wine, alcohol content ranges from 3% to 7% (v/v) for fermented palm sap and 20%–40% (v/v), when distilled (Atputharajah et al., 1986; Bennett et al., 1998). Natural fermentation of coconut palm by wild yeast produces ethanol contents much below the theoretical yield (Bazirake, 2008; Liyanage et al., 1981). The alcohol content of toddy can however be increased by inhibiting the growth of non-fermenting microorganisms.

The presence of *Saccharomyces marxianus*, *S. exiguous*, and *Candida* has been recorded from samples of toddy. A total of 166 isolates of yeasts and 39 isolates of bacteria have been identified. Seventeen species of yeasts belonging to eight genera have also been recorded. The largest number of isolates (72%) belonged to genera *Candida*, *Pichia* and *Saccharomyces* while *Saccharomyces chevalieri* was the most dominant yeast species and accounted for 35% of the total isolates. Seven genera of bacteria have been isolated and the predominant genera was *Bacillus*. Others included *Enterobacter*, *Leuconostoc*, *Micrococcus*, and *Lactobacillus* (Atputharajah et al., 1986).

The palm sap fermentation involved alcoholic–lactic–acetic acid fermentation, in the presence of mainly yeasts and LAB (Borse et al., 2007; Law et al., 2011) but *Saccharomyces* spp. present in the natural fermented palm sap were important for the formation of the characteristic aroma of the palm wine (Aidoo et al., 2005). *S. cerevisiae* and *Schizo pombe* were found to be the dominant yeast species (Odunfa and Oyewole, 1998), while other yeast types (Aidoo et al., 2005) include *Saccharomyces chevalieri*, *Saccharomyces exiguus*, *Candida* spp., *Pichia* spp., and *Saccharomycodes ludwigii* present in the coconut palm wine (*toddy*) (Atacador-Ramos, 1996). *Saccharomyces cerevisiae*, *S. pombe*, *Kodamaea ohmeri*, and *Hanseniaspora occidentalis* were reported as the major ethanol producers in toddy (Aidoo et al., 2005; Joshi et al., 1999).

LAB and other bacteria, such as *Lactobacillus plantarum*, *L. mesenteroides*, *Acetobacter* spp., and *Zymomonas mobilis*, are also present, reportedly originating from the palm tree, the gaurd employed for sap collection and fermentation, or the tapping equipment (Jespersen et al., 1994; Law et al., 2011).

Volatiles from (i) fresh, (ii) clarified, and (iii) fermented coconut sap *neera* were isolated by a simultaneous distillation and solvent extraction method using a Likens–Nikerson apparatus and subjected to GC–MS analysis for identification of chemical constituents (Uzochukwu et al., 1998). Twenty-one compounds, constituting more than 98% of the volatiles from fresh *neera* were characterized. Typical major flavor compounds found in volatiles of fresh *neera* were ethyl lactate, phenyl ethyl alcohol, ethyl lactate, 3-hydroxy-2-pentanone, farnesol, 2-methyl tetrahydrofuran, and tetradecanone. Clarified *neera* contained lower quantities of volatiles, in which 13 compounds, constituting more than 97%, were identified. However, the typical flavor components retained were ethyl lactate, phenyl ethyl alcohol, 1-hexanol, 2-methyl tetrahydrofuran, 3-hydroxy-2-pentanone, and 2-hydroxy-3-pentanone. Fermented *neera* contained a greater quantity of volatiles, in which 12 compounds, representing more than 95% of the volatiles, were characterized. Ethyl lactate, phenyl ethyl alcohol, and farnesol were among the seven compounds retained from fresh *neera*. The astringency and harsh note of the fermented *neera* could be due to the increased amounts of acids (19.0 mg/L), such as palmitoleic acid and dodecanoic acid, along with higher concentrations of ethyl alcohol and ethyl esters (Borse et al., 2007).

The quality of the fermented toddy can be improved by using the pure cultures inoculation method (Sanchez, 1979). During the fermentation process, yeast sediments collect at the bottom of the container after a few hours which produce a typical yeasty odor. The composition and quality of palm sap are greatly affected by the location, weather, time, and duration of tapping (Borse et al., 2007; Law et al., 2011).

5.3.2 Lactic Acid Fermented Food

5.3.2.1 Fermented Milk Products The fermented product "*dahi*" was a chance discovery and later on it was discovered that it is fermented by LAB namely, *Leuconostoc*, *Streptococcus*, *Lactobacillus* (Rao et al., 2005; Sarkar, 2008).

5.3.2.2 Curd or Dahi   *Dahi*, a most popular fermented dairy product prepared traditionally throughout India, is obtained by lactic acid fermentation of cow or buffalo or mixed milk (Rao et al., 2005) through the application of a single or mixed strain of LAB accompanied by alcoholic fermentation by yeast (Raju and Pal, 2011). Other lactic acid fermented milk products include *mohi, gheu, dudh, churpi, churipi, chhu, somar, philu,* and *kalari* (Dewan and Tamang, 2007). *Shrikhand* and a large variety of cheeses are also lactic acid fermented products. The single most important quality characteristic of such products is lactic acid production and low pH. The most important characteristic of raw material on which the final quality depends are the fats, SNF, and microbial quality. The lactic acid fermented products have probiotic activity and, thus, are considered very useful. Conventional toxicology and safety evaluation alone is of limited value in the safety evaluation of probiotic organisms. In case of new products, it is necessary to prepare the products that are at least as safe as conventional counterparts, regardless of potential health benefits. The safety record of the test probiotic culture is excellent, and the adverse effects of eating this culture are almost negligible, indicating no health risk posed by consuming these microorganisms.

5.3.2.3 Shrikhand   The overall acceptability of *shrikhand* depends on its quality, which in turn is decided by its chemical composition, microbiological attributes, consistency, and sensory profile. A good quality *shrikhand* should be free from any sign of free fat or syrup separation or uneven color distribution, and should have a clean, pleasant, sweetish sour flavor and be free from any off flavors. Its consistency, it should typically be a uniform semi-solid with a smooth texture without any sign of graininess. The optimum acidity should be around 1.0%–1.1% expressed as lactic acid (Desai and Salunkhe, 1986).

It was found that 5%–6% fat in finished *shrikhand* gives the desired smoothness to the product (Aneja, 2002). Laboratory made *shrikhand* by Tamime and Robinson (1999) contained on an average 5.16% fat; 6.59% protein; 1.68% reducing sugar; 39.37% non-reducing sugar and 42.24% moisture. *Shrikhand* containing about 6% fat; 35%–40% moisture and 40% sugar was highly preferable with respect to sensory profile and consistency of the product (Desai et al., 1985; Miyani, 1962; Patel and Chakraborty, 1985; Reddy et al., 1984) reported that *shrikhand* of superior quality can be prepared from buffalo milk (6% fat and 9% SNF) treated with sodium citrate (0.25%), while Desai et al. (1985) found that homogenization of milk improved the organoleptic quality attributes of *shrikhand*. The most desirable *shrikhand* has 6% fat, 41% sugar, and 40% moisture (Patel and Abd-Ed-salam, 1986; Patel and Chakraborty, 1988). An acceptable quality *shrikhand* was made using *Streptococcus cremoris* as the starter culture with the addition of 40% sugar and 20% cream (40% fat). Aneja (2002) studied the effects of fat content ranging from 1% to 12% on the quality of *shrikhand* and found that 5%–6% fat in a finished product gives the desired smoothness to the product. De and Patel (1991) analysed market samples of *shrikhand* and documented average composition of the market *shrikhand*, as given in Table 5.8.

Table 5.8 Average Composition of *Shrikhand*

PARAMETERS	CONCENTRATION
Moisture (%)	34.48–35.66
Protein (%)	5.33–6.13
Reducing sugars (%)	1.56–2.18
Sucrose (%)	55.55–58.67
Fat (%)	1.93–5.6

Source: Adapted from De, A. and Patel, R.S. 1991. *Cult Dairy Prod J* 25(2): 21.

The market samples of *shrikhand* examined for microbiological quality showed wide variation in count of various groups of microorganisms. *Shrikhand* contained 2.4×10^4 to 1.6×10^6 total viable counts/g, 0–930 coliforms/g, $0–1 \times 10^4$/g psychrotrophs, and $0–1.8 \times 10^5$/g yeast and molds. All samples were however, free of enteric pathogens (Borate et al., 2011).

5.3.3 Fermented Fruit and Vegetable Products

5.3.3.1 Gundruk and Sinki
Gundruk is one of the most common and highly preferred fermented dry vegetable product indigenous to Nepal. It is primarily valued for its uniquely appetizing flavor and is mostly used in the preparation of curry, soup, chutney, and other local delicacies. All Nepalese, irrespective of wealth, status, relish *gundruk*. Thus, it has an important bearing on the Nepalese diet. During fermentation of *gundruk* and *sinki*, common indigenous non-salted fermented vegetable products, sp. *Lactobacillus* and *Pediococcus* produce lactic acid and acetic acid (Tamang and Sarkar, 1993). Unlike sauerkraut and pickles, *gundruk* is used as a condiment to enhance the overall flavor of the meal. The microflora of Nepalese pickles *gundruk* was investigated (Karki et al., 1983a). Depending upon the substrate, the acidity (% as lactic acid) of *gundruk* is also variable. The acidity of mustard and cauliflower *gundruk* has been reported to be 0.48% and 4.5% (as lactic acid on dry basis), respectively. The degree and direction of all the twenty amino acids varied with the type of vegetable used (Karki et al., 1983b) and solar cabinet dried food items were superior to those which were sun dried when evaluated in terms of taste, color, and mold counts, and retain much vitamin A (Whitfield, 2000).

The moisture content of *gundruk* can be brought down to 6.6% in a solar cabinet drier in 6 h and to 9.9% moisture in 11 h in the sun, respectively. The sensory quality of solar dried *gundruk* was superior ($p < 0.05$) to the sun dried in terms of color, aroma, taste and overall quality. This methodology can make *gundruk* of almost consistant quality, attaining higher acidity in a week's time, and solar drying can maintain hygienic condition and retain better sensory quality.

The effect of cabbage starter on the sensory quality attributes of *gundruk* (Karki et al., 1986), blanched and unblanched cabbage, was examined and the results are shown in Tables 5.9 and 5.10. The flavor of *gundruk* (Karki et al., 1983a) was enhanced

Table 5.9   Effect of Cabbage Starter on Sensory Quality Attributes of *Gundruk*s on 7th and 8th Days

QUALITY ATTRIBUTES	MEAN SENSORY SCORES							
	TREATMENTS							
	GT1	GT2	GT3	GT4	GT5	GT6	GT7	GT8
Color	7.3 (0.48)	7 (0.47)	7.8(0.63)a	7.5 (0.53)ab	7.2 (0.42)bc	7.4 (0.52)ab	7.4 (0.52)abc	7.2 (0.42)bc
Aroma	6.8 (0.42)abc	7.1 (0.57)abc	7.3 (0.48)a	7.2 (0.42)abc	6.9 (0.57)abc	6.7 (0.48)c	7.3 (0.48)ab	7 (0.47)abc
Taste	7.1 (0.57)cde	6.8 (0.42)e	8.4 (0.52)a	8.2 (0.42)a	7 (0.67)cde	7.4 (0.52)bcd	7.7 (0.48)b	7.5 (0.53)bc
Overall Acceptability	6.7 (0.67)d	6.9 (0.57)cd	8.5 (0.53)	7.8 (0.42)a	6.2 (0.63)	6.9 (0.32)cd	7.4 (0.52)ab	7.3 (0.48)bc

Source: From Karki, T. et al., 1983a. *Nippon Shokuin Kogyo Gakkaishi* 30: 357–367. With permission.
Mean separation within rows by DMRT, 5%. Values in parentheses are SD (standard deviation).
Mean in rows bearing similar notations are not significantly different.
The various *Gundruk* treatments coded as gt1-radish, gt2-rayo, gt3-radish + 10%cabbage, gt4-rayo + 10%cabbage, gt5-radsh, gt6-rayo, gt7-radish + 10%cabbage, gt8-rayo + 10%cabbage.

Table 5.10   Effect of Cabbage Starter on Sensory Quality Attributes of Blanched and Unblanched *Gundruk*s on 7th and 8th Days

QUALITY ATTRIBUTES	MEAN SENSORY SCORES					
	TREATMENTS					
	GT1	GBT2	GUT3	GUT4	GBT5	GBT6
Color	6.2 (1.15)b	6.35 (1.27)b	7.15 (1.18)ab	6.5 (1.24)ab	5.95 (1.50)b	5.7 (1.22)b
Aroma	6.9 (1.25)ab	5.6 (1.35)cde	7.01 (1.12)a	6.15 (1.46)bc	4.5 (1.43)	5.75 (1.62)cd
Taste	6.24 (1.3)abc	4.71 (2.02)de	6.94 (1.39)a	6.82 (1.07)ab	4.29 (1.79)	5.35 (1.62)cd
Overall acceptability	6.53 (1.18)abc	5.47 (1.81)de	7.41 (1.0)a	6.18 (1.74)bcd	5.01 (1.625)e	5.35 (1.62)ce

Source: From Karki, T. et al. 1983a. *Nippon Shokuin Kogyo Gakkaishi* 30: 357–367. With permission.
Mean separation within rows by DMRT, 5%.
Values in parentheses are SD (standard deviation).
Mean in rows bearing similar notations are not significantly different.
The various *Gundruk* treatments are coded as Gt1-radish/8days, Gbt2-radish blanched, Gut3-radish unbalanced + 10%cabbage, Gut4-radish unblanched + 10%cabbage, Gbt5-radish = 10%cabbage, Gbt6-radish = 10%cabbage.

by the addition of cabbage starter due to desirable lactic acid production, and seventh day *gundruk* was the best. *Gundruk* made of unblanched radish with 10% cabbage starter fermented for seven days had superior overall quality. Cabbage starter proved an important and reliable source of LAB that helped in increasing the acidity level in a very short time, thus enhancing the flavor of *gundruk* (Karki et al., 1986).

5.3.3.2 Palm Vinegar   Joint FAO/WHO Food Standards Programme (1987) (www.codexalimentarius.net) has defined vinegar as a liquid fit for human consumption, produced from a suitable raw material of agricultural origin, containing starch, sugars, or both, by the process of double fermentation, alcoholic, and acetous, and containing a specified amount of acetic acid (Joshi and Thakur, 2000). Vinegar must contain not less than 4% acetic acid. The quality of palm vinegar, however, varies across the producers, depending upon individual production techniques. The final product differences are based on personal hygiene, sanitary facilities, harvesting condition, fermentation conditions, post-fermentation treatment (pasteurization temperature and time), and storage conditions. The quality standards for cider vinegar are given in Table 5.2, so products like palm vinegar should be evaluated for physical (color, turbidity, pH, and TSS) chemical (total acidity, acetic acid and sugars—total sugars and reducing sugar concentration—and ethyl acetate) properties, sensorial profile, aroma intensity, general impression, pungent sensation, and richness in aroma, as quality parameters of palm vinegars are of different types.

5.3.4 Meat and Meat Products

The quality of meat is influenced by the conditions of storage, and can certainly be controlled by conditions such as temperature and packaging. Fermented sausages are generally shelf-stable products, but refrigeration greatly prolongs their shelf-life and maintains their optimum quality. At various stages in the marketing and distribution channels, refrigeration of fermented sausages needs to be resorted to maintain chemical and sensory quality (Wani and Sharma, 1999). Various physicochemical and microbiological quality parameters of the stored samples were evaluated at 15-day intervals; the moisture content was $45.96\% \pm 0.38\%$, fat content was $15.95\% \pm 0.49\%$, protein content was $31.83\% \pm 0.48\%$, pH was 4.80 ± 0.03, salt content was $-3.30\% \pm 0.04\%$, must:protein ratio was 1.45 ± 0.05 and weight loss was $25.73\% \pm 0.13\%$. The microbial counts of fermented mutton sausages were significantly affected by the storage period. The total aerobic mesophilic count and lactic acid bacterial (LAB) count increased substantially at each storage interval. While the total counts increased from about log 7 at day 0 to log 7.6/g on day 120, the LAB increased from log 6.7 to log 7.5/g during the same period. The yeast and mold counts also tended to increase ($p < 0.05$) during the storage period. Thus, polypropylene-packaged fermented mutton sausages could be stored upto 105 days at $5 \pm 1°C$ and maintain their physicochemical quality and microbial stability without any loss of overall acceptability (Mir and Sharma, 2004).

5.3.5 Fermented Fish and Shrimp Products

The quality of fermented fish and shrimp products is measured by the total crude protein concentration, salt content, and TSS, whereas the product safety is determined based on the concentration of total volatile nitrogen, histamine, 3-monochloropropane-1,2-diol (3-MCPD), and microbiological load (www.food technology.w.nz).

A fish sauce, regardless of the local name and the origin of production, should contain total crude protein and salt content of not less than 5% and 15%, respectively. A high quality fish sauce should contain TSS of above 37°B. Total volatile based-nitrogen (TVBN) content in fermented fish and shrimp products has been considered as an important quality indicator of the products. TVBN shows the increased breakdown of protein due to the enzymatic action of bacteria which produces amines that reduce the nutritional value of products (Kerr et al., 2002). If TVBN content level reaches 30–35 mg/100 g, products are considered spoiled and if the level is 40 mg/100 g and above, products are unsafe for consumption (Gill, 1990). TVBN content increases with increasing temperature and storage period.

Salt concentration of more than 15% in fermented products lowers TVBN formation due to the inhibition of spoilage bacteria such as *Salmonella putrefaciens*, *Photobacterium phosphoreum*, and *Vibrio aceae*, but at such a salt concentration, halotolerant hydrophilic which plays a role in fermentation, are able to grow (Gram and Huss, 1996; Karacam et al., 2002; Yildirim, 2009). The taste of fish sauce is classified as '*umami*' to characterize the unique flavor compound which produces the most acceptable aroma and taste (Mizutani et al., 1992).

5.4 Microbial Profile of Indigenous Fermented Foods

In the indigenous fermented foods, the molds, yeasts, and bacteria play very important role. In the alcoholic beverages production, the yeast are involved to the greatest extent.

Lactic acid fermentation occurs during the preparation of a wide variety of foods made from raw materials of plant and animal origins (Tanasupawat et al., 2013). In fermented vegetables, cereals, fish, and meat products (Tamang, 2010), the homofermentative, strains of *L. plantarum*, *L. curvatus*, *L. sake*, *Pediococcus acidilactici*, and *P. pentosaceus*, and the heterofermentative strains of *L. fermentum*, *L. brevis*, *L. divergens*, and *L. sanfrancisco* produce DL-lactic acid (Dworkin et al., 2006). The other lactic acid producing strains, *L. casei*, *L. carnis*, *L. amylophilus*, *Lactococcus lactis*, *S. lactis*, *S. thermophilus*, *Enterococcus faecium*, *E. cecorum*, *E. durans*, and D-lactic acid producing strains, *L. delbrueckii*, *L. bulgaricus*, *L. coryniformis*, *L. fructosus*, *Leu. mesenteroides*, *Leu. lactis*, *Leu. citreum*, and *Leu. fallax* were isolated (Dworkin et al., 2006; Shaw and Harding, 1989; Tolonen et al., 2004). Yeasts distributed in fermented products give several beneficial effects. They produce alcohol and gas that contribute to food preservation, besides producing flavor and taste (Nout, 2003). The production of alcohol improves the aroma of the product and a certain concentration also makes the substrate unsuitable for spoilage-causing microorganisms (Nout, 2003). The effect is increased by the presence of lactic acid produced by the LAB bacteria. Some yeast strains—*Saccharomycopsis fibuligera*, and *Pichia anomala*—show strong phosphatase and peptidase activity (Tamang, 2009a,b, 2010). The mold strains *Rhizopus* spp. have a major role in saccharifying starch, while *Saccharomycopsis fibuligera* supports this activity. They degrade starch and produce glucose, and *Saccharomyces* strains grow rapidly to produce ethanol (Thapa, 2001).

5.4.1 Fermented Vegetables

The populations of LAB in Himalayan fermented vegetables, *gundruk*, *sinki*, *goyang*, *inziangsang* (the liquid form is called *ziang dui*), *khalpi*, *ekung*, *eup*, *hirring*, *mesu, soibum, soidon,* and *soijim* ranged from 6.0 ± 0.2 log CFU/g per sample to 7.9 ± 0.1 log CFU/g per sample, while the yeast and aerobic mesophilic bacteria are up to 5.1 ± 0.6 log CFU/g per sample and 8.1 ± 0.5 log CFU/g per sample, respectively (Tamang, 2010). Homofermentative strains include *Lactobacillus plantarum, L. casei, Pediococcus acidilactici, P. pentosaceus, Lactococcus lactis, Enterococcus faecium,* and *L. curvatus,* while heterofermentative strains of *L. fermentum, L. brevis, Leuconostoc fallax, Leuconostoc citreum,* and *Leuconostoc lactis* were isolated. Halophilic LAB and *Tetragenococcus halophilus* strains were isolated from *ekung,* while *L. plantarum, L. brevis, P. pentosaceou, Lactococcus lactis, Bacillus circulans, B. firmus, B. sphaericus,* and *B. subtilis* were found in *Tuaithur* (Tamang et al., 2012) (Table 5.11). The yeasts *Candida, Saccharomyces,* and *Torulopsis* spp. were also isolated (Tamang and Fleet, 2009). LAB are dominant microflora in Himalayan fermented vegetable products; however, some yeasts—*Pichia, Candida, Saccharomyces,* and *Rhodotorula*—have been reported in *sinki, goyang, khalpi, ekung, hirring,* and *mesu* (Tamang, 2006). The yeasts might appear during storage and thus, cause spoilage in the products.

Table 5.11   Microbial Flora of Fermented Vegetable Products

NAME (RAW MATERIALS)	MICROORGANISMS	REFERENCES
Gundruk	*L. fermentum, L. plantarum*	Karki et al. (1983a,b)
Leaves	*L. casei, P. pentosaceus*	Tamang et al. (2005)
Sinki	*L. plantarum, L. brevis*	Tamang and Sarkar (1993)
	L. caseii, Leuconostoc fallax	Tamang et al. (2005)
Goyang (green vegetable)	*L. plantarum, L. brevis, Lactobacillus lactis, E. faecium, Candida* spp.	Tamang and Tamang (2007)
Inziangsang (mustard leaves)	*L. plantarum, L. brevis*	Tamang et al. (2005)
Khalpi or khaipi (cucumber)	*Leuconostoc fallax*	
Mesu (bamboo shoot)	*L. plantarum, L. brevis*	Tamang and Sarkar (1996)
	L. curvatus, Leuconostoc citreum, P. pentosaceus	Tamang et al. (2008)
Soibum (bamboo shoot)	*L. plantarum, L. brevis*	Giri and Janmey (1987); Tamang et al. (2012)
	L. coryniformis, L. delbrueckii	Tamang et al. (2008)
	Leuconostoc fallax, Leuconostoc lactis	Sarangthem and Singh (2003)
	Leuconostoc mesenteroides, E. durans	
	Streptococcus lactis, B. subtilis, B. licheniformis	
	B. coagulans, M. luteus, Candida spp., *Saccharomyces, Torulopsis* spp.	
Soidon (bamboo shoot) and *Soijim*	*L. brevis, Leuconostoc fallax*	Tamang et al. (2008)
Bamboo shoot tips	*Leuconostoc lactis*	
Ekung	*L. plantarum, L. brevis*	Tamang and Tamang (2009)
Hirring (bamboo shoot tips)	*L. plantarum, Lac. lactis*	Tamang and Tamang (2009)
Lung-siej (bamboo shoot tips)	*Lactobacillus, Pediococcus*	Tamang (2010)

5.4.2 Fermented Legumes and Black Gram Food

Fermented soy bean foods of the Himalayas include *kinema, hawaijar, tungrymbai, aakhone, bekang, peruyyan,* and fermented black gram called *maseura. B. subtilis* strains were the dominant bacterial flora (10^8 CFU/g) in the products (Jeyaram et al., 2008b, 2009; Tamang, 1992). *B. cereus, B. licheniformis, B. mycoides, B. pumilus, B. laterosporus, Bacillus* spp., *E. faecium, E. durans,* along with yeasts *C. parapsilosis, Geotrichum candidum, S. cerevisiae, Pichia burtonii,* and *C. castelli,* including LAB, *L. fermentum, P. acidilactici,* and *L. salivarius,* were distributed in the products. *Staphylococcus aureus, S. sciuri,* and *Alcaligenes* spp. were also isolated (Table 5.12).

5.4.3 Fermented Milk Products

LAB (10^8 CFU/g) were the dominant populations of Himalayan fermented milk products *dahi, mohi, gheu, chhurpi, dudh chhurpi, chhu, somar, philu,* and *kalari* (Dewan and Tamang, 2007; Tamang and Fleet, 2009), especially the rod-shaped LAB, *L. bifermentans, L. alimentarius, L. paracasei, L. farciminis, L. plantarum, L. curvatus, L. fermentum, L. kefir, L. hilgardii,* and *L. salivarius* (Dewan, 2002; Mikelsaar et al., 2002). Yeast isolates—strains of *Saccharomycopsis* spp., and *Candida* spp.—and enterobacteriaceae are distributed in these products (Table 5.13). In fermented milk (Tamang, 2010), homofermentative strains, *L. alimentarius, L. paracasei, Lac. lactis, L. farciminis, L. plantarum, L. curvatus, L. salivarius, E. faecium,* and heterofermentative strains, *L. fermentum, L. brevis, L. kefir, L. bifermentans, L. hilgardii, Leu. mesenteroides, W. confuse,* are distributed, which differ from the yogurt and related starter cultures *S. thermophilus, L. delbrueckii, L. bulgaricus,* and *L. acidophilus* (Aneja, 2002).

Table 5.12 Microbial Flora of Fermented Legumes and Black Gram Products

NAME (RAW MATERIALS)	MICROORGANISMS	REFERENCES
Kinema	*B. subtilis, E. faecium, C. parapsilosis*	Tamang (1992)
	Geotrichum candidum, LAB	Sarkar et al. (1994)
Hawaijar (soybean)	*B. subtilis, B. cereus*	Jeyaram et al. (2008b)
	B. licheniformis, S. aureus S. sciuri, Alcaligenes spp.	Tamang et al. (2009)
Tungrymbai, Aakhone	*B. subtilis, Bacillus* spp.	Tamang et al. (2009)
Bekang, Peruyyan (soybean)	*B. subtilis, Bacillus* spp.	
Maseura (black gram)	*B. subtilis, B. mycoides B. pumilus, B. laterosporus L. fermentum, P. acidilactici, L. salivarius, E. durans Saccharomyces cerevisiae Pichia burtonii Candida castelli*	Chettri and Tamang (2008)

Table 5.13 Microflora of Milk Products

NAME (RAW MATERIALS)	MICROORGANISMS	REFERENCES
Dahi (cow milk),	*L. bifermentans, L. alimentarius*	Dewan and Tamang (2007); Tamang and Fleet (2009)
Mohi (cow milk)	*L. paracasei, L. lactis* *Saccharomycopsis* spp., *Candida* spp.	
Gheu (cow milk)	*L. lactis*	Dewan (2002)
Chhurpi (hard variety) and *dudh chhurpi* (cow milk)	*L. farciminis, L. paracasei, W. confusa, L. bifermentans*	Dewan (2002)
Chhurpi (soft variety) (cow milk)	*L. plantarum, L. curvatus, L. fermentum, L. paracasei L. alimentarius, L. kefir L. hilgardii, E. faecium, L. mesenteroides*	Tamang et al. (2000)
Chhu (cow/yak milk)	*L. farciminis, L. brevis L. alimentarius, L. salivarius, L. lactis Saccharomycopsis* spp., *Candida* spp.	Dewan and Tamang (2007)
Somar (cow/yak milk)	*L. paracasei, L. lactis*	Mikelsaar et al. (2002)
Philu (cow/yak milk)	*L. paracasei, L. bifermentans, E. faecium*	Dewan and Tamang (2007)

Microflora of *srikhand* is similar to *dahi*, mostly consisting of *S. salivarius* ssp. *Thermophilus* and *L. delbrueckii* ssp. *bulgaricus*. Apart from these types, potential pathogens may also gain access into the product due to unhygienic practices in its preparation (Mårtenssona et al., 2002).

5.4.4 Fermented Cereal Foods

LAB (above 10^8 CFU/g) are the dominant populations of Himalayan fermented cereals, selroti (Yonzan, 2007) the count of 10^5 to 10^6 CFU/g were found in *jalebi* (Batra, 1986). They are *L. curvatus, P. pentosaceus, L. fermentum, L. buchneri, L. bulgaricus, S. lactis, S. thermophilus, L. mesenteroides, E. faecalis, E. faecium*, and *Lactococcus* spp., as in Table 5.5 (Tamang, 2009a,b). The yeasts, *Saccharomyces cerevisiae, S. kluyveri, S. bayanus, Debaromyces hansenii, Pichia burtonii, Zygosaccharomyces rouxii* (Tamang et al., 2012), and *Hansenula anomala* were isolated (Batra and Millner, 1976; Yonzan, 2007). The microflora of fermented cereal foods is depicted in Table 5.14.

Salmonella (detection limit, 1 cell (25 g)–1) occurred in 12 samples of *wadi, idli* and papad, but was absent in the other three products. *Clostridium perfringens* (detection limit, 10 CFU g–1) and *Shigella* (detection limit, 1 cell (25 g)–1) were not detected. Thus, these foods were manufactured using poor-quality starting materials, processed under unhygienic conditions, or/and temperature-abused during transportation and storage (Roy et al., 2007).

Changes in *warries* fermentation affect the quality of a product (Soni and Sandhu, 1989, 1999). Aerobic bacteria, Bacillus strains, especially B. subtilis, in fermented legumes, *kinema*, are distributed in raw soybeans (Tamang, 1992) and play the main

Table 5.14   Microflora of Fermented Cereal Foods

NAME (RAW MATERIALS)	MICROORGANISMS	REFERENCES
Selroti (rice-wheat flour-milk)	*L. mesenteroides*	Yonzan (2007)
	E. faecium, P. pentosaceus, L. curvatus	
	Saccharomyces cerevisiae, S. kluyveri	
	Debaromyces hansenii, Pichia burtonii	
	Zygosaccharomyces rouxii	
Jalebi (wheat flour)	*L. fermentum*	Batra and Millner (1976)
	L. buchneri	Soni and Sandhu (1990)
	L. bulgaricus, S. lactis, S. thermophilus	Tamang (2009a,b)
	E. faecalis, Saccharomyces bayanus, S. cerevisiae	
	Hansenula anomala	
Nan (wheat flour)	*S. kluyveri*	Batra (1986)
Siddu (wheat)	*LAB, yeasts*	
Chilra (buckwheat)	*Lactobacillus, Leuconostoc, Schizosaccharomyces,*	Kanwar et al. (2007)
	Lactococcus, Saccharomyces, Debaryomyces	
Marchu (wheat flour)	*LAB, yeasts*	
Bhaturu (wheat flour)	*Lactobacillus, Leuconostoc, Schizosaccharomyces,*	Kanwar et al. (2007)
	Lactococcus, Saccharomyces, Debaryomyces	

role in fermentation. *Bacillus* strains showed strong activities of peptidase and phosphatase, but had weak lipase and esterase activity (Tamang, 2010). The fermentation of *warries* is brought about by natural microflora. The distribution of contaminants such as *B. cereus* and *Staphylococcus aureus* strains, including enterobacteriaceae, in some fermented vegetables, fermented milk, and fermented fish products might have been introduced during the handling of the raw materials (Dewan, 2002; Tamang, 2006; Thapa, 2002). The salting of raw materials at the initial stage can inhibit the growth of pathogens. However, the food borne pathogens, *B. cereus*, *Listeria* spp., and *Salmonella* spp. have not been isolated in fermented cereal, seroti (Yonzan, 2007) or in meat products (Rai, 2008; Tamang, 2009a,b).

5.4.5 Fermented Fish Products

5.4.5.1 Fermented Fish   During the preparation of Ngathuchar in Assam, India, hill stream, fishes are collected, washed, and rubbed with salt and dried in the sun for 4–7 days, or kept above the *machang* near the fire place. The dried fish product is stored at room temperature for 3–4 months before consumption (Tamang, 2009a,b). The presence of *Lactococcus plantarum*, *Leuconostoc mesenteroides*, *Lactobacillus plantarum*, *Bacillus subtilis*, and *Candida* sp. in different fermented fish product was documented. The pH of all these products was slightly acidic in nature, due to the pre-dominance of LAB flora. The antagonistic properties of the strains, isolated from fish products of the different villages of the NC Hills, were tested against the indicator strains (*Listeria monocytogenes* DSM 20600, *Bacillus cereus* CCM 2010, and *Enterococcus faecium* DSM 20477).

Table 5.15   Microflora of Fermented Fish Products

NAME (RAW MATERIALS)	MICROORGANISMS	REFERENCES
Suka ko maacha, Gnuchi (river fish)	*Lactobacillus lactis*	Thapa et al. (2006)
Sidra, Sukuti (fish)	*Lactobacillus plantarum*	
	Leuconostoc mesenteroides	Hurtado et al. (2012)
	E. faecium	
	E. faecalis	
	P. pentosaceus	
	Candida chiropterorum	
	C. bombicola	
	Saccharomycopsis spp.	
Ngari, Hentak, Tungtap (fish)	*Lactobacillus lactis*	Thapa et al. (2004)
	Lactobacillus plantarum	Tamang et al. (2012)
	E. faecium	
	L. fructosus	Tamang et al. (2012)
	L. amylophilus	
	L. coryniformis	
	L. plantarum	
	B. subtilis	
	B. pumilus	
	Micrococcus spp.	
	Candida spp.	Tamang et al. (2012)
	Sacchromycopsis spp.	

LAB, bacilli, and micrococci, including yeasts, were distributed 70%, 15%, 5% and 10% in the fish products, respectively (Thapa, 2002). LAB populations of *suka ko maacha, gnuchi, sidra, sukuti, ngari, hentak,* and *tungtap* ranged from 4.6.0 to 7.3 log CFU/g per sample, while the yeast, spore forming bacteria and aerobic mesophilic bacteria were up to 3.2 log CFU/g, 4.2 log CFU/g, and 7.6 log CFU/g, respectively (Thapa et al., 2006). *L. plantarum, L. coryniformis, L. amylophilus, L. fructosus, P. pentosaceus, L. lactis, L. plantarum, L. mesenteroides, E. faecium,* and *E. faecalis* were the LAB that were isolated (Table 5.15).

5.4.6 Fermented Meat Products

The LAB flora of *lang kargyong, yak kargyong, faak kargyong* (Tamang, 2009a,b), *lang satchu, yak satchu, suka ko masu, chartayshya,* and *arjia* ranged from 6.0 ± 0.1 to 7.8 ± 0.1 log CFU/g per sample. They were *L. sake, L. casei, L. plantarum, L. brevis, L. divergens, L. carnis, L. sanfrancisco, L. curvatus, P. pentosaceus, L mesenteroides,* and *E. faecium* strains (Maria et al., 2010) (Table 5.16). The bacilli, micrococcaceae, aerobic mesophilic bacteria, yeast, and molds were 1.5 ± 0.3 to 3.8 ± 0.1 log CFU/g, 4.9 ± 0.1 to 6.6 ± 0.1 log CFU/g, 6.2 ± 0.1 to 9.0 ± 0.1 log CFU/g, 4.4 ± 0.1 –5. 7 ± 0.1 log CFU/g, and up to 3.5 ± 0.1 log CFU/g, respectively (Rai, 2008).

Table 5.16   Microflora of Fermented Meat Products

NAME (RAW MATERIALS)	MICROORGANISMS	REFERENCES
Lang kargyong (beef)	*L. sake, L. divergens*	Rai (2008)
	L. carnis, L. sanfrancisco	
	L. curvatus	
	L. mesenteroides	
	E. faecium, B. subtilis	
	B. mycoides, B. thuringiensis	
	S. aureus, Micrococcus spp.	
	Debaryomyces hansenii	
	Pichia anomala	
Yak kargyong (yak)	*L. casei, L. plantarum*	Tamang (2010)
	L. sake, L. divergens	
	L. carnis, L. sanfrancisco	
	L. curvatus, L. mesenteroides	Tamang et al. (2012)
	E. faecium, B. subtilis	
	B. mycoides, Debaryomyces pseudopolymorphus	
Faak kargyong (pork)	*L. brevis, L. carnis*	Rai (2008)
	L. plantarum, L. mesenteroides	
	E. faecium, B. subtilis	
	B. mycoides, B. licheniformis	Maria et al. (2010)
	S. aureus, Micrococcus spp.	
	Debaryomyces polymorphus	
	Candida famata	
Lang satchu (beef meat)	*P. pentosaceus, L. casei*	Rai (2008)
	L. carnis, E. faecium	
	B. subtilis, B. mycoides	Tamang (2009a,b)
	B. lentus, S. aureus	
	Micrococcus spp.	
	Debaryomyces hansenii	Maria et al. (2010)
	Pichia anomala	
Yak satchu (yak meat)	*P. pentosaceus, B. subtilis*	Rai (2008)
	B. mycoides, B. licheniformis	
	S. aureus	
	Micrococcus spp.	Tamang et al. (2012)
	Debaryomyces polymorphus	
Suka ko masu (buffalo meat)	*L. canis, L. plantarum*	Rai (2008)
	E. faecium, B. subtilis	
	B. mycoides, B. thuringiensis	
	S. aureus, Micrococcus spp.	Tamang (2009a,b)
	Debaryomyces hansenii	
	Pichia famata	
Chartayshya (chevon, goat meat)	*L. divergens, E. faecium*	Rai (2008)
	P. pentosaceus, B. subtilis	
	B. mycoides, B. thuringiensis	
	S. aureus, Micrococcus spp.	Tamang (2009a,b)
		(*Continued*)

Table 5.16 (Continued)   Microflora of Fermented Meat Products

NAME (RAW MATERIALS)	MICROORGANISMS	REFERENCES
	Debaryomyces hansenii	
	Pichia burtonii	
Jamma/geema (chevon, goat meat)	*L. sanfrancisco, L. divergens*	Rai (2008)
	Leuconostoc mesenteroides	
	P. pentosaceus, E. cecorum	
	E. faecium, B. subtilis	
	B. sphaericus, S. aureus	Tamang (2009a,b)
	Micrococcus spp.	
	Debaryomyces hansenii	
	C. albicans	
Arjia (chevon, goat meat)	*P. pentosaceus, E. faecium*	Rai (2008)
	B. subtilis, B. mycoides	
	B. thuringiensis, S. aureus	Tamang (2009a,b)
	Micrococcus spp.	
	Debaryomyces hansenii	
	Cryptococcus humicola	

5.4.7 Alcoholic Fermented Beverages

The microbial profile of indigenous alcoholic beverage *Sur* is discussed here as an example.

5.4.7.1 Sur   Most of the collected *sur* samples showed the microbiological contamination. The total plate counts of the various collected *sur* samples are given in Table 5.17. The maximum CFU/mL were observed in the Kangra 2 and Sirmaur 1 samples, whereas, the Kullu 2 and Mandi JN2 samples showed no microflora.

5.4.8 Alcoholic Fermented Beverages and Starters

The starter culture *marcha* is used to prepare alcoholic beverages from starchy substrates. This starter culture is a mixed starter contained molds, yeasts, and LAB; 10^6 CFU/g, 10^8 CFU/g, and 10^7 CFU/g, respectively (Tamang, 1992). *Mucor* and *Rhizopus* were dominant mycelia fungi, and the yeasts, *Saccharomyces cerevisiae*, *Saccharomycopsis fibuligera*, *Saccharomycopsis capsularis*, *P. anomala*, *P. burtonii*, and *Candida versatilis* were isolated (Tamang et al., 2007a; Tsuyoshi et al., 2005). The other starter cultures, *manapu*, *mana*, and *hamei* contained *Saccharomyces cerevisiae*, *Candida* spp., *Pichia* spp., *Trichosporon* sp., *Torulaspora delbrueckii*, *Mucor* spp., *Rhizopus* spp., *Aspergillus oryzae*, and LAB (Endo and Okada, 2008; Jeyaram et al., 2008a; Nikkuni et al., 1995; Shrestha et al., 2002; Tamang et al., 2007a; Thakur et al., 2004) (Table 5.18). Alcoholic fermented beverages like *kodo ko jaanr* or *chyang*, *chyang*, *bhaati jaanr*, *poko*,

Table 5.17   Total Plate Count of Collected *sur* Samples

SAMPLE	LOG CFU/ML
KANGRA 1	3.18
KANGRA 2	4.18
KANGRA 3	4.12
KANGRA 4	3.48
KANGRA 5	2.70
KANGRA 6	3.08
Kullu 1	3.67
Kullu 2	Nil
Kullu 3	2.30
Kullu 4	3.48
Mandi 1	3.68
Mandi 2	4.21
Mandi 3	3.70
Mandi 4	2.48
Mandi 5	3.43
Mandi 6	3.70
Mandi 7	3.51
Mandi 8	3.41
Mandi J N 1	Nil
Mandi J N 2	3.72
Mandi J N 3	3.65
Sirmaur 1	4.18

Source: From Sharma, A. 2013. *Preparation and Evaluation of Sur Production in Himachal Pradesh.* MSc thesis, Dr. Y.S. Parmar University of Horticulture and Forestry, Nauni, Solan, HP. With permission.

makai ko jaanr, gahoon ko jaanr, atingba, and *zutho* or *zhuchu* contained *Mucor* spp., *Rhizopus* spp., *Candida* spp., *Pichia* spp., *Saccharomyces cerevisiae, Saccharomycopsis fibuligera,* and LAB (Tamang, 2009a,b) (Table 5.19).

An inoculum used frequently in the preparation of *Chhang* is *phab, phaff,* or *phapus* which was analysed for microflora (Table 5.20), and *Candida krusei* and *Saccharomyces cerevisiae* (two different strains) were isolated and identified (Bhatia et al., 1977).

Table 5.18   List of Microorganisms Isolated from the Inoculum of Various Fermented Foods and Beverages of Himachal Pradesh

SOURCES OF ISOLATES	MICROORGANISMS ISOLATED AND IDENTIFIED
Chfca "Treh"	*Saccharomyces cerevisiae*
Bhatura "Malera"	*S. cerevisiae, Lactobacillus* sp.
Borhe dough	*S. cerevisiae*
Siddu/Khobli "Malera"	*S. cerevisiae, Candida valida*
Phab	*S. cerevisiae, Bacillus* sp., *Actinomycetes*

Source: Adapted from Thakur, N., Savitri, and Bhalla, T.C. 2004. *Indian J Trad Know* 3(3): 325–335.

Table 5.19   List of Microorganisms Isolated from Fermented Beverages of Himachal Pradesh

FERMENTED BEVERAGE	MICROORGANISMS ISOLATED
Sur	Saccharomyces cerevisiae, Zygosaccharomyces bisporus
Lugri	S. cerveisiae, Leuconostoc
Angoori	S. cerevisiae
Daru	S. cerevisiae, Candida famata, Candida valida, Kluveromyces
Chhang	Candida cocoi, S. cerevisiae

Source: Adapted from Thakur, N., Savitri, and Bhalla, T.C. 2004. *Indian J Trad Know* 3(3): 325–335; Kanwar, S.S. et al. 2007. *Indian J Trad Know* 6: 42–45.

Commonly encountered microorganisms in the inocula of fermented foods and beverages are *Saccharomyces cerevisiae*, *Candida* sp., *Zygosacharomyces bisporus*, and *Kluveromyces thermotolerance* (Table 5.20). The predominant bacteria in the fermented products included species of *Leuconostoc* and *Lactobacillus* (Thakur et al., 2004). Microbiological analysis of the starter culture revealed a mixed microflora consisting

Table 5.20   Microflora of Starters and Fermented Alcoholic Beverages

NAME (RAW MATERIALS)	MICROORGANISMS	REFERENCES
Marcha (starter, rice, wild herbs, spices)	Mucor circinelloides	Tamang (1992)
	M. hiemalis	Tamang and Sarkar (1995); Tamang (2009a,b)
	Rhizopus chinensis	Thapa (2001)
	R. stolonifer	Tsuyoshi et al. (2005)
	S. fibuligera	Tamang et al. (2007a)
	Saccharomycopsis capsularis	Tamang and Fleet (2009)
	P. anomala, P. burtonii	
	Saccharomyces cerevisiae	
Manapu (mixed starter, rice-wheat, herbs)	Saccharomyces cerevisiae	Shrestha et al. (2002)
	Candida versatilis	Tamang (2009a,b)
	Rhizopus spp.	
	P. pentosaceus	
Mana (starter, wheat, herbs)	Aspergillus oryzae	Nikkuni et al. (1996)
	Rhizopus spp.	Shrestha et al. (2002)
Hamei (starter, rice wild herbs)	Mucor spp.	Tamang et al. (2007a)
	Rhizopus spp.	Jeyaram et al. (2008b)
	S. cerevisiae	
	Pichia anomala	
	P. guilliermondii	Tamang (2009a,b)
	P. fabianii	
	Trichosporon sp.	
	C. tropicalis	
	C. parapsilosis	
	C. Montana	

(Continued)

Table 5.20 (*Continued*)   Microflora of Starters and Fermented Alcoholic Beverages

NAME (RAW MATERIALS)	MICROORGANISMS	REFERENCES
	Torulaspora delbrueckii	
	P. pentosaceus	
	L. brevis	
ALCOHOLIC FOOD BEVERAGES		
Kodo ko jaanr or chyang (finger millet)	*Mucor circinelloides*	Thapa (2001)
	Rhizopus chinensis	Thapa and Tamang (2004)
	Pichia anomala	
	S. cerevisiae	
	C. glabrata	Tamang et al. (2012)
	S. fibuligera	
	P. pentosaceus	
	L. bifermentans	
Chyang (finger millet, barley)	*S. cerevisiae*	Batra (1986)
	S. uvarum	
Bhaati jaanr (rice)	*Mucor circinelloides*	Thapa and Tamang (2006)
	M. hiemalis	
	Rhizopus chinensis	
	R. stolonifer	
	S. fibuligera	Tamang (2009a,b, 2012)
	S. capsularis	
	Pichia anomala	
	P. burtonii	
	S.cerevisiae	
	S. bayanus	
	Candida glabrata	
	P. pentosaceus	
	L. bifermentans	
	L. brevis	
Poko (rice)	*Rhizopus* spp.	Shrestha et al. (2002)
	S.cerevisiae	
	C. versatilis	
	P. pentosaceus	Tamang (2012)
Makai ko jaanr (maize)	*Mucor circinelloides*	Thapa (2001)
Gahoon ko jaanr (wheat)	*M. hiemalis*	
	Rhizopus chinensis	
	R. stolonifer	
	S. fibuligera	Tamang (2009a,b, 2012)
	Pichia anomala	
	P. burtonii	
	S. cerevisiae	
	C. glabrata	
	P. pentosaceus	
	L. bifermentans	
	L. brevis	
Atingba (rice)	*Same as hamei*	
Zutho or zhuchu (rice)	*S. cerevisiae*	Teramoto et al. (2002)

of yeasts, actinomycetes, and *Bacillus* species (Thakur et al., 2004). Since no stringent microbiological precautions are observed during the preparation and formulation of *Phab*. These variations are expected since it not only contains fermentative microorganisms but also has saccharolytic and acidogenic microorganisms, which might be responsible for the release of fermentable sugars from starch-based raw materials, like rice, in *Chhang* preparation (Thakur et al., 2004).

5.5 Spoilage, Shelf-Life, and Safety of Indigenous Fermented Products

Spoilage, consequently shelf-life of any food product and its safety is intimately interconnected. From safety point of view, the plant or animal based raw material may contain toxin or is formed by contamination by pests and or microorganisms. These aspects have been discussed in this section.

5.5.1 Causes of Spoilage and Shelf-Life

5.5.1.1 Causes of Spoilage A food is considered spoiled when a discriminating consumer refuses to accept it (Frazier and Westhoff, 1978). In other words, decay or decomposition of an undesirable nature that makes the food unfit for consumption is known as spoilage (Joshi et al., 2000). Spoilage may be caused by one or more of the following reasons: growth or activity of microorganisms; insect food pests; action of the enzymes of the plant or animal food; purely chemical reactions (which are not catalyzed by enzymes of the tissues or of microorganisms); and physical changes caused by freezing, burning, drying, pressure, etc.

5.5.1.2 Deterioration of Processed Foods The principal mechanisms involved in the deterioration of processed foods includes microbiological spoilage sometimes accompanied by the growth of pathogens, development of chemical and enzymatic activity causing lipid breakdown, color, odor, flavor, and textural changes, and moisture and/or other vapor migration producing changes in texture, water activity, and flavor.

Different processing variables that affect these mechanisms and which can be used to control deterioration include: (1) moisture and water activity (www.Medallionlabs. com); (2) pH; (3) heat treatments; (4) emulsifier systems; (5) preservatives and additives; and (6) packaging.

Amongst various fermented foods, alcoholic beverages are important products. Alcoholic beverages like wine can have defects from both microbial and nonmicrobial causes, where the latter include spoilage due to metals and their salts, and enzymes and materials used for clarification (Joshi, 1997, 2006b; Joshi et al., 2000). Microbial spoilage occurs only if either the wine making procedure is faulty or the wine is handled improperly during storage. In a beverage if alcoholic content is less than 15%, spoilage by LAB takes place (Frazier and Westhoff, 1978; Joshi, 1997). Growth of any bacteria or yeast in wines can cause an undesirable cloudiness, or any acetic or heterofermentative LAB growth in wine may increase volatile acidity. Indigenous fermented alcoholic beverages like *chhang* and palm wine get spoiled if kept for more than 3–4 days at normal temperature. However, when the product is distilled it does not spoil microbiologically. Some of the spoilages caused by microbes in wine are listed in Table 5.21.

Various types of breads, if stored for more time, get contaminated by fungi on the surface. The common fungi include *Penicillium*, *Aspergillus*, *Rhizopus*, and *Mucor*, while bacteria like *Bacillus* can grow and spoil loaves of bread. The lactic acid fermented vegetables like sauerkraut can be easily spoiled by the surface growth of yeasts like *Pichia*, *Rhodotorula*, etc. Pickles can also be spoiled by fungi if there is insufficient salt, acid, or spices (Frazier and Westhoff, 1978; Joshi et al., 2000). Milk products like *dahi* usually get spoiled after 24 h unless refrigerated. Increase in acidity makes the *dahi* unacceptable even if not spoiled microbiologically. Vinegar is usually not spoiled microbiologically, except when not pasteurized (Joshi and Sharma, 2009). Other products like *shrikhand* are spoiled in a week or so if kept without refrigeration.

Table 5.21   Summary of Spoilage of Wine by Microbes

TYPE OF SPOILAGE	MICROORGANISMS INVOLVED	TRAITS
SPOILAGE BY YEASTS		
Ester taints	*Hanseniaspora uvarum, Hansenula anomala, Brettanomycas*	High concentration of esters and associated aroma
Flowers of yeast (film formation)	*Candida, Metschnikowia, Pichia* and *Hansenula* as part of film and oxidative yeast, cause severe taints	Growth of film yeasts on wine surface Oxidized taste of wine
Spoilage by refermentation	*Zygosaccharomyces bailli*	Turbidity and sediment in wine, reduced acidity
SPOILAGE BY BACTERIA		
Vinegary taints	*Gluconobacter oxydams, Acetobacter pasteurians, Acetobacter acetii*	Intense smell of acetic acid Wines with low volatile acidity but high ethyl acetate
Ropiness	*Streptococcus mucilaginous* var. *vini, Pediococcus cerevisiae, Leuconostoc* sp.	Slimy, viscous oily wine High volatile acidity Ropiness starts from bottom of tank
Mannitol taints	Some strains of homofermentative lactic acid bacteria	Sliminess with diacetyl taste, vinegary estery taste, high mannitol, acetic acid and lactic acid contents
Mousiness	Lactic acid bacteria and *Brettanomyces* sp.	Occurs in low acid wines with insufficient SO_2 Wine smells like mouse or urine or acetamide

Source: Adapted from Joshi, V.K., Sandhu, D.K., and Thakur, W.S. 1999. *Biotechnology: Food Fermentation.* Vol. II. Educational Publishers and Distributors, Kerala, India. pp 647–744.

5.5.1.3 Factors Affecting Spoilage Spoilage caused by microorganisms and enzymes depends upon the kind and number of agents present and depends upon the environment. The kind and number of microorganisms present on/in the food will be influenced by the nature and extent of contamination, previous opportunities for certain kinds of growth, and any pretreatments that the food has received. Associations of microorganisms are involved in the spoilage or fermentation of most kind of food. The environment, however, should determine which of the different kinds of microorganisms present in the food will outgrow the others, causing characteristic changes or spoilage. The various factors involved have been discussed earlier (Joshi et al., 2000).

5.5.2 Shelf-Life of Indigenous Fermented Products

The modern food industry has developed and expanded because of its ability to deliver a wide variety of high quality food products to the consumers (www.foodtechnology. co). It has achieved this by building stability into the products through processing, packaging, and additives that enable the foods to remain fresh and wholesome throughout the distribution process (www.Medallionlabs.com).

In general, the shelf-life of any cultured milk product mainly depends on the initial milk quality, the heat treatment given, the starter culture used for fermentation, the conditions of incubation, the handling care during manufacture, prior to packaging, and the storage conditions of the product during distribution as well as by the consumers. Microbial tests are conducted to determine the number of specific starter bacteria used for fermenting milk, and also the contaminating microbes. For the assessment of proper hygiene, it is worthwhile to detect the presence of coliforms, staphylococci, and enterococci. Plastic cups and lids are the main packaging materials used commercially. This aspect is illustrated by taking an example of indigenous fermented food. *Shrikhand* has a longer shelf-life than other cultured milk products due to low moisture and higher sugar content (Patel and Chakraborty, 1988). The keeping quality of *shrikhand* has been found to be about 12–14 days under refrigeration, after which mold growth and increase in acidity take place (Gandhi and Jain, 1977). It was found that *shrikhand* stored at $10 \pm 3°C$ developed an off flavor and an unpleasant odor in about 40 days, whereas that stored at 37°C spoiled within a period of one week. The storage life was found to be 50 days at $7 \pm 2°C$ (Deasi et al., 1985); at 10°C it would about 42 days, but only 2–3 days at 30°C. However, a shelf-life of more than 30 days at ~10°C was reported by Desai (1983). Aidoo et al. (2005) reported the shelf-life of *shrikhand* to be about 35–40 days at 8°C and 2–3 days at 30°C, and that this could be improved by 20% by heating the product to 70°C for 2 minutes, or by the addition of 0.05% potassium sorbate.

5.5.3 Naturally Occurring Toxins of Plant Origin

5.5.3.1 Cyanogenic Compounds in Cassava Cassava is an important staple food crop (Westby et al., 1997) and the roots contain the cyanogenic glucosides linamarin and lotaustralin which can be hydrolysed to the corresponding ketone and glucose by the endogenous enzyme linamarase when cellular damage occurs (De Bruijn, 1973; Nartey, 1978). Cyanohydrins break down non-enzymatically at a rate dependent upon the pH and temperature (Cooke, 1978), with their stability increasing at acidic pH values. A second enzyme, hydroxynitrile lyase, may also contribute to cyanohydrins breakdown. Dietary cyanide from cassava has been implicated in a number of health problems. It is reasonable to consider that cyanide exposure from insufficiently processed cassava can cause acute poisoning (Bokanga et al., 1994; Oirschot et al., 2012; Westby et al., 1997). There is also strong evidence though for a causal link between cyanide and the paralytic disease konzo (Tylleskar, 1994) and tropical atoxic neuropathy (Osuntokun, 1994).

5.5.3.2 Other Antinutritional Factors   The main antinutritional compounds of interest in plant foods are phytate, tannins, saponins, oxalates, lectins, and enzyme inhibitors (Westby et al., 1997). The concentrations of these antinutritional factors are reduced during the preparation of fermented foods, but in general the mechanisms involved and the roles of microorganisms are poorly understood. There is a need for a search for suitable strains of bacteria, molds or yeasts that can eliminate antinutritional and toxic components in foods during fermentation (Reddy and Pierson, 1994; Westby et al., 1997). The role of fermentation to select for or introduce microorganisms that have the ability to reduce antinutritional factors has been further elaborated in Chapter 13 of this text.

5.5.3.3 Ethyl Carbamate   Ethyl carbamate (also called urethane) is a chemical compound with the molecular formula $C_3H_7NO_2$ first prepared in the nineteenth century (www.Medlibrary.com). Structurally, it is an ester of carbamic acid. Despite its common name, it is not a component of polyurethanes (Westby et al., 1997). Ethyl carbamate is a carcinogen that can be produced in some fermented foods by the reaction of naturally occurring urea and carbamyl phosphate with ethanol (Hasegawa et al., 1990; Ough, 1976). Urethane (ethyl carbamate) is formed naturally in wines during the yeast fermentation of fruit juice (Figure 5.3).

Most yeast-fermented alcoholic beverages contain traces of ethyl carbamate (15–12 ppm) (www.Medlibrary.com), however, other foods and beverages prepared by means of fermentation also have ethyl carbamate (Haddon et al., 1994, www. Medlibrary.com). Quantities of both ethyl carbamate and methyl carbamate have also

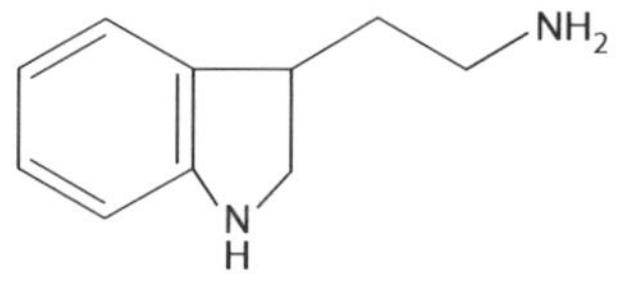

Figure 5.3   Structure of ethyl carbamate ($C_3H_7NO_2$).

been found in wines, sake, beer, brandy, whiskey, and other fermented alcoholic beverages (www.Medlibrary.com). Alcoholic beverages, particularly certain stone-fruit spirits and whiskies, tend to have much higher concentrations of ethyl carbamate. Heating (e.g., cooking) the beverage increases the ethyl carbamate content. It also has a tendency to accumulate in the human body from a number of daily dietary sources, for example, alcohols, bread, fermented grain products, soy sauce, orange juice, and commonly consumed foods such as *kimchi*, soybean paste, buns, crackers, and bean curd, along with wine, sake, and plum wine (www.Medlibrary.com). Determination of ethyl carbamate in alcoholic beverages and soy sauce by gas chromatography shows that distilled spirits contained 50–330 ng EC/g (ppb), fortified wine 40–160 ppb, table wine 10–50 ppb, and soy sauce 15–70 ppb (Weber and Sharypov, 2009). It has been found to be a possible carcinogen in humans. According to IARC, it has been found to be a mutagenic in several assays. As per the group of experts from WHO/FAO (JECFA), this compound is of concern for consumers of alcoholic beverages, and measures to reduce its levels have to be continued. Limits have been set in regulations (e.g., for Switzerland, 1 mg/L in a specific category of alcoholic drink). Nout (1994) studied the occurrence of ethyl carbamate in *kenkey* and found its concentration to be insignificant, most likely due to inadequate levels of ethanol and/or precursors such as citrullen, arginine, or urea in the product.

5.5.3.4 3-MCPD (3-monochloropropane-1,2-diol)   3-MCPD is a carcinogen contaminant resulting from the reaction of chloride with lipids in food. 3-MCPD is not produced naturally during the fermentation of fish sauces, but is a well-known contaminant of acid-hydrolyzed vegetable protein (acid-HVP) (Afiza et al., 2007; JFSSG, 1999). Contamination of the sauces with 3-MCPD occurs if acid-HVP is added as a savory ingredient to the sauces, or hydrochloric acid is used for fish/shrimp hydrolysis in the production of sauces (Afiza et al., 2007; Wong et al., 2006). There has been no scientific report on the occurrence of 3-MCPD in fish sauces, but a consumer association reported that an imported fish sauce available in the Malaysian market contained as much as 213 ppb 3-MCPD (Idris, 2007), more than 20 times the permitted level of 10 ppb (FAC, 1999). Out of 421 samples of soy and oyster sauces analysed for 3-MCPD level, 45 samples (10.7%) had 3-MCPD levels of over 20 ppb (Wong et al., 2006). It has been found that *budu* is free from 3-MCPD, but the histamine content is slightly high (50 mg/100 g) due to the high salt content of the unprocessed *budu*, as well as unhygienic processing procedure that encourages microbial enzyme reaction on the histidine in fish mixtures (Rosma et al., 2009). Thus, 3-MCPD levels should be determined for all protein-rich sauces, namely soy, oyster, fish sauces, and shrimp pastes.

Fish sauce is made by a fermentation process, but its usage seems limited due to health and safety issues, as most of it, including *budu*, is produced under unhygienic conditions with extremely high salt content. Thus, it has the potential to contain certain amount of histamines or 3-MCPD, as the spoilage causing organisms of fermented fish products can decarboxylase the histidine of fish mixture via microbial reaction as described earlier. These chemical contaminants will pose serious health problems when consumed in large amounts.

5.5.4 Toxins of Biological Origin

5.5.4.1 Biogenic Amines   Biogenic amines of toxicological significance include histamine, putrescine, cadaverine, tyramine, 2-phenylethylamine, and tryptamine (Stratton et al., 1991) (Table 5.22). They are formed by microbial decarboxylases, the enzymes elaborated by the LAB and *Enterobacteriaceae* (Westby et al., 1997). Histamine is the most widely studied biogenic amine which is heat stable and survives normal cooking temperatures. Investigation of biogenic amines in *kenkey*, a lactic fermented maize product from Ghana, showed that the total amine levels were very low (60 ppm), but were increased tenfold on the addition of cowpeas (Nout, 1994). Both red and white cowpeas were added to obtain products with increased protein content. The increase was attributed to the addition of precursor amino acids in cowpeas and the possible presence of polyphenols in the seed coats, which could have inhibited decarboxylases. Prolonged cooking of *kenkey*, however, contributed only marginally to lowering the amine levels (Nout, 1994).

Table 5.22   List and Structure of Some Important Biogenic Amines Formed during Fermentation

COMPOUND NAME	STRUTURE
Histamine	
Putrescine	
Cadaverine	
Tyramine	
2-Phenylethylamine	
Tryptamine	

Source: Adapted from Shukla, S., Kim, J.K., and Kim, M. 2011. *Soybean and Health.* InTech. ISBN: 978-307-535-8.

Scromboid toxins or histamines intoxication is the consequence of bacterial action on scromboid fish such as Tuna or mackerel. The onset of symptoms takes from minutes to hours, and includes nausea, vomiting, diarrhea, cramps, flushing, headache, and burning mouth. Histamine has also been found in wines which can cause serious pathological symptoms if its content increases to more than 10 mg/L. A *Pediococcus* strain has been found to be capable of this reaction. There is little information on histamine production in indigenous fermented foods, but the average concentration of histamines was found to be 3.4 mg/L in red wines and 1.1 mg/L in white wines.

5.5.4.2 N-Nitrosamines Nitrosamines (Table 5.22) are the potential carcinogenic compounds mostly found in beers. The concentration of nitrosamines in beers ranged from 0 to 1.6 ppb, while in India, high concentrations of 3.6 ppb are observed. However, there is no information on indigenous fermented foods for nitrosamines.

5.5.4.3 Mycotoxins and Other Toxins

5.5.4.3.1 Mycotoxins Microbial toxins include components of a microbial cell or product of its metabolism which, upon contact or ingestion by a living organism, lead to toxic manifestation. These compounds can be produced by a broad group of microorganisms, but bacteria and fungal toxins are of major concern in the safety of foods (Table 5.23).

Mycotoxins are the most commonly occurring toxins of microbial origin in cereals and legumes used as raw materials for fermented foods (Westby et al., 1997). Chemically, aflatoxins are highly fluorescent and oxygenated heterocyclic compounds, characterized by a caumarin nucleus fused to a bifuran moiety, and either a cyclic pentenone ring or 6-membered lactone. These are secondary metabolites that are formed during the growth of certain fungi and are some of the most potent toxic compounds known to man (Bol and Smith, 1989; Coker, 1995; Zaika and Buchanan, 1987). Along with four major aflatoxins, several of its metabolites (M1, M2, M4, B2a, G^, etc.) have been identified and characterized. The risk of contamination by mycotoxins of raw materials used in fermented foods is a serious food safety hazard. The production of fermented foods usually involves various processing steps, such as cleaning, soaking, milling, dehusking, and cooking which can contribute to the reduction

Table 5.23   Some Important Mycotoxins Producing Fungi and Their Toxicity

MYCOTOXIN	PRODUCER FUNGUS	TOXICITY
Aflatoxin	*A. flavus, A. parasiticus*	Carcinogenic, Mutagenic, Teratogenic
Ochratoxin	*A. ochraceous*	Nephrotoxic
Citrinin	*P. viridicatum, P. citrinum*	Nephrotoxin
Patulin	*P. urticae, P. patulum*	Sarcoma
Ergot alkaloids	*Claviceps* sp.	Ergotism

Source: Adapted from Sharma, A. 1999. *Biotechnology: Food Fermentation.* Educational Publishers and Distributors, New Delhi, pp. 321–344.

in the contamination of the final products and should not be ignored when considering the efficiency of a whole processing procedure. Influence of grain quality, heat and processing time on the reduction of aflatoxin in fermented food has been described earlier (Adegoke et al., 1974) while ability of various fungi to degrade mycotoxin has been documented (Kanittha, 1990). Aflatoxine B1 (AFB_1) present in food has been found to be excreted as aflatoxin M (AFM) in milk, notwithstanding processing steps like chilling, separation, pasteurization, and boiling. The toxins get concentrated in concentrated/dried dairy products, like *khoa*, *chhana*, condensed milks, dried milks, infant formula, etc. Behavior of aflatoxin during manufacture and storage of yoghurt has been documented (Blanco et al., 1993; Govasis et al., 2002; Wiseman and Marth, 1983) while the changes in Aflatoxin M1 during manufacture and storage of pickled cheese have also been determined (Deveci, 2007), and bread (El Banna and Scott, 1983). Molecular approaches to reduce aflatoxin have also been made (Bhatnagar et al., 1995). Reduction of mycotoxins using fermentation is discussed in detail in Chapter 13 of this text.

5.5.4.3.2 Ochratoxin Ochratoxin A is produced by the fungi *Aspergillus ochraceous* and *Penicillium verucosum* (Campbell-Platt, 1987; Madhyastha et al., 1990; Milanez and Leitão, 1996). Commodities contaminated with the molds are wheat, barley, oats, corn, dry beans, peanuts, cheese, and coffee. Pathological changes may involve tubular necrosis and carcinoma of the kidneys and mild liver damage, affecting swine, duckling, and chickens as well as humans.

5.5.4.3.3 Patulin Patulin is a mycotoxin produced by a number of organisms involved in food spoilage, such as *Penicillium expansum* (Joshi et al., 2013). It is toxic to higher animals, carcinogenic when administered subcutaneously to rats, and teratogenic to chicken embryos. It can however, be degraded during alcoholic fermentation of apple juice (Stinson et al., 1978) (see Chapter 13 of this text for more details).

5.5.5 Food Safety Issues and Concerns

A large number of microorganisms are involved in the production of fermented products, therefore, the safety of fermented foods resulting from the probable production of toxic factors by microorganisms is a matter of public health concern (Padmaja and George, 1999). Thus, the safety of such products assumes great importance due to the possible presence of pathogenic bacteria, viruses, parasites, or even mold spores on the ingredients, equipment, or the hands of processing personnel (Steinkraus, 1983; 1997; Wani and Sharma, 1999). The contamination of *Aspergillus flavus* in traditional starter cultures for rice wine and soybean sauce has caused severe food poisoning incidences (Lee and Lee, 2002). Fermented foods generally have a very good safety record, even in the developing world where the foods are manufactured by people without any formal training in microbiology or chemistry. These foods are often made in unhygienic and contaminated environments (Steinkraus, 1997), but most of the traditional fermentation methods have their own inbuilt safeguard mechanisms. Food poisoning microorganisms contaminating the vegetables and other raw materials such, as *kimchi* have been found killed within a week of beginning of fermentation, mainly due to the production of acids and bacteriocins (Lee, 1997). A large amount of nitrate and secondary amines in vegetable products are also reduced by fermentation (Lee, 1990). In the Japanese food industry, the aflatoxin-forming ability of 238 strains of *Aspergillus* was studied and it was observed that 52 strains produced florescent compounds, but none produced aflatoxins (Matsuura et al., 1970). Industrial *koji* molds from Japan were also analysed, but did not show the presence of any aflatoxins (Murakami et al., 1968).

Alkaline fermented foods are generally recognized as safe (Steinkraus, 1997). Indeed, to produce alkaline fermented foods, the raw material undergoes a long cooking period prior to fermentation, which ensures almost complete elimination of nonspore-forming pathogenic bacteria. However, there is a risk of recontamination during subsequent handling, and spore forming pathogenic bacteria like *B. cereus* can easily survive the cooking process and resume growth during fermentation (Ouoba et al., 2008b; Parkouda et al., 2010; Thorsen et al., 2010). Nevertheless, fermented foods and condiments are often cooked during the preparation of various dishes where they are used, giving extra safety margins to the products, although there are some exceptions; for example, the emetic toxin cereulide produced by some strains of *B. cereus* is heat stable, tolerates extreme pH (2–11) values, and is not degraded by digestive enzymes like pepsin and trypsin (Shinagawa et al., 1995, 1996). Consequently, there is a risk of food poisoning associated with dishes prepared using alkaline fermented foods or condiments, even when cooked. Thus, steps should be taken to avoid or limit the growth of *B. cereus*.

At the same time, various *Bacillus* spp. are known to produce a wide arsenal of antimicrobial substances, including bacteriocins, polyketide, peptide, and lipopeptide antibiotics, some of which are active against *B. cereus* (Abriouel et al., 2011; Kabore et al., 2012; Zimmerman et al., 1986). During *kinema* production, the capacity of *B. subtilis* to reduce the growth of *B. cereus* has been shown to prevent the production of enterotoxin. Further, *B. subtilis* strains isolated from *Maari* produce bacteriocins which are active against Gram-positive pathogens such as *L. monocytogenes* and *B. cereus*. Nout et al. (1998), Kabore et al. (2012), Guo et al. (2006) and Ouoba et al. (2007b) have also found that *B. subtilis* and *B. pumilus* isolated from *Soumbala* have antimicrobial properties against both Gram-positive and -negative bacteria (Wacher et al., 2010), including *Micrococcus luteus*, *Staphylococcus aureus*, *Bacillus cereus*, *Enterococcus facium*, *Listeria monocytogenes*, *Escherichia coli*, *Salmonella typhimurium*, *Shigella dysenteriae*, and *Yersinis enterocolitica*, as well as ochratoxigenic molds such as *Aspergillus ochraceus*. During fermentation, elimination or reduction of antinutritional factors is also observed, making the food more edible (Odunfa, 1988; Ouoba et al., 2007a; Sarkar et al., 1997a; Steinkraus, 1996). Similarly, LAB isolated from fermented vegetables produces bacteriocin and exihibits antimicrobial activity against pathogenic bacteria like *E.coli* and *Bacillus* (Joshi et al., 2006), which could also be a factor contributing to the better shelf life and safety of indigenous fermented foods.

5.6 Improving the Safety of Indigenous Fermented Foods

The safety of fermented foods can be improved by various means, such as using quality raw materials, using unique, detoxifying starter cultures, maintaining proper hygienic standards in the processing environment, using proper packaging (Ijabadeniyi, 2004; Ijabadeniyi and Omoya, 2006), and, finally, implementing behavior-based food safety management (Yiannas, 2009).

5.6.1 Using Quality Raw Materials

By choosing suitable raw materials for the production of indigenous fermented foods, the quality of such products can be improved considerably as raw materials of poor quality cannot give rise to acceptable finished products. If the raw material has already started fermenting, it will not make a quality product, even when inoculated with a desirable microorganism. Other ingredients, such as water, and more importantly the equipment, must be clean. Water quality is essential to make quality processed products, and food processing requires a good supply of clean, potable water for cleaning equipment, cooling filled containers, and, sometimes, as an ingredient (Fellows et al., 1995). Such water should be free from coliform bacteria. Suitable methods need to be employed to purify the water to be used for the production of indigenous fermented food products. Chlorination is an effective method of sanitizing water, and is recommended for the food industry.

5.6.2 Using Starter Cultures

Microorganisms responsible for effecting desirable changes in the food during fermentation can be selected and subjected to genetic improvement geared toward maximizing desirable quality attributes in the food and limiting any undesirable attributes. Using appropriate starter cultures is advantageous due to the competitive role of microorganisms and their metabolites in preventing the growth and metabolism of unwanted microorganisms (Holzapfel, 1995; Ijabadeniyi and Omoya, 2006). According to Nout (1994), the selection of starter cultures may be achieved by both conventional selection and mutation, or by recombinant-DNA techniques, to obtain increased levels of safety.

5.6.3 Maintaining Proper Hygiene Standards and Packaging

During the preparation of fermented foods, the maintenance of proper hygiene standards in the food processing environment should be a top priority. The raw materials for use must be fresh and without any microbial contamination. For example, moldy cereals cannot be used for the preparation of fermented cereal product (Nout, 1994). It is equally essential that the workers in a food processing factory must remain clean, and wash their hands with detergent before commencing the preparation of indigenous fermented foods. Not only this, the workers must also be healthy, without any communicable disease. The equipment to be used for the preparation of such foods must also be thoroughly washed and cleaned. After the preparation of fermented foods, it is necessary for them to be properly packaged for marketing and consumption (Ijabadeniyi and Omoya, 2006).

5.6.4 Application of Hazard Analysis Critical Control Point (HACCP)

Generally in the food industry, approaches based on good manufacturing practice (GMP) are being largely replaced by application of the Hazard Analysis Critical Control Point (HACCP) concept. It improves on traditional practices by introducing a more systematic, rule-based approach for applying our knowledge of food microbiology to quality. The same concept can also be adopted with physical and chemical factors affecting food safety (Adams and Moss, 1996), and since small-scale fermented foods are an integral part of the global food supply, it will not be out of place for the HACCP concept to be used in the quality assurance and improvement of safety for small-scale fermented foods. The implementation of the HACCP system for the preparation of indigenous fermented foods should be implemented striclty in the different countries of South Asia, which will ensure safe and hygienically prepared foods (Quevedo, 1997; WHO, 1993).

5.6.5 Implementation of a Behavior-Based Food Safety Management

Present food safety management systems may not be sufficient to ensure the safety of indigenous fermented foods. Achieving food safety success in this changing environment involves going beyond traditional training, testing, and inspectional approaches to managing risks (Greenstreet Berman et al., 2012; Yiannas, 2009). A better understanding of organizational culture and the human dimensions of food safety is needed (Greenstreet Berman et al., 2012). Essentially, food processors should know and adhere to behaviors that will keep fermented foods safe (Moy et al., 2010). Five key food behaviors that will help ensure safety of foods are: (1) keep food clean; (2) cook thoroughly; (3) separate raw and cooked food; (4) keep food at safe temperatures, either hot or cold; and (5) choose foods for safety.

5.7 Summary and Future Prospects

The traditional methods of preparing indigenous foods are no doubt simple, low cost, and use locally available raw material, but the techniques used are labor intensive, time consuming, and have low productivity, and success depends on employing good manufacturing practices. It is also apparent that some of the products are not of a good standard, or have doubtful safety. For example, alcoholic beverages which, as naturally fermented alcoholic beverages, have variable alcohol content along with far greater volatile acids than prescribed, indicate the lack of standardized technology in their production. The presence of coliform bacteria in the Naturally Fermented Beverages (NFB) beverages also indicate the unhygienic conditions in which such beverages are prepared. Such beverages have some other undesirable quality characteristics, including high levels of methanol. Similar instances for other indigenous fermented foods can be cited. In addition, because of the primitive technology used in the production of fermented foods, the role of water quality and the pesticides/herbicides used and their high residues in the foods cannot be ignored. Therefore, the simple technology used in the production of the indigenous foods fails to stand up in the face of such challenges. There are no checks in respect of the quality of the raw materials, of the control of the fermentation process, or, finally, of the quality of the finished products.

Developments in science and technology, especially microbiology, biochemistry, and genetic engineering, show that considerable improvements can be made in the quality of indigenous products with respect to both shelf-life, safety and marketability. For example, to improve the quality of such beverages, the use of pure cultures of wine yeast rather than the mixed bacterial and yeast cultures of natural fermentation need to be encouraged. In addition, the use of preservative like potassium metabisulphite, of controlled distillation practices, and the manipulation of the conditions of alcoholic fermentation or other food fermentation to ensure adequate quality of the final fermented food should be advocated. At the same time, there is much to be learnt about the role of individual microorganisms and standardization of their levels by artificial inoculation into the ingredients, along with the optimization of various physico-chemical factors, to improve the quality of the indigenous fermented foods. The application of simple methods, such as moisture content in solid state fermentation, can select the desired microorganism to improve the quality of the food products.

Fermentations are brought about by the microflora that come from the raw materials and the environment. There is also the possibility of the development of various undesirable microorganisms in natural fermentation, and the production of toxic substances, such as mycotoxins or patulin, by certain species, which is of serious concern. Research should be directed toward identifying and applying novel food processing techniques (e.g., irradiation and high pressure) to improve the quality and safety of indigenous fermented foods.

Fermentation is known to play a significant role in improving food safety with respect to toxic compounds in raw materials, so it can be used with potential benefits. Not only this, an understanding of the dynamics of food fermentation and the mechanisms by which they can be controlled, can be applied to reduce toxins through the promotion of specific groups of microorganisms, or even by the production of antimicrobial substances by groups of microorganisms, which can help in the safety of food fermentation. Future research and development work should focus on several areas, such as understanding the mechanisms by which toxic substances are reduced during fermentation to maximize their potential application. In brief, the strategy should be to produce and popularize indigenous fermented foods by the application of science and technology to increase product quality, safety, and nutrition.

References

Abriouel, H., Benomar, N., Lucas, R., and Gálvez, A. 2011. Culture-independent study of the diversity of microbial populations in brines during fermentation of naturally-fermented Aloreña green table olives. *Int J Food Microbiol* 144: 487–496.

Adams, M.R. and Moss, M.O. 1996. Fermented and microbial foods. In: *Food Microbiology*. Adams, M.R. and Moss, M.O. (Eds.), 1st edn. New Age International Publishers, New Delhi, pp. 252–302.

Adegoke, G.O., Otumu, E.J., and Akanni, A.O. 1994. Influence of grain quality, heat, and processing time on the reduction of aflatoxin B1 levels in "tuwo" and "ogi": Two cereal-based products. *Harz Foods Hwnan Nutr* 45: 113–117.

Afiza T.S., Rosma A., Faradila B., Wan Nadiah, W.A., and Ibrahim, C.O. 2007. Quality index of kalantan unprocessed "budu." In: *Proceedings of the 9th Symposium of the Malaysian Society of Applied Biology, Exploring the Source of life as a catalyst for Technological Advancement*, pp. 344–347. Penang: University Sains Malaysia.

Aidoo, K.E., Nout, M.J.R., Sarkar, P.K., and Arnold, J.P. 2005. *Occurrence and Function of Yeasts in Asian Origin and History of Beer and Brewing: From Prehistoric Times to the Beginning of Brewing Science and Technology*. Reprint Edition by Beer Books, Cleveland, OH. ISBN 0-9662084-1-2.

Amerine, M.A., Berg, H.W., Kunkee, R.E., Ough, E.S., Singleton, V.L., and Webb, A.D. 1980. *The Technology of Wine Making*. 4th edn. AVI Publishing Co., Inc., Westport, CT.

Aneja, R.P. 2002. *Technology of Indian Milk Products*. Dairy India Publication, New Delhi.

Anon. http://www.bestmedicalhealth.com/c/yersiniosis

Anon. http://thedailyomnivore.net/category/food/

Anon. http://www.crcpress.com/

Arogba, S.S., Ademola, A., and Elum, M. 1995. The effect of solvent treatment on the chemical composition and organoleptic acceptability of traditional condiments from Nigeria plant foods. *Hum Nutr* 48: 31–38.

Atacador-Ramos, M. 1996. Indigenous fermented foods in which ethanol is a major product. In: *Handbook of Indigenous Fermented Foods*. Steinkraus, K.H. (Ed.), Marcel Dekker, New York, NY, pp. 363–508.

Atputharajah, J.D., Widanapathirana, S., and Samarajeewa, U. 1986. Microbiology and biochemistry of natural fermentation of coconut palm sap. *Food Microbiol* 3: 273–280.

Batra, L.R. 1986. Microbiology of fermented cereals and grains legumes of India and vicinity. In: *Indigenous Fermented Food of Non-Western Origin*. Hesseltine, C.W. and Wang, H.L. (Eds.), J. Cramer, Berlin, pp. 85–104.

Batra, L.R. and Millner P.D. 1976. Asian fermented foods and beverages. *Dev Ind Microbiol* 17: 117–128.

Bazirake, G.H. 2008. *The Effect of Enzymatic Processing on Banana Juice and Wine*. Stellenbosch University Institute for Wine Biotechnology, South Africa.

Bennett, L.A., Campillo, C., Chandrashekar, C.R., and Gureje, O. 1998. Alcoholic beverage consumption in India, Mexico, and Nigeria: A cross-cultural comparison. *Alcohol Health Res World* 22: 2r43–252.

Bhatia, A.K., Singh, R.P., and Atal, C.K. 1977. *Chhang*—Fermented beverage of Himalayan folk. *Ind Food Packer* 1(4): 34–38.

Bhatnagar, D., Cleveland, T., Linz, J., and Payne, G. 1995. Molecular biology to eliminate aflatoxins, *INFORM* 6:262.

Blanco, J.L., Carrion, B.A., Liria, N., Diaz, S., Garcia, M.E., Domineuez, L., and Suarez, C.l. 1993. Behaviour of aflatoxins during manufacture and storage of yoghurt. *Milchtvissenschafi* 48: 385–387.

Bokanga, M., Essers, A.J.A., Rosling, H., Poulter, N.H., and Tewe, O. 1994. Summary and recommendations, International workshop on cassava safety. *Acta Hortic* 375: 11–19.

Bol, J. and Smith, J.E. 1989. Biotransformation of aflatoxin. *Food Biotechnol* 3: 127–144.

Borate, A.J., Gubbawars, G., Shelke, R.R., and Chavan, S.D. 2011. Studies on keeping quality of *Shrikhand* prepared from buffalo milk blended with soymilk. *Food Sci Res J* 2(2): 205–210.

Borse, B.B., Rao, L.J.M., Ramalakshmia, K., and Raghavana, B. 2007. Chemical composition of volatiles from coconut sap (*neera*) and effect of processing. *Food Chem* 101(3): 877–880.

Campbell-Platt, G. 1987. Fermented foods of the world—A dictionary and guide. *Butterworths Lond* 294: 134–142.

Chettri, R. and Tamang, J.P. 2008. Microbiological evaluation of *maseura*, an ethnic fermented legume-based condiment of Sikkim. *J Hill Res* 21: 1–7.

Christensen, H.E. 1973. *The Toxic Substances*. List Natl. Inst for Occupational Safety and Health. Washington, DC, 534 pp.

Coker, R.D. 1995. Controlling mycotoxin in oilseeds and oilseed cakes. *Chem Ind.* April, 260–264.

Cooke, R.D. 1978. An enzymatic assay for the total cyanide content of cassava (*Manihot esculenta* Crantz). *J Sci Food Agric* 29: 345–352.

De, A. and Patel, R.S. 1991. Technology of *shrikhand* powder. *Cult Dairy Prod J* 25(2): 21.

De Bruijn, G.H. 1973. The cyanogenic character of cassava (*Manihot esculenta*). In: *Chronic Cassava Toxicity*. Nestel, B.L. and MacIntyre, R. (Eds.), International Development Research Centre, Ottawa, Canada, pp. 43–48.

Desai, B.B. and Salunkhe, D.K. 1986. Dhokla and Khaman. In: *Legume Based Fermented Foods*. Reddy, N.R., Pierson, M.D., and Salunkhe, D.K. (Eds.), CRC Press, Boca Raton, FL pp. 161–171.

Desai, H.K., Vyas, S.H., and Upadhyay, K.G. 1985. Influence of homogenization of milk on the quality of Chakka and *Shrikhand*. *Indian J Dairy Sci* 38(2): 102–106.

Deveci, O. 2007. Changes in the concentration of aflatoxin M1 during manufacture and storage of White Pickled cheese. *Food Control* 18(9): 1103–1107.

Dewan, S. 2002. Microbiological evaluation of indigenous fermented milk products of the Sikkim Himalayas. PhD thesis, Food Microbiology Laboratory, Sikkim Government College (under North Bengal University), Gangtok, India.

Dewan, S. and Tamang, J.P. 2007. Dominant lactic acid bacteria and their technological properties isolated from the Himalayan ethnic fermented milk products. *Int J Gen Mol Microbiol* 92: 343–352.

Dworkin, M., Falkow, E., Rosenberg, K.H., and Schliefer, E. 2006. *Organic Acid and Solvent Production*. Springer, New York, NY, pp. 511–756.

El Banna, A.A. and Scott, P.M. 1983. Fate of mycotoxins during processing of foodstuffs. I. Aflatoxin Bl during making of Egyptian bread. *Journal of Food Protection* 46(4): 301–304.

Endo, A. and Okada, S. 2008. Reclassification of the genus *Leuconostoc* and proposals of *Fructobacillus fructosus* gen. nov., comb. nov., *Fructobacillus durionis* comb. nov., *Fructobacillus ficulneus* comb. nov. and *Fructobacillus pseudoficulneus* comb. nov. *Int J Syst Evol Microbiol* 58(9): 2195–2205.

Fellows, P., Axtell, B., and Dillon, M. 1995. *Quality Assurance for Small-Scale Rural Food Industries*. FAO Agricultural Services Bulletin, Rome, Italy, 120 pp.

Fleet, G.H. 1990. Growth of yeasts during wine fermentations. *J Wine Res* 1(3): 133–142.

Food Advisory Committee (FAC). 1999. Recent developments on 3-MCPD in food and food ingredients. FAC Press Release 5/99.

Fowles, G. 1989. The complete home wine maker. *New Scientist*, September, 38.

Frazier, W.C. and Westhoff, D.C. 1978. *Food Microbiology*. 3rd edition. Tata McGraw Hill Pub. Co., New Delhi, pp. 17–34.

Gandhi, N.K. and Jain, S.C. 1977. A study on the development of a new high protein formulated food using buffalo milk. *J FD Sci Technol* 14(4): 156.

Gill, T.A. 1990. Objective analysis of seafood quality. *Food Rev Int* 6: 681–714.

Giri, S.S. and Janmey, L.S. 1987. Microbial and chemical contents of the fermented bamboo shoots in soibum. *Front Bot* 1: 89–100.

Gould, W.A. 1977. *Food Quality Assurance*. AVI Publishing Co., Westport, CT.

Govaris, A., Roussi, V., Koidis, A., and Botsoglou, N. 2002. Distribution and stability of aflatoxin M1 during production and storage of yoghurt. *Food Addit Contam* 19(11): 1043–1050.

Gram, L. and Huss, H.H. 1996. Microbiology spoilage of fish and fish product. *Int J Food Microbiol* 33: 121–137.

Greenstreet Berman (Firm), Wright, M., Leach, P., Palmer, G., Great Britain. Food Standards Agency 2012. *A Tool to Diagnose Culture in Food Business Operators*. Greenstreet Berman Ltd., London.

Guo, X.H., Li, D.F., Lu, W.Q., Piao, X.S., and Chen, X.L. 2006. Screening of *Bacillus* strains as potential probiotics and subsequent confirmation of the *in vivo* effectiveness of *Bacillus subtilis* MA139 in pigs. *Anton Leeuw* 90: 139–146.

Haddon, W.F., Mancini, M.I., Mclaren, M., Effio, A., Harden, L.A, Egre, R.I., and Bradford, J.L. 1994. Occurrence of ethyl carbamate (urethane) in U.S. and Canadian breads: measurements by gas chromatography–mass spectrometry. *Cereal Chem* 71(2): 207–215.

Hasegawa, Y., Nakamura, Y., Tonogai, Y., Terasawa, S., Ito, Y., and Uchiyama, M. 1990. Determination of ethyl carbamate in various fermented foods by selected ion monitoring. *J Food Prot* 53: 1058–1061.

Holzapfel, W. 1995. Use of starter cultures in fermentation on a household scale. *Background Paper Prepared for WHO/FAO Workshop: 'Assessment of fermentation'*, Pretoria, South Africa.

http://www.codexalimentarius.net/download/report/433/al83_19e.pdf-FAO REPORT

http://www.foodtechnology.co.nz/packaging

http://www.medallion lab.com

http://www.niscair.res.in/ScienceCommunication/ResearchJournals/rejour/ijtk/ijtk2k4/ijtk_jul 04.asp

Hurtado, A., Reguant, C., Bordons, A., and Rozès, N. 2012. Lactic acid bacteria from fermented table olives. *Food Microbiol* 31(1): 1–8.

Idris, M. 2007. Test all sauces for cancer-causing agent. *New Straits Times*. 29-1-2007.

Ijabadeniyi, A.O. 2004. *Effect of methods of processing on the microbiological and nutrient composition of "ogi" produced from three maize varieties*. M.Tech thesis.

Ijabadeniyi, A.O. 2007. Microbiological safety of gari, lafun and ogiri in Akure metropolis, Nigeria. *Afr J Biotechnol* 6(22): 2633–2635.

Ijabadeniyi, A.O. and Omoya, F.O. 2006. Safety of small-scale food fermentations in developing countries. In: *13th IUFOST World Congress of Food Science and Technology. Food is Life*. Nantes, France, September 17–21, pp. 1833–1845.

Jespersen, L., Halm, M., Kpodo, K., and Jakobsen, M. 1994. Significance of yeasts and moulds occurring in maize dough fermentation for 'kenkey' production. Int *J Food Microbiol* 24: 239–248.

Jeyaram, K., Mohendro Singh, Capece, A., and Romano, P. 2008a. Molecular identification of yeast species associated with *"Hamei"*—A traditional starter used for rice wine production in Manipur, India. *Int J Food Microbiol* 124: 115–125.

Jeyaram, K., Singh, W.M., Premarani T., Devi R.A., Chanu K.S., Talukdar N.C., and Singh M.R. 2008b. Molecular identification of dominant microflora associated with hawaijar: A traditional fermented soybean (*Glycine max* {L}) food of Manipur, India. *Int J Food Microbiol* 122: 259–268.

Jeyaram, K., Singh T.A., Romi, W., Devi, A.R., Singh, W.M., Dayanidhi, H., Singh, N.R., and Tamang, J.P. 2009. Traditional fermented foods of Manipur. *Indian J Trad know* 8(1): 115–121.

JFSSG (Joint MAFF/DH Food Safety and Standards Group) 1999. Survey of 3-monochloropropane-1,2-diol (3-MCPD) in acid hydrolyzed vegetable protein. *Food Surveillance Information Sheet No.* 181.

Joint FAO/WHO Food Standards Programme. 1987. www.codexalimentarius.net

Joshi, V.K. 1997. *Fruit Wines*. 2nd edn. Directorate of Extension Education, Dr. Y.S. Parmar University of Horticulture and Forestry, Nauni, Solan, HP, 255 p.

Joshi, V.K. 2006a. *Sensory Science: Principles and Applications in Evaluation of Food*. Agro-Tech Publishers, Udaipur, 527 p.

Joshi, V.K. 2006b. Food fermentation: Role, significance and emerging trends. In: *Microbiology: Applications and Current Trends*. Trivedi, P.C (Ed.), Pointer Publisher, Jaipur, pp. 1–35.

Joshi, V.K., Attri, D., Singh, T.K., and Absol, G.S. 2011. Fruit wines: Production technology. *Handbook of Enology: Principles, Practices and Recent Innovations*. Vol. III. Joshi, V.K. (Ed.), Asia-Tech Publisher and Distributors, New Delhi, pp. 1177–122.

Joshi, V.K. and Bhushan, S. 2000. Sensory evaluation of fruits, vegetables and their products. In: *Postharvest Technology of Fruits and Vegetables*. Verma, L.R. and Joshi, V.K. (Eds.), The Indus Publ., New Delhi, pp. 286–336.

Joshi, V.K., Bhutani, V.P., and Sharma, R.C. 1990. Effect of dilution and addition of nitrogen source on chemical, mineral and sensory qualities of wild apricot wine. *Am J Enol Vitic* 41(3): 229–231.

Joshi, V.K., Kaushal, N.K., and Nevedita, S. 2000. Spoilage of fruits vegetables and their products. In: *Postharvest Technology of Fruits and Vegetables*. Verma, L.R. and Joshi, V.K. (Eds.), The Indus Publ., New Delhi, pp. 235–285.

Joshi, V.K., Kumar, A., and Thakur, N.S. 2015. Technology and status of *Sur* production in Himachal Pradesh: A critical appraisal. *Indian J Trad Know* (in press).

Joshi, V.K., Lakhanpal, P., and Kumar, V. 2013. Occurrence of Patulin its dietary intake through consumption of apple and apple products and methods of its removal. *Int J Food Ferment Technol* 3(1): 15–32.

Joshi, V.K. and Pandey, A. 1999. *Biotechnology: Food Fermentation Microbiology, Biochemistry and Technology*. Vols. I and II. Educational Publishers and Distributors, Kerala, India.

Joshi, V.K. and Sandhu, D.K. 2000. Quality evaluation of naturally fermented alcoholic beverages, microbiological examination of source of fermentation and ethanolic productivity of the isolates. *Acta Aliment* 29(4): 323–334.

Joshi, V.K., Sandhu, D.K., and Thakur, N.S. 1999. Fruit based alcoholic beverages. In: *Biotechnology: Food Fermentation*. Vol. II. Joshi, V.K. and Pandey, A. (Eds.), Educational Publishers and Distributors, New Delhi, pp. 647–744.

Joshi, V.K. and Sharma, S. 2009. Cider vinegar: microbiology, technology and quality. In: *Vinegars of the World*. Solieri, L. and Giudiet, P. (Eds.), Springer-Verlag, Italy, pp. 197–207.

Joshi, V.K., Sharma, S., Bhushan, S., and Attri, D. 2004. Fruit based alcoholic beverages. In: *Concise Encyclopedia of Bioresource Technology*. Pandey (Ed.), Haworth Inc., New York, NY.

Joshi, V.K., Sharma, S., and Rana, N. 2006. Production, purification, stability and efficiency of bacteriocin from the isolate of natural lactic acid fermentation of vegetables. *Food Technol Biotechnol* 44(3): 435–439.

Joshi, V.K. and Singh, R.S. 2012. *Food Biotechnology*. IK Publisher, New Delhi, pp. 952.

Joshi, V.K. and Thakur, N.S. 2000. Vinegar: Composition and production. In: *Postharvest Technology of Fruits and Vegetables*. Verma, L.R. and Joshi, V.K. (Eds.), The Indus Publ., New Delhi, pp. 1128–1170.

Kabore, D., Thorsen, L., Nielsen, D.S., Berner, T.S., Sawadogo-Lingani, H., Diawara, B., Dicko, M.H., and Jakobsen, M. 2012. Bacteriocin formation by dominant aerobic spore-formers isolated from traditional *maari*. *Int J Food Microbiol* 154: 10–18.

Kanittha, S. 1990. Kan khat luak chuara ti salai san allatoxin [Degradation of aflatoxin by selected molds]. Report, Kasetsart University, Bangkok, Thailand.

Kanwar, S.S., Gupta, M.K., Katoch, C., Kumar, R., and Kanwar, P. 2007. Traditional foods of Lahaul and Spiti area of Himachal Pradesh. *Indian J Trad Know* 6: 42–45.

Karacam, H., Kutlu, S., and Kose, S. 2002. Effect of salt concentrations and temperature on the quality and shelf-life of brined anchovies. *Int J Food Sci Technol* 37: 19–28.

Karki, T., Kofi, H., Hayashi, K., and Kozaki, M. 1983b. Chemical changes occurring during *Gundruk* fermentation. Part II—1 amino acids. *Lebensm-Wiss Tecrtol* 16(3): 180–183.

Karki, T., Okada, S., Baba, T., Itoh, H., and Kozaki, M. 1983a. Studies on the microflora of Nepalese pickles *gundruk*. *Nippon Shokuin Kogyo Gakkaishi* 30: 357–367.

Karki, T.B., Itoh, H., Nikkuni, P., and Kozaki, M. 1986. Improvement of gundruk processing by selected lactic strains. *Nippon Skokuhin Kogyo Gakhaishi*, 33(10), 734–739.

Kerr, M., Lawichi, P., Aguirre, S., and Rayner, C. 2002. *Effect of Storage Conditions on Histamine Formation in Fresh and Canned Tuna*. State Chemistry Laboratory-Food Safety Unit, Department of Human Service, pp. 5–20

Law, S., Bakar, A., Hashim, F., and Hamid, A. 2011. Popular fermented foods and beverages in Southeast Asia. *Int Food Res J* 18: 475–484.

Lazos, E., Aggelousis, G., and Bratakos, M. 1993. The fermentation of *trahanas*: A milk-wheat flour combination. *Plant Foods Hum Nutr* 44(1): 45–62.

Lee, C.H. 1990. Cereal fermentation in African Countries. In: *Fermented Cereals: A Global Perspective*. Haard, N.F., Odunfa, S.A., and Lee, C., Quintero-Ramirez, R., and Warcher-Radarte, C. (Eds.). FAO Agricultural Services Bulletin: 66–67.

Lee, C.H. 1997. Lactic acid fermented foods and their benefits in Asia. *Food Control* 8: 259–269.

Lee, C.H. and Lee, S.S. 2002. Cereal fermentation by fungi. *Appl Myco Biotechnol* 2: 151–170.

Lee, C.Y., Robinson, W.R., Van buren, J.P., Acree, T.E., and Stuews, G.S. 1975. Methanol in wines in relation to processing and variety. *J Eno AVWC* 26: 184.

Lisle, D.B., Richards, C.P., and Wardleworth, S. 1978. The identification of distilled alcoholic beverages. *J Inst Brew* 84: 93.

Liyanage, A.W., Hettiarachchi, M.R., and Jayatissa, P.M. 1981. Yeasts of coconut and polmyrah palm wines of Srilanka. *J Food Sci Technol* 186: 256.

Madhyastha, S.M., Marquardt, R.R., Frolich, A.A., Platford, G., and Abramson, D. 1990. Effect of different cereal and oil seeds substrates on the production of toxins by *Aspergillus alubaccu* and *P verruesarum*. *J Agric Food Chem* 38: 1506.

Maria, M., Theron, J.F., and Rykers, L. 2010. Microbial organic acid producers. Chapter 4. *Organic Acids and Food Preservation*. CRC Press, Boca Raton, FL, pp. 97–116.

Mårtenssona, O., Oste, R., and Holst, O. 2002. The effect of yoghurt culture on the survival of probiotic bacteria in oat-based, non-dairy products. *Food Res Int* 35(8): 775–784.

Matsuura, S., Manabe, M., and Sato, T. 1970. Toxic microorganisms. In: *Proc 1968 US-Japan Conference on Toxic Microorganisms*, US Govt. Print Office, Washington, DC.

Mikelsaar, M., Annuk, H., Shchepetova, J., Mändar, R., Sepp, E., and Björkstén, B. 2002. Intestinal Lactobacilli of Estonian and Swedish Children. *Microbiol Ecol Health Dis* 14(2): 75–80.

Milanez, T. and Leitão, M. 1996. The effect of cooking on ochratoxin A content of beans, variety "Carioca." *Food Addit Contam* 13(1): 89–93.

Mir, S. and Sharma, N. 2004. *Effect of Refrigerated Storage on the Quality of Fermented Mutton Sausages*. Division of Livestock Products Technology, FVSc&AH, SKUAST-K, Shuhama, Srinagar 190006, Kashmir, J&K, India; National Dairy Research Institute, Karnal, Haryana, India.

Miyani, R. 1962. MSc thesis, Gujarat Agricultural University, Anand Gujarat.

Mizutani, T., Kimizuka, A., Ruddle, K., and Ishige, N. 1992. Chemical components of fermented fish products. *J Food Compos Anal* 5: 152–159.

Moy, G., Han, F., and Chen, J. 2010. Ensuring and promoting food safety during the 2008, Beijing Olympics. *Foodborne Pathog Dis* 7(8): 981–983.

Murakami, H.S., Takase, S., and Kuwabara, K. 1968. Non-productivity of aflatoxin by Japanese industrial strains of *Aspergillus* II. Production of fluorescent substances in rice *koji* and their identification by absorption. *J Gen Appl Microbiol* 14: 97.

Nartey, F. 1978. *Cassava—Cyanogenesis, Ultrastructure and Seed Germination*. Munksgaard, Copenhagen.

Nikkuni, S., Karki, T.B., Vilku, K.S., Suzuki, T., Shindoh, K., Suzuki, C., and Okada, N. 1995. Mineral and amino acid contents of kinema, a fermented soybean food prepared in Nepal. *Food Sci Technol Int* 1: 107–111.

Nout, M.J.R. 1994. Fermented foods and food safety. *Food Res Int* 27: 291–298.

Nout, M.J.R. 2003. Traditional fermented products from Africa, Latin America and Asia. In: *Yeasts in Food: Beneficial and Detrimental Aspects*. Boekhourt, T. and Robert, V. (Eds.), GmbH and Co., Behr's Verlag, pp. 451–473.

Nout, M.J.R., Bakshi, D., and Sarkar, P.K. 1998. Microbiological safety of kinema, a fermented soya bean food. *Food Control* 9: 357–362.

Odunfa, S.A. 1988. Review: African fermented foods: From art to science. *Mircen J App Microbioland Biotechnol.* 4(3): 259–273.

Odunfa, S.A. and Oyewole, O.B. 1998. African fermented foods. In: *Microbiology of Fermented Foods*. Vol. 2, Wood, B.J.B. (Ed.), Blackie Academic and Professional, London, pp. 713–752.

Oirschot, Q., Cornelius, E., Amjad, M., Rees, D., Westby, A., Rees, D., Cheema, M., Tomlins, K., and Farrell, G. 2012. *Tropical root crops*. In: *Crop Post-Harvest: Science and Technology*. Rees, D., Farrell, G., and Orchard, J. (Eds.). Wiley-Blackwell, Chichester, U.K.

Osuntokun, B.O. 1994. Chronic cyanide intoxication of dietary origin and a degenerative neuropathy in Nigerians. *Acta Hortic* 375: 311–321.

Ough, G.H. 1976. Ethyl carbamate in fermented beverages and foods. *J Agric Food Chem* 24: 323–331.

Ouoba, L.I.I., Diawara, B., Christensen, T., Mikkelsen, J.D., and Jakobsen, M. 2007a. Degradation of polysaccharides and nondigestible oligosaccharides by *Bacillus subtilis* and *Bacillus pumilus* isolated from soumbala, a fermented African locust bean *Parkia biglobosa*, food condiment. *Euro Food Res Technol* 224: 689–694.

Ouoba, L.I.I., Diawara, B., Jespersen, L., and Jakobsen, M. 2007b. Antimicrobial activity of *Bacillus subtilis* and *Bacillus pumilus* during the fermentation of African locust bean *Parkia biglobosa*, for soumbala production. *J Appl Microbiol* 102: 963–970.

Ouoba, L.I.I., Thorsen, L., and Varnam, A.H. 2008b. Enterotoxins and emetic toxins production by *Bacillus cereus* and other species of *Bacillus* isolated from soumbala and bikalga, African alkaline fermented food condiments. *Int J Food Microbiol* 124: 224–230.

Padmaja, G. and George, M. 1999. Oriental fermented foods; biotechnological approaches. In: *Food Processing: Biotechnological Applications*. Marwaha, S.S. and Arora, J.K. (Eds.), Asia Tech Publishers Inc., New Delhi, pp. 143–189.

Parkouda, C., Thorsen, L., Compaore, C.S., Nielsen, D.S., Tano-Debrah, K., Jensen, J.S., Diawara, B., and Jakobsen, M. 2010. Microorganisms associated with maari, a Baobab seed fermented product. *Int J Food Microbiol* 142: 292–301.

Patel, R.S. and Abd-Ed-salam, M.H. 1986. *Shrikhand* and Indian analong of western quarg. *Cult Dairy Product J* 21(1): 6–7.

Patel, R.S. and Chakraborty, B.K. 1985. Reduction of curd-forming period in *Shrikhand* manufacturing process. *Le Lait* 65(1): 55–64.

Patel, R.S. and Chakraborty, B.K. 1988. *Shrikhand*: A review. *Indian J Dairy Sci* 41: 1–6.

Postel, W. and Adam, L. 1979. Gas chromatographic characterization of brandy cognac and Armagnac I study of methods. *Dia Brand Wein Wirtschaft* 2: 404–409.

Postel, W. and Adam, L. 1980. Gas chromatographic characterization of brandy, Cognac and Armagnac II content of volatile subrtancis. *Die Brand Wein Wistschoft*, 2: 154.

Quevedo, F. 1997. Food safety and fermented foods in selected Latin American countries. *Food Control* 8: 299–302.

Rai, A. 2008. *Microbiology of Traditional Meat Products of Sikkim and Kumaun Himalaya.* PhD thesis, Food Microbiology Laboratory, Sikkim Government College and Department of Botany, Kumaun University, Nainital, India.

Raju, P.N. and Pal, D. 2011. Effect of bulking agents on the quality of artificially sweetened misti dahi (caramel colored sweetened yoghurt) prepared from reduced fat buffalo milk. *LWT—Food Sci Technol* 44(9): 1835–1843.

Ranganna, S. 1986. *Handbook of Analysis and Quality Control for Fruit and Vegetable Products.* 2nd edn. Tata Mcgraw Hill Publishing Co., New Delhi.

Rao, E.R., Varadaraj, M.C., and Vijayendra, S.V.N. 2005. Fermentation biotechnology of traditional foods of the Indian subcontinent. Chapter 3.18. In: *Food Biotechnology.* Kalidas, S., Gopinadhan, P., Anthony, P., and Robert, E.L. (Eds.), 2nd edn. CRC Press, Boca Raton, FL.

Reddy, K.K., Pasha Ali, M., Rao, B.V., and Jagannadha, R.T. 1984. Studies on production and quality of shrikhand from buffalo milk. *Indian J Dairy Sci* 37: 293–296.

Reddy, N.R. and Pierson, M.D. 1994. Reduction in antinutritional toxic components in plant foods by fermentation. *Food Res Int* 27: 281–290.

Rosma, A., Afiza, T.S., Wan Nadiah, W.A., Liong, M.T., and Gulam, R.R.A. 2009. Short communication microbiological, histamine and 3-MCPD contents of Malaysian unprocessed "budu." *Int Food Res J* 16: 589–594.

Roy, A., Moktan, B., and Sarkar, P.K. 2007. Microbiological quality of legume-based traditional fermented food marketed in West Bengal, India. *Food Control* 18: 1405–1411.

Sanchez, P.C. 1979. The prospects of fruit wine production in the Philippines. *J Crop Sci* 4: 183–190.

Sarangthem, K. and Singh, T.N. 2003. Microbial bioconversion of metabolites from fermented succulent bamboo shoots into phytosterols. *Curr Sci* 84: 1544–1547.

Sarkar, P.K., Jones, L.J., Craven, G.S., and Somerset, S.M. 1997a. Oligosaccharide profiles of soybeans during kinema production. *Lett Appl Microbiol* 24: 337–339.

Sarkar, S. 2006. Shelf-life extension of cultured milk products. *Nutr Food Sci* 36(1): 24–31.

Sarkar, S. 2008. Innovations in Indian fermented milk products—A review. *Food Biotechnol* 22(1): 78–97.

Saxena, S. 2000. Alcohol problems and responses: Challenges for India. *J Substance Use* 5(1): 62–70.

Sekar, S. and Mariappan, S. 2005. Usage of traditional fermented products by Indian rural folks and IPR. *Indian J Trad Know* 6: 111–120.

Sharma, A. 1999. Microbial toxins. In: *Biotechnology: Food Fermentation.* V.K. Joshi and A. Pandey (Eds.), Educational Publishers and Distributors, New Delhi, pp. 321–344.

Sharma, A. 2013. *Preparation and Evaluation of Sur in Himachal Pradesh.* MSc thesis, Dr. Y.S. Parmar University of Horticulture and Forestry, Nauni, Solan, HP.

Sharma, P.C. 1986. *Home Wine Making.* Govt. of H. P. Publication (HP), Shimla, India.

Shaw, B.G. and Harding, C.D. 1989. *Leuconostoc gelidum* sp. nov. from chill-stored meats. *Int J Syst Bacteriol* 39: 217.

Shinagawa, K., Konuma, H., Sekita, H., and Sugii, S. 1995. Emesis of rhesus monkeys induced by intragastric administration with the HEp-2 vacuolation factor cereulide, produced by *Bacillus cereus. FEMS Microbiol Lett* 130: 87–90.

Shinagawa, K., Ueno, Y., Hu, D., Ueda, S., and Sugii, S. 1996. Mouse lethal activity of a hep-2 vacuolation factor, cereulide, produced by *Bacillus cereus* isolated from vomiting-type food poisoning. *J Vet Med Sci* 58: 1027–1029.

Shrestha, H., Nand, K., and Rati, E.R. 2002. Microbiological profile of *murcha* starter and physicochemical characteristics of *poko,* a rice based traditional food products of Nepal. *Food Biotechnol* 16: 1–15.

Shukla, S., Kim, J.K., and Kim, M. 2011. Occurrence of biogenic amines in soybean food products. In: Soybean and Health. El-Shemy, H. (Ed.), InTech. ISBN: 978-307-535-8.

Soni, S.K. and Sandhu, D.K. 1989. Biochemical and nutritional changes associated with Indian Punjabi warri fermentation. *J Fd Sci Technol* 27: 82.

Soni, S.K. and Sandhu, D.K. 1990. Indian fermented foods: Microbiological and biochemical aspects. *Indian J Microbiol* 30: 130–157.

Soni, S.K. and Sandhu, O.K. 1999. Fermented cereal products In: *Food Fermentation: Microbiology, Biochemistry and Technology.* Joshi, V.K. and Ashok, P. (Eds.), Educational Publisher and Distributor, New Delhi, pp. 895–948.

Steinkraus, K.H. 1983. *Handbook of Indigenous Fermented Foods.* Marcel Dekker Inc., New Delhi.

Steinkraus, K.H. 1996. Chinese *sufu.* In: *Handbook of Indigenous Fermented Foods.* Steinkraus, K.H. (Ed.), Marcel Dekker Inc., New York, NY, pp. 633–641.

Steinkraus, K.H. 1997. Classification of fermented foods: Worldwide review of household fermentation technique. *Food Control* 8: 311–317.

Stinson, E.E., Osman, S.F., Huhtanen, C.N., and Bills, D.D. 1978. Disappearance of Patulin during alcoholic fermentation of apple juice. *Appl Environ Microbiol* 36(4): 620–622.

Stratton, D., Hutkins, J.E., and Taylor, S.L. 1991. Biogenic amines in cheese and other fermented foods—A review. *J Food Prot* 54: 460–470.

Tamang, B. 2006. *Role of Lactic Acid Bacteria in Fermentation and Biopreservation of Traditional Vegetable Products.* PhD thesis, Food Microbiology Laboratory, Sikkim Government College, North Bengal University.

Tamang, B. and Tamang, J.P. 2007. Role of lactic acid bacteria and their functional properties in Goyang, a fermented leafy vegetable product of the Sherpas. *J Hill Res* 20(2): 53–61.

Tamang, B., Tamang, J.P., Schillinger, U., Franz, C.M.A.P., Gores, M., and Holzapfel, W.H. 2008. Phenotypic and genotypic identification of lactic acid bacteria isolated from ethnic fermented tender bamboo shoots of North East India. *Indian J Trad Know* 121: 35–40.

Tamang, J.P. 1992. *Studies on the Microflora of Some Traditional Foods of the Darjeeling Hill and Sikkim.* PhD thesis, North Bengal University, Darjeeling.

Tamang, J.P. 2009a. Ethnic starters and alcoholic beverages. Chapter 8. *Himalayan Fermented Foods. Microbiology, Nutrition, and Ethnic Values.* CRC Press, Boca Raton, FL.

Tamang, J.P. 2009b. Fermented cereals. Chapter 5. *Himalayan Fermented Foods: Microbiology, Nutrition, and Ethnic Values.* CRC Press, Boca Raton, FL, pp. 117–138.

Tamang, J.P. 2010. *Himalayan Fermented Foods: Microbiology, Nutrition and Ethnic Values, 295.* CRC Press, Taylor & Francis Group, New York.

Tamang, J.P. and Sarkar, P.K. 1993. *Sinki:* A traditional lactic acid fermented radish taproot product. *J Gen Appl Microbiol* 39: 395–408.

Tamang, J.P. and Sarkar, P.K. 1995. Microbiology of *murcha*—An amylolytic fermentation starter. *Microbios* 81: 115–122.

Tamang, J.P. and Sarkar, P.K. 1996. Microbiology of *mesu,* a traditional fermented bamboo shoot product. *Int J Food Microbiol* 29: 49–58.

Tamang, J.P., Chettri, R., and Sharma, R.M. 2009. Indigenous knowledge of Northeast women on production of ethnic fermented soybean foods. *Indian J Trad Know* 8: 122–126.

Tamang, J.P., Dewan, S., Tamang, B., Rai, A., Schillinger, U., and Holzapfel, W.H. 2007a. Lactic acid bacteria in *Hamei* and *Marcha* of North East India. *Indian J Microbiol* 47(2): 119–125.

Tamang, J.P., Dewan, S., Thapa, S., Olasupo, N.A., Schillinger, U., and Holzapfel, W.H. 2000. Identification and enzymatic profiles of predominant lactic acid bacteria isolated from soft-variety *chhurpi,* a traditional cheese typical of the Sikkim Himalayas. *Food Biotechnol* 14: 99–112.

Tamang, J.P. and Fleet, G.H. 2009. Yeasts diversity in fermented foods and beverages. In: *Yeast Biotechnology: Diversity and Applications.* Satyanarayana, T. and Kunze, G. (Eds.), Springer, Netherlands, pp. 169–198.

Tamang, J.P., Tamang, B., Schillinger, U., Franz, C.M.A.P., Gores, M., and Holzapfel, W.H. 2005. Identification of predominant lactic acid bacteria isolated from traditional fermented vegetable products of the Eastern Himalayas. *Int J Food Microbiol* 105: 347–356.

Tamang, J.P., Tamang, N., Thapa, S., Devan, S., Tamang, B., Yonzan, H., Rai, A., Chettri, R., Chakrabarty, J., and Kharel, M. 2012. Microorganisms and nutritional value of ethnic fermented foods and alcoholic beverages of North East India. *Indian J Trad Know* 11(1): 7–25.

Tamime, A.Y. and Robinson, R.K. 1999. *Yoghurt: Science and Technology.* Woodhead, Cambridge, U.K.

Tanasupawat, S., Visessanguan, W., and Boziaris, B. 2013. Fish fermentation. In: *Seafood Processing: Technology, Quality and Safety.* Boziaris, I.S. (Ed.), Wiley-Blackwell, Chichester, U.K., 508 pp.

Teramoto, Y., Yoshida, S., and Ueda, S. 2002. Characteristics of rice beer (zutho) and a yeast isolated from the fermented product in Nagaland, India. *World J Microbiol Biotechnol* 18: 813–816.

Tesfaye, W., Morales, M.L., Parrilla, M.C., and Troncoso, A.M. 2002. Wine vinegar: Technology, authenticity and quality evaluation. *Trends Food Sci Technol* 13(1): 12–21.

Thakur, N., Savitri, and Bhalla, T.C. 2004. Characterizations of some fermented foods and beverages of Himachal Pradesh. *Indian J Trad Know* 3(3): 325–335.

Thapa, N. 2002. *Studies of Microbial Diversity Associated with Some Fish Products of the Eastern Himalayas.* PhD thesis, Food Microbiology Laboratory, Sikkim Government College North Bengal University. India.

Thapa, N., Pal, J., and Tamang, J.P. 2004. Microbial diversity in ngari, hentak and tungtap, fermented fish products of Northeast India. *World J Microbiol Biotechnol* 20: 599–607.

Thapa, N., Pal, J., and Tamang, J.P. 2006. Phenotypic identification and technological properties of lactic acid bacteria isolated from traditionally processed fish products of the Eastern Himalayas. *Int J Food Microbiol* 107: 33–38.

Thapa, S. 2001. *Microbiology and Biochemical Studies of Indigenous Fermented Cereal-Based Beverages of the Sikkim Himalayas.* PhD thesis, Food Microbiology Laboratory, Sikkim Government College (under North Bengal University), 190 pp.

Thorsen, L., Azokpota, P., Hansen, B.M., Hounhouigan, D.J., and Jakobsen, M. 2010. Identification, genetic diversity and cereulide producing ability of *Bacillus cereus* group strains isolated from Beninese traditional fermented food condiments. *Int J Food Microbiol* 142: 247–250.

Tolonen, M., Rajaniemi, S., Pihlava, J., Johansson, T., Saris, P., and Ryhänen, E. 2004. Formation of nisin, plant-derived biomolecules and antimicrobial activity in starter culture fermentations of sauerkraut. *Food Microbiol* 21(2): 167–179.

Tsuyoshi, N., Fudou, R., Yamanaka, S., Kozaki, M., Tamang, N., Thapa, S., and Tamang, J.P. 2005. Identification of yeast strains isolated from marcha in Sikkim, a microbial starter for amylolytic fermentation. *Int J Food Microbiol* 99: 135–146.

Tylleskar, T. 1994. The association between cassava and the paralytic disease konzo. *Acta Horticul* 175: 321–331.

Uzochukwu, S., Balogh, E., Tucknot, O.G., Lewis, M.J., and Ngoddy, P.O. 1999. Role of palm wine yeasts and bacteria in palm wine aroma. *J Food Sci Technol* 36: 301–304.

Valyasevi, R. and Rolle, R.S. 2002. An overview of small-scale food fermentation technologies in developing countries with special reference to Thailand: Scope for their improvement. *Int J Food Microbiol* 75: 231–239.

Wacher, C., Díaz-Ruiz, G., and Tamang, J.P. 2010. Fermented vegetable products. Chapter 5. In: *Fermented Foods and Beverages of the World.* Tamang, J.P. and Kasipathy, K. (Eds.), CRC Press, Boca Raton, FL.

Wani, S.A. and Sharma, B.D. 1999. Fermented meat products. In: *Biotechnology: Food Fermentation, Microbiology, Biochemistry and Technology.* Vol. II. Joshi, V.K. and Pandey, A. (Eds.), Educational Publishers and Distributors, New Delhi, pp. 971–1002.

Weber, J.V. and Sharypov, V.I. 2009. Ethyl carbamate in foods and beverages—A review. *Climate Change, Intercropping, Pest Control and Beneficial Microorganisms. Sutainable Agriculture Reviews.* Vol. 2, Springer Science + Business Media B.V., Dordrecht, pp. 429–452.

Westby, A., Reilly, A., and Bainbridge, Z. 1997. Review of the effect of fermentation on naturally occurring toxins in fermented food safety. Nout, M.J.R. and Motarjemi, Y. (Eds.). *Food Control* 8(5–6): 329–339.

Whitfield, D.E. 2000. Solar drying systems and the internet: Important resources to improve food. In: *Solar Drying Presented at International Conference on Solar Cooking*, Klmborly, South Africa, 26–29th November.

Wiseman, D.W. and Marth, E.E. 1983. Behaviour of toxin, in yogurt, buttermilk and kefir. *J Food Prot* 46: 115–118.

Wong, K.O., Cheong, Y.H., and Seah, H.L. 2006. 3-Monochloropropane-1, 2-diol (3-MCPD) in soy and oyster sauce: Occurrence and dietary intake assessment. *Food Control* 17: 408–418.

World Health Organization. 1993. Application of the Hazard Analysis Critical Control Point (HACCP) system for the improvement of food safety. *WHO-supported Case Studies on Food Prepared in Homes, at Street Vending Operations, and in Cottage Industries.* Food Safety Unit, WHO, Geneva.

Yiannas, F. 2009. Food safety culture. In: *Creating a Behavior-Based Food Safety Management System.* Doyle, M.P. (Ed.), Springer, New York, p. 95.

Yildirim, G. 2009. *Effect of storage time on olive oil quality.* Msc thesis, Graduate School of Engineering and Sciences of Izmir Institute of Technology, Turkey.

Yonzan, Y. 2007. *Studies on Selroti, a Traditional Fermented Rice Product of the Sikkim Himalayas: Microbiological and Biochemical Aspects.* PhD thesis, Food Microbiology Laboratory, Sikkim Government College, North Bengal University.

Zaika, L.L. and Buchanan, R.L. 1987. Review of compounds affecting the biosynthesis or bioregulation of aflatoxin. *J. Food Protect.* 50: 691.

Zambonelli, C., Chiavari, C., Benevelli, M., and Coloretti, F. 2002. Effects of lactic acid bacteria autolysis on sensorial characteristics of fermented foods. *Food Technol Biotechnol* 40(4): 347–351.

Zimmerman, S.B, Schwartz, C.D., Monaghan, R.L., Pelak, B.A., Weissberger, B., Gilfillan, E.C., Mochales, S. et al. 1986. Difficidin and oxydifficidin: Novel broad spectrum antibacterial antibiotics produced by *Bacillus subtilis.* I. Production, taxonomy and antibacterial activity. *J Antibiol* 12: 1677–1681.

6

HEALTH-RELATED ISSUES AND INDIGENOUS FERMENTED PRODUCTS

ADITI SOURABH, AMIT KUMAR RAI,
ARJUN CHAUHAN, KUMARSWAMI JEYARAM,
MALAI TAWEECHOTIPATR, PARMJIT S. PANESAR,
RAKESH SHARMA, REEBA PANESAR, S.S. KANWAR,
SOHINI WALIA, SOMBOON TANASUPAWAT,
SWATI SOOD, V.K. JOSHI, VANDANA BALI,
VANDITA CHAUHAN, AND VIKAS KUMAR

Contents

6.1 Introduction

Fermented food products are essential dietary components for most people of South Asian countries, and are consumed in various forms as beverages, main dishes, or condiments, at affordable costs (Campbell-Platt, 1994). Man's food and the fermentative activities of microorganisms have been intimately connected with each other since the beginning of human civilization, with respect to the production of fermented foods and beverages (Joshi and Pandey 1999; Joshi et al., 1999a; Blandino et al., 2003). Many of the indigenous fermented foods are prepared by process of solid substrate fermentation, either by natural microflora or by adding starter cultures involving molds, yeasts, or bacteria, including the Himalayas (Tamang, 1998). Several indigenous fermented foods are associated with therapeutic values such as in the folk medicines. (Anonymous FAO, 1997) in the ayurvedic system of medicine, extensive use is made of indigenous fermented foods, and such foods can be modified and future stratergies geared up to develop more of the foods with nutraceutical or functional and probiotic properties using fermentation technology. It is therefore, imperative to know the functional components of indigenous fermented foods and their effects on human health, especially the theraputic disease-combating effects, their bioactive molecules, mechanism of action, etc. Food fermentation has been reported to impart antimicrobial properties to a food product, thus making it safe to eat (Dewan, 2002; Thapa et al., 2006). All these aspects, along with other relevant issues, are discussed in this chapter.

6.2 Health and Beneficial Effects of Indigenous Fermented Foods

Indigenous fermented foods have emerged not only as a source of nutrition but also as functional and probiotic foods, and this has stimulated the production of these foods and beverages at industrial scale (Chavan and Kadam, 1989; Campbell-Platt, 1994; Steinkraus, 1996; Lee, 1997). In South Asia, several fermented foods are produced and consumed, and the diversity of these foods has bean discussed in Chapter 2 of this text. These include alcoholic beverages, lactic acid fermented foods, acetic acid fermented foods and alkaline fermented foods (see Chapters 1 and 3 of this text). The healthful aspects of some of the fermented food has been described in this section.

6.2.1 Alcoholic Beverages

Apart from consumption in social and religious functions, alcoholic beverages have traditional medicinal uses among the ethnic groups of South Asia. Irregular menstrual cycle, infertility, obesity, loss of appetite, and under-nourishment are some of the problems that are traditionally regulated using *Yu*, a traditional fermented food of Manipur (India) (Singh and Singh, 2006). Massaging the parts of the body also involves combining certain plant species with *Yu*. Parts of the plants *Pogostemon purpurascens* Dalz., *Coriandrum sativum* L., *Cynodon dactylon* Pers., *Oryza sativa* L, *Vangueria spinosa* Roxb., *Mussaenda frondosa* L., and *Holmskioldia sanguinea* Retz. are mixed in different combinations with a specified volume of alcohol for massaging the sore body parts. In Sekmai and Phayeng village in Imphal West District (India), the alcohol is smeared on the face and body as a beauty care product (Singh and Singh, 2006). Some of the therapeutic effects of alcoholic contents in ayurvedic medicines have been discussed in the next section of this text.

A variety of alcoholic beverages, including different types of wine and brandy are prepared (Joshi et al., 1999b; Joshi et al., 2011; Reddy et al., 2011). The consumption of alcoholic beverages in moderation have been associated with several benefits (Joshi, 2012) apart from antimicrobial activity, glucose tolerance factors etc. The presence of phenolics in wines red wine is of paramount importance and has profound implications in human health, due to their recently discovered role as an antioxidant (Joshi and Sharma, 2004; Joshi et al., 2009a), which protect the heart from cardiovascular diseases (Bission et al., 1995; Stockley, 2011). Phenolics have been found in other wines also (Joshi and Sharma, 2004). Antioxidant-rich apple and strawberry wines have been made by adding spices like aonla and ginger, known to possess medicinal and antimicrobial properties, and the amount of phenols added by the herbs and spices was almost double (Joshi et al., 2009a). Due to the excellent effect of honey on digestion and metabolism it is used to make wine called mead, and fruit mead are made from fruits like apple, plum, and pear (Joshi et al., 1990, 2009b). Various indigenous alcoholic beverages are made by using various herbs and spices, similar to the wines prepared commercially, called vermouth. Vermouth (spiced wine) produced commercially from grapes, can also be made from sand pears, plums, and apples (Joshi et al., 1991; Joshi and Sandhu, 2000; Panesar et al., 2009, 2011). Consumption of vermouth can combat heart disease and other ailments due to the contribution of phenolic compounds from the spices, herbs, and fruit (Panesar et al., 2009, 2011). It also protects the human body from free radical attack and increased high density lipoprotein (HDL) levels (Joshi et al., 1999a,b; Stockely, 2011). Phytoalexins like resveratrol have been found in grapes and have cancer chemopreventive activity besides antioxidant and anti-inflammatory activities (Joshi and Devi, 2009; Das et al., 2012). Thus, the beverages containing such components would be useful from health point of view as several indigenous alcoholic beverages are fruit based, completely or partly.

The excessive consumption, however, has been associated with several problems also (Alves and Herbert, 2011). The ingested alcohol is rapidly absorbed in the gastro-intestinal tract (stomach and intestines) and other parts of body, and has a pronounced effect on the brain, exerting a depressant action. Judgment, self-criticism, the inhibitions learned from earlier experience, are depressed first, and the loss of this regulatory control gives rise to excitement. As a result, the drinker gradually becomes less alert, loses awareness of his environment, becomes dim and hazy, muscular coordination deteriorates, and sleep is facilitated. Besides, excessive alcohol consumption leads to cirrhosis-cancer of the liver (Alves and Herbert, 2011).

Handia, an indigenous alcoholic fermented beverage and its concentrate, prepared from parboiled rice by the ethnic tribes (Santal, Sabar, Bhumij, Paroja, Kondh, Kolh, Mundari, Jung, etc.) spread among the eastern region specially in Orissa and West Bengal, India, was evaluated for antioxidant activity in terms of total antioxidant, reducing capacity, free radical scavenging, and metal chelating activities (Roy et al., 2012). Both *Handia* and its concentrate exhibited strong antioxidant activity, reducing capacity, and free radical scavenging activity.

6.2.2 Acetic Acid Fermented Products

Vinegars are produced by acetic acid fermentation including fruit juices and have been documented in biblical times (Joshi and Thakur, 2000). The use of vinegar to fight infections and other acute conditions dates back to Hippocrates (460–377 bc, the father of modern medicine) who recommended vinegar for treatment of ulcerations and for the treatments of sores. The use of vinegar has been found useful in cardiovascular disorders, tumor, blood glucose etc. as summarised earlier (Joshi, 2012). Apple cider vinegar has been claimed to treat a number of different human health conditions, such as arthritis, high blood pressure, all types of skin problems, chronic fatigue, and heart burn, besides giving an antiglycemic effect (Joshi and Sharma, 2009). Similarly, black plum (*Jamun*) (*Syzygium cumini* L) vinegar is associated with beneficial effect. It is prepared and consumed in many rural areas in India.

6.2.3 Lactic Acid Fermented Foods

Among the various indigenous fermented foods, lactic acid fermented foods are of high significance due to the probiotic activity of the lactic acid bacteria (LAB) that cause this fermentation. Thus, the lactic acid bacteria isolated from spontaneously fermented foods have potential probiotic attributes (Kalui et al., 2010). The functional properties of the probiotic bacteria present in fermented foods include antimutagenic and anticarcinogenic properties, serum cholesterol level reducing properties, lactose metabolism improvement properties, and immune system stimulation properties (Shah, 2003). Such foods with probiotic bacteria are marketed, and play a very significant role as carriers of probiotics (Heller, 2001; Rivera-Espinoza and Gallardo-Navarro, 2010). Consumers have therefore developed a keen interest in the health aspects of such foods, due to specific physiological functions of health relevance (Katina, 2005; Katina et al., 2012).

From the time of Mechnikoff, fermented foods have been studied for their role in enhancing immunity and for healthy living. Consequently, the knowledge of health benefits associated with fermented foods is constantly growing (Dirar, 1993). The possible health benefits from the consumption of indigenous fermented foods include enhanced lactose digestion by individuals with lactose intolerance (Salwa et al., 2004); prevention/treatment of acute rotavirus and antibiotic induced diarrhea; stimulation of intestinal immunity; stabilization of Crohn's disease; stimulation of intestinal peristalsis; improvement of balance between microbial populations (probiotic foods contain live microorganisms to actively enhance the health of consumers by improving the balance of microflora in the gut when ingested; Golden and Gorbach, 1992; Sanders et al., 2003; Salwa et al., 2004); inhibition of food-spoilage-causing bacteria and certain pathogens by lowering the pH of the food (Hammes and Tichaczek, 1994; Battcock and Azam-Ali, 1998; Sekar and Mariappan, 2007); decreased fecal enzyme activity; inhibition of some form of cancers (substances in fermented foods protect from the development of cancer; Frohlich et al., 1997); reduction in serum cholesterol level; reduction in hypertension; and the production of antibiotics and bacteriocins (Wood and Hodge, 1985; Nout, 1995; Adams and Nicolaides, 1997; Battcock and Azam-Ali, 1998) by certain lactic acid bacteria (e.g., *Lactobacillus acidophilus*) and molds.

Lactic acid bacteria, besides fermenting the food, also impart their health and nutritional benefits to the food. They produce various compounds, such as organic acids, diacetyl, hydrogen peroxide, and bacteriocins or bactericidal proteins during lactic acid fermentation (Sharma et al., 2003; Joshi and Sharma, 2012; Bhushan et al., 2013). These components are not only desirable for their effects on food taste, smell, color and texture, but they also inhibit undesirable microflora, thus preventing spoilage, extending shelf life, and inhibiting pathogenic organisms (Rattanachaikunsopon and Phumkhachorn, 2010). These beneficial microorganisms produce numerous helpful enzymes and antibiotic and anticarcinogenic substances, and bacteriocins (Joshi et al., 2006). The fermentation process concentrates the antioxidant and immune-stimulating effects of crucifer vegetables, like cabbage, and also the various bioactive compounds, such as α-tocopherol, β-carotene, vitamins, selenium, or phenolic compounds, etc., contributed by these vegetables (Tamang, 2010). Studies have shown that S-methylmethionine, isothiocyanates, and indoles present in fermented cabbage extract or sauerkraut reduces the risk of stomach cancer, and that they are also antagonistic towards various microorganisms (Wacher et al., 2010). Some of the lactic acid bacteria, their bacteriocin type and their inhibitory spectrum, are given in Table 6.1.

Table 6.1   Lactic Acid Bacteria, their Bacteriocin Type and Inhibitory Spectrum

LACTIC ACID BACTERIA	BACTERIOCIN TYPE	INHIBITORY SPECTRUM
Lactococcus lactis subsp. *lactis*	Nisin	Many Gram-positive bacteria; *Bacillus, Clostridium, Listeria, Micrococcus* etc.
Lactococcus lactis subsp. *cremoris*	Diplococcin	Wide range of Gram-positive bacteria
Bacillus subtilis	Subtilin	It has inhibitory spectrum similar to Nisin
Pediococcus acidilactici	Pediocin1 Pediocin AcH	Some LAB and *Listeria, Clostridium sp.*
Pediococcus pentosaceus	Pediocin A	Some LAB and *Listeria, Staphyllococci* and *Clostridium botulinum*
Lactobacillus reuterii	Reuterin	Gram-positive and -negative bacteria, yeast and fungi
Lactobacillus acidophilus	Lactococin/Acidolin/Lactacin	Narrow-spectrum antibiotic
Lactobacillus plantarum	Lactolin/Plantaricin	Related strains of LAB; *L. brevis, L. acidophilus* etc.
Bifidobacterium sp.	Bifidin	*Bifidobacteria, E. coli* and related strains

Source: Adapted from Sharma, R. et al. 2003. *Bev Food World*, 30(7): 36–40; Joshi, V.K. and Sharma, S. 2012. *International Journal of Food and Fermentation Technology*, 2(1): 1–12.

6.2.3.1 Probiotics   The first report of consumption of fermented milks has been estimated to be over 2000 years ago, but the work of Metchinkoff in 1907 boosted the interest in this area (Das et al., 2012). The word "probiotic" is translated from the Greek, meaning "for life." So a probiotic can be defined as a "live microbial food supplement which, when administered in adequate amounts, beneficially affects the host animal by improving its intestinal microbial balance" (Fuller, 1989; Das et al., 2012). In the present functional food era, probiotic products are becoming more popular than ever. According to Das et al. (2012), probiotics generally include the following categories of bacteria:

- Lactobacilli such as *L. acidophilus, L. casei, L. brevis,* and *L. delbrueckii* subsp. *bulgaricus.*
- Gram-positive cocci such as *Lactococcus lactis,* and *Streptococcus salivarius* subsp. *thermophilus.*
- Bifidobacteria such as *B. bifidun, B. infantis, B. longum,* and *B. thermophilum.*

Lactic acid bacteria with probiotic activity isolated from fermented foods from South Asia are given in Table 6.2. Probiotics are generally used to treat gastrointestinal (GI) conditions such as lactose intolerance, acute diarrhea, and antibiotic-associated GI side effects. Probiotics are non-pathogenic, non-toxic, resistant to gastric acid, and adhere to gut epithelial tissues producing antibacterial substances (Das et al., 2012). There is evidence that administration of probiotics decreases the risk of systemic conditions, such as allergy, asthma, cancer, and several other infections, including those of the ear and the urinary tract.

Fermented dairy products are the most typical food for probiotic bacteria, whereas the other fermented and non-fermented products such as fruits, vegetables, cereals, *soja,*

Table 6.2 Lactic Acid Bacteria Isolated from South Asian Fermented Foods and their Probiotic, Antimicrobial, and Related Activities

LACTIC ACID SPECIES	FERMENTED PRODUCT	PROBIOTIC AND RELATED ACTIVITIES	REFERENCE
LAB	*Idli*	Reduce the mutagenicity of aflatoxins in the *Salmonella* mutagenicity	Thyagaraja and Hosono (1993)
Leu. paramesenteroides and *L. casei*	*Idli*	Binding activity towards the amino acid pyrolysates	Thyagaraja and Hosono (1994)
LAB	*Chhurpi*	Showed peptidases, esterase-lipases and high hydrophobicity	Tamang et al. (2000)
Lactobacillus plantarum	*Kanjika*	Production of Vitamin B12	Madhu et al. (2010)
Pediococcus sp. GS4	*Khadi*	Antioxidant potential	Sukumar and Ghosh (2011)
W. paramcscntcroidcs, *L. plantarum* and *L. fermentum*	*Kallappam batter, koozh and mor kuzhambu*	Antibacterial activity toward *Salmonella typhi*, *Vibrio parahaemolyticus* and *Listeria monocytogenes*	Kumar et al. (2010)
L. plantarum AS1	*Kallappam*	Bind to cultured human intestinal cell line HT-29 and inhibit cell attachment by *Vibrio parahaemolyticus*	Kumar et al. (2011)
L. plantarum AS1	*Kallappam*	Suppress 1,2-Dimethyl hydrazine (DMH)-induced colorectal cancer in male wistar rats	Kumar et al. (2012)
LAB	*Ngari, hentak* and *tungtap*	Provided a high degree of hydrophobicity	Thapa et al. (2004)
L. pobuzihii strains	*Tungtap*	Bacteriocinogenic activity against *Salmonella typhi* MTCC 733, *Bacillus cereus* MTCC 430, *Klebsiella pneumoniae* MTCC 109, *Escherichia coli* MTCC 118 and *Bacillus licheniformis* MTCC 429	Rapsang et al. (2011)
E. faecalis, Leuconostoc sp.	Curd, pickle	Bacteriocin like inhibitory substance (BLIS)	Mallesha et al. (2010)
Pediococcus sp. GS4	*Khadi*	Antibacterial ability against *Staphylococcus aureus* ATCC 25923, *Escherichia coli* ATCC 25922 and *Pseudomonas aeruginosa* ATCC 25619	Sukumar and Ghosh (2010)
Lactobacillus spp.	Curd, *idli* batter, *dosa* batter	Antibacterial activity against *Staphylococcus aureus*	Bhattacharya and Das (2010)
LAB	*Dahi, mohi, chhurpi, somar, philu* and *shyow*	*Enterobacter agglomerans, Enterobacter cloacae, Klebsiella pneumonia*	Dewan and Tamang (2007)
L. plantarum Lp9	Buffalo milk	Exhibited high resistance against low pH and bile, and possessed antibacterial, antioxidative and cholesterol lowering properties	Kaushik et al. (2009)

LAB = Lactic acid bacteria.

and meat may possibly be probiotic foods (Rivera-Espinoza and Gallardo-Navarro, 2010). Different fermented milk products, such as yoghurt, have been used as carrier foods for probiotic microorganisms, and the consumers can take in large amounts of probiotic cells for the their therapeutic effect (Lourens-Hattingh and Viljoen, 2001). Many *yogurts* or *dahi* for that purpose, contain lower amounts of lactose than milk, and are helpful in the digestion of lactose (Martini et al., 1987; Oskar et al., 2004; Panesar, 2011). During fermentation, some of the lactose (milk' sugar) is also converted into lactic acid by starter cultures that may produce the enzyme lactase, which hydrolyses lactose. Good quality curd contains live bacteria that provide a host of health benefits, such as longer life, and may fortify the immune system, particularly in people, so increasing the resistance to immune-related diseases. Eating curd may also help to prevent vaginal yeast infections. The fermentation of milk makes a large amount of phosphorus and calcium available to the digestive system by their precipitation in the lower intestines due to the acidic condition induced by *Lactobacillus* sp. However, the consumption of sour milk also results in the increased efficiency of the human body to cope up with a sudden influx of lactic acid in the system. *Dahi* in its different forms such as *lassi*, *kadhi*, *shrikhand*, etc., also significantly improves the nutritive value of an average Indian diet. It is also attractive due to providing value-added ingredients in a beverage system, such as yogurt drinks fortified with ingredients such as prebiotics and probiotics (Allgeyer et al., 2010). Traditionally, probiotics have been added to yoghurt and other fermented food products, usually together with a yoghurt starter culture of *S. thermophilus* and *Lactobacillus delbrueckii* subsp. *bulgaricus*, as probiotic strains, on their own, are slow fermenters, possibly due to poor functioning or complete lack of proteolytic systems.

The conventional yoghurt starter bacteria, *L. delbrueckii* subsp. *bulgaricus* and *Streptococcus thermophilus*, have been found to lack the ability to survive passage through the intestinal tract and consequently do not play a role in the human gut (Gilliland, 1989). At the same time, production of organic acids by the probiotics lowers the pH and alters the oxidation–reduction potential in the intestine, resulting in antimicrobial action that prevents the growth of any pathogenic contaminants (Micanel et al., 1997; Lourens-Hattingh and Viljoen, 2001). Viable LAB suppress the colonization and subsequent proliferation of pathogens either directly or through production of anti-bacterial substances, thus preventing various gastrointestinal infections and antibiotic-induced diarrhea (Gorbach, 2000; Culligan et al., 2009; Panesar, 2011). Himalayan ethnic fermented milk products have been found appropriate for the alleviation of lactose intolerance (Tamang et al., 2000; Dewan and Tamang, 2006, 2007). The consumption of fermented foods containing viable cells of *L. acidophilus* decreases the conversion of procarcinogens to carcinogens, thereby, activating the immune system of consumers (LeBlanc and Perdigon, 2005). *L. bulgaricus* have antimutagenic effects due to their ability to bind with the heterocyclic amines, which are carcinogenic substances formed in cooked meat (Panesar, 2011). The probiotic strain of *L. acidophilus* La-5 produces conjugated linoleic acid (CLA), an anticarcinogenic agent (Macouzet et al., 2009). Thus, the consumption of probiotic food can prevent cancer.

The functional probiotic food products should fulfill some essential prerequisite conditions before being labelled as "probiotic," such as those laid down by a task force constituted by ICMR (Indian Council of Medical Research), for the evaluation of probiotics in food in India (Evaluation of probiotics in food, Ganguly et al., 2011). It is established that fermented dairy products are the most typical food for probiotic bacteria. LAB from traditional fermented foods with antimicrobial activities should also be evaluated for probiotic potential. Kurmann and Rasic (1991) suggested that to achieve optimal potential therapeutic effects, the number of probiotic organisms in a probiotic product should meet a suggested minimum of $>10^6$ CFU/mL. Depending upon the species, and even strains, this criterion is refered to as the "therapeutic minimum" in the literature (Davis et al., 1971; Rybka and Kailasapathy, 1995).

The gastrointestinal (GI) tract of man represents an ecosystem of the highest complexity, and it is colonized by approximately 10^{14} microbial cells per gram (Luckey and Floch, 1972). The immense metabolic potential of the intestinal microflora suggests a strong regulatory effect on body functions, especially in the colon, where the largest concentration of upto 5×10^{11} bacterial cells per gram of more than 400 species is found (Holzapfel and Schillinger, 2002). These , have been divided roughly into three groups on the basis of metabolic activity such as lactic acid bacteria, anaerobes and aerobes, or nonpathogens, including *Bifidobacterium, Lactobacillus,* and *Streptococcus* (including *Enterococcus*); the putrefactive bacteria, including *Clostridium perfringens, Clostridium* spp., *Bacteroides, Peptostreptococci, Veillonella, Escherichia coli, Staphylococcus,* and *Pseudomonas aeruginosa,* and other bacteria, including *Eubacterium, Ruminococcus, Megasphaera, Mitsuokella, C. butyricum,* and *Candida* (Moore et al., 1978; Mitsuoka, 2002). The intestinal microflora is effected by the consumption of probiotic food.

LAB, *L. paramesenteroides,* and *L. casei* strains isolated from *idli* reduced the mutagenicity of aflatoxins and also showed binding activity towards the amino acid pyrolysates (Thyagaraja and Hosono, 1993, 1994). LAB from *chhurpi* showed peptidases, esterase-lipases and high hydrophobicity (Tamang et al., 2000). The *L. plantarum* strain from *kanjika* produced Vitamin B12 (Madhu et al., 2010). *Pediococcus* sp. GS4 from *khadi* showed antioxidant potential (Sukumar and Ghosh, 2011). *W. paramesenteroides, L. plantarum* and *L. fermentum* strains from *kallappam batter, koozh,* and *mor kuzhambu* showed antibacterial activity toward *Salmonella typhi, Vibrio parahaemolyticus,* and *Listeria monocytogenes* (Kumar et al., 2010). *L. plantarum* AS1 from *kallappam* could bind to cultured human intestinal cell line HT-29 and inhibit cell attachment by *Vibrio parahaemolyticus* (Kumar et al., 2011), including suppressing 1,2-dimethyl hydrazine (DMH)-induced colorectal cancer in male wistar rats (Kumar et al., 2012). LAB from *ngari, hentak,* and *tungtap* provided a high degree of hydrophobicity (Thapa et al., 2004). *L. pobuzihii* strains from *tungtap* showed bacteriocinogenic activity against *Salmonella typhi* MTCC 733, *Bacillus cereus* MTCC 430, *Klebsiella pneumoniae* MTCC 109, *Escherichia coli* MTCC 118 *Bacillus licheniformis* MTCC 429 (Rapsang et al., 2011). Strains of *E. faecalis* and *Leuconostoc* sp. isolated from curd and pickle produced bacteriocin-like inhibitory substance (BLIS) (Mallesha et al., 2010). *Pediococcus* sp. GS4 isolated from *khadi* showed antibacterial ability against *Staphylococcus aureus* ATCC 25923, *Escherichia coli* ATCC 25922, and *Pseudomonas aeruginosa* ATCC 25619 (Sukumar and Ghosh, 2010). Jeyaram et al. (2008) developed fibrinolytic fermented soybean using *Bacillus subtilis* MH10B5 (MTCC 5481) as a starter culture. *Lactobacillus* spp. strains from *curd, idli batter, and dosa batter* showed antibacterial activity against *Staphylococcus aureus* and diarrheagenic pathogens (Bhattacharya and Das, 2010; Shruthy et al., 2011). LAB from *dahi, mohi, chhurpi, somar, philu,* and *shyow* inhibited *Enterobacter agglomerans, Enterobacter cloacae,* and *Klebsiella pneumonia* (Dewan and Tamang, 2007). *L. plantarum* strain Lp9 has been found to exhibit high resistance against low pH and bile, and possesses antibacterial, antioxidative, and cholesterol-lowering properties. Thus, there is lot of potential for developing indigenous functional food or nutraceuticals (Kaushik et al., 2009).

Some of the species of LAB can be regarded as protective cultures, and most of the species involved in fermented foods do not pose any health risk, and thus, are designated as "GRAS" (generally recognized as safe) organisms (Adams and Nicolaides, 1997; Nout, 2001; Hansen, 2002).

6.2.3.2 Prebiotics   Prebiotics are the non-digestible or low-digestible food ingredients that benefit the host organism by selectively stimulating the growth or activity of one or a limited number of probiotic bacteria in the colon and they can be considered as an alternative to probiotics. Fermentable carbohydrates, like oligosaccharides ard inulin, and its hydrolysates and oligofructanes, which are not digested or poorly digested in the small intestine, stimulate, preferentially the growth of bifidobacteria and some Gram-positive bacteria belonging to the probiotic bacteria administered to humans. Such complex carbohydrates in the small intestine of the lower gut are available for some colonic bacteria, but are not utilized by the majority of the bacteria present in the colon. The metabolism of carbohydrates mainly give rise to shortchained fatty acids, such as acetate, butyrate, and propionate, used by the host organism as an energy source (Grajek et al., 2005). Vegetables like chicory roots, banana, tomato, and allium are rich in fructo-oligosaccharides, whereas oligosaccharides such as raffinose and stachyose are found in beans and peas. A daily intake of 520 g of inulin oligosachharides promote the growth of bifidobacteria. Combined mixtures of probiotics and prebiotics are often used for their synergic effects on food products, and are called *Synbiotics* (Das et al., 2012).

6.3 Indigenous Fermented Food in Health and Combating Diseases

6.3.1 Indigenous Fermented Foods in Health

There are many traditional fermented food products with medicinal properties and this aspect has been summarized here.

6.3.1.1 Fermented Milk Products   Milk and other fermented milk products are a rich source of calcium, thereby, decreasing the risk of osteoporosis, colon cancer, and hypertension. Curd (*dahi*), a traditional dietary component among Indians, has the benefit of wide acceptability, feasibility, cost effectiveness, and no known side effects (Saran, 2004). Exopolysaccharide (EPS) producing non-ropy strain of lactic acid bacteria (LAB) is found in *dahi* (Vijayendra et al., 2008a). Though tasteless, the EPS from LAB increases the residence time of milk products in the mouth and the perception of taste (Duboc and Mollet, 2001). In addition, EPS remains for a longer time in the gastro-intestinal tract, thus, enhancing the colonization by probiotic bacteria (German et al., 1999). Other benefits such as antitumor effect (Kitazawa et al., 1998), cholesterol lowering ability (Pigeon et al., 2002), and immuno-stimulatory activity (Chabot et al., 2001) have also been attributed to EPS produced by LAB (Vijayendra et al., 2008b).

In India, fermented milk products, like *lassi*, are made of mixed culture by spontaneous fermentation, and contain a number of microorganisms which are beneficial for human health. Fermented milk containing *L. casei* and Indian *dahi* help in combating acute diarrhea (Agarwal and Bhasin, 2002). Fermentation of food with certain strains of lactic acid bacteria help in the conversion of aflatoxin to less toxic derivatives (Croci et al., 1995). Fermentation using different cultures of *Lactobacillus acidophilus*, *S. thermophilus*, and *Lactococcus lactis* reduce 30%–58% of aflatoxin M1 in mixed milk. Lower rates of colon cancer among the consumers of fermented dairy products have also been documented (Ogueke et al., 2010). Recently, some yogurt products have been reformulated to include *L. acidophilus* and *Bifidobacterium* strains (AB-cultures), in addition to the conventional yogurt starter. Fruit-yogurt-like fermented milk products with living probiotic bacteria significantly shorten the duration of antibiotic-associated diarrhea and improve gastrointestinal complaints. Fermented milk is a matrix suitable for probiotic bacteria, and reduced *H. pylori* infection and urease activity (Lin et al., 2011); those with *L. johnsonii* La1 improved intestinal microflora and blood phagocytic activity, while *Bifidobacterium animalis* Bb-12 increased intestinal microflora and intestinal IgA (Fukushima et al., 1998; Galdeano et al., 2007). The consumption of fermented milk containing living *B. animalis* DN-173 010 improved the total colonic transit time (CTT) in humans (Bouvier et al., 2001). Milk fermented by *L. helveticus* containing valyl-prolyl-proline and isoleucyl-prolyl-proline provided a non-pharmacological approach to the management of hypertension (Nakamura, 2004).

Whey—a product of cheese processing—contains many unique components with broad antibacterial properties, including immunoglobulin (Igs), lactoferrin (Lf), lactoperoxidase (Lp), glycomacropeptides (GMP), and sphingolipids. Lactoferrin has been found to exhibit immune-modulating activity through both antimicrobial and antitoxin activity, besides providing protection against viruses such as hepatitis, cytomegalovirus, and influenza (Harper, 2000). These compounds have been observed to survive passage through the stomach and small intestine, and to arrive as intact proteins in the large intestine, where they exert their biological effects (Warny et al., 1999). Whey also has various therapeutics properties (Kar and Misra, 1999). Whey has also been studied as a carrier of probiotic against diarrhea (Goyal and Gandhi, 2008).

The bacterium *L. plantarum* has antioxidant and anticancer properties, thus making it a probiotic to be used in the fermentation of dairy products (Kumar et al., 2011). *L. acidophilus* in acidophilus milk reduces cholesterol, thus acting as a dietary adjunct. Trials in some human subjects have demonstrated that fermented dairy foods can produce a modest reduction in total and low density lipoproteins (LDL) cholesterol levels in those individuals with normal levels to begin with (Boris et al., 2007). Probiotics are also used in the treatment of *H. pylori* infections, which cause peptic ulcers in adults, when used in combination with standard medical treatment (Srividya and Vishnuvarthan, 2011; Matsushima and Takagi, 2012). In the colon and plasma of cancer-bearing animals, an increased level of lipid peroxide products and increased activity of antioxidant and marker enzymes was observed when fed with fermented milk. The probiotics strain *L. plantarum* AS1 was isolated from the South Indian fermented food, *Kallappam*, and has been shown to modulate the development of DMH-induced rat colon carcinogenesis through an antioxidant dependent mechanism (Kumar et al., 2012) Antioxidant activities have been reported in many ethnic fermented soybean foods, such as *chungkokjang* (Shon et al., 2007; Tamang et al., 2009) and *Pazhaya saadam*, a fermented rice dish from South India. It is rich in all natural nutritional supplements, and contains a number of microbes, contributing vitamin B complex and K vitamins, and preventing constipation problems due to its fiber content, as well as helping intestinal disorders, including dueodenal ulcers, infectious colitis, ischemic colitis, radiation colitis, ulcerative colitis, Crohn's disease, diverticular disease, hemorrhoids, irritable bowel syndrome (IBS), celiac disease (Gobbetti et al., 2005), *Candida* infections, etc.

6.3.1.2 Cereal-Legume Products   *Dhokla* and *idli* are popular breakfast in India. The antioxidant activities of cereal and legume mixed batter as influenced by process parameters during perparation of *dhokla* and *idli*, traditional steamed pancakes, was determined (Moktan et al., 2011). Fermentation enhanced more than 2.5-fold TPC (total phenolic content) and more than 125% reducing activity in the fermented batters or their steamed products.

6.3.1.3 Alcoholic Beverages   The fermented coconut palm wine called *Toddy* also has health benefits, as it is said to act as a sedative, a mild laxative, help with good eye sight, and as a tonic during recovery from diseases, including chicken pox (Steinkraus, 1996). In order to gain strength, ailing persons and postnatal women in the Himalayas consume *bhaati jaanr* extract (a fermented rice beverage) and *kodo ko jaanr* (a fermented finger millet product) due to their high calorific content (Tamang et al., 2012).

6.3.1.4 Fruits and Vegetable Products   In the northern Himalayas, a fermented leafy vegetable product called *gundruk* and a fermented radish tap-root (*sinki*) have large amounts of lactic acid, ascorbic acid, carotene, and dietary fiber, which have anticarcinogenic effects.

6.3.1.5 Fermented Fish Products It is belived by the people of tribal communities that *Hukoti* is a pain killer as well as is a local therapeutic to cure malaria. Similarly, *Hentak* is offered to mothers in confinement and patients in convalscence.

6.3.2 Indigenous Food as Functional Food

Functional foods promote health by improving wellbeing (mental and physical conditioning) and reducing the risk of diseases, and have useful effects in bioregulation, including central action, peripheral action, appetite, and absorption; biodefense, including the suppression of allergies and immune-stimulation; and prophylaxis against hypertension, diabetes, cancer, hypercholesterolemia, anaemia, and platelet aggregation (Mata et al., 1996; Sharma et al., 2012). These foods arc also considered to prevent geriatric diseases through a reduction of reactive oxygen species (ROS) (Mitsuoka, 2002). Functional foods have been classified into three groups, probiotics, prebiotics and biogenics, based on their mechanisms of action (Mitsuoka, 2002), as shown in Figure 6.1. Definitions of probiotics and prebiotics have already been discussed in an earlier sections 6.2.3.1 and 6.3.1 of this chapter, and are described elsewhere also (Sohail et al., 2012). The main therapeutic benefits attributed to consumption of probioitcs are summarized in Table 6.3.

6.3.3 Ayurvedic Context of Indigenous Fermented Foods

There are many types of indigenous ayurvedic medicines, including capsules, powders, infusions, decotions, infused oils, ointments, and creams/lotions, which are made by fermentation and have added herbals infusions/extracts (Sekar, 2007).

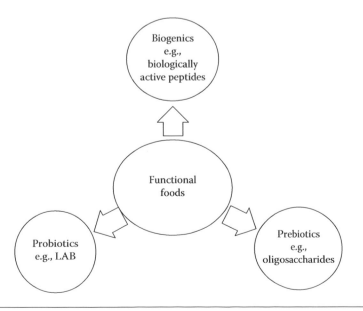

Figure 6.1 Classification of functional foods.

Table 6.3 Therapeutic Benefits from the Consumption of Probiotics

EFFECT	REMARKS	REFERENCES
Enhancement of the immune system	Lactic acid bacteria play an important role and function in the host's immunoprotective system by increasing specific and nonspecific mechanisms to have an antitumor effect	Sekar and Mariappan (2007), Schiffrin et al. (1995), Rafter (2004)
Reduction of lactose intolerance	*L. acidophilus* and *B. bifidum* produced β-ᴅ-galactosidase supported certain people who had the abdominal discomfort in the intestine due to the inability to digest lactose adequately	Kim and Gilliland (1983)
Reduction in serum cholesterol levels	The consumption of fermented milk significantly reduced serum cholesterol Role in lowering serum cholesterol by feeding of bifidobacteria may involve HMG-CoA reductase.	Mann and Spoerry (1974), Gilliland (1989), Homma (1988)
Antibiotic-induced diarrheal disease	The patients receiving *Saccharomyces boulardii* who developed antibiotic-associated diarrhea were only 7.2% whereas in placebo it was 14.6%.	McFarland et al. (1995)
Rotavirus diarrhea	Probiotic microorganisms have been demonstrated as a treatment for rotavirus infection	Isolauri et al. (1991), Kaila (1992)
Food borne pathogen-associated with gastrointestinal infections	Several probiotics lactobacilli have exhibited antagonistic to *Bacillus cereus, E. coli, S. aureus, Yersinia enterocolitica* and *Listeria in vitro*	Chaveerach et al. (2004); Batdorj et al. (2007); Zhang and Beynan (1993)
Helicobacter pylori gastroenteritis	Oral consumption of specific strain of lactobacilli may be effective in mediating human protection against *H. pylori* infections	Marteau et al. (2002); Wang et al. (2004), Yang et al. (2012)
Role in inflammatory bowel diseases	*L. acidophilus* combined with *B. animalis* subsp. *lactis* can reduce clinical symptom in IBD patients and *L. rhamnosus* GG showed significant improvements of clinical outcomes in children with Crohn's disease	Guandalini (2002)
Anticarcinogenic activity	Antitumor activity of probiotic is attributed to the inhibition of carcinogens and/or procarcinogens, inhibition of bacteria that convert procarcinogens to carcinogens	Gilliland (1989), Gorbach et al. (1987), Lourens-Hattingh and Viljoen (2001)
Antibacterial and antimold activity	LAB can inhibit the growth of other spoilage causing bacteria by producing bacteriocins or other substances like antibiotic reutramycin, etc.	Todorov (2008); Corsetti et al. (2000)

Indigenous alcoholic beverages are well known therapeutics with ayurvedic properties, if taken according to the prescription, in the proper dose, along with a balanced diet (Dash and Kashyap, 1980; Sekar, 2007). No proof is required to advocate the fact that indigenous people, for example, tribal people, are more hale and hearty, and possess an immense knowledge of the environment, ayurveda and healing, and the properties of flora and fauna, and their usage and management.

Indigenous fermented foods have been prepared commonly by using local food crops and other biological resources in India (Roy et al., 2004), but the nature and type of the raw materials vary from region-to-region. They can be classified as sugar based, fruit based, cereal based, and cereal and/or fruit based with/without herbs. Asavas and Arishthas are very important ayurvedic products. A number of alcoholic beverages, generally called *Madya*, are essential component of this system of medicine, as they have a tremendous effect on human health. Such beverages are grouped into seven major categories, based on the nature of raw materials used and the nature of the fermentation, such as *sura, sukta, sidhu,* etc., while *Ira, madira, hala,* and *halavallabha* are synonyms of *Madya* (Sekar, 2007). All kinds *of Madya* are hot in potency, slightly sweet, bitter, pungent in taste, slightly astringent, and sour. These aggravate *pitta*, mitigate *vata* and *kapha*, cause purgation, are digested quickly, create dryness, are nonviscid, improve digestion, helps in better taste and are beneficial to those having loss of sleep or excess sleep for both lean and stout persons (Sekar, 2007).

The effect of such ayuervedic medicine details about it is illustrated by taking an example madya here. The details of its preparation have been described in Chapter 9 of this text. The upper portion of *madya* is light in nature and is called *prasanna* (Sekar, 2007). The portion below it is relatively denser, and is called *kadambari* or Kadamba (Dash and Kashyap, 1980). The portion below *kadambari* is called *jagala*, while the portion at the bottom region of the container, is called *medaka*. The lowest layer of *medaka* containing the paste of drugs is called *vakkasa*. The material used for initiating fermentation of alcoholic drinks is called *kinva* or *surabija* (it is a microbial inoculum). If *kinva* is not matured, then it is called *madhulaka*. It is present in improperly fermented *madya* (Dash and Kashyap, 1980; Patwardhan et al., 2004), *Prasanna* alleviates vomiting, pain in the heart and the abdomen, *kapha, vata,* constipation, and hardness of bowels. It also cures *anaha* (flatulence), *gulma* (gastritis), *arsas* (piles), *chardi* (vomiting) and *arocaka* (anorexia) (Dash and Kashyap, 1980). *Kadambari, a* type of alcoholic drink, stimulates digestion, cures *anaha* (flatulence), pain in the heart and pelvic region, and colic pain. It is highly aphrodisiac, alleviates *vata*, and is a laxative. *Jagala* alleviates *kapha*, cures constipation, *sopha* (oedema), *arsas* (piles), and *grahani* (irritable bowel syndrome). It is hot, carminative, and strength-promoting. It cures *ksut* (morbid hunger), *trsna* (morbid thirst), and *aruci* (anorexia), and is beneficial for colic, dysentery, *borborygmi* (sound of flatus in intestine), and constipation (Valiathan, 2003). *Jagala* is digestive, reduces oedema, and alleviates dysentery, gurgling sound in bowels, piles, *vata*, and consumption (Sekar, 2007). It is applied externally for oedema. *Bakkasah* and *jagala* are free from liquid and consist only of yeast and drugs. It is *vata* aggravating, an appetizer, a laxative, a diuretic, and is non-slimy. Freshly prepared alcoholic drinks (*nava madya*) are *abhisyandi* (which obstructs channels of circulation). It alleviates all the three *doshas*, but is a laxative. It is neither good for the heart nor tasty (*virasa*), and causes a burning sensation and produces a putrid smell. It is *visada* (nonslimy), heavy, and difficult to digest. But *Purana madya*, the same alcoholic drink, after preserving for a long time is relished and cures *krmi* (parasitic infection) and aggravation of *kapha* as well as *vata*. It is a cardiac tonic, fragrant, endowed with good qualities, cleanses and opens the channels of circulation and improves appetite (Dash and Kashyap, 1980).

6.3.4 Bioactive Peptides as Functional Foods

6.3.4.1 Bioactive Compounds   Bioactive substances or compounds are food components which can affect biological processes and thus, influence the body's functions or condition and, ultimately, health. Bioactive compounds are to be considered "bioactive" only if, as a dietary component, they can impart a measurable biological effect at a physiologically realistic level, and, secondly, that the "bioactivity" measured should have the potential (at least) to affect health in a beneficial way, thus excluding from this definition potentially damaging effects (such as toxicity, allergenicity, and mutagenicity), which are undoubtedly a reflection of "bioactivity" in its broadest sense (Moller et al., 2008a,b). Bioactive peptides may affect many different biological functions:

- Improved uptake of minerals (calcium and iron) (casein-phospho-peptide)
- ACE-inhibition (casokinin)
- Gastrointestinal mobility (casomorphin)
- Antithrombotic agents (casoplatelin)
- Enzyme secretion and acidification of the stomach
- Antimicrobial properties (casecidins, lactoferricin)
- Better utilization of protein
- Improved appetite and growth

There are many naturally occurring bioactive compounds from foods which possess important properties which are helpful to slow down disease progression and inhibit pathophysiological mechanisms, as well as suppressing the activities of pathogenic molecules. Among these bioactive compounds, proteins and peptides play significant roles (Hartmann and Meisel, 2007) and are, thus, gaining importance as nutraceuticals and functional, and healthful foods, besides having antidisease characteristics (antiproliferative, antimutagenic, anti-inflammatory, anticancerous, or antioxidative) that are manifested in many diseases, including cancer, diabetes, and inflammatory disorders.

Bioactive peptides, also known as functional peptides, are food-derived peptides that, in addition to their nutritional value, exert a physiological effect in the body (Hettiarachchy, 2012). These have a defined sequence of amino acids which are inactive within the original protein, but which display specific properties once they are released by enzymatic hydrolysis (Vermeirssen et al., 2004). They normally remain dormant until they are acted upon by specific proteases. Gastrointestinal proteolytic enzymes release the peptides, and the small fraction released and absorbed is sufficient to impart a biological function. These are fragments of proteins, upon digestion using specific proteolytic enzymes or fermentation, that impart different positive functions or have several benefits that influence human health.

6.3.4.2 Bioactive Peptides   Bioactive peptides derived from food proteins during fermentation exert various physiological functions beyond their nutritional value. These peptides are inactive within the sequence of the parent protein, and are released by enzymatic hydrolysis during fermentation or on gastrointestinal digestion. The size of the peptide formed on hydrolysis may range from 2 to 20 amino acids, and the activity is dependent on its size, amino acid content, and their composition. Depending on their amino acid sequence, they exhibit various properties, including antioxidant (Erdmann et al., 2008), antimicrobial (Haque and Chand, 2008), immunomodulatory (Vinderola et al., 2007a,b), antithrombic (Erdmann et al., 2008), hypocholesterimic (Hartmann and Meisel, 2007), and antihypertensive (Phelan and Kerins, 2011). There are several foods derived bioactive peptides (Korhonen and Pihlanto, 2001; Niels et al., 2008). Fermented milk and soybean products are one of the richest sources of bioactive peptides (Gill and Sutherland, 2000), and these have been purified and characterized (Choi et al., 2012). Some fermented fish products, such as blue mussels (Jung et al., 2005) and shrimp paste (Peralta et al., 2008) have also been studied for bioactive peptides. How they are produced and function has been reviewed (Korhonen and Pihlanto, 2006). These peptides improve the functional properties of the foods and can also act as a natural alternative to various synthetic drugs.

6.3.4.3 Occurrence of Bioactive Peptides in Dairy Products   It is now well established that bioactive peptides can be generated during milk fermentation using the starter cultures, traditionally employed by the dairy industry (Gill and Sutherland, 2000). Consequently, peptides with various bioactivities are found in the end-products, such as various varieties of cheese and fermented milk (Gobbetti et al., 2007). These traditional dairy products may, under certain conditions, carry specific health effects when ingested as a part of the daily diet. The occurrence of various peptides in different fermented milk products has been established and listed in Table 6.4. A great variety of peptides are formed during cheese ripening, many of which have been shown to exert biological activities (Haque and Chand, 2006). Not only this, secondary proteolysis during cheese ripening may lead to the formation of other bioactive peptides, and the occurrence of bioactivity appears to be dependent on the ripening stage of the cheese. Meisel et al. (2006) detected higher ACE-inhibitory activities in medium-aged *gouda* cheese than in either fresh or long-term ripened cheese (Haque et al., 2008). Thus, the concentration of active peptides in cheese increases with maturation, but starts to

Table 6.4   Bioactive Peptides Identified from Milk Products

PRODUCT	ORIGIN	BIO-FUNCTIONAL ROLE
Cheddar cheese	α_{s1}- and β-casein fragments	Several phosphopeptides with a range of properties including the ability to bind and solubilize minerals
Italian cheeses: Mozzarella, Crescenza, Italico, Gorgonzola	β-CN f (58–72)	ACE inhibitory
Yoghurt-type products	α_{s1}-, β- and κ-CN fragments	ACE inhibitory
Gouda cheese	α_{s1}-CN f (1–9), β -CN f (60–68)	ACE inhibitory
Festivo cheese	α_{s1}-CN f (1–9), f (1–7), f (1–6)	ACE inhibitory
Emmental cheese	α_{s1}- and β -casein fragments	Immunostimulatory, mineral binding and solubilizing, antimicrobial
Manchego cheese	Ovine α_{s1}-, α_{s2}- and β-casein fragments	ACE inhibitory
Sour milk	β-CN f (74–76, f (84–86), κ-CN f (108–111)	ACE inhibitory/Antihypertensive
Dahi	Ser-Lys-Val-Tyr-Pro	ACE inhibitory

Source: Adapted from Korhonen, H.J. 2009. *Bioactive Components in Milk and Dairy Products.* Wiley-Blackwell, Oxford; Fatih, Y. 2010. *Amino Acids, Oligopeptides, Polypeptides, and Proteins. Advances in Food Biochemistry.* Taylor & Francis Group, LLC. pp. 51–100; Haque, E. and Chand, R. 2006. Milk protein derived bioactive peptides. [Online]. http://www.dairyscience.info/exploitation-of-anti-microbial-roteins/111-milk-protein-derived-bioactive-peptides.html?showall=1.

α_{s1}-CN = α_{s1}-casein, β-CN = β-casein, κ-CN = κ-casein.

decline when proteolysis exceeds a certain level. Accordingly, ACE-inhibitory activity was low in products having a low degree of proteolysis, such as yoghurt, fresh cheese, and quark (Korhonen and Pihlanto, 2007).

6.3.4.4 *Effects of Bioactive Peptides*

6.3.4.4.1 Biological Functions of Bioactive Peptides   Bioactive peptides present in foods may help to reduce the worldwide epidemic of chronic diseases. Functional proteins and peptides are now an important category of compounds within the nutraceutical food sector. These may affect many different types of biological functions (Moller et al., 2008a,b; Liu and Pan, 2011) such as improved uptake of minerals (Calcium and iron) (Casein-phospho-peptide), angiotensin I converting enzyme (ACE)-inhibition (casokinin), gastrointestinal mobility (casomorphin), antithrombotic agents (caso-platelin), enzyme secretion and acidification of the stomach, antimicrobial properties (casecidins, lactoferricin), better utilization of protein, and improved appetite and growth.

Generally, the peptides are rapidly metabolized to the constituent amino acids by brush border membrane peptidases after oral dosing, but the absorption and bioavailability remains very low (Jakala and Vapaatalo, 2010). They may be absorbed *via* carrier-mediated transport or paracellular diffusion (Shimizu, 2004). Upon oral administration, bioactive peptides may affect the major body systems namely, the cardiovascular, digestive, immune, and nervous systems, depending upon their amino acid sequence (Figure 6.2).

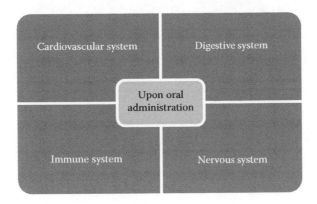

Figure 6.2 Effect of oral administration of bioactive peptides on major parts of the body.

Numerous sources of bioactivity peptides have been described as peptides released from dietary proteins by enzymatic proteolysis, including opiates, antithrombotics, and antihypertensives, and with properties including immunomodulating, antilipemic, osteoprotective, antioxidative, antimicrobial, ileum contracting, anticariogenic, and growth promoting. Specific peptides can have one or more different biological functions (Moller et al., 2008a,b), depending upon their inherent amino acid composition and sequence. The size of active sequences may vary from two to twenty amino acid residues, and many peptides are known to show multifunctional properties.

6.3.4.4.1.1 Angiotensin I Converting Enzyme (ACE) Inhibitors   Angiotensin I converting enzyme (ACE) is known to play a key physiological role in controlling the blood pressure by the renin angiotensin pathway (Li et al., 2005; Takenaka, 2011), as shown in Figure 6.3. Protein-rich fermented foods are a good source of ACE

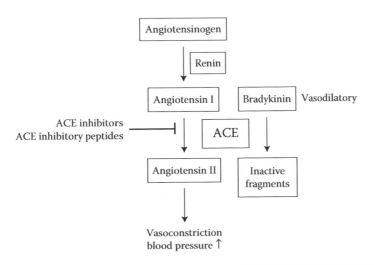

Figure 6.3 A diagrammatic view of the action of the renin–angiotensin system. (From Erdmann, K., Cheung, B.W.Y., and Schroder, H. 2008. *J Nutr Biochem*, 19: 643–654. With permission.)

inhibitory peptides, which are released during fermentation from the native protein. ACE inhibitory peptides inhibit the activity of ACE and prevent the conversion of angiotensin I to angiotensin II, a potent vasoconstrictor. Fermented soybean products are a good source of ACE inhibitory peptides. These peptides have also been reported in many other traditional Asian fermented soy foods, such as *douchi* (Zhang and Delworth, 2005; Mejia and Lumen, 2006), soy sauce (Kinoshita et al., 1993), Korean soy paste, *Tempeh* (Gibbs et al., 2004), and *Natto* (Okamoto et al., 1995). ACE inhibitory peptides have to be absorbed through the small intestine, and reach the cardiovascular system in an active form. Within the renin–angiotensin system, ACE catalyses the conversion from angiotensin I (Ang I) to angiotensin II (Ang II), a hormone that cause vasoconstriction and, subsequently, in an increase of blood pressure (Moller et al., 2008a,b).

ACE inhibitors function by maintaining the balance between the associated vasoconstrictive and salt-retentive property of Ang II with the vasodilatory effect of bradykinin. The balance is maintained by decreasing the production of Ang II and reducing the degradation of bradykinin. Thus, ACE-inhibitiors lower hypertension and are believed to prevent cardiovascular diseases (Ryan et al., 2011). Various fermented milk products have been studied for ACE inhibitory properties, including cheese (Singh et al., 1997; Ong et al., 2007; Moller et al., 2008a,b; Meira et al., 2012), sour milk (Nakamura et al., 1995), *dahi* (Ashar and Chand, 2004), sheep's milk yoghurt (Papadimitriou et al., 2007), fermented camel milk (Mosleliishad et al., 2013), and fermented goat's milk (Minervini et al., 2009). In hypertensive patients, milk fermented with *Lactobacillus helveticus* LBK-16H was able to reduce blood pressure in a long-term study (Seppo et al., 2003). Angiotensin I converting enzyme (ACE) inhibitory activity of fermented *dahi* batter as influenced by fermentation emolition was determined (Rajwani and Ananthanarayan, 2013). The activity was influenced by time of fermentation, proportion of ingredient temperature of fermentation and addition evtrinsic protease. There are two more peptides (Tyr-Pro and Lys-Val-Leu-Pro-Val-Pro-Gln) shown to possess ACE inhibitory properties in spontaneous hypertensive rats. The peptides were purified and characterized from fermented milk (Maeno et al., 1994; Yamamoto et al., 1999). Fermented fish sauce in many Asian countries shows ACE-inhibitory activity, and the peptide was isolated from fermented blue mussel sauce. It was purified and characterized (Je et al., 2005), and oral administration of the same significantly reduced blood pressure in spontaneous hypertensive rats. Thus, protein rich fermented foods can become a potential source of ACE-inhibitory peptides having the ability to control hypertension.

ACE inhibitory peptides were first discovered in snake venom (Ferreira et al., 1970) and, since then, numerous synthetic ACE inhibitors have been produced, with captopril being the most common. However, captopril and other synthetic ACE inhibitors have been found to have various side-effects, such as coughing, taste disturbances, and skin rashes (Vermeirssen et al., 2002; Ryan et al., 2011). These side effects, coupled with a high hypertensive patient population, has contributed to the ongoing search for food-derived antihypertensive peptides for exploitation as antihypertensive agents in functional foods and nutraceuticals (Lopez et al., 2006; Ryan et al., 2011). The cardiovascular effects of LVV-hemorphin-7, a member of the family of fragments from the β-chain of human or bovine hemoglobin, were studied in conscious spontaneously hypertensive rats (SHR) and Wistar-Kyoto (WKY) rats by radiotelemetry. Intraperitoneal injection of hemorphin in a dose of 100 μg/kg significantly decreased blood pressure in SHR, whereas a negligible effect was seen in normotensive WKY rats. Blood pressure changes were accompanied by a reduction of heart rate. It is concluded that a direct effect of LVV-hemorphin-7 on blood pressure was demonstrated in SHR. These biologically active peptides appear to be involved in blood pressure regulation, especially in hypertensive rats, but the precise mechanism has not been elucidated (Cejka et al., 2004).

6.3.4.4.1.2 Anticholesterol Properties   Hypercholesterolemia is the most significant risk factor for cardiovascular diseases in many countries. Cholesterol-lowering effects have been associated with consumption of whey (Nagaoka et al., 1992; Zhang and Beynan, 1993), soya (Hori et al., 2001), and fish protein (Wergedahl et al., 2004). The exact mechanism behind the anticholesterol properties of these proteins or peptides derived from fermentation, however, is not known. But the effect of the consumuption of probiotic yogurt on serum cholesterol level in mildly to moderately hypercholesterolemic subjects was to cause significant decrease in serum total cholesterol (p < 0.05) during the clinical trials (Jafari et al., 2009). Yogurt prepared with buffalo milk and soy milk also shows plasma cholesterol lowering properties in rats fed on a cholesterol-enriched diet.

6.3.4.4.1.3 Antithrombotic Properties   Fermented foods have antithrombic properties, either due to bioactive molecules or enzymes formed during fermentation which have the ability to reduce thrombus formation. The antithrombotic effect can be by: (i) limiting the migration or aggregation of platelets; (ii) limiting the ability of the blood to clot; or (iii) dissolving the clots after they have been formed. Different fermented foods reduced platelet aggregation and dissolved fibrin clots by fibrinolytic enzymes (Pais et al., 2006). Fermented foods with fibrinolytic and thrombolytic activity, therefore, are in demand to prevent rapidly emerging cardiovascular diseases.

6.3.4.4.1.4 Inhibiting Platelet Aggregation   There are different bioactive components in different fermented foods which have the ability to inhibit platelets aggregation. The main component in fermented soybean, genestein, reduced platelet aggregation and was able to inhibit collagen and TXA2 analog-induced platelet aggregation (Nakashima et al., 1991; McNicol, 1993). Daidzein is another soybean flavanoid which inhibited platelets aggregation induced by collagen and TXA2 analogs (Gottstein et al., 2003). Genistein and daidzein are the major isoflavones which are present in higher concentrations in fermented soybean compared to unfermented cooked soybean (Ozaki et al., 1993). In addition, peptides derived from milk protein in fermented milk products have also been shown to possess antithrombic properties (Phelan and Kerins, 2011).

6.3.4.4.1.5 Antidiabetic Properties   Diabetes mellitus is a complex metabolic disorder which is increasing globally and affects 7% of the world's adult population (IDF, 2008). Some fermented foods are rich in flavonoids which at physiological concentration are comparable to clinically used antidiabetic drugs. Anti-α-glucosidase inhibitors prevent type II diabetes by combining with α-glucosidase and blocking the uptake of postprandial blood glucose (Holman and Jones, 1998; Chen et al., 2007). Accordingly, α-glucosidase inhibitors from naturally occurring food sources are becoming more popular as the synthetic inhibitors used presently cause hepatic disorders and other gastrointestinal symptoms (Murai et al., 2002). The high antidiabetic property of fermented soybean could be due to the changes that occur in the composition of isoflavones and peptides during the fermentation process. The incidence of type-2 diabetes is lower in Asian countries in comparison to Western countries and this might be correlated with the consumption of fermented soybean products (Kwon et al., 2010). Aqueous extract of fermented rice with *Monascus purpureus* has exhibited antidiabetic properties in streptozotocin-induced diabetic rats. Fermentation improves the antidiabetic properties by improving the bioavailability of the bioactive component in comparison to the nonfermented products (Kwon et al., 2010).

6.3.4.4.1.6 Regulation of the Immune System   Bioactive peptides regulate the immune system in man (Korhonen and Pihlanto, 2007). The immunomodulatory properties of the peptides are shown in Table 6.5. Results of some of the studies on this aspect are presented in Table 6.6.

6.3.4.4.1.7 Antioxidant Properties   Free radicals and reactive oxygen species play a major role in causing various diseases such as cancer, inflammation, ageing, and the toxicity of numerous compounds. A constant supply of natural antioxidants through proper diet to compensate for the losses of the defense system against oxidative stress and other metabolic disorders is necessary. Many fermented products have antioxidant properties (Jung et al., 2005; Moktan et al., 2008; Meira et al., 2012). Fermented soybean is one of the most popular foods of many Asian countries and is rich in isoflavones and peptides (Wang et al., 2007; Kwak et al., 2009). The improvement in the antioxidant properties of fermented soyabean products takes place mainly due to free polyphenols and peptides, which are formed by enzymatic hydrolysis during fermentation due to the conversion of glycone (the bound form) of isoflavone to aglycone (the free form) (Cho et al., 2011). Fermentation improves the antioxidant properties of many Asian fermented foods, such as Indian *Kinema*, fermented with *Bacillus subtilis* (Moktan et al., 2008).

The peptide fraction obtained from fermented camel and cow's milk by *Lactobacillus rhamnosus* PTCC 1637 shows antioxidant properties (Mosleliishad

Table 6.5 Function of Immunomodulatory Peptides

- Immunomodulatory peptides act on the immune system and cell proliferation responses.
- Several milk proteins such as whole casein, whole whey protein, lactoferrin, lactoperoxidase, milk growth factors and milk immunoglobulin G. have been reported to modulate lymphocyte proliferation *in vitro* (Moller et al., 2008a,b).
- Milk feeding facilitates passive immunity in neonates and, thus, contributes to their protection against harmful environmental pathogens (Moller et al., 2008a,b).
- Some milk factors like lactoferrin and lactoperoxidase exhibit their immunomodulating actions only in their isolated form (Moller et al., 2008a,b).
- Some immunomodulatory peptides are multifunctional peptides and may modulate cell proliferation by interacting with opioid receptors.
- These peptides can modulate the proliferation of human lymphocytes, down-regulate the production of certain cytokines and stimulate the phagocytic activities of macrophages and they can regulate the development of the immune system in newborn infants (Korhonen and Pihlanto, 2007).
- Lactoferrin, an iron binding glycoprotein which occurs in the milk of all mammals, shows diverse effects on the host defense system, such as antimicrobial action and various immunomodulating effects (Moller et al., 2008a,b).
- Casein, hydrolysated by *Lactobacillus* GG and digestive enzymes (pepsin and trypsin), to yield compounds having both stimulating and suppressing effects on lymphocyte proliferation (Sutas et al., 1996).
- A proline-rich polypeptide (PRP) isolated from ovine colostrum is known for its immunoregulatory properties.
- Protein rich peptides activates cytokine production by murine macrophages induces growth and differentiation of resting B-lymphocytes (Julius et al., 1988; Moller et al., 2008a,b).
- Tuftsin, a tetrapeptide (TKPR), is generated by the digestion of heavy chain Fc region of bovine and human immunoglobulin G with endopeptidase and leukokininase and has leucocyte chemotaxis and phagocyte motility, enhancement of phagocyte oxidative metabolism and antigen processing, and increase in monocyte- and NK cell-mediated tumor cell cytotoxicity (Werner et al., 1986).
- The impact of the response of specific antibodies to Helicobacter, enriched from the colostrum of hyperimmunized cows, on *Helicobacter felis*-infected Balbic mice (Marnila et al., 2003) reduced the colonization of gastric antrum by *H. felis* in comparison to a control group (Marnila et al., 2003).

et al., 2013). Antioxidant peptides have also been isolated from other products, such as cheese (Meira et al., 2012), yogurt (Perna et al., 2014), fermented fish products, such as fish sauce (Harada et al., 2003; Michihata, 2003), and fermented shrimp paste (Peralta et al., 2005). The antioxidant properties of peptides are mainly due to their composition, structure, and hydrophobicity. Amino acids which exhibit anti-oxidant properties are tyrosine, tryptophan, methionine, lysine, cysteine, and histidine, but the total activity is dependent upon the presence of proper amino acids and their correct positioning in the peptide (Sarmadi and Ismail, 2010). Consumption of fermented products with antioxidant properties has the potential to protect the body against oxidative stress, which can be a preventive measure for various degenerative diseases.

6.3.4.5.2 Groups of Bioactive Peptide Depending on their functionality, there are several bioactive peptides, performing various functions, as outlined earlier also. (Figure 6.4).

6.3.4.5.2.1 Opioids The first bioactive peptide released from food proteins, that is, bovine β-casomorphin-7 (YPFPGPI), an opioid peptide from a casein hydrolysate, was isolated during late 1970. Unlike endomorphins found in human beings, exorphins

Table 6.6   Bioactive Peptides and the Immune System

- Milk protein hydrolyzates and peptides derived from caseins and whey proteins can enhance immune cell functions, measured as lymphocyte proliferation, antibody synthesis and cytokine regulation (Gill et al., 2000)
- Peptides released during milk fermentation with lactic acid bacteria have been found to modulate the proliferation of human lymphocytes, to down-regulate the production of certain cytokines and to stimulate the phagocytic activities of macrophages (Meisel and FitzGerald, 2003)
- Immunomodulatory milk peptides however may alleviate allergic reactions in atopic humans and enhance mucosal immunity in the gastrointestinal tract (Korhonen and Pihlanto, 2003)
- Immunomodulatory peptides may regulate the development of the immune system in newborn infants
- Commercial whey protein isolates contain immunomodulating peptides which can be released by enzymatic digestion (Mercier et al., 2004)
- Immunopeptides formed during milk fermentation have also been shown to contribute to the antitumor effects observed with fermented milks (Marnila et al., 2003)
- CPPs have been shown to exert cytomodulatory effects that inhibit cancer cell growth or stimulate the activity of immunocompetent cells and neonatal intestinal cells (Meisel and FitzGerald, 2003)
- GMP and its derivatives exhibited a range of immunomodulatory functions, like immune-suppressive effects on the production of IgG antibodies
- It was concluded based on mouse model studies, that peptides released by bacterial proteolysis might have important implications in modulation of the host's immune response impacting the inhibition of tumor development (Marnila et al., 2003)

such as casomorphins from casein are found in other milk proteins (β-lactoglobulin, α-lactalbumin, lactoferrin), in cereal proteins, such as wheat (gluten, gliadin) and also barley (hordein, avenin, secalin, zein) and rice (albumin), in vegetables, such as soybeans (α-protein), spinach (rubisco protein), and in meat/poultry (albumin, haemoglobin, γ-globulin) and eggs (ovalbumin). Homologous sequences have also been identified in both human and goat milk. Casomorphins are found in hydrolyzed casein, fermented milk products, and in cheese. The quantities found in cheese are, however, significantly less than 1 mg/kg. Exorphins primarily affect the intestinal lumen and mucosa by regulating gastro-intestinal motility, as well as gastric and pancreatic secretions.

6.3.4.5.2.2 Mineral-binding Peptides   Like mineral-binding proteins, casein-derived phosphopeptides, reported as caseinophosphopeptides (CPP), have shown mineral-binding properties related to the presence of the phosphorylated serine residues, which can form salts with minerals such as calcium (Phelan et al., 2009). These peptides are involved both in the re-mineralization of tooth enamel and in the increased absorption and bioavailability of calcium and other minerals (zinc, copper, manganese, iron) in the intestine (Bouhallab et al., 2002).

6.3.4.5.2.3 Antioxidative Peptides   Bioactive peptides have an antioxidative effect. A study of the presence of antioxidative peptides, derived from hydrolyzed food proteins such as caseins, whey proteins, soybean, rice bran, quinoa seed protein, buckwheat protein, egg-yolk protein, porcine myofibrillar proteins and aquatic by-products proteins, has been carried out (Pihlanto, 2006), and they were found to be effective: against enzymatic and non-enzymatic peroxidation of lipids and essential

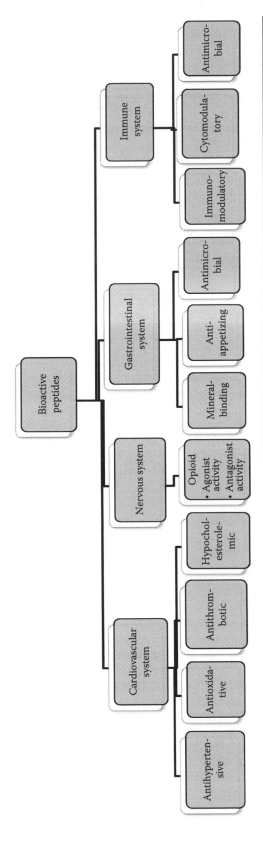

Figure 6.4 Milk based peptides and their biological functions in human body. (Adapted from Korhonen, H. and Pihlento, A. 2006. *International Dairy Journal* 16: 945–960).

Table 6.7   ACE Inhibitory Activity (IC50) of Lentil Protein Hydrolyzates Obtained by Trypsin, Pancreatin, and Chymotrypsin

TIME (H)	TRYPSIN	PANCREATIN	CHYMOTRYPSIN
0.5	2.204	0.069	4.452
1.5	0.45	2.057	3.875
2.5	2.134	2.242	2.971
3.5	2.167	2.249	2.879

Note: Inhibitor peptide concentration in mg protein/mL required to inhibit 50% ACE.

fatty acids; as free radical scavengers; in metal ions chelation; and in adduct formation (Conde et al., 2013), besides bradykinin potentiating effect (Perpeto et al., 2003). The inhibition of oxidative processes is of particular importance for the survival of cells in an organism. The pepsin hydrolysate obtained showed higher antioxidant activity, while those obtained with pancreatin had higher antihypertensive activity (IC50 = 0.069 mg/mL). It was, therefore, concluded that lentil protein concentrate and hydrolyzates represent an alternative for bioactive peptides and functional food production (Table 6.7).

6.3.4.5.2.4 Antihypertensive Peptides   Out of the various peptides, those with antihypertensive peptide activity, particularly those which inhibit ACE, have been investigated (Vercruysse et al., 2005; Ryan et al., 2011).

6.3.4.4.2.5 Immunomodulatory Peptides   Various peptides have shown immunomodulatory properties. Bioactive peptides help in the regulation of different body systems.

6.3.5 Bioactive Ingredients in Fermented Foods and Their Transformation during Fermentation

6.3.5.1 Benefits of Fermentation and Biogenics   Some South Asian fermented foods and their bioactive components and health benefits are shown in the Table 6.8.

The most beneficial outcome of fermentation is the enrichment of the fermented product with vitamins, essential amino acids, and bioactive components (peptides, free polyphenols). In addition, the fermentation also helps to increase the bioavailability of minearals and reduce several antinutritional and toxic components (oxalic acid, trypsin inhibitors, phytic acids). The bioactive components in fermented foods

Table 6.8   Some South Asian Fermented Foods, Their Bioactive Components, and Their Health Benefits

FERMENTED FOOD	COUNTRY	FERMENTING MICROORGANISM	BIOACTIVE COMPONENTS	FUNCTIONAL PROPERTIES	REFERENCE
Dahi	India	Lactic acid bacteria	Peptides	Antidiabetic probiotic	Yadav et al. (2007)
Kinema	India	*Bacillius subtilis*	Polyphenols	Antioxidant	Moktan et al. (2008)

produced by microbes during fermentation include bacteriocin, exopolysaccharides, and enzymes. The central role played by microorganisms in fermentation, especially LAB, is now widely acknowledged. It is an accepted fact that these microorganisms exert beneficial effects through two mechanisms: firstly, the direct effects of the live microbial cells, also known as the "probiotic effect" (Fuller, 1989), and secondly, the indirect effects during fermentation, where the microbes act as cell factories for the generation of secondary metabolites with health-promoting properties. For the latter function, the term "biogenics" has been coined. Among the most important biogenic compounds in fermented milk are bioactive peptides (Mitsuoka, 2000), released from milk proteins by members of the genera *Lactobacillus* (Lb.), *Bifidobacterium*, and other LAB. There are several bioactive components present in different fermented foods, which are being discussed in the subsequent sections.

6.3.5.2 Polyphenols/Flavones Polyphenols are one of the abundant micronutrients available in various fermented foods (*Kinema, Natto, Douche*, etc.) that have been well studied for their role in the prevention of degenerative diseases such as cancer and cardiovascular disease (Gottstein et al., 2003; Cho et al., 2011). Their beneficial effects, however, depend on the amount consumed and on their bioavailability at the specific site of action. Free polyphenol content was found to increase in various fermented soybean products such as *Kinema* (Moktan et al., 2008), *Cheonggukjang* (Cho et al., 2011), and *Tempe* (Nout and Kiers, 2005). Increase in free phenolic acids and a ferullic acid also took place in rye bran fermentation using *Saccharomyces cerevisiae* (Katina et al., 2007, 2012). Sourdough fermentation of whole gram, oats, and barley using lactic acid bacteria increased the concentration of phenolic acids more than five fold (Hole et al., 2012). The higher bioavailability of polyphenols in fermented foods is beneficial for wide spectrum of biological activities and for the prevention of cardiovascular disease (Fraga et al., 2010). The main isoflavones in soy protein extracts have been found to include genistein, daidzein, glycitein, and their naturally occurring glycosides (genistin, daidzin and glycitin), and malonyl glycosides (Cho et al., 2011).

6.3.5.3 Proteins Proteins are one of the major components in different fermented foods (soybean, milk, meat) having nutritional and biofunctional properties. The biofunctional properties are due to enzymes (fibrinolytic enzymes) or peptides (bacteriocin) produced by the starter culture or due to the peptide formed due to hydrolysis of food protein.

6.3.5.4 Therapeutic Enzymes Fibrinolytic enzymes are the enzymes that are used to dissolve fibrin clots, and they have been found in various traditional Asian fermented foods. In most cases, fermented soybean and fish products were found to harbor attractive fibrinolytic enzymes (Kotb, 2012). Fifty-four traditional fermented foods of Northeast India were recently screened for fibrinolytic activity. Among these, fermented soybean products and fermented fish products had fibrinolytic activity.

The fibrin zymogram pattern generated from these fibrinolytic foods revealed the microbial origin of fibrinolytic activity in fermented soybean products and the endogenous origin in fish products (Singh et al., 2013). The *Bacillus subtilis* strain MH10B5 (MTCC 5481) isolated from *Hawaijar* (Jeyaram et al., 2008), a traditional fermented soybean food of Manipur, India, with high fibrinolytic activity and superior fermentative parameters, was selected for fibrinolytic fermented soybean production. The solid state fermentation (SSF) yielded up to 5000 Kilo plasmin units/kg of soybean substrate, which is 30 fold higher than submerged fermentation, using 0.5% casein broth.

6.3.5.5 Carbohydrate   Carbohydrates constitute a major component in various cereal-based fermented foods such as *Idli* and *Meso* (Blandino et al., 2003). Apart from the food itself, the microbes involved in the fermentation process, such as LAB and *Bacillus subtilis*, also produce exopolysaccharides, which have been shown to possess antioxidant and antitumor properties (Abdel-Fattah et al., 2012). Besides the technological properties, exopolysaccharide produced by LAB exhibit various biofunctional characteristics, including anticancer (Kitazawa et al., 1998), cholesterol lowering (Pigeon et al., 2002), and immunomodulatory properties (Chabot et al., 2001). Exopolysaccharide producing LAB have also been isolated from cheese (Kojic et al., 1992), Indian *Dahi* (Vijayendra et al., 2008b), etc. The presence of these exopolysaccharides produced by the starter culture in fermented foods enhance the biofunctionality of the products.

6.3.5.6 Lipids   Lipids are one of the important components of fermented foods, and determine both the functionality and the sensory properties. Lipid content and their fatty acid composition can affect a wide range of quality attributes. The presence of n-3 poly unsaturated fatty acids (PUFA) can improve the biofunctional properties of the fermented product, and this depends on the raw material used for fermentation. Yoghurt made from sheep milk has higher n-3 fatty acids than cow's milk (Serafeimidou et al., 2013). In the case of fermented organic milk, higher amounts of trans-octadecenoic acid (C18:1, 1.6 times) and polyunsaturated fatty acids, including cis-9 trans-11, C18:2 conjugated linoleic (CLA, 1.4 times), and α-linolenic acids (ALA, 1.6 times), than conventional fermented milks has been documented (Florence et al., 2012; Bang et al., 2014). Fermented soybeans that are consumed in many Asian countries contain the essential fatty acid linoleic acid (Sarkar et al., 1996), which has a preventive effect on cardiovascular disease. Higher amounts of eicosapentaenoic and docosahexaenoic acids present in fermented shrimp paste did not significantly change during fermentation for almost a year (Peralta et al., 2008).

6.3.5.7 Addition of Beneficial Compounds   During the fermentation process, the starter culture produces various beneficial compounds—such as enzymes, peptides, organic acids, esters, and alcohols—which affect the flavor, nutritional attributes, and functional properties of the final products. Organic acids affect the flavor and preserve the food for many years (Caplice and Fitzgerald, 1999). The enzymes, such as protease, amylase, etc., are known to improve the digestibility of the fermented product. LAB in various fermented foods are reported to produce antibacterial peptides (bacteriocin) having preservative properties (Ross et al., 2002), and exopolysaccharides have immunomodulatory properties (Xu et al., 2005). The health promoting effects ascribed to fermented dairy products and probiotic bacteria arise not only from the bacteria themselves, but also from the metabolites produced by them during milk fermentation (Erdmann et al., 2008; Haque and Chand, 2008). Exopolysaccharides produced during milk fermentation by LAB and peptides derived during fermentation by proteolytic starters in protein rich foods are potential modulators of various regulatory processes in the body (Kitazawa et al., 2000; Choi et al., 2012).

6.3.5.8 Reduction of Antinutritional Factors   Raw beans contain a significant amount of antinutritional factors such as phytic acids, trypsin, and amylase inhibitors. During fermentation, the starter culture produces enzymes (e.g., phytase) which degrades the antinutritional factors (e.g., phytic acid) present in the raw material. The fermentation of soybean, cowpea, and groundbean using *Rhizopus oligosporus* has resulted in the reduction of phytic acid, which is an antinutritional factor (Egounlety and Aworh, 2003) that reduces the bioavailability of essential minerals. The fermentation of *Garcinia* using *Saccharomyces cerevisiae* reduced oxalic acid (an antinutritional factor) in the final fermented beverage (Rai et al., 2010) and increased aspartic and glutamic acid. Traditional fermentation of pearl millet for the preparation of *Iohah* (fermented bread) significantly reduced trypsin and amylase inhibitors and phytic acid content (Osman, 2004). It is apparent that fermentation not only improves bioactivity but also reduces antinutritional factors in raw materials.

6.3.6 Pediatric Health Implications of Fermented Foods

It is estimated that by 2025, of five million deaths of children younger than five years in the developing world, about 97% will be due to infectious diseases, among which diarrhea continues to play a prominent role (Thapar and Sanderson, 2004). A number of fermented food items contain probiotic microorganisms which exist in normal human gut microflora (*L. acidophilus*, *L. casei*, *Bifidobacterium bifidum*, *B. breve*, etc.), and, thus, they can help in improving intestinal microbial balance and overcoming pathogenic microbes. Probiotics decrease both the frequency and duration of diarrhea, in malnourished and dehydrated children in particular (Szajewska and Mrukowicz 2001; Van Neil et al., 2002). Fermented milk is an excellent source of nutrients, such as calcium, protein, phosphorus, and riboflavin, which improves and increases lactose digestibility. Consumption of fermented milk products builds up optimal bone mass in children both before and during early puberty, thereby helping to prevent osteoporosis and fractures during childhood (Abrams and Stuff, 1994). It has a probiotic effect, which protects infants from gastrointestinal infections. It improves iron status by improving absorption of iron and the bioavailability of iron from other foods. Milk protein (casein) and cheese protect against tooth decay. Calcium and phosphorus present in fermented milk products like cheese have also been found to promote remineralization and prevent demineralization of teeth, and to have anticarcinogenic effects as well (Jenkins, 1990). In addition, the production of lactic acid and other organic acids during fermentation reduces stomach pH and enhances iron absorption. Whey gives distinct advantages in the treatment of diarrhea due to the presence of amino acids, vitamins, and minerals, which are lost in deficiency diseases and malnutrition. *Rokshi*, a fermented drink in Sikkim, is used for the treatment of children to alleviate colds and coughs (Singh and Jain 1995). *Idli*, being easily digestible, is used as food for infants and invalids (Steinkraus, 1996).

6.4 Summary and Future Prospects

Various fermented food products, especially milk products, contain many health-promoting components. Fermented milk and milk products thus, constitute a balanced nutritional diet, especially for children, pregnant and lactating mothers, and elderly people. An increased intake of fermented milk may increase the level of calcium, which can reduce premenstrual syndrome (PMS) symptoms and lower the risk of several chronic diseases (such as cancer, hypertension and osteoporosis) in children and adults. Similarly, indigenous fermented foods have been associated with health benefits. It is well established that Ayurveda an ancient system of Indian medicine is known to prepare and prescribe a large profile of beverages with varying level alcohol with therapeutic properties. It is indicated that such alcoholic drinks, taken according to the prescribed dose, at the proper time, along with wholesome food, and according to the capacity of the individual, produce effects like ambrosia. When used inappropriately, they can cause diseases and work as a poison. It is important to apply modern science to understand the traditional systems of medicine such as Ayurveda, which requires a multifaceted approach of research for validation. Further, assessment of the quality of beverages in terms of ethyl alcohol, volatile acid content, fixed acid content, sugar content, and testing for methyl alcohol, etc., have to be performed in order to use them medicinally. Several probiotic bacteria of human origin are being exploited commercially, but a wide variety of indigenous probiotics are still underutilized. This could be a potential area of research.

A great focus is being made on bioactive peptides, as they play a very crucial roles in human body as antihypertensive, antioxidant, antimicrobial or immunomodulates, thus have great potential in combating several diseases. Indigenous fermented foods are the major sources, reflecting the great usefulness of these foods. More research in this direction is the need of the hour.

Indigenous fermented foods have been prepared and consumed in South Asia from times immemorial, yet the scientific research into the role they play in health and human well-being is quite recent. In the future, more in-depth studies could reveal the nutritional and therapeutic role of these products. Needless to say, their toxicity or adverse effects on health should also be forth coming. Due to several health-promoting benefits, there is an ample scope for promotion of these age-old foods in the global market as bio-foods or functional foods. Multi-institutional collaborative research can lead to standardization of the fermented functional food products and an increase in their shelf-life. Accordingly, a database should be developed listing all the fermented foods available in the region, along with their place of origin, and production, raw materials used, microorganisms involved, nutritional value, and the cost involved. The traditional fermented foods of South Asia need to be explored so as to conserve the microbial resources and to exploit the prophylactic and therapeutic potential of indigenous probiotic bacteria for their further development into commercial functional foods. The future should see the large-scale production of indigenous fermented foods, including those from fruits and vegetables, which are known to be protective foods. Research and development work in the analytical field will greatly assist in identifying any toxic components in even the lowest concentrations. It is expected that, in future, most indigenous fermented foods from South Asia will be available as safe, nutritious, and therapeutic products for the betterment and well-being of mankind.

References

Abdel-Fattah, A.M., Gamal-Eldeen, A.M., Helmy, W.A., and Esawy, M.A. 2012. Antitumor and antioxidant activities of levan and its derivative from the isolate *Bacillus subtilis* NRC1aza. *Carbohydr Poly*, 89: 314–322.

Abrams, S.A. and Stuff, J.E. 1994. Calcium metabolism in girls: Current dietary intakes lead to low rates of calcium absorption and retention during puberty. *Am J Clin Nutr*, 60: 739.

Adams, M.R. and Nicolaides, L. 1997. Review of the sensitivity of different foodborne pathogens to fermentation. *Food Cont*, 8: 227–239.

Agarwal, K.N. and Bhasin, S.K. 2002. Feasibility studies to control acute diarrhea in children by feeding fermented milk preparations actimel and Indian dahi. *Eur J Clin Nutr*, 56 (suppl 4): S56–S59.

Allgeyer, L.C., Miller, M.J., and Lee, S.Y. 2010. Drivers of liking for yogurt drinks with prebiotics and probiotics. *J Food Sci*, 75(4): S212–S219.

Alves, A. and Herbert, P. 2011. Toxicological aspects related to wine consumption. In *Handbook of Enology*, Vol. II. Joshi, V.K., ed. Asia Tech Publishers, Inc., New Delhi. pp. 209–233.

Anonymous dairy science. http://www.dairyscience.info/index.php/probiotics/110wheyprobiotics.html?showall=1

Anonymous FAO. 1997. *Chapter 1: Fermented Fruits and Vegetables—A Global Perspective. The Benefits of Fermenting Fruits and Vegetables*. FAO, Rome.

Ashar, M.N. and Chand, R. 2004. Fermented milk containing ACE-inhibitory peptides reduces blood pressure in middle aged hypertensive subjects. *Milchwissciiseliat*, 59: 363–366.

Bang, M., Oh, S., Lim, K.-S., Kim, Y., and Oh, S. 2014. The involvement of ATPase activity in the acid tolerance of Lactobacillus rhamnosus strain GG. *Int J Dairy Technol*, 67: 229–236.

Batdorj, B., Trinetta, V., Dalgalarrondo, M., Prévost, H., Dousset, X., Ivanova, I., Haertlé, T., and Chobert, J.M. 2007. Isolation, taxonomic identification and hydrogen peroxide production by *Lactobacillus delbrueckii* subsp. *lactis* T31, isolated from Mongolian yoghurt: Inhibitory activity on food-borne pathogens. *J Appl Microbiol*, 103: 584–593.

Battcock, M. and Azam-Ali, S. 1998. Fermented frutis and vegetables—A global perspective. FAO Agricultural Services Bulletin No. 134.

Bhattacharya, S. and Das, J. 2010. Study of physical and cultural parameters on the bacteriocins produced by lactic acid bacteria isolated from traditional Indian fermented foods. *Am J Food Technol*, 5(2): 111–120.

Bhushan, S., Sharma, S., Joshi, V.K., and Abrol, G. 2013. Bio-preservation of minimally processed fruits and vegetables using lactic acid bacteria and bacteriocin. In: *Food Processing and Preservation*. Bakshi A.K., Joshi, V.K., Vaidya, D., and Sharma, S., Eds. Jagmander Book Agency, New Delhi. pp. 355–374.

Bission, L.F., Butzke, C.E., and Ebeler, S.E. 1995. The role of moderate consumption in health and human nutrition. *Am. J. Enol. Vitic*, 46(4): 449–455.

Blandino, A., Al-Aseeri, M.E., Pandiella, S.S., Cantero, D., and Webb, C. 2003. Cereal-based fermented foods and beverages. *Food Res Int*, 36: 527–543.

Boris, H., Catherine, N., and Florent, L. 2007. Effect of low-fat, fermented milk enriched with plant sterols on serum lipid profile and oxidative stress in moderate hypercholesterolemia. *Am J Clin Nutr*, 86: 790–796.

Bouhallab, S., Vladas, C., Nabil, A., François, B., Dominique, N., Pierre, A., Jean-Louis, M., and Dominique, B. 2002. Influence of various phosphopeptides of caseins on iron absorption. *J Agri Food Chem*, 50(24): 7127–7130.

Bouvier, M., Meance, S., Bouley, C., Berta, J.-L., and Grimaud, J.-C. 2001. Effects of consumption of milk fermented by the probiotic strain *Bifidobacterium animalis* DN-173 010 on colonic transit times in healthy humans. *Biosci Microflora*, 20(2): 43–48.

Campbell-Platt, G. 1994. Fermented foods—A world perspective. *Food Res Int*, 27: 253–257.

Caplice, E. and Fitzgerald, G.F. 1999. Food fermentations: Role of microorganisms in food production and preservation. *Int. J. Food Microbiol*, 50: 131–149.

Cejka, J.J., Zelezná, B., Velek, J., Zicha, J., and Kuneš, J. 2004. LVV-hemorphin-7 lowers blood pressure in spontaneously hypertensive rats: Radiotelemetry study. *Physiol Res*, 53(6): 603–607.

Chabot, S., Yu, H.L., De Leseleuc, L. et al. 2001. Exopolysaccharide from *Lactobacillus rhamnosus* RW-9595M stimulate TNF, IL-6 and IL-12 in human and mouse cultured immunocompetent cells and IFN-g in mouse splenocytes. *Lait*, 81: 683–697.

Chavan, J.K. and Kadam, S.S. 1989. Nutritional improvement of cereals by fermentation. *CRC Crit Rev Food Sci Nutr*, 28: 349–400.

Chaveerach, P., Lipman, L.J.A., and van Knapen, F. 2004. Antagonistic activities of several bacteria on *in vitro* growth of 10 strains of *Campylobacter jejuni/coli*. *Int J Food Microbiol*, 90: 43–50.

Chen, J., Cheng, Y.Q., Yamaki, K., and Li, L.T. 2007. Anti-α-glucosidase activity of Chinese traditionally fermented soybean (douche). *Food Chem*, 103: 1091–1096.

Cho, K.M., Lee, J.H., Yun, H.D., Ahn, B.Y., Kim, H., and Seo, W.T. 2011. Changes of phytochemical constituents (isoflavones, flavanols, phenolic acids) during cheonggukjang soybean fermentation using potential prebiotic *Bacillus subtilis* CS90. *J Food Comp Anal*, 24: 402–410.

Choi, J., Sabikhi, L., Hassan, A., and Anand, S. 2012. Bioactive peptides in dairy products. *Int J Dairy Technol*, 65(1): 1–12.

Conde, E., Moure, A., Domínguez, H., and Parajó, J.C. 2013. Extraction of natural antioxidants from plant food. In: *Separation, Extraction and Concentration Processes in the Food, Beverage and Nutraceutical Industries*. Rizvi, S.S.H., Ed. Woodhead Publishing Limited, Cambridge, UK, pp. 506–594.

Corsetti, A., Gobbetti, M., and De Marco, B. 2000. Combined effect of sourdough lactic acid bacteria and additives on bread firmness and staling. *J Agric Food Chem*, 48: 3044–3051.

Croci, L., Toti, L., di Pasquale, S., Mirdglia, M., Brera, C., and de Dominicis, R. 1995. Lactic acid bacteria influence in variability of production and biotransformation of aflatoxins from *Aspergillus parasiticus*. *Rivista di Scienza dell'Alimentazione*, 24: 59–66.

Culligan, E.P., Colin, H., and Sleator, R.D. 2009. Probiotics and gastrointestinal disease: Successes, problems and future prospects. *Gut Pathog*, 1: 19.

Das, L., Bhaumik, E., Raychaudhuri, U., and Chakraborty, R. 2012. Role of nutraceuticals in human health. *J Food Sci Technol*, 49(2): 173–183.

Dash, V.B. and Kashyap, V.L. 1980. *Materia Medica of Ayurveda*. Concept Publishing Co., New Delhi. pp. 187–203.

Davis, J.G., Ashton, T.R., and McCaskill, M. 1971. Enumeration and viability of *Lactobacillus bulgaricus* and *Streptococcus thermophilus* in yogurt. *Dairy Ind*, 36: 569–573.

Dewan, S. 2002. Microbiological evaluation of indigenous fermented milk products of the Sikkim Himalayas. PhD thesis, Food Microbiology Laboratory, Sikkim Government College (under North Bengal University), Gangtok, India.

Dewan, S. and Tamang, J.P. 2006. Microbial and analytical characterization of *Chhu*, a traditional fermented milk product of the Sikkim Himalayas. *J Sci Ind Res*, 65: 747–752.

Dewan, S. and Tamang, J.P. 2007. Dominant lactic acid bacteria and their technological properties isolated from the Himalayan ethnic fermented milk products. *Antonie van Leeuwenhoek*, 92(3): 343–352.

Dirar, H.A. 1993. *The Indigenous Fermented Foods of the Sudan: A Study of African Food and Nutrition*. CAB International, Wallingford.

Duboc, P. and Mollet, B. 2001. Applications of exopolysaccharides in dairy industry. *Int Dairy J*, 11: 759–768.

Egounlety, M. and Aworh, O.C. 2003. Effect of soaking, dehulling, cooking and fermentation with *Rhizopus oligosporus* on the oligosaccharides, trypsin inhibitor, phytic acid and tannins of soybean (*Glycine max* Merr.), cowpea (*Vigna unguicidata* L. Warp) and groundbean (*Macrotyloma geocarpa* Harms). *J Food Eng*, 56: 249–254.

Erdmann, K., Cheung, B.W.Y., and Schroder, H. 2008. The possible roles of food-derived bioactive peptides in reducing the risk of cardiovascular disease. *J Nutr Biochem*, 19: 643–654.

Fatih, Y. 2010. *Amino Acids, Oligopeptides, Polypeptides, and Proteins. Advances in Food Biochemistry*. Taylor & Francis, Boca Raton, pp. 51–100.

Ferreira, S.H., Bartelt, D.C., and Greene, L.J. 1970. Isolation of bradykinin-potentiating peptides from Bothrops *jararaca venom. Biochemistry*, 9: 2583–2593.

Florence, A.C., Béal, C., Silva, R.C., Bogsan, C.S., Pilleggi, A.L., Gioielli, L.A., Oliveira, M.N. 2012. Fatty acid profile, trans-octadecenoic, α-linolenic and conjugated linoleic acid contents differing in certified organic and conventional probiotic fermented milks. *Food Chem*, 135(4): 2207–2214.

Fraga, C.G., Galleano, M., Verstraeten, S.V., and Oteiza, P.I. 2010. Basic biochemical mechanism behind the health benefits of polyphenols. *Mol Aspects Med*, 31: 435–445.

Frohlich, R.H., Kunze, M., and Kiefer, I. 1997. Cancer preventive impact of naturally occurring non-nutritive constituents in food. *Acta Med Austriaca*, 24: 108–113.

Fukushima, Y., Kawata, Y., Hara, H., Terada, A., and Mitsuoka, T. 1998. Effect of a probiotic formula on intestinal immunoglobulin A production in healthy children. *Int J Food Microbiol*, 42: 39–44.

Fuller, R. 1989. Probiotics in man and animal. *J Appl Bacteriol*, 66: 365–378.

Galdeano, C.M., de LeBlanc, M.G., Vinderola, M.E., and Bibas Bonet, G.P.A. 2007. Proposed model: Mechanisms of immunomodulation induced by probiotic bacteria. *Clin Vaccine Immunol*, 14(5): 485–492.

Ganguly, N.K., Bhattacharya, S.K., Sesikeran, B. et al. 2011. Guidelines for evaluation of probiotics in food. *Indian J Med Res*, 134(1): 22–25.

German, B., Schiffrin, E., Reniero, R., Mollet, B., Pfeifer, A., and Neeser, J.R. 1999. The development of functional foods. Lessons from the gut. *Trends Biotechnol*, 17: 492–499.

Gibbs, B.F., Zougmanb, A., Massea, R., and Mulligann, C. 2004. Production and characterization of bioactive peptides from soy hydrolysate and soy-fermented food. *Food Res Int*, 37: 123–131.

Gill, H.S., Doull, F., Rutherford, K.J., and Cross, M.L. 2000. Immunooregulatory peptides in bovine milk. *Br J Nutr*, 84: SIII-7.

Gilliland, S.E. 1989. Acidophilus milk products: A review of potential benefits to consumers. *J Dairy Sci*, 72: 2483–2494.

Gobbetti, M., De Angelis, M., Corsetti, A., and Di Cagno, R. 2005. Biochemistry and physiology of sourdough lactic acid bacteria. *Trends Food Sci Technol*, 16: 57–69.

Gobbetti, M., Rizzello, C.G., Di Cagno, R., and De Angelis, M. 2007. Sourdough lactobacilli and celiac disease. *Food Microbiol*, 24: 187–196.

Golden, B.R. and Gorbach, S.L. 1992. Probiotics for humans. In: *Probiotics: The Scientific Basis*. Fuller, R. Ed. Chapman & Hall, London. pp. 366–368.

Gorbach, S.L. 2000. Probiotics and gastrointestinal health. *Am J Gastroenterol*, 95: S2–S4.

Gorbach, S.L., Chang, T.W., and Goldin, B. 1987. Successful treatment of relapsing *Clostridium difficile colitis* with *Lactobacillus* GG. *Lancet*, 2: 1519.

Gottstein, N., Ewins, B.A., Eccleston, C. et al. 2003. Effect of genistein and daidzein on platelet aggregation, monocyte and endothelial function. *Br J Nutr*, 89: 607–616.

Goyal, N. and Gandhi, D.N. 2008. Whey, a carrier of probiotics against diarrhea. [Online]. http://www.dairyscience.info/index.php/probiotics/110wheyprobiotics.html?showall=1. Accessed August 31, 2014.

Grajek, W., Olejnik, A., and Sip, A. 2005. Probiotics, prebiotics and antioxidants as functional foods. *Acta Biochimica Polonica*, 52 (3): 665–671.

Guandalini, S. 2002. Use of *Lactobacillus*-GG in paediatric Crohn's disease. *Dig Liver Dis*, 34(suppl 2): S63–S65.

Hammes, W.P. and Tichaczek, P.S. 1994. The potential of lactic acid bacteria for the production of safe and wholesome food. *Z Lebensm Unters Forsch*, 198(30): 193–201.

Hansen, E.B. 2002. Commercial bacterial starter cultures for fermented foods of the future. *Int J Food Microbiol*, 78: 119–131.

Harada, K., Okano, C., Kadoguchi, H., Okubo, Y., Ando, M., and Kitao, S. 2003. Peroxyl radical scavenging capability of fish sauces measured by the chemiluminescence method. *Int J Mol Med*, 12: 621–625.

Harper, W.J. 2000. *Biological Properties of Whey Componenets: A Review*. The Americal Dairy Products Institute, Chicago, IL.

Hartmann, R. and Meisel, H. 2007. Food-derived peptides with biological activity: From research to food applications. *Curr Opin Biotechnol*, 18: 163–169.

Hata, Y., Yamamoto, M., Ohni, M., Nakajima, K., Nakamura, Y., and Takano, T. 1996. A placebo-controlled study of the effect of sour milk on blood pressure in hypertensive subjects. *Am J Clin Nutr*, 64(5): 767–771.

Haque, E. and Chand, R. 2006. Milk protein derived bioactive peptides. [Online]. http://www.dairyscience.info/exploitation-of-anti-microbial-roteins/111-milk-protein-derived-bioactive-peptides.html?showall=1.

Haque, E. and Chand, R. 2008. Antihypertensive and antimicrobial bioactive peptides from milk protein. *Eur J Food Res Technol*, 227: 7–15.

Haque, E., Chand, R., and Kapila, S. 2008. Biofunctional properties of bioactive peptides of milk origin. *Food Rev Int*, 25(1): 28–43. doi: 10.1080/87559120802458198.

Heller, K.J. 2001. Probiotic bacteria in fermented foods: Product characteristics and starter organisms. *Am J Clin Nutr*, 73: 374–379.

Hettiarachchy, N.S. 2012. *Bioactive Food Proteins and Peptides-Applications in Human Health*. CRC Press, Taylor & Francis Group, Boca Raton, FL. p. 315.

Hole, A.S., Rud, I., Grimmer, S., Sigl, S., Narvhus, J., and Sahlstrøm, S. 2012. Improved bioavailability of dietary phenolics in whole grain barley and oat groat following fermentation with probiotic *Lactobacillus acidophilus*, *Lactobacillus johnsonii*, and *Lactobacillus reuteri*. *J Agri Food Chem*, 60: 6369–6375.

Holman, D.J. and Jones, R.E. 1998. Longitudinal analysis of deciduous tooth emergence II: Parametric survival analysis in Bangladeshi, Guatemalan, Japanese and Javanese children. *Am J Phys Anthropol*, 105(2): 209–230.

Holzapfel, W.H. and Schillinger, U. 2002. Introduction to pre- and probiotics. *Food Res Int*, 35(2): 109–116.

Homma, N. 1988. Bifidobacteria as a resistance factor in human beings. *Bifidobacteria Microflora*, 7: 35.

Hori, G., Wang, M.F., Chan, Y.C. et al. 2001. Soy protein hydrolysate with bound phospholipids reduces serum cholesterol level in hypercholesterolemic adult male volunteers. *Bioscience Biotechnology and Biochemistry*, 65: 72–78.

IDF. 2008. International diabetes federation. *Diabetes Atlas*. http://www.eatlas.idf.org.

Isolauri, E., Juntunen, M., Rautanen, T., Sillanaukee, P., and Koivula, T. 1991. A human *Lactobacillus* strain (*Lactobacillus casei* sp. strain GG) promotes recovery from acute diarrhea in children. *Pediatrics*, 88: 90–97.

Jafari, A., Larijani, A., and Majd, B.A. 2009. Ann NutrMetab: Cholesterol-lowering effect of probiotic yogurt in comparison with ordinary yogurt in mildly to moderately hypercholesterolemic subjects. *Alter Med Rev* 54(1): 22–27.

Jakala, P. and Vapaatalo, H. 2010. Antihypertensive peptides from milk proteins. *Pharmaceuticals*, 3: 251–272.

Je, J.Y., Park, J.Y., Jung, W.K., Park, P.J., and Kim, S.K. 2005. Isolation of angiotensin I converting enzyme (ACE) inhibitor from fermented oyster sauce, *Crassostrea gigas*. *Food Chem*, 90: 809–814.

Jenkins, G.N. 1990. Cheese as a protection against dental caries. *Nutr Q*, 13: 33.

Jeyaram, K., Mohendro Singh, W., Premarani, T. et al. 2008. Molecular identification of dominant microflora associated with 'Hawaijar'—A traditional fermented soybean (Glycine max (L.)) food of Manipur, India. *Int J Food Microbiol*, 122: 259–268.

Joshi, V.K. 2012. Health benefits and therapeutic value of fermented foods. In *Health Foods: Concept, Technology and Scope*. Gupta, R.K., Bansal, S., and Mangal, M., Eds. Biotech, New Delhi. pp. 183–235.

Joshi, V.K., Attri, B.L., Gupta, J.K., and Chopra, S.K. 1990. Comparative fermentation behaviour, physico-chemical characteristics of fruit honey wines. *Indian J Hort*, 47(1): 49–54.

Joshi, V.K., Attri, B.L., and Mahajan, B.V.C. 1991. Studies on the preparation and evaluation of vermouth from plum. *J Food Sci Technol*, 28(3): 138–141.

Joshi, V.K., Bhutani, V.P., and Thakur, N.K. 1999a. Composition and nutrition of fermented products. In *Biotechnology: Food Fermentation*, Vol. I. Joshi, V.K. and Pandey, A., Eds. Educational Publishers and Distributors, New Delhi. pp. 259–320.

Joshi, V.K., Devender, A., Tuhin Kumar, S., and Ghanshyam, A. 2011. Fruit wines: Production technology. In *Handbook of Enology*, Vol. II. Joshi, V.K., Ed. Asia Tech Publishers, Inc., New Delhi. pp. 1177–1221.

Joshi, V.K. and Devi, P.M. 2009. Resvertatrol: Importance, role, contents in wine and factors influencing its production. *Proc. Nati. Acad. Sci. India*, Sect. B, 79:111.

Joshi, V.K. and Pandey, A. 1999. Biotechnology: Food fermentation. In *Biotechnology: Food Fermentation*, Vol. I. Joshi, V.K. and Pandey, A., Eds. Educational Publishers and Distributors, New Delhi. pp. 1–24.

Joshi, V.K. and Sandhu, D.K. 2000. Influence of ethanol concentration, addition of spices extract and level of sweetness on physico-chemical characteristics and sensory quality of apple vermouth. *Brazilian Arch Biol Technol*, 43(5): 537–545.

Joshi, V.K., Sandhu, D.K., and Thakur, N.S. 1999b. Fruit based alcoholic beverages. In *Biotechnology: Food Fermentation*, Vol. II. Joshi, V.K. and Pandey, A., Eds. Educational Publishers and Distributors, New Delhi. pp. 647–744.

Joshi, V.K. and Sharma, S. 2004. Importance, nutritive value and medicinal contribution of wines. *Bever Food World*, 2: 41–45.

Joshi, V.K. and Sharma, S. 2009. *Cider Vinegar: Microbiology, Technology and Quality*. Solieri, L. and Gludiet, P., Eds. Vinegars of the World. Springer-Verlag, Italy. pp. 197–207.

Joshi, V.K. and Sharma, S. 2012. A panorama of lactic acid bacterial fermentation of vegetables—A review. *Int. J. Food Ferment. Technol*, 2(1): 1–12.

Joshi, V.K., Sharma, S., Devi, P.M., and Bhardwaj, J.C. 2009b. Effect of initial sugar concentration on the physico-chemical and sensory qualities of plum wine. *J North East Foods*, 8(1,2): 1–7.

Joshi, V.K., Sharma, S., and Thakur, N.S. 2011. Lactic acid fermented foods. In: *Food Biotechnology: Principles and Practices*. Joshi, V.K. and Singh, R.S. Eds. IK International Publishing House, New Delhi. pp. 375–415.

Joshi, V.K., Sharma, S., John, S., Kaushal, B.B.L., and Rana, N. 2009a. Prepration of antioxidant rich apple and strawberry wines. *Proc Natl Acad Sci India, Sect B*, 79(IV): 415–420.

Joshi, V.K., Sharma, S., and Rana, N. 2006. Production, purification, stability and efficiency of bacteriocin from the isolate of natural lactic acid fermentation of vegetables. *Food Technol Biotechnol*, 4(3): 435–439.

Joshi, V.K. and Thakur, S. 2000. Lactic acid fermented beverage. In *Postharvest Technology of Fruits and Vegetables*. Verma, L.R. and Joshi, V.K., Eds. The Indus Publishers, New Delhi. pp. 1102–1127.

Julius, M.H., Janusz, M., and Lisowski, J. 1988. A colostral protein that induces the growth and differentiation of resting B lymphocytes. *J Immunol*, 140: 1366–1371.

Jung, W.K., Rajapakse, N., and Kim, S.K. 2005. Antioxidative activity of a low molecular weight peptide derived from the sauce of fermented blue mussel, *Mylillus edulis. Eur Food Res Technol*, 220: 535–539.

Kaila, M., Isolauri, E., Soppi, E., Virtanen, E., Laine, S., and Arvilommi, H. 1992. Enhancement of the circulating antibody secreting cell response in human diarrhea by a human *Lactobacillus* strain. *Pediatr Res*, 32: 141–144.

Kalui, C.M., Julius, M.M., and Kutima, P.M. 2010. Probiotic potential of spontaneously fermented cereal based foods—A review. *Afr J Biotechnol*, 9(17): 2490–2498.

Kar, T. and Misra, A.K. 1999. Therapeutic properties of whey used as fermented drink. *Rev de Microbiol*, 30: 163–169.

Katina, K. 2005. Sourdough: A tool for the improved flavour, texture and shelf-life of wheat bread. *VTT Publications*, 569: 1–92.

Katina, K., Juvonen, R., and Laitila, A. 2012. Fermented wheat bran as a functional ingredient in baking. *Cereal Chem*, 89: 126–134.

Katina, K., Laitila, A., and Jovonen, R. 2007. Bran fermentation as a means to enhance technological properties and bioactivity of rye. *Food Microbiol*, 24: 175–186.

Kaushik, J.K., Kumar, A., Duary, R.K., Mohanty, A.K., Grover, S., and Batish, V.K. 2009. Functional and probiotic attributes of an indigenous isolate of *Lactobacillus plantarum*. *PLoS ONE*, 4(12): e8099. doi:10.1371/journal.pone.0008099

Kim, H.S. and Gilliland, S. 1983. *Lactobacillus acidophilus* as a dietary adjunct for milk to aid lactose digestion in humans. *J Dairy Sci*, 66: 959–966.

Kinoshita, E., Yamakoshi, J., and Kikuchi, M. 1993. Purification and identification or an angiotensin I-converting enzyme inhibitor from soy sauce. *Biosci Biotech Biochem*, 57: 1107.

Kitazawa, H., Harata, T., Uemura, J., Saito, T., Kaneko, T., and Itoh, T. 1998. Phosphate group requirement for mitogenic activation of lymphocytes by an extracellular phosphopolysaccharide from *Lactobacillus delbrueckii* spp. *bulgaricus. Int J Food Microbiol*, 40: 169–175.

Kitazawa, H., Ishii, Y., Uemura, J., Kawai, Y., Sailo, T., Kaneko, T., Noda, K., and Itoh, T. 2000. Augmentation of macrophage functions by an extracellular phosphopolysaccharidc from *Lactobacillus delbrueckii* ssp. *bulgaricus. Food Microbiol*, 17: 109–118.

Kojic, M., Vujcic, M., Banina, A., Cocconcelli, P., Cerning, J., and Topisirovic, L. 1992. Analysis of exopolysaccharide production by *Lactobacillus casei* CG11, isolated from cheese. *Appl Environ Microbiol*, 58: 4086–4088.

Korhonen, H.J. 2009. Bioactive components in bovine milk. In: *Bioactive Components in Milk and Dairy Products*. Park, Y.W., Ed. Wiley-Blackwell, Oxford.

Korhonen, H. and Pihlanto, A. 2001. Food-derived bioactive peptides—Opportunities for designing future foods. *Curr Pharm Des*, 9: 1297–1308.

Korhonen, H. and Pihlanto, A. 2003. Food-derived bioactive peptides-opportunities for designing future foods. *Curr Pharm Des*, 9: 1297–1308.

Korhonen, H. and Pihlanto, A. 2006. Review bioactive peptides: Production and functionality. *Int Dairy J*, 16(9): 945–960.

Korhonen, H. and Pihlanto, A. 2007. Bioactive peptides from food proteins. In: *Handbook of Food Products Manufacturing*. Hui, Y.H. Ed. John Wiley & Sons, Inc., Hoboken, NJ. doi:10.1002/9780470113554.ch46

Kotb, E. 2012. Fibrinolytic bacterial enzymes with thrombolytic activity. In: *Fibrinolytic Bacterial Enzymes with Thrombolytic Activity*. Springer, Berlin, Heidelberg. pp. 1–74.

Kumar, R.S., Kanmani, P., Yuvaraj, N., Paari, K.A., Pattukumar, V., and Arul, V. 2011. *Lactobacillus plantarum* AS1 binds to cultured human intestinal cell line HT-29 and inhibits cell attachment by enterovirulent bacterium. *Vibrio parahaemolyticus, Lett Appl Microbiol*, 53: 481–487.

Kumar, R.S., Kanmani, P., Yuvaraj, N., Paari, K.A., Pattukumar, V., Thirunavukkarasu, C., and Arul, V. 2012. *Lactobacillus plantarum* AS1 isolated from south Indian fermented food kallappam suppress 1,2-dimethyl hydrazine (DMH)-induced colorectal cancer in male wistar rats. *Appl Biochem Biotechnol*, 166: 620–631.

Kumar, R.S., Varman, D.R., Kanmani, P., Yuvaraj, N., Paari, K.A., Pattukumar, V., and Arul, V. 2010. Isolation, characterization and identification of a potential probiont from south Indian fermented foods (*Kallappam, Koozh* and *Mor Kuzhambu*) and its use as biopreservative. *Probiotics Antimicro Prot*, 2(3): 145–151

Kurmann, J.A. and Rasic, J.L. 1991. The health potential of products containing bifidobacteria. In: *Therapeutic Properties of Fermented Milks*. Robinson, R.K., Ed. Elsevier Applied Food Science Series, London. pp. 117–158.

Kwak, S.Y., Seo, H.S., and Lee, Y.S. 2009. Synergistic antioxidative activities of hydroxyl-cinnamyl-peptides. *J Peptide Sci*, 15: 634–641.

Kwon, Y., Shen, W.L., Shim, H.S., and Montell, C. 2010. Fine thermotactic discrimination between the optimal and slightly cooler temperatures via a TRPV channel in chordotonal neurons. *J Neurosci*, 30(31): 10465–10471.

LeBlanc, A. and Perdigon, G. 2005. Reduction of b-Glucuronidase and nitroreductase activity by yoghurt in a murine colon cancer model. *Biocell*, 29(1): 15–24.

Lee, C.H. 1997. Lactic acid fermented foods and their benefits in Asia. *Food Control*, 8: 259–269.

Li, N., Zimpelmann, J., Cheng, K., Wilkins, J.A., and Burns, K.D. 2005. The role of angiotensin converting enzyme 2 in the generation of angiotensin 1–7 by rat proximal tubules. *Am J Physiol*, 288(2): F353–F362.

Lin, W.H., Wu, C.R., Fang, T.J., Guo, J.T., Huang, S.Y., Lee, M.S., and Yang, H.L. 2011. Anti-*Helicobacter pylori* activity of fermented milk with lactic acid bacteria. *J Sci Food Agric*, 91(8): 1424–1431.

Liu, C.F. and Pan, T.M. 2011. *Beneficial Effects of Bioactive Peptides Derived from Soybean on Human Health and Their Production by Genetic Engineering, Soybean and Health*. Prof. Hany El-Shemy, Ed., ISBN: 978-953-307-535-8, In Tech. http://www.intechopen.com/books/soybean-andhealth/beneficial-effects-of-bioactive-peptides-derived-from-soybean-on-human-health-and-their-production-b.

Lopez Fandino, R., Otte, J., and Van Camp, J. 2006. Physiological, chemical and technological aspects of milk-protein-derived peptides with antihypertensive and ACE-inhibitory activity. *Int Dairy J*, 16: 1277–1293.

Lourens-Hattingh, A., and Viljoen, B.C. 2001. Yogurt as probiotic carrier food—Review. *Inter Dairy J*, 11: 1–17.

Luckey, T.D. and Floch, M.H. 1972. Introduction to intestinal microecology. *Am J Clin Nutr*, 25: 1291–1295.

Macouzet, M., Lee, B.H., and Robert, N. 2009. Production of conjugated linoleic acid by probiotic *Lactobacillus acidophilus* La-5. *J Appl Microbiol*, 106: 1886–1891.

Madhu, A.N., Giribhattanavar, P., Narayan, M.S., and Prapulla, S.G. 2010. Probiotic lactic acid bacterium from kanjika as a potential source of vitamin B12: Evidence from LC-MS, immunological and microbiological techniques. *Biotechnol Lett*, 32: 503–506.

Maeno, M., Ong, R.C., Suzuki, A., Ueno, N., and Kuug, H.F. 1994. A truncated bone morphogenetic protein 4 receptor alters the fate of ventral mesoderm to dorsal mesoderm: Roles of animal pole tissue in the development of ventral mesoderm. *Proc Natl Acad Sci USA*, 91: 10260–10264.

Mallesha, R.S., Selvakumar, D., and Jagannath, J.H. 2010. Isolation and identification of lactic acid bacteria from raw and fermented products and their antibacterial activity. *Rec Res Sci Tech*, 2: 42–46.

Mann, S.V. and Spoerry, Y. 1974. Studies of a surfactant and cholesterolemia in the Maasai. *Am J Clin Nutr*, 27: 464–470.

Marnila, P., Rokka, S., Rehnberg-Laiho, L., Karkkainen, P., Kosunen, T.U., Rautelin, H., Hanninen, M.L., Syvaoja, E.L., and Korhonen, H. 2003. Prevention and suppression of *Helicobacter felis* infection in mice using colostral preparation with specific antibodies. *Helicobacter*, 8: 192–201.

Marteau, P., Seksik, P., and Jain, R. 2002. Probiotics and intestinal health effects: a clinical perspective. *Br J Nutr*, 88: S51–S57.

Martini, M.C., Bollweg, G.L., Levitt, M.D., and Savaiano, D.A. 1987. Lactose digestion by yogurt β-galactosidase: Influence of pH and microbial cell integrity. *Am J Clin Nutr*, 45: 432–436.

Matsushima, M. and Takagi, A. 2012. Is it effective? to "How to use it?": The era has changed in probiotics and functional food products against *Helicobacter pylori* infection. *J Gastroenterol Hepatol*, 27: 888–892.

McFarland, L.V., Surawicz, C.M., Greenberg, R.N., Elmer, G.W., Moyer, K.A., Melcher, S.A., Bowen, K.E., and Cox, J.L. 1995. Prevention of beta-lactamase associated diarrhea by *Saccharomyces boulardii* compared with placebo. *Am J Gastroenterol*, 90: 439–448.

McNicol, A. 1993. The effects of genistein on platelet function are due to thromboxane receptor antagonism rather than inhibition of tyrosine kinase. *Prostaglandins Leukot Essent Fatty Acids* 48: 379–384.

Meira, Q.G.S., Barbosa, I.M., Athayde, A.J.A.A., Siqueira-Junior, J.P., and Souza, E.L. 2012. Influence of temperature and surface kind on biofilm formation by *Staphylococcus aureus* from food-contact surfaces and sensitivity to sanitizers. *Food Control*, 25: 469–475.

Meisel, H. and FitzGerald, R.J. 2003. Biofunctional peptides from milk proteins: Mineral binding and cytomodulatory effects. *Curr Pharm Des*, 9: 1289–1295.

Meisel, H., Walsh, D.J., Murray, B., and FitzGerald, R.J. 2006. In *ACE Inhibitory Peptides, Nutraceutical Proteins and Peptides in Health and Disease*. Mine, Y., and Shahidi, F., Eds. Taylor & Francis, Boca Raton, FL.

Mejia, E. and de Lumen, B.O. 2006. Soybean bioactive peptides: A new horizon in preventing chronic diseases. *Sex Reprod Meno*, 4(2): 91–95.

Mercier, A., Hidalgo, R.Y., and Hamel, J.-F. 2004. Aquaculture of the Galapagos sea cucumber *Isostichopus fuscus*. In: *Advances in Sea Cucumber Aquaculture and Management*. Lovatelli, A., Conand, C., Purcell, S., Uthicke, S., Hamel, J.-F., and Mercier, A. Eds. FAO, Rome.

Micanel, N., Haynes, I.N., and Playne, M.J. 1997. Viability of probiotic cultures in commercial Australian yogurts. *Aust J Dairy Technol*, 52: 24–27.

Michihata, T. 2003. Components of fish sauce Ishiru in Nolo peninsula and its possibilities as a functional food. *Foods Ingred J*, 208: 683–602.

Minervini, F., Bilancia, M.T., Siragusa, S., Gobbetti, M., and Caponio, F. 2009. Fermented goats' milk produced with selected multiple starters as a potentially functional food. *Food Microbiol*, 26: 559–564.

Mitsuoka, T. 2000. Significance of dietary modulation of intestinal flora and intestinal environment. *Biosci Microflora*, 19(1): 15–25.

Mitsuoka, T. 2002. Prebiotics and intestinal flora. *Biosci Microflora*, 21(1): 3–12.

Moktan, B., Saha, J., and Sarkar, P.K. 2008. Antioxidant activities of soybean as affected by Bacillus fermentation to kinema. *Food Res Int*, 41: 586–593.

Moktan, B., Roy, A., and Sarkar, P.K. 2011. Antioxidant activities of cereal-legume mixed batters as influenced by process parameters during preparation of dhokla and idli, traditional steamed pancakes. *Int. J. Food Sci. Nutr*, 62(4): 360–369.

Moller, P.N., Scholz-Ahrens, K.E., Nils, R., and Schrezenmeir, J. 2008a. Bioactive peptides and proteins from foods: Indication for health effects. *Eur J Nutr*, 2: 1–12.

Moller, P.N., Scholz-Ahrens, K.E., Nils, R., and Schrezenmeir, J. 2008b. Bioactive peptides and proteins from foods: Indication for health effects. *Eur J Nutr*, 47(4): 171–182.

Moore, W.E.C., Cato, E.P., and Holdeman, L.V. 1978. Some recent concepts in intestinal bacteriology. *Am J Clin Nutr*, 31: S33–S42.

Moslehishad, M., Mirdamadi, S., Ehsani, M.R., Ezzatpanah, H., and Moosavi-Movahedi, A.A. 2013. The proteolytic activity of selected lactic acid bacteria in fermenting cow's and camel's milk and the resultant sensory characteristics of the products. *Int J Dairy Technol*, 29: 82–87.

Murai, A., Iwamura, K., Takada, M., Ogawa, K., Usui, T., and Okumura, J. 2002. Control of postprandial hyperglycaemia by galactosyl maltobionolactone and its novel anti-amylase effect in mice. *Life Sci*, 71: 1405–1415.

Nagaoka, S., Kanamaru, Y., Kojima, T., and Kuwata, T. 1992. Comparative studies on serum cholesterol lowering action of whey protein and soybean protein in rats. *Biosci Biotechnol Biochem*, 56: 1484–1485.

Nakamura, T. et al. 1995. Cloning and characterization of the *Saccharomyces cerevisiae* SVS1 gene which encodes a serine- and threonine-rich protein required for vanadate resistance. *Gene*, 165(1): 25–29.

Nakamura, Y. 2004. Studies on anti-hypertensive peptides in milk fermented with *Lactobacillus helveticus*. *Biosci Microflora*, 23(4): 131–138.

Nakashima, S., Koike, T., and Nozawa, Y. 1991. Genistein a protein tyrosine kinase inhibitor, inhibits thromboxane A2-mediated human platelet responses. *Mol Pharmacol*, 39: 475–480.

Niels, P.M., Scholz-Ahrens, K.E., Nils, R., and Schrezenmeir, J. 2008. Bioactive peptides and proteins from foods: indication for health effects. *Eur J Nutr*, 47(4): 171–182.

Nout, M.J.R. 1995. Fungal interactions in food fermentations. *Can J Bot*, 73: 1291–1300.

Nout, M.J.R. 2001. Fermented foods and their production. In: *Fermentation and Food Safety*. Adams, M.R. and Nout, M.J.R., Eds. Aspen Publishers Inc., Gaithersburg, Maryland. pp. 1–30.

Nout, M.J.R. and Kiers, J.L. 2005. Tempe fermentation, innovation and functionality: Update into the third millennium. *J Appl Microbiol*, 98: 789–805.

Ogueke, C.C., Owuamanam, C.I., Ihediohanma, N.C., and Iwouno, J.O. 2010. Probiotics and prebiotics: Unfolding prospects for better human health. *Pak J Nutr*, 9(9): 833–843.

Okamoto, M., Ogawa, H., Yoshinaga, Y., Kusunoki, T., and Odawara, O. 1995. Behavior of key elements in Pd for the solid state nuclear phenomena occurred in heavy water electrolysis. *Proc 1CCF4*, 3: 14.

Ong, L., Henriksson, A., and Shah, N.P. 2007. Development of probiotic cheddar cheese containing *Lactobacillus acidophilus*, *Lb. casei*, *Lb. paracasei* and *Bifidobacterium* spp. and the influence of these bacteria on proteolytic pattern and production of organic acid. *Int Dairy J*, 16, 446–456.

Oskar, A., Nikbin, M.S., and Russell, R.M. 2004. Yogurt and gut function. *Am J Clin Nutr*, 80(2): 245–256.

Osman, M.A. 2004. Changes, in sorghum enzyme inhibitors, phytic acid, tannins, and *in vitro* protein digestibility occurring during Khamir (local bread) fermentation. *Food Chem*, 88: 129–134.

Ozaki, Y., Yatomi, Y., Jumai, Y., and Kume, S. 1993. Eiiects of genistein, a tyrosine kinase inhibitor, on platelet functions: Genistein attenuates thrombin-induced CaCa²⁺ mobilization in human platelets by affecting polyphosphoinositide turnover. *Biochem Pharmacol*, 46: 395–403.

Pais, E., Alexy, T., Holsworth, R.E., and Meiselman, H.J. 2006. Effects of nattokinase, a pro-fibrinolytic enzyme, on red blood cell aggregation and whole blood viscosity. *Clin Hemorheol Microcirc*, 35: 139–142.

Panesar, P.S. 2011. Fermented dairy products: Starter cultures and potential nutritional benefits. *Food Nutr Sci*, 2: 47–51.

Panesar, P.S., Joshi, V.K., Panesar, R., and Abrol, G.S. 2011. Vermouth: Technology of production and quality characteristics. In: *Advances in Food and Nutritional Research*, Taylor, S. Ed. vol. 63. Elsevier, Inc., London. pp. 253–271.

Panesar, S.P., Narender Kumar, Marwaha, S.S., and Joshi, V.K. 2009. Vermouth production technology—An overview. *Nat Product Radiant*, 8: 334–341.

Papadimitriou, C.G., Vafopoulo-Mastrojiannaki, A., Viera Silva, S., Gomes, A.M., Malcata, F.X., and Alichanidis, E. 2007. Identification of peptides in traditional and probiotic sheep milk yoghurt with angiotensin I-converting enzyme (ACE)-inhibitory activity. *Food Chem*, 15: 647–656.

Patwardhan, B., Vaidya, A.O.B., and Chorghade, M. 2004. Ayurveda and natural products drug discovery. *Cur. Sci*, 86(6): 789–799.

Peralta, E., Hatate, H., Watanabe, D., Kawabe, D., Murata, H., and Kama, Y. 2005. Antioxidative activity of Philippine salt-fermented shrimp paste and variation of its contents during fermentalion. *Oleo Sci*, 54: 553–558.

Peralta, E.M., Hatate, H., and Kawabe, D. 2008. Improving antioxidant activity and nutritional components of Philippine salt-fermented shrimp paste through prolonged fermentation. *Food Chem*, 111: 72–77.

Perna, A., Intaglietta, I., Simonetti, A., and Gambacorta, E., 2014. Antioxidant activity of yogurt made from milk characterized by different casein haplotypes and fortified with chestnut and sulla honeys. *J. Dairy Sci.* 97: 1–9.

Perpetuo, E.A., Juliano, L., and Lebrun, I. 2003. Biochemical and pharmacological aspects of two bradykinin-potentiating peptides from tryptic hydrolysis of casein. *J Prot Chem*, 22: 601–606.

Phelan, M., Aherne, A., and FitzGerald, R.J. 2009. Casein-derived bioactive peptides: Biological effects, industrial uses, safety aspects and regulatory status. *Int Dairy J*, 19: 643–654.

Phelan, M. and Kerins, D. 2011. The potential role of milk derived peptides in cardiovascular diseases. *Food Funct*, 2: 153–167.

Pigeon, R.M., Cuesta, E.P., and Gilliland, S.E. 2002. Binding of free bile acids by cells of yoghurt starter culture bacteria. *J Dairy Sci*, 85: 2705–2710.

Pihlanto, A. 2006. Antioxidative peptides derived from milk proteins. *Int Dairy J*, 16: 1306–1314.

Rafter, J. 2004. The effects of probiotics on colon cancer development. *Nutr Res Rev*, 17: 277–284. doi:10.1079/NRR200484.

Rai, A.K., Prakash, M., and Anu Appaiah, K.A. 2010. Production of Garcinia wine: Changes in biochemical parameters, organic acids and free sugars during fermentation of Garcinia must. *Int J Food Sci Technol*, 45: 1330–1336.

Rajwani, H.G. and Ananthanarayan, L. 2013. Angiotensin-I converting enzyme (ACE) inhibitory activity of fermented idli batter as influenced by various parameters prevailing during fermentation. *Int. J. Food Ferment. Technol*, 3(1): 71–77.

Rapsang, G.F., Kumar, R., and Joshi, S.R. 2011. Identification of *Lactobacillus pobuzihii* from *tungtap*: A traditionally fermented fish food, and analysis of its bacteriocinogenic potential. *Afr J Biotechnol*, 10(57): 12237–12243.

Rattanachaikunsopon, P. and Phumkhachorn, P. 2010. Lactic acid bacteria: Their antimicrobial compounds and their uses in food production. *Ann Biol Res*, 1(4): 218–228.

Reddy, L.V.A., Joshi, V.K., and Reddy, O.V.S. 2011. Utilization of tropical fruits for wine production with special emphasis on mango (*Mangifera indica* L.) wine. *MicroSust Agri Biotechnol*, 30:679–710.

Rivera-Espinoza, Y. and Gallardo-Navarro, Y. 2010. Non-dairy probiotic products. *Food Microbiol*, 27(1): 1–11.

Ross, R.P., Morgan, S., and Hill, C. 2002. Preservation and fermentation: Past, present and future. *Int J Food Microbiol*, 79: 3–16.

Roy, A., Khanra, K., Mishra, A., and Bhattacharya, N. 2012. General analysis and antioxidant study of traditional drink Handia, its concentrate and volatiles. *Adv. Life Sci. Appl*, 1(3): 54–57.

Roy, B., Kala, P., Farooquee, N., and Majila, B.S. 2004. Indigenous fermented food and beverages: A potential for economic development of the high altitude societies in Uttaranchal. *J Hum Ecol*, 15(1): 45–49.

Ryan, J., Thomas, R., Paul, R., Declan, B., Gerald, F.F., and Stanton, C. 2011. Bioactive peptides from muscle sources: Meat and fish. *Nutrients*, 3(9): 765–791.

Rybka, S. and Kailasapathy, K. 1995. The survival of culture bacteria in fresh and freeze-dried AB yoghurts. *Aust J Dairy Technol*, 50: 51–57.

Salwa, A.A., Galal, E.A., and Elewa, N.A. 2004. Carrot yoghurt: Sensory, chemical, microbiological properties and consumer acceptance. *Pak J Nutr*, 3(6): 322–330.

Sanders, M.E., Morelli, L., and Tompkins, T.A. 2003. Sporeformers as human probiotics: *Bacillus, Sporolactobacillus*, and *Brevibacillus. Comp Rev Food Sci Food Saf*, 2: 101–110.

Saran, S. 2004. Use of fermented foods to combat stunting and failure to thrive: Background of the study. *Nutrition*, 20: 577–578.

Sarkar, P.K., Jones, L.J., Gore, W., Craven, G.S., and Somerset, S.M. 1996. Changes in soya bean lipid profiles during Kinema production. *J Sci Food Agri*, 71(3): 321–328.

Sarmadi, B.H. and Ismail, A. 2010. Antioxidative peptides from food proteins: A review. *Peptides*, 31: 1949–1956.

Sekar, S. 2007. Traditional alcoholic beverages from Ayurveda and their role on human health. *Indian J Trad Know*, 6(1): 144–149.

Sekar, S. and Mariappan, S. 2007. Usage of traditional fermented products by Indian rural folks and IPR. *Indian J Trad Know*, 6(1): 111–120.

Seppo, L., Jauhiainen, T., Poussa, T., and Korpela, R. 2003. A fermented milk high in bioactive peptides has a blood pressure-lowering effect in hypertensive subjects. *Am J Clin Nutr*, 77: 326–330.

Serafeimidou, A., Zlatanos, S., Kritikos, G., and Tourianis, A. 2013. Change of fatty acid profile, including conjugated linoleic acid (CLA) content, during refrigerated storage of yogurt made of cow and sheep milk. *J Food Comp Anal* 31: 24–30.

Shah, N. 2003. *Bifidobacterium* spp.: Applications in fermented milks. In: *Encyclopedia of Dairy Sciences*. Roginski, H., Fuquay, J.W., and Fox, P.F., Eds. Academic Press, London. pp. 147–149.

Sharma, R., Sharma, S.K., Kumar, R., and Kamboj, P. 2003. Food preservation of biological origin—An overview. *Bev Food World*, 30(7): 36–40.

Sharma, R., Joshi, V.K., and Abrol, G.S. 2012. Fermented fruit and vegetable products as functional foods—An overview. *Indian Food Packer*, 66(4): 45–53.

Shimizu, M. 2004. Food-derived peptides and intestinal functions. *BioFactors*, 21: 43–47.

Shon, M.-Y., Lee, J., Choi, J.-H., Choi, S.-Y., Nam, S.-H., Seo, K.-I., Lee, S.-W., Sung, N.-J., and Park, S.-K. 2007. Antioxidant and free radical scavenging activity of methanol extract of chungkukjang. *J Food Compos Anal*, 20: 113–118.

Shruthy, V.V., Pavithra, M., and Ghose, A. 2011. Probiotic potential among lactic acid bacteria isolated from curd. *IJRAP*, 2(2): 602–609.

Singh, H.B., Sharma, B.R., and Pradhan, B. 1995. Ethnobotanical observation on the preparation of millet beer in Sikkim State, India. In: *Ethnobotany of Medicinal Plants of Indian Subcontinent*. Maheshwari, J.K., Ed., Scientific Publishers, Jodhpur. pp. 580–582.

Singh, P.K. and Singh, K.I. 2006. Traditional alcoholic beverage, Yu of Meitei communities of Manipur. *Indian J Trad Know*, 5(2): 184–190.

Singh, T.A., Devi, K.R., Ahmed, G., and Jeyaram, K. 2013. Microbial and endogenous origin of fibrinolytic activity in traditional fermented foods of Northeast India. *Food Res Int*. 55: 356–362. doi:10.1016/j.foodres.11.028.

Singh, T.K., Fox, P.F., and Healy, A. 1997. Isolation and identification of further peptides in the diafiltration retentate of the water-soluble fraction of Cheddar cheese. *J Dairy Res*, 64: 433–443.

Sohail, M.U., Hume, M.E., Byrd, J.A., Nisbet, D., Ijaz J., Sohail, A.A., Shabbir, M.Z., and Rehman, H. 2012. Effect of supplementation of prebiotic mannan-oligosaccharides and probiotic mixture on growth performance of broilers subjected to chronic heat stress. *Poultr Sci*, 91(9): 2235–2240.

Srividya, A.R. and Vishnuvarthan, V.J. 2011. Probiotic: A rational approach to use Probiotic as medicine. *Int J Pharm Fron Res*, 1(1): 126–134.

Steinkraus, K.H. 1996. *Handbook of Indigenous Fermented Foods*, 2nd edition. Marcel Dekker, New York, NY.

Stockely, C.S. 2011. Therapeutic value of wine: Clinical and Scientific prospective. In: *Handbook of Enology Principles, Practices and Recent Innovations*, vol II, Joshi, V.K. Ed. Asiatech Publishers, New Delhi. pp. 146–208.

Sukumar, G., and Ghosh, A.R. 2010. *Pediococcus* spp.—A potential probiotic isolated from Khadi (an Indian fermented food) and identified by 16S rDNA sequence analysis. *Afr. J. Food Sci*, 4(9): 597–602.

Sukumar, G. and Ghosh, A.R. 2011. Anti-oxidative potential of probiotic bacteria from Indian fermented food. *Int J Res Ayurveda Phar*, 2(3): 23–30.

Sutas, Y., Soppi, E., Korhonen, H., Syvaoja, E.L., Saxelin, M., Rokka, T., and Isolauri, E. 1996. Suppression of lymphocyte proliferation *in vitro* by bovine caseins hydrolyzed with *Lactobacillus casei* GG-derived enzymes. *J Allergy Clin Immunol*, 98: 216–224.

Szajewska, H. and Mrukowicz, J.Z. 2001. Probiotics in the treatment and prevention of acute infectious diarrhea in infants and children: A systematic review of published randomized, double-blind, placebo-controlled trials. *J Pediatr Gastroenterol Nutr*, 33(suppl 2): S17–S25.

Takenaka, T. 2011. Properties of angiotensin I-converting enzyme (ACE). In: *Soybean—Biochemistry, Chemistry and Physiology*. Tzi-Bun Ng, Ed. InTech, Rijeka, Croatia. pp. 21–30.

Tamang, J.P. 1998. Role of microorganisms in traditional fermented foods. *Indian Food Ind*, 17(3): 162–167.

Tamang, J.P. 2010. Diversity of fermented foods. In: *Fermented Foods and Beverages of the World*. Tamang, J.P. and Kailasapathy, K., Eds. CRC Press, Boca Raton. pp. 41–84.

Tamang, J.P., Chettri, R., and Sharma, R.M. 2009. Indigenous knowledge on North-East women on production of ethnic fermented soybean foods. *Indian J Trad Know*, 8(1): 122–126.

Tamang, J.P., Dewan, S., Thapa, S., Olasupo, N.A., Schillinger, U., and Holzapfel, W.H. 2000. Identification and enzymatic profiles of predominant lactic acid bacteria isolated from soft-variety *chhurpi*, a traditional cheese typical of the Sikkim Himalayas. *Food Biotechnol*, 14(1–2): 99–112.

Tamang, J.P., Tamang, N., Thapa, S., Dewan, S., Tamang, B., Yonzan, H., Rai, A.K., Chettri, R., Chakrabarty, J., and Kharel, N. 2012. Microorganisms and nutritional value of ethnic fermented foods and alcoholic beverages of North East India. *Indian J Trad Know*, 11(1): 7–11.

Thapa, N., Pal, J., and Tamang, J.P. 2004. Microbial diversity in Ngari, Hentak and Tungtap, fermented fish products of North-East India. *World J Microbiol Biotechnol*, 20(6): 599–607.

Thapa, N., Pal, J., and Tamang, J.P. 2006. Phenotypic identification and technological properties of lactic acid bacteria isolated from traditionally processed fish products of the Eastern Himalayas. *Int J Food Microbiol*, 107: 33–38.

Thapar, N. and Sanderson, I.R. 2004. Diarrhoea in children: An interface between developing and developed countries. *Lancet*, 363: 641–653.

Thyagaraja, N. and Hosono, A. 1993. Antimutagenicity of lactic acid bacteria from 'Idli' against food-related mutagens. *J Food Prot*, 56: 1061–1066.

Thyagaraja, N. and Hosono, A. 1994. Binding properties of lactic acid bacteria from 'Idli' towards food-borne mutagens. *Food Chem Toxic*, 32: 805–809.

Todorov, S.D. 2008. Bacteriocin production by Lactobacillus plantarum AMA-K isolated from Amasi, a Zimbabwean fermented milk product and study of adsorption of bacteriocin AMA-K to *Listeria* spp. *Braz J Microbiol*, 38: 178–187.

Valiathan, M.S. 2003. *The Legacy of Caraka*. Orient Longman Private Ltd, Hyderabad. pp. 125–127.

Van Niel, C.W., Feudtner, C., Garrison, M.M., and Christakis, D.A. 2002. *Lactobacillus* therapy for acute infectious diarrhea in children: A meta-analysis. *Pediatrics*, 109: 678–684.

Vercruysse, L., Van Camp, J., and Smagghe, G. 2005. ACE inhibitory peptides derived from enzymatic hydrolysates of animal muscle protein: A review. *J Agric Food Chem*, 53: 8106–8115.

Vermeirssen, V., Van Camp, J., and Verstraete, W. 2002. Optimisation and validation of an angiotensin-converting enzyme inhibition assay for the screening of bioactive peptides. *J Biochem Biophys Methods*, 51: 75–87.

Vermeirssen, V., van Camp, J., and Verstraete, W. 2004. Bioavailability of angiotensin I-converting enzyme inhibitory peptides. *Br J Nut*, 92: 357–366.

Vijayendra, S.V.N., Palanivel, G., Mahadevamma, S., and Tharanathan, R.N. 2008a. Physico-chemical characterization of an exopolysaccharide produced by a non-ropy strain of *Leuconostoc* sp. CFR 2181 isolated from *dahi*, an Indian traditional lactic fermented milk product. *Carbohyd Poly*, 72(2): 300–307.

Vijayendra, S.V.N., Palanivel, G., and Tharanathan, R.N. 2008b. Physico-chemical characterization of an exopolysaccharide produced by a non-ropy strain of *Leuconostoc* sp. CFR 2181 isolated from *dahi*, an Indian traditional lactic fermented milk product. *Carbohyd Poly*, 72: 300–307.

Vinderola, G., Matar, C., and Perdigón, G. 2007b. Milk fermented by *Lactobacillus helveticus* R389 and its non-bacterial fraction confer enhanced protection against *Salmonella enteritidis* serovar. *Typhimurium* infection in mice. *Immunobiology*, 212: 107–118.

Vinderola, G., Matar, C., Palacios, J., and Perdigón, G. 2007a. Mucosal immunomodulation by the non-bacterial fraction of milk fermented by *Lactobacillus helveticus* R389. *Int. J. Food Microbiol*, 115: 180–186.

Wacher, C., Díaz-Ruiz, G., and Tamang, J.P. 2010. Fermented vegetable products. In: *Fermented Foods and Beverages of the World*. Tamang, J.P. and Kailasapathy, K., Eds. CRC Press, Boca Raton. pp. 149–190.

Wang, K.Y., Li, S.N., Liu, C.S., Perng, D.S., Su, Y.C., Wu, D.C., Jan, C.M., Lai, C.H., Wang, T.N., and Wang, W.M. 2004. Effects of ingesting Lactobacillus- and Bifidobacterium-containing yogurt in subjects with colonized Helicobacter pylori. *Am J Clin Nutr*, 80(3): 737–741.

Wang, X., Auler Augusto, S., Edwards, R.L., Hai, C., Emi, I., Yongjin, W., Xinggong, K., and Maniko, S. 2007. Millennial-scale precipitation changes in southern Brazil over the past 90,000 years. *Geophys Res Lett*, 34: L23701.

Warny, M., Fatima, A., and Bosrwick, E.F. 1999. Bovine immunoglobulin concentrate *Clostridium difficile* retains *C. difficile* toxin neutralizing activity after passage through the human stomach and small intestine. *Gut*, 44(2): 212–217.

Wergedahl, H., Liaset, B., Gudbrandsen, O.A., Lied, E., Espe, M., Muna, Z., Mork, S., and Berge, R. K. 2004. Fish protein hydrolysate reduces plasma total cholesterol, increases the proportion of HDL cholesterol, and lowers acyl-CoA:cholesterol acyltransferase activity in liver of Zucker rats. *J. Nutr*, 134: 1320–1327.

Werner, G.H., Floch, F., Migliore-Samour, D., and Jolles, P. 1986. Immunomodulating peptides. *Experientia*, 42: 521–531.

Wood, B.J.B. and Hodge, M.M. 1985. Yeast-lactic acid bacteria interaction. In: *Microbiology of Fermented Foods*. Wood, B.J.B., Ed. Elsevier Applied Science Publishers, England, UK. pp. 263–293.

Xu, Q., Yajima, T., Li, W., Saito, K., Ohshima, Y., and Yoshikai, Y. 2005. Levan (β-2, 6-fructan), a major fraction of fermented soybean mucilage, displays immunostimulating properties via Toll-like receptor 4 signalling: induction of interleukin-12 production and suppression of T-helper type 2 response and immunoglobulin E production. *Clin Exp Allergy*, 36: 94–101.

Yadav, H., Jain, S., and Sinha, P.R. 2007. Production of free fatty acids and conjugated linoleic acid in probiotic dahi containing *Lactobacillus acidophilus* and *Lactobacillus casei* during fermentation and storage. *Int Dairy J*, 17: 1006–1010.

Yamamoto, N., Macno, M., and Takano, T. 1999. Purification and characterization of an antihypertensive peptide from a yogurt-like product fermented by *Lactobacillus helveticus* CPN4. *Dairy Set*, 82: 1388–1393.

Yang, Y.J., Chuang, C.C., Yang, H.B., Lu, C.C., and Sheu, B.S. 2012. *Lactobacillus acidophilus* ameliorates *H. pylori*-induced gastric inflammation by inactivating the Smad7 and NF-κB pathways. *BMC Microbiol*, 12: 38.

Yuan, G.F. Chen, X.E., and Li, D. 2014. Conjugated linolenic acids and their bioactivities: A review. *Food Funct*, 5(7): 1360–1368. doi:10.1039/c4fo00037d.

Zhang, R. and Delworth, T.L. 2005. Simulated tropical response to a substantial weakening of the Atlantic thermohaline circulation. *J Clim*, 18(12): 1853–1860.

Zhang, X. and Beynan, A.C. 1993. Lowering effect of dietary milk—Whey protein v casein on plasma and liver cholesterol concentration in rats. *Br J Nutr*, 70: 139–146.

7

CEREAL-BASED NON-ALCOHOLIC INDIGENOUS FERMENTED FOODS

ANILA KUMARI, ANITA PANDEY, ANUP RAJ, ANUPAMA GUPTA, ARINDAM ROY, B.L. ATTRI, BHANU NEOPANY, BIJOY MOKTAN, C.K. SUNIL, CHETNA JANVEJA, DEV RAJ, DORJEY ANGCHOK, GHULAM MUEEN-UD-DIN, GITANJALI VYAS, JAHANGIR KABIR, KONCHOK TARGAIS, L.V.A. REDDY, LOK MAN S. PALNI, MANAS R. SWAIN, NIVEDITA SHARMA, PRABHAT K. NEMA, RAMAN SONI, RAMESH C. RAY, RUPESH S. CHAVAN, S.S. THORAT, S.K. SONI, SAVITRI, SHRADDHA R. CHAVAN, SWATI KAPOOR, TEK CHAND BHALLA, TSERING STOBDAN, AND V.K. JOSHI

Contents

7.1 Introduction

7.1.1 Indigenous Cereal Fermented Foods

The production of indigenous fermented foods and beverages pre-dates historical records (Joshi and Pandey, 1999; Steinkraus, 1996; Tamang, 2010a,b). Such foodstuffs are widely consumed around the world, contributing approximately 20% of the total food consumption (Kwon, 1994). Many of the indigenous fermented products are valued for their taste, flavor, and aroma (FAO, 1999, ch 1, 2), and, among the indigenous fermented foods, cereal-based varieties occupy an important place in the cuisine of many countries and regions, including South Asia (FAO, 1999, ch 1, 2 Adams and Moss, 1996).

The countries of South Asia exhibit climatic, ethnic, and religious diversities, resulting in the production of a number of traditional fermented foods and beverages, especially cereal-based foods. Some of these are used as beverages and breakfasts or snack foods, while others are consumed as staples (FAO, 1999, ch 2). Large quantities of acid-leavened bread and pancakes are consumed daily in South Asian countries such as India, Sri Lanka, Pakistan, and Nepal, *idli*, *dosa*, *kulcha*, *khaman*, *vada*, and *dhokla* are produced primarily in India and Sri Lanka, while *jalebis* are consumed throughout India, Nepal, and Pakistan (FAO, 1999, ch 2). Fermented cereals are particularly important as weaning foods for infants and as dietary staples for adults. Fermented foods and beverages have almost always featured in India's national cuisine, not only this a significant amount of diversity prevails in the food habits of people living in different states of the country (Savitri and Bhalla, 2013), which, in turn, has given rise to a large variety of traditional fermented foods (Aachary et al., 2011; Soni and Sandhu, 1990).

The skills of food preservation existed even in the native peoples, and this knowledge was propagated orally (Farnworth, 2008; NPCS Board, 2012; Prajapati, 2003) as is apparent in several states of India. Some products, such as *bhaturu*, *siddu*, *chilra*, *marchu*, *manna*, *dosha*, *pinni*, *seera*, etc., are unique to district of kullu, Mandi, and Bilaspur of Indian states like Himachal Pradesh (Figure 7.1). In the Indian tribal districts of Lahaul and Spiti, and Kinnaur, a large variety of fermented foods is prepared daily, during special occasions, or for consumption during journeys. Traditional starter cultures such as "*malera*" and "*treh*" are used as inocula in the preparation of

Figure 7.1   Some cereal-based fermented foods (*chilra*, *marchu*, *tchog*, *bhatooru*, *siddu*, and *seera*) prepared in Himachal Pradesh. (Copyright Dr. T.C. Bhalla and Dr. Savitri.)

these foods, while natural fermentation (i.e., without the addition of inoculum, as microorganisms present in the raw materials carry out fermentation) is used in the production of *seera, sepubari, borhe,* etc. (Thakur et al., 2004).

The people of Orissa (a state in the east of India) also have a tradition of relishing a variety of cakes, locally called *pithas,* produced from the fermentation of cereal–legume batters and made during various festivals and rituals (Roy et al., 2007). These products include *chitou, arisa, chakuli, chhunchipatra, endure, munha,* and *poda pitha.* These *pithas* are unknown to the scientific community and require scholarly study—neverthless are known to be delicious and easily digestible. Not only this, these are also suitable for ailing persons, pre-or post-natal women, and children, and are consumed by all the communities, irrespective of caste and creed (Roy et al., 2007). These foods form an important component of the diet of the tribal people in Arunachal Pradesh in northeast India, too, and are known to play a vital role in their traditional lifestyle as well (Tiwari and Mahanta, 2007). In addition, *paddu* is one of the famous fermented, shallow fat-fried products of south India, where it is generally consumed as a breakfast or snack item. Overall, the fermented foods and beverages of India have been reviewed earlier and discussed extensively (Joshi and Pandey, 1999; Padmaja and George, 1999).

Wheat is used for making flours, such as maida and semolina, that are ultimately utilized in the preparation of a variety of food products, including fermented foods (Figure 7.2). Rural women in Maharashtra (western India), during hot summers, using their indigenous knowledge, prepare the very popular *papadi, kurdai,* and *sandage* from wheat, sorghum, and *ragi* (finger millet) cereal crops. However, a large variation in production techniques of various fermented foods exists from region-to-region. This knowledge has been collected from the literature (e.g., Joshi et al., 2012; Padmaja and George, 1999; Kanwar et al., 2007; Tamang et al., 1988, 2012); Thakur et al., 2004) and from the results of survey carried out and documented. Similar examples of other South Asian countries can also be cited (Vijaylakshmi, 2005). Many of the conventional food products are made only at home (Soni and Arora, 2000) for household consumption, although, as a source of income, many women also cook such foods at home or on a place setting in small wooden carts to sell to the wider local population.

Fish and rice are the staple foods in Bangladeshi cuisine, but with a small portion used to make various indigenous fermented foods, including fermented beverages made from rice. A wide variety of traditional fermented products are available in Bangladesh, including *modhubhat* (a fermented foodstuff prepared from germinated rice powder and boiled rice), *kanjibhat* (fermented rice), *pantabhat* (fermented cooked rice), *zilapi* (a savory product made from fermented Bengal gram), *vapa pitha* (fermented snacks), and *bundia* (fermented Bengal gram) (Batra, 1981; Das et al., 2012; Sankaran, 1998; Steinkraus, 1996). Similarly, fermented foods from cereals are prepared and consumed in pakistan, another country of South Asia. (Parveen and Hafiz, 2003).

(a)

(b)

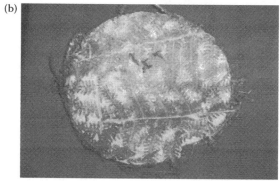

Figure 7.2   (a) Some of the cereal-based fermented foods. (b) Murcha the starter culture. (Adapted from Soni, S.K. and Sandhu, D.K. 1999. *Biotechnology: Food Fermentation, Microbiology, Biochemistry and Technology*, Vol II. Educational Publishers and Distributors, New Delhi, pp. 895–949.)

7.1.2 Cereal Fermentation

Worldwide, cereal grains constitute a major source of dietary nutrients, and are the principal source of energy, carbohydrates, protein, B-complex vitamins, vitamin E, fiber, and some micronutrients in the diets of many poor people (Belton and Taylor, 2004; Arendt and Zannini, 2013). Indeed, cereal grains were one of man's earliest sources of nutrition (Matz, 1971).

One way of processing cereal grains into food is through fermentation (Adeyemi and Beckley, 1986; Akingbala et al., 1981; Akinrele, 1970; Fields et al., 1981), which produces a variety of fermented foods. They can be fermented individually or in combination with some other legumes, such as Bengal gram and soya beans. Only two cereals, wheat and rye, are suitable for use in the preparation of leavened bread. The fermentation of cereals increases their relative nutritive value and available lysine (an important amino acid). Furthermore, bacterial fermentations involving proteolytic activity increase the biological availability of essential amino acids more than yeast fermentations, which mainly degrade carbohydrates while starch and fiber tend to decrease (FAO, 1999, ch 2). It does not however alter the mineral content of the product, but the hydrolysis of chelating agents such as phytic acid during fermentation improves the bioavailability of minerals while changes in the vitamin content of cereals with fermentation vary according to the fermentation process, and the raw material used. B group vitamins generally show an increase in fermentation, thus fermentation may be the most simple and economical way of improving their nutritional value, sensory properties, and functional qualities (Pereira et al., 2002; Al-Aseeri et al., 2003; FAO, 1999, ch 2) (For more details on changes during fermentation, see Chapter 3).

Unlike fruit and milk fermentations, cereal fermentation requires a saccharification process, which is not accomplished easily. A primitive method of cereal saccharification was achieved by chewing the raw cereals and spitting them into a vessel so as to allow saccharification to occur through the action of salivary amylase. It can also be accomplished through the malting process, which occurs naturally through the wet damage of cereals during storage, and is used frequently for beer making. But, in Asia, instead of the malting process, fermentation starters prepared from the growth of molds on raw or cooked cereals is more commonly practiced (FAO, 1999, ch 3). So use of starter cultures in fermentation at house hold level has been found useful (Holzapfel, 1997).

Murcha is a traditional starter used to produce several beverages by the indigenous people of the Darjeeling hills (Sekar and Mariappan, 2007; Tamang and Sarkar, 1995). Tamang et al. (1988) illustrated the sequential addition of ingredients in *murcha*. Rice is soaked in water overnight (10–12 h), and then placed in a large wooden mortar. A few pieces of the root of *sweto-chitu* (*Plumbago zelanica* L.), a few leaves of *vimsen pathe* (*Buddleja asiatica* Lour), and certain spices such as ginger and chilli are blended using a heavy wooden pestle (Tamang et al., 1988). Water is added to make a thick paste, which is kneaded to give a flattened cake of 2–5 cm in diameter. Old *murcha* is powdered and sprinkled over the new cakes, which are then, wrapped in fern fronds (Figure 7.2b). The cakes are placed above kitchen oven over a straw-covered bamboo floor and covered successively with fresh fronds, dried fronds, straw, or simply with rice straw, and, finally, sackcloth. At a temperature of 25–35°C, drying takes normally 1–3 days. The cakes are dried in sun, and remain active for several months at room temperature (Sekar and Mariappan, 2007). The microflora of *murcha* was reported to contain *P. pentasaceus*, *Saccharomycopsis fibuligera*, *Pichia anomala*, *Mucor circinelloides*, and *R. chinensis*. Among the molds, *S. fibuligera* has amylolytic activity (Sekar and Mariappan, 2007; Tamang and Sarkar, 1995).

A typical cereal fermentation with molds and other microorganisms converts unpalatable carbohydrates of low digestibility and proteins into palatable sugars and amino acids, respectively, with a high conversion efficiency (FAO, 1999, ch 3).

7.2 Bread Types and Production

7.2.1 Bread

7.2.1.1 Types of Bread

There is considerable variation in bread types, differentiated according to a number of criteria as follows:

- *Leavening*: Breads may be unleavened, such as *chapatti*; leavened with yeast, such as pan breads; chemically leavened, such as Irish soda bread; or leavened with bacteria, such as salt-rising bread.
- *Formulation*: Many breads around the world are made with lean formulas comprising wheat flour, water, yeast, and salt, while others, such as certain premium breads, also contain large amounts of additional ingredients, such as fat, eggs, and milk.
- *Shape and size*: A virtually endless variety of shapes and sizes have evolved, partly because some breads are baked in pans while others are baked in open hearths.
- *Specific volume*: This factor can vary from the low-density, high-volume, white, pan breads to the dense, dark breads.
- *Crust characteristics*: Some breads, such as Vienna loaves, have thin crispy crusts, while others, such as German pumpernickel breads have thick crusts. Colors can range from quite light to very dark, too.
- *Crumb characteristics*: Continuous mix breads in the United States have a very fine, uniform crumb structure, while other breads, such as French baguettes, have coarse, irregular structures. Some Middle Eastern breads have little crumb structure at all, being comprised largely of crust because of the high baking temperatures that are used.

7.2.1.2 Role of Ingredients in Bread Making

Production of bread requires key ingredients such as flour, yeast, and water. Each ingredient has its own significance, both in conventional as well as in sourdough-based product manufacture (Chavan and Chavan, 2011). Flour is the most important ingredient in bread making as it gives the specific characteristics to the bakery products. It consists of protein, starch and other carbohydrates, ash, fibers, lipids, water, and small amounts of vitamins, minerals, and enzymes. Wheat flour is the most common flour used. Gluten formed during the preparation of dough forms a cohesive and elastic network, which gives wheat its functional properties (Chavan and Chavan, 2011). *Saccharomyces cerevisiae* is the most common yeast used in bread making. Yeast cells metabolize fermentable sugars (glucose, fructose, sucrose, and maltose) under anaerobic conditions producing carbon dioxide, which acts as a leavening agent. Water is necessary for the formation of dough and is responsible for its moisture. It is used for the dissolution of salt and sugars, and assists in the dispersion of yeast cells. Sugars are normally used by yeast during the early stages of fermentation. Later, more sugars are released for gas production by the action of enzymes in the flour. Salt strengthens the gluten and controls the action of yeast, and, therefore, controls the loaf volume and improves flavor (Chavan and Chavan, 2011). Salt also favors the action of amylases, helping to maintain a supply of maltose as food for the yeast. Lipids are an optional ingredient in bread, but can improve dough handling and crumb appearance and contribute to the product flavor (Chavan and Kadam, 1989; Giannou et al., 2003).

7.2.1.3 Production of Bread Bread is one of the oldest of the prepared foods, dating back to the Neolithic era (Panesar and Marwaha, 2013), and is one of the most commonly used baked foods and has traditionally been an important factor in human nutrition. In many countries, bread consumption accounts for a large part of the diet. The earlier breads produced were cooked versions of a grain paste, made from ground cereal grains and water. Bread was first baked in Egypt more than 6000 years ago, and, from that time and place, it traveled West, varying in ingredients, flavor, texture, and form. European and Middle Eastern immigrants brought recipes for their native breads that were handed down to second- and third-generation populations. It is known to have been produced in Roman times, and is now baked in all regions of the world, including South Asian countries.

Bread is often made from a wheat flour dough that is inoculated with yeast, allowed dough to rise, and is then, baked in an oven (Figure 7.3). Owing to its high levels of gluten, which gives the dough sponginess and elasticity, wheat is the most common grain used in the preparation of bread, though it is also made from the flours of rye, barley, maize, and oats, *albeit* usually in combination with wheat flour.

The protein of the wheat flour is unique, in that, when the flour is mixed with water in certain proportions, the protein forms an elastic colloidal mass, or dough, that can hold gas and forms a spongy structure, when baked (Panesar and Marwaha, 2013). The bread dough is usually leavened with bread yeasts by fermenting the sugars in the dough and producing mainly carbon dioxide and alcohol. However, the activity of gas-forming microorganisms such as wild yeasts, coliform bacteria, saccharolytic

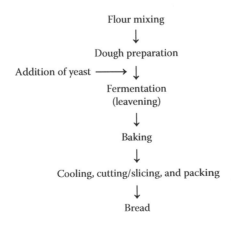

Figure 7.3 Flowchart for the preparation of bread.

Clostridium species, and heterofermentative lactic acid bacteria (LAB) can also be used instead of bread yeasts for leavening (NPCS Board, 2012).

The baking process is the last and most important step in the production of bread. Through the agency of heat, an unpalatable dough mass is transformed into a light, porous, easily digestible, and very appetizing product. The biological activities that take place in the dough are stopped by baking, with the destruction of microorganisms and enzymes present. The unstable colloidal system is then, stabilized through leavening (NPCS Board, 2012). The basic characteristics of the native starch and proteins are drastically changed. At the same time, new flavoring substances are formed, including caramelized sugars, pyrodextrins, and melanoidins, which give the bread its desirable and distinctive sensory properties. The type of bread most widely produced and consumed in different parts of the world is white pan bread.

The sponge starter mixture, consisting about 65% of the total flour, plus a portion of the total dough, water, yeast, and ammonium salts to stimulate the yeasts, and other salts, including $KBrO_3$, KIO_3, CaO_2, and $(NH_4)_2S_2O_8$, are added to improve the dough's characteristics. The dough is prepared and discharged into a trough (as described below), where it undergoes a fermentation period of 4.5 h in a controlled environment (Panesar and Marwaha, 2013). From a starting temperature of about 25°C, the final temperature increases by approximately 6°C due to the exothermic reactions brought about by the yeast activity. At the end of the fermentation, the sponge mixture is transferred into a dough mixer. The balance of the flour is also placed into the mixer, along with the water and other ingredients. This is mixed until the dough is transformed from a sticky, wet-appearing mixture into a smooth, cohesive dough, characterized by a glossy sheen (Panesar and Marwaha, 2013). As *S. cerevisiae* develops carbon dioxide, the gas diffuses into previously formed gas vesicles and is retained in the dough, due to the unique nature of gluten. The mixed dough is placed into troughs and allowed to rest for 20–30 min, during which the dough recovers from the mechanical stress, relaxes, and is better able to undergo the remaining processing stages.

Dividing the dough is the next stage: the dough is cut into pieces of the desired weight by a machine that volumetrically divides the pieces and discharges them into cylinders, which are then fed to bread pans (Panesar and Marwaha, 2013). Pans containing the dough pieces are placed in fermentation units called proof boxes for the last fermentation period prior to baking. The environment in these units is typically maintained at 35–43°C at a relative humidity of 80%–95%. The dough pieces expand in the pans to the desired volume, a process usually requiring approximately 60 min (Panesar and Marwaha, 2013). The proofed loaves are then placed in an oven for baking. Gas within the dough expands; steam and alcohol vapors also contribute to this expansion. Enzymes are active until the bread reaches about 75°C. At this temperature, the starch gelatinizes and the dough structure is set. With the bread surface temperature reaching 130–140°C, sugars and soluble proteins react chemically to produce an attractive crust color. The center of the loaf however, does not exceed 100°C. Remaining stages in the bread-making process include cooling of the baked bread, slicing, wrapping, and its distribution to stores for sale to the consumer.

In a research conducted in Pakistan, four different Pakistani wheat varieties—namely, Inqulab-91, Auqab-2000, Iqbal 2000, and AS-2002—were evaluated for their physicochemical and sensory characteristics to determine their suitability for use in the production of flat bread (*naan*). On the basis of above cited characteristics, AS-2002 was found to be the most suitable for use in the production of flat breads (Mueen-ud-din et al., 2007).

7.2.2 Sourdough Bread

Sourdough bread is usually made from a mixture of rye and wheat flour, containing up to 80% of wheat flour. Its acid taste may be obtained by adding mixtures of lactic acid and acetic acid (Edema, 2011). Such breads are usually leavened with yeast and the action of LAB. Sourdough starter cultures are also available commercially, and these can be used to inoculate the doughs. Such cultures usually contain mixtures of LAB (Padmaja and George, 1999; Splcher and Schroder, 1978). Ordinary, commercially available baker's yeast (*S. cerevisiae*) typically contains 10,000 to several million LAB per gram, and its use will result in a souring of the dough if fermentation is carried out for prolonged periods of time and at relatively elevated temperatures. The characteristic yeast of this process is *Torulopsis holmii* and the bacterium producing both lactic acid and acetic acid is *Lactobacillus sanfrancisco* (Sugihara et al., 1971). The sponge starter contains only water, flour, and microorganisms, and requires about 8 h at 25°C for full development. The final pH value may be as low as 3.9. About 20% of the starter sponge is used in the dough.

7.2.2.1 Sourdough Fermentation

Bread and its production techniques—converting cereal flours into an attractive, palatable, and digestible food—differ widely around the world. For example, the characteristics that are desirable for leavened wheat breads are high volume, a soft and elastic crumb structure, a good shelf life, and the microbiological safety of the product (Cauvain, 2003; Chavan and Chavan, 2011; Chavan and Kadam, 1989). Although yeast fermented breads are widely consumed in the Asian region, leavened bread type foods are not traditional staples of the region. Other types of breads are prepared primarily by acid fermentation of rice flour dough, which is similar to the Indian *idli*, except for the fact that it does not contain any legumes (FAO, 1999, ch 3).

Most bread types are manufactured by using a straight dough process in which all of the ingredients are combined in one step, and the dough is baked after proofing (Chavan and Chavan, 2011). Alternatively, though, bread can be made from sourdough with or without addition of a starter culture in the form of well-defined pure strains of microorganisms or pieces of previously fermented dough (used in back-slopping). Sourdough is the foremost cereal fermentation performed in a variety of technologies

Table 7.1   Role of Sourdough LAB in Wheat Fermentation

- Creates an optimum pH for the activity of endogenous aspartic proteinases (Thiele et al., 2002) that improves textural changes (Clarke et al., 2004) and flavor (Gaggiano et al., 2007; Gobbetti et al., 2005; Hammes and Ganzle, 1998)
- Contributes directly to the bread flavor, especially through the synthesis of acetic acid (Gobbetti et al., 2005)
- Increases the loaf volume (Corsetti et al., 1998)
- Delays starch retrogradation and bread firming (Corsetti et al., 2000)
- Inhibits ropiness by spore-forming bacteria (Gobbetti et al., 2005)

LAB = Lactic Acid Bacteria.

with almost any cereal (Catzeddu et al., 2006; Chavan and Chavan, 2011; De Vuyst et al., 2002; Moroni et al., 2009; Valcheva et al., 2006).

Spontaneous sourdough fermentation is one of the oldest known cereal fermentations, and has a well-established role in improving the flavor and structure of bread (Arendt and Bello, 2008). Sourdough is a dough whose microorganisms, mainly LAB and yeasts, originate from sourdough or a sourdough starter and are metabolically active or need to be activated (Di Cagno et al., 2002; Ferchichi et al., 2008; Coda et al. 2014). The main function of LAB and yeast is to leaven the dough to produce a more gaseous substance, and, as such, a more aerated bread (Decock and Cappelle, 2005). It can be performed as a firm dough or as a liquid suspension of flour in water. Traditional sourdough is simply a piece of dough from the previous baking that is mixed with flour, salt, and water to make the bread dough (Jacob, 1997). In wheat fermentation, sourdough LAB plays several roles, as shown in Table 7.1.

7.2.2.2 Classification of Sourdough   Sourdoughs are classified into three types: Type I is a sourdough that is restarted using a part of the previous fermentation (Rollán et al., 2010; Katina et al., 2005), Type II is an industrial type in which adapted strains are used to start the fermentation, and Type III is a dried type of sourdough that is often used by industrial bakeries as the quality is constant and there are no longer end-product variations, due to the sourdough's freshly produced factor (Böcker et al., 1995). Types II and III require the addition of baker's yeast as a leavening agent, whereas Type I sourdoughs does not require such an addition (Mueen-ud-din, 2009). Types I, II, and III consist of obligate and facultative heterofermentative and obligate homofermentative species of LAB. Lyophilized strains of *Lactobacillus delbrueckii*, *Lactobacillus fructivorans*, *Lactobacillus plantarum*, and *Lactobacillus brevis* are very well established as sourdough LAB (Corsetti et al., 2003; Hammes and Ganzle, 1998).

7.2.2.3 Sourdough Processing and Its Effects   As well as *S. cerevisiae*, beer yeast has historically been used for dough leavening (Chavan and Chavan, 2011; Kulp and Lorenz, 2003; Spicher and Stephan, 1999), while, today, sourdough is typically employed in the manufacture of breads, cakes, and crackers (Mueen-ud-din, 2009; Ottogalli et al.,

1996). Its distinctive characteristic is mainly due to its microflora, principally represented by LAB and yeasts: during fermentation, biochemical changes occur in the carbohydrate and protein components of the flour due to microbial and indigenous enzyme activity (Chavan and Chavan, 2011; Spicher, 1983).

Specifically, in the sourdough manufacturing process, fermentation is of vital importance as it affects the dough rheology at two levels: in the sourdough itself and in bread dough containing sourdough. The effects may be ascribed to the metabolic activities of resident LAB: lactic fermentation, proteolysis, synthesis of volatile compounds, antimold, and antiropiness are among the most important activities taking place during sourdough fermentation (Gobbetti et al., 1999; Hammes and Ganzle, 1998). Endogenous factors present in the cereal products (carbohydrates, nitrogen sources, minerals, lipids and free fatty acids, and enzyme activities) and process parameters (temperature, dough yield, oxygen, fermentation time, and the number of sourdough propagation steps) influence the microflora of sourdough and, consequently, the features of leavened baked goods (Hammes and Ganzle, 1998).

7.2.2.4 Microbiota of Sourdough Sourdough is a very complex ecosystem owing to its microbial composition and their interactive effects during bread-making processes, as well as the presence of different ingredients (Gobbetti et al., 1999). LAB dominates the microbiota of sourdough, occurring at numbers in excess of 10^8 colony-forming units (CFU) per gram (g), and may co-exist or possibly be symbiotic with yeasts (Gobbetti et al., 1999; Vogel et al., 2002). With few exceptions, the majority of species from sourdough fall within one of the four genera: namely, *Lactobacillus*, *Pediococcus*, *Leuconostoc*, and *Weissella*. *Lactobacillus* is the dominant genera and more than 23 species are used, classified into obligately homofermentative (*L. acidifarinae*, *L. brevis*, *L. buchneri*, (Corsetti and Settanni, 2007), *L. fermentum*, *L. fructivorans*, *L. frumenti*, *L. hilgardii*, *L. panis*, *L. pontis*, *L. reuteri*, *L. rossiae*, *L. sanfranciscensis*, *L. siliginis*, *L. spicheri*, *L. zymae*), facultatively heterofermentative (*L. plantarum*, *L. pentosus*, *L. alimentarius*, *L. paralimentarius*, *L. casei*), and obligately heterofermentative strains (*L. amylovorus*, *L. acidophilus*, *L. delbrueckii* sp. *delbrueckii*, *L. farciminis*, *L. mindensis*, *L. crispatus*, *L. johnsonii*, *L. amylolyticus*).

Yeasts present in sourdough mainly belong to 20 species, with some of the well-known ones being *Candida milleri*, *Candida holmii*, *Saccharomyces exiguous*, and *S. cerevisiae*, *Pichia anomala* as *Hansenula anomala*, *Saturnispora saitoi* as *Pichia saitoi*, *Torulaspora delbrueckii*, *Debaryomyces hansenii*, and *Pichia membranifaciens* (Foschino and Galli, 1997; Gobbetti et al., 1995; Hammes and Ganzle, 1998; Mueen-ud-din, 2009; Succi et al., 2003). Factors such as dough hydration, the level and type of cereal used, the leavening temperature, and the sourdough maintenance temperature control the number and type of yeast in dough (Gobbetti et al., 1994).

Various acid-leavened breads and noodles prepared in the South Asia are described in Table 7.2.

Table 7.2 Some Common Acid-Leavened Breads, Noodles, Porridges, Pastes, and Condiments

PRODUCT NAME	COUNTRY OF USE	MAJOR INGREDIENTS	MICROORGANISM	APPEARANCE AND USAGE
Idli	South India, Sri Lanka	Rice grits, black gram powder	*L. mesenteroides, S. faecalis, T. candida, T. pullulans*	Steamed cake
Dosa	India	Rice flour, black gram powder	*L. mesenteroides, S. faecalis, T. candida, T. pullulans*	Griddled pancake
Dhokla	India	Rice, Bengal gram	*L. mesenteroides, S. faecalis, T. candida, T. pullulans*	Steamed cake
Jalebis	India, Nepal, Pakistan	Wheat flour	*S. bayanus*	Pretzel-like confection
Adai	India	Rice, black gram, Bengal gram, red gram	LAB	Griddled pancake
Bhatura	India	Wheat flour	LAB (Tamang, 2012)	Deep-fried bread
Nan	India, Pakistan, Afghanistan, Iran	Wheat flour	LAB, *Saccharomyces cerevisiae*	Baked, leavened flat bread
Kulcha	India, Pakistan	Wheat flour	LAB, yeast	Baked, leavened flat bread
Appam	India	Rice	*LAB*	Steamed cake
Ambali	India	Ragi millet flour, rice	*Leuconostic mesenteroides, Lactobacillus fermentum, Streptococcus faecalis*	Porridge
Rabadi	India, Pakistan	Maize flour	*Pediococcus acidilactis, Bacillus* sp., *Micrococcus* sp.	Thick sauce
Punjabi warri	India	Black gram, spices	*Leuconostoc mesenteroides* (Panesar and Marwaha, 2013), *Streptococcus faecalis, Lactobacillus fermentum, Saccharomyces cerevisiae* (Panesar and Marwaha, 2013), *Pichia membranaefaciens, Trichosporon beigelii,* and *Hansenula anomala*	Spicy, hollow, brittle, friable balls used as condiment
Papad	India	Black gram, spices, rice, potato	*Leoconostoc mesenteroides, Streptococcus faecalis, Saccharomyces cerevesiae*	Thin, crisp, circular tortilla-like wafers
Bhalle	India	Black gram	LAB, yeasts	Deep-fried spongy balls
Vadai	India, Sri Lanka	Black gram, lentil, potato	LAB, yeats	Deep-fried wheel-shaped spongy with hole in the middle

Source: Adapted from FAO, 1999. Cereal fermentations in countries of the Asia-Pacific region. Chapter 3. In: Haard, N.F. et al. (Eds.) *Fermented Cereals: A Global Perspective.* FAO Agricultural Services Bulletins, 138. http://www.fao.org/docrep/x2184e/x2184e09.htm. and the literature cited in the table.

7.2.3 Rye Bread

Rye bread is typically made from rye flour with or without a starter or sour. The old method of preparing sour bread depends upon the bacteria naturally present in a mixture of rye flour and water. The mixture is allowed to ferment for 5–10 h, then more flour and water are added, and the fermentation is continued for an additional 5 or 6 h. Half of the sour is incorporated in the sponge or dough for the bread and the rest is carried over to start a new sour. This sour has been modified by some bakers through the addition of yeast and acetic acid bacteria from cultured buttermilk or buttermilk to a sour that was made afresh daily. Modern methods involve the addition of considerable amounts of cultures of acid-forming bacteria to the dough mass to be used as a sour, and controlling the time of fermentation between 18 and 24 h and the incubation temperature around 25°C. An excessively high incubation temperature—for example, 32–35°C—supports the growth of undesirable gas formers, such as coliform and butyric bacteria. Some bakers prefer to use low-temperature LAB, whereas others use high-temperature lactics and adjust the incubation temperature, accordingly. The growth of heterofermentative LAB is considered desirable. The starter imparts a desired sour flavor to the rye bread. Several variants of rye bread are also available, and two of these are discussed here.

7.2.3.1 Straight Rye Bread

Simple, all-rye bread can be made using a sourdough starter and rye meal. It will not rise as high as a wheat bread, but will be more moist with a substantially longer preservation time. Such breads are often known as "black breads," from their darker color compared to that of wheat breads (enhanced by long baking times, creating Maillard reactions in the crumb) (Medlibrary.com encyclopedia). All-rye breads can have very long shelf-life, measured in months rather than days, and are popular as storage rations for long boat trips and outdoor expeditions. Such breads are usually sliced very thinly because of their density, sometimes only a few millimeters thick, and are sold pre-sliced (Medlibrary.org.com).

7.2.3.2 Crisp Rye Flat Breads

There are three different types of rye crisp breads: yeast fermented, sourdough fermented, and cold bread crisp bread (Medlibrary.org.com). The third type, the so-called "cold bread crisp bread," is essentially a type of hardtack that is baked without the addition of any leavening (Åkerström and Carlson, 1936; Medlibrary.org.com). The dough attains the correct texture through a foaming process in which air is incorporated into the cooled dough, which also leads to the almost white color of the finished bread. Crisp bread owes its long shelf-life to its very low water content (5%–7%).

7.3 Indigenous Breads of Ladakh

Breads are made by the people in Ladakh (India) from wheat flour (*paqphey*) as well as barley flour (*narjen*, which translates to "uncooked barley"), or a mixture of the two.

Furthermore, this region's breads are also sometimes made from pea or lentil flour; often, the pea and wheat elements are mixed and ground into flour, which makes it more nourishing and palatable. Before baking powder was available, *pul* (local soda) from the Nubra valley was used to prepare these breads. The *pul* is mixed with the flour to prepare the dough by mixing water or any other diluting agent, such as *chhang*, curd, yogurt, whey, local tea, etc., the consistency of which depends on the type of bread to be made. The dough is left overnight to ferment, and then the breads are prepared. Some of these local breads are described here.

7.3.1 Tagi Khambir or Skyurchuk (Browned Sourdough Bread)

This bread (*tagi* is the general name for bread) is eaten throughout Ladakh. *Skyur* means "sour" and *chuk* means "mixed with." So *Skyurchuk* is the name by which the bread is generally known in the villages; *tagi khambir* is its cosmopolitan name. In traditional kitchens (Figure 7.4), wheat flour is kneaded into dough after adding baking powder or local soda (*pul*) and a starter for initiating the fermentation. It is then, kept overnight for fermentation. Fermented dough is rolled into small balls and flattened to form a thick round bread. The dough will first be cooked on a cast iron plate placed over one of the cooking holes in the stove, under which is the fire of wood or cow dung. Then, it will be finished-off on the ambers inside. It is half baked on a flat, thick, stone surface or iron griddle and then, put directly on cinder for roasting. The final product is thick and soft. It is called *skyurchuk, thuk-tuk, tagi khambir*, or simply *tagi*. Starter may be buttermilk (*tara*), *chhang* (local beer), or back-slop. Readymade yeast available in the market is also used nowadays. The quantity of starter and incubation time varies according to the sourness required.

Tagi khambir is served with butter, vegetables, curd, or tea. This type of bread is fermented with LAB and yeasts. CO_2 produced by yeast leavens the bread, producing anaerobic conditions, and baking produces a dry surface resistant to invasion by organisms. Baking also destroys many of the microbes present in the bread itself.

Figure 7.4 Traditional kitchen.

7.3.2 Tagi Buskhuruk (Puffed Unleavened Bread)

Buskhuruk means "unleavened bread." The dough used is the same as for *tagi shrabmo* (similar to *chappati*), but the balls with which it is made are thicker and smaller in size than those used for *tagi shrabmo*. They are cooked briefly, as above, but are allowed to puff up, and then, if possible, put in the ashes of stove or fire. Traditionally, it is baked on a plate placed on a cow dung fire, covered with another plate and more dung.

7.3.3 Tagi Thalkhuruk (Bread Uncovered and Baked in Ashes)

This bread is lighter than *tagi thalshrak* (covered and backed in ashes, Figure 7.5) and is, therefore, good for people who are not well (chiefly, those with poor digestion). The dough is made in the same way, but the bread is cooked while uncovered on a plate, and on the ashes—*thalkhuruk.*

7.3.4 Tagi Tain-Tain

It is soft to eat and resembles a *dosa*, except that it is thicker, and is served with milk or tea. A batter of wheat flour is mixed with a small amount of starter (back-slop) and is kept overnight for fermentation. The fermented batter is spread over a layer of hot cooking oil for frying in a griddle (*tawa*). Once one side is cooked, it is turned over to fry the other side. Both sides are cooked on moderate heat. *Tagi tain-tain* is also known as *poli* in Kargil.

7.3.5 Tagi Kiseer/Giziri

It is made as per *tagi tain-tain* except that wheat flour is substituted for buckwheat flour (Mir and Mir, 2000). The rest of the procedure is the same as described for other types of bread.

Figure 7.5 Tagi thalshrak.

7.3.6 Skien

Skien is made by making a dough of flour and sheeps' fat, which is given the shape of an ibex horn, and then, baked on charcoal. When it is eaten, it is broken up and the bread along with the fat is heated in a bowl and mixed together. It is made especially during *Losar* (Ladakhi New Year, which generally falls in December–January), when it is decorated on the kitchen shelf (Figure 7.6) and is presented to close relatives and neighbors.

7.3.7 Kaptsey/Makhori

It is a thick round bread made from wheat flour. Its edge is turned, twisted, and pressed to give a woven appearance, then cooked in a medium oil; sugar may be added to taste. This bread is presented when a girl child is born. In *Baltis* and *Shin* tribes (in Kargil district, India), it is also presented by the groom's family to the bride's family during the betrothal and marriage ceremonies. At *Losar*, the *Dard* or *Shin* tribals of Drass stuff the *makhori* with animal fat and send it as a present to close female relatives, like daughters, sisters, and close cousins.

7.3.8 Tagi Tsabkhur *UK (Ground, Sprouted Wheat Bread)*

This is easy to cook and the taste is very sweet. Making the flour from the sprouted wheat or barley is a somewhat complicated process, but it is interesting. *Tsabkhuruk* means "sprouted grain" and "cooked while covered" (as opposed to *thalkhuruk*, which means "not sprouted" and "uncovered"). It is said to be good for pregnant women.

7.3.9 Sephe Tagi *UK (Freshly Sprouted Wheat Bread)*

This type of bread is made in spring time in the Zanskar valley of Ladakh. The grains are sprouted as above, but are used fresh when the sprouts are about 1 cm long. The sprouts are crushed in a pestle and mortar, and then mixed with water. This mixture is then, added to the flour to make moist, loose dough. After this, the dough

Figure 7.6 Skien placed on kitchen shelf.

is cooked with a little fat, or dry on a heated stone or griddle. It swells, like *tagi kham-bir*, and is eaten split with a little fat or butter added inside the loaf.

7.3.10 Khura (Sweet Deep-Fried Biscuits)

This is made especially on the occasion of *Losar*. When a married girl visits her parent's home during *Losar*, she will carry with her a plate of *khura* (Figures 7.7 through 7.9) for the family members. In the old days, local flour was used, and this may still

Figure 7.7 Khura.

Figure 7.8 *Phaps*: starter culture.

Figure 7.9 Spreading of boiled grains on *khol-chari*. (Courtesy of Targais, K., Stobdan, T. and Mundra, S. 2012. *Indian Journal of Traditional Knowledge* 11(1): 190–193.)

be in use in some of the villages. But, nowadays, people often use rice flour to achieve an even finer consistency. The technique for giving shape to this biscuit is quite tricky. They can be salty, when salt solution is used instead of the syrup.

7.3.11 Markhour

Molten butter and well-beaten egg white are added to wheat flour, kneaded to a hard consistency by adding small quantity of water, and made into breads of about 2.5 cm thickness and 12–15 cm diameter. Egg yolk is spread on the upper surface, giving it a shiny appearance. A motif on the upper surface can also be added. The bread is placed between two iron plates and then, baked by embedding them in cinders.

7.4 Bread-like Products

A large number of other bread-like, leavened, and baked foods are also produced (Gotcheva et al., 2001) and are consumed in most of the countries of the world, including South Asian countries, as a staple food. But considerable variations again exist in the type of foods from country-to-country, and among regions within a given country (Soni and Marwaha, 2003). Some of these are discussed here, and have been listed in Table 7.2.

7.4.1 Kulcha (Bread-Like, Cereal, Non-Alcoholic Products)

Kulcha is a type of an Indian flat bread made from refined wheat flour (Figure 7.10) (Medlibrary.org.com). It is particularly popular in India and Pakistan, and is usually eaten with chickpea curry or with vegetables and meat as, a staple diet. Refined wheat flour, sugar, a portion of a previously fermented batch, and water are kneaded to form a smooth dough, which is then, baked in an oven until golden brown. The dough required for its preparation is usually made by the straight dough method described above. White wheat flour, sugar, a portion of the dough from a previously

Figure 7.10   Kulcha—an Indian flat bread.

fermented batch, and water are mixed (100:1:10:50) at room temperature, kneaded by hand, and allowed to ferment overnight in earthen pots. Sometimes, gram flour (*Cicer aerietinum* L.) is also added (at the 20% level). After fermentation, the dough is made into balls, placed on a smooth surface, lightly sprinkled with flour, and flattened by hand to disks of 15–20 cm in diameter. The smoothly flattened dough is then, transferred onto a metallic baking tray, or scooped in a special oven made of clay. The baking is completed in 30–40 min (Soni and Sandhu, 1999). In some areas of Pakistan, the dough is raised with a low level of inoculum, but the fermentation is prolonged, which gives characteristic flavor and taste to the product. In some parts of India, a commercial baker's yeast as well as some chemical leaveners are also added to raise the dough.

7.4.2 Bhatura/Bhatooru

Bhatura is a leavened, deep-fried bread (Figure 7.11) made from white wheat flour that is consumed as a breakfast food and as a snack with *masala chana* in northern India (Joshi, 2005; Panesar and Marwaha, 2013). The dough is prepared by mixing white wheat flour with water and kneading for 20 min at room temperature. The dough is supplemented with 1% common salt, inoculated with a portion of previous batch ripe batter, and kept in the open to undergo fermentation for 8–14 h. The dough is then, made into balls, which is then, flattened by hand, using vegetable oil, to 12–18 cm in diameter before being deep fried in oil (Soni and Sandhu, 1999).

 Bhatooru (*sumkeshi roti*, *tungi roti*) is an indigenous leavened bread, or *roti*, and makes up the staple diet of rural populations living in the unindustrialized areas of Kullu, Mandi, Kangra, Chamba, and Shimla districts of Himachal in India (Savitri and Bhalla, 2013; Thakur et al., 2004). The rural migrants in urban areas also prepare *bhatooru*.

Figure 7.11   Bhaturu—a leavened deep fried bread.

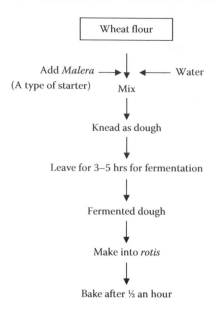

Figure 7.12   Flowchart for *Bhatooru* preparation (Practiced in Kullu district of Himachal Pradesh, India).

A flow diagram illustrating the preparation of *bhatooru* is shown in Figure 7.12. It is traditionally made by using wheat flour or sometimes barley flour (Joshi, 2005). Inoculum, *malera* (a traditional inoculum used in the preparation of some fermented foods; mainly consisting of LAB and yeasts, it is a portion of previously leftover fermented dough), along with water is added and kneaded as dough, and then, left for 3–5 h for fermentation. The fermented dough is then, made into *roti*, spread onto cotton/woolen sheet (*pattu*, etc.) and covered to allow further fermentation to complete, and then, baked.

Bhatooru is served with vegetables, *dal*, or curry for routine meals, and deep fried on festive occasions (Savitri and Bhalla, 2013). Sometimes, the *bhatooru* are stuffed with spices and a salted paste of black gram, and then, deep fried to make *bedvin roti* or *kachauri*.

Suhalu/Babroo is a traditional fermented snack food prepared during special occasions, such as birthdays, marriages, or other such celebrations, in Himachal Pradesh in India. Wheat flour, sugar, water (2:1), and *malera* (inoculum) are mixed together and made into a semi-solid paste that is kept for fermentation at 25–30°C for 3–4 h. After fermentation, the slurry is cooked as flat pancakes with oil. This foodstuff is popular in Mandi, Suket and other areas of Himachal Pradesh (Thakur et al., 2004).

Bedvin roti is a modified form of *bhatura* as they are stuffed with a mixed paste of spices, *dal* (black gram), poppy seeds, or walnut. They can be either simple baked or deep fried. The fried versions are called "*kachoris*" in the Mandi district of Himachal Pradesh. They are consumed as breakfast or as a snack food with tea.

7.4.3 Naan

This is leavened flat bread with a central pouch (Panesar and Marwaha, 2013). It is consumed as a staple food in north India, Pakistan, Afghanistan, and Iran. *Naan* is a kind of *tannor* bread; several names, like *tannori, tandour, khubz*, and *naan*, are given to essentially the same product in some countries of South Asia (Mueen-ud-din et al., 2007). It is prepared by mixing white wheat flour with sugar, salt, a small portion of previously fermented dough, and water to form a smooth dough (100:2:10:50). *Naan* is made from a finer granulated flour than that used for *chapattis* because the finer the granulation, the more rapid is the process of fermentation (Mueen-ud-din, 2009; Qarooni, 1996). The dough is left for 12–24 h, formed into balls, and flattened (FAO, 1999, ch 2). The smoothly flattened dough is slapped into the inner wall of the clay-clad brick oven, called a *tandoor*, where it sticks while baking until the dough is puffed up and becomes light brown in color.

Butter *naan* is prepared by coating the fermented flattened dough with butter before baking. Yeast (*S. cerevisiae*) and LAB are involved in the fermentation of the dough (Blandino et al., 2003; Sandhu et al., 1986), which is then, kneaded by hand at room temperature. Sometimes, the dough is made by the sponge-and-dough method with yogurt added as an inoculum along with baker's yeast and some chemical leaveners. The flattened dough is then, transferred onto a circular pad of cotton cloth and plastered onto the inner wall of a *tandoor*. Before baking, some spices, such as caraway seeds (*Carum carvi*), coriander (*Coriandrum sativum*), and *kalaunji* (*Bochanania lauzan*), are sprinkled onto it. The finished product is obtained by baking at temperature 120–150°C for 10–15 min (Soni and Sandhu, 1999).

As noted earlier, yeast (*S. cerevisiae*) and LAB are involved in the fermentation process (Panesar and Marwaha, 2013). New dough used for making *naan* contains 10^5 yeast CFU/g and 10^2 LAB CFU/g, compared to the respective counts of 10^8 and 10^9 for ripe fermented dough (FAO, 1999, ch 3).

7.4.4 Marchu

Marchu/marquee/poltu/pole is a fermented and deep-fried *rotii* prepared especially during the tribal festivals and marriages in the Lahaul Spiti and Kinnaur districts of Himachal Pradesh. They are prepared during festivals, and religious and marriage ceremonies as a snack/breakfast food to be taken with tea (Figure 7.13) (Thakur et al., 2004). It is also customary for a daughter to take *marchu* with her, whenever she visits her maternal home from that of her in-laws, or vice versa (Thakur et al., 2004).

The major ingredient for *marchu*, prepared during the local festivals (*phagli, halda*) in Lahaul valley, is wheat flour. The inoculum used for fermentation is *malera*, or previously naturally fermented leftover dough (Thakur et al., 2004). It is allowed to ferment for 6–8 h during summers, or overnight during winters at an ambient temperature.

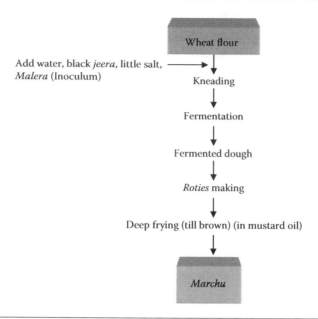

Figure 7.13   Flowsheet for the preparation of *Marchu*.

The dough is made into small balls and then, flattened (Thakur et al., 2004). The *marchu* are made on a wooden base with carving to give designed imprints on them, and then, deep fried in oil (Thakur et al., 2004). When they are shaped like folded spirals, they are called *doshu*. They can be sweet or salty, and are consumed with tea/milk by the people of Himachal Pradesh. The *marchu* is somewhat similar to the *bhatura* prepared in northern India (Punjab), which is also a wheat-based, fermented, deep-fried bread.

7.4.5 Doli ki Roti

This is an indigenous, fermented, wheat-based bread prepared by the Indian Punjabi communities who migrated from western Pakistan at the time of the partition of India. The fermentation of the batter is carried out in an earthen pot called a *"doli"*; thus, the name of bread is *doli ki roti*. Traditionally, the inocula of this bread was made in the temples and distributed to the local people so that they could prepare the bread on special occasions like fasts, when people used to eat *"basa"* food (food prepared on the previous day, and not the fresh hot food prepared on the day of a fast). The spices that are added impart flavor and also have antimicrobial properties, thereby restricting the growth of pathogenic microflora during fermentation. A combination of cereal and legumes in this *roti* improves the protein quality of a vegetarian meal and is also nutritious due to the cumulative effect of germination and fermentation (Bhatia and Khetarpaul, 2012). It contains less amount of phytic acid and higher availability of dietary essential minerals (Bhatia and Khetarpaul, 2002). The process for preparation of *doli ki roti* is depicted in Figure 7.14.

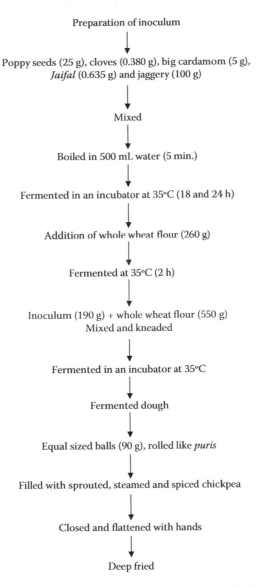

Preparation of inoculum

↓

Poppy seeds (25 g), cloves (0.380 g), big cardamom (5 g), *Jaifal* (0.635 g) and jaggery (100 g)

↓

Mixed

↓

Boiled in 500 mL water (5 min.)

↓

Fermented in an incubator at 35°C (18 and 24 h)

↓

Addition of whole wheat flour (260 g)

↓

Fermented at 35°C (2 h)

↓

Inoculum (190 g) + whole wheat flour (550 g)
Mixed and kneaded

↓

Fermented in an incubator at 35°C

↓

Fermented dough

↓

Equal sized balls (90 g), rolled like *puris*

↓

Filled with sprouted, steamed and spiced chickpea

↓

Closed and flattened with hands

↓

Deep fried

Figure 7.14 Flow sheet for preparation of "*Doli ki Roti.*" (From Bhatia, A. and Khetarpaul, N. 2012. *Indian Journal of Traditional Knowledge* 11(1): 109–113. With permission.)

7.5 Cereal-Based Fermented Products

7.5.1 Siddu

Siddu is a traditionally fermented, steam-cooked, oval- or disk-shaped dish prepared as a delicacy in rural areas of the Mandi Kullu, Bilaspur, and Shimla districts (Joshi, 2005) (Thakur et al., 2004). It is also called *khobli* in the Shimla district. Wheat flour, spices, and a mixed paste of opium seeds/walnut/black gram are its main ingredients (Thakur et al., 2004). The inoculum used in its preparation is *malera* (previously naturally fermented leftover dough). The process is illustrated in Figure 7.15.

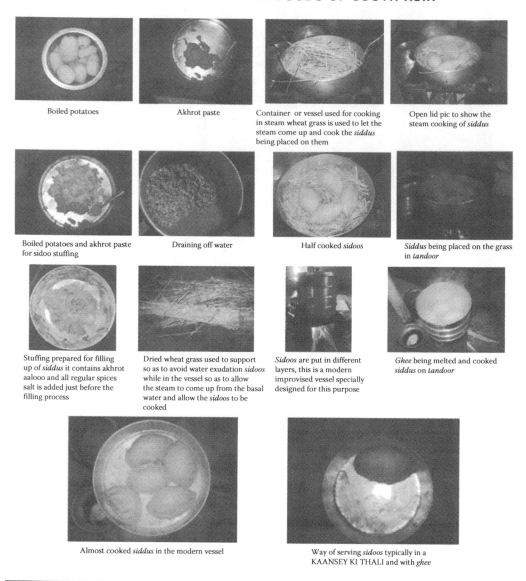

Figure 7.15   Pictorial representation of the process of *siddu* preparation and method of serving. (Courtesy of Dr. Arjun Chauhan.)

Today, yeast powder is also used for fermentation (Thakur et al., 2004). The complete process of *siddu* preparation has been summarized as a flow sheet in Figure 7.16. It is served hot with *desi ghee* or chutney. Village cooperatives (*mahila mandals*) put up stalls at local fairs to sell *siddu*, while restaurants in Kullu, Manali, and Shimla have also included them on their menu. *Siddu* is very popular among the rural as well as urban populations.

7.5.2 Appam/Kallappam/Vellayappam/Hopper

This is a type of fermented food made in south India, especially in Kerala and Sri Lanka. For its preparation, rice is soaked in water for 3–6 h and then, ground to

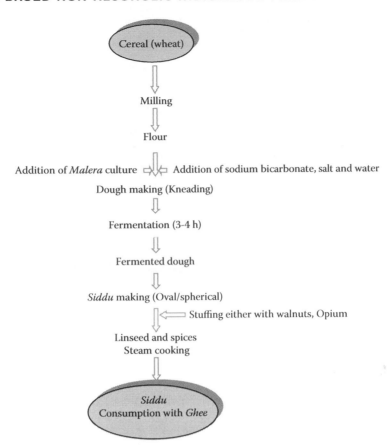

Figure 7.16 Flowsheet for the preparation of *siddu*. (Adapted from Thakur, N., Savitri and Bhalla, T.C. 2004. *Indian Journal of Traditional Knowledge* 3(3): 325–335.)

a fine paste. Sugar and yeast/"*toddy*" (a palm wine) and water are added to the paste to form a free-flowing batter, which is fermented for 4–6 h. The fermented batter is then, poured into moulds and cooked by steaming to form *idli*-like pancakes (Figure 7.17). *Appam* is also known as *kallappam* or *vella appam* in various regions of south India, and as *appa* in Sri Lanka. Using the modern method, the *toddy* is replaced by

Figure 7.17 Appam—idli-like pancake.

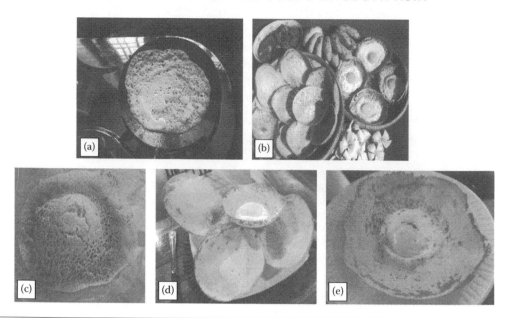

Figure 7.18   (a) Hopper being cooked, (b) hopper while serving, (c) a "common" or plain Hopper, (d) kallappam, and (e) egg hopper.

yeast and the pH is controlled, which gives the product a shelf life of 2–4 days. These are available along with other bakery products in a packaged form in Kerala (India).

"Hopper" *or appa* is made with fermented rice batter and coconut milk. It is very popular in Sri Lanka, where it is commonly referred to by its anglicized name of "hopper" (Figure 7.18); it is called *chitau pitha* in Oriya, *paddu* or *gulle eriyappa* in Kodava, *appa* in Sinhala, and *arpone* in Burmese. It is eaten most frequently for breakfast or dinner, and can be of several types. For example, plain hoppers (*vella appam*) are bowl-shaped thin pancakes made from a fermented rice flour batter using rice, yeast, salt, and a little sugar. After the mixture has stood for a couple of hours, it can be fried in the *appachatti* with a little oil. *Vella appam* are fairly neutral in taste and mostly served with some spicy condiment or curry. Another form of *appam* is one in which "*kallu*" (Malayalam/Tamil), or *toddy*, is added to the fresh batter to kick-start the fermentation. Egg hoppers are like a plain hoppers, but with an egg broken into the pancake as it cooks. Sometimes, pancake batter is simply spruced up with coconut milk and a splash of *toddy* (Sri Lankan palm wine). The unique part is that hoppers are cooked in small, wok-like, rounded pans so the dough cooks thick and soft on the bottom, and thin and crunchy around the edges.

7.5.3 Jalebis

Pretzel-like, syrup-filled confections called *jalebis* are prepared from deep-fried, fermented, wheat flour dough (Figure 7.19). They are consumed hot as a confection food throughout India, Nepal, and Pakistan, and eaten as a snack particularly on festive occasions (Soni and Sandhu, 1999). To prepare them, fine wheat/rice flour is mixed

Figure 7.19 Jalebis—pretzel like syrup filled confection.

with curd, water, and a small portion of a previous batch to form a thick batter. This is then, allowed to ferment at room temperature for 8–12 h. *L. fermentum*, *L. lactis*, *L. buchneri*, and *Saccharomyces bayanus* have been isolated as fermenting organisms from wheat/rice flour batter during the preparation of *jalebis*. Mixed fermentation is by LAB, *L. fermentum*, *L. lactis*, and *L. buchneri*, with *S. lactis* and *Streptococcus faecalis* producing lactic acid, and lowering the pH from an initial 4.4 to 3.3 (Panesar and Marwaha, 2013). The yeast *S. cerevisiae* is also active, producing CO_2 and causing about a 10% increase in volume during fermentation. Both amino nitrogen and total sugars were found to decrease during fermentation (Batra, 1981; Ramakrishnan, 1977; Steinkraus, 1996).

The fermented batter is squeezed through a cloth with a hole about 4 mm in diameter and deposited as continuous spirals into heated oil or ghee (160–180°C) (Panesar and Marwaha, 2013). In about a minute, these spirals become light brown, and are removed from the frying pan using a sieve-like spatula (5–7 cm). The excess fat is drained away and the *jalebi* is immediately immersed into a sugar syrup (70% sugar) of about the consistency of honey for 2–3 min, and then, removed (Panesar and Marwaha, 2013). Often, rose water or other flavoring agents, such as *kewda* water (*Pandarws tectahus*), and orange food-grade color are added to the syrup to flavor, and impart an attractive color to the *jalebis* (Batra, 1981).

Another method of making *jalebis* recommends that maida and buttermilk are mixed to make a smooth batter (with water, if necessary). This is left overnight at room temperature. The next morning, it is mixed thoroughly with the remaining maida, saffron color is added to it, and the mixture is put to one side for 2–3 h. Water is brought to the boil and sugar is added, allowing some water to evaporate while stirring. Once the sugar syrup begins to thicken up, some liquid is taken between the fingers to see if a thread is being formed or not. Once it is done, it is removed from heat and allowed to cool. Then, oil is heated in a deep frying pan and the batter is poured in concentric rings using a cake decorating pipe, or squeezed through a thick cloth with a hole of 4 mm in diameter at its center. These are fried until crisp and golden brown, drained, dipped in pre-prepared sugar syrup for about 30 seconds, and then, taken out. *Jalebis* are served on a platter, and dusted with cardamom powder and powdered sugar.

7.5.4 Wanndu

Wandu is a traditional Sri Lankan confection made using rice flour, thick coconut milk, ground *jaggery* (a hard treacle)/sugar, coconut water, and baking soda. Salt is added to taste and *kenda kola* is also used. First, the rice flour is mixed with coconut water and left covered for 4 h. A small quantity of coconut milk is then, added and mixed with the *jaggery*, baking soda, and salt to form a thick batter. Next, the mixture is left covered for another 2 h. Kanda leaf (*Macaranga peltata*) is rolled to make a cup (*gotta*), with the two edges bound using a toothpick. The thick batter is poured into the cup and steamed until a crack in the center is observed. It is served with grated coconut mixed with treacle.

7.5.5 Koozhu

Koozhu is a cereal-based, fermented, breakfast food of the Tamil Nadu state of India that is made using sorghum, pearl millet, little millet, and foxtail millet flour. One part of millet flour is mixed with two parts of water and the mixture is allowed to ferment overnight to sour; the fermented batter is then, cooked on a medium-low heat until most of the water is evaporated and the *ragi* (millet) is well cooked. A small amount of cooked rice is added to the cooked *ragi*, which is again cooked for some time; cooking is then, stopped and the dish is allowed to cool. The preparation is mixed with half a cup of yogurt and salt to taste (FAO, 1999, ch 3). It is also consumed with side dishes such as black-eyed beans curry (*Karakozhambu* in the Tamil language), dried fish curry (*karuvaatu kozhambu* in Tamil), or fish curry.

7.5.6 Mangalore Bonda

Mangalore bonda is fermented *maida* (fine wheat powder) made into small balls and deep fried. It is usually consumed hot at tea time. For its preparation, maida is mixed with water and a small quantity of *dahi* to a thick consistency. Various spices and additives, such as green chillies (cut into small pieces), coconut gratings, and cumin seeds, are added, and the batter is then, allowed to ferment for 4 h. The acidity developed during the fermentation stage adds to the final taste of the product.

7.5.7 Chilra

Chilra, or *lawr* in Lahaul, is more or less like *dosa*, but differs slightly in terms of ingredients and shape (Joshi, 2005). It is an indigenous fermented pancake used as

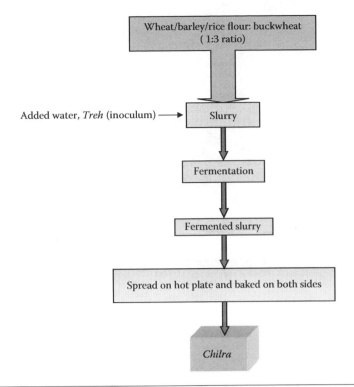

Figure 7.20 Flow sheet for preparation of *Chilra*.

a staple food, and is mainly prepared from wheat/barley and buckwheat flour (1:3) with the inoculum *treh*. The traditional, bucket-shaped, wooden vessel used for fermentation is called a "*lwarenza*" (Tamang, 2009a,b). Once prepared, the fermented mixture is spread onto a hot plate and baked on both sides to make the *chilra*, which is served with coriander, chutney, potato, and mutton soup in the Lahaul valley as a popular snack food. It is also often prepared for marriage ceremonies and local festivals (Figure 7.20) (Thakur et al., 2004). *Chilra* is enjoyed with milk, too, in the Mandi district, while sweet *chilra* is made and consumed in the Kangra district of Himachal Pradesh.

7.5.8 Jhan Chhang

In Himachal Pradesh, during the winter season, a cereal-based fermented foodstuff called *jhan chhang* is enjoyed (FAO, 1999, ch 3). For its preparation, barley grains are washed, ground into a paste, and made into a slurry, which is cooked for 2 h in an open vessel or for 15–30 min in a pressure cooker. The cooked slurry is cooled, then *phab* (a fermentation starter) is added to it, and it is wrapped in a woolen cloth. This mixture is kept at 25–30°C for 2–3 days to complete the fermentation, a process that is helped by the microorganisms present in the *phab* (Figure 7.21). The end product is then, directly consumed as a snack food (Kanwar et al., 2007).

Figure 7.21 A pictorial representation of traditional preparation of *Jhan chhang.*

7.5.9 Tchog

Tchog is a hard, solid dough in the form of a ball prepared during Buddhist religious ceremonies in Lahaul Spiti (Himachal Pradesh), India. It is made by mixing *sattu* (roasted barley flour), *chhang, jaggery,* and *ghee.*

7.5.10 Kurdai

These are fermented cereal products prepared and consumed in rural areas of Maharashtra (India). *Kurdai* is a made from fermented and cooked wheat gruel. The cooked, thick, wheat gruel/extract is taken in a small quantity and pressed through a machine with a plate at the bottom that has number of small holes (Figure 7.22). The cooked gruel comes out in the form of thin circular strands, which are dried in circular forms known as *kurdai* (Figure 7.23).

7.5.11 Wheat Papadis

Papadis are prepared from fermented and cooked wheat gruel. The wheat is soaked in water for 3–4 days at an ambient temperature (30–35°C). The water is changed daily, and, at the end of the process, it is drained using a bamboo screen. The wheat is pulverized either in a crusher or traditionally by using stone *pata* and *warwata.* The mixture is then, washed with water to separate out the bran, and the bran-free milky extract is allowed to settle the colloidal matter for 2 days. The water is drained and the settled residue is mixed with water, then homogenized and cooked. The cooked, thin,

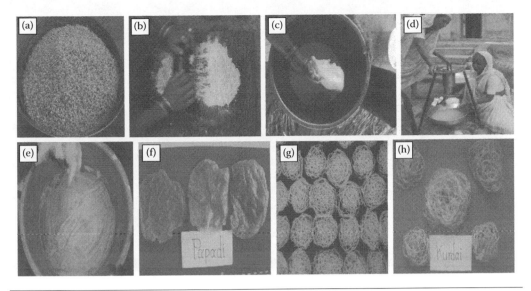

Figure 7.22 (a) Bamboo *chhalni*. (b) Crushing the soaked grains for taking the extract. (c) Residue of extract, mixing with water. (d) Making kurdai. (e) Cooked extract of wheat. (f) Dried *papadi*. (g) Drying of kurdai. (h) Dried *kudai*.

hot wheat gruel/extract is taken in small quantities and spread onto a clean, wet cloth on a circular, wooden plank (*palput*) into thin circles using one's fingers (and using cold water for dipping the fingers). These flat, circular shapes are then transferred onto a high-density polyethylene or cloth sheet for drying in the sunlight. *Papadis* are dried completely, which takes 8–10 h during the summer. After drying, they turn into hard structures that, on breaking, give a cracking sound. They are carefully staked one above the other and stored in small containers made up of *nirgudi* sticks with a cover (*kaning*), and are used throughout the year as snack food.

They are served in either roasted or fried form with *jaggery* or peanuts, and are very popular in both rural and urban areas. They are prepared in summer and used throughout the year, but are more popular during the rainy season. *Papadis* can also be prepared by using locally available sorghum, with slight variation in the preparation process (Figure 7.24); see *bajara*, *ragi* (Figure 7.25), etc.

7.5.12 Meethe (Sweetened) Dahi Vade

This is a sweet dish popular among the Muslim community of Gujarat (India). A dough of refined wheat flour with curd is made, and allowed to ferment for 2 h. Small balls from this dough are flattened and fried in oil or *ghee* and then, dipped in a sugar syrup.

7.5.13 Malpuda

A batter of wheat flour, sugar, and egg is allowed to ferment for 4–5 h, and then, poured as small flat pieces into hot *ghee* or oil and fried.

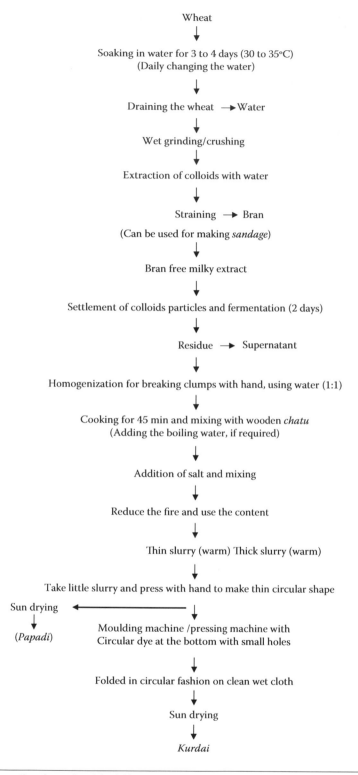

Figure 7.23 Preparation of papadi and *kurdai* from wheat.

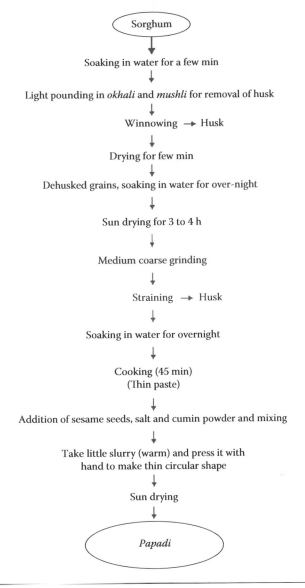

Figure 7.24 Preparation of *papadi* from sorghum.

7.5.14 Dahi *Karamba*

This is similar to the curd rice of South India. This mixture of rice and curd is added with sugar and *ghee* and garnished with dried fruits. This food product is popular among the Muslims of Gujarat (India).

7.6 Cereal- and Legume-Based Products

Soybeans, black gram, mung beans, and Bengal grams are the main legumes used in the preparation of a variety of fermented foods in different parts of the world (Panesar and Marwaha, 2013). They are fermented separately or in combination with cereals

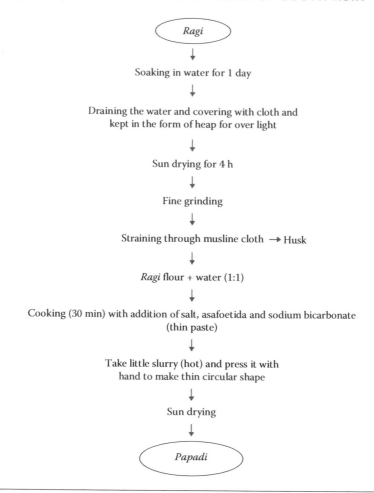

Figure 7.25   Preparation of *papadi* from *ragi*.

(Soni and Sandhu, 1999). There are a number of important cereal- and legume-based fermented foods prepared and consumed in India (Al-Aseeri et al., 2003). A fermented, thick suspension made of a blend of rice (*Oryza sativum*) and dehulled black gram (*Phaseolus mungo*) is used in several traditional fermented foods including such as *idli* and *dosa* which are very popular in India and Sri Lanka (Sands and Hankin, 1974; Batsa, 1981; Joshi et al., 2012; Steinkraus, 1996).

7.6.1 Papads

Papad (*papadam*) is another important condiment or savory food popular in India, Pakistan, and Bangladesh (Aidoo et al., 2006). Known as *papad* in northern India, *pappadam* in Malayalam, *happala* in Kannada, and *appalam* in Tamil, it is a thin, crisp, Indian preparation sometimes described as a cracker. This thin, usually circular, tortilla- or wafer-like product (Figure 7.26) is commonly served with curry, eaten by itself as a snack, or served as an accompaniment to a main meal in India (Aidoo et al., 2006). It can also be consumed with toppings, such as chopped onions, chutney,

Figure 7.26 Papad.

or other condiments. In some parts of India, it is served as the final item in a meal (Wikipedia the free encyclopedia). They are typically made from the flour or paste derived from either lentils, chickpeas, black gram (*urad* flour), rice, or potato (Anon, TNAU Agriteck Portal, 2014). It is normally consumed in toasted or fried form.

Papads are made of legumes, cereals, or starchy crop flour (Sankaran, 1998); they are typically made from the flour or paste derived from either lentils, chickpeas, black gram (*urad* flour), rice, or potato (Anon, TNAU Agriteck Portal, 2014). For their preparation, black gram flour, water, sodium bicarbonate, asafetida, and refined oil are mixed well, with the dough created after adding water. The dough is prepared and spiced as with *warries* (see below) but without fenugreek and ginger. It is kneaded well, divided into balls of about 15 g each, and each ball is pressed to a 0.8 cm thickness with a 8–9 cm diameter. The *papads* are stacked and stored overnight, conditioned, and dried under the sun for 10 min to reduce their moisture content by 10%–12.0%. They are then, packed in polyethylene pouches. Various steps involved in the processing of *papad* are outlined in Figure 7.27.

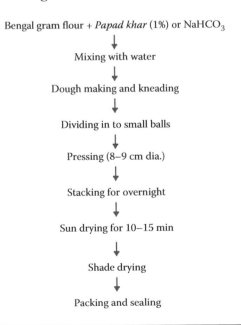

Bengal gram flour + *Papad khar* (1%) or $NaHCO_3$
↓
Mixing with water
↓
Dough making and kneading
↓
Dividing in to small balls
↓
Pressing (8–9 cm dia.)
↓
Stacking for overnight
↓
Sun drying for 10–15 min
↓
Shade drying
↓
Packing and sealing

Figure 7.27 Preparation of *Papad*.

Papads making under controlled conditions has evolved into a cottage- or small-scale industry. Here, black gram flour, or a blend of black gram with Bengal gram, lentil (*Lens culinaris*), red gram, or green gram (*Vigna radiata*) flour, is hand-kneaded with a small quantity of peanut oil, common salt (about 8%, w/w), *papad khar* (saltworts produced by burning a variety of plant species, or from very alkaline deposits in the soil), and water, and then pounded into a stiff paste. The dough (sometimes with a back-slop and spices added) is left to ferment for 1–6 h. The fermented dough is shaped into small balls, which are rolled into thin, circular, flat sheets (10–24 cm diameter, 0.2–1.2 mm thick) and, generally, dried in shade to remove 12% to 17% (w/w) moisture content (Wikipedia the free encyclopedia). *Candida krusei* and *S. cerevisiae* are involved in the preparation of *papad* (Shurpalekar, 1986).

Similar to *papad*, the noodle is a snack product that has been developed by fermentation and subsequent drying. The preparation process consists of soaking wheat kernels in water in 1:3 ratio for 1–2 h at room temperature. The steep water is discarded, and the fermented grains are blended with additional water to form a thin slurry. The slurry is then, strained through a muslin cloth and fermented for 8–10 h. The clear liquid is discarded, and the thick paste is mixed with water in a 1:1 ratio, cooked to a thick paste, extruded through a hand extruder into circular noodles, and sun-dried. The product is stored, and consumed after deep-fat frying.

7.6.2 Millet-Based Kambarq Coozhi/Ambali

It is known by various names, including *Kambarq coozhi/Ambali*. It is generally made by fermenting millet or *ragi* (*Eleusine coracana*) and pearl millets, and is consumed as a liquid product, especially in the summer.

Ambali (Figure 7.28) is also called as *ambil* in some areas. It is a traditional fermented product that is very popular in the interiors of Maharashtra (Ramakrishnan, 1977), as well as in the small villages and towns of Tamil Nadu (India). For its preparation, the

Figure 7.28 *Ambali.*

millet flour is mixed with water thoroughly to form a thick batter, which is allowed to ferment in earthen pots for 14–16 h. The fermented batter is sometimes cooked with (3/4) cooked rice and diluted with buttermilk or water. Salt is added to taste. It is drunk alongside some side dishes, such as raw mango, chilli, fried vegetables, etc. (Ramakrishnan, 1979), and is consumed in large quantities in the summer and considered good for one's health; the millet flour and rice are fermented, and are an easily digestible food. Lactic acid fermentation, involving the heterofermentative bacteria *L. mesenteroids*, produces CO_2 aerating batter, while the homofermentative *L. fermentum* (1.6×10^9/g) and *S. faecalies* (1.6×10^9/g) produce only lactic acid, which gives the characteristic flavor as well as other minor components to the product. Vitamins are also synthesized during the fermentation stage. The pH decreases from 6.4 to 4.0, and the volume increases by about 20% as a result of CO_2 production (Ramakrishnan, 1979). In an alternate method, the fermented batter is added to partially cooked rice with continuous stirring until the rice is cooked completely. Sour milk is then added, and the *ambali* is ready to eat.

7.6.3 Adai

Adai is a famous dish of south India that is made of rice and split legumes, like *dosa*, and is eaten with sauce, chutney, or sugar. Ten parts rice and three parts each of black gram *dhal* and red gram *dhal* are cleaned well and soaked in water for 2–3 h, then coarsely ground and left to ferment for 2–3 h in the summer or overnight during winters, after the addition of some salt (Panesar and Marwaha, 2013). Following fermentation, 1.5 parts grated coconut, 1–2 parts finely chopped onions, and 0.6 parts chilli powder or chopped green chillies are added with enough water to make a batter. About 40 g batter is then, spread on an oily hot plate (about 10 mm thick), browned on both sides, and served hot with coconut chutney or chutney powder with oil (Soni and Arora, 2000; Soni and Marwaha, 2003).

According to another method, 10 parts of rice and 6 parts of each of the legumes are soaked in water for 3–4 h, then ground coarsely with garlic, red chillies, cumin seeds, and asafoetida, and left to ferment overnight, or for 2–3 h in summer. Fried chopped onions, chillies, curry leaves, and coconut leaves are seasoned, added to the batter, mixed well to a thick consistency, and fried like *dosa* on a *tawa*. After one side is cooked, it is turned over and cooked until becoming brown. *Adai* is probably the oldest lactic-acid fermented product.

7.6.4 Handawa

Handwa is a popular fermented and baked food of Gujarat (western India) made out of a combination of cereal and pulse flours. Since the major food groups—as represented by cereals, pulses, fruits, and vegetables—are included in its preparation, it has scope to meet the nutritional needs of various populations. Almost 85% of the region's families consume *handwa* in the form of a "major meal," accompanied by oil, tea, buttermilk, tomato ketchup, or chutney. The most popular mix (used by 30% of the families), comprises approximately 60% rice, 20% red gram dal flour, and 20% Bengal gram dal flour. Lactic-acid fermentation of the batter prepared using this mix was more popular (75% families) than that using natural fermentation.

Handwas can be made with mix mesh size ranging from 25 to 75, using 75 mL of water per 50 g mix to prepare the batter. The fermentation of the batter produces gas for the porous texture, acidity, and flavor that are typical of the foodstuff, and which is achieved by the addition of lactic cultures in the form of a curd or natural starter developed by auto-fermentation. The titratable acidity rises steadily as the fermentation progresses from 6 to 24 h: at the end of 24 h, an initial titratable acidity of 0.2% can increase to 1.8% lactic acid, while increasing the inoculum from 1% to 10% has only a slight effect on the increasing the rate of acid production. Varying the duration of fermentation only slightly affects the cell size, cell distribution, crumb tenderness, and moistness of the *handwas*. On the other hand, flavor and overall acceptability were found to be significantly associated with the extent of fermentation. Although 10–14 h of fermentation results in the production of *handwas* with desirable acidity, a 12 h fermentation period using 2.5% of a natural culture developed in a laboratory using a back-slopping method per 100 g mix, incubated at 31.5°C (±1.5°C), results in producing *handwas* with the best sensory qualities. In tests, the baking of *handwa* batter standardized for a sand bath oven involved 90 min of total baking time, keeping the flame high for the initial 10 min and reducing it for the rest of the baking stage. These *handwas* could be stored for 18 days at room temperature and for 5 months at refrigerated temperatures without significantly affecting the microbial quality.

Furthermore, a variety of pulses, including broad beans, moth beans, black gram dals, lentils, and peas, can be incorporated in standard *handwa* at upto 25% without any significant change in their sensory qualities. Note that wheat and pearl millet can be incorporated only up to 10%, whereas maize and sorghum can be incorporated at upto 50%. Although wheat, due to its gluten content, is a crucial ingredient for breads, cakes, and similar products, its excessive use in *handwa* can give rise to higher acidity and moisture retention, affecting the quality of *handwa*. To some extent, the addition of millets along with the wheat can counteract the undesirable effects of the latter. However, a combination of three cereals, such as wheat (20%), *jowar* (40%), and rice (40%), made *handwas* comparable to the controls in tests. The typical *handwa* mix, having rice and germinated powders of red gram and Bengal gram with 15% less batter moisture, produced *handwas* that were comparable to the controls and also reduced the baking time by 20 min. Vegetables such as string beans, cluster beans, and bottle gourd can also be added at a 30% level without significantly affecting the specific volume or sensory qualities of *handwa*. A combination of 10 min steaming followed by 50 min baking reduces the overall cooking time from 90 to 60 min without significantly affecting the physical and sensory qualities of *handwa*.

7.6.5 Idli

An *idli* is a sour, steamed product made through the fermentation of rice and black gram, and is among the most popular of the cereal-and legume-based foods of India and Sri Lanka (Figure 7.29) (Vijaylakshmi, 2005). *Idlis* and the process of steaming were known in India as early as 700 (common era) (www.bhavnakitchen.com, 2013), with the steaming process probably introduced from Indonesia at some point between 800 and 1200 (common era). *Idlis* have been certainly been enjoyed in south India since 1100 (common era) (Ramakrishnan, 1979), and the credit for introducing them into the country's general diet rightly goes to the housewives of south India, particularly Tamil Nadu.

Each *idli* is a spongy, acid-leavened, steamed, whitish pancake (2–3 inches in diameter) of uniform porosity and physical stability with a characteristic taste and flavor. The microorganisms involved in their fermentation are mostly Lactic acid bacteria and some yeasts. (Iyer et al., 2013; Soni and Sandhu, 1999; Soni and Arora, 2000). They are of particular interest because of their high degree of acceptability as a food in south India, the protection they offer against food poisoning and the transmission of pathogenic organisms (due to its high acidity), They can be prepared using different combinations of cereal grains and acids to produce acid-leavened bread or pancake products (FAO, 1999, ch 2).

As noted, *idlis* are made involving the bacterial fermentation of a thick batter of carefully washed and soaked rice and black gram *dhal* (Rhee et al., 2011). Bacteria such as *Leuconostoc mesenteroides, Lactobacillus coryneformis, L. delbrueckii, L. fermentum, L. lactis, S. faecalis, S. cerevisiae, D. hansenii, H. anomala, Torulopsis candida, Trichosporon beigelii,* and *Pediococcus cerevisiae* (FAO, 1999, ch 3; Soni and Arora, 2000) are involved in their fermentation. The rice and black gram cotyledons are washed several times

Figure 7.29 *Idlis* and *Wade.* (a) *Idlis,* (b) *Idlis,* as served, and (c and d) *wade* as served.

with clean water to remove residual dirt, dust particles, and surface microorganisms, and are soaked separately (Reddy et al., 1982). The proportion of rice (*O. sativa*) to black gram (*P. mungo*) used in *idli* fermentation ranges from 1:4 to 4:1. Black gram is *dhal* soaked in water overnight, or for about 8 h, and ground into a smooth paste. The rice is washed, drained, ground coarsely, and then, mixed with the black gram paste to form a batter. The amount of water required to prepare a batter of a desirable and uniform consistency varies from 1:5 to 2:2 times the dry weight of the ingredients (Rao et al., 2005; Soni, 1987; Sandhu and Soni, 1989; Wang and Hesseltine, 1982). Salt and spices are added to the batter, and it is set aside in a warm place for 8–9 h or overnight for fermentation. The batter is then, allowed to ferment for 18–30 h by means of natural microflora, or by inoculating with fermented batter from a previous batch after the addition of approximately 1% salt. It is allowed to ferment overnight, until it expands to about 2.5 times its original volume. The fermentation is carried out in a closed container to avoid contamination and to keep insects away from the batter.

To cook them, a tree holding trays above the level of boiling water in a pot is used, with the pot covered for 10–25 min, depending upon size (www.bhavnaskitchen.com, 2013). The molds used (Figure 7.30) are perforated to allow the *idlis* to be cooked evenly. Historically, when *idli* mold cooking plates were not popular or widely available, the thick *idli* batter was poured onto a cloth tightly tied onto the mouth of a concave, deep cooking pan or *tawa* half filled with water. A heavy lid used to be placed on the pan, and the pot kept on the boil until the batter was cooked into *idli*. This was often a large *idli*, depending on the circumference of the pan. It was then cut into bite-sized pieces and served with *sambhar* (curry) or coconut *chutney* (Soni and Sandhu, 1999; Soni and Marwaha 2003; Panesar and Marwaha, 2013). Today, the product is sometimes salted and flavored with seasoning such as chillies. A variety of *idlis* are available, based on the raw materials used, such as *Kanchivaramidli*, Soy *idli*, Rava *idli*, and Malligae *idli*.

Figure 7.30 *Idli* preparation equipment.

In terms of taste, *idlis* are soft and spongy with a desirable sour flavor and eaten with coconut chutney, pickles, lentil curry, or chutney powder with a mixture of roasted and coarsely ground seasonings, such as sesame seeds, coriander seeds, red chillies, and split legumes. These preparations are consumed for breakfast with a spicy *sambar* and coconut *chutney*. Mixtures of crushed dry spices such as *milagai podi* are the preferred condiment for *idlis* eaten on the go. A variant of *idli* known as *sanna* is very popular among the people of Goa and other Konkani people.

Prepackaged mixes of *idli* allow for almost instant versions of the foodstuff; however, the additional health benefits of the fermentation process will be lacking. The plain rice and black lentil *idli* continues to be a particularly popular version, but they may also incorporate a variety of extra ingredients, savory or sweet. Mustard seeds, fresh chilli peppers, black pepper, cumin, coriander seed and its fresh leaf form (*cilantro*), fenugreek seeds, curry leaves, fresh ginger root, sesame seeds, nuts, garlic, scallions, coconut, and the unrefined sugar *jaggery* are all options. Filled *idlis* contain small amounts of *chutneys*, *sambars*, or sauces placed inside before steaming. *Idlis* are sometimes steamed in a wrapping of leaves such as banana leaves or jackfruit leaves.

Attempts have been made to substitute black gram *dhal* with other legumes, such as the common bean (Steinkraus et al., 1967), soya bean (Akolkar and Parekh, 1983; Panesar and Marwaha, 2013; Ramakrishnan et al., 1976; Sandhu and Soni, 1989) great northern bean (Sathe and Salunkhe, 1981), and mung bean (Sandhu and Soni, 1989). A flow chart depicting *idli* production is presented in Figure 7.31.

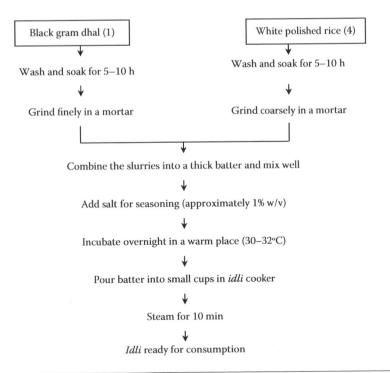

Figure 7.31   Flowchart for the production of traditional Indian *idli*. (Adapted from Panesar, P.S. and Marwaha, S.S. 2013. *Biotechnology in Agriculture and Food Processing: Opportunities and Challenges*. CRC Press, Boca Raton, FL, pp. 217–217.)

Several attempts to improve the *idli* fermentation process through the standardization of various factors involved have been made (Soni, 1987; Sandhu and Soni, 1989). For example, it has been found that *idli* batter fermentation rate increases with a rise in temperature, but the batter ripening time decreases from 16 h at 31°C to 10 h at 41°C. Organoleptically, *idlis* made from batter fermented at 31°C are soft, retaining their full and characteristic aroma, while those fermented at 41°C, though soft, lack the characteristic aroma (Rao et al., 2005). The fortification of *idli* batters with glucose at 1% level has a beneficial effect on gas production and leavening during fermentation.

During the preparation stage, after draining the water, the rice and the black gram are ground independently for 2–3 min in a blender, with the occasional addition of water during the process of paste making in a heavy stone grinding vessel (Al-Aseeri et al., 2003). The rice is coarsely ground while the black gram is finely ground. Then, the rice and the black gram batters are mixed together (2:1 ratio) with an addition of 0.8%–1% salt and allowed to ferment overnight at room temperature (about 30°C) (Reddy et al., 1982), depending upon the proportion of ingredients used and temperature (Steinkraus et al., 1967; Venkatasubbaiah et al., 1984; Yajurvedi, 1980).

The large-scale production of *idli* carried out in batch compartmental steaming units is labor intensive and has limited capacity. Equally, with the growing demands for breakfast foods, *idlis* are being consumed on an increasing scale in many Indian institutions, such as in the army, on railways, and in industrial canteens. In order to meet this demand, numerous studies have been carried out on the development of continuous units for the production of *idli* (Blandino et al., 2003; Murthy et al., 1994; Nagaraju and Manohar, 2000; Sridevi et al., 2010). *Idli* batter samples were prepared using lactic starter cultures such as *P. pentosaceus* (Pp), *Enterococcus faecium* MTCC 5153 (Ef), and *Ent. faecium* (IB2 Ef-IB2), individually, along with the yeast culture, *Candida versatilis* (Cv). *Idli* batter prepared using Ef and Ef-IB2 cultures gave better results, when evaluated for the rise in batter volume (80 mL), level of CO_2 production (23.8%), titratable acidity (2.4%–3.5%; lactic acid), and pH (4.3–4.4). The storage stability of batter made with selected starter cultures was determined by analysing the *idlis* prepared using the batter stored for 1 and 5 days for texture, nutrient composition, and sensory quality. Slight variations in the results were seen among the *idlis* of different combinations of cultures, but these results were better than those for the *idlis* made using naturally fermented *idli* batter. Sensory profile of *idlis* prepared using starter cultures showed a higher score (3.9–4.4) compared to the control (3.6) for overall acceptability (Sridevi et al., 2010). Murthy et al. (1994) developed an automatic *idli*-making unit that produces 1200 *idlis* per hour. The unit consists of an automatic *idli* batter depositor, a special *idli* pan conveyor, steam chamber, and an *idli* scooping system. The automatic batter depositor of the continuous *idli*-making unit works on a gravity flow system (Aachary et al., 2011).

Black gram is the main ingredient responsible for the characteristic texture of *idli* as observed. Earlier the surface-active proteins and polysaccharides of black gram are well conditioned to retain a large volume of gases to give a soft and fluffy texture (Susheelamma and Rao, 1978), but, after a certain period of fermentation, the batter starts collapsing, and, with further days of storage, there is whey separation, resulting in *idlis* with a very hard texture. Specifically, batter without stabilization or preservatives, even in refrigerated storage, has a limited shelf-life of few days, after which there is collapse in batter volume causing whey separation and consequently, very hard *idlis* (Nisha et al., 2005). Studies into the preservation of *idli* batter found a combination of 7.5 ppm nisin and 2000 ppm potassium sorbate to give good quality *idlis*, after 10 days of storage of batter at room temperature (28–30°C) and 30 days in refrigerated (4–8°C) storage.

Furthermore, milk proteins, such as whey proteins and casein, can act as good surface-active agents, because protein, in addition to lowering the interfacial tension, can form a continuous viscoelastic film (Damodaran and Paraf, 1997; Nisha et al., 2005). The polysaccharide in black gram is characterized as an arabinogalactan, which stabilizes the soft porous texture of *idli*. Arabinogalactan and guar gum are speculated to have similarity in their primary structure (Susheelamma and Rao, 1978). The stabilization of foam is very important with respect to *idli* texture (Glicksman, 1986; Nisha et al., 2005), and an addition of 0.1% pre-swollen xanthan to instant *idli* batter has been found to result in a desirable *idli* texture (Thakur, 1995).

Reduction in the fermentation time of the *idli* batter is of great commercial significance, too, for large-scale *idli* production, and this can be potentially achieved through the addition of enzymes externally. To achieve this, an exogenous source of α-amylase enzyme, 5, 15, and 25 U per 100 g batter, was added to the *idli* batter, which was allowed to ferment. The fermentation time was reduced from 14 to 8 h used conventionally, and the sensory attributes of the final product were also successfully maintained (Iyer and Ananthanarayan, 2008).

7.6.6 Dosa

Dosa or *thosai* is one of the most popular traditional fermented foods of south India as well as Sri Lanka (Soni, 1987; Soni and Sandhu, 1999; Vijaylakshmi, 2005). It is a thin, crisp, fried pancake, and a tasty and attractive staple foodstuff eaten as a breakfast product (Figure 7.32), but increasingly popular as a snack food throughout India (Soni and Arora, 2000). *Dosa* or *thosai* is also a popular breakfast food in Sri Lanka.

It is prepared from naturally fermented rice (*O. sativa*) and black gram (*P. mungo*). Rice and black gram *dhal* (1:1) are soaked separately for 4–6 h in water, and then ground to a fine paste and mixed together to form a free-running slurry (Vijayendra et al., 2010). This is then, allowed to ferment overnight naturally for 8–20 h after the addition of salt (FAO, 1999, ch 2). Following fermentation, the batter is quickly fried as a thin, crisp pancake and eaten directly. *L. mesenteroides*, *S. faecalis*, *T. candida*, and *Trichosporon pullulans* are the most common microorganisms associated with *dosa* fermentation (Al-Aseeri et al., 2003; Steinkraus, 1996). The microbiological, physical,

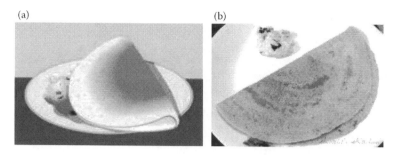

Figure 7.32   (a) Dosa (b) Thosia.

and biochemical changes in *dosa* during fermentation as well as its nutritive value are quite similar to those of *idli* (Chavan and Kadam, 1989; Purushothaman et al., 1993; Ramakrishnan, 1993; Sands and Hankin, 1974; Shortt, 1998). However, more rice flour is used in *dosa* batter than in *idli* batter (Sandhu and Soni, 1988).

Traditional *dosa* batter fermentation has revealed the occurrence and role of several bacteria alone or in combination with yeasts in bringing about various biochemical changes (Sandhu et al., 1986; Soni et al., 1986). The frying of fermented thin batter (a layer of 1–5 mm) is done on a thick iron plate smeared with little oil or fat. A sol to gel transformation occurs during the heating, and, within a few minutes, a circular, semi-soft to crisp product resembling a pancake, ready for consumption, is obtained (Bhattacharya and Bhat, 1997; Das et al., 2012). It is eaten when hot with special *dhal*, coconut chutney, or *sambhar*. *Dosa* batter is very similar to *idli* batter, except that both the rice and the black gram are finely ground. The batter is also thinner than for the *idli* (Al-Aseeri et al., 2003). Also, the fermented suspension, instead of being steamed, is heated with a little oil on a flat plate.

When the *dosa* batter is made without inoculation, it is necessary that the rice and legume be soaked ground with water and incubated at room temperature. The utilization of a hot soak and hot grinding destroys the microorganisms essential for fermentation. Fermentation time varies from 14 to 24 h. The inoculum is the microorganisms developing during the initial soaking and then, during the night fermentation (Mukherjee et al., 1965). The optimum temperature is 25–35°C. India's Central Food Technological Research Institute recommends the addition of buttermilk. *L. mesenteroids* and *S. faecalis* are developed during soaking, and continue to multiply, following grinding reaching more than $1 \times 10^9/g$ (Rhee et al., 2011). These organisms are known to be present on the ingredients, and, generally, it is unnecessary to add inoculum. The addition of yeast contributes to leavening, flavor, and an enhanced content of thiamin (Soni and Sandhu, 1990; Tamang and Fleet, 2009). The recommended step-by-step method is given below:

- Soak 3 parts rice and 1 part black gram in water for 4–6 h at 27–30°C.
- Grind to a fine paste and mix both to get a free-running batter.

- Add 1% of salt to the batter and allow to ferment overnight (10–16 h). The natural microflora of both the rice and *dhal* will act as a starter and bring about the fermentation.
- Pour about 50–80 mL of batter into a hot, oiled pan and fry until crisp and browned on one side, then turn to the second side, cook well, and roll.
- Serve this crispy and roasted *dosa* with either coconut chutney or with a lentil and vegetable curry (Soni et al., 1985).

7.6.7 Adai Dosa

This foodstuff is made using rice, Bengal gram, red gram, black gram, and green gram (Panesar and Marwaha, 2013). The legumes and rice are soaked in water for 2–3 h and then, ground into a coarse batter and left to ferment for 2–3 h in summer and overnight during winter, after the addition of some salt. After fermentation, coriander leaves, chopped onions, curry leaves, asafoetida (*hing*), and grated coconut are added to the batter. A small amount of batter is spread onto an oily plate and toasted on a low flame, and then, served hot with coconut chutney and *jaggery*. *Pediococcus* sp., *Streptococcus* sp., and *Leuconostoc* sp. have been found to be associated with *adai dosa* fermentation (Chavan and Kadam, 1989).

7.7 Legume-Based Fermented Products

7.7.1 Punjabi Warri/Wadiyan

Warries, in the form of spicy (Yadav and Khetarpaul, 1994), hollow, brittle, fried balls of 5–8 cm in diameter, are very popular in northern India, especially in Amritsar city (Punjab). They are used as an accompaniment in cooking with vegetables, legumes, or rice (Panesar and Marwaha, 2013). Dehulled black gram (*P. mungo*) grains, after washing and soaking overnight in water, are ground to a paste. The paste is then spiced with asafoetida (*Ferula foetida*, 0.5%–1.0%), caraway (*C. carvi*, 0.5%–1.0%), cardamom (*Elettaria cardamomum*, 1.0%), cloves (*Syzygium aromaticum*, 0.4%–0.5%), fenugreek (*Trigonella foenumgraecum*, 1.0%), ginger (*Zingiber officinale*, 8.0%), and red pepper (*Capsicum annum*, 1.0%), and molded into small balls that are set aside to undergo fermentation and drying in open air for 4–8 days.

7.7.2 Bhallae

Bhallae are snack foods prepared from unspiced black gram paste, similar to *warri* but without spices and fermented for 12–18 h (Panesar and Marwaha, 2013).

The paste is pan-or deep-fried and consumed as a snack (Figure 7.33) as prepared, or following a brief soak in spiced tamarind water or curd, in various parts of Pakistan and India.

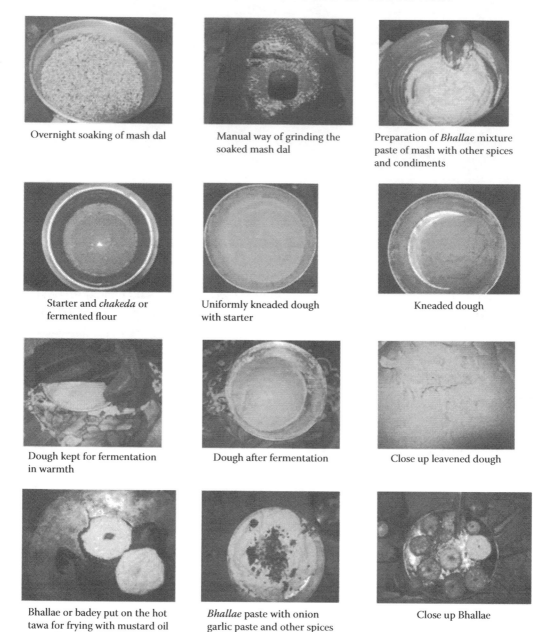

Overnight soaking of mash dal

Manual way of grinding the soaked mash dal

Preparation of *Bhallae* mixture paste of mash with other spices and condiments

Starter and *chakeda* or fermented flour

Uniformly kneaded dough with starter

Kneaded dough

Dough kept for fermentation in warmth

Dough after fermentation

Close up leavened dough

Bhallae or badey put on the hot tawa for frying with mustard oil typically

Bhallae paste with onion garlic paste and other spices

Close up Bhallae

Figure 7.33   The pictorial representation of preparation of Bhallae.

The microflora involved in *bhallae* fermentation generally comprises yeasts and bacteria. *L. fermentum* and *L. mesenteroides* are the predominant bacteria (Panesar and Marwaha, 2013) with *Bacillus subtilis*, *S. faecalis*, *Flavobacter*, and *Achromobacter*, while *Kluyveromyces marxianus* and *T. beigelii* are the most predominant yeast species, followed by *Trichosporon pullulans*, *H. anomala*, *D. hansenii*, *Pichia membranaefaciens*, *S. cerevisiae*, and *Candida curvata*. However, the number and proportion of the microbial types varies with the season, with summers favoring bacteria and winters favoring yeasts (Hammes et al., 2005; Soni, 1987; Soni and Arora, 2000; Soni and Sandhu, 1990).

7.7.3 Vada/Vadai

Vada—also known as *wada*, *vade*, *vadai*, or *bara*—is a savory fritter-type snack from south India (Figure 7.34) (http://www.absoluteastronomy.com, 2013). It is a traditional food known from antiquity (www.Medlibrary.com 2012). Although they are commonly prepared at home, *vadas* are also a typical street food in the Indian subcontinent and in Sri Lanka. They can vary in shape and size, but are usually either doughnut- or disk-shaped and between 5 cm and 8 cm across. They are made from black gram, lentil, gram flour, or potato. They are also known as *ulundhu vadai*. *Vadas* are typically and traditionally served along with a main course such as *dosa*, *idli*, or *pongal* (www.simplyindianrecipes.com, 2012), and are preferably eaten freshly fried, while still hot and crunchy, and served with a variety of dips, ranging from *sambar* to *chutney* to curd.

Vadas are made from black gram based on an unspiced legume paste and deep fried in oil, as with *bhallae*, but the dough is comparatively finer and the incubation period is 18–24 h (Panesar and Marwaha, 2013). To prepare *vada*, legume beans are soaked in water for 4–6 h, then ground to a fine paste, and left to ferment at ambient temperature for few hours, after the addition of salt, chillies, and sometimes onions (Panesar and Marwaha, 2013). The paste is then made into balls and deep fried in vegetable oil, and consumed as a snack or staple food in various parts of India (Shurtleff and Aoyagi, 2012). *Ulundhu vadai* are thus, marvelously tasty little fritters made from daal, combined with incredible spices, and deep fried to crunchy perfection.

7.7.4 Khandvi/Panori/Patudi

This is a condiment, prepared from a thin batter of gram flour with buttermilk and salt, of Gujarat. The batter is cooked to a soft thick batter and then, spread into thin

Figure 7.34 *Vada—a savory fritter type snack.*

layers on a greased plate, cut into strips, and formed into a roll. These pieces are garnished with cracked mustard and coriander leaves.

7.8 Rice-Based Fermented Extract Product

This type of fermentation is practiced widely in the households in Kerala and in some parts of Tamil Nadu and the Andhra Pradesh states of India (Tamang, 2009a,b; Vijaylakshmi, 2005). In it, the rice is cooked, then cooled to ambient temperature; water is added, mixed gently, and left for fermentation overnight. The next morning, the water is decanted to yield a fermented rice extract, which is mixed with buttermilk and salt, and directly consumed as a beverage. Alternatively, the extract is used for cooking vegetables. The leftover rice, following the decantation of the fermented rice extract, is also consumed, after mixing it with *dahi* and salt (Ramakrishnan, 1979). The microorganisms isolated from this rice fermentation were *S. faecalis* (2.7×10^7/g), *Paediococcus acidolactice* (2.7×10^7/g), *Bacillus* sp. (1.6×10^8/g), *Microbacterium flavum* (1.1×10^8/g), *Candida tropicalis*, *Candida gulliermondii*, *H. anomala*, and *Geotrichum candidum*, and the pH of extracted decreased from 6.1 to 5.7 in 16 h of fermentation.

7.8.1 Sonti Annam

This is essentially an incipiently fermented, cooked rice product that is popular in the Nellore district of Andhra Pradesh (India). The microorganisms involved are unknown yeasts and the mold *Rhizopus sonti*. The product, with a characteristic pleasant aroma and taste, is considered to have cooling and refreshing effects and, therefore, is consumed mainly in summer months.

In its preparation, the cooked rice is allowed to cool to ambient temperature, mixed with the inoculum, and allowed to ferment at room temperature (Gadaga et al., 1999). The inoculum from a previous batch of fermentation is employed. A pleasing aroma develops when the rice has fermented for 1–2 days, and the product is consumed at this stage. Further, the continuation of the fermentation stage leads to the disintegration and liquefaction of the cooked rice grains to produce a yellowish-colored alcoholic liquor that is mildly intoxicating. For inoculum preparation, the *Sonti annam* from a previous batch is absorbed on the paddy husk, made into the form of small balls, and used as such or sun-dried when not to be used immediately. The microorganisms present here are those present in *dahi*.

7.8.2 Adirasams

Adirasams are a deep-fried product made from rice flour and *jaggery*. *Jaggery* is dissolved in water, strained, and boiled until thick; then *ginjly* (sesame) seeds and coconut scrapings are added to it, mixed well, and then removed from the fire. Rice flour is added to the syrup mixture, again mixed thoroughly, and cooked on a slow flame until the consistency of the mixture comes to be a string when taken in a spoon. This is allowed to stand at room temperature for a day, and then, small balls are made and flattened with greasy hands, then fried in hot oil until golden brown. The microorganisms involved are probably those encountered in *idli* batter.

7.8.3 Selroti

Selroti is a popular, fermented, rice-based, spongy, pretzel-like, deep-fried food consumed in Sikkim and the Darjeeling hills in India, Nepal, and Bhutan (Das et al., 2012; Tamang et al., 1988; Tamang, 2009a, 2010a,b, 2012). It is served as a confectionery item during religious festivals and special occasions. It is mostly prepared at home, to the extent of 75% while the proportion of market purchases is estimated to be about 15% (Yonzan and Tamang, 2009). Its per capita consumption is 8 g/day, and the average annual production per household has been found to be 18.5 kg.

In its preparation, a local variety of rice (*O. sativa*) is sorted, washed, and soaked in cold water overnight (6–8 h) (Das et al., 2012). The water is then, decanted from the rice using a bamboo-constructed sieve and spread over a woven tray, and then dried for 1 h (Tamang, 2009a,b). The pre-soaked rice is pounded into a coarse powder using a wooden mortar and pestle. Larger particles of pounded rice flour are separated from the rest through a winnowing process using a bamboo tray (Das et al., 2012). Then, the rice flour is mixed with nearly 25% refined wheat (*Triticum aestivum*) flour, 25% sugar, 10% butter or fresh cream, and 2.5% spices/condiments, including large cardamom (*Amomum subulatum*), cloves (*S. aromaticum*), coconut (*Cocus nucifera*), fennel (*Foeniculum vulgare*), nutmeg (*Myristica fragrans*), cinnamon (*Cinnamomum zeylanicum*), and small cardamom (*E. cardamomum*), which are added to the rice flour and mixed thoroughly.

Milk or water is added and the mixture is kneaded into a soft dough, and ultimately, into a batter with an easy flow. It is then, left to ferment naturally at an ambient temperature (20–28°C) for 2–4 h during summer and at 10–18°C for 6–8 h during winter (Yonzan and Tamang, 2010). The fermented batter is squeezed and deposited as a continuous ring into hot cooking oil and fried until golden brown, then removed from the oil and drained using a poker and served as a confectionary bread (Yonzan and Tamang, 2010). The traditional method of *selroti* preparation is shown in Figure 7.35.

The microorganisms involved in the fermentation have been identified as bacteria such as *Leuconostoc mersenteroides*, *E. faecium*, *P. pentosaceus*, and *Lactobacillus curvatus* and yeasts (Tamang et al., 2012) such as *S. cerevisiae*, *Saccharomyces kluyveri*, *D. hansenii*, *Pichia burtonii*, and *Zygosaccharomyces rouxii* (Yonzan and Tamang, 2010).

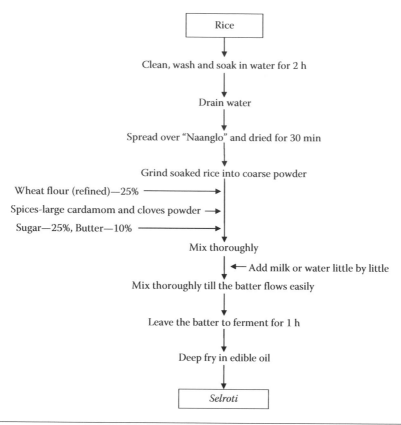

Figure 7.35 Traditional method for the preparation of *Selroti*. (Adapted from Yonzan, H. and Tamang, J.P. 2010. *Food Biotechnology* 24(3): 227–247.)

7.8.4 Fermented Rice

Fermented rice is another popular, easily digestible breakfast food made from rice in Tamil Nadu. To make it, water is added to cooked rice and allowed to ferment overnight at about 30–32°C. With the addition of a little salt, it can be consumed directly along with curd. *S. faecalis*, *P. acidilactici*, *Bacillus* sp., and *M. flavum* have been found to be involved in the fermentation of rice (Ramakrishnan, 1977).

7.8.5 Sez

Sez is a traditional semi-fermented food used by the Bhotiya community in Uttaranchal (India) (Das et al., 2012). The method of its preparation is similar to that for *jann chhang* (an alcoholic beverage) except that rice is used as a substrate and it is mostly used as a snack food (Sekar and Mariappan, 2007). Historically, it was a delicacy and used to be prepared only during certain festivals. *Sez* is made by cooking the rice and by adding *balam* starter powder to the cooked rice. Fermentation is carried out for 2–3 days at an ambient temperature, and the fermented rice can be consumed as it is. Commonly, *sez* is extracted during the preparation of rice *jann* (an alcoholic beverage), described in Chapter 9 of this text.

7.8.6 Pantha Bhat

This is a lightly fermented rice-based dish popular in rural areas, where it is served for breakfast with salt, onion, and chilli. It is a cold and wet food is suitable for summer mornings. There are many variations of the dish, but a common one is made by soaking cooked rice in water overnight. Care must, however, be taken to cover the dish during long soaking to avoid contamination with microorganism and insects (Wood, 1985). *Pantha bhat* contains a small amount of alcohol, which is formed during fermentation (Nasrullah, 2003). The fermentation improves the bioavailability of minerals such as iron and zinc as a result of the phytic acid hydrolysis, and increases the content of riboflavin and vitamin B (Ruel, 2001). *S. faecalis, P. acidilactici, Bacillus* sp., and *M. flavum* have been isolated from fermented rice. During fermentation, the pH decreases from 6.1 to 5.7 in 16 h, but there is no change in volume, amino nitrogen, or free sugar content (Steinkraus, 1996).

7.8.7 Khoblu

It is a traditional fermented food of the Mandi Suket area of Himachal Pradesh (India). Cooked rice is further cooked in buttermilk. During the cooking process, wheat flour *pedas* are also put in buttermilk mixtures and cooking is continued until the *pedas* are cooked properly. Cooked *pedas* are consumed along with the buttermilk rice mixture (Sharma and Singh, 2012; Thakur et al., 2004).

7.8.8 Water Rice (Pakahla)

Pakahla is a partially fermented, homemade, non-alcoholic, rice product. Depending on the quality of the rice and additive, ingredients such as spices and curd, which are added to the rice in water, the *pakahla* can be classified as *saja pakhaka* (fresh water rice), *basi pakhala* (stale water rice), *jeera pakala* (spice water rice), *dahi pakhala* (curd water rice), and so on. Water rice is locally called *pakahla*, which is an Odiya term and is derived from the Pali word *pakhalita* as well as from the Sanskrit word *prakshalana*, which means "washed to wash." *Pakahla* is widely eaten, particularly by Odiya people, during the summer season, where the average temperature is more than $40 \pm 2°C$ with relative humidity of $80\% \pm 5\%$. Eating *pakahla* in this season imparts a soothing effect to the human body. Today, the *pakahla* recipe has also been introduced to the 3- and 5-star hotel menus of Odisha to attract the guests who are interested in traditional ethnic cuisines.

In *pakahla* preparation, rice is boiled fully and the excess boiled water is removed from it (FAO, 1999, ch 2). The drained rice is allowed to cool down to room temperature, and ordinary water is added to it so that the rice remains submerged completely. Then, the excess water-mixed rice is allowed to ferment at room temperature for 8–12 h. After the incubation period, the fermented rice with water, locally called *torani*, is served. It is popularly served with roasted vegetables such as potato (*Solanum tuberosum* L.) and *brinjal* or eggplant (*Solanum melongena* L.), and is often fried with small fish. *Pakahla* is sometimes co-fermented or mixed with curd, cucumber (*Cucumis sativus* L.), cumin (*Cumin* L.) seeds, fried onion (*Allium cepa* L.), and mint (*Mentha longifolia*) leaves. Preliminary studies have shown that *pakahla* contains several yeast (*Saccharomyces* sp.) and lactobacilli, which bring the typical aroma and sourness to the foodstuff.

7.9 Cereal and Legume Batter Fermented Cakes

7.9.1 Dhokla

Dhokla is not only a fermented spongy cake consumed as breakfast or snack food in various states of western India, but also a fermented food commonly identified with Gujarat (Tamang and Fleet, 2009). The ingredients—namely, Bengal gram cotyledons and polished white rice—are washed with water to remove any adhering dust and surface microorganisms, and soaked in water for 5–10 h, separately (Reddy et al., 1982). Rice and black gram *dhal*, in the proportion of 3:1, is mixed with one part curd and water to prepare the dough for the *dhokla* preparation. The soaked ingredients are ground in a stone mortar separately to obtain a fine paste and then, mixed together with salt (approximately 1.0%, w/w). The combined batter is incubated in a warm place (30–32°C) overnight (12–14 h), and chopped green fenugreek leaves are added to the fermented batter for an enhanced taste (FAO, 1999, ch 3). The mixture is poured into a greased tray and steamed in an open condition (i.e., rather than in a covered *idli* steamer) for 15–20 min (Joshi et al., 1989; Panesar and Marwaha, 2013). After cooling, the *dhokla* is cut into squares and seasoned with coriander leaves, green chillies, mustard, and asafoetida (Figure 7.36). The entire process is also shown diagrammatically in Figure 7.37.

Bacteria belonging to *L. fermentum*, *L. lactis*, *L. delbrueckii*, and *L. mesenteroides* and yeast belonging to *Hansenula silvicola* from fermenting batter have been generally

Figure 7.36 Dhokla—a fermented spongy cake.

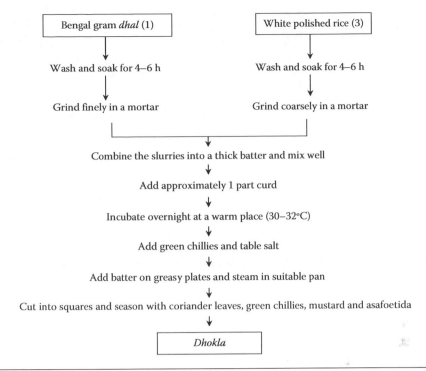

Figure 7.37 Flow chart for the production of traditional Indian *Dhokla*.

associated with *dhokla* fermentation (Joshi et al., 1989). Other microorganisms, such as *L. fermentum*, *L. mesenteroides*, *Pichia silvicola*, *S. faecalis*, *Torulopsis* sp., *Candida* sp., and *T. pullulans*, are also associated with this fermentation (Soni and Arora, 2000). The characteristic flavor of this foodstuff could be due to the acetoin and less volatile fatty acids, acetic acid, propionic acid, isobutyric and isovaleric acid, which show a marked rise during fermentation (Panesar and Marwaha, 2013; Wiseblatt and Kohn, 1960).

Bhat dhokla is another variety of *dhokla* prepared by the people of the Saurashtra region. In it, rice is mixed with gram flour and curd and allowed to ferment for 2 h before being steamed (Panesar and Marwaha, 2013).

7.9.2 Chakuli

Chakuli is a fermented fried pancake prepared from varying proportions of rice (*O. sativa*) and black gram (*P. mungo* L.). It resembles *dosa* and is a round, fried pancake that is generally a little bit thicker but smaller in diameter (Figure 7.38). *Chakuli*, which resembles *dosa* (a popular South Indian fermented food), is eaten in Odisha daily as a snack food. Its ratio of ingredients is chosen according to the availability of ingredients and consumer preferences, although 2:1 is the most common ratio (FAO, 1999, ch 2 and ch 3). A little amount of boiled rice may be added to make the product softer. Alternatively, black gram may be substituted with juice of jackfruit (*Artocarpus*

Figure 7.38 Steps in the preparation of legume-based traditional fermented food. (Modified from Roy, A., Moktan, B., and Sarkar, P.K. 2007. *Indian Journal of Traditional Knowledge* 6(1): 12–16.)

heterophyllus Lam.) or palmyra palm (*Borassus flabellifer* L.), fruit during summer. A small amount of *jaggery* may also be added to make it sweet, which is occasionally supplemented with the pulp of jackfruits (*A. hetrophyllus* Lam.), or Palmyra palm (*B. flabellifer* L.), or sweet potato (*Ipomoea batatas* L.), or mahula (*Madhuca latifolia* L.) flower paste, depending upon the availability.

For its preparation, the rice is washed, soaked for 6 h, decanted, and briefly sundried (Figure 7.38a and b) (FAO, 1999, ch 2). Next, the dried rice grains are pounded in an iron or wooden mortar and then, sieved to obtain a fine powder (Figure 7.38c and d). Black gram is soaked until the seed coats becomes easily removable by applying a gentle pressure. The grains are then, rubbed by hand to loosen the seed coats, which are allowed to float away. The black gram is then, made into a smooth paste using a stone grinder (Figure 7.38e). The paste is beaten repeatedly by hand, with a little amount of water added and mixed with rice powder, as well as an appropriate amount of lukewarm water and salt (Figure 7.38f). The batter is left to ferment under cover for 4–5 h during summer (12–15 h during winter) (Figure 7.38g). Then, the fermented batter is fried over a hot greased pan to round-shaped flat cakes (Figure 7.38h). Spices such as ginger (*Z. officinale* L.), onion (*A. cepa* L.), and black pepper (*Piper nigrum*) powder are sometimes added at the time of frying (*pitha*). *Chakuli* is served with a variety of side dishes, including *sambar*, sugar, *jaggery*, tea, milk, vegetable curry, and mutton, or even without any side dish in the Orissa state of India (Roy et al., 2007). The complete process of *chakuli* making is shown in Figure 7.39.

Different varieties of *chakuli* include *saru chakuli*, *budha chakuli*, and *jau chakuli*. *Saru chakuli* is a thinner variety, while a thicker consistency of the batter with added *jaggery*, *paneer*, and grated coconut makes *budha chakuli*. The name *budha chakuli* has a link with the Buddhist spiritual leader Gautama Buddha, giving an idea of its origin. In some regions, fresh juices of jackfruit (*A. heterophyllus* Lam.) or palmyra palm (*B. flabellifer* L.) fruit is added to the batter after fermentation to render a fruity flavor (Roy et al., 2007). Parboiled rice is cooked to make boiled rice and then, incubated overnight. This fermented rice is mixed with black gram paste to make the batter of *jau chakuli*.

Chakuli tastes best when consumed hot, but has shelf-life of one day only. It is prepared during all the festivals, including *satha sasthi*, *raja*, *prathama astami*, and *bataosha*.

7.9.3 Chhuchipatra Pitha

The batter of *chhuchipatra pitha* is prepared (Figure 7.38) as with *chakuli* by grinding (Figure 7.38a and b) and fermenting (Figure 7.38g) the same ratio of raw ingredients. The fermented batter is then, spread thinly over a hot greased pan both horizontally and vertically using a spatula. The cake is then filled at the center with grated coconut, *dahi-chhana* (curd), and sugar before folding in a square shape to fry (Figure 7.40h–q). Due to its sugar content, it has a shelf-life a little greater than that of *chakuli*, and is usually taken without any accompaniment. *Chhuchipatra pitha* is prepared during the

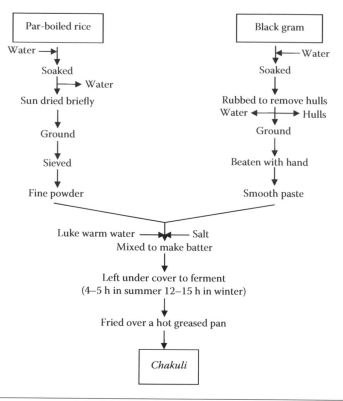

Figure 7.39 Flowsheet for the preparation of *chakuli*. (From Roy, A. Moktan, B., and Sarkar, P.K. 2007. *Indian Journal of Traditional Knowledge* 6(1): 12–16. With permission.)

bataosha festival in Orissa, India. The entire process of preparing it is shown in Figure 7.39.

The fermented batter, as is done for *chakuli* preparation, is flattened over an oil-greased pan as a thin layer and filled with grated coconut, curd, and sugar in its center; the pancake is then, folded in a square shape to be fried (Roy et al., 2007). It has a shelf-life of two days, and is usually consumed without any accompaniment due to its sweet taste. The art of preparing this *pitha* is gradually vanishing, as it is somewhat cumbersome. However, it is still regularly prepared during the *raja sankranti* (summer) festival in rural Odisha.

7.9.4 Chitou

Chitou is a fried cake. The batter is prepared (Figure 7.40) and fermented as in *chakuli*. Sugar and grated coconut are mixed with the fermented batter. A special earthen mould or deep bowl is smeared with oil, filled with the batter mix, and covered with a lid. The junction is closed with a wet cloth and water is sprinkled intermittently. It is fried on a low heat. *Chitou* is consumed fresh and hot with curry, sugar, curd, or tea. It has a shelf-life of one day, and is prepared for popular festivals, like *maker sankranti* and *chitou-amabasa*.

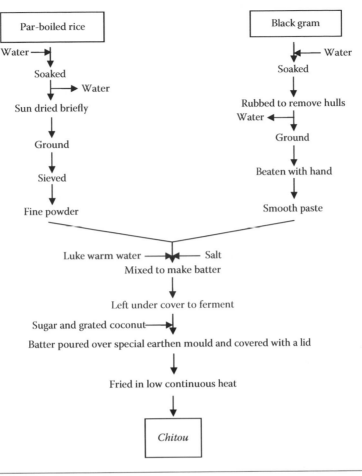

Figure 7.40 Flow sheet for the preparation of *chitou*.

7.9.4.1 Chitou Pitha   *Chitou* is prepared by mixing the fermented rice batter (without any legume*)* with sugar and grated coconut (*C. nucifera* L.) (Roy et al., 2007). It is then, taken in a special earthen mould or in deep bowl and covered with a lid. The junction is closed with a wet cloth and water is sprinkled intermittently. It is fried on a low heat. Although it has shelf-life of one day, *chitou* is delicious when taken fresh and hot. Generally, it is taken with curry, sugar, curd, or milk. *Chitou* is prepared in popular harvest festivals, like *maker sankranti* and *chitou amabasya*. Preparation of the *chitou pitha* is given in a flow sheet in Figure 7.41.

7.9.5 Arish Pitha

Arisa pitha (Figure 7.42) is made by creating a thick batter prepared from rice flour, locally called *jantani*, which is then, made into small, semi-flat, round shapes and fried in *ghee* or oil until the color changes to golden brown (http://bestfoodrecepies.in/arisa-pitha-orissa-recipe/2014). During the preparation stage, raw rice is submerged in water for 8–12 h. Then, the water-soaked rice is allowed to dry under shade, and ground into

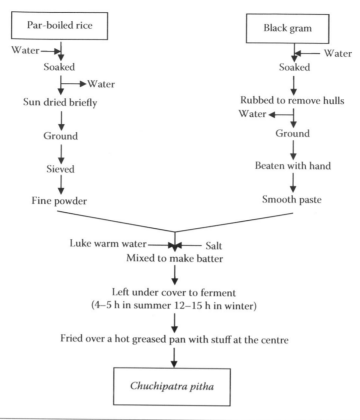

Figure 7.41   Flowsheet for the preparation of *chuchipatra pitha*. (From Roy, A., Moktan, B., and Sarkar, P.K. 2007. *Indian Journal of Traditional Knowledge* 6(1): 12–16. With permission.)

Figure 7.42   *Arisa pitha.*

a fine flour with the help of a mechanical grinder. The flour is then, used for the *jantani* preparation. The shelf-life of this *arisa pitha* varies from six months to a year.

7.9.6 Enduri Pitha

Enduri pitha is a flavored and steamed cake. The preparation procedure (Figure 7.43) of the batter of *enduri pitha* is similar to that for *chakuli*. The fermented batter,

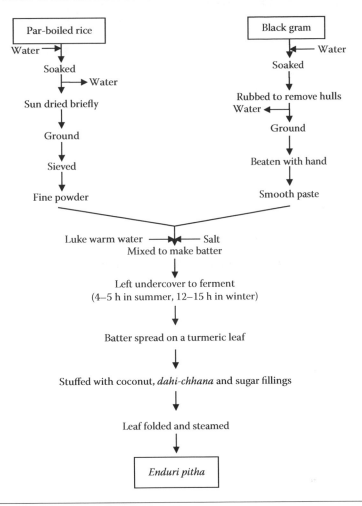

Figure 7.43 Flow sheet for the preparation of *enduri pitha*.

a little thicker than that of *chakuli*, is taken in a turmeric leaf, then stuffed with grated coconut, *dahi-chhana*, and sugar fillings, and folded through the mid-vein. The traditional method of steaming is done by heating an earthen pitcher (*handi*) filled with water. The mouth of the bowl is tied with a piece of cloth, keeping a shallow cavity to put the batter-filled leaves over there which is then, covered with an earthen lid to capture the steam. Its shelf-life is also two days, and it may be consumed without any side dish due to its sweet taste and appealing aroma (FAO, 1999, ch 3; Roy et al., 2007).

It is prepared customarily in the state of Odisha during the occasion of *Prathamstami*, a festival to celebrate the wellness of the eldest child in the family. *Enduri pitha* is a traditional delicacy, and also shows how to make use of the medicinal properties of the turmeric plant. Many Ayurvedic physicians have observed that eating the extract of turmeric leaves through this traditional food in winter helps in the strengthening of the immune system (Roy et al., 2007).

7.9.7 Munha Pitha

Munha pitha is a cereal- and pulse-based fermented food of Orissa that is mainly prepared during festival seasons, particularly during the *raja* (harvest) festival in June. It is a steamed, spongy cake similar to *idli* (Figure 7.44). It is also made from parboiled rice powder and black gram paste in a 3:1 ratio. The two ingredients are mixed with salt and fermented to develop a pleasant acidic flavor with definite leavening. The consistency of the batter should be thicker than that of *chakuli* (FAO, 1999, ch 3). Sugar or *jaggery*, minced coconut, raisins, and cashew (*Anacardium occidentale* L.) nuts are added to the fermented batter for a delicate taste. Steam is generated as with the *enduri pitha* preparation process, and the thick batter is poured over a cloth (Figure 7.44). An inverted, empty *handi* is suitable to capture the steam. The continuously generated steam cooks the material, and completion of cooking is checked by inserting a sharp object through the center of the batter mass and observing if the batter

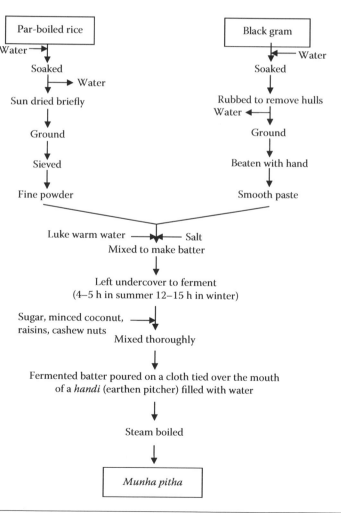

Figure 7.44   Flow sheet for the preparation of *munha pitha*.

has stuck to the surface. No adherence of batter to the object indicates completion of cooking even at the center. A good quality *munha pitha* becomes spongy, like *idli* (another popular South Indian food item), and it is served by cutting into pieces and then, consumed without any accompaniment or with sugar or curry. The shelf-life of this *pitha* is one day only (Roy et al., 2007).

7.9.8 Podo Pitha

Podo pitha (Figure 7.45) is, a baked cake. It is slow-cooked *pitha*. During its prepara-tion, fermented batter (fermented rice and black gram, similar to that used in *chakuli*) is mixed with minced coconut, raisins, cashew nuts, and sugar. The method of prepa-ration (Figure 7.46) of the batter is similar to that of *enduri pitha* in its composition and fermentation. In the preparation of *podo pitha*, rice and black gram *dal* are soaked overnight, separately. Both of the ingredients are ground into a thick batter. After the addition of common salt, it is allowed to ferment for 2–4 h until the inside is soft and white.

The fermented batter mixture is then packed in sal (*Shorea robusta* C.F. Gaertn.) or banana (*Musa paradisiaca* L.) leaves. The packets are then, covered all over with hot charcoal in an earthen oven to bake in a low but continuous heat for 5–10 h (Roy et al., 2007). Completion of cooking is checked with a sharp object, as described in the cooking stage for *munha pitha* (Roy et al., 2007). The *pitha* is cut into pieces and served. It has a shelf-life of 2–3 days. It is consumed like *munha pitha*. Its serving and mode of consumption are similar to those for *enduri pitha*. *Raja* and *bijoya dashami* are two important festivals during which *podo pitha* is prepared (Figure 7.47).

7.9.9 Sandan

Sandan is a snack or breakfast food of Maharashtra and Gujarat (India) that is pre-pared by mixing jackfruit juice into coarsely ground rice flour, followed by over-night fermentation and the addition of *jaggery*, and then, steaming like *dhoklas*. The

Figure 7.45 Podo pitha—a baked pitha.

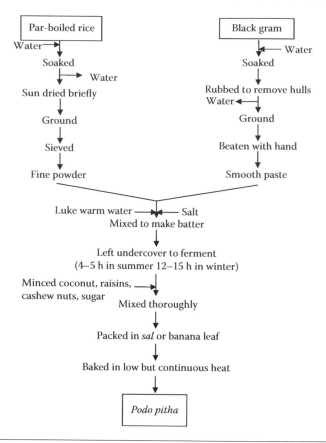

Figure 7.46 Flow sheet for the preparation of *podo pitha*.

fermentation process is being followed in some households in the preparation of this foodstuffs, while being totally neglected in others.

7.9.10 Sandage

These are prepared from the coarse ground cereals with bran after fermentation. The bran residues obtained after the preparation of *papadi* and *kurdai* are also used in the preparation of *sandage* (Figures 7.48 and 7.49). These are rich in crude and dietary fibers. The process of preparation of *sandage* is outlined in Figure 7.49. They are generally used as vegetables, after cooking with spices, while the pearl millet (*bajra*) *sandage* can be roasted with little oil and eaten with peanuts as a snack.

7.9.11 Anarse

This is an authentic Maharashtrian (Maharashtra, India) snack especially prepared on festive occasions such as *ganesh chaturthi*, *diwali*, etc. It is also prepared and consumed in the Gujarat state of India. Although *anarse* are simple to make, the snack normally requires at least 7 days for preparation (Figure 7.50).

Figure 7.47 Pictorial representation of preparation of *Pida Pitha*.

During *anarse* preparation (Figure 7.51), the rice is cleaned, washed, and soaked in enough water for 3–4 days during winter and 2 days during summer at ambient temperature, changing the water daily. The water is then, decanted using a strainer/bamboo *chalni* and the rice is spread over a clean cloth, and partially dried in shade for 1–2 h. It is then, pounded into a fine powder in a mortar and pestle, known as *okhali* (stone) and *musali* (wooden), or in a mixer to a smooth powder. It is sieved through fine mesh and then mixed with sugar/*jaggery* in the ratio of 2:1. This mixture is kneaded to a hard dough without adding water, shaped into big balls (Figure 7.50), and kept in a clean, sealed container for 4–5 days in a refrigerator, or can be used fresh or as and when required for making *anarse*. To make it, the balls are broken and mixed with milk/banana/curd (full fat) and made into dough. The dough is divided into small balls and flattened into a medium, thick, circular shape using moist hands

Figure 7.48   Sandage.

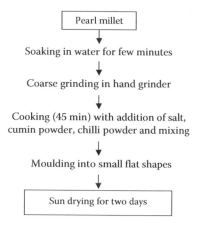

Figure 7.49   Preparation of *sandage* from pearl millet.

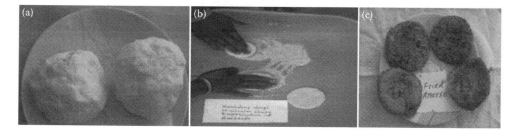

Figure 7.50   (a) Rice flour and sugar balls after fermentation for making *anarse*; (b) Pressing the *anarse* in aniseed; (c) Fried *anarse*.

to make each about a 3- to 3.5-inch size diameter circle in poppy seed (Figure 7.50) taken in a bamboo tray (*supali*). Do not make the *anarse* too thin, as this will make it very hard.

This circular shape is finally fried in hot oil on a slow heat until golden brown, and is drained from the hot oil using a poker, locally called a *zaraya*, or a spatula (Das et al., 2012). Frying *anarse* is done over a low flame or medium-low heat because of its tendency to break if fried too quickly. Once golden, they are allowed to drain

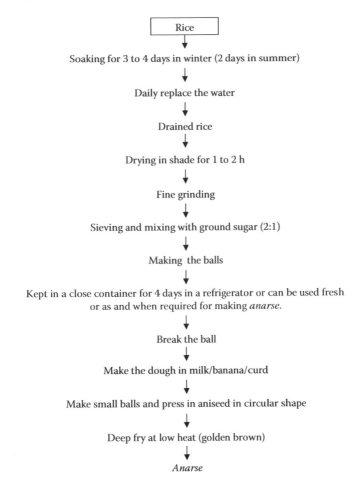

Figure 7.51 Process of making *anarse* from rice.

and cool, usually for up to 5 or 6 h, until they are crispy. Do not turn the poppy seed-side down in the frying pan because it will make the poppy seeds come out and oil will start burning soon. It is a porous, textured, sweet product (Figure 7.51). The traditionally accepted quality criterion for this foodstuff is a light, fluffy, golden brown-colored product with a honeycomb structure and a crisp, chewy texture.

7.9.12 Seera/Kheera

This is a nutritious, easily digestible snack food made in the Bilaspur, Kangra, Hamirpur, Chamba, Mandi, Bilaspur, and Kullu districts of Himachal Pradesh (India). In the preparation of *seera*, wheat grains are soaked in water for 2–3 days to allow natural fermentation to occur (Thakur et al., 2004). After fermentation, the grains are ground and steeping is done, to allow the starch grains and some proteins to settle, and the bran is removed (Thakur et al., 2004). The settled solids are then, sun-dried and the dried material is called *seera/kheera* (Thakur et al., 2004). This dried material is made into a slurry mixture by soaking in water, which is then, poured into hot *ghee*, sugar is then added, and it is then cooked and served as a sweet dish/snack. It holds special significance to the village people of Bilaspur, where, during droughts, *seera* is offered to *Jal Devta* (god of water) for rain. People suffering from jaundice or hepatitis are also given *seera*. It is used during fast, too. Village-based cooperatives market *seera* in packets. It is prepared occasionally or offered to the guests as a sweet dish in the rural or urban areas of Himachal Pradesh (Thakur et al., 2004). The microflora isolated from *seera* mainly comprise of *S. cerevisiae*, *Cryptococcus laurentii*, and *T. delbrueckii* among the yeasts and *Lactobacillus amylovorus*, *Cellulomonas* sp., *Staphylococcus sciuri*, *Weisella cibaria*, *Bacillus* sp., *Leuconostoc* sp., and *Enterobacter sakazakii* among the bacteria (Savitri et al., 2012).

7.9.13 Bagpinni/Jagcha

It is a cereal-based traditional fermented food of Himachal Pradesh, prepared from barley and *chhang*/buttermilk (FAO, 1999, ch 3). For the preparation of *bagpinni*, roasted barley flour (*sattu*) is mixed with buttermilk or *chhang* (a traditional beverage see Chapter 9 of this text) and kneaded into dough, which is finally, given the shape of a ball with a cavity in the center. This cavity is filled with *ghee*/butter and consumed during long journey (Thakur et al., 2004).

7.10 Cereal-Based Fermented Beverages

7.10.1 Ambil

This is a fermented product of *ragi* (*Eleusine coracana* L. Gaertn.) and rice. In earlier times, it was used as a breakfast food, and is generally regarded as providing energy and being good for one's health overall. In the villages of Andhra Pradesh, it is particularly utilized by workers and laborers, in contexts where energy is needed. Sometimes, it is consumed with buttermilk, green chilli, onions, and pickles, which can give it a better taste (Ramakrishnan, 1979). Today, *ambil* (Figure 7.52) is typically enjoyed as an appetizer. It is a common product used during *yalamusha* in large quantities in Maharashtra.

To prepare it, *ragi* flour is made into a thick batter with water and allowed to ferment for 14 h in an earthen pot. The fermented *ragi* batter is then, added to cooked rice, stirring to avoid lump formation. It is allowed to cool, mixed with sour buttermilk, and consumed. Before serving, though, it is mixed with spices (e.g., ginger, garlic, salt, chilli powder, coriander, cumin).

The LAB present in buttermilk degrade the starch of sorghum into simple sugars, resulting in a decrease of pH from 6.0 to 5.4 in 5–6 h, while the volume increases by about 20%, indicating CO_2 production (Das et al., 2012; Ramakrishnan, 1979). The fact that CO_2 is produced may be due to the heterofermentative LAB. It is reported

Figure 7.52 Ambil.

that the *ambila* fermented product of *ragi* contains *L. mesenteroides*, *L. fermentum*, and *S. faecalis* microorganisms and *Saccharomyces* sp. (Ramakrishnan, 1979).

7.10.2 Kali

Kali is one of the indigenous fermented rice products of Andhra Pradesh, India. To the leftover cooked rice, water is added and the mixture is allowed to ferment overnight. It is generally prepared in the summer season because of the high temperatures in the nights. People in Andhra Pradesh believed that *kali* cools the body and is very good for one's health (Sekar and Mariappan, 2007). The microorganisms involved in its fermentation are bacteria such as *L. mesenteroides* and *L. fermentum Beij* and yeasts such as *Hansnula*, *Candida* and *Saccharomyceae* sp. (Aidoo et al., 2006).

7.11 Summary and Future Prospects

A large variety of cereal-based fermented foods are prepared and consumed throughout the countries of South Asia, as discussed in this chapter. Indigenous fermented cereal foods show a great variety, meeting the tastes of local populations and depending upon the availability of the foodstuffs' raw ingredients, which illustrates the best use of available resources. These cereal-fermented products have better nutritional values than their raw materials, and have an improved shelf-life. The microflora responsible for the fermentation in many cases includes strains of LAB, yeast, and fungi. Single or mixed cereals sometimes mixed with other pulses are used, and the final texture of each product can vary according to the processing method and the fermentation conditions.

Traditionally, fermented foods, including cereal-based fermented foods, have formed an integral part of various rituals of many communities in different countries of South Asia. The preparation procedures of these fermented foods are complex and time consuming, involving soaking, grinding, incubating, and frying or steaming. Due to this, while once undertaken regularly, this traditional skill set is now mainly practiced by rural women during specific festivals. Likewise, there is a wide variety of cereal-based fermented products that have not received the scientific attention they deserve. As with other indigenous fermented foods, rapid urbanization and modernization have affected the time-tested traditional technologies for the preparation of fermented foods. If academics continue to ignore this important issue, due to modern dietary habits and changing lifestyle patterns, in times to come, our indigenous traditional knowledge will vanish before it can be validated and improved. Thus, the traditional products, such as those discussed in this chapter and across the book, should be promoted, due to their comparatively enhanced nutritive value and sensory quality. The benefits and risks associated with specific indigenous fermented cereals; elucidating the contributions of microorganisms, enzymes, and other cereal constituents in the fermentation process; developing starter cultures, unique microbial strains for nutritional improvement and detoxification; and testing new cereal varieties for their suitability as fermentation substrates, all need to be looked into in our future research and development programmes.

Indigenous fermented foods and beverages also offer an opportunity for scaling up to make presently household-confined fermented food products available to a much wider population. Modern science and technology have the potential to facilitate the transforming of the home-scale art of preparation of fermented foods into small- to medium-sized industry applications, leading to the production of better and a consistent quality of product with an improved shelf-life and enhanced nutritional value to meet the demand of present and future consumers. Sourdough technology, although a traditional process, when combined with the modern manufacturing techniques could yield healthier end products for consumers (Holzapfel, 1997). Natural preservatives such as LAB bacteriocins or other such antimicrobial products have shown capabilities for inhibiting food-borne pathogens and/or food-spoilage bacteria, including *Listeria monocytogenes*, *B. subtilis*, and *Staphylococcus aureus*, and their use as food additives or in the application of the producer strains as starter or protective cultures could significantly contribute to the production of safer indigenous fermented food products from cereals in the future.

References

Aachary, A.A., Gobinath, D., and Prapulla, S.G. 2011. Short chain xylo-oligosaccharides: A potential prebiotic used to improve batter fermentation and its effect on the quality attributes of *idli*, a cereal–legume-based Indian traditional food. *International Journal of Food Science and Technology* 46(7): 1346–1355.

Adams, M.R. and Moss, M.O. 1996. Fermented and microbial foods. In: Adams, M.R. and Moss, M.O. (Eds.) *Food Microbiology*, 1st edn. New Age International Publishers, New Delhi, pp. 252–302.

Adeyemi, I.A. and Beckley, O. 1986. Effect of period of maize fermentation and souring on chemical properties and amylograph pasting viscosities of ogi. *Cereal Science* 4: 353–360.

Aidoo, K., Nout, R., and Sarkar, P. 2006. Occurrence and function of yeasts in Asian indigenous fermented foods. *FEMS Yeast Research* 6(1): 30–39.

Åkerström, J. and Carlson, G. (trans.) 1936. *The Princesses Cook Book*. Albert Bonnier Publishing, New York.

Akingbala, J.O., Rooney, K.W., and Faubion, J.M. 1981. A laboratory procedure for the preparation of ogi: A Nigerian fermented food. *Food Science* 45(5): 1523–1526.

Akinrele, I.A. 1970. Fermentation studies on maize during the preparation of traditional African starch-cake food. *Journal of the Science of Food and Agriculture* 21: 619–625.

Akolkar, P.N. and Parekh, L.J. 1983. Nutritive value of soy idli. *Journal of Food Science and Technology* 20: 1–4.

Al-Aseeri, M., Pandiella, S., Blandino, A., Canter, D., and Webb, C. 2003. Cereal-based fermented foods and beverages. *Food Research International* 36: 527–543.

Anon, 2014. TNAU (Tamil Nadu Agricultural University) Agritech Portal. 2014. http://agritech.tnau.ac.in/postharvest/pht_pulses_valueaddtn.html.

Arendt, E. and Bello, F. 2008. Functional cereal products for those with gluten intolerance. In: Hamekar, B.R. (Ed.) *Technology of Functional Cereal Products*. Cambridge, U.K.: Woodhead Publishing Ltd., pp. 446–475.

Arendt, E. and Zannini, E. 2013. *Cereal Grains for the Food and Beverage Industries. Woodhead Publishing Series in Food Science, Technology and Nutrition*, pp. 1–66, 67.

Batra, L.R. 1981. Fermented cereals and gram legumes of India and vicinity. In: Moo Young, M. and Robinson, C.W. (Eds.) *Advances in Biotechnology*. Pergamon Press, Toronto, pp. 547–554.

Belton, P.S. and Taylor, J.R.N. 2004. Sorghum and millets: Protein sources for Africa. *Trends in Food Science and Technology* 15: 94–98.

Bhatia, A. and Khetarpaul, N. 2002. Effect of fermentation on phytic acid and *in vitro* availability of calcium and iron of 'Doli ki roti'—An indigenously fermented Indian bread. *Ecology of Food and Nutrition* 41(3): 243–253.

Bhatia, A. and Khetarpaul, N. 2012. *Doli ki roti*—An indigenously fermented Indian bread: Cumulative effect of germination and fermentation on bioavailability of minerals. *Indian Journal of Traditional Knowledge* 11(1): 109–113.

Bhattacharya, S. and Bhat, K.K. 1997. Rheological properties of *Dosa* batter. *Journal of Food Engineering* 34: 429–443.

Blandino, A., Al-Aseeri, M., Pandiella, S.S., Cantero, D., and Webb, C. 2003. Cereal-based fermented foods and beverages. *Food Research International* 36: 527–543.

Bőcker, G., Stolz, P., and Hammes, W.P. 1995. Neue Erkenntnisse zum Őkosystem Sauerteig und zur Physiologie des Sauerteig-Typischen Stämme *Lactobacillus sanfrancisco* und *Lactobacillus pontis. Getreide, Mehl und Backwaren* 49: 370–374.

Catzeddu, P., Mura, E., Parente, E., Sanna, M., and Farris, G.A. 2006. Molecular characterization of lactic acid bacteria from sourdough breads produced in Sardinia (Italy) and multivariate statistical analyses of results. *Systematic and Applied Microbiology* 29: 138–144.

Cauvain, S.P. 2003. Breadmaking: An overview. In: Cauvain, S.P. (Ed.), *Breadmaking: Improving Quality*. Woodhead Publishing, Cambridge, pp. 8–28.

Chavan, R.S. and Chavan, S.R. 2011. Sourdough technology—A traditional way for wholesome foods: A review. *Comprehensive Reviews in Food Science and Food Safety* 10(3): 169–182.

Chavan, J.K. and Kadam, S.S. 1989. Nutritional improvement of cereals by fermentation. *CRC Critical Reviews in Food Science and Nutrition* 28: 349–400.

Clarke, C.I., Schober, T.J., Dockery, P.Ő., Sullican, K., and Arendt, E.K. 2004. Wheat sourdough fermentation: Effects of time and acidification on fundamental rheological properties. *Cereal Chemistry* 81: 409–417.

Coda, R., Cagno, R.D., Gobetti, M., and Rizello, C. 2014. Sourdough lactic acid bacteria: Exploration of non-wheat cereal-based fermentation. *Food Microbiology* 37: 51–58.

Corsetti, A., Angelis, M.D., Dellaglio, F., Paparella, A., Fox, P.F., and Settanni, L. 2003. Characterization of sourdough lactic acid bacteria based on genotypic and cell-wall protein analyses. *Journal of Applied Microbiology* 94(4): 641–654.

Corsetti, A., Gobbetti, M., Balestrieri, F., Paoletti, F., Russi, L., and Rassi, J. 1998. Sourdough lactic acid bacteria effects on bread firmness and staling. *Journal of Food Science* 63: 347–351.

Corsetti, A., Gobbetti, M., and De Marco, B. 2000. Combined effect of sourdough lactic acid bacteria and additives on bread firmness and staling. *Journal of Agricultural and Food Chemistry* 48: 3044–3051.

Corsetti, A. and Settanni, L. 2007. Lactobacilli in sourdough fermentation. *Food Research International* 40(5): 539–558.

Damodaran S, Paraf A. 1997. Protein-stabilized Foams and Emulsions. In: Damodaran S. and Paraf A., editors. Food proteins and their applications. CRC Press Taylor and Francis group, Roca, Baton, FL. p 57–105.

Das, A., Raychaudhuri, U., and Chakraborty, R. 2012. Cereal-based functional food of Indian subcontinent: A review. *Journal of Food Science and Technology* 49(6): 665–672.

De Vuyst, L., Schrijvers, V., Paramithiotis, S., Hoste, B., Vancanneyt, M., and Swings, J. 2002. The biodiversity of lactic acid bacteria in Greek traditional wheat sourdoughs is reflected in both composition and metabolite formation. *Applied and Environmental Microbiology* 68: 6059–6069.

Decock, P. and Cappelle, S. 2005. Bread technology and sourdough technology. *Trends in Food Science and Technology* 16: 113–120.

Di Cagno, R., de Angelis, M., Lavermicocca, P., De Vincenzi, M., Giovannini, C., Faccia, M., and Gobbetti, M. 2002. Proteolysis by sourdough lactic acid bacteria: Effect on wheat flour protein fractions and gliadin peptides involved in human cereal intolerance. *Applied and Environmental Microbiology* 68: 623–633.

Edema, O.M. 2011. A modified sourdough procedure for non-wheat bread from maize meal. *Food and Bioprocess Technology* 4(7): 1264–1272.

Farnworth, E.R. 2008. Handbook of fermented functional foods, 2nd edn. In: Mazza, G. (Ed.) *Functional Foods and Nutraceuticals Series*. Taylor & Francis, CRC Press, Boca Raton, FL.

Ferchichi, M., Valcheva, R., Oheix, N., Kabadjova, P., Prévost, H., Onno, B., and Dousset, X. 2008. Rapid investigation of French sourdough microbiota by restriction fragment length polymorphism of the 16S-23S rRNA gene intergenic spacer region. *World Journal of Microbiology and Biotechnology* 24(11): 2425–2434.

Fields, M.L., Hamad, A.M., and Smith, D.K. 1981. Natural lactic acid fermentation of corn meal. *Journal of Food Science* 46: 900–902.

Food and Agriculture Organization of the United Nations (FAO) 1999. Cereals: Rationale for fermentation. Chapter 1. In: Haard, N.F., Odunfa, S.A., Lee, C.-H., Quintero-Ramírez, R., Lorence-Quinones, A. and Wacher, R.C. (Eds.) *Fermented Cereals: A Global Perspective*. FAO Agricultural Services Bulletins, 138. http://www.fao.org/docrep/x2184e/x2184e09.htm.

Food and Agriculture Organization of the United Nations (FAO) 1999. Cereal fermentations in African countries. Chapter 2. In: Haard, N.F., Odunfa, S.A., Lee, C.-H., Quintero-Ramírez, R., Lorence-Quinones, A. and Wacher, R.C. (Eds.) *Fermented Cereals: A Global Perspective*. FAO Agricultural Services Bulletins, 138. http://www.fao.org/docrep/x2184e/x2184e07.htm.

Food and Agriculture Organization of the United Nations (FAO) 1999. Cereal fermentations in countries of the Asia-Pacific region. Chapter 3. In: Haard, N.F., Odunfa, S.A., Lee, C.-H., Quintero-Ramírez, R., Lorence-Quinones, A. and Wacher, R.C. (Eds.) *Fermented Cereals: A Global Perspective*. FAO Agricultural Services Bulletins, 138. http://www.fao.org/docrep/x2184e/x2184e09.htm.

Foschino, R. and Galli, A. 1997. Italian style of life: Pane, amore, lievito naturale. *Tecnologie Alimentari* 1: 42–59.

Gadaga, T., Mutukumira, A., Narvhusb, J., and Feresu, S. 1999. A review of traditional fermented foods and beverages of Zimbabwe. *International Journal of Food Microbiology* 53(1): 1–11.

Gaggiano, M., Cagno, R.D., Angelisa, M.D., Arnault, P., Tossut, P., Fox, P., and Gobbetti, M. 2007. Defined multi-species semi-liquid ready-to-use sourdough starter. *Food Microbiology* 24(1): 15–24.

Giannou, V., Kessoglou, V., and Tzia, C. 2003. Quality and safety characteristics of bread made from frozen dough. *Trends in Food Science and Technology* 14: 99–108.

Glicksman, M. 1986. *Food Hydrocolloids*, Vol. III. CRC Press, Boca Raton, FL.

Gobbetti, M., Corsetti, A., and De Vincenzi, S. 1995. The sourdough microflora. Characterization of homofermentative lactic acid bacteria based on acidification kinetics and impedance tests. *Italian Journal of Food Science* 2: 91–102.

Gobbetti, M., Corsetti, A., Rossi, J., La Rosa, F., and De Vincenzi, S. 1994. Identification and clustering of lactic acid bacteria and yeasts from wheat sourdoughs of central Italy. *Italian Journal of Food Science* 1: 85–94.

Gobbetti, M., De Angelis, M., Arnault, P., Tossut, P., Corsetti, A., and Lavermicocca, P. 1999. Added pentosans in breadmaking: Fermentations of derived pentoses by sourdough lactic acid bacteria. *Food Microbiology* 16: 409–418.

Gobbetti, M., De Angelis, M., Corsetti, A., and Di Cagno, R. 2005. Biochemistry and physiology of sourdough lactic acid bacteria. *Trends in Food Science and Technology* 16: 57–69.

Gotcheva, G., Pandiella, S., Angelov, A., Roshkova, Z., and Webb, C. 2001. Monitoring the fermentation of the traditional Bulgarian beverage Boza. *International Journal of Food Science and Technology* 36(2): 129–134.

Hammes, W.P., Brandt, M.J., Francis, K.L., Rosenheim, J., Seitter, M.F., and Vogelmann, S. 2005. Microbial ecology of cereal fermentations. *Trends in Food Science Technology* 16(1–3): 4–11.

Hammes, W.P. and Ganzle, M.G. 1998. Sourdough bread and related products. In: Wood, B.J.B., (Ed.) *Microbiology of Fermented Foods*, 2nd edn, Vol 1. Blackie Academic and Professional, London, pp. 199–216.

Holzapfel, W. 1997. Use of starter cultures in fermentation on a household scale. *Food Control* 8: 241–258.

http://bestfoodrecepies.in/arisa-pitha-orissa-recipe/2014.

http://www.absoluteastronomy.com.2013.

http://www.simplyindianrecipes.com.2012.

Iyer, B., Singhal, R., and Ananthanarayan, L. 2013. Characterization and *in vitro* probiotic evaluation of lactic acid bacteria isolated from *idli* batter. *Journal of Food Science and Technology* 50: 1114–1121.

Iyer, B.K. and Ananthanarayan, L. 2008. Effect of α-amylase addition on fermentation of idli – A popular south Indian cereal–legume-based snack food. *LWT – Food Science and Technology* 41: 1053–1059.

Jacob, H.E. 1997. *Six Thousand years of Bread: Its Holy and Unholy History*. Lyons and Burford, New York, NY, 27 pp.

Joshi, N., Godbole, S.H., and Kanekar, P. 1989. Microbial and biochemical changes during dhokla fermentation with special reference to flavor compounds. *Journal of Food Science and Technology* 26: 113–115.

Joshi, V.K. and Pandey, A. 1999. Biotechnology: Food fermentation. In: Joshi, V.K. and Pandey, A. (Eds.) *Biotechnology: Food Fermentation, Microbiology, Biochemistry and Technology*, Vol. 1. Educational Publishers and Distributors, New Delhi, pp. 1–24.

Joshi, V.K. 2005. Traditional fermented foods and beverages of North India—An overview In: Proceedings of first work shop and strategic meeting of South Asian Network on fermented foods. May 26–27, 2005, Anand, Gujarat, India, pp. 59–73.

Joshi, V.K., Garg, V. and Absol, G.S. 2012. Indigenous fermented foods. In: *Food Biotechnology*. Joshi, V.K. and Singh, R.S. (Eds). International Publishers and Distributors, New Delhi, pp. 337–373.

Kanwar, S.S., Gupta, M.K., Katoch, C., Kumar, R., and Kanwar, P. 2007. Traditional fermented food of Lahaul and Spiti area of Himachal Pradesh. *Indian Journal of Traditional Knowledge* 6(1): 42–45.

Katina, K., Arendt, E., Liukkonen, K.H., Autio, K., Flander, L., and Poutanen, K. 2005. Potential of sour dough for healthier cereal products. *Trends in Food Science and Technology* 16(1–3): 104–112.

Kulp, K. and Lorenz, K. 2003. *Handbook of Dough Fermentations*. Marcel Dekker, Inc., New Delhi, pp. 1–21.

Kwon, T.W. 1994. The role of fermentation technology for the world food supply. In: Lee, C.H., Adler, N. and Barwald, G., (Eds.) *Lactic Acid Fermentation of Non-Dairy Food and Beverages*. Ham Lim Won, Seoul, South Korea, pp. 187–193.

Matz, S.A. 1971. *Cereal Science*. AVI Publishing Co., Westport, CT, 241 pp.

Medlibrary org.com. Encyclopedia. 2012.

Mir, M.S. and Mir, A.A. 2000. Ethnic foods of Ladakh (J&K) India. In: Sharma, J.P. and Mir, A.A. (Eds.) *Dynamics of Cold Arid Agriculture*. Kalyani Publication, New Delhi, pp. 297–306.

Moroni, A., Bello, F., and Arendt, E. 2009. Sourdough in gluten-free bread making: An ancient technology to solve a novel issue? *Food Microbiology* 26(7): 676–684.

Mueen-ud-din, G. 2009. Effect of wheat flour extraction rates on physicochemical characteristics of sourdough flat bread. PhD thesis, Faisalabad University, Pakistan.

Mueen-ud-din, G., Rehman, S., Anjum, F.M., and Nawaz, H. 2007. Quality of flat bread (*Naan*) from Pakistan wheat varieties. *Pakistan Journal of Agricultural Sciences* 44(1): 171–175.

Mukherjee, S.K., Albury, M.N., Pederson, C.S., van Veen, A.G., and Steinkraus, K.H. 1965. Role of *Leuconostoc mesenteroides* in leavening of *idli*: A fermented food of India. *Applied Microbiology* 13: 227–231.

Murthy, C.T., Nagaraju, V.D., Rao, P.N.S., and Subba Rao, V.N. 1994. A device for continuous production of idli. Indian Patent No. 950/DEL/94.

Nagaraju, V. and Manohar, B. 2000. Rheology and particle size changes during *Idli* fermentation. *Journal of Food Engineering* 43(3): 167–171.

Nasrullah, S. 2003. Liberalizing alcohol policy. *Daily Star (Bangladesh)* pp. 8–15.

Nisha, P., Ananthanarayan, L., and Singhal, R. 2005. Effect of stabilizers on stabilization of idli (traditional South Indian food) batter during storage. *Food Hydrocolloids* 19(2): 179–186.

NPCS (NIIR Project Consultancy Services) Board. 2012. *Modern Technology on Food Preservation*, 2nd edn. Asia Pacific Business Press, Inc., New Delhi, 528 pp.

Ottogalli, G., Galli, A., and Foschino, R. 1996. Italian bakery products obtained with sourdough: Characterization of the typical microflora. *Advances in Food Sciences* 18: 131–144.

Oyewole, O.B. 1997. Lactic fermented foods in Africa and their benefits. *Food Control* 8(5–6): 289–297.

Padmaja, G. and George, M. 1999. Oriental fermented foods: Biotechnological approaches. In: Marwaha, S.S. and Arora, J.K. (Eds.) *Food Processing: Biotechnological Applications*. Asia Tech Publishers Inc., New Delhi, pp. 143–189.

Panesar, P.S. and Marwaha, S.S. 2013. *Biotechnology in Agriculture and Food Processing: Opportunities and Challenges*. CRC Press, Boca Raton, FL, pp. 217–217.

Parveen, S. and Hafiz, F. 2003. Fermented cereal from indigenous raw materials. *Pakistan Journal of Nutrition* 2(5): 289–291.

Pereira, M., Jacobs, D., and Pins, J. 2002. Effect of whole grain on insulin sensitivity in overweight hyperinsulinemic adults. *American Journal of Clinical Nutrition* 75: 848–855.

Prajapati, J.B. 2003. Fermented foods of India. In: *Proceedings of International Seminar and Workshop on Social Well Being*, Anand, India, pp. 1–4.

Purushothaman, D., Dhanapal, N., and Rangaswami, G. 1993. Indian idli, dosa, dhokla, khaman and related fermentations. In: Steinkraus, K.H., (Ed.) *Handbook of Indigenous Fermented Foods*. Marcel Dekker, New York, NY, pp. 149–165.

Qarooni, J. 1996. *Flat Bread Technology*. New York, NY: Chapman and Hall, pp. 99–101.

Ramakrishnan, C.V. 1977. The use of fermented foods in India. In: *Symposium on Indigenous Fermented Foods*, Bangkok, Thailand.

Ramakrishnan, C.V. 1979. Studies on Indian fermented foods. *Baroda Journal of Nutrition* 6: 1–7.

Ramakrishnan, C.V. 1993. Indian idli, dosa, dhokla, khaman, and related fermentations. In: Steinkraus K.H., Ed., *Handbook of Indigenous Fermented Foods*. Marcel Dekker, New York, NY, pp. 149–165.

Ramakrishnan, C.V., Parekh, L.J., Akolkar, P.N., Rao, G.S., and Bhandari, S.D. 1976. Studies on soy *idli* fermentation. *Plant Foods for Man* 2(1–2): 15–33.

Rao, E.R., Varadaraj, M. C. and Vijayendra, S.V.N. 2005. Fermentation biotechnology of traditional foods of the Indian subcontinent. Chapter 3.18. In: Shetty, K., Paliyath, G., Pometto, A., and Levin, R.E. (Eds.) *Food Biotechnology*, 2nd edn. CRC Press, Boca Raton, FL, pp. 1759–1794.

Reddy, R., Sathe, S., Pierson, D., and Salunkhe, D. 1982. *Idli*, an Indian fermented food: A review. *Journal of Food Quality* 5(2): 89–101.

Rhee, S.R., Lee, J., and Lee, C. 2011. Importance of lactic acid bacteria in Asian fermented foods. *Microbial Cell Factories* 10(Suppl 1): S1–S5.

Rollán, G., Gerez, C.L., Dallagnol, A.M., Torino, M., and Font, G. 2010. Update in bread fermentation by lactic acid bacteria. In: Mendez, A. (Ed.) *Current Research, Technology and Education Topics in Applied Microbiology and Microbial Biotechnology*. IBA, Bucharest, pp. 1168–1174.

Roy, A. Moktan, B., and Sarkar, P.K. 2007. Traditional technology in preparing legume-based fermented foods in Orissa. *Indian Journal of Traditional Knowledge* 6(1): 12–16.

Ruel, M.T. 2001. *Can Food-Based Strategies Help Reduce Vitamin A and Iron Deficiencies?* International Food Policy Research Institute, Washington, D.C.

Sandhu, D.K. and Soni, S.K. 1988. Optimization of physicochemical parameters for Indian dosa batter fermentation. *Biotechnology Letters* 10(4): 277–282.

Sandhu, D.K. and Soni, S.K. 1989. Microflora associated with Indian *Punjabi warri* fermentation. *Journal of Food Science and Technology*, 26: 21–25.

Sandhu, D.K., Soni, S.K., and Vikhu, K.S. 1986. Distribution and role of yeast in Indian fermented foods. *Proceedings of the National Symposium on Yeast Biotechnology, Haryana Agriculture University*, Hissar, December, pp. 142–148.

Sands, D.C. and Hankin, L. 1974. Selecting lysine-excreting mutants of *Lactobacilli* for use in food and feed enrichment. *Journal of Applied Microbiology* 28: 523–534.

Sankaran, R. 1998. Fermented foods of the Indian subcontinent. In: Wood, B.J.B., Ed., *Microbiology of Fermented Foods*, Vol. II. Blackie Academic and Professional, London, pp. 753–789.

Sathe, S.K and Salunkhe, D,K. (1981), Functional Properties of the Great Northern Bean (Phaseolus vulgaris L.) Proteins: Emulsion, Foaming, Viscosity, and Gelation Properties. *Journal of Food Science*, 46: 71–81. doi: 10.1111/j.1365-2621.1981.tb14533.x

Savitri and Bhalla, T.C. 2013. Characterization of *bhatooru*, a traditional fermented food of Himachal Pradesh: Microbiological and biochemical aspects. *3 Biotech* 3(3): 247–254.

Savitri, Tahkur, N., Kumar, D., and Bhalla, T.C. 2012. Microbiological and biochemical characterization of seera: A traditional fermented food of Himachal Pradesh. *International Journal of Food and Fermentation Technology* 2(1): 49–56.

Sekar, S. and Mariappan, S. 2007. Usage of traditional fermented products by Indian rural folks and IPR. *Indian Journal of Traditional Knowledge* 6(1): 111–120.

Sharma, N. and Singh, A. 2012. An insight into traditional foods of North-Western area of Himachal Pradesh. *Indian Journal of Traditional Knowledge* 11(1): 58–65.

Shortt, C. 1998. Living it up for dinner. *Chemistry and Industry* 8: 300–303.

Shurpalekar, S.R. 1986. Papads. In: Reddy, N.R., Pierson, M.D. and Salunkhe, D.K. (Eds.) *Legume-Based Fermented Foods*. CRC Press, Boca Raton, FL, pp. 191–217.

Shurtleff, W. and Aoyagi, A. 2012. *History of Uncommon Fermented Soyfoods*. Soyinfo Center Lafayette, CA.

Soni, S.K. 1987. Studies on some Indian fermented foods: Microbiological and biochemical aspects. PhD thesis, Guru Nanak Dev University, Amritsar, India.

Soni, S.K. and Arora, J.K. 2000. Indian fermented foods: Biotechnological approaches. In: Marwaha, S.S. and Arora, J.K. (Eds.) *Food Processing: Biotechnological Applications. 1.* New Delhi: Asiatech Publishers Inc., New Delhi,pp. 143–190.

Soni, S.K. and Marwaha, S.S. 2003. Cereal products: Biotechnological approaches for their production. In: Marwaha, S.S. and Arora, J.K. (Eds.) *Biotechnology: Strategies in Agro-Processing.* Asiatech Publishers Pvt. Ltd. New Delhi, India, pp. 236–266.

Soni, S.K. and Sandhu, D.K. 1990. Indian fermented foods: Microbiological and biochemical aspects. *Indian Journal of Microbiology* 30(2): 135–137.

Soni, S.K. and Sandhu, D.K. 1999. Fermented cereal products. In: Joshi, V.K. and Pandey, A. (Eds.) *Biotechnology: Food Fermentation, Microbiology, Biochemistry and Technology*, Vol II. Educational Publishers and Distributors, New Delhi, pp. 895–949.

Soni, S.K., Sandhu, D.K., and Vilkhu, K.S. 1985. Studies on dosa – An indigenous Indian fermented food: Some biochemical changes accompanying fermentation. *Food Microbiology* 2: 175–181.

Soni, S.K., Sandhu, D.K., Vilkhu, K.S., and Kamra, N. 1986. Microbiological studies on *dosa* fermentation. *Food Microbiology*, 3: 45–53.

Spicher, G. 1983. Baked goods. In: Rehm, J.H. and Reed, G. (Eds.) *Biotechnology.* Verlag Chemie, Weinheim, pp. 1–80.

Spicher, G. and Schroder, R. 1978. The microflora of sourdough IV. The species of rod shaped lactic acid bacteria of the genus *Lactobacillus* occurring in sourdoughs. *Zeitschrift für Lebensmittel-Untersuchung und -Forschung* 167: 342–354.

Spicher, G. and Stephan, H. 1999. *Handbuch sauerteig, biologie, biochemie, technologie*, 5th edn. Hamburg: Behr's Verlag.

Sridevi, J., Halami, P.M., and Vijayendra, S.V.N. 2010. Selection of starter cultures for idli batter fermentation and their effect on quality of idlis. *Journal of Food Science and Technology* 47: 557–563.

Steinkraus, K.H. 1996. *Handbook of Indigenous Fermented Foods.* Marcel Dekker, New York, NY, p. 352.

Steinkraus, K.H., Van veen, A.G., and Thiebeau, D.B. 1967. Studies on *Idli*: An Indian fermented black gram food. *Food Technology* 21: 916–919.

Succi, M., Reale, A., Andrighetto, C., Lombardi, A., Sorrentino, E., and Coppola, R. 2003. Presence of yeasts in Southern Italian sourdoughs from *Triticum aestivum* flour. *FEMS Microbiology Letters* 225: 143–148.

Sugihara, T.F., Kline, L., and Miller, N.W. 1971. Microorganisms of the San Francisco sourdough bread process. I. Yeasts responsible for the leavening action. *Applied Microbiology* 21: 456.

Susheelamma, N.S. and Rao, M.V.L. 1978. Isolation and characterization of arabino-galactan from black gram (*Phaseolus mungo*). *Journal of Agricultural and Food Chemistry* 26: 1434–1437.

Tamang, J.P. (Ed.) 2009a. Fermented cereals. Chapter 5. In: *Himalayan Fermented Food Microbiology, Nutrition, and Ethnic Values.* J.P. Tamang (Ed) CRC Press, Boca Raton, FL.

Tamang, J.P. (Ed.) 2009b. Ethnic starters and alcoholic beverages. Chapter 8. In: *Himalayan Fermented Foods: Microbiology, Nutrition, and Ethnic Values.* CRC Press, Boca Raton, FL.

Tamang, J.P. 2010a. *Himalayan Fermented Foods: Microbiology, Nutrition and Ethnic Value.* CRC Press, Taylor & Francis Group, New York, NY.

Tamang, J.P. 2010b. Diversity of fermented foods. In: Tamang, J.P. and Kailasapathy, K. (Eds.) *Fermented Foods and Beverages of the World.* CRC Press, Boca Raton, pp. 41–84.

Tamang, J.P. 2012. Plant-based fermented foods and beverages of Asia. Chapter 4. In: Hui, Y.H. and Evranuz, Ö.E. (Eds.) *Plant-Based Fermented Foods.* CRC Press, Boca Raton, FL, pp. 49–90.

Tamang, J.P. and Fleet, G.H. 2009. Yeasts diversity in fermented foods and beverages. In: Satyanarayana, T. and Kunze, G. (Eds.) *Yeast Biotechnology: Diversity and Applications.* Springer, Berlin, Germany, pp 169–198.

Tamang, J.P. and Sarkar, P.K. 1995. Microflora of murcha; an amylolytic fermentation starter. *Microbios* 81: 327.

Tamang, J.P., Sarkar, P.K., and Hesseltine, C.W. 1988. Traditional fermented foods and beverages of Darjeeling and Sikkim—A review. *Journal of the Science of Food and Agriculture* 44(4): 375–385.

Tamang, J.P., Tamang, N., Thapa, S., Devan, S., Tamang, B., Yonzan, H., Rai, A. et al. 2012. Microorganisms and nutritional value of ethnic fermented foods and alcoholic beverages of North East India. *Indian Journal of Traditional Knowledge* 11(1): 7–25.

Targais, K., Stobdan, T. and Mundra, S. 2012. *Chhang* – A barley based alcoholic beverage of Ladakh, India. *Indian Journal of Traditional Knowledge* 11(1): 190–193.

Thakur, S. 1995. Effect of xanthan on textural properties of *idli* (traditional south Indian foods). *Food Hydrocolloids* 9: 141–145.

Thakur, N., Savitri and Bhalla, T.C. 2004. Characterizations of some fermented foods and beverages of Himachal Pradesh. *Indian Journal of Traditional Knowledge* 3(3): 325–335.

Thiele, C., Ganzle, M.G., and Vogel, R.F. 2002. Contribution of sourdough *Lactobacilli*, yeast, and cereal enzymes to the generation of amino acids in dough relevant for bread flavour. *Cereal Chemistry* 79: 45–51.

Tiwari, S.C. and Mahanta, D. 2007. Ethnological observations on fermented food products of certain tribes of Arunachal Pradesh. *Indian Journal of Traditional Knowledge* 6: 106–110.

Valcheva, R., Ferchichi, M., Korakli, M., Ivanova, I., Ganzle, M.G., and Vogel, R.F. 2006. *Lactobacillus nantensis* sp. nov. isolated from French wheat sourdough. *International Journal of Systematic and Evolutionary Microbiology* 56: 587–591.

Venkatasubbaiah, P., Dwarakanath, C.T., and Murthy, V.S. 1984. Microbiological and physicochemical changes in idli batter during fermentation. *Journal of Food Science and Technology* 21: 59–62.

Vijayendra, S., Rajashree, K., and Halami, P. 2010. Characterization of a heat stable antilisterial bacteriocin produced by vancomycin sensitive *Enterococcus faecium* isolated from *idli* batter. *Indian Journal of Microbiology* 50(2): 243–246.

Vijaylakshmi, G. 2005. Fermented foods of southern India and Srilanka. Fermented food, health status and social well-being. *First Workshop and Strategic Meeting of the Swedish South Asian Network on Fermented Foods* held on Anand, India, pp. 1–11.

Vogel, R.F., Ehrmann, M.A., and Ganzle, M.G. 2002. Development and potential of starter lactobacilli resulting from exploration of the sourdough ecosystem. *Antonie van Leeuwenhoek* 81: 631–638.

Wang, H.L. and Hesseltine, C.W. 1982. Oriental fermented foods. In: Reed, G. (Ed.) *Prescott and Dunn's Industrial Microbiology.* AVI Publishing Co. Inc., Westport, CT.

Wikipedia The free encyclopedia.

Wiseblatt, L. and Kohn, F.W. 1960. Some volatile aromatic compounds in fresh bread. *Cereal Chemistry* 37: 55–56.

Wood, B.J.B. 1985. *Microbiology of Fermented Foods.* Springer, Berlin Heidelberg New York.

www.bhavnaskitchen.com. 2013.

Yadav, S. and Khetarpaul, N. 1994. Indigenous legume fermentation: Effect on some antinutrients and *in-vitro* digestibility of starch and protein. *Food Chemistry* 50(4): 403–406.

Yajurvedi, R.P. 1980. Microbiology of idli fermentation. *Indian Food Packer* 34: 33.

Yonzan, H. and Tamang, J.P. 2009. Traditional processing of Selroti: A cereal-based ethnic fermented foods of the Nepalis. *Indian Journal of Traditional Knowledge* 8(1): 110–114.

Yonzan, H. and Tamang, J.P. 2010. Microbiology and nutritional value of *Selroti*, an ethnic fermented cereal food of the Himalayas. *Food Biotechnology* 24(3): 227–247.

8

INDIGENOUS FERMENTED FOODS INVOLVING ACID FERMENTATION

A.K. SENAPATI, ANITA PANDEY,
ANTON ANN, ANOOP RAJ, ANUPAMA GUPTA,
ARUP JYOTI DAS, B. RENUKA, BHANU NEOPANY,
DEV RAJ, DORJEY ANGCHOK, FOOK YEE CHYE,
GITANJALI VYAS, J.P. PRAJAPATI,
JAHANGIR KABIR, JARUWAN MANEESRI,
K.S. SANDHU, KHENG YUEN SIM,
KONCHOK TARGAIS, L.V.A. REDDY,
LAXMIKANT S. BADWAIK, LOK MAN S. PALNI,
M. PREEMA DEVI, MANAS R. SWAIN,
MD. SHAHEED REZA, NIVEDITA SHARMA,
PALLAB KUMAR BORAH, RAMESH C. RAY,
S.G. PRAPULLA, S.V. PINTO, SANKAR CHANDRA
DEKA, SOMESH SHARMA, SURESH KUMAR,
TSERING STOBDAN, AND V.K. JOSHI

Contents

8.1 Introduction

"Fermentation" is employed in the production of foods through the application of microorganisms or their enzymes (Geis, 2006). Such foods are an intricate part of the diet of people all over the world, and it is the diversity of the raw materials used as substrates, the methods of preparation, and the sensory qualities of the finished products that contribute to the popularity of fermented foods (Beuchat, 1995; Stein Kraus, 1996). As one come across more and more of fermented foods, one begins to learn more about the eating habits of various cultures. Out of the lactic acid fermented foods, lactic acid fermented milk was undoubtedly the first product made and consumed by man. Out of the various approaches to fermentation, lactic acid fermentation, using natural microflora or lactic acid bacterial (LAB) cultures, is employed throughout the world, in conjunction with chemical preservation, using salt and acid to preserve various foods such as milk, cereals, meat, and fruits and vegetables (Thokchom and Joshi, 2012). Historically speaking the Chinese were the first to ferment vegetables, as evidenced by the fact that they had prepared such vegetables at the time of building the Great Wall of China (Pederson, 1971). The basic mechanism of the preservation of foods is the production of acid, chiefly by LAB, which lowers the pH to a level at which most of the spoilage-causing microorganisms cannot grow, and, thus, the food is preserved (Pederson, 1971; Frazier and Westhoff, 1998; Joshi and Thakur, 2000). In addition, today, lactic acid fermentation is being used more as a taste diversification tool than as a method of preservation. It is extensively used in the cuisines of various countries in South Asia, and thus, the importance of lactic acid is being stressed in developed countries more as a method through which to prepare food with therapeutic value than as a food preservation method. The indigenous, lactic acid-fermented foods made in different countries of South Asia include *gundruk, sinki, dahi, paneer, shoidon, soijin, ngaree,* and *shrikhand*. The lactic acid fermented foods include those from fruits and vegetable, milk, cereal and pulses, meat and similar products. In the modern era, the lactic acid fermented products especially milk products have assumed a great importance due to the probiotic effect, they imparts to the consumers. This chapter focuses on various aspects connected with the production of indigenous fermented foods from fruits and vegetables, and milk made using lactic acid fermentation while those connected with meat and fish are described in Chapter 11 of this text.

8.2 Lactic Acid-Fermented Fruits and Vegetables Products

8.2.1 Vegetable Fermentation for Preservation

Fresh vegetables cannot be kept for long, and so, over the centuries, humankind has developed ways and means of preserving many of them. It has been employed to make products like sauerkraut kimchi, several pickles including olive pickle, *kanji, gundruk, sinki, khalpi,* etc. all over the world, while the last four are made in South Asia (Sekar and Mariappan, 2007; Joshi et al., 2012). Lactic acid fermentation is one of the earliest methods used to preserve vegetables such as carrots, radishes, cucumbers, and turnips, in order to provide dietary variety in the form of diversified products (Joshi and Sharma 2009; 2012; Sharma et al., 2012; Bhushan et al., 2013). Methods for producing several vegetable based products have been standardized and the products are prepared and consumed as a routine diet (Karkri, 1986; Montet et al., 1999; Joshi and Thakur, 2000; Pandya et al., 2006; Joshi, 2015). It is one of the tools used to develop new, fermented, vegetable-based products after the natural or inoculated fermentation of vegetables. The LAB fermentation of carrots (Asiatic type) and radishes (Chinese pink) with a salt concentration of 2.5%, mustard of 2.0%, a temperature of 26°C, and a sequential culture was found to provide the best product (Joshi et al., 2003, 2008). However, a LAB fermentation of cucumber, with a salt concentration of 3.0%, mustard of 2.0%, a temperature of 32°C, and a sequential culture produced products of the highest quality. Thus, sequential culture fermentation is considered to hold promise for the production of fermented vegetables with better quality attributes, similar to those produced by natural fermentation. It also has the advantage over natural fermentation with respect to controlled fermentation.

8.2.1.1 Factors Affecting the Lactic Acid Fermentation of Fruit and Vegetables   To preserve a vegetable with lactic acid fermentation, the choice of technique is determined by a number of factors. Specifically, microorganisms, salt concentration, temperature, chemical additives, the amount of fermentable carbohydrates in the vegetables, and the availability of nutrients in the brine are all known to affect the lactic acid fermentation (Pederson, 1971; Joshi et al., 1993; Joshi and Sharma, 2012), and several of these factors are discussed in the subsequent sections.

8.2.1.1.1 Microorganisms   LAB are one of the important microorganisms in food fermentation, and have been shown by serological techniques and 16S ribosomal RNA cataloging to be phylogenetically related and to share a number of common features (Adams and Moss, 1996). The LAB produce lactic acid as the major end product of the fermentation of carbohydrates (Battcock and Azmi-ali, 1998). These bacteria are within the genera of *Lactobacillus*, *Streptococcus*, *Pediococcus*, and *Leuconostoc*, and include *Lactobacillus brevis*, *Lactobacillus plantarum*, *Leuconostoc mesenteroides*, *Streptococcus faecalis*, and *Pediococcus cerevisiae* (Gibbs, 1987). More details on this area can be found in Chapter 3 of this text and for more detail on microbial ecology, see Daeschal et al. (1987).

Lactic fermentation is a natural process brought about by the LAB present in the raw food, such as vegetables, or those derived from a starter culture (Pederson, 1971; Motarjemi and Nout, 1996; Sagarika and Pradeepa, 2003). It is initiated by heterofermentative LAB—namely, *Leuconostoc mesenteroides*—followed by homolactic bacteria, such as *Lactobacillus brevis*, followed by next stage of fermentation, which is dominated by homofermentative LAB, selectively favored by the complete lack of oxygen, lowered pH, and an elevated salt content (Sankaran, 1998). Species of the genera *Streptococcus* and *Leuconostoc* produce the least acid; next are the heterofermentative species of *Lactobacillus*, which produce intermediate amounts of acid; followed by the *Pediococcus*; and, lastly, the homofermentors of the *Lactobacillus* species, which produce the most acid (Battcock and Azmi-ali, 1998; Anonymous, FAO, 1997). The initial population and the growth rate of microorganisms, as well as salt and acid tolerances, are important factors that influence the sequential development of various LAB in most of the vegetable fermentations. A reduction of pH and a removal of carbohydrates by fermentation are the primary preserving actions that these bacteria provide. But they are also capable of producing inhibitory substances, such as nisin, that are antagonistic toward other microorganisms (Nettles and Barefoot, 1993; Adams and Moss, 1996; Eijsink et al., 1998; Joshi and Thakur, 2000; Silva et al., 2002; Mindy et al., 2005). More of such details are given in Chapters 1, 3, and 6 of this text.

The fermentation of vegetables is difficult to control, primarily due to variations in shape, the large number and types of naturally occurring microorganisms, and the variability in their nutrient content (Buckenhuskes, 1993). One promising approach might be the application of defined starter cultures, which are capable of growing rapidly and are highly competitive under environmental conditions used to make fermented products (Buckenhuskes, 2001; Li, 2003). The LAB tolerate high salt concentrations, which give them an advantage over other less salt-tolerant species and allows the LAB to produce acid that inhibits the growth of undesirable microorganisms (Anonymous, FAO, 1997). *Leuconostoc* is noted for its high salt tolerance, and, for this reason, initiates the majority of lactic acid fermentations (Battcock and Azmi-ali, 1998). Mushroom in brine inoculated with LAB produced more acid than did uninoculated mushroom. *Lactobacillus plantarum* produced the highest amount of acid, followed by *Streptococcus lactis* (Joshi et al., 1996). In cucumber and green olive fermentations, too, *Lactobacillus plantarum* produced the highest acid (Etchells et al., 1966). Similar, attempt with fermentation bacteria has been attempted in low salt cucumber brine (Chavasit et al., 1991).

8.2.1.1.2 Temperature Temperature affects the growth and activity of all living cells, and microbial cells are no exception. At high temperatures, microorganisms are destroyed, while, at low temperatures, their rate of activity is decreased or suspended. Accordingly, they are classified into three distinct categories: psychrophiles, mesophiles, and thermophiles. LAB are mesophilic and work best in a temperature range of 18–22°C (Battcock and Azmi-ali, 1998). Temperature is a critical factor for producing high-quality fermented vegetables as it affects the acidification rate of the vegetable and promotes the growth of a single microbial species, giving it a competitive edge over other species. Influence of salt and temperature has aho been determined in radish (Sharma and Joshi, 2007).

8.2.1.1.3 Salt Concentration There are three methods—high-salt brine salting, low-salt brine salting, and dry salting used in lactic acid fermentation (Pederson, 1971). Dry salting is employed by adding dry salt to vegetables with a high water content or by soaking others in brine. However, dry salt has to be added regularly to successive layers of the product to avoid high local salt concentrations that could promote yeast development instead of LAB. Any variety of common salt is suitable as long as it is pure, as impurities or additives cause problems—such as those of iron and magnesium, that result in the blackening of vegetables and imparting a bitter taste (Anonymous, FAO, 1997), while carbonates in pickles are known to cause a soft texture (Lal et al., 1986; Anonymous, FAO, 1997; Battcock and Azmi-ali, 1998). In principle, the vegetables can also be fermented without the addition of sodium chloride (NaCl) (Fleming et al., 1975), though it is often a very important ingredient for many reasons, with a major one being its contribution to flavor, as well as its role in supporting the development of anaerobic conditions on fermentation vessels. The amount used, however, depends on the particular vegetable and on consumer preferences (Buckner et al., 1993).

In the pickling industry, salt has historically been used for directing the fermentation of cucumbers, radishes, and carrots (Thompson et al., 1979; Hudson and Buescher, 1985; Fleming et al., 1987; Mcfeeters et al., 1989). The complex changes that occur in vegetable fermentation are produced by the growth of a sequence of LAB. The growth of each species depends upon its initial presence on the vegetable, the sugar and salt concentrations, and the temperature. Salt concentration in vegetable fermentation can range from 20 to 80 g/L during fermentation, and up to 160 g/L in some stored vegetables (Cheigha et al., 1994). A high salt concentration inhibits the growth of unwanted microorganisms and induces the plasmolysis of plant cells, thus promoting anaerobiosis in the medium, which is, however, more effective in finally cut and shredded plant material. Whole or low water-content vegetables are therefore, fermented in highly salted brine. Plasmolysis also promotes the growth of LAB by releasing nutrients contained in the plant cells (Cheigha et al., 1994). The dominant bacteria at the outset of fermentation are resistant to high salt concentration. Due to the progressive acidification of the medium, acid tolerant bacteria are present by the final stages of the process. A sequence of dominant flora during the course of lactic acid fermentation has been identified: successively, *Pediococcus, Leuconostoc*, and *Lactobacillus* (Hubert and Dupuy, 1994). Most vegetables can be fermented at 12.5–20° salometer salt. However, at a higher salt concentration of about 60° salometer, the lactic acid fermentation ceases to function (Vaughn, 1985; Anonymous, FAO, 1997; Battcock and Azmi-ali, 1998).

8.2.1.1.4 Growth Stimulators Many foods contain the ingredients essential for the growth of even the most fastidious LAB. Their nutritional requirements are particularly complex and exacting, and, in addition to an energy source, a variety of essential growth factors must be made available (Vokbeck et al., 1963; Pederson, 1971). Some of the ingredients, added more or less empirically to lactic acid-fermented vegetables or fruits, seem to enhance the development of lactic flora, which generally requires very special conditions to proliferate. The LAB depend primarily on plant sugars for their growth (Montet et al., 1999), but, in some vegetables with low nutrient contents, such as turnip and cucumber, the addition of sugar promotes bacterial growth, thereby accelerating lactic acid fermentation. For example, *jaggery* (*gur*) is added during the lactic acid fermentation of sweet turnip. Adding 1–2 g of sucrose or *jaggery* to 100 g of hand-shredded turnip (mixed with 6 g of salt and 6 g of mustard seed) has been found to increase acid production during lactic acid fermentation (Anand and Das, 1971; Rao et al., 2006), but, at a high sucrose concentration (25 g), the fermentation was found to slow down. Addition of garlic in kimchi fermentation has been made to study its effect on growth of microorganisms during kimchi fermentation (Cho et al., 1988).

8.2.1.1.5 Other Additives Many ingredients apart from salt can be used in the preparation of lactic acid-fermented fruits and vegetables. They have three main functions: as a source of nutrients (sugars, mineral salts, and vitamins) for the fermentation-causing microorganisms, to help restrict the growth of unwanted bacteria (either through a regulatory effect on the pH level or by producing inhibitory substances), and, in the case of spices, to have a final flavor-determining role in the fermented vegetables (Montet et al., 1999).

8.2.1.1.5.1 Whey Whey is often added in the traditional lactic acid -fermentation processes for vegetables due to its high lactose content, a potential energy substrate for lactic bacteria. It is highly recommended for the preparation of slightly salted sauerkraut (0.2%–0.3% salt, w/w of brine; Aubert, 1985), or for fermenting low-nutrient vegetables (Schoneck, 1988).

8.2.1.1.5.2 Spices Spices or aromatic herbs are added to most of the lactic-acid fruit and vegetable fermentation preparations to improve the flavor of the finished products. Aromatic compounds in these spices (mainly terpenes and polyphenols) often have an antimicrobial effect, which means that they can have a selective role in the development of bacteria during fermentation (Laencina et al., 1985; Rao et al., 2006).

8.2.1.1.5.3 Mustard Seed Mustard seeds are also of interest as they contain allyl isothiocyanate, a volatile aromatic compound with antibacterial and antifungal properties. Ground mustard seeds or oil are widely used in traditional fermented vegetable preparations in India (Rao et al., 2006). Adding 6–10 g of mustard seed powder to turnip (100 g) with a salt content of 6 g increased lactic acid levels during fermentation (Anand and Das, 1971). Increasing the proportion of mustard from 1% to 2% with different salt concentrations (2%–12%) progressively increased the rate of lactic acid production and suppressed the spoilage in cauliflower slurry (Sethi and Anand,

1984). The maximum acidity was developed in 4% salt when used either alone or along with 2.0% mustard. Mustard has a selective preservative action, allowing LAB to grow and suppress the surface yeast (Rao et al., 2006). Among the different varieties of mustards tried, *Brassica nigra* (Banarsi rye) was the best, followed by *B. juncea* (Rye), *B. campestris* var. *sarson*, or *B. campstris* var. *Toria*. Mustard, being an oil source, is an important condiment and used in various food products, and has advantage of not only imparting flavor and taste to the food products but also providing a preservative action (Anand and Johar, 1957). The addition of mustard powder to pickles of turnip, cabbage, and cucumber has been found to increase the rate of lactic fermentation (Anand and Das, 1971; Mikki, 1971; Sethi and Anand, 1972; Rao et al., 2006). Using response surface methodology, optimization of lactic acid fermentation of radish and carrot has been optimized with respect to salt, additives (including mustard seeds) and growth stimulation (Chauhan and Joshi, 2014; Joshi et al., 2014).

8.2.2 Indigenous Fermented Vegetable Products

Many different forms of vegetables are fermented and preserved by the tribal people of the Indian state of Sikkim. These products are similar to *sauerkraut* (Europe), *kimchi* (Korea), *oncom* (Indonesia), *tsukemono* (Japan), *suan cai* (China), and *atchara* (Philippines) (Pederson and Albury, 1969; Mikky, 1971; Banwart, 1981; Kumari et al., 1993; Lee, 2009). Some common fermented vegetable products of this northeast region of India are listed in Table 8.1.

8.2.2.1 Sinki
This is a form of non-salted, fermented, radish (*Raphanus sativus* L.) taproot that is consumed by the Nepalese in Sikkim (Figure 8.1). It is made during winter when the weather is least humid and when there is an ample supply of this vegetable (Tamang et al., 1988; Tamang and Sarkar, 1993; Sekar and Mariappan, 2007). It is consumed by all ethnic people of Darjeeling (Steinkraus, 1996). The method of preparation is similar to that of *gundruk* except that the substrate is the taproots and its fermentation takes 30–40 days (Tamang et al., 1988).

To prepare *sinki,* fresh taproots of radish are cleaned by washing, wilted by sundrying for 1–2 days until they become soft, and then shredded, dipped in lukewarm water, squeezed, and placed tightly into an earthen jar with the help of a heavy wooden pestle. The jar is sealed with an earthen lid and covered with radish leaves. It is then,

Table 8.1 Traditional Lactic Acid-Fermented Vegetables of the Sikkim Himalayas

PRODUCT	SUBSTRATE	FERMENTATION TIME (DAYS)	TEXTURE AND USE
Gundruk	Leafy vegetable	7–15	Dried, sour; soup/pickle
Sinki	Radish tap root	25–30	Dried, sour; soup/pickle
Khalpi	Cucumber	5–7	Sour; pickle
Mesu	Bamboo shoots	7–10	Sour; pickle

Source: Adapted from Tamang, J.P. et al. 2010. *Himalayan Study Monographs,* No. 11, pp. 177–185.

Figure 8.1   *Sinki*—a non-salted fermented radish.

kept in a warm and dry place for 15–30 days (Tamang and Sarkar, 1993; Rao et al., 2006; Toshirou and Tamang, 2010).

Alternatively, a pit of about 1 m in depth and diameter is dug in a dry place. It is cleaned and dried by lighting a fire. The ash is removed and the sides are plastered with mud while still hot. It is then, covered on all sides with dried leaves of bamboo, banana, or radish. The shredded roots are pressed tightly into this pit, then covered with dried leaves and weighed with heavy stones or wooden planks. The top is then plastered with mud or cow dung and left to ferment for a period of 30–40 days (Tamang et al., 2012). Afterwards, the fermented mass is taken out, cut into small pieces, and sun-dried for 3–5 days (Rao et al., 2006). This product can be kept for two years or more at room temperature by exposing it to sunlight periodically (Tamang and Sarkar, 1993; Tamang and Tamang, 2009).

Sinki is a naturally fermented product where *Lactobacillus fermentum* initiates the fermentation, followed by *Lactobacillus brevis* and *Lactobacillus plantarum*. During fermentation, its pH drops from 6.7 to 3.3, with an increase in acidity (Tamang and Sarkar, 1993). Others have reported *L. plantarum*, *L. brevis*, *L. casei*, and *Leuconostoc fallax* as the microorganisms involved in *sinki* fermentation (Tamang and Sarkar, 1993; Tamang et al., 2005).

With its highly acidic flavor, *sinki* is typically used as a base for soup and as a pickle. The soup is made by soaking the *sinki* in water for about 10 min, squeezing out the liquid, and frying the foodstuff along with chopped onion, tomatoes, green chillies, and salt. The soup is served hot along with main meals. It is said to be a good appetizer, and people use it as a remedy for indigestion (Tamang and Sarkar, 1993). Meanwhile, the pickle is prepared by soaking *sinki* in water, squeezing it, and then mixing it with salt, mustard oil, onion, and green chillies.

8.2.2.2 Goyang *Goyang* is an ethnic, fermented, acidic, vegetable foodstuff (Figure 8.2) prepared from *magane-saag* (Carmen et al., 2010). During the rainy season, the leaves of the wild, edible plant *magane-saag* are in good supply. They are collected, washed, chopped into pieces, and squeezed to remove the excess water. The leaves are tightly pressed into a bamboo basket lined with 2–3 layers of fig (*Ficus carica*) leaves. The top of the basket is also covered with fig leaves, and it is left to ferment at room temperature (15–25°C) for a period of 1 month (Carmen et al., 2010). After this stage, the *goyang* is transferred to an airtight container, where it can be stored for a period of 2–3 months. The product can also be made into balls and sun-dried, which increases its shelf-life. The different microorganisms involved in its fermentation have been reported to be *L. plantarum*, *L. brevis*, *Lactococcus lactis*, *Enterococcus facium*, *P. pentosaceus*, and the yeast *Candida* spp. (Tamang and Tamang, 2007; Tamang et al., 2012). The Sherpa tribe belonging to Sikkim (India) prepares this fermented product from leaves of the wild plant (Tamang and Tamang, 2009). Locally, *goyang* is boiled with beef or yak meat and noodles and made into a thick, soup-like dish called *thupka* (Tamang and Tamang, 2009).

8.2.2.3 Khalpi *Khalpi* is a traditional fermented cucumber product of Sikkim. To prepare it, mature and ripened cucumbers (*Cucumis sativus L.*) are cut into fixed sizes and sun-dried for 2 days (Carmen et al., 2010). They are then put into bamboo vessels called "*dhungroo*" and sealed. Fermentation is allowed to take place for 4–7 days at room temperature. The product can be stored for about a week in an airtight container. The microorganisms involved are *L. plantarum*, *L. brevis*, and *Lactococcus fallax* (Carmen et al., 2010; Tamang and Tamang, 2010). It is generally made for home consumption by the Nepalese Brahmins belonging to the Bahunand Chettri castes. It is consumed as a pickle after mixing with mustard oil, chillies, and salt (Carmen et al., 2010; Tamang and Tamang, 2009).

Figure 8.2 *Goyang*—an ethnic fermented acidic vegetable product.

8.2.2.4 Gundruk Gundruk is one of the most common and highly preferred of the fermented dry vegetables, indigenous to Nepal (Tamang, 2010). It is primarily valued for its uniquely appetizing flavor, and is used in the preparation of curry, soup, *chutney*, and other local delicacies (Yonzan and Tamang, 2010). All Nepalese people, irrespective of wealth status, relish *gundruk*. It is commonly used as a condiment to enhance the overall flavors of meals, and a small amount of *gundruk* can make a bland diet much more appealing. It is traditionally used by different ethnic group of Darjeeling.

The substrate used for preparation of *gundruk* is usually ryo leaf (*Brassica campestris* L. var. *cumifolia* Roxb.), or radish (*Raphanus sativus* L.), or cauliflower (*Brassica oleracia* var. *botrytis* L.). The most common raw materials used for preparing *gundruk* in Nepal are the leaves of *Brassica* species vegetables, such as mustard (*Brassica compestris* L.), rayo (*Brassica juncea* L.) or broad leaf mustard, cauliflower (*Brassica Oleracea* L. var. *botrytis* L), and radish (*Raphanus sativus* L.). The leaves of the wild plant *Arisaema* (*Arisaema utile Hook.f. ex Schoot–Dhokaya*) are also used for making *gundruk* by some people in villages in central Nepal (Manandhar, 1998). The methods of preparing *gundruk* differ slightly, according to region and ethnic communities (Karki and Horsford, 1986). In general, the leaves are wilted, shredded, placed in earthenwares or a pit to ferment for 7–9 days, and then sun-dried.

Specifically, ryo leaves are washed and dried slightly in the sun to soften and for the removal of moisture. The dried leaves are crushed lightly and packed into a polyethylene bag, which is then tied with a thread or rope. Next, the bag is buried in a matured cow-dung heap. After 15–20 days, when the fermentation is complete, the substrate is taken out and sun-dried for a few days (2–4 days). It is then, cut into pieces in preparation for its consumption as *gundruk*. The process is depicted as a flow chart in Figure 8.3, and also shown pictorially in Figure 8.4.

A study has indicated that the degree and direction of all the 20 amino acids varies with the type of vegetable used (Karki et al., 1983). The taste, flavour, and acidity of *gundruk* are due to the synergistic action of three lactic strains, *L. cellobiosus*, *P. pentosaceus*, and *L. plantarum*, with the preservation of product quality (Karki and Horsford, 1986). Depending upon the substrate, the acidity (percentage as lactic acid) of *gundruk* is also variable. The acidity of mustard and cauliflower *gundruk* has been reported to be 0.48% and 4.5% (as lactic acid on a dry weight basis), respectively. According to Whitfield (2000), solar-dried food items were superior to those that were sun-dried when evaluated in terms of taste, color, and mold counts, and retained much of the vitamin A. The sensory quality of solar-dried *gundruk* was found to be superior to that of the sun-dried version in terms of color, aroma, taste, and overall quality, thus solar drying helps in retaining most of the characteristics of *gundruk* that are important from a sensory perspective. The acidity level of *gundruk* can be increased through the addition of fermenting cabbage as a starter; organoleptically, the seventh day-fermented *gundruk* was of superior quality.

Leafy Vegetables

↓

Wilting (1–2 days)

↓

Shredding

↓

Tight packing of leaves in earthern pots

↓

Covering of leaves with warm water and straw

↓

Fermentation

↓

Drying

Figure 8.3 Flow chart of the preparation of *gundruk*. (Adapted from Joshi, V.K. et al. 2003. *International Seminar and Workshop on Fermented Foods, Health Status and Social Well-Being*, 2003, Anand, India, pp. 24–28.)

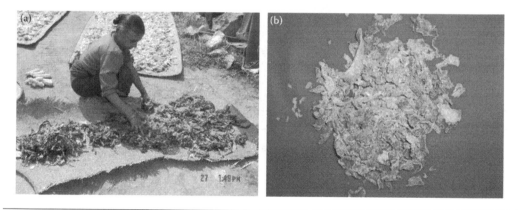

Figure 8.4 (a) Drying of *gundruk*; (b) Dried *gundruk*. (Courtesy of Anonymous, Carrying Capacity Study of Teesta Basin in Sikkim Edible Wild Plants and Ethnic Fermented Foods. Vol. VIII Biological Environment—Food Resources. Centre for Interdisciplinary Studies of Mountain and Hill Environment (CISMHE) University of Delhi, Delhi.)

8.2.3 *Traditional Pickles*

Pickling is one of the oldest methods of food preservation and has been practiced widely. It is also an important consumer product, with different varieties of pickles produced in different countries of South Asia, including India (Anonymous, Pickles making). They are made through the natural fermentation of fruits and vegetables, and, besides having nutritional value, pickles also act as a food accompaniments and palatability enhancers (Joshi and Bhat, 2000; Savitri and Bhalla, 2007).

The notable versions made in India are mango, drumstick, *brinzal*, *gongura*, and *mirchi* pickle. In Orissa and Andhra Pradesh (India), people consume various pickles as a part of their daily diet, along with staple foodstuffs such as rice, *chapathi*, bread, *samosa*, *upma*, and so on. In both the states, particularly in the tribal areas, the availability of the raw ingredients for making pickles is very high. Thus, pickling is practiced widely in tribal areas (Sekar and Mariappan, 2007; Anonymous, Pickles making).

Pickles of vegetables such as *lingri* (fern), bottle gourd, *kachnar*, *ghandoli*, *tardi*, *maslam*, *brinjal*, some species of mushroom, and fruits such as pear, peach, plum, *bidana*, *galgal* (a type of lemon), etc., are prepared and consumed in various parts of the Himalayan region (Savitri and Bhalla, 2007). In the lower parts of Himachal Pradesh, pickled mangoes, *galgal*, lime, and so on are packed in earthern pots, covered with wooden lids, and sterilized by the fumes generated from burning red chillies along with asafoetida (*Heeng*) and a little mustard oil. The antimicrobial properties of the fumes of red chillies, mustard oil, and asafoetida sterilize the containers and increase the shelf-life of the final product. In the preparation of pickles, fully matured fresh vegetables and fruits are washed and cut into standard sizes; salt, *mirchi*, oil, and other ingredients are added in the necessary quantities. The preparation of all pickles is basically similar (Lal et al., 1986; Joshi and Bhat, 2000). The addition of salt to the pickles restricts the growth of gram-negative bacteria and enhances the growth of LAB. In all types of pickles, the dominant microorganisms are *Leuconostoc mesenteroids*, *L. pallax*, and *Lactobacillus plantarum*, while *Staphylococcus aureus*, *Saccharomyces cerevisiae*, and *A. niger* may be present when a pickle is spoiled.

8.2.3.1 Mango Pickle A fermented mango pickle known as *achar* is very popular throughout India, including Murshidabad and Malda, the mango production district of west Bengal, and many producing areas of northern India and Andhra Pradesh. Fully developed, under-ripe, and tart varieties of mango are used. The fruits are washed and then, sliced with a stainless steel knife, discarding the stones. (There is also a widely practiced method in which the stone is cut, too, and the content except for stony outer layer is eaten.) The slices are placed in brine of 2%–3%, to prevent the blackening of the cut surface. Then, the slices are taken out and mixed with salt powder in a glazed jar. The freshly prepared pickles are cured for a week, during which process the jars are kept in the sun for 4–5 days. Then, the slices are mixed with other spices and smeared with a little rapeseed oil. The pickles are then, packed into a glass jar and covered with a thin layer of oil; the pickle is ready in about 2–3 weeks (Lal et al., 1986; Joshi and Bhat, 2000). A flow chart depicting the preparation of mango pickle is given in Figure 8.5. The list of ingredients for mango pickle encompasses mango slices (0.90 kg), common salt (226 g), ground fenugreek (113 g), ground nigella (28 g), turmeric powder (28 g), red chilli powder (28 g), black pepper (28 g), and fennel (28 g).

8.2.3.2 Cabbage and Carrot Pickle The pickling of vegetables, especially cabbage, radish, turnip, and carrot, is a common method of preserving the surplus produced during

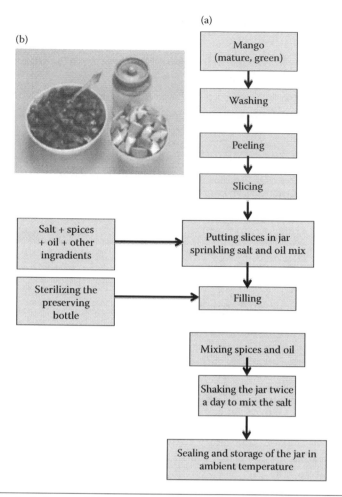

Figure 8.5 (a) Flow chart of mango pickle preparation. (b) Mango pickle, mango slices, and a ceramic container.

the summer months for consuming during the winter season when the availability as well as affordability of vegetables becomes an issue, especially in hilly areas. Shredded cabbage and chopped carrot are thoroughly mixed with mustard oil and many spices and condiments, such as mustard, salt, fennel, chili powder, black pepper, cumin, and cardamom. This concoction is tightly packed in a wide-mouthed glass bottle and kept in the sun for seasoning for about 15 days. Thereafter, the bottle is filled with mustard oil so as to remove any air left in it. It is kept for consumption during the long winter season without getting spoiled (Lal et al., 1986; Joshi and Bhat, 2000).

8.2.4 Acid-Fermented Fruit and Vegetable-Based Products

The preparation of fermented foods using LAB has been shown to improve their acceptability, as fermented vegetables as such do not find much acceptability with Indian consumers. Lactic acid fermentation is employed to prepare various products such as *kanji*, fermented cucumber extract, fermented carrot-, radish-, and cucumber-based appetizers, sauce, ready-to-serve (RTS) chutney, sauces, and other products (Sharma et al., 2008; Joshi and Sharma, 2010; Joshi et al., 2011a,b; Joshi and Singh, 2012; Sharma et al., 2012).

8.2.4.1 Fermented Vegetable-Based Ready-to-Serve Drinks   Different drinks have been prepared using lactic acid-fermented radish, carrot, and cucumber blended with different pulps of fruits. Such drinks were developed as per the flow chart given in Figure 8.6. Among the different RTS drinks, the carrot-based drink blended with apricot has been found to be the best. Specifically, based on the physico-chemical characteristics and sensory evaluation scores of different attributes, the carrot and

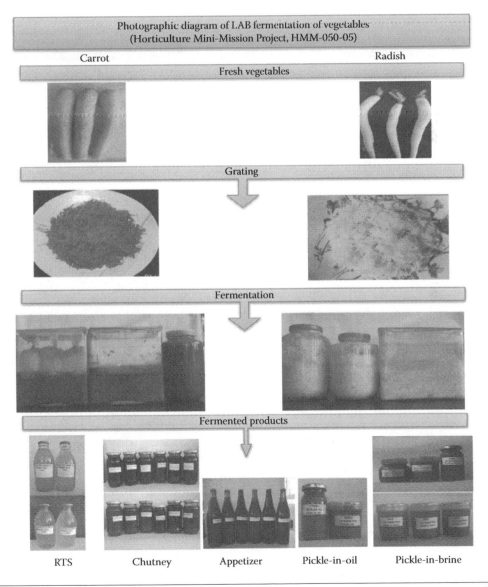

Figure 8.6   Diagrammatic representation of lactic acid-fermented vegetable products. (Adapted from Dr. V.K. Joshi, Mini Mission Project of the Indian Council of Agricultural Research (ICAR), 2013.)

mango RTS drink preparation of 40% fermented carrot + 60% mango was ranked the best and therefore, considered the most suitable blend for fermented RTS drinks. Likewise, among the radish-based RTS drinks, those containing 30% fermented radish pulp + 70% mango were pronounced the best. Finally, on the basis of physicochemical and sensory characteristics of various fermented cucumber-based products, the RTS drink prepared with 20% fermented cucumber + 80% mango pulp was adjudged as the best (Sharma et al., 2008).

8.2.4.2 Lactic Acid–Fermented Appetizers   In another attempt, lactic acid-fermented carrot and different blends of fruit pulps were used to prepare appetizers (Figure 8.6), and the appetizer containing 20% radish + 10% apricot was rated the best. Similarly, on the basis of the physico-chemical and sensory characteristics of various lactic-fermented cucumber-based products, an appetizer containing 10% fermented cucumber + 20% apricot pulp was considered to be the best (Joshi et al., 2011a).

8.2.4.3 Lactic Acid–Fermented Sauces   Different combinations of fermented vegetables such as carrot, radish, and cucumber with pear and mango pulps were made separately and processed into sauces (Figure 8.6; Joshi and Sharma, 2010). Here, the sauces prepared with 25% radish + 75% pear and 50% cucumber + 50% pear were preferred to others.

8.2.4.4 Ready-to-Serve Chutney   Different combinations of fermented vegetables such as carrot, radish, and cucumber with *amchoor* (dried powder of green mango) and *anardana* (seeds of wild pomegranate) powder were made separately to prepare instant chutney (Figure 8.6; Joshi and Sharma, 2010). In the study, all of the instant chutney powders prepared using different combinations of dried and powdered fermented vegetables, and with acidulant added in the ratio of 1:1, were evaluated (Joshi et al., 2011b). The results showed that the instant chutney prepared with fermented carrot + *anardana* (1:1 ratio) ranked the best. Consistent with the trends of various sensory attributes, the overall acceptability score confirmed the suitability of 50% carrot + 50% *anardana* for fermented carrot-based instant chutney, while the instant chutney prepared with fermented cucumber + *anardana* (1:1 ratio) was adjudged to be the best (Joshi et al., 2011b).

8.2.5 Lactic Acid–Fermented Pickles

8.2.5.1 Carrot   Fermentation using LAB, either natural or inoculated, is also carried out traditionally (Buckenhuskes, 1993). In tests, carrots were fermented in brine with different salt concentration, and it was found that fermentation of carrot in 2% brine at 30 ± 2°C for six days yielded an acceptable product (Ramdas and Kulkarni, 1987). The brine containing 2% and 4% salt had a faster drop in pH than did those with 6%, 8%, and 10% salt, and the fermentation was completed in 6–7 days, whereas, in case of 6%, 8%, and 10% brine, it took 3–4 days more to develop maximum acidity. Based on sensory evaluation, the product with 2% brine, 0.78% acid, 3.38 pH, and a final microbial count of colony-forming units (CFU) of 67×10^8 CFU/mL was identified as the best.

In carrot fermentation, the acidity development was fastest at 37°C, but the product had a "cooked" flavor, while the acidity development was slower at 22°C than that at room temperature (i.e., 30°C) (Ramdas and Kulkarni, 1987). Thus, the fermentation of carrots at 30 ± 2°C for six days yielded an acceptable product. LAB fermentation in a sequential manner using *Lactobacillus*, *Streptococcus*, and *Pediococcus* gave the highest acid production at a salt concentration of 2.5% and a temperature of 26°C (Joshi et al., 2009). A generalized flow chart for the lactic acid fermentation of radish, carrot, and cucumber fermentation followed in these studies is depicted in Figure 8.7.

Batches of pickled cucumbers using the pure culture of *Lactobacillus planta-rum* and/or *Pediococcus cerevisiae* were prepared, and the best results were given by mixed cultures (Leon et al., 1975). The effect of inoculation (*Lactobacillus plantarum* alone, and mixed with *Lactobacillus brevis*) and a brine purge with a N_2 and $CaCl_2$

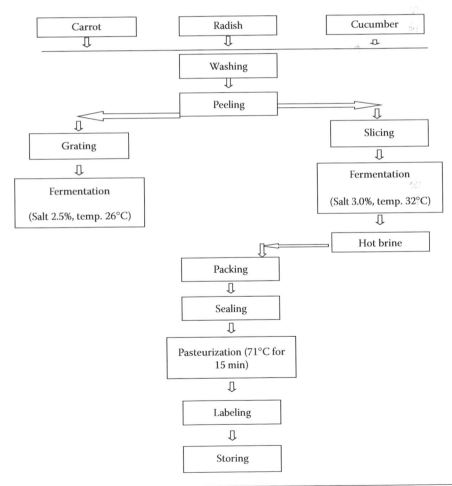

Figure 8.7   Generalized flow chart of the production of lactic acid-fermented pickles.

addition on cucumber fermentation and on the product quality has also been studied (Rodrigo et al., 1985), in which large-sized, fresh cucumber fruits (8–11 cm) in a low salt brine (4.7% NaCl) at 17°C were fermented. In the inoculated samples, fermentation was completed in 11–21 days; however, non-inoculated samples took 35 days and non-inoculated samples with 10.7% NaCl had hardly begun fermentation on the 87th day. Brine purge helped to avoid bloater formation. Model studies carried out to evaluate the fermentative bacteria (*Bifidobacterium bifidium, Lactobacillus casei, Lactobacillus plantarum, Lactococcus diacetylactis, Leuconostoc mesenteroides, Leuconostoc oenos, Pediococcus pentosaceus* and a mixed culture of *Propionibacterium shermanii* and *Pediococcus pentosaceus*) in low-salt (2.5% NaCl) cucumber-juice brine at 22–26°C for 39 days resulted in the formation of constituents such as acetic acid, acetoin, ethanol, lactic acid, mannitol, and propionic acid (Chavasit et al., 1991).

An initial salt concentration of 5.0%–7.5% at an equilibrated level has been found to be optimum for the fermentation of pickling cucumbers (Suresh et al., 1997), and a good-quality brine-stock pickling cucumber can be obtained by fermentation in a 7.5% equilibrated concentration of NaCl solution for 4–5 days, initially, and then raising the salt level to 14%. The controlled fermentation of cucumbers reduced the salt requirements in the brine, formerly used to enable the naturally low count of lactic acid-forming bacteria to develop, and reduced the incidence of bloater formation (Etchells et al., 1975; Andres, 1977; Chavasit et al., 1991). Model studies have been carried out to evaluate the effect of fermentative bacteria in low-salt cucumber brine (2.5% NaCl), and found that the percentage of sugar fermented ranged from 16.2 to 87.7 (Chavasit et al., 1991).

8.2.5.2 Cucumber Salt is used to provide a selective environment for the LAB responsible for desirable changes in the quality of the product (Fleming and McFeeters, 1981; Fleming, 1982; Naewbanij et al., 1986) by either inhibiting the degradative enzymes or preventing the growth of microorganisms that produce such enzymes (Fleming et al., 1978; Mcfeeters et al., 1989). Both the concentration and chemical composition of the brine play a regulating role in the sequence of microorganisms that occur in natural cucumber fermentation resulting in the growth of LAB. Initiation of cucumber fermentation is carried out in many instances by *Leuconostoc mesenteroides*, particularly at lower salt concentrations and temperatures (Pederson and Albury, 1950). But, at higher temperatures and salt concentrations, fermentation is often initiated by *Streptococcus faecalis, Pediococcus cerevisiae, Lactobacillus plantarum*, and, to a lesser extent, *Lactobacillus brevis*. The latter two species are responsible for high acidity levels.

8.2.5.3 Olives Olives are also cultivated in India, where they are made into a pickle-in-brine. *Pediococcus cerevisiae* play a major role in these vegetable fermentations, along with *Lactobacillus plantarum*, particularly those in brine and with the extraction of oil. The fruit is intensely bitter and cannot be consumed direct from harvesting.

The brining of olives is thus, carried out, prior to which the olives are "de-bittered" using a sodium hydroxide solution followed by a neutralization process (Pederson, 1971). In olive pickling, the spoilage bacteria present in the primary stage are eliminated due to acid production by LAB, mostly *Streptococcus*, *Pediococcus*, and *Leuconostoc* (Vaughn, 1975). The intermediate stage features *Lactobacillus* as the dominating microflora. In the last stage, as the acid content is increased (0.8%–1.0%), pH decreases. In an early study, Cruess (1948) made use of starter cultures for the preparation of Spanish-type green olives, which increased the rate of acid production and reduced gas pocket formation and other types of spoilage (Vaughn et al., 1943). Details of the process for olive fermentation are shown in Figure 8.8a and b.

Lactic acid-fermented and brined olives are not liked by most of the Indian population, so further experiments were conducted, and it was found that olives, after lactic fermentation and brining, when converted into special pickles have large acceptance.

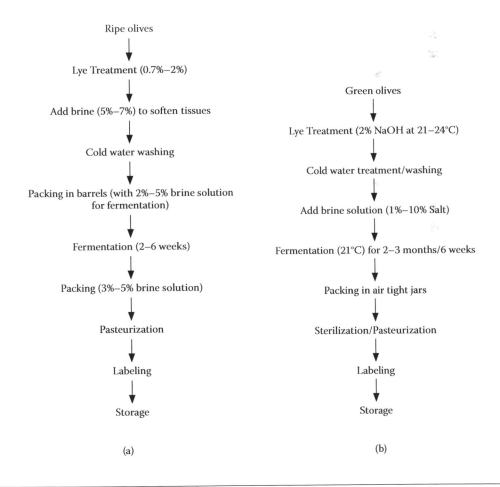

(a)   (b)

Figure 8.8   (a) Flow chart for the prepartion of olive in brine pickle (ripe olives); (b) Flow chart for the preparation of olives in brine (green pickles). (Adapted from Joshi, V.K., Sharma, S., and Thakur, N.S. 2003. *International Seminar and Workshop on Fermented Foods, Health Status and Social Well-Being* held on 2003 at Anand, India, pp. 24–28.)

A recipe for this product has also been developed accordingly (Joshi and Bhat, 2000; Joshi et al., 2005).

8.2.6 Vegetable-Based Fermented Foods

8.2.6.1 Anishi This is a fermented cake made from the leaves of the *Colocasia* plant exclusively by the Ao Naga tribe of northeast India. Its preparation involves the packing of the *Colocasia* leaves in gunny (natural fiber) bags, or, alternatively, wrapped in banana leaves for about 3–4 days or until they become yellow. Then, the leaves are pounded into pastes and made into cakes. These cakes are then, wrapped in banana leaves and kept under the ash near the fireplace or exposed to the sunlight until they are completely dried and become hard (Jamir and Deb, 2014).

8.2.6.2 Hungrii This is a fermented product, prepared from the leaves of the mustard plant (*Brassica* spp.), that is popular with the Rengma Naga tribe of Nagaland (India). Fresh mustard leaves are taken and sun-dried (Jamir and Deb, 2014). A pit is dug out and banana leaves are laid at the bottom of the pit (Figure 8.9). The dried leaves are then, wrapped in the banana leaves and left in the pit, covered with soil, for about 15–18 days (Figure 8.10).

8.2.6.3 Tsutuocie *Tsutuocie* is made from cucumber fruits and leaves mostly by the Angami Naga tribe of Nagaland (Jamir and Deb, 2014). Matured cucumber fruits are preferred, and these are cut into pieces. The leaves are first washed and then, shredded

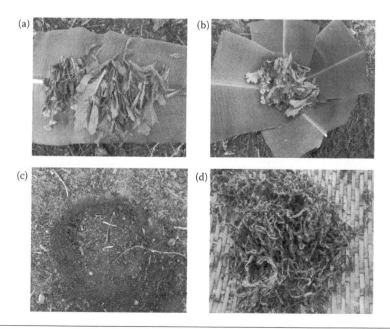

Figure 8.9 Different stages of *hungrii* preparation. (a) Fresh mustard leaves; (b) leaves being wrapped in banana leaves; (c) buried in the ground; (d) the fermented product, *hungrii*. (Adapted from Jamir, B. and Deb, C.R. 2014. *Int. J. Food. Ferment. Technol.* 4(2): 121–127.)

Leaves of *Brassica* sp.

↓

Sun dried and wrapped in banana leaves

↓

Pit made in the ground where leaves are buried

↓

Fermentation for 15–18 days

↓

Hungrii

Figure 8.10 Flow chart for preparation of *hungrii*. (From Jamir, B. and Deb, C.R. 2014. *Int. J. Food. Ferment. Technol.* 4(2): 121–127. With permission.)

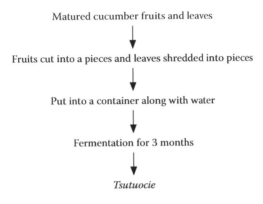

Matured cucumber fruits and leaves

↓

Fruits cut into a pieces and leaves shredded into pieces

↓

Put into a container along with water

↓

Fermentation for 3 months

↓

Tsutuocie

Figure 8.11 Flow chart for the preparation of *tsutuocie* (Angami).

into pieces by hand. The mixture of the fruit and the leaves is then, put into a container with a sufficient amount of water. It is then, allowed to undergo fermentation for about 3 months (Figure 8.11), following which it becomes a thick, sluggish, green paste used as a condiment during the preparation of meat and chutney. *Tsutuocie* has a shelf-life of over a year.

8.3 Beverages Made by Using Lactic Acid Fermentation

Lactic acid fermentation is employed to prepare various products such as *kanji*, fermented grape juice, fermented nut milk, "sogurt" from soybeans, fermented cucumber extract, fermented beverages from wheat and maize (Bucker et al., 1979; Gobbetti and Rossi, 1989; Cheng et al., 1990; Naewbanij et al., 1990; Takagi et al., 1990), and many other similar products. Some of these have been produced since ancient times using indigenous skills, and are described here.

8.3.1 Kanji

This fermented carrot-based and colorful beverage is made and consumed throughout India. In the traditional method of preparation, its fermentation is carried out using the natural microflora. Due to this method, as well as variations in the raw ingredients, the quality of this product varies. Therefore, a method for the preparation and preservation of this beverage from fermented black carrots has been standardized and documented (Sethi, 1990). Among the treatments, the addition of 3% salt, 1.0% mustard, 0.015% sodium benzoate, and 0.01% potassium metabisulfite to a 1:1 carrot:water mixture has been recommended for the preparation of this fermented beverage. The preservatives are added to the *kanji* after fermentation for its preservation. The shelf-life of the fermented juice could be increased either by addition of 1.0% mustard or by the use of chemical preservatives, as noted earlier. Thus, the addition of mustard along with preservatives has a complementary role for extending the shelf-life of this fermented juice up to 20 weeks.

8.3.2 Soybean-Based Products: "Sogurt"

The soybean is employed to produce a milk-like product called "soymilk." Soybeans (comprising 42% protein and 20% fat) are inexpensive proteins and an excellent source of energy, but soymilk is commonly characterized as having a beany, grassy, or soy flavor. However, this can be improved by lactic acid fermentation (Joshi and Thakur, 2000), which is employed to prepare beverages in which lactic acid is the major constituent. It was found that soymilk-based yogurt ("sogurt") could offer several distinct nutritional advantages over the milk-based yogurt to the consumer; namely, reduced levels of cholesterol, saturated fat, and lactose (Lee et al., 1990; Ruei and Lin, 2000).

The various operations involved in sogurt production have been standardized, and are described here. Soybeans are cleaned, washed, drained, and then, ground. The soy flour is de-fatted to remove substrates for lipoxygenase. De-fatted soy flour is mixed with water (1:7.6) for 5 min in a blender, autoclaved at 121°C for 15 min, and then gelatin (0.5%) and lactose (0% or 2%) are added. After cooling the soymilk to 45°C, inoculum (2% *L. casei* and 2% *S. thermophilus*) and coagulant (0.15% calcium acetate) are added. The mixture is incubated at 30°C for 18 h for the growth of the flavor-producing bacteria *L. casei*, and then, at 45°C for 16.5 h for the growth of the acid-producing bacteria *S. thermophilus*. The sogurts are then, packed and stored at 4–5°C (Cheng et al., 1990).

8.3.3 Fermented Milk

Fermented milks are manufactured throughout the world, and, while around 400 generic names are applied to both the traditional and the industrialized products, the list may in fact only include a few varieties (Kurmann, 1984). Such products are made and consumed extensively in several different countries of South Asia. These have also been a part of the staple diet of those living in the mountainous regions of Bhutan and Nepal as well as the Nepalese-origin ethnic communities living in different parts of the world (Thapa et al., 2003). They have been defined as products prepared from milk—be it whole, partially or fully skimmed, concentrated milk, or milk substituted from partially or fully skimmed dried milk, homogenized or unhomogenized, pasteurized, sterilized, and fermented by specific organisms (as per International Dairy Federation definitions, 1969; Batish et al., 1999; Robinson and Tamine, 2006). See Wattiaux (2012) for composition of milk. Some of fermented milks and byproducts made and consumed in South Asian countries are discussed here.

8.3.3.1 Buttermilk or Chhaas (Lassi)   "Buttermilk" is defined as the aqueous phase released during the churning of cream in butter manufacturing (Corredig and Dalgleish, 1997; Sodini et al., 2006; Morin et al., 2007). It includes a wide range of milk-fat byproducts with various compositions according to the raw materials used, pre-treatment conditions, and butter-making process (Vanderghem et al., 2010). In the Indian subcontinent, buttermilk is the liquid left after extracting butter from churned yogurt (*dahi*). This is a traditional buttermilk that is still common in many Indo-Pakistani households. In southern India and most areas of the Punjab, buttermilk with added water, sugar and/or salt, asafoetida, and curry leaves is available at stalls during festivals (Anonymous, Wikipedia, lassi). It is produced as an indigenous fermented product, and also represents the successful industrialization of indigenous technology with improvements made scientifically.

Lassi is a much-diluted verison of buttermilk that is a part of the regular diet of the people of Saurashtra, Gujrat, as well as those from other states of India and regions in South Asia. The people of the Saurashtra regions call this type of buttermilk "*tikhari chash.*" As a byproduct of *deshi* (traditional) butter prepared in households, it is known as *chhas* or *matha* (Anonymous, Wikipedia, lassi). The term "*lassi*" is also used in some parts of northern India to refer to a cold refreshing beverage obtained by blending *dahi* with water and sugar. Indeed, it is appreciated throughout India as a beverage for its palatability and as thirst-quenching/refreshing drink, aside from its therapeutic values. The palatability and wholesomeness of the product, however, depends on the quality of curd churned (using a churner; Figure 8.12) and the temperature of churning. A curd obtained by fermentation with contaminants is highly sour or "off" flavored and produces a *lassi* unfit for human consumption (Anonymous, Wikipedia, lassi; Gandhi, 2002).

The various steps involved in the preparation of *lassi*, as practiced in northern states of India, are described in Figure 8.13. The lactic microflora commonly associated with *lassi* are *Lactobacillus lactis*, *Lactobacillus delbrucki*, *L. acidophilus*, *Lueconostoc mesenteroides*, *L.diacetyllactis*, *L. helveticus*, and *S. thermophilus*. To get a good-quality *lassi*, a restricted fermentation with cultured microorganisms with an ability for enhanced diacetyle production, should be used (Gandhi, 1989).

At an industrial scale, buttermilk is also produced by culturing skim or low-fat milk with a lactic acid-production culture. It is separated into two layers, then allowed to

Figure 8.12   A typical churner used for the separation of buttermilk from butter. (Courtesy of Dr. V.K. Joshi.)

stand for some time. The upper layer is made of whey and the second, heavier layer is made of casein that has curdled into fine lumps. Buttermilk is a low-cost product available in large quantities as a byproduct of the dairy industry, and is also considered to be an important foodstuff due to its composition of proteins and polar lipids from the milk-fat globule membrane (Vanderghem et al., 2008). It can also be produced using

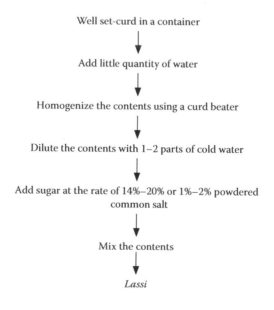

Figure 8.13   General steps to prepare *lassi*. (Adapted from Anonymous, Wikipedia, lassi http://www.agriinfo.in/?page=topic&superid=9&topicid=725.)

LAB: specifically, cultured buttermilk is made by adding LAB (*Streptococcus lactis*) to milk when an exclusive unit for better milk production is set up.

8.3.3.1.1 Cultured Buttermilk   Commercially available cultured buttermilk comprises pasteurized milk, homogenized (1% or 2% fat), and inoculated with a culture of LAB to simulate the action of naturally occurring bacteria (Figure 8.13). In the dairy industry, buttermilk is used in several ways, such as in cheese making (Batish et al., 1999), in the formulation of ice cream (Sodini et al., 2006) or yogurt (Trachoo and Mistry, 1998), or in the manufacture of recombined milks (Singh and Tokley, 1990). Skim milk is preferred in the preparation of buttermilk, and it should be free from any antibiotics and inhibitory substances. The milk should be filtered or clarified to remove the visible dust particles, which may alter the texture and aesthetic properties of the final product. The customary method of buttermilk production is shown in Figure 8.14.

In the preparation of buttermilk, the milk should be preheated to about 35–40°C to ease the filtration carried out at the next stage. Following filtration, the milk is pasteurized at a temperature range of 82–88°C for about 30 min and then, cooled to

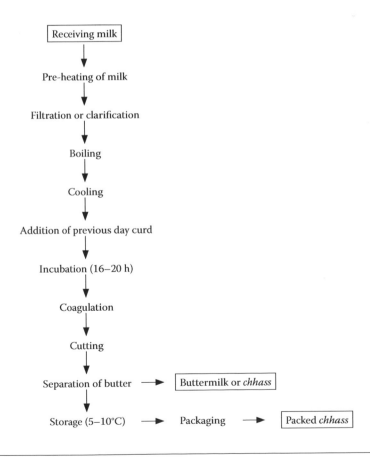

Figure 8.14   Traditional method of buttermilk production.

22°C. Starter culture is then, added to the pasteurized milk at a rate of 1%–2%. The most frequently used culture is *Lactococcus lactis* ssp. *lactis*, as well as *L. lactis* ssp. *cremoris*, and the mixture is incubated at 20–25°C for 16–20 h for the development of coagulation and so an acidity of between 0.80% and 0.85% is reached (Goswami et al., 2011). Then, the curd or coagulant formed is churned by slow agitation to give a smooth consistency. The released fat aggregates into a solid lipid matrix forming butter (Vanderghem et al., 2008). The product is then, cooled to 5–10°C. After churning (Figure 8.12), the butter layer is separated and the remaining product, the buttermilk, is packed and then, stored at 5°C. At the time of packaging, the acidity of the buttermilk should be 0.85%–0.90%. A rich flavor is imparted to the buttermilk through the addition of cream, which also helps in decreasing the viscosity. The latter can, however, be increased by adding stabilizers, non-fat dry milk, or polysaccharide-producing cultures (Goswami et al., 2011). Both lactic-acid producing and flavor-producing strains are used for culturing. Salt is be added to the finished product to further improve its flavour.

The lactic microflora commonly associated with buttermilk are *Lactobacillus lactis*, *Lactobacillus delbrucki*, *L. acidophilus*, *Lueconostoc mesenteroides*, *L.diacetyllactis*, *L. helveticus*, and *S. thermophilus*. To get good-quality buttermilk, restricted fermentation with cultured microorganisms such as enhanced acetaldehyde and diacetyle-producing organisms is carried out.

8.3.3.1.2 Tara *(Buttermilk)* Pastoralism is the mainstay of the people of the Ladakh region of India who live in the area's elevated regions of Changthang and Zanskar. Milk is processed in a number of ways here, and many of its products are unique to Ladakh. Milching animals reared in the region include sheep, goat, cow, *dzomo*, and *dri*; a *dzomo* is the female of a crossbreed of cow and yak while a *dri* is a female yak.

Milk (*oma*) is boiled, allowed to cool, and then, inoculated with the previous batch of buttermilk and incubated overnight in a warm place to form a curd (known as "*jho*" in Leh). In the Changthang and Zanskar areas of Ladakh, the curd (*jho*) is shaken vigorously in a bag made of goatskin to separate the butter from the buttermilk. In other parts of the region, however, buttermilk is made by churning curd in a special wooden vessel (*zem*) made of juniper wood. In Kargil (India), the nomenclature for these products is different again: they call milk "*orjen*," curd "*oma*," and buttermilk "*derba*."

8.4 Fermented Milk Products Other than Beverages

8.4.1 Fermented Milk

Milk can be consumed as fluid milk, but is also used as a raw material for the preparation of different products such as fermented milk and other foodstuffs (Batish et al., 1999; Chandan, 2013), such as flavored milk beverages, ice cream, cheese, *ghee*, and many more. It may have been a simple accident when people first experienced the taste of fermented food, so the first fermentation might have been started with the storage of surplus milk that resulted in a fermented product the next day (Vedamuthu, 1991; Steinkraus 1996). In any case, after drying, fermentation is considered to be the oldest method of food preservation. Over the time, people have realized the sensory, nutritional, and therapeutic value of fermented foods and drinks, thus making them more popular (Farnworth, 2008). In Asia, indigenous fermented milk products were known for their better preservation quality, stability, digestibility, and other beneficial attributes. These products have also been observed to play a role in preventing various disorders, such as obesity, osteoporosis, dental caries, poor gastrointestinal health, cardiovascular diseases, hypertension, colon and rectal cancer, bone ailments, eye diseases, aging, etc. (Sharma and Rajput, 2006). Fermented milk and milk products are produced and consumed (Law et al., 2011) because, although it is a highly perishable commodity, milk is an important food for the majority of the population and is the only food suitable for babies. Because of its low acidity and its nutrients, bacteria grow more easily and quickly in it, thus necessitating milk's conversion into a shelf-stable product such as a fermented food. Fermented milk can, for example, be stored for about 20 days, compared to the fresh milk, which can often be stored for only a day or even less. Furthermore, milk and milk products have been associated with therapeutic values, including extending the longevity of its consumers (Granato et al., 2010) and probiotic activity (see Chapter 6 for more information).

In some South Asian countries, Himalayan fermented yak milks have been reported to have probiotic properties (Tamang et al., 2000). Lactic acid bacteria processing technological properties have been isolated from indigenous fermented milk products of Sikkim in India (Tamang et al., 2004). Here, as elsewhere, the preservation of milk is achieved by fermentation (increasing acidity), in addition to other methods of preservation or the separation of the components to produce foods such as cheese or butter (Anonymous, Wikipedia, milk). There are several fermented products manufactured such as *dahi*, cheese, buttermilk, *shrikhand*, etc. with different flavors. The ethnic peoples of the Himalayan regions of India, Nepal, Bhutan, and China consume a large variety of indigenous fermented milk products made from milk of cows and yaks. These lesser-known ethnic fermented foods are *dahi*, *mohi*, *chhurpi*, *somar*, *philu*, and *shyow* (Dewan and Tamang, 2007), which are produced by natural fermentation. However, the use of different microorganisms has led to the development of a wide range of additional indigenous fermented milk products, such as yogurt, *shrikhand*, *lassi*, *kefir*, *koumiss*, *yakult*, *laben*, and so on.

Initially, the souring of the milk was done by natural fermentation, but, as many communities began to acquire a taste for sour milk, techniques were developed to ensure the process of souring taking place. The constant use of the same vessels or the addition of fresh milk to an ongoing fermentation might have given rise to the gradual evolution of locally popular products (Tamime and Robinson, 1988). Many countries in South Asia, being tropical in climate, might have embarked some time ago in the preparation of sour milk, recognizing it as being more stable and containing high-quality nutrients compared to fresh milk, hence its early popularity in these regions (Mann, 1977). The milk from at least eight species of domesticated mammals—including cows, buffalo, sheep, goats, horses, yaks, and zebras—is known to be used in the preparation of fermented milk products (Tamime and Robinson, 1988; Lowe and Fox, 2012). Purified bacteriocin Basicin can be used to improve the shelf-life of dairy products (Sharma et al., 2011).

8.4.2 Indigenous Fermented Milk Products

A considerable portion of milk in India is utilized for the manufacture of indigenous fermented milk products such as clarified butter (*ghee*), heat-desiccated milk (*khoa*), acid-coagulated milk (*chhana*), Indian cheese (*paneer*), and a varieties of sweets. Heat and acid coagulation, heat desiccation, separation, and fermentation are some of the different processing steps involved in the preparation of three indigenous fermented milk products: *dahi* (curd), *shrikhand* (sweetened concentrated curd), and *lassi* (stirred curd), which may be considered as equivalent to Western yogurt, *quarg*, and stirred yogurt, respectively (Sekar and Mariappan, 2007; Sarkar, 2008).

8.4.2.1 Dahi (*Sanskrit*: Dadhi)

Dahi is a well-known, indigenous, commercialized successfully, and marketed product. It is a fermented milk product made from the milk of cows, buffalo, or a mixed milk, and it is widely consumed in most South Asian countries, including the Himalayan region, either plain, sugared, or salted (Farnworth, 2008; Tamang et al., 2010). It can be consumed as a sweet or savory drink, or as a dessert containing sugar, spices, fruits, nuts, etc., either as a part of a daily diet or as a refreshing beverage. It has acidic taste, and is yellowish creamy-white or creamy-white in color with a smooth, firm, and glossy surface. In India, it is also known as *dadhi*.

It is a product made from a household level to an industrial scale. According to FSSAI regulations (Food Safety and Standard authority of India, 2011), "*dahi* is a curd as a semi-solid product obtained from pasteurized or boiled milk by souring (natural or otherwise), using a harmless lactic acid or other bacterial cultures. It may contain added cane sugar, but retains the same minimum percentage of milkfat and milk solids-not-fat (SNF) as the milk from which it is prepared." Where *dahi* or curd is sold or offered for sale without any indication of its class of milk, the standards prescribed for *dahi* prepared from buffalo milk would apply (Vats, 2013).

8.4.2.1.1 Classification of Dahi

Dahi is made in different varieties, according to region-specific tastes, and can have several uses as

- *dahi* used for direct consumption and the production of *desi* butter
- for the production of *chakka, shrikhand*, and *lassi*
- *dahi* prepared from whole milk, skim milk, standard milk, or special milk

- *dahi* prepared with the addition of sugar and fruits
- the acidity of normal *dahi* is less than 0.7%, while the acidity of sour *dahi* is more than 0.7%

There are two types of *dahi* prevalent in India for direct consumption: a sweet and mildly acidic variety with a pleasant flavour, and a sour variety with a sharp, acidic flavour.

8.4.2.1.2 The Traditional and Standardized Method of Preparing Dahi   The indigenous method of *dahi* preparation is depicted in Figure 8.15.

8.4.2.1.3 Standardized Method of Dahi *Making*   *Dahi* is ideally made from buffalo's milk, which produces a thick-bodied product due to its high SNF content, which should range between 11% and 13%. The increased protein content in the mix gives a custard-like, thick consistency after fermentation. The higher milk solids also prevent the product from syneresis, which is a very common defect. *Dahi* prepared from whole milk contains, in terms of percentages, fat (about 5%–8%), protein (3.2%–3.4%), lactose (4.6%–5.2%), ash (0.70%–0.72%), and a titratable acidity of 0.60%–0.80%. Skim milk (9% SNF, 0.05% fat) is heated to 90°C for 15 sec in a high-temperature–short-time pasteurizer, cooled to 30°C, and inoculated with 0.25%–0.50% *dahi* culture of mixed strains (Kilara and Chandan, 2013). The milk is boiled and sometimes concentrated before the addition of the starter.

The inocula used in *dahi* preparation are a mixture of *Lactobacillus* and *Streptococci* spp. Commonly, a small quantity of the curd from a previous fermentation is used as an inoculum. A mixed culture containing *Lactococcus lactis* ssp. *Lactis*, *Lactococcus lactis* ssp.

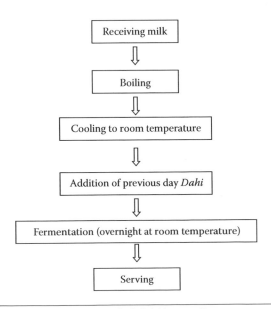

Figure 8.15   Flow chart depicting the indigenous method of *dahi* preparation.

diacetilactis/Leuconostoc, and *Lactococcus lactis* ssp. *cremoris* in the ratio of 1:1:1 may be used. The other commonly associated lactic microflora in the inoculum are *Lactobacillus lactis*, *Lactobacillus delbrucki*, *L. acidophilus*, *L. helveticus*, and *S. thermophilus*. The quality of the *dahi* may vary with the type of starter culture used (Masud et al., 1991; Tamime et al., 2006). *Dahi* of an acceptable quality can be obtained with the application of acid-producing as well as flavor- (primarily diacetyl) producing microorganisms and adopting a two-stage fermentation. Moreover, the inclusion of certain probiotic and beneficial bacteria for further enhancement of the dietetic properties of traditional *dahi* can be undertaken. The application of bio-preservatives and thermization (mild heat treatment) may be recommended for the shelf-life improvement in order to extend market reach of the foodstuff (Gandhi, 1989; Sarkar, 2008). After eight hours of inoculation, the required acidity (0.8%–1.0% lactic acid) should be achieved, and the curd is ready for further processing.

8.4.2.1.4 Commercial Manufacturing Process of Dahi   Good-quality fresh milk of cows, buffalo, or a mix of sources is received and preheated to 35–40°C, following filtration and clarification to remove the impurities. Then, the milk is standardized to adjust the fat solid ratio: the fat is adjusted to 2.5% with 10% SNF to improve the texture of the product. The milk is heated to 90°C for 15 min, cooled, and then inoculated with *L. acidophilus*, *L. casei*, and *Lactococcus lactis* var. *diacetylactis*. It is then, transferred to containers of the required capacity and incubated at 37°C for 12–14 h (Elliot et al., 2007). During incubation, the acidity of the milk reaches 0.7%–1% and curd formation takes place. The cooling of the curd is done to 12°C, and it is stored at about 5°C, before being prepared for sale. Attempt to use thermal processing of *dahi* to improve shelf-life has also been made (Aziz, 1985).

8.4.2.1.5 Mishti Dahi   *Mishti dahi* is a fermented milk product with a creamish to light brown color, sweet acidic taste, firm body, smooth texture, and pleasant aroma. Analogous varieties of *dahi* known as *mishti doi*, *mishti dahi*, *lal dahi* (red *dahi*), or *payo-dhi* in the eastern region of the Indian subcontinent are also very popular. The product is commonly sold in earthen pots of varying sizes, and served chilled. It is made by adding 6.0%–6.5% sugar to the boiled milk. Artificial color, caramel, and *jaggery* may also be added. The milk is cooled to 40–45°C and incubated for 12–15 h. The sweetened, concentrated form of *dahi* consumed in Bengal is known as *mishti doi* or *sweet dahi*.

8.4.2.2 Raita   *Raita* is a product that is prepared using *dahi* as its base material. No standardized method for the manufacture of *raita* is available since its composition varies from region-to-region, depending on local consumers' preferences. In general, though, for preparing *raita*, stirred *dahi* is sweet-salted according to the taste. Crumbs are then, mixed in (at 10%–15% by weight), which give the product a good body and texture. Sometimes, grated cucumber or mixed fruit pieces are also added to give the product an enhanced flavor. *Raita* is generally served fresh.

8.4.2.3 Dahi Karamba   This is similar to the curd rice of south India, and is a mixture of rice and curd with added sugar and *ghee* and a garnish of dried fruits. This food product is popular among the Muslim populations of Gujarat, India.

8.4.2.4 Other Dahi-Based Fermented Foods   In the state of Gujrat, some of the well-known foods that have either been fermented as a stage of processing or that include fermented ingredients are as follows:

- *shahi gatte*—a culinary preparation in which strips of gram flour are first steam cooked and then, cooked in a mixture of curd and spices;
- *dahi vada*—flattened and fried *vadas* are prepared from de-skinned black gram and then, dipped in beaten fresh curd and seasoned with spices;
- *ghevar*—a sweet dish prepared by fermenting the slurry of refined flour before frying it and dipping it in sugar syrup.

8.4.2.5 Curd/Yogurt   Curd, or yogurt, is a fermented dairy product made by fermentation with bacterial cultures added to the milk (Figure 8.16). Manufacter of yogurt has been described in detail (Chandan, 2006). The bacteria used to make yogurt are known as "yogurt cultures." The fermentation of lactose by these bacteria produces lactic acid, which acts on milk protein to give yogurt its texture and characteristic tang (Rasic, 1987; Tamine and Robinson, 1999, 2007). Worldwide, cow's milk (the protein within which is mainly casein) is most commonly used to make yogurt. Milk from water buffalo, goats, ewes, mares, camels, and yaks, though, is also used in various parts of the world.

Yogurt is produced using a culture of *Lactobacillus delbrueckii* ssp. *Bulgaricus* and *Streptococcus thermophilus* bacteria (Tamine et al., 2006). The method is the same as that employed to make *dahi*. It is a tasty fermented drink that is nutritious and easily digestible. It is popular in Nepal, where it is served as both an appetizer and a dessert. Locally called *dahi*, it is a part of the Nepali culture, used in local festivals, marriage ceremonies, parties, religious occasions, family gatherings, and so on. The most famous type of Nepalese yogurt is called *juju dhau*, and it orginates from the city

Figure 8.16   Yogurt/curd.

of Bhaktapur (Anonymous, wiki yogurt). These days yogurt drink with prebiotic and probiotic is getting increased popularity (Allgeyer et al., 2010). Using soybean yogurt-like product sogurt has been made successfully (Cheng et al., 1990).

8.4.2.6 Labo (*Cottage Cheese*) *Tara* (buttermilk) is boiled and then, allowed to cool for 10–20 min This results in a separation of solid from liquid, which is filtrated out. The liquid (whey) element is known as *chhurkhu* and solid (cottage cheese) as *labo*. The latter is eaten fresh with bread or *kholak*, usually after adding sugar to it. *Chashrul* is a kind of semi-solid preparation made by mixing *labo*, *phey*, and salt tea, and served hot. *Chhurkhu* is thought to provide instant energy, and is therefore often consumed by farmers after they finish their work in the fields (Angchok et al., 2009). In Changthang, *chhurkhu* is mixed with *nyasphey* to make a cake that is consumed during travel (Figure 8.17). This type of cheese is also one of the most valuable and popular of the indigenous fermented dairy products, and its status is increasing throughout the world due to its taste, improved digestibility, nutritional value, and functional properties.

8.4.2.6.1 Burnt Sweet Cheese (Chhenapoda) *Chhenapoda* (Figure 8.18)—literally, "burnt sweet cheese"—is a special type of cheese-based sweetmeat that had its origin in Odisha during the twentieth century, and has proved popular in the state as well as elsewhere. It is made of well-kneaded homemade cottage cheese, sugar, cashew nuts, and raisins, and is baked slowly for several hours in an oven until the mixture turns brown. *Chhenapoda* is the only well-known Indian dessert for which the flavor is predominantly derived from the caramelization of sugar. The shelflife of *chhenapoda* is usually 3–4 days (Anonymous, Wikipedia).

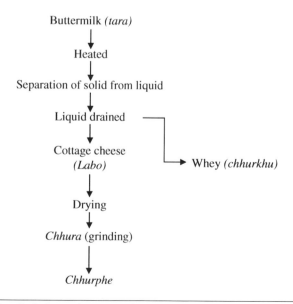

Figure 8.17 Traditional method of preparing cottage cheese and its derivatives.

Figure 8.18 *Chhenapoda* (Burnt sweet cheese).

8.4.2.6.2 Rasgullas The utilization of dairy byproducts in the production of new food products has a twofold purpose. First, the nutritional and the functional virtues of the milk constituents can be made use of for enhancing the overall quality of these new products, and, second, the disposal of the byproducts in this way reduces pollution problems and enhances the economy of the dairy-processing operation over-all. *Rasgullas* are fermented cheese curd kneaded to a dough-like consistency and rolled into balls. The balls are dipped in sugar syrup, and then, eaten (Anonymous, Wikipedia).

8.4.2.6.3 Chhura (*Dried Cottage Cheese*) *Labo* is a perishable commodity that cannot be stored, and so it is dried to increase its shelf-life. The dried product is known as *chhura* (Figure 8.19), while its powdered form is known as *chhurphe*. *Chhura* is one of the ingredients of *thukpa*. On cooking, it becomes soft, which means it is then, easily chewed.

Thuth is prepared by mixing *chhurphe*, *maar*, and sugar. Generally, it is consumed in the form of a cake, which can be of different shapes and size. It can be stored during winters, and served especially during *losar* (Ladakhi New Year). In *Changthang*

Figure 8.19 *Chhura* (dried cottage cheese) kept in a bowl.

(India), *thuth* is almost always used with *paq* (a kind of *kholak*) plus one slice of frozen liver and two slices of frozen meat. It is also common in Zanskar.

8.4.2.7 Shrikhand   Shrikhand is a staple dessert made out of fermented milk. "Shrikhand" means a product obtained from *chakka* or skim milk *chakka* to which milkfat is added (FSSAI, 2011). Fruits, nuts, sugar, cardamom, saffron, and other spices are added, but the final product should not contain any added coloring or artificial flavoring substances.

It is a semi-solid, sweetish–sour, fermented milk dessert that is very similar to flavored "*quarg*" available in Germany (Kilara and Chandan, 2013). In addition, it is served as a "curry" with *pooris* as a breakfast item or as a snack, or is served at parties either as a dessert or in place of *sorbet* between courses to cleanse the taste-buds. It is a popular dessert of various states of India, including Gujarat, Rajasthan, Punjab, Maharashtra, and Karnataka, and is an important part of festive occasions. It is a part of many north Indian wedding feasts, too. Aptly, it is also prepared to celebrate the birth of Lord Krishna, the divine protector of cowherds (Anonymous, Wikipedia yogurt).

Shrikhand get its name from *ksheer* ("milk" in Sanskrit) and *qand* ("sweet" in Persian; as per Wikipedia shrikhand). The exact origin of the foodstuff is unknown, but western India is attributed with the first historical mention of the dish. The legend states that traveling herdsmen used to hang curd or yogurt overnight to make it easier to carry while traveling. The thick yogurt that used to be collected the next day was then, mixed with sugar and nuts to make it palatable for consumption during the long journey (Anonymous, Wikipedia shrikhand). Historically, it was made by Halwai (confectioners or sweet-makers), but today is produced mechanically as a continuous process. Milk is fermented with LAB, then whey from the curd is removed, and sugar, flavouring, and spices are added to create the final mixture (Patel and Chakraborty, 1988). Like *dahi*, *shrikand* can be manufactured by traditional methods as well as by commercial or standardized methods, as detailed in the following sections.

8.4.2.7.1 Traditional Preparation   Traditionally, *shrikhand* is prepared following several steps as follows:

- The milk is heated and then, cooled to room temperature (0.5%).
- The preparation of curd or *dahi* by culturing milk with a natural starter is then, undertaken.
- A firm curd is obtained, and then, transferred into a muslin cloth and hung for 12–18 h to remove the whey.
- The *chakka*, or solid mass obtained, is mixed with specified amounts of sugar, color and flavoring materials, and spices, and then, blended to a smooth and homogenous consistency.
- The pulps of fruits—apple, mango, papaya, banana, guava, and *sapota*—are introduced through blending (Bardale et al., 1986; Dadarwal et al., 2005).

• Cocoa powder with or without papaya pulp is also used (Vagdalkar et al., 2002).

8.4.2.7.2 Factors Affecting Production   With the development of the dairy industry, *shrikhand* production has been industrialized, too. This process basically consists of the centrifugal separation of the whey from the curd and the mechanical mixing of the *chakka*, sugar, and spices, rendering the entire process hygienic as well as labor and time saving. Before the details of industrial method are described in further detail, though, it is pertinent to first discuss various factors involved in the production of *shrikhand*; specifically, the type of milk used and the use of additives.

8.4.2.7.2.1 Type of Milk   The type of milk and its quality, and composition affect the quality of the *chakka* and the resultant *shrikhand*. Different types of milk for *shrikhand* manufacture have been researched. In one study, whole milk was heated to 85°C for 15 min, cooled to 30°C, inoculated with 2% *Streptococcus lactis*, and incubated for 18 h at 30°C. The curd obtained was drained in a muslin cloth for 8 h and the resultant *chakka* was used to make *shrikhand* through the addition of 40% sugar. It was found that a 35%–37% fat loss occurred during *chakka* manufacture, while the fat loss in whey was between 1.23% and 2.1%, and overall loss was 6%–9.8%.

Using standardized buffalo milk for *chakka* making, yields a smooth and fine-flavored *chakka* as well as *shrikhand*, although it also gives higher fat losses in the whey. Upadhyay and Dave (1977) and Rao et al. (1987) recommended the use of buffalo skim milk to prepare *chakka* and the addition of "plastic cream" (cream that has been centrifuged at high speed causing it to form an oil-in-water emulsion) (70%–80% milkfat) to *chakka* to adjust desirable fat levels in the final *shrikhand*.

The use of skim milk not only prevents the fat losses but also helps with faster moisture expulsion from the curd. Alternatively, the use of reconstituted skim milk has also been suggested for *shrikhand* making (Patel and Chakraborty, 1985a,b). As noted, the use of skim milk for *chakka* manufacture, with addition of cream during *shrikhand* manufacture, has reduced fat losses (Rao et al., 1987). The best quality *shrikhand* was considered to have been made from cowmilk and buffalo milk in a ratio of 1:1 (% w/w), which imparted a smooth texture and a firm, soft body (Ghatak and Dutta, 1998). Patel and Chakraborty (1985c) studied different sources of milk solids for *shrikhand* making, and reconstituted skim milk (40%–43% total solids) was found to be an effective alternative to fresh skim milk. Buttermilk was the least favorable for *shrikhand* manufacture, while concentrated milk had poor sensory appeal, due to lack of acidity and presence of saltiness.

8.4.2.7.2.2 Homogenization   *Chakka* made from skim milk was considered to be relatively rough and to lack characteristic aroma. In contrast, *chakka* obtained from whole milk was judged superior in both consistency and aroma, and it has also been found that the limitations of high fat loss in whey can be overcome by the homogenization of milk. Desai (1983) prepared *dahi* from standardized (4% fat), homogenized (100 kg/cm² at 60–65°C) milk (to which sodium alginate or gelatin was added at the rate of 0.2%, w/v). In this and related research, the use of homogenized milk was found to significantly improve the flavor, body/texture, and total sensory scores of *shrikhand*, compared to that made with un-homogenized milk (Desai et al., 1985).

8.4.2.7.2.3 Additives   Different additives have been utilized in milk to improve the sensory profile, reduce fat losses, and increase the yield of *shrikhand*. Kadam et al. (1984) reported that the *chakka* produced from cowmilk (containing 2% skim milk powder and 0.2% sodium alginate) was similar to that obtained from buffalo milk. Reddy et al. (1984) prepared *shrikhand* from buffalo milk treated with 0.25% sodium citrate, which increased the moisture content but reduced the fat losses. Ashwagandha powder has also been added in preparation of *shrikhand* (Landge et al., 2010). The addition of diacetyl at 10 ppm improved the flavor of *shrikhand* made with a yogurt starter. However, when cardamom was used for flavoring the *shrikhand*, the effect of added diacetyl was not perceptible, and was therefore, considered unnecessary (Patel and Chakraborty, 1985b). The addition of 5% sour whey concentrate in *chakka* (both from cow's milk) increased the yield of *shrikhand* by 5% over that achieved by the traditional method without bringing about any change in the physical attributes of the *shrikhand* (Giram et al., 2001).

8.4.2.7.2.4 *Chakka*   As *chakka* is the base material for *shrikhand* making, its organoleptic quality and chemical composition greatly influence the consistency, composition, and flavor attributes of the *shrikhand*. Various quality characteristics have been described in the literature (Desai and Gupta, 1986; Pandya et al., 2006; FSSAI, 2011). The quantity of sugar to be added depends on the acidity of *chakka* (Aneja et al., 1977; Upadhyay and Dave, 1977; Tamime and Robinson, 2007) and has been reported that the yield of *chakka* depends upon its moisture content and the amount of coagulable protein in curd. Kadam et al. (1984) found that the yield of *chakka* from cow and buffalo milk was 22.60% and 29.81%, respectively. Patel and Chakraborty (1988) made *chakka* from buffalo skim milk, with a minimum of 23.5% total solids, 14.1% total proteins, 0.35% soluble proteins, 3.1%–3.2% reducing sugar, and 2.2% titrable acidity as lactic acid (LA) When different milks, such as buffalo, cow, and goat, were used in *shrikhand* making, the yield of *chakka* was the highest with the buffalo milk (26.2%) and lowest with the goat milk (24.0%). The total solids recovery of *chakka* was 54.7% for buffalo milk, 43.8% for goat milk, and 39.9% for cowmilk (Subramanian et al., 1997).

8.4.2.7.3 *Mechanized Preparation*   *Shrikhand* preparation involves (Kilara and Chandan, 2013):

- The formation of curd by culturing boiled and cooled milk with *dahi* from a previous batch or using a lactic starter culture

- The preparation of *chakka* (the base material for *shrikhand*) by tying the curd/*dahi* in a cloth bag and hanging it for about 6–8 h, or using mechanical means such as a basket centrifuge or a continuous curd separator for draining most of the whey from *dahi*
- The addition of sugar and/or cream (80% fat) in the case of skim milk *chakka*
- Mixing *chakka* with approximately an equal amount of sugar, and then, kneading manually, or in a planetary mixer to obtain a homogeneous consistency
- The addition of flavors, fruits, colors, etc., as desired
- Post-production heat treatment (optional)
- Packaging in a suitable container, followed by cooling and storing the product under refrigeration

The method is simple and widely practiced. The different types of *shrikhand* available include plain *shrikhand* and fruit *shrikhand*, but it is perhaps the quantity and quality of these blending materials, the method of blending, and so on that will ultimately determine the compositional, microbiological, and sensory characteristics of the resultant product.

Chakka/maska is made by separating the whey from the *dahi*. The removal of whey from curd to obtain *chakka* (Figure 8.20) is very important step in *shrikhand* production, as it influences the body and texture characteristics of the finished product. Traditionally, the removal of whey from curd is done by hanging/tying curd in cloth bags for 6–8 h or more until the draining has apparently ceased (Puntambekar, 1968; Ingle and Joglekar, 1974; Gandhi and Jain, 1977; Upadhyay and Dave, 1977; Kadam et al., 1984; Reddy et al., 1984; Rao et al., 1986). In the industrial method, a 28-inch diameter basket centrifuge at 1100 rpm used to be employed that produced 80 kg of

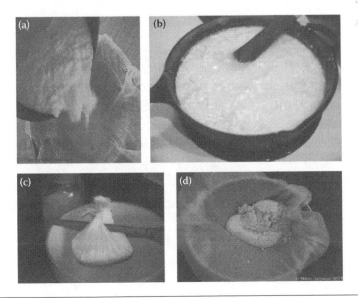

Figure 8.20 *Chakka*, the base material for *shrikhand*. (a) Curd; (b) Separation of whey; (c) Complete separation of whey; and (d) *Chakka*.

curd per hour. Now, the process has been further upgraded by using a *quarg* separator with a capacity of 2500 kg of curd per hour.

The processing conditions for the manufacture of *shrikhand* have been standardized. It is prepared by adding sugar at the rate of 80% of the amount of *chakka*, and mixed in a planetary mixer. The required amount of plastic cream (80% fat) is added along with the sugar to the *chakka* so as to give at least 8.5% fat in the finished product on a dry matter basis (Chandan, 2006).

Upadhyay and Dave (1977) suggested that the best quality *shrikhand*, with a homogeneous consistency, can be achieved through the addition of an equal quantity of sugar along with plastic cream (70%–80% fat) or unsalted white butter (a calculated amount) to skim milk *chakka*, and then, kneading the mixture on a stainless steel wire mesh. The end product can then, be blended with the desired flavor, color, fruits, and nuts. Aneja et al. (1977) employed a planetary mixer for mixing sugar, cream (80% fat), and cardamom with the *chakka*. The fat content of the *shrikhand* was adjusted to 5% through the addition of an 80% fat pasteurized cream. Upadhyay (1981) mixed *chakka*, sugar (80% by weight of the *chakka*), and plastic cream (calculated on 5% fat in final product), and kneaded it over a clean, dry, and sanitized stainless steel wire mesh to obtain a smooth consistency of *shrikhand*.

Miyani (1982) kneaded *chakka* over a cleaned and sanitized stainless steel wire mesh to obtain uniformity, and then, mixed it with crystalline sugar and pasteurized cream (70% fat). The admixture was then, mechanically mixed through the application of 60 (to-and-fro) strokes with the help of a stainless steel ladle. Desai (1983) prepared *shrikhand* by kneading *chakka* over a cleaned and sanitized stainless steel wire mesh screen (30 meshes) to obtain uniformity, and then mixed it with sugar (80% by weight of chakka), and allowed it to stand for about 30 min to ensure the proper dissolution of the sugar. The mixture was then, manually stirred using a stainless steel ladle, giving it about 60 (to-and-fro) strokes. Patel and Chakraborty (1985a,b,c) prepared *shrikhand* by mixing calculated quantities of *chakka*, sugar, and cream (77%–80% fat) in a planetary mixer at 30–35 rpm for 30 min, to get 41% sugar and 6% fat in the finished product. In general, the use of a planetary mixer has been recommended for the industrial method of *shrikhand* manufacture, for mixing and blending the *chakka* and other ingredients (Aneja et al., 1977; Patel and Abd-El-Salam, 1986).

Response surface analysis of data revealed that the *shrikhand* containing fat and sugar at the levels of 2%–4% and 30%–35%, respectively, had lower acceptability, whereas that containing 7%–9% fat and 33%–39% sugar had higher acceptability. The most desirable combination of fat and sugar levels in *shrikhand* that give maximum acceptability were found to be 8% fat and 36% sugar (Nalawade et al., 1998). Among the different levels of sugars used, *shrikhand* (made by separated buffalo milk) using *Streptococcus thermophillus* and *Lactobacillus delbruckii* sub. *bulgaricus* prepared using raftilose (4%) and sugar (12.5%) was rated as the most acceptable by sensory panels (Singh and Jha, 2005). *Shrikhand* containing ~6% fat, 35%–40% moisture, and 40% sugar was found to be highly preferable with respect to the sensory profile and consistency of the product. Regarding the latter, smoother product preparation needs a chosen rate of acid development in the curd, or else the product becomes grainy.

The introduction of semi-mechanized lines for the manufacture of *shrikhand* on an industrial scale has brought about a revolution in its manufacture that is being employed successfully and now covers a large market share. Nonetheless, it has been observed that traditionally made *shrikhand* has superior sensory attributes than those of mechanically made *shrikhand*.

The quality of the *shrikhand* of various treatments was evaluated at refrigeration temperature at 0-, 10-, and 20-day storage periods. Good-quality *shrikhand* can be prepared by fortifying it with 10%–20% mango pulp, and this can be stored acceptably for up to 10 days without any preservatives at refrigeration temperature. Apple was also found to be a promising fruit that can be used for fortification at a 10% level, but pineapple and *sapota* pulp were not suitable.

8.4.2.7.4 Shelf-Life and Post-Production Heat Treatment   In general, the shelflife of any cultured milk product is determined by the initial milk's quality, the heat treatment given to the milk, the starter culture used, the conditions of incubation and handling during the manufacture and until packaging, and the storage conditions of the product. Plastic cups with lids are the main packaging materials used commercially (Chandan and Shahani, 2001). *Shrikhand* has longer shelf-life than other cultured milk products due to its low moisture and higher sugar content (Patel and Chakraborty, 1988). The preservation quality of *shrikhand* ranges between 12 and 14 days under refrigeration (Gandhi and Jain, 1977), after which mold growth and increases in acidity take place. *Shrikhand* stored at $10 \pm 3°C$ was found to develop "off" flavors and an unpleasant odor in about 40 days, whereas *shrikhand* stored at 37°C got spoiled within a period of one week (Sharma and Zariwala, 1980). Upadhyay (1981) reported the storage life of *shrikhand* to be 50 days at $7 \pm 2°C$, while Patel (1982) found the preservation quality of *shrikhand* at 10°C to be about 42 days, and 2–3 days only at 30°C, whereas it was 30 days at ~10°C according to Desai (1983). Patel and Abd-El-Salam (1986) and Pandya et al. (2006) reported the shelf-life of *shrikhand* to be about 35–40 days at 8°C, and 2–3 days at 30°C. Moreover, the shelflife of this product can be improved by 20% by heating it at 70°C for 2 min or through the addition of 0.05% potassium sorbate. For chemical and microbiological quality of market *shrikhand* see the literature cited (Upadhyay et al., 1975a,b).

Strawberry *shrikhand* prepared with 15% pulp and 40% sugar levels was found to be the best of all the treatments, including control for sensory quality, and was classified as "liked very much." The addition of strawberry pulp significantly improved the flavor of *shrikhand* without adversely affecting other important sensory attributes. Strawberry pulp, when incorporated in the *shrikhand* (up to the extent of 15% of *chakka*), improved its flavor characteristics with only a slight (7%) rise in the cost of production.

The use of post-production heat treatment to prolong the shelf-life of cultured milks and their products is based on its destructive effect on microorganisms and enzymes (Rao et al., 2006; Tamime and Robinson, 2007). Such a heat-treated product is more stable than untreated against biochemical and organoleptic changes during storage, besides being the most economical method for extending shelf-life. *Shrikhand* was given post-production heat treatments—55°C/30 min, 60°C/20 min, 65°C/10 min, 70°C/5 min, and 75°C/2 min—and the product was evaluated for its quality and shelf-life in storage at ambient as well as refrigerated temperatures. The post-production heat treatment of *shrikhand* had no adverse effects on its chemical composition, titratable acidity, pH, precipitated protein, syneresis, or sensory characteristics, but had a marked destructive effect on the groups of microorganisms (acid producers, proteolytic, lipolytic, yeasts, and fungi) studied, depending on the severity of the heat treatment and the type of microorganisms involved (Prajapati et al., 1991). The unthermized product had a shelf-life of 5 days at ambient temperature (35–37°C), while thermized *shrikhand* (in the range of 65–75°C) remained acceptable upto 15 days at an ambient temperature of 35–37°C (Mital and Garg, 1992; Prajapati et al., 1992). At refrigerated temperatures (8–10°C), unthermized *shrikhand* samples became unacceptable after 45 days and thermized (55°C/30 min) on the 70th day of storage, whereas thermized (60°C/20 min, 65°C/10 min, 70°C/5 min, and 75°C/2 min) versions had a storage life of more than 70 days (Prajapati et al., 1993).

8.4.2.7.5 Prebiotic- and Probiotic-Enriched Shrikhand   Several attempts have been made to incorporate different additives into *shrikhand* to diversify the food product and attract a wider range of consumers. The incorporation of probiotic organisms (Geetha et al., 2003), the pulp of fruits such as apple, mango, papaya, banana, guava, and *sapota* (Bardale et al., 1986; Dadarwal et al., 2005), ashwagandha (Landge et al., 2010) and of cocoa powder with and without papaya pulp (Vagdalkar et al., 2002) to *shrikhand* have all been researched.

Shrikhand, as a fermented milk-based dessert enriched with fructooligosaccharides (FOS), a low-calorie prebiotic, and probiotic *Enterococcus faecium* CFR 3002 has successfully been made (Patel and Chakraborty, 1982, 1985a,b,c,d; Tamime and Robinson, 1988). No colors and flavors were added to the prepared product. The process for the preparation of *shrikhand* enriched with prebiotic FOS and probiotic *Enterococcus faecium* CFR 3002 is shown in the Figure 8.21. No significant differences were observed in the overall sensory-quality scores of the *shrikhand* enriched with FOS and blended, but the scores for the sucrose-sweetened *shrikhand* were slightly higher those achieved for the FOS-enriched version (Vijayendra and Gupta, 2011). The viability of the probiotics was >140 × 10^6 CFU/100 g during 60 days of storage. It had a sweetish–sour taste, typical of *shrikhand*.

It has been shown that the incorporation of prebiotics can increase the mouthfeel attributes of the product while providing an adequate sweetness level (Guggisberg et al., 2009). *Shrikhand* enriched with FOS and blend have obtained fairy good score when compared with sucrose sweetened *shrikhand*. The *L* a* b** values (Figure 8.22) of *shrikhand* enriched with prebiotic FOS and a blend with sucrose along with probiotic

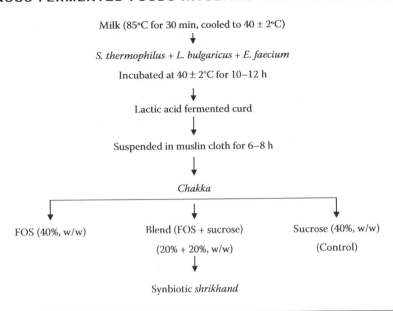

Milk (85°C for 30 min, cooled to 40 ± 2°C)
↓
S. thermophilus + *L. bulgaricus* + *E. faecium*

Incubated at 40 ± 2°C for 10–12 h
↓
Lactic acid fermented curd
↓
Suspended in muslin cloth for 6–8 h
↓
Chakka

FOS (40%, w/w)   Blend (FOS + sucrose)   Sucrose (40%, w/w)
(20% + 20%, w/w)   (Control)

Synbiotic *shrikhand*

Figure 8.21 Flow chart of the process for the preparation of prebiotic- and probiotic-enriched *shrikhand*.

E. faecium CFR 3002 were measured immediately after its preparation ("0" day), and after storage for 20, 40, and 60 days at refrigerated temperatures (4 ± 2°C; Renuka et al., 2010). As can be seen from Figure 8.22, the L^* and b^* values of *shrikhand* sweetened with FOS and a blend of FOS and sucrose decreased gradually, with an increase during the storage period (60 days). The replacement of sucrose either fully or partially with prebiotic FOS along with *E. faecium* CFR 3002 did not affect the color of the final product. Thus, the prebiotic *shrikhand* with FOS can drive preference and present additional health benefits, beyond traditional ingredients (Allgeyer et al., 2010). It is currently used in a wide range of foods (Renuka et al., 2010) and beverages (Renuka et al., 2009), as it doesn't interact with any commonly used food ingredients.

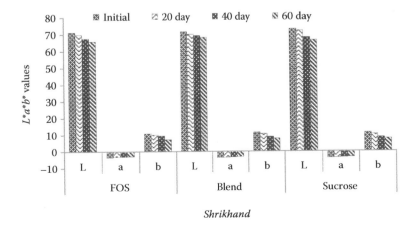

Figure 8.22 Color of *shrikhand* enriched with blended prebiotic FOS, sucrose, and probiotic *E. faecium* CFR 3002. (Copyright Parpulla and Renuka.)

Figure 8.23 Scanning electron micrograph of *shrikhand* enriched with (a) FOS; (b) sucrose; and (c) a blend of FOS–sucrose. (Copyright Parpulla and Renuka)

The structure of foods greatly affects the texture and appearance of an end product. In particular, their microstructure has a major impact on the texture and other physical properties of acid milk gels (Haque and Kayanush, 2002). The incorporation of FOS (Figure 8.23a) might have a different effect on the *shrikhand* gel-network formation than would sucrose (Figure 8.23b). Scanning electron microscopy observations revealed that, in the presence of either FOS or a blend of FOS and sucrose (Figure 8.23c), the *shrikhand* matrix was predominantly casein micelles arranged in double longitudinal polymers along with clusters in some spots. The differences in the microstructures, however, did not have any bearing on the palatability or the acceptability of the FOS-enriched *shrikhand*.

The survival of total probiotics in *shrikhand* enriched with FOS and in a blend of FOS and sucrose along with probiotic *E. faecium* CFR 3002 are given in Table 8.2 (Renuka et al., 2010). No change in the colony counts of total probiotics was observed for 20 days of storage at 4 ± 2°C. There was a slight decrease in the viable count of probiotics from 40 days and at the end of storage period (60 days), but the population of all the probiotics remained at ≥140 × 10^6 CFU/g. This is well above the levels suggested for providing therapeutic benefits (Boylston et al., 2004; Michael and Kailasapathy, 2006). All of the three types of *shrikhand* showed similar trends with only a slight

Table 8.2 Total Probiotics* of *Shrikhand* Enriched with Prebiotic FOS, a Blend of FOS and Sucrose, and Sucrose along with Probiotic *E. faecium* CFR 3002 Stored at Refrigerated Temperatures (4 ± 2°C)

STORAGE PERIOD (DAYS)	TOTAL PROBIOTICS (CFU/100 G)		
	FOS	BLEND	SUCROSE
0	>300 × 10^6	>300 × 10^6	>300 × 10^6
20	>300 × 10^6	>300 × 10^6	>300 × 10^6
40	281 × 10^6	274 × 10^6	256 × 10^6
60	172 × 10^6	148 × 10^6	141 × 10^6

Source: Adapted from Renuka, B. et al. 2010. *J Texture Studies* 41(4): 594–610.
* Total count of probiotic microorganisms *E. faecium, S. thermophilus, L. bulgaricus.*
CFU = colony forming units.

decrease in viable probiotics, confirming that the probiotic cultures remained viable in the product until the end of storage (60 days), which is satisfactory for *shrikhand* to be claimed as "probiotic *shrikhand*." Relatedly, Desai et al. (2004) and Ozer et al. (2005) have reported an improvement in the survival of probiotics in yogurt in the presence of inulin (Ramchandran and Shah, 2010).

The foregoing results showed that *shrikhand* enriched with prebiotic FOS and probiotic *E. faecium* did not present any significant change with respect to their sensory attributes, such as color, taste, texture, flavor, mouthfeel, and overall acceptability. Nor any significant differences were observed between the overall quality scores of the *shrikhand* enriched with FOS or the blend and sucrose, though the scores of sucrose-sweetened *shrikhand* were slightly higher (Basappa et al., 1997).

Shrikhand has a long shelf-life compared with other cultured milk products, and has been reported to be stable for >30 days (Sarkar et al., 1996b). The analysis of freshly prepared *shrikhand* showed no growth (plate count, yeast/molds and coliform) at "0" day. However, the number of CFU increased (20–25×10^2 CFU/100 g) at the end of storage period (60 days). The presence of a low number of CFU in the *shrikhand* prepared with replacement of FOS is associated with the rather rare ability of microorganisms to metabolize FOS, compared with the microbial utilization of sucrose (Winkelhausen and Kuzmanova, 1998). In view of these observations, the symbiotic *shrikhand* enriched with FOS along with probiotics are not only microbiologically safe but their shelf-life could be much longer, too.

8.4.2.8 Kadhi Kadhi is a popular culinary food item prepared from *dahi* in several states of India, including Gujarat. No standardized method for the manufacture of *kadhi* is available, since its composition varies from region to region and depends on consumer preferences. It exhibits a mildly acidic character and a specific cooked flavor. The milk solids content in *kadhi* varies from 6% to 8%, while other solids have been reported in the range of 6%–7% (Aneja et al., 2002).

It is prepared by adding 5%–8% Bengal gram flour (*besan*) to *chhash* made from *dahi* by adding equal amount of water. After mixing thoroughly, the combination is heated to boiling using a slow flame with continuous stirring. At this stage, salt, chopped ginger, chopped green chillies, turmeric, and cinnamon powder are added, and boiling is continued for 10–15 min on a slow flame. The product can be seasoned by heating groundnut oil in a pan; cumin seeds, mustard seeds, a few curry leaves, and asafoetida are added until they splutter, and the seasoning is then, poured over the *kadhi*. It is generally light yellowish brown in color with a pourable, thick consistency like soup, and is eaten hot on its own or with rice, *pulav*, or *chapatties*.

8.4.2.8.1 Pakore ki Kadhi To the *kadhi*, fried balls prepared from a batter of gram flour are added, and the resultant foodstuff is called *pakore ki kadhi* (Rai et al., 2012).

8.4.2.9 Butter Butter (maar) is one of the important fermented dairy products (Rangappa and Achaya, 1974). It is used in small quantities as table butter, while the

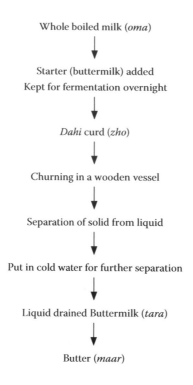

Whole boiled milk (*oma*)

↓

Starter (buttermilk) added
Kept for fermentation overnight

↓

Dahi curd (*zho*)

↓

Churning in a wooden vessel

↓

Separation of solid from liquid

↓

Put in cold water for further separation

↓

Liquid drained Buttermilk (*tara*)

↓

Butter (*maar*)

Figure 8.24   Traditional method of butter (*maar*) preparation.

major portion is used in the preparation of *ghee* and clarified butterfat. In its preparation (Figure 8.24) from fresh milk, first, curd is made, which is churned with a wooden or metal churner, during which process the butter rises to the top and is removed. Indian farmers' knowledge regarding this process is traditional and handed down from ancient times like the Vedas (Soni et al., 2013). Today, following the establishment of the cooperative movement in Gujarat (India), butter is prepared commercially on many dairy farms in large quantities.

The most predominant microorganisms present in butter are *lactobacilli*, followed by *Streptococci*, coliforms, and aerobic spore formers. Yeasts and molds are also present in considerable numbers and contribute to the product's acidity and colors, as well as its off flavor (Soni et al., 2013). *S.lactis* ssp. *diaacetylactis* produces the flavor compounds diacetyl, acetaldehyde, acetoin, and carbon dioxide (Gandhi, 1989).

8.4.2.9.1 Maar (*butter*)   Butter in known as "*maar*" in Ladakh (India). After churning the curd, the butter is separated from buttermilk by filtering through a cotton or muslin cloth. The solid portion is dipped into cold water to further separate the liquid portion from the butter. Butter thus, prepared is considered very pure and preferred for all religious purposes. Butter in the Zanskar region of Ladakh (India) is packed tightly in a bag stitched of goatskin, and exported throughout Ladakh (Figure 8.25). Molten butter is known as "*maarkhu*" and is used to light up lamps in monasteries, used as cooking oil, and for making *kholak* (Figure 8.26).

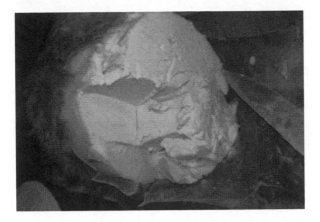

Figure 8.25   Butter packed in goatskin brought from Zanskar for sale in Leh market in India.

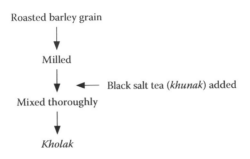

Figure 8.26   Traditional method of *kholak* preparation.

8.4.2.10 Ghee   *Ghee* is the Indian name for clarified butterfat and is mainly made from cow, buffalo, sheep, or goat milk. There is no need for cold storage for this product as it has a good shelf-life at room temperature. It forms an essential part of ceremonial religious offerings among the Hindu populations.

It is made by heating butter or churned curd at 100–140 rpm in a churner (Figure 8.12) for 5–10 min, which removes the water through evaporation. It is then clarified using a muslin cloth and cooled to solidify. *Ghee* prepared from buffalo milk has a white color, but, due to the presence of carotenoids, *ghee* prepared from cowmilk is yellow. All the microorganisms present in curd and butter are killed during heating process, so the *ghee* is free of microorganisms (Campbell-Platt, 1987; Aneja et al., 2002) and, hence, stable.

8.4.3 Fermented Milk Products of Sikkim

Consumption of milk and milk products is a part of the dietary culture of ethnic people living in the Sikkim Himalayas. Cow's milk is popular, but, in the high altitudes (>2100 m), mostly in north Sikkim, yak rearing is a more common practice for milk as well as meat, skin, and hair. The production of ethnic fermented milk products is

Table 8.3   Traditional Fermented Foods and Beverages of the Sikkim Himalayas

PRODUCT	SUBSTRATE	FERMENTATION TIME	TEXTURE AND USE
Chhurpi (soft)	Cow/Yak milk	1 day	Soft mass, cheese-like; curry/pickle
Chhurpi (hard)	Cow/Yak milk	2–3 days	Hard mass; masticator
Chhu/Sheden	Cow/Yak milk	~7 days	Soft mass, strong-flavored; curry
Philu	Cow/Yak milk	5–7 days	Cream; fried curry with butter
Somar	Cow/Yak milk	~1 month	Soft paste, strong flavor; condiment
Dahi	Cowmilk	10–12 h	Curd; savory
Gheu	Cow/Yak milk		Butter
Mohi	Cowmilk	10–12 h	Buttermilk

Source: Adapted from Tamang, J.P. et al. 2007. *Journal of Hill Research* 20(1): 1–37.

mainly confined to unorganized or informal sectors as well as to the individual house-hold level in Sikkim (Table 8.3; Anonymous, volume viii—Biological Environment).

8.4.3.1 Chhu   *Chhu* or *sheden* is an indigenous, cheese-like, fermented yak- or cow-milk product consumed mostly by the Bhutias and Lepchas in Sikkim. *Shyow* (curd) is prepared from boiled or unboiled milk of yak or cow. It is churned in a bamboo or wooden vessel, with the addition of warm or cold water to produce *maa* and *kachhu*. *Kachhu* is cooked for 15 min until a soft, whitish mass is formed. This mass is sieved out and put inside a muslin cloth, which is hung by a string to drain out the remaining whey. The product is called *chhu*. *Chhu* is placed in a closed vessel and kept for several days to months to ferment the product further, after which it is consumed (Anonymous, volume viii—Biological Environment).

The microorganisms used in fermentation are *Lactobacillus farciminis*, *Lactobacillus brevis*, *Lactobacillus alimentarius*, and *Lactococcus lactis* ssp. *cremoris*. Initially, it has a rubbery texture with slightly sour taste when fresh; after further fermentation, it becomes creamish- to pale yellow-colored and develops a strong flavor. It is made into a curry by cooking it in *maa* along with onions, tomato, and chillies, and is consumed with boiled rice. Soup prepared from strong-flavored *chhu* is also consumed by the Bhutia. It has a sour taste with a strong aroma, and is enjoyed as an appetizer (Anonymous, volume viii—Biological Environment).

8.4.3.2 Philuk   *Philuk* is a typical, indigenous, fermented, butter-like milk product made from cow or yak milk, with an inconsistent semi-solid texture. For its preparation, fresh milk, collected in cylindrical bamboo vessels (locally called "*dzydung*" by the Bhutia) or in wooden vessels (called "*yadung*"), is slowly swirled around the walls of these vessels by rotating them for a few minutes. Sometimes, a thick mesh of dried creeper is kept inside the vessel to increase the surface area onto which the *philu* can adhere. Thus, a creamy mass sticks to the walls of the vessel and around the creeper. The milk is then, poured off and utilized elsewhere. The vessel is kept in an upside-down position to drain out the remaining liquid (Anonymous, volume viii—Biological Environment). This process is repeated daily for about

6–7 days until a thick, white, cream layer is formed on the vessel walls and the creeper surface. This soft mass, as *philuk*, is scraped off and stored in a dry place for later consumption.

Microorganisms conducting the fermentation include: *Lactobacillus casei* ssp. *casei, Lactobacillus bifermentans* and *Enterococcus faecium* (Anonymous, volume viii—Biological Environment.

Philu obtained from yak milk has a creamy white color with an inconsistent semi-solid texture. It is commonly eaten by the Bhutias and Sherpa of Sikkim, in the northeast of the region. The Sherpa call it *"philuk." Philu* is consumed as a side dish with rice and is cooked with butter and a little salt.

Philuk is a high-priced traditional milk product sold in local markets in the Sikkim Himalayas, where the rural people are dependent on it for their livelihood. In north Sikkim, it too is produced mostly from yak milk and is consumed at the household level by the Bhutias.

8.4.3.3 Chhurpi   Fermented milk is the key ingredient of this product also. In its traditional preparation, *dahi* (curd) is churned in a bamboo or wooden vessel, with the addition of warm or cold water, to produce *gheu* and *mohi*. *Mohi* (buttermilk) is cooked for about 15 min until a soft, whitish mass is formed. This mass is sieved out and put inside a muslin cloth, which is hung by a string to drain out the remaining whey (Anonymous, volume viii—Biological Environment).

In its fermentation, *Lactobacillus plantarum, Lactobacillus curvatus, Lactobacillus fermentum, Lactobacillus paracasei* ssp. *pseudoplantarum, Lactobacillus alimentarius, Lactobacillus kefir, Lactobacillus hilgardii, Enterococcus faecium* and *Leuconostoc mesenteroides* are involved (Tamang et al., 2000; Dewan, 2002; Anonymous, volume viii—Biological Environment).

A soft variety of *chhurpi* is another cheese-like fermented milk product (Figure 8.27). It has a rubbery texture with a slightly sour taste and an excellent aroma when it is

Figure 8.27   *Chhurpi* curry. (Courtesy of Carrying capacity study of Teesta basin in Sikkim. *Edible Wild Plants and Ethnic Fermented Foods*, Vol. VIII *Biological Environment-Food Resources*.)

fresh. Soft *chhurpi* is used to prepare various dishes, and its popularity is increasing as it provides a different taste to dishes. It is consumed as an excellent source of protein and as a substitute for vegetables. It is also made into a curry by cooking it in oil along with onions, tomato, and chillies. The curry is also prepared with edible ferns, locally called "*sauney ningro*" (*Diplaziumpolypodiodes*) and "*kali ningro*" (*Diplazium* sp.), and eaten with rice. In addition, it can be used to prepare "*achar,*" or pickle, by mixing it with chopped cucumber, radish, chillies, etc. (Tamang et al., 1988). Soup prepared from soft *chhurpi* can be consumed as a substitute for *dal* along with rice. *Chhurpi* is sold in all local markets by rural women, who pack it in the leaves of the fig plant and then tie it loosely with straw. One kilogram of *chhurpi* costs about Rs. 60/- or more (Anonymous, volume viii—Biological Environment).

8.5 Indigenous Fermented Foods from Bamboo

8.5.1 Bamboo Shoot

Bamboo is a grass (subfamily *Bambusoideae*, family *Poaceae*) that grows in the tropical and subtropical climates of Asia (Lu et al., 2005). Major species include *Dendrocalamus strictus* in India. In several states of northeast including Assam, fermented bamboo shoot is an important part of the traditional cuisine, and some of the edible species that are suitable for processing and available in Assam include *Dendrocalemus giganteus* (*Worra*), *D. hamiltonii* (*Kako*), *D. strictus* (*Lathi bans*), *Melocanna bambusoides* (*Tarai, Arten*), *Bambusa balcooa* (*Bhaluka*), *B. tulda* (*Jati*), *B. polymorpha* (*Jama betwa, Betwa*), *B. nutans* (*Kotoha*), and *B. pallida* (*Bijuli, Bakhal*). Among these, *B. balcooa*, *B. tulda*, and *D. hamiltonii* are mostly used for preparing fermented bamboo shoots especially in Assam (Kar and Borthakur, 2007).

The shoots or sprouts are the edible parts of such bamboo species, and are consumed either raw or processed because of their exotic taste, flavour, and medicinal value. Raw bamboo shoots are crisp, tender, but bitter tasting and so cannot be eaten directly and are hard to digest. For consumption, the shoots are sold in various processed shapes, are available in fresh, dried, fermented, pickled, and canned forms (Nimachow et al., 2010; Anonymous-Bambooshoots), and can be preserved for two years, as long as they are not damaged after harvesting. Due to a high enzymatic action in bamboo shoots, they are only harvested during the months of May, June, and July, and the local people have access to the shoots only for this period, so food processing and preservation methods are necessary to ensure that these products are available throughout the year. Processing is different for each type of bamboo shoot, and traditional fermentation techniques have proved to be suitable.

The preservation of bamboo shoots through fermentation refers, as elsewhere, to extending the storage life (up to a year or more) and enhancing the safety of these foods using natural microflora and their antibacterial compounds. The people of Assam (India) have been using lactic acid fermentation of bamboo shoots to enhance their shelf-life without the aid of modern methods of processing, like refrigeration,

Figure 8.28 Bamboo shoot fermentation. (a) Young bamboo shoots; (b) peeled bamboo shoots; (c) sliced bamboo shoots; (d) slices filling earthen pots; (e) and (f) fermentation in earthen pots.

canning, and vacuum packaging. The fermented shoots are used in local cuisines, as medicine, and in pickle making. These techniques of bamboo shoot fermentation have been perfected over hundreds of years based on trial and error, while scant scientific research has been published regarding the biochemistry or microbiology of the processes. Traditionally, bamboo shoot fermentation (Figure 8.28) is done in earthen pots, and the process of this fermentation. The resulting products, include *khoria*, *poka khorisa*, *khorisa pani*, *kahudi*, and *miyamikhri*.

8.5.2 Khorisa

Khorisa is an ethnic, fermented, bamboo shoot product of Assam (India), and is made during the monsoons, when the bamboo shoots are available. To prepare *khorisa*, the shoots are harvested and then, the outer surface is peeled off and the white inner part is used. These are washed and hammermilled in a traditional, wooden, husking pedal called a *dheki* that breaks down the bamboo shoots into a mash of pulp. The pulp is then, packed inside earthen pots, which are smoked prior to packing. In some regions, small dried pieces of *Garcinia pendaculata* Roxb., locally known as *bor-thekera*, are also mixed in along with the bamboo shoot pulp, as an acidifier (Jeyaram et al., 2009). Additionally, dried chillies are placed inside the earthen pots, plus a small amount of water. All the ingredients are mixed and mildly pressed into the earthen pots, and then, the mouth is tied with banana leaves (Figure 8.29). The entire system is made factitively anaerobic. The mixture is then allowed to ferment naturally for a period of 4–12 days, depending upon shoot species, region, and locality. A mild acidic taste

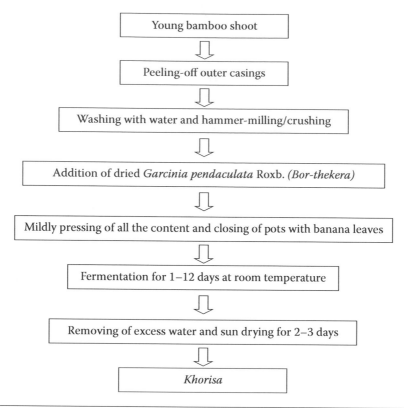

Figure 8.29 Method of preparation of *khorisa*. (Adapted from Tamang, B., and Tamang, J.P. 2009. *Indian J Trad Know* 8(1): 89–95.)

and sour smell indicates the completion of fermentation, and the entire pulp is removed from the pots. The excess water is removed by pressing and the fermentation material is sun-dried for 2–3 days. When the moisture content has reduced substantially and the product becomes crispy, it is stored in jars for further use. This method of *khorisa* preparation is somewhat similar to that for *soidon*, a fermented bamboo shoot product of Manipur *Lactobacillus* species are mainly responsible for the fermentation of bamboo shoots (Tamang and Tamang, 2009), as described here.

Khorisa is used in traditional meals including fish, meat, and sweets, and dishes cooked with it are popular appetizers among the indigenous population (Figure 8.30a). Scientific studies of this traditional foodstuff are not known, but it is observed to be popularly consumed by the tribal people as an important component of their diet. It has high nutritional potential as well as medicinal value, due to presence of a useful amount of phenolic compounds with high antioxidant properties.

8.5.3 Poka Khorisa

Poka khorisa/Khorisa tenga is also an ethnic, fermented, bamboo-shoot foodstuff of Assam (India). It is whitish in color with a faint aroma and sour taste. However, it is not dry and crispy like *khorisa*, and has more moisture. The smell and taste of *poka*

Figure 8.30 Fermented bamboo shoot products of Assam. (a) *Khorisa*; (b) *poka khorisa*; (c) *khorisa pani*; (d) *kahudi*; (e) *miyamikhri*.

khorisa is particularly appealing for the indigenous population of Assam. In its preparation, locally grown, young, edible bamboo shoots of *Bhaluka baah* (*Bambusa balcooa*) and *Kako baah* (*Dendrocalamus hamiltonii*) are defoliated, hammermilled, mixed with dried *Garcinia pedunculata* Roxb. (*Bor-thekera*) fruit and dried chillies, and then, packed inside pre-smoked earthen pots and pressed mildly, as with *khorisa*. Next, the mouth of the earthen pot is tied with banana leaves, and the pot is left to ferment anaerobically for 4–12 days. Completion of fermentation is indicated by the typical *poka khorisa* smell. The pulp is taken out and the excess water is removed by pressing, and then the solid fermented product is stored in jars. It is used in various dishes, pickle making, and as a medicine. It is also mixed with edible oils, chillies, and salt (pickled) and can be kept in closed containers for up to two years. The non-pickled fermented *poka khorisa* can be kept in closed jars for more than a year. Like *khorisa*, a group of *Lactobacillus* species are mainly responsible for the *poka khorisa* fermentation stage (Figure 8.30b).

8.5.4 Khorisa Pani

Khorisa pani is another ethnic fermented product of Assam, but it is not solid like *khorisa* or *poka khorisa*. Instead, it is a liquid, and has a sour acidic taste, similar to that of *poka khorisa* (Figure 8.30c). When bamboo shoots are fermented in earthen pots, this sour liquid is produced. When the fermentation process is complete, the liquid is collected in bottles and is later used in making curry, meat, sour fish curry, etc. It has a shelflife of not more than 7 days, so must be consumed within a week and any remaining liquid should be discarded. The product is associated with medicinal properties, and is administered to children suffering from measles or chickenpox as it is believed to help hasten the de-pigmentation of pox marks.

8.5.5 Kahudi

Kahudi is one of the traditional, fermented, bamboo-shoot products mainly consumed by the people of the river island of Majuli in Assam (India). Mustard seeds are kept submerged in *khorisa pani*, the sour liquid that is collected after the fermentation of *poka khorisa*. These seeds are kept submerged for 3–4 days in the liquid, and are then, taken out and sun-dried for a day, and then, mixed with *khorisa pani* again and blended into a paste (Figure 8.30d). The paste is then, transferred to a vessel, and can be stored and consumed for up to 6 months.

8.5.6 Miyamikhri

Miyamikhri is a traditional, fermented, bamboo-shoot product mainly consumed by the tribes of the North Cachar Hills district of Assam (India). Young edible bamboo shoots are defoliated and rendered into small pieces, which are wrapped in banana leaves and then, put into an earthen pot to ferment for about 4–5 days (Figure 8.30e). When the typical *miyamikhri* smells is noted, the foodstuff is shifted to a glass vessel. The local people use it for up to a year as a pickle, or mix it with curry (Chakrabarty et al., 2009).

8.5.7 Mesu

For the indigenous peoples of the eastern Himalayan regions, including the Darjeeling Hills, *mesu* is a common food (Sekar and Mariappan, 2007). Young edible shoots of *Choya bans* (*Dendrocalamus hamiltonii*), *karati bans* (*Bambusa tulda*), and *bhalu bans* (*Dendrocalamus sikkimensis*) are finely chopped and traditionally, put into a bamboo vessel (hollow bamboo) tightly packed in an airtight environment (Tamang et al., 1988; Tamang et al., 2012; Figure 8.31). The top of the vessel is covered tightly with the leaves of bamboo or other wild plants and left to ferment under natural anaerobic conditions for 7–15 days at an ambient temperature (20–25°C). The fermentation of *mesu* is initiated by *Pediococcus pentosaceus*, followed by *Lactobacillus brevis*, and, finally, dominated by *L. plantrum*. Studies have recorded the microorganisms involved to include *Lactobacillus curvatus*, *Lactobacillus plantarum*, *Lactobacillus brevis*, *Leuconostoc citreum* and *Pediococcus pentosaceus* (Tamang et al., 2008; Tamang, 2009). The pH declines from 6.4 to 3.8 due to increase in titratable acidity from 0.04% to 0.95% (Sekar and Mariappan, 2007). Completion of fermentation is indicated by the typical *mesu* flavor and taste (Tamang and Tamang, 2009).

Young bamboo shoot

↓

Peel-off outer inedible casings

↓

Chop the inner part; wash with water, cut into pieces

↓

Press tightly into a fresh bamboo-vessel, made air tight

↓

Keep the container in an upside down position

↓

Ferment for 7–12 days

↓

Mesu

Figure 8.31 Traditional method for preparation of *mesu*. (Adapted from Tamang, J.P. 2009. *Himalayan Fermented Foods: Microbiology, Nutrition and Ethnic Value*, CRC Press, Taylor & Francis Group, New York.)

Mesu is prepared in the months of June to September. A very common pickle is produced from it, and it is also used as a base for curry. For the pickle, it is mixed with edible oil, chillies, and salt and is kept in a closed jar for several months without refrigeration. *Mesu* is also stored in a green bamboo vessel, loosely capped by fig plant leaves, and tied by straw, in which form it is commonly sold during the rainy season in the local markets of Darjeeling Hills and Sikkim by Limboo women (Tamang, 2009).

Mesu products are low in fat and cholesterol, but have potassium, carbohydrates, and dietary fibers. Many nutritious and active materials (vitamins and amino acids) and antioxidants (flavones, phenols, and steroids) have been systematically analysed, compared, and reported (Nirmala et al., 2008; Choudhury et al., 2012).

8.5.8 Fermented Bamboo Shoots

Soaking, as an approach to food fermentation, can be employed to prepare both sour and salted pickles. There are two methods. In the first, soaking in a 5%–8% saline solution is completed, followed by fermentation for 3–5 days or until the taste of lactic acid becomes evident. The second method includes soaking in vinegar or mixing with sugar, salt, and spices for a good taste and flavor. The sour pickled bamboo shoot version can also be made by soaking the food in an acid solution. Also, the quantity of salt can be decided upon according to a ratio of food to salt. In another variation, the bamboo shoot is soaked in a concentrate of saline solution of about 20%–25%, and this process can preserve salted pickle for longer than the sour pickle.

8.5.8.1 Soibum

Fermented bamboo-shoot products are a traditional food called "*soibon*" in the Manipur state of India (Jeyaram et al., 2010). This is a non-salted, acidic, fermented bamboo-shoot product which made using a traditional liquid starter called "*soidon mahi*," and prepared by mixing the acidic juice extract of about 1–1.5 kg of *Garcinia pedunculata* Roxb. fruits with 10–15 L of rice (*Oryza sativa* Lin.) are washed with water, and then mixed with the succulent bamboo shoot tips of *Schizostachyum capitatum* Munro for fermentation (Rao et al., 2006). The fermented juice, as a starter, can be kept for a year and used by dilution with water in a ratio of 1:1 (Jeyaram et al., 2010).

The succulent bamboo shoots of *B. balcooa*, *D. strictus*, and *Melocana baccifera* are used in this dish. The outer casing of the tender tips of the shoots are removed and cut, as with other, similar product preparation (only the inner white portion is used for fermentation). These are immersed into the traditional starter dilutant and kept for fermentation (Jeyaram et al., 2009), which is carried out in an earthen pot for 20 days (Tamang et al., 2008). Predominant LAB associated with fermented bamboo-shoot products in India have been identified as *Lactobacillus brevis*, *Lactobacillus plantarum*, *Lactobacillus curvatus*, *Pediococcus pentosaceus*, *Leuconostoc mesenteroides* ssp. *mesenteroides*, *Leuc. fallax*, *Leuc. lactis*, *Leuc. cireum*, and *Enterococcus durans* (Tamang et al., 2008).

Soibum is eaten as a pickle and as a curry mixed with fermented fish. There are many ways for the preparation of fermented bamboo shoot that are employed in Manipur (Devi and Kumar, 2012), including those detailed here.

8.5.8.1.1 Inside a Pit

In this method, a pit is first dug and then, a basket is made using bamboo and in the shape of the pit into which it is to be placed. Care should be taken to slightly incline the bamboo basket while placing so as to allow the flow of the water produced by the bamboo shoots during fermentation. The water collected from this process can also be preserved, and is used again in subsequent, new fermentations of bamboo shoots.

Next, wild *colocasia* leaves are put in and around the pit in a thick layer of about 2–3 inches. In fact, today, instead of wild *colocasia* leaves, plastic sheets are often used, with holes provided in the bottom to allow the drainage of water (Devi and Kumar, 2012). The bamboo shoots are cut into longitudinal shreds and kept in an airtight condition. The fermented bamboo shoots are ready for sale, or for making curry, within 3–5 days, and are locally known as *soibum* (Figure 8.32). This kind of fermented bamboo shoot can be kept for relatively long periods (that is, for a month or more) if kept airtight. The degree of sourness of product shows a rapid increase in the initial stage of fermentation, but, as it reaches a peak point at about 7–10 days, it started to decrease.

This method of preparing fermented bamboo shoots is followed in almost all of the hill districts of Manipur. Bamboo varieties such as *saneibi*, *nath*, *unal*, *longa wa*, *meiribob*, and *ooii* are used for this purpose. Care is taken to avoid varieties such as *utang* and *khok* as they are not eatable. "*Sanaibi*" is deemed to be the best bamboo variety for use in the preparation of fermented shoots, followed by the "*nath*" variety,

Figure 8.32 (a) *Soibum*; (b) a vendor selling *soibum*.

while the "*unal*" variety is thought to give a better texture and appearance (Devi and Kumar, 2012).

8.5.8.1.2 In an Earthen Pot   This method of the fermenting bamboo shoot is almost the same as above, with the only difference being that, instead of fermenting the shoots inside a pit, an earthen pot is used in which a hole is made at the bottom for the drainage of excess water. During fermentation in an earthen pot, some people prefer to add *heiboong* (*Garcinia anomala*) to enhance the fermentation, and also to achieve a sourer taste in the final product (Devi and Kumar, 2012).

8.5.8.1.3 Open Condition   In the "open condition," wild *colocasia* leaves are used in thick layers. The sliced bamboo shoots are placed on top of this, and are again covered with the leaves. Fermentation is allowed to take place in this way, with no other additions (Devi and Kumar, 2012).

8.5.8.1.4 Water-Dipped Nath   Here, a special bamboo variety is used: the *nath* variety, which is locally known as *nath ki soibum*. This method is commonly used by the people of the Bishempur district of Manipur. As the *nath* bamboo is very small and long, it is sliced lengthwise into about 2–3 cm pieces, and then, placed in a container (plastic buckets are generally employed these days) into which water is poured—just enough to dip the contents. This is covered and kept until it is sold. The fermented bamboo shoots are ready to sell after three days. The *nath* variety gives very tasty fermented bamboo shoots, but they can't be stored for very long (Devi and Kumar, 2012).

8.5.8.1.5 Dried Usoi   For this approach, any edible bamboo shoot variety is used. All the shoots are sliced into small pieces, boiled in water, and then dried in a bamboo tray under the sun after draining off the excess water. The dried bamboo shoots are packed in plastic sheets and used in off-seasons or for longer-distance selling (Devi and Kumar, 2012).

8.5.8.1.6 Fermented and Dried Soibum   In this method, after the completion of the normal fermentation of the bamboo shoots, they are dried either under the sun or in the top of the fire. The Tankhul people of Manipur use a special variety of bamboo shoot that is very small and long, locally known as *ngathan*. It gives a twisted appearance after drying, just like noodles (Figure 8.34a; Devi and Kumar, 2012).

8.5.8.1.7 Fermented Bamboo Shoot Pickle   *Soibum* (fermented or dried) is consumed by most of the people of Manipur, irrespective of caste or tribe. It is eaten raw with fermented fish or in boiled and other cooked form with any meat, fish, or vegetables. It represents an important food element of all of the festivals observed by the Manipuries. Currently, in some small-scale industries, the bamboo shoots are blanched, once sliced into small pieces, in hot water to reduce their enzymatic activity, and are treated with potassium metabisulphide (KMS) (1%) for 10 min. After this, the foodstuff is sun-dried, packed in an airtight container, and sent for sale (Devi and Kumar, 2012).

8.5.9 Karadi

There are many indigenous, forest-based, fermented products, and *karadi* is one such product. This fermented *Bambusa arundinacea* L. shoot is locally known in the districts of undivided Ganjam, Kalahandi, Koraput, and Sambalpur. It is usually produced during June–September, when bamboo shoots first sprout.

   The tips of young bamboos are collected, sliced into pieces, and then, dipped in water for a day to be fermented. During the process of the fermentation, the bitterness of the bamboo shoot is reduced, or washed out in the water. *Karadi* is cooked along with locally available vegetables such as *taro* (*Colocasia*), cucumber (*Cucumis sativus* L), potato (*Solanum tuberosum* L.), *brinjal* (*Solanum melongena* L.), and so on. It is sometimes pounded and sun-dried, with the powdered form locally known as *handua*, which is cooked as a curry throughout the year (Panda and Padhy, 2007).

8.5.10 Bastanga

Bastanga is made from succulent bamboo shoots (*Dendrocalamus hamiltonii, Bambusa tulda*) mostly by the Lotha Naga tribe, and named *Rhujuk* in Lotha dialect. Young shoots are harvested and their sheaths are removed until only the soft, white part of the shoot remains. The shoots are then, cut into small pieces and pressed tightly into bamboo baskets covered with banana leaves. A hole is made in the middle so as to let the juice drain out. The preparation is kept in such a manner for about 2–3 weeks, until the bamboo shoots are completely drained of their juice. The fermented bamboo

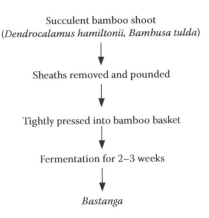

Succulent bamboo shoot
(*Dendrocalamus hamiltonii, Bambusa tulda*)

↓

Sheaths removed and pounded

↓

Tightly pressed into bamboo basket

↓

Fermentation for 2–3 weeks

↓

Bastanga

Figure 8.33   Flow chart of preparation of *bastanga*/rhujuk (Lotha). (From Jamir, B. and Deb, C.R. 2014. *Int. J. Food. Ferment. Technol.* 4(2): 121–127. With permission.)

shoots are then, dried (Jamir and Deb, 2014). Different grades of dried bamboo shoots are obtained, depending on the way they are cut (Figures 8.33 and 8.34a,b). The juice can also be stored for a year.

8.6 Summary and Future Prospective

Acid fermentation has been employed from times immemorial to prepare a large diversity of food products with a correspondingly wide range of flavors, including milk and

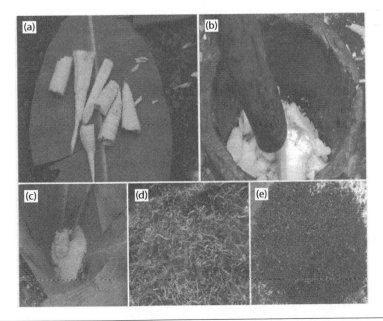

Figure 8.34   Different stages of *bastanga* preparation. (a) The succulent bamboo shoots; (b) pounded in traditional mortar and pestel; (c) *bastanga* in bamboo basket; (d) and (e) different grades of dried bamboo shoots. (From Jamir, B. and Deb, C.R. 2014. *Int. J. Food. Ferment. Technol.* 4(2): 121–127. With permission.)

milk-based products, fruits and vegetables, forest products, combinations of cereals, etc. Undoubtedly, the exact methods of preparation do vary from country-to-country, but the common component is the production of acid—mainly, lactic acid. Lactic acid fermentation as a method of preservation is employed using indigenous skills, and has proved to be safe and sustainable approach to food preparation and storage, despite an apparent or historical lack of microbiological, chemical, and toxicological grounding.

The interest in traditionally fermented foods has greatly increased in recent years, because more emphasis is being placed on plant materials as human foods. Rising costs of and demand for animal protein is another reason for this trend. Lactic acid-fermented food products have also been recommended as dietary additions because of their beneficial effects on health. Vegans and some vegetarians avoid dairy products because of ethical, dietary, environmental, political, and religious concerns, so there is a need to conduct more research on indigenous fermented products, including on the microorganisms involved, to meet the requirements of these consumers.

It is well established that the market for fermented milks is booming, especially probiotics and those with particular added ingredients as they provide a mechanism to preserve a perishable item like "milk" for longer periods. *Shrikhand* is one of the major indigenous fermented milk products, and it is palatable and has a characteristic taste. It also offers a great potential to serve as a vehicle for probiotic cultures and prebiotics in the wider human diet, with the advantage of being accepted and enjoyed by all age groups. Moreover, people suffering from lactose intolerance can consume fermented products. The value-added *shrikhand*, as well as similar products, represents a highly successful research venture with further commercial potential. Given these viewpoints, lactic acid-fermented beverages assume a still greater significance.

An increasing demand for traditional dairy products presents an immense opportunity for organized dairies in the countries of South Asia to modernize and scale up their production. From *burfi* to *kulfi*, *kalakand* to *shrikhand*, and *gulabjamun* to *chumchum* extends the enticing world of Indian milk delicacies. From ancient times, many of these processes have largely remained unchanged, being in the hands of the *Halwais*, or traditional sweetmeat makers, who formed the core of this localized industry, and, therefore, there are opportunities to develop further and to introduce suitable, mechanically controlled methods to commercialize these products.

Fermented bamboo products are associated with many health benefits, too, and yet detailed information about these indigenous fermentation techniques are often scarce outside of the region in which they are produced. The fermentation processes described above for *khorisa*, *poka khorisa*, *khorisa pani*, *kahudi*, and *miyamikhri* are the traditional methods through which the shelflife of the product is prolonged for more than a year, without refrigeration, and can be consumed when fresh bamboo shoots are not available.

Fermentation is undoubtedly a great way to preserve food, and yet there are no preservatives added to these products, a feature that is of great significance. Fermented bamboo products are very good appetizers, and are used to aid digestion.

Research findings on the microbial preservation of bamboo shoots and the use of indigenous lactic acid-producing bacteria can be applied in the preservation of many other edible substances in different parts of the South Asia, where such traditional knowledge is lacking.

The traditional acid-fermented products described in this chapter are generally prepared at the household level, but demand for them is increasing due to changes in lifestyle, customs, and consumers' health awareness. Hence, there is a strong need to prepare these products on commercial scale in order to meet the greater demand for them. In the future, it is expected that scientific and technological inputs will help to retain and enhance the value of fermentation as a way of preserving food and developing new products. Such research input should help to elucidate the role of lactic acid-fermenting microorganisms, their biochemistry, probiotic effect, and their therapeutic values for the health of consumers at large.

References

Adams, M.R. and Moss, M.O. 1996. Fermented and microbial foods. In: *Food Microbiology*. New Age International (P) Ltd. Publishers, New Delhi. pp. 252–302.

Allgeyer, L.C., Miller, M.J., and Lee, Y. 2010. Drivers of liking for yogurt drinks with prebiotics and probiotics. *J Food Sci* 75(4): S212–S219.

Anand, J.C. and Das, L. 1971. Effects of condiments on lactic fermentation in sweet turnip pickle. *J Food Sci Technol* 8(3): 143.

Anand, J.C. and Johar, D.S. 1957. Effect of condiments on control of *Aspergillus niger* in mango pickle. *J Sci Indus Res* 16: 370.

Andres, C. 1977. Controlled fermentation reduces salt usage, improves pickle quality and profitability. *Food Process.* 38(12): 50.

Aneja, R.P., Mathur, B.N., Chandan, R.C., and Benerjee, A.K. 2002. *Technology of Indian Milk Products*. Dairy India Publication, New Delhi.

Aneja, R.P., Vyas, M.N., Nanda, K., and Thareja, V.K. 1977. Development of an industrial process for the manufacture of *Shrikhand*. *J Food Sci Technol* 14(4): 159–163.

Angchok, D., Dwivedi, S.K., and Zhmad, Z. 2009. Traditional foods and beverages of Ladakh. *Indian J Trad Know* 8(4): 551–558.

Anonymous, bamboo shoots. http://www.meeting-buyers-suppliers-food-sector.eu/search.php?phrase=Used

Anonymous, FAO, 1997. Fermented frutis and vegetables. A global perspective: Bacterial fermentations. FAO, Publications, Rome.

Anonymous, Pickles making NSIC http: //www.nsicpartners.com/schemes/documents/projprofiles/PICKLES%20MAKING.pdf

Anonymous, shrikhand http://ifood.tv/indian/shrikhand/about

Anonymous, Volume VIII *Biological Environment—Food Resources*. Centre for Inter-disciplinary Studies of Mountain & Hill Environment (CISMHE), University of Delhi, Delhi.

Anonymous, wiki http://en.wikipedia.org/wiki/Buttermilk

Anonymous, wiki yogurt http://en.wikipedia.thelinks.com.pl/wiki/Yogurt

Anonymous, Wikipedia, lassi. http://www.agriinfo.in/?page=topic&superid=9&topicid=725

Anonymous, Wikipedia, milk. Although cow's milk is the most popular in many countries, milk can be obtained from many different sources, GCSE/Design & Technology/Food Technology, 2003-08-19.

Anonymous Wikipedia http://en.wikipedia.org/wiki/Chhena_Poda

Aubert, C. 1985. Les aliments fermentes traditionnels: Une richesse meconnue. *Collection Les Vrais Aliments d' Aujourd'hui et de Demain*. Terre Vivante, Paris, p. 263.

Aziz, T. 1985. Thermal processing of dahi to improve its keeping quality. *Ind J Nutr Diet* 22: 80–87.

Banwart, G.J. 1981. Sauerkraut. In: *Basic Food Microbiology*, AVI, Westport, CT, p. 454.

Bardale, P.S., Waghmare, P.S., Zanzad, D.M., and Khedkar, D.M. 1986. The preparation of *Shrikhand* like product from skim milk chakka by fortifying with fruit pulps. *Indian J Dairy Sci* 39: 480–483.

Basappa, S.C., Somashekar, D., and Agrawal, R. 1997. Nutritional composition of fermented ragi (channg) by phab and defined starter cultures as compared. *Int J Food Sci* 48(5): 313–319.

Batish, V.K., Grover, S., Pattaink, P., and Ahmed, N. 1999. Fermented milk products. In: *Biotchnology: Food Fermentation, Microbiology, Biochemistry and Technology*, Joshi, V.K. and Pandey, A (eds.), vol. II. Educational Publication & Distributors, New Delhi, pp. 781–864.

Battcock, M. and Azmi-ali, S. 1998. Fermented fruits and vegetables. In: *Fermented Fruits and Vegetables. A Global Perspective*. FAO Agricultural services bulletin No 134, Food and Agriculture Organization of the United Nations, Italy, Rome.

Beuchat, L.R. 1995. Indigenous fermented foods. In: *Biotechnology*, Rehm, H.-J., Reed, G., Puhler, A., and Studler, P. (eds.), 2nd edition, vol. 9, Chapter 13. VCH Publication, Weinheim. pp. 505–559.

Bhushan, S., Sharma, S., Joshi, V.K., and Abrol, G. 2013. Bio-preservation of minimally processed fruits and vegetables using lactic acid bacteria and bacteriocin. In: Bakshi, A.K., Joshi, V.K., Vaidya, D., and Sharma, S. (eds.), *Food Processing and Preservation* Jagmander Book Agency, New Delhi, pp. 355–374.

Boylston, T.D., Vinderola, C.G., Ghoddusi, H.B., and Reinheimer, J.A. 2004. Incorporation of bifidobacteria into cheeses: Challenges and rewards. *Int Dairy J* 14: 375–387.

Buckenhuskes, H.J. 1993. Selection criteria for lactic acid bacteria to be used as starter cultures for various food commodities. *FEMS Microbiol Rev* 12: 253–272.

Buckenhuskes, H.J. 2001. Fermented vegetables. In: *Food Microbiology: Fundamentals and Frontiers*, Doyle, M.P., Beuchat, L.R., and Montiville, T.J. (eds.), 2nd edition. ASM Press, Washington, DC. pp. 665–679.

Bucker, E.R., Mitchell, J.H., and Johnson, M.G. 1979. Lactic fermentation of Peanut milk. *J Food Sci* 44(5): 1534–1538.

Buckner, H., Salmen, A., Amar, A., and Buckenhuskes, H.J. 1993. Fermented dairy like products from advanced soy proteins. *Proc Euro Food Chem* 7: 215–222.

Campbell-Platt, G. 1987. Fermented Foods of the World—A Dictionary and Guide. Butterworths, London.

Carmen, W., Gloria, D.-R., and Tamang, J. 2010. Fermented vegetable products. In: *Fermented Foods and Beverages of the World*, Tamang, J.P., and Kailasapathy, K. (eds.), CRC Press, Taylor & Francis, Boca Raton, FL. pp. 149–190.

Chakrabarty, J., Sharma, G.D., and Tamang, J.P. 2009. Substrate utilisation in traditional fermentation technology practiced by tribes of North Cachar hills district of Assam. *Assam Univ J Sci Technol: Biol Sci* 4(1): 66–72.

Chandan, R.C. 2006. History and consumption trends. In: *Manufacturing Yogurt and Fermented Milks*. Wiley-Blackwell, Oxford, UK, pp. 3–15.

Chandan, R.C. 2013. History and consumption trends. In: *Manufacturing Yogurt and Fermented Milks*, Kilara, A. and Chandan, R.C. (eds.), Wiley-Blackwell. pp. 3–17.

Chandan, R.C. and Shahani, K.M. 2001. Other fermented dairy products. In *Biotechnology*. Rehm, H.-J. and Reed, G. (eds.), 2nd edition. Wiley-VCH Verlag GmbH, Weinheim. doi:10.1002/9783527620999.ch10j

Chauhan, A. and Joshi, V.K. 2014. Application of response surface methodology (RSM) in optimization of lactic acid fermentation of carrot: Influence of addition of salt, additives and growth stimulators. *Int. J. Curr. Res.* 6(1): 15–21.

Chavasit, V., Hudson, J.M., Torres, J.A., and Daeschel, M.A. 1991. Evaluation of fermentative bacteria in a model low salt cucumber juice brine. *J Food Sci* 56(2): 462–465.

Cheigha, H.S., Parka, K.Y., and Leeb, C.Y. 1994. Biochemical, microbiological, and nutritional aspects of kimchi (Korean fermented vegetable products). *Crit Rev Food Sci Nutr* 34(2): 175–203.

Cheng, Y.J., Thompson, L.D., and Brittin. H.C. 1990. Sogurt, a yogurt-like soybean product: Development and properties. *J Food Sci* 55(4): 1178–1179.

Cho, N.C., Jhon, D.Y., Shin, M.S., Hong, Y.H., and Lim, H.S. 1988. Effect of garlic concentration on growth of microorganisms during kimchi fermentation. *Korean J Food Sci Technol* 20: 231.

Choudhury, D., Sahu, J.K., and Sharma, G.D. 2012. Bamboo shoot: Microbiology, biochemistry and technology of fermentation—A review. *Indian J Trad Know* 11(2): 242–249.

Corredig, M. and Dalgleish, D.G. 1997. Isolates from industrial buttermilk: Emulsifying properties of materials derived from the milk fat globule membrane. *J Agric Food Chem* 45: 4595–4600.

Cruess, W.V. 1948. Spoilage and discolouration of sauerkraut. In: *Commercial Fruit and Vegetables Products*. 4th edition. McGraw-Hill Book Co., New York, NY. p. 724.

Dadarwal, R.B., Beniwal, B.S., and Singh, R. 2005. Process standardization for preparation of fruit flavoured *shrikhand. J Food Sci Technol* 42(1): 22–26.

Daeschal, M.A., Andersson, R.E., and Fleming, H.P. 1987. Microbial ecology of fermenting plant materials. *FEMS Microbiol Rev* 46: 357–367.

Desai, A.R., Powell, I.B., and Shah, N.P. 2004. Survival and activity of probiotic *lactobacilli* in skim milk containing prebiotics. *J Food Sci* 69(3): 57–60.

Desai, H.K. 1983. Evaluation of Effects of Homogenization of Milk and Addition of Stabilizers on the Quality of Shrikhand. M.Sc. thesis, Gujarat Agril. Uni. S.K. Nagar Gujarat.

Desai, H.K. and Gupta, S.K. 1986. Sensory evaluation of *Shrikhand. Dairy Guide* 8(12): 33.

Desai, H.K., Vyas, S.H., and Upadhyay, K.G. 1985. Influence of homogenization of milk on the quality of *chakka* and *Shrikhand. Indian J Dairy Sci* 38(2): 102–106.

Devi, P. and Kumar, S.P. 2012. Traditional, ethnic and fermented foods of different tribes of Manipur. *Indian J Trad Know* 1(1): 70–77.

Dewan, S. 2002. *Microbiological Evaluation of Indigenous Fermented Milk Products of the Sikkim Himalayas*. PhD thesis. Food Microbiology Laboratory, Sikkim Government College, Gangtok.

Dewan, S. and Tamang, J.P. 2007. Dominant lactic acid bacteria and their technological properties isolated from the Himalayan ethnic fermented milk products. *Antonie van Leeuwenhoek Int J Gen Mol Microbiol* 92(3): 343–352.

Eijsink, V.G.H., Skeeie, M., Middelhoven, P.H., Brurberg, M.B., and Nes, I.F. 1998. Comparative studies of class IIa bacteriocins of lactic acid bacteria. *Appl Environ Microbiol* 64: 3275–3281.

Elliot, R., Elmer, R., and Marth, H. 2007. Incidence and behavior of *Listeria monocytogenes* in cheese and other fermented dairy products. In: *Listeria, Listeriosis, and Food Safety*, Elliot, R., Elmer, R., and Marth, H. (eds.), 3rd edition. Food Science and Technology Series, p. 896.

Etchells, J.L., Costilow, R.N., Anderson, T.E., Bell, T.A., and Fleming, H.P. 1966. Pure culture fermentation of green olives. *Appl. Microbiol.* 14: 1027–1041.

Etchells, J.L., Fleming, H.P., Hortz, L.H., Bell, T.A., and Monroe, R.S. 1975. Factors influencing bloater formation in brined cucumbers during controlled fermentation. *J Food Sci* 40(3): 569–575.

Farnworth, E.R. 2008. *Handbook of Fermented Functional Foods*. 2nd edition, CRC Press Taylor & Francis Group, Boca Raton, FL.

Fleming, H.P. 1982. Fermented vegetables. In: *Economic Microbiology: Fermented Foods*. Rose, A.H. (ed.), Academic Press, New York, NY. pp. 227–258.

Fleming, H.P., Etchells, J.L., Thompson, R.L., and Bell, T.A. 1975. Purging of CO_2 from cucumber brines to reduce bloater damage. *J Food Sci* 40(6): 1304–1310.

Fleming, H.P. and McFeeters, R.F. 1981. Use of microbial cultures: Vegetable products. *Food Technol* 35(1): 84–88.

Fleming, H.P., McFeeters, R.F., and Thompson, R.L. 1987. Effects of sodium chloride concentration on firmness retention of cucumbers fermented and stored with calcium chloride. *J. Food Sci.* 52: 653–657.

Fleming, H.P., Thompson, R.L., Bell, T.A., and Hontz, L.H. 1978. Controlled fermentation of sliced cucumbers. *J. Food Sci.* 43: 888.

Frazier, W.C. and Westhoff, D.C. 1998. *Food Microbiology.* 7th edition. Tata Mcgraw Hill Publi. Co., New Delhi.

FSSAI. 2011. Food Safety and Standards Act, India.

Gandhi, D.N. 1989. *Dahi* and Acidophilus milk. *Indian Dairyman* 41: 323–327.

Gandhi, D.N. 2002. Potential application of lactic acid bacteria for the development of fermented milk products and in bioprocessing of whey. *Indian Dairyman* 54(12): 64–67.

Gandhi, N.K. and Jain, S.C. 1977. A study on the development of a new high protein formulated food using buffalo milk. *J Food Sci Techno* 14(4): 156.

Geetha, V.V., Sarma, K.S., Reddy, V.P., Reddy, Y.K., Moorthy, P.R.S., and Kumar, S. 2003. Physico-chemical properties and sensory attributes of probiotic *shrikhand. Indian J Dairy Biosci* 14: 58–60.

Geis, A. 2006. Genetic engineering of bacteria used in food fermentation. In: *Genetically Engineered Food: Methods and Detection,* 2nd edition. Heller, K.J. (ed.), Wiley-VCH Verlag GmbH & Co., KGaA, Weinheim. doi:10.1002/9783527609468.ch5

Ghatak, P.K. and Dutta, S. 1998. Effect of admixing of cow and buffalo milks on compositional and sensory qualities of *shrikhand. Indian J Nutr Diet* 35(2): 43–46.

Gibbs, P.A. 1987. Novel uses for lactic acid fermentation in food preservation. *J Appl Bacteriol Symp Suppl* 515–585.

Giram, S.D., Barbind, R.P., Pawar, V.D., Sakhale, B.K., and Agarkar, B.S. 2001. Studies of fortification of sour whey concentrate in *chakka* for preparation of *Shrikhand. J Food Sci Technol* 38(3): 294–295.

Girdhari Lal, G., Siddappa, S., Tandon, G.L. 1986. *Preservation of Fruits and Vegetables.* Indian Council of Agricultural Research, New Delhi, 207pp.

Gobbetti, M. and Rossi, J. 1989. Wheat and maize flours as media for lactic acid beverages. *Annals Fac Agr Univ Perugia* 3: 377.

Goswami, P., Singh, S., and Sharma, K.P. 2011 Fermented Milk Products—Nature's Blessings, http://www.pfionline.com/index.php/columns/value-addition/146-fermenenter-milk-products

Granato, D., Branco, G.F., Cruz, A.G., Faria, J.A.F., and Shah, N.P. 2010. Probiotic dairy products as functional foods. *Compr Rev Food Sci Food Safety* 9(5): 455–470.

Guggisberg, D., Cuthbert-Steven, J., Piccinali, P., Bütikofer, U., and Eberhard, P. 2009. Rheological, microstructural and sensory characterization of low-fat and whole milk set yogurt as influenced by inulin addition. *Int. Dairy J.* 19(2): 107–115.

Haque, Z. and Kayanush, A. 2002. Effect of sweeteners on the microstructure of yogurt. *Food Sci Technol Res* 8: 21–23.

Hubert, J.C. and Dupuy, P. 1994. Conservation des fruits et legumes. In: *Bacteries Lactiques: Aspects Fondamentaux et Technologiques.* de Roissart, H. and Luquet, F.M. (eds.), 11: p. 233.

Hudson, J.M. and Buescher, R.W. 1985. Pectic substances and firmness of cucumber pickles as influenced by CaCl2, NaCl and brine storage. *J. Food Biochem* 9: 211–215.

Ingle, U.M. and Joglekar, N.V. 1974. Effect of fat and sugar variations on the acceptability of *Shrikhand* preparation. *J Food Sci Technol* 11(4): 189.

Jamir, B. and Deb, C.R. 2014. Studies on some fermented foods and beverages of Nagaland, India. *Int J Food Ferment Technol* 4(2): 121–127.

Jeyaram, K., Romi, W., Singh, A., Devi, A.R., and Devi, S. 2010. Bacterial species associated with traditional starter cultures used for fermented bamboo shoot production in Manipur state of India. *Int J Food Microbiol* 143(1–2): 1–8.

Jeyaram, K., Singh, T.A., Romi, W., and Devi, A.R. 2009. Traditional fermented foods of Manipur. *Indian J Trad Know* 8(1): 115–121.

Joshi, M.U., Sarkar, A., Singhal Rekha, S., and Pandit, A.B. 2009. Optimizing the formulation and processing conditions of *Gulab Jamun*: A statistical design. *Int J Food Prop* 12(1): 114–128.

Joshi, V.K. and Bhat, A. 2000. Pickles: Technology of its preparation. In: *Postharvest Technology of Fruits and Vegetables*. Verma, L.R. and Joshi, V.K. (eds.), vol. 2. The Indus Publ. New Delhi. pp. 777–820.

Joshi, V.K., Chauhan, A., Devi Sarita, and Kumar Vikas. 2014. Application of response surface methodology in optimization of lactic acid fermentation of radish: Effect of addition of salt, additives and growth stimulators. *J. Food Sci. Technol.* (submitted). DOI: 10.1007/s13197-014-1570-9.

Joshi, V.K., Chauhan, S.K., and Sharma, R. 1993. Preservation of fruits and vegetables by lactic acid fermentation. *Bev Food World* 19(1): 9.

Joshi, V.K., Garg, V., and Abrol, G.S. 2012. Indigenous fermented foods. In: Joshi, V.K. and Singh, R.S. (eds.), *Food Biotechnology: Principles and Practices*. IK International Publishing House, New Delhi, pp. 337–373.

Joshi, V.K., Kaur, M., and Thakur, N.S. 1996. Lactic acid fermentation of mushroom (*Agaricus bisporus*) for preservation and preparation of sauce. *Acta Alimentaria* 25(1): 1–11.

Joshi, V.K. and Sharma, S. 2009. Lactic acid fermentation of radish for shelf-stability and pickling. *Nat. Prod. Rad.* 8(1): 19–24.

Joshi, V.K. and Sharma, S. 2010. Preparation and evaluation of sauces from lactic acid fermented vegetables. *J. Food Sci. Technol.* 47(2): 214–218.

Joshi, V.K. and Sharma, S. 2012. A panorama of lactic acid bacterial fermentation of vegetables—Review. *Int. J. Food Fermentation Technol* 2(1): 1–12.

Joshi, V.K., Sharma, S., and Neerja, R. 2011a. Preparation and evaluation of appetizers from lactic acid fermented vegetables. *J. Hill Agric.* 2(1): 20–27.

Joshi, V.K., Sharma, S., and Thakur, N.S. 2003. Technolgy of fermented fruits and vegetables. In: *International Seminar and Workshop on Fermented Foods, Health Status and Social Well-Being* held on Nov 2003 at Anand, India, pp. 24–28.

Joshi, V.K., Somesh, S., Bhushan, S., Sharma, R., and Singh, R.P. 2005. Preparation and evaluation of olive pickles (Indian style): Selection of suitable cultivar and recipe. *Process. Food Ind.* 8(9): 14–16.

Joshi, V.K., Somesh, S., Chauhan, A., and Thakur, N.S. 2011b. Preparation and evaluation of instant chutney mix from lactic acid fermented vegetables. *Int. J. Food Ferment. Technol* 1(2): 201–209.

Joshi, V.K. and Singh, R.S. 2012. *Food Biotechnology: Principles and Practices*. I.K. International Publ. House (P), New Delhi.

Joshi, V.K. 2015. *Preparation and Evaluation of Lactic Acid Fermented Fruits and Vegetables—A Practical Approach*. Department of Food Science and Technology, Dr. YSP University of Horticulture and Forestry, Nauni, Solan, Himachal Pradesh, India, pp. 1–66.

Joshi, V.K. and Thakur, S. 2000. Lactic acid fermented beverages. In: *Postharvest Technology of Fruits and Vegetables*, Verma, L.R. and Joshi, V.K. (eds.), Vol. II. Indus Publ. Co., New Delhi. p. 1102.

Kadam, S.J., Bhosale, D.N., and Chavan, I.G. 1984. Studies on preparation of *chakka* from cow milk. *J Food Sci Technol* 21(3): 180.

Kar, A. and Borthakur, S.K. 2007. Wild vegetables sold in local markets of *Karbi Anglong*, Assam. *Indian J Trad Know* 7(1): 169–172.

Karki, C.B. and Horsford, R.M. 1986. Epidemic of wheat leaf blight in Nepal caused by *Pyrenophora trilici-repentis Nepalese J. Agric.* 17: 69–74.

Karki, T., Okada, S., Baba, T., Itoh, H., and Kozaki, M., 1983. Studies on the microflora of Nepalese pickles gundruk. *Nippon Shokuhin Kogyo Gakkaishi* 30: 357–367.

Karkri, T.B. 1986. Fermented vegetables. In: *Concise Handbook of Indigenous Fermented Foods in the ASCA Countries*, Saono, S., Hull R.R., and Dhamcharee, B. (eds.). Govt. of Australia, Canberra, 157 p.

Kilara, A. and Chandan, R.C. 2013. Greek-style yogurt and related products. In: *Manufacturing Yogurt and Fermented Milks*. Kilara, A. and Chandan, R.C. (eds.), Wiley-Blackwell. pp. 297–315.

Kumari, A., Kalia, M., and Attri, S. 1993. Studies on chemical and organoleptic evaluation of cabbage and its product 'Sauerkraut'. *Indian Food Packer* 47(2): 11–14.

Kurmann, J.A. 1984. Aspects of the production of fermented milks. IDF Bulletein No. 178. Brussels: Int Dairy Federation. pp. 16–26.

Laencina, J., Guzman, G., Guevara, L.A., and Flores, J. 1985. Accion de los aceites esenciales sobre levaduras de la fermentacion de aceitunas (*Olea europaea* L.). *Essenze Derivati Agrumari* 27: 35.

Lal, G., Siddappa, G.S., and Tondon, G.L. 1986. Chutneys, sauces and pickles. In: *Preservation of Fruits and Vegetables*, ICAR, New Delhi. pp. 235–269.

Landge, U.B., Pawar, B.K., and Choudhari, D.M. 2010. Preparation of shrikhand using Ashwagandha powder as additive. *J Dairy Foods Home Sci* 40(2): 79–84.

Law, S.V., Abu Bakar, F., Mat Hashim, D., and Abdul Hamid, A. 2011. Popular fermented foods and beverages in Southeast Asia. *Int Food Res J* 18: 475–484.

Lee, C.H. 2009. Food biotechnology. In: *Food Science and Technology*. Campbell-Platt, G. (ed.), CRC Press, Boca Raton, FL. pp. 85–95.

Lee, I.S., Kim, S.H., Kang, K.H., and Rawer, W.D. 1990. *Dairy Sci. Abstr.* 52: 398.

Leon, R., Cabrera, S.S. and Rolz, C. 1975. Controlled fermentation of cucumbers by bacterial cultures. *Rav Latino Microbiol* 17(1): 33–37.

Li, K.-Y. 2003. Fermentation: Principles and microorganisms. Chapter 9. In: *Handbook of Vegetable Preservation and Processing*. Hui, Y.H., Ghazala, S., Graham, D.M., Murrell, K.D., and Nip, W.-K. (eds.), CRC Press, Boca Raton, FL.

Lu, B.Y., Wu, X., Tie, X., Zhang, Y., and Zhang, Y. 2005. Toxicology and safety of antioxidant of bamboo leaves. Part I: Acute and subchronic toxicity studies on antioxidant of bamboo leaves. *Food Chem Toxicol* 43(5): 783–792.

Manandhar, N.P. 1998. The preparation of *Gundruk* in Nepal: A sustainable rural industry? *Ind Know Dev Monit* 6(3).

Mann, C.V. 1977. A factor of yoghurt which lowers cholesteremia in man. *Atherosclensis* 26: 335–340.

Masud, T., Sultana, K., and Shah, M.A. 1991. Incidence of lactic acid bacteria isolated from indigenous *Dahi*. *Australian J Anim Sci* 4: 329–331.

McFeeters, R.F., Senterh, M.M., and Fleming, P. 1989. Softening effects of monovalent cations in acidified cucumber mesocarp tissue. *J Food Sci* 54: 366.

Michael, P. and Kailasapathy, K. 2006. Viability of commercial probiotic cultures (*L. acidophilus*, *Bifidobacterium* sp., *L. casei*, *L. paracasei* and *L. rhamnosus*) in cheddar cheese. *Int J Food Microbiol* 108(2): 276–280.

Mikki, M.S. 1971. *Studies on sauerkraut*. M.Sc. Thesis, Division of Horticulture and Fruit Technology, IARI, New Delhi.

Mindy, M.B., Alejandro, A., and Divya, J. 2005. Control of food borne bacterial pathogens in animals and animal products through microbial antagonism, Chapter 3. In: *Food Biotechnology*. Shetty, K., Paliyath, G., Pometto, A., and Levin, R.E. (eds.), CRC Press, Boca Raton, FL.

Mital, B.K. and Garg, S.K. 1992. Acidophilus milk products: Manufacture and therapeutics. *Food Rev Int* 8(3): 347–389.

Miyani, R.V. 1982. Effect of Different Levels of Moisture, Sugar and Fat on Consistency and Acceptability of Shrikhand. MSc thesis, Agril. Uni. S.K. Nagar. Gujarat.

Montet, D., Loiseau, G., Zakhia, N., and Mouquet, C. 1999. Fermented fruits and vegetables. In: *Biotechnology: Food Fermentation*. Joshi, V.K. and Pandey, A. (eds.), vol. II. Edu. Publ. Distri., New Delhi. pp. 951–969.

Morin, P., Britten, M., Jiménez-Flores, R., and Poulio, Y. 2007. Microfiltration of buttermilk and washed cream buttermilk for concentration of milk fat globule membrane components. *J Dairy Sci* 90: 2132–2140.

Motarjemi, Y. and Nout, M.J.R. 1996. Food fermentation: A safety and nutritional assessment. *Bull World Health Org* 74: 553–559.

Naewbanij, J.O., Stone, M.B., and Chambers, E. 1990. *Lactobacillus plantarum* and *E. cloacae* growth in cucumber extracts containing various salts. *J Food Sci* 55(6): 1634.

Naewbanij, J.O., Stone, M.B., and Fung, D.Y.C. 1986, Growth of *Lactobacillus plantarum* in cucumber extract containing various chloride salts. *J Food Sci* 51: 1257–1259. doi:10.1111/j.1365-2621.1986.tb13099.

Nalawade, J.S., Patil, G.R., Sontakke, A.T., and Hassan-Bin, A. 1998. Effect of compositional variables on sensory quality and consistency of *shrikhand*. *J Food Sci Technol* 35(4): 310–313.

Nettles, C.G. and Barefoot, S.F. 1993. Biochemical and genetic characteristics of bacteriocins of food associated lactic acid bacteria. *J Food Prot* 56: 338–356.

Neuwhoff, M. 1969. Variation in quality. In: *Cole crops*. Leonard-Hill, London. p. 267.

Nimachow, G., Rawat, J.S., and Dai, O. 2010. Prospects of bamboo shoot processing in northeast India. *Curr Sci* 98(3): 288–289.

Nirmala, C., Sharma, M.L., and David, E.A. 2008.Comparative study of nutrient components of freshly harvested, fermented and canned bamboo shoots of *Dendrocalamus giganteus* Munro. *J Am Bamboo Soc* 21(1): 41–47.

Ozer, D., Akin, S., and Ozer, B. 2005. Effect of inulin and lactulose on survival of *Lactobacillus acidophilus* LA-5 and *Bifidobacterium bifidum* BB-02 in acidophilus-bifidus yogurt. *Food Sci Technol Int* 11: 19–24.

Panda, T. and Padhy, R. 2007. Sustainable food habits of the hill-dwelling *Khanda* tribe in Kalahandi distrit of Orissa. *Indian J Trad Know* 6(1): 103–105.

Pandya, A.J., Mohamed, M., and Khan, H. 2006. Traditional Indian dairy products. In: *Handbook of Milk of Non-Bovine Mammals* Park, Y.W. and Haenlein, G.F.W. (eds.), Blackwell Publishing Professional, Ames, IA. doi:10.1002/9780470999738.ch10

Patel, R.S. 1982. *Process Alteration in Shrikhand Technology*. PhD thesis, Kurukshetra University, Haryana, India.

Patel, R.S. and Abd-El-Salam. 1986. *Shrikhand*—An Indian analogue of Western quarg. *Cult Dairy Prod J* 21(1): 6–7.

Patel, R.S. and Chakraborty, B.K. 1982. Process alterations in shrikhand technology. PhD thesis, Kurukshetra University, India.

Patel, R.S. and Chakraborty, B.K. 1985a. Reduction of curd-forming period in *Shrikhand* manufacturing process. *Dairy Sci Technol* 65, 647–648.

Patel, R.S. and Chakraborty, B.K. 1985b. Factors affecting the consistency and sensory properties of *Shrikhand*. *Egyptian J Dairy Sci* 12: 73–78.

Patel, R.S. and Chakraborty, B.K. 1985c. Use of different sources of milk solids for *Shrikhand* making. *Egyptian J Dairy Sci* 13: 79–84.

Patel, R.S. and Chakraborty, B.K. 1988. *Shrikhand*: A review. *Ind J Dairy Sci* 41: 126.

Pederson, C.S. 1971. *Microbiology of Food Fermentation*. AVI Publishing Co. Inc., Westport, CT. pp. 108–152.

Pederson, C.S. and Albury, M.N. 1950. The effect of temperature upon bacteriological and chemical changes in fermenting cucumbers. *Agr Expt Sta Bull*. p. 744.

Pederson, C.S. and Albury, M.N. 1969. The sauerkraut fermentation. *Agr Sta Bull*. p. 824.

Prajapati, J.P., Upadhyay, K.G., and Desai, H.K. 1991. Study on influence of post production heat treatment on quality of fresh *shrikhand. J Food Sci Technol* 28(6): 365–367.

Prajapati, J.P., Upadhyay, K.G., and Desai, H.K. 1992. Comparative quality appraisal of heated shirkhand stored at ambient temperature. *Austr. J. Dairy Technol.* 47: 18–22.

Prajapati, J.P., Upadhyay, K.G., and Desai, H.K. 1993. Quality appraisal of heated shrikhand stored at refrigerated temperature. *Cultured Dairy Prod. J.* 28(2): 14–17.

Puntambekar, P.M. 1968. Studies on Levels of Fat and Sugar on the Quality of Shrikhand and Estimation of Fat by Modifying the Gerber Test for Milk. MSc thesis, Sardar Patel University.V.V. Nagar, Ahmedabad, India.

Rai, S., Mehrotra, S., Dhingra, D., Prasad, M., and Suneetha, V. 2012. Preparation of curd in the presence of easily available prebiotic sources and study of their effect on physiochemical, sensory and microbiological properties of the curd. *Int J Pharma Sci Rev Res* 17(1): 40–43.

Ramchandran, L. and Shah, N.P. 2010. Characterization of functional, biochemical and textural properties of synbiotic low-fat yogurts during refrigerated storage. *LWT—Food Sci Technol* 43(5): 819–827.

Ramdas, A.R. and Kulkarni, P.R. 1987. Fermentative preservation of carrots. *Indian Food Packer* 4(5): 40–48.

Rangappa, K.S. and Achaya, K.T. 1974. *Indian Dairy Products*, 2nd edition. Asia Publishing House, Bombay, New Delhi. pp. 119–124.

Rao, D.R., Alhajali, A., and Chawan, C.B. 1987. Nutritional, sensory and microbiological qualities of Labneh made from goat milk and cow milk. *J. Food Sci.* 52: 1228–1230. DOI: 10.1111/j.1365-2621.1987.tb14049.x.

Rao, H.G.R., Thygraj, N., and Puranik, D.B. 1986. Standardised methods for preparation of *Shrikhand*—A popular fermented milk product. *Dairy Guide*, 8(11): 35.

Rao, R.E., Vijayendra, S.V.N., and Varadaraj, M.C. 2006. Fermentation biotechnology of traditional foods 25 of the Indian subcontinent. In: *Food Biotechnology*, Shetty, K., Paliyath, G., Pometto, A. and Levin, R.E. (eds.), 2nd edition. CRC Press Taylor & Francis, Boca Raton, FL. pp. 1759–1794.

Rasic, J.L. 1987. Yogurt and yogurt cheese manufacture. *Cult Dairy Prod J* 22: 6–8.

Reddy, K.K., Ali, M.P., Rao, B.V., and Rao, T.J. 1984. Studies on the production and quality of *shrikhand* from buffalo milk. *Indian J Dairy Sci* 37(4): 293–296.

Renuka, B., Kulkarni, S.G., Vijayanand, P., and Prapulla, S.G. 2009. Fructooligosaccharide fortification of selected fruit juice beverages: Effect on the quality characteristics. *LWT—Food Sci Technol* 42(5): 1031–1033.

Renuka, B., Prakash, M., and Prapulla, S.G. 2010. Fructooligosaccharides based low calorie *gulab jamun*: Studies on the texture, microstructure and sensory attributes. *J Texture Stud* 41(4): 594–610.

Robinson, R.K. and Tamime, A.Y. 2006. Types of fermented milks. In: *Fermented Milks*. Tamime, A. (ed.), Blackwell Publishing Ltd, Oxford. doi:10.1002/97 80470995501.ch1

Rodrigo, M., Alvarruiz, A., Villa, R., and Feria, A. 1985. Lactic acid controlled fermentation of fresh cucumbers and product quality. *Rev Agro Technol Alimen* 25(1): 104–116.

Ruei, L.J. and Lin, C.W. 2000. Production of Kefir from soymilk with or without added glucose, lactose, or sucrose. *J Food Sci* 65(4), 716–719.

Sagarika, E. and Pradeepa, J. 2003. Fermented foods of Sri Lanka and Maldives. In: *International Seminar and Workshop on Fermented Foods, Health Status and Social Well Being*, held at Anand, India, on November 13–14.

Sankaran, R. 1998. Fermented foods of the Indian sub-continent. In: *Microbiology of Fermented Foods*, Wood, B.J.B. (ed.), 2nd edition. Blackie Academic and Professional, London.

Sarkar, S. 2008. Innovations in fermented milk products—A review. *Food Biotechnol* 22(1): 78–97.

Sarkar, S., Kuila, R.K., and Misra, A.K. 1996b. Effect of incorporation of Gelodan™ SB 253 (Stabilizer- cum- preservative) and nisin on the microbiological quality of *shrikhand*. *Indian J Dairy Sci* 49: 176–184.

Savitri and Bhalla, T.C. 2007. Traditional foods and beverages of Himachal Pradesh. *Indian J Trad Know* 6(1): 17–24.

Schoneck, A. 1988. Des crudites toute l'annee: les legumes Lacto- fermentes. *Collection Les Vrais Aliments d' Aujourd'hui et de Demain*. Terre Vivante, Paris, p. 96.

Sekar, S. and Mariappan, S. 2007. Usage of traditional fermented products by Indian rural folks and IPR. *Indian J Trad Know* 6(1): 111–120.

Sethi, V. 1990. Lactic fermentation of black carrot juice for spiced beverage. *Indian Food Packer* 44(3): 7–12.

Sethi, V. and Anand, J.C. 1972. Effect of mustard on curing of longmelons for pickles. *Pros Hort* 3(1): 11.

Sethi, V. and Anand, J.C. 1984. Effect of mustard and its components on the fermentation of cauliflower. *Indian Food Packer* 38(4): 41–46.

Sharma, N. Gupta, A., and Gautam, N. 2011. Application of purified bacteriocin—Basicin produced from *Bacillus* sp. AG1 to improve safety and shelf life of dairy food products. *Int J Food Ferment Technol* 1(1): 119–127.

Sharma, R. and Rajput, Y.S. 2006. Therapeutic potential of milk and milk products. *Indian Dairyman* 58: 70–80.

Sharma, S. and Joshi, V.K. 2007. Influence of temperature and salt concentration on lactic acid fermentation of radish (*Raphnus sativus*). *J food Sci Technol* 44: 611–614.

Sharma, S., Joshi, V.K., and Lal Kaushal, B.B. 2008. Preparation of ready-to-serve drink from lactic acid fermented vegetables. *Indian Food Packer* 62(3): 17–21.

Sharma, S., Joshi, V.K., and Thakur, N.S. 2012. Lactic acid fermented foods. In: Joshi, V.K. and Singh, R.S. (eds.), *Food Biotechnology: Principles and Practices*. IK International Publishing House, New Delhi, pp. 375–415.

Sharma, U.P. and Zariwala, I.T. 1980. Deterioration of *shrikhand* during storage. *Indian J Dairy Sci* 33(2): 223–231.

Silva, J., Carvalho, A.S., Teixeira, P., and Gibbs, P.A. 2002.Bacteriocin production by spray—Dried lactic acid bacteria. *Lett Appl Microbiol* 34: 77–81.

Singh, H. and Tokley, R.P. 1990. Effects of preheat treatments and buttermilk addition on the seasonal variations in the heat stability of recombined evaporated milk and reconstituted concentrated milk. *Aust. J. Dairy Technol.* 45: 10–16.

Singh, R. and Jha, Y.K. 2005. Effect of sugar replacers on sensory attributes, biochemical changes on shelf life of *shrikhand*. *J Food Sci Technol* 42(2): 199–202.

Sodini, I., Morin, P., Olaki, A., and Jimenez-Flores, R. 2006. Compositional and functional properties of buttermilk: A comparision between sweet, sour and whey buttermilk. *J Dairy Sci* 89: 525–536.

Soni, S., Soni, R., and Janveja, C. 2013. Production of fermented foods. In: *Biotechnology in Agriculture and Food Processing Opportunities and Challenges*, Panesar, P.S. and Marwaha, S.S. (eds.), CRC Press, Taylor & Francis group, Boca Raton, FL. p. 637.

Steinkraus, K.H. 1996. *Handbook of Indigenous Fermented Books*. Marcel Dekker, New York, NY.

Subramonian, S.B., Naresh, Kumar, C., Narasimhan, R., Shanmugam, A.M., and Mohamedhabibulla Khan, M. 1997. Selection of level and type of 'LAB' starter in the preparation of dietetic *shrikhand*. *J Food Sci Technol* 34(4): 340–342.

Suresh, E.R., Yakashekhara, E., and Ethiraj, S. 1997. Preparation of brine stock pickling cucumbers by fermentation. *Indian Food Packer,* 51(3): 13.

Takagi, K., Toyoda, M., Saito, Y., Niwa, M., and Morimoto, H. 1990. Composition of fermented grape juice continuously produced by immobilized *Lactobacillus casei*. *J. Food Sci.* 55: 455–457.

Tamang, B. and Tamang, J.P. 2007. Role of lactic acid bacteria and their functional properties in *Goyang*, a fermented leafy vegetable product of the *Sherpas*. *J Hill Res* 20(2): 53–61.

Tamang, B. and Tamang, J.P. 2009. Traditional knowledge of biopreservation of perishable vegetables and bamboo shoots in Northeast India as food resources. *Indian J Trad Know* 8(1): 89–95.

Tamang, B. and Tamang, J.P. 2010. In *situ* fermentation dynamics during production *of gundruk* and *khalpi*, ethnic fermented vegetables products of the Himalayas. *Indian Journal of Microbiology* 50(Suppl 1): 93–98.

Tamang, B., Tamang, J.P., Schillinger, U., Franz, C.M.A.P., Gores, M., and Holzapfel, W.H. 2008. Phenotypic and genotypic identification of lactic acid bacteria isolated form ethnic fermented tender bamboo shoots of Northeast India. *Int J Food Microbiol* 121: 35–40.

Tamang, J.P. 2010. Himalayan Fermented Foods: Microbiology, Nutrition and Ethnic Value. CRC Press, Taylor & Francis Group, New York, NY.

Tamang, J.P., Dewan, S., Thapa, S., Olasupo, N.A., Schillinger, U., and Holzapfel, W.H. 2000. Identification and enzymatic profiles of predominant lactic acid bacteria isolated from soft-variety *chhurpi*, a traditional cheese typical of the Sikkim Himalayas. *Food Biotechnol* 14(1&2): 99–112.

Tamang, J.P., Dewan, S., and Holzapfel, W.H. 2004. Technological properties of predominant lactic acid bacteria isolated from indigenous fermented milk products of Sikkim in India. In: Proceeding Abstract of the 19th International ICFMH Symposium Food Micro 2004 on *"New Tools for Improving Microbial Food Safety and Quality: Biotechnology and Molecular Biology Approaches,"* organized by University of Ljubljana, Slovenian Microbiological Society and *International Committee for Food Microbiology and Hygiene*, pp. 12–16.

Tamang, J.P., Okumiya, K., and Yasuyuki, K. 2010. Cultural adaptation of the Himalayan ethnic foods with special reference to Sikkim, Arunachal Pradesh and Ladakh. In: *Himalayan Study Monographs, No.* 11, pp. 177–185.

Tamang, J.P. and Sarkar, P.K. 1993. *Sinki*: A traditional lactic acid fermented radish tap root product. *J Gen Appl Microbiol* 39: 395–408.

Tamang, J.P., Sarkar, P.K., and Hesseltine, C.W. 1988. Traditional fermented foods and beverages of Darjeeling and Sikkim—A review. *J Sci Food Agri* 44: 375–385.

Tamang, J.P., Tamang, B., Schillinger, U., Franz, C.M., Gores, M., and Holzapfel, W.I I. 2005. Identification of predominant lactic acid bacteria isolated from traditionally fermented vegetable products of the Eastern Himalayas. *Int. J. Food Microbiol.*, 105: 347–356.

Tamang, J.P., Tamang, N., Thapa, S., Dewan, S., Tamang, B., Yonzan, H., Rai, A.K., Chettri, R., Chakrabarty, J., and Khare, N. 2012. Microorganisms and nutritional value of ethnic fermented foods and alcoholic, beverages of North East India. *Indian J Trad Know* 11(I): 7–25.

Tamang, J.P., Thapa, N., Rai, B., Thapa, S., and Yonzan, H.S. 2007. Food consumption in Sikkim with special reference to traditional fermented foods and beverages: A micro-level survey. *J Hill Res Supply Issue* 20(1): 1–37.

Tamang, J.P. 2009. *Himalayan Fermented Foods: Microbiology, Nutrition and Ethnic Value.* CRC Press, Taylor & Francis Group, New York.

Tamime, A.Y. and Robinson, R.K. 1988. Fermented milks and their future trends. Part II. Technological aspects. *J Dairy Res* 55(2): 281–307.

Tamime, A.Y. and Robinson, R.K. 1999. *Yogurt: Science and Technology*. 2nd edition. CRC Press, Boca Raton, FL.

Tamime, A.Y. and Robinson, R.K. 2007. Traditional and recent developments. In: *Yogurt Production and Related Products, Tamime and Robinson's Yogurt*, A volume in Woodhead Publishing Series in Food Science, Technology and Nutrition. Woodhead Publishing. Cambridge, U.K., pp. 348–467.

Tamime, A.Y., Skriver, A., and Nilsson, L.-E. 2006. Starter cultures. In *Fermented Milks*. Tamime, A. (ed.), Blackwell Publishing, Oxford. doi:10.1002/9780470 995501.ch2

Thapa, T.B., Jagat Bhadur, K.C., Gyamtsho, P., Karki, D.B., Rai, B.K., and Limbu, D.K. 2003. Fermented foods of Nepal and Bhutan. In: *Proceedings of International Seminar and Workshop on Fermented Foods, Health Status and Social, Well-Being*, held at Anand, India, pp. 13–14.

Thokchom, S. and Joshi, S.R. 2012. Microbial and chemical changes during preparation in the traditionally fermented soybean product *Tungrymbai* of ethnic tribes of Meghalaya. *Indian J Trad Know* 11(1): 139–142.

Thompson, R.L., Fleming, H.P., and Monroe, R.J. 1979. Effects of storage conditions on firmness of brined cucumbers. *J Food Sci* 44: 843–846.

Toshirou, N. and Tamang, J.P. 2010. Fermented legumes: Soybean and non-soybean products. In: *Fermented Foods and Beverages of the World*. CRC Press, Taylor & Francis, Boca Raton, FL. pp. 191–224.

Trachoo, N. and Mistry, V.V. 1998. Application of ultra filtered sweet buttermilk and sweet buttermilk powder in the manufacture of non fat and low fat yogurts. *J Dairy Sci* 81: 3163–3171.

Upadhyay, K.G. and Dave, J.M. 1977. *Shrikhand* and its technology. *Indian Dairyman* 29: 487–490.

Upadhyay, K.G., Vyas, S.H., Dave, J.M., and Thakar, P.N. 1975a. Studies on chemical composition of market samples of *shrikhand. J Food Sci Technol* 12: 190–194.

Upadhyay, K.G., Vyas, S.H., Dave, J.M., and Thakar, P.N. 1975b. Studies on microbiological quality of market s*hrikhand. Indian J Dairy Sci* 28: 147–149.

Upadhyay, S.M. 1981. Assessing the Suitability of Different Microbiological and Chemical Tests as Keeping Quality Tests for Shrikhand. MSc thesis, Gujarat Agricultural University S.K. Nagar.

Vagdalkar, A.A., Chavan, B.R., Morkile, V.M., Thalkari, B.T., and Landage, S.N. 2002. A study on preparation of *Shrikhand* by using cocoa powder and papaya pulp. *Indian Dairyman* 54: 49–51.

Vanderghem, C., Blecker, C., Danthine, S., Deroanne, C., Haubruge, E., Guillonneau, F., De Pauw, E., and Francis, F. 2008. Proteome analysis of the bovine milk fat globule: Enhancement of membrane purification. *Int. Dairy J.* 18(9): 885–893.

Vanderghem, C., Bodson, P., Danthine, S., Paquot, M., Deroanne, C., and Blecker, C. 2010. Milk fat globule membrane and buttermilks: From composition to valorization. *Biotechnol Agron Soc Environ* 14(3): 485–500.

Vats, P. 2013. *Dahi or curd*, foodsafetyhelpline.com March 18.

Vaughn, R.H. 1975. Lactic acid fermentation of olives with special reference to California conditions. In: Carr, J.G., Cutting, C.V., and Whiting, G.C. (eds.), *Lactic Acid Bacteria in Beverages and Food*. Academic Press, New York.

Vaughn, R.H. 1985. Microbiology of vegetable fermentations. In: Wood, B. (ed.), *Microbiology of Fermented Foods*, Vol. 1. Elsevier, Amsterdam.

Vaughn, R.H., Doughlas, H.C., and Gilliland, J.R. 1943. *Production of Spanish Type Green Olives*. Bull. Univ. of California, Agr. Exp. Sta. Berkeley. p. 678.

Vedamuthu, E.R. 1991. The yogurt story-past, present and future. *Dairy Food Environ Sanit* 7: 371–374.

Vijayendra, S.V.N. and Gupta, R.C. 2011. Assessment of probiotic and sensory properties of dahi and yoghurt prepared using bulk freeze-dried cultures in buffalo milk. *Ann Microbiol* 62(3): 939–947.

Vokbeck, M.L., Albury, M.N., Mattick, L.R., Lee, F.A., and Pederson, C.S. 1963. Lipid alterations during the fermentation of vegetables by the lactic acid bacteria. *J Food Sci* 28(5): 495–502.

Wattiaux, M.A. 2012. *Milk Composition and Nutritional Value*. Babcock Institute for International Dairy Research and Development Dairy Essentials, University of Wisconsin, MA.

Whitfield, D.E. 2000. Solar drying systems and the Internet: Imporatant resources to improve food. In: *Solar Drying Presented at International Conference on Solar Cooking* Kimberly-South Africa, November 26–29.

Winkelhausen, E. and Kuzmanova, S. 1998. Microbial conversion of D-xylose to xylitol. *J Ferm Bioeng* 86: 1–14.

Yonzan, H. and Tamang, J.P. 2010. Microbiology and nutritive value of selroti, an ethnic fermented food of the Nepalis. *Indian J. Tradit. Knowl.* 8(1): 110–114.

9

INDIGENOUS ALCOHOLIC BEVERAGES OF SOUTH ASIA

ANILA KUMARI, ANITA PANDEY, ANTON ANN,
ANUP RAJ, ANUPAMA GUPTA, ARJUN CHAUHAN,
ARUN SHARMA, ARUP JYOTI DAS,
ASHWANI KUMAR, B.L. ATTRI, BHANU NEOPANY,
CHAMGONGLIU PANMEI, DEEPA H. DIWEDI,
DORJEY ANGCHOK, FOOK YEE CHYE,
GEORGE F. RAPSANG, GITANJALI VYAS,
GURU ARIBAM SHANTIBALA DEVI, IDAHUN BAREH,
JAHANGIR KABIR, JAYASREE CHAKRABARTY,
KHENG YUEN SIM, KONCHOK TARGAIS,
KUNZES ANGMO, L.V.A. REDDY, LOK MAN S. PALNI,
M. PREEMA DEVI, MANAS R. SWAIN,
MONIKA, NAVEEN KUMAR, NEELIMA GARG,
NINGTHOUJAM SANJOY SINGH, NIVEDITA SHARMA,
PREETI YADAV, RAMESH C. RAY, S.S. THORAT,
SANKAR CHANDRA DEKA, SATYENDRA GAUTAM,
SAVITRI, SHARMILA THOKCHOM, S.R. JOSHI,
SURESH KUMAR, SUSHMA KHOMDRAM,
TEK CHAND BHALLA, TSERING STOBDAN,
V.K. JOSHI, VANDITA CHAUHAN,
AND VIDHAN JAISWAL

Contents

9.1 Introduction

One of the most important fermented food products are the alcoholic beverages, and a number of such beverages have been prepared and consumed by mankind since ancient times in most of the world's cultures (Borgstrom, 1968; Joshi and Pandey, 1999; Sekar and Mariappan, 2007; Tamang, 2008; Joshi et al., 2011a; Kanwar et al., 2011). Traditional alcohol brewing has been a home-based industry, mostly carried out by rural women using indigenous knowledge of fermentation. The common knowledge of making fermented alcoholic beverages is believed to have originated from the Yunnan–Guizhou provinces of China.

9.1.1 Types of Alcoholic Beverages

A variety of indigenous alcoholic beverages is produced in South Asia by many tribes using fruits, cereals, grains, etc. These beverages can be classified into various types such as wine, beer, brandy, whiskey, etc. based on the use of raw material, processing steps, alcoholic fermentation (by yeast, principally *S. cerevisiae*), distillation of fermented materials, and post-distillation processing (Bluhm, 1995; Franz et al., 2011). Among the alcoholic beverages, beer and wine are produced by the fermentation of sugar or starch, while those made by fermentation followed by distillation have a greater quantity of alcohol and are known as spirits. Beer is one of the world's oldest and most widely consumed alcoholic beverages (mostly 4%–6% alcohol by volume, ABV) made by brewing and fermenting starches derived from cereal grains, malted barley, etc. Hops are the female flower clusters of a hop species, *Humulus lupulus*, and are used primarily as a flavoring agent, imparting a bitter, tangy flavor (Lichine, 1987; Arnold, 2005; Nelson, 2005). Wine, with alcohol content of 9%–16% (ABV), is produced primarily from grapes, but other fruits are also used (Anonymous Wikipedia; Vine, 1983; Joshi, 1997; Joshi et al., 1999; Verma and Joshi, 2000; Joshi et al., 2011b; Joshi et al., 2012a) When wine is distilled it produces brandy. Wine consumption is associated with protection in cardio vascular disease, due to phenolic compounds (Joshi et al., 2004; Joshi, 1997) as phenolic compounds have been associated with protection from several diseases including cardiovascular diseases (Arts et al., 2000), which are found in several foods including wines.

9.1.2 Alcoholic Beverages in Health and as Therapeutic Beverage

These beverages are inexpensive, palatable, safe, and nutritious. They serve as a source of calories, are important for celebrations and ritual ceremonies, and are even incorporated in various food preparations. For example, *chhang* is supposed to provide energy and refreshment (Targais et al., 2012). People used to drink palm wine in the early morning on an empty stomach, before breakfast, for health benefits. The wine had a special place in traditional celebrations and ceremonies such as marriages and festivals. *Toddy* is believed to be beneficial for the health and eyesight, and is sedative for the consumers. The medicinal importance of alcoholic beverages is also discussed in "Ayurveda" in which they are called "madya." Wine is also a mild laxative beverage, relieving constipation (Sekar and Mariappan, 2007). Various positive health effects have been attributed to the consumption of fermented alcoholic beverage such as rice beer, being effective against insomnia, headache and body ache, inflammation of body parts, diarrhoea and urinary problems, expelling worms, and as a treatment for cholera (Samati and Begum, 2007; Deka and Sharma, 2010). Many more such examples can be cited from other countries of south Asia (see Chapter 6 of this text for more information).

9.1.3 Alcoholic Fermentation

Natural or spontaneous fermentation is normally carried out to produce indigenous alcoholic beverages. Such fermentations may contain a mixed microflora, often a lactic–alcoholic–acetic one, resulting in higher acidity in the final product (see Chapter 3 for more information). Besides sugar-rich material such as fruits, alcoholic fermented foods/beverages can also be produced after starch hydrolysis by amylolytic moulds, and the fermentation carried out by yeasts and flavor enhancing lactic acid bacteria. The combination of acids and alcohol *via* microbial reaction during fermentation eventually enhances the aroma of the end product. In traditional forms of alcoholic beverages, it is difficult to control the product quality, and hygienic conditions during processing are generally lacking (Joshi and Sandhu, 2000; Joshi, 2005). Besides, knowledge of significance of the presence of a number of metabolites and their chemical interactions, health relevant constituents, and contaminants in these alcoholic beverages and foods (Law et al., 2011) is essential. For more details, see Chapter 6 of this text.

9.1.4 Consumption of Alcoholic Beverages and Culture

Generally, the consumption of alcoholic beverages is socially accepted by people around the world. Fermented alcoholic beverages of different forms have been consumed by the tribal peoples of North-East India since times immemorial (Tamang et al., 2012). More than 250 types of ethnic fermented foods and alcoholic beverages are produced and consumed in North East India (Tamang, 2010; Tamang et al., 2012). Rice beer preparation and consumption is very common among many tribal peoples of north-eastern states of India for ages (Ghosh and Das, 2004; Jeyaram et al., 2008). The consumption of alcoholic beverages throughout South Asian countries has its roots in many cultural and religious practices of the people and has practically no ill effects upon the health of the hard working population. The consumption of such beverages has strong ritual importance among the ethnic people in the Sikkim Himalayas, where social activities require the provision and consumption of appreciable quantities of alcohol (Tamang et al., 1996). *Losar* (the New Year of the *Kinnauris* and the *Ladhakis*) and *phagli* (the New Year of the *Lahaulis*) are major events for the preparation of *chhang/lugri*, which is also associated with all farm activities and every auspicious occasion and celebration among Buddhists in Ladakh. It is offered to guests, priests, and even dieties. When offered to someone held in high esteem, the drink is referred to as *skyems*. *Chapskyen* is the brass pot used to serve *chhang*, and *donskyok* is the cup/bowl to drink it from. In the tantric system of Buddhism, one of the most popular forms practiced in Ladakh, *chhang* is an indispensable part of the rituals. It is looked upon as *ambrosia* and is integrated into the socio-cultural fabric of Buddhist Ladakh. A proposal of marriage from the groom's family is invariably associated with an offer of *chhang*, acceptance of which by the bride's family conveys their assent to the request. During the marriage ceremony, and on some special occasions, guests are welcomed by women in traditional dresses holding decorated *Chapskyen* filled with *chhang*. At the time of childbirth, relatives and friends bring *chhang* along with other gifts for the family.

Judima is an integral part of ritual for the *Dimasa*. During religious festivals, freshly prepared *judima* is offered to the family gods and goddesses (Chakravarty et al., 2013).

During birth, a drop of *judima* is administered to the lips of the new born baby to protect from any evil force. *Judima* and *juharo* are essential also to solemnize marriage ceremonies. Traditionally, the newly wedded bride visits her parents' house once in a year, and when she returns to her husband's house she carries *judima*. The traditional festival *busudima* of the Dimasa is celebrated with freshly prepared *judima* (Chakravarty et al., 2013). In death ceremonies also, freshly prepared *judima* is offered to the dead persons and also to the ancestors. Without *judima* no celebration or religious ceremony is complete, and guests are also served *judima*. Like other indigenous fermented foods, the native skills for preparing *judima* have been passed from one generation to another. The role of indigenous women in the North Cachar Hills district of Assam in the production and marketing of ethnic fermented foods and beverages has been highlighted. (Chakrabarty et al., 2014).

In this chapter, we focus on the methods employed in the production of indigenous fermented alcoholic beverages from various substrates, using the indigenous technology and knowledge.

9.2 Diversity of Alcoholic Beverages of South Asia

Many types of alcoholic beverages are found in South Asian regions such as: (1) rice and cereal wine, (2) palm wine, (3) distilled spirit from rice, cereal, or palm wine, (4) different wines using fruits or their combinations with *jaggery*, (5) flower-based wines, and (6) distilled alcoholic beverages (Thakur et al., 2004). Raw materials used for brewing methods differ accordingly. For many centuries, the people of Himalayan regions prepared and consumed more than forty varieties of common as well as lesser-known indigenous fermented foods and beverages (Tamang, 2001). The high altitude Himalayan region is characterized by diverse ethnic groups, who have developed their own cultures for fermentation based on available natural resources, giving rise to a cultural diversity at par with the high level of biological diversity found in the region. Amongst the inhabitants of the high Himalayas, the making and use of fermented foods and beverages using local food crops and other biological resources is very common (Roy et al., 2004). In the North Eastern region of India, several alcoholic beverages (Table 9.1) are traditionally made and consumed, and even marketed locally (Tamang, 2001). Even preparation of traditional fermeted foods and beverages including alcoholic beverages is a means of livelihood (Elain and Danilo, 2012) in several countries of South Asia. It remains an important household-cum-societal drinking activity associated with religious ceremonies among different ethnic tribal groups in North East India (Sharma and Mazumdar, 1980).

The *Bhotiya* tribes of Uttarakhand produce *jann* and *daru* as the common fermented drinks of the community. Among the fermented foods, is the semi-fermented rice called *sez*, which is taken as a light snack by these communities (Roy et al., 2004). Rice wine from glutinous rice is also a popular traditional alcoholic beverage, manufactured under non-sterile condition at home scale using traditional solid flat rice cakes as a starter (Jeyaram et al., 2008). Manipur, one of the states in the North

Table 9.1   Ethnic Alcoholic Beverages of north east India, along with their substrates, starters and consumers

ALCOHOLIC BEVERAGE	SUBSTRATE	AMYLOLYTIC STARTER	CONSUMERS/REGION
Kodo ka jaanr	Finger millet	*Marcha*	*Gorkha*
Bhaati ka jaanr	Rice	*Marcha*	*Gorkha*
Makai ka jaanr	Maize	*Marcha*	*Gorkha*
Gahoon ka jaanr	Wheat	*Marcha*	*Gorkha*
Faapar ka jaanr	Buckwheat	*Marcha*	*Gorkha*
Jao ka jaanr	Barley	*Marcha*	*Gorkha*
Poko	Rice	*Manapu*	*Gorkha*
Raksi	Cereal	*Marcha*	*Gorkha*
Arak	Cereal	*Phab*	*Bhutia, Tibetan*
Chayang	Finger millet/barley	*Phab*	*Bhutia, Tibetan*
Chee	Finger millet/barley	*Buth*	*Lepcha*
Atingba	Rice	*Hamai*	*Meitei*
Yu	Rice	*Hamai*	*Meitei*
Jou	Rice	*Khekhrii*	*Naga*
Zutho	Rice	*Khekhrii*	*Naga*
Duizou	Red rice	*Khekhrii*	*Naga*
Nchiange	Red rice	*Khekhrii*	*Naga*
Ruhi	Rice	*Khekhrii*	*Naga*
Madhu	Rice	*Yeast, Moulds*	*Naga*
Dekuijao	Rice	*Nduhi*	*Naga*
Apong	Rice	*Ipoh*	*Monpa, Apatani, Nishi, Adi*
Pona	Rice	*Ipoh*	*Monpa, Apatani, Nishi, Adi*
Ennong	Rice, paddy husk	*Ipoh*	*Monpa, Apatani, Nishi, Adi*
Oh	Rice-millet	*Ipoh*	*Monpa, Apatani, Nishi, Adi*
Themsing	Finger millet/barley	*Ipoh/ Siye*	*Monpa, Apatani, Nishi, Adi*
Mingri	Maize/rice/barley	*Phab*	*Monpa, sherdukpen, memba*
Lohpani	Maize/rice/barley	*Phab*	*Monpa, khamba*
Aara	Cereals	*Phab*	*Monpa, khamba*
Kiad-lieh	Rice	*Thiat*	*Khasi*
Judima	Rice	*Humao*	*Dimasa*
Juharo	Rice	*Humao*	*Dimasa*
Zu	Rice	*Humao*	*Dimasa*
Juhning	Rice	*Chol*	*Hrangkhol*

Source: Adapted from Tamang, J.P. et al. 2012. *Indian Journal of Traditional Knowledge*, 11(9): 7–25.

Eastern India is inhabited by different communities such as Meitei, Nagas, Kukis and Meitei Pangals, of Mongoloid and Indo-Aryan stock. Some ethnic groups within the major tribal groups of the Nagas and Kukis still brew alcoholic beverages. These products are similar to *shaosingiju* and *laochao* from China, *sake* from Japan, *chongju* and *takju* from Korea, *brem bali*, *tape-ketan*, and *tapuy* from Indonesia, *khaomak* from Thailand, and *tapai pulul* from Malaysia (Lee, 2009; Das et al., 2014). The high altitude residents of Uttaranchal state, comprised of two regions, Gharwal and Kumaon, make the local fermented beverage made from barley "*chakti*" in Dharchula, "*daru*" in Munsyari, and "*chang*" in Chimoli and Uttarakashi (Roy et al., 2004). In Sri Lanka, *toddy* is a traditional alcoholic beverage prepared and consumed widely. There are similar products from other countries, like those from Nepal, such as *rakshi*. Based on diversity and consumption, it can be concluded that there is a large scope of indigenous fermented especially alcoholic beverages for production at commercial scale for extensive marketing, as observed earlier also (Tamang, 2011).

9.3 Ethnic Starter Cultures Used in Alcoholic Beverages Preparation

Microbial inocula, in the form of dry powders or hard balls known as starters, are used to carry out the fermentation process (Tsuyoshi et al., 2005). These starter preparations have a variety of names, such as *marcha* or *murch*, in the Himalayan regions of India, Nepal, and Bhutan. Various ethnic cultures used in different parts of India are listed in Table 9.2. In Asia, three major types of starter are commercially produced (Takeuchi et al., 2006; Tamang and Fleet, 2009), as described here.

First type: Here the starter is a mixed culture of mycelial or filamentous molds, yeasts, and lactic acid bacteria (LAB), along with rice or wheat as the base, in the form of dry, flattened or rounded balls of various sizes (Tamang, 2009a,b). It is made by inoculation with a previous starter. This mixed flora is allowed to develop for a short time, then dried, and used to make either alcoholic or fermented foods from starchy materials. Some of these ethnic starters are called *ragi* and *marcha* (Tamang, 2010).

Second type: This type of starter is a combination of *Aspergillus oryzae* and *A. sojae*, and is called *koji*. The term *koji* is Japanese word, meaning naturally, spontaneously, or artificially moulded cereals and beans (Yokotsuka, 1991; Tamang, 2012).

Third type: Whole-wheat flour, with its associated flora, is moistened, made into large compact cakes and then, incubated to select certain desirable microorganisms. This type of starter contains yeasts and filamentous molds, and is mostly used for alcohol production (Tamang, 2010).

The various starters frequently employed in various countries of South Asia are described in the subsequent sections.

9.3.1 Ragi

Ragi is an ethnic starter in the form of dry and flat cakes (Tamang, 2012). To produce it, rice or millet or cassava or other starchy bases are milled, mixed with herbs and spices, roasted together, and then, sieved. The mixture is mixed with water and 2%–4% powder of old *ragi*, mixed thoroughly, shaped into balls, and is fermented at 25–30°C for 72 h, in a humid environment. Fermented balls are sun dried and used as a starter in the preparation of alcoholic beverages and drinks similar to that in Indonesia (Saono et al., 1974).

9.3.2 Koji

Koji is a mould starter culture used to produce alcoholic beverages. During *koji* preparation, rice is steamed, cooled on bamboo trays, stacked with gaps of about 10 cm in

Table 9.2 Ethnic Starters of India Used in Production of Alcoholic Beverages

ETHNIC STARTER	SUBSTRATE	APPEARANCE	BEVERAGES	ORGANISMS	PLACE	REFERENCES
Bakhar	Ginger and some plant materials and rice flour	Dry cakes	Pachwai	Rhizopus, Mucor and yeast	North India	Hutchinson et al. (1925)
Marcha	Rice, wild herbs, spices	Dry mixed starter	Bhaati Jaanr, Kodo ko jaanr and Raksi	Molds, yeasts, LAB	Darjeeling and Sikkim	Tamang et al. (2007)
Hamei	Rice, wild herbs	Dry mixed starter	Atingbai	Molds, yeasts, LAB	Eastern Himalaya	Jeyaram et al. (2009) and Tamang et al. (2007)
Phab	Rice, wild herbs	Dry mixed starter	Themsing	Yeast	Arunachal Pradesh	Singh et al. (2007)
Ipoh	Rice, wild herbs	Dry mixed starter	Apong, Ennog	Yeast	Arunachal Pradesh	Tiwari and Mahanta (2007)
Mod pitha	Rice and 31 plant materials	Small round cakes	Sujen	–	Assam and Arunachal Pradesh	Deori et al. (2007)
Vekur pitha	Rice, wild herbs	Dry plate-disc-shaped	Ahom	Yeast	Assam	Saikai et al. (2007)
Thiat	Rice, wild herbs	Dry cake	Kiad	Yeast	Meghalaya	Samati and Begum (2007)
Ranu dabai	Rice, herbal plants	Dry small tablets	Jhara or Harhia.	Mold	West Bengal	Sekar and Mariappan (2007)
Keem	Barley, plants part	Dry cakes	Soor	–	Uttaranchal	Rana et al. (2004)
Balam	Wheat, herbs and spices	Dry balls	Jann, daru	–	Uttaranchal	Sekar and Mariappan (2007)
Dhehli	Barley, 36 wild herbs	Dry brick shaped	Sura	Mold, yeast	Himachal Pradesh	Thakur et al. (2004)
Phab	Barley, herbs	Dry, small ball, cakes	Chhang, aarak	Mold, yeast	Himachal Pradesh, Ladakh, Tibet	Bhatia et al. (1977); Thakur et al. (2004)

Source: Adapted from Tamang, J.P. 2010. *Fermented Foods and Beverages of the World*. CRC Press, Taylor & Francis, Boca Raton, pp. 85–125.

between to allow air circulation, inoculated with 0.1% mold spores, and incubated at 23–25°C. The rise in temperature due to the growth of mold is kept within the range of 35–45°C by stirring and turning the *koji* from the top to the bottom on trays at about 20–40 h intervals; it is normally fermented for three days or till the mycelium grows and spreads, but does not sporulate (Lotong, 1985; Tamang, 2010).

9.3.3 Mana

Mana is a granular-type starter prepared in Nepal from wheat flakes (Tamang, 2010). To prepare it, wheat grains are soaked in water overnight, steamed for 30 min, transferred to a bamboo basket, drained, and ground into a lump. The floor is cleaned, straw is spread on the ground, and the wheat lump is placed over it, covered with paddy straw or a straw mat, and fermented for 6–7 days, after which a green mold appears on the wheat grains. This *mana* is then, dried in the sun and stored (Tamang, 2010) (Table 9.3).

9.3.4 Dhehli

Dhehli is a herbal mixed starter used to produce a traditional beverage of Himachal Pradesh called *sura* or *sur*. It is prepared as an annual community effort, in which elderly people go to the forests on the 20th day of *Bhadrapada* month (usually 5th or 6th of September) and collect approximately 36 fresh herbs (Thakur et al., 2004; Tamang, 2010; Sharma, 2013; Joshi et al., 2015). Some of the important herbs used in *dhehli* preparation are *Pistacia integerrima* (*kkakar shinga*), *Solanum xanthocarpum* (*katari*), *Clitoria ternatea* (*kkayal*), *Aegel marmelos* (*bhel*), *Viola cinerea* (*banaksa*), *Cannabis sativa* (*bhang*), *Trachyspermum copticum* (*ajwain*), *Micromeria biflora* (*chharbara*), *Spiranthes australis* (*bakarshingha*), *Saussurea* sp. (*bbacha*), *Bupleurum lanceolatum* (*nimla*), *Drosera lunata* (*oshtori*), *Salvia* sp. (*kotugha*), *Arisaema helleborifolium* (*chidi ri chun*), and *Fragaria* sp. (*dudlukori*) (Thakur et al., 2004). The collected herbs are crushed in a stone with a large conical cavity (*ukhal*), using a wooden bar (*mussal*) and the extract, as well as the plant biomass, are added to roasted barley flour and roughly kneaded. This dough is put into a wooden mold, to give the shape of a brick, that is dried, and is called *dhehli* (Thakur et al., 2004; Tamang, 2009; Sharma, 2013).

9.3.5 Phab

Phab is the traditional inoculum used for the preparation of traditional alcoholic beverages in Ladakh and the hilly areas of Kullu and Kinnaur in Himachal Pradesh

Table 9.3 Ethnic Starters of the South Asia Employed in Production of Alcoholic Beverages

ETHNIC STARTER	SUBSTRATE	APPEARANCE	PRODUCTS	ORGANISMS	COUNTRY
Marcha	Rice, wild herbs, spices	Dry, mixed starter	*Kodo ko jaanr, bhaati, jaanr, gahoon ko, jaanr, makai ko, jaanr, raksi*	Molds, yeasts, LAB	India, Nepal
Mana	Wheat, herbs	Dry, granulated starter	Alcoholic drinks	*A. oryzae*	Nepal

Source: Adapted from Tamang, J.P. and Kailasapathy, K. 2010. *Fermented Foods and Beverages of the World*. CRC Press, Taylor & Francis, Boca Raton, FL; Tamang, J.P. 2010. *Fermented Foods and Beverages of the World*. CRC Press, Taylor & Francis, Boca Raton, FL, pp. 85–125.

(Joshi and Sandhu, 2000; Thakur et al., 2004; Tamang, 2010). Traditionally, it is prepared only in the Nubra valley and is supplied throughout Ladakh as a source of yeast for making *chhang*. The ingredients of *phab* include ground roasted barley (*nasphey*), powdered black pepper (*phoarulu*), dried ginger (*che'ga*) powder, crushed paddy (*so'a*), wild herbs/medicinal plants (*mansa*), and *phab* made earlier as the starter. All these ingredients are kneaded and rolled into small balls which are then, spread over a bed of branches of the wild shrub, *burtse* (*Artemisia* sp.), and also covered over with *burtse* branches. It is then, covered with a blanket, allowed to ferment for 2–3 days, dried, and stored or sold (personal communication with Padma Yangzom, a lady resident of Nubra). *Phab* is reported to remain potent for about a year. Microbiological analysis of the *phab* starter culture has revealed a mixed microflora consisting of yeasts, actinomycetes, and bacillus species (Thakur et al., 2004). The commonly encountered organisms in *Phab* are *Saccharomyces cerevisiae* and *Candida* species, and *Zygosaccharomyces bisporus* and *Kluveromyces thermo–tolerances* (Thakur et al., 2004). The predominant bacteria in the fermented products includes species of *Leuconostoc* and *Lactobacillus*. Though not much information on the microbiological aspect of *chhang* is available, three strains of yeast species in *phab* have been reported, which include one strain of *Candida krusei* and two strains of *Saccharomyces uvarum* (Bhatia et al., 1977).

9.3.6 Keem

Keem is a starter cake from the Tons valley located in Garhwal Himalaya of Uttaranchal, used for the preparation of a beverage called *soor*. For *keem* preparation, the fresh twigs of *Cannabis sativa*, leaves of *Sapindus mukorossi*, and approximately 40 other herbal plant materials are dried in the shade, powered, and mixed with barley flour. The desired quantity of the above dried flour is mixed with a sufficient quantity of *jayaras* (an infusion prepared from the finely chopped leaves and tender parts of *Melica azedarach*, *Zanthoxylum armatum*, *Leucas lanata*, and *Dicliptera roxburghiana*) in a large container, left overnight, and then, made into a round cake. The cakes are stacked on a bed made from the tender shoots of *Cannabis sativa* and *Pinus roxburghii*, layered alternately between the cakes, in a closed room, and left undisturbed for 24 days. On the 25th day, the cakes are turned upside down and again kept undisturbed for 12 days. Then, they are dried in the sun or open air and stored (Rana et al., 2004; Sekar and Mariappan, 2007).

9.3.7 Balam or Balma

Balam is a wheat-based starter used in the fermentation of the two common traditional beverages *jaan* and *daru* by the Bhotiya community of Uttaranchal Himalaya (Das and Pandey, 2007). It is also made in Kumaun, and is known as *balma* in the Garhwal regions of Uttarakhand. The traditional knowledge for making the starter culture was obtained from the grazier community, known as *anwals*, of the alpine regions. First, the raw wheat is washed in water and then, sun dried; then it is ground into flour and roasted over a fire, but removed before it becomes brown in color. The roasted flour is then, mixed with spices like *Cinnamomum zeylanicum*, *Amomum subulatum*, and *Piper longum*, and the leaves of wild chillies and the seeds of *Ficus religiosa* (Roy et al., 2004). To this mixture, old *balam* powder is also added and thoroughly mixed with the required quantity of water. The mixture is kneaded into a thick paste, and then, rolled into balls of the required size, using the hands. These balls (Figure 9.1a) are then, dried in the shade and stored for future use (Roy et al., 2004; Das and Pandey, 2007). Various ingredients used in the preparation of balam are shown in Table 9.4.

Balam starter collected from the *Johar* valley of Pithoragarh district has been analysed for microbial characteristics (Das and Pandey, 2007). The bacterial and fungal cultures, including yeasts, were obtained from the prescribed media (Figure 9.1(b)), while maximum counts of bacteria (21.3×10^5 cells/g sample) were recorded on nutrient agar and maximum yeast colonies (26.2×10^5 cells/g sample) were recorded on malt extract agar. Microscopic examination of distinct colonies revealed two types of bacteria and three types of yeasts (Figure 9.1(d–f)). After biochemical characterization, the yeast colonies were assigned to *Saccharmycopsis fibuligera*, *Kluveromyces maxianus*, and *Saccharomyces* sp. Two of these yeast cultures, *K. maxianus* and *S. fibuligera*, have been given the accession numbers MTCC 7313 and MTCC 7314, respectively, by the Microbial Type Culture Collection and Gene Bank, Institute of Microbial Technology, Chandigarh, India. The two types of rod-shaped Gram +ve

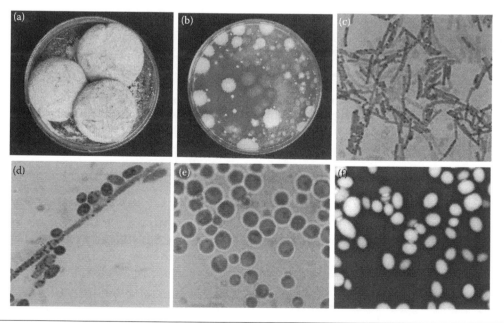

Figure 9.1 (a) The starter culture (*balam* balls), (b) yeast and bacterial colonies on malt extract agar, (c) the bacilli, (d) *Saccharmycopsis fibuligera*, (e) *Kluveromyces maxianus*, and (f) *Saccharomyces* sp. (Bar = 2 μm).

Table 9.4 Ingredient Required in the Making of *Balam* or *Balma*

NAME OF THE INGREDIENT	QUANTITY
Wheat flour	1 kg
Clove powder	5–10 g
Cardamom powder	5–10 g
Pepper powder	20–30 g
Old balma powder	40–60 g
Mirchi powder	2–3 g
Pipal seeds (*Ficus religiosa*)	3–4 g

Source: From Roy, B. et al. 2004. *Journal of Human Ecology*, 15(1): 45–49. With permission.

bacteria isolated from wheat balls were assigned to the genus *Bacillus*. The importance of various species of *Bacillus*, mainly *B. subtilis* and *B. cereus*, from fermented foods has been well documented (Roy et al., 2007).

9.3.8 Marcha or Murcha

Marcha is dry, round-to-flattened, creamy-white-to-dusty-white, solid ball amylolytic culture used for the production of various indigenous alcoholic beverages, such as *jaanr*, in the Darjeeling hills and Sikkim in India, Nepal, and Bhutan (Tamang and Sarkar, 1996; Tsuyoshi et al., 2005; Dung et al., 2006; Tamang et al., 2007; Jeyaram et al., 2008; Tamang, 2012; Krishna and Baishya, 2013). In its preparation, glutinous rice is soaked in water for 6–8 h and crushed in a foot-driven heavy wooden mortar and pestle. To 1 kg of ground rice, are added the roots of *Plumbago zeylanica* L., 2.5 g; the leaves of *Buddleja asiatica* Lour, 1.2 g; the flowers of *Vernonia cinerea* (L.) Less, 1.2 g; ginger, 5.0 g; dried red chilli, 1.2 g; and some previously prepared *marcha* as the mother culture, 10.0 g (Tamang, 2010). The mixture is kneaded into flat cakes of varying sizes and shapes and placed individually on a bed of fresh fronds of the ferns (*Glaphylopteriolopsis erubescens*) on a platform of bamboo strips suspended in the kitchen below the ceiling, which is then, covered with dry ferns and jute bags. These cakes are left to ferment for 1–3 days before being sun-dried for 2–3 days, after which they can be stored in a dry place for more than a year (Tamang, et al., 1996; Thapa, 2002; Tsuyoshi et al., 2004, 2005). *Marcha* is similar to the *phab* produced in the Nubra valley in Ladakh and Manali in Himachal Pradesh (Tsuyoshi et al., 2005).

9.3.9 Hamei

Hamei is a round or flattened starter from Manipur, used to prepare a rice-based beverage locally called *atingba* and a distilled clear liquor called *yu*. It is mostly prepared in the summer (May–July) and stored for use during the year (Tamang, 2010).

During its production, local varieties of raw rice, either without soaking or soaked for 30 min and then dried, are crushed and mixed with the powdered bark of *yangli* (*Albizia myriophylla*) and a pinch of previously prepared powdered *hamei*. To this mixture, boiled water in 1:1 is added to make a paste, which is hand pressed into flat discs of different shapes and sizes. The pressed cakes of *hamei* are kept over paddy husk in a bamboo basket for 2–3 days at room temperature and then sun dried for 2–3 days (Tamang et al., 2007; Jeyaram et al., 2009; Tamang, 2010).

9.3.10 Ranu Dabai

It is a starter from West Bengal used for the preparation of the beverage *jhara* or *harhia*. To prepare *ranu dabai*, rice is washed with clean water and the decanted water is stored. Twelve fresh herbal plant materials are chopped, ground thoroughly along with the rice and some old *ranu dabai* using wooden husking machine called a *dhiki*. The gound mixture is made into paste, using the decanted rice water. Clean gunny bags are spread on the floor inside shaded rooms, and the paste is made into tablets by hand and arranged in rows on the gunny bags, where they are left for 40–60 min to dry slightly (Sekar and Mariappan, 2007). They are then, arranged in a single layer in a large bamboo basket, covered with straw, and another layer of tablets placed on top. This process is repeated until the basket is filled. Finally, a large amount of straw is placed on top and the entire basket is covered by polythene sheets or gunny bags and stored in dark warm place for 2–3 days. During this time a layer of fungal mycelia develops on the tablets, which are then, dried in the sun for 7–8 days and stored for future use (Sekar and Mariappan, 2007).

9.3.11 Humao

Humao is the traditionally prepared mixed amylolytic dough inocula used as a starter for the production of various indigenous alcoholic beverages in North Cachar Hills district of Assam. *Humao* is a dry, flat- rounded and oval, creamy white to dusty white, solid starter ranging from 1.4 cm to 10.7 cm in diameter with the weight ranging from 20 g to 25 g. Dimasa calls it *Humao*, Hrangkhol calls it *Chol* and Jeme Naga calls it *Nduhi*.

During the traditional method of preparation, bark of *Albizia myriophylla* Benth. (Family -Mimosaceae) is used. Sticky rice was soaked for 10–12 h at room temperature and there after it was crushed with bark of *Albizia myriophylla* and 1%–2% of previously prepared starter in the form of powder. The mixture is then made into paste by adding water and kneaded into flat rounded and oval cakes of varying sizes and kept it for one to three days at 25–30°C and sun dried . The traditional method of preparation of *Humao* is protected as hereditary trade and passes from mother to daughters (Chakrabarty et al., 2013).

9.4 Alcoholic Beverages Production where Sugars are the Main Fermentable Carbohydrates

9.4.1 Daru/Ghanti/Liquor

The alcoholic drink made using the flowers of *Dhataki* (*Woodfordia fruitions* (L.) Kurz), water, and *gur* (*jaggery*) is called a *gauda* (Sekar, 2007). It is said to promote the power of digestion, complexation, and strength. *Gauda* eases passage of flatus and feces and thus, improves appetite (Sekar, 2007).

Daru is a *jaggery*-based traditionally fermented beverage produced and consumed in rural areas of Shimla, Kullu, Kinnaur, and other regions of Himachal Pradesh (India). Also called *chakti* in the Kullu valley (Thakur et al., 2004), it is prepared by adding the locally available inoculum known as *phab* to a mixture of *gur* and water (Figure 9.2). Babool wood, locally called *kikar* (*Acacia nilotica*), is added to give taste and aroma to it. The mixture is then, heated and left to simmering heat for 4–5 days to carry out the fermentation (Thakur et al., 2004). After the completion of fermentation, the mixture is filtered through a cloth and the filtrate is collected as *daru* (Thakur et al., 2003). In a few places, the product is made by distillation of *jaggery* fermented liquor, which is carried out by traditional distillation method, as described below:

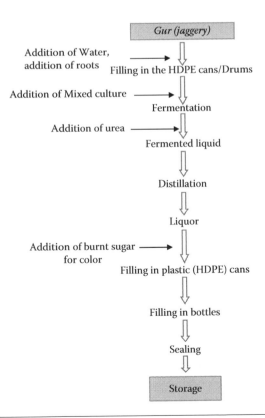

Figure 9.2 Flowsheet of production of *Daru/ghanti/liquor.*

Figure 9.3   Traditional distillation system being practised in rural areas of Kinaur, HP (India). (Courtesy of Dr. Bharati Negi.)

The traditional distillation set for *daru* consists of 3 parts:

- *Parar* (a saucepan like container with a flat bottom)
- *Jokhal* (a flat wooden device having an elongated channel with a hole at the center)
- *Tal* (a simple cooking vessel)

A similar distillation apparatus, as described above, is used by the tribal people of Himachal Pradesh and Utteranchal (Joshi and Sandhu 2000; Roy et al., 2004). *Daru* collected in the first 3–4 bottles during the process of distillation contain a very high percentage of alcohol, whereas the final few bottles have a low alcohol content (*piskani*); the bottles in between have a moderate amount of alcohol, and are rated good for consumption (Roy et al., 2004).

The pictorial representation of the production *ghanti* in Kinnaur area of Himachal Pradesh is shown in Figure 9.3.

9.4.2 Sugar Cane-Based Wines/Alcoholic Beverages

The drink *Jann* is also prepared using fruits like apple, banana, pumpkin, and orange, instead of cereals, using *balam* as the starter; the procedure used is essentially similar to that described earlier. Cane sugar is used to increase the sugar content prior to fermentation. Apples (for example) are first cut into pieces and then, mixed with *balam* powder for fermentation (Roy et al., 2004). Bananas are used without removing the outer skin. The preparation of *jann* from pumpkin is slightly different: a small hole is made in a large-sized pumpkin in such a way that the cut piece can be fitted back into the hole. First the seeds and loose tissues of the pumpkin are removed through the opening and boiled rice or other substrate mixed with *balam* powder is poured into the empty hollow pumpkin (Roy et al., 2004). It is then, re-sealed placing the cut piece back in the hole. The process of fermentation takes place inside, and the rice, along with the inner soft tissue of the fruit, is fermented, yielding *jann* in due course.

9.4.3 Fruit-Based Fermented Beverage

The different Naga tribes usually prepare different kinds of beverages from fruits like Naga apple (*Docynia indica*), passion fruit (*Passiflora edulis*), plum (*Prunus* sp.), and gooseberry (*Phyllanthus emblica*). The steps involved in preparation of fermented beverages are shown in Figure 9.4. The fruits are collected and boiled after removing the seeds (Jamir and Deb, 2014). The boiled fruit pulp is soaked in sugar syrup for ~1–2 week for fermentation. The fermented product is taken as a beverage.

9.4.4 Mead/Honey Wines

Honey was the only concentrated sugar widely available in pre-historic times, and it has been used from times immemorial to prepare the wine called "mead" (Joshi et al., 1990). It is nutritious, containing many nutrients required by man, and has an excellent effect on digestion and metabolism (Ioyrish, 1974; Lee et al., 1985; Joshi et al., 1999). It has been suggested that a half empty honey pot left in the rain, might have been the source of the first alcoholic drink (Brothwell and Brothwell, 1969). Available archaeological evidence for the production of mead dates back to 700 BC. Pottery vessels containing mixtures of mead, rice, and other fruits, with the organic compounds of fermentation, have been found in Northern China (Anonymous http://www.absoluteastronomy.com/topics/Mead). The first known description of mead is found in the Rigveda dating back to 1700–1100 BC (Gupta and Sharma, 2009).

Honey contains about 80 different substances, but 95%–99% of the total solids are sugars. The color, flavor and aroma are important quality characteristics of honey from the consumer's point of view, as well as for mead preparation. Honey of a light yellow or creamy fawn color is preferred, as dark honey contains more pollen and, consequently, have more growth factors than the light honey and, therefore, ferments more rapidly, which tends to make strongly flavored, unpleasant mead (Foster, 1967; Joshi et al., 1990; Steinkraus, 1996). The other factors which effect the preparation of alcoholic beverages are the microorganisms, additives, the fermentation and maturation procedures. The alcohol content of mead can vary between 7% and 22%. By varying the proportions of honey and water, and the point where fermentation is stopped, different types of mead can be prepared.

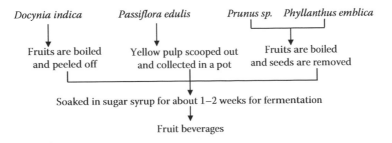

Figure 9.4 Flow chart of preparation of different fruit beverages. (From Jamir, B. and Deb, C.R. 2014. *International Journal of Food and Fermentation Technology*, 4(2): 121–127. With permission.)

9.4.4.1 Addition of Additives to Honey to Make Mead   Fruit juices, salts, and acids have been used as additives to stimulate fermentation and to improve the flavor of mead (Fabian, 1935). Honey contains no acids or tannins (Foster, 1967) so honey drinks lack body and are too sweet. Fruit juices can be added to to the mead must to contribute acid and growth factors for yeast (Joshi et al., 1990; Joshi et al., 1999), along with sufficient tannins. The flavor and aroma of hops are quite popular, but hops also contribute tannins, which can precipitate proteins, thus causing cloudiness (Prescott and Dunn, 1959). Since honey is deficient in nitrogen, minerals, and growth factors that stimulate yeast growth and fermentation, the addition of nitrogenous and phosphorus sources to the diluted honey is essential (Filipello and Marsh, 1941; Joshi et al., 1999). This problem can be overcome by adding 250 mg diammonium phosphate and potassium bitartarate 250 mg/L. To adjust acidity, 1.875 g of tartaric acid or 1.750 g/L of citric acid are usually added (Maugenet, 1964). To prevent the growth of lactic acid bacteria, 25–50 mg sulphurous anhydride or 50–100 mg/L potassium metabisulphite can be added.

9.4.4.2 Method of Preparation   There are several strains of yeast, and two of the best yeasts for this purpose are Maury and Vierke yeasts. To prepare mead, honey is diluted with water to a concentration of 22°B (Joshi et al., 1999). To each litre of this diluted honey solution, additives such as 5 g citric acid, 1.5 g diammonium monohydrogen, 1 g potassium bitartarate, 0.25 magnesium chloride, and 100 ppm SO_2 are added. Juice/pulp from apple, plum and pear with honey were also used to prepare the musts, keeping blending ratio of 8:5:3 for pulp/juice, water and honey. Out of various fruit meads, apple based was rated the best (Joshi et al., 1990, 1999; Joshi et al., 2011b). To start the fermentation process, a 3%–5% active yeast culture of *Saccharomyces cerevisiae* is added. The fermentation process is continued till total soluble solids (TSS) are stabilized. The mead or fruit mead is then siphoned, racked, and matured (Amerine et al., 1967). It is pasteurized or sterile bottled. The mead can also be fortified with high proof brandy. The entire process of mead making is shown in Figure 9.5.

9.4.5 Production of Wine/Fruit-Based Alcoholic Beverages

In South Asia, fermented fruit-based alcoholic beverages like wines are quite popular, because in this area many kinds of fruits with high sugar concentration are abundant. Fruits such as grapes, apples, plums, pineapples, custard apples, passion fruits, bananas, etc., are all used for fermentation (Joshi et al., 2011). The fruit low in sugar content is enriched with extra sugar to increase the fermentable sugar levels. The fruit is thoroughly, mixed with the starter after adding the additives and kept in a metallic container for some days. Fermented liquids are then taken out and consumed. The fermented mixture is consumed usually without filtering. Tropical fruits such as guava, mango, and pawpaw also contain fermentable sugars with levels varying from 10% to 20% (Kordylas, 1992) and are, therefore, also used to prepare wine.

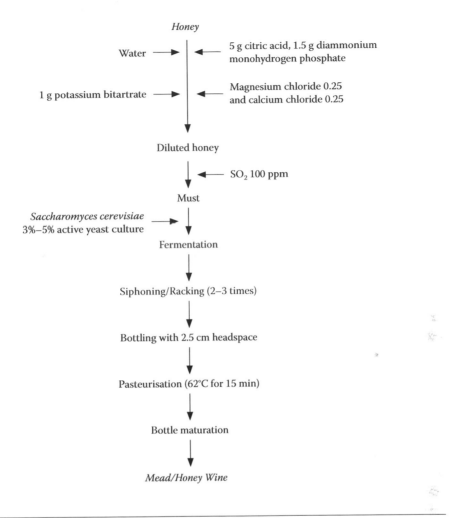

Figure 9.5   Flow sheet for mead preparation. (Adapted from Joshi, V.K. et al. 1990. *Indian Journal of Horticulture*, 47(1): 49–54.)

9.4.5.1 Chulli   This is a popular fruit-based distilled alcoholic beverage, commonly made by the people of Kinnaur, Himachal Pradesh (India) from wild apricots (Joshi et al., 1990; Joshi, 1997). The method of preparation is very crude and is the result of natural fermentation, followed by distillation of the wine (Joshi et al., 1990; Reddy et al., 2012). It is a traditional fermented beverage of the tribal hill people and the process of production consists of collecting the wild apricots, crushing them, adding water and *phab*, and then, warming the mixture (Thakur et al., 2004). It is allowed to ferment for 2–3 days and then, filtered followed by fermentation for 10–15 days. The flow sheet of preparation of the distilled alcoholic beverage, *chulli*, as practiced traditionally, is shown in Figures 9.6.

A method for the preparation of wine from wild apricot has been developed which consists of diluting the pulp in the ratios of 1:2, adding DAHP @ 0.1% and 0.5% pectinol, and fermenting with *Saccharomyces cerevisiae* (Joshi et al., 1990, 2012). The

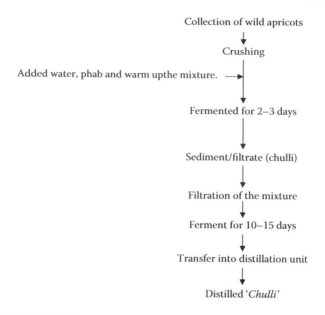

Figure 9.6   Flow diagram for *chulli* beverage by the traditional method employed.

process is depicted in Figure 9.7. Further, with the increase in the dilution level, the rate of fermentation, alcohol content, and pH of the wines increase, whereas a decrease in titratable and volatile acidity, phenols, TSS, color values and K, Na, Ca, Mg, Zn, Fe, Mn, and Cu takes place. Addition of DAHP at the rate of 0.1% enhanced the rate of fermentation. The filtrate collected is known as *chulli*. The wine from 1:2 diluted pulp is rated as the best (Joshi et al., 1990, 2012; Reddy et al., 2012). Like all other traditional beverages made in Kinnaur, it has its own sociocultural importance, and is consumed by the people of Kinnaur to help cope with the adverse dry, cold conditions of region. This alcoholic drink is served during local festivals, fairs, and marriage ceremonies (Thakur et al., 2004).

9.4.5.2 Angoori   This is an alcoholic beverage prepared locally in the *Ribba* region of Kinnaur, Himachal Pradesh. It is made from black and green grapes. A variety of grapes are grown in this district and are used to produce alcoholic beverages, as a part of their culture. In the preparation of *angoori*, *chholtu* red and white varieties of grape are generally used (Figure 9.8). The preparation of the beverage consists of picking the grapes, usually red or green, which are then crushed, and sugar and water are added to increase the rate of fermentation and the yield of alcohol. After this, the locally available inoculum, *phab*, is added. The fermentation is allowed to take place for about 15 days, after which is the mixture is filtered and distilled. *Angoori* is also known as *kinnauri*, which the most popular traditional fermented beverage consumed during local festivals and marriage ceremonies by the tribal people of Kinnaur. It has the same social-cultural importance as *chulli*. A brief outline of method of *angoori* preparation is shown in Figure 9.9.

Figure 9.7 Pictorial representation of wine preparation from wild apricot (*chulli*) (a) wild apricot fruit, (b) must preparation (c) fermentation, and (d) prepared wine in wine glasses. (Courtesy of Dr. G.S. Abrol.)

9.4.5.3 Rguntshang (Fermented Grape Drink)   Grapes are also grown in the lower valleys of Ladakh, in the Da-Hanu areas, where the *Brokpa*, or *Dards*, live. A similar process as described earlier is used to make alcoholic beverage. This drink, prepared from grapes, is mildly alcoholic.

Figure 9.8 Grape varieties grown in Kinnaur HP (India) and used for *Angoori* preparation (Courtesy of Ms Raj Kumri Negi.)

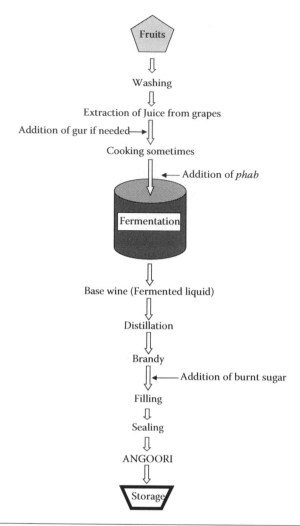

Figure 9.9 Flowsheet of production of brandy (fruit liquor)/*Angoori* (grapes). (Adapted from Joshi, V.K. 2005. Traditional fermented foods and beverages of northern India an overview. In: *Proceedings of First Workshop and Strategic Meeting of South Asian Network on Fermented Foods*, May 26–27, Anand, Gujarat, India, pp. 59–73.)

9.4.5.4 Banana Wine   Wines are also prepared from banana (Joshi, 1997; Joshi et al., 2011). Preparation of alcohol from banana is somewhat different in Andro villages, predominantly in the Meitei community in the Imphal East district. To the ripe bananas *hamei* a type of inocula is added, and mixed thoroughly in a metallic container. The mash is kept for 3–4 days and allowed to ferment. The fermented liquor is collected by filtering through a clean cloth, which results in the production of delicious banana wine. If too much *hamei* is used, the taste becomes too bitter for consumption. Another kind of alcoholic beverage prepared from banana, commonly consumed by the Naga tribes of Manipur, is also known as banana wine. Which is prepared by fermenting the bananas in a closed container with a little water. No inoculum is added to conduct the fermentation. The wine is ready to drink after three days, and should be consumed within 1–2 days after opening the cover (Devi and Kumar, 2012).

9.4.5.5 Apricot Wine Apricot is a delicious fruit grown in many parts of hilly temperate countries, including India, and has great promise for conversion into wine (Reddy et al., 2012). The preparation of apricot wine from the New Castle variety and from wild apricots (*chulli*) with better physico-chemical and sensory qualities has been reported (Joshi et al., 1990, 2004; Joshi and Sharma, 1994). For the preparation of apricot wine from the New Castle variety, the extraction of pulp is either by the hot method (Method I) or by the addition of enzymes and water to the fruits (Method II) (Joshi and Sharma, 1994; Joshi et al., 2012). The wine prepared from the latter method has higher titratable acidity, K, Na, and Fe content but lower phenolic contents. The effect of initial sugar levels on the quality of apricot wine was also determined by Joshi and Sharma (1994). It was found that dilution of pulp in the ratio of 1:1 with water, the addition of DAHP at 0.1%, and raising the TSS to 30°B instead of 24°B made wine of superior quality. The higher sensory quality of wine with 30°B could be attributed to balanced acid/sugar/alcohol content, besides the production of a lower amount of volatile acidity due to the higher initial sugar concentration of the must (30°B) (Joshi and Sharma, 1994).

9.4.5.6 Indian Jackfruit Wine Jackfruit (*Artocarpus heterophyllus*) is an important crop of India, Burma, China, Sri Lanka, etc., and is quite popular in Eastern and Southern India, where it is cultivated widely in Kerala, Karnataka, Andhra Pradesh, Tamil Nadu, West Bengal, Maharashtra, Assam, and the Andaman and Nicobar islands. The fruit is a good source of sugars, proteins, and also flavoring compounds. The fruit pulp segments yield 53%–60% of juice by weight with 12–12.4°B TSS. Sucrose content is 4%–9%, reducing sugars are 4.5%–5.2%, and pectin content is about 3.5%–4%. So the production of jackfruit wine is practiced in Nagaland, Tripura, and other Eastern hilly areas of India by fermentation of jackfruit pulp (*Atrocarpus heterophyllus*). It produces a characteristic pungent aroma and flavor, and is used by the people of the eastern hilly areas of India and Sri Lanka (Dahiya and Prabhu, 1977; Sekar and Mariappan, 2007).

Ripe fruits are used for the production of the wine. The skin of the ripe fruit is peeled and the seeds are removed, and the pulp is soaked in water (Sekar and Mariappan, 2007). The pulp is ground to extract the juice, which is collected in earthen ware/pots. A little water is added to the pots along with fermented wine inoculum from a previous batch. The pots are covered with banana leaves and allowed to ferment at 18–30°C for about a week (Figure 9.10). The liquid is then, decanted and drunk as a wine. During fermentation, the pH of the wine reaches a value of 3.5–3.8, suggesting that an acidic fermentation has also taken place at the same time as the alcoholic fermentation. Finally, the alcohol content reaches up to 7%–8% (v/v) within a fortnight (Steinkraus, 1996; Ward and Ray, 2006; Anonymous, FAO). The yeasts involved in the fermentation resemble *Endomycopsis* (Steinkraus, 1996).

9.4.5.7 Fenny/Feni Cashew was introduced into Goa by the Portuguese government for the purpose of containing soil erosion. Goa is the only state in India where the

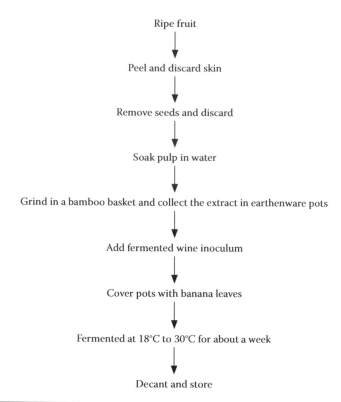

Ripe fruit

↓

Peel and discard skin

↓

Remove seeds and discard

↓

Soak pulp in water

↓

Grind in a bamboo basket and collect the extract in earthenware pots

↓

Add fermented wine inoculum

↓

Cover pots with banana leaves

↓

Fermented at 18°C to 30°C for about a week

↓

Decant and store

Figure 9.10 Flow sheet for production of wine from Indian jackfruit. (From Anonymous FAO http://www.fao.org/docrep/x0560e/x0560e09.htm. With permission.)

fruit is utilized to produce an alcoholic drink known as *Urrack*, which, on further concentration, results in *feni*, a traditional exotic spirit, which is an essence of the Goan lifestyle, synonymous with Goa just as Scotch is synonymous with Scotland, Tequila with Mexico, or Champagne with France. There is much romance and ritual to the growing, harvesting, fermentation, and distillation of *feni* (Anonymous, FAO). Recently *feni* has been given a GI registration and has thus, become the patent product of Goa.

The traditional *feni* manufacturing process is very crude and quite unhygienic. The efficiency of fermentation and the quality of the product is also quite low. It is estimated that about 3.5 kg of cashew apples by the traditional process gives about 1 L of juice. Cashew apples are manually crushed into a cake, called a *coimbi*, in a rock which is shaped like a basin, with an outlet for the juice. A huge boulder is then, placed on the top of it. The final quota of juice which trickles out in a clean form is called *neero*, and allowed to ferment. After three days of fermentation, the product is distilled (Sekar and Mariappan, 2007; Osho, 1995). The process is shown diagrammatically in Figure 9.11.

The traditional method of distilling cashew *Feni* on the hill is equally interesting to watch. The cashew juice is put in a big pot called a *bhann* (cauldron) (Figure 9.12) which serves as a closed boiler. It is connected to a smaller pot called a *launni*

Figure 9.11   Flow diagram for *fenny* preparation.

by means of a conduit that serves as a receiver or collector. The juice in the big pot is then, boiled by burning firewood under it. As the process of vaporization and distillation goes on the concentrated liquid collects in the smaller pot. The pressure in the receiver is kept in check by pouring cold water on it, frequently, with a wooden ladle. The first stage of processing may be done on a big fire, but the later stage of distillation has to be done on a slow fire to keep the pressure and heat under control. The process of distilling *feni* with such apparatus, locally called *bhatti*, takes about a total of 8 h. One can tell from a distance that *feni* is being distilled since the surrounding area is filled with its aroma (Anonymous wiki).

In the distillation, three products are obtained. The first distillation produces *urrac*, the second distillation produces *cazulo*, and the third distillation produces cashew *fenny*, with an alcohol content up to 30%–40% (v/v) (Mohanty et al., 2006; Sekar and Mariappan, 2007). The *urrac*, which is the product of the first distillation, is light and can be consumed neat (without dilution). Its strength ranges between 14 and 16 grao. However, when consumed in excess, *urrac* intoxicates the mind like any other alcoholic beverage. It can be consumed with orange or lemon juice also. The *fenni* produced from the third distillation has a long shelf-life. Since *cazulo* is no longer made, *feni* is produced after the second distillation (Anonymous Feni_(liquor)).

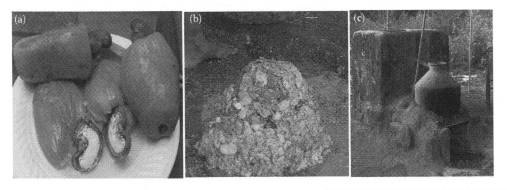

Figure 9.12   (a) Cashew apple, (b) *Coimbi* and (c) traditional distillation unit (*bhann*). (Adapted from Anonymous http:// en.wikipedia.org/wiki/Feni_(liquor))

Feni's strength is its exotic nature and its traditional Goan roots, as well as its compatibility with Indian cuisine and its versatility as a cocktail base. Its weaknesses, however, are that first-time consumers of *feni* and non-drinkers alike find the smell of *feni*, which is an integral part of it, unpleasant, though it is loved by experienced *feni* consumers. The excise law prevalent in Goa governs the strength of *feni*, and the maximum permitted strength is between 25% proof and 75% proof, equivalent to 42.8% (v/v). The local measure is known as a *grao* (Cartier scale). The strength of *feni* distilled by the traditional method ranges between 20 and 24 grao.

9.4.5.8 Neera *(Palm Nectar) and* Toddy/*Palm Wine*   *Neera*, also called sweet *Toddy* or Palm Nectar, is a sap extracted from the inflorescence of various species of *Toddy* palms. It is sweet, oyster white, and translucent like water (Anonymous medlibrary. org/medwiki/Neera). *Neera* is a non-alcoholic and unfermented beverage which, upon fermentation, is converted into *toddy* (Anonymous, medwiki). *Neera* is derived from incising the flower clusters of the coconut. It is widely consumed in South Asian countries including India and Sri Lanka. In India, on an average, the yield of *neera* is about 18 L per spadix for a tapping period of about one month. Trees, which yield a large number of nuts, also yield plenty of sap/*neera*. On an average, the yield of *neera* per palm per day is about 1.5 L. The composition and quality of *neera* has been found to vary from place-to-place, and with time and duration of tapping, so it is difficult to obtain *neera* of a consistent quality (Chandrashekar, 2007).

Toddy and palm wine are alcoholic beverages that are made by fermenting the sugary sap from various palms, such as palmyra and coconut palms (Verma and Joshi, 2000; Rundel, 2002). In India and other South Asia countries, coconut palms and palmyra palms such as the *Arecaceae* and *Borassus* are preferred. It is mainly produced from the lala palm (*Hyphaene coriacea*) by cutting the stem and collecting the sap (Anonymous, wiki). The fermented beverage is called "*panam culloo*" in Sri Lanka, "*tuba*" in Malaysia, *Toddy* in India, and Bangladesh (Lee and Fujio, 1999; Law et al., 2011). In India, there are three types of *toddy*: "sendi," obtained from the wild date palm (*Phoenix sylvestris*); "*tari*," from the palmyra palm (*Borassus flabellifer*) and date palm (*Phoenix dactylifera*); and "*nareli* (*nira*)," from the coconut palm (*Cocos nucifera*) (Batra and Millner, 1974; Aidoo et al., 2006). The wild date (*Phoenix sylvestris*), coconut palm (*Cocos nucifera*), and palmyra palm (*Borassus flabellifer*), etc., are also frequently used for this purpose (Shamala and Sreekantiah, 1988). The kithul palm grows in the wild, in forest covers, in fields, and in rain-forest clearings and is a species of indigenous flowering plant in the palm family from Sri Lanka and India (Law et al., 2011). Virtually, any sugary plant sap can be converted into an alcoholic beverage—it just needs the correct yeasts, temperature, and processing conditions. In brief, *toddy* refers to the fermented flower sap from the coconut palm (*Cocus neusifera*), while palm wine refers to the fermented sap collected from the trunk of other palms including Raphia (*Raphia hookeri* or *R. vinifera*) and the Jelluga (*Caryota palm*) (Anonymous, *toddy* and palm wine). The sweetened *toddy* is called *Karuppany* which is delicious and a good source of vitamin B complex. *Neera* is highly susceptible to natural fermentation at ambient temperature, and after fermentation it is transformed into *toddy* (Anomymous, *toddy*-and-palm-wine). The fermentation is conducted, after collection, by natural yeasts in the air and residual yeasts in the collecting container. The inside of the pot is lined with a little lime (calcium oxide or calcium hydroxide) to prevent it from fermenting before it is tied to the spathe.

Palm *toddy* is a generally sweetish, heavy, milky white, vigorously effervescent, alcoholic beverage (Sekar and Mariappan, 2007). Actually, it is a suspension of living microorganisms in fermenting sap. Alcohol content is 4%, when consumed. In Sri Lanka, fermented *toddy* is consumed fresh, bottled, and pasteurized or distilled to produce the spirit known as *Arrack* (Atputharajah, 1986).

9.4.5.8.1 Technology of Toddy *Making*   The method of *toddy* making is shown in Figure 9.13 and depicted pictorially in Figure 9.14, while sap collection, distillation, and the *toddy* itself is shown in Figure 9.15.

9.4.5.8.2 Collection of Sap   In the traditional method, local manufacturers used bamboo tubes, which covered with many lactic acid bacteria and yeasts, to collect the sap oozing from the palm (Aidoo et al., 2006; Law et al., 2011). Before blossoming, when the flower is just budding, a coconut tree climber would reach the bud and attach a clay pot (Figure 9.15). A small hole of about 15 cm is made in front of the bud so that the white milk-like juice starts to ooze out and is collected in the pot. The sap obtained from the tender inflorescence of the coconut palm (*Cocos nucifera* Linn.) is the major raw material in the fermentation industry in Sri Lanka. In the evening the mouth of the pot is covered with a cotton cloth to protect it from the outside

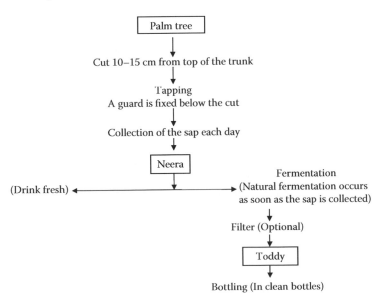

Figure 9.13   Flow diagram for making *neera* and *toddy*.

Figure 9.14   Palm/*toddy* wine collection and *toddy* collector in Andhra Pradesh. (a) Coconut tree, (b) *Toddy* wine in earthen pot, (c) Palm wine collected in glass beaker, (d) Palm tree, (e) Instruments used to collect palm/*toddy* wine, and (f) *Toddy* collector). (Adapted from Half yearly report on Standardization of selected ethnic fermented foods and beverages by rationalization of Indigenous Knowledge, 30031.)

environment. The sap is transported to the collection centers every 24 h. The collected juice starts fermenting immediately. In a day or two, the alcoholic strength is like that of wine, but is not too strong (Anonymous, India wine).

Packaging and Storage Toddy is not usually packaged and is sold immediately or transferred to a refrigerator to extend its shelf-life for 1 or 2 days. Packaging is usually required to keep the product clean and to transport it for its relatively short shelf-life.

Figure 9.15   (a) *Toddy* sap collecting from the *toddy Palmyra* tree, (b) *toddy* distillation unit, and (c) *toddy* wine.

Clean glass or plastic bottles should be used. The product should be kept in a cool place away from the direct sunlight.

9.4.5.9 Palm Wine Fresh palm juice is a sweet, clear, dirty brown/colorless juice containing 10%–12% sugar, while the palm sap has a sugar content of 14%–18%, and is concentrated to make syrup for later fermentation. As the yeasts multiply, it becomes eventually milky white and opalescent (Anonymous, FAO). The sap is not heated and the wine is an excellent substrate for microbial growth, but proper hygienic collection procedures are followed to prevent contaminating bacteria from competing with the yeast and producing acid instead of alcohol (Sekar and Mariappan, 2007). The tapper (Figure 9.16) usually changes the pot twice daily (Grimwood and Ashman, 1975). Tapping of another popular palm, the Nipah palm (*Nypa fruticans*), is, however, slightly different from the coconut; here the sap is collected from the mature fruit stalk after cutting away the almost fully grown fruit head. During tapping, a bamboo container or plastic bag is fastened at the sliced end for sap collection (Figure 9.17). The stalks grow from the ground, so climbing is not necessary for tapping neera sap. Collection of sap from the palmyra palm is done by cutting the panicles grown at the head of the tree with a very sharp sickle or knife. The inflorescence axis of these palms when tapped, can secrete about 4 L of sap per day, with maximum production in April and May when the plants are in full bloom (Steinkraus, 1985). The sap from the *kithul* palm sap is extracted from the young inflorescence in the same way as other palms. A thin slice is cut from the end of the inflorescence axis each time a collection is made and replaced with a fresh pot (Atputharajah et al., 1986; Law et al., 2011). This process is carried out for a period of about two months. A flow chart for the preparation of palm wine by natural fermentation is given in Figure 9.17.

Figure 9.16 (a) A *toddy* tapper collecting the sap, (b) with a container, (c) making an incision, (d) incision making and collecting tools, and (e) sap in a container.

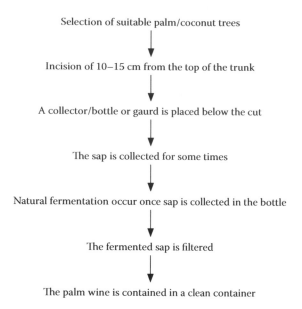

Selection of suitable palm/coconut trees

↓

Incision of 10–15 cm from the top of the trunk

↓

A collector/bottle or gaurd is placed below the cut

↓

The sap is collected for some times

↓

Natural fermentation occur once sap is collected in the bottle

↓

The fermented sap is filtered

↓

The palm wine is contained in a clean container

Figure 9.17 Natural fermentation of palm wine. (Adapted from Anonymous how to pedia, http://en.howtopedia.org/wiki/How_to_Make_Palm_Wine)

The fermentation starts as soon as the sap flows into the pitcher due to yeasts that are present in the sap and those that are added from the previous batch of *toddy* as described earlier also. Sugar in the sap is converted to alcohol which helps to preserve the product. Half of the total sugars are fermented during the first 24 h of fermentation. If allowed to continue to ferment for more than a day, it starts turning into vinegar. The main control points are extraction of a high yield of palm sap without excessive contamination by spoilage microorganisms, and proper storage to allow the natural fermentation to take place. The freshly cut sap is generally a dirty brown sweet liquid having 10%–18% (w/w) sugar (depending on the palm being used), which after fermentation results in the formation of a product containing as much as 9% (v/v) ethanol (Joshi et al., 1999; Aidoo et al., 2006). Tapped palm saps left to spontaneously ferment promotes the proliferation of the yeasts and bacteria present. The natural fermentation converts the sap into palm wine with a mildly alcoholic flavor, with a sweet taste, vigorous effervescence, and a milky white color, as it contains a suspension of numerous bacteria and yeasts (Law et al., 2011).

Acetobacter species were earlier isolated from palm wine (Faparusi and Bassir, 1972; Faparusi, 1973; Okafar, 1975) and from the immature spadix of palm trees (Faparusi, 1973). *A. aceti* subsp. *xylinium* was present on the leaflets of the palm tree and in the surrounding air, and *Gluconobacter oxydans* subsp. *suboxydans* was found on the floret of palm trees (Faparusi, 1973); in the tap holes and in palm sap (Faparusi, 1974). Yeasts belonging to the genera *Saccharomyces*, *Pichia*, and *Candida* are predominant isolates in the fermentation of coconut palm sap, in which *Saccharomyces chevalieri* is the most dominant species, which accounted for 35% of the total isolates (Atputharajah et al., 1986), while the yeasts that are responsible for fermenting wild date palm sap into *toddy* include *Saccharomyces cerevisiae* and *Schizosaccharomyces pombe* (Shamala and Sreekantiah, 1988). *A. Aceti, A. rancens,* and *A. suboxydans* are predominant acetic acid bacteria that have been isolated from palm wine. Other bacteria such as *Lactobacillus* sp., *Pediococcus* sp., *Leuconostoc dextranicum,* and *Leuconostoc* sp. have also been found in palm wine (Atputharajah et al., 1986; Shamala and Sreekantiah, 1988). These bacteria produced high amounts of acetic acid from alcohol. There are also other compounds that can influence the wine quality (Drysdale and Fleet, 1989). Some polysaccharides such as cellulose, levan, and dextran are also reportedly produced by these bacteria (Hibbert and Barsha, 1931; Hehre and Hammilton, 1953). Further, the yeast species from the genus of *Saccharomyces, Candida,* and *Zygosaccharomyces,* and lactic acid bacteria (LAB), such as *Lactobacillus, Leuconostoc, Acetobacter, Pediococcus,* and *Micrococcus* spp., are also involved in palm wine fermentation (Shamala and Sreekantiah, 1988; Stringini et al., 2009; Marshell and Danilo, 2012). Changes during fermentation in various physico-chemical characterics are depicted in Table 9.5 (Anonymous, *toddy* and palm wine).

The quality of the final wine is determined by the conditions used for the collection of the sap. Often the collecting gourd is not washed between the collections and consequently residual yeasts in the gourd quickly begin the fermentation. This is beneficial as it prevents the growth of bacteria which can spoil the sap (Grimwood and Ashman, 1975). The sap is also collected from the cut flowers of the palm tree or by slicing-off the tip of an unopened flower. Flowers can be used until they cease to provide sap or become infected (Anonymous, FAO). A small amount of *toddy* from the previous day's fermentation should be left in the pot to start the next fermentation. An alternate method is the felling of the entire tree; where this is practiced, a fire is sometimes lit at the cut end to facilitate the collection of sap.

9.4.5.10 Distilled Toddy Arrack   *Toddy* can be distilled to make a brandy-like spirit known as *arrack* in Sri Lanka. But distillation requires a special license (Marshall and Mejia, 2011). Palm wine may also be distilled to prepare a stronger drink, which goes

Table 9.5   Change in Physico-Chemical Characterics of *Toddy* at Different Fermentation Intervals

SAMPLE NO	pH			TITRABLE ACIDITY (%)			BRIX (SUCROSE)		
	FRESH	24 H	48 H	FRESH	24 H	48 H	FRESH	24 H	48 H
T1	3.50	3.46	3.22	0.68	1.82	2.40	15.0	13.0	11.6
T2	3.52	3.50	3.46	0.46	1.20	1.32	12.0	11.5	9.4
T3	3.62	3.55	3.43	0.75	1.11	1.088	15.2	14.0	9.7
T4	3.71	3.58	3.54	0.73	0.98	1.425	15.4	12.0	10.0

Source: From Springer Science+Business Media: *Microorganisms in Sustainable Agriculture and Biotechnology,* Utilization of tropical fruits for wine production with special emphasis on mango (*Mangifera indica* L.) wine. 2012, pp. 679–710, Reddy, L.V.A., Joshi, V.K., and Reddy, O.V.S.

by different names, depending on the region (e.g., *arrack*, village gin, *charayam*, and country whiskey).

9.4.5.11 Coconut Toddy and Nectar   The chief product of the palmyra is the sweet sap obtained by tapping the tip of the inflorescence, as is done with the other sugar palms, and, to a lesser extent, with the coconut palm. The sap flows for 5–6 months—200 days in Ceylon—each male spadix producing 4–5 L/day; the female gives 50% more than the male. The *toddy* ferments naturally (Figure 9.17) within a few hours of sunrise, and is locally popular as a beverage; it is distilled to produce the alcoholic liquor called palm wine, *arrack*, or *arak*. The sap can be reduced in volume by boiling to create a sweet syrup or candy, as in the Maldives.

9.4.6 Mnazi

Mnazi (coconut *toddy*) is a whitish, effervescent, acidic, alcoholic beverage (Swings and De Ley, 1977) made by fermentation. The process of getting the sap is depicted in Figure 9.18 while the flow sheet of the process is shown in Figure 9.19. In the first step, the sugar of the sap is fermented to ethanol within 8–12 h by yeasts and LAB, thus, creating a highly suitable medium for the development of acetic acid bacteria (Ou Tait and Lombrechs, 2002). *Acetobacter* sp. perfer ethanol as carbon source and usually dominate during the later stages of wine fermentation (Joyeux et al., 1984). During fermentation, the acetic acid bacteria (Kersters et al., 2006) appear after 2–3 days, utilizing glucose and/or sucrose that might be present in the earlier stages of the *mnazi* fermentation (Okafar, 1975). *Acetobacter* and *Gluconobacter* were isolated and were found to be responsible for the spoilage of *mnazi* (Blackwood et al., 1969; Joyeux et al., 1984). *Acetobacter* usually dominates at later stages of fermentation (Du Toit and Lambrechts, 2002). *Acetobacter* and *Gluconobacter* both show positive growth at 25, 30, and 40°C and also at pH 7.0 and 4.5, while there was no growth at 45°C, pH 2.5 and 8.5 (Blackwood et al., 1969). The coconut palm wine is sweet, but will turn sour easily due to lactic acid fermentation if kept at ambient temperature. *Tuak*, on the other hand, is another traditional palm wine that is derived from *nipah* palm (*Nypa fruticans*). The container is covered by a layer of plastic to ensure fermentation is carried out smoothly on site (Figure 9.19).

Figure 9.18   Process of getting *nipah* palm sap.

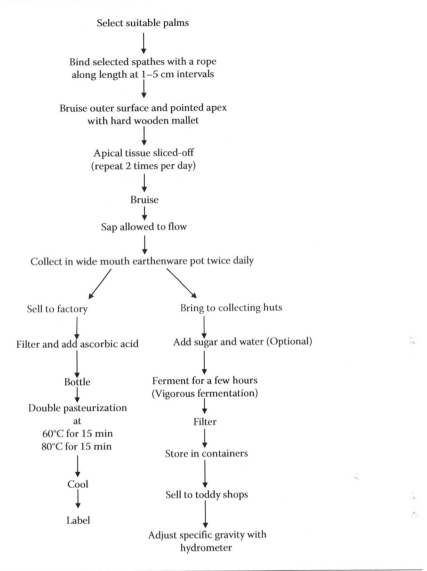

Figure 9.19 Flow sheet for *mnazi* prepartion.

9.4.7 Sara

Sara is a distilled liquor and is a very popular alcohol beverage/drink in the villages of Andhra Pradesh (India) which is also called *Naatusarai* (country *sara*). It is prepared for use in festive seasons like *Sankranthi*, *Ugadi*, and *Vijayadasimi*. It contains a high alcohol concentration of around 20%–25% (v/v).

9.4.8 Flower-Based Alcoholic Beverages

9.4.8.1 Mahua Liquor (Mahua Daaru) The *Mahua* tree (*Madhuca indica*) belongs to *Sapotaceae* family and bears flowers (Figure 9.20a and b) from March to May. The *Mahua* or *Kalpavriksha* (*Bassia latifolia*), is the most highly valued tree among

Figure 9.20   (a) *Mahua* tree bearing flowers, and (b) *Mahua* flowers (*Bassia latifolia*).

tribal communities, and every part is utilized in central India. It is known by different vernacular names such as *Mahua*, *Mohwa*, or *Mauwa* in Hindi, *Mahwa*, *Maul*, or *Mahula* in Bengali, *Mahwa* or *Mohwra* in Marathi, *Mahuda* in Gujrati, *Ippa* in Telugu, *Illupei* or *Elupa* in Tamil, *Hippe* in Kannada, *Poonam* or *Ilupa* in Malyalam, and *Mahula*, *Moha*, or *Madgi* in Oriya. As a consequence, *mahua* trees are some of the largest, oldest, and most common in the fields and forests. *Mahua* flowers are rich in total (64.60%) and reducing (56.30%) sugars, which include sucrose, maltose, glucose, fructose, arabinose, and rhamnose (Belavady and Balasubramanian, 1959; Patil et al., 2010). Some biochemical parameters of *mahua* flower juice are given in Table 9.6.

The flowers are fermented to make country wine, which is also distilled to make liquor. Tribals use indigenous technology for fermenting *mahua* flowers to make quality fermented beverages. About 5–6 bottles of liquor can be made from 2 to 3 kg of dried flowers by mixing and heating palm sugar, urea, and water with the flowers. The liquor is made under most unhygienic conditions, even using prohibited additives, which sometime results in a poisonous product. Redistilled and carefully prepared liquors however are of good quality. *Mahua* liquor is grouped under the category of country liquor, and hence the Government of India has imposed legal restrictions on transactions of *mahua* flowers away from their area of production. In Madhya Pradesh, *mahua* liquor is a substantial revenue-earner, for the state and for tribals alike. Nearly 70%–80% of *mahua* flowers collected from the forest by tribal people are used for the production of liquor. It is a very popular alcoholic beverage in the tribal regions of India.

Table 9.6   Biochemical Composition of *Mahua* Flower Juice

PARAMETERS	*MAHUA* JUICE
T.S.S (°B)	13.0
Acidity (%)	0.11
Ascorbic Acid (mg/100 mL)	3.15
Tannins (%)	0.11
Reducing sugar (g%)	1.04

Source: From Yadav, P., Garg, N., and Diwedi, D.H. 2009. *Natural Product Radiance*, 8(4): 406–418. With permission.

9.4.8.2 Mahua *Wine* For preparation of indigenous *mahua* liquor, fermentation is carried out under ambient temperature conditions. In Northern India, the temperature during summer rises as high as 40°C and if the wine is fermented at this temperature, the quality will be far inferior. Wine fermented at 16°C has the highest content of alcohol (9.9%) and ascorbic acid (0.9 mg %) and higher sensory scores. But lower temperature such as 13°C decreases both the fermentation and the yeast growth rates (Beltran et al., 2007). Sensory evaluation reveals that tannin addition is not required in the preparation of *mahua* wines (Sener et al., 2007). Yeast extract addition during fermentation improves the sensory acceptability of *mahua* wine (Yadav et al., 2009) as is the case with other wines (Vine, 1983; Jackson, 1994; Osho, 2005). The addition of flavorings to the *mahua* wine, namely, lemon, cinnamon, raw mango, and mint, improves the aroma and acceptability over the control (Yadav et al., 2009). Wine with lemon is superior to cinnamon treated wine. Citronellal, geranial, and 1,8-cineole are the major flavor producing compounds of lemon. A comparison of the HPLC profile of *mahua* wines with or without lemon grass is made in Figures 9.21a and b.

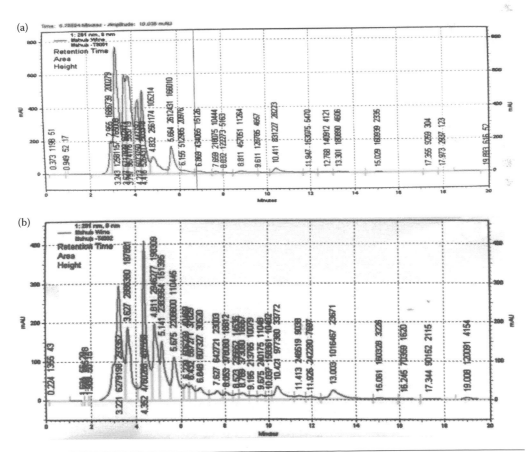

Figure 9.21 (a) HPLC analysis of control *mahua* flowerwine, and (b) HPLC analysis of lemon treated *mahua* flower wine. (From Yadav, P., Garg, N., and Dwivedi, D. 2009. *Natural Product Radience*, 8(4): 406–418. With permission.)

9.4.8.3 Mahua Vermouth   Vermouth is an aromatized wine having added sugar, roots, herbs, spices, and flowers (Amerine et al., 1980). It contains ethyl alcohol, sugar, acid, tannins, aldehyde, esters, amino acids, vitamins, anthocyanins, fatty acids, and minor constituents like flavoring compounds (Joshi, 1997). The additives do not boost the alcohol content, but they do improve the flavor of the wine. Vermouth can have between 15% and 19% alcohol content by volume (Amerine et al., 1980; Joshi et al., 1999). The protocol for the development of *mahua* vermouth has been developed (Yadav et al., 2012) is described in Figure 9.22. Since *mahua* flowers have a burnt starchy flavor, herbs have been used to mask the same and replaced it with a sweet spicy flavor. No microbial growth was observed in *mahua* vermouth after one year of aging. The sensory results clearly reflect the improvement in its aroma and acceptability during storage. Vermouth prepared from *mahua* has 18.4% alcohol and 1.26 mg/100 g tannin content compared to base wine where it is 10.50% and 0.32%, respectively (Effect of location of cultivar, fermentation temperature and additives on the physico-chemical and sensory qualities on Mahua (Madhuca Indica J.F. Gmel) wine preparation *Nat. Product Radiance*, 8(4): 406–418). Non-significant changes were observed in the biochemical parameters during storage. HPLC analysis indicates the presence of phenolics compounds, namely, gallic acid (122.59 mg/100 g), caffeic acid (10.76 mg/100 g), p-coumaric acid (4.18 mg/100 g), and kaempferol (282.16 mg/100 g), as well as the flavanols (+)-catechins (35.31 mg/100 g) and (–)-epi-catechin (39.51 mg/100 g). Polyphenol content of food products are considered as of high significance, being ant, oxidants (Arts et al., 2000). Wines made from different germ plasms growing around Lucknow were found to produce no significant difference in terms of biochemical and sensory quality of the product. Sweet *mahua* wines can be prepared by maintaining a sugar acid blend of 11.6°B; 0.6% results in optimum organoleptic quality.

9.4.8.4 Mahua Daaru   The flowers are collected, dried, and used for making *mahua daaru*, in which the alcohol content ranges from 20% to 40%. It is considered to be inexpensive and is the most popular beverage consumed among the *Rajputs* of North-Western India (Bennett, 1999). The *mahua* flowers are mixed with water (1:4) and

Figure 9.22   Assembly for distillation of *mahua daaru.*

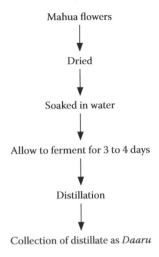

Mahua flowers

↓

Dried

↓

Soaked in water

↓

Allow to ferment for 3 to 4 days

↓

Distillation

↓

Collection of distillate as *Daaru*

Figure 9.23   Process for making *mahua daaru.*

are allowed to ferment in a container for 2–3 days. In the fermentation medium, *navshadar* (Ammonium chloride) is exclusively added, and sometimes *jaggery* is also added as needed, generally in the ratio of (2:1, flowers:*jaggery*). Some people also add black pepper to make *daaru* slightly hot. After fermentation, the fermented mixture is kept for distillation in a special apparatus that has a container (Aluminum pateela) above which a wood structure (*Badgi*) is kept to collect the distillate, and this is covered with a aluminum container with cold water for condensation of the ethanol vapors (Figure 9.22). The cold water is replaced when it becomes hot. The distillate is collected in a bottle. It contains around 45%–60% alcohol. The yield of *daaru* is about 300–400 mL/kg dried flowers. The entire process of *mahua daaru* preparation is depicted in Figure 9.23.

9.4.9 Mahuli (Country Liquor)

Mahuli is made from distilled fermented *mahula* flowers, commonly prepared by the tribal peoples in Odisha (India). The method of fermentation is similar to the earlier method of *mahua darru* preparation, but with variations in the distillation. Usually, in the preparation of *mahuli*, the flowers are thoroughly washed in water and submerged in plastic drums or tanks for a period of 4 days with the addition of *bakhar* (syn. *ranu*) (Figure 9.24a) (Dhal et al., 2010). The list of herbal and other plant parts used in the production of *bakhar* is given in Table 9.7. Fermented *mahula* flower mass is distilled in a metallic (aluminum) container by keeping another earthen pot on the top of the first container in a reverse manner. The joints of the two vessels are sealed by using sticky pond mud. A metallic pipe is connected to the upper earthen vessel, which passes through water and opens to a collecting vessel. The lower metallic container with

(a)

(b)

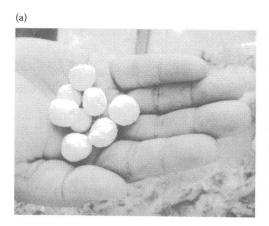

Figure 9.24   (a) *Bakhar* tablets. (b) Small scale traditional mahuli distillation process from fermented mahula flower. Dotted white arrow: metallic pipe connected to the upper vessel, Black arrow: mud water, through which metallic pipe passes and it helps in condensing vapors to liquid, White arrow: condensed *Mahuli* collecting vessel.

fermented *mahua* flower mass is heated at low temperature with a wood fire (Behera et al., 2012). Finally, the vapors are condensed in a metallic pipe and collected in a collecting vessel. The traditional distillation process is shown graphically in Figure 9.24b, and that used in industry in Figure 9.25. The alcohol (ethanol) concentration in the distillates varies from 30% to 40%. The distillate produced from *mahula* flowers alone (10 kg) and mahula + sugar cane molasses (4 kg mahula flowers and 6 kg molasses) are 6.5 and 9 L/batch, respectively (Behera et al., 2012). The distillate is diluted to approximately 10%–15% alcohol (ethanol) and is consumed as a "country liquor."

9.5 Cereal-Based Indigenous Alcoholic Beverages

Fermentation of cereal grains to produce a wide variety of alcoholic beverages has been a practice for a long time in various South Asian countries. Rice wine is one such popular alcoholic beverage (Law et al., 2011).

9.5.1 Kalei, Yu or Wanglei

Types of traditional alcohol produced by different communities of Manipur are depicted in Figure 9.26. *Kalei*, *yu*, or *wanglei* is the traditional alcoholic beverage of the *Meitei* community and is also produced by other tribal communities of Manipur, India. There are local variations in the production and enhancement techniques. The fermentation techniques employed in Andro, a Meitei village in Imphal East district and Langthabal, a Kabui village in the Imphal West district are profiled here.

9.5.1.1 Traditional Way of Yu Fermentation (Andro Type)

9.5.1.1.1 Yu   The commonly consumed alcoholic beverage in Manipur, known as *yu*, is prepared from rice. Any kind of rice can be used in the preparation of alcoholic

Table 9.7 Phytotherapeutic Plants Used in Preparation of *Bakhar*

NAME OF THE PLANT	LOCAL NAME	FAMILY	PARTS USED	MEDICINAL USES/PROPERTIES
Asparagus racemosus Willd	Kader	Liliaceae	Roots	Roots used for curing fever, nutritive tonic
Cissampelos pareira L	Andia kidula	Menispermaceae	Roots	Roots used to increase milk in lactating mother
Clerodendrum serratum (L) Moon, Cat. Pl	Saram lutur	Verbenaceaer	Roots	Roots used in rheumatism
Dipteracanthus suffruticosus (Roxb.) Voigt	Ranuran	Acanthaceae	Roots	Roots used in renal problems
Elephantopus scaber L.	Hadem ran	Asteraceae	Roots	Roots used in diarrhoea, dysentery
Gardenia gummifera L.f.	Bhurlu	Rubiaceae	Young shoot	Gums of young shoot used in nervous disorder of children
Holarrhena pubescens (Buch Ham) Wall ex. G. Don	Hat	Apocyanaceae	Bark	Bark used in amoebic dysentery and diarrhea
Homalium nepalense Benth J Linn.	Danmari	Flacourtiaceae	Bark	Bark juice is used to cure colic
Lygodium flexuosum (Linn.) Sw	Nanjam rehed, Aliz tukah	Lygodiaceae	Roots	Fresh roots used in eczema
Madhuca indica Gmel.	Matkam	Sapotaceae	Seeds, Leaves & Bark	Mahula oil obtained from seeds is used in rheumatism, leaves and bark used in diabetes
Ochna obtusata DC. Var. pumila (Buch Ham. Ex DC) Kanis	Ot champa	Ochnaceae	Roots	Roots used as antidote in snake bite
Orthosiphon rubicundus (D. Don) Benth	Khara ranu ran	Lamiaceae	Roots tuber	Roots tubers used to cure colic
Polygala crotalarioides Buch Ham ex DC	Lilkathi	Polygalaceae	Bark	Used for cough
Phoenix acaulis Buch. Ham ex Roxb.	Kitah	Arecaceae	Roots	Laxative
Rauvolfia serpentina (L.) Benth. Ex Kurz	God	Apocyanaceae	Roots	Roots used to relieve from nervous disorders
Smilax macrophylla Roxb	Atkir	Smilaceae	Roots, Stems	Roots used for urinary complaints, stems used as tooth brush
Woodfordia fruticosa (L.) Kurz	Iche	Lythraceae	Flowers	Dried flowers are used as astringent
Xantolis tomentosa (Roxb.) Rafin.	Dhumodhur	Sapotaceae	Fruits	Fruits contain a thermostable anticholeric principal

Source: From Behera et al. 2012. Traditional knowledge on the utilization of mahula (*Madhuca latifolia* L.) flowers by the santhal tribe in similipal biosphere reserve, Odisha, India: A survey. *Annals of Tropical Research* (in press). With permission.

beverages, but the *Tankhul* tribe use only sticky rice. In some alcoholic beverages, a yeast inoculum is required to conduct the fermentation (Devi and Kumar, 2012). Most of the traditional alcoholic fermentation processes in Manipur utilize a starter culture, as is observed in other communities of North East India. Various communities of North East India, even though following identical protocols for fermentation of

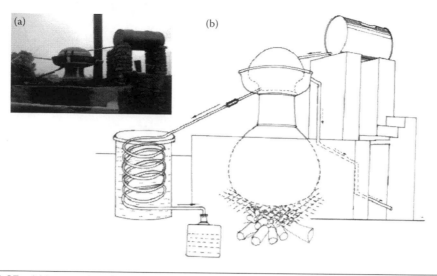

Figure 9.25   (a) Large-scale traditional *mahuli* distillation plant (b) Outer sketch of large scale distillation plant.

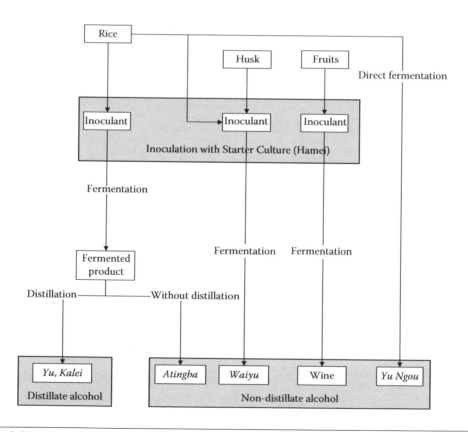

Figure 9.26   Types of traditional alcohol produced by different communities of Manipur.

Table 9.8 Plants Associated with Starter Culture Preparation in the Alcoholic Fermentation in North East India

COMMUNITY	STATE	STARTER	PLANT USED	TRADITIONAL ALCOHOL
Adivasi	Assam	Dabai	*Persicaria hydropiper* (L.) Delarbre *Nicotiana tabacum* L.	Haria
Ahom	Assam	Xajar Pitha	*Datura metel* L. *Zanthoxylum nitidum* (Roxb) DC. *Clerodendrum infortunatum* L. *Adenanthera* sp. *Solanum viarum* Dunal	Xajpani
Bodo	Assam	Emao	*Artocarpus integer* (Thunb.) Merr. *Clerodendrum infortunatum* L. *Persicaria hydropiper* (L.) Delarbre	Zu
Karbi	Assam	Thap	*Dryopteris* sp.	Arak
Dimasa	Assam	Humao	*Oldenlandia diffusa* (Willd.) Roxb. *Solanum ferox* L. *Glycyrrhiza glabra* L., *Albizia myriophylla* Benth	Zu, Judima
Mishing	Assam	Apong kusure	*Artocarpus integer* (Thunb.) Merr. *Ananas comosus* (L.) Merr. *Saccharum officinarum* L. *Psidium guajava* L. *Musa balbisiana* Colla *Capsicum annuum* L. *Persicaria hydropiper* (L.) Delarbre *Piper nigrum* L.	Apong
Apatani	Arunachal Pradesh	Epo	*Eleusine coracana* (L.) Gaertn. *Saurauia roxburghii* Wall.	Chu
Meitei	Manipur	*Hamei*	*Clerodendrum glandulosum* Lindl. *Dryopteris* sp. *Albizia myriophylla* Benth.	Yu, Kalei
Kabui	Manipur	*Hamei*	*Albizia myriophylla* Benth. *Albizia kalkora* (Roxb.) Prain	Zou
Nepali	Sikkim	Marcha	*Calotropis gigantea* (L.) Dryand. *Clerodendrum infortunatum* L.	Jarr
Angami	Nagaland	Yei	*Oryza sativa* L.	Peyazu
Khasi	Meghalaya		*Cheilocostus speciosus* (J.Koenig) C.D. Specht	U-Phandieng
Zeme	Assam	Nduhi	*Glycyrrhiza glabra* L.	Nduijao

Sources: Compiled from different sources: Tanti, B. et al. 2010. *Indian Journal of Traditional Knowledge*, 9(3): 463–466; Chakrabarty, J., Sharma, G.D., and Tamang, J.P. 2009. *Assam University Journal of Science & Technology: Biological Sciences*, 4(1): 66–72; Tsuyoshi, N. et al. 2004. *Journal of Applied Microbiology*, 97(3): 647–655; Hodson, T.C. 1999. *The Meitheis*. LPS Publications, New Delhi; Singh, P.K. and Singh, K.I. 2006. *Indian Journal of Traditional Knowledge*, 5(2): 184–190; Borthakur, S.K. 1990. *Contributions to Ethnobotany of India*. Jain, S.K. (ed.), Scientific Publisher, Jodhpur, India, pp. 277–283.

alcohol, use different types of plant species in the starter culture preparation process (Tanti et al., 2010) as shown in Table 9.8.

9.5.1.1.2 Preparation of Yeast (Hamei) The quality of rice beer is partly dependent on the quality of yeast used. It is called *hamei* by Kabui tribes and *chamri* by the *Tankhul* tribes. Alcohol fermentation in Manipur uses solid *hamei* as the starter for enhancing the quality and quantity of the liquor. *Hamei* is a flat rice cake (Figure 9.27), similar

Figure 9.27 Starter (*hamei*).

to the *ragi* of Indonesia, *budob* of the Phillipines, *chu* of China, *naruk* of Korea, and *marchu* of Sikkim, that has been traditionally used for the preparation of rice wine (Tsuyoshi et al., 2005). In the olden days, there were reports of the use of tree bark extract and fern leaves during indigenous rice beer fermentation by the Meiteis in Manipur (Hodson, 1999). Now-a-days, *hamei* is prepared mainly from *Albizia myriophylla* and *Albizia kalkora* plants, collectively known as *Yangli*, along with the flour of unpolished rice.

For the preparation of *Hamei* (Figures 9.28 and 9.29), finely ground rice powder, where the rice has been previously soaked in water for 2–3 h, is thoroughly mixed with the powdered bark of *Yanglei* (*Albizia myriophyla*). The mixture is kept in a large vessel and water is added slowly till the mixture becomes a paste with the required consistency. The paste is then, spread on a bamboo mat or banana leaves and made into small cubes or tablets. The prepared tablets are sun dried until completely dry. *Hamei* can be stored in a cool, dry place for over a year. For 1 kg rice, around 8–10 g *Yanglei* (*Albizia myriophyla*) is added.

Three different kinds of alcoholic beverages are prepared and consumed (Devi and Kumar, 2012), as described here.

Yu angouba: To prepare *yu angouba*, the rice is soaked in water for around 2–3 h, along with some germinated paddy. For 1 kg rice, around 100 g germinated paddy is added. After this, the water is drained out and the soaked rice is crushed to a powder with the help of a mortar. In another vessel, water is boiled and the crushed rice is added to this boiled water, with continuous stirring, until it is cool. It is then, covered with a muslin cloth and kept for 2–3 days without any disturbance. During these days, foam forms and a typical flavor and odor is released, indicating that the *yu angouba* is now ready to consume (Devi and Kumar, 2012). The tankhul tribe of Manipur (India) use a typical pot which has the shape of a conical flask, and after pouring the content it is sealed with a mixture of cow dung and ashes. This tribe uses only paddy and not rice for the preparation of *yu angouba*, and it is known as *khor* in their local dialect. *Yu angouba* cannot be stored for a long period of time—the maximum is about 7 days. This kind of alcoholic beverage, if

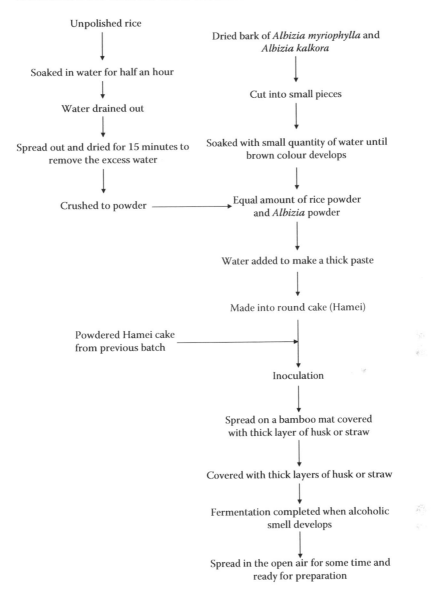

Figure 9.28 Traditional method of *hamei* preparation.

consumed within 2–3 days, is said to be beneficial for the body. It has been compared to drinking of milk, but in a limited quantity—that is, 500 mL at the maximum.

Yu in *Kabui* or *acham* in *Tankhul* is prepared from *Atingba*. *Atingba* is poured into an aluminum pot and is cooked in a low flame. Above the pot, an aluminum funnel is placed and from this a pipe is connected to the other part of the pot. This pipe is used for collecting *yu*. The pot is covered tightly with an aluminum plate (Figure 9.30). On top of this, another aluminum pot is placed containing cold water. All the connecting points are sealed properly with cowdung paste. Distillation continues until all the alcohol present is distilled out. This can be checked by dipping a small stick into the boiling *Atingba* and igniting it—if it produces a green flame, then there is still some alcohol content left

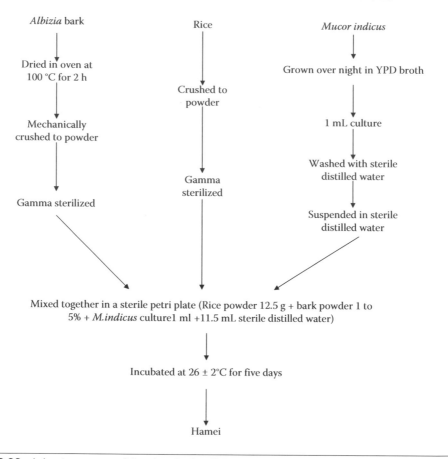

Figure 9.29 Laboratory process of *hamei* production. (Adapted from Panmei, C. 2004. *Studies on the role of Albizia spp. in the fermentation process of different rice cultivars.* PhD thesis, Manipur University, India.).

Figure 9.30 *Yu* making process.

(Devi and Kumar, 2012). The remaining content after the extraction of *yu* is used as a pig feed. This type of alcoholic beverage is very strong due to greater alcoholic content than the other beverages described earlier (Devi and Kumar, 2012).

A bag of hulled rice is soaked in water overnight and the next morning, the rice is taken out and drained. After draining, the wet rice is exposed to hot steam and the rice is put inside a container with a large hole at the bottom plugged with straw. The container is again placed as a lid of another larger container with boiling water. When the rice is properly cooked, it is thoroughly washed in water and excess water is drained-off. The washed rice is dried by spreading it on a plastic sheet or bamboo mat. The thoroughly dried rice is mixed with the starter culture. Then, the mixture is put inside an earthen container and covered with a gunny bag and kept for four days. After that, pond water is added to the container. A second addition of water is made after two days, and then, the mixture is boiled. The vapors are made to pass through a pipe and allowed to condense and the condensed drops are collected. The liquid thus, collected constitute the *kalei* (*yu*) and can be used. Care is taken while handling the cooked rice; it can be handled with bare hands before adding water, but handling is not allowed after the addition of water, as this reportedly produces a sour taste. An acidic taste also develops if the washing with water is not complete, or there is a lack of proper draining of the water. The first drops collected form a special liquid called *machin*. This distillate has a high percentage of alcohol and its use is considered to be harmful to the health of consumers. The *machin* is usually mixed with distillate collected later to produce a balanced beverage, which reportedly has medicinal value, such as for massaging sore muscles. The flow sheet for the entire process is shown in Figure 9.31.

There are many variations in the traditional process of *yu* fermentation. One of these is typified in a Kabui village in Langthabal, where the addition of a starter culture is made after cooking. In this preparation, native rice varieties are used for cooking. After proper cooking, the rice is cooled down and then, mixed with the starter culture. The mixture is put inside a metal container and stored for 4–5 days. While storing, the mixture is sprinkled with water at regular intervals. Then, water is added to the mixture in the ratio of 1.4–1.5 L of water per kg of rice and stored for another day. On the next day, the whole material is transferred to another large container covered with a lid with a pipe on the side. The mixture is then, boiled but the lid is kept cool by placing a pot of cold water above it. After some time, alcohol drips down through the pipe. It is somewhat cloudy in nature; clear alcohol is made by filteration using a cotton plug or cotton cloth.

9.5.1.2 Laboratory Production of Yu   *Hamsei* cake produced using a laboratory method (Figure 9.29) was used for the production of *yu* in laboratory conditions to standardize the process (Figure 9.30). Instead of hand pounding, the rice was dehusked in a grinder and the husks were separated by an air blower. Fifty grams of rice was cooked in 5 volumes of water, cooled, and mixed with a single cake. The mixture was transferred into a sterile volumetric flask, plugged with cotton, wrapped with aluminium foil, and

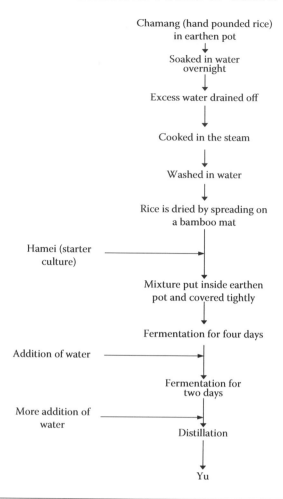

Chamang (hand pounded rice)
in earthen pot
↓
Soaked in water
overnight
↓
Excess water drained off
↓
Cooked in the steam
↓
Washed in water
↓
Rice is dried by spreading on
a bamboo mat

Hamei (starter
culture) ─────────────→
↓
Mixture put inside earthen
pot and covered tightly
↓
Fermentation for four days

Addition of water ─────────────→
↓
Fermentation for
two days

More addition of
water ─────────────→
Distillation
↓
Yu

Figure 9.31   Flowchart of *yu* fermentation (Andro Style).

incubated for five days at a temperature of $26 \pm 2°C$. After incubation, the fermented mash was transferred into a bottle and centrifuged at 6000 g for 20 min. The supernatant was double distilled at 78°C for the laboratory production of *yu*. The optimum amount of water was found to be 1:5 (w/v). There was no significant difference in alcohol production between two types of *Albizzia* species, namely *Albizzia myriophylla* and *Albizzia kalkora*. Alcohol yield depended on the quality of rice used for *hamei* preparation, setting of fermentation broth, amount of water used, as well as quantity of *hamei* the inoculum used.

9.5.1.2.1 Atingba   This is an alcoholic beverage traditionally made from glutinous rice without distillation. In this type of alcoholic beverage, rice is cooked and spread in a container or in a tray made of bamboo. *Hamei* is mixed properly along with the cooked rice. The whole content is transferred to a vessel and a small amount of water is added. Then, it is covered with a muslin cloth. Heat is released for 2–3 days (Coleman et al., 2007; Devi and Kumar, 2012). Water is again poured to reduce the heat. *Atingba* is ready to drink 6–8 h after adding the water, but this gives a very light drink. Hot water is added to the mixture during the winter, while cold water is used in the summer. A proper *atingba* is formed after 4–5 days of fermentation during summer and after 7–8 days in winter. This kind of alcoholic beverage can be consumed for only 1–2 days after fermentation, but can be kept for around 1–2 months to be used for preparing *Yu*. The *tankhul* tribe call this type of wine as *patso* (Devi and Kumar, 2012).

9.5.2 Non-Distillate Alcohols

9.5.2.1 Waiyu   *Waiyu* is a mild alcohol traditionally prepared from rice and husk. It is produced by many ethnic groups in Manipur for societal and religious purposes. *Waiyu* preparation methods in Andro (Meitei) and Langthabal (Kabui) villages are described here (Figure 9.32).

9.5.2.2 Traditional Fermentation of Waiyu *(Andro Type)*   Before fermentation, hand pounded rice is mixed with the husk and soaked in water for one day. On the next day, the mixture is taken out allowing excess water to drain away. Then, the materials are heated using steam. When the materials soften, they are spread on a wide mat and mixed with the starter after optimal cooling. The mixture containing starter is put inside an earthen pot and steamed. Then, the whole set up is covered with gunny bag and stored. After a week of storage, the materials are transferred to another pot and sufficient water (about 2.5 L of water/kg of rice) is added and kept for 1 h. Then, a process called *Yu Thaba* is started, where the liquid is taken out through a pipe. It could be called siphoning or racking in terms of enology the science of wine making. The process does not involve distillation. *Waiyu* is a mild alcohol that can also be consumed even by women. The alcohol is taken as a delicious drink among indigenous people, and also used in religious ceremonies.

9.5.2.3 Traditional Fermentation of Waiyu *(Langthabal Type)*   Hand pounded rice is properly washed in water and dried. Then, it is cooked by steaming. Cooked rice is spread in a bamboo mat and cooled down. Husk is mixed with the cooked rice in 1:4 ratio after adding some starter (*hamei*). The mixture is put inside an earthen container and closed air tight for storing. The whole set up is stored for 3–4 days in the summer, or 3–6 days in the winter. After fermenting for these days, water is added and kept for 2–3 h. The fermented yellowish liquid is drawn through a pipe with a filter. *Waiyu* is a very popular drink among the Maring, a tribal community in Manipur. Any visitor or guest can drink the *waiyu* by using a tube from a large container.

9.5.3 Jhara

Rice beer known as *jhara* or *harhia* is extremely popular among the tribal inhabitants in different district of West Bengal, particularly the Oraon and Santhal workers in the

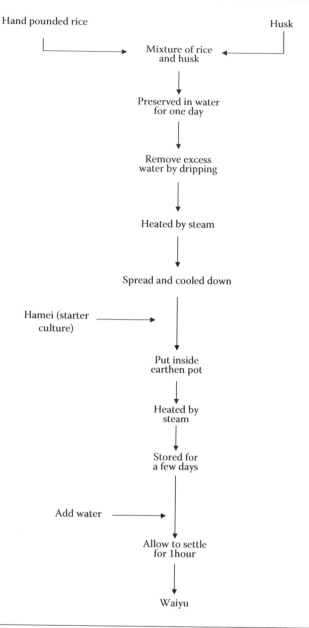

Figure 9.32 Flowchart of *waiyu* fermentation.

Terai Tea Gardens (Ghosh and Das, 2004a,b; Sekar and Mariappan, 2007). In the preparation of *jhara*, various herbal plants and their parts are used, which impart sweetness, bitter flavors, and color, and also act as preservatives. In addition to the use of five core plants (*Oryza sativa*, *Coccinia grandis*, *Plumbago zeylanica*, *Vernonia cinerea*, and *Clerodendrum viscosum*), tribals also use quite a few more plants to modify the taste and color of *jhara* (Ghosh and Das 2004a,b; Sekar and Mariappan, 2007). *Ranu dabai* is used as a starter mixture for the preparation of *jhara*. The fermented liquid is diluted in the proportion 55 L/kg with water before consumption (Sekar and Mariappan, 2007).

9.5.4 Jhar or Jnard

Jhar or *Jnard* is a common fermented beverage traditionally prepared and consumed by various ethnic groups of the Darjeeling district of West Bengal. *Jhar* is made from wheat, although various other substrates such as rice, maize, *bajra* (millet), etc., are used. The beverage is known by specific names according to the substrate used; from rice it is known as *vate jhar*, from maize it is known as *makai jhar*, etc. *Jhar* is also known as *tchang* by the Tibetans, and some ethnic people of Darjeeling which is fermented product of millet (Tamang et al., 1988). The manufacturing process using wheat involves various steps, which are shown in Figure 9.33. The method of preparation of other *jhar* from ingredients like rice (*vathi jhar*) and maize (*makai jhar*) is more or less the same.

9.5.5 Jannr/Jaand

Jannr/jaand is a slightly acidic, alcoholic traditional beverage of Nepal and Tibet (Tamang et al., 1988). *Kodo ko jaanr* is the most common fermented alcoholic beverage prepared from the dry seeds of finger millet (*Eleusine coracana*), locally called *kodo* in the Eastern Himalayan regions of the Darjeeling hills and Sikkim in India, Nepal, and Bhutan. *Jaanr* is the common name for all alcoholic beverages in Nepal; Tibetans call it *minchaa chhyaang* and the Lepcha call it *mong chee* (Thapa and Tamang, 2004). *Jaanr* is an alcoholic beverage prepared by the people of Nepal, Tibet, Bhutan, and Eastern India. Traditional alcoholic beverages constitute an integral part of dietary culture among the ethnic peoples in the Himalayas (Bhalla et al., 2004). It is exclusively produced from locally grown cereal grains, and the product is named according to the substrate used. The major raw material used for the preparation of *kodo ko jaanr/chyang/chee*, *bhaati jaanr*, *makai ko jaanr*, *gahoon ko jaanr*, *jao ko jaanr*, and *faapar ko jaanr* are finger millet, rice, maize, wheat, barley, and buckwheat, respectively (Table 9.9). The common method of *jaanr* preparation is summarized in Figures 9.34 and 9.35 (Tamang, 2009a and b). *Toongba* is the traditional vessel used to drink *Jannr/Jaand*.

9.5.5.1 Marcha Preparation

Marcha or *Murcha* is an amylolytic starter usually used to produce the alcoholic beverage *jaanr*. The microflora of *marcha* consists of filamentous molds such as *Mucor circinelloides* and *Rhizopus chinensis*, yeasts such as *Saccharomycopsis fibuligera* and *Pichia anomala*, and bacteria such as *Pediococcus pentosaceus* (Tamang and Sarkar, 1996).

9.5.5.1.1 Method of Preparation

A traditionally prepared mixed inocula or starter called *marcha*, is used for the preparation of *jaanr*. To prepare *marcha*, glutinous rice is soaked in water for 6–8 h and pounded in a foot-driven heavy wooden mortar and pestle. To 1 kg of the ground rice, are added the roots of *Plumbago zeylanica*

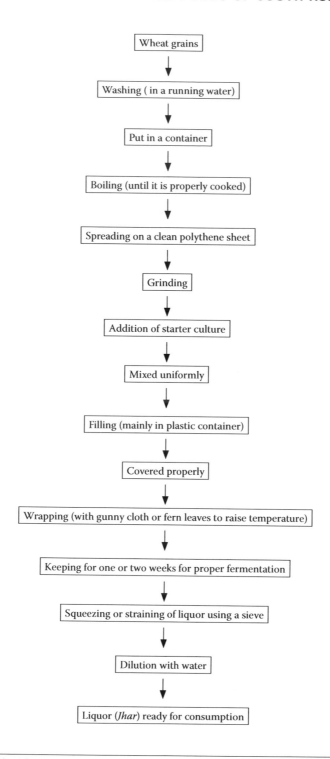

Figure 9.33 Flow sheet for preparation of *jhar*.

Table 9.9   Different *Jaanr* Products with Different Uses

PRODUCT	SUBSTRATE	NATURE AND USE
Makai ko jaanr	Maize	Mild-alcoholic, sweet-sour, beverage
Gahoon ko Jaanr	Wheat	Mild-alcoholic, slightly acidic, beverage
Simal tarul ko Jaanr	Cassava tuber	Mild-alcoholic, sweet-sour, food beverage
Jao ko Jaanr	Barley	Mild-alcoholic, slightly acidic, beverage
Faapar ko jaanr	Buck wheat	Mild-alcoholic, slightly acidic, beverage

Source: Adapted from Thapa, S. and Tamang, J. 2004. *Food Microbiology*, 21(5): 617–622; Tamang, J.P. et al. 2007. *Journal of Hill Research*, 20(1): 1–37.

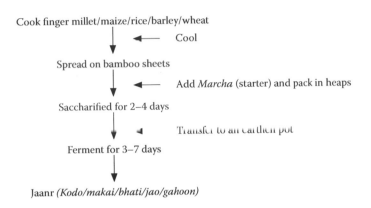

Figure 9.34   Flow diagram for *jaanr* preparation. (Adapted from Tamang, J.P. 2009. Ethnic starters and alcoholic beverages. Chapter 8. In: *Himalayan Fermented Foods: Microbiology, Nutrition and Ethnic Values*. CRC Press, Taylor & Francis, Boca Raton, FL, pp. 187–228.)

Figure 9.35   Pictures showing (a) rice (raw) and *bhaati jaanr* and (b) drinking *jaanr* with *toongbaa* and (c) traditional vessel *toongbaa* used for drinking *jaanr*.

L. (2.5 g), the leaves of *Buddleja asiatica* Lour (1.2 g), the flowers of *Vernonia cinere* Less (1.2 g), *Gingiber officinale* (5.0 g), red dry chilli (1.2 g), and previously prepared *marcha* (10.0 g). The mixture is kneaded into flat cakes, placed on a bamboo mat lined with fresh fronds of ferns (*Glaphylopteriolopsis erubescens* [Wall ex Hook.] Ching) and covered with dry ferns and jute bags (Tsuyoshi et al., 2005). This is placed above the kitchen and allowed to ferment for 1–3 days. They are then sun dried for 2–3 days, and can then be stored in a dry place for more than a year (Tsuyoshi et al., 2005).

Good quality *jaanr* has a pleasant sweet aroma blended with mild alcohol (Tamang et al., 1988; Tsuyoshi et al., 2005).

Kodo ka Jannr: *Kodo ka jannr* is a traditional mild alcoholic beverage prepared from the seeds of finger millets in the Eastern Himalayas (Thapa and Tamang, 2004). During the preparation of *kodo ka janar*, dry seeds of finger millet are cleaned, washed, and cooked for about 30 min in an open cooker, excess water is drained off, and the cooked seeds are spread on a mat made up of bamboo, locally called *mandra*, for cooling. After this, 2% of a mixed starter culture of dry, powdered, *marcha* (Tamang et al., 1996) is sprinkled over cooked seeds, mixed thoroughly and packed into a bamboo basket lined with fresh fern (*Thelypterise rubescens*), locally called *thadre unioon*, or banana leaves. Different microorganisms have been associated with *mascha*, such as *Mucor cicinelloides*, *Rhizopus chinensis*, *R. stolonifer* var. *lyococcus*, *Saccharomyces cerevisiae*, *S. bayanus*, *Hansenula anomala*, *Pediococcus pentosaceus*, *Lactobacillus* sp., *Candida glabrata*, *Saccharomycopsis capsularis*, *S. fibuligera*, *Pichia burtonii*, *P. anomala*, *Cryptococcus* sp., *Trichosporon* sp., *Debaryomyces* sp., *Kluyveromyces* sp., *Myxozyma* sp., *Bullera* sp., *Rhodotorula* sp., and *Tremella* sp. (Tamang and Sarkar, 1995; Tsuyoshi et al., 2005; Sekar and Mariappan, 2007; Tiwari and Mahanta, 2007). It is covered with sack cloth and kept for 2–4 days at room temperature for saccharification. During saccharification a sweet aroma is emitted, and the saccharified mass of finger millet is transferred into an earthen pot or into a specially made bamboo basket called a *septu*, and made air-tight and fermented for 3–4 days during summer (5–7 days in winter) at room temperature. About 200–500 g of the fermented mass is put into a vessel called a *toongbaa* and lukewarm water is added up to the rim. After about 10–15 min, the milky white extract of *jaanr* can be sipped through a narrow bamboo straw called a *pipsing*, which has a hole in the side near the bottom to avoid the passing of grits (Figure 9.36). Water can be added 2–3 times after drinking the extract (Singh and Jain, 1995; Tamang et al., 1996). Good quality *jannr* has a sweet taste with a mild alcoholic flavor (Tamang, 2012).

The drink constitutes an integral part of dietary culture and religious beliefs among the ethnic people in Sikkim (Tamang et al., 1996; Sekar and Mariappan, 2007).

Tongba: The ingredient for its preparation is pearl millet (*Pennisetum glaucum*). The process of preparation starts the same as that of *jhar* (wheat), and the subsequent steps are shown in Figure 9.33. The liquor is a strained into in a cylindrical container, traditionally a bamboo vessel, and after about 10 min, the beverage is ready to drink. It is consumed normally by sucking through a small pipe of bamboo.

9.5.6 *Alcohol Without Starter Culture:* Yu Ngou (Jou Ngou)

Though most of the alcoholic beverages are produced by using starter culture, *yu ygou* (*zou ngou* in Kabui) is prepared without a starter culture. In this type of beverage,

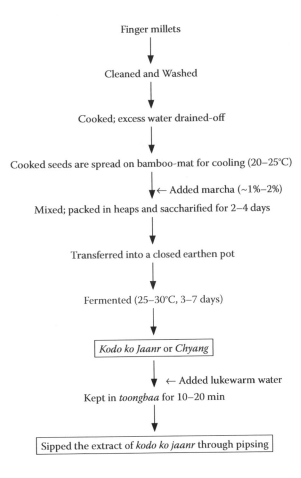

Finger millets

↓

Cleaned and Washed

↓

Cooked; excess water drained-off

↓

Cooked seeds are spread on bamboo-mat for cooling (20–25°C)

↓ ← Added marcha (~1%–2%)

Mixed; packed in heaps and saccharified for 2–4 days

↓

Transferred into a closed earthen pot

↓

Fermented (25–30°C, 3–7 days)

↓

| Kodo ko Jaanr or Chyang |

↓ ← Added lukewarm water

Kept in *toongbaa* for 10–20 min

↓

| Sipped the extract of *kodo ko jaanr* through pipsing |

Figure 9.36   Flow sheet of *kodo ko jaanr* preparation in East Sikkim. (Adapted from Carrying capacity study of Teesta basin in Sikkim, edible wild plants and ethnic fermented foods. *Biological Environment: Food Resources* Vol. III.)

sun-dried sprouted paddy is used which is mixed thoroughly with washed and dried hulled rice. The mixture is powdered in a traditional pestle and mortar. Only the powder is used—the larger pieces are separated by sieving. Hot boiling water is added to the powder and thoroughly mixed by stirring. The liquid is then, allowed to cool at room temperature and ferment for some days—about 4–5 days in summer, but up to 10 days in winter. This type of alcoholic fermentation does not require any starter culture. The fermented liquid is milky in nature and also forms a curd-like coagulum. As the liquid contains starchy material, it can also be a source of energy. Local people consume this beverage after hard work such as agricultural activity and shifting cultivation. *Yu ngou* can also be prepared from aromatic rice and used in religious activities, and in rites of passage of the tribal communities. Another beverage, called *ban*, is made from aromatic rice (*chak-hao*), which, due to its high alcoholic concentration, is highly intoxicating. Sprouted paddy called *nappok* is important for preparing this type of alcohol. It is a tradition to keep *nappok* for every household.

9.5.7 Rice Wines

Rice (*Oryza sativa*) wines are popular alcoholic beverages in the Asian countries and are made from the waxy or glutinous rice, the round and short grains of which are ideal for this purpose. *Tapuy* is a sweet and acidic alcoholic rice wine, whereas *ruounep* is a turbid suspension of pink–red color with some residual sugar and containing 8%–14% (w/v) alcohol (Aidoo et al., 2006; Law et al., 2011). The preparation of rice wine varies according to location and traditional practices. Generally, rice wine fermentation can be done by the submerged and solid state processes. The submerged process involving saccharification of rice to liberate sugar which is converted to ethanol by the submerged fermentation of yeast in a liquid medium occurs with plenty of free water (Ray and Ward, 2006).

The general procedure to produce rice wine includes washing the rice, immersing it in water for a certain period of time followed by steaming. The steamed rice are then, spread out on cloth for cooling to room temperature (Sanchez et al., 1988). A starter culture is then, added a rate of at 1% (w/w) of raw rice and mixed well. The starter culture normally contains *Rhizopus*, which have the ability to hydrolyse amylase, and *Sacharomycopsis*, which promotes alcohol fermentation. The inoculated glutinous rice is incubated in trays covered with paper (Chiang et al., 2006; Law et al., 2011). After two days at room temperature (25–30°C), the product is transferred to a fermentation jar with a water seal to allow the rice to ferment for two weeks. The fermented mass is squeezed using cheese cloth to collect the alcoholic liquor, and the residue is discarded. The wine is allowed to stand for 1–3 months in dark, cool place to prevent discoloration. After maturation, the wine can be bottled and pasteurized at 65–70°C for 20 min. Approximately, 1 L of rice wine can be harvested from 1 kg of rice.

Solid state fermentation is the process where microbial growth and product formation occurs on the surface of solid materials (Aidoo et al., 2006; Law et al., 2011). Glutinous rice is first soaked in water overnight and cooked by steaming. The steamed rice is cooled, mixed with dry starter culture, and incubated at room temperature for 1–2 days. The inoculated rice paste is then transferred into an earthenware jar, covered, and tightly sealed. The jar is left at room temperature to ferment for more than a week. The fermented content is diluted with water for drinking (Aidoo et al., 2006; Chuenchomrat, et al., 2008). Mixed cultures are commonly used for fermentation of carbohydrate-rich substrates into alcoholic beverages. Hesseltine et al. (1988) and Law et al. (2011) stated that amylolytic starters typically combine fungi, yeasts, and bacteria. Mixed cultures are used instead of sequential fermentation in amylolytic starters, where moulds are able to degrade the starch and the yeasts are for alocoholic fermentation (Aidoo et al., 2006). Rice wine preparation involves alcoholic fermentation using mainly *Saccharomycopsis* spp. and *S. fibuligera* yeasts, which have high ethanol producing capacity and amylolytic activity (Limtong et al., 2002; Law et al., 2011).

9.5.7.1 Tapai

The *tapai* is considered as a ceremonial wine, served during special occasions, such as weddings, and large celebrations, such as a bountiful harvest festival

(*Pesta keamatan*). Since the native people can produce *tapai* in their homes, it is also drunk on a daily basis. Similar other products are also found around the Asian region, including the *ruou de* (Vietnam), *yakju* (Korea), and *mi jiu* (China) (Anonymous, rice wine).

The main substrates for producing *tapai* are rice or glutinous white rice and a starter culture (*sasad tapai*) to initiate fermentation (Figure 9.37). The glutinous rice is cleaned and washed thoroughly prior to immersing in water for 1 h before cooking in a pot (Figure 9.37). The cooked glutinous rice is spread out on a clean cloth to cool to room temperature. At this stage, the powdered starter cake, at approximately 1.0%–1.5%, is sprinkled on the cooled glutinous rice, followed by continuous mixing. The traditional starters normally contain *Rhizopus* and *Saccharomyces* spp. that initiate fermentation. The mixture is then, transferred to an earthenware jar or *tajau*, and left open for one day at room temperature before the lid is sealed (Chiang et al., 2006). The fermenting mass is stirred from time-to-time in order to keep the surface moist. The *tapai* is ready to harvest after 3–4 weeks of fermentation. Matured *tapai* has a strong alcoholic aroma and a sweet taste, and is a clear yellow liquid. The liquid is drawn off into smaller bottle or containers for consumption. A simplified process flow diagram for making *tapai* is shown in Figure 9.38.

9.5.7.2 Kiad A popular local liquor known as *kiad* accompanies every religious festival and ceremony of the Jaintia tribes in India. It plays an important role in Jaintia

Figure 9.37 Preparation of *tapai* and its consumption in the locale.

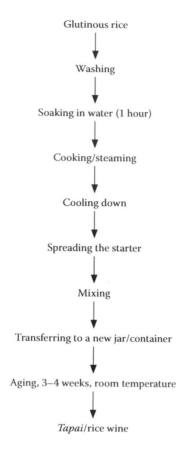

Glutinous rice
↓
Washing
↓
Soaking in water (1 hour)
↓
Cooking/steaming
↓
Cooling down
↓
Spreading the starter
↓
Mixing
↓
Transferring to a new jar/container
↓
Aging, 3–4 weeks, room temperature
↓
Tapai/rice wine

Figure 9.38   Process flow of making *tapai*/ rice wine.

socio-cultural life, and daily consumption in small amounts is considered a remedy for ailments such as urinary trouble and dysentery (Anonymous, wiki). *Kiad* is brewed and distilled in local breweries called *patas*. A local red rice (*kho so*) or white rice may also be used to prepare *kiad*. However, the local brown millet (*Pennisetum typhoides*) is widely used these days. Millet is one of the major cereal crops in the semi-arid regions of Asia. The process of preparing *kiad* can be divided into two parts: preparing the natural yeast, *thiat* (Figure 9.39), and brewing the liquor (Figure 9.40) (Jaiswal, 2010).

9.5.7.3 Natural Yeast (Thiat)   The leaves of *Amomum aromaticum* Roxb (*khaw-iang*) are collected, washed, cleaned, and sun dried. A handful of these dried leaves are ground into powder using a mortar (*thlong*) and a pestle (*Surai*), both made of hardwood. 1–2 kg of washed local red rice *Oryza sativa* (*kho so*) is soaked in spring water (*um pohliew*) and ground to a thick sticky paste using *thlong* and *surai* until a damp powder is obtained. The powdered *khaw-iang* leaves are mixed with the rice powder, some spring water (*um pohliew*) is added to get a thick paste, and the mass is transferred to a cone-shaped basket (*khrie*) made of bamboo. Round cakes of 4–5 cm

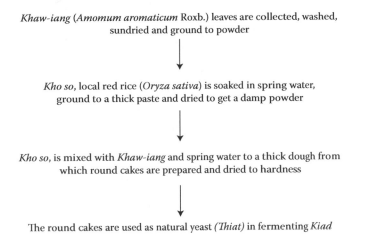

Khaw-iang (*Amomum aromaticum* Roxb.) leaves are collected, washed, sundried and ground to powder

↓

Kho so, local red rice (*Oryza sativa*) is soaked in spring water, ground to a thick paste and dried to get a damp powder

↓

Kho so, is mixed with *Khaw-iang* and spring water to a thick dough from which round cakes are prepared and dried to hardness

↓

The round cakes are used as natural yeast (*Thiat*) in fermenting *Kiad*

Figure 9.39 Natural yeast (*thiat*) preparation.

in diameter and 0.8–1 cm in thickness are prepared from this dough, transferred to a round basket (*Malieng*) made of bamboo, (Jeyaram et al., 2008), covered by banana leaves (*Sla-pashor*), and exposed to sunlight or tied about 1.5 m above the fireplace for drying until the cakes harden (Samati and Begum, 2007). These dried cakes (Figure 9.41a) known as the *thiat* are used as yeast inocula (Samati and Begum, 2007; Jaiswal, 2010; Das and Deka, 2012).

9.5.7.4 Kiad Brewing   Local millet grains (*Pennisetum typhoides*) 4–5 kg, are soaked in spring water (*um pohliew*) and cooked for an hour in a metallic vessel (*khiaw–heh*) on wood fire (Figure 9.41b). During cooking, it is repeatedly stirred with a spoon (*siang-ja*) to avoid burning or over cooking. The cooked millet is taken out, cooled, and dried by spreading on a wooden board or in a round bamboo basket (*malieng*). Natural yeast (*thiat*) cake is finely crushed by hand and mixed with the cooked millet. 1–2 *thiat* cakes are needed per kg of millet. The millet-yeast mixture is blended to uniformity by hand and transferred to a cone-shaped basket (*shang*) made of bamboo and left overnight. The next day, a yellowish white liquid starts dripping. The liquid can be extracted as a

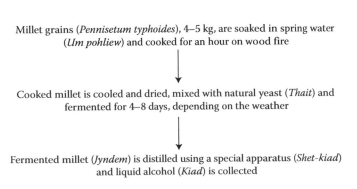

Millet grains (*Pennisetum typhoides*), 4–5 kg, are soaked in spring water (*Um pohliew*) and cooked for an hour on wood fire

↓

Cooked millet is cooled and dried, mixed with natural yeast (*Thait*) and fermented for 4–8 days, depending on the weather

↓

Fermented millet (*Jyndem*) is distilled using a special apparatus (*Shet-kiad*) and liquid alcohol (*Kiad*) is collected

Figure 9.40 The process of brewing *kiad*.

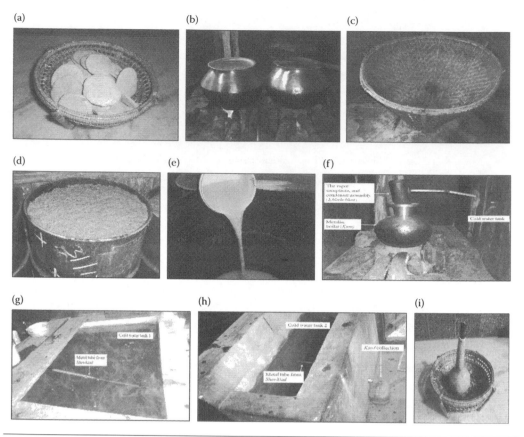

Figure 9.41 Pictorial representation of brewing *kiad*. (a) Natural yeast (*thiat*) used as starter culture in brewing *kiad*. (b) Millet grains (*Pennisetum typhoides*) being cooked on wood-fire. (c) The cone-shaped bamboo basket (*shang*). (d) The fermenting millet (*jyndem*) in metallic drums. (e) The fermented millet (*jyndem*) liquid ready for distillation. (f) The *jyndem* being distilled using a distillation apparatus (*shet-kiad*). (g) The alcohol vapors from *shet-kiad* are condensed using cold water. (h) The condensed vapor from *shet-kiad* is collected as *kiad* liquor. (i) The traditional drinking vessel (*klong-skoo*) is used to serve *kiad*.

beer, which is locally known as *sadhiar*. The fermenting millet (*jyndem*) is transferred from the *shang* into metallic drums and left to ferment (for 4–5 days in summer; 6–8 days in winter), depending on the surrounding temperature.

Basically, *kiad* is a distilled beverage. A special set of apparatus (*shet-kiad*) is used to distill *kiad* from the fermented millet (*jyndem*). The s*het-kiad* is a crude distillation assembly consisting of a metallic boiler (*kum*), and a condenser and metal tubing assembly (*khlieh-bhot*). Fermented *jyndem* is mixed with spring water and transferred to the *kum* and boiled over a wood fire. The vapor receptacle, condenser, and tubing assembly (*khlieh-bhot*) is attached to the *kum* and the joint is made airtight by wrapping with a cloth (*khats*) or clay (*kew byrtha*) to prevent the leakage of vapor. The vapor arising from boiling *jyndem* passes through the metal tubing that is routed through cold water tanks, and the condensed vapors, in the form of liquid alcohol *kiad*, is collected in an appropriate container (Figure 9.41a to h). The residual mass left after distillation is called *ja-jyndem*, which is used as fodder for cattle and pigs.

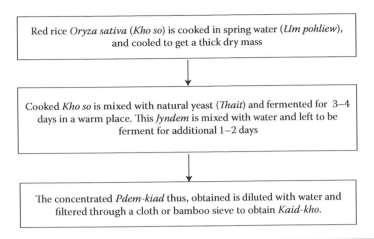

Figure 9.42 The process of brewing *kiad-kho.*

Preparation of *kiad-kho:* The *jaintia* rice beer (*kiad-kho*) is prepared with some modifications of the *kiad* fermentation process, without distillation (Figure 9.42). The natural yeast (*thiat*) is prepared as outlined in the *kiad* preparation process. Local red rice is soaked and cooked in spring water in a metallic vessel over a wood fire. The cooked rice is cooled, dried, and mixed with crushed *thiat* (Anonymous, wiki). The mixture is transferred to a cone-shaped bamboo basket (*shang*), covered with banana leaves (*sla-pashor*) and a piece of cloth and tied with a thread (Tamang, 2012). The contents of the *shang* are left to ferment in a warm place for 3–4 days.

When a yellowish white liquid starts dripping from the fermenting mixture (*jyndem*), the *shang* is opened, and its contents are transferred to a metallic vessel. Some spring water is added to the mixture and set aside to ferment for an additional 1–2 days. The concentrated *pdem-kiad* thus, obtained is mixed with spring water and filtered through a cloth or bamboo sieve. The liquid filtrate (*kiad-kho*) is consumed as a rice beer. The alcohol content of *kiad-kho* is lower than *kiad.*

9.5.7.5 Bhaati Jaanr   This is a Himalayan sweet–sour, mildly alcoholic beverage, in the form of a paste, prepared from rice, and consumed as a staple food (Tamang, 2010). Glutinous rice is steamed, spread, on a bamboo mat for cooling, and 2%–4% of powdered *marcha* is sprinkled over the cooked rice, mixed well, and kept in a vessel or an earthen pot for 1–2 days at room temperature. After saccharification, the vessel is made air tight and fermented for 2–3 days (in summer; 7–8 days in winter). *Bhaati jaanr* is made into a thick paste by stirring the fermented mass with the help of a hand-driven wooden or bamboo stirrer. It is consumed directly as a food. Occasionally, *bhaati jaanr* is stored in an earthen ware crock for 6–9 days, and a thick yellowish-white supernatant liquor, locally called *nigaar*, is collected at the bottom of the vessel. *Nigaar* is drunk directly with or without the addition of water. *Bhaati jaanr* is an inexpensive, high-calory staple food consumed as a beverage by post-natal women and old people in the villages, who believe that it helps them to regain their strength (Tamang, 2010).

9.5.7.6 Poko   *Poko* is an ethnic, mildly alcoholic beverage prepared from rice in Nepal, which is similar to *bhaati jaanr* (Tamang, 2010). It is believed that *poko* promotes good health, nourishes the body, and provides vigor and stamina. In the traditional method of *poko* production, rice is soaked overnight, steamed until cooked and sticky, and spread to cool on a clean floor at room temperature (Neema et al., 2003). Powdered *manapu* is sprinkled on the cooked rice, mixed well, and packed in earthen vessels, covered, and allowed to ferment at room temperature for 2–5 days. The sticky rice is transformed to a creamy white, soft juicy mass that is sweet–sour, with a mildly alcoholic and aromatic flavor, and is consumed as a dessert. *Rhizopus, Saccharomyces cerevisiae, Candida versatilis,* and *Pediococcus pentosaceus* are present in *poko* (Shrestha et al., 2002; Tamang, 2010).

9.5.8 Rice Beer

Combined with water and wild yeast, rice is fermented and a crude beer is made. From crude beginnings, the people of Arunachal Pradesh (India) have come a long way in the production of beer, especially with respect to the use of more scientific and technologically sound methods of production and quality control; the rice beer known as *opo* is no exception. It is made from rice, maize, and sometimes tapioca also. Some rice grains broken during normal milling, are used in beer brewing, along with the whole grain. Different methods to the manufacture rice beer are used, and a flow chart for the preparation of *opo* is given in Figures 9.43 and 9.45.

Cleaned whole or broken rice is boiled until its consistency is the same as that of cooked rice. Only potable water is used in its preparation. Cleaned rice husk is taken separately, heaped, and burnt or smoked until it becomes brown or slightly blackish. Boiled rice is now spread on a mat and the burnt rice husk is sprinkled on it and then, mixed thoroughly. While mixing, yeast (*epo*) is added in a proper proportion, and the mixture, locally known as *pon*, is then, kneaded by hand before again spreading it uniformly on a mat, at a thickness of not more than 5–6 cm. The mixture, after drying, is collected and kept wrapped in *oko* or *okkam* leaves to make packets. These packets (about 6 kg) are stacked in a room well-enclosed on all sides, where they are allowed to ferment for 10–30 days. The longer the fermentation period, the better the production of filtrate. The packets are then opened and once again mixed thoroughly. For large scale production at local level, an apparatus is prepared, as shown in Figure 9.44. On a raised bamboo platform (4)—about 10 feet high—a bamboo funnel and basket called *obades* and *sades* are fitted. The bamboo is split into strips throughout its length, except at one end of the bamboo funnel (3); strips are joined by bamboo rings at a suitable gap (Neema et al., 2003). Banana leaves cover these gaps and the bamboo cone is filled with *pon* (2). Drinking water is boiled and sprayed on top of the *pon* (1) in small quantities. After initial absorption of the water by the *pon* the liquid oozes out at the bottom, which is the purest filtrate, rich in alcohol, and with long shelf life. As the process goes on, the concentration of the filtrate decreases and contamination

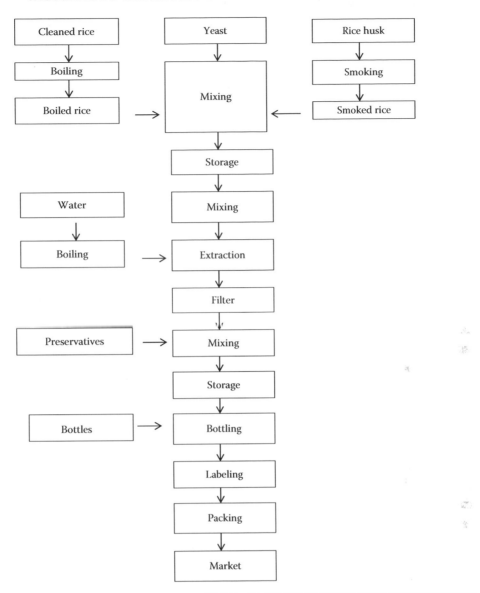

Figure 9.43 Process flow chart for the preparation of *opo*. (From Neema, P.K., Kario, T., and Singh, A.K. 2003. Opo—The rice beer. In: *International Seminar and workshop on Fermented Foods, Health Status and Well Being Held at Anand on 13–14th November, 2003*, pp. 59–61. With permission.)

can occur. The filtrate is collected in a bucket (5) and stored for some time to allow the large impurities to settle at the bottom. It may then, be filtered to remove suspended particles, especially carbon particles. However, the tribes of Arunachal Pradesh like to consume the unfiltered beer as it is, with suspended carbon particles, which generally amounts up to 0.05%–0.50% of the total weight. The *opo* is consumed in bamboo bottles known as *podu*. After sometime, the rice beer is ready and is transferred to bottle guard vessels or put in a basket vessel called a *yaphong*.

Highly concentrated filtrate can be stored for about a month, but it has to be consumed sooner if more water is added to it, as the shelf-life depends on the amount

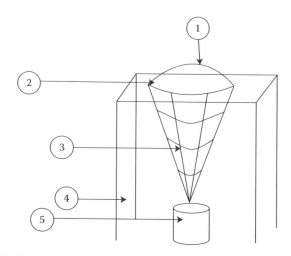

Figure 9.44 Set up of *opo* extraction from *pon*: (1) Top of *pon*, (2) *Pon*, (3) Bamboo funnel, (4) Raised platform, (5) Bucket. (From Neema, P.K., Kario, T., and Singh, A.K. 2003. *International Seminar and workshop on Fermented Foods, Health Status and Well Being Held at Anand on 13–14th November, 2003*, pp. 59–61. With permission.)

of water in the pure filtrate. However, adding proper preservatives can enhance the shelf-life. The product is put into bottles and, after labeling is sold.

9.5.8.1 Traditional Method of Rice Beer Preparation A field survey was carried out in the villages and rural areas of the states of Assam, Nagaland, and Arunachal Pradesh during the month of December 2010 (Das et al., 2014) and, based on the information available, the prevalence of traditional methods of rice beer preparation was documented. The information was collected from the producers predominantly involved in the process of making rice beer. The women in all the communities visited were mostly involved and they were asked about their practices for preparation, for example, making of starter cakes, along with the plants and their parts added, the fermentation procedure, and the duration and uses of the beverage. Some of the nearby fields and forests were also visited, and, with local help, the available plant samples were collected and identified at the Department of Agronomy, Assam Agricultural University, Jorhat, Assam and Department of Botany, Darrang College, Tezpur, Assam (India).

9.5.8.2 Methodology The methodology of fermentation carried out by different tribes was reported to be almost the same, except that difference came from the different types of plant species used in starter culture preparation. In beer preparation, rice (either glutinous or non-glutinous) were half-cooked and allowed to cool and then, mixed with powdered starter cakes and spread out for some time. The mixture was then, kept in an earthen pot, the mouth sealed, and kept in a closed room for a period of 3–5 days for fermentation. After this, some water used to be added to the fermented mass and left for about 10 min. The mass was then, strained and the liquid obtained was the rice beer.

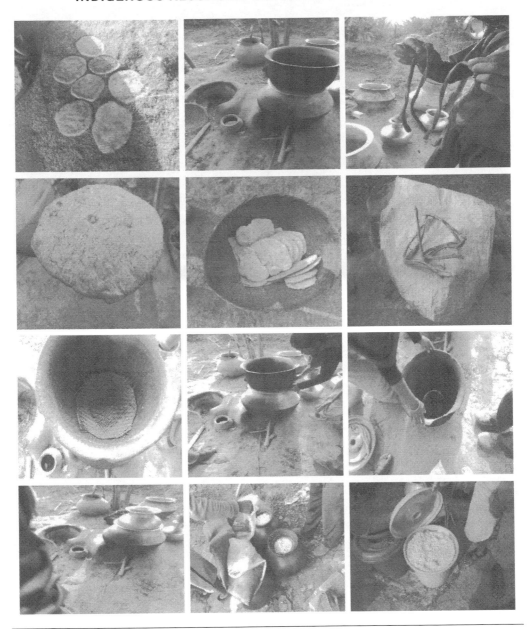

Figure 9.45 Process for *opo* preparation—a pictorial representation.

Angkur—Bodos of Assam (India): The rice beer made by the Bodos is known as *jou bishi*, and the starter cakes are known as *angkur* (Krishna and Baishya 2013). For preparing *angkur*, different plant materials are used, but the most common species are the leaves of Agarsita (*Xanthium strumarium*) and Dongphang rakhep (*Scoparia dulcis*) and either the roots or leaves of lokhunath (*Clerodendrum viscosum*). These plants are first washed properly and allowed to dry in the air. Rice grains are soaked in water for about 5 h and then, mixed with the plants and ground together in a wooden mortar with a pestle (this set of apparatus is called *wayai*). A dough is made by adding a little water to the mixture, which is then made into round cakes of about 5.5 cm diameter and 0.5–1 cm thickness and covered with dry mixture powder and allowed to dry for a period of 3–4 days by covering with *gigab* (paddy straw).

Thap—Karbis of Assam: The *Karbis* prepare a traditional alcoholic beverage called *hor-alank* and the starter culture is known as *thap* (Krishana and Baishya, 2013). For preparing *thap*, rice is soaked in water for one day and then, mixed with the leaves of *marthu* (*Croton joufra*), *janphong* (*Artocarpus heterophyllus*), *jockan* (*Phiogocanthus thysiflows*), *hisou-kehou* (*Solanum viarum*), and the bark of the *themra* (*Acacia pennata*) plant. The mixture is ground together in a wooden mortar (*longo*) with a pestle called (*lingpum*) in order to make a paste. This paste is then, made into small flat cakes of about 6 cm in diameter and 0.5 cm in thickness. These are overlaid with powder of previous *thaps* and kept in a bamboo sieve called an *ingkrung* and dried for about three days under the sun or above the fire (Krishana and Baishya, 2013).

Vekur pitha—Ahoms of Assam: The Ahoms call their rice beer *xaj pani* or *koloh pani*. The starter cake is known as *vekur pitha* and consists of the leaves of *banjaluk* (*Oldenlandia corymbosa*), *kopou lota* (*Lygodium* sp.) *horuminimuni* (*Hydrocotyle sibthorpioides*), *bormanmunii* (*Centella asiatica*), and *tubuki Iota* (*Cissampelos pareira*), and seeds of *jaluk* (*Piper nigrum*). All these are washed and dried well and then, ground in an wooden mortar (*ural*) with a pestle, and mixed with ground rice and a little water in a vessel and made into a paste. From this, oval-shaped balls of about 4.5 cm × 3 cm are made and placed on *kol pat* (banana (*Musa* sp.) leaves) and dried, either in the sun or over the fireplace, taking care not to get them too close to the fire. After a period of about five days, they become hard and are ready for use.

Apop pitha—Misings of Assam: The rice beer prepared by the Misings is known as *apong* and the starter cake is called as *apop pitha* (Krishana and Baishya, 2013). The different leaves needed for preparing *apop pitha* are of the plants *bormanimuni* (*Centella asiatica*), *sorumanimuni* (*Hydrocotyle sibthorpioides*), *banjaluk* (*Oldenlandia corymbosa*), *kuhiar* (*Saccharum officinarum*), *dhapattita* (*Clerodendrum viscosum*), *bhilongoni* (*Cyclosorus exlensa*), bam kolmou (*Ipoemea* sp.), *senikuthi* (*Scoparia dulcis*), *lai jabori* (*Drymeria cordata*), *jalokia* (*Capsicum annuum*), *anaras* (*Ananas comosus*), and *kopou dhekia* (*Lygodium flexuosum*). All these leaves are cleaned and dried by placing on a bamboo mat called an *opoh*. Soaked rice and the leaves are ground separately in a *kipar* (wooden grinder) and they are mixed together in a vessel with a little water. From the dough, oval-shaped balls of about 6 cm × 3 cm are made and dried in the sun (Krishana and Baishya, 2013).

Perok kushi—Deoris of Assam: The indigenous rice beer of the Deoris is known as *sujen*. The starter material is known as *perok kushi*, and the plant materials used are the leaves of bhatar *duamali* (*Jasminum sambac*), *thok thok* (*Cinnamomum byolgha*), *tesmuri* (*Zanthoxylum hamiltonianum*), *zing zing* (*Lygodium flexuosum*), *zuuro* (*Acanthus leucostychys*), *bhilongoni* (*Cyclosorus exlensa*), and *sotiona* (*Alstonia scholaris*), the roots of *dubusiring* (*Alpinia malaccensis*), and the stem and rhizome of the plant *Tomlakhoti* (*Costus speciosus*). These are washed and cut into small pieces and then, ground in a wooden grinder called a *dheki*. The mixture is then, soaked in water in a vessel until the water becomes colored. The whole mixture is added to ground rice in a vessel so as to make dough. Round balls of about 4 cm diameter are made and dried, either in the sunlight or over the fireplace by placing on a bamboo mat called as *aaphey*. After drying, they are stored in bamboo containers (*kula*), the inside and mouth of which is covered with *kher* (paddy straw).

Piazu—Angamis/Aos of Nagaland: The brew made by the Angami tribe is known as *zutho*, which is called *litchumsu* by the Ao tribe. This starter material used in the preparation of *zutho* is sprouted rice, known as *piazu*. For preparing *piazu*, un-hulled rice is first soaked in water for a period of about 3–4 days, after which, some of the water is drained out and the grains are allowed to germinate for about a week, depending on the prevailing temperature. After drying in the air, the sprouted grains are pounded in a wooden mortar with a pestle to obtain the powdered *piazu*.

Siiyeh—Adi-Calos of Arunachal Pradesh: The rice beer made by the Adi-Galo tribe is called *opo* and the starter cake is known as *siiyeh* or *opop*. For preparing *opop*, the leaves and bark of the plants *dhapat* (*Clerodendron viscosum*) and *Lohpohi* (*Veronia* sp.) are washed, sun dried, and made into powder, which is then mixed with powdered rice and a little bit of previously prepared *opo* in order to make a paste. From this, flat cakes of about 10–11 cm diameter are made and placed upon bamboo mats which are kept in the hearth for about 3–4 days, when the cakes become hardened. Other details are the same, as described earlier.

Jann: Jann is a traditional drink having low alcohol concentration consumed by the Bhotiyas community of Uttarakhand in India, during the winter. It is prepared from different types of cereals like rice (*Oryza sativa*), wheat (*Triticum aestivum*), *jau* (*Hordeum vulgare*), *koni* (*Setaria italica*), *china* (*Panicum miliaceum*), *oowa* (*Hordeum himalayens*), and *chuwa* (*Amaranthus paniculatus*) (Roy et al., 2004). The best quality *jann* made from local millet, is known as *koni* (*Setaria italica*), and is judged by its taste (sweetness), smell, and strength. Rice *jann* is commonly prepared by the community (Roy et al., 2004). The process of *jann* preparation is summarized in Figure 9.46.

Rice and apples are the most common cereal and fruit used in preparation. For preparing the rice *jann*, rice is boiled for about half an hour, so that it becomes soft. Excess water is drained-off and the remaining material (cooked rice) is spread over a flat container. After cooling, the cooked rice is mixed with the starter culture (*balam*) in the proportion of 5 kg rice to 40 g starter, to allow the process of fermentation. The mouth of the container is usually sealed with a piece of cloth. The mixture is then, kept in an airtight container in a dark and warm place for fermentation. In cold conditions, the rate of fermentation is slower than in warm, but yields good quality *jann*. Its preparation usually takes one week; the period can however, be stretched to get an improved quality product. After completion of the fermentation in about one week, the product is filtered. It has been proved that longer the period of fermentation, the greater the reduction in phytic acid content, while the availability of *in-vitro* minerals increases (Bhatia and Khetarpaul, 2012). The filtrate is a white liquid, which is thrown away or is used as animal feed. Earlier, when the Bhotiyas used to migrate

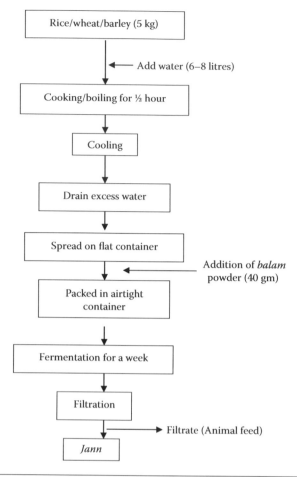

Figure 9.46 Traditional method of preparation of *jann* in the Kumaon Hills. (From Roy, B. et al. 2004. *Journal of Human Ecology,* 15(1): 45–49. With permission.)

to their winter settlement in the lower valleys, before the migration they prepared the *jann* and left it to ferment. For the six months of winter their entire settlement used to be buried/submerged under snow, and as a result of the internal heat generated by the pressure of the ice from the top, the *jann* fermentation was slow but steady. On their return in summer, the people found their *jann* ready to drink, and *jann* produced in this way is considered to be the best in quality (Roy et al., 2004). The entire process has been depicted in a flow diagram in Figure 9.46. The preparation of *jann* is the same from other cereals like *koni*, wheat, *oowa*, *chuwa*, and *cheena*. Only in the case of barley are the seeds partially ground before boiling, to give a quick fermentation and optimum yield.

The traditional starter used in the preparation of fermented beverages is called *balam* in the Kumaon and *balma* in the Garhwal regions as discribed in an earlier section also. It is made up of wheat mixed with a number of herbs and spices. A brief method of *balam* preparation is summarized in Figure 9.47.

Figure 9.47   Flow diagram for *balam* preparation.

Spices—*Dalchini* (*Chinnamomum zeylanicum*), *elachi* (*Amomum subulatum*), *kalimirch* (*Piper longum*), the leaves of *mirchi-ghash* (wild chillies), and the seeds of *pipal* (*Ficus religiosa*).

Judima or *zu* (Figure 9.48) is a mildly-alcoholic beverage, with a distinct sweet aroma, prepared from steamed glutinous rice (Chakravarthy et al., 2013). Different ethnic groups call it by their own dialect, such as *judima* (by the Dimasa), *juhning*

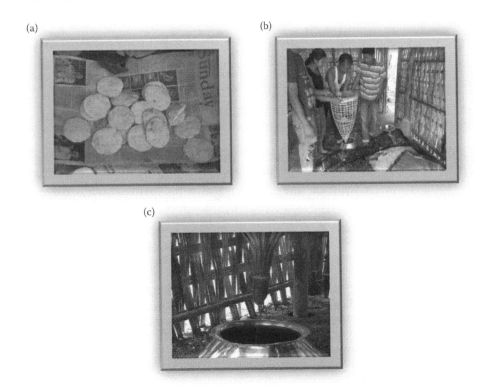

Figure 9.48   (a) Humao (starter), (b) fermented rice is transferred into *khulu,* (c) judima collected in an empty vessel or hundy which is kept below the *khulu.*

(Hrangkhol), *jeme naga* (*Deuijao* and *Nduijao*), and *remalu baitui* and *bumong baitui* (Baite). The rice beer of the Dimasas is called *judima*, and the starter cake is called *umhu* or *humao*, which is a mixture of rice and the bark of the *thempra* (*Acacia pennata*) plant. The bark is cut into small pieces and dried in the sun. Rice is soaked in water until softened, followed by grinding in a wooden pestle and mortar, called a *rimin*, along with the bark of the *thempra* plant. A little water is added so as to make a paste which is then, made into cakes of appropriate sizes and allowed to dry for a week (Figure 9.48a).

The Dimasa prefer sticky rice (*Oryza sativa* L., local cultivars *Bairon/Maiju-walao/ Maijuwalao-gedeba/Maiju-hadi*) for a better quality of *judima* preparation, or sometimes, if sticky rice is not available, they mix 80% ordinary rice with 20% of sticky rice. During the traditional method of preparation of *judima*, the sticky rice is cleaned, washed, cooked, and excess water is drained out. The cooked rice is spread over bamboo mats or banana leaves (in a layer 6–8 inches thick) for cooling (nowadays polythene sheet is also used) and then ~1% of *humao* is mixed in thoroughly, and a little water is added. The fermenting mass is placed in fresh banana leaves and covered by more banana leaves or polythene sheet and fermented for 1–2 days (in summer, or 3–4 days in winter) at room temperature. When a sweet aroma is emitted (saccharification), the fermented mass is transferred in a *khulu* (triangular shape bamboo cone) (Figure 9.48b) covered with banana leaves, or a hollow earthen pot. An empty vessel, or *hundy* (Figure 9.48c), is kept below the *khulu* and the juices are collected in it (it takes 1–2 days), and called *judima*. After collection of the *judima*, the remaining rice, called *jugap*, is distilled to make high-alcohol-content liquor, locally called *juharo*. *Judima* is similar to the *bhaati jaanr* of the Himalayan regions of India, Nepal, and Bhutan (Chakravarty et al., 2013). *Judima/zu* is drunk with or without water. After the celebration of traditional rituals and festivals, *judima* is served with meat.

Juharo, the distilled liquor, is more alcoholic and acidic in taste. It is a traditional diet for women in villages who believe that it helps them to regain their strength. The best quality of Judima is pineapple juice colored or yellowish-red in color. The color of the product however depends on the quality of rice. Besides home consumption, *judima* is sold in the local market of NC (North Cachar) Hills by some people who are economically dependent upon *judima*. It costs Rs. 45–50 per 750 mL bottle.

Zutho: Zutho (rice beer) is a traditional alcoholic beverage prepared from rice (*Oryza sativa* L.), named according to the *Angami* "Naga" dialect (Jamir and Deb, 2014). To prepare *zutho*, polished rice grain is soaked in water for about 2 h and the excess water is drained off (Figure 9.49). It is then allowed to air dry and is pounded into powder. For the other part of the preparation, unhulled rice grains are soaked in water for about 2–3 days and allowed to germinate. The germinated grains are sun dried and made into powder. The polished rice powder and germinated grain powder are mixed in a ratio of 10:3 and made into paste by mixing slowly with boiling water. The mixture is then allowed to cool and incubate at room temperature for 4–5 days. Fermentation is completed after about 4–5 days. The first fermented product in its

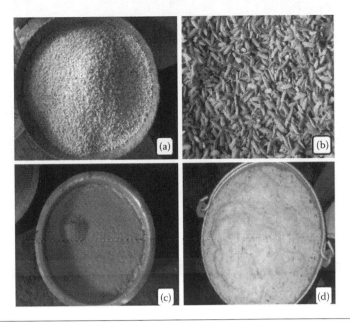

Figure 9.49 Steps involves in *zutho* preparation. (a) Polished rice, (b) unhulled germinated rice grains, (c) mixed powder of (a, b) under fermentation, and (d) *zutho* ready-to-serve.

pure form is called *thoutshe,* and when *thoutshe* is diluted with water it is called *zutho.* It is consumed as a popular alcoholic beverage in Nagaland.

9.5.9 Soor/Sura/Sur

Soor/Sura/Sur is a local alcoholic beverage made up of cereals and fruits rich in sugars by the natives of Uttarakhand and Himachal Pradesh in India (Rana et al., 2004; Bhalla et al., 2004; Sharma, 2013). People drink *soor* to cope up with the adverse climatic condition and also at ceremonies and festivals. Barley, rice, finger millet, apples, pears, apricots, and peaches are the common raw material used for the preparation of *soor,* which contains 35%–40% alcohol (Rana et al., 2004). It is made by using a starter culture material called *keem.* It is mainly prepared during the rainy season by the villagers. It is made up of different plant species (Rana et al., 2007). A brief method of *soor* preparation is summarized in Figure 9.50.

The distilled *soor* collected is further classified on the basis of its alcohol content as:

- Super—1st hour of distillation
- Good—after 2 h of distillation
- Moderate—after 3 h of distillation

It is an indigenous alcoholic beverage of Himachal Pradesh (India), produced mainly in the Kullu, Mandi, and Kangra districts, from the alcoholic fermentation of locally grown finger millet (*Elusine coracana*) (Joshi et al., 2012b; Sharma, 2013; Joshi et al., 2015). Starch is the main carbohydrate present in finger millet (*ragi*) and it constitutes of 59.4%–70.2% of total carbohydrates (Nirmala et al., 2000). In addition,

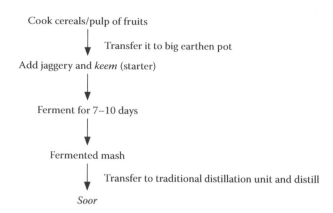

Cook cereals/pulp of fruits

↓ Transfer it to big earthen pot

Add jaggery and *keem* (starter)

↓

Ferment for 7–10 days

↓

Fermented mash

↓ Transfer to traditional distillation unit and distill

Soor

Figure 9.50   Flow diagram for *soor* preparation.

these have polyphenols or tannins known to have antioxidant properties. *Dhehli* is the inoculum/starter used in the fermentation of *sura* (Figure 9.51) which is prepared by the elderly people of the Kullu valley during the rainy season (*Bhadpadra* month). *Dhehli* (Figure 9.52) is a bioactive mixture of 36 different herbs and roasted barley flour (*sattu*) made in the shape of bricks, which imparts additional flavor (Thakur et al., 2004; Sharma, 2013).

Sur production involves the grinding of finger millet to flour, which is kneaded with water to form dough. After the preparation of the dough, it is left for 10 days in a container for natural fermentation, which leads to the hydrolysis of starch and the conversion of dough into slurry. This slurry is spread on a hot plate, and half-baked *roties* are

(a)   (b)

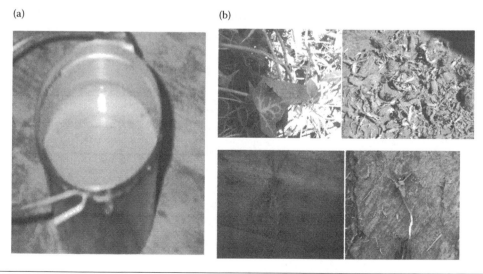

Figure 9.51   Pictures of (a) *sura* and (b) few herbs used in preparation of *dhehli*. (Courtesy of Sharma, 2013. *Preparation and Evaluation of Sur production in Himachal Pradesh.* Thesis, Dr. Y.S. Parmar University of Horticulture and Forestry. Nauni, Solan. HP, India; Ashwani, K. 2013. *Preparation and Evaluation of Sur production in Himachal Pradesh.* Thesis, Dr. Y.S. Parmar University of Horticulture and Forestry. Nauni, Solan. HP, India.)

Figure 9.52 (a) Pictorial view of *dhehli* (b) Method used in maintaining the temperature of the fermenting pot in Himachal Pradesh. (Courtesy of Ashwani, K. 2013. *Preparation and Evaluation of Sur production in Himachal Pradesh.* Thesis, Dr. Y.S. Parmar University of Horticulture and Forestry. Nauni, Solan. HP, India.)

made (Thakur et al., 2004; Ashwani, 2013). These *roties* are cut into pieces, cooled, and powdered *dheli* and water are mixed and put into an earthen pot. However, in some cases, jaggery syrup is also added so as to increase the intial sugar content to produce more alcohol in the final product (Ashwani, 2013; Joshi et al., 2015). This mixture is fermented for 8–10 days by covering the pot and the end product obtained is called *sura*. *Sura* obtained by this method is further filtered in some regions, whereas in other areas it is consumed without filtration. Fermentation involves hydrolysis of starch and conversion of the resulting fermentable sugars to alcohol by natural microflora. A brief overview of *sura* preparation is shown in Figure 9.53.

9.5.10 Chhang/LugriChakhti/Jhol

Chhang is a popular indigenous alcoholic beverage of the hill folk in the Western Himalayas, made by fermenting barley with 4.61% alcohol, 1.68% dry extract, and 65.94 meq total acids, of which volatile acids are 3.18 meq. It is an indigenous beer made in the tribal belt of Lahaul and Spiti, Kullu and Kangra. *Chhang* is also called *jhol* and *chakti* in Kullu. Huskless *sherokh* variety of barley, locally called as *grim* is used in its preparation (Bhatia et al., 1977; Targais, 2012). In Ladakh, *chhang* (Skyems) is the general word for alcoholic beverages, but here it refers to the barley drink. No good party is without it, although nowadays there seems to be a tendency, among the men at least, to favor hard liquor such as rum or whisky. The preparation of *chhang* involves solid-state fermentation, as no additional water is added to the ingredients—namely cooked rice/wheat/barley and *phab* (the traditional inoculum) or *phapus* (Figure 9.54) (Thakur et al., 2004). Sufficient water is used for absorption. The *Ladakhi* word for yeast (starter culture) is *phabs* and mostly comes from the Nubra valley, but also can be found in the market in nugget form (Figure 9.54). There are different strengths of the beverages, which are carefully tasted by one of the older ladies and evaluated. The traditional vessel, made of metal or stone, used to store *chhang*, is called an *uthi* in Lahaul. The *chhang/lugri* is a very popular fermented beverage which is served to

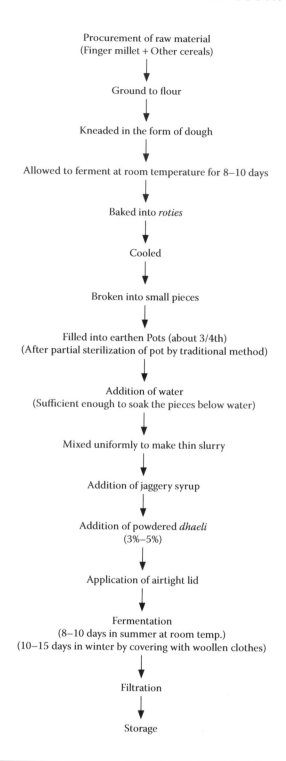

Procurement of raw material
(Finger millet + Other cereals)

↓

Ground to flour

↓

Kneaded in the form of dough

↓

Allowed to ferment at room temperature for 8–10 days

↓

Baked into *roties*

↓

Cooled

↓

Broken into small pieces

↓

Filled into earthen Pots (about 3/4th)
(After partial sterilization of pot by traditional method)

↓

Addition of water
(Sufficient enough to soak the pieces below water)

↓

Mixed uniformly to make thin slurry

↓

Addition of jaggery syrup

↓

Addition of powdered *dhaeli*
(3%–5%)

↓

Application of airtight lid

↓

Fermentation
(8–10 days in summer at room temp.)
(10–15 days in winter by covering with woollen clothes)

↓

Filtration

↓

Storage

Figure 9.53 Generalized process for *sur* production as per the information gathered from various regions. (Adapted from Ashwani, K. 2013. *Prepartion and Evaluation of Sur production in Himachal Pradesh*. Thesis, Dr. Y.S. Parmar University of Horticulture and Forestry. Nauni, Solan. HP, India.)

Major steps followed for production of *sur* in laboratory by traditional method

Cleaning of grains   Milled flour   Kneaded flour

Flour covered with natural micro-flora   Baking into roties   Broken pieces of *roties*

Sterilization of pot by smoking   Filling of pieces into pot   Fermentation

Figure 9.54 Pictorial representation of different steps in *sur* preparation. (Courtesy of Ashwani, K. 2013. *Preparation and Evaluation of Sur production in Himachal Pradesh.* Thesis, Dr. Y.S. Parmar University of Horticulture and Forestry. Nauni, Solan. HP, India.)

guests during *Phagli* (traditional New Year of the Lahulis) and at marriage ceremonies. As a sign of honor, guests are served *chhang* with a small piece of butter on the beautiful brass pot known as a *chhabskyen* (Figure 9.54) from which *chhang* is usually served, or one is put in the glass (Thakur et al., 2004)). The distilled form of *chhang* is known as *sura* in the Lahaul valley.

9.5.10.1 The Preparation of Chhang   The different steps involved in the preparation of *chhang* are shown in Figure 9.55.

9.5.10.1.1 Cleaning and Boiling of Barley Grains   Substrates for fermentation during the preparation of *chhang* may be wheat grain, rice, or grapes instead of barley, and

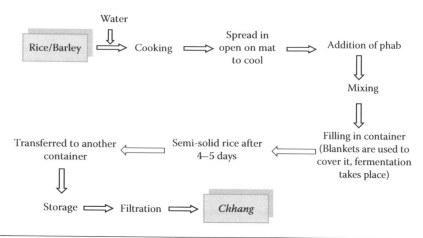

Figure 9.55 Flow diagram of *chhang* making.

the products in the respective cases are known as *tochhang, dechhang,* or *rguntshang,* in that order (Angchok et al., 2009; Targais et al., 2012). Cleaned grains are boiled in a large vessel. The grain quantity is measured in terms of *bho,* a wooden mug-shaped container. One *bho* contains 1–2 kg of grains, depending on the size. On an average, 5 kg of grains and 10 L of water are used for the boiling process. Once the contents start boiling, the fire is kept at low intensity until the grains have absorbed all of the water. After cooking, the grains are spread on a blanket for about 10 min, just to drain off excess water. The boiled grains, locally known as *lhums,* are spread on a canvas known as a *khol–char* made of woven Yak's fur. The whole content is mixed with a goat or sheep's scapula at regular interval till the temperature drops to 28–32°C. The end point of cooling process is checked by touching the *lhums* to the cheek (Targais et al., 2012).

9.5.10.1.2 Adding Starter Culture On cooling, the locally available starter culture, in the form of nuggets known as *phabs,* weighing 1.3–1.7 g, is crushed to powder and spread uniformly over the boiled grains and mixed thoroughly (Figure 9.56). Generally 3 g *phabs* are sufficient for 5 kg of grains during summer, while up to 7 are needed during the winter (Targais et al., 2012), depending on the quality of the culture. After mixing in the starter culture, the inoculated grains are transferred into a cloth bag, known as a *chhang-sgey,* and placed in a heap of wheat or barley straw. A large flat stone weighing 2–4 kg is placed on top of the straw heap to increase the temperature around the contents in the bag, while in some cases the container is kept inside a straw store, locally known as a *sabrak* or *phukrak,* so as to insulate it from the cold. Usually, it takes two days for the starter culture to grow during summer, and five days during winter. The end point for the growth of the starter culture is checked by a combination of fermentation smell and wetness of the grains. In severe winter when the starter culture takes a longer time to grow, a heated fist-sized spherical stone known, as a *chhang-rdo,* is placed in the middle of the contents in the bag and

Figure 9.56 Pictures showing (a) *phab*, (b) copper kettle (*chapskan*) used to serve *chhang*, (c) *jau* (barley) *chhang* and (d) rice *chhang*.

kept for an additional day. The temperature of the stone and its placement is an art and, depending on experience and mastery of the art, the taste of the *chhang* either improves or deteriorates. It is also believed that placing an iron object on top of the bag generally improves the taste of the beverage (Targais et al., 2012).

9.5.10.1.3 Fermentation The incubation period and number of *phabs* added as a starter, determines the strength of the *chhang*. After 2–3 days of growth of the starter culture, the contents of the bag is transferred into an earthen pot. The mouth of the pot is closed with a spherical stone wrapped in clean cloth and made air tight with ash clay. The pot is placed in the heap of straw or kept at ambient conditions, usually for 7 days, depending on the ambient temperature. The seven day duration is called the *chhang-zak* or *lhums-zak*, meaning that the *chhang* should be ready within this period. Depending on the temperature, the duration ranges from three days in summer to 10 days in winter. If a strong beverage with high alcohol content is required, the duration is further extended (Targais et al., 2012). Completion of fermentation is judged almost invariably by an old woman of the household by tasting the fermented grains (*lhum*). The fermented grains (*lhum*) are then transferred to a wooden cask (*zem*) and then rinsed with water and drained to collect the *chhang* (Figure 9.57). *Chhang* is generally stored in a *zangboo* (a copper pot). The fermented barley grains left after extraction are locally called b*angma*, which are generally fed to animals or used to prepare *bangphe*. In other cases, the fermented mass is transferred to a narrow-mouthed stone, metal, or plastic jar and filled up to the brim. The vessel is plugged tightly with cloth and covered with mud plaster to provide airtight sealing (Tamang, 2009a,b).

A minimum storage period of five days is considered essential to enhance the aroma and alcohol content of the beverage.

9.5.10.1.4 Extraction and Filtration   After completing the fermentation process, the contents are taken out of the pot and transferred into a cylindrical wooden drum called a *zem* (Figure 9.57). Water is added till the fermented content is submerged. After 2–5 h, the first filtrate is taken out from the hole made in the lower part of the wooden drum. It is sieved through a *tsagma*, a sieve made from straw. The filtrate is called *machu*, meaning concentrate, or *tang-po*, meaning first. After the first filtrate is taken out of the drum, water is again added till the fermented grains are submerged. The filtrate is again taken after 2–5 h and is called *nyis-pa*, meaning second. The process is continued, and the third and fourth filtrates are called *sum-pa* and *gyi-pa*, respectively. The alcohol content of the first filtrate is the highest, and reduces in the subsequent steps. Based on taste and alcoholic content of each filtrate, a blend is made by mixing the filtrates. The beverage is put into a *chhabskyen* (Figure 9.57) and served. The final blended product contains 5%–7% alcohol, has pH 3.6–3.8, and acidity of 0.55%–0.65% (Targais et al., 2012). Unused *chhang* can be stored for further use. A small amount of fermented grains

Figure 9.57   (a) Traditional kitchen. (b) *Tagi thalshrak.* (c) *Skien* (placed on kitchen shelf). (d) *Khura.* (e) *Phabs*: starter culture. (f) *Phabs* (indigenous yeast inoculants) prepared in Nubra valley of Ladakh. (g) Spreading of boiled grains on *kholchari.* (h) *Zem*: wooden drum for extraction and filtration of *chhang.* ((a–e,g–h) Adapted from Targais, K. et al. 2012. *Indian Journal of Traditional Knowledge,* 11(1): 190–193, (f) Courtesy of Dr. Anup Raj.)

can be crushed in water and the suspension produced, known as *boza*, is drunk (Targais et al., 2012). A similar process is carried out to produce *chhang* in other areas.

9.5.10.2 Chhang Kholak   Chhang is locally barley made fermented drink. *Phey* is added to *chhang* to make *chhang kholak*. It has a sour taste and mostly preferred by travelers.

9.5.10.2.1 Distillation of Chhang   The *chhang* prepared can also be distilled by heating in a copper container with an inlet pipe for the circulation of cold water, an outlet to collect the distilled liquor, and a lid for evaporation. The distilled product is directly consumed or stored for a longer period (Kanwar et al., 2011).

Raksi: Raksi (commonly known as rice wine) is a traditional distilled alcoholic beverage made in Bhutan, Bangladesh, and Nepal. It is prepared from rice, barley, finger millet, and maize (Singh et al., 2007), with different grains producing different flavors. Besides, these grains, potato and cassava are also used. It tastes somewhat like Japanese *sake. Raksi* is commonly used in different religious rituals and social events. It is an ethnic Himalayan alcoholic drink with a characteristics aroma, distilled from the traditional fermented cereal beverages such as *Kodo ko jaans/bhaati* jaanr, *makai ka* jaanr, *gahoon ka jaanr* etc. (Kozaki et al., 2000). *Raksi* is a common term in Nepali meaning any alcoholic drink (Tamang, 2009).

Das-chhang (fermented rice brew): This is a new introduction to this region, brought about by the migrating laborers from Nepal, and they generally call it *rakshi*. Though rice is not grown in Ladakh, *rakshi* is prepared from the rice available in the local market. *Rakshi* or *Rokshi* is also a fermented product prepared from wheat, rice, or *jowar*/sorghum (*Sorghum bicolor*). This drink is restricted to the Nepali laborers and is not prepared or consumed by local people. A brief method of *raksi* preparation, as shown in Figures 9.58 and 9.59, is described here.

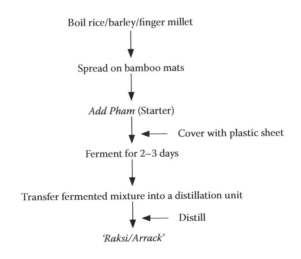

Figure 9.58 Flow diagram for *raksi/arrack* preparation.

Figure 9.59 Traditional distillery unit for *raksi*.

Steps of preparation:

1. Wheat grains are washed well with running water and kept in a container.
2. These are then, boiled with water until properly cooked.
3. The boiled material is then, spread on clean polyethylene.
4. A starter culture (*murcha*) is added to the material after grinding. The starter culture (*pham*) @ 5–10 g/kg of wheat is added. *Pham* is the dry yeast cake prepared from rice paste and *S. khasianum* leaves.
5. The material is mixed properly with the starter culture.
6. It is then, put (mostly) into a plastic containers and covered properly.
7. The containers are wrapped with gunny bags or fern leaves to raise the temperature inside.
8. The containers are kept for 1–2 weeks in this condition to complete the fermentation.
9. The fermented product derived at this stage is placed at the bottom of a large metallic chamber.
10. Inside the metallic chamber above the fermented product, a wooden tool is placed. A wide shallow metallic container is kept on the tool. This is connected to a metallic pipe outside the metallic chamber to siphon out the fermented liquor.
11. Another metallic container is placed at the top of the large container and tightly fitted with cloth and the container is filled with cold water.
12. The large metallic container is heated. The liquor of the fermented product boils and evaporates, moves up and get condensed when it comes in contact with the bottom of the container filled with cold water, placed at top. When the water of the container at top becomes hot it is replaced with cold water.

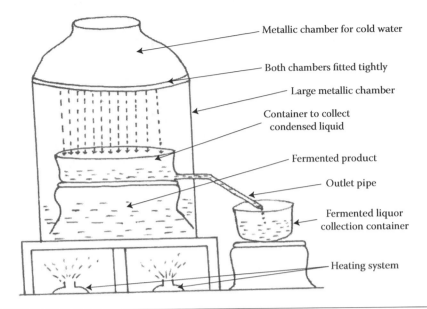

Metallic chamber for cold water

Both chambers fitted tightly

Large metallic chamber

Container to collect
condensed liquid

Fermented product

Outlet pipe

Fermented liquor
collection container

Heating system

Figure 9.60 Metallic chamber for preparation of *rakshi*.

13. The condensed vapor drops in the form of liquid into the container placed on the tool. The liquid is collected through a pipe drop-wise externally in a container (Figure 9.60). *Raksi* contains about 20%–40% (v/v) of alcohol.

9.6 Preparation of *Daru*

Daru is a distilled liquor containing ethyl alcohol at a much higher concentration than other alcoholic beverages. Rice and *jaggery* are the common substrates used for its preparation, though other cereals like *koni*, *chuwa*, *oowa*, and wheat are also used (Roy et al., 2004). The taste of *daru* does not vary according to the type of substrate used. The most commonly available and cost-effective materials used in the preparation of *daru* are rice (5 kg) and *jaggery* (2 kg). Cooked rice on becoming cool is mixed with the powder of *balam*, the proportion of *balam* powder required in preparation of *daru* being much more than is required in *jaan* preparation (100 g/5 kg rice). The mixture is then kept in an airtight container for fermentation, preferably in a warm place. After about a week of fermentation, when the mixture is in a semi-liquid condition, it is distilled in a distillation vessel. The distillate is called *daru*, which is collected in bottles (Roy et al., 2004).

The traditional distillation method is still practised in various countries of the region. The indigenous distillation set, which is quite simple, has three parts—the *parar*, the *jokhal*, and the *tal*. The *parar* is a big saucepan like container with flat bottom; the *jokhal* is a flat wooden device like a dish, having an elongated channel with a hole at the centre; and the *tal* is a simple cooking vessel, but the neck of *tal* and bottom of the *parar* are of such a size that they hold the *jokhal* perfectly (Roy et al., 2004). The rest

(a)   (b)

Figure 9.61   (a) Covering the fermented vessel. (b) Distillation of *daaru*. (Courtesy of Dr. Nehal A. Farroquee.)

of the procedure is the same for any distilled product. A small quantity of turmeric is hung at the mouth of the distillation vessel to give a bright yellow color to the product. Traditionally, *daru* is graded into three categories, as described earlier. The complete process for the preparation of *daru* is shown in Figures 9.60 and 9.61.

The distillation unit used (Roy et al., 2004), is a unique example of use of locally available material in a traditional manner (Figure 9.61).

9.7 Production of *Jann*, *Sez*, and *Daru* from a Common Cycle of Fermentation

All the three categories of fermented beverages and foods can be prepared from a common fermentation cycle only when rice is used as the substrate. Rice is first cooked in boiled water for half an hour or until soft and edible. It is then, kept in a flat container to drain-off the excess water and also to cool down (Roy et al., 2004). This boiled rice is mixed with *balam* powder and the mixture is kept in an airtight container, preferably in a dark and cool place. The container is traditionally an earthenware or wooden vessel, although nowadays plastic is also used. After one or two days of fermentation, the *sez* is ready for consumption. The required quantity of *sez* can be removed at this stage. After that, the container is again kept airtight for another five to ten days for further fermentation, to produce *jaan*. The *jaan* is removed by filtration, that is, by passing the contents through a sieve or a piece of cloth. In the remaining mixture, *jaggery* and fresh balls of *balam* at a ratio of one ball per kg *jaggery*, is added for the preparation of *daru*. The complete process for the preparation of *jann*, *sez*, and *daru* is shown in Figure 9.62 (Roy et al., 2004).

9.8 Ayurvedic Context of Indigenous Alcoholic Fermented Foods

9.8.1 *Tonic Wines*

Tonic wines are medicinally useful, increase vitality, and improve digestion. These wines are made by steeping medicinal herbs in wine for several weeks (Singh et al., 1999; Sekar, 2007). A simple and effective way to make a tonic wine is in a jar or a ceramic vat with a tap at the base to enable the wine to be drawn off without disturbing

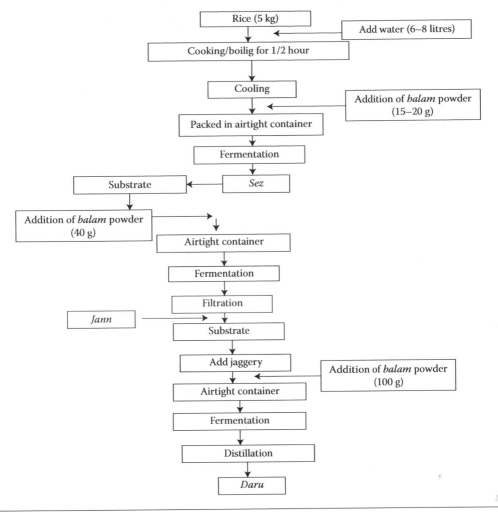

Figure 9.62 Traditional method of preparation of *jann*, *sez*, and *daru* in the Kumaon Hills.

the herbs. Wine can be added periodically to keep the herbs covered, although, in time, this will reduce the tonic effectiveness of the wine. If exposed to the air, the herbs may go moldy due to the unwanted growth of fungi, making the remedy not only ineffective but also unsafe to consume (Dhiman, 2004; Sekar, 2007).

The upper portion of *madya*, which is light in nature, is called *prasanna*. The portion below that, which is relatively denser, is called *kadambari*. The portion below *kadambari* is called *jagala*. The portion, which is at the bottom region of the container, is called *medaka*. The lowest layer of *medaka* containing the paste of drugs is called *vakkasa*. The material that is used for initiating fermentation of alcoholic drinks is called *kinva* or *surabija* (this is the microbial inocula). If *kinva* is not matured, then it is called *madhulaka*. It is present in improperly fermented *madya*.

Prasanna: This supernatant clear portion of alcoholic preparation alleviates vomiting, anorexia, pain in the heart and abdomen, *kapha*, *vata*, piles, constipation, and hardness of bowels (Sharma, 2004; Sekar, 2007). *Prasanna* cures *anaha* (flatulence), *gulma* (gastritis), *arsas* (piles), *chardi* (vomiting) and *arocaka* (anorexia) (Dash and Kashyap, 1980; Sekar, 2007).

Kadambari: The lower thick portion of wine is called *kadambari* (Dash and Kashyap, 1980). *Kadambari* type of alcoholic drink is a digestive stimulant. It cures *anaha* (flatulence), pain in the heart and pelvic region, and colic pain. It is heavy, aphrodisiac, alleviator of *vata*, and a laxative.

Jagala and *bakkasahi:* This is constipative. It cures *sopha* (oedema), *arsas* (piles), and *grahani* (irritable bowel syndrome). It is hot, carminative and strength promoting. It cures *ksut* (morbid hunger), *trsna* (morbid thirst), and *aruci* (anorexia). It is beneficial for colic, dysentery, *borborygmi* (sound of flatus in intestine), and constipation (Valiathan, 2003; Sekar, 2007). *Jagala* is a digestive, produces oedema, and alleviates dysentery, gurgling sound in bowels, piles, *vata* and consumption. In oedema, it is applied externally. *Bakkasah* are *jagala* are free from liquid, consisting only of yeast and drugs. It is *vata* aggravating, an appetizer, a laxative, a diuretic, and nonslimy.

Medaka: This is sweet, strength promoting, *stambhana* (which increases the power of retention), cooling and heavy (Dash and Kashyap, 1980; Sekar, 2007).

Vakkasa: This portion from which alcohol is taken out is constipative and it aggravates *vata* (Patwardhan et al., 2004; Sekar, 2007).

Kinvaka: This alleviates *vata*. It is not good for the heart and is difficult to digest.

Madhulaka: This is unctuous, constipative, aggravates *kapha*, and is difficult to digest (Dash and Kashyap, 1980, 2002; Sekar, 2007).

9.8.2 Fresh and Stored Wines and Their Properties

Freshly prepared alcoholic drinks (*nava madya*) are *abhisyandi* (obstruct channels of circulation). It alleviates all three *doshas* and is a laxative. It is not good for the heart and not tasteful (*virasa*). It causes a burning sensation and produces a putrid smell. It is *visada* (non-slimy), heavy, and difficult to digest. It is *navam*, which means immature in terms of taste and clarity, or of less than a year (Sharma, 2004; Sekar, 2007). They should not be used by persons who are having purgation (or have had purgative therapy) and who are hungry (Murthy, 1994; 1998; Sekar, 2007).

Purana madya, the same alcoholic drink, when preserved for a long time and used, is relishing. It cures *krmi* (parasitic infection) and the aggravation of *kapha*, as well as *vata*. It is a cardiac tonic, fragrant, endowed with good qualities and light. It cleanses the channels of circulation, improves the appetite, and opens the body channels (Dash and Kashyap 1980; Sekar, 2007). Old wine is an appetizer, much relishing, anthelmintic, and pacifies *vata* and *kapha*, and promotes digestion. It is consumed when older than one year (Murthy, 1998; Sekar, 2007).

9.8.2.1 Asavas and Arishthas

Ayurveda comprises of various types of medicines, including the fermented forms, namely *arishtas* (fermented decoctions) and *asavas* (fermented infusions). These are regarded as valuable therapeutics due to their efficacy and desirable features. The bulk of knowledge on these fermented medicines however, remains undocumented, unrecognized, and unvalidated. The fundamental concepts in the design of *arishtas* and *asavas*, with representative examples, are briefly summarized. The preparation, fermentation, storage, and usage of these products are also summarized earlier (Roy et al., 2004; Sekar and Mariappan, 2008) and briefly described.

Arishtas and *asavas* are self-generated herbal fermentations of the traditional Ayurvedic system. They are alcoholic medicaments prepared by allowing herbal juices or their decoctions to undergo fermentation with the addition of sugars. *Arishtas* are made with decoctions of herbs in boiling water, while *asavas* are prepared by directly using fresh herbal juices. Fermentation of both the preparations is brought about by the addition of a source of sugar with *dhatuki* (*Woodjordia fruticosa* Kurz) flowers (Murthy, 1998). They are moderately alcoholic and sweetish in taste, and with a slight acidity and an agreeable aroma.

Sugarcane is produced in several South Asian countries, especially India and Pakistan, and its juice is used both for the production of wines and for distilled liquors. Sugarcane wines have also been described in *ayurveda* as *sidhu* (*sita rasa* or *sita rasika*, *paversa sidhu*, *sasyaka*, and *aksiakah-sidhu*). Those made out of uncooked materials are called *sita rasa* or *sita rasika*. Unboiled sugarcane juice is used in their preparation, but this type of *sidhu* is inferior in quality. It improves digestion, voice, and complexion, combats swelling, abdominal disorders, and piles, and is useful for slimming. The product made from boiled sugarcane juice is known as *Pakvarasa sidhu*. It aggravates *vata* and *pitta* and diseases of *kapha*, obesity, dropsy (an excessive accumulation of fluid in any of the tissues or cavities of the body), enlargement of the abdomen, and hemorrhoids. *Sidhu* is better than *sita rasa*. *Pakavarasa sidhu* promotes good voice, digestive power, and serves as a cardiac tonic; it is unctuous, and an appetizer (Sekar, 2007).

Sasyaka: This *sidhu*, prepared using boiled sugarcane juice fermented with *dhataki* and after a certain period of fermentation with little *ghee* and *jaggery* added, is known as *sasyaka* (Sekar, 2007).

Aksikah-sidhu: *Aksikah-sidhu* is a decoction of *bibhitaka* (*Terminalia bellirica* (Gaertn), Roxb.) and *jaggery*, and processed with *dhataki* flowers. It alleviates anemia, is astringent, sweet, *pitta* alleviating, and is a blood purifier (Sekar, 2007).

9.8.3 Method of Preparation

The basic equipments required for preparation of the *ayurveda* products are an earthenware pot sufficiently large and strong with a glazed exterior, or a glazed porcelain jar of suitable size; a lid of the correct size to close the vessel; a cloth ribbon to seal the vessel; a paddle like stirrer; a clean cloth of fine and strong texture for filtering; and a vessel to keep the juices or boil the herbs (Shastri, 1968; Mishra et al. 2012). The major ingredients are herbs from which the extract or decoction is made. The herbs yield compounds which are of pharmacological and therapeutic significance, and the names of the medicines are derived from these herbs, denoting their importance. The herbs, besides contributing to the flavor, have their own pharmacological action. The fermentation initiator is an inoculum, while sugars are also required in the medium for fermentation. Analysis of the components of typical *arishta* and *asava* could support and provide insights and serve as the basis of designing them. In *Asokarishta*, the main herb is *asoka* (*Saraca asoca* De Wilde). Those contributing to flavor are; *Cuminum cyminum* L., *Santalum album* L., and *Zingiber officinale* Roscoe. *Woodfordia fruticosa* (L.) Kurz, is a fermentation initiator, and honey and *jaggery* are a source of sugar. Similarly, in *Kanakusava*: *Kanaka* (*Datura metel* L.) is the main herb, while *Piper longum* L. and *Zingiber officinale* Roscoe contribute to the flavor (Mishra et al., 2012).

9.8.4 *Process of Preparation*

The proportion of the different ingredients are, water 32 *seers* (or 1024 *tolas*, 1 *tola* = 12 g), treacle or *jaggery* 12 and half *seers* (or 400 *tolas*), and honey 6 *seers* (or 200 *tolas*), medicinal substances (such as roots, leaves, or bark, etc., of plants cut into pieces), 1 *seer* (or 40 *tolas*), in powder or decoction farm. The basic herbs from which the extract is prepared are first cleaned and rinsed in water to get rid of dirt (Kushwaha and Karanjekar, 2011). In the case of fresh plants, they are cleaned, pulverized and pressed for collection of the juice. If the dried plant is needed in the preparation of *asava*, it is coarsely crushed and added to water to which the prescribed quantities of honey, *jaggery*, and/or sugar are added. If it is an *arishta*, a decoction is obtained by boiling the herb/plant in the specified volume of water (Kushwaha and Karanjekar, 2011). The water used should, however, be clean, clear, and potable. When the extracts are obtained, the sugar (cane sugar), *jaggery*, and/or honey are added and completely dissolved. Sometimes, any one or more of these sugary substances are omitted if so directed by the recipe, but those used should be pure. The *jaggery* to be added should be very old (*prapurana*) because fresh *jaggery* aggravates *kapha* and suppresses the power of digestion. The flavoring agents are coarsely powdered and added to the sweetened extract. Too fine powder of the flavoring agent is undesirable as it causes sedimentation in the prepared medicine and its filtration is difficult. In *asavas*, the *avapa* (drugs which are added in powder form at the end) should be one tenth in quantity and honey should be three fourths of quantity of *jaggery* (Kushwaha and Karanjekar, 2011).

The earthen pot or jar intended for fermenting the medicine is tested for weak spots and cracks and similarly, a lid is also chosen. The earthen pot is prepared from the soft mud collected from the silt in the bank of a river or lake (Kushwaha and Karanjekar, 2011). The pot should be greasy, thick, light, and smooth and be free from holes or cracks and be homogenous. An echo should come out from inside of the jar. Its circumference in the middle should be 42 *angulas* (I *angula* = 5 inches) and its height should be 43 *angulas*. Its wall should be one *angulas* in thickness and compact. In shape, the pot should be like the fruit of *bakula* (*Mimusops elengi* L.). The internal surfaces of the pot and the lid should be wiped with a clean dry cloth and cow ghee should be smeared on this surface to prevent oozing of the contents when pouring. Glazed porcelain ware may also be used instead of earthenware. At large scale, the fermentation is carried out in huge wooden vats with wooden covers to make it them tight. The filtration is carried out by electric filter presses which efficiently separate the suspended particles and isolate clear medicine. The powdering, grinding, and mixing are done by mills, pulverizers, and mixing machines. The decoctions are made in large steam-jacketed boilers, heated by superheated steam under pressure.

9.8.4.1 Inoculum When the pot or the jar is ready, the sweetened and flavored drug extract is poured into it up to three quarters of the capacity, and one quarter space is kept for the fermenting liquid to rise up due to frothing and evolving a large amount of carbon dioxide. Otherwise, the medium may damage the container and flow out. Then, the inoculum is added to initiate fermentation. The process of fermentation necessitates the presence of yeasts. The yeast inoculum comes from the *dhataki* flowers, which contain a wild species of yeast in the dry nectariferous region (Vohra and Satyanarayana, 2001; 2004). The presence of tannins both in hops and in these flowers favors a suitable environment for yeast growth. After addition of the flowers, the contents are stirred well to distribute the inoculum. Apart from the *dhataki* flowers, other ingredients like honey and raisins (gum) are also added. Alternatively, the inoculum of yeasts can come either from *mahua* (*Madhuca tongifolia* Macbr.) flowers, the honey, or the raisins used to initiate the process of fermentation. The yeasts multiply rapidly by division in a short time. Finally, the vessel should be closed and sealed with a long ribbon of cloth smeared with clay on one surface. While sealing, the blank surface of the ribbon should line the rim of the vessel and lid and the clay side should be external. The vessel is then, placed in a dark place without much circulation of air, such as a grain store, buried in a heap of grain, or in a pit in the soil. A soft packing of straw should be provided around the vessel to prevent breakage.

9.8.4.2 The Fermentation Process During the autumn and summer seasons, fermentation takes place in 6 days, but in winter, it takes about 10 days, while in the rainy season and spring, eight days are sufficient (Kushwaha and Karanjekar, 2011). The fermentation vessel is left undisturbed for a month and then, opened. The medicinal extract is filtered, and if needed, allowed further sedimentation time. It is then, allowed to stand for a few more days and again filtered to separate the sediment. In the usual practice, 7–10 days are enough in the hot tropical climate, whereas a long period of 30 days is allowed in cool temperate climate, when biological activity of yeast is low (Joshi et al., 2011b; Santhosh et al., 2012).

9.8.4.3 Storage and Usage The filtered medicine is stored in lightly stoppererd glass bottles and taken for use, whenever necessary. The *asava* and *arishta* remain good for any length of time, and, actually, the medicinal value of the preparation is said to increase with time. If *asava* or *arishta* show any sign of mold development at any stage, they should be rejected as they become unfit for use. *Asava* or *arishta* are mixed with an equal volume of water before consumption. The container should be kept well closed, as these sweet medicines attract flies.

9.8.4.4 Doses   The dose of both the *asava* and *arishta* is one *pala* (48 mL approx.). *Arishta* is better than *asava*, as the former is light because of boiling. It is evident that

Ayurveda constitutes a profile of alcoholic beverages with therapeutic properties. Such alcoholic drinks, taken according to the prescribed proper dose, at the proper time, along with wholesome food, and according to the capacity of the individual, produce effects like *ambrosia*. When used inappropriately, it causes diseases and works as a poison. The dose taken should not cause intoxicated movement of the eyeball.

9.9 Summary and Future Prospects

The alcoholic beverages, such as wine, beer, brandy, other distilled berverages are known to play a significant role in the socio-cultural life of the local people of South Asian countries, and their production serves as a source of income, thus, supporting the livelihood of the ethnic tribes. The production and consumption of the alcoholic beverages is an ancient practice. There is a large diversity in the type of these beverage as is the diversity in production technology. Traditionally made alcoholic beverages have enhanced flavor and aroma, bionutrients, increased digestibility, and exert health-promoting benefits due to antioxidant activity and antimicrobial effects. The alcoholic fermentation also increases the shelf life, reduce volume and cooking time, and increases nutritive value of the fermented food over the non-fermented ingredients. Traditionally made alcoholic beverages are also a source of bioactive compounds, like angiotensin, which reduces the blood pressure, and serve as a social lubricant. The technology of the production of indigenous alcoholic beverages also reveals several sound practices, similar to the scientific principles discovered later on and being applied in the production of such beverages, at present. No doubt, in the absence of scientific knowledge, the indigenous knowledge was the outcome of trial and error. However, the traditional system is being ignored. Attempts to apply modern science universally without regard for traditional knowledge systems have reduced sustainable resource management as well as eroding the biological diversity. There is a need to have an integrated approach, where the traditional knowledge is reinforced with modern scientific tools, especially with respect to the safety and nutrition of such beverages.

A critical review of the process of rice beer preparation followed by different ethnic tribes residing in different states of North-East India is more or less similar to that followed at the alcoholic beverage industrial scale. The only difference is the ingredients, in the form of the different parts of various plants species used by the tribal people, based on their availability. The local brews, such as rice beer, have very significant roles in the culture and traditions of the tribal people residing in a particular part of the country. The preparation of rice beer is considered sacred by all the tribes, and the consumption of a moderate amount of alcohol in the form of rice beer gives some relaxation to the hard working population of these states and, practically, has no side effects on their health. In addition, besides imparting sensory qualities, like taste, color, flavor and sweetness to the beer, the different plants used in the preparation of starter cultures are also known to have many medicinal properties and are a source of the general microflora present in the starter cakes. The quality of the starter culture used in production of such beverages, is said to be dependent on the variety of plant parts used, and also on the maintenance of proper sanitary conditions, as is the case with the variety of rice.

Chhang is an indispensable fermented beverage in Ladakh and several areas of Himachal Pradesh, and no social activity is complete without it. Traditional method of alcoholic beverage production from barley is a well established process practiced for centuries, but introduction of new alcoholic beverages made by distillation have posed a threat to it and other such beverages. The time-tested method of *chhang* preparation needs to be preserved to maintain the cultural identity of the people of Ladakh. Wherever possible, improvements made by the application of science, of course, could be incorporated. It is cultural necessity which has helped conserve these traditional fermented food and beverages.

The traditional foods and beverages are indicative of the simplicity of the ways of life of the people, and the impact of the surrounding environment in the high altitude cold-arid regions such as of Ladakh. Another example is the traditions maintained by various communities of the Manipuris who have a strong system of ethics. There should be focused efforts to promote traditional food systems within rural communities. Methods of fermentation, knowledge of microorganisms, and instrumental methods of quality control could be defined to improve the processing and production of indigenous alcoholic beverages.

Ayurveda has had the vision to exploit the fermentation technique for a wide variety of therapeutic purposes since long ago. The approach was simple, cautious, and pragmatic, with minimum intervention, though several processes lack a scientific base. They have been continued from ancient times, when there was no knowledge of microbiology, molecular biology, chemistry, or engineering. Thus, it is important to apply modern science to understand the traditional systems of human life. But the tradition of Ayurveda requires multifaceted approach of research for validation. For example, the medicinal properties of the beverages can be studied by modern scientific tools. The microorganisms involved in these biomedical fermentations, the chemistry of fermentation, and the associated biotransformation processes, can be studied in detail so as to validate and improve the great living tradition of Ayurveda. Assessment of the beverages in terms of various quality characteristics ethyl alcohol, volatile acid content, fixed acid content, and sugar content, and testing for methyl alcohol, etc., could be performed in order to use them medicinally. There is a dise need to thoroughly understand the therapeutic aspects of the alcoholic beverages and more efforts made to impart such properties to these beverages. Research on traditional knowledge should be undertaken with the aim of preserving indigenous knowledge for the benefit of the future, and documentation of these products and technologies is also needed at the present time.

References

Aidoo, K.E, Nout, M.J.R., and Sarkar, P.K. 2006. Occurrence and function of yeasts in Asian indigenous fermented foods. *FEMS Yeast Research*, 6: 30–39.

Amerine, M.A.; Berg, H.W., and Cruess, W.V. 1967. *The Technology of Wine Making*. AvI Publishing Company Inc., Westport, CT.

Amerine, M.A., Berg, H.W., Kunkee, R.E., Qugh, C.S., Singleton, V.L. and Webb, A.D. 1980. *The Technology of Wine Making*, 4th edn. AVI Publishing Co., Inc. Westport, CT.

Angchok, D., Dwivedi, S.K., and Ahmad, Z. 2009. Traditional foods and beverages of Ladakh. *Indian Journal of Traditional Knowledge*, 8(4): 551–558.

Anonymous FAO http://www.fao.org/docrep/x0560e/x0560e09.htm

Anonymous Feni http://www.goafeni.com/aboutdrink.htm

Anonymous how to pedia,http://en.howtopedia.org/wiki/How_to_Make_Palm_Wine

Anonymous http://en.wikipedia.org/wiki/Feni_(liquor)

Anonymous http://en.wikipedia.org/wiki/Palm_wine

Anonymous http://www.absoluteastronomy.com/topics/Coconut

Anonymous http://www.absoluteastronomy.com/topics/Mead

Anonymous Indian wine http://indianwine.com/wine*toddy*.htm

Anonymous med wikihttp://medlibrary.org/medwiki/Neera

Anonymous palm wine http://www.pinoybisnes.com/food-business-ideas/how-to-make-*toddy*-palm-wine

Anonymous rice wine-http://hubpages.com/hub/tapuy-filipino-rice-wine-how-to-make

Anonymous *toddy* and-palm-wine http://practicalaction.org/*toddy*-and-palm-wine

Anonymous wiki-http://en.wikipedia.org/wiki/Alcohol_beverage

Arnold, J.P. 2005. Beer Books, Cleveland, OH. ISBN 0-9662084-1-2.

Arts, I.C.W., Putte, B., and Hollman, P.C.H. 2000. Catechin contents of food commonly consumed in the Netherlands, 2. Tea, wine, fruit juices and chocolate, milk. *Journal of Agricultural and Food Chemistry*, 48(5): 1752–1757.

Atputharajah, J.D., Widanapathirana, S., and Samarajeewa, U. 1986. Microbiology and biochemistry of natural fermentation of coconut palm sap. *Food Microbiology*, 3 (4): 273–280.

Atputharajah, J.D, Widanapathirana, S., and Samarajeeva, U. 1986. Microbiology and biochemistry of natural fermentation of coconut palm sap. *Food Microbiology*, 3(4): 273–280.

Batra, L.R. and Millner, P.D. 1974. Some Asian fermented foods and beverages and associated fungi. *Mycologia*, 66: 942–950.

Behera, S., Ray, R.C., Swain, M.R., Mohanty, R.C., and Biswal, A.K. 2012. Traditional knowledge on the utilization of mahula (*Madhuca latifolia* L.) flowers by the santhal tribe in similipal biosphere reserve, Odisha, India: A survey. *Annals of Tropical Research* (in press).

Belavady, B. and Balasubramanian, S.C. 1959. Nutritive value of some Indian fruits and vegetables. *Indian Journal of Agricultural Science*, 29(2/3): 151–163.

Beltran, G., Rozas, N., Mas, A., and Guillamom, J. 2007. Effect of low temperature fermentation on yeast nitrogen metabolism. *World Journal of Microbiology and Biotechnology*, 23(6): 809–815.

Bennett, P. 1999. Understanding responses to risk: Some basic findings. In: *Risk Communication and Public Health*. Bennett, P. and Calman, K. (eds.), Oxford University Press, Oxford, pp. 3–19.

Bhatia, A. and Khetarpaul, N. 2012. 'Doli Ki Roti'—An indigenously fermented Indian bread: Cumulative effect of germination and fermentation on bioavailability of minerals. *IJTK*, 11(1): 109–113.

Bhatia, A.K., Singh, R.P., and Atal, C.K. 1977. *Chhang*: The fermented beverage of Himalayan folk. *Indian Food Packer*, July–August: 1–8.

Blackwood, A.C., Guimberteau, G., and Penaud, E. 1969. Sur les bactéries acétiques isolées de raisins. *Comptes rendus hebdomadaires des séances de l'Académie des Sciences Série D.*, 269: 802–804.

Bluhm, L. 1995. Distilled beverages. In: Biotechnology. Food and feed production with composition of Tapai, A Sabah's fermented beverage. *Malaysian Journal of Microbiology*.

Borgstrom, G. 1968. *Principles of Food Science, Food Microbiology and Biochemistry*. Macmillan, New York, NY.

Borthakur, S.K. 1990. Studies in ethnobotany of the Karbis (Mikirs), Plant masticatories and dyestuffs. In: *Contributions to Ethnobotany of India*. Jain, S.K. (ed.), Scientific Publisher, Jodhpur, India, pp. 277–283.

Brothwell, D. and Brothwell, P. 1969. *Food in Antiquity: A Survey of the Diet of Early Peoples*. Expanded Edition 1998. John Hopkins University Press, Baltimore, MD. ISBN 0-8018-5740-6. pp. 283.

Chandrashekar, M.V. 2007. Neera Board to bring out value-added products. http://www.fnbnews.com April 29, 2012.

Chakravarty, J., Sharma, G.D., and Tamang, J.P. 2009. Substrate utilization in traditional fermentation technology practiced by Tribes of North Cachar Hills district of Assam. *Assam University Journal of Science & Technology: Biological Sciences*, 4(1): 66–72.

Chakravarty, J., Sharma, G.D., and Tamang, J.P. 2013. Indigenous technology for food processing by the tribes of *Dima* Masao (North Cachar Hills) district of Assam for social security. In: Das Gupta, D. (Ed.), Food and Environmental Foods Security Impertives of Indigenous Knowledge System. *Agrobios*, Jaipur, India, pp. 32–45.

Chakrabarty, J, Sharma, G.D., and Tamang, J.P. 2014. Traditional technology and product characterization of some lesser-known ethnic fermented foods and beverages of North Cachar Hills District of Assam. *Indian Journal of Traditional Knowledge*, 13(4): 706–715.

Chiang, Y.W., Chye, F.Y., and Ismail, Mohd. 2006. A microbial diversity and proximate composition of Tapai, a Sabah's fermented beverage. *Malaysian Journal of Microbiology*, 2(1): 1–6.

Chuenchomrat, P., Assavanig, A., and Lertsiri, S. 2008. Volatile flavor compounds analysis of solid state fermented Thai rice wine (Ou). *Science Asia*, 34: 199–206.

Coleman, M.C., Fish, R., and Block, D.E. 2007. Temperature-dependant kinetic model for nitrogen–limited wine fermentations. *Applied and Environmental Microbiology*, 73(18): 5875–5884.

Dahiya, D.S. and Prabhu, K.A. 1977. Indian jackfruit wine. *Symposium on Indigenous Fermented Food*, Bangkok, Thailand.

Das, A.J., Khawas, P., Tatsuro, M., and Deka, S.C. 2014. HPLC and GC-MS analyses of organic acids, carbohydrates, amino acids and volatile aromatic compounds in some varieties of rice beer from northeast India: HPLC and GC-MS analyses of rice beer. *Journal of the Institute of Brewing*, 120(3): 244–252.

Das, A.J. and Deka, S.C. 2012. Fermented Foods and Beverages of the North East India. *International Food Research J.*, 19(2): 377–392.

Das, C.P. and Pandey, A. 2007. Fermentation of traditional beverages prepared by Bhotiya community of Uttaranchal Himalaya. *Indian Journal of Traditional Knowledge*, 6(1): 136–140.

Dash, V.B. and Kashyap, V.L. 1980. *Materia Medica of Ayurveda*. Concept Publishing Company, New Delhi, pp. 187–203.

Dash, V.B. and Kashyap, V.L. 2002. *Latro-chemistry of Ayurveda*. Concept Publishing Company, New Delhi , pp. 69–79.

Deka, D. and Sharma, G.C. 2010. Traditionally used herbs in the preparation of rice-beer by the *Rabha* tribe of Goalpara district, Assam. *Indian Journal of Traditional Knowledge*, 9: 459–462.

Deori, C., Begum, S.S., and Mao, A.A. 2007. Ethnobotany of *sujen*: A local beer of *Deori* tribe of Assam. *Indian Journal of Traditional Knowledge*, 6: 121–125.

Devi, P. and Kumar, S.P. 2012. Traditional, ethnic and fermented foods of different tribes of Manipur. *Indian Journal of Traditional Knowledge*, 11(1): 70–77.

Dhal, N.K., Pattanaik, C., and Reddy, C.S. 2010. *Bakhar* starch fermentation-a common tribe practice in Orissa. *Indian Journal of Traditional Knowledge*, 9(2): 279–281.

Dhiman, A.K. 2004. *Common Drug Plants and Ayurvedic Remedies*. Reference Press, New Delhi.

Drysdale, G.S. and Fleet, G.H. 1985. Acetic acid bacteria in some Australian wines. *Food Technology of Australia*, 37: 17–20.

Dung, N.T.P., Rombouts, F.M., and Nout, M.J.R. 2006. Functionality of selected strains of moulds and yeasts from Vietnamese rice wine starters. *Food Microbiology*, 23: 331–340.

Du Toit, W.J. and Lambrechts, M.G. 2002. The enumeration and identification of acetic acid bacteria from South African red wine fermentations. *International Journal of Food Microbiology*, 74: 57–64.

Elaine, M. and Danilo, M. 2012. Traditional fermented food and beverage for improved livelihoods. *FAO Diversification Booklets* (21): 86.

Fabian, F.W. 1935. The use of honey in making fermented drinks. *Fruit Production: Journal of Food Manufacture*, 14: 363–366.

Faparusi, S.I. 1973. Origin of initial microflora of palm wine from oil palm trees (*Elascis guineensis*). *Journal of Applied Bacteriology*, 36: 559–565.

Faparusi, S.I. 1974. Microorganisms from oil palm tree (*Elascis quineensis*) tap holes. *Food Science*, 39: 755–757.

Faparusi, S.I. and Bassir, O. 1972. Factors affecting the quality of palm wine. 2. Period of storage. *West African Journal of Biological and Applied Chemistry*, 15: 24–28.

Filipello, F. and Marsh, G.L. 1941. Honey wine. *Fruit Production: Journal of Food Manufacture*, 41: 78–79.

Foster, C. 1967. *Home Wine Making*. Wardlock Co., London.

Franz, C.M. Huch, M., Abrioue, H., Holzapfel, W., and Galvez, A. 2011. Enterococci as probiotics and their implications in food safety. *International Journal of Food Microbiology.*, 151, 125.

Ghosh, C. and Das, A.P. 2004. Preperation of rice beer by the tribal inhabitants of tea gardens in Terai of West Bengal. *Indian Journal of Traditional Knowledge*, 3: 374–382.

Grimwood, B.E. and Ashman, F. 1975. *Coconut Palm Production: Their Process in Developing Countries*. Food and Agriculture Organization, Rome, pp. 189–190.

Guide to mead by Michael Faul (c), 1991–2005, http://ittd.com/articles/216/mcad/.

Gupta, J.K. and Sharma, R. 2009. Production technology and quality characteristics of mead and fruit-honey wines: A review. *Natural Product Radiance*, 8(4): 345–355.

Hesseltine, C.W., Rogers, R., and Winarno, F.G. 1988. Microbiological studies on amylolytic oriental fermentation starters. *Mycopathologia*, 101: 141–155.

Hibbert, H. and Barsha, J. 1931. Studies on reactions relating to carbohydrates and polysaccharides: xxxix. Structure of the cellulose synthesized by the action of *Acetobacter xylinus* on glucose. *Canadian Journal of Research*, 5(5): 580–591. 10.1139/cjr31-096.

Hodson, T.C. 1999. *The Meitheis*. LPS Publications, New Delhi.

Hutchinson, C.M. and Ram Ayyar, C.S. 1925. *Bakhar*: The Indian rice beer ferment. *Memoirs of the Department of Agriculture in India: Bacteriological Series*, 1: 137–168.

Ioyrish, N. 1974. *Bees and People*. MIR Publishers, Moscow, 213.

Jackson, R.S. 1994. *Wine Science: Principles and Applications*. Academic Press, San Diego, CA, p. 592.

Jaiswal, V. 2010. Culture and ethnobotany of jaintia tribal community of Meghalaya, Northeast India—A mini review. *Indian Journal of Traditional Knowledge*, 9: 38–44.

Jamir, B. and Deb, C.R. 2014. Studies on some fermented foods and beverages of Nagaland, India. *International Journal of Food and Fermentation Technology*, 4(2): 121–127.

Jeyaram, K., Singh W.M., Capece, A., and Romano, P. 2008. Molecular identification of yeast species associated with 'Hamei'—A traditional starter used for rice wine production in Manipur, India. *International Journal of Food Microbiology*, 124: 115–125.

Jeyaram, K., Singh, T.A., Romi, W., Devi, A.R., Singh, W.M., Dayanidhi, H., Singh, N.R., and Tamang, J.P. 2009. Traditional fermented foods of Manipur. *Indian Journal of Traditional Knowledge*, 8(1): 115–121.

Joshi, V.K. 1997. *Fruit Wines*. 2nd edn. Directorate of Extension Education. Dr. Y.S. Parmar University of Horticulture and Forestry, Nauni, Solan, HP, p. 255.

Joshi, V.K. 2005. Traditional fermented foods and beverages of northern India an overview. In: *Proceedings of First Workshop and Strategic Meeting of South Asian Network on Fermented Foods*, May 26–27, 2005, Anand, Gujarat, India, pp. 59–73.

Joshi, V.K., Attri, B.L., Gupta, J.K., and Chopra, S.K. 1990. Comparative fermentation behaviour, physico-chemical characteristics of fruit honey wines. *Indian Journal of Horticulture*, 47(1): 49–54.

Joshi, V.K. Attri, D., Singh, T.K., and Abrol, G. 2011b. Fruit wines: Production technology. In: *Handbook of Enology*. Joshi, V.K. (ed.), Vol. III. Asia-Tech Publishers, New Delhi, pp. 1177–1221.

Joshi, V.K., Garg, V., and Abrol, G.S. 2012b. Indigenous fermented foods. In: *Food Biotechnology: Principles and Practices*. Joshi, V.K. and Singh, R.S. (eds.), IK International Publishing House, New Delhi, pp. 337–373.

Joshi, V.K., Kumar, A., and Thakur, N.S. 2015. Technology and production of *sur* in Himachal Pradesh. *Indian Journal of Traditional Knowledge* (in press).

Joshi, V.K. and Pandey, A. 1999. Biotechnology: Food fermentation. In: *Biotechnology: Food Fermentation*. Joshi, V.K. and Pandey, A. (eds.), Vol. I. Educational Publishers and Distributors, New Delhi, pp. 1–24.

Joshi, V.K. and Sandhu, D.K. 2000. Quality evaluation of naturally fermented alcoholic beverages, microbiological examination of source of fermentation and ethanolic productivity of the isolates. *Acta Alimentaria*, 29(4): 323–334.

Joshi, V.K, Sandhu, D.K., and Thakur, N.S. 1999. Fruit based alcoholic beverages. In: *Biotechnology: Food Fermentation: Microbiology, Biochemistry and Technology*. Joshi, V.K. and Pandey, A. (eds.), Vol. II. Educational Publishing and Distributors, New Delhi, pp. 647–744.

Joshi, V.K. and Sharma, S.K. 1994. Effect of method of must preparation and initial sugar levels on the quality of apricot wine. *Research on Industry*, 39(4): 255–257.

Joshi, V.K., Sharma, R., and Abrol, G.S. 2012a. Stone fruit: Wine and brandy. In: *Handbook of Plant-Based Fermented Food and Beverage Technology*. Hui, Y.H. and Özgül, E. (eds.), CRC Press, Taylor & Francis, Boca Raton, FL, pp. 273–306.

Joshi, V.K., Sharma, S., Shashi, B., and Attri, D. 2004. Fruit based alcoholic beveregaes. In: *Concise Encyclopedia of Bioresource Technology*. Pandey, A. (ed.), Haworth Inc., New York, NY.

Joshi, V.K., Thakur, N.S., Bhatt, A., and Garg, C. 2011a. Wine and brandy: A perspective. In: *Handbook of Enology*. Joshi, V.K. (ed.), Vol. 1 Asia-Tech Publishers, Inc., New Delhi, pp. 3–45.

Joyeux, A., Lafon-Lafourcade, S., and Ribéreau-Gayon, P. 1984. Evolution of acetic acid bacteria during fermentation and storage of wine. *Applied and Environmental Microbiology*, 48: 153–156.

Kanwar, S.S., Gupta, M.K., Katoch, C., and Kanwar, P. 2011. Cereal based traditional alcoholic beverages of Lahaul and spiti area of Himachal Pradesh. *Indian Journal of Traditional Knowledge*, 10(2): 251–257.

Kersters, K., Lisdiyanti, P., Kazuo, K., and Jean, S. 2006. The Family Acetobacteraceae: The Genera *Acetobacter, Acidomonas, Asaia, Gluconacetobacter, Gluconobacter, and Kozakia.* (Chapter 3.1.8). In: *The Prokaryotes.* Martin, D. and Stanley, F. (eds.), Springer, New York, NY, pp. 163–200.

Kordylas, J.M. 1992. Biotechnology for production of fruits, wines and alcohol. In: *Application of Biotechnology to Traditional Foods.* Board of Science and Technology for International Development (BOSTID), Washington, DC, p. 188.

Kozaki, M., Tamang, J.P., Katoka, J., Yamanaka, S. and Yoshida, S. 2000. Cereal wine (*Jaans*) and distilled wine (*raksi*) in Sikkim. *J. Brewing Society of Japan,* 95(2): 115–122.

Krishna, B. and Baishya, K. 2013. Ethno medicinal value of various plants used in the preparation of traditional rice beer by different tribes of Assam, India, *Drug Invention Today,* 5(4): 335.

Kushwaha, R. and Karanjekar, S. 2011. Standardization of *Ashwagandharishta* formulation by TLC method. *International Journal of Chem.Tech. Research,* 3(3): 1033–1036.

Law, S.V., Abu Bakar, F., Mat Hashim, D., and Abdul Hamid, A. 2011. MiniReview: Popular fermented foods and beverages in Southeast Asia. *International Food Research J.,* 18: 475–484.

Lee, C.H. 2009. Food biotechnology. In: *Food Science and Technology.* Campbell-Platt, G. (ed.), John Wiley and Sons, New York, NY, pp. 85–95.

Lee, C.Y., Smith, N.L., Kimc, R.W., and Morse, R.A. 1985. Source of the honey protein responsible for apple juice clarification. *Journal of Apicultural Research,* 24: 190–194.

Lee, A.C. and Fujio, Y. 1999. Microflora of banh men, a fermentation starter from Vietnam. *World Journal of Microbiology Biotechnology,* 15(1): 51–55.

Lichine, Alexis. 1987. *Alexis Lichine's New Encyclopedia of Wines & Spirits,* 5th edn. Alfred A. Knopf, New York, NY, pp. 707–709.

Limtong, S., Sintara, S., Suwanarit, P., and Lotong, N. 2002. Yeast diversity in Thai traditional alcoholic starter. *Kasetsart Journal Natural Sciences,* 36: 149–158.

Lotong, N. 1985. *Koji.* In: *Microbiology of Fermented Food.* Wood, B.J.B. (ed.), Elsevier Applied Science Publishers, London, U.K., pp. 237–270.

Marshall, E. and Mejia, D. 2011. Traditional fermented food and beverages for improved livelihoods. Rural Infrastructure and Agro-Industries Division Food and Agriculture Organization of the United Nations, Rome. http://www.fao.org/docrep/015/i2477e/i2477e00.pdf

Maugenet, J. 1964. Hydromel. *Annals Abeille,* 7: 165–179.

Mishra, A.K., Singh, R., Mishra, K.K., and Pathak, M.K. 2012. Quality assessment of different marketed brands of ashokarishta: An ayurvedic formulation. *International Journal of Pharmacy and Pharmaceutical Sciences,* 4: 506–508.

Mohanty, S., Ray, P., Swain, M.R., and Ray, R.C. 2006. Fermentation of cashew (*Anacardium occidentale* L.) apple into wine. *Journal of Food Processing and Preservation,* 30: 314–322.

Murthy, S.K.R. 1994. *Astanga Hrdayam.* Vol. 1, Krishnadas Academy, Varanasi, pp. 68–73.

Murthy, S.K.R. 1998. *Rhavaprakasa.* Vol. I, Krishnadas Academy, Varanasi, pp. 479–484.

Neema, P.K., Kario, T., and Singh, A.K. 2003. Opo—The rice beer. In: *International Seminar and workshop on Fermented Foods, Health Status and Well Being held at Anand on 13–14th November, 2003,* pp. 59–61.

Nelson, Max. 2005. *The Barbarian's Beverage: A History of Beer in Ancient Europe.* books.google. co.uk. ISBN 978-0-415-31121-2.

Nirmala, C., David, E., and Sharma, M.L. 2000. Changes in nutrient components during ageing of emerging juvenile bamboo shoots. *International Journal of Food and Nutritional Sciences,* 58: 345–352.

Okafar, N. 1975. Microbiology of Nigerian palm wine with particular reference to bacteria. *Journal of Applied Bacteriology,* 38: 81–88.

Osho, A. 1995. Evaluation of cashew apple juice for single cell protein and wine production. *Nahrung,* 39: 521–529.

Osho, A. 2005. Ethanol and sugar tolerance of wine yeasts isolated from fermenting cashew apple juice. *African Journal of Biotechnology*, 4(7): 660–662.

Panmei, C. 2004. *Studies on the role of Albizia spp. in the fermentation process of different rice cultivars*. PhD thesis, Manipur University, India.

Patil, S.V., Salunkhe, R.B., Patil, C.D., Patil, D.M., and Salunke, B.K. 2010. Bioflocculant exopolysaccharide production by *Azotobacter indicus* using flower extract of *Madhuca latifolia* L. *Applied Biochemistry and Biotechnology*, 162(4): 1095–1108.

Patwardhan, B., Vaidya, A.D.B., and Chorghade, M. 2004. Ayurveda and natural products drug discovery. *Current Science*, 86: 789–799.

Prescott, S.C. and Dunn, C.G. 1959. eds. *Industrial Microbiology*. McGraw Hill, New York, NY.

Rana, J.C., Pradeep, K., and Verma, V.D. 2007. Naturally occurring wild relatives of temperate fruits in western Himalayan region of India analysis. *Biodiversity Conservation*, 16: 3963–3991.

Rana, T.S., Datt, B., and Rao, R.R. 2004. *Soor*: A traditional alcoholic beverage in tons valley, Garhwal Himalaya. *Indian Journal of Traditional Knowledge*, 3(1): 59–65.

Ray, R.C. and Ward, O.P. 2006. Post harvest microbial biotechnology of tropical root and tuber crops. In: Ray, R.C. and Ward, O.P. (eds.). *Microbial Biotechnology in Horticulture*, Vol. 1. Science Publishers, Enfield, New Hampshire, pp. 345–396.

Ray, R.C. and Ward, O.P. 2006. Microbial biotechnology in horticulture—An overview. In: *Microbial Biotechnology in Horticulture*. Ray, R.C. and Ward, O.P. (eds.). Vol. 1. Science Publishers, Enfield, NH. (in press).

Reddy, L.V.A., Joshi, V.K., and Reddy, O.V.S. 2012. Utilization of tropical fruits for wine production with special emphasis on mango (*Mangifera indica* L.) wine. In: *Microorganisms in Sustainable Agriculture and Biotechnology*. Satyanarayana, T., Johri, B.N., and Prakash, A. (eds.), Springer, Germany, pp. 679–710.

Roy, A., Moktan, B., and Sarkar, P.K. 2007. Characteristics of *Bacillus cereus* isolates from legume-based Indian fermented foods. *Food Control*, 18: 1555–1564.

Roy, B., Kala, C.P., Farooquee, N.A., and Majila, B.S. 2004. Indigenous fermented food and beverages: A potential for economic development of the high altitude societies in Uttaranchal. *Journal of Human Ecology*, 15(1): 45–49.

Rundel, P.W. 2002. The Chilean wine palm. *Mildred E. Mathias Bot. Garden Newslett.*, 5(4): 1.

Saikai, B., Tag, H., and Das, A.K. 2007. Ethnobotany of foods and beverages among the rural farmers of Tai Ahom of North Lakhimpur district, Asom. *Indian Journal of Traditional Knowledge*, 6: 126–132.

Samati, H. and Begum, S.S. 2007. *Kiad*—A popular local liquor of *Pnar* tribe of Jaintia hills district, Meghalaya. *Indian Journal of Traditional Knowledge*, 6(1): 133–135.

Sanchez, P.C., Julianno, B.O., Laude, V.T., and Perez, C.M. 1988. Nonwaxy rice for *tapuy* (rice wine) production. *Cereal Chemistry*, 65: 240–243.

Santhosh, B., Jadar, P.G., and Nageswara, Rao. 2012. Kanji: An ayurvedic fermentative preparation. *International Research Journal of Pharmacy*, 3(1): 154–155.

Saono, S., Gandjar, I., Basuki, T., and Karsono, H. 1974. Mycoflora of ragi and some other traditional fermented foods of Indonesia. *Annales Bogorienses*, 4: 187–204.

Sekar, S. 2007. Traditional alcoholic beverages from Ayurveda and their role on human health. *Indian Journal of Traditional Knowledge*, 6(1): 144–149.

Sekar, S. and Mariappan, S. 2007. Usage of traditional fermented products by Indian rural folks and IPR. *Indian Journal of Traditional Knowledge*, 6(1): 111–120.

Sekar, S. and Mariappan, S. 2008. Traditionally fermerted biomedicines, arishtas asovas from Ayurveda. *Indian Journal of Traditional Knowledge*, 7(4): 548–556.

Sener, A., Canbas, A., and Unal, M. 2007. The effect of fermentation temperature on the growth kinetics of wine yeast species. *Turkish Journal of Agriculture and Forestry Sciences*, 31(5): 349–354.

Shamala, T.R. and Sreekantiah, K.R. 1988. Use of wheat bran as a nutritive supplement for the production of ethanol by *Zymomonas mobilis*. *Journal of Applied Bacteriology*, 65: 433–436.

Sharma, A.K. 2013. *Prepartion and Evaluation of Sur production in Himachal Pradesh*. Thesis, Dr. Y.S. Parmar University of Horticulture and Forestry, Nauni, Solan.

Sharma, P.V. 2004. *Susrutasamhita* (*Chaukhambha Visvabharati* Varanasi), 449–459.

Sharma, T.C. and Mazumdar, D.N. 1980. *Eastern Himalayas: A Study on Anthropology and Tribalism*. Cosmo Publications, New Delhi.

Shastri, M.V. 1968. *Vaidya Yoga Ratnavali*. IMPCOPS, Madras, 6–10.

Shrestha, H., Nand, K., and Rati, E.R. 2002. Microbiological profile of murcha starters and physico-chemical characteristics of poko, rice based traditional fermented food product of Nepal. *Food Biotechnology*, 16(1): 1–15.

Singh, H.B. and Jain, A. 1995. Ethnobotanical observation on the preparation of millet beer in Sikkim state, India. In: *Ethnobotany and Medicinal Plants of Indian Subcontinent*. Maheshwari, J.K. (ed.), Scientific Publisher, Jodhpur, pp. 577–579.

Singh, I.S., Srivastava, A.K., and Singh, V. 1999. Improvement of some under utilizes fruits through selection. *Journal of Applied Horticulture*, 1(1): 34–37.

Singh, P.K. and Singh, K.I. 2006. Traditional alcoholic beverage, Yu of Meitei communities of Manipur. *Indian Journal of Traditional Knowledge*, 5(2): 184–190.

Singh, R.K., Singh, A., and Sureja, A.K. 2007. Traditional foods of *Monpa* tribes of West Kameng, Arunachal Pradesh. *Indian Journal of Traditional Knowledge*, 6: 25–36.

Steinkraus, K.H. 1985. Bio-enrichment: Production of vitamins in fermented foods. In: Wood, B.J.B. (ed.), *Microbiology of Fermented Foods*. Elsevier Applied Science, London, pp. 323–344.

Steinkraus, K.H. 1996. *Handbook of Indigenous Fermented Foods*. Marcel Dekker, New York, NY, pp. 149–165.

Stringini, M., Comitini, F., Taccari, M., and Ciani, M. 2009. Yeast diversity during tapping and fermentation of palm wine from Cameroon. *Food Microbiology*, 26(4): 415–420.

Swings, J. and De Ley, J. 1977. The biology of *Zymomonas*. *Microbiol. Mol. Biol. Rev.*, 41(1): 1–46.

Takeuchi, A., Shimizu-ibuka, A., Yoshitaka, N., Kiyoshi, M., Okada, S., Tokue, C., and Arai, S. 2006. Purification and characterization of an α-amylase of *Pichia burtonii* isolated from the traditional starter "Murcha" in Nepal. *Bioscience, Biotechnology, and Biochemistry*, 70(12): 3019–3024.

Tamang, J.P. 2001. Food culture in the Eastern Himalayas. *Journal of Himalayan Research and Cultural Foundation*, 5: 107–118.

Tamang, J.P. 2009. Ethnic starters and alcoholic beverages, Chapter 8. In: *Himalayan Fermented Foods: Microbiology, Nutrition and Ethnic Values*. CRC Press, Tayler & Francis, Boca Raton, FL, pp. 187–228.

Tamang, J.P. 2010. Diversity of fermented beverages and alcoholic drinks. In: *Fermented Foods and Beverages of the World*. CRC Press, Tayler & Francis, Boca Raton, FL, pp. 85–125.

Tamang, J.P. 2011. *The 12th Asaan Food Conference on Prospects of Asian Fermented Foods in Global Markets*. 16–18 June, 2011. BITEC Bangna, Bangkok, Thailand.

Tamang, J.P. 2012. Plant-based fermented foods and beverages of Asia. In: *Handbook of Plant-Based Fermented Food and Beverage Technology*, 2nd edn. Hui, Y.H. and Özgül, E. (eds.), CRC Press, Tayler & Francis, Boca Raton, FL, p. 821.

Tamang, J.P., Dewan, S., Tamang, B., Rai, A., Schillinger, U., and Holzapfel, W.H. 2007. Lactic acid bacteria in *Hamei* and Marcha of North East India. *Indian Journal of Microbiology*, 47(2): 119–125.

Tamang, J.P. and Fleet, G.H. 2009. Yeasts diversity in fermented foods and beverages. In: *Yeasts Biotechnology: Diversity and Applications*. Satyanarayana, T. and Kunze, G. (eds.), Springer, New York, NY, pp. 169–198.

Tamang, J.P. and Nikkuri, S. 1998. Effect of temperatures during pure culture fermentation of Kinema. *World J. Microbiol. Biotechnol.*, 14: 847–850.

Tamang, J.P. and Sarkar, P.K. 1995. Microflora of murcha: An amylolytic fermentation starter. *Microbios*, 81: 115–122.

Tamang, J.P. and Sarkar, P.K. 1996. Microbiology of *mesu*, a traditional fermented bamboo shoot product. *International Journal of Applied Microbiology*, 29(1): 49–58.

Tamang, J.P., Sarkar, P.K. and Messelfine, D.N. 1998. Traditional Fermented Foods and Beverages of Darjeeling and Sikkim–A review. *J. Sci. Food Agric.*, 44: 375–385.

Tamang, J.P., Tamang, N., Thapa, S., Dewan, S., Tamang, B. Yonzan, H., Rai, A.K., Chettri, R., Chakrabarty, J., and Kharel, N. 2012. Microorganisms and nutritional value of ethnic fermented foods and alcoholic beverages of North East India. *Indian Journal of Traditional Knowledge*, 11(9): 7–25.

Tamang, J.P., Thapa, S., Tamang, N., and Rai, B. 1996. Indigenous fermented food beverages of Darjeeling hills and Sikkim: Process and product characterization. *Journal of Hill Research*, 9(2): 401–411. http://www.fnbnews.com/article/detnew.asp?articleid=20651&sectionid=3

Tamang, J.P., Thapa, N., Rai, B., Thapa, S., Yonzan, H., Dewan, S., Tamang, B., Sharma, R., Rai, A., and Chettri, R. 2007. Food consumption in Sikkim with special reference to traditional fermented foods and beverages: A microlevel survey. *Journal of Hill Research*, 20(1): 1–37.

Tanti, B., Gurung, L. Sarma, H.K., and Buragohain, A.K. 2010. Ethnobotany of starter cultures used in alcohol fermentation by a few ethnic tribes of Northeast India. *Indian Journal of Traditional Knowledge*, 9(3): 463–466.

Targais, K. 2012. Chhang—A barley based alcoholic beverage of Ladakh, India. *Indian Journal of Traditional Knowledge*, 11(1): 190–193.

Thakur, N., Kumar, D., Savitri, and Bhalla, T.C. 2003. Traditional fermented foods and beverages of Himachal Pradesh. *Invention Intelligence*, July–August: 173–178.

Thakur, N., Savitri, and Bhalla, T.C. 2004. Characterization of some traditional foods and beverages of Himachal Pradesh. *Indian Journal of Traditional Knowledge*, 3(3): 325–335.

Thapa, N. 2002. *Studies of microbial diversity associated with some fish products of the Eastern Himalayas.* PhD thesis, North Bengal University. India.

Thapa, S. and Tamang, J. 2004. Product characterization of *kodo ko jaanr*: Fermented finger millet beverage of the Himalayas. *Food Microbiology*, 21(5): 617–622.

Tiwari, S.C. and Mahanta, D. 2007. Ethnological observations fermented food products of certain tribes of Arunachal Pradesh. *Indian Journal of Traditional Knowledge*, 6(1): 106–110.

Tsuyoshi, N., Fudou, R., Yamanaka, S., Kozaki, M., Tamang, N., Thapa, S., and Tamang, J.P. 2004. Identification of yeast strains isolated from marcha in Sikkim, a microbial starter for amylolactic fermentation. *Journal of Applied Microbiology*, 97(3): 647–655.

Tsuyoshi, N., Fudou, R., Yamanaka, S., Kozaki, M., Tamang, N., Thapa, S., and Tamang, J.P. 2005. Identification of yeast stains isolated from marcha in Sikkim, a microbial starter for amylolytic fermentation. *International Journal of Food Microbiology*, 99(2): 135–146.

Valiathan, M.S. 2003. *The Legacy of Caraka.* Orient Longman Private Ltd., Hyderabad, pp. 125–127.

Verma, L.R. and Joshi, V.K. eds. 2000. *Postharvest Technology of Fruits and Vegetables.*, Vols. 1 and 2. The Indus Publ, New Delhi. p. 1242.

Vine, R.P. 1981. *Commercial Winemaking.* AVI Publishing Co. Inc., Westport, CT, pp. 1–481.

Vohra, A. and Satyanarayana, T. 2001. Phytase production by the yeast *Pichia unomala. Biotechnology Letters*, 23: 551–554.

Vohra, A. and Satyanarayana, T. 2004. A cost effective cane molasses medium for enhanced cell bound phytase production by *Pichia anomala. Journal of Applied Microbiology*, 97: 471–476.

Yadav, P., Garg, N., and Dwivedi, D. 2012. Preparation and evaluation of *mahua* (*Bassia latifolia*) vermouth. *International Journal of Food Fermentation Technology*, 2(1): 57–61.

Yadav, P., Garg, N., and Diwedi, D.H. 2009. Effect of location of cultivar, fermentation temperature and additives on the physico-chemical and sensory qualities on *Mahua* (*Madhuca indica* J.F. Gmel.) wine preparation. *Natural Product Radience*, 8(4): 406–418.

Yokotsuka, T. 1991. Nonproteinaceous fermented foods and condiments prepared with *koji* molds. In: *Handbook of Applied Mycology*. Arora, D.K., Mukerji, K.G., and Marth, E.H. (eds.), Marcel Dekker, Inc., New York, NY, pp. 293–328.

10

ACETIC ACID FERMENTED FOOD PRODUCTS

AHMAD ROSMA, ABU HASSAN SITI NADIAH, ANUP RAJ, SOMBOON TANASUPAWAT, SOMESH SHARMA, AND V.K. JOSHI

Contents

10.1 Introduction

10.1.1 Vinegar

The history of vinegar production dates back to around 2000 BC (Solieri and Giudici, 2009). It is not considered as a "food" but as a condiment (Cruess, 1958), and is made by the transformation of an alcoholic product into acetic acid. There are two types of vinegars (Joshi and Thakur, 2000); synthetic and brewed vinegar. Synthetic vinegar is directly prepared from synthetic acetic acid with the addition of water, colored using caramel, whereas brewed vinegars are prepared from fruit juices, e.g. apples, grapes, oranges, pears, etc.; starchy materials, e.g., malted cereals; spirit or alcohol; dilute denatured ethyl alcohol; and sugar, such as syrups, molasses, honey, maple skimmings, etc. It is one of several fermented foods prepared and used by early man (Hesseltine, 1965; Steinkraus, 1996). Even today, it is consumed in large quantities all over the world. Vinegar is used as a flavoring agent, as a preservative and, in some countries, also as a healthy drink (Solieri and Giudici, 2009).

Vinegar is made by a two-step fermentation process involving yeasts (Aidoo et al., 2006) as the first agent for alcoholic fermentation, followed by acetic acid fermentation by acetic acid bacteria (Adrian et al., 1995). It is an inexpensive product, and its production requires low-cost raw materials, such as substandard fruit, seasonal agricultural surpluses, by products from food processing, and fruit waste (Solieri and Giudici, 2009).

The manufacturing process of vinegar has developed from an uncontrolled, naturally occurring sequence of fermentation to a carefully monitored and automated fermentation. Consumer requirements demand industrial fermentation systems capable of producing a large amount of vinegar with reliable controls and optimum conditions for acetic acid bacteria fermentation (DeOry et al., 1999). Many techniques have been developed to improve industrial production of vinegar, such as to increase the speed of the transformation of ethanol into acetic acid by the acetic acid bacteria (Tesfaye et al., 2002). The most common technology in the vinegar industry is based on submerged culture (Hromatka et al., 1951), with diverse technical modifications which try to improve the general fermentation conditions (aeration, stirring, heating, etc.; San Chiang Tan, 2005).

Earlier, it was manufactured from the juice or sap of the date palm, date or raisin wine, and from beer also. Theoretically, 1 g of glucose will produce 0.67 g of acetic acid but, as this figure is never achieved in practice, at least 2% (w/v) sugar is required for every 1% (w/v) acetic acid in the final product. In most cases, the raw material used in vinegar production contains sufficient nutrients to support the growth and metabolism of acetic acid bacteria (AAB), but low-sugar juices can be supplemented with extra sugars or concentrated by evaporation or reverse osmosis prior to alcoholic fermentation (Igbinadolor, 2009).

10.1.2 Historical Aspects of Vinegar

English people consumed beer and ale, so their first vinegars were obtained by the souring of these beverages rather than from wine. The earliest product was simply ale converted into vinegar by long exposure to the air. Since wine making dates back to atleast 10,000 years, it can be safely assumed that the existence of vinegar is at least that old (Duddington, 1961). Since the factors involved in the conversion of wine to vinegar were not known at that time, the spoilage of wine was a common occurrence.

There are also many records which refer to vinegar, attesting to its importance in ancient civilizations, and the earliest are those from Babylon about 5000 bc. Some of the historical events in the production of vinegar are summarized in Table 10.1.

During Biblical times, vinegar became well known and was mainly used as a refreshing and energizing drink, often added to water, or as a condiment for adding flavor to other foods (Stefane and Marooka, 2009). It was also used in preserving the fresh food items and, consequently, extending the time period when these foods were available for consumption (Hubner, 1927; Liu et al., 2004). Vinegar manufactured by the Romans when mixed with water or sometimes with water and eggs was called "Posca," a common drink of the Roman soldiers and in the slums. Both Greeks and

Table 10.1 Historical Events in Vinegar Production and Usage

- The word "*Vinegar*" has been in use in the English language since the fourteenth century, when it arrived in the British Isles, from the French word *vinaigre,* which simply means "sour wine," and that came in turn from the Latin *vinum acre,* "sour wine" or, more commonly, *vinum acetum,* "wine vinegar"
- The word *acetum,* meaning "vinegar" in the most proper sense, is derived from the verb *acere,* meaning "to become pungent, go sour," and is similar to the ancient Greek[*akme*], "spike"
- Looking to another ancient culture, the Hebrew word for "vinegar" in the Old Testament was *koe-metz* (also transliterated as *chomets, hometz*), meaning "pungent" or "fermented" (Stefano and Marooka, 2009). So since ancient times, the words indicating vinegar were always associated with the idea of being pungent, strong-tasting, and sharp, or, more simply, the idea of being acid. Notwithstanding these, very little is known about the origin of vinegar (Stefano and Marooka, 2009)
- Of all the people of ancient times, the Egyptians were probably the first to discover and use true vinegar, as they had been brewing beer from barley, wheat, or millet since the origin of their civilization (Stefano and Marooka, 2009)
- Vinegar is the world's oldest cooking ingredient and food preservation method. Its use can be traced back over 10,000 years (Vinegar Institute, 2006). In fact, flavored vinegars have been manufactured and sold for almost 5000 years (San Chiang Tan, 2005)
- Persia (now Iran), according to evidence from many sources, is the oldest winemaking country in the world. Large urns and vases have been excavated from the kitchen area of a mud-brick building in Hajji Firuz Tepe, a Neolithic village in Iran's Northern Zagros Mountains (Stefano and Marooka, 2009)
- The jars dating back to 6000 BC were found to have a peculiar reddish-yellow coating. Using infrared spectrometry, liquid chromatography, and a wet chemical test, a team from the University of Pennsylvania Museum found that the coating contained calcium salts from tartaric acid, which occurs naturally in large amounts only in grapes (McGovern et al., 1996). The jars were possibly used to store wine and vinegar
- The processing methods used in prehistoric and ancient times were certainly crude, and the hot, dry climate of the desert might have encouraged a quick fermentation, rapidly turning grape juice into an indeterminate alcoholic–acidic beverage (Stefano and Marooka, 2009)
- The acetic fermentation resulting in vinegar, although not entirely welcomed by the early winemakers, who used to control fermentation by adding clay as a stopper, to prevent the wine—or other alcoholic beverage—from turning into vinegar (Stefano and Marooka, 2009)
- It is, however, still uncertain whether the resin from the terebinth tree *(Pistacia terebinthus),* a renowned natural preservative, was deliberately used for the control of fermentation, or simply as an ingredient for adding flavor and taste to drinks (Stefano and Marooka, 2009). In ancient Greece, Hippocrates of Kos (460–377 BC), father of modern medicine, was the first to study the human digestive system and fix the principles of dietology, because he considered every food as a cause of illness or good health. Following his theory, he prescribed vinegar as the main remedy against a great number of diseases, including the common cold and cough (Flandrin et al., 2000)
- In the third century BC, the Greek philosopher Theophrastus of Eressos (370–285 BC) described how vinegar acted on metals to produce pigments useful in art including *white lead* and *verdigris,* a green mixture of copper salts such as copper acetate (Stefano and Marooka, 2009)
- An important ancient source, where vinegar is quite often mentioned, is the Bible, both in the Old and in the New Testament (Stefano and Marooka, 2009)
- There is an evidence of *Saccharomyces cerevisiae* fermentation in ancient wine (Cavalieri et al., 2003).

Romans extensively used vinegar in cooking, preserving lentils, and green olives using vinegar, oil, and spices. Pickling came into increased popularity in Rome during the years between Cato and Columella (Clark and Goldblith, 1974).

10.1.3 Vinegar: Definitions and Legislation

Vinegar production is regulated through an extensive set of statutes, and the definition of vinegar itself differs from country-to-country. FAO/WHO defines vinegar as any liquid, fit for human consumption, produced exclusively from suitable raw materials containing starch and/or sugars by the process of two fermentations, first alcoholic and then, acetous. The residual ethanol content must be less than 0.5% in wine vinegar and less than 1% in other vinegars (Joint FAO/WHO Food Standards Programme, 1998). In the USA, the Food and Drug Administration (FDA) requires that vinegar products must possess a minimum of 4% acidity (FDA, 2007). European countries have regional standards for the vinegar produced or sold in the area. Unlike the USA, the EU has established thresholds for both acidity and ethanol content. In general, the definition "vinegar" is used for the products having a minimum of 5% (w/v) acidity and a maximum of 0.5% (v/v) ethanol. Wine vinegar is exclusively obtained by acetous fermentation of wine and must have a minimum of 6% acidity (w/v) and a maximum of 1.5% (v/v) of ethanol (Regulation (EC) No. 1493/1999; Solieri and Giudici, 2009).

In China, the term "vinegar" is used to indicate both fermented and artificial vinegars, according to the Chinese National Standard definitions (CNS14834, N5239) (Chinese National Standard, 2005). Considering the different laws on vinegar, it is clear that acidity and residual ethanol are the two main parameters used to establish an all-encompassing vinegar classification (Table 10.2). The parameters change on the basis of raw materials used, the microorganisms involved in the fermentation process, the technology employed, but mainly on the basis of culture and "vinegar lore" (Solieri and Giudici, 2009).

The strength of vinegar earlier was expressed in grains, which vary from country-to-country. These grains were once literally equivalent to the number of grains of soda required to neutralize a fixed volume of vinegar. But it is better to describe vinegar strength directly in terms of their per cent acetic acid content. Nevertheless, the strength of different vinegars in different countries in terms of grains is: USA 1 grain equals to 0.1%; U.K: 0.25%; Canada: 0.6 (%) acetic acid (w/v). Based on the production, cider vinegar is classified into low strength and high strength; low strength is

Table 10.2 Acidity and Residual Ethanol Content in Several Vinegars

VINEGAR	ACIDITY (% W/V)	ETHANOL (% V/V)
Malt vinegar	4.3–5.9	–
Cider vinegar	3.9–9.0	0.03
Wine vinegar	4.4–7.4	0.05–0.3
(semi-continuous process)	(8–14)	–
Rice vinegar	4.2–4.5	0.68
Chinese rice vinegar	6.8–10.9	–
Cashew vinegar	4.62	0.13
Coconut water vinegar	8.28	0.42
Mango vinegar	4.92	0.35
Sherry vinegar	7.0	–
Pineapple vinegar	5.34	0.67

Source: Adapted from Springer Science+Business Media: *Vinegars of the World*, 2009, pp. 1–16, Solieri, L. and Giudici, P. (eds).

cider vinegar produced from cider of a total concentration (acid % by weight plus alcohol % by volume) of less than 8%–9%, and high strength is cider vinegar with more than 9% and upto 13% total concentration (Joshi and Sharma, 2009).

10.1.4 Uses of Vinegar

10.1.4.1 Food Uses   Fruits, vegetables, meat, and fish, are preserved by pickling in vinegar, usually with the addition of spices, herbs, and pepper, while melons, cucumbers, pumpkins, peaches, apricots, apples, and pears are preserved by cutting them into pieces and pickling them in fruit vinegar to which spices and herbs are added (Joshi and Thakur, 1999). From the early days of agriculture until today, mankind has always used vinegar for different purposes: as a condiment, as a pickling or preserving agent, as a disinfectant, as a cleansing agent, and as a beverage, with virtually no exceptions for all the cultures in the world. It is always present in old-fashioned attics or in the cheapest supermarkets, and it serves the same purposes today as it did when mankind first discovered it (Stefano and Marooka, 2009). On both sides of the Atlantic, there is an increasing interest in cider vinegar as a health food, to be taken internally as a remedy for a variety of complaints, ranging from digestive disorders to arthritis, in both man and animal (Jarvis, 1959). Some recent studies have indicated the health related properties of vinegars, such as the dietary acetic acid has been reported to reduce serum cholesterol and triglycerols in rats when fed cholesterol rich diet (Fushimi et al., 2006). In another study, the extract of "kurosu" vinegar inhibited the proliferation of human cancer (Nanda et al., 2004), also proved to be anti-hypertensive (Nishidai et al., 2001) and suppressed lipid peroxidation in mouse skin (Ohigashi, 2000) as was the case with rice vinegar (Sugiyama et al., 2008).

10.1.4.2 Non-Food Uses   Vinegar was the first antibiotic known to man, a bacterial product antagonistic to other microorganisms. Hippocrates (father of medicine), applied it to his patients in 400 BC (Hippocrates translated by Adams, 1849), while Pliny listed several maladies which were claimed to be relieved by vinegar or its by-products. Medicinal vinegars and their several formulations, such as antiseptic vinegar, antiscorbutic vinegar, camphorated vinegar, colchicum vinegar, squill vinegar, black drop vinegar, and syrup of vinegar have also been documented (Dussance, 1871). It was also employed with considerable success to prevent an incipient attack of scurvy among soldiers, as well as for the treatment of wounds during the First World War and in the American Civil War. Formulations of apple cider vinegar have been used to treat lameners, ivy poison, shingles, night sweats, burns, varicose veins, impetigo, and ringworm (Jarvis, 1959). In Ireland, cider vinegar as a component of animal feed, is regarded as a coat conditioner for horses and as a treatment for mastitis in dairy cattle (Lea, 1988). However, vinegar finds little use in modern systems of medicine. Nevertheless, it is interesting to note that some *Acetobacter* spp. produce biologically active compounds such *Acetobacter rancens*, which produces a compound proved lethal to yeasts (Gilliland and Lacey, 1966). Another biologically active material in vinegar that inhibits the growth of human tumor cells *in vitro* has also been documented (Adams, 1978; Webb, 1983; Tamir and Eskin, 2005). Vinegar also inhibits tumor growth in mice injected with Ehrlich's tumor fluid (Lanzani and Pecile, 1960).

The solvent effects of vinegar were well understood by ancient people. Olives were sprinkled with vinegar to release them from their pits (Low, 1901). Vinegar was used in early times and even today as a cosmetic aid as a beautifier (King et al., 1956). Sometimes, women have been known to finish a hair shampoo with a vinegar rinse (Fowler and Schwartz, 1973).

10.1.4.3 Vinegar in Food Processing The attributes of vinegar which made it valuable in the past and also today are its preservative actions and flavor-enhancing properties. It can also be used for a number of other purposes, e.g., salad dressing, a flavor for cooking, etc. French cooking has long been noted for the quality of its sauces, and many of these are flavored with vinegar. Both the British and Americans use it on fish and chips (Gogui, 1856). Vinegar has been described as a wholesome food condiment. Seasoned vinegars prepared by infusion of cider or white vinegar with herbs, spices, shallots, or garlic oil and for condiment purposes are employed extensively in food preparations (Crawford, 1929). The addition of ginger and garlic extract to apple vinegar gives successful preparations of apple ginger and apple garlic vinegars of standard quality with additional potential benefits (Kocher et al., 2007). Since vinegar can reduce the pH of certain food products to a level which inhibits microbial growth, including some spore forming bacteria, it is used as an additive of food products, such as pickles, pickled vegetables, spiced fruits, salad dressings, mayonnaises, mustards, ketchups and other tomato products, relishes, fish products, barbecued poultry, marinated and pickled meats, breads, sauces, cheese, dressings, and soft drinks.

10.1.5 Geographical Distribution and Origin in Asia

A number of vinegars are produced in South Asian countries, as listed in Table 10.3. Very little however is known about the origins of vinegar in the history of Asia, but

Table 10.3 Some of the Vinegars of South Asia, the Raw Materials Used, Intermediate Products, and Name of the Vinegar

RAW MATERIAL	INTERMEDIATE	VINEGAR NAME	GEOGRAPHICAL DISTRIBUTION
Palm sap	Palm wine (toddy, tari, tuack, tuba)	Palm vinegar, toddy vinegar	India, Sri Lanka
Apple	Cider	Cider vinegar	India
Sugarcane	Fermented sugarcane juice	Cane vinegar	India
Coconut	Fermented coconut water	Coconut water vinegar	Sri Lanka, India
Mango	Fermented mango juice	Mango vinegar	India
Mulberry	Fermented mulberry juice	Mulberry vinegar	India, Asia
Honey	Diluted honey wine	Honey vinegar	India

Source: Adapted from Springer Science+Business Media: *Vinegars of the World*, 2009, pp. 1–16, Solieri, L. and Giudici, P. (eds.); Joshi, V.K. and Thakur, N.S. 2000. *Postharvest Technology of Fruits and Vegetables*. Indus Publishing Co., New Delhi.

the earliest records are found in China at the time of the Zhou Dynasty, 0021–22, in texts such as *Tso Chuan* and *Mo Tzu*, which refer to a seasoning called *liu*, which has usually been interpreted as vinegar. During the Zhou Dynasty, a condiment is said to have been obtained from plums and salt that became very popular (Fong, 2000). It is beyond doubt that in China, vinegar soon became a very important ingredient in the preparation of all sorts of dishes. Sour, along with sweet, salty, pungent, and bitter, is one of the "five flavors" of classical Chinese cooking (Bo, 1988a,b; Simoons, 1991; Stefano and Marooka, 2009).

There is also a Chinese traditional painting called *The vinegar tasters* that depicts three men dipping their fingers into a vat of vinegar and tasting it (Stefano and Marooka, 2009). Coconut vinegar, made from the sap or "toddy" of the coconut palm, is produced and used extensively in South Asian countries, including India. It is a cloudy white liquid, has a particularly sharp, acidic taste (4% acetic acid) with a slightly yeasty flavor. Cane sugar vinegar, made from sugarcane juice, is also popular in rural India, and ranges in color from dark yellow to golden brown with a mellow flavor. When it is produced by the traditional static fermentation method, it requires about a month to complete (Kozaki and Sanchez, 1974; Murooka et al., 2009). In Table 10.3, some of the vinegars of South Asia are listed, along with the details of raw materials used, etc.

10.2 Types of Vinegar

Vinegar varieties vary greatly from country-to-country. Some of the most popular vinegars are described here (Joshi and Thakur, 2000; San Chiang Tan, 2005):

- *Balsamic vinegar* is brown in color with a sweet–sour flavor and is an aromatic, aged type of vinegar (Giudici et al., 2006). It is made from the white Trebbiano grape and aged in barrels of various woods.
- *Canesugar vinegar* is made from fermented sugar cane juice and has a very mild, rich–sweet flavor (Anonymous, 2008).
- *Cider vinegar* is made from apples and is the most popular vinegar used for cooking (Joshi and Sharma, 2009).
- *Coconut vinegar* is low in acidity, with a musty flavor and a unique aftertaste (San Chiang Tan, 2005).
- *Fruit vinegars* are made from fruit wines, such as apple, blackcurrant, raspberry, and quince, usually without any additional flavoring (Joshi and Thakur, 2000). Typically, the flavors of the original fruits remain in the final product.
- *Distilled vinegar* is harsh vinegar made from grains and is usually colorless (San Chiang Tan, 2005).
- *Flavored vinegar.* Popular fruit-flavored vinegars include those infused with whole raspberries, blueberries, or figs. Vinegar may also be flavored with herbs and spices.

- *Jamun sirka*, a vinegar produced from the *jamun* fruit in India, is considered to be medicinally valuable for stomach, spleen, and diabetic ailments (Joshi and Thakur, 2000).
- *Palm vinegar* is made from the fermented sap from flower clusters of the nipa palm.
- *Malt vinegar* is made from fermented barley and grain mash, and flavored with woods such as beech or birch (San Chiang Tan, 2005).
- *Wine vinegar* can be made from white, red, or rose wine. These vinegars make the best salad dressings (San Chiang Tan, 2005).

10.3 Vinegars: Raw Materials

Most vinegars have a plant origin, with two exceptions: those produced from whey and from honey. Whey, which is the milk serum residue of the cheese-making process, is rich in lactose and/or its corresponding hydrolysed sugars, galactose and glucose, depending on the cheese-making technology employed. The other raw material used for vinegar production include fruit juices, like apple juice, jamun juice, mango juice, grape juice, cane sugar juice, coconut water, and palm sap (Solieri and Giudici, 2009).

The chemical composition of the raw material exerts a strong selective pressure on microorganisms and determines the dominant species involved in acetification (Solieri and Giudici, 2009). The critical steps in vinegar production are the alcoholic fermentation of raw materials fermentable sugary solutions followed by acetic acid fermentation. The various steps include slicing and/or crushing of fruits like apple to obtain fruit juice (Joshi and Sharma, 2009), enzymatic digestion of starch in cereals, as well as cooking and steaming in some cases. In general, fruits require less preparation than grains, though the grains are more easily stored and preserved, and consequently their use is independent of the harvest. Fruits are highly perishable, rich in water, and need to be processed very quickly; in some conditions, such as at high temperatures or in the case of damaged fruits, the processing has to be done immediately after harvest (Solieri and Giudici, 2009).

10.4 Microbiology of Vinegar Production

Production of vinegar involves two types of fermentation involving two different types of microorganisms (Hesseltine, 1983). The first is the alcoholic fermentation due to the yeast *Saccharomyces cerevisiae*. In natural fermentation, several yeasts are involved but, finally, it is *Saccharomyces cerevisiae* that carry out the fermentation (Aidoo et al., 2006). The yeast is elliptical in shape and divides by budding, and is known for high fermentation capacity. It has capacity to convert glucose into ethyl alcohol through glycolysis. Ethyl alcohol and carbon dioxide are produced during this fermentation. Most times, the natural yeast flora in the juice containers/air is able to conduct the alcoholic fermentation, but on an industrial scale the yeast is grown in the laboratory and inoculated to get the predictable results, i.e., ethanol production.

Acetobacter strains are usually involved in vinegar production in the second fermentation. A pellicle formation on various liquids occurs, which gave "mother of vinegar" the name Mycoderma, i.e., muciliginous skin (Persoon, 1822). On examination, this was found to contain cells resembling those of yeast. The microbial nature of vinegar was recognized first by Conner and Allgeier in 1976 when it was observed that the mother of vinegar contained minute dot like organisms arranged in chains, and they were the first to isolate acetic acid bacteria from vinegar, although he called the organism *Ulvina aceti*—an algae.

The acetic acid fermentation is a biological process whose inception and maintenance is the result of the metabolism of a living organism *Mycoderma aceti* (Pasteur, 1868). The role of *Acetobacter* (*M. aceti*) in the conversion of ethanol to acetic acid was, however, not immediately and universally accepted. The ethanolic product obtained is converted by the activity of acetic acid bacteria, *Acetobacter* and *Gluconobacter* (Palacios et al., 2002; Solieri and Giudici, 2009) into acetic acid.

Different microorganisms develop during the "let alone" or "Orleans process. *Acetobacter* strains are widely distributed in nature and have been isolated from many sources, including vinegar generators, beer, wine, yeast, wort, apple juice, cane sugar juice, flowers, vegetables, and many fruits (Wegst and Lingens, 1983). Different species of *Acetobacter* were first distinguished, and three species, later named *Bacterium aceti*, *Bacterium pasteurianum* and *Bacterium kutzingianum* were isolated. All the important strains or species that have later been described from time-to-time (Pasteur, 1868) have been divided into 10 species, falling into four groups on the basis of their biochemical properties: peroxydans, oxydans, mesoxydans and sub-oxydans (Sharma and Joshi, 2005).

Most of the *Acetobacter* cultures are believed to be a mixture of two or more species derived by spontaneous mutation (Shimwell, 1959). Strains of *Acetobacter*, other than *A.* aceti, which have been used commercially or isolated from vinegar generators are referred to as *Bacterium curvum*, now called *A. curvum* (Henneberg, 1926); *Bacterium schuzanbachii* called *A. schiizenbachii* (Gamova-Kayukova, 1950), *A. acidophilum* (Yamamato et al., 1959), and *A. rancens* (Suomalainen et al., 1965; Matthias et al., 1989).

Until recently, strain differentiation for similar species involved in traditional static fermentation had not been possible as no method was available. But techniques such as the enterobacterial repetitive intergenic consensus-polymerase chain reaction (ERIC-PCR) method (Nuswantara et al., 1997), random amplified polymorphic DNA (RAPD) finger printing analysis (Track et al., 1997), and repetitive extragenic palindromic (REP) element-PCR methods (Nuswantara et al., 1997) have been used for the taxonomic grouping of bacteria, including acetic acid bacteria (Murooka et al., 2009).

Acetic acid bacteria responsible for acetic fermentation are strictly aerobic, Gram–ve, with a strong oxidizing effect on ethanol, and have been classified in two genera, *Acetobacter* and *Gluconobacter*, belonging to the *Acetobacteriaceae* family (Gillis and De Ley, 1980; Saeki, 1993; Yukphan et al., 2009). These bacteria are very resistant to acetic acid concentrations and are even active upto 2.0% acidity (Anonymous, 1971).

Acetobacter is capable of oxidizing ethanol not only to acetic acid but also totally to carbon dioxide when ethanol is used for vinegar production (Willems et al., 2007). Strains of *A. aceti* are generally used in modern submerged culture fermentors, while *A. xylinum* was usually used in the traditional Orleans (barrel) process.

In actual practice, most of the vinegar manufacturers probably do not select particular bacterial strains; rather, fermentation conditions (air, volume, and temperature) are experimentally optimized, and whatever bacteria in the feedstock or air proliferate under these conditions are selected for the fermentation.

Nanda et al. (2001) isolated acetic acid bacterial strains from samples obtained during commercial *komesu* production in Japan in the traditional static method. They used the ERIC-PCR and RAPD methods. These fermentations had not been inoculated with a pure culture, since the vinegar production began in 1907. A total of 126 isolates were divided into groups A and B on the basis of DNA finger printing analyses. The 16S ribosomal DNA sequences of strains belonging to each group showed similarities of more than 99% with *A. pasteurianus*, and Group A strains overwhelmingly dominated all stages of fermentation. It was concluded that the appropriate strains of acetic acid bacteria had spontaneously established nearly pure cultures, after a century of *komesu* fermentation (Murooka et al., 2009). Besides, other techniques such as RAPD-PCR profiling of *Acetobactor* have also been employed (Track et al., 1997).

10.5 Metabolism and Biochemistry of Vinegar Production

10.5.1 Nutrition and Metabolism of Acetobacter

Acetobacter strains used for vinegar production, under proper conditions of high aeration rate and high acidity, convert ethanol rapidly, with high yields, to acetic acid. Ammonia serves as a source of nitrogen for most of the strains, while neither vitamins nor growth promoters are required by *A. aceti* in laboratory cultures (Gossele et al., 1983). Nevertheless, it is a common practice in the industry to add autolysed yeast or pantothenate to the diluted ethanol for the production of white distilled vinegar (Hildebrandt et al., 1961; Nickol et al., 1964; Trcek et al., 2006; López, 2012).

10.5.1.1 Metabolism of Carbon *Acetobacter aceti* can convert about 80% of the consumed glucose into gluconate, and use the remaining 20% as a source of carbon and energy (De Ley, 1959, 1961). Glucose is metabolized by two independent, competing oxidative pathways, one leads to the formation of gluconate, and the other to the hexokinase pathway followed by the hexose monophosphate cycle. Mannose, galactose, xylose, l-arabinose, and ribose are oxidized into the corresponding sugar acids. They are generally, not considered as carbon sources for the growth of most acetic acid bacteria. The mannitol, fructose, mannose, galactose, xylose, glycerol, erythritol, sodium-d lactate, ethanol, and sodium acetate are oxidized by the strain of *A. aceti* (De Ley, 1961). The absence of the glycolytic cycle in *Acetobacter* is now well recognized. *A. suboxydans* does not contain the enzymes of the TCA cycle (King and Cheldelin,

1952). The presence of all these enzymes in strains of *A. pasteurianum* and *A. aceti* has been demonstrated (Rao, 1955; King et al., 1956).

The carbohydrate metabolism of 20 *Acetobacter* strains belonging to the *oxydans,* *mesoxydans,* and *suboxydans* groups reveals that the enzymes of the pentose and hexose monophosphate oxidative cycle are present in all the groups (Stouthamer, 1959). The TCA cycle was present in the *oxydans* and *mesoxydans* strains but was absent in the *suboxydans* group (Stouthamer, 1959, De Ley, 1961). The presence of enzymes involved in the TCA cycle in the crude extract of an organism for commercial vinegar production has been established, showing oxygen consumption in the presence of β- hydroxy butyrate, citrate, ketoglutarate, succinate, melate, oxaloacetate, and pyruvate (Nakayama, 1960).

10.5.1.2 Metabolism of Nitrogen

Acetobacter grows well when lactate is provided as a source of energy and carbon (Rainbow and Mitson, 1953; Brown and Rainbow, 1956). The bacterium grows readily on a substrate containing ammonium sulphate as a nitrogen source and lactate as a source of energy (lactophile) and carbon. Glutamate, proline, aspartate, and alanine are rapidly assimilated by lactophiles acid-hydrolysed casein as a substrate.

10.5.2 Mechanism of Biological Conversion of Ethanol to Acetic Acid

It was recognized by Rosier in 1786 that air was absorbed by wine in the process of ethanol being converted to vinegar, and further shown by Lavosier that oxygen was the only compound of the air involved in the conversion (Lafar, 1898). Dobereiner (cited in Mitchell, 1926) found that ethanol reacted with oxygen (not with CO_2) to yield water and acetic acid. Acetaldehyde was found to be an intermediate in the formation of acetic acid. The first step in the conversion of ethanol to acetic acid was the formation of acetaldehyde (Henneberg, 1897) according to the equation:

$$\text{(I)} \quad \underset{\text{Ethanol}}{CH_3CH_2OH + O} \xrightarrow{\text{Alcohol dehydrogenase}} \underset{\text{Acetaldehyde}}{CH_3CHO + H_2O}$$

The second step is the formation of acetic acid from acetaldehyde. The latter first reacts with water to yield hydrated acetaldehyde, which in turn is oxidized or dehydrogenated (properly) to yield acetic acid, in accordance with the following equation:

$$\text{(II)} \quad \underset{\text{Acetaldehyde}}{CH_3CHO + H_2O} \xrightarrow{\text{Aldehyde dehydrogenase}} CH_3 - \underset{\underset{OH}{|}}{\overset{\overset{H}{|}}{C}} - OH + [0] \longrightarrow \underset{\text{Acetic acid}}{CH_3COOH + H_2O}$$

In the second mechanism, acetaldehyde is converted to acetic acid. Two molecules of acetaldehyde formed as in reaction (II) may react with each other in the so called Cannizzaro reaction to yield acetic acid and ethanol, as shown below:

$$(\text{III}) \quad CH_3CHO + CH_3CHO \longrightarrow CH_3COOH + CH_3CH_2OH$$

$$\text{Acetaldehyde} \qquad\qquad \text{Acetic acid} \quad \text{Ethanol}$$

The ethanol then, undergoes re-oxidation according to reaction (1) and the cycle is repeated until all the ethanol is converted into acetic acid.

10.5.3 Behavior of Acetobacter during Vinegar Fermentation

Acetobacter is extremely sensitive to the oxygen level during fermentation. The utilization of oxygen in the air stream can be raised to the point where only 4 vol % residual oxygen remains in the effluent gas without any adverse effects on the fermentation; thus, a proper use of air is quite important, as the volatility of ethanol and acetic acid is very high. But the use of pure oxygen or highly oxygen-enriched air is known to damage the acetic acid bacteria (Raspor and Goranovic, 2008). These bacteria are also known to be damaged if vinegar fermentation is conducted out to the point where all of the ethanol has been oxidized and/or the addition of fresh ethanol/cider has been delayed (Ebner et al., 1996). However, the fermentation is not affected if the temperature changes every 2 h between 32°C and 26°C, but the cells of *Acetobacter* are damaged when the duration of cooling is greater, or the temperature is higher, with higher concentration of acetic acid. There is no lag phase of *Acetobacter* at all, if the time of pumping out of vinegar and the method of adding fresh mash is chosen correctly. Changes in the concentration will have a detrimental effect on the fermentation (Ebner, 1982). Over-oxidation of acetic acid to CO_2 and H_2O is undesirable. It is well known that, if it starts once, it proceeds simultaneously with the oxidation of ethanol to acetic acid. It can be recognized by lower yields. It is important to make sure that the fermentation does not proceed to the point where the ethanol is used up, to avoid over-oxidation (Callejón et al., 2009).

10.6 Technology of Vinegar Production

In its production, initially, yeasts ferment the natural sugars in the raw materials to alcohol, followed by acetic acid bacteria (*Acetobacter*) that convert the alcohol to acetic acid. General, basic, safe food operating principles, such as good agricultural practices (GAP), good manufacturing practices (GMP), and good hygiene practices (GHP) should be followed in all the steps in vinegar production. After acetification, there is no real danger of spoilage, since acetic acid has strong antibacterial activity at low pH, though prior to fermentation the raw materials can be spoiled microbiologically, and even toxins, like aflatoxin, can be produced. Different microbial species are involved at various stages of the fermentation process, as illustrated in Table 10.4. The great

Table 10.4   Main Microorganisms Involved in Vinegar Production in South Asia

VINEGARS	YEASTS	LAB	AAB	REFERENCES
Beer/malt vinegar	*Saccharomyces Sensu-stricto*	*Lb. brevis. Lb. buchneri, P. damnosus*	*A. cerevisiae, Ga. sacchari*	White (1970); Greenshields (1975a,b); Fleet (1998); Cleenwerck et al. (2002)
Coconut water vinegar	*Saccharomyces* spp.	–	*A. aceti*	Steinkraus (1996)
Nata de coco	*Saccharomyces* spp.	–	*Ga. xylinus*	Steinkraus (1996); Iguchi et al. (2004)
Fruit vinegars	*S. cerevisiae, Candida* spp.	–	*A. aceti, A.pasteurianus*	Maldonado et al. (1975); Uchimura et al. (1991); Suenaga et al. (1993)
Honey vinegar	*S. cerevisiae*	*Lactobacillus*	*Acetobacter* spp.	Adams and Nielsen (1963)
	Zygosaccharomyces spp., *Torulopsis* spp.	*Streptococcus, Leuconostoc* and *Pediococcus* spp.	*Gluconacetobacter* spp.	Snowdon and Cliver (1996) Ilha et al. (2000); Bahiru et al. (2006)
Palm wine vinegar	*S. cerevisiae, S. uvarum,*	*Lb. plantarum*	*Acetobacter* spp.	Okafor (1975)
	C.utilis, C. tropicalis.	*Lc. mesenteroides*	*Zymomonas mobilis*	Uzochukwu et al. (1999)
	Schizo. pombe, K. lactis			Ezeronye and Okerentugba (2001); Amoa-Awua et al.(2007)
Whey vinegar	*K. marxianus*	Nd	*Ga. liquefaciens, A. pasteurianus*	Parrondo et al. (2003)
Wine vinegar	*S. cerevisiae*	Nd	*Ga. europaeus* *Ga. oboediens* *A. pomorum* *Ga. intermedius* *Ga. entanii*	Sievers et al. (1992) Sokollek et al. (1998) Boesch et al. (1998) Schuller et al.(2000)

Source: With kind permission from Springer Science+Business Media: *Vinegars of the World*, 2009, pp. 1–16, Solieri, L. and Giudici, P. (eds).

nd, not determined—not detected; A., *Acetobacter*; B., *Brettanomyces*; C., *Candida*; Ga., *Gluconacetobacter*; K., *Kluyveromyces;* Lb., *Lactobacillus:* Lc, *Leuconostoc;* P., *Pediococcus;* S., *Saccharomyces;* S'odes, *Saccharomycodes;* Schizo., *Schizosacharomyces;* Z., *Zygosaccharomyces.*

microbial diversity reflects the variety of raw materials, of the physico-chemical characteristics, sugar sources, and processes, as well as environmental conditions, e.g., temperature, pH, water activity). After raw material preparation, fermentation plays a key role in vinegar production (Solieri and Giudici, 2009).

Among the yeasts, *Saccharomyces cerevisiae* is the most widespread species in fruit and vegetable vinegars; the lactose-fermenting yeast *Kluyveromyces marxianus* is the species responsible for whey fermentation; and a physical association of yeasts, LAB (lactic acid bacteria) and AAB (acetic acid bacteria) is involved in the fermentation of *kombucha.* Even though there are now, ten generally recognized genera of AAB the majority of the species detected in vinegars belong to the genera *Acetobacter* and *Glucobacter*. However, it is likely that several of the species and genera involved in vinegar production have not yet been described because of the difficulties in cultivating acetic acid bacteria.

10.6.1 Spontaneous Fermentation

In most of the cases, fermentation is induced spontaneously, sometimes by back-slopping, or by the addition of starter cultures. The latter is the usual practice at industrial scale. In spontaneous fermentation, the raw material is processed and the changed environmental conditions encourage the most appropriate indigenous microflora. The more stringent the growth conditions are, the greater becomes the selective pressure exerted on the indigenous microorganisms. In a very acidic and sugary environment, such as some fruit juices, only yeast, LAB, and AAB can grow. Spontaneous fermentation is only suitable for small-scale production and that too for specific juices (Solieri and Giudici, 2009). The method is, however, difficult to control, and there is a great risk of spoilage. In most such fermentations, a microbial succession takes place, and quite often LAB and yeasts dominate initially, which consume sugars and produce lactic acid and ethanol, respectively. The acid and alcohol inhibit the growth of many bacterial species, determining prolongation of the shelf-life of the product.

Molds mainly grow aerobically and, therefore, their occurrence is limited to specific production steps or to crops before and after the harvest (Solieri and Giudici, 2009). Nevertheless, molds are of great concern for safety, as some of the genera and species produce aflatoxin. It is, therefore, of utmost importance that the molds used for starch hydrolysis of grains should be Generally Recognized As Safe (GRAS).

10.6.2 Back-Slopping Fermentation

In back-slopping, a part of a previously fermented batch is used to inoculate a new batch, and it increases the initial number of desirable microorganisms and ensures a more reliable and faster process than spontaneous fermentation. It is a primitive precursor of the starter-culture method, because the best-adapted species succeed over the indigenous populations (De Vuyst, 2000). Back-slopping enhances the growth of useful yeasts, while inhibiting the growth of pathogenic microorganisms, and the laborious and time-consuming starter selection process is avoided. It is particularly useful for inoculating AAB cultures, which are very fastidious microorganisms, needing special case to produce true starter cultures. In the semi-continuous submerged acetification process, at least one-third of the vinegar is left in the fermenter to inoculate the new wine, whereas in surface-layer fermentation a physical transplant of the AAB film can easily be done in order to preserve the integrity of the cell layer. Some times, a tool is used to take out the film of acetic acid becteria. The tool used consists of a handle and mesh, through which liquid oozes but bacterial film is retained without breakage, and is transferred to the barrel containing the ethanolic solution. Use of this procedure, implants the acetic acid bacteria in a proper way (Solieri and Giudici, 2009).

10.6.3 Starter Culture Fermentation

A starter culture is a microbial preparation of a large number of cells of a microorganism (in some cases more than one), which is added to the raw material to produce a fermented food by accelerating and guiding its fermentation process (Leroy and De Vuyst, 2004). The practice of pure culture inoculation was originally elaborated by Robert Koch for bacteria (Ranieri et al., 2003), each microbial colony of which is made up of cells that all originate from the same single cell and, therefore, can be relied upon to produce the desired biochemical reactions (Solieri and Giudici, 2009).

In some Asian vinegars, a mixed starter culture of undefined molds and yeasts, called *koji*, is used to saccharify and ferment rice and cereals. However, *koji* cannot be considered to be a true starter, as its exact microbial composition is not known. True starter cultures of oenological *S. cerevisiae* strains, selected for winemaking, are used for producing the alcoholic bases for vinegars, such as beer, wine, and cider vinegar. Regarding acetous fermentation, the application of starter cultures is still not practised on a large scale, as the AAB are nutritionally demanding microorganisms, and are difficult to cultivate and maintain in laboratory media, nor can they be preserved as a dried starter. The second major reason is that vinegar is generally an inexpensive commodity and, therefore, its manufacture does not warrant an expensive starter culture selection (Solieri and Giudici, 2009).

10.6.4 Ethanolic Fermentation

The initial fermentation is alcoholic fermentation in which *Saccharomyces cerevisiae*, either from the raw materials, the equipment, or the environment conducts the fermentation. At industrial scale, it is an inoculated fermentation. In this fermentation, glucose is converted into ethyl alcohol, which is then, converted into acetic acid in the subsequent fermentation. The alcoholic fermentation is effected by the sugar level, the nitrogen level, and the inoculum level.

10.6.5 Acetification

Vinegar is produced by either fast or slow acetic acid fermentation processes. For the quick method, the liquid is oxygenated by agitation and the bacterial culture is submerged, permitting rapid fermentation. Slow methods are generally used for the production of traditional wine vinegars, where the culture of acetic acid bacteria (AAB) grows on the surface of the liquid and fermentation proceeds slowly over the course of weeks or months. The longer fermentation period allows for the accumulation of a non-toxic slime composed of yeast and acetic acid bacteria, known as the "mother of vinegar." Vinegar "eels" (the nematode *Turbatrix aceti*) feed on these organisms and occur in naturally fermenting vinegar. Most manufacturers filter and pasteurize their product before bottling to prevent these organisms from growth (San Chiang Tan, 2005). Vinegar production ranges from traditional methods employing wood casks and surface culture to submerged fermentation in acetators.

Two processes used for making vinegar include the Orleans process (which is also known as the slow process), the generator process (the quick process), and the submerged culture process (San Chiang Tan, 2005). Generally, static fermentation is less costly and is employed in traditional vinegar production as an indigenous fermented product. In terms of plant investment, it is cheap, and the quality of the product is also good, as a long time is taken to complete the fermentation, and it does not require strict sterilization measures or a pure culture to start the vinegar fermentation. The acetic acid bacteria consist of two genera, *Acetobacter* and *Gluconobacter*, but strains of *Acetobacter* are generally involved in vinegar production (Sokollek et al., 1998). Both quick and submerged culture processes are used for commercial vinegar production today. Acetic acid yield from fermented sugar is approximately 40%, with the remaining sugar metabolites either lost to volatilization or converted into other compounds (San Chiang Tan, 2005). Acid yield improvements can be made using high rates of aeration during continuous production (Ghommidn et al., 1986). Vinegar also requires packaging; intermediate bulk containers and tanks should be manufactured of stainless steel, glass, or plastic material resistant to corrosion (Solieri and Giudici, 2009).

10.7 Production Technology of Vinegars of South Asia

10.7.1 Cider Vinegar

The souring of apple cider to yield vinegar (Sharma and Joshi, 2005) was an ancient art, which probably descended the biblical period. To prepare cider vinegar, culled, undersized, sub-standard or unmarketable surplus apples are frequently used, although the raw material may differ depending on whether the manufacturer is producing other vinegars, processed apple products, apple juice, or fermented apple cider (Lea, 1988; Joshi and Sharma, 2009). The juice is extracted by grating followed by pressing (Bereau International due Travail, 1990). The water used for the preparation of mashes/juices must be clear, colorless, odorless, and with no sediment or suspended particles. It can be demineralized followed by the addition of specific minerals (Joshi and Sharma, 2009). It must, however, be bacteriologically clean. Most natural raw materials do not require the addition of extra-nutrients, though cider is usually lower in nitrogenous compounds, so ammonium phosphate @ 1.0% is added (Ebner, 1982; Ebner et al., 2000). Most manufacturers set side specific batches for vinegar production at an early stage. Cider vinegar like other vinegars production, involves two basic fermetative processes: (i) alcoholic fermentation, and (ii) acetic acid fermentation. These fermentations take place in proper sequence (Figure 10.1).

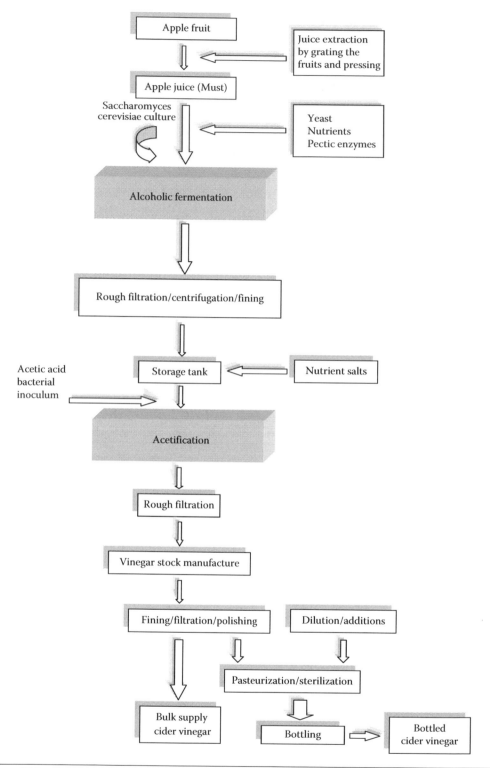

Figure 10.1   Flow chart for cider vinegar production. (Adapted from Springer Science+Business Media: *Vinegars of the World*, Cider vinegar: Microbiology, technology and quality, 2009, pp. 197–207, Joshi, V.K. and Sharma, S.)

10.7.1.1 Alcoholic Fermentation The first is alcoholic fermentation, mainly carried out by yeast *Saccharomyces cerevisiae*, either by pure culture inoculation or by the natural process of fermentation. The process can be represented by a simplified equation:

$$C_6H_{12}O_6 \longrightarrow 2CO_2 + 2C_2H_5\text{-}OH + 55 \text{ Kcal} \quad\text{————}\quad \text{(IV)}$$

Glucose   Carbon dioxide   Ethyl alcohol

In the process, ethyl alcohol is not the only product, but small amounts of other compounds like glycerol, succinic acid, amyl alcohol, propyl alcohol etc. are also produced in the fermentation which is anerobic. Apple juice normally contains about 10%–13% sugar. Apple juice concentrate may contain as much as 60%–80% sugar, so is diluted to the desired concentration. In either case, the juice is fermented with the wine yeast *Saccharomyces cerevisiae*, as in the wine or cider preparation. Jeong et al. (1999) and Seo et al. (2001) reported an optimum fermentation time of 5 days and 67.32 h, respectively, for apple juice. There is also a practice which involves natural fermentation. In any case, when the fermentation is complete, the yeast and fruit pulp settles to form a compact mass at the bottom of the tank, from which the fermented liquid is separated (Board, 2003). The clear liquid is stored in vessels for conversion into vinegar through the process of acetic acid fermentation (Joshi and Sharma, 2009).

10.7.1.2 Acetous Fermentation The second fermentation is acetic acid fermentation. It is an oxidative fermentation carried out by acetic acid bacteria such as *Acetobacter aceti*. In traditional vinegar production, a pure culture of acetic acid bacteria is not used, due to the greater efficiency of mixed cultures. The oxidation reaction is shown in the following equation:

$$C_2H_5OH + O_2 \xrightarrow[\text{Bacteria}]{\text{Acetic acid}} CH_3COOH + H_2O + 116 \text{ Kcal}$$

Ethyl alcohol   Acetic acid

The optimum temperature of fermentation is 26°C, which is achieved by the heat generated in the process. The fermentation can be achieved either by a slow or quick method. For acetic acid fermentation, the fermented liquid is adjusted to 7%–8% alcohol. Mother of vinegar containing *Acetobacter* or acetic acid bacteria, is then added in the ratio of 1:5 of alcoholic fermented juice to hasten the process of acetification and to check the growth of undesirable bacteria (Board, 2003).

For production of vinegar in South Asian countries, the temperature is a big constraint. A normal temperature for the acetification process is 30°C, and the optimal temperature for growth of AAB is 25–30°C (Holt et al., 1994), while in many South Asian countries, for most of the time, the temperature remains above the optimum temperature. The fermentation temperature for vinegar production falls within a narrow range of temperature, so an increase of 2–3°C causes a serious deterioration in both the acetification ratio and efficiency (Adachi et al., 2003). In submerged cultures, large amounts of heat are generated, so that cooling costs become rather high

(Adachi et al., 2003). These factors need to be considered for vinegar production at small scale or at the household level. The regions where cider vinegar is produced is normally cold, so such problem of temperature is not encountered in most of the countries of South Asia.

10.7.1.2.1 Slow Process   This process takes a long time and involves the slow oxidation of ethanol. The juice kept in the barrels is allowed to undergo both alcoholic and acetic fermentations slowly over time. The bung hole of the barrel is covered with a piece of cloth to screen-off the dust and flies, and the barrel is placed in a damp but warm place (Joshi and Sharma, 2009). It takes about 5–6 months to complete the whole alcoholic and acetous fermentation to form the vinegar from the juice (Figure 10.2). However, Mendonca et al. (2002), Kocher et al. (2003), and Joshi and Sharma (2009) reported a wide variation in fermentation time of 16–60 days for batch vinegar fermentations, using "NRRL746" and "NRRL 1036" strains. The main drawbacks of this process are that the alcoholic fermentation is often incomplete, the acetic fermentation is very slow, and the yield is low, with the production of inferior quality vinegar (Figure 10.2).

10.7.1.2.2 Orleans Process   It is one of the oldest commercial vinegar processes which is named after the French city (Mitchell, 1926; Wustenfeld, 1930; Trcek, 2005) and is an improvement of the old fielding process which was used extensively for cider

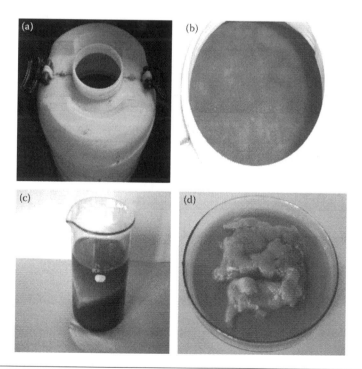

Figure 10.2   Depiction of acetic acid fermentation and acetic acid bacteria in a mat. (a) Acetous fermentation in barrel. (b) Top view of mother of vinegar. (c) Mat of vinegar cultured. Mat of acetic acid bacteria. (From Dr. V.K. Joshi, Fermentation Technology Laboratory; Dr. Yspuhf, Nauni Solan, H.P., India.).

vinegar production in earlier times. The efficiency of this process is very low compared to the present standards (77%–84%) (Wustenfeld, 1930; Joslyn, 1970). This process converts ethanol to acetic acid at a rate of about 1% per week. The diagrammatic illustration of containers with mother of vinegar is shown in Figure 10.3.

In this process, a barrel is partially (3/4th) filled with the fermented juice and inoculated with mother of vinegar, and the barrel is placed on its side. On either sides of the barrel, just above the level of the juice, in addition to the bung hole, two holes, each about 2.5 cm in diameter, are made. These three holes are covered with wire gauze to screen out insects, vinegar flies, etc. The barrel is kept in shady and warm place at 21–27°C, and the acetic acid fermentation is allowed to proceed till the acid content reaches its maximum strength, normally in about 3 months. Then, about 3/4th of the vinegar is taken out and an equal amount of fermented alcoholic juice is added for further vinegar production. The process can be repeated once every 3–4 months. The film of vinegar bacteria (Figure 10.2) on the surface of the fermenting juice should not be disturbed, otherwise the broken film will sink to the bottom and in the absence of air, the nutrients will be extracted without producing any vinegar. Breaking and sinking of the film is avoided as described earlier when the vinegar is withdrawn. Vinegar produced by the slow Orleans matures during the fermentation, and is clear and superior in quality (Board, 2003). The objectionable features of this method are the gradual filling of the barrel with slime or mother of vinegar, and the slow rate of reaction (Mitchell, 1926).

10.7.1.2.3 Quick Process   In quick processes like the generator processes, the alcoholic liquid is in motion. This process is applied mostly to the production of vinegar from spirit (alcohol). In it, ethanol is converted to acetic acid at a rate of about 1% per day, which is faster than the Orelans process. Fruit liquors are well supplemented with food for the vinegar bacteria, but to maintain active vinegar bacteria in the generator

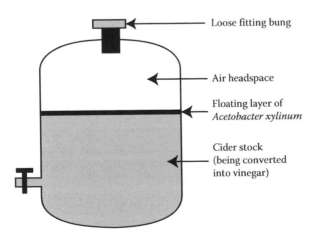

Loose fitting bung

Air headspace

Floating layer of *Acetobacter xylinum*

Cider stock (being converted into vinegar)

Figure 10.3   A barrel with cider stock, floating layers of acetic acid bacteria, illustrating the Orleans process. (Adapted from Joshi, V.K. and Thakur, N.S. 2000. *Postharvest Technology of Fruits and Vegetables.* Indus Publishing Co., New Delhi.)

methods using alcohol denatured with ethyl acetate or vinegar, it must be supplemented with a combination of organic and inorganic compounds, known as vinegar food (Joshi and Thakur, 2000). Combinations of substances such as dibasic ammonium phosphate, urea asparagins, peptones, yeast extract, glucose, malt, starch, dextrins, salts, and other substances are normally used.

The credit for constructing the first generator using packing and applying the trickling principle for vinegar production goes to the Dutch technologist Boerhaave (Boerhaave, 1732). Materials such as pumice, branches of vines, grape stems, corn cobs, etc., are used for packing the generators. Schiizenbach introduced the use of a vat instead of a cask for the acetification process, and provided mechanical means for the repeated distribution of the acid liquid over the packing (Mitchell, 1926). Holes for ventilation near the bottom of the generator were provided and the "German Process" was, thus, invented (Wustenfeld, 1930).

The process needs an additional supply of oxygen, which is made available for the bacteria by increasing the surface area for the action of the bacterial culture, which increases the rate of fermentation (Joshi and Sharma, 2009). The equipment used in this process is known as an Upright Generator (Board, 2003). This process involves the movement of the alcohlolic liquid during the process of acetification. This liquid is trickled down over the surface on which films of the vinegar bacteria have grown and to which a plentiful supply of air is provided.

Currently, the generator method is used to produce cider vinegar (Figure 10.4). The upright generator in its simplest form is a cylindrical tank that comes in different sizes and is usually made of wood (Vyas and Joshi, 1986). The interior of this, is divided into three parts:

1. Upper section: In this part, the alcoholic liquid is introduced
2. Large middle section: Here the liquid is allowed to trickle down over beechwood shavings, corn cobs, charcoal, coke, or some other material that will provide a large total surface area yet not settled into a compact mass.
3. Bottom section: This section is for the vinegar collection.

The alcoholic liquid is put in at the top through an automatic feed trough or a sprinkling device (sparger) and trickles down over the shavings or other material on which a slimy growth of acetic acid bacteria is developed, and the bacteria oxidize the alcohol to acetic acid (Joshi and Thakur, 2000). Air enters through the false bottom of the middle section, and after becoming warm, it is exhausted through a vent above (Kersters, 2006). Being an exothermic process, it is necessary to control the temperature below 30°C, which is achieved by using cooling coils, by adjusting the rate of alcoholic liquid input and feed air, and also by cooling the alcoholic liquid before it enters the generator or by cooling the partially acetified liquid that is returned to the top from the bottom section of the tank for further action (Bamforth, 2006). It is essential that the slime of vinegar bacteria be established before vinegar can be made from a new generator. To build up bacterial growth on the shavings, the alcoholic

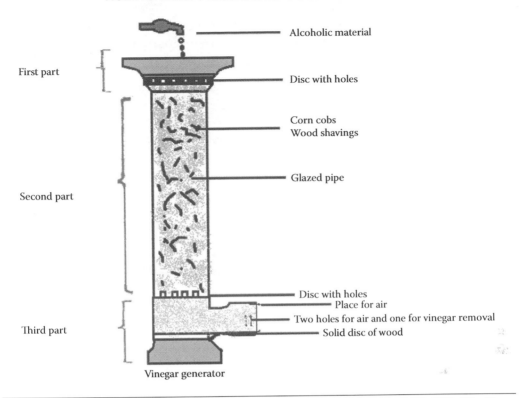

Alcoholic material

First part

Disc with holes

Corn cobs
Wood shavings

Glazed pipe

Second part

Disc with holes
Place for air
Two holes for air and one for vinegar removal
Solid disc of wood

Third part

Vinegar generator

Figure 10.4   Diagrammatic sketch of vinegar generator. (Adapted from Vyas, K.K. and Joshi, V.K. 1986. *Cider Vinegar* (Hindi). Dept. of Horticulture, Navbahar, Shimla, H.P., India.)

liquid, acidified with vinegar is slowly trickled through the generator, and then recirculated. Vinegar can be made by a single run of the alcoholic liquid through the generator. If insufficient acid is produced at first, or too much alcohol is left in the vinegar, then the vinegar collected at the bottom may be recirculated through the generator to get the required acid level.

10.7.1.3 Modern Processes of Vinegar Production: Significant improvements have taken place in vinegar production as a result of modern equipment and pure culture techniques, as detailed below.

10.7.1.3.1 The Circulating Generator   Introduction of the Frings circulating generator was a major improvement in the quick vinegar production process (Frings, 1932; Frings, 1937; Stefano and Marooka, 2009), though the process is similar in principle to the earlier described methods (Prescott and Dunn, 1959), it has many advantages over them. The Frings generator is a large, cylindrical, air tight tank which is equipped with a sprinkler at the top and has cooling coils in the lower part of the middle section and facilities to enable recirculation of the vinegar from the bottom collection chamber through the system. Such generators are equipped with various automatic controls and thus, give high yields of acetic acid and leave little residual alcohol.

10.7.1.3.2 Submerged Cultures: Frings Acetator   In the preparation of pure cultures of *Acetobacter*, the effect of inoculum amount, concentration of ethanol, forced aeration rate, temperature, stimulants such as activated salts, and the action of extraneous bacteria on submerged acetification have been reviewed earlier (Joshi and Thakur, 2000). Cider vinegar production has now been superceded by continuously aerated submerged-culture fermenters without packing materials. The submerged culture for vinegar production has also been extensively studied and reported (Tait and Ford, 1987; Sethi and Maini, 1999). Vinegar production in the acetator takes place in either low strength or high strength fermentation. The submerged culture technique in use today is the result of several investigations (Hromatka et al., 1951; Hromatka et al., 1953; Hromatka and Ebner, 1955). Frings-types submerged fermentation and tower fermenters have many advantages, such as better conversion, can be continuous or semi-continuous, have no slimy strain of *A. xylinum* residue, and can be operated with 2%–10% alcohol.

There has been a continuing interest in improving the design and production efficiencies of the process, which resulted in the development of the Frings acetator. This automatic fermentor makes possible shorter production cycles and higher yields than those obtained with the circulating generator. It also eliminated much of the unexplained variation in production ratio and efficiencies which are inherent in the circulating generator.

Most fermenters (acetifiers) are now of the Frings type (Figure 10.6) in which fresh air is being entrained and distributed continuously to the turbulent feedstock and *Acetobacter* mixture by a vortex stirrer mounted at the bottom of the fermentation tank (Ebner, 1982; Heikefelt, 2011), where the growth of the bacteria is considerable (from 8 to 10 cells/mL). For continuous removal of foam, a foam breaker is incorporated in the head of the tank. When in operation, some of the finished vinegar is drawn-off periodically and fresh cider is added to the fermenter. There is a device known as alkograph attached to the fermenter, which monitors the conversion of alcohol to acetic acid continuously. The system is designed in a way to automatically discharge vinegar and take in a fresh charge of cider when the alcohol level in the fermenter falls below 0.1%–0.2% (Lea, 1988). Operation of the acetator to produce as much as 13% acetic acid has also been described (Duddington, 1961; Ebner, 1967). A self-priming acetator which operates without a compressor has also been documented (Ebner, 1967; Ebner et al., 1967). The kinetics of submerged acetic fermentation and the behavior of *Acetobacter rancens* towards oxygen has also been documented (Mori et al., 1970). A constant supply of air (capacity 100–250 L/min per 1000 L of liquid) is an essential requirement for the submerged culture fermenters, and even a one-minute stoppage in air supply is sufficient to kill most of the *Acetobacter* cells in submerged culture units. Both the Frings and the Greenshields acetifiers are operated between 30°C and 35°C and require a cooling water flow under normal conditions, although this may be temporarily halted as new alcoholic liquid is added and the temperature falls. The presence of SO_2 or other contaminants in the air supply may also have an inhibitory effect on the *Acetobacter*. In most of the plants, the alcoholic liquid stock is fortified with nutrients before acetification, as described earlier. High alcohol liquid stock may be blended at an early stage with 10% of the fresh vinegar in a holding tank and stored for a month.

10.7.1.3.3 The Cavitator The success of the Frings acetator prompted others to develop a less expensive model, which is believed to be the first continuous submerged culture carried out for the manufacture of vinegar. The cavitator, however, had actually been developed for the biological oxidation of sewage (Burgoon et al., 1960). Despite the attempts made to use the cavitator for vinegar production on a commercial scale, technical difficulties have forced its abandonment.

10.7.1.3.4 The Tower Fermenter A novel improvement in aeration systems applied to vinegar production has been made (Greenshields and Smith, 1971) which uses a tower fermenter (Figure 10.5) or the column fermenter, consisting of a column fitted at the bottom with a perforated plate.

This acetifier is distinguished by having no moving parts whatsoever, and depends entirely on the agitation and aeration created by air pumped through a sintered glass disk at the bottom of the tower. The tower can be made of polypropylene reinforced fiberglass, and has a length:diameter ratio of about 12:1. Cider (fermented apple juice) trickles in slowly at the bottom and finished vinegar is taken out at the top so this is a truly continuous system (Greenshields, 1978; Heikefelt, 2011).

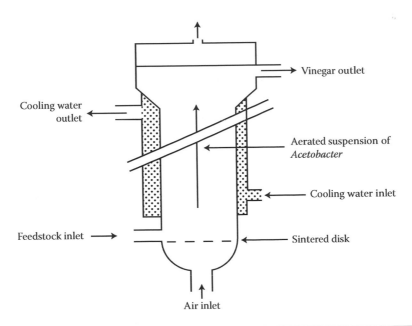

Figure 10.5 A diagrammatic view of tower-type generator used in vinegar production. (Adapted from Lea, A.G.H. 1988. *Processed Apple Products*, Van Nostrand Reinhold, New York, NY, pp. 279–301.)

10.7.1.4 Maturation/Aging   After the completion of acetification, the vinegar is transferred to wooden or stainless steel tanks for storage and maturation. A rough filtration is carried out at this stage. During storage, the harsh flavor of vinegar changes to a more pleasant aroma and bouquet, probably due to the oxidation of vinegar brought about by air entering through the pores in the wood. Acetic acid may also react with alcohol during storage to form ethyl acetate, which has a fruity flavor. In the case of cider vinegars, color is generally due to oxidation of polyphenols during processing, but in vinegars made from apple juice concentrate it is believed to be due to Maillard reactions (Joshi and Sharma, 2009).

10.7.1.5 Clarification and Packaging   The vinegar is made sparkling clear before bottling by clarification and bottling. A well-aged vinegar is rough filtered through a sheet or filter to give a bright product, which is diluted with water to the required strength (Lea, 1988; Joshi and Sharma, 2009).

10.7.1.6 Fining   Fining reduces the total load of suspended material (mainly *Acetobacter*) and reduces the polyphenol levels, and is an alternative to rough filtration (Joshi and Sharma, 2009). Fining is achieved by the addition of gelatin followed by a slurry of bentonite, and the mixture is stirred well and then, allowed to settle for at least a week before racking (Lea, 1988).

10.7.1.7 Pasteurization/Sterile Bottling   After aging and clarification, the vinegar is pasteurized to stop any spoilage. Pasteurization is caused out at 65–85°C by passing through a plate heat exchanger, and hot filled into glass bottles, or cooled before filling into flasks or plastic bottles (Joshi and Sharma, 2009). It can also be flash pasteurized by passing it through a tubular heat exchanger at 66°C. Bottled vinegar can also be pasteurized by immersing the bottles in hot water until the vinegar inside attains a temperature of 60°C (Joshi and Sharma, 2009). Cold sterile bottling through a microporous membrane has been used as an alternative to pasteurization. Nevertheless, pasteurization is the only effective procedure to prevent post-bottling haze formation from non microbiological causes, where the elimination of O_2 is also necessary. Nitrogen sparging is sometimes used at the filling tank reservoir.

10.7.1.8 Additives   At the bottling stage, additives are added to cider vinegar to prevent excessive browning, non-microbiological haze formation, and inhibition of bacterial growth. Sulphur dioxide (SO_2) is the most effective of all, as it has antimicrobial properties, is most effective at low pH, inhibits enzymes, and has antioxidative and carbonyl-binding functions. Ascorbic acid is however, considered to be less effective than SO_2 unless used in very high quantity (>250 ppm), when it acts as an antioxidant. Pectin and gum Arabic have also been added to stabilize against haze formation (Joshi and Thakur, 2000). Iron, copper, oxygen, procyanidine, and oxidized ascorbic acid should be reduced to the lowest possible levels to avoid post-bottling coloring of the vinegar.

10.7.1.9 Ultrafilteration An alternative method of stabilizing vinegars has recently been promoted—ultrafiltration—which could replace normal filtration and sterilization procedures. To minimize the chances of oxidation and bacterial reinfection, it should, however, be done immediately before bottling.

10.7.2 Palm Vinegar

Traditionally, alcoholic fermentation of the sugary sap of various palms (tribe *Cocoineae*, family *Palmae*) is carried out throughout the tropics to produce palm vinegar (Igbinadolor, 2009). Palm vinegar is a liquid substance consisting mainly of water and acetic acid produced through the slow fermentation of palm sap over the course of weeks or months (Anonymous, 1995). The traditional method of producing palm vinegar consists of spontaneous fermentation without the addition of a mother culture. The details are given here.

10.7.2.1 Collection Method for Palm Sap The palm tree is very abundant in tropical regions. Palm trees such as coconut palm (*Cocos mucifera*), nypa palm (*Nypa fructicans*), palmyra palm (*Borassus flabellifer*) and kithul palm are prevalent in various countries of Asia (Stanton and Owens, 2002). Sap of the nypa palm is obtained by tapping the stalk of the mature fruit head (Figure 10.6a). Prior to tapping, the stalk is massaged vigorously in order to encourage the flow of the sap. It is then, cut with a sharp sickle or knife, and the sap is collected into a plastic container (Figure 10.6b). The tapper usually changes the container twice daily to obtain fresh sap. Palm sap is a sweet whitish liquid that gradually turns milky as a result of the growth of microorganisms which contaminate the sap as it oozes out of the tree, causing a spontaneous fermentation (Igbinadolor, 2009).

Figure 10.6 (a) Cutting the stalk by a sharp sickle or knife lets the sap trickle out to be collected in a plastic container. (b) Tapping procedure of the stalk of the mature fruit head to collect sap of the nypa palm. (Courtesy of Dr. Rosma.)

The methods of procuring the unfermented sap from palm trees vary according to the tree and the locality. One method includes the felling of the oil palm tree (*Elaeis guineensis*) and the collecting of the sap from a cut on the stem, or by cutting the terminal bud. The palm wine produced is called "down wine," and differs from other palm wines because of its high content of ethanol, methanol, and propanol (Ayernor and Matthews, 1971). In another method of tapping, an incision is made at the base of the immature male inflorescence after removing the bracts, and is left to dry for 2 days. After this, the hole is reopened and the sap is collected in a gourd. It is done twice daily for 2–3 weeks. It is a method which spares the tree and produces a wine that commands a high price (Igbinadolor, 2009).

The outer spathes (flowering sheaths) of some palms, such as coconut or palmyra, yield toddy or sugary sap that can be converted into alcoholic beverages, such as wine and vinegar (Mozingo, 1989). The palms most frequently tapped for sap are raphia palms (*Raphia hookeri* and *Raphia vinifera*) and the oil palm (*Elaeis guineese*). The sugar content of the palm sap is between 10% and 14% (w/v), making it an ideal substrate for yeast growth (González, 2009). Generally, the pH of fresh nypa sap is approximately 7. The initial microbial load of log 7–8 cfu/mL (aerobic count) and log 4–5 cfu/mL (yeasts and molds) in fresh sap plays a major role in fermentation sap.

The composition of fresh nypa palm sap after 6 h of tapping the stalk of the fruit head is given in Table 10.5.

As the sap drips from the tapping hole, it is contaminated by microorganisms from the bark. During the collection, the sugary sap is highly susceptible to spontaneous yeast–lactic acid fermentation and the process is rapid in Sun light (Borse et al., 2007). Due to this contamination, it ferments rapidly, losing its sweet taste and becomes sour and milky-white within 24 h (Uraih and Izuagbe, 1990; Igbinadolor, 2009). The origin of microorganisms, microbiology of palm wine and factors affecting the quality of palm wine have also been examined (Faparusi, 1973; Faparusi and Bassir, 1971, 1972).

Table 10.5 Physico-Chemical Composition and Microbiological Count of Fresh Nypa Sap

COMPOSITION	VALUE
pH	5.33–6.32
Total soluble solids (°B)	14.6–15.7
Sucrose (g/L)	130.8–170.1
Glucose (g/L)	4.7–11.7
Fructose (g/L)	9.9–18.6
Ethanol (%)	0.02–0.10
Acetic acid (%)	0.6–1.0
Total aerobic count (cfu/mL)	Log 7–8
Yeast and mold count (cfu/mL)	Log 4.5–5
Lactic, succinic and propionic acids	Trace amount

10.7.2.2 Fermentation The first fermentation is alcoholic fermentation mainly to produce alcoholic liquid or palm wine. Palm sap fermentation involves alcoholic–lactic–acetic acid fermentation by environmental yeasts and LAB (Okafor, 1975, 1978; Law et al., 2011; Ciani et al., 2012). The fermenting microorganisms are dominated by yeast particulary *Saccharomyces cerevisiae* and lactic acid bacteria. In general, lactic acid bacteria produce very little ethanol while it is the yeast produce ethanol as a major end product in metabolism (Lindsay, 1996). It is carried out in a HDPE cylindrical container, measuring 90 cm (height) × 60 cm (diameter) (Gupta and Kushwaha, 2011), which is covered with a piece of cloth to prevent any insects or dirt from getting inside the container while still allowing in air for acetification process. The trees after cutting are allowed to remain as such for two weeks and the sap is collect for upto 8 weeks (Hartley, 1984). The sap is left to ferment by itself for about 2 months at a temperature between 25°C and 32°C. After fermentation is complete, the nypa vinegar is filtered with a sieve to get clear vinegar liquid. The vinegar is then, manually bottled in plastic bottles and is ready to be sold. In another method, the sap is spontaneously fermented by yeast, mainly *Saccharomyces cerevisiae* (Amoa-Awua et al., 2007), in the sap collectors during tapping, representing a method of continuous production. The alcoholic fermentation of the sap is usually spontaneous and starts a few hours after sap collection. It takes around 2 days to complete the fermentation, reaching ethanol concentration of above 7% (v/v) (Amoa-Awua et al., 2007). In alcoholic fermentation, *Saccharomyces* spp. present in the natural fermentation palm sap are important as it contributes to the fermentation of aroma (Aidoo et al., 2006). *S. cerevisiae* and *Schizo pombe* have reported to be the dominant yeast-species (Odunfa and Oyewole, 1998). See Atputharajah et al. (1986) for more details on microbiology of palm fermentation. The second fermentation or acetification also takes place spontaneously. The acetification process takes about 4 days to be completed, and the acetic acid concentration is around 4% (v/v) (Battcock and Azam-Ali, 1998; Gonzalez, 2009). Changes in the physico-chemical and sensory characteristics are summarized in Table 10.6.

Table 10.6 Changes in the Taste, Alcoholic Content, Sugar, and Acid Content during Spontaneous Oxidation of Palm Wine

CHARACTERISTICS	0 H	3 H	6–12 H	24 H
Taste	Fresh, sweet	Sweet	Slightly sour	Very sour
Alcohol (%)	3.78	4.84	6.32	6.70
Sucrose (%)	6.80	4.18	1.48	0.35
Acetic acid (%)	0.49	0.54	0.57	0.69

Source: Modified from Chinnarasa, E. 1968. *The Preservation and Bottling of Palm Wine.* Research Report No. 38. Federal Institute of Industrial Research, Nigeria; With kind permission from Springer Science+Business Media: *Vinegars of the World*, Other tropical fruit vinegars, 2009, pp. 261–270, Igbinadolor, R.O.

H = Hours after collection at 28–30°C.

10.7.2.3 Quality of Palm Vinegar The final composition of palm vinegar is influenced by the raw material employed, as well as by the technology used (Tesfaye et al., 2002).

10.7.3 Coconut Water Vinegar

Coconut trees are the most extensively grown nut trees, and provide the inhabitants with the raw materials for their basic needs such as food, drinks, furniture, etc. The Coconut-based technologies require very simple, locally available, materials and equipment, and are quite easy to perform (Banzon and Velasco, 1982). These factors qualify coconut water a very appropriate primary product for bioconversion into vinegar in rural areas, as no sophisticated apparatus or equipment is required. Coconut water is a waste produced in copra making process or in desiccated coconut factories. Thus, vinegar production gives an added value to its use and also provides an extra source of income for coconut farming communities (Punchihewa, 1997; Gonzalez, 2009). See Borse et al. (2007) for chemical composition of coconut sap (*Neera*).

Coconut water vinegar is produced by allowing coconut water filtered through a cloth to undergo alcoholic fermentation and subsequent, acetification at ambient temperature, followed by adjusting the sugar content to 162 g/L by the addition of refined sugar (normal sugar content of coconut water is around 1% w/v). The liquid is then, pasteurized and, once cooled, it is inoculated with a source of yeast, such as active dry yeast, for an alcoholic fermentation, taking 5–7 days. Normally, a spontaneous alcoholic fermentation takes place without the pasteurization step. After fermentation, the alcoholic coconut water is transferred to other containers, usually made of plastic or stainless steel, and "mother of vinegar" or a starter culture of AAB is added. The container is not fully filled to allow head space for aeration, as AAB are strictly aerobic microorganisms, thus ensuring a higher effectiveness of the process. The mixture is stirred and covered with a cloth. The acetification process takes about 7 days (Punchihewa, 1997). The vinegar is clarified with egg albumin (2 egg white/10 L vinegar) (Truong, 1987). After addition, the vinegar is heated, allowed to stand, and then decanted. The clarified vinegar is aged for 3–4 weeks prior to bottling and pasteurization. The vinegar so treated remained stable for a year and is accepted by consumers for different uses in India. The final product contains between 3% and 4% (v/v) acetic acid and is an indispensable commodity in many households (Sanchez, 1990). Vinegar production from coconut water is also a source of income for the producers (Batugal and Ramanatha Rao, 1998).

10.7.4 Mango Vinegar

The mango tree (*Mangifera indica*) is appreciated in many countries for its fruit (Rey et al., 2006). The production is concentrated mainly in Asia, more precisely in India. The production volumes of this fruit are high, and large quantities are often wasted

(Akubor, 1996). Those culled mango fruits can be used for the production of fermented products, like vinegar, which is produced from mango pulp by alcoholic fermentation as a first step and acetification as a second step (Gonzalez, 2009). Figure 10.7 outlines the general unit operations for the preparation of vinegar from fruits including mango. The pulp is usually diluted with water (in a pulp:water ratio of 1:5) and filtered through a filter cloth to obtain the juice (Garg et al., 1995), which normally contains 3.6% (w/v) of total sugars. Therefore, more sugar is added to the filtered juice, up to

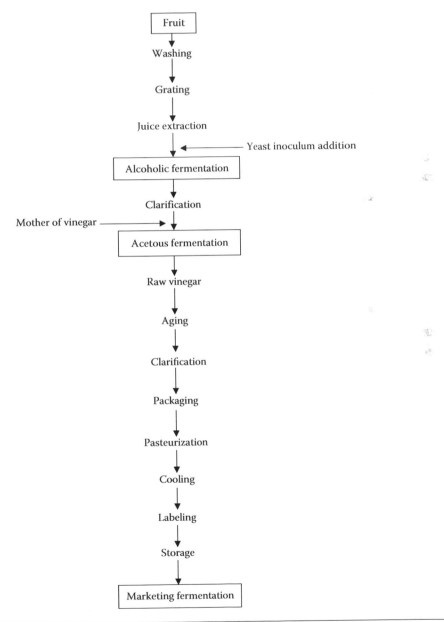

Figure 10.7 Unit operations for preparation of fruit vinegar. (Adapted from Joshi, V.K. and Thakur, N.S. 2000. *Postharvest Technology of Fruits and Vegetables.* Indus Publishing Co., New Delhi.)

approximately 200 g/L to obtain sufficient conversion of sugar into ethanol. The juice is pasteurized and the fermentation is carried out by the addition of a starter culture of AAB, and lasts for 14 days, yielding around 8% (v/v) of ethanol (Akubor, 1996). If the fermentation continues for up to 3 weeks, the ethanol concentration can increase up to 9% (v/v), probably due to the additional conversion of malic acid into ethanol (Kunkee and Amerine, 1970; Akubor, 1996). The mango wine is then transferred to other containers for the acetification process. "Mother of vinegar" is added as a starter. Another technique used for this second process is the use of immobilized cells of AAB on wood shavings (Garg et al., 1995). In this technique the immobilized microorganism is *Acetobacter aceti*, together with *Acetobacter pasteurianus*—the most common AAB found in acetification processes. *Acetobacter lovaniensis* has also been isolated from mangoes (Lisdiyanti et al., 2001). It could be a potential starter for mango vinegar production, and mango vinegar has an acidity of around 5.3% (v/v) as acetic acid (Gonzalez, 2009).

The fermentable sugar adhering to the fruit processing waste materials is an ideal substrate for fermentation. Peel and stones from mango waste, 20%–30% of the fruit, have been made into vinegar. Direct fermentation of this waste yields more alcohol than fermentation of cold or hot water extract. However, the alcoholic content being less (2.5%–3.5%) is raised to 5%, either by fortification with alcohol or by conducting fermentation by addition of cane sugar, for vinegar production (Gonzalez, 2009). The basic process for mango vinegar production is, however, the same as outlined earlier. The batch type process takes 3–12 days for completion. Good quality vinegar with 4.5%–5.0% acetic acid can be obtained, with a characteristic mango flavor. Direct fermentation can be adopted as a standard method for fermentation, as the hot or cold extraction method yields thick pulpy slurry which is difficult to ferment (Ethiraj and Suresh, 1992).

10.7.5 Honey Vinegar

Honey has traditionally been used mainly to produce alcoholic beverage, and the yeasts commonly carrying out the alcoholic fermentation are osmophiles, such as *Zygosaccharomyces*. Vinegar is produced from honey wine by acetification with AAB. Its production, however, is hardly feasible economically, mainly due to the time required and the cost of honey. It is commonly produced for home consumption only (Krell, 1996; Gonzalez, 2009).

10.7.6 Plum Vinegar

Surplus and substandard plums can be processed into quality vinegar by fermentation (Grewal and Tewari, 1990). To prepare vinegar, alcoholic fermentation of plum varieties *Alubokhara* and *Katruchak* is carried out by using *Saccharomyces cerevisiae* var. *ellipsoideus*, and the maximum alcoholic yields obtained were 10.26%–10.37% (v/v), respectively. Using *Acetobacter aceti*, the wine was converted into vinegar. The maximum fermentation efficiency was obtained by using inocula at the rate of 10% (v/v), 1.5% (w/v) initial acidity, and 6.0% ethanol concentration at a temperature of 30°C. Initial ethanol concentration has a large influence on acetic acid fermentation. Other details of vinegar production are similar to other fruit vinegar production.

10.7.7 Banana Vinegar

Bananas are a widespread fruit and have a short shelf-life, with a rapid rate of deterioration (Akubor et al., 2003). Fermenting banana juice is, therefore, considered to be an attractive way of utilizing culled and over-ripe bananas to make products with economic benefits. The main fermented products of these include vinegar. The various steps in the production of banana vinegar include washing, peeling, and slicing the fruits. The slices are blended and boiled with water and then, filtered (Akubor et al., 2003). Prior to the alcoholic fermentation, sugar is usually added. The next steps of the vinegar production process are similar to other vinegars, as described above (Gonzalez, 2009). The detailed method is described in a flow sheet in (Figure 10.8).

10.7.8 Pineapple Vinegar

To make pineapple vinegar, pineapple peel is passed through a crusher, followed by pressing in a hydraulic press; the juice is ameliorated to 10°B with a sugar source and mixed with a 0.05% ammonium phosphate solution which is mixed with a starter culture made with sterilized juice inoculated with Brewer's yeast. The acetification is carried out using two types of reactors. The fermented juice mixed with mother of vinegar (3:1) is transferred to a generator and allowed to undergo acetification at 34°C. After 4 days, the liquid is siphoned-off (Satyavati et al., 1972). To improve the traditional process of vinegar production, several solutions have been proposed, such as fine filtration of fermented fruit juice (Kubota et al., 1988; Kubota, 1989), use of a membrane recycling bioreactor for both ethanol and acetic fermentation (Mehaia and Cheryan, 1991), and application of immobilized *Acetobacter* for acetification.

10.7.9 Date Vinegar

The lack of acidity of dates makes it less attractive for production of vinegar (Barreveld, 1993). For its production, water is added to the dates and the mixture is boiled in the case of a non-spontaneous alcoholic fermentation (Ali and Dirar, 1984). The acetification of the date wine is carried out by adding *"mother of vinegar"* to the wine (Gonzalez, 2009). Fermentation of dates is done using both batch and continuous membrane reactors. Using a membrane recycle bioreactor with 150 g/L cell density, the sugar present is completely consumed and an ethanol content of 68 g/L (eight times more than conventional batch processing) can be obtained (Mehaia and Cheryan, 1991). The final vinegar commonly has 4%–5% (v/v) acidity as acetic acid (Gonzalez, 2009).

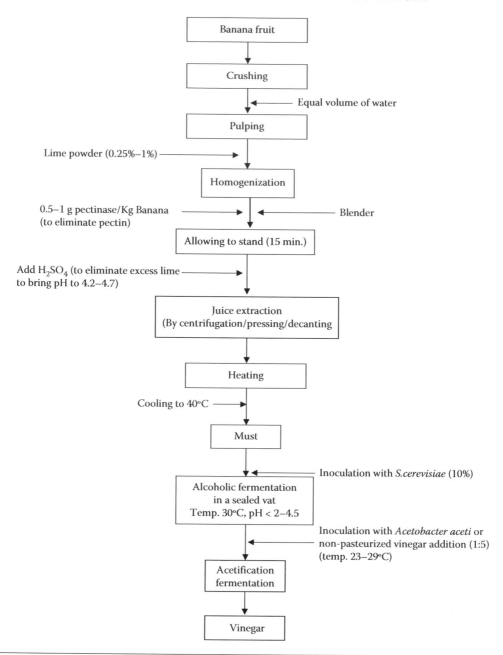

Figure 10.8 Flow diagram for production of banana vinegar. (Adapted from Joshi, V.K. and Thakur, N.S. 2000. *Postharvest Technology of Fruits and Vegetables.* Indus Publishing Co., New Delhi; from Springer Science+Business Media: *Vinegars of the World*, Other tropical fruit vinegars, 2009, pp. 261–270, Igbinadolor, R.O.)

10.7.10 Bamboo Vinegar

Bamboo vinegar is obtained, as in the case of palm vinegar, from the sap of the plant. The process of obtaining the sap—and so the vinegar—is, therefore, very similar to that used in the production of palm vinegar. The main difference is that the alcoholic fermentation of bamboo sap is extremely fast, taking 5–12 h to obtain bamboo wine.

The bamboo wine is converted to vinegar by natural fermentation by acetic acid bacteria. The speed of the fermentation confers an added value to bamboo vinegar production (Gonzalez, 2009).

10.7.11 Cocoa Vinegar

Cocoa mucilage (sweatings) is a pale yellowish liquid, which is a waste by-product of the cocoa industry. It is derived from the breakdown product of the mucilage (pulp) surrounding the fresh cocoa beans of the tree *Theobroma cacao*, and constitutes about 10% of the weight of the cocoa fruit (Adams et al., 1982). Cocoa sweatings are a by-product obtained from traditional cocoa fermentation, obtained by the activity of pectolytic enzymes which are secreted by some microorganisms involved in the fermentation process (Ansah and Dzogbefia, 1990). It is rich in soluble sugars and pectin, besides other components (Table 10.7) suitable for producing alcoholic drinks, vinegar, etc. (Opeke, 2005). Cocoa mucilage is free of alkaloids and other toxic substances (Igbinadolor, 2009).

The initial pH of the pulp is approximately 3.6 due to its high citric acid content, which favors the growth of yeasts, together with a low level of oxygen. Different yeast species contaminate cocoa pulp, mainly *Kloeckera apiculata*, *Kluyveromyces marxianus*, *Saccharomyces cerevisiae*, *Pichia fermentans*, *Lodderomyces elongisporus*, and *Candida bombi*. During the fermentation process, *Saccharomyces cerevisiae* is the most dominant yeast species due to its ability to tolerate high ethanol concentrations. Therefore, *Saccharomyces cerevisiae* is the yeast most frequently used in fermentation of most fruit juices (Prashant and Rajendra, 1989). The succession of different yeast species active during alcoholic fermentation of cocoa pulp is summarized in Figure 10.9 (Igbinadolor, 2009).

Alcoholic fermentation of cocoa mucilage is carried out in relatively simple vessels, such as open vats of wood or concrete, or earthenware pots, and without any form of

Table 10.7 Chemical Composition of Cocoa Mucilage

COMPOSITION	PERCENTAGE
Water	79.2–84.2
Dry substances	15.8–20.8
Nonvolatile acids	0.77–1.50
Volatile acids	0.02–0.04
Glucose	5.32–11.60
Sucrose	0.11–0.90
Pectin	5.00–6.90
Protein	0.42–0.50
Ash	0.40–0.50

Source: From Adams, M.R. et al. 1982. *Agric Wastes*, 4, 225–229. With permission.

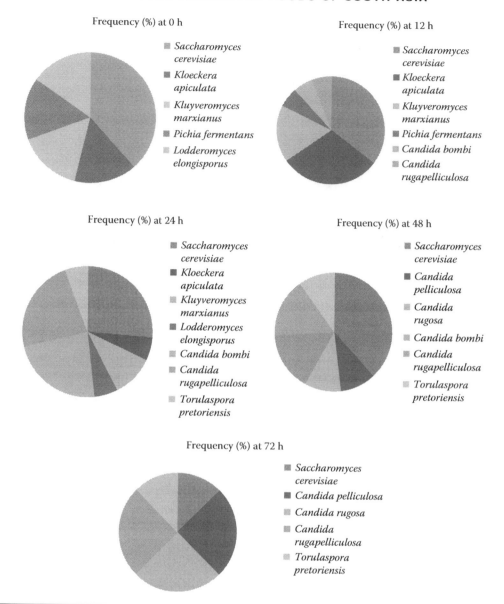

Figure 10.9 Frequencies (%) of species isolated during cocoa fermentation. (With kind permission from Springer Science+Business Media: Other Tropical Fruit Vinegars, *Vinegars of the World*, 2009, pp. 261–270, Igbinadolor, R.O.)

temperature control, particularly when small vessels are used and the ambient temperature is suitable for yeast growth (25–30°C). No starter cultures are generally used. The alcoholic fermentation generally ends within 48–72 h. When fermentation is complete, the alcoholic cocoa liquid may be centrifuged to remove yeast cells and then, mixed with a proportion of suitable "seed" vinegar or "mother of vinegar," which is generally a portion of a previous successful acetification. The AAB involved in vinegar production belong mainly to the genera *Acetobacter* and *Gluconobacter* (Sievers and Swings, 2005; Igbinadolor, 2009).

10.7.12 Cashew Vinegar

The cashew tree *(Anacardium occidentale)* is a medium-sized fruit-bearing tree, and its fruit consists of a kidney-shaped nut and a pseudoapple with a brilliant yellow or red skin color. The cashew apple is five to ten times as heavy as the nut when ripe and contains 85% juice with 10% sugar, most of which is invert sugar (Ohler, 1979).

Cashew juice is obtained by removing the nut from the apples, cutting the fruits into small pieces, and squeezing them by hand or with a machine. The juice is often clarified by filtering through a sieve. A large number of different yeasts can colonize the cashew juice due to its low pH (3.8–4.0). So it is pasteurized and fermentation is started by inoculating it with a desired yeast starter, such as *Saccharomyces cerevisiae*. After fermentation, back-slopping with seed vinegar harvested from the previous batch is carried out to produce cashew vinegar.

The clarification can be effected by filtration or by fining. Generally, filtration is preferred, as it reduces the bacterial population and removes vinegar eels, if present (Cruess, 1958; Frazier and Westhoff, 1978; Adams et al., 1982). In the fining method, clarifying agents such as casein, gelatin, bentonite, sodium alginate, and isinglass are mixed with the cashew vinegar and the mixture is allowed to stand until clear (Prescott and Dunn, 1959). The clarified cashew vinegar is pasteurized. The bottles should be completely filled and tightly capped or corked with treated corks to prevent the entry of air. The temperature and time of pasteurization vary according to microbial contamination and conditions of filling. Generally, a temperature of 60–66°C for about 30 minutes is adequate for cashew vinegar pasteurization (Igbinadolor, 2009).

10.7.13 Sugarcane Juice Vinegar

Natural vinegar can be prepared from a variety of substrates having fermentable sugar, such as sugarcane juice, which is available in many states of India, including Uttar Pradesh, Punjab, Himachal Pradesh, and Haryana (Joshi and Thakur, 2000). In these states, at the village level, sugarcane juice is fermented to make sugarcane juice vinegar for home consumption. However, it is not produced at any industrial scale. The process is similar to other vinegar production, employing alcoholic and acetous fermentation. The former is carried out by yeast and latter by acetic acid bacteria. Mostly the fermentation is carried out by natural microflora. Evaluation of the efficiency of different species and isolates showed *Acetobacter aceti* to produce the highest amount of acetic acid (4.5% w/w) in sugarcane vinegar production in 69 days under constant conditions (Tewari et al., 1991; Vegas et al., 2010). The yield of acetic acid was improved by adjusting the pH of mash to 3.5 (Lee et al., 1999). However, the addition of ammonium sulphate, ammonium phosphate, and peptone inhibited vinegar fermentation.

Vinegar made by the traditional method generally develops *Acetobacter xylinum*, vinegar eels, etc., which contaminate and spoil the product. Pasteurized clarified cane-juice fermentation with *Sacharomyces cerevisiae* var. *ellipsoideus* and *Acetobacter aceti* in earthen pitchers under uncontrolled conditions yielded 8.3% (v/v) acetic acid, and the vinegar produced was found to be commercially acceptable (Tewari et al., 1991).

10.8 Socioeconomic Aspects of Vinegar Production

In the South Asian countries, there is no organized production of vinegar at industrial scale. Most of the production is undertaken at home scale and for local consumption. However, vinegar production is a small industry in the overall economy of industrialized countries (Adams, 1998). Global shares of the different kinds of vinegar during 2005 are compared in Figure 10.10 (Vinegar Institute, 2006; Solieri and Giudici, 2009). The production statistics indicate that vinegar production in South Asian countries, if taken up as an industrial venture, would have large potential for revenue generation. Besides, it can save a lot of fruit from wastage, as at present 25%–30% of such production is wasted after harvest in South Asian countries, including India (Verma and Joshi, 2000). It can easily be developed as a small-scale industry initially. Going by the utility of vinegar, it should not pose any problem for marketing. In developing countries or South Asian countries, where food preservation and technology options are limited, vinegar is an important agent for preserving fresh fruit and vegetables from rapid deterioration, especially in the tropics, where the environmental conditions accelerate food spoilage. Developing and improving small-scale vinegar production, and food fermentation technologies in general, are one of the goals of the FAO (Anonymous, 1995; FAO, 1998; Solieri and Giudici, 2009).

10.9 Summary and Future Projection

The development of agriculture in South Asian Countries is creating significant quantities of surplus food products that small farms are not able to utilize. These wastes are an excellent raw material for the production of vinegar, but economic investment in factories for such processes has been very slow. The availability of a wide variety

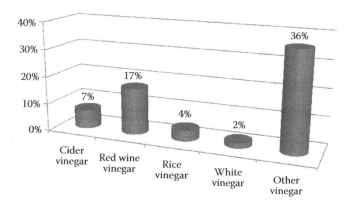

Figure 10.10   Global shares (%) of different vinegar types in 2005. (From Vinegar Institute. 2006. Market Trends. Available at www.vcrsatilevinegar.org/markeltrends.html; With kind permission from Springer Science+Business Media: *Vinegars of the World*, 2009, pp. 1–16, Solieri, L. and Giudici, P. (eds).)

of foods for biotransformation into vinegar is advantageous for the farmers (fitting-in well with their main agricultural activities) and helps in the conservation of natural resources. However, the large number of different raw materials used makes it impossible to introduce common techniques and processes to the producers, which increases the cost of vinegar production.

There are several technological factors to take into account for vinegar production in South Asian countries, such as room temperature, species of AAB used, amount of water, etc. The fermentation temperature for vinegar production falls within a narrow range, so an increase of 2–3°C causes a serious deterioration in both the acetification ratio and efficiency. Not only that; in submerged cultures, large amounts of heat are generated, so that cooling costs become rather high. The need for cooling of the *bioreactors* (which is normally performed with ordinary water) increases the overall economic cost of vinegar production. Another problem relates to the AAB as reliable starter cultures of AAB can accelerate the fermentation process and improve control over it, as well as prevent contamination with unwanted substances. In turn, this increases the production of high-quality vinegars safe for human consumption. There is need for proper solutions to all these problems.

The possibility of standardization of the production of such vinegars, as well as stimulating economic development, could be very important when specialized products from particular raw materials are produced. Such value-added products represent an attractive niche market, both because of the economic benefits and also, in particular, from a gastronomical point of view. Vinegar is an important ingredient in a variety of commercial products, such as ketchup, mayonnaise, sauces, and mustards. Therefore, local vinegar production, particularly if it uses agricultural waste products, should be encouraged.

References

Adachi, O., Moonmangmee, D., Toyama, H., Yamada, M., Shiagawa, E., and Matsuchita, K. 2003. New developments in oxidative fermentation. *Appl Microbiol Biotechnol*, 60, 643–653.

Adams, F. 1849. Hippocrates. *Genuine Works of Hippocrates* (transl. by F. Adams). Lydenham Society, London, Vol. I, p. 301.

Adams, M.R. 1978. Small scale vinegar production from banana. *Trop Sci*, 20, 11–19.

Adams, M.R. 1998. Vinegar. In: Wood, B.J.B. (ed.) *Microbiology of Fermented Food*. Blackie Academic and Professional, London, pp. 1–44.

Adams, M.R., Drugan, J., Glossop, E.J., and Twiddy, D.R. 1982. Cocoa sweatings: An effluent of potent value. *Agric Wastes*, 4, 225–229.

Adams, S.L. and Nielsen, G.V. 1963. Beverage from honey and process for making it. United States Patent Office, 3.100.706.

Adrian, J., Potus, J., and Frangc, R. 1995. *La Science alimentaire de A a Z. Collection Technique et Documentation*, Lavoisier, Paris, p. 477.

Aidoo, K.E., Nout, M.J.R., and Sarkar, P.K. 2006. Occurrence and function of yeasts in Asian indigenous fermented foods. *FEMS Yeast Res*, 6, 30–39.

Akinwale, T.O. 2000. Extraction of Pulp from Fresh Cocoa Beans for Wine Production. Anni Report of Cocoa Research Institute of Nigeria (CRIN).

Akubor, P.I. 1996. The suitability of African bush mango juice for wine production. *Plant Foods Hum Nutr*, 49, 213–219.

Akubor, P.I., Obio, S.O., Nwadomere, K.A., and Obiomah, E. 2003. Production and quality evaluation of banana wine. *Plant Foods Hum Nutr*, 58, 1–6.

Ali, M.Z. and Dirar, H.A. 1984. A microbiological study of Sudanese date wines. *J Food Sci*, 49, 459–460.

Amoa-Awua, W.K., Sampson, E., and Tano-Debrah, K. 2007. Growth of yeasts, lactic acid and acetic acid bacteria in palm wine during tapping and fermentation from felled oil palm (*Elaeis guinecnsis*) in Ghana. *J Appl Microbiol*, 102, 599–606.

Anonymous. 1971. *Coconut Vinegar*. Cylon Ind. Dev. Board, Ind. Prospect Rep., p. 25.

Anonymous. 1995. http://www.answers.com/topic/vinegar

Anonymous. 2008. http://gleez.com/articles/food/all-about-vinegar

Ansah, F. and Dzogbefia, V.P. 1990. The role of microbial and endogenous pectolytic enzymes in cocoa fermentation. *J Univ Sci Technol Kumasi Ghana*, 10, 69–74.

Atputharajah, J.D., Widanapathiranat, S., and Samarajeewa, U. 1986. Microbiology and biochemistry of natural fermentation of coconut palm sap. *J Food Microbiol*, 3, 273–280.

Ayernor, G.K.S. and Matthews, J.S. 1971. The sap of the palm *Elaeis guineensis Jacq* as raw material for alcoholic fermentation in Ghana. *Trop Sci*, 13, 71–83.

Bahiru, B., Mchari, T., and Ashenafi, M. 2006. Yeast and lactic acid flora of tej, an indigenous Ethiopian honey wine: Variations within and between production units. *Food Microbiol*, 23, 277–282.

Bamforth C.W. Ed. 2006. *New Brewing Technologies: Setting the Scene*. Woodhead Publishing Limited, Cambridge, Vol 126, pp. 1–9.

Banzon, J.A. and Velasco, J.R. 1982. *Coconut: Production and Utilisation*. Coconut Research and Development Foundation, Pasig, Metro Manila.

Barreveld, W.H. 1993. *Date Palm Products*. FAO Agricultural Services Bulletin no. 101. Food and Agriculture Organization of the United Nations, Rome.

Battcock, M. and Azam-Ali, S. 1998. *Fermented Fruits and Vegetables: A Global Perspective*. FAO Services Bulletin no. 134. Food and Agriculture Organization of the United Nations, Rome.

Batugal, P.A. and Ramanatha Rao, V. 1998. *Coconut Breeding*. Proceedings: Standardization of Coconut Breeding Research Techniques, 20–25 June 1994, Port Bouet, Cote d'Ivoire. Serdang: International Plant Genetic Resources Institute—Regional Office for Asia, the Pacific and Oceania.

Bereau International due, Travail. 1990. Conservation des fruitsapetite echclle. Serie Technologique, dossier technique no.14, Geneva.

Bo, T.A. 1988a. Brewing techniques for Chinese vinegars: 1. *Jyozo Kyokai Schi*, 83, 462–471 (in Japanese).

Bo, T.A. 1988b. Brewing techniques for Chinese vinegars: 2. *Jyozo Kyokai Schi*, 83, 534–542 (in Japanese).

Board NIIR. 2003. *The Complete Technology Book on Processing, Dehydration, Canning, Preservation of Fruits and Vegetables*. National Institute of Industrial Research, p. 862.

Boerhaave, H. 1732. Glementa Chemicae, Lugduni Batavorum, 2, 179 (cited by Lafar (1898) and by Mitchell, 1926).

Boesch, C., Track, J., Sievers, M., and Teuber, M. 1998. *Acetobacter intermedius*, sp. *nov. Syst Appl Microbiol*, 21, 220–229.

Borse, B.B., Rao, L.J.M., Ramalakshmi, K., and Raghavan, B. 2007. Chemical composition of volatiles from coconut sap ("neera") and effect of processing. *Food Chem*, 101, 877–880.

Brown, G.D. and Rainbow, C. 1956. Nutritional patterns in acetic acid bacteria. *J Gen Microbiol*, 15, 61–69.

Burgoon, D.W., Ciabattari, E.J., and Yeomans, C. 1960. *Mixing Apparatus*. US Patent 2,966,345 Dec 27.

Callejón, R.M., Tesfaye, W., Torija, M.J., Mas, A., Troncoso, A.M., and Morales, M.L. 2009. Volatile compounds in red wine vinegars obtained by submerged and surface acetification in different woods. *Food Chem*, 113(4), 1252–1259.

Cavalieri, D., McGovern, P.E., Hartl, D.L., Mortimer, R., and Polsinelli, M. 2003. Evidence for *S. cerevisiae* fermentation in ancient wine. *J Mol Evol*, 57, 226–232.

Chinese National Standard (CNS). 2005. Edible Vinegar. No. 14834, N52.

Chinnarasa, E. 1968. *The Preservation and Bottling of Palm Wine*. Research Report No. 38. Federal Institute of Industrial Research, Nigeria.

Ciani, M., Stringini, M., and Comitini, F. 2012. Palm wine. In: *Handbook of Plant-based Fermented Food and Beverage Technology*. Y. H. Hui and E.O. Evranuz (eds). 2nd edition, CRC Press, Boca Raton, FL, pp. 631–638.

Clark, J. and Goldblith, S.A. 1974. *Processing and Manufacturing of Food in Ancient Rome*, contribution No. 2327. Department in Nutrition and Food Science, Massachusetts Institute of Technology, Cambridge.

Cleenwerck, I., Vandemeulebroccke, K., Janssens, D., and Swings, J. 2002. Re-examination of the genus Acetobacter, with descriptions of *Acetobacter cerevisiae* sp. *nov.* and *Acetobacter malorum* sp.*nov*. *Int J Syst Evol Microbiol*, 52, 1551–1558.

Conner, H.A. and Allgeier, R.J. 1976. Vinegar: Its history and development. *Adv Appl Microbiol*, 20, 81–133.

Crawford, S. L. 1929. Fruit prod. *J Am Vinegar Ind*, 9, 80–86 (cited by Conner and Allgeier, 1976).

Cruess, W.V. 1958. *Commercial Fruit and vegetable Products: A Textbook for Student, Investigator and Manufacturer*, 4th edition, McGraw-Hill, New York, NY.

De Ley, J. 1959. On the formation of acetoin by *Acetobacter*. *J Gen Microbiol*, 21, 352–365.

De Ley, J. 1961. Comparative carbohydrate metabolism and a proposal for a phylogenetic relationship of the acetic acid bacteria. *J Gen Microbiol*, 24, 31–50.

De Ory, L., Romero, L.E., and Cantero, D. 1999. Maximum yield of acetic acid fermenter. *Bioproc Eng*, 21, 187–190.

De Vero, L., Gala, E., Gullo, M., Solieri, L., Landi, S. and Giudici, P. 2006. Application of denaturing gradient gel electrophoresis (DGGE) analysis to evaluate acetic acid bacteria in traditional balsamic vinegar. *Food Microbiol*, 23, 809–813.

De Vuyst, L. 2000. Technology aspects related to the application of functional starter cultures. *Food Technol Biotechnol*, 38, 105–112.

Duddington, C.L. 1961. *Microorganisms as Allies: The Industrial Use of Fungi and Bacteria*. Macmillan, New York, NY.

Dussance, H. 1871. *A General Treatise on the Manufacture of Vinegar*. Baird, Philadelphia, PA.

Ebner, H. 1967. Latest development in the technical verification of the submerged vinegar fermentat Zentralbl Bakteriol, Parasitenkd, Infektionskr. *Hyg Abt 1: Orig Suppl*, 2, 65.

Ebner, H. 1982. Vinegar. In: Reed, G., Prescott, S.C. and Dunn, C.G. (eds) *Industrial Microbiology*. 4th edition. AVI Publishing Co., Westport, CT, pp. 802–834.

Ebner, H., Follmann, H. and Sellmer, S. 1996. *Vinegar, Ullmann's Encyclopedia of Industrial Chemistry*. Wiley-VCH Verlag GmbH & Co. KgaA, Weinheim, Vol A27, pp. 403–418.

Ebner, H., Pohl, K., and Enenkel, A. 1967. Self-priming aerator and mechanical defaomer for microbiological processes. *Biotechnol Bioeng*, 9, 357–364.

Ethiraj, S. and Suresh, E.R. 1992. Studies on the utilization of mango processing waste for production of vinegar. *Food Sci Technol Abstract*, 24, 10.

Ezeronye, O.U. and Okerentugba, P.O. 2001. Genetic and physiological variants of yeast selected from palm wine. *Mycopathologia*, 159, 85–89.

FAO, 1998. *Fermented Fruits and Vegetables: A Global Perspective* [available at www.fao.org].

Faparusi, S.I. 1973. Origin of initial microflora of palm wine from oil palm trees (*Elaeis guineensis*). *J Appl Bacteriol*, 36, 559–565.

Faparusi, S.I. and Bassir, O. 1971. Microbiology of fermenting palm wine. *J Food Sci Technol*, 8, 206.

Faparusi, S.I. and Bassir, O. 1972. Factors affecting the quality of palm wine. I. Period of tapping a palm tree. *W Afr J Biol Appl Chem*, 15, 17–23.

Flandrin, J.L., Montanari, M., and Sonnenfeld, A. 2000. *Food: A Culinary History*. Columbia University Press, New York, NY.

Fleet, G.H. 1998. The microbiology of alcoholic beverages. In: Wood, B.J.B. (ed.) *Microbiology of Fermented Food*. Blackie Academic and Professional, London, Vol 1, pp. 217–262.

Fong, S. 2000. Chinese vinegar. *Flav Fort*, 7, 5–24.

Food and Drug Administration (FDA). 2007. FDA/ORA Compliance Policy Guides, Sec. 525.825 Vinegar, Definitions: Adulteration with Vinegar Eels (CPG 7109.22) available at www.fda.gov/ora/compliance_ref/cpg/cpgfod/cpg525–825.htmll.

Fowler, G.R. and Schwartz, M.H. 1973. In: Preece, W.E. (ed.) *Encyclopedi Britannica*. William Benton, IL, Vol 6, pp. 657–659.

Frazier, W.C. and Westhoff, D.C. 1978. *Food Microbiology*. McGraw-Hill, New York, NY.

Frings, H. 1932. *Manufacture of Vinegar*. U.S. Patent.1880, 3811.

Frings, H. 1937. US Patent. 2,094,592.

Fushimi, T., Suruga, K., Oshima, Y., Fukiharu, M., Tsukamoto, Y., and Goda, T. 2006. Dietary acetic acid reduces serum cholesterol and triacylglycerols in rats fed a cholesterol-rich diet. *Br J Nutr* 95, 916–924.

Gamova-Kayukova, N.I. 1950. *Mikrobiologi*, 29, 137–148 (cited by Conner and Allgeier, 1976).

Garg, N., Tandor, D.K., and Kalra, S.K. 1995. Production of mango vinegar by immobilized *Acetobacter aceti. J Food Sci Tehcnol*, 32, 216–218.

Ghommidn, C., Cutayar, J.M., and Navdiro, J.M. 1986. A study of acetic acid production by immobilized *Acetobacter* cells oxygen transfer. *Biotechnol Bioeng Lett*, 8, 13.

Gilliland, R.B. and Lacey, J.P. 1966. *J Inst Brew*, London, 72, 291 (cited by Joshi and Thakur, 2000).

Gillis, M. and De Ley, J. 1980. Intra and intergenetic similarities of the ribonucleic acid cistrons of *Acetobacter* and *Gluconobacter. Int J Syst Bacteriol*, 30, 7.

Giudici, P., Gullo, M., Solieri, L., De Vero, L., Landi, S., Pulvirenti, A., and Ranieri, S. 2006. Le fermentazioni dell'aceto balsamico tradizionale.Stati du Luogo Diabasis, Reggio Emilia, Italy.

Gogui, A. 1856. Secrets de la cuisine francaise. Libraire de L. Hachette et Cie, Paris.

González, Á. 2009. Vinegars from tropical Africa. In: Solieri, L. and Gludici, P. (eds) *Vinegars of the World*. Springer-Verlag, Italy.

Gossele, F., Swings, J., Kersters, K, Pauwels, P., and De-Ley, J. 1983. Numerical analysis of phenotypic features and protein gel electrophoregrams of a wide variety of acetobacter strains. Proposal for the improvement of the taxonomy of the genus acetobacter beijerinck 1898, 215, Systematic. *Appl Microbiol*, 4(3), 338–368.

Greenshields, R.N. 1975a. Malt vinegar manufacture: I. *Brewer*, 61, 295–298.

Greenshields, R.N. 1975b. Malt vinegar manufacture: II. *Brewer*, 61, 401–407.

Greenshields, R.N. 1978. Acetic acid-Vinegar. In: Rose, A.H. (ed.) *Economic Microbiology. Primary Products of Metabolism*. Academic Press, New York, NY, Vol. 2, pp. 121–186.

Greenshields, R.N. and Smith, E.L. 1971. Tower fermentation systems and their applications. *Chem Eng*, 249, 182–190.

Grewal, H.S. and Tewari, H.K. 1990. Studies on vinegar production from plums. *J Res PAU*, 27, 272.

Gupta, N. and Kushwaha, H. 2011. Date palm as a source of bioethanol producing microorganisms. In: Jain, S.M., Jameel, M., Al-Khayri, and Dennis V.J. (eds.) *Date Palm Biotechnology*. Springer Publishers, Dordrecht, Heidelberg, New York, London, pp. 711–721.

Hartley, C.W.S. 1984. The Oil Palm. *Elaeis Guineensis Jacq*, 2nd edn. Longman Group.

Heikefelt, C. 2011. Chemical and sensory analyses of juice, cider and vinegar produced from different apple cultivars. Självständigt arbete vid LTJ-fakulteten, SLUDegree project in the Horticultural Science Programme, 30 ECTS.

Henneberg, W. 1897. Zentralbl. Bakteriol, Parasitenkd. Infektionskr Abst, 23, 933.

Henneberg, W. 1926. Handbook der Garungsbakteriologie—Berlin. (cited by Shimwell 1954).

Hesseltine, C.W. 1965. A millennium of fungi, food and fermentation. *Mycologia*, 57, 149–197.

Hesseltine, C.W. 1983. Microbiology of oriental fermented foods. *Ann Rev Microbiol*, 37, 575–601.

Hildebrandt, F.M., Nickol, G.B., Dukowicz, M., and Conner, H.A. 1961. *Vinegar*. Newsletter No. 33, US Ind. Chem. Chem Co., New York, NY.

Holt, L.J.M., Krieg, N.R., Sneath, P.H.A., Staley, J.Y., and Williams, S.T. 1994. Genus *Acetobacter Gluconobacter*. In: *Bergey's Manual of Determinative Bacteriology*, 9th edition. Williai Wilkins, Baltimore, MD, pp. 71–84.

Hromatka, O. and Ebner, H. 1955. Method for the preparation of vinegar acids by oxidative fermentation of alcohols. US Patent, 2707, 683.

Hromatka, O., Ebner, H., and Csoklich, C. 1951. Investigations of vinegar IV. About the influence of a total interruption of the aeration. *Enzymologia*, 15, 134–153.

Hromatka, O., Kastner, G., and Ebner, H. 1953. Investigations of vinegar about the influence of temperature and total concentration on the submerged fermentation. *Enzymologia*, 15, 337–350 (German).

Hubner, E. 1927. *Dtsch Essigind*, 31, 28 (cited by Joshi and Thakur, 2000).

Igbinadolor, R.O. 2009. Other tropical fruit vinegars. In: Solieri, L. and Gludici, P. (eds) *Vinegars of the World*. Springer-Verlag, Italy, pp. 261–270.

Iguchi, M., Yamanaka, S., and Budhiono, A. 2004. Bacterial cellulose, a masterpiece of nature's arts. *J Mater Sci*, 35, 261–270.

Ilha, E.C., Sant'Anna, E., Torres, R.C., Claudia Porto, A., and Meinert, E.M. 2000. Utilization of bee (*Apis mellifera*) honey for vinegar production at laboratory scale. *Acta Cient Venez*, 51, 231–235.

Jarvis, D.C. 1959. *Folk Medicine: A Vermont Doctors Guide to Good Health*. Hanry & Holt, New York, NY.

Jeong, Y.J., Seo, J.H., Lee, G.D., Park, N.Y., and Choi, T.H. 1999. The quality comparison of apple vinegar by two stages fermentation with commercial apple vinegar. *J Korean Soc Food Sci Nutr*, 28, 353–358.

Joshi, V.K. and Sharma, S. 2009. Cider vinegar: Microbiology, technology and quality. In: Solieri, L and Gludici, P. (eds) *Vinegars of the World*. Springer-Verlag, Italy, pp. 197–207.

Joshi, V.K. and Thakur, N.S. 2000. Vinegar: Composition and production. In: Verma, L.R. and Joshi, V.K. (eds) *Postharvest Technology of Fruits and Vegetables*. Indus Publishing Co., New Delhi, pp. 1128–1170.

Joslyn, M.A. 1970. Vinegar. In: *Kirk-Othmer Encyclopedia of Chemical Technology*. 2nd edn. John Wiley & Sons, New York, NY, Vol 21, pp. 254–269.

Kersters, K., Vos Paul, D., Gillis, M., Swings, J., Vandamme, P., and Stackebrandt, E. 2006. Introduction to the Proteobacteria. The family Acetobacteraceae: The genera *Acetobacter, Acidomonas, Asaia, Gluconacetobacter, Gluconobacter*, and *Kozakia, The Prokaryotes*, pp. 3–37.

King, T.E. and Cheldelin, V.H. 1952. Phosphorilative and non-phosphorilative oxidation in *Acetobacter suboxydans. J Biol Chem* 198, 135–141.

King, T.E., Kawasaki, E.H., and Cheldelin, V.H. 1956. Tricarboxylic acid cycle activity in *Acetobacter pasteurianum. J Bacteriol*, 72, 418–421.

Kocher, G.S., Kalra, K.L., and Tewari, H.K. 2003. Production of vinegar from cane juice. In: *Proceedings of Symposium on Food and Nutritional Security: Technological Interventions and Genetic options* (In late arrivals), Sept 18–19, HPKV, Palampur, India.

Kocher, G.S., Singh, R., and Kalra, K.L. 2007. Preparation of value added vinegar using apple juice. *J Food Sci Technol*, 44, 226–227.

Kozaki, M. and Sanchez, P.C. 1974. Fermented foods in Philippines. *Shokuhin Yoki*, 15, 66–79 (in Japanese).

Krell, R. 1996. *Value-added Products from Beekeeping*. FAO Agricultural Bulletin no. 124 and Agriculture Organization of the United Nations, Rome.

Kubota, T. 1989. *Vinegar production* (from fermented fruit or cereals) US Patent 4770881; Tamanoi, Osaka, Japan.

Kubota, T., Kato, H., Tanifiri, S., and Matsuda, H. 1988. *Processing for Production Vinegar*. US Patent 477081; Tamanoi vinegar, Osaka, Japan.

Kunkee, R.E. and Amerine, M.A. 1970. Yeasts in wine making. In: Rose, A.H. and Harrison, J.S. (eds) *Yeasts*. Academic Press, London, Vol. 3, pp. 6–71.

Lafar, F. 1898. *Technical Mycology*.Griffin, London, Vol. 1.

Lanzani, G.A. and Pecile, A. 1960. Activity of a substance extracted from the fermentation products of *Acetobacter aceti* on Ehrlich's ascites tumour cells. *Nature*, 185, 175–176.

Law, S.V., Abu Bakar, F., Mat Hashim, D., and Abdul Hamid, A. 2011. Mini review popular fermented foods and beverages in Southeast Asia. *Int Food Res J*, 18, 475–484.

Lea, A.G.H. 1988. Cider Vinegar. In: Downing, D.L. (ed.) *Processed Apple Products*, Van Nosand Reinhold, New York, NY, pp. 279–301.

Lee, F.L. Liang, S., and Chan, T.W. 1999. A thermotolerant and high acetic acid-producing bacterium Acetobacter sp. I14–2. *J Appl Microbiol*, 86(1), 55–62.

Leroy, F. and De Vuyst, L. 2004. Lactic acid bacteria as functional starter cultures for the food fermentation industry. *Trends Food Sci Technol*, 15, 67–78.

Lindsay, R.C. 1996. Flavors. In: Fenemma, O.R. (ed.) *Food Chemistry*. Marcel Dekker, Inc., New York, NY, pp. 723–765.

Lisdiyanti, P., Kawasaki, H., Seki, T., Yamada, Y., Uchimura, T., and Komagata, K. 2001. Identification of *Acetobacter* strains isolated from Indonesian sources and proposals of *Acetobacter syzgii*sp.nov., *Acetobacter cibinogenesis* sp. *nov* and *Acetobacter orientalis* sp. *nov*. *J Gen Appl Microbiol*, 47, 119–131.

Low, I. 1901. In: Singer, I. (ed.) *The Jewish Encyclopedia*. Funk and Wagnalls, New York, NY, Vol 12, p. 439.

Maldonado, O., Rol, C., and Schneider de Cabrera, S. 1975. Wine and vinegar from tropical fruits. *J Food Sci*, 40, 262–265.

Matthias, K., Wolfgang, S., Follmann, W., Heinrich, T., and Hans, G. 1989. Isolation and classification of acetic acid bacteria from high percentage vinegar fermentations. *Appl Microbiol Biotechnol*, 30(1), 47–52.

McGovern, P.E., Glusker, D.L., Exner, L.J., and Voigt, M.M. 1996. Neolithic resinated wine. *Nature*, 381, 480–481.

Mehaia, M.A. and Cheryan, M. 1991. Fermentation of date extracts to ethanol and vinegar in batch and continuos membrane reactors. *Enzyme Microbiol Technol*, 13, 257.

Mendonca, C.R.B., Granada, G.G., Rosa, V.P., and Zambiazi, R.C. 2002. Alternative vinegars: Physical, sensory and chemical characteristics. *Aliments Nutr*, 13, 35–47.

Mitchell, C.A. 1926. *Vinegar: Its Manufacture and Examination*, 2nd edition. Griffin, London.

Mori, A., Kunno, N., and Terni, G. 1970. Kinetic studies on submerged vinegar fermentation. I. Behaviour of *Acetobacter rancens* cells toward dissolved oxygen. *J Ferment Technol*, 48, 203–212 (in Japanese).

Mozingo, H.N. 1989. Palm. In: Holland, D.T. (ed) *The Encyclopedia Americana International Edition*. Grolier Incorporated, Danbury, CT, Vol 21, pp. 319–332.

Murooka, Y., Nanda, K., and Yamashita, M. 2009. Rice vinegars. In: *Vinegars of the World*. Springer-Verlag, Italia, pp. 121–133.

Nakayama, T. 1960. Studies on acetic acid bacteria. II. Intracellular distribution of enzymes related to acetic acid fermentation and some properties of a highly purified TPN-dependent aldehyde dehydrogenase. *J Biochem*, 48, 812–830.

Naknean, P., Nathathai J., and Titada, Y. 2013. Influence of clarifying agents on the quality of pasteurised palmyra palm sap (*Borassus flabellifer* Linn.). *Int J Food Sci Technol*, 494, 1175.

Nanda, K., Miyoshi, N., Nakamura, Y., Shimoji, Y., Tamura, Y., Nishikawa, Y., Uenakai, K., Kohno, H., and Tanaka, T. 2004. Extract of vinegar "Kurosu" from unpolished rice inhibits the proliferation of human cancer cells. *J Exp Clin Cancer Res*, 23, 69–75.

Nanda, K., Taniguchi, M., Ujike, S., Ishihara, N., Mori, H., Ono, H., and Murooka, Y. 2001. Characterization of acetic acid bacteria in traditional acetic acid fermentation of rice vinegar (komesu) and unpolished rice vinegar (kurosu) production in Japan. *Appl Environ Microbiol*, 67, 986–990.

Nickol, G.B., Conner, H.A., Dukowiez, M., and Hildebrandt, F.M. 1964. *Vinegar*. Newsletter, No. 43. US Ind. Chem. Co., New Delhi.

Nishidai, S., Nakamura, K., Torikai, K., Yamamoto, M., Ishihara, N., Mori, H., Nishikawa, Y. et al. 2001. Antihypertensive effect of kurosu extract, a traditional vinegar produced from unpolished rice, in the SHR rats. *Nippon Shokuhin Kagaku Kogaku Kaisi*, 48, 73–75 (in Japanese).

Nuswantara, S., Fujie, M., Sukiman, H., Yamashita, M., Yamada, T., and Murooka, Y. 1997. Phylogeny of bacterial symbionts of the leguminous tree *Acacia mangium*. *J Ferment Bioeng*, 84, 511–518.

Odunfa, S.A. and Oyewole, O.B. 1998. African fermented food. In: Wood, B.J.B. (ed.) *Microbiology of Fermented Food*, Blackie Academic and Profession, London, Vol 2, pp. 713–752.

Ohigashi, H. 2000. Kurosu, a traditional vinegar produced from unpolished rice, suppresses lipid peroxidation *in vitro* and in mouse skin. *Biosci Biotechnol Biochem*, 64, 1909–1914.

Ohler, J.G. 1979. *Cashew*. Royal Tropical Institute, Amsterdam.

Okafor, N. 1975. Microbiology of Nigerian palm wine with particular reference to bacteria. *J Appl Bacteriol*, 38, 81–88.

Okafor, N. 1978. Microbiology and biochemistry of oil palm wine. *Adv Appl Microbiol*, 24, 237–254.

Opeke, L.K. 2005. *Tropical Commodity Tree Crops*. 2nd edition, Spectrum Books Ltd, Ibadan.

Palacios, V., Manuel, V., Ildefonso, C., and Luis, P. 2002. Chemical and biochemical transformations during the industrial process of sherry vinegar aging. *J Agric Food Chem*, 50(15), 4221–4225.

Parrondo, J., Herrero, M., Garcia, L.A., and Diaz, M. 2003. Production of vinegar from whey. *J Inst Brew*, 109, 356–358.

Pasteur, L. 1868. *Etudes sur le vinaigre, sa fabrication, ses maladies, moyens de les prévenir*. Masson, Paris.

Persoon, C.H. 1822. *Mycol Eur*, 42(1), 96 (cited by Conner and Allgeier, 1976).

Prashant, M. and Rajendra, P. 1989. Relationship between ethanol tolerance and film of *Saccharomyces cerevisae*. *Appl Microbiol Biotechnol*, 30, 294–329.

Prescott, S.C. and Dunn C.G. 1959. *Industrial Microbiology*. McGraw-Hill, Kogakusha, Tokyo.

Punchihewa, P.G. 1997. *Status on the Coconut Industry*. Proceedings: Workshop on Coconut Biotechnology, Merida, Mexico.

Rainbow, C. and Mitson, G.W. 1953. Nutritional requirement of acetic acid bacteria. *J Gen Microbiol*, 9, 371–375.

Ranieri, S., Zambonclli, C., and Kaneko, Y. 2003. Saccharomyces sensu stricto: Systematics, genetic diversity and evolution. *J Biosci Bioeng*, 96, 1–9.

Rao, M.R.R. 1955. Pyruvate and Acetate metabolism in *Acetobater aceti* and *Acetobacter suboxydans*. Thesis Urbana, Illinois (cited by De Ley 1961).

Raspor, P. and Goranovic, D. 2008. Biotechnological applications of acetic acid bacteria. *Crit Rev Biotechnol*, 28(2), 101–124.

Rey, J.Y., Diallo, T.M., Vanniere, M., Didier, C., Ketia, C., and Sangare, M. 2006. The mango in french-speaking west Africa. *Fruits*, 61, 281–289.

Saeki, A. 1993. Application of *Gluconobacter oxydans* sub-sp. sphaericus IFO 12467 to vinegar production. *J Ferm Bioeng*, 75(3), 232–234.

Sakakibara, S., Yamauchi, T., Oshima, Y., Tsukamoto, Y., and Kadowaki, T. 2006. Acetic acid activates hepatic AMPK and reduces hyperglycemia in diabetic KK-A (y) mice. *Biochem Biophys Ros Commun*, 344, 597–604.

San Chiang Tan, B.S. 2005. Vinegar fermentation. A Thesis submitted to the Graduate Faculty of the Louisiana State University Agricultural and Mechanical College in Partial fulfilment of the requirements for the degree of Master of Science in the Department of Food Science University of Louisiana at Lafayette.

Sanchez, P.C. 1990. Vinegar. In: Coconut Research and Development Foundation (ed.) *Coconut as Food*. Philippine Inc Publication, Manila, pp. 151–161.

Satyavati, V.K., Bhat, V.G., Varkey, G., and Mookerji, K.K. 1972. Preparation of vinegar from pineapple waste. *Indian Food Packer*, 26, 50.

Schuller, G., Hertel, C., and Hammes, W.P. 2000. *Gluconacetobacter entanii* sp. *nov.*, isolated from submerged high-acid industrial vinegar fermentations. *Int J Syst Evol Microbiol*, 50, 2013–2020.

Seo, J.H., Lee G.D., and Jeong, Y.L. 2001. Optimization of the vinegar fermentation using concentrated apple juice. *J Korean Soc Food Sci Nutr*, 30, 460–465.

Sethi, V. and Maini, S.B. 1999. Production of organic acids. In: Joshi, V.K. and Pandey, A. (eds) *Biotechnology Food Fermentation: Microbiology, Biochemistry and Technology*. Educational Publishers & Distributors, Kerala, Vol 2.

Sharma, R.C. and Joshi, V.K. 2005. The technology of apple processing. In: Chadha, K.L. and Awasthi, R.P. (eds) *The Apple*. Malhotra Publ. Co., New Delhi.

Shimoji, Y., Tamura,Y., Nakamura, Y., Nanda, K., Nishidai, S., Nishikawa, Y., Ishihara, N., Uenakai, K., and Ohigashi, H. 2002. Isolation and identification of DPPH radical scavenging compounds in Kurosu (Japanese unpolished rice vinegar). *J Agric Food Chem*, 50, 6501–6503.

Shimwell, J.L. 1959. A re-assessment of the genus *Acetobacter*. *Antonie Van Leeuwenhoek*, 25, 49–67.

Sievers, M. and Swings, J. 2005. Family II acetobacteriaceae gills and deLoy 1980. In: Garrity, G.M., Winters, M., and Searles, D.B. (eds) *Bergey's Manual of Systematic Bacteriology*. 2nd edition. Springer, New York, NY, Vol 2, pp. 41–48.

Sievers, M., Sellmer, S., and Teuber, M. 1992. *Acetobacter europaeus* sp. *nov.*, a main component of industrial vinegar fermenters in Central Europe. *Syst Appl Microbiol*, 15, 386–392.

Simoons, F.J. 1991. *Food in China: A Cultural and Historical Inquiry*. CRC Press, Boca Raton, FL.

Snowdon, J.A. and Cliver, D.O. 1996. Microorganisms in honey. *Int J Food Microbiol*, 31, 1–26.

Sokollek, S.J., Hertel, C., and Hammes, W.P. 1998. Description of *Acetobacter oboediens* sp. *nov.* and *Acetobacter pomorum* sp. *nov.*, two new species isolated from industrial vinegar fermentations. *Int J Syst Bacteriol*, 48, 935–940.

Solieri, L. and Giudici, P. (eds) 2009. *Vinegars of the World*. Springer-Verlag, Italia, pp. 1–16.

Stanton, W.R. and Owens, J.D. 2002. *Fermented Foods-Fermentation of the Far East*. Encyclopedia of Food Sciences and Nutrition, pp. 2344–2355.

Stefano, M. and Marooka, Y. 2009. Vinegars through the ages. In: Solieri, L. and Gludici, P. (eds) *Vinegars of the World*. Springer-Verlag, Italy, pp. 17–39.

Steinkraus, K.H. 1996. *Handbook of Indigenous Fermented Foods*. 2nd edition. Marcel Dekker, New York, NY.

Stouthamer, A.H. 1959. Oxidative possibilities in the catalase positive *Acetobacter* species. *Antonie van Leeuwenhoek*, 25(1), 241–264.

Suenaga, H., Furuta, M., Ohta, S., Yamagushi, T., and Yamashita, S. 1993. Persimmon vinegar production from brewed persimmon using a horizontal rotory drum reactor. *J Jpn SocFood Sci Technol*, 40, 225–227.

Sugiyama, S., Kishi, M., Fushimi, T., Oshima, Y., and Kaga, T. 2008. Hypotensive effect and safety of brown rice vinegar with high concentration of GABA on mild hypertensive subjects. *Jpn Pharmacol Ther*, 36, 429–444 (in Japanese).

Suomalainen, H., Keranen, A.J.A., and Kangasperko, J.1965. *J Inst Brew* London, 71, 41.

Tait, A.H. and Ford, C.A. 1987. US Patent, 181999.

Tamir, S. and Eskin Michael, N.A. 2005. *Dictionary of Nutraceuticals and Functional Foods*. CRC Press, Boca Raton, FL, 520 p.

Teoh, A.L., Heard, G., and Cox, J. 2004. Yeast ecology of kombucha fermentation. *Int J Food Microbiol*, 95, 119–126.

Tesfaye, W., Morales, M.L., Garca-Parrilla, M.C., and Troncoso, A.M. 2002. Wine vinegar: Technology, authenticity and quality evaluation. *Trends Food Sci Technol*, 13(1), 12–21.

Tewari, H.K., Marwaha, S.S., Gupta, A., and Khanna, P.K. 1991. Quality vinegar production from juice of sugar cane cv. *Coj-64*. *J Res Punjab Agric Univ*, 28, 77.

Track, J., Ramus, J., and Raspor, P. 1997. Phenotypic characterization and RAPD-PCR profiling of *Acetobacter* sp. isolated from spirit vinegar production. *Food Technol Biotechnol*, 35, 63–67.

Trcek, J., Toyama, H., Czuba, J., Misiewicz, A., and Matsushita, K. 2006. Correlation between acetic acid resistance and characteristics of PQQ-dependent ADH in acetic acid bacteria. *Appl Microbiol Biotechnol*, 70(4), 366–373.

Trcek, J. 2005. Quick identification of acetic acid bacteria based on nucleotide sequences of the 16S-23S rDNA internal transcribed spacer region and of the PQQ-dependent alcohol dehydrogenase gene. *Syst Appl Microbiol*, 28(8), 735–745.

Truong, Van Denard Marquez, M.E. 1987. Handling of coconut water and clarification of coco vinegar for small scale production. *Ann Trop Res*, 9, 13.

Uchimura, T., Niimura, Y., Ohara, N., and Kozaki, M. 1991. Microorganisms in Luck pang used for vinegar production in Thailand: Microorganisms in Chinese starters from Asia: 5. *J Brew Soc Jpn*, 86, 62–67.

Uraih, N. and Izuagbe, Y.S. 1990. *Public Health, Food, and Industrial Microbiology*. Uniben Press, Nigeria.

Uzochukwu, S., Balogh, E., Tucknot, O.G., Lewis, M.J., and Ngoddy, P.O. 1999. Role of palm wine yeasts and bacteria in palm wine aroma. *J Food Sci Technol*, 36, 301–304.

Vegas, C., Mateo, E., González, Á., Jara, C., Guillamón, J.M., Poblet, M.J., Torija, M., and Mas, A. 2010. Population dynamics of acetic acid bacteria during traditional wine vinegar production. *Int J Food Microbiol*, 138(1–2), 130–136.

Verma, L.R. and Joshi, V.K. 2000. (eds) *Postharvest Technology of Fruits and Vegetables*. Indus Publishing Co., New Delhi. 1194 p.

Vinegar Institute. 2006. Market Trends (available at www.vcrsatilevinegar.org/markeltrends.html)

Vyas, K.K. and Joshi, V.K. 1986. *Cider Vinegar* (Hindi). Dept. of Horticulture, Navbahar, Shimla (H.P) India.

Webb, A.D. 1983. Vinegar. In: *Kirk-Othmer Encyclopedia of Chemical Technology*, 3rd edition. John Wiley & Sons, New York, NY, Vol 23, pp. 753–757.

Wegst, W. and Lingens, F. 1983. Bacterial degradation of ochratoxin A. *FEMS Microbiol Lett*, 17(1–3), 341–344.

Wei, X. 2001. A milestone of condiments administration with industrial standard. *Chin Condiment*, 1, 3–8.

White, J. 1970. Malt vinegar manufacture. *Proc Biochem*, 5, 54–56.

Willems, A., Bassirou, N., Ilse, C., Katrien, E., Robin, D-D., Amadou, T., Guiro, S., Tefanie, V.T., and Phillipe, T. 2007. Acetobacter senegalensis sp. nov., a thermotolerant acetic acid bacterium isolated in Senegal (sub-Saharan Africa) from mango fruit (*Mangifera indica* L.). *IJSEM*, 57(7), 1576–1581.

Wustenfeld, H. 1930. *Lehrbuch der Essigfabrikation*. Parey, Berlin.

Yamamato, T., Lanzani, G.A., and Cernuschl, E. 1959. *Clin Ostet Ginecol*, 61, 35 (cited by Asai 1968).

Yukphan, P., Taweesak, M., Yuki, M., Mai, T., Mika, K., Wanchern, P., Somboon, T. et al. 2009. Ameyamaea chiangmaiensis gen. nov., sp. nov., an Acetic Acid Bacterium in the Proteobacteria. *Biosci Biotechnol Biochem*, 73(10), 2156–2162.

11

INDIGENOUS FERMENTED FOODS

Fermented Meat Products, Fish and Fish Products, Alkaline Fermented Foods, Tea, and Other Related Products

AHMAD ROSMA, AMARJIT SINGH,
ANTON ANN, ANUP RAJ, ANUPAMA GUPTA,
AVANISH KUMAR, B.M.K.S. THILAKARATHNE,
BHANU NEOPANY, DORJEY ANGCHOK,
FOOK YEE CHYE, GEORGE F. RAPSANG,
GITANJALI VYAS, ISHIGE NAOMICHI,
JAHANGIR KABIR, K. LAKSHMI BALA,
KENNETH RUDDLE, KHENG YUEN SIM,
KONCHOK TARGAIS, M. PREEMA DEVI,
MD. SHAHEED REZA, NIVEDITA SHARMA,
RANENDRA KUMAR MAJUMDAR,
S.R. JOSHI, SHARMILA THOKCHOM,
SOMBOON TANASUPAWAT, SURESH KUMAR,
TSERING STOBDAN, V.K. JOSHI,
VIKAS KUMAR, AND W.A. WAN

Contents

11.1 Introduction

Food fermentation constitutes one of the oldest forms of food preparation and preservation technologies which not only prevents spoilage but also improves the physico-chemical characteristics and nutritional quality, leading to favorable sensory qualities of foods. Indeed, fermentation to produce the indigenous fermented foods is a part of ethno-knowledge (Dirar, 1993) and is now receiving far greater attention than ever to produce functional foods with therapeutic values. A great upsurge in investigating methods of production and quality evaluation, especially the role in well being of man, has been made in various parts of the world.

11.1.1 Fermented Meat and Meat Products

Amongst the fermented foods, the meat and meat products play a very important role in human nutrition (Ahmad and Srivastava, 2007). Being highly perishable, their storage and marketing demands a considerable amount of energy input in the form of refrigeration and freezing, which are costly and scarce in countries of South Asia (Ahmad and Srivastava, 2007; Venugopal, 2005). Meat fermentation as a means of preservation, and the use of microbial cultures as a natural preservative, have been practised from ancient time. This is particularly true of the areas like Himachal Pradesh, and states of Northeast in India, where meat is salted and kept for long time at very low temperature, where products are naturally dried and are preserved. Interestingly, these products have a long track record of safety. Apparently, the sub-zero temperature of these regions has contributed to the safety of the products. In an environment where nothing but snow can be seen for months together, fermented and dried products from meat have remained a significant contributor to the nutrition of the people inhabiting these places. Thus, it could be an excellent means of combating malnutrition in developing countries. Even in India, the need for development of low pH meat products to circumvent the non-availability of refrigeration has been stressed earlier (Anon, 1988; Sharma, 1987).

11.1.2 Fermented Fish and Fish Products

In some South Asian countries, animal protein in the human diet is supplied by fish, which is consumed daily, either as a salty side dish or a condiment that facilitates rice consumption. Fermented products serve this purpose more appropriately, since they are simple to produce and cook, have a long shelf-life, and impart *umami* and a salty taste to vegetable dishes (Kimizuka et al., 1992; Mizutani et al., 1987). *Umami* is a category recognized by the Japanese as the taste of glutamic acid (Ishii and O'Mahony, 1987). Much of the seasonally available surplus fish is preserved by fermentation.

The term "fermented fish products" is used in this chapter to describe the products of freshwater and marine finfish, shell-fish, and crustaceans that are processed with salt to cause fermentation and, thereby, prevent putrefaction (Tamang et al., 2012; Tanasupawat and Visessanguan, 2013). Fermented fish products such as *shidal* (Assam, India), *nya sode* (Bhutan), *pedah* (Indonesia), *jadi* (Sri Lanka), among others do not fit into the category of intentionally fermented products (Ruddle and Ishige, 2005). The *liquamen* or *garam* of Imperial Rome was fermented intentionally, and is of the same type as Asian fermented fish products (Corcoran, 1963; Gamer, 1987; Grimal and Monod, 1952).

11.1.3 Alkaline Fermented Foods

Similar to the meat and meat products are the alkaline fermented foods that serve as a protein source and condiment. Alkaline fermented foods including *natto, kinema*, and *thua–nao* have traditionally been produced and consumed in several parts of the world. These foods form an important part of the diet in many countries including those of South Asia. Alkaline fermentations make otherwise inedible foods edible, enhance flavor and nutritional value, decrease toxicity, preserve the food, decrease cooking times and energy requirements, and, in general, bring diversity to the kinds of foods and beverages available (Chukeatirote et al., 2010; Parkouda et al., 2009; Sarkar and Nout, 2014; Steinkraus, 1998). It is a process during which the pH of the substrate increases to alkaline values as high as 9 (Amadi et al., 1999; Omafuvbe et al., 2000; Tamang and Sarkar, 1996) due to the enzymatic hydrolysis of proteins from the raw material into peptides, amino acids, and ammonia (Kiers et al., 2000), and/or due to alkali-treatment during their production (Parkouda et al., 2009; Wang and Fung, 1996). In several countries including those of Asia, the traditional diets of the majority of people rely largely on starchy staples which are rich in calories but poor in other nutrients, especially proteins, and the alkaline fermented foods provide a protein rich supplement to the diet (Achi, 2005; Dakwa et al., 2005; Parkouda et al., 2010).

11.1.4 Fermented Food Condiments

Fermented food condiments also form a significant component of the diets of many people in developing countries (Parkouda et al., 2010; Tamang, 2010a,b). Most of the indigenous alkaline-fermented foods are derived from the seeds or leaves of various wild trees, and cultivated plants (Parkouda et al., 2009; Steinkraus, 2004).

11.1.5 Soy Based Products

The diversity of indigenous fermented meat and meat products or protein rich products from soybeans is very high in countries of South Asia. These products, especially the processes used to produce them, have the potential to inspire researchers and producers in other parts of the World (Guizani and Mothershaw, 2005) to attempt similar products. Apart from *natto*, which is industrially produced, the production of other such indigenous fermented foods is still carried out at household scale.

The current health concerns to decrease dietary fat intake may make indigenous fermented foods more appealing, due to their inherent low-fat characteristics. Advanced fermentation technologies and exploration of new research areas may create new food items, and enhance the desirable qualities of indigenous fermented foods. And in doing so it may offer convenience, safety, and variety, and also provide an appropriate marketing base, employment, profits to entrepreneurs, and good returns to the producers.

Efforts have been made to document the knowledge of the ethnic people of the Himalayas on the production of indigenous meat products. The aim of the present chapter is to give an overview of the main indigenous alkaline fermented foods and meat and meat-like products, their production, fish and fish products, and the technological and functional properties.

11.2 Traditional Acid Fermented Meat Products

Lerstner (1986) and Zeuthen (1995) have given a good account of the history of meat preservation through fermentation, and fermented sausage is one of the oldest processed foods known to man. A distinctive category of sausage products has been established for well over 2000 years (Smith, 1988). Fermented sausages probably originated in the Mediterranean. The Romans knew that ground meat with added salt, sugar and spices turned into a palatable product with a long shelf-life, if prepared and ripened properly.

11.2.1 Fermented Meat Products of Himalaya

Ethnic people of the Himalayas are mostly non-vegetarian (www.shodhganga. inflibve.net.ac-2014), and meat remains a part of the daily diet for many ethnic people dwelling in the Himalayan regions of India (Tamang, 2009a,b,c,d,e; Tamang, 2013). Raw meat becomes spoiled at high ambient temperatures within a few hours due to its high moisture and protein content, and it is mostly dried or smoked or fermented to prolong the shelf-life (Dzudie et al., 2003; Rantsiou and Cocolin, 2006).

11.2.1.1 Uttarakhand   Several meat based preparations, made through drying, smoking, or fermentation of meat are popular in various Himalayan states (Tamang, 2010a).

Table 11.1   Traditional Meat Products of Uttarakhand (India)

MEAT PRODUCT	DESCRIPTION	MICROORGANISMS INVOLVED
Chartayshya	Goat (*chevon*) meat prepared from red meat mixed with salt, eaten as curry component	Species of *Enterococcus, Pediococcus* and *Weissella*
Jamma/Geema	Goat (*chevon*) meat prepared from red meat mixed with finger millet (*Eleusine coracana*), wild pepper (*Zanthozylum* sp.), chilli powder and salt, eaten as cooked sausage or as a curry component	Species of *Enterococcus, Leuconostoc, Pediococcus*
Arjia	Similar to *jamma* with slight modification, eaten as deep fried sausage or as a curry component	Species of *Enterococcus* and *Pediococcus*

Source: Adapted from Oki, K. et al. 2011. *Food Microbiol* 28(7): 1308–1315.

In Table 11.1, a list of traditional meat products of Uttarakhand along with description is given. Some of these products, like *chartayshya, jamma*, and *arjia* are from high altitude areas of Uttarakhand, (India) which have been studied in detail for the microflora associated with these products (Rai et al., 2009, 2010). The diversity of lactic acid bacteria in these products has been enumerated through cultural and molecular tools (Namugumya and Muyanja, 2009). The microorganisms involved in the fermentation belonged mainly to the species of *Enterococcus, Leuconostoc, Pediococcus,* and *Weissella.* All these nutritive products are prepared from red meat mixed with salt, spices, and/or animal blood (Rai et al., 2009, 2010) (Figure 11.1).

11.2.1.2 North Eastern States   *Kheuri, satchu,* and *kargyong* are popular products in Sikkim, other Northeastern states, Tibet, and Bhutan, and are prepared from yak/ beef/pork meat (Rai et al., 2009). *Suko ko masu* and *chilu* are dried/smoked meat products mostly made from yak/beef/lamb; *chilu* is a stored animal-fat-based product (Rai et al., 2009).

11.2.1.2.1 Kargyong   *Kargyong* is an indigenous sausage-like meat product of the Northeast Region, specifically of Sikkim, prepared from meat (Tamang et al., 2012). Meat (yak/beef/pork) along with its fat is chopped finely and combined with crushed

(a)   (b)   (c)

Figure 11.1   Meat products of Uttarakhand: (a) *Chartayshya*, (b) *Jamma*, and (c) *Arjia*. (Photo reproduced with permission from the thesis (Tamang, 2009c). "Microbiology of traditional meat products of Sikkim and Kumaun Himalaya," from the work of the scholar and his supervisors.)

garlic, ginger, salt, and a little water. The prepared mixture is stuffed into segments of gastro-intestinal tract of animal, locally called *gyuma*, which is used as a casing, 3–4 cm in diameter and 40–60 cm long. One end of the casing is tied up with the rope, and the other end is sealed after stuffing, and the sausage is boiled for 20–30 min. When cooked, the sausages are taken out and hung on bamboo strips above the kitchen oven (*chullah*) for smoking for 10–15 days (Rai et al., 2009). *Kargyong* is boiled for 10–15 min, then sliced and fried in edible oil, with onion, tomato, powdered or ground chillies, and salt, and is made into curry (Tamang, 2009a,b,c,d,e). It is also consumed as fried sausage with distilled liquor. In its production *Lactobacillus sake*, *L. divergens*, *L. carnis*, *L. sanfrasisco*, *L. curvatus*, *Leuco. mesenteroides*, *Enterococcus faecium*, *Bacillus cereus*, and *Micrococcus. Debarmyces hansenii*, and *Pichia anomala* are involved (Tamang et al., 2012).

11.2.1.2.2 Satchu   *Satchu* is an ethnic dried or smoked meat (beef/pork/yak) product of the Northeast Himalayas (Sikkim) (Tamang et al., 2012). Red yak meat or pork is sliced into several strands of about 60–90 cm and is mixed thoroughly with turmeric powder, edible oil or butter, and salt. The meat strands are hung on bamboo strips or wooden sticks and are kept in the open air in a corridor of the house, or are smoked above the kitchen oven, for 10–15 days (Rai et al., 2009; Tamang et al., 2012). In fermentation, *Pediococcus pentosaceous*, *Lactobacillus casei*, *Lactobacillus carnis*, *Enterococcus faecium*, *Bacillus subtilis*, *B. mycoides*, *B. lentus*, *S. aureus*, *Micrococus* and *Debaromyces hansenii*, and *Pichia anomala* are involved (Tamang et al., 2012). *Satchu* is made into curry by washing and briefly soaking in water and then, frying in yak/cow butter with chopped garlic, ginger, chillis, and salt. A thick gravy is made, which is eaten with boiled or baked potatoes (Tamang, 2009a,b,c,d,e).

11.2.1.2.3 Suka Ko Masu   This is a dried or smoked meat product consumed by the non-vegetarian *Gorkha* of Sikkim (Tamang et al., 2012) which is made by cutting the red buffalo meat into strips up to 25–30 cm long and mixing with turmeric powder, mustard oil, and salt (Tamang, 2009a,b,c,d,e). The meat strips are hung on bamboo and kept above the earthen kitchen and smoked for from 7–10 days to several weeks. After complete drying, the smoked meat product is called *suka ko masu* or *seakua*, and it can be stored at room temperature for several weeks (Tamang, 2009a,b,c,d,e). It is washed and soaked in lukewarm water, excess water is squeezed out, and it is fried in heated mustard oil, with chopped onion, ginger, chilli powder and salt. Coriander leaves are sprinkled over the curry and it is eaten with boiled rice. In its preparation *L. carnis*, *L. plantarum*, *Enterococcus faecium*, *Bacillus subtilis*, *B. mycoides*, *B. thuringiensis*, *S. aureus*, *Micrococcus*, and the yeasts *Debarmyces hansenii* and *Pichia burtoii* (Rai et al., 2010; Tamang, 2009a,b,c,d,e) are involved. But regular consumption of meat is expensive for the majority of the rural people (Tamang, 2009a,b,c,d,e). So they slaughter domestic animals usually only on special occasions, festivals, and marriages. After the ceremony, the meat is cooked and eaten. The remaining flesh is preserved by smoking to make *suka ko masu* for future consumption (Rai et al., 2009).

11.2.1.3 Fermenetd Meat Products of Ladakh Ladakhi people are mostly non-vegetarians. Meat of goat (*ra*), sheep (*luk*), yak, cow (*balang*), or *zho* (a crossbred of cow and yak) is used for making various indigenous meat products. Animals are slaughtered every year just before the onset of winter, at the time when animals are in best condition. Rather than by conventional method of butchery, animals in Ladakh are slaughtered by asphyxiation which is done for their religious dispensation and also to save blood—one of the ingredients of some meat products—from being wasted.

Some of the meat is consumed fresh and the rest is frozen or dried. Freezing and drying takes place in the ambient environment which is arid and cold at the time of slaughter. The raw meat is dipped in salt water before drying. Salt crystals are hygroscopic and absorb some of the water from the meat, preserving the meat surfaces by keeping them dry, as dry meat surfaces inhibit the growth of bacteria and moulds, hence, increasing the shelf-life. Slight oxidation of the meat fats contributes to the typical flavor of dried meat.

11.2.1.3.1 Sharjen The air-dried meat is very dry and hard, and has a distinctive flavor. Some of this dried meat is eaten as it is, cutting or tearing the strips into smaller pieces. Otherwise it is consumed after cooking, for which there are two main methods. One is to roast it by burying the meat in the dung fuelled hearth (*chullah*), until the meat smells fragrant. It is then taken out, cleaned, cut into pieces, and eaten. The other method is to soak the dried meat for several hours and then, boil it in water. Salt and condiments are usually not added. This product is known as *sharjen*.

11.2.1.3.2 Kheuri A product similar to *kheuri* prepared in Sikkim (Rai et al., 2009) is also prepared in Ladakh. Chopped meat and viscera (including liver, lungs, etc.) of the slaughtered animals are put into a cleaned, empty sheep's stomach. The stomach opening is later stitched. It is then, allowed to freeze in the natural "cold storage." Before consuming, it is put in warm water for some time for thawing. The desired portion of the meat is then, taken out and cooked.

11.2.1.3.3 Thakpol Ground roasted barley (*nyasphey*), animal fat (*tsilu*), comminuted meat, blood, and salt are kneaded into a dough. From this paste, bun type bread is made, which is cooked slowly on hot ash for about 2–3 h. The blood used for *thakpol* and other products is collected from the thoracic cavity of the dead animal after removing the lungs and heart.

11.2.1.3.4 Gyuma This is a sausage-like product prepared traditionally by the ethnic communities living in the highlands of Ladakh. *Nyasphey* or wheat flour is mixed with chopped meat, fat, and salt. Water is added to this paste to make it thin enough to be poured into the casings of small animal intestines. Sometimes, fresh blood is also added to the mixture before stuffing the casing. Both ends of the tube are tied before placing it into water for boiling for 10–20 min. The stuffed intestine is pierced at intervals to avoid bursting during boiling. It is eaten either at the time it is made, or after storing it by hanging indoors. The sausage may also be roasted on the top of the stove fuelled by yak dung. It is consumed with or without frying. Deep fried *gyuma* is generally, served with *chhang*.

11.2.1.3.5 Nangchi This is meat sausage usually encased in large intestine. The stuffing is composed of meat, fat, and fresh animal blood. The meat and fat are chopped into pieces and salt, condiments, and blood are added before the mixture is stuffed into a large intestine. The filled casing is tied, pricked and boiled. It is normally consumed fresh. Meat sausage can be stored for about a month, and is quite similar to the *arjia* made in the *Kumaun* Himalayas.

11.2.2 Chartayshya—A Traditional Meat Product of Kumaon

11.2.2.1 Chartayshya *Chartayshya* is a traditional goat meat product of the Kumaon Himalayas, consumed by *Bhotiya* of Dharchula and Munsiary in the district of Pithoragarh (Tamang, 2009c,d). Red goat meat is cut into small pieces of 3–4 cm, mixed with salt, sewed in a long thread and hung in an open air corridor of the house for 15–20 days (Figure 11.2) (Tamang, 2009c). It can be kept at room temperature for several weeks for future consumption. In Western Nepal, a similar product called *sukha sikhar* is prepared from chevon (Tamang, 2009c). Curry is made by frying in edible oil with tomato, ginger, garlic, onion, and salt. This is the most delicious meat item of the Kumaon Himalayas (Tamang, 2009c,d). The ethnic people of the Kumaon Himalayas prepare *chartayshya* curry especially during the *kolatch* festival (worshipping ancestral spirits) and offered to the ancestors before eating.

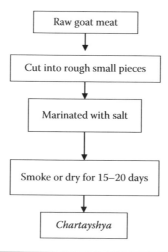

Figure 11.2 Traditional method of preparation of *Chartayshya* in the Kumaon Hills. (Adapted from Tamang, J.P. 2009c. *Himalayan Fermented Foods: Microbiology, Nutrition, and Ethnic Values.* CRC Press, Boca Raton, FL, Chapter 7, pp. 161–186.)

11.2.2.2 Geema/Jamma   *Geema/Jamma* is a traditional fermented chevon sausage of the Kumaon Himalayas (Tamang, 2009a,b,c,d,e). Red goat meat is chopped into fine pieces, ground finger millet (*Eleusine coracana*), wild pepper, locally called *Timbur* (*Zanthoxylum* sp.), chilli powder, and salt are added and mixed. A small amount of fresh animal blood is also added (Oki et al., 2011). The meat mixture is made semi-liquid by adding water, and then it is stuffed into the small goat intestine of about 2–3 cm in diameter and 100–120 cm length with the help of a funnel and tied at both the ends. It is pricked randomly to prevent bursting while boiling. After boiling for 15–20 min the stuffed intestines are smoked above the kitchen oven for 15–20 days. It is consumed as a curry by mixing with onion, garlic, ginger, tomato, and salt (Tamang, 2009c). It is also deep fried and eaten with local alcoholic beverages. Sometimes, *geema* may be eaten as cooked sausage (Figure 11.3).

11.2.2.3 Arjia   *Arjia* is also a sausage-like product made from goat meat in the Kumaon Himalayas and consumed by the *Bhotiya*. The method of preparation of *arija* is similar to *jamma* (Tamang, 2009c). However, in *arija* preparation, a mixture of chopped lungs of goat, salt, chilli powder, *timbur* (*Zanthoxylum* sp.) and fresh animal blood are stuffed into the large goat intestine, instead of small, and boiled for 15–20 min. As in *jamma*, pricking of the stuffed large intestine is necessary to prevent bursting while boiling. It is dried for 15–20 days or smoked above the kitchen oven (Tamang, 2009a,b,c,d). *Arija* is consumed as a curry or deep fried sausage along with a main meal. The flow diagram of *arija* preparation is given in Figure 11.4.

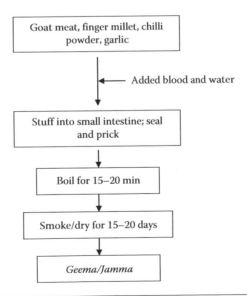

Figure 11.3   Traditional method of preparation of *geema* or *jamma* in the Kumaon Hills. (Adapted from Tamang, J.P. 2009c. *Himalayan Fermented Foods: Microbiology, Nutrition, and Ethnic Values.* CRC Press, Boca Raton, FL, Chapter 7, pp. 161–186.)

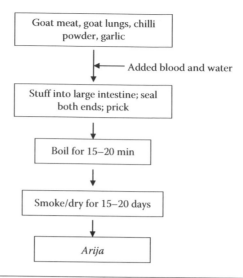

Figure 11.4   Traditional method of preparation of *arija* in the Kumaon Hills.

11.2.3 Meat Based Fermented Food

11.2.3.1 Jangpangnatsu "Japangangnatsu" is a fermented product made from crab (*Scylla* sp.), named according to the *Ao Naga* dialect (Jamir and Deb, 2014). To prepare it, crabs are first washed thoroughly and shredded into pieces and the hard shell pieces removed (Figures 11.5 and 11.6). The shreds are then, mixed with ground black *til* (*Sesamum orientale* L.) and wrapped in banana leaves or *Phrynium pubinerve* leaves or kept in a pot. After 3–4 days of keeping near warm or the fire place, the fermentation is complete.

11.2.3.2 Jang Kap "Jang kap" is made from buffalo skin, named according to the *Ao Naga* dialect. The skin is separated from the flesh completely and stacked in a tin or pot with a tight covering. It is kept for about a week to allow the fermentation process to complete. Afterwards, all the hairs are completely scrapped-off and it is either dried in the sun or near the fire place. People usually pressure cook and consume *Jang kap* as it becomes hard after it is dried (Jamir and Deb, 2014) (Figure 11.7).

Figure 11.5   Flow chart of preparation of *Jangpangnatsu*.

Figure 11.6 Different stages of *Jangpangnatsu* preparation. (a) The shredded pieces of crab. (b) Black til (*Sesamum orientale* L.) fry. (c) Mixture of a and b and made paste. (d) The fermented product *Jangpangnatsu*. (Adapted from Jamir, D. and Deb, C.R. 2014. *Int J Food Ferment Technol* 4(2): 121–127.)

Skin of buffalo
↓
Stacked in a tin or pot
↓
Fermentation for 1 week
↓
Hairs removed and fermented product dried near the fire place or sun
↓
Jang kap

Figure 11.7 Flow chart of preparation of *Jang kap* (*Ao*). (From Jamir, B. and Deb, C.R. 2014. *Int J Food Ferment Technol*, 4(2): 121–127. With permission.)

11.2.3.3 Fermented Pork Fat   Pork fat is fermented and used as a condiment in preparation of vegetables and curries by almost all the *Naga* tribes (Jamir and Deb, 2014). Pork fat is cut into small pieces and boiled. It is then, put into bamboo containers, the mouths of which are sealed with banana leaves. The fermentation process in completed in about 1 week or so (Figure 11.8).

11.3 Fermented Fish Products

There are several general reviews on fermented fish products published by the Indo-Pacific Fisheries Council (1967), the Tropical Products Institute (1982) and Steinkraus (1983a,b, 1985), which outline the main processing techniques for fermented fish products and describe the basic chemical components of the principal products. In addition, scattered reports (e.g., Lee, 2009) contain basic chemical analyses from some

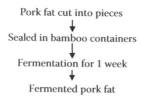

Pork fat cut into pieces
↓
Sealed in bamboo containers
↓
Fermentation for 1 week
↓
Fermented pork fat

Figure 11.8   Flowchart for the preparation of fermented pork fat. (From Jamir, B. and Deb, C.R. 2014. *Int J Food Ferment Technol*, 4(2): 121–127. With permission.)

other countries. Prior to the work of Ishige (1986) and Ruddle and Ishige (2005), these products were virtually ignored in Western publications. During 1982–1985, Ishige and Ruddle conducted a comprehensive field survey on the fermented fish products industry, from the catching of the raw materials to their culinary use in several countries: Bangladesh, Cambodia, China, India, Indonesia, Japan, Korea, Malaysia, Myanmar, the Philippines, Taiwan, Thailand, and Vietnam (Ishige and Ruddle, 1987, 1990; Ruddle and Ishige, 2005; Shio, 2006).

A wide range of fermented fish foods is produced in Asia, so a strict classification by product type must be limited to individual countries or linguistic groups (Salampessy et al., 2010; Tanasupawat and Visessanguam, 2013). Therefore, a simple generic classification (Figure 11.9) based on both the nature of the final product and the method used to prepare it has been employed. The prototypical product is probably highly salted fish, which in Japan is known as *shiokara*. Since there are no succinct equivalent English terms for these products, simple Japanese terms are used throughout this chapter. The product of combining fish and salt that preserves the shape of the original raw fish material is termed *shiokara*. This can be comminuted to *shiokara* paste, which has a condiment-like character. If no vegetable ingredients are added, the salt fish mixture yields fish sauce, a liquid used as a pure condiment. If cooked vegetable ingredients are added to the fish and salt mixture, it becomes *narezushi*.

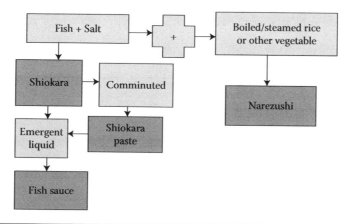

Figure 11.9   Generic classification of fermented fish products. (Adapted from Ruddle, K. and Ishige, N. 2005. *Fermented Fish Products in East Asia*. International Resources Management Institute, Hong Kong.)

11.3.1 Traditional Fermented Fish Products of India

11.3.1.1 Fermented Fish Products of Northeast India The Northeastern Region of India is known for its vast natural resources, and it is a cauldron of different people and cultures. Lying deep in the lap of the easternmost Himalayan hills in the North-Eastern part of India, it is connected to the rest of India by a strip of land only 20 km wide (www.newseducation4india.com). The region comprises of States like Arunachal Pradesh, Nagaland, Manipur, Tripura, Mizoram, Meghalaya, Assam, and Sikkim (www.indianetzone.com). The fish fermentation technology in the Northeastern states of India has evolved by necessity rather than choice. The region, having the highest rainfall area of the world, does not provide a congenial environment for simple sun drying of fish. People used to preserve fish for use in lean periods by drying in the sun (Muzaddadi and Basu, 2012), but such drying was prolonged due to the very humid atmosphere and frequent rainfall, particularly during the peak fishing seasons (i.e., from May to September). The northeast, being abundant in low lying areas where accumulation of water during rainy season, offers an ideal habitat for the breeding of weed fishes, such as *Puntius* spp., "*Darkina*" (*Esomusdanricus*), and "*Mola*" (*Amblypharyngodon mola*) (Muzaddadi and Basu, 2012). The clever fishermen, therefore, were in search of a method through which they could preserve the heavy catches of such less-valued weed fishes for consumption and sale in the dry seasons (November–April) when there was a scarcity of fresh fish in the markets (Muzaddadi and Basu, 2012). Moreover, due to non-availability of ice and of good road communications, these perishable products could not be transported to distant markets where they could get a higher price than they could earn in the village markets. The food habit of the rice-eater is to eat the rather tasteless rice mixed with little morsels of produce with strong flavors. In this situation, fermented fish could be an ideal product to cater this need. The Northeast region has many fermented fish products, such as *shidal, ngari, hentak, lonailish, tungtap*, etc. (Tamang 2009d; Tamang, 2013). Fermented fish contributes to a regular menu, especially in the diets of tribal people (Baishya and Deka, 2009). Unlike the salted fermented fish products of Southeast Asia, the salt-free fermented fish products of the Northeast indicate the fact that the technology of salt-free fermentation originated long before the man started using salt (sodium chloride). Later on, salt was used for fermentation, as is practised in the case of *Lonailish*. Preparation of such fermented fish products is simple, but most of the critical steps are optimized by experiences to get a quality product. Traditional practices are still in use today, due to lack of proper scientific intervention and standardization of methods.

11.3.2 Traditional Fermented Fish Products of Bangladesh

Fish and rice are the staple foods in Bangladeshi cuisine, and it is said that these two food items are what make a Bengali (*Machh-e-Bhat-e-Bangali*). The majority of these foods are consumed freshly cooked, while a minor portion is used to prepare various indigenous fermented foods, including fermented beverages from rice and different fermented fishery products. The former are largely consumed by the tribal villagers in the Chittagong hill tracts, while the latter are consumed in different parts of the country, including Mymensingh, Comilla, Sylhet, Rangpur, Gazipur, and Chittagong.

11.3.3 Fermented Fish Products of Sri Lanka and the Maldives

Sri Lanka possesses a 1560 km long coastline and an exclusive economic zone of around 20,000 km, and the fisheries sector contributes 2.5% to the national economy of Sri Lanka (Anon, 2000). It provides food and employment to thousands of rural people. Fish accounts for nearly 70% of protein intake of both rural and urban populations. One of the problems identified in the fisheries sector is the large post-harvest loss through spoilage due to poor handling. It has been estimated that 25% of catch goes waste due to poor preservation. Besides human consumption, most of the fish caught in inland plantation sectors is converted to salted dried fish and other products. All the fermented fishery products of Sri Lanka are salt-based, depending on the product and variety of fish used as raw material. The fish species with a high fat content, are very prone to rancidity and quickly spoil (Southgate, 1984), so most of the fatty fish are traditionally preserved by curing and fermentation. Generally, cured fermented fish products have an appealing flavor and good shelf-life. Many such products, like *jadi*, salted fish, smoked fish, Maldive fish, pickled fish, and *ambulthiyal*, are popular among most of the Sri Lankan populations. Today, fish sausages, fish sauces, and other fish products are arriving in the commercial markets. *Jaadi* and pickled and salted fish sauces are a useful low cost fish preservation method, particularly when these are plentiful, there is no sunshine, and there are difficulties of cold storage, transport, and processing.

11.4 Technology of Fish Product Preparation

Some fermented fish products do not originate from South Asian countries—India, Pakistan, Sri Lanka, Nepal, Bhutan, Bangladesh, Afghanistan, and the Maldives—but are described briefly here since they represent distinct typical products with basic technology used in the countries of South Asia.

11.4.1 Fish Shiokaras

Shiokara is made by mixing fish and salt to yield a fermented product that preserves either in whole or in part the original shape of the raw fish. This is probably the prototypical form, from which all other variants of *shiokara* developed. *Shiokara* is used mostly as a side dish, and is important in several cuisines, including Lower Myanmar. Fish *shiokara* is commonly dried and comminuted into a paste by pounding or grating, so that it can be eaten easily. It has a condiment-like character and can be dissolved easily in water, and then, used as either a soup stock or for dipping.

Sometimes, a small quantity of boiled rice is also added. *Shiokara* products contain salt plus other materials as taste enhancers. Flavor arises mainly from chemical reaction.

11.4.1.1 The Generic Production Process   Despite considerable local variations, there are many common aspects to the production process employed, wherever *shiokara* is made.

11.4.1.2 Preliminary Processing   Before the main processing, several preliminary tasks are commonly performed on the fish. The principal among these are washing, draining, or drying, beheading, evisceration, scaling, and deboning. Sometimes, the fish is washed prior to the main processing (IPFC 1967), as in Myanmar (Maxwell, 1904; Scott, 1910) and the Philippines, and the cleaned fish may then, be either drained or sun-dried prior to the main processing (Ruddle and Ishige, 2005). Other preliminary steps include the soaking of the fish overnight, until swollen, or the separation of larger fish for special handling in the first stages of the main processing, as in parts of Myanmar (Ruddle and Ishige, 2005; Scott, 1910). In general, these steps are performed only when either larger fish are used, as at Theingangyam Township, in Myanmar (Ruddle and Ishige, 2005), or when whole fish *shiokara* is made, as at Payagyi Village, in Myanmar (Scott, 1910). Larger fish may sometimes be softened prior to fermentation by trampling (Ruddle and Ishige, 2005). Fish are scaled by trampling in the Tharrawaddy District of Myanmar (Anon, 1920), which also functions to soften them.

11.4.1.3 Main Processing
   11.4.1.3.1 Mixing Salt and Fish   This step is universal, although both the amount of salt used and number of times it is done, vary considerably. Fish may be salted up to 4 times, the amount of salt applied ranging from 10% to 100% of the fish used, by weight. Additional salt may be added by covering the top of the fermentation vessel with a salt layer to protect against maggots (Ruddle and Ishige, 2005).

11.4.1.4 Sun-Drying   Sun drying of the salted fish is common, but not universal. It may be done from one to three times, and for up to four days. The sequencing of this step varies considerably; it may follow comminution into *shiokara* paste (IPFC, 1967) or it may follow preliminary salting, as in Myanmar (Anon, 1920; Maxwell, 1904; Scott, 1910).

11.4.1.5 Comminuting   Fish must be comminuted to make *shiokara* paste. This paste is made only in Myanmar, where fish is comminuted, usually using a mortar and pestle, from 1 to 3 times (Anon, 1920; IPFC, 1967; Maxwell, 1904; Scott, 1910), or using a foot-operated pounder. Repeated comminution is done to ensure an even texture and blending of the final product, particularly when several fish species are being used together (Ruddle and Ishige, 2005).

11.4.1.6 Fermentation   The total fermentation time ranges from 3 days to 12 months. A variety of fermentation containers is employed, including ceramic or glass jars, wooden vats, and concrete vats (Ruddle and Ishige, 2005).

11.4.1.7 Compression   In some instances, the fish is compressed with weights. Most commonly, this is done during the main period of fermentation. Less commonly, the salted fish is compressed as a part of the preliminary processing, before the main fermentation period begins, as in the preparation of *tangtha ngapi* (Anon, 1920) and *shiokara* made from mudfish (Scott, 1910), in Myanmar, when the ingredients are compressed prior to sun-drying and packing into fermentation containers.

11.4.1.8 Draining of the Emergent Liquid   Only relatively rarely is the liquid that emerges from the fermenting fish drained-off and used as a fish sauce (Ruddle and Ishige, 2005).

11.4.1.9 Mixing of Additives   In Thailand, materials are added to the *shiokara* during the production process such as rice bran (Ruddle and Ishige, 2005).

11.4.1.10 Packing   A variety of packing containers is used in the sale of *shiokara*, depending upon the nature of the market. Large cans are used in the Philippines and Thailand, where the product is sold wholesale, whereas a range of jars and bottles are used in retailing. The principal fish *shiokara* in Thailand is known locally as *pla-ra* and is produced throughout the year at home and as a cottage industry, mainly in Northeast Thailand (Figure 11.10). In that region, it is known colloquially as *pla daek* (Ruddle and Ishige, 2005). The equipment used in the production process is simple, obtained locally either cheaply or at no cost. The main expense is for fermentation jars; Chinese-made pickled egg jars of 150 kg, 100 kg and 50 kg capacity have been re-cycled for the purpose. The mouths of the fermenting jars are covered with tight fitting woven bamboo lids. Stone weights are placed over the lids to compress the fish. The fermenting contents of the jars are stirred periodically with a stirrer made from an old oar. Re-cycled metal containers for edible oil, each with a 25-kg capacity, are used to pack the fermented fish for direct retail marketing (Ruddle and Ishige, 2005).

11.4.2 Shrimp Shiokara

Shrimp *shiokara* is widely consumed throughout continental and insular Southeast Asia, and generally in preference to fish *shiokara*, as an indispensable condiment in local and national cuisines.

11.4.2.1 The Generic Production Process   The processes locally used to make shrimp *shiokara* vary widely in their details. Many Industrial- and cottage industry-scale operations (full processors) undertake both preliminary and main processing. Others (industrial

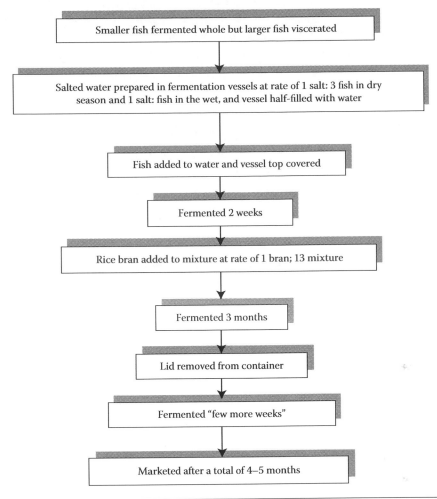

Figure 11.10 The production process of *pla ra* in a small-scale factory at Taman Ubon Rat, Khon Kaen District, and Northeast Thailand. (From Ruddle, K. and Ishige, N. 2005. *Fermented Fish Products in East Asia*. International Resources Management Institute, Hong Kong. With permission.)

re-processors) obtain partly processed shrimp, which have been partially salted and may be crudely comminuted and partly fermented by household or cottage-industry producers, for re-processing and refining to commercially acceptable standards. There is considerable variation in the degree of complexity as well as sophistication of the processes used at the household level.

11.4.2.2 Preliminary Processing

11.4.2.2.1 Pre-Processing The shrimp are washed in seawater as the first stage of processing. Some processors sort the catch, which is done by some Malaysian producers, prior to washing (IPFC, 1967). Shrimp paste producers at Mohishkali Island, Bangladesh, sort the shrimp for fermentation into those caught near the sea surface and those taken near the bottom (Ruddle and Ishige, 2005). A task performed assiduously during preliminary processing, particularly where the intention is to prepare the highest quality shrimp *shiokara*, is the removal of unwanted material from among the shrimp, including juvenile finfish, crabs, seaweed, and small pieces of inorganic material. The larger pieces are also commonly removed when making the lower grades. This is done during the sun-drying process at Choupaldandi Village, in Bangladesh (Ruddle and Ishige, 2005).

11.4.2.3 Mixing Salt and Shrimp There is considerable variation throughout the region in the ratio of shrimp and salt used, and the number of times and stages of the process that salt is added to the shrimp. The salted types are said to have a shelf-life of only 6 months, whereas the unsalted types may be kept for up to 2 years without spoiling. Slightly salty types are made with a shrimp-salt mixture of around 25:1 to 20:1. Moderately salty shrimp *shiokara*, which has a shrimp salt ratio ranging from about 10:1 to 5:1, occurs more widely, including Choupaldandi Village, Bangladesh (Ruddle and Ishige, 2005). Of the 21 processors examined by Ruddle and Ishige (2005) who produced salted types, all but two mixed salt with the shrimp only once. Exceptions occur in the sophisticated process employed sometimes at Choupaldandi Village, Bangladesh, depending on the weather conditions prevailing (Ruddle and Ishige, 2005). The point during the processing at which the salt and shrimp are mixed also varies considerably depending upon the locality. Only occasionally are shrimp salted on board the catching vessel prior to being landed. At Choupaldandi Village, for example, shrimp is salted at sea when rough conditions preclude rapid landing of the catch (Ruddle and Ishige, 2005). This is done to prevent both putrescence and insect infestation. Salting at sea is also undoubtedly done elsewhere, under similar circumstances.

Under normal conditions, and elsewhere, the shrimp are mixed with salt after or during preliminary processing and before the material is left to ferment. Elsewhere the timing is somewhat different. For example, the materials may be mixed after the first sun-drying.

11.4.2.4 Main Processing

11.4.2.4.1 Sun-Drying Both the total time and the number of times that the shrimp is sun-dried also vary considerably, although sun-drying is done in almost all the cases. Sun-drying is performed on an average of two times during processing, although this ranges from no times to as many as eight. The average total time that the salted shrimp is sun-dried is 2–3 days. However, total drying time can vary considerably at any location, depending on the degree of cloudiness and average ambient air temperature (Ruddle and Ishige, 2005). Most commonly, sun-drying is done between comminuting stages, as at Choupaldandi Village, Bangladesh (Ruddle and Ishige, 2005).

11.4.2.5 Drainage Relatively few production processes drain the material at any stage. The shrimp is drained for 1 day prior to comminuting or is drained-off overnight.

After the first comminuting, it is drained twice, once during each fermentation period (Ruddle and Ishige, 2005); it is also drained overnight twice, once after salting and again after the first comminuting, while at other places the mixture is drained during fermentation, and the liquid used to make fish sauce (Ruddle and Ishige, 2005) or sometimes boiled and returned later to the comminuted solids (IPFC, 1967), or blended with the solid to produce a *shiokara* paste (Lu'ong, 1981).

11.4.2.6 Comminuting　In most production processes, the shrimp are comminuted to a paste, either by using a mortar and pestle, or trampling by foot, or mechanically, or with either a powered or manual grinder or mincer. Grinders or mincers are used only at small-scale factory or industrial levels, whereas only traditional techniques are used in household and cottage industry processing. Where comminuting is practiced, frequency ranges from 1 to 6 times, with an average of 2 (Ruddle and Ishige, 2005).

Most commonly, multiple comminuting stages follow periods of sun-drying, as at Choupaldandi Village, Bangladesh (Ruddle and Ishige, 2005); comminuting may be done at the same time as the shrimp is salted, or it may follow salting, and sometimes drainage also (IPFC, 1967; Ruddle and Ishige, 2005).

11.4.2.7 Fermentation　Deliberate fermentation periods, in contrast to those that occur while raw materials are either in transit or in storage, occur in most production processes. There is also a considerable variation in the length of the fermentation period, from 1 day to 6 months at various places. Average fermentation time, however, is about a month. In general, the soft or wet, lightly or unsalted shrimp, *shiokara* pastes are fermented for only a few days. Those with a higher salt-to-shrimp ratio, and where a solid final product is desired, are fermented for a longer period (Ruddle and Ishige, 2005).

The containers used for fermentation range from none at all to specialized vats or concrete tanks. In Bangladesh, the salted shrimp is simply heaped in the shade, without being put into a container. However, at the industrial level of production, specialized wooden vats or concrete tanks may be used. At the cottage industry level, salted shrimp is generally fermented in large earthenware or ceramic jars, or in bamboo baskets.

11.4.2.8 Artificial Coloring Agents　Artificial coloring agents are added to shrimp *shiokara* in some cases in Indonesia and in all of the Philippines (Ruddle and Ishige, 2005).

11.4.2.9 Blending　Blending is done either to soften hard pastes, with the addition of liquids that emerge out during the production process, or to blend *shiokara* of different qualities. The purpose of the blending done in some communities in Bangladesh is to produce a uniform grade of *shiokara* from disparate qualities of fresh shrimp— those caught in the bottom waters which have ingested large amounts of detritus, and cleaner shrimp taken near the surface (Ruddle and Ishige, 2005).

11.4.2.10 Adulteration   Apart from the admixture of by-catch, particularly anchovy, in the lower grades of "shrimp" *shiokara*, little adulteration of the product appears to be practiced.

11.4.2.11 Shaping   The finished paste is sent to the market in a variety of individual containers. Where shaped, it is generally in the form of a block, which may range in size from the 2 kg blocks sold to middlemen to precisely measured 500 g blocks produced by small-scale factories; however, sausage shapes are favored (Ruddle and Ishige, 2005).

11.4.2.12 Packing   Traditionally, most shrimp *shiokara* has been marketed loose, being sent to market in bamboo baskets. This is often still the case. Where packaging is used, it ranges from banana or other leaves, to plastic packaging, as found on the products of cottage- and small-scale industry (Ruddle and Ishige, 2005).

11.4.2.13 Typical Example   A shrimp *shiokara* paste, locally called *nappi*, is produced in coastal villages in the Cox's Bazaar District of Southern Bangladesh by the Rakhiyane people, a minority population of Arakanese origin, where shrimp *shiokara* production is undertaken as a community-wide cottage industry (Ruddle and Ishige, 2005). Most of the 80 households in Choupaldandi, for example, make *nappi*, the combined production of which was 40 tonnes in 1983. Traditionally, men fish and do some of the *shiokara*-making tasks, but women do much of the preliminary processing, sun-drying, and comminuting. The peak shrimping season is September–January, when *shiokara* production also peaks. However, since some shrimp can be caught the year-round, some *nappi* is also produced throughout the year.

The tools used are simple, and, except for the set bag net (*behundi*) for catching shrimp and various metal containers, are mostly made locally from readily available materials. They comprise bamboo baskets for unloading the shrimp from the boats, bamboo mats for sun-drying the shrimp, mortars and pestles for comminuting, bamboo packing and transport baskets, and leaves for lining the basket (Ruddle and Ishige, 2005).

Unlike most other parts of Asia, in southern Bangladesh the shrimp used to make *shiokara* are mostly not planktonic, but juveniles of *Penaeus indicus*, *Metapenaeus monoceros*, and *Perapaeniopsis stylifera* (Tanasupawat and Visessanguan, 2013). They are caught either directly by the *nappi*-making household or purchased from fishers. Fresh shrimp is purchased in 40 kg lots at a price based on the prevailing whole-sale price of the finished *nappi*. Only local salt made by solar evaporation is used to make *nappi* (Ruddle and Ishige, 2005). The *nappi* production process is shown in Figure 11.11. After unloading, and, if necessary, sprinkling with salt, the shrimp are spread on a bamboo mat to sun-dry for a day. Next, the material is comminuted to a coarse paste using a mortar and pestle. More salt is added during this procedure. The crushed shrimp are then, spread again on a bamboo mat to sun-dry further for a day. Following this, the material is returned to the mortar and pestle, slightly moistened

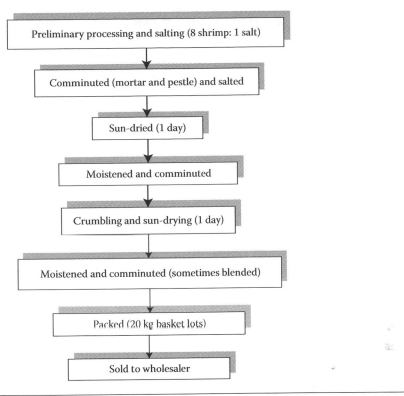

Figure 11.11 The production process of shrimp shiokara (*nappi*) at Choupaldandi Village, Cox's Bazaar District, Bangladesh. (Adapted from Ruddle, K. and Ishige, N. 2005. *Fermented Fish Products in East Asia*. International Resources Management Institute, Hong Kong.)

with freshwater, and then, pounded for about 20 min to reduce it to a fine, paste-like consistency. On the third morning, the paste is separated into small lumps, and again spread on the bamboo mat to sun-dry for a day. On the 4th morning, the material is returned to the mortar, sprinkled with freshwater, and pounded. This procedure results in a highly comminuted, smooth shrimp *shiokara*.

During the bright, warm weather that is ideal for sun-drying, the shrimp are sprinkled with salt as soon as they are dumped on the bamboo drying mat. To prevent insect infestation or putrescence, the shrimp may be salted at sea, when rough conditions preclude prompt landing. Similarly, during cloudy weather, when drying conditions are poor, the shrimp are salted prior to the first sun drying. Extraneous matter, the by-catch, of small finfish and crabs, is carefully picked out and, generally, discarded.

The total rate of salting varies considerably, according to village. At Choupaldandi, only a lightly salted *shiokara*, with a shrimp–salt ratio of 40:1–1.5, is produced, whereas the saltier product of Mohiskhali Island is made at a ratio of 8:1.

At this stage, in the specialized *shiokara*-making Mudichara Hamlet, of the Gorakghata Union, on Mohiskhali Island, *nappi* of different shades of a brown–pink–purple range are blended together to produce a product of uniform color. Since the lighter *nappi* has a higher economic value, the darker shades are blended with it to produce an average color that is more readily marketable. The rate of turbidity in the bay waters affects both the color and taste of the *nappi*; the higher the rate, the darker and poorer-tasting is the *nappi*. It is because shrimp in such water ingest a higher rate of mud and detritus (Ruddle and Ishige, 2005). Immediately after the third comminuting, market intermediaries pack the *shiokara* in 20-kg lots for sale and transport. Bamboo baskets lined with leaves are used for this purpose.

11.4.3 Shidal

Shidal is a salt-free fermented fish product indigenous to the Northeast sector of India. It has several local names, and is popularly called *"seedal," "sepaa," "hidal," "verma,"* and *"shidal"* in Tripura, Assam, Mizoram, Arunachal Pradesh, and Nagaland states of India. In particular, Assam and Tripura are the major producers of *shidal* amongst the Northeast states of India. The technology of its processing is very old and originated in the erstwhile undivided India (now Bangladesh) and is believed to have come into existence at least before the British Era in the Northeastern states of India, that is before 1824 (Muzaddadi and Basu, 2012). Indian history reveals that the people of this region did not know the use of salt before its introduction by the British (Muzaddadi and Basu, 2012). Even after the British Era, salt was treated as a highly valued and scarce commodity, and as an alternative of salt people used a substance known as *"khar,"* made from banana or papaya plants (Muzaddadi and Basu, 2012). Thus, people could not afford salt for fish preservation, and this may be one of the reasons for preserving fish in this unique way, without using salt. The plains of Bangladesh and adjoining Northeastern sector of India are famous for their "beel fisheries", which are weed infested shallow water bodies that dry up fully or partially during winter. They are an excellent habitat for weed fish, such as *Puntius* sp., that propagate naturally with the beginning of the rainy season and form a good fishery when the water level starts decreasing with the onset of winter. The reason for the exclusive use of *Puntius* sp. for *shidal* production possibly lies with the huge availability of this fish, particularly in the post-monsoon period, and probably this necessitated the evolution of this cheaper technology of fish preservation for their use in lean periods.

Shidal is very popular due to its strong flavor, and the product is usually prepared from *Puntius* sp. It is solid, compressed, and pasty. The shape of the fish remains almost unchanged except for a little disintegration near the belly and caudal portion. The color of the best quality product is dull white, which gradually becomes slight brown to deep brown on continuous exposure to air. A strong odor permeates the air in and around the storage, and gives the area a characteristic smell of *shidal*. Its quality, however, deteriorates very quickly after breaking the seal of the container and exposure to air (Majumdar and Basu, 2009). Presently, the production of *shidal* is confined to particular districts of the states of Assam (Nagaon, Dhubri, Goal para, and Kachar districts), Manipur (Imphal city), and Tripura (West Tripura district). *Shidal* is usually produced in the months of November–February. Dried salt-free *Puntius* species are usually used for preparation of *shidal*. The different steps in shidal preparation are described here.

11.4.3.1 Matka and Its Oil Processing   "*Matka*" is the local name of the pear shaped earthen container used for fermentation of fish (Figure 11.12). Although m*atkas* of different sizes are in use, the most common size has a neck diameter of 8 inches, a middle, expanded diameter of 24 inches, and a height 36 inches. The capacity of a *matka* usually ranges from 1 to 40 kg. Since *matkas* are made of earth they are breakable, but can be used for several batches of fish until they break.

The best quality "*matkas*" are made from very fine black soil, as these absorb much less oil during processing and provide much less air permeability. For similar reasons, producers believe that the older the *matkas*, the better the product quality, and the less the cost of production. Before use, *matkas* are smeared with oil for the reasons which have been described earlier. Oil extracted from *Puntius* is generally preferred by fishers and commercial producers if it is available in plenty. In the case of large scale production of *shidal*, vegetable oil, especially mustard oil, is preferred. Oil is smeared in both inner and outer walls of the *matka* followed by drying in the sun. The oil smearing and subsequent drying process is continued for 7–10 days for new *matkas*, until they become fully saturated with oil and unable to absorb any more; they are now ready for packing with fish (Muzaddadi and Basu, 2012). In the case of re-used *matkas*, 2–5 days of oil smearing and subsequent drying is required.

11.4.3.2 Water Soaking and Drying of Dried Fish   The dried *Puntius* is cleaned and sorted and further dried in the sun for 3–5 days to remove moisture from the fish to maximum possible extent, and also to drive away the maggots, if any, although those with signs of maggot infestation are not used for *shidal* production. Drying is followed by water soaking of the fish in porous bamboo baskets, usually for 10–15 minutes, preferably in the running water. The absorption of water is greater and quicker due to

(a)   (b)

Figure 11.12   (a) Traditional oil-processed mutka. (b) Traditional mutka packed up. (From Muzaddadi, A.U. and Mahanta, P. 2013. *Afr J Microbiol Res* 7(13): pp. 1086–1097. With permission.)

the previous drying of fish. After water soaking, fish are spread on bamboo mats or on a cement floor in the shade over night for drying. The evening hours is the best time for water soaking, because the subsequent drying of water soaked fish for 8–10 hours passes without any problem from flies and birds. This step is a very critical step for yield of a good quality *shidal* and for determining the total fermentation period. The duration of water soaking and subsequent drying, is determined by previous experience, depending upon the quality, period of fermentation, and shelf-life of the end product required. After water soaking and drying, the fish become soft textured with a dry surface and are ready for packing into the *matka*.

11.4.3.3 Filling of Matka   Before filling, the oil processed *matka* is placed in the ground by digging a hole in such a way that one third of the belly remains buried in the ground. The dug out soil is gathered around the underground portion of the belly, and *matka* is fixed firmly, ensuring that it stands exactly vertically so that it can withstand the pressure during the filling and compaction of the fish (Muzaddad and Basu, 2012). Clean gunny bags are spread surrounding the *matka* to avoid any raw material getting contaminated with the soil underneath. After fixing the *matka* in the ground, the partially dried fish are spread in a layer and uniform pressure is applied with bare hands or feet (in the case of large-mouthed *matkas*) (Muzaddad and Basu, 2012). Once the layer is tightly packed, subsequent layers are added in a similar manner util the layers reach near the neck. Sometimes, a wooden stick is used along with the hands or feet to make an almost air tight packing. About 35–37 kg of dried *punti* is required to fill one 40 kg capacity *matka*.

11.4.3.4 Sealing of Filled Matka   Once the matka is filled to the neck, it is primarily sealed with a cover paste which is made from a dust of dry fish. The thickness of the cover layer is 2–2.5 inch. Then, either any broad leaf or newspaper is placed over the wet seal of cover paste and finally, the *matka* is sealed by a layer of wet mud made from clay soil. This layer is often checked for any crack and is repaired immediately by wet mud again. The filled *matkas* are lifted to the surface and left undisturbed in the shade for maturation. The usual period of maturation is 4–6 months, but this may be extended to a year. About 40–42 kg *shidal* is obtained from each *matka*. The entire process of shidal making is shown in Figure 11.13.

11.4.4 Sidol

Sidol is a semi-smoked fish product popular in the lower Assam region of Assam State, especially amongst the *Rajbangshi* communities. The product is of the ready-to-eat variety, and can be consumed with main rice dishes. There is no offensive smell associated with the product. In this process, heat from burning of wood is used to dry the fish because of continuous rain during the months of monsoon in the Northeast region of India. The fish used for *Sidol* production includes indigenous trash fishes such as

Sort uniform sized fish (Raw material dried *Puntius* sp.)

↓

Dry under bright sunlight for 2–3 days

↓

Wash in river water taking in bamboo basket

↓

Dip in river water for 5–15 min depending on the texture of the fish

↓

Spread on a shaded floor and dry overnight

↓

Smear fish with vegetable oil/fish oil and dry under sun

↓

Smear earthen pots (*mutka*) with vegetable oil/fish oil in both sides and dry under sun

↓

Repeat the above step for 5–10 days until *mutka*s become saturated with oil

↓

Bury the *mutka* into the ground (up to the belly) to fix tightly with ground keeping upright and spread gunny bags surrounding it

↓

Spread a thin layer of fish inside *mutka* and press fish with bare foot in standing position so that no air gaps remain between fish

↓

Repeat the above step until *mutka* is packed up to the neck

↓

Fill *mutka* from the neck to the rim with fish-paste made from broken pieces of fish after grinding them in a wooden mortar

↓

Cover mouth with a sheet of tree leaf/polythene and keep it in a sheltered place for 1–2 days

↓

Seal the mouth of *mutka* with clay prepared from fine soil and store for incubation

↓

Promptly mend the cracks on the seal which are likely to appear on subsequent days of storage and keep on mending until no more cracks appear on the seal

↓

Store for incubation undisturbed for 4 – 6 months depending on the texture of raw material

↓

Remove the clay-seal and fish-paste carefully and discard the seal

↓

Take out ready *shidal*

Figure 11.13 Flowsheet of the traditional method of *shidal* preparation from dry *Puntius* sp. in Tripura state. (From Muzaddadi, A.U. and Mahanta, P. 2013. *Afr J Microbiol Res* 7(13): pp. 1086–1097. With permission.)

Puntius and *Mystus* species. The peak season of production is the monsoon, that is June–October. Fresh fish are collected from the markets or landing centres and dressed by removing scales, viscera, etc. Dressed fish are washed thoroughly and placed over a bamboo meshed tray to drain out the water. This is followed by placing the moist fish over a bamboo platform 3–4 feet high with a meshed top. A smokeless fire is provided at the bottom of the platform to heat the fish until it is dried completely. The dried fish are then, mixed with different spices and macerated along with the petiole of aroid plants (*Colocassia*) to make a homogenous paste. The plant ingredients enhance the stickiness of the product. Small balls are then, made from this paste, wrapped tightly with leaves of aroid plants, and put in an earthen container (pot) which is closed properly with a polythene sheet. The filled pot is left undisturbed in a dark place for fermentation. After 4 days, the product is ready to eat. A small preparation however, is needed before consumption. A few balls of *Sidol* are taken out from the container and baked for sometime. Then, a small quantity of salt and mustard oil is added and made into paste and consumed as "*chutney*." For long-term storage, the wrapped balls of *Sidol* are dried under direct sunlight periodically—at least once in a month—and again put in a container. During this time, the earthen containers are smeared with a little mustard oil, which is believed to act as a preservative. With this practice of periodical drying, the product can be preserved for 2–3 years without any spoilage.

11.4.5 Fermented Fish Pastes

In most Asian countries, people rely heavily on fish or shrimp paste in their daily diets. Unfortunately, the numbers of these marine products are declining due to global warming, as well as pollution of the oceans. So people have devised their own way to sustain the availability of fish or shrimp products to fulfill their protein intake, such as by transforming the fish or shrimp into paste-like products *via* fermentation, for extended storage as well as to impart unique sensory qualities to the product. Fermented fish paste is one of the diverse fish products which serve to enhance the flavor of dishes (Tamang, 2010a). The product has vernacular names such as *hentak*, *ngari*, and *tungtap* in India (Baishya and Deka, 2009). The method of their preparation is similar, except for the use of different fish species and the preparation steps, according to respective ethnic preferences.

11.4.6 Ngari

Ngari has been valued by the people of the Manipur district for decades for its enormous value, and it is one of the essential ingredients of every household in the area, due to its taste, therapeutic properties, and strong appetizing nature (Figure 11.14). Because of these qualities, its value as a food ingredient has spread into other states of the Northeastern region, and even into Myanmar. Considering the importance of *ngari* to the people of Manipur, sun-dried *Setipinna* species were recently introduced as a raw material for its

Figure 11.14   *Ngari*—A fermented fish product.

preparation. It is consumed daily as a side dish called "Iromba" (mixed with potatoes, chillies, etc.) with cooked rice by people from all communities of Manipur. For the preparation of *ngari* (fermented fish), a typical small type of fish, locally called *phabou nga*, is used (Devi and Suresh, 2012; Tamang et al., 2012). For the Manipuris, the daily meal is not complete without *ngari*, which is eaten either in the form of chutney (*iromba or morok metpa*) or as an ingredient in other curries (Baishya and Deka, 2009).

The production of *ngari* is, however, restricted to the valley region of Imphal (Manipur) and surrounding areas during the months of October–January. The preparation procedure of *Ngari* starts with the collection of locally available punti fish. Sun-dried, salt-free punti fish (*Puntius sophore*) are also imported from the Brahmaputra valley of Assam and Bangladesh. However, in commercial production of Ngari in bulk by major entrepreneurs, the dried *Puntius* fishes are collected from more distant states, such as Andhra Pradesh and West Bengal, in dried form. The local fisherman catch these fishes from the wetlands or pats, and also as by-catch from local ponds. The processing industry is absent in Manipur, so the processing and marketing of this product is confined to some rich household, which, along with some hired workers, run the businesses. *Ngari* production technology can be of two major types; one is the indigenous method, and other one is a modified commercial operation.

The *phabou nga* (*Puntius* sp.) is washed thoroughly with water and sun-dried until it becomes crispy (Devi and Suresh, 2012). Crushing of the fish head should be done properly with the help of a hammer. A special vessel is used for the preparation of *ngari*, where mustard oil is spread and then, the dried fish is placed in layers. The container is sealed after filling and, to make it air tight, sand is placed on the top, and fermentation is allowed to occur naturally. It takes about 3–6 months to mature and to be ready to be eaten, which is judged by a typical odor.

11.4.6.1 Ngari chaphu The fermentation jar traditionally used for the purpose of *ngari* production in Manipur is locally known as *ngari chaphu*, which is a round-bottomed, narrow necked, earthen pot—called "*kharung*" in some localities of Manipur—specially designed for the purpose (Devi and Suresh, 2012). For the preparation of *ngari*, jars of varied sizes can be used which can hold from 15 to 75 kg of fish. The thickness of these pots is about 1.5–2.5 cm and they are brick red in color in the initial phase. The product's quality is enhanced as the pots are used repeatedly for many years, and as the older *chaphus* become more air-tight thus, giving better conditions for fish fermentation.

Pre-processing of chapus: Before using them, *chaphus* require pre-processing; this is achieved by smearing the inner walls of the *chaphus* with oil and drying them in the sun. The oils used may be any vegetable oils, like sunflower, mustard, or palm oil. Traditionally, either vegetable or fish oil is used, depending upon availability and cost. In some places, fish oil is incorporated in varying amounts, as it provides some essential *n*–3 fatty acids. In commercial production, the use of vegetable oil reduces the cost. When the oil is absorbed and finally fully dried, the process is repeated (Muzaddadi and Basu, 2012). The process is continued for 1–2 weeks until the *chaphus* become fully saturated with the oil on the inner wall. It is economical to use old and used *chapaus*, as newly used *chaphus* absorb more oil and thus, increase the cost of production, and are more prone to the fish sticking to the inner walls. The earthen jars are tightly bound with wires to facilitate easy handling and to avoid leakage and breakage.

Covering paste: This is used mainly for the purpose of covering the mouth of the *chaphus* and is prepared from fish powder or even from some trash fishes and from *Puntius* that are discarded during the drying process. Fine powder is made by drying in sunlight followed by grinding. This is followed by moistening using a little water to make a dough-like paste which is used to seal the mouth of the *chaphus*.

Covering leaf: These are the temporary covers used to wrap the covering paste in the mouth of *chaphus*. Any broad leaf can be used for this purpose. In Manipur, the most commonly used cover leaves are those of *Musa paradacea*, although others, like bottle gourd and catesu leaves can also be used; newspaper can also be an ideal replacement.

Sealing the mouth: Finally, the mouth of the *chaphus* is sealed tightly using a thick mud paste made from fine soil. Even old rug sacs are used after making them solid and compact.

11.4.6.2 Indigenous Method This is more economical and practiced more frequently in the fishing seasons when *Puntius* fish are available as by-catch in many village fish ponds. In this process, the fish are collected from local fish ponds or wetlands (*pats*), and whole fish are used in *ngari* preparation. The fish are washed thoroughly and only intact fish are used for drying. When fish are semi-dried, they are rubbed with a little mustard oil or fish oil and again dried for 3–4 days in sunlight. While rubbing with the oil, a little salt is also added, which is believed to facilitate the absorption of oil deep into the fish's body. However, the amount of oil added should be optimized, as it determines the texture of the final product; when it is too low the texture of the product becomes very firm and is unpopular. The rest of the procedure is as described earlier.

After keeping for a period of 3–6 months, the *chaphus* are ready; this is known locally as "*chaphu kaiba*" in Manipur. The final product is removed carefully, layer by layer, and packed in smaller containers or pots. However, care must be taken while handling the product, and "*chaphu kaiba*" is usually done in a dry day to prevent fungal infestation that might occur in moist conditions.

11.4.6.3 Modified Commercial Method The fish are sun dried for 4–5 days to reduce the available moisture to a minimum and to chase out any maggots, if present. The dried fish are soaked in water for a period of 10–15 minutes, and the absorption of water is rapid due to the previous drying. The water soaking process is usually carried out in the evening so that the next step of partial drying for 8–10 h is done at night time to avoid bird predation. The partially dried fish are of soft textured with a dry surface. The fish are then, smeared with vegetable oil followed by another drying under the sun for about an hour. The fish are then, trampled or rolled by rollers to make them soft and tender.

The oil smeared, partially dried fish are taken in a big tumbler, and the air present is removed by continuous piercing with rod or bamboo stick into the heap of fish. Usually, the processors trample the product layer-by-layer so as to make it air tight. The fish now become ready for packing in the *chaphus*. The pre-processed *chaphus* are buried in pits excavated earlier and layered with moist sacks with half or one-third of the belly underground to fix it tightly. Initially, 5–6 kg of fish is put in each *chaphus* followed by pressing with the feet until wet liquid is released from the fish, and the process is repeated until the containers are filled with fish up to the neck. It is then sealed, first with a covering of paste and then, with wet mud on top of a layer of leaves, and kept for maturation for 3–6 months. After maturation, the top layer of fish in the *chaphus* (locally called as "*phumai*") is removed and the rest is *ngari*. The shelf-life of *ngari* is about 12–18 months.

11.4.7 Hentak

Hentak is a fermented fish paste of Manipur, India (Baishya and Deka, 2009). It is a different kind of fermented dry fish mixture, with added plant ingredients such as "*Hongoo*" or "*Paangkhok*" (*Colocasis* sp.), although the petioles of aroid plant (*Alocasia macrorhiza*) are mostly used. The plant ingredients give taste and also help in the fermentation of the dry fish. Since times immemorial, the preparation of *Hentak* has been traditionally practiced in Manipur, but there is no commercialization of it. Nevertheless, the home-made *Hentak* is tasty and is offered to mothers in confinement and patients in convalescence for its nutritional value. The principal aim behind the production of *Hentak* is to preserve the animal protein along with vegetable ingredients for a longer duration without any spoilage. It is consumed as curry as well as a condiment with boiled rice.

Various dried fish like *Puntius* spp. (Ngakha); Mola (*Amblypharyngodon mola*) (Mukaangaa); Darkina (*Esomus danricus*) (Agashaang), etc., are used for the production of Hentak. The dried fish of any variety is washed thoroughly and allowed to dry completely by exposing them under the sun continuously for a few days till the fish easily crumble. The completely dried fish are ground into fine powder in a grinder. In some places, "Shoombal" (a manually operated wooden grinder) is used for grinding. The petioles of aroid plants are finely chopped, washed thoroughly, and dried in the sun for about an hour. The dried vegetable and powder of dried fish are taken in a 1:1 ratio and made in to a homogenous paste in a grinder. The texture of the paste is most important for its quality and shelf-life. The texture should not be too soft nor too hard. Optimum texture quality can be achieved by adding adequate quantity of vegetable component or fish powder. The texture becomes softer if more of the vegetable component is used, harder when fish powder is used. Small balls, not bigger than an egg, are made from this paste and are kept in an air-tight earthen container and allowed to undergo fermentation for four days. After fermentation, the balls are taken out and kneaded uniformly and again made into small balls. These balls are again kept for another six days in the same air-tight container. The balls are then, taken out of the container and kneaded well with the addition of a little mustard oil, and small balls are made out of the paste and againstored in the same ait-tight container for 10 days. The texture of the product is checked for hardness periodically. The fermented small balls are then, wrapped in clean banana leaves and put in the earthen container which is made air-tight with a paste made from clay and raw cow dung in a 1:1 ratio. After a storage period of six months, the product is ready for consumption. For consumption, these balls are crushed and again made into paste with a little water.

11.4.8 Sukati Machh

It is a powder of smoked fish and is popular in some communities (Hazarika, Bora, Kalita, Rajbangshi, Mohanta, etc.) of Assam State. It is used to make soup with chilli and garlic. There is a belief prevalent amongst the users that it is helpful for the prevention of malaria. Any indigenous trash fish can be used to prepare *Sukati Machh*. In this process, fish is preserved in powder form by the combined preservative effect of salt and smoke, generated by burning wood (Tamang et al., 2012). In its preparation, fish (e.g., *Harpodon nehereus* Hamilton) is collected, washed, rubbed with salt, and dried in the sun for 4–7 days; it is stored for 3–4 months (Thapa et al., 2006).

11.4.8.1 Sakuti   To prepare *sakuti*, any variety of fresh trash fish is dressed by removing scales and viscera, and washed thoroughly with clean quality water (Figure 11.15). The cleaned fish are then, immersed in brine (6%–10% solution of sodium chloride) for 10–15 minutes and then, dried in the sun for about 30 minutes. The partially dried fish are then, placed over a wooden sieve (mesh size is according to the size of the fish), which is placed over a fire to expose the fish to the smoke and heat produced by the burning wood. After 4–5 hours (depending upon the size of the fish and previous

Figure 11.15   *Sukuti.*

experience), the fish are turned upside down so that both sides are exposed to the smoke. Then, after complete drying, the smoked fish are ground with the help of an electrical grinder or "Gaiyl-Chekai" (a traditional, wooden manually operated grinder) to a powdery form. This fish powder is then, stored in air tight containers and kept in a dry place. The product has a long shelf-life and is a very popular sun-dried fish product in the cuisine of the Gorkha (Tamang et al., 2012). It is consumed in pickles, soups, and curries, and is also commonly sold in local markets.

11.4.9 Kharang and Khyrwong/Kha Pyndong

Kharang and *Khyrwong/Kha Pyndong* are the products of age-old traditional technology of processing fish which is still practiced by the people of Umladkur and Thangbuli villages of the Jaintia hills district of Meghalaya, India. They are indigenous smoked fish products, and several dishes, including curry, chutney, etc., are prepared from them. When there is sufficient catch, the excess fish is processed and preserved for use in the lean period. During the off-season, fish imported from other states are also used for the production. Usually, the fish of bigger varieties, like Chocolate Mahseer (*Acrossocheilus hexagonalepis*), Rohu, Mrigal, Grass carp, etc., are preferred for the production of *Kharang* and *Khyrwong/Kha Pyndong*.

A special type of small hut or smoke house is employed for the purpose. The roof of the hut is covered with leaves/grass (locally known as *Tynriew*), its floor is muddy having a traditional furnace in the centre of the room. The door is closed during smoking. Wooden frames are fitted surrounding the furnace in such a way that flame can flow through the centre of the frame. The fish are pre-processed before smoking. They are first washed properly and their viscera are removed. When bigger fish of more than 500 g–1.0 kg are used, the product is known as *Kharang* and in case of smaller size (less than 500 g) fish, it is called as *Khyrwong/Kha Pyndong*. After thorough washing of dressed fish with plenty of water, a single fish, in the case of *Kharang*, or two fish at a time, in the case of *Khyrwong/Kha Pyndong*, are pierced by bamboo sticks. The bamboo stick is pushed through the mouth of the fish through the belly up to the caudal peduncle region. The fish are now ready and arranged in the wooden frame and the smoking process starts by burning wood in the furnace. Bigger fish (*Kharang*) are arranged near the flame. Fish are placed upside down in such a way that the head and belly dries up first, and then, gradually rolled-over/turned in the frame, using the protruded bamboo sticks, for uniform smoking and drying. When the fish become brown in color, they are removed from the wooden frame and allowed to cool for storage. The smoked fish are then, properly arranged in bamboo baskets, covered, and sent to the market for sale. Different items are traditionally prepared from this smoked fish, such as soup, curry, chutney, fish balls, etc.

11.4.9.1 Lonailish

Lonailish (Figure 11.16) is a salt fermented fish product, made exclusively from Indian shad (*Tenualosa ilisha*), a high fat fish (fat content of adult

Figure 11.16   Dry salted *Hilsa*.

hilsa ranges from 14% to 25%). It is a very popular product and is widely consumed in Northeast India and Bangladesh, mainly due to its typical flavor, aroma, and texture (www.ifr.ac.uk-2011). Tripura (India) is the major producer of *lonailish* amongst the Northeast States, and the product is exported to other neighboring states. The fish (*Tenualosa ilisha*) are washed properly with potable water and are then, descaled. The tail and head portions are removed leaving the gut inside. The fish are cut diagonally in such a way that the steak/chunk has more flesh exposed than the skin. The thickness of the steaks are generally 1.5–2.0 cm. Each of the fish steak is rolled thoroughly in salt (fish to salt ratio is 4:1) and kept in a bamboo basket (locally called a "tukri"), layer after layer, with flesh side down. Some salt is also put at the bottom of the basket before the fish (Majumdar et al., 2006). Besides, salt is sprinkled between each layer and above the top layer. The filled basket is covered with a black polythene sheet so as to avoid the entry of light. The baskets are stored in a dark place. The brine formed is allowed to drain. The fish steaks are kept in a dry salted condition for 48 h. A considerable amount of moisture content of fish is removed during this process, and the color of the flesh becomes dull white and the texture becomes somewhat tough.

The salt cured steaks are then, packed in a container. The containers used for this fermented product are the empty tin containers (cap. 18 L) used for cooking oil. Packing of cured *hilsa* is done layer after layer and compacted uniformly by hand after adding each layer until the layers reach to 2–3 cm below the top. The salted *hilsa* steaks are shaked well before packing to remove adhering salt. Then, cold saturated brine is poured slowly in the container over the fish to fill the voids between the steaks and maintain a level of brine about 2–3 cm above the fish (Kakati and Goswami, 2013). The saturated brine solution is boiled one day before packing and cooled overnight. The container is closed with a lid followed by their stacking in a dark room, and left undisturbed for 4–6 months for fermentation. The *lonailish* tin (fermenting container) can be kept over a year, but the marketing starts after a few months of maturation. The fish remains submerged until sold. Once a container is opened, the materials are sold part by part by taking fish from out the brine solution. If the fish is exposed to air and light for few hours, the pink red color changes to grayish black, which is unacceptable.

Sliced *hilsa* is about 1.50–2.00 cm in thickness with the texture remaining firm and the flesh not easily separing out from the bone (Majumdar and Basu, 2009). It has a characteristic strong aroma mixed with some sweet, fruity, and acidic notes, along with some saltiness. The strong odor spreads in the air during its storage and gives the area a characteristic smell of *lonailish*. It is kept immersed in saturated brine until consumption (www.ifr.ac.uk-2011). The uniqueness of this product lies in the fact that, despite the presence of salt and metals coming from the container used for fermentation, the rancidity of this highly-unsaturated fatty-acid containing fish is kept under control, and is not manifested as long as the fish steaks are kept immersed in the fermenting brine.

There are some advantages of the traditional method of preparation of *lonailish*. The period from June to September is the peak time of preparation of *lonailish*, when *hilsa* is caught in large quantities during its upward migration for breeding (Majumdar et al., 2006). During this time, the fish has a very high fat content—up to 19%–22% (Majumdar et al., 2006). A diagonal cut appears to be helpful in providing more surface area of fish flesh exposed to salt as well as halotolerant bacterial action. Loss of a significant amount of water from the fish as self-brine during dry salting prevents dilution of saturated brine during the ripening stage (Majumdar et al., 2006). Boiling of brine is done firstly to prepare a saturated solution and secondly to destroy undesirable microorganisms present in the salt that might contaminate the product. During maturation, the texture of the fish changes and the ripened product attains the texture of cooked meat, and the enzymatically hydrolysed fat enhances the flavor.

11.4.9.2 Hukoti   *Hukoti* is a traditional indigenous dry fish product, made by different tribal communities of upper Assam (India). Fatty fishes such as *Puntius* spp. are normally preferred. Fishes are first dressed to remove scales, fins, and intestines. Dressed fishes are mixed with salt and turmeric in no fixed ratio, and kept overnight in a container covered with a plastic sheet so as to prevent any fly infestation. The next day fishes are washed and again mixed with salt and turmeric, and spread uniformly on a perforated bamboo tray (locally called as "*chelani*") and sun dried for 3–4 days, or dried over kitchen fire by placing on a rack made of wood or bamboo, so as to make the texture tough to facilitate subsequent grinding operation. Dried fishes are later ground in wooden domestic grinders (locally called as "*dekhi*"). The powdered dry fish are retrieved from the grinder and sieved using a locally made bamboo sieve to remove extraneous matter like scales, bones, etc. and get homogenous material. The powder so obtained is further ground, and at this time the stem of de-skinned *Colocasia* spp. (cut into pieces) are mixed with the powdered fish. About 100 g of *Colocasia* stem is mixed with 1 kg of dry fish powder. Occasionally, the leaves of a shrubby plant *Euphorbia ligularia* Roxb. (commonly known as "leafy spurge" or "milk hedge," and locally known as *Siju*) is also added, along with *Colocasia*. Approximately, 100 g of Siju leaves per kg of dry fish powder is mixed and ground properly. The villagers, believe that these ingredients are incorporated to increase the adhesiveness of the prepared mixture. Some communities also add a particular type of chilli (*Capsicum* chinense or *C. frutescens*, and ginger, garlic, etc.) during grinding to suit their own taste and preferences. After grinding, the resultant paste becomes green in color. The paste is then, stuffed into locally available matured bamboo cylinders (approx. 2.5 feet in length to accommodate nearly 2 kg of the paste). After filling, the bamboo cylinder is sealed with dry banana leaves and then, with moistened clay to make it airtight. The sealed bamboo cylinders are then, placed on a bamboo or wooden rack fitted over the kitchen for drying. The usual drying time requirement is 10–15 minutes, 4–5 times a day, for a period of about 3 months. After this period, "*hukoti*" is ready for consumption and is sold in the local markets. It is consumed by frying in oil alone or with other ingredients/vegetables. Usually cooked "*hukoti*" is consumed with rice. According to the popular belief of the tribal communities, *Hukoti*, is used as a pain-killer as well as a local therapeutic to cure malaria.

11.4.10 Smoked Fish Products of Northeast States of India

Amongst the Northeast states of India, Manipur is famous for different varieties of smoked fish. Smoked *Esomus danricus*, *Puntius sophore*, *Mystus bleekeri*, *Amblypharyngodon mola*, *Notopterus notopterus* and *Glossogobius giuris* are some of the important smoked fishes of Manipur. The smoking of medium size fish is done by spreading the fish on a wire tray and then, exposing to flame briefly to burn the skin. The process is repeated after turning the fishes upside down. In the case of small size fish, they are simply spread on a wire tray. Then they are exposed to smoke from burning paddy husk from a distance of about 30 cm below for about 2–3 h at 70–80°C.

11.4.10.1 Numsing   This is a traditional fish product developed by the "*Mising*" community of Assam, India. *Numsing* is a semi-dried and semi-smoked paste like product prepared by mixing fish, petioles of arum (*Alocasia macrorrhiza*), and spices. Small cheap fish species are preferably used for this product. The fish are first dressed to remove gills, scales, and intestines and then, washed with clean water. The edible part of the arum is sorted out, peeled, and sliced. The fish are dried on a specially made bamboo rack fixed at a height of 3–4 feet over a fire. The fish are spread on it and are heated with a nearly-smokeless fire that is created by burning dry bamboo. It is usually done during the night hours and is continued with a low flame until the fish become moderately hard. Arum slices are sun dried for a day.

The flame-dried fish and sun-dried arum slices are mixed together in the ratio of 4:1 with other spices such as red pepper and green chilli, ginger, and garlic are also used, occasionally, for taste, but without salt. The mixture is ground in locally-made foot-operated wooden grinders (*Dhekee*). The ground mixture is stuffed compactly in a bamboo cylinders, keeping a headspace of about 10–15 cm for subsequent sealing. The bamboo cylinder is the single internode of immature bamboo. The filled bamboo cylinders are sealed with previously washed raw leaves of bladder fern (genus *Cystopteris* sp.). Finally, the bamboo cylinders are sealed with moist clay soil so as to prevent entry of air or insects. The sealed bamboo cylinders are then, hung about 2–3 feet above the traditional mud-oven in the outdoor kitchen where the *Mising* people cook food. After allowing a fermentation period of about 30 days in the summer, with intermittent heating, the product becomes ready for consumption. The shelf-life of the product is about 2–3 years. *Numsing* is usually consumed along with vegetables.

11.4.10.2 Gnuchi   *Gnuchi* is a traditional smoked fish product of the *Lepcha* community of Sikkim (Tamang et al., 2012). Fish (*Schizothorax richardsonii, Labeo dero, Acrossocheilus* spp., *Channa* sp.) are collected from the river, kept on a big bamboo tray to drain-off water, degutted, and mixed with salt and turmeric powder. The bigger fish are selected and spread in an upside down manner on *"sarhang"* and kept above the earthen oven in the kitchen. The small sized fishes are hung on a bamboo strip above the earthen oven and kept for 10–14 days. *Gnuchi* will keep well at room temperature for a period of 2–3 months and is eaten as a curry.

11.4.10.3 Suka ko Maacha   Traditionally, smoked fish *Suka ko Maacha* is a product of the *Gorkha* community of Sikkim. The hill river fish "dothay asala" (*Schizothorax richardsonii* Gray) and "chuchay asala" (*Schizothorax progastus* McClelland) are collected in a bamboo basket from the river or streams, and are degutted, washed, and mixed with salt and turmeric powder (Tamang et al., 2012). Degutted fishes are hooked in a bamboo-made string and are hung above the earthen oven in the kitchen for 7–10 days. *Suka ko Maacha* can be preserved for a period of 4–6 months and is eaten as a curry.

11.4.10.4 Sidra   *Sidra* is a sun-dried fish product commonly consumed by the *Gorkha* community of Sikkim (Tamang et al., 2012). Fish (*Puntius sarana* Hamilton) is collected, washed, dried under the sun for 4–7 days and is stored at room temperature for 3–4 months. *Sidra* pickle is a popular cuisine (Thapa et al., 2006).

11.4.10.5 Khainga   Preparation of the dry salted *Khainga* (Figure 11.17) fish preparation is similar to the dry salted *Hilsa* preparation. The catch is usually made from the sea or Chilika Lake. This fermented fish has its own flavor. The fermentation period is usually 4–6 months and, like *hilsa*, the fermented fishes are dried in the shade for 10–15 days. The shelf-life of *khainga* dry fish is about 12–18 months.

11.4.11 Non-Salted Dry Fish and Prawns

Rice field fishes and fresh water pencil fishes *Nannostromus beckfordi* are used widely for preparation of dried fish. The fish are harvested in plenty during the month of June and are washed properly and then, sun-dried in open, or some time directly on, gunny

Figure 11.17   Dry salted Khainga (*Mugil cephalus*).

bags. In some cases, the washed fishes are submerged in turmeric water for a few hours and then, sun dried. The sun drying of the fishes is continued for 20–25 days in direct sun light. The non-salted fishes are stored in gunny bags or bamboo baskets for further use; the shelf-life of these types of non-salted dried fish is short, in most cases less than six months.

Prawns and shrimps are preserved in a similar ways to small fish. They are washed, sun dried in the open and graded into size and then, packed and sealed for transporting to market (http:cismhe.org; 2006).

11.4.11.1 Fermentation of Fish by Naga Tribes   Fish is fermented by Naga tribes, and in a very different way by the "Lotha" tribe, and this is still a favorite food item of the locality (Mao and Odyuo, 2007). Small fish are used whole, whereas bigger ones are cut into smaller pieces. The fish is washed put inside a bamboo (normally in *Dendrocalamus hamiltonii*), tightly plugged with leaves and kept over the fire place for fermentation (Mao and Odyuo, 2007). Within a few days, the fish is fermented and ready for use as a taste-enhancer for vegetable curry. The flavor of fermented fish is of prime importance. However, the shelf-life of this fermented fish is only one month, as it spoils gradually and becomes unpalatable. Fermented fish products are mostly consumed as condiments, although, low-salt fermented fish products can be consumed in large quantities too.

11.4.11.2 Tungtap   *Tungtap* is a fermented fish product prepared by the ethnic *Khasi* and *Jaintia* tribes of Meghalaya (India) (Figure 11.18) (Tamang, 2009a,b,c,e). Its preparation follows the principal methods of fish preservation such as fermentation, salting, drying, and smoking (Deshpande, 2002; Rapsang and Joshi, 2012; Rapsang et al., 2011). However, it is different from the *hentak* of Manipur (India), *nam-pla* of Thailand, *nuoc-mam* of Vietnam, *sushi* of Japan, and *patis* of the Philippines, which are either fish pastes or fish sauce (Toyokawa et al., 2010). *Tungtap*, on the other hand,

Figure 11.18 The pictorial view of *Tungtap* preparation.

retains the original shape but is softened by salting of the fish prior to fermentation. It is usually consumed as a pickle or taste enhancer. Although there are some variations in the process of its preparation between the *"Khasi"* and *"Jaintia"* communities, the basic process is detailed here. Trash fish like *Puntius*, *Danio* sp., etc. are commonly used. The fish are first eviscerated and sun dried for 3–4 days (Figure 11.18). Then, the dried fish are mixed with salt, fresh fish and spices, and put in air-tight earthen containers for fermentation. After 3–7 days at room temperature (18–28°C), the semi-fermented fish is taken out and roasted in a dry pan and then, ground by hand to a fine paste with chillies, onion, mint leaves, turmeric powder, and salt (Figure 11.19). This paste serves as an important side dish on the local cuisine of the *"Khasi"* people.

11.4.11.3 Chepa Shutki *Chepa shutki* is a home processed semi-fermented food, prepared from a small sized fish especially small carp and minnows, like *Puntius sophore*, *P. stigma*, *P. sarana*, and *P. ticto*, by artisanal fishermen, where the fish still retains its original form with a cured texture and aroma when the fermentation is

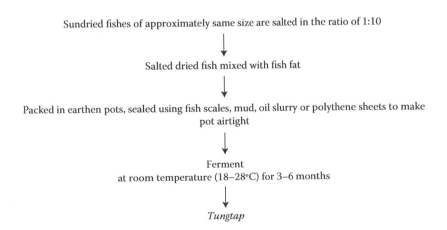

Figure 11.19 Flow sheet for the preparation of *Tungtap*.

completed. Semi-fermentation is one of the most important methods of preserving fish in Bangladesh, where it serves as an important source of protein, vitamins, and minerals for its highly dense, economically disadvantaged, malnourished people. The calcium, phosphorus, magnesium, and iron contents of *chepa shutki* have been found to be higher than those of similar kinds of Japanese processed fish (Khanum et al., 1999), and it is regarded as a high-quality protein food. It is a common food commodity in many districts of Bangladesh, including Mymensingh, Netrokona, Kishorgonj, Brahmanbaria, Comilla, Sylhet, Sunamgonj, Rangpur, and Gazipur. The particular taste and flavor of *chepa shutki* is the cause of its acceptability and popularity among Bangladeshis (Mansur, 2007). A large number of people in Bangladesh are engaged in the production and marketing of *chepa shutki*, and it plays a vital role in the socio-economic conditions of the people of the region.

11.4.11.3.1 Technology of Processing of Chepa Shutki *Chepa shutki* is mainly produced in the winter season because of the availability of the raw materials and favorable weather conditions. Dried small fishes of *Puntius* sp. are the major raw material which are usually collected from the dry fish processors or wholesalers. Other fish species used include *Hairfin anchovy* or "faisha" (*Setipinna taty*) and coromandel ilisha or "choukya" (*Ilisha filigera*). Production of *chepa shutki* starts in October and it continues until March. The various steps involved in the production have been sub-divided into two major stages: (i) production of dried small indigenous species (SIS); and (ii) production of *chepa shutki* from dried SIS (Figure 11.20).

i. Processing technique of dried SIS

Collection of raw material: Raw materials are collected in the fresh condition, usually from landing centers or from fishermen or middlemen. Usually, the small-scale fishermen bring their catch at the landing center using traditional bamboo baskets, to the pre-selected buyers or middlemen, locally called "*Mohajon*". These "*mohajon*" give

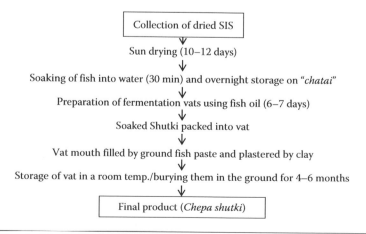

Figure 11.20 Flow sheet for the preparation of *Chepa shutki*.

loans to the poor fishermen and the fishermen become bound to sell their catch to them alone. At the peak season of harvesting, all the catch is not sold in the market, and there are no adequate facilities, such as cold storage/icing facilities, to preserve the raw fish. At that time, the dried product processors collect the raw fish and, in many cases, producers use poor quality raw materials for the production of dried SIS.

Dressing and gutting: After collection, the raw materials are gutted. Generally, village women are involved in dressing and gutting.

Salting: After gutting of the fish, salting is done (optional) to protect fish from flies, insects, or their larval infestation. Sometime, salt is used to give extra weight to the fish. Generally, 125 g of salt is used for 1 kg fish.

Washing: Next day after salting, the processed SIS are kept overnight at room temperature and washed with river or spring water to remove salt and other particles. After washing, the fishes are ready for drying.

Sun drying: After washing, small fishes are dried directly in the sun by spreading them on bamboo "*chatai*" or in the open field. Depending on the weather condition, drying normally takes 2–3 days.

Pre-packing treatment and storage: After drying, the products are graded according to their size and stored in bamboo baskets or under mats at room temperature until marketing.

ii. Processing technique of *chepa shutki* from dried SIS

Collection of dried SIS: For the preparation of *chepa shutki*, the processors collect dried SIS from *shutki* processing centers. The dried SIS is carried by van, rickshaw or bicycle. Some processors also collect the *shutki* from the wholesale and retail market.

Sun drying and pre-treatment before fermentation: After collection, they again sun dry the *shutki* for 10–12 days and during this period they also allow the fish to absorb adequate moisture from the night fogs in winter.

Preparation of fermentation vats: At this time, the processors prepare large-sized clay vats using fish oil. These vats have a round shaped wide-mouth with a capacity of 50 kg fermented fish (http://hrf.org, November 6, 2013). The inside of the vat is polished adequately with fish oil and sun dried for 3–4 days before packing with the dried fish. The vats, thus dried and matured, do not allow any air or moisture to pass through the pores of the earthen walls.

Washing and soaking of fish with water: After drying of *shutki* for 10–12 days, they are washed for at least 30 min using ground water to remove dust from the fish body, and are kept overnight on a bamboo "*chatai*" or basket to absorb more water. When the fish flesh become soft and slimy they are ready for packing into the prepared vats.

Filling of vats with fish and burying the vats in the ground: The vat is then packed with fish tightly by pressing with hands and feet until filled up to the rim. The remaining parts in the mouth of the vats are then filled by pastes of ground *shutki*. Then the mouth of the vat is covered with polythene and plastered with a heavy layer of clay to create anaerobic conditions and to accelerate the fermentation process. The sealed vat is kept at room temperature or often buried in the ground to assure optimal temperature for a period of 4–6 months (Figure 11.21), after which it is ready for consumption.

11.4.11.4 Nga-Pi   *Nga-pi* is a fermented fish product traditionally produced in Cox's Bazar, Barguna, Patuakhali, and Chittagong districts of Bangladesh. Mainly the *Rakhaings* people prepare—and eat and sell—this product from marine fish (including their fry and fingerlings) and shrimp such as *Mysid* sp. and *Acetes* sp. There are large communities of the *Rakhaings* living in different parts of Bangladesh, estimated to be more than 300,000 (Sein et al., 2001), and they have brought this traditional fishery product from Myanmar. *Nga-pi* (fermented fish paste) is made by pounding or grinding fish or shrimp with 20%–25% salt and partially drying the mixture in the sun for 3–4 days (Tanasupawa and Visessanguan, 2013). The mixture is stored by pressing it tightly into earthen jars or concrete vats and leaving for 3–6 months to mature. *Nganpyaye* (fish sauce) is obtained as a by-product of the fermentation of *nga-pi* (Tyn, 1996, 2004). It is a blackish pounded dough with a strong pungent flavor, eaten mainly as a condiment with curry and rice (Nowsad, 2007). A small amount of *nga-pi* is dissolved in a cup of water and the watery extract is used as a flavor enhancer in curry (Figure 11.22).

11.4.11.4.1 Processing Technology   Typically *nga-pi* is prepared from small *Acetes* and *Mysid* shrimps, which are pounded to a paste with a proportion of added salt (http:bfrf. org, Nov 6, 2013). The paste is then, subjected to several alternate sun-drying and fermentation processes, before being matured in an air-tight container. In *Rakhaing* villages, the whole process is done traditionally, where a lot of chances for contamination

Figure 11.21   Pictorial representation of process for the preparation of *Chepa shutki* in Bangladesh. (a) Collection of dried SIS, (b) sun drying for 10–12 days, (c) shutki soaked into water, (d) fermentation vat soaking with fish oil, (e) soaked fish put into vat at heavy pressure, (f) vat mouth covered by a layer of clay, (g) vat buried in ground and kept at room temp. for 4–6 month, and (h) final product (*Chepa shutki*).

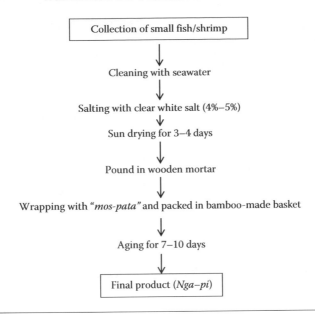

Figure 11.22 Flow sheet for the preparation of *nga-pi*.

and deterioration of the raw materials and products exist. The major steps involved in the processing of *nga-pi* at Chowfalldandi in Cox's Bazar district are described here.

Collection of raw material: *Nga-pi* processors collect raw materials either from fishermen, from landing centers or from middlemen. During the peak harvesting season, processors collect the raw fish at comparatively cheap prices. Small shrimp harvested from the Moheskhali channel or from the shallow continental shelf near the shore on a day-to-day basis are used for *nga-pi* preparation. The catch is neither iced nor salted on board the vessel. Generally, smaller sized shrimp of *Acetes* sp. and *Mysid* sp. are used. However, other fish species, including various small marine fish, their fry, and fingerlings, various shrimps, and even sea snails and molluscs, are utilized. *Acetes* shrimps are pinkish in color with comparatively bigger size, with a harder shell, and are locally called "ming," whereas *Mysid* shrimps are very small in size with a milky-white soft body, locally called "maishi." According to the villagers, "ming" can produce the best quality *nga-pi*. Sometimes fish and crabs also remain in the mixture. The sorting of bigger shrimp and fish larvae, small fish, other shrimps, sea snails, and other molluscs, however, is not done prior to drying.

Salting and sun drying: Unsalted small shrimps are carried by bamboo baskets from the boats. A small amount of salt is spread over the bamboo "*chatai*" and the *shrimps* are spread in thin layers over the mats (http://bfrf.org, Nov 6, 2013). The salt treatment given varies with the type and quality of shrimp. Generally, shrimps on the mat are dried in the sun for 3–4 days.

Pounding in wooden mortar: Semi-dried shrimp is ground in a wooden mortar with salt the following night (http://bfrf.org, Nov 6, 2013). The amount of salt varies from 1 to 2 kg per 40 kg shrimp (2.5%–5%) (http://bfrf.org, Nov 6, 2013). On the following day, the salt-ground paste is dried in the sun on the mat for the whole day and

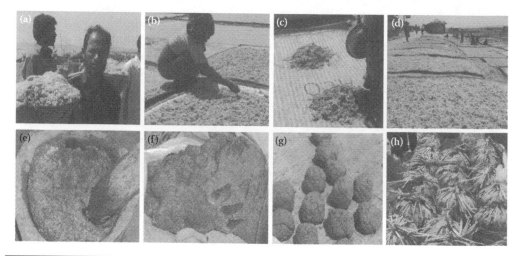

Figure 11.23 Processing technique of *nga-pi* in Bangladesh. (a) Collection of raw shrimp, (b) sorting of big shrimp and crabs, (c) salting of raw materials, (d) drying on chatai for 3–4 days, (e) pounding in wooden mortar, (f) pounded *nga-pi*, (g) round shaped *nga-pi*, and (h) warping with leaves for aging for 7–10 days.

ground again at night. Salt is not added during the second grinding. Similarly, the product is dried for the 3rd day and ground, finally, into paste at night with no more incorporation of salt.

Wrapping with leaves: The final paste has a deep-grayish to blackish appearance. No additional color is incorporated. The final paste is shaped into blocks or balls and wrapped in the large leaves of a wild hill tree called "mos-pata." These tree leaves are very thick and can absorb the water released from the product due to the salt action. Wrapped *nga-pi* is packaged in bamboo baskets.

Aging: 20 kg of the dough is packed in one bamboo basket and kept for 7–10 days for aging. The product is generally sold within a week, or sometimes stored for a longer time to get a better price (http://bfrf.org, Nov 6, 2013). According to the processors, *nga-pi* produced in this way has a shelf-life of 6 months (Figure 11.23).

11.4.12 *Fermented Fish with Salt and Carbohydrate*

There are many kinds of fermented fish with salt and carbohydrate, such as *pla-ra* (fish, salt, and roasted rice powder), *pla-som* (whole fish, salt, cooked rice, and garlic), *pla-chao* (fish, salt, and khao-mak), *som-fak* (minced fish, salt, cooked rice, and garlic), and *pla-chom* (fish, salt, garlic, and roasted rice powder) (Tanasupawat and Visersanguan, 2013). The fish used are mainly fresh water fish, though freshwater shrimp. The rice is a source of carbohydrate for the microorganisms involved in the fermentation, mainly LAB, which give a characteristic flavor to the products. The roasted rice also gives a brown color. There are two types of fermented rice used *khao-mak*, which is a mold, and yeast fermented and *ang-kak* rice which is fermented using the mold *Monascus purpureus* which gives a red colour in *pla-paeng daeng* (Phithakpol et al., 1995). A few typical products with their methods of preparation are briefly described here.

11.4.12.1 Pla-ra Pla-ra is made from freshwater fish (*Crossocheilus* sp. soi, *Channa striatus* Chorn, *Cyclocheilichthys* sp. takok, *Labiobarbus leptocheilus* soi, *Puntius gonionotus* ta-pian, *Trichogaster* sp. kra-dee and is fermented with salt and roasted rice/paddy or unroasted rice bran for 6–10 months (Figures 11.24 and 11.25) (Tanasupawat and Visersanguan, 2013).

11.4.12.2 Pla-som Pla-som is made from fresh water fish (Puntius gonionotus ta-pian) with salt and minced cooked rice, fermented for 5–7 days (Figure 11.25).

11.4.13 Sun Dried, Salted Dried, and Salted Fish

The most simple curing practice used in Sri Lanka is sun drying. The drying method varies with the size of fish. Typically, small fish varieties such as anchovies, pony fish, sardines, and silver bellies are used for sun drying (Bramsnaes, 1965) while salt drying is used for larger fishes and those which are moderately fatty. The salt proportion varies from 1:3 to 1:10, depending upon the size of the fish. In smaller fish, their ventral side is split open, gills and intestines removed and cleaned by washing. Large fishes are split dorso-ventrally, gills and intestine are removed, washed, cleaned and spread out at the time of drying. Salting is a common fish-curing practice in the rainy season. After 6–7 months, at ambient temperature, in the above manner fish are

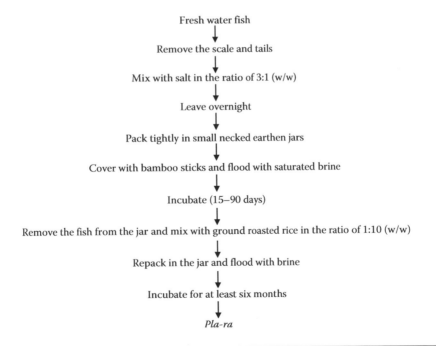

Figure 11.24 Flow sheet of *pla-ra* production. (Adapted from Appleton, J., Briggs, C., and Rhatigan, J. 1978. *Pieces of Eight: The Rites, Roles, and Style of the Dean by Eight Who Have Been There*. Portland, OR: NASPA Institute of Research and Development; Tanasupawat, S. and Visessanguan, W. *Seafood Processing: Technology, Quality and Safety*. 2013. Copyright Wiley-VCH Verlag GmbH & Co. KGaA.)

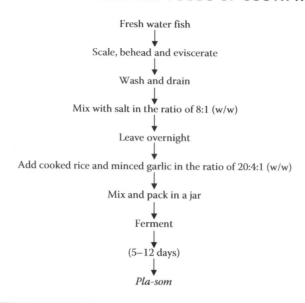

Fresh water fish

↓

Scale, behead and eviscerate

↓

Wash and drain

↓

Mix with salt in the ratio of 8:1 (w/w)

↓

Leave overnight

↓

Add cooked rice and minced garlic in the ratio of 20:4:1 (w/w)

↓

Mix and pack in a jar

↓

Ferment

↓

(5–12 days)

↓

Pla-som

Figure 11.25 Flow sheet of *pla-som* production. (Adapted from Appleton, J., Briggs, C., and Rhatigan, J. 1978. *Pieces of Eight: The Rites, Roles, and Style of the Dean by Eight Who Have Been There.* Portland, OR: NASPA Institute of Research and Development.)

matured for dipping in low concentration salt solution or oil and canned or vacuum packed.

11.4.13.1 Maldive Fish This is lightly salted, smoked, and sun dried to very low moisture content, giving hard "*katsuboshi*" like products, widely used as flavoring agents in most of the local preparations. The characteristic flavors of various products have been shown to be at least partly due to the growth of molds on them during the curing process. *Aspergillus flavus* plays an important role in development of flavor compounds during the curing process.

11.4.13.2 Jaadi (Pickle Cure) This is prepared by the Colombo cure or Pickle cure method practiced in Sri Lanka. Gutted fish are mixed with salt in a ratio of 3 parts of fish to one part of salt in large cement tanks, and pieces of dried pods of *goraka* or tamarind are placed on the fish. *Goraka* improves the process by lowering the pH below 4. A significant improvement in the quality of the product is obtained by replacing tamarind with 5% acetic acid (vinegar) in the ratio of one part of acid to 20 parts of fish. It can be kept for a year without spoilage.

11.4.13.3 Ambulthiyal The traditional tuna fish curry in Sri Lanka is known as *ambulthiya*. Its speciality is that it has a shelf-life of around seven days at ambient temperature in clay pots. It is prepared using fish, salt, pepper, and *goraka* (Malabar tamarind, *Garcinia cambogia*). Ready-to-eat products are becoming popular in Sri Lanka by using retort pouch sterilization, which increases the shelf-life to 180 days at ambient temperature.

11.4.14 Fish Sauces

To produce fish sauces, the production process used to make *shiokara* is prolonged and is usually more elaborate than *shiokara*. Fish sauce is a condiment widely used, where production processes range from simple procedures to satisfy household needs to large-scale factories that employ industrial techniques.

11.4.14.1 Fish Sauce Preparation of fish sauce is a process that extends the shelf-life of a valuable commodity and provides a condiment that is widely varied and popular. Fish sauce is the liquid extract of the flesh of fish made by the process of prolonged salting and fermentation. It is made from small fish, like sardines and anchovies, and molluscs, especially squid. Fish pastes are more important than fish sauces as a source of protein. Fish pastes include partially dried products and the residues from fish sauce production.

11.4.14.2 The Generic Process Like the other types of fermented products already described, there are many variations in the processes used to make fermented fish and shrimp sauces, depending upon the ingredients and processes as described here.

- Salting rate
  The ratio of fish to salt used varies considerably. For fin fish, rates vary from a low of 10 fish to 1 salt to a high of 1:1 (Ruddle and Ishige, 2005).
- Grades of sauce produced
  Household and cottage industry-level producers mostly make just a single grade of sauce; industrial (including small-scale industrial) producers manufacture two to four grades—most produce three (Ruddle and Ishige, 2005).
- Fermentation vessels used
  Household and cottage industry producers use either ceramic jars, wooden barrels, or, occasionally, lengths of concrete pipe, as fermentation vessels. Traditional industrial producers use large wooden vats, whereas newer factories or those with modernized facilities have installed concrete or cement tanks and drainage channels (Ruddle and Ishige, 2005).
- Compression
  The fish and salt mixture is sometimes compressed during the fermentation process at all scales of production (Ruddle and Ishige, 2005).
- Fermentation period
  Fermentation periods vary considerably, and mainly according to the fish species used, grade of sauce prepared, and climate. Apart from fish liver sauce in Cambodia, which is processed in 8–10 days, fermentation periods of first grade sauces are from two months in Myanmar to 36 months in cooler areas (Ruddle and Ishige, 2005).
- Blending

Most small producers do not blend the different grades of sauces that they manufacture. Grades are blended by industrial-level producers in Myanmar and in other countries (Ruddle and Ishige, 2005).

- Additions to the residue

  To prepare lower grades of sauce, some producers add materials to the residue of fish and salt remaining after the first grade sauce has been drawn-off. These additions usually consist of either just salt, brine, or salted water, or almost "*instant*" sauce is made by adding *mieki*, acetic acid, and sometimes monosodium glutamate and sugar, in addition to the salted water. After such ingredients have been added, the mixture is fermented again (Ruddle and Ishige, 2005).

- Additions to the sauce

  Some producers add ingredients like flavor and/or color enhancers to their sauces, such as palm sugar (Cambodia), sugar (Thailand), caramelized sugar (Vietnam), tamarind or tamarind water (Malaysia), and red dye (Malaysia) (Ruddle and Ishige, 2005).

- Exposure to sunlight

  About half the producers remove the covers from fermentation containers each day to expose the fermenting mixture to sunlight during the later stages of production (Ruddle and Ishige, 2005).

- Boiling the sauce

  About half the producers boil the sauce drawn-off (Ruddle and Ishige, 2005).

- Filtering the sauce

  Somewhat more than 50% of the producers filter the fish sauce before packing it. Most use a simple cloth filter, but other materials used include cloth with paper or with charcoal, sea sand, or an aluminum screen (Ruddle and Ishige, 2005).

- Packing the final product

  Regardless of scale of production or whether for the wholesale or retail market, most manufacturers bottle their sauce. At the industrial scale, only three factories distributed sauce in plastic containers (Ruddle and Ishige, 2005).

- Marketing sauce

  All the industrial- and small-scale industrial producers sampled sell their sauce both wholesale and retail. Only a large-scale factory in Cambodia sold just wholesale. In contrast, cottage industry and household producers were just retailers (Ruddle and Ishige, 2005).

- Use of the final residue

  In some cases, a residue remains after the last of the sauce has been drawn-off. This is generally used either as fertilizer or animal feed (Ruddle and Ishige, 2005).

11.4.14.3 Fish Sauce (nam-pla)   Fish sauce is a traditional liquid product made from brackish water and sea water fish and fresh water fish (*Corica argentius, siu-kaew, C.*

Figure 11.26 Flow sheet of fish sauce production. (Adapted from Phithakpol, B. et al. 1995. *The Traditional Fermented Foods of Thailand.* Institute of Food Research and Product Development, Kasetsart University, Bangkok, p. 11.)

soborna, Stolephorus indicus, S. commersonii, sai-ton and *Clupeoides* sp. *ka-tak*) or the other kinds of fish, with high salt content and fermented for 5–18 months (Tanasupawat and Visessanguan, 2013). The product is clear, yellow or brownish yellow in color, with a salty and fishy aroma (Tanasupawat and Visessanguan, 2013). This hydrolyzed fish condiment is produced in many countries, with different names, such as *nam–pla* (Thailand, Loas), *yu–lu* (China), *colombo-cure* (India and Pakistan), *Jeot-kalor aekjeot* (Korea) and *bu–du* (Malaysia) (Phithakpol et al., 1995; Wood, 1998). The process of preparation of fish sauce is shown in Figure 11.26.

11.5 Alkaline Fermented Foods

11.5.1 Fermented Soybean Products

11.5.1.1 Tungrymbai *Tungrymbai* is an ethnic fermented soybean food prepared by the indigenous *Khasi* and *Jaintia* tribes of Meghalaya, India. It is a sticky, slightly alkaline food that has unique flavor and texture (Thokchom and Joshi, 2012). It serves as an important source of protein in the local diet of the people. This fermented food, however, has little identity with the original bean and is typical of the region. *Tungrymbai* is similar to other fermented soybean foods of North East India, namely *hawaijar* of Manipur, *bekang* of Mizoram, *aakhone* of Nagaland, *peruyaan* of Arunachal Pradesh, and *kinema* of Sikkim. It is also similar to *natto* of Japan, *chungkok-jang* of Korea, *thua-nao* of Thailand, *pe-poke* of Myanmar, and *dou-chi* of China (Tamang, 2003; Tamang et al., 2012).

11.5.1.1.1 Preparation of Tungrymbai   *Tungrymbai* is prepared from the local variety of soybean seeds, *Glycine max* (L.) Merrill (Thokchom and Joshi, 2012). The seeds are washed and boiled for about 45 minutes to 2 hours until softened. The excess water is drained-off and the boiled soybean seeds are allowed to cool (25–30°C) (Thokchom and Joshi, 2012). The seeds are transferred to a bamboo basket lined with fresh leaves of *Pyrnium pubinerve* Bl. (Marantaceae) locally called "*slamet*" in which 2–3 hot charcoals are placed and are fully wrapped with the leaves. The basket is placed inside a jute bag and kept for fermentation at 25–30°C for 3–4 days, preferably near a fireplace (Sohliya et al., 2009; Thokchom and Joshi, 2012). The fermented soybeans are crushed lightly in a wooden mortar and pestle to obtain the final product, *tungrymbai* (Figure 11.27).

11.5.1.2 Kinema   *Kinema* is a popular product in Darjeeling, particularly among the Lepchas. *Kinema* is produced by the traditional alkaline fermentation of soybeans, constitute an important contribution to the diet of people living in the Northeastern states of India, as well as in Nepal and Bhutan, who use it in local dishes as a seasoning agent and as a low-cost source of protein (Moktan et al., 2011; Sarkar and Nout, 2014; Sarkar et al., 1994, 1997; Sekar and Mariappan, 2007; Tamang, 2010b; Tamang and Nikkuni, 1996, 1998). Kinema serves as a major source of protein in the diet of the people of this region as it has about 48% (dry weight basis) of crude protein and free amino acids which account to approximately 26% of total amino acid content (Sarkar et al., 1994, 1997).

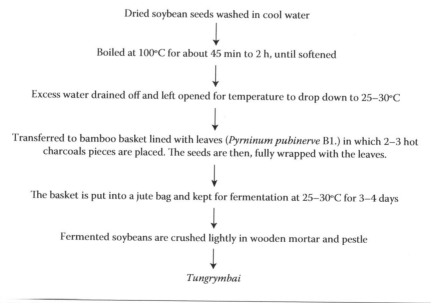

Dried soybean seeds washed in cool water

↓

Boiled at 100°C for about 45 min to 2 h, until softened

↓

Excess water drained off and left opened for temperature to drop down to 25–30°C

↓

Transferred to bamboo basket lined with leaves (*Pyrninum pubinerve* Bl.) in which 2–3 hot charcoals pieces are placed. The seeds are then, fully wrapped with the leaves.

↓

The basket is put into a jute bag and kept for fermentation at 25–30°C for 3–4 days

↓

Fermented soybeans are crushed lightly in wooden mortar and pestle

↓

Tungrymbai

Figure 11.27   Flow diagram of *tungrymbai* production.

11.5.1.2.1 Method of Preparation Kinema (Figure 11.29) is produced by cleaning the soybeans (*Glycine max* L.) (often yellow seeded), washing, soaking overnight (12–20 h), boiling until softened, and crushing into grits. The cotyledons are crushed by hand with a wooden ladle. The crushed material is scattered on a flat surface, preferably on a polyethylene sheet. A small amount of yeast is added and mixed with the soybean mixture. The mixture is then, wrapped tightly with banana or other wild plant leaves (*Glaphylopteriolopsis erubescens*), preventing air from entering. Wrapped in fresh leaves and sackcloth, the grits are left to ferment for 1–3 days at 25–35°C until the beans are covered with a stringy, mucilaginous coating, and the typical kinema flavor, dominated by that of ammonia, appears (Moktan et al., 2011; Sarkar and Tamang 1995; Sarkar et al., 2002; Tamang et al., 2009d). Kinema has a pH between 7.9 and 8.5. During its production, a small amount of firewood ash is optionally added to the crushed beans (Sarkar et al., 2002). Fresh kinema can be kept for 2–3 days during summer and up to 1 week during winter (Dahal et al., 2005). A flow-sheet for the preparation kinema is shown in Figure 11.28.

For food preparation, fresh kinema is fried in oil with other ingredients such as chopped onion, tomatoes, and turmeric powder, and eaten with rice and vegetables. Tamang et al. (1996) and Tamang (1998) have produced kinema of acceptable quality with a high level of soluble proteins using *Bacillus subtilis* as a starter culture for controlled fermentation at 40°C for 20 h, followed by maturation at 5°C for 1 day. Tamang and Holzapfel (1999) developed a ready-to-use cost-effective and easy to handle pulverized starter using *B. subtilis* KK2:B10 strain for small-scale kinema productions in the Himalayas. In addition to its valuable contribution of protein to the diet in the Eastern Himalayas, kinema has also been found to be a rich source of some B vitamins (Sarkar et al., 1998) and minerals, and also have antioxidative properties (Moktan et al., 2011, Tamang et al., 2009d), besides a large amount of Group B saponin content, which has health promoting benefits (Tamang et al., 2012).

Many kinema-like sticky bacilli-fermented soybean-based products are consumed in the Northeastern states of India bordering with Bhutan, China, and Myanmar, which include *Hawaijar* in Manipur, *Bekang* in Mizoram, *Peruyaan* in Arunachal Pradesh, *Aakhone* in Nagaland, and *Tungrymbai* in Meghalaya (Sarkar and Nout, 2014; Tamang, 2010b; Tamang et al., 2009d).

There are other varieties of kinema prepared in Sikkim (http:cismhe.org). In one such method, the grits of cooked soybeans along with other ingredients are packed in a bamboo basket which is covered with a jute bag and left to ferment naturally at ambient temperature (which is about 20–35°C in Sikkim). The whole mass is then placed over an earthen oven and left for 1–2 days, after which the completion of fermentation is indicated by the appearance of white viscous mass on the soybeans and the release of an ammoniacal odor (Tamang, 2009a,b,c,d,e). Its shelf-life is 2–3 days in summer and 5–7 days in winter (Tamang et al., 2009d). The prepared kinema is shown in Figure 11.29. The indigenous source of microorganisms can be traced to the practice of not cleaning the mortar and pestle to preserve and supplement the

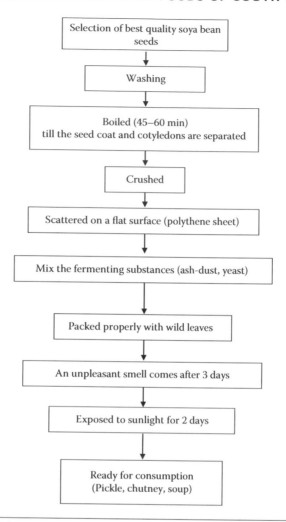

Selection of best quality soya bean seeds

↓

Washing

↓

Boiled (45–60 min)
till the seed coat and cotyledons are separated

↓

Crushed

↓

Scattered on a flat surface (polythene sheet)

↓

Mix the fermenting substances (ash-dust, yeast)

↓

Packed properly with wild leaves

↓

An unpleasant smell comes after 3 days

↓

Exposed to sunlight for 2 days

↓

Ready for consumption
(Pickle, chutney, soup)

Figure 11.28   Flow sheet for preparation of kinema.

Figure 11.29   *Kinema*. (Courtesy of Carrying capacity study of teesta basin in Sikkim edible wild plants and ethnic fermented foods Vol. VIII Biological Environment: Food Resources) (From www.cismhe.org)

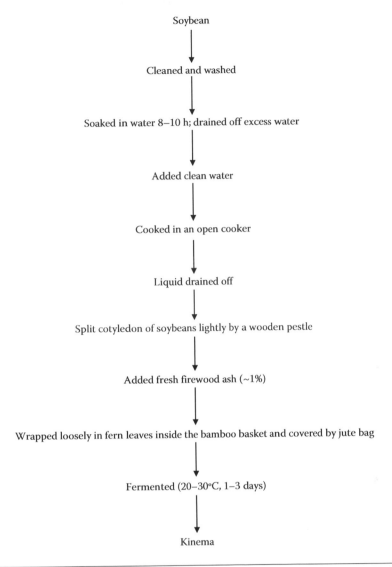

Figure 11.30 *Kinema* production in Sikkim by Limboo women. (Adapted from Tamang, J.P. et al. 2000. *Food Biotechnol* 14 (1–2): 99–112.; http:cismhe.org.com.)

microorganisms for spontaneous fermentation without the use of a starter cultures (Tamang, 2003) (http:cismhe.org). A little water is added to it to make a thick gravy, which is cooked for 5–7 minutes (Tamang, 2009d).

It is produced exclusively by Nepali *Sikkim* women (Figure 11.30). The soya bean is locally known as *bhatmas* and the varieties used are "yellow cultivar" and "dark brown cultivar." It is produced individually or at the household level, and is sold in the local markets. The skill of production of this delicacy has been protected as a hereditary right and passed from one generation to another (Tamang, 2009d).

11.5.1.3 Hawaijar   It is an indigenous traditional fermented soybean with characteristic flavor and stickiness (Devi and Suresh, 2012; Sarkar and Nout, 2014). It is consumed commonly in the local diet as a low cost source of high protein food and plays an economical, social and cultural role in Manipur, India.

11.5.1.3.1 Traditional Method of Preparation In the traditional method of *Hawaijar* preparation, medium and small sized soybean (*Glycine max* L.) seeds are cleaned and sorted (Devi and Suresh, 2012; Tamang, 2009d, Tamang, 2012). The graded soybean seeds are put in water, where the water level should be twice that of the seeds, and left overnight. The seeds are then, washed properly 2–3 times in running water and then, cooked, either in a pressure cooker or by conventional means (Devi and Suresh, 2012; Sarkar and Nout, 2014). The solid portion, that is, the cooked soya bean. which is neither too soft nor too hard, is placed in a bamboo basket after draining. The beans are rinsed until they are non-greasy, and any remaining water is drained-off completely and the contents are turned upside down once or twice. The drained water portion was said to be useful in washing clothes in old times, and it is believed to help in curing T.B. (Tuberclosis) and also to be good for women.

A cloth, folded 2 or 3 times, is placed in a coarse bamboo basket. A thick layer of *Ficus hispida* leaves, locally known as *"Asse heibong"* or banana (*Musa* spp.) leaves or paddy is placed on the cloth (Devi and Suresh, 2012; Sarkar and Nout, 2014). The cooked soybeans are then, placed in alternate layers. At the end, another cloth which is folded 2–3 times, is placed and then, the whole content is tied tightly with another cloth, which should now be air-tight. Paddy husk/straw provides favorable conditions for the growth of naturally occurring microorganism for fermentation to take place. Moreover, paddy husk/straw absorbs some of the ammonia odor produced during fermentation and improves the characteristic taste and flavor of *hawaijar* (Devi and Suresh, 2012; Tamang, 2009a). The fermentation of *hawaija*r takes place above room temperature, so it is maintained between $30 \pm 1°C$ and $40 \pm 1°C$ by exposing the basket to sun during day time or near an earthen oven at night. The *hawaijar* is ready in 3 days during summer, but during winter it takes 5 days. The basket is preferably covered with gunny bag to prevent the escape of heat from inside the basket during fermentation. *Hawaijar* is traditionally prepared from soybeans fermented with naturally occurring *Bacillus subtilis* (Jeyaram et al., 2008). In order to add more flavor, it is kept near the fireplace as long as possible. In olden times, rice husks were used instead of cloth (Devi and Suresh, 2012). *Hawaijar* can be eaten raw, with salt and chilli, or cooked. The preparation of *hawaijar* by natural fermentation leads to variations in quality due to varying methodologies, fermentation times, and temperature of incubation, due to uncontrolled environmental conditions, often leading to unsuccessful fermentation and a poor quality product.

The traditional *hawaijar* (Figure 11.31) is characterized by its alkalinity (pH 8.0–8.2), stickiness, and pungent odor. The preparation of *hawaijar* is very simple, similar to that of Japanese *"Itohiki–Natto"* (the whole soya bean is used for fermentation). But in *"Kinema"* (another Indian fermented soya bean), it is dehulled and cracked into pieces before fermentation. Unlike *"Kinema"* there is no addition of firewood ash

Figure 11.31 *Hawaijar.*

during *hawaijar* production (Devi and Suresh, 2012). The entire process of *hawaijar* preparation is shown in Figure 11.32 as a flow sheet, and depicted pictorially in Figure 11.33.

A good quality *Hawaijar* has a very strong ammoniacal smell and produces lots of spider-web-like mucus strings when the beans are pulled apart. It is most commonly eaten with rice as a side dish. It may also accompany other dishes or be fried. The perceived flavor of *hawaijar* can differ from person-to-person. *Hawaijar* has a moisture content of about 80%, because of which its shelf-life is only 2–3 days at ambient

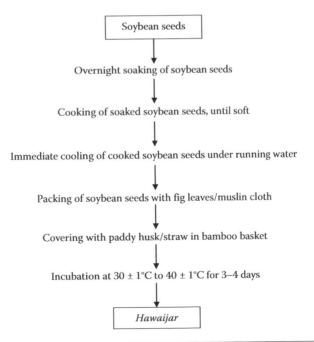

Figure 11.32 Flowchart for *hawaijar* production. (Adapted from Devi, P. and Suresh, P. 2012. *Indian J Trad Knowl* 11(1): 70–77.)

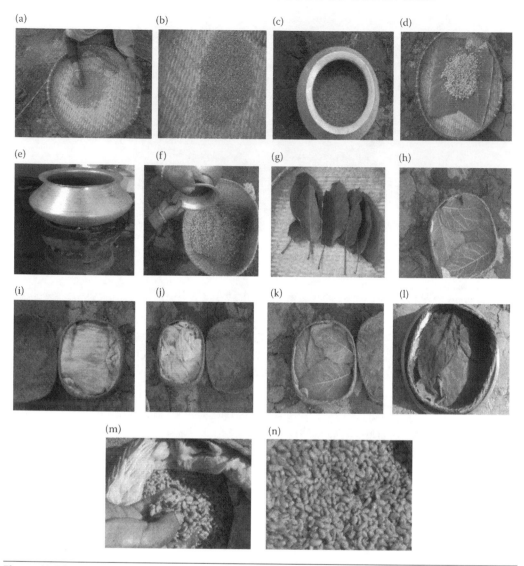

Figure 11.33 Pictorial representation of preparation of *hawaijar.* (a) Cleaning of soybean seeds, (b) cleaned soybean seeds, (c) soaking, (d) soaked soybean seeds, (e) cooking until soft, (f) cooling of cooked soybean, (g) fig leaves for packaging of cooked soybean for fermentation, (h) lining woven bamboo basket with fig leaves, (i) after 4 days of fermentation, (j) cooked soybean packed in muslin cloth, (k) final covering of soybean packed in muslin cloth with fig leaves, (l) after 4 days of fermentation, (m) fermented soybean inside muslin cloth, and (n) final product, *Hawaijar* (fermented soybean).

conditions, although this can be extended under low temperature storage using polystyrene containers as a packaging material.

11.5.1.3.2 Dried Hawaijar There are two types of dried *hawaijar* commonly prepared by the people of Manipur, India. The first method involves the common or traditional technique of making *hawaijar* (Devi and Suresh, 2012). After one week of preparation, salt is mixed with the *hawaijar*. The whole contents are poured into a bamboo (any bamboo with a longer internode, bigger hole and a thin outer part) where it is sealed with the help of bamboo leaves and tied very tightly with a plastic

Figure 11.34   Dried *hawaijar.*

sheath. This bamboo is then, placed on top of the fireplace in the kitchen and is kept for a week. The *hawaijar* obtained from this method has less smell but a better taste, and can be kept for a longer period of time. The second method is to dry the *hawaijar* in direct sunlight to make it moisture free (Figure 11.34). This method has been practiced recently by the people of Manipur (India).

11.5.1.3.3 Pickled Hawaijar   The fermented soya bean that is the *hawaijar* is fried in oil along with some spices (*masala*) and salt to taste and then, put into a plastic or glass jar along with the excess oil and sealed. It can be stored for a long time.

11.5.1.4 Axone/Akhone   *Axone* is a fermented soybean (*Glycine max* L.) product, named according to the "*Sema Naga*" dialect. Soya bean are cooked and packed in a bamboo basket with the base lined with leaves of *Ficus* species and covered by the same on top. It is kept near the fireplace for about 3–4 days for the fermentation to complete (Jamir and Deb, 2014). Most people go for longer fermentation to reduce the strong smell of the fermented product. The fermented soya bean is then, made into a paste and wrapped in banana leaves or *Phrynium pubinerve* leaves and kept for another 3–4 days near the fireplace (Figures 11.35 and 11.36).

11.6 Tea

Tea is the most common drink in the world, including countries of the South Asia, and is served to welcome guests as a refreshing beverage in modern homes as a sign of hospitality (http://mag.com). It is brewed and served with milk and sugar, or the leaves are boiled with milk, water, spices, and sugar. On railway stations, trains, and street corners, sweet milky tea is poured from hot kettles into disposable cups or mugs in most parts of India. The tea plant belongs to the genus *Camellia* (*Camellia sinensis*). Fermentation is one of the steps in tea processing. In South Asia, the tea producing countries are India, Sri Lanka, and Burma. The earstwhile East India Company

Soybean seeds boiled
(3–4 h in pot or 1 h in pressure cooker)

↓

Excess water drained-off
packed in bamboo basket lined with leaves of *Ficus* spp.
at bottom and covered at top

↓

Keep near the fire place for 3–4 days

↓

Make into a paste, wrapped and
kept near the fire place again for about 3–4 days

↓

Axone

Figure 11.35   Flow chart for preparation of *Axone*.

Figure 11.36   Different stages of *Axone/Akhuni* preparation. (a) The boiled soybean seeds, (b) the paste is wrapped in leaves, (c) the semi-fermented product, and (d) the matured *Axone*. (From Jamir, B. and Deb, C.R. 2014. *Int J Food Ferment Technol* 4(2): 121–127. With permission.)

developed the tea industry in India and now in India and Sri Lanka it is one of the finest agro-based industries, that earns substantial foreign exchange, besides providing employment.

11.6.1 Types of Tea and Its Processing

Conventional teas are: (a) totally fermented black tea, (b) raw or unfermented green tea, and (c) partially fermented Oolong (red and yellow) tea; non-conventional tea products are instant tea (cold-and hot-soluble), flavored tea, and de-caffeinated tea (Srikantayya, 2003).

11.6.1.1 Black Tea   Out of various teas, from the indigenous fermented product point of view, white, red, and black teas are most significant (Srikantayya, 2003) and, therefore, are described here (Figure 11.37). Black tea is made from the young leaves and unopened buds of the tea plant. The major steps involved in the manufacture of black tea are plucking, withering, leaf distortion, fermentation, firing, grading, packing, and storage.

11.6.1.2 Red Tea   Red tea is a fermented tea and its production is based on a controlled combination of enzymic and thermo-chemical processes, where the rate of the enzymic process is slower than that in black tea manufacture (Srikantayya, 2003). It has a pleasant, mild, and astringent taste, and a strong, stable aroma. The various steps involved in its manufacture are plucking, withering, rolling, roasting, sorting, and, finally, firing (Bokuchava and Skobeleva, 1980).

11.6.1.3 White Tea   White tea is a fermented type of fresh leaves: one bud, and one to two leaves with profuse hairs, or only buds, from the first plucking of spring tea, with a moderate content of polyphenols. To prepare it the bud and leaves are withered, fired to obtain raw tea, sorted, and packed as white tea (Srikantayya, 2003). The infusion of white tea is light orange–yellow, the infused leaf is open and mixed, the aroma is fresh but pure, and the taste is plain (Bokuchava and Skobeleva, 1980).

11.6.1.4 Oolong Teas   Oolong teas are the semi-fermented teas that are partially fermented before drying to preserve the natural flavors (http://www.energymanagement training.com). The process of producing Oolong tea begins with picking of the two leaves and a bud, generally early in the morning. The leaves are then, partially dried indoors to promote fermentation. When the leaves start turning red—at a stage, when 30% of the leaves are red, and the rest 70% are green—the leaves are rubbed repeatedly by hand, or mechanically, to generate flavor and aroma, and finally, dried over charcoal and blended.

Figure 11.37   Different types of fermented teas. (a) black tea, (b) white tea, and (c) red tea. (Adapted from Srikantayya, N. 2003. *Handbook of Postharvest Technology: Cereals, Fruits, Vegetables, Tea, and Spices.* CRC Press, Boca Raton, FL, pp. 741–778.)

11.6.2 Technology of Production of Tea

11.6.2.1 Harvesting   The fresh green tea leaves are usually harvested by hand, at intervals of 7–14 days, throughout the year. Generally, the rapidly growing shoot tips down to about the second or third unfolded leaf are plucked and used (Srikantayya, 2003). A long plucking interval or use of clones that deviate from the desirable leaf standard, lowers the levels of theaflavin, caffeine and volatile flavor compounds, and impart a poor sensory quality (sweet, flowery, and grassy aromas).

11.6.2.2 Withering   The withering step makes freshly plucked tea leaves undergo certain biochemical and physiological changes that assist the further processing steps—rolling and fermentation (Srikantayya, 2003). The physiological and biochemical changes that occur in the living tea leaf continue, but withering alters the pattern and rate of these changes, which ensures the quality of black tea (Srikantayya, 2003).

The various chemical and biochemical changes occurring during withering can be summarized as:

- Increase in amino acids, simple carbohydrates, and caffeine levels
- Maximal activity of polyphenol oxidase
- Loss of pectinase activity
- Breakdown of chlorophyll

Withering is carried out with leaves spread in thin layers (0.3–0.7 kg/m²) in trays/open loft system/rooms on the upper story of the factory in the traditional processing of tea, and varies from 16 to 20 h, depending upon the condition of the leaves and the requirement of the tea to be made. Hot air is blown from the bottom of the withering tray/trough system. The moisture in the leaves is evaporated by air, thus causing drying. Low altitude factories use ambient air and factories at high elevation use hot air for drying. Short withering periods (12 h) and low temperature (10–15°C) produce tea with good flavor, whereas longer withering time (20–30 h), forced withering, and high temperatures (25–30°C) have a good effect on color, but adversely effect the chemical and flavor qualities of tea (Srikantayya, 2003). Different withering techniques used include drum, tunnel, trough, and continuous withering systems. Trough withering is the most popular method, as it has low costs of construction and maintenance (Anonymous, 1974).

11.6.2.3 Leaf Distortion/Rolling for Manufacture of Orthodox Tea   After the withering process, the leaf is distorted by rolling or cutting. Conventional processing of leaf requires rolling in order to produce orthodox teas, namely, black tea and green tea (Nagalakshmi et al., 2003). Leaf distortion is not carried out immediately after plucking. The rolling technique brings out the juice from the leaf and twists the leaf. A roller consists of a circular table, a cylindrical box or jacket, and a cap to apply pressure. The leaf is bruised and twisted, then broken into small pieces by increased pressure and sifted by green leaf sifters. The remaining bulk is rolled, and the duration of each roll varies from 15 to 60 minutes, and the number of rolls varies from 2 to 3, in normal practice, depending upon the degree of wither, type of tea required, roller charge and speed, rolling conditions, and temperature (Anonymous, 1974).

11.6.2.4 Leaf Distortion for Non-Wither Teas   In the case of non-wither teas, a number of leaf distorting versatile modern machines are used, such as Legg-cut, CTC crushing, tearing, curling, and Rotorvane, singly or in combination (Srikantayya, 2003). The purpose is intensive maceration of the tea leaf to ensure rapid and complete fermentation. The CTC machine consists of two engraved metal rollers running close together, one at 70 rpm and the other at 700 rpm. The soft withered leaf is cut, torn, or rolled in the small gap between the serrated surfaces of the rollers. Machines used for leaf distortion include the Triturator, Ceylon continuous tea processing machine, Tocklai continuous roller (TCR), Barbora leaf conditioner (BLC), and the USSR continuous rolling expresser (Anonymous, 1974).

11.6.2.5 Fermentation   It is the main step in black tea processing, most important for the necessary chemical and biochemical changes (Nagalakshmi et al., 2003). The process is initiated at the onset of leaf maceration and is allowed to continue under ambient conditions. The green leaf after rolling and sifting in the case of orthodox tea or macerated leaf (CTC type), is spread in thin layers 5–8 cm deep on the factory floor or on racked trays in a fermentation room. Temperature control and air diffusion are facilitated by using humidifiers or cool air. Fermentation is carried out for between 45 minutes and 3 hours, depending on the nature of leaf, the maceration techniques, ambient temperatures, and requirement of the tea to be made. The temperature employed varies between 24°C and 27°C, although low temperatures (15–25°C) improve flavor.

At the end of fermentation, leaf color changes from green to coppery red, along with the development of a pleasant characteristic aroma. The completion of fermentation is determined by the skill of the tea maker or by instrumental techniques. It can also be assessed by measuring the theaflavin and thearubigin content, which are formed in the ratio of 1:10, under ideal conditions of fermentation. Tannin decreases during this period, from 20% in green tea leaf to 10%–12% in fermented tea and, therefore, is another measure of fermentation, which is terminated by the firing step. Modern developments in fermentation technology such as skip, trough, and drum continuous fermenting systems, have merits such as controlled optimal temperatures, reduced cost, lowered floor space requirement, and improved briskness in tea liquor (Anonymous, 1974). During fermentation, monomeric flavonoids (flavan 3-ols or tea catechins) present in tea leaves are transformed to polymeric theaflavin and thearubigin by oxidation occurring during fermentation. The development of the distinctive color, decreased bitterness and astringency, and characteristic flavor are derived from the fermentation process, giving fermented teas a marked distinction from nonfermented green tea. The formation of aroma compounds in tea is summarized in Table 11.2.

Table 11.2   Formation of Aroma Compounds in Tea

- The oxidized catechin reacts with the precursor molecules present in green tea and produces the volatile compounds in black tea.
- Oxidation of amino acids, carotenes, and unsaturated lipids leads to the formation of aroma compounds during the fermentation period aldehydes of amino acids formed as a result of Strecker's degradation reaction are of significant importance.
- Ionones, terpene alcohols, terpene aldehydes, and their oxidation products such as theaspirone and dihydroactinidiole result from oxidation of the carotenoids present in the flush by quinones.
- More than 638 aroma compounds in tea have been identified.
- The most important of aroma components are terpenes, terpene alcohols, lactones, ketones, esters, and spiro compounds.

Source: Adapted from Srikantayya, N. 2003. *Handbook of Postharvest Technology: Cereals, Fruits, Vegetables, Tea, and Spices.* CRC Press, Boca Raton, FL, pp. 741–778.

Aroma compounds of some of the world renowned teas with unique flavor characteristics have also been identified: India's Darjeeling tea has linalool, linalool oxides I and II, and geraniol; Sri Lanka's Uva tea contains methylsalicylate, linalool, and linalool oxides (Flament, 1989; Sanderson and Graham, 1973; Takeo and Mahantha, 1983).

11.6.2.6 Grading and Storage   Finally, tea is winnowed to remove stalky material and sieved to obtain different grades, based on particle size. Chemical changes take place during the storage of finished tea products, which tend to lose all the residual greenness and harshness within a few weeks' time (Nagalakshmi et al., 2003). If kept in a cool place and protected from moisture and oxygen, tea remains sound and full of flavor for more than a year (Anonymous, 1974).

11.6.3 New Manufacturing Technology

The traditional technology of tea manufacture has certain disadvantages: (a) undamaged tea leaf tissues (20%–25%) during rolling or CTC method, which in turn produce a nonuniform rate of oxidation that results in high losses of polyphenols; (b) loss of 70%–80% of essential oils in the firing stage, which reduces the aroma; and (c) rapid aging of tea, resulting in loss of quality. Thermal treatment of under-fermented tea is carried out in a factory or storehouse to 40°C, particularly in China and India, to eliminate the grassy odor and coarse taste in unfermented teas, mainly due to polyphenols, catechins, and other constituents present in high proportions. Besides this, it also causes isomerization and epimerization of catechins, degradation of chlorophyll, and synthesis of aldehydes and essential oils that improve the flavor quality of manufactured tea (Yamanishi, 1978).

11.7 Summary and Future Prospectives

It is shown from the review of indigenous products made from meat, fish, alkaline fermented products, production of tea, etc., that throughout South Asia, these products play a significant role both in the diet and the socio-economic life of the people of these countries. There is large diversity in these products; even the names of some similar products are different, though there may be very small variations in their production methods.

Meat and meat products play a crucial role in human nutrition. But being highly perishable, their storage and marketing require immediate refrigeration and freezing, which are costly and scarce in various South Asian countries. The development of simple technologies for the manufacture of shelf-stable ready-to-eat meat products is, thus, urgently needed. Fermentation of meat as a means of preservation, and the use of microbial cultures, therefore, appear to be a suitable method of natural preservation. The present world-wide research on fermented meat products is focused on the reduction of microbial contamination, identification and characterization of LAB native to meat, accelerated fermentation processes, bacteriocinogenic/probiotic/protective/genetically engineered starter cultures, flavor, lipolysis, proteolysis, lipid oxidation inhibition, additives, preservatives, biogenic amines, etc. It is pretty certain that future research on fermented meat will focus more on making the production process safer and the end product healthful for consumers.

Similarly to other fermented foods, the microbiology and biochemistry of meat and meat-products, alkaline fermented food, fermented fish, etc., are poorly understood and need more elaboration. The safety of these products, especially, needs considerable attention. The use of starter cultures, along with exemplary hygiene in respect of meat production facilities, and close monitoring using sophisticated techniques, could lead to immense improvement in the quality of these products. Through advanced fermentation technologies it may be possible to enhance the desirable qualities of indigenous fermented foods, including alkaline fermented foods.

In some South Asian countries fermented fish products are easily made and consumed, using simple techniques, requiring only a few, commonplace ingredients. Fermented fish products are indispensable for economically poorer populations in South Asia, who consume them daily in relatively large amounts. As a rule, fermented fish products are added to vegetables and eaten with rice, so they serve mainly as a salty and *umami* condiment to help in the consumption of large quantities of rice. But as the household incomes improve, consumption decreases in favor of either delicious fermented fish products, or increasingly commercialized products such as fish sauce, that displace the coarser, traditional village items, like fish paste. Despite variations of detail, the ingredients, production techniques, and culinary uses for fermented fish products are similar throughout the region of South Asia. Among desirable changes is the reduced use of high-salt-content fermented fish products, which are a serious health concern that need to be addressed by researchers. Developments in machinery, new techniques, and the art of texturizing protein products have changed, and as a result the spectrum of ingredients that can be texturized into useable end products has increased greatly and the knowledge can be applied to the indigenous fermented foods and thus, holds promise for the future.

In India, the tea industry has developed, but in some areas, like Kangra in Himachal Pradesh, the production of herbal tea is declining, and measures to revive it need to be taken, as the tea due to medicinal/therapeutic values being attached is getting increased recognition.

The technology of indigenous fermented food production is relatively simple, and most of the products are for house-hold consumption only. Out of all the products in South Asia, commercialization has only taken place in the case of tea, so there is ample scope for commercialization of production technology for other indigenous fermented products, to serve as a tool for the economic development of these countries.

References

Achi, O.K. 2005. The potential for upgrading traditional fermented foods through biotechnology. *Afric J Biotechnol* 4(5): 375–380.

Ahmad, S. and Srivastava, P.K. 2007. Quality and shelf life evaluation of fermented sausages of buffalo meat with different levels of heart and fat. *Meat Sci* 75(4): 603–609.

Amadi, K., Nwana, E.J.C., and Otubu, J.A.M. 1999. Effect of thyroxine on the contractile responses of the Vas deference to prostaglandin E2. *Arch Androl* 42: 55–62.

Anon. 1988, FAO Meat Quality. http://www.fao.org/docrep/t0562e/T0562E02.htm

Anon. 1920. *Burma Gazetteer, Tharrawaddy District*. Office of the Superintendent of Government Printing, Rangoon, Burma.

Anon. 2000. Sri Lanka Fisheries Yearbook 1999. Socio-Economic and Market Research Division. National Aquatic Resource Research and Development Agency, Sri Lanka.

Anon. 1974. *J. Workhoven, FAO Bulletin*, Vol. 26. Rome.

Appleton, J., Briggs, C., and Rhatigan, J. 1978. *Pieces of Eight: The Rites, Roles, and Style of the Dean by Eight Who Have Been There*. NASPA Institute of Research and Development, Portland, OR.

Baishya, D. and Deka, M. 2009. *Fish Fermentation Traditional to Modern Approaches*. New India Publishing Agency, Pitam Pura, New Delhi, pp. 1–123.

Bokuchava, M.A. and Skobeleva, N.I. 1980. The biochemistry of tea manufacture. *CRC Crit Rev Food Sci Nutr* 12: 303.

Bramsnaes, F. 1965. In: Borgstrom, G. (Ed.), *Fish as Food*, Vol. 4. Academic Press, New York and London.

Chukeatirote, E., Dajanta, K., and Apichartsrangkoon, A. 2010. Thua nao, indigenous Thai fermented soybean: A review. *J Biol Sci* 10: 581–583.

Corcoran, T.H. 1963. Roman fish sauces. *Classic J* 58(5): 204–221.

Dahal, R.N., Karki, T.B., Swamylingappa, B., Li, Q., and Gu, G. 2005. Traditional foods and beverages of Nepal—a review. *Food Rev Int* 21(1): 1–25.

Dakwa, S., Sakyi-Dawson, E., Diako, C., Annan, N.T., and Amoa-Awua, W.K. 2005. Effect on the fermentation of soybeans into dawadawa (soydawadawa). *Int J Food Microbiol* 104: 69–82.

Deshpande, S.S. 2002. Toxicology in foods. Part II. In: *Handbook of Food Toxicology*. CRC Press, Boca Raton, FL. doi:10.1201/9780203908969.pt2.

Devi, P. and Suresh, P. 2012. Traditional, ethnic and fermented foods of different tribes of Manipur. *Indian J Trad Knowl* 11(1): 70–77.

Dirar, H.A. 1993. *The Indigenous Fermented Foods of the Sudan. A Study in African Food and Nutrition*. CABS International, Wallingford, pp. xvii + 552.

Dzudie, T., Bouba, M., Mbofung, C.M., and Scher, J. 2003. Effect of salt dose on the quality of dry smoked beef. *Ital J Food Sci* 15: 433.

Flament, L. 1989. Coffee, cocoa, and tea. *Food Rev Int* 5: 317–414.

Gamer, G. 1987. Antike anlagen zur Fischverarbeitung. In: *Hispanien und Mauretanien*. Antike Welt. *Zeitschr Archaol Kulturgeschichte* 18(2): 19–28.

Grimal, P. and Monod, T. 1952. Sur la Véritable Nature de "Garum." *Rev Étud Anciennes* 54: 27–38.

Guizani, N. and Mothershaw, A. 2005. Fermentation. Chapter 63. In: *Handbook of Food Science, Technology, and Engineering*. Vol. 4. Hui, Y.H. (Ed.), CRC Press, Boca Raton, FL. doi:10.1201/b15995-7010.1201/b15995-70.

http://bfrf.org/site/site/wp-content/uploads/2012/06/Chapter-13b.pdf

http://cismhe.org/cc/Vol-VIII_Food%20Resources.pdf8

http://news.education4india.com/464/intel-to-teacher-training/

http://shodhganga.inflibnet.ac.in/bitstream/10603/5613/19/19_references.pdf

http://www.docstoc.com/docs/74239291/Handbook-of-Fermented-Functional-Foods-2nd-edn

http://www.energymanagertraining.com/tea/pdf/Tea%20Types%20and%20Production%20Proccss006.pdf

http://www.ifr.ac.uk/SFC/Total%20Food%202006%20Abstract%20Book.pdf

http://www.indianetzone.com/37/north-east_indian_tribes.htm

IPFC (Indo-Pacific Fisheries Council). 1967. *Fish Processing in the Indo Pacific Area, Regional Studies No. 4*. IPFC, Bangkok.

Ishige, N. 1986. Narezushi in Asia: A study of fermented aquatic products (2). *Bull Natl Museum Ethnol* 11(3): 603–668.

Ishige, N. and Ruddle, K. 1987. Gyosho in Southeast Asia—A study of fermented aquatic products. *Bull Natl MusEthnol* 12(2): 235–314.

Ishige, N. and Ruddle, K. 1990. Gyosho to Narezushi no Kenkyu (Research on fermented fish products and Narezushi). *J Food Technol* 21, 55–60.

Ishii, R. and O'Mahony, M. 1987. Defining a taste by a single standard: Aspects of salty and umami tastes. *J Food Sci* 52: 1405–1409.

Jamir, B. and Deb, C.R. 2014. Studies on some fermented foods and beverages of Nagaland, India. *Int J Food Ferment Technol* 4(2): 121–127.

Jeyaram, K., Singh, W.K., Premarani, T., Devi, A.R., Chanu, K.S., Talukda, N.C. and Rohinikumar, M.R. 2008. Molecular identification of dominant microflora associated with "Hawaijar"—A traditional fermented soybean (Glycine max L.) food of Manipur, India. *Int J Food Microbiol* 122(3): 259–268.

Kakati, B.P. and Goswami, U.C. 2013. Microorganisms and the nutritive value of traditional fermented fish products of Northeast India. *Glob J Biosci Biotechnol* B.2(1): 124–127.

Khanum, H., Islam, N.M., and Nahar, N.K. 1999. Intestinal nematode infestation among children of lower income group employees in Dhaka city. *Bangladesh J Zool* 27(2): 177–183.

Kiers, J.L., Van Laekan, A.E.A., Rombouts, F.M., and Nout, M.J.R. 2000. *In vitro* digestibility of *Bacillus* fermented soya bean. *Int Food Microbiol* 60: 163–169.

Kimizuka, A., Mizutani, T., Ruddle, K., and Ishige, N. 1992. Chemical Tanzanian food product. *Int J Food Microbiol* 56: 179–190.

Lee, C.H. 2009. Food biotechnology. In: *Food Science and Technology*. Campbell-Platt, G. (Ed.), Wiley-Blackwell, West Sussex, U.K., pp. 85–113.

Lerstner, L. 1986. Allgemeines Uber Rohwurst. *Fleischwirtschaft* 66: 290–300.

Lu'ong, H.D. 1981. *Che Bien Tu Ca Va Hai San Khac (Some Products Prepared from Fish and Other Marine Resources)*. Tran Phu, Ho Chi Minh City (in Vietnamese).

Majumdar, R.K. and Basu, S. 2009. Evaluation of the influence of inert atmosphere packaging on the quality of salt-fermented Indian shad. *Int J Food Sci Technol* 44(12): 2554–2560.

Majumdar, R.K., Basu, S., and Nayak, B.B. 2006. Studies on the biochemical changes during fermentation of salt-fermented Indian Shad (*Tenualosa ilisha*). *J Aquat Food Prod Technol* 15(1): 53–69.

Mansur, M.A. 2007. A review of different aspects of fish fermentation in Bangladesh. *Bangladesh J Prog Sci Technol* 5: 185–190.

Mao, A.A. and Odyuo, N. 2007. Traditional fermented foods of the Naga tribes of Northeastern, India. *Indian J Trad Knowl* 6(1): 37–41.

Maxwell, 1904. Report on inland fisheries and sea fisheries in the thongwa, Myaungma, and Bassein districts and reports on the turtle-banks of the Irrawaddy Division. Office of the Superintendent of Govt. Printing, Rangoon, Burma.

Mizutani, T., Kimizuka, A., Ruddle, K., and Ishige, N. 1987. A chemical analysis of fermented fish products and discussion of fermented flavors in Asian cuisines, a study of fermented fish products. *Bull Natl Mus Ethnol* 12(3): 801–864.

Moktan, B., Roy, A., and Sarkar, P.K. 2011. Antioxidant activities of cereal-legume mixed batters as influenced by process parameters during preparation of dhokla and idli, traditional steamed pancakes. *Int J Food Sci Nut* 62: 360–369.

Muzaddadi, A.U. and Basu, S. 2012. A traditional fermented fishery product of North-east India. *Indian J Trad Know* 11(2): 322–328.

Muzaddadi, A.U. and Mahanta, P. 2013. Effects of salt, sugar and starter culture on fermentation and sensory properties in Shidal (a fermented fish product). *Afr J Microbiol Res* 7(13): pp. 1086–1097.

Nagalakshmi, D., Sastry, V.R.B., and Pawde, A. 2003. Rumen fermentation patterns and nutrient digestion in lambs fed cottonseed meal supplemental diets. *Anim Feed Sci Technol* 103(1–4): 1–4. Doi: 10.1016/S0377-8401(02)00140-2.

Namugumya, B.S. and Muyanja, C.M.B.K. 2009. Traditional processing, microbiological, physiochemical and sensory characteristics of Kwete, a Ugandan fermented maize based beverage. Report. *Afr J Food Agric Nutr Dev* June: 9.

Nowsad, A.K.M.A. 2007. *Participatory Training of Trainers: A New Approach Applied in Fish Processing*. Bangladesh Fisheries Research Forum, Bangladesh, p. 328.

Oki, K., Rai, A.K., Sato, S., Watanabe, W., and Tamang, J.P. 2011. Lactic acid bacteria isolated from ethnic preserved meat products of the Western Himalayas. *Food Microbiol* 28(7): 1308–1315.

Omafuvbe, B.O., Shonukan, O.O., and Abiose, S.H. 2000. Microbiological and biochemical changes in the traditional fermentation of soybean for soydaddawa—A Nigerian food condiment. *Food Microbiol* 17: 469–474.

Parkouda, C., Nielsen, D.S., Azokpota, P., Ouoba, L.II, Amoa-Awua, W.K., Thorsen, L., Hounhouigan, J.D. et al. 2009. The microbiology of alkaline-fermentation of indigenous seeds used as food condiments in Africa and Asia. *Crit Rev Microbiol* 35: 139–156.

Parkouda, C., Thorsen, D., Compaoré, C.S., Nielsen, D.N., Tano-Debrah, K., Jensen, J.S., Diwara, B., and Jakobsen, M. 2010. Microorganisms associated with Maari, a Baobab seed fermented product. *Int J Food Microbiol* 142(3): 292–301.

Phithakpol, B., Varanyanond, W., Reungmaneepaitoon, S., and Wood, H. 1995. *The Traditional Fermented Foods of Thailand*. Institute of Food Research and Product Development, Kasetsart University, Bangkok, pp. 157.

Rai, A.K., Palni, U., and Tamang, J.P. 2009. Traditional knowledge of the Himalayan people on production of indigenous meat products. *Indian J Trad Know* 8(1): 104–109.

Rai, A.K., Tamang, J.P., and Palni, U. 2010. Nutritional value of lesser-known ethnic meat products of the Himalayas. *J Hill Res* 23(1–2): 22–25.

Rantsiou, K. and Cocolin, L. 2006. New developments in the study of the microbiota of naturally fermented sausages as determined by molecular methods. A review. *Int J Food Microbiol* 108: 255.

Rapsang, G.F. and Joshi, S.R. 2012. Bacterial diversity associated with Tungtap, an ethnic traditionally fermented fish product of Meghalaya. *Ind J Trad Knowl* 11(1): 134–138.

Rapsang, G.F, Kumar, R., and Joshi, S.R. 2011. Identification of *Lactobacillus pobuzihii* from tungtap: A traditionally fermented fish food and analysis of its bacteriocinogenic potential. *Afr J Biotechnol* 10(57): 12237–12243.

Ruddle, K. and Ishige, N. 2005. *Fermented Fish Products in East Asia*. International Resources Management Institute, Hong Kong.

Salampessy, J., Kailasapathy, K., and Thapa, N. 2010. Fermented fish products. Chapter 10. In: *Fermented Foods and Beverages of the World*. Tamang, J.P. and Kailasapathy, K. (Eds.), CRC Press, Boca Raton, FL, pp. 289–307.

Sanderson, G.W. and Graham, H.N. 1973. On the formation of black tea aroma. *J Agric Food Chem* 21: 576–585.

Sarkar, P.K., Hasenack, B., and Nout, M.J.R. 2002. Diversity and functionality of *Bacillus* and related genera isolated from spontaneously fermented soybeans (Indian Kinema) and locust beans (African Soumbala). *Int J Food Microbiol* 77(3): 175–186.

Sarkar, P.K., Tamang, J.P., Cook, P.E., and Owen, J.D. 1994. Kinema—A traditional soybean fermented food: Proximate composition and microflora. *Food Microbiol* 11: 47–55.

Sarkar, P.K., Tamang, J.P. 1995. Changes in the microbiol profile and proximate composition during natural and controlled fermentations of soybeans to produce kinema. *Food Microbiol*, 12: 317–325.

Sarkar, P.K., Morrison, E., Tingee, U., Sommerset, S.M., and Craven, G.S. 1998. B-group vitamins and mineral contents of soybean during kinema production. *J Sci Food Agric* 78: 498–502.

Sarkar, P.K., Jones, L.J., Craven, G.S., Somerset, S.M., and Palmer, C. 1997. Amino acid profiles of kinema, a soybean-fermented food. *Food Chem* 59(1): 69–75.

Sarkar, P.K. and Nout, M.J.R. (Eds.), 2014. *Handbook of Indigenous Foods Involving Alkaline Fermentation*. CRC Press, Boca Raton, FL.

Sarkar, P.K. and Tamang, J.P. 1995. Changes in the microbial profile and proximate composition during natural and controlled fermentations of soybeans to produce kinema. *Food Microbiol* 12: 317–325.

Scott, J.G. 1910. *The Burman: His Life and Notions*. School of Food Science and Nutrition.

Sein, M.T., Aggamedha, B.M., and Kawthanlla, U.M. 2001. The Rakhaing Welfare Society. Cox's Bazar, Bangladesh. Shrimps during long storage at normal room temperatures. *Sea Food*.

Sekar, S. and Mariappan, S. 2007. Usage of traditional fermented products by Indian rural folks and IPR. *Indian J Trad Know* 6(1): 111–120.

Sharma, N. 1987. Processed meat products and their potential in 2000 AD. In: *Advances in Meat Research*. Khot, J.B., Sherikar, A.T., Jayarao, B.M., Pillai (Eds.), Red and Blue Cross Publisher, Bombay.

Shio, S. 2006. Fermented fish products in East Asia: IRMI Research Study 1, Kenneth Ruddle, and Naomichi Ishige. International Resources Management Institute, Hong Kong (2005). *Trends Food Sci Technol* 17(11): 626.

Smith, D.R. 1988. Sausages—A food of myth, mystery and marvel. *Food Techno Aust* 40(2): 51–56.

Sohliya, I., Joshi, S.R., Bhagobaty, R.K., and Kumar, R. 2009. Tungrymbai—Traditional fermented soybean food of the ethnic tribes of Meghalaya. *Ind J Trad Know* 8(4): 559–561.

Southgate, H. 1984. *Meat Fish, Eggs and Novel Proteins on Fish Nutrition*. Churchill Livingstone, New York/London, pp. 363–371.

Srikantayya, N. 2003. Tea: An appraisal of processing methods and products. Chapter 25. In: *Handbook of Postharvest Technology: Cereals, Fruits, Vegetables, Tea, and Spices*. Hosahalli, S., Ramaswamy, G.S., Vijaya, R., Amalendu, C., and Arun, S.M. (Eds.), CRC Press, Boca Raton, FL, pp. 741–778.

Steinkraus, K.H. 1983a. Handbook of indigenous fermented foods. *Food Contrib* 8: 311–317.

Steinkraus, K.H. 1983b. Traditional food fermentations as industrial resources. *Acta Biotechnol* 43(1): 3–12.

Steinkraus, K.H. 1985. Indigenous fermented food technologies for small scale industries. *Food Nutr Bull* 7(2): 21–27.

Steinkraus, K.H. 2004. *Industrialization of Indigenous Fermented Foods*, 2nd ed. CRC Press, Boca Raton, FL.

Steinkraus, K.H. 1998. Bio-enrichment: Production of vitamins in fermented foods. In: *Microbiology of Fermented Foods*. Wood, J.B. (Ed.), Blackie Academic and Professional, London, pp. 603–619.

Takeo, T. and Mahantha, P.K. 1983. Comparison of black tea aromas of orthodox and CTC tea and of black teas made from different varieties. *J Sci Food Agric* 34: 307.

Tamang, J.P. 1998. Role of microorganisms in traditional fermented foods. *Indian Food Ind* 17(3): 162–167.

Tamang, J.P. 2003. Native microorganisms in the fermentation of Kinema. *Int J Food Microbiol* 43: 127–130.

Tamang, J.P. 2009a. *Himalayan Fermented Foods: Microbiology, Nutrition, and Ethnic Values*. CRC Press, Boca Raton, FL, pp. 255–284.

Tamang, J.P. 2009b. Antiquity and ethnic values. Chapter 9. In: *Himalayan Fermented Foods: Microbiology, Nutrition, and Ethnic Values*. CRC Press, Boca Raton, FL, pp. 229–246.

Tamang, J.P. 2009c. Ethnic meat products, Chapter 7. In: *Himalayan Fermented Foods: Microbiology, Nutrition, and Ethnic Values*. CRC Press, Boca Raton, FL, pp. 161–186.

Tamang, J.P. 2009d. Fermented legumes. Chapter 3. In: *Himalayan Fermented Foods: Microbiology, Nutrition, and Ethnic Values*. CRC Press, Boca Raton, FL, pp. 65–94.

Tamang, J.P. 2009e. Ethnic fish products. Chapter 6. In: *Himalayan Fermented Foods: Microbiology, Nutrition, and Ethnic Values*. CRC Press, Boca Raton, FL, pp. 139–160.

Tamang, J.P. 2010. Himalyan fermented foods: Microbiology, Nutrition and Ethone value. CRC press, Taylor and Francis group, New York.

Tamang, J.P. 2010a. *Himalayan Fermented Foods: Microbiology, Nutrition and Ethnic Value*. CRC Press, Taylor & Francis Group, New York.

Tamang, J.P. 2010b. Diversity of fermented foods. In: Tamang, J.P. and Kailashpathy, K. (eds.), *Fermented Foods and Beverages of the Word*. CRC Press, Taylor & Francis Group, Boca Raton, FL, pp. 41–84.

Tamang, J.P. 2012. Plant-based fermented foods and beverages of Asia. Chapter 4. In: *Plant-Based Fermented Foods and Beverages*. Hui, Y.H. and Özgül, E. (Eds.), CRC Press, Boca Raton, FL, pp. 49–90.

Tamang, J.P. 2013. Animal-based fermented foods of Asia. Chapter 4. In: *Animal-Based Fermented Foods*. Hui, Y.H. and Özgül, E. (Eds.), CRC Press, Boca Raton, FL, pp. 61–67.

Tamang, J.P., Dewan, S., Thapa, S., Olasupo, N.A., Schillinger, U., and Holzapfel, W.H. 2000. Identification and enzymatic profiles of predominant lactic acidbacteria isolated from soft-variety chhurpi, a traditional cheese typical of the Sikkim Himalayas. *Food Biotechnol* 14 (1–2): 99–112.

Tamang, J.P. and Holzapfel, W.H. 1999. Biochemical identification techniques-modern techniques: Microfloras of fermented foods. In: *Encyclopedia of Food Microbiology*. Robinson, R.K., Batt, C.A., and Patel, P.D. (Eds.), Academic Press, London, pp. 249–252.

Tamang, J.P. and Nikkuni, S. 1996. Selection of starter culture for production of kinema, fermented soybean foods of the Himalaya. *World J Microbiol Biotechnol* 12(6): 629–635.

Tamang, J.P. and Nikkuni, S. 1998. Effect of temperatures during pure culture fermentation of kinema. *World J Microbiol Biotechnol* 14: 847–850.

Tamang, J.P. and Sarkar, P.K. 1996. Microbiology of *mesu*, a traditional fermented bamboo shoot product. *Int J Food Microbiol* 29(1): 49–58.

Tamang, J.P, Tamang, N., Thapa, S., Devan, S., Tamang, B., Yonzan, H., Rai, A., Chettri, R., Chakrabarty, J., and Kharel, M. 2012. Microorganisms and nutritional value of Ethnic fermented foods and alcoholic beverages of Northeast India. *Indian J Trad Know* 11(1): 7–25.

Tamang, J.P., Thapa, S., Tamang, N., and Rai, B. 1996. Indigenous fermented food beverages of Darjeeling hills and Sikkim: Process and product characterization. *J Hill Res* 9(2): 401–411.

Tanasupawat, S., Namwong, S., Kudo, T., and Itoh, T. 2007. *Piscibacillus salipiscarius* gen. nov., sp. nov., a moderately halophilic bacterium from fermented fish (pla-ra) in Thailand. *Int J Syst Evol Microbiol* 57: 1413–1417.

Tanasupawat, S., Namwong, S., Kudo, T. and Itoh, T. 2008–2009. Identification of halophilic bacteria from fish sauce (nam-pla) in Thailand. *J Culture Collections* 6, 69–75.

Tanasupawat, S., Thongsanit, J., Thawai, C., Lee, K.C., and Lee, J.S. 2011. *Pisciglobus halotoleransgen.* nov., sp. nov., isolated from fish sauce in Thailand. *Int J Syst Evol Microbiol* 61: 1688–1692.

Tanasupawat, S. and Visessanguan, W. 2013. Fish fermentation. In: *Seafood Processing: Technology, Quality and Safety.* Boziaris, I.S. (Ed.), John Wiley & Sons, New York, NY.

Thapa, N., Pal, J., and Tamang, J.P. 2006. Microbial diversity in *ngari, hentak* and *tungtap,* fermented fish products of Northeast India. *World J Microbiol Biotechnol* 26(6): 599–607.

Thokchom, S and Joshi, S.R. 2012. Microbial and chemical changes during preparation in the traditionally fermented soybean product Tungrymbai of ethnic tribes of Meghalaya. *Indian J Trad Know* 11(1): 139–142.

Toyokawa, Y., Takahara, H., Reungsang, A., Fukuta, M., Hachimine, Y., Tachibana, S., and Yasuda, M. 2010. Purification and characterization of a halotolerant serine proteinase from thermotolerant Bacillus licheniformis RKK-04 isolated from Thai fish sauce. *Appl Microbiol Biotechnol* 86(6): 1867–1875.

TPI (Tropical Products Institute). 1982. *Fermented Fish Products: A Review. Fish Handling, Preservation and Processing in the Tropics, Part 2*: 18–22, TPI, London.

Tyn, M.T. 1996. Trends of fermented fish technology in Burma. In: *Fish Fermentation Technology.* Lee, C.-H., Steinkraus, K.H., and Alan Reilly, P.J. (Eds.), United Nations University, New York, NY, pp. 129–153.

Tyn, M.T. 2004. Industrialization of Myanmar fish paste and sauce fermentation. In: *Industrialization of Indigenous Fermented Foods.* 2nd edn., revised and expanded. Marcel Dekker, New York, NY, pp. 737–759.

Venugopal, V. (Ed.), 2005. Value addition of freshwater and aquacultured fishery products through quick freezing, retortable packaging and cook-chilling. Chapter 12. In: *Seafood Processing.* CRC Press, Boca Raton, FL, pp. 341–376.

Wang, J. and Fung, D.Y. 1996. Alkaline-fermented foods: A review with emphasis on pidan fermentation. *Crit Rev Microbiol* 22: 101–138.

Wood, B.J.B. 1998. *Microbiology of Fermented Foods.* Vol.1, Blackie Academic & Professional, London.

Yamanishi, T. 1978. In: *Flavor of Foods and Beverages, Chemistry and Tech.* Charalambous, G. and Inglett, G.E. (Eds.), p. 305.

Zeuthen, P. 1995. Historical aspects of meat fermentation. In: *Fermented Meats.* Campbell-Platt, G. and Cook, P.E. (Eds), Blackie, Glasgow, pp. 39–52.

12

Technology of Mushroom Production and Its Postharvest Technology

B.C. SUMAN, DHARMESH GUPTA, P.K. KHANNA, REENA CHANDEL, SHAMMI KAPOOR, AND V.K. JOSHI

Contents

12.1 Introduction

Mushrooms have been a part of fungal diversity for about 300 million years. They represent a highly specialized group of fungi endowed with the ability to degrade and bio-convert a variety of inedible plant wastes into a useful form of food. Apart from their culinary, nutritional, and health benefits, mushrooms are now recognized as a food contributing to amelioration of protein malnutrition for the ever increasing population. It is a "macrofungus" with a distinctive fruiting body, which can either be epigeous or hypogeous and large enough to be seen with naked eye and to be picked by hand (Chang and Miles, 1992). Thus, mushrooms need not be basidiomycetes, nor aerial, or fleshy, or edible but can be ascomycetes, grow underground, have a non-fleshy texture, and need not be edible.

To the ancient Romans, they were "the foods of the Gods" resulting from bolts of lightning thrown to the earth by Jupiter during thunder storms; the Egyptians considered them as "a gift from the God Osiris"; while the Chinese viewed them as "the elixir of life." Throughout history, many cultures, including oriental ones, have built-up a practical knowledge of which mushrooms were suitable to eat and those that were poisonous, while other mushrooms might have profound health-promoting benefits (Hobbs, 1995).

12.1.1 Mushrooms as a Fermented Food Product

In nature, saprophytic mushrooms are able to obtain their nutrients from plant residues and dead wood, while cultivated mushrooms are generally grown on fermented or composted plant residues. The process of fermenting the substrates to produce a medium from dead organic materials suitable for the production of these mushrooms is called composting. Since mushrooms are grown on substrates produced from fermentation of agri-residues and are produced through solid state fermentation like some of the fermented foods, they are considered to be one of the oldest fermented products.

12.1.2 History of Mushroom Cultivation

Historically, mushrooms were gathered from the wild for consumption and for medicinal use. China has been the main source of much early cultivation of mushrooms, for example *Auricularia auricula* (600 AD), *Flammulina velutipes* (800 AD), *Lentinula edodes* (1000 AD), and *Tremella fuciformis* (1800 AD). *Agaricus bisporus* was first cultivated in France in 1600 BC, while *Pleurotus ostreatus* was first grown in the USA in 1900. Mushrooms nowadays are popularly known as functional foods (Liu and Wang, 2009) and have been recognized as an alternate source of good quality protein and produce, with production of the highest quantity of protein per unit area from agro-wastes (Dama et al., 2010). They also provide potential for generating employment, improving the economic status of growers, help in checking pollution, and earn foreign exchange (Rai and Arumuganathan, 2005). There are about 20 varieties of mushroom being cultivated throughout the world for food. In India, only white button mushrooms (*Agaricus bisporus*), oyster mushrooms (*Pleurotus* spp.), paddy straw mushrooms (*Volvariella volvacea*), and milky mushrooms (*Calocybe indica*) are grown commercially, with white button mushrooms contributing to 90% of the total production. Mushroom cultivation now spans more than 100 countries of the world, but it is only during the last 2–3 decades that we have witnessed major expansions in basic research and practical knowledge leading to the creation of a major worldwide mushroom industry (Chang and Miles, 1989). Innovative technological developments in the mushroom industry during the last 10–15 years have seen increasing production capacities and marketing of culinary and medicinal mushrooms from the Asian countries. In the developed countries of the European Union (EU) and the USA, it has attained the status of a hi-tech industry, with higher levels of mechanization and automation. In spite of these factors, the cost of mushroom production in these countries has gone up, leading to stagnation in mushroom production, although it has opened new opportunities to the third world countries to capitalize on due to widening gaps between demand and supply. Contributions from developing countries such as India, Poland, Hungary, and Vietnam have been steadily increasing and, as a result, the World production of edible mushrooms is estimated at about 12 million tons, but the annual growth rate is still above 8% (Rai and Arumuganathan, 2008a). In India, current production has crossed the one lakh ton mark, with an annual growth rate of above 15%.

12.1.3 Production of Mushroom in South Asia

Mushroom production represents an important opportunity for developing countries, particularly in Asia. In Asia's mushroom industry much attention has been paid to specialty mushrooms. There are several countries in South Asia, like Nepal, Sri Lanka, India, Pakistan, and Bhutan (Khanna and Sharma, 2013), contributing significantly to the pool of mushrooms production.

12.2 Mushroom Types

Mushroom in general can be placed under the following categories:

- Edible mushrooms
- Specialty mushrooms
- Medicinal mushrooms
- Mycorrhizal mushrooms
- Poisonous mushrooms

12.2.1 Edible Mushrooms

Edibility may be defined by criteria that include absence of poisonous effects on humans and desirable taste and aroma. Edible mushrooms are consumed by humans

Figure 12.1   Some specialty mushrooms: (a) *Lentinula edodes*, (b) *Auricularia polytricha*, (c) *Pleurotus* spp., (d) *Flammulina velutipes*.

for their nutritional and, occasionally, medicinal value. These fungal species are either harvested wild or cultivated. Some of the examples of commercially cultivated edible mushrooms include *Agaricus bisporus*, *Pleurotus* sp., *V. volvacea*, and *Calocybe indica*.

12.2.2 Specialty Mushrooms

These mushrooms represent an array of species with diverse flavors, textures, and colors that have documented food and medicinal values. Specialty mushrooms differ significantly from common button mushroom in appearance and the substrate in which they are grown. Some examples of specialty mushroom (Figure 12.1) include *Lentinula edodes* (shiitake), *Auricularia polytricha* (black ear mushroom), *Pleurotus* spp. (oyster), *Flammulina velutipes* (enokitake), *Tremella fuciformis* (snow fungus), *Hericium erinaceous* (pimpom), *Stropharia rugosoannulata* (wine cap), and *Pholiota squarrosa* (shaggy scaly cap). Specialty mushrooms are grown commercially in many South East Asian countries and are frequently used in Japanese, Korean, and Chinese cuisine for soups and salads, both in fresh and dried form.

12.2.3 Medicinal Mushrooms

Medicinal mushrooms have become even more widely used as traditional medicinal ingredients for the treatment of various diseases and related health problems, largely due to the increased ability to produce the mushrooms by artificial methods. While much attention has been focused on various immunological and anticancer properties of these mushrooms, they also offer other potentially important therapeutic properties, including antioxidants, antihypertensive, cholesterol-lowering, liver protection, antifibrotic, anti-inflammatory, antidiabetic, antiviral, and antimicrobial. Some of the most important medicinal mushrooms include: *Ganoderma lucidum*, *Lentinula edodes*, *Auricularia auricula*, *Hericium erinaceus*, *Grifola frondosa*, *Flammulina velutipes*, *Trametes (Coriolus) versicolor*, and *Cordyceps sinensis*.

12.2.4 Mycorrhizal Mushrooms

These fungi live in the soil in symbiotic association with the roots of vascular plants of different tree species in woodlands and in forest ecosystems. Many of the mycorrhizal species of mushrooms are edible. Around 5000 species of fungi are known to form ectomycorrhizae with about 2000 woody hosts that belong to Hymenomycetes in Basidiomycotina, while some are Gasteromycetes and possibly Ascomycetes. These are predominantly in the genera *Amanita*, *Boletus*, *Tricholoma*, *Russula*, *Laccaria*, *Lactarius*, *Cantharellus*, *Cortinarius*, etc. These mushroom species cannot be cultivated indoors like other cultivable mushrooms. Outdoor/field cultivation technology has been developed for *Boletus granulatus* Linn. ex Fr., *B. edulis* Schaeff. ex Fr., *Tricholoma matsutake* (Ito et Inai) Sing., *Tuber melanosporum* Vitt (Perigold black truffle), and *T. magnatum* Pico ex Vitt.

12.2.5 Poisonous Mushrooms

Of the many thousands of mushroom species in the world, many have been associated with fatalities while others have been found to contain some toxins. By far, the majority of mushroom poisonings are not fatal, but the majority of fatal poisonings are attributable to the *Amanita phalloides* mushroom. Poisonous mushrooms contain a variety of different toxins that can differ markedly in toxicity. Symptoms of mushroom poisoning may vary from gastric upset to life-threatening organ failure, resulting in death. Serious symptoms do not always occur immediately after eating; often not until the toxin attacks the kidneys or liver, sometimes days or weeks later.

12.3 Wild Edible Mushrooms of South Asia

In India and Pakistan, *Morchella conica* and *M. esculenta* dominate the market for wild edible fungi. Almost all the morels are exported to Europe. The collectors get between one-half and two-thirds of the export price (Sabra and Walter, 2001; Prasad et al., 2002). Other species includes *Agaricus augustus* (Karim Nawaz, 2011), *A. compestris*, *A. placomyces*, *A. rodmani*, *A. cilvaticus*, *A. silvicola*, *Armilaria mellea*, *Cantharellus cibarius*, *Craterellus cornucopioides*, *Flammulina velutipes*, *Macrolepiota procera*, *Morchella angusticeps*, *Podaxis pistillaris*, *Termitomyces clypeatus*, *T. eurhizus*, *T. heimii*, *T. microcarpus*, *T. radicatus*, and *T. striatus* (FAO, 2004). Nepal is rich in biodiversity of mushrooms because of variation in topography, climate, and latitudinal changes that occur within a short distance. Nepalese mycoflora includes 585 genera and 1822 species. These wild species include various fungi of economic importance (edible 110 sp, poisonous 48 sp, medicinal 17 sp, mycorrhizal 50 sp, decorative 12 sp, and parasitic 970 sp. (Adhikari, 1995)). Among the ascomycetes, *morels* (mainly

Morchella conica) are the only wild edible fungi which are exported, in larger quantities. They are mainly found in pine forests between 2000 and 3500 m, and are most abundant in the western part of Nepal. *Morels* (*Morchella* spp.) are the most sought after and costly edible mushrooms in the world, and include several species that differ in head color and ecological characteristics (Khan and Khan, 2011). The morels possess an excellent flavor and are highly appreciated for their culinary aspects and gastronomical delights. They are polymorphic with respect to head shape, stalk-to-head ratio, immature and mature color, taste, and edibility (Weber, 1997; Kuo, 2005; Kellner, 2009). The polysaccharides from *Morchella esculenta* fruit bodies have several medicinal properties, including antitumor effects, immunoregulatory properties, fatigue resistance, and antiviral effects (Wasser, 2002; Rotaoll et al., 2005; Nitha and Janardhan, 2008).

The fruit bodies of *Morchella* species grow on soil rich in organic matter, in coniferous, oak, and deodar forest, usually appearing during the spring season in March and April when the snow begins to melt and there is some precipitation. Species of *Morchella* have rarely been recorded/collected during the rainy season when other mushrooms appear abundantly. Different species of *Morchella* namely *Morchella angusticeps* Peck., *M. conica* Pers., *M. crassipes* (vent.) Pers. Ex. Fr., *M. deliciosa* Fr., *M. esculenta* (L.) pers. Ex. Fr. and *M. semilibera* DC. ex. Fr. (Syn.) have been reported from different parts of India, but are chiefly found in the north-western Himalayas. *M. esculenta* was also recorded in the Kodaikanal Hills (Western Ghats) in Tamil Nadu, South India for the first time in 2006 (Kaviyarasan et al., 2006).

Among the basidiomycetes, 18 edibles species have been described according to area. In Kathmandu valley, among the Clavariales, only *Ramaria botrytis* and *Clavulinopsis fusiformis* are found to be edible. In Lumle, *Cantharellus cibarius*, *Grifola frondosa*, *Laccaria laccata*, *Lactarius volemus*, *Laetiporus sulphureus*, and *Termitomyces clypeatus* are good for eating. Similarly, *Hericium erinaceus*, *Oudemansiella radicata*, *Ramaria flaccid*, and *Russula chloroides* are considered good to eat, whereas *Auricularia auricula-judae*, *Clavulinopsis fusiformis*, *Exobasidium butleri*, and *Lactarius piperatus* are considered to be not so tasty or good for eating (Adhikari et al., 2005). Some of the very popular wild edible mushrooms of Bhutan include *Matsutake* spp., *Chanterelle* spp., *Shimeji* spp., *Rozites caperatus*, Shiitake, Oyster spp., *Auricularia* spp., and *Ramaria* spp.

The larger Basidiomycetes are presently the best known group in Sri Lanka, with records of 513 species of agarics in 50 genera, such as *Psalliota* (*Agaricus*) 35 spp., *Pleurotus* 12 spp., *Marasmius* 53 spp., *Lentinus* 14 spp., and *Hygrophorus* 27 spp. A comprehensive *Agaric Flora of Sri Lanka* was published by Pegler in 1983.

12.4 Solid State Fermentation for Mushroom Production

The technology for production of mushrooms basically involves solid-state fermentation (Pandey and Ramachandran, 2005) at three different stages, that is, composting, spawn manufacture, and the growth of the mushrooms themselves on the moist substrate.

12.4.1 Composting

The major aim of composting in mushroom cultivation (Pandey and Ramachandran, 2005) is to produce a selective growth medium that renders successful mycelia colonization and fruiting body formation. Practically, composting is accomplished by piling up the substrates for a period, during which various changes take place so that the composted substrate is quite different from the starting material (Chang, 1991; Chang and Miles, 2004). Natural composts are derived from substrates like horse manure or chicken manure, while synthetic composts have been made from almost every type of agro-industrial waste products and residues, such as wheat straw, paddy straw, barley straw, rice bran, saw dust, banana, maize stover, tannery waste, wool waste, sugarcane bagasse, and horse chestnut wood.

The moisture content of the compost plays a major role in maintaining the ultimate sporophore quality (Pandey and Ramachandran, 2005). Water content in the range of 55%–70% is required for the best growth of most of the mushrooms (Flegg, 1960, 1962; Nigam et al., 2003). A temperature in the range of 22–27°C is most suitable for growth and rapid colonization of the mycelia (Carey and Connor, 1991). Nutrient status and pH, along with the salinity and ammonia content of the compost significantly influences the quality of the mushroom crop. The compost preparation must have a nitrogen content of approximately 2%, carbon to nitrogen ratio of 17:1, a reserve of fats, oils, minerals, vitamins, and viable but dormant microbial biomass (Carey and Connor, 1991). The pH should be in the range of 6–7.5, which is optimum for the growth of most of the mushrooms (Allison and Kneebone, 1962). A high concentration of soluble salts however, retards the fruiting of mushrooms. An excess of sodium, potassium, and ammonium ions in the compost is also detrimental to the growth of mycelia (Carey and Connor, 1991).

12.4.2 Spawn Preparation

Mushroom spawn is the inoculum of the desired mushroom strain in a very active form (Chang, 1991). Spawn production involves the growth of mycelia of mushrooms on cereal grains (Pandey and Ramachandran, 2005). Different cereal grains are autoclaved at 121–135°C for 2–3 h to get the best conditions. Before inoculation, the grain substrate is maintained at a moisture content of 40%–50% to favor maximum mycelial growth and a good shelf-life. Fungal strains maintained on Potato Dextrose Agar (PDA) are inoculated under a laminar flow, where strict sterility is maintained. The inoculated cereals grains are incubated at 25–30°C. Mechanical shaking of the containers is known to encourage homogenous growth of mycelia. Transportation of spawn should be done under a conditioned atmosphere, otherwise temperature fluctuation diminishes the quality (Zardrazil et al., 1992).

12.4.3 Mushroom Development

The requirements for fruiting body formation (mushroom development) differ from those required for mycelia development. The fruiting bodies need a lower temperature than mycelia culture (Chang, 1991). Accumulation of carbon dioxide favors good mycelia growth, but it is detrimental for fruiting body formation. Hence, proper ventilation of accumulated carbon dioxide should be taken care of. Once the fruiting bodies are formed, harvesting is based upon the type of species and to some extent on market demand.

12.5 Production Technology of Mushroom

Major Phases of Mushroom Cultivation

Mushroom farming is a complex business, which requires precision. The major steps of mushroom cultivation are (Chang and Chiu, 1992; Chang 1998):

- Selection of an acceptable mushroom species
- Selection of a good quality fruiting culture
- Development of robust spawn
- Preparation of selective substrate/compost
- Care of mycelial (spawn) running
- Management of fruiting/mushroom development
- Careful mushroom harvesting

12.5.1 Selection of Acceptable Mushroom Species/Strains

Before any decision to cultivate a particular mushroom is made, it is important to determine if that species possesses sensory qualities acceptable to consumers, if suitable substrates for cultivation are plentiful, and if environmental requirements for growth and fruiting can be met without employing excessively costly systems of mechanical control. A "fruiting culture" is defined as a culture with the genetic capacity to form fruiting bodies under suitable growth conditions. The stock culture which is selected should be acceptable in terms of yield, flavor, texture, fruiting time, etc.

12.5.2 Selecting a Good Quality Fruiting Culture

1. Sources of the cultures
   a. Tissue culture.
      i. Healthy mushrooms should be chosen either in the later button or egg stage
      ii. Should be cleaned with 75% alcohol
      iii. Should be split in half by cutting it longitudinally

iv. Should be cultured as a routine procedure at incubation temperatures ranging between 25°C and 34°C for ten days, depending on the mushroom type

v. Transfer the mycelium to spawn substrate to make spawn

b. Spore culture.

i. Transfer individual spores singly to a test tube or petri dish and allow to develop and germinate into mycelium

ii. Single-spore isolates from homothallic mushrooms, for example, *Volvariella volvacea* (primary homothallism) or *Agaricus bisporus* (secondary homothallism), can be used as fruiting culture to make spawn

iii. When single-spore isolates are from heterothallic mushrooms, for example, *Lentinula edodes*, *Pleurotus sajor-caju*, and *Ganoderma lucidum*, then they cannot form fruiting cultures and cannot make spawn, requiring mating with a compatible single-spore isolate.

iv. After mating, they form a dikaryon/fruiting culture and they can then, be used to make spawn.

c. Pure culture from other laboratories.

i. Obtain a test tube culture from a research laboratory which is already tested for their production characteristics and guaranteed to be pure.

d. Cultures from another source.

i. Cultures may also be grown from spawn obtained from another source

ii. A piece of the spawn is aseptically transferred to agar slants and incubated properly

iii. It is however, risky because the number of transfers that the spawn culture has undergone is rarely known

2. Culture media

Mushrooms grow on a variety of cultural media and on different agar formulas, both natural and synthetic, depending on the organism to be cultivated and the purpose of the cultivation.

a. PDA (potato dextrose agar) is the simplest and the most popular medium for growing mycelia of most of the cultivated mushrooms.

b. A ready-made MEA (malt extract agar) powder is also available commercially.

The strain selection and maintenance of new and improved strains of commercial mushroom species, namely *Agaricus bisporus*, *Pleurotus* species, and *Volvariella volvacea*, is a continuous process and the easiest approach to better production. Strain selection and improvement results in better sporophore quality for various market outlets, less shrinkage in processing, resistance to pests and pathogens, and better keeping quality, etc. Several studies have been conducted on this aspect, as briefly reviewed here.

Figure 12.2 Strain U$_3$ of *Agaricus bisporus*. (Courtesy of Dr. Dharmesh Gupta.)

12.5.2.1 Button Mushroom (Agaricus bisporus)   A strain improvement program in button mushrooms in India started in 1973 when Seth (1973) procured 19 strains of *A. bisporus* from different countries and evaluated them at the Mushroom Research Laboratory of the Department of Mycology and Plant Pathology, College of Agriculture, Solan (HP). Strain S-11 of *A. bisporus*, after proper evaluation on long method of compost, gave the highest yield, with all the desirable morphological characteristics. This strain remained the most sought after strain for cultivation till the introduction of strain U$_3$.

The strains S-11 and S-310 were also found to be good yielders in both long and short compost methods (Mehta and Bhandal, 1989). However, in India, the introduction of exotic strains continued till the late 1980s, even though by that time the necessity of a comprehensive breeding program through (i) hybridization and (ii) genetic manipulation started getting attention from Indian scientists (Dhar and Kaul, 1987). However, the situation did not improve further, except that some single spore isolates showing improved performance were found to perform well during multi location testing under diverse growing conditions and variable substrate formulation. Kumar and Sharma (1997) introduced hybrid strain U$_3$ from Holland after proper evaluation under Indian conditions, and it was found to be the highest yielding among the existing stock (Figure 12.2).

Suman (2001) developed the A-16 single spore isolate of button mushroom *A. bisporus*, which recorded a significant yield of 23.5 kg/100 kg compost in the long compost method, which was not only 40% higher than the S-11 parent, but was also comparable to U$_3$. Other important characteristics of this isolate were its white color, round shape, and better keeping quality.

Singh and Kamal (2011) isolated 132 single spore isolates of which 24 were found non-fruiting, whereas 11 were fertile. The variability was further analysed, and selected single spore isolates were subjected to RAPD and gene sequencing. A significant breakthrough in hybrid development of *Agaricus bisporus* was achieved at the

Figure 12.3   Newly developed hybrid-4 of *Agaricus bisporus*. (Courtesy of Dr. Manju Sharma.)

Figure 12.4   Fruiting body of *Pleurotus djamor*. (Courtesy of Dr. Dharmesh Gupta.)

Dr Y.S. Parmar University of Horticulture and Forestry in 2012 by developing seven hybrids using single spore isolates and the hyphal fusion technique. The formation of hybrids was further confirmed by utilizing RAPD markers and gene sequencing. Among these, hybrid-4 was found superior, giving an outstanding average yield of more than 26 kg/100 kg compost with all the desirable morphological characteristics (Figure 12.3) in a cropping period of 60 days (Sharma, 2012).

12.5.2.2 Oyster Mushroom (Pleurotus spp.)   Among the various oyster mushroom species, including *Pleurotus florida*, *P. sapidus*, *P. cornucopiae*, *P. eryngii*, *P. eous*, *P. djamor* (Figure 12.4), and *P. flabellatus* (Figure 12.5), *P. sajor-caju*, introduced by Jandaik and Kapoor (1974) in 1974 is the most important commercial species. Very few attempts have been made to improve the available strains, either by conventional methods or through the use of modern biotechnological techniques. Jandaik (1987)

Figure 12.5 *P. flabellatus* on wheat straw.

described the potential of *P. sajor-caju* for breeding improved strains. Pandey and Tiwari (1989, 1994) came across a natural sporeless mutant, a much desirable trait for oyster mushroom due to the problem of spore allergy. Among the very few contributions made on these lines, the work of Ghosh and Chakraborty (1991) on dikaryotization and on the production of protoplasts and their regeneration in a bid to obtain mutants of *P. florida* are worth a mention. Also, findings have indicated the possibility of raising high yielding hybrid strains by monokaryon isolates raised from diverse germ plasm. Several methods used for strain improvement in *Pleurotus* spp. include selection, hybridization, etc.

12.5.2.3 Straw Mushroom (Volvariella spp.)   More than 100 species of *Volvariella* have been recorded from various parts of the globe, but only 13 species are known from India (Singh et al., 2011), out of which three are under cultivation, namely, *Volvariella volvacea*, *V. diplasia*, and *V. esculenta*. Very limited efforts so far have been made to improve the existing strains of these two species. *Volvariella* is a primary homothallic fungus with most of its monosporus isolates being self sterile, making hybridization a difficult proposition (Chang and Chu, 1969). Evaluation of the existing cultures to identify the highest yielding strains of *V. diplasia* and *V. volvacea* have not yielded tangible results (Garcha et al., 1987). The other possibility to obtain an improved strain is through mutation, but that too has limited chances of success due to the multinuclear nature of its mycelium, which tends to mask the effect of mutation. Kapoor (1987) has, therefore, emphasized the need to give more thrust to breeding of *V. volvacea*, with the primary objectives of attempting somatic hybridization through protoplast fusion. Out of many important species, the cultivation technology of *V. volvacea* has been commercially exploited. Studies on the sexuality, genetics, and transformation in this mushroom need to be explored to obtain improved strains giving consistent yields (Sodhi and Kapoor, 2007).

12.5.3 Development of Robust Spawn

The medium through which the mycelium of a fruiting culture has grown and which serves as the inoculum of "seed" for the substrate in mushroom cultivation is called the "mushroom spawn." Failure to achieve a satisfactory harvest may often be traced to unsatisfactory spawn used. Consideration must also be given to the nature of the spawn substrate, since this influences the rapidity of growth in the spawn medium as well as the rate of mycelial growth and filling of the beds following inoculation.

12.5.3.1 Mushroom Spawn Preparation

a. Spawning means seeding of the compost. The word "spawn" is derived from an old French word, *espandre*, meaning to spread out or expand, which was earlier derived from the Latin, *expandere*, meaning to spread. It is also defined by Webster's Dictionary as "the mycelium of fungi, especially of mushrooms grown to be eaten, used for propagation." In the mushroom industry, spawn is a substrate into which a mushroom mycelium has impregnated and developed, and can be used as a seed in propagation for mushroom production.

b. Spawn substrates: A number of materials, mostly agricultural wastes, can be used to prepare mushroom spawn, such as chopped rice straw, sawdust, water hyacinth leaves, used tea leaves, cotton wastes, and lotus seed husks. Cereal grains (wheat, rye, or sorghum) are used as mother spawn, and agricultural wastes as the planting spawn substrates in most laboratories. The mother spawn is used to inoculate the final spawn container in which the planting spawn will be produced. The planting spawn is used to inoculate the compost/substrate for mushroom production.

c. Preparation of mother spawn: Wheat grains are soaked in water for 12 h or overnight. Dead seeds or those that float on water are carefully removed. The grains are then washed again and boiled in water for at least 10–15 min until they expand but are not quite broken. The grains are drained and allowed to cool. Precipitated chalk (1.5% on wet basis) is added to the grains. The grains are then, loosely packed in bottles (2/3 full) and are plugged with cotton wool or covered by a double-layer of aluminium foil and are sterilised in an autoclave at 22 psi for 2 h. The bottles are cooled prior to inoculation.

Four methods of spawning have frequently been used with grain spawn (Shandilya et al, 1974; Jain and Singh, 1982). These are (a) *Spot spawning*—planting of lumps of grain spawn about 5 cm below the surface of compost about 20–25 cm apart, (b) *Surface spawning*—spawn is spread at the top surface of the compost and slightly mixed into the upper layer of the compost, (c) *Layer spawning*—the compost is successively filled with about 5 cm depth and spawned in 2–3 layers, and (d) *Thorough spawning*—the spawn is mixed in the compost and then, filled in the container. Each kind of spawning technique has its own merits. The one layered surface spawning gives better results for early sown crops, while the layered technique is suitable for late spawning.

Thorough spawning is preferred for cultivation in polythene bags in low temperature conditions or controlled temperature mushroom houses. It has been observed that, irrespective of the method of spawning (Khanna and Sharma, 2013), a spawn rate of 0.5%–0.75% fresh weight of compost is the optimum dose of spawn for button mushrooms. A lower rate of spawning results in slow growth of mycelium and increases the possibility of infestation by competitors/pests, whereas higher doses of spawn not only increase the cost of production but also results in increased temperature of the compost in the bed, which is detrimental for the developing mycelium. "Super spawning" or "active mycelium" spawning involves spawning of mushroom compost using colonized compost free from contaminants as inoculum for the fresh compost. Some growers opt for a shake up of the compost 1 week after spawning which results in a quicker spawn run by redistribution of inoculum to increase the loci for spread of mushroom mycelium and dispersal of ammonia in case any residual ammonia is left at the time of casing. During the spawn run, the temperature in the growing room is maintained at 23–24°C with 80%–85% humidity, and a high CO_2 concentration is allowed to build up by keeping the fresh air ventilation closed. Under such conditions, it takes about 12–14 days for the spawn to colonize the compost. The spawn run period, however, depends upon the quality and moisture content of compost, the strain used, and the environmental conditions (in case of seasonal growing).

1. Casing soil and Casing

   Casing is a 3–4 cm thick layer of soil applied on top of the spawn run compost, and it is a prerequisite for fructification of *A. bisporus*. A variety of casing materials are used the world over, and lot of variation in yield of button mushrooms has been observed, indicating the role of several intrinsic qualities of the casing material. The texture, bulk density, water holding capacity, pH, and electrical conductivity are the major physical and chemical factors that influence the yield. Although casing is an indispensable step for button mushroom production, less attention has been devoted to this aspect as compared to other facets. Peat, available in abundance in the west is a widely used material for casing due to its unique abilities to adsorb and release water quickly, enough porosity for air exchange, and its availability. However, the occurrence of peat is limited to only the Kashmir valley in India. Other alternative mixtures are used with success in India such as:

   - Mixture of 2 years old FYM (Farm Yard Mannure) and 2-year-old spent compost (1:1)
   - Garden soil and sand mixture (4:1 by volume)
   - Decomposed FYM and loam soil (1:1 by volume)

   The casing materials may harbor many pests and pathogens. Before these are used by the growers, they are treated to selectively kill the harmful flora without affecting the useful ones. The industrial growers, having the facilities to do so, pasteurize the casing mixture using steam in bulk chambers. For steam pasteurization, the casing soil is treated with a jet of pressure steam in a closed chamber at 62–65°C for 6 h. Alternatively, the seasonal growers also use chemical disinfection by making use of a 4% solution of formaldehyde, which has been found quite effective.

12.5.3.2 Mushroom Spawn Handling

a. *Maintenance of spawn quality.* Mushroom spawn, whether prepared as a family home project or on industrial scale using modern equipment, should be in excellent condition when delivered to growers. Spawn of most mushrooms can be refrigerated, but it should be warmed to normal room temperature before it is used as an inoculum or for planting.

Vigorous growth of the planting spawn is a prerequisite to good growth and yield. If the spawn is not vigorous, the mushroom mycelium will be overgrown by competitor organisms. If it is vigorous, it will overcome many of the competitive organisms and produce more mushrooms. When purchasing spawn, ask the spawn maker how long the spawn can be kept before planting. Old spawn is not acceptable because its vigor may have decreased. Buyers or users should know the "expiry" date of the spawn.

Spawn are maintained by controlling environmental conditions during the spawn run, that is, air temperature (23 ± 1°C), humidity (90%), and high CO_2 (1500 ppm) are maintained in the growing room during the first week of casing, after which the temperature is lowered to 15–17°C with fresh ventilation (30%) for 1–2 h which reduces CO_2 level (300–1000 ppm) and humidity to 85%. In the seasonal growing rooms, the temperature is also lowered by an additional spray of water, the circulation of fresh cool air during the evening/night, and maintenance of high humidity. The optimal environmental parameters during various stages of cropping are given in Table 12.1.

The pin heads of button mushroom start appearing after about 1 week, and at this time, fresh air circulation in the room is further increased (4 air changes/h) to allow the pin heads to develop into full buttons in the next 4–5 days. The crop appears in flushes and generally about 5 flushes are harvested in a cropping period of 45–50 days. Controlled units with mechanized control over temperature, humidity, and CO_2 prefer to have a shorter cropping

Table 12.1 Optimum Environmental Parameters during Various Stages of Cropping

PARAMETERS	SPAWN RUN	CASE RUN	PINNING	CROPPING
Air temperature	23 ± 1°C	23 ± 1°C	15–17°C	15–17°C
Compost temperature	24–25°C	24–25°C	16–18°C	16–18°C
CO_2 concentration	About 15,000 ppm	10–15,000 ppm	300–1000 ppm	300–800 ppm
Relative humidity	90%–95%	90%–95%	About 85%	About 85%
Air changes	Nil	Nil	Fresh air (30%)	4 changes per hour
	No fresh air (100% recirculation)		Recirculation (70%)	

period of about 30–35 days. The yields of up to 20%–25% of compost from short method compost under controlled growing conditions and 10%–15% from long method compost under seasonal cultivation, can be obtained.

b. *Spawn quantities.* The quantity of the spawn used does not directly affect yield. However, the use of more spawn may reduce the effect of competitive organisms present in the planting substrates. So, 2%–4% of spawn is suggested for inoculation into the spawning substrate. Once the container is opened, the spawn should be used in its entirety. Unused and opened bottles or bags of spawn, however, can be kept in the refrigerator for 2–6 days, as long as they are not contaminated during storage.

12.5.4 Making a Composted Substrate

Compost preparation is an aerobic process and is the most important part of *A. bisporus* production. Today, the white button mushroom is cultivated on compost prepared by a traditional method, known as the long method of composting (LMC), on the short method of compost (SMC), or on compost prepared by an accelerated method, also known as the rapid composting method (indoor compost).

12.5.4.1 Long Method of Composting (LMC)   It is the oldest method for the preparation of compost. Compost prepared by such a method, besides taking more time (18–24 days), gives a lower yield (Khanna and Sharma, 2013), as it is prone to attack by many pests and diseases. The initial nitrogen-level of the composting pile in various formulations is kept at around 1.5% of the dry weight, and a specific turning schedule is followed, with the addition of various supplements. This method has been replaced by the SMC (Short Mehtod of Composting) in developed countries, but still, in India, bulk of white button mushroom is produced on this type of compost. Given below is the detail of one of the most prevalent methods of LMC procedures recommended by the Punjab Agricultural University, Ludhiana, for the seasonal growers of the plains of Punjab and adjoining areas, to prepare wheat straw based synthetic compost (Garcha and Khanna, 2003).

Wheat straw is spread over the platform and wetting of the straw is started by sprinkling water and turning manually with forks until sufficient water is absorbed (Figure 12.6). Wetting is continued for 48 h and the straw attains a moisture level of around 75%. Fertilizers and wheat bran are mixed separately, moistened, and kept covered with a gunny bag or polythene sheet for 24 h. Subsequently, this mixture is broadcasted on the wetted straw, mixed with forks, and made into a heap (Day 0). The wooden/iron boards are arranged in the form of a rectangular block and the entire mixture is filled into it by simultaneous pressing from the sides and on top. The process is repeated until a compost pile of rectangular shape is formed. Normally, a pile of $5' \times 5' \times 5'$ ($l \times b \times h$) is achieved, starting with 300 kg of base material (Khanna and Sharma, 2013). If the quantity is more, the boards are moved lengthwise to accommodate it, so that the width and height of the pile remain the same. Too compact a pile is not desirable as anaerobic conditions can occur, the temperature of the pile will rise beyond 75°C, and secondary fermentation will lead to the production of undesirable substances. Leaching of water from the pile is avoided by carefully pouring it back on top of the pile.

12.5.4.2 Turning Schedule   Seven turnings are given to the composting pile by following the schedule of turning at 4, 8, 12, 15, 18, 21, and 24 days, along with the mixing-in of the remaining ingredients during these turnings. The turning procedure can be accomplished either manually or using a low-cost fabricated turner to achieve mechanical bruising of straw—saving labor, giving uniform decomposition, and increased processing of the material. The objective of the turning is to have equal amount of aeration in each portion of the pile so that uniform decomposition of the straw in the preparation of selective substrate is achieved. In the stack, the core portion is fermented at its peak due to the temperature build up, while the rest of the portion exposed to air is either not fermenting at all or is fermenting improperly. During each turning, one foot portion of the pile from the four sides and from the top are removed, kept separately, and the moisture level is adjusted with clean water. The rest of the pile is dismantled. A new stack is then, made using the wooden/iron moulds, where the side and top portions of the pile material are placed at the centre, while the bottom and core portions come to the top and the sides. In this way, pile of Day 0 is turned inside out on the first turning. This procedure of turning is repeated at each turning and some additives or supplements are added, as given in Table 12.2.

The prepared compost is checked for the smell of ammonia and can be kept for 1–2 days more if it is detected (Khanna and Sharma, 2013). This substrate is now ready for spawning and filling. The whole composting process is accomplished in 26 days (–2, 0, 4, 8, 12, 15, 18, 21, 24 days), and a dry matter of about 35% is lost. If paddy straw is to be used along with wheat straw (50:50, w/w), its addition to the composting pile is recommended on the 8th day of turning, but care should be taken to keep the total moisture content of the pile between 62% and 65%. The compost prepared using the Long Method of Composting (LMC) should be dark brown in color, free from ammonia smell, around 70% moisture, pliable (non-greasy and non-sticky), pH between 7.0 and 8.0, and should be free from insects and nematodes.

Improvements in the long method of composting have been carried out in the direction of (i) introducing different base materials and combinations, (ii) the use

Table 12.2 Details of Turning and Additives or Supply Additives during Composting

Day 4 (1st turning)	Molasses are added
Day 8 (2nd turning)	Plain turning is given and moisture level in the side and top portion is adjusted, if required
Day 12 (3rd turning)	At this turning required quantity of gypsum is added
Day 15 (4th turning)	Plain turning with no addition of any supplement is given
Day 18 (5th turning)	Turning given and required quantity of Furadon 3G added
Day 21 (6th turning)	No addition of ingredient is given during this turning
Day 24 (7th turning)	The pile is dismantled and cooled

Table 12.3   Some Compost Formulations Used in India

	LONG METHOD OF COMPOSTING (LMC)		SHORT METHOD OF COMPOSTING (SMC)	
DMR, Solan	Wheat and paddy (1:1)	300 kg	Wheat straw	300 kg
	CAN	9 kg	Wheat bran	15 kg
	Urea	25 kg	Chicken manure	125 kg
	Wheat bran	20 kg	BHC (10%)	125 g
	Gypsum	20 kg	Gypsum	20 kg
PAU, Ludhiana	Wheat and paddy (1:1)	300 kg	Wheat and paddy (1:1)	300 kg
	CAN	9 kg	Chicken manure	60 kg
	Urea	3 kg	CAN	6 kg
	Superphosphate	3 kg	Superphosphate	3 kg
	Muriate of potash	3 kg	Muriate of potash	3 kg
	Wheat bran	15 kg	Wheat bran	15 kg
	Gypsum	30 kg	Gypsum	30 kg
	BHC (5%)	250 g	Paddy straw	300 kg
	Furadan 3G	150 g	Chicken manure	150 kg
IIHR, Bangalore	Paddy straw	150 kg	Wheat bran	12.5 kg
	Maize stalks	150 kg	Gypsum	10 kg
	Ammonium sulfate	9 kg		
	Superphosphate	9 kg		
	Urea	4 kg		
	Rice bran	50 kg		
	Cotton seed meal	15 kg		
	Gypsum	12 kg		
	Calcium carbonate	10 kg		
Mushroom Research Laboratory, Solan	Wheat straw	1000 kg	Wheat straw	1000 kg
	CAN	30 kg	Chicken manure	400 kg
	Superphosphate	25 kg	Brewer's manure	72 kg
	Urea	12 kg	Urea	14.5 kg
	Sulfate of potash	10 kg	Gypsum	30 kg
	Wheat bran	100 kg		
	Molasses	16.6 L		
	Gypsum	100 kg		
RRL, Srinagar	Wheat straw or paddy straw	300 kg		
	Molasses	12 kg		
	Urea	4.5 kg		
	Wheat bran	50 kg		
	Muriate of potash	2 kg		
	Cotton seed meal	15 kg		
	Gypsum	15 kg		

of alternative ingredients to improve productivity of the substrate, (iii) reduction in the period of composting, and (iv) low cost pasteurization without the use of steam (chemical and solar pasteurization) to eliminate the pathogens and pests in the compost. Based on these studies, a number of compost formulations Table 12.3 have been made and are in use.

12.5.4.3 Short Method of Composting (SMC)   Around 1915, growers in the USA found that if prepared compost is kept in shelves in growing rooms and subjected to high temperature (around 60°C) for sometime, it gives a higher and consistent yield (Singh et al., 2011). This process was later termed as "sweating out," and it laid down the foundation of pasteurization of compost. Based on this principle, Sinden and Hauser (1950, 1953) proposed a new method of composting where pasteurization became its integral part. The first step in standardizing the SMC in India, involving both outdoor and indoor phases, with wheat straw as the base material, along with chicken manure, was taken in 1976 (Shandilya, 1976; Hayes and Shandilya, 1977). This method was initially based on tray pasteurization by peak heating, but very soon the bulk pasteurization concept was introduced (Tunney, 1980) in the country. Work on Short Method of Composting (SMC) has been done in India on very few aspects, such as microbial succession during different stages (Shandilya, 1976), paddy straws based SMC (Shandilya, 1989), chemical characteristics, and co-relation studies on compost quantity and yield of mushrooms (Shandilya et al., 1987). The method consists of two phases:

Phase I—Outdoor composting for 10–12 days.

Phase II—Under defined conditions, it involves pasteurization and conditioning of compost inside an insulated room by free circulation of air. This phase lasts for around 7 days.

Phase I (Outdoor composting): This phase starts with the wetting of the wheat straw and chicken manure thoroughly till they absorb sufficient moisture (around 75%). A stack is made out of the materials. After 2 days, the stack is broken, water is added to the dry portions, and again a stack is made. Prewetting and mixing of ingredients is a must before starting the actual composting procedure on Day 0 and the stacks made during this process are wide with a low height of 3–4 feet. On Day 0, the stack is again broken, urea and wheat bran are added, along with water, if required, and a high aerobic stack is again made. Three to four turnings are given after every 2 days to the compost, which can be done manually or by compost turners. Gypsum is added at the third turning. A high temperature (70–80°C) is built up in the core region (Khanna and Sharma, 2013), which results in the charring of ingredients, which gives a distinct brown color to the compost. The temperature around the core region ranges between 50°C and 60°C, which harbors huge population of thermophilic microflora that serves as the inoculum for the whole pile during turnings to bring about decomposition of the straw ingredients. On Days 8–10, the compost is ready for pasteurization.

Phase II: This phase is generally performed in a pasteurization tunnel, in bulk. It can be divided into two stages: conditioning and pasteurization.

12.5.4.4 Conditioning   During this stage, the whole of the compost mass is brought to a temperature range optimum for the growth of thermophilic flora (45–52°C). In this process, ammonia gets fixed in a lignin–humus complex, and the excess of ammonia is released into the atmosphere. An optimum temperature range, at which these microflora grow and ammonia generation takes place, is close to 45°C. Compost should not be conditioned below 40°C, as some mesophilic fungi may be produced at this temperature, rendering the compost unsuitable for mushroom growth. An oxygen concentration above 10% is supplied to the compost mass in order to maintain aerobic conditions. Both pasteurization and conditioning make the compost most selective for the growth of button mushrooms.

12.5.4.5 Pasteurization This can be done soon after tunnel/room filling, or after a few days (Singh et al., 2011). Pasteurization is required to kill or inactivate harmful organisms. The compost is pasteurized properly if it is kept at 59°C for 4–6 h. A temperature above 60°C is harmful, as this temperature may kill all kinds of organisms, including thermophilic fungi, which are essential for governing the phase II of composting. The activity of the compost is a very important factor during phase II. If it is active (green compost), its temperature rises to 57°C due to self-generation of heat; otherwise steam through a boiler is supplied.

The phase II of composting is accomplished either in trays by peak heating them, or in a bulk chamber by maintaining a very systematic and controlled regime of temperature, time, aeration, ventilation, and ammonia level. All these parameters are monitored and controlled using sensors and probes inside the area. Detailed configuration of a bulk pasteurization room or tunnel and peak heating room required to produce about 20–22 tonnes of compost has been described by Vijay and Gupta (1995). A lot of empirical data has been generated on the various parameters, like levels of aeration/ventilation, temperature, and duration, and level of various gases in the tunnel for obtaining a highly productive compost mixture. During phase II, about 25%–30% of the compost weight is lost, so as to produce 20 tonnes of compost from the tunnel (36′ × 9′ × 12′) or peak heating room (24′ × 16′ × 8′) to accommodate 250 standard trays, roughly 28 tonnes of compost should be used, for which one may start with 12 tonnes of raw material, for which a composting yard of 100′ × 40′ is large enough.

12.5.5 Aerated Phase/Composting

Traditionally, compost for button mushroom growing is prepared partly outdoor/partly indoors, and this results in environmental pollution. A need is also felt to monitor and control the entire process of composting exclusively selective for mushroom mycelium. For that, it has to be carried out indoors, where all parameters can be kept controlled. So, indoor composting is carried out in especially designed chambers with forced circulation. The ingredients are thoroughly wetted and mixed, spread over the composting yard (8–10′ height) to increase the bulk density of ingredients. The mixed ingredients are made into a high aerobic heap and left for 48 h. The material is then, transferred to a phase I bunker where a temperature of 70–75°C is obtained through the introduction of air. After 3 days of partial fermentation in the phase I bunker, the entire compost mass is transferred to another bunker for another 3–4 days under the same set of conditions. Once phase I is over, compost is transferred to the phase II tunnel for the usual phase II operation, as described for SMC, which is to be completed in 6–7 days. It is desirable not to allow a build up of temperature beyond 75°C. Using indoor composting, it has been possible to obtain 20–22 kg of mushrooms from 100 kg compost in 4–6 weeks of cropping.

12.5.6 Growing Systems for Mushrooms

Three systems of mushroom growing are practiced:

1. *Tray growing* is used mainly by medium and large growers. This is the major method used for the growing of mushrooms in Australia. Trays are made of wood, usually 1 m × 2 m × 0.3 m in size, fastened using stainless steel fittings.
2. *Bag growing* is becoming increasingly popular with small-to-medium growers and new entrants into the mushroom industry. This system requires a smaller capital outlay than the tray system. Bag growing offers advantages in pest and disease control by allowing fast and easy removal of infected bags. However, these advantages are offset by the larger labor input per kg output required by this method than either of the other two systems.
3. *Shelf-growing* offers large savings in labor costs, but this is offset by very large capital set-up costs.

Of paramount importance is the level of expertise required in both the management and growing methods in mushroom growing. If skills in either area are lacking, production can vary greatly between crops, regardless of the growing system used.

12.5.7 Substrate Supplementation

The agricultural base materials, like cereal straws, maize stalks, and sugarcane bagasse, are supplemented with other materials with nitrogen and carbohydrate sources in order to meet the required C/N ratio to start the fermentation process.

These materials can be animal manure; nitrogen fertilizers like urea, ammonium sulfate and calcium nitrate; carbohydrate rich nutrients like molasses, brewers grain, potato waste, apple, grape pomace, etc.; or concentrated meals like seed meals of cotton, soya, mustard, castor, etc.

12.5.8 Spent Mushroom Substrate (SMS)

Spent mushroom substrate (SMS) normally contains 1.9:0.4:2.4%, N–P–K before weathering and 1.9:0.6:1.0, N–P–K after weathering for 8–16 months. Nitrogen and phosphorus do not leach out during weathering but potassium is lost in significant amount during weathering (Gupta et al., 2004). SMS contains much less heavy metals than sewerage sludge, which precludes its classification as a hazardous substance. Weathering causes a slow decrease in the organic matter contents (volatile solids) and leads to different characteristics of weathered SMS because of on-going microbial activity (Beyer, 1999). The SMS obtained from different sources usually has conductance in the range of 1.9–8.3 mm cm^{-1}. However, the chloride content in fresh SMS varies from 1.5 to 7.5 kg t^{-1} (Gerrits, 1987) and only about 294 ppm in a well-rotted SMS (Chong and Wickware, 1989).

There are contradictory reports regarding the pH of fresh and weathered Spent Mushroom Substrate (SMS). According to Wuest and Fahy (1991), SMS has an initial pH of around 7.28, which increases during weathering, while, Devonald (1987) reported pH of the fresh SMS in the range of 7.01–8.04.

The volume of SMS also decreases (shrinkage) over a period of time. The fresh SMS obtained from various sources varies in density: 0.198 g cm^{-3}, with a range of 0.15–0.24 g cm^{-3} in United Kingdom (Devonald, 1987) and 0.24–0.62 g cm^{-3} in USA.

12.6 Production Technologies of Commercial Mushrooms

12.6.1 Agaricus bisporus

In India, *Agaricus bisporus* (the white button mushroom) contributes about 90% of the country's total production, as against its global share of about 40%. The method of cultivation of mushroom was recorded as early as 300 BC. In India, commercial production of white button mushrooms was initiated in the hilly regions of India like Chail (Himachal Pradesh), Kashmir, and Ooty (Tamil Nadu). Mushroom cultivation slowly spread to the North Western plains of India (seasonal crop during winter). It requires a temperature of 14–18°C during cropping, hence its cultivation has become popular in the cooler hilly regions of India. Unlike other cultivated species, its cultivation needs more technical skill, which can be acquired by training. Today, white button mushrooms are cultivated on compost prepared either by the traditional method known as the long method of composting (LMC) or on the short method of composting (SMC) (Sinden and Hauser, 1953) or on a compost prepared by an accelerated method or by a rapid composting method (indoor compost). The various steps involved in its cultivation are illustrated in Figure 12.6.

12.6.1.1 Compost and Composting Compost is the substrate used for the cultivation of *A. bisporus* which provides all the nutrients required for the growth and reproduction of mushroom fungus. Composting is a multistep process by a consortium of microorganisms, especially fungi, bacteria, and actinomycetes, that inhabit the compost mass in a well defined succession for creation of the selective substrate for button mushroom cultivation. Traditionally, in India the compost is prepared using a base material consisting of cereal straw and/or horse/chicken manure mixed with various

Figure 12.6 Different steps in cultivation of button mushroom. (a) Wetting of straw for composting; (b) Mixing of fertilizers before stacking; (c) Preparation of stack during composting; (d) Spawning of ready compost; (e) Bag/shelf system of growing; (f) Tray system of growing; (g) Trays ready for casing; (h) Preparation of casing mixture; (i) Appearance of pin heads of mushrooms and (j) Mushrooms ready for harvest. (Courtesy of Dr. S. Kapoor.)

supplements/nutrients. The choice of a particular nutrient is affected by its cost as well as its availability. Each category of these nutrients has a specific role to play during the process of composting. The majority of the seasonal growers in India do not have compost pasteurization facilities and still use the long method of composting.

12.6.1.2 Compost Formulations A large number of formulations are available for growers based on the cost and availability of raw materials in the particular region (Table 12.3). These are used with some modifications at the growers' level by most of the seasonal growers in the plains of the Northern States and in Himachal Pradesh for preparing compost, using the LMC. The use of chicken manure/horse manure in these formulations is recommended where pasteurization facilities are available.

A widely adopted objective in the computation of formulations is to achieve some form of balance, especially between carbon and nitrogen nutrients (C:N ratio), which favors the establishment of a correct sequence of dominance in microbial succession to maintain an overall active fermentation process. A mixture containing 1.4%–1.5% N of dry matter is the target at the beginning of the composting which provides compost with C/N of 16–18.1 and N content between 1.75% and 2.0% at the completion of the process.

12.6.2 Pleurotus spp.

The *Pleurotus* mushroom is generally called the oyster mushroom because the pileus or cap is shell-like, spatulate, and the stipe eccentric or lateral (Chang and Miles, 2004). It is ranked third in global production, contributing about 14% of the total (Shukla and Jaitly, 2011). *P. ostreatus* was first cultivated in the USA in 1900, and some other species, such as *P. sajor-caju*, were initially cultivated in India after the late 1940s. It belongs to the family Tricholomataceae/Pleurotaceae and has about 40 well recognized species, out of which 12 species are cultivated in different parts of India (Ahmed et al., 2009). Some of the important cultivated species include *Pleurotus ostreatus*, *P. sajor-caju*, *P. florida*, *P. cystidiosus*, *P. cornucopiae*, *P. pulmonarius*, *P. tuber-regium*, *P. citrinopileatus*, and *P. flabellatus* (Madhav and Uday, 2007). *Pleurotus* spp. are a rich source of proteins, minerals (Ca, P, Fe, K and Na), and vitamins C and B complex (thiamin, riboflavin, folic acid and niacin). They are consumed for their nutritive as well as medicinal values (Agahar-Murugkar and Subbulakshmi, 2005), such as a high potassium to sodium ratio, which make them an ideal food for patients suffering from hypertension and heart disease (Patil et al., 2010).

The *Pleurotus* mushroom is popularly grown in several states of India such as Orissa, Karnataka, Maharashtra, Andhra Pradesh, Madhya Pradesh, West Bengal, and in the North-Eastern States of Meghalaya, Tripura, Manipur, Mizoram, and Assam (Singh et al., 2011; Upadhyay, 2011). Various species are cultivated in both temperate and subtropical climates. *Pleurotus* species are efficient lignin degraders, so can grow on a wide variety of agricultural wastes (most hardwoods, wood by-products

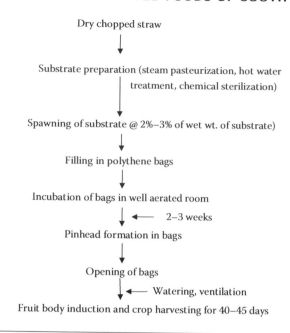

Dry chopped straw

↓

Substrate preparation (steam pasteurization, hot water treatment, chemical sterilization)

↓

Spawning of substrate @ 2%–3% of wet wt. of substrate)

↓

Filling in polythene bags

↓

Incubation of bags in well aerated room

↓ ← 2–3 weeks

Pinhead formation in bags

↓

Opening of bags

↓ ← Watering, ventilation

Fruit body induction and crop harvesting for 40–45 days

Figure 12.7   Cultivation process of *Pleurotus* spp.

such as sawdust, paper, pulp sludge, all the cereal straws, corn, and corn cobs, coffee residues, such as coffee grounds, hulls, stalks, and leaves, banana fronds, and waste cotton, often enclosed in plastic bags and bottles) with broad adaptability to varied agro-climatic conditions (Jandaik and Goyal, 1995). Its cultivation can be carried out on unfermented pasteurized/ unpasteurized substrates in polyethylene bags with a high biological efficiency. The basic steps for cultivation involved are shown in Figure 12.7. Fresh mushrooms should be packed in perforated polythene bags for marketing. They can also be sun dried by spreading on a cotton cloth in bright sunlight or diffused light. The dried fruiting bodies with 2%–4% moisture can be stored for 3–4 months after proper sealing.

12.6.3 *Lentinula edodes*

Lentinula edodes commonly called shiitake is the second largest proportion of mushroom cultivation (accounting for 25% of the production) in the world after *Agaricus bisporus* (Singh et al., 2011; Chang, 2005) and is liked by consumers for its unique flavor and medicinal properties. Its proven pharmacological properties, especially the polysaccharide, lentinan (Bisen et al., 2010), eritadenine, shiitake mushroom mycelium, and culture media extracts (LEM, LAP and KS-2) which act as antibiotics and anticarcinogenics, and also have some antiviral compounds that have been isolated intracellularly (fruiting body and mycelia) and extracellularly (culture media) (Bisen et al., 2010). In the early 1980s, Pegler (1983) assigned shiitake to the genus *Lentinula* and currently, there are six species that are generally recognized in the genus *Lentinula*, three (*L. edodes*, *L. lateritia* [Berk.] Pegler, and *L. novaezelandiae*

[Stev.] Pegler) are of Asia–Australasian distribution, while the remaining three (*L. boryana* (Berk. & Mont.) Pegler, *L. guarapiensis* (Speg.) Pegler, and *L. raphanica* (Murrill) Mata and R.H. Petersen) are distributed in the Americas (Nicholson et al., 2009). Today, shiitake cultivation is widely practiced, not only in Southeast Asia, but also in North America and Europe, Australia, and New Zealand (Oei, 1996; Romanens, 2001), making its cultivation a global industry. In India, cultivation of this mushroom was first reported at Solan in Himachal Pradesh (Shukla, 1994).

Shiitake mushrooms develop as saprophytes on tree logs and form fruiting bodies at low temperature (15–20°C) and high humidity levels (Maki et al., 2001). It grows in nature on the wood of broad leaf trees, mainly oak and chestnut, and on dead tree trunks or stumps. In India, it is cultivated on an artificial medium: saw dust and wheat straw supplemented with wheat and rice bran (Thakur and Sharma, 1992). The substrate formula standardized for *L. edodes* production involves sawdust or wheat straw supplemented with wheat bran, gypsum and lime and, commercially, polypropylene bags filled with substrate are used. The basic steps involved are shown in Figure 12.8.

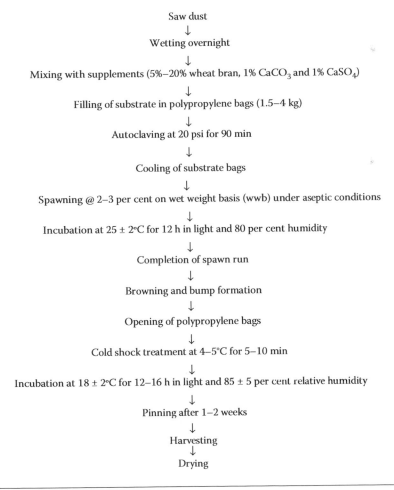

Saw dust
↓
Wetting overnight
↓
Mixing with supplements (5%–20% wheat bran, 1% $CaCO_3$ and 1% $CaSO_4$)
↓
Filling of substrate in polypropylene bags (1.5–4 kg)
↓
Autoclaving at 20 psi for 90 min
↓
Cooling of substrate bags
↓
Spawning @ 2–3 per cent on wet weight basis (wwb) under aseptic conditions
↓
Incubation at 25 ± 2°C for 12 h in light and 80 per cent humidity
↓
Completion of spawn run
↓
Browning and bump formation
↓
Opening of polypropylene bags
↓
Cold shock treatment at 4–5°C for 5–10 min
↓
Incubation at 18 ± 2°C for 12–16 h in light and 85 ± 5 per cent relative humidity
↓
Pinning after 1–2 weeks
↓
Harvesting
↓
Drying

Figure 12.8 Cultivation process of *L. edodes*.

12.6.4 Auricularia spp.

Long popular in China as a food and medicine, *Auricularia* is reported to be the first mushroom cultivated by humans (Chang and Miles, 1987; 2004) and is the fourth most popular mushroom variety in the world. It has several medicinal functions, such as promoting blood circulation, treating hemorrhoids, and having analgesic, antitumor, immuno-stimulating, hypolipidaemic and hypocholesterolaemic effects (Dai et al., 2009; 2010). This jelly fungus is known as Jew's ear (a contraction of Judas' ear). It is worldwide in its distribution in temperate and tropical regions but extensive cultivation takes place in China, Taiwan, Thailand, and the Philippines. It has tremendous scope of cultivation in a temperature range of 20–30°C. Its cultivation technology has also been standardized in India but it has not become popular. The two species of *Auricularia* cultivated worldwide include *A. polytricha* and *A. auricula-judae*. *A. polytricha* is a widely distributed tropical and subtropical species and is darker, thicker and long haired, while *A. auricula-judae* is a temperate species that occurs occasionally in the subtropics (Chang and Miles, 2004) and is thin and light in color. This mushroom can be grown both on wood logs as well as supplemented sawdust, and cereal straw in plastic bags. The basic steps involved in its production are shown in Figure 12.9. An average yield of 1.0–1.4 kg fresh mushroom per kg dry straw in 3–4 flushes can be obtained. Traditionally, *Auricularia* cultivation is carried out on artificial logs. The basic steps include preparation of artificial log, autoclaving/sterilization, inoculation, incubation, fruiting body formation and harvest. Proper hygienic conditions are necessary because

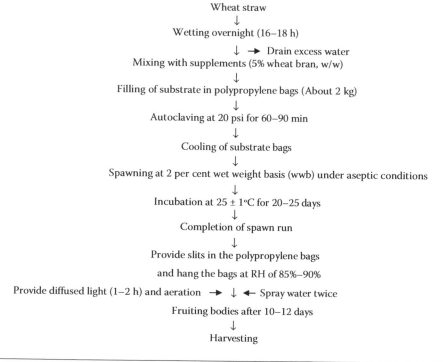

Wheat straw
↓
Wetting overnight (16–18 h)
↓ → Drain excess water
Mixing with supplements (5% wheat bran, w/w)
↓
Filling of substrate in polypropylene bags (About 2 kg)
↓
Autoclaving at 20 psi for 60–90 min
↓
Cooling of substrate bags
↓
Spawning at 2 per cent wet weight basis (wwb) under aseptic conditions
↓
Incubation at 25 ± 1°C for 20–25 days
↓
Completion of spawn run
↓
Provide slits in the polypropylene bags
and hang the bags at RH of 85%–90%
Provide diffused light (1–2 h) and aeration → ↓ ← Spray water twice
Fruiting bodies after 10–12 days
↓
Harvesting

Figure 12.9 Cultivation process of *Auricularia*, (wwb = wet weight basis).

addition of supplements may attract moulds in the absence of proper sterilization. Presence of high level of carbon dioxide and absence of light results in the development of abnormal fruiting bodies. Under very humid conditions, cobweb has been found to attack this mushroom.

12.6.5 *Agrocybe aegerita*

Commonly known as the black poplar mushroom, it grows mostly on poplar and willow wood from spring to autumn. It has a unique flavor and good nutritive and medicinal value, but low biological efficiency. It belongs to the white rot fungi and is a medium-sized agaric, having a very open and convex cap, almost flat, of 3–10 cm in diameter. Underneath, it has numerous whitish radial plates adherent to the foot, later turning to a brownish-gray color, and light elliptic spores. It possesses a variety of bioactive secondary metabolites (Parera and Li, 2011), such as indole derivatives, with free radical scavenging activity, cylindan, with anticancer activity, and also agrocybenine, with antifungal activity. It can be grown on barley straw, wheat straw, rice straw, orange peels, maize stalks, grape stalks, rice husks, and sawdust. Its cultivation technology has also been standardized in India, and the basic steps involved are depicted in Figures 12.10 and 12.11. On an average a bag of 2 kg yields about

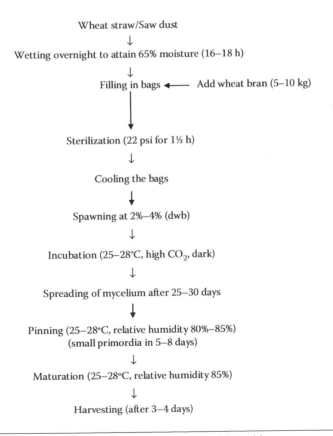

Wheat straw/Saw dust
↓
Wetting overnight to attain 65% moisture (16–18 h)
↓
Filling in bags ◄──── Add wheat bran (5–10 kg)

Sterilization (22 psi for 1½ h)
↓
Cooling the bags
↓
Spawning at 2%–4% (dwb)
↓
Incubation (25–28°C, high CO_2, dark)
↓
Spreading of mycelium after 25–30 days
↓
Pinning (25–28°C, relative humidity 80%–85%)
(small primordia in 5–8 days)
↓
Maturation (25–28°C, relative humidity 85%)
↓
Harvesting (after 3–4 days)

Figure 12.10 Cultivation process of *Agrocybe aegerita*, (dwb = dry weight basis).

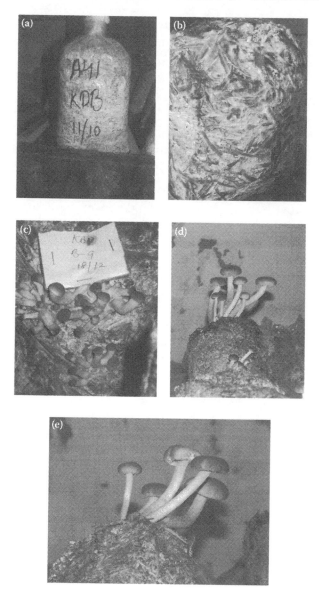

Figure 12.11   Pictorial view of different steps in cultivation of *Agrocybe aegerita* mushroom. (a) Sterilized and inoculated bag; (b) Spawn impregnated bag; (c) Primordia development; (d) Developing fruiting bodies; and (e) Fruit bodies ready for harvest. (Courtesy of Dr. S. Kapoor.)

500–600 fresh mushrooms. The fruiting bodies can be stored for 10–12 days in a refrigerator at 4°C. The fruit bodies are easily sun dried, with dry matter of 8%–12%.

12.6.6 *Flammulina velutipes or Enokitake*

This is known as the winter mushroom, velvet foot or velvet stem, golden needle mushroom, and enoki, *Flammulina velutipes* was cultivated as early as 800 AD in China and today it is being cultivated all over the world (Upadhyay and Singh, 2011). Species of *Flammulina* have been reported to occur naturally on various deciduous tree species, namely poplar (*Populus* spp.), willows (*Salix* spp.), elms (*Ulmus* spp.), plum (*Prunus* spp.), maple (*Acer* spp.), and birch (*Betula* spp.), as a parasite and later as a saprophyte, growing on the trunks or stumps of these broadleafed trees from the end of autumn to early spring. Fruiting bodies appear after a few months at temperatures between −2°C and 14°C, under low intensity winter light (Poppe, 1974; Zadrazil, 1999) on wounded or weakened trees. This mushroom is particularly known for its taste and preventive as well as curative properties for liver diseases and gastroenteric ulcers. The cultivation process is depicted in Figure 12.12. The fruiting bodies starts in the dark, but light is necessary for further development. For the proper growth of fruit bodies, the temperature is lowered from 8–12°C to 3–5°C during maturation, which encourages stiff, white and drier fruit bodies. When the fruit bodies are 13–14 cm long they are harvested and packed. The first flush usually amounts to 200–240 g/bag and second flush yields 160–180 g/bag. The cultivation of *Flammulina* takes about 50–60 days from the initial fruiting to the crop.

12.6.7 Calocybe indica

The milky mushroom, a potentially new species to the world mushroom growers, is a tropical mushroom which belongs to the class basidiomycetes, order Agaricales, and family Tricholomataceae. The three species of *Calocybe* namely, *C. ionides*, *C. carnea*, and *C. gambosa* are collected from forests but there is no report as yet of their domestication (Dickinson and Lucas, 1979). *C. indica* was reported for the first time from West Bengal in India in 1974 (Purkayastha and Chandra, 1974) but its edibility was only confirmed later on (Purkaystha and Chandra, 1976). It is well appreciated due to its large-sized milky white sporophores, simple production technology, and low capital investment. Moreover, its ability to grow at a temperature above 30°C makes

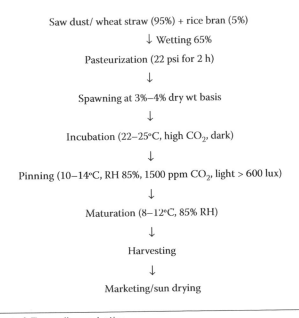

Saw dust/ wheat straw (95%) + rice bran (5%)

↓ Wetting 65%

Pasteurization (22 psi for 2 h)

↓

Spawning at 3%–4% dry wt basis

↓

Incubation (22–25°C, high CO_2, dark)

↓

Pinning (10–14°C, RH 85%, 1500 ppm CO_2, light > 600 lux)

↓

Maturation (8–12°C, 85% RH)

↓

Harvesting

↓

Marketing/sun drying

Figure 12.12   Flow chart of *Flammulina* production.

it suitable for cultivation in the sub-tropical environment, particularly during the summer months, or as a seasonal crop where such a favorable tropical climate is available for a few months in a year. Its cultivation has gained momentum in several states of India, such as Andhra Pradesh, Goa, Karnataka, Kerala, Maharashtra, Tamil Nadu, Punjab, and Uttar Pradesh. It can be grown easily on substrates containing lignin, cellulose, and hemicelluloses such as straw of paddy, wheat, ragi, maize, cotton stalks, sugarcane bagasse, cotton, and juice wastes, dehulled maize cobs, and tea and coffee waste. The basic steps involved in cultivation are depicted in Figures 12.13 and 12.14. After 10–15 days of casing, mushrooms started appearing in flushes (30–35°C, 80%–85% relative humidity). Harvesting is by gentle twisting of the fruit bodies.

12.6.8 *Volvariella volvacea*

This edible mushroom is also known as the paddy straw or Chinese mushroom of the tropics and subtropics. First cultivated in China as early as in 1822, around 1932–35, it was introduced into other Asian countries. Only three species of the straw mushroom, *V. volvacea*, *V. esculanta*, and *V. diplasia* are under artificial cultivation. It derives its name from 'Volva' which means a wrapper, that completely envelops the main fruiting body during the young stage. The fruiting body formation starts with tiny clusters of white hyphal aggregates called primordial, which are followed by several morphological stages in the fruit body development process (Chang and Yau, 1971).

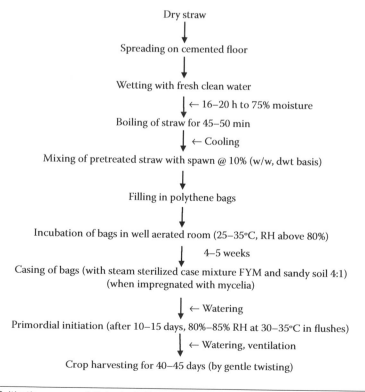

Figure 12.13 Cultivation process of *Calocybe indica*.

Figure 12.14 Different steps in cultivation of milky mushroom. (a) Wetting of straw; (b) Spawning of bags; (c) Spawn impregnated bags; (d) Preparation of casing soil; (e) Casing of bags; (f) Fruiting body development; and (g) Fruit bodies ready for harvest. (Courtesy of Dr. S. Kapoor.)

The successive stages are called the "button," "egg," "elongation," "mature" stages, respectively. Differentiation can be seen first at the "button" stage. At maturity, the buttons enlarge and umbrella-like fruit bodies emerge after the rupture of the volva (Li, 1982). The traditional method of cultivation is on paddy straw, both indoors and outdoors, where biological efficiency of 15%–20% is obtained.

Sophisticated indoor technology is recommended for industrial-scale production of this mushroom, but low-cost technology appropriate for rural development is normally practiced at the community level both for indoor and outdoor cultivation. The basic steps involved in cultivation are: preparation of paddy straw bundles (90 cm × 15 cm) tied with gunny threads. These bundles are soaked overnight in water in a tank to attain 70%–75% moisture. Beds are made by placing 4 layers of bundles, each layer comprising of 5 bundles. While laying beds, the distance between the beds should be 2–2.5 feet which is used as a working space. These beds are spawned at 1.5% of dry weight of paddy straw, and the whole bed is covered with a polythene sheet which increases both the temperature and humidity. After 4–5 days of spawning, the sheet is removed and the bed is sprinkled with water. A temperature of 28–32°C and relative humidity of 80% are maintained. Mushrooms start appearing after 7–9 days of spawning and continue for the next 20 days (Figure 12.15).

Figure 12.15   Different steps in cultivation of paddy straw mushroom. (a) Making paddy straw bundles; (b) Wetting of bundles; (c) Laying of beds; (d) Fruit bodies ready for harvest. (Courtesy of Dr. S. Kapoor.)

12.7 Nutritional Quality and Functional Properties

White button mushrooms (*Agaricus bisporus*) are a valuable source of several micro-nutrients in a low energy, nutrient-dense food. Bano (1976) suggested that the food value of mushrooms lies between meat and vegetables. They are also low in sodium and calories, but are excellent source of proteins, vitamins, and minerals (Prakash and Tejaswin, 1991; Ghosh and Singh, 1995). Gruen and Wong (1982) found that the edible mushrooms were highly nutritional and compared favorably with meat, egg, and milk food sources. The composition of white button mushroom (*Agaricus bisporus*) and nutritive value has been reviewed extensively (Sharma and Vaidya, 2011). The composition of different mushrooms is given in Table 12.4.

Table 12.4   Nutritive Value of Different Mushrooms (Dry Weight Basis g/100 g)

MUSHROOM	CARBOHYDRATES	FIBER	PROTEIN	FAT	ASH	ENERGY (Kcal)
Agaricus bisporus	46.17	20.90	33.48	3.10	5.70	499
Pleurotus sajor-caju	63.40	48.60	19.23	2.70	6.32	412
Lentinula edodes	47.60	28.80	32.93	3.73	5.20	387
Pleurotus ostreatus	57.60	8.70	30.40	2.20	9.80	265
Volvariella volvacea	54.80	5.50	37.50	2.60	1.10	305
Calocybe indica	64.26	3.40	17.69	4.10	7.43	391
Flammulina velutipes	73.10	3.70	17.60	1.90	7.40	378
Auricularia auricular	82.80	19.80	4.20	8.30	4.70	351

Source: From Stamets, P. 2005. *Mycelium Running: How Mushrooms Can Help Save the World.* Ten Speed Press, Berkeley, CA. (*A. bisporus, P. sajor-caju, L. edodes*); FAO, U.N. 1972. *Food Composition Table for Use in East Asia.* FAO, Rome. p. 334 (*P. ostreatus, V. volvacea*); Doshi, A. and Sharma, S.S. 1995. *Advances in Horticulture.* 13, MPH, New Delhi. (*C. indica*); Crisan, E.V. and Sands, A. 1978. *The Biology and Cultivation of Edible Mushroom.* Academic Press, New York, NY, pp. 137–168 (*F. velutipes* and *Auricularia* spp.).

12.7.1 Moisture Content

In the proximate analysis of the mushrooms, moisture is a variable component (Pardo et al., 2002) and is significantly affected by environmental factors such as temperature and relative humidity during growth and storage, as well as by the relative amount of metabolic water which may be produced or utilized during storage (Crisan and Sands, 1978). Moisture content of most of the fresh mushrooms varies between 85% and 95% (Crisan and Sands, 1978).

12.7.2 Carbohydrates and Fiber

The carbohydrate content of mushrooms represents the bulk of the fruiting body, accounting for 50%–65% of dry weight. Dominating carbohydrates present in mushroom are raffinose, sucrose, glucose, fructose, and xylose (Singh and Singh, 2002). Water soluble polysaccharides of mushrooms have antitumor properties (Yoshioka et al., 1975). Free sugars amount to about 11%, but *Coprinus atramentarius* (Bull.: Fr.) has 24% of carbohydrate on a dry weight basis (Florezak et al., 2004). Mannitol, called mushroom sugar, constitutes about 80% of the total free sugars, hence it is dominant (Tseng and Mau, 1999; Wannet et al., 2000). Fresh mushrooms contain 0.9% mannitol, 0.28% reducing sugar, 0.59% glycogen, and 0.91% hemicellulose (Florezak et al., 2004).

The glycogen is a reserve polysaccharide and its content varies between 5% and 10% of dry matter. Chitin accounts for up to 80%–90% of dry matter in mushroom cell walls and is responsible for high protein values (Crisan and Sands, 1978). Mushrooms have glycogen and chitin as main polysaccharides, occurring in animals but not in plants. They have β-glucans considered significant due to their health-positive effects. Various polysaccharides of mushroom are regarded as a potential source of prebiotics.

All the mushrooms have a very high fiber content. The soluble fiber content ranges between 4% and 9% and insoluble fiber between 22% and 30% in mushrooms (Kalac, 2009). Fiber content in *Pleurotus* species ranges between 0.7% and 1.3% on fresh weight basis, and is the highest in *P. membranaceus* (Rai et al., 1988). On a dry weight basis, *Pleurotus* species are reported to contain 7.5%–27.6% fiber (Bano and Rajarathanam, 1982), against 10.4% in *Agaricus bisporus* (Crisan and Sands, 1978).

12.7.3 Proteins

The protein content of mushrooms depends on the composition of the substratum, the size of pileus, the harvest time, and the species (Bano and Rajarathnam, 1982). On a dry matter basis, the protein content of mushrooms varies between 19g/100g and 39g/100 g (Weaver et al., 1977; Breene, 1990). The dry matter is mainly represented by proteins (Aletor, 1995; Florezak and Lasota, 1995; Chang and Buswell, 1996). Protein in *A. bisporus* mycelium ranges from 32% to 42% of dry weight, but Abou-Heilah et al. (1987) found 46.5% protein on a dry weight basis (dwb) in *A. bisporus*. Purkayastha and Chandra (1976) found 14%–27% crude protein (dwb) in *A. bisporus, Lentinus subnudus,*

Calocybe indica, and *Volvariella volvacea.* With respect to crude protein, mushrooms rank below animal meats but well above most of other foods, including milk (Chang, 1980). Mushrooms are very useful for vegetarians because they contain some essential amino acids which are found in animal proteins. Rai and Saxena (1989) observed a decrease in the protein content of mushrooms on storage. The total nitrogen content of dry mushrooms is contributed by protein amino acids, which also reveals that crude protein is 79% compared with 100% for an ideal protein (Friedman, 1996).

Mushroom protein contains all the nine essential amino acids required by man (lysine, methionine, tryptophane, threonine, valine, leucine, isoleucine, histidine, and phenylalanine) which must be present simultaneously and in the correct relative amounts for protein synthesis to occur. The most abundant essential amino acid in mushrooms is lysine, while tryptophan and methionine are present in low amounts (Table 12.5). The most abundant amino acid in *A. bisporus* is the non-essential amino acid glutamic acid, which occurs both as a free amino acid and integrated in proteins.

12.7.4 Fat

Mushrooms have very low fat content in comparison to the carbohydrates and proteins. The fat present in mushroom fruiting bodies are dominated by unsaturated fatty acids. Total fat content in *A. bisporus* was reported to be 1.66–2.2 g/100 g (dwb) (Maggioni et al., 1968). The crude fat of mushrooms has representatives of all classes of lipid compounds, including free fatty acids, monoglycerides, diglycerides, triglycerides, sterols, sterol esters, and phospholipids. Most common fatty acids in *A. bisporus* are palmitic, stearic, oleic, and linoleic acids, but the linoleic acid is nutritionally the most desirable polyunsaturated fatty acid predominant, with 70% of fatty acid content of neutral lipid fraction and 90% of polar lipid fraction (Holtz and Schisler, 1971), it is comparable to that found in safflower oil. Hugaes (1962) observed that mushrooms are rich in linolenic acid, which is an essential fatty acid. In 100 g fresh matter of *A. bisporus* (Lange) Sing and *Pleurotus ostreatus* (Jacq: Fr.) Kumm, the contents of fatty compounds were found to

Table 12.5   Essential Amino Acids (% crude protein) in Edible Mushrooms

AMINO ACID	*AGARICUS BISPORUS*	*PLEUROTUS SAJOR-CAJU*	*VOLVARIELLA VOLVACEA*
Leucine	7.5	7.0	4.5
Isoleucine	4.5	4.4	3.4
Valine	2.5	5.3	5.4
Tryptophan	2.0	1.2	1.5
Lysine	9.1	5.7	7.1
Threonine	5.5	5.0	3.5
Phenyl alanine	4.2	5.0	2.6
Methionine	0.9	1.8	1.1
Histidine	2.7	2.2	3.8

Source: Adapted from Rai, R.D. 2003. *Fungal Biotechnology in Agricultural, Food and Environmental Applications.* CRC Press, Boca Raton, FL.

be 0.3 and 0.4 g, respectively (Manzi et al., 2001), and 2 g and 1.8 g (dwb), respectively (Shah et al., 1997).

Lipid content in *A. bisporus* and *V. volvacea* has been found to be higher than that in other edible mushrooms. 72% of total fatty acids in *L. edodes*, 70% in *V. volvacea*, 69% in *A. bisporus*, and 63% in *P. sajor-caju* is linoleic acid.

Mushrooms have a high proportion of unsaturated fatty acids, especially linoleic acid, with no cholesterol, and are, therefore, regarded as 'health food'. Among the sterols, ergosterol has been found most abundant, and cholesterol is non-existent in mushrooms. Most common sterols found in mushrooms were identified as provitamin D_2 (ergosterol), provitamin D_4 (22-dihydro-ergosterol) and g-ergosterol (Huang et al., 1985). Ultraviolet radiation converts ergosterols present in mushrooms to Vitamin D in the human body.

Triglycerides are found to be the main component (58.72%–63.95%), while free fatty acids and hydrocarbons are also present in significant amounts. Partial glycerides, free sterols, and sterol esters are other components of the non-polar lipid fraction, which are present in relatively low concentrations (Khanna and Garcha, 1983). Analysis of phospholipid and glycolipid contents of the polar lipid fraction indicates that phosphatidyl choline (PC) contributes more than 50% in all the species. The other phospholipids detected in high concentration are phosphatidyl ethanolamine (PE) and phosphatidyl inositol (PI). However, the lyso forms of PC, PE and PS (phosphatidyl serine) are present in low concentration. Khanna and Garcha (1983) identified two glycolipids (monogalactosyl diglyceride and digalactosyl diglyceride) in mushrooms.

12.7.5 Vitamins

Mushrooms are one of the best sources of vitamins especially Vitamin B (Breene, 1990; Mattila et al., 1994, 2000; Zrodlowski, 1995; Chang and Buswell, 1996). Manning (1985) have given comprehensive data on the vitamin content of mushrooms and some vegetables. The B-complex group vitamins include Vitamin B1 (Thiamin), Vitamin B_2 (Riboflavin), niacin, and biotin (Chang and Miles 1989). *A. bisporus* contains very low levels of vitamin A, vitamin D, vitamin E, and thiamin (Anderson and Fellers, 1942). Levels of vitamin D_2 are also low, inspite of relatively high concentrations of the precursor ergosterol.

Mushrooms also have vitamin C. Rai and Saxena (1989) found 8, 4 and 3 mg vitamin C per 100 g fresh weight in *A. bisporus*, *P. sajor-caju*, and *P. ostreatus*. The vitamin content of mushrooms, along with the daily requirement, is given in Table 12.6.

Table 12.6   Major Vitamin Contents in Mushrooms

MAJOR VITAMINS	DAILY REQUIREMENT	MUSHROOMS
Thiamin (B-1)	1.4 mg	4.8–8.9 mg
Riboflavin (B-2)	1.5 mg	3.7–4.7 mg
Niacin	18.2 mg	42–108 mg

Source: Adapted from Rai, R.D. and Saxena, S. 1989. *Mushroom J. Trop.* 9: 43–46.

12.7.6 Mineral Constituents

The fruiting bodies of mushrooms are characterized by a high level of mineral elements, which include major minerals namely, K, P, Na, Ca, and Mg, and minor elements like Cu, Zn, Fe, Mo, and Cd (Bano et al., 1981; Bano and Rajarathanum, 1982; Chang, 1982). K, P, Na, and Mg constitute about 56%–70% of the total ash content of the mushrooms (Li and Chang, 1982), while potassium alone forms 45% of the total ash. Abou-Heilah et al. (1987) found that the content of potassium and sodium in *A. bisporous* was 300 and 28.2 ppm, respectively. Varo et al. (1980) reported that *A. bisporus* contains Ca (0.04 g), Mg (0.16 g), P (0.75 g), Fe (7.8 g), Cu (9.4 mg), Mn (0.833 mg), and Zn (8.6 mg) per kg fresh weight. Mushrooms have been found to accumulate heavy metals like cadmium, lead, arsenic, copper, nickel, silver, chromium, and mercury (Schmitt and Sticher, 1991; Mejstric and Lepsova, 1993; Wondratschek and Roder, 1993; Kalac and Svoboda, 2000; Issilogglu et al., 2001; Svoboda et al., 2001; Malinowska, 2004). Apart from *L. edodes*, mushrooms are rather poor in calcium. Potassium and phosphorus are the main constituents of ash of *Pleurotus* species (Bano and Rajarathnam, 1982). Iron is present in appreciable amounts in *Pleurotus* species (Table 12.7). About one-third of the total iron in mushrooms is in the available form (Anderson and Fellers, 1942). Na and Ca are present in approximately equal concentrations in all the mushrooms except *L. edodes*, in which Ca is present in an especially large amount. The low Na concentration in mushrooms is suitable for people suffering from high blood pressure. The mineral content of various *Pleurotus sp.* is shown in Table 12.7. Of all the heavy metals, the Zn content was the highest in all species of *Pleurotus*. All edible mushrooms are good sources of selenium, which is of high significance as an anticancer substance to reduce the risk of prostate cancer. Bano et al. (1981) have also assessed the minerals and heavy metal contents in *Pleurotus* species.

12.7.7 Antimicrobial, Antioxidant and other Functional Properties

Experimental evidence has revealed that mushrooms are a rich source of many biologically active components which offer health benefits and protection against degenerative diseases (Barros et al., 2008). These include lectins, terpenoids, beta-glucans, ascorbic acid, tocopherols, carboxylic acids, and various dietary fibers (Parslew et al., 1999; Wasser and Weis, 1999; Mau et al., 2001; Wasser, 2002). These have been found effective in

Table 12.7   Mineral Content of *Pleurotus* Mushrooms

SPECIES	K	P	Mg	Na	Ca	Fe	Cd	Zn	Cu	Pb	Hg
			(Mg/100 g DRY WT.)						(ppm)		
P. sajor-caju	2240	565	156	256	40	2.8	–	2.6	0.5	–	–
P. eous	4570	1410	242	78	23	9.0	0.4	82.7	17.8	1.5	0
P. flabellatus	3760	1550	292	75	24	12.4	0.5	58.6	21.9	1.5	0
P. florida	4660	1850	192	62	24	18.4	0.5	11.5	15.8	1.5	0

Source: Adapted from Bano Z. et al. 1981. *Mushrooms News Lett Trop* 1: 6–10.

cholesterol reduction, against cancer stress, insomnia, asthma, allergies, and diabetes (Bahl, 1983). The high amount of protein in mushrooms are useful in the fight against protein malnutrition. Mushrooms as functional foods are used as nutrient supplements to enhance immunity. Due to low starch content and low cholesterol, they suit diabetic and heart patients. One-third of the iron in mushrooms is in the available form. Their polysaccharide content is used as an anticancer drug. They have even been used to combat HIV effectively (King, 1993; Nanba, 1993). Further, compounds from mushrooms possess antifungal, antibacterial, antioxidant, and antiviral properties, and have been used as insecticides and nematicides. Many species of mushrooms have been found to be highly potent immune enhancers, potentiating animal and human immunity against cancer (Wasser and Weis, 1999; Kidd, 2000; Feng et al., 2001).

12.7.7.1 Antioxidant Components Phenols such as tocopherols, BHT, and gallate are found in mushrooms and are known to be effective antioxidants. Total flavonoids in *Agaricus bisporus* cap were found to be 2.173 ± 0.007 and 1.533 ± 0.005, which might account for ferrous ion chelation and superoxide scavenging activities (Babu and Rao, 2011). Flavonoid and phenolic compounds are potent, water soluble free radical scavengers, which prevent oxidative cell damage (Marja and Anu, 1999; Okwu, 2004; Olajide et al., 2013). The presence of phenols in *A. bisporus* make this macrofungus a good option for formulating antioxidant products. Tyrosinase from *A. bisporus* is an antioxidant (Shi et al., 2002).

12.7.7.2 Antimicrobial Activity The antimicrobial activity of *A. bisporus* has been attributed to the presence of essential bioactive components. Catechin, the phenolic component, has exhibited antimicrobial, antioxidant, anticancer, and antiallergy properties (Baise et al., 2002; Shimamura et al., 2007; Yaltirak etal., 2009). Caffeic acid and rutin have also shown antimicrobial activity (Baise et al., 2002). Gallic acid, which is a bioactive compound and is widely present in plants (Li et al., 2000; Yaltirak et al., 2009), is a strong natural antioxidant, as well as having anti-inflammatory, antitumor, antibacterial, and antifungal activity (Kroes et al., 1992; Li et al., 2000; Miki et al., 2001; Panizzi et al., 2002).

12.8 Postharvest Technology of Mushrooms

12.8.1 Postharvest Shelf-life Extension of Fresh Mushrooms

In general, mushrooms have very high metabolic activity, especially respiration, and so have a very short postharvest life. Shelf-life of mushrooms is mainly dependent upon harvesting time and storage. To extend the postharvest life, several methods have been developed (Sharma and Vaidya, 2011), which are summarized in this section.

12.8.1.1 Harvesting, Pre-cooling, Packaging, and Transportation Mushrooms are generally harvested after 3 weeks of casing. Button mushrooms are generally harvested when the cap size is 30–45 mm in diameter. After harvesting, precooling of the produce, packaging, and preservation, proper transportation of the produce is normally carried out.

Washing of mushrooms: The washing of mushrooms prior to packing is very important for enhancing the shelf-life and extending the marketing period (Maini et al., 1987). Washing of fresh mushroom in water containing sodium sulfite solutions did reduce bacterial counts and improved initial appearance, but more rapid bacterial growth and browning occur during subsequent storage compared to the unwashed controls (Guthrie and Bellman, 1989).

Pre-cooling: Pre-cooling of mushrooms is a major traditional application of vacuum cooling, the porous structure and higher moisture content of mushroom making it possible (Frost et al., 1989; Zheng and Sun, 2005). The advantage of vacuum cooling is equivalent to a prolonged shelf-life of 24 h after 102 h storage (Burton et al., 1987). The influence of vacuum cooling on mushroom quality however, reveals no significant difference of product quality between vacuum and conventional cooling if the mushrooms are stored at 5°C after cooling (Frost et al., 1989). Enzymatic browning of mushroom caps caused by the enzyme polyphenol oxidase is a major criterion of mushroom quality, but pre and post packaging of similar types of mushrooms reduces activity of the polyphenol oxidase and lowers the incidence of browning (Gormley and MacCanna, 1967). Vacuum cooling of mushrooms results in around 3.6% of weight loss, which is higher than the 2% for air blast chilling (Wang and Sun, 2001). However, during storage, vacuum cooled mushroom experienced less weight loss than air blast cooled ones (Sun, 1999a) which helps to compensate for the cooling loss. Pre-wetting mushroom prior to vacuum cooling has been demonstrated to be an effective method to increase product yield (Sun, 1999b).

Packaging: Mushrooms are sensitive to desiccation and drought; consequently a suitable package is very important during storage. These are usually packed in polypropylene bags of 250–500 g capacities. These small packs are stacked in large containers for transportation (Sethi and Anand, 1976, 1984–85). Saxena and Rai (1988) stored button mushrooms in polypropylene bags of less than 100 gauge thickness with perforations having a vent area of about 5%, and documented the exacerbated veil-opening, browning, and reduction in weight during the storage at 15°C and found that the mushrooms were best preserved in nonperforated bags kept at 5°C. Storage in polystyrene or pulp-board punnets for transporting long distances, instead of using polythene bags, is preferred. Dhar (1992) also found that fruiting bodies of summer white button mushrooms (*Agaricus bitorquis*) could be stored without a significant loss of quality for 6 days at 15°C in non-perforated packs without any chemical treatment or washing in water. Use of oriented polypropylene (OPP) film can double the storage period compared to PE film, and maintains mushroom quality for at least 2 days at 18°C (De la Plaza et al., 1995).

Transportation: The positive effects of pre-cooling and packing will be partially neutralized if the product after pre-cooling is stored and transported in a hot environment. So mushrooms need a complete cool-chain for storage and transportation. To achieve this objective, mushrooms are cooled, during short distances under ambient conditions, in polypacks stacked in small wooden cases or boxes with sufficient crushed ice in polypacks (overwrapped in paper), by the small growers. For transport of large quantities for long distances, refrigerated trucks, though costly, are indispensable (Rai and Arumuganathan, 2008b).

Modified Atmosphere Packaging (MAP): During postharvest storage, the main processes related to mushroom deterioration are related to the development of sporophore, such as breaking of the veil, elongation of the stipe, opening of the pileus, expansion of gill-tissue, and spore formation (Lopez-Briones et al., 1992; Braaksma et al., 1994; Eastwood and Burton, 2002), along with browning of the cap and gills and a general loss of appearance. So these negative quality characteristics limit the shelf-life of this commodity. MAP technology for extending the shelf-life of button mushroom has been tried. The incorporation of a small area of a microporous film into the overwrapping film creates a modified atmosphere in mushroom punnets with moderately low but acceptable oxygen levels. It delays mushroom development and reduces browning and incidence of visible symptoms of diseases on mushrooms at 18°C storage (Burton, 1988).

Nichols and Hammond (1975) packed mushrooms in six different types of films, stored them at two different temperatures (2°C and 18°C) and evaluated the effect of modified atmospheres on the quality. It was found that at 2°C, equilibrium was established roughly after 24 h at 4%–10% CO_2 and 11%–17% O_2; but the mean concentration was dependent on the type of film. At 18°C, equilibrium was established at 8%–15% CO_2 and 1%–2% O_2 and thus, film need to be chosen according to the storage temperature of the product. The poor keeping quality of mushroom was mainly attributed to enzyme and microbial activity. An atmosphere containing 5% CO_2 with or without 1% O_2 level prevented cap opening of mushroom for up to five weeks at 0°C (Murr and Morris, 1974). Nichols and Hammond (1975) reported that 1–4 pinholes (1 mm diameter) in overwrapping PVC film could control the degree of modification of the atmosphere. The optimum conditions for retaining the most acceptable color and appearance of mushrooms were to store them in perforated plastic packs at 4–7°C and 40%–50% relative humidity (Bush and Cook, 1976). To prevent anaerobiosis, the overwrap films are perforated to keep the oxygen levels above 4% (Nichols, 1985). Button mushroom at the ambient (18°C) and lower storage temperatures (10°C and 2°C) with a modified atmosphere delayed the postharvest development of mushrooms (Burton and Twyning, 1989).

Ozone treatment (100 mg/h) of mushrooms prior to packaging increased the external browning and reduced the internal browning rates (Escriche et al., 2001; Thompson, 2007). Ozone treatment exhibited no significant differences in terms of texture, maturity index, and weight loss of mushrooms.

12.8.2 Processing and Preservation of Mushrooms

12.8.2.1 Fresh Sliced Mushroom   Demand of sliced mushroom is growing because of increased consumer demand for fresh ready-to-use products and convenience foods. The process for sliced whole mushrooms includes soaking of mushrooms for 10 min in the solution of citric acid or hydrogen peroxide, followed by slicing, packing, and storage at 4°C for a period of 19 days. Both the treatments reduced the bacterial count and improved the keeping quality of the sliced mushroom as compared to water soaked slices (Brennan et al., 2000).

12.8.2.2 Steeping Preservation   Fresh mushrooms are washed and blanched in 0.05% of potassium metabi-sulphite (KMS) for 5 min. After draining, they are washed with cold water for 4–5 times and then, put into bottles or cans. Hot brine of 18%–20% NaCl and 0.1% citric acid is added, followed by proper lidding and storage at room temperature. This method of steeping preservation is found to be good for storing mushrooms for up to 3 months (NRCM, Annual Report-2000–01). Water-only blanched mushrooms had a yellow color, whereas whiteness was maintained in the treated mushrooms (Pruthi et al., 1984).

12.8.2.3 Freezing   Individual quick freezing (IQF) is another popular processing method followed in large industrial units. In this method, the raw material is washed at processing units after receipt from farms, and then, the mushrooms are inspected, sliced, and graded according to quality. After that, blanched and water-cooled mushrooms are subjected to tunnel freezing. At this stage, the mushrooms are cooled in a system having a temperature of around −40°C until the core areas of the mushroom pieces reach a temperature of around −18°C. Subsequently, they are packed in multi-layer poly-bags and stored in a cold storage at −20°C to −35°C. Vacuum freeze drying (VFD) is another development in mushroom processing, where the original shape, quality, color, size, texture, and freshness properties of the treated produce are retained. It involves the cooling of the mushrooms to −40°C, when the moisture present in the mushrooms is converted to tiny ice molecules which are further directly sublimed into vapor when subjected to vacuum, with a slight rise in temperature, resulting in a dried end product.

Mushrooms are freeze dried at −20°C and the moisture is removed by sublimation at a very low vacuum (0.012 mbar) for 12–16 h, and the dried mushrooms had superior flavor and appearance, but were brittle (Kapoor, 1989). The appearance of freeze dried mushrooms is very similar to fresh mushrooms, but as the product is brittle, it has to be packed in sturdy packaging or cushion-packs flushed with nitrogen for better keeping qualities (Saxena and Rai, 1990). The product can be stored up to 6 months without any change in its quality and appearance. Being a very costly and energy-intensive process, the venture depends upon the demand and price for such products. Freeze-drying has been carried out by slicing mushrooms and immersing them in a solution of 0.05% KMS and 2% salt for about 30 min. The pretreated mushrooms are then blanched in boiling water for 2 min, followed by cooling. The product is frozen at −22°C for 1 min. The frozen mushrooms are dried to a moisture content of 3% in a freeze drier and packed under vacuum (Kannaiyan and Ramsamy, 1980). The drying characteristics for fluidized bed button mushrooms have been determined and have been found that the quality of the dehydrated mushrooms is significantly influenced by pretreatments as well as temperature; the samples treated with 1% KMS, 0.2% citric acid and 3% salt solution and dried at 50°C gave satisfactory results (Singh et al., 2001).

12.8.2.4 Canning The canned mushrooms can be stored for longer periods up to a year and most of the international trade in mushrooms is done in this form. The canning of mushroom involves several unit operations, such as cleaning, blanching, filling, sterilization, cooling, labeling, and packaging (Figure 12.16). To produce good quality canned mushroom, they should be processed as soon as possible after the harvest, but where delay is inevitable, they can be stored at 4–5°C until processed. The mushrooms with a stem length of one cm are preferred and are canned whole, sliced, and as stems-and-pieces, as per market demand (Beelman and Edwards, 1989). Longitudinal (mushroom shape) slicing is common (Mudahar and Bains, 1982; Pruthi et al., 1984).

A brine solution with 2% common salt, 1% sugar, and 0.05% citric acid for filling the cans for better results is recommended (Azad et al., 1987). Adsule et al. (1983) used tomato juice in place of brine solution for canning of mushrooms; unlike brine solutions, there is no need to add citric acid to tomato juice to lower the pH of the filling medium, and the nutrients of the mushrooms can be retained in the tomato

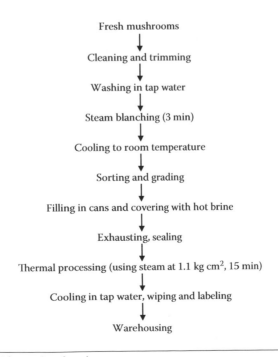

Figure 12.16 Flow chart for canning of mushroom.

juice. Arumuganathan et al. (2004) obtained improved quality of canned button mushroom when the mushrooms were pre-treated with ethylene diamine tetra acetic acid (EDTA).

Shrinkage losses in canned mushrooms have been a serious problem and can be reduced by the vacuum treatment of the fresh mushrooms cutting the blanching time, and also prehydration treatments. The levels of vacuum achieved and the quality of the mushrooms has affected shrinkage losses of 5%–10% (Steinbuch, 1978). To increase the drained weight (reducing shrinkage) of canned products agar-agar, methyl cellulose, carboxy methyl cellulose, pectin, and pectin–calcium chloride have been used by various researchers (Singh et al., 1982). Steam blanching for low loss in drained weight has been documented (Kapoor, 1989). A method to determine the shrinkage of mushrooms during processing based on a liquid displacement method has been developed (Konanayakam et al., 1987), consisting of immersing the sample in a glass container with an over-flow spout. Water treated with a surfactant was used as the displacement liquid in the glass container.

12.8.2.5 Pickling and Preparation of Chutney/Sauce Pickling of mushroom is also a popular method of preservation (Figure 12.17). It is a more economically viable method of preservation during surplus periods. Joshi et al. (1991, 1996) reported the preparation of sweet chutney and mushroom sauce from edible mushrooms having a shelf-life of over a year, with better sensory qualities. Pickle prepared from paddy straw mushrooms had better quality and could be stored upto 1 year in bottles (Saxena and Rai, 1990).

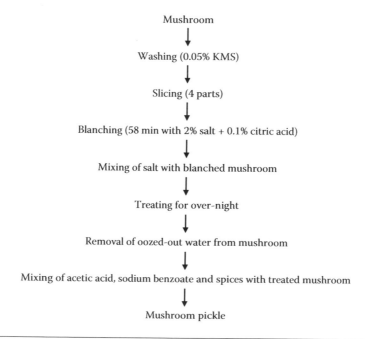

Figure 12.17   Flow chart for mushroom pickle. (From Saxena, S. and Rai, R.D. 1990. *Postharvest Technology of Mushrooms.* Technical Bulletin No. 2: NRCM, Solan.)

12.8.2.6 Drying The removal of moisture from any commodity prevents its spoilage. Mushrooms that containing about 90% moisture at the time of harvesting, can be dried to a moisture level below 10%–12%. At a drying temperature of 55–60°C any insects and microbes on the mushrooms are killed in a few hours, and gives a dehydrated final product with a lower moisture content and a longer shelf-life. The temperature, moisture of the mushrooms, and humidity of the air affect the color of the dried product (Yapar et al., 1990). Dehydrated mushrooms are used as an important ingredient in several food formulations, such as instant soups, pasta, snack seasonings, casseroles, and meat and rice dishes (Tuley, 1996; Gothandapani et al., 1997). They can be further ground into flour to make value-added products like noodles, soup, and *tikki*.

The water blanching of the mushroom *Agaricus bisporus* for 5 min, along with 0.5% citric acid, 0.1% KMS and 125 ppm EDTA to improve the color and texture of mushroom slices is recommended. Longitudinally sliced and blanched button mushrooms, when dried at 60°C for 5 h, had a drying ratio of 10.8:1 and rehydration ratio of 2.78, as against cross slit mushrooms, with a drying time of 8 h, drying ratio of 10.9:1, and rehydration ratio of 2.80 (Pruthi et al., 1984). The blanching of both button and oyster mushrooms in boiling water for 1 min and treating in a solution containing 0.1% citric acid and 0.25% KMS for 15 min at room temperature reduced the browning index and the activation energy values of both button and oyster mushrooms (Arora et al., 2003).

Drying of mushrooms in a mechanical dehydrator was found to be the fastest because of the high air temperature and forced air circulation (Katiyar, 1985). Mean dehydration time was 8.4 h in a mechanical dryer compared to 16.8 hours needed in sun drying. However, Kumar (1992) dehydrated *Agaricus bisporus* for 9 h at 60 ± 2°C to a constant weight. Lidhoo and Agrawal (2006) dried white button mushroom in a hot air oven and observed that minimum browning index was recorded at 65°C, with a rehydration ratio of 2.9. Another development in drying of mushrooms is osmotic dehydration of the button mushrooms in a continuously-circulated contacting reactor, which showed 1% NaCl as the optimum (Yang and Le Maguer, 1992), but pretreatments of the mushrooms in high concentrations of sucrose, followed by high salt concentration, was the most effective method to remove water and load salt to further decrease water activity in the mushrooms. Osmosis using 15% brine solution removed about 35% of initial moisture in 1 h (Kar and Gupta, 2001).

12.8.2.7 Irradiation Radiation preservation or "cold sterilization" is the method where the mushrooms can be preserved without any marked change in their natural characteristics. Low dosages of γ-radiation reduce microbial contamination and extend the shelf-life of mushrooms. However, irradiation should be done immediately after harvest for optimum benefits (Hamid et al., 2013). Various types of beneficial effects of radiation have been observed in preserving button mushrooms (Staden, 1967; Campbell et al., 1968; Wahid and Kovacs, 1980; Lescane, 1984; Roy and Bahl, 1984a), such as to delay maturation (development of cap, stalk, gill, and spore), and reduce the loss of water, color, flavor, texture, and, finally, quality. Cobalt 60 (Co^{60}) has been used as a common source of γ-rays. A dose of 400 krad gave whiter button mushrooms than control (without radiation) when the atmospheric temperature during growth and subsequent, handling was slightly lower than 20°C (Roy and Bahl, 1984b). A dose of 10 kGy (kilo Gray) completely destroys the microorganisms, while enhancing the shelf-life of *Agaricus bisporus* up to a period of 10 days has been achieved by application of gamma rays close to 2 kGy and storage at 10°C (Lescane, 1984). Irradiation reduces the incidence of fungal and bacterial infection and also retards the breakdown of mannitol and trehalose, as well as decreasing the flavor components in irradiated mushrooms. Amino acids in fresh mushrooms were better preserved by γ-irradiation, and irradiation at low levels proved better than irradiation levels of 1 and 2 kGy (Roy and Bahl, 1984a). The effect of electron-beam irradiation on the quality of white button mushroom showed that irradiation levels above 0.5 kGy prevented microbe-induced browning (Koorapati et al., 2004).

12.9 Marketing of Mushrooms

Fresh mushrooms have a very short life cycle, cannot be transported long distances without a refrigerated transport facility, and are sold in local markets near the production areas. Of all the mushrooms cultivated, the button mushroom holds the main attraction for consumers. Paddy straw mushrooms (*Volvariella volvacaea*) are famous for specific bite and a watery texture. In India, it is widely cultivated in Orissa, where high temperatures predominate. Its color ranges from white to dark brown. It is available in fresh, canned, and dried form, mainly for overseas markets. The different species of *Pleurotus* (Oyster Mushroom) have a very broad temperature range of growth, and have considerable variations in taste and appearance. Most of the species with beautiful colors break easily and require careful handling and packing. Marketing potential of dried oyster mushrooms is still limited, though the taste becomes stronger after drying. There is a ready market for dried shiitake mushrooms (*Lentinula edodes*). A good quality mushroom brings relatively high prices. It has a specific taste which becomes much stronger after drying.

Marketing of mushrooms in India is not well organized. It is the simple system of producers selling directly to retailers or even to consumers, which has its own limitations. The production of mushrooms, especially of white button mushrooms, in India has gone up in the recent years, but it has also exacerbated its marketing problems. The bulk of India's mushroom production emanates from Himachal Pradesh, Jammu, Kashmir, Punjab, and Haryana. There is not yet much of a market for processed foods in India, and the maximum sale, as in most countries of South Asia, is of fresh mushrooms. Only a fraction of the produce is canned and marketed in India.

The major problems related to marketing of mushroom have been identified as no proper agency to purchase the mushroom, the low price of mushroom, the large distance to transport the produce, and malpractices by the purchasing agencies (Singh and Singh, 2006). It has been suggested that farmers' cooperative marketing societies must be promoted to take care of surplus mushrooms. Being a highly perishable crop and prone to high temperature deterioration, it needs a marketing infrastructure, such as cold storage facilities, which are very expensive but of immense value. The canning and processing units for preservation of surplus mushrooms needs to be established, as there is going to be a good demand for the processed mushrooms. Per capita consumption of mushrooms in India is less than 50 g. Mushroom product diversification also has great scope in the near future.

12.10 Summary and Future Prospects

The operations essential to successful cultivation of mushrooms involve: selection of mushroom spores or strains, maintenance of mycelial cultures, development of spawn or inoculum, preparation of growing medium, inoculation, colonization, and crop management for optimum production. South Asia has tremendous potential for mushroom production, and all commercial edible and medicinal mushrooms can be grown. Different countries have a wide range of agro-climatic conditions that can be exploited for production of different varieties of mushrooms all the year round. However, some obstacles to increased commercialization of mushrooms by small scale farmers need to be addressed with adequate solution.

Commercialization of new developments in mushroom production and processing requires links between agricultural research and farmers. There is an increasing demand for quality mushrooms and their products at competitive rate, both in domestic and export markets. There is a need to develop technology to grow other mushrooms with better nutrients, shelf-life, and/or sensory qualities. The dense and undernourished population of South Asian countries also provides a huge domestic market. Though growth of the mushroom industry will depend on increasing and widening the domestic market in coming years, the export market will also be equally attractive. However, being a nutritious and health-promoting commodity, both the producer and the consumer need to enhance production and marketing strategies to give remunerative prices to the growers.

Mushrooms, being highly perishable, force the producer to preserve and process them. Preservation is essential to make them available throughout the year, to retain maximum nutrients, texture, and flavor, and to increase per capita consumption in developing countries. Value added products from mushrooms are a promising enterprise that will not only cater to protein and micronutrient requirements, but will also enable the consumers to live a healthy life. Presently, the mushroom products available are pickles, chutneys, nuggets, papads and fast food items like burgers, cutlets, pizza, and bakery products (biscuits, bread, cakes), etc. It is, therefore, imperative to produce quality processed products at affordable price. It is also important to commercially utilize the spent mushroom substrates for further economies in the process, and for total recycling of agrowastes.

References

Abou-Heilah, A.N., Kasionalsim, M.Y., and Khaliel, A.S. 1987. Chemical composition of the fruiting bodies of *Agaricus bisporus*. *Int J Exp Bot* 47: 64–68.

Adhikari, M.K. 1995. Mycodiversity in Nepal, a glimpse. *Bull Nat Hist Mus (Nepal)* 3–4B(1–4): 4–6.

Adhikari, M.K., Devkota, S., and Tiwari, R.D. 2005. Ethnomycolgical knowledge on uses of wild mushrooms in Western and Central Nepal. *Our Nat* 3: 13–19.

Adsule, P.G., Onkaraya, H., Tewari, R.P., and Girija, V. 1983. Tomato juice as a new canning medium for button mushroom (*Agaricus bisporus* (Lange) Sing.). *Mushroom J* 124: 143–145.

Agahar-Murugkar, D. and Subbulakshmi, G. 2005. Nutritional value of edible wild mushrooms collected from the Khasi hills of Meghalaya. *Food Chem* 89: 599–603.

Ahmed, S.A., Kadam, J.A., Mane, V.P., Patil, S.S., and Baig, M.M.V. 2009. Biological efficiency and nutritional contents of *Pleurotus florida* (Mont.) singer cultivated on different agro-wastes. *Nat Sci* 7: 44–48.

Aletor, V.A. 1995. Compositional studies on edible tropical species of mushrooms. *Food Chem* 54: 265–268.

Allison, W.H. and Kneebone, L.R. 1962. Influence of compost pH and casing soil pH on mushroom production. *Mushroom Sci* 5: 81–90.

Anderson, E.E. and Fellers, C.R. 1942. The food value of mushrooms *(A. campestris)*. *Proc Am Soc Hortic Sci* 41: 301.

Arora, S., Shivhare, U.S., Ahmed, J., and Raghavan, G.S.V. 2003. Drying kinetics of *Agaricus bisporus* and *Pleurotus florida* mushrooms. *Trans ASAE* 46(3): 721–724.

Arumuganathan, T., Rai, R.D., Chandrasekar, V., and Hemakar, A. K. 2004. Studies on canning of button mushroom, *Agaricus bisporus* for improved quality. *Mushroom Res* 12(2): 117–120.

Azad, K.C., Srivastava, M.P., Singh, R.C., and Sharma, P.C. 1987. Commercial preservation of mushrooms—1. A technical profile of canning and its economics. *Indian J Mushrooms* XII-XIII: 21–29.

Babu, D.R. and Rao, G.N. 2011. Antioxidant properties and electrochemical behavior of cultivated commercial Indian edible mushrooms. *J Food Sci Technol* doi:10.1007/s13197-011-0338-8.

Bahl, N. 1983. Medicinal value of edible fungi. In: *Proceeding of the International Conference on Science and Cultivation Technology of Edible Fungi*. Indian Mushroom Science II, pp. 203–209.

Baise, H.P., Walker, T.S., Stermitz, F.R., Hufbauer, R.S., and Vivanco, J.M. 2002. Enantiometric dependent phytotoxic and antimicrobial activity of catechin: A rhizose-creted racemic mixture from *Centaurea maculosa* (spotted knapweed). *Plant Physiol* 128: 1127–1135.

Bano, Z. 1976. Nutritive value of Indian mushrooms and medicinal practices. *Ecol Bot* 31: 367–371.

Bano, Z., Bhagya, S., and Srinivasan, K.S. 1981. Essential amino acid composition and proximate analysis of Mushroom, *Pleurotus florida*. *Mushrooms News Lett Trop* 1: 6–10.

Bano, Z. and Rajarathanam, S. 1982. *Pleurotus* mushrooms as a nutritious food. In: *Tropical Mushrooms—Biological Nature and Cultivation Methods*. Chang, S.T. and Quimio, T.H. (Eds.) The Chinese University Press, Hong Kong, pp. 363–382.

Barros, L., Falcao, S., Baptista, P., Freire, C., Vilas-Boas, M., and Ferreira, I.C.F.R. 2008. Antioxidant activity of *Agaricus* sp. mushrooms by chemical, biochemical and electrochemical assays. *Food Chem* 111: 61–66.

Beelman, R.B. and Edwards, C.E. 1989. Variability in the protein content and canned product yield of four important processing strains of the cultivated mushrooms (*Agaricus bisporus*). *Mushroom News* 37(7): 17–26.

Beyer, D.M. 1999. *Spent Mushroom Substrate Fact Sheet*. http:/mushroomspawn.cas.psu.edu/spent.htm.

Bisen, P.S., Baghel. R.K., Sanodiya, B.S., Thakur, G.S., and Prasad, G.B. 2010. *Lentinus edodes*: A macrofungus with pharmacological activities. *Curr Med Chem* 17(22): 2419–2430.

Braaksma, A., Schaap, D.J., de Vrije, T., Jongen, W.M.F., and Woltering, E.J. 1994. Ageing of mushroom (*Agaricus bisporus*) under post-harvest conditions. *Postharvest Biol Technol* 4: 99–110.

Breene, W.M. 1990. Nutritional and medicinal value of specialty mushrooms. *J Food Protect* 53: 883–894.

Brennan, M., Le Port, G., and Gormley, R. 2000. Post-harvest treatment with citric acid or hydrogen peroxide to extend the shelf life of fresh sliced mushrooms. *Food Sci Technol* 33(4): 285–289.

Burton, K.S. 1988. Effect of pre-and post-harvest development on mushroom tyrosinase. *J Hortic Sci* 63(2): 255–260.

Burton, K.S., Frost, C.E., and Atkey, P.T. 1987. Effect of vacuum cooling on mushroom browning. *Int J Food Sci Technol* 22: 599–606.

Burton, K.S. and Twyning, R.V. 1989. Extending mushroom storage life by combined modified atmosphere packaging and cooling. *Acta Hortic* 258: 565–569.

Bush, P. and Cook, D.J. 1976. The purchase and domestic storage of pre-packed mushrooms. *Mushroom J* 39: 76, 78–80.

Carey, A.T. and Connor, T.P.O. 1991. Influence of husbandry factors on the quality of fresh mushrooms (*Agaricus bisporus*). In: *Science and Cultivation of Edible Fungi*. Vol. 2. Maher, M.J. (Ed.) Balkana, Rotterdam, pp. 673–682.

Campbell, J.D., Stothers, S., Vaisey, M., and Berck, B. 1968. Gamma radiation influence on the storage and nutritional quality of mushroom. *J Food Sci* 3: 540–542.

Chang, S.T. 1980. Mushroom as human food. *Biol Sci* 30: 339–401.

Chang, S.T. 1982. Prospects for mushroom protein in developing countries. In: *Tropical Mushroom—Biological Nature and Cultivation Methods*, Chang, S.T. and Quimio, T.H. (Eds.) Chinese University Press, Hong Kong, pp. 463–473.

Chang, S.T. 1991. Cultivated mushrooms. In: *Handbook of Applied Mycology, Foods and Feeds*. Vol. 3. Arora, D.K., Mukerji, K.G., and Marth, E.H. (Eds.) Marcel Dekker, Inc., New York, NY, pp. 221–240.

Chang, S.T. 1998. Development of novel agroscience industries based on bioconversion technology. In: *Frontiers in Biology: The Challenges of Biodiversity*, Chou, C.H. and Shao, K.T. (Ed.) Academia Sinica, Taipei, pp. 217–222.

Chang, S.T. 2005. Witnessing the development of the mushroom industry in China. In: *Proceedings of the 5th International Conference on Mushroom Biology and Mushroom Products*. Tan et al. (Eds.) *Acta Edulis Fungi* 12, pp. 3–19.

Chang, S.T. and Buswell, J.A. 1996. Mushroom nutriceuticals. *World J Microbiol Biotechnol* 12: 473–476.

Chang, S.T. and Chiu, S.W. 1992. Mushroom production—An economic measure in maintenance of food security. In: *Biotechnology: Economic and Social Aspects*. Da Silva, E.J., Ratledge, C., and Sasson, A. (Eds.) Cambridge University Press, Cambridge, U.K., pp. 110–141.

Chang, S.T. and Chu, S.S. 1969. Nuclear behavior in basidium of *Volvariella volvacea*. *Cytologia* 34: 293–299.

Chang, S.T. and Miles, P.G. 1987. Historical record of the early cultivation of *Lentinus* in China. *Mushroom J Trop* 7: 31–37.

Chang, S.T. and Miles, P.G. 1989. *Edible Mushrooms and their Cultivation*. CRC Press, Boca Raton, FL, pp. 345.

Chang, S.T. and Miles, P.G. 1991. Recent trends in world production of cultivated edible mushrooms. *Mushroom J* 503: 15–18.

Chang, S.T. and Miles, P.G. 1992. Mushroom biology—A new discipline. *The Mycologist* 6: 64–65.

Chang, S.T. and Phillip, G.M. 2004. Volvariella—A high temperature cultivated mushroom. *Mushrooms Cultivation, Nutritive Value, Medicinal Effect and Environmental Impact.* CRC Press, Boca Raton, FL, pp. 277–304.

Chang, S.T. and Yau, C.K. 1971. *Volvariella volvacea and its life history. Ann J Bot* 58: 552–561.

Chong, C. and Wickware, M. 1989. Mushroom compost trial at Canavonda Nursery. *Hortic Rev* 7(6): 10–11, 13.

Crisan, E.V. and Sands, A. 1978. Nutritive value. In: *The Biology and Cultivation of Edible Mushroom.* Chang, S.T. and Hayes, W.A. (Eds.). Academic Press, New York, NY, pp. 137–168.

Dai, Y.C., Yang, Z.L., Cui, B.K., Yu, C.J., and Zhou, L.W. 2009. Species diversity and utilization of medicinal mushrooms and fungi in China. *Int J Med Mushrooms* 11: 287–302.

Dai, Y.C., Zhou, L.W., Yang, Z.L., Wen, H.A., Bau, T., and Li, T.H. 2010. A revised checklist of edible fungi in China. *Mycosystema* 29: 1–21.

Dama, C.L., Sunil, K., Brijesh, K.M., Kunj, B.S., Sudha, M., and Anila, D. 2010. Antioxidative enzymatic profile of mushrooms stored at low temperature. *J Food Sci Technol* 47(6): 650–655.

Dhar, A.K. and Kaul, T.N. 1987. Genetics and improvement of mushroom crops. *Indian Mushroom Sci* 2: 339–347.

De la Plaza, J.L., Alique, R., Zamorano, J.P., Calvo, M.L., and Navarro, M.J. 1995. Effect of the high permeability to O_2 on the quality changes and shelf-life of fresh mushrooms stored under modified atmosphere packaging. *Mushroom Sci* 14(2): 709–716.

Devonald, V.G. 1987. Spent mushroom compost, a possible growing medium ingredient. In: *Compost: Production, Quality and Use.* Elsevier AppliedScience, London, pp. 785–791.

Dhar, B.L. 1992. Postharvest storage of white button mushroom *Agaricus bitorquis. Mushroom Res* 1: 127–130.

Dickinson, C. and Lucas, J. 1979. *The Encyclopedia of Mushrooms.* Lucea Orbis Publishing, London.

Doshi, A. and Sharma, S.S. 1995. Cultivation of white summer mushroom. In: *Advances in Horticulture.* Chadha, K.L. and Sharma, S.R. (Eds.). 13, MPH, New Delhi.

Eastwood, D. and Burton, K. 2002. Mushrooms—A matter of choice and spoiling onself. *Microbiol Int* 2: 7–14.

Escriche, I., Serra, J.A., Gomez, M., and Galotto, M.J. 2001. Effect of ozone treatment and storage temperature on physicochemical properties of mushrooms (*Agaricus bisporus*). *Food Sci Technol Int* 7(3): 251–258.

FAO, U.N. 1972. *Food Composition Table for Use in East Asia.* FAO, Rome. pp. 334.

FAO, U.N. 2004. *Wild Edible Fungi: A Global Overview of Their Use and Importance to People. Non-Wood Forest Products*, Vol. 17. Rome.

Feng, W., Nagai, J., and Ikekawa, T. 2001. A clinical pilot study of EEM for advanced cancer treatment with EEM for improvement of cachexia and immune function compared with MPA. *Biotherapy* 15: 691–696.

Flegg, P.B. 1960. *Mushroom Composts and Composting: A Review of Literature.* Rep Glasshouse Crops Res Inst.

Flegg, P.B. 1962. The development of mycelial strands in relation to fruiting of the cultivated mushroom (*Agaricus bisporus*). *Mushroom Sci* 5: 300–313.

Florezak, J., Karmnska, A., and Wedzisz, A. 2004. Comparision of the chemical contents of the selected wild growing mushrooms. *Bromatol Chem Toksykol* 37: 365–371.

Florezak, J. and Lasota, W. 1995. Cadmium uptake and binding by artificially cultivated cultivated (*Pleurotus ostreatus*). *Bromatol Chem Toksykol* 28: 17–23.

Friedman, M. 1996. Nutritional value of proteins from different food sources: Review. *J Agric Food Chem* 44: 6–29.

Frost, C.E., Burton, K.S., and Atkey, P.T. 1989. A fresh look at cooling mushroom. *Mushroom J* 193: 23–29.

Garcha, H.S. and Khanna, P.K. 2003. *Mushroom Cultivation*. Punjab Agricultural University, Ludhiana, India, pp. 60.

Garcha, H.S., Sodhi, H.S., and Khanna, P.K. 1987. Evaluating strains of paddy straw mushroom (*Volvariella* spp.) in India. In: *Cultivating Edible Fungi*. (Wuest, P.N., Royse, D.T., and Beelman, R.B. (Eds.). Elsevier Science Publishers, the Netherlands, pp 101–108.

Gerrits, J.P.G. 1987. Compost for mushroom production and its subsequent use for soilimprovement. In: de Bertoldi, M., Ferranti, M.P., Hermite, P.L., and Zucconi, F. (Eds.). Elsevier Sci., London, pp. 431–439.

Ghosh, N. and Chakraborty, D.K. 1991. Studies on evolving new strains of *Pleurotus sajor-caju* by selective dikaryotization. *Adv Mushroom Sci* 28–32.

Ghosh, S. and Singh, S. 1995. Utilization of whey for the manufacture of ready-to-serve mushroom soup. *Mushroom Res* 4: 23–26.

Gormley, T.R. and MacCanna, C. 1967. Pre-packaging and shelf life of mushroom. *Irish J Agric Res* 6: 255–265.

Gothandapani, L., Parvathi, K., and Kennady, Z.J. 1997. Evaluation of different methods of drying on the quality of oyster mushroom (*Pleurotus* sp.). *Dry Technol* 15(6–8): 1995–2004.

Gruen, V.E.C. and Wong, H.X. 1982. Immunodulatory and antitumour activities of a polysaccharide–peptide complex from a mycelial culture of *Trichoderma* sp. *Sciences* 57: 269–281.

Gupta, P., Indurani, C., Ahlawat, O.P., Vijay, B., and Mediratta, V. 2004. Physicochemical properties of spent mushroom substrates of *Agaricus bisporus*. *Mushroom Res* 13(2): 84–94.

Guthrie, B.D. and Bellman, R.B. 1989. Control of bacterial deterioration in fresh washed mushrooms. *Mushroom Sci* 12(2): 689–699.

Hamid, A., Seyed, M.K., and Mohammad, A.S. 2013. Deterioration and some of applied preservation techniques for common mushrooms (*Agaricus bisporus*, followed by *Lentinus edodes*, *Pleurotus* spp.). *J Microbiol Biotechnol Food Sci* 2(6): 2398–2402.

Hayes, W.A. and Shandilya, T.R. 1977. Casing soil and compost substrate used in artificial culture of *Agaricus bisporus* (Lange) Sing. The cultivated mushroom. *Indian J Mycol Plant Pathol* 7: 5–10.

Hobbs, C. 1995. *Medicinal Mushrooms*. Botanica Press, Santa Cruz, CA.

Holtz, R.B. and Schisler, L.C. 1971. Lipid metabolism of *Agaricus bisporus* (Lange) Sing.: I. Analysis of sporophore and mycelial lipids. *Lipids* 6: 176–180.

Huang, B.H., Yung, K.H., and Chang, S.T. 1985. The sterol composition of *Volvariella volvacea* and other edible mushrooms. *Mycologia* 77: 959–63.

Hugaes, D.H. 1962. Preliminary characterization of the lipid constituents of the cultivated mushroom *Agaricus campestris*. *Mushroom Sci* 5: 540–546.

Issilogglu, M., Yilmaz, F., and Merdivan, M. 2001. Concentrations of trace elements in wild edible mushrooms. *Food Chem* 73: 163–175.

Jain, V.B. and Singh, S.P. 1982. Effect of spawning method on the yield of *Agaricus brunnescens* Peck. *Prog Hortic* 14(4): 246–248.

Jandaik, C.L. 1987. Breeding potential of *Pleurotus* species in India with special reference to *Pleurotus sajor-caju*. *Indian Mushroom Sci* 2: 355–360.

Jandaik, C.L. and Goyal, S.P. 1995. Farm and farming of oyster mushroom (*Pleurotus* sp.). In: *Mushroom Production Technology*. Singh, R.P. and Chaube, H.S. (Eds.). G. B. Pant Univ. Agril. and Tech., Pantnagar, India, pp. 72–78.

Jandaik, C.L. and Kapoor, J.N. 1974. Studies on cultivation of *Pleurotus sajor-caju*. *Mushroom Sci* 9(1): 667–672.

Joshi, V.K., Seth, P.K., Sharma, R.C., and Sharma, R. 1991. Standardization of a method for the preparation of sweet chutney from edible mushrooms *Agaricus bisporus*. *Indian Food Packer* 45(2): 39–43.

Joshi, V.K., Mohinder, K., and Thakur, N.S. 1996. Lactic acid fermentation of mushroom (*Agaricus bisporus*) for preservation and preparation of sauce. *Acta Aliment* 25(1): 1–13.

Kalac, P. 2009. Chemical composition and nutritional value of European species of wild growing mushrooms: A review. *Food Chem* 113: 9–16.

Kalac, P. and Svoboda, L. 2000. A review of trace element concentrations in edible mushrooms. *Food Chem* 69: 273–281.

Kannaiyan, S. and Ramaswamy, K. 1980. *A Handbook of Edible Mushrooms*. Todays & Tomorrow's Printers and Publishers, New Delhi, pp. 44–50.

Kapoor, J.N. 1987. Cytology, sexuality and breeding of *Volvariella volvacea*. *Indian Mushroom Sci* 2: 361–363.

Kapoor, J.N. 1989. *Mushroom Cultivation*. ICAR, New Delhi, p. 15.

Kar, A. and Gupta, D.K. 2001. Osmotic dehydration characteristics of button mushrooms. *J Food Sci Technol* 38(4): 357–257.

Karim, N. 2011. *Desert Truffle Mushrooms in Spate Irrigation Areas*. Practical Notes No. 14. Spate Irrigation.

Katiyar, R.C. 1985. *Evaluation of drying characteristics and storage behaviour of cultivated mushroom (Agaricus bisporus (Lange) Sing.)*. MSc thesis. Dept. of Pomology and Fruit Technology. HPKV College of Agriculture, Solan, India.

Kaviyarasan, V., Kumar, M., Siva, R., and Natarajan, K. 2006. *Morchella esculenta*—a new record from south India. *Mushroom Res* 15(1): 87–88.

Kellner, H. 2009. *Morel Project*. http://haraldkellner.com/html/morel_project.html.

Khan, A.A. and Khan, J. 2011. Market survey of useful plants in the mountain region of Abbottabad District Pakistan. *World Appl Sci J* 14(4): 510–513.

Khanna, P.K. and Garcha, H.S. 1983. Lipid composition of *Pleurotus* spp. (dhingri). *Taiwan Mushrooms* 7(1): 18–23.

Khanna, P.K. and Shivani, S. 2013. Production of mushrooms. In: *Biotechnology in Agriculture and Food Processing: Opportunities and Challenges*. Pamesar, P.S. and Marwaha, S.S. (Eds). CRC Press, Taylor and Francis, Boca, Raton FL. pp. 509–555.

Kidd, P.M. 2000. The use of mushroom glucans and proteoglycans in cancer therapy. *Altern Med Rev* 5: 4–27.

King, T.A. 1993. Mushrooms, the ultimate health food but little research in U.S. to prove it. *Mushroom News* 41: 29–46.

Konanayakam, M., Sastry, S.K., and Anantheswaran, R.C. 1987. A method to determine shrinkage of mushrooms during processing. *J Food Sci Technol* 24(5): 257–258.

Koorapati, A., Foley, D., Pilling, R., and Prakash A. 2004. Electron-beam irradiation preserves the quality of white button mushroom (*Agaricus bisporus*) slices. *J Food Sci* 69(1): 25–29.

Kroes, B.H., Van den Berg, A.J.J., Quarles van, H.C., Ufford, H.D., and Labadie, RP. 1992. Anti-inflamatory of gallic acid. *Planta Med* 58: 499–504.

Kumar, A. 1992. *Studies on storage and dehydration of white button mushroom, Agaricus bisporus*. MSc thesis, Dept. of Postharvest Technology, UHF, Solan, India.

Kumar, S. and Sharma, V.P. 1997. Evaluation of various strains of *Agaricus bisporus* (Lange) Imbach for commercial cultivation under H.P. conditions. *Mushroom Res* 6: 55–68.

Kuo, M. 2005. *Morels*. The University of Michigan Press, Ann Arbor, MI, p. 205.

Lescane, C. 1984. Extension of mushroom (*Agaricus bisporus*) shelf life by gamma radiation. *Post Harvest Biol Technol* 4: 255–260.

Li, G.S.F. 1982. Morphology of *Volvariella volvacea*. In: *Tropical Mushrooms: Biological Nature and Cultivation Methods*. Chang, S.T. and Quimio, T.H. (Eds.). Chinese University Press, Hong Kong.

Li, A.S., Bandy, B., Tsang, S.S., and Davison, A.J. 2000. DNA-breaking versus DNA-protecting activity of four phenolic compounds *in vitro*. *Free Radic Res* 33: 551–566.

Li, G.S.F. and Chang, S.T. 1982. Nutritive value of *Volvariella volvacea*. In: *Tropical Mushrooms-Biological Nature and Cultivation Methods*. Chang, S.T. and Quimio, T.H. (Eds.). The Chinese University Press, Hong Kong, pp. 199–219.

Lidhoo, C.K. and Agrawal, Y.C. 2006. Hot-air over drying characteristics of button mushroom-safe drying temperature. *Mushroom Res* 15(1): 59–62.

Liu, G.Q. and Wang, X.L. 2009. Selection of a culture medium for reducing costs and intracellular polysaccharide production by *Agaricus blazei* AB2003. *Food Technol Biotechnol* 47: 210–214.

Lopez-Briones, G., Varoquaux, P., Yves, C., Bouquant, J., Bureau, G., and Pascat, B. 1992. Storage of common mushroom under controlled atmospheres. *Int J Food Sci Technol* 28: 57–68.

Madhav, P. and Uday, K.W. 2007. *Abstract of I.G.K.V. Theses.* Nehru Library, Indira Gandhi Krishi Vishwavidyalaya, Raipur, Chattisgarh.

Maggioni, A., Passera, C., Renosto, F., and Benetti, E. 1968. Composition of cultivated mushrooms (*Agaricus bisporous*) during the growing cycle as affected by the nitrogen source in compositing. *J Agric Chem* 16: 517–519.

Maini, S.B., Sethi, V., Diwan, B., and Munjal, R.L. 1987. Pre-treating mushrooms to enhance their shelf life and marketability. *Ind Mushroom Sci* 2: 215.

Maki, C.S., Teixeira, F.F., Paiva, E., and Paccola-Meirelles, L.D. 2001. Analysis of genetic variability in *Lentinula edodes* through mycelial responses to different abiotic conditions and RAPD molecular markers. *Braz J Microbiol* 32: 170–175.

Malinowska, E., Szefer, P., and Faradays, J. 2004. Metals bioaccumulation by bay Bolete, *Xerocomos badius* from selected sites. *Pol Food Chem* 84: 405–416.

Manning, K. 1985. Food value and chemical composition. *The Biology and Technology of the Cultivated Mushroom.* Flegg, P.B., Spencer, D.M., and Wood, D.A. (Eds.). John Willey and Sons, New York, NY, pp. 221–230.

Manzi, P.A., Aguzzi, A., and Pizzoferrato, L. 2007. Nutritional value of mushrooms widely consumed in Italy. *Food Chem* 73: 321–325.

Marja, P.K. and Anu, I.H. 1999. Activity of plant extracts containing phenolic compounds. *J Agric Food Chem* 47: 3954–3962.

Mattila, P.K., Konko, M., Eurola, J., Pihlava, J., Astola, L., Vahteristo, V., Hietaniemi, J., Kumpulainen, N., Valtonen, V., and Piironen, V. 2000. Contents of vitamins, mineral elements and some phenolic compounds in the cultivated mushrooms. *J Agric Food Chem* 49: 2343–2348.

Mattila, P.H., Piironen, V.I., Uusi, R., and Koivistoinen, P.E. 1994. Vit. D contents in edible mushrooms. *J Agric Food Chem* 42: 2449–2453.

Mau, J.L., Chao, G.R., and Wu, K.T. 2001. Antioxidant properties of methanolic extracts from several ear mushrooms. *J Agric Food Chem* 49: 5461–5467.

Mehta, K. and Bhandal, M.S. 1989. Mycelial growth variation of six *Pleurotus* spp. at different temperatures. *Ind J Mushroom* 14: 64–65.

Mejstric, V. and Lepsova, A. 1993. Applicability of fungi to the monitoring of environmental pollution by heavy metals. In: *Plants as Biomonitors.* Markert, B. (Ed.). VCH Weinheim, Germany, pp. 365–378.

Miki, K.R., Yasushi, R., Yunmo, L. et al. 2001. Antitumor effect of gallic acid on LL-2 lung cancer cells transplanted in mice. *Anti-Cancer Drugs* 12: 847–852.

Mudahar, G.S. and Bains, G.S. 1982. Pre-treatment effect on quality of dehydrated *Agaricus bisporus* mushrooms. *Indian Food Packer* 28(5): 19–27.

Murr, D.P. and Morris, L.L. 1974. Influence of O_2 and CO_2 on O-diphenol oxidase activity in mushrooms. *J Am Soc Hortic Sci* 99: 272–277.

NRCM Annual Report—2000–2001. Published by National Research Centre for Mushroom, Solan, HP, p. 45.

Nanba, H. 1993. Maitake mushroom the king mushroom. *Mushroom News* 41: 22–25.

Nitha, B. and Janardhan, K.K. 2008. Aqueous-ethanolic extract of morel mushroom mycelium *Morchella esculenta*, protects cisplatin and gentamycin induced nephrotoxicity in mice. *Food Chem Toxicol* 46: 3193–3199.

Nichols, R. 1985. Post harvest physiology and storage. In: *The Biology and Cultivation of Cultivated Mushroom*. Flegg, P.B., Spencer, D.M., and Wood, D.A. (Eds.). John Wiley and Sons Ltd., New York, NY, pp. 195–210.

Nichols, R. and Hammond, J.B.W. 1975. Storage of mushroom in prepacks. The effect of changes of carbon dioxide and oxygen on quality. *J Food Sci Agric* 24: 1371–1381.

Nicholson, M.S., Bunyard, B.A., and Royse, D.J. 2009. Phylogenetic implications of restriction maps of the intergenic regions flanking the 5S ribosomal RNA gene of *Lentinula* species. *Fungi* 2(4): 48–57.

Nigam, P., Tim, R., and Dalel, S. 2003. Solid state fermentation-An overview. In: *Handbook of Fungal Biotechnology*. Dilip, K.A. (Ed.). CRC Press, Boca Raton, FL.

Oei, P. 1996. *Mushroom Cultivation (With Emphasis on Techniques for Developing Countries)*. Tool Publications, Leiden, the Netherlands, pp. 93–204.

Okwu, D.E. 2004. Phytochemical and vitamin content of indigenous species of southern Nigeria. *J Sustain Agric Environ* 6: 30–37.

Olajide. O.B., Fadimu, O.Y., Osaguona, P.O., and Saliman, M.I. 2013. Ethnobotanical and phytochemical studies of some selected species of leguminoseae of northern Nigeria: A study of Borgu local government area, Niger state, Nigeria. *Int J Sci Nat* 4(3): 546–551.

Pardo Arturo, J., de Arturo, J., and Emilio Pardo, J. 2002. Production, characterization and evaluation of composted vine shoots as a casing soil additive for mushroom cultivation. *Biol Agricul Hortic: Int J Sustain Prod Syst* 19(4): 377–391.

Pandey, A. and Ramachandran, S. 2005. Process development in solid state fermentation for food applications. In: *Food Biotechnology*. 2nd edn. Shetty, K., Paliyath, G., Pometto, A., and Robert, E.L. (Eds.) CRC Press, Boca Raton, FL.

Pandey, M. and Tiwari, R.P. 1989. A natural sporeless variant of *Pleurotus*. *Ind J Mushroom* 15: 25–28.

Pandey, M. and Tiwari, R.P. 1994. Strategies for selection and breeding of edible mushroom. *Adv Mushroom Biotechnology*. Nair, M.C., Gokulapaln, C. and Das, L. (Eds.). Scientific Publishers, Jodhpur, pp. 61–68.

Panizzi, L., Caponi, C., Catalona, S., Cioni, P.L. and Morelli, I. 2002. *In vitro* antimicrobial activity of extracts and isolated constituents of *Rubus ulmifolius*. *J Ethanopharmacol* 79: 165–168.

Parslew, R., Jones, K.T., Rhodes, J.M., and Sharpe, G.R. 1999. The antiproliferative effect of lectin from the edible mushroom (*Agaricus bisporus*) on human keratinocytes: Preliminary studies on its use in psoriasis. *Br J Dermatol* 140: 56–60.

Patil, S.S., Ahmed, S.A., Telang, S.M., and Baig, M.M.V. 2010. The nutritional value of *Pleurotus ostreatus* (Jacq.:FR) Kumm cultivated on different lignocellulosic agrowastes. *Innovat Roman Food Biotechnol* 7: 66–76.

Pegler, D. 1983. The genus *Lentinula* (Tricholomataceae tribe Collybieae). *Sydowia* 36: 227–239.

Perera, P.K. and Li, Y. 2011. Mushrooms as a functional food mediator in preventing and ameliorating diabetes. *Funct Foods Health Dis* 1(4): 161–171.

Poppe, J. 1974. *Collybia velutipes* as tree wound parasite and as cultivated mushroom. *Mededelingen Fakulteit Landbouwwetenschappen-Gent* 39: 957–970.

Prakash, J.D. and Tejaswin, M.S. 1991. Mushroom for many uses. *Food Digest* 22(1): 84–91.

Prasad, P., Chauhan, K., Kandari, L.S., Maikhuri, R.K., Purohit, A., Bhatt, R.P., and Rao, K.S. 2002. *Morchella esculenta*: Need for scientific intervention for its cultivation in Central Himalaya. *Curr Sci* 82: 1098–1100.

Pruthi, J.S., Manan, J.K., Raina, B.L., and Teotia, M.S. 1984. Improvement in whiteness and extension of shelf-life of fresh and processed mushrooms (*Agaricus bisporus* and *Volvariella volvacea*). *Indian Food Packer* 38(2): 55–63.

Purkayastha, R.P. 1976. A new method of cultivation of *Calocybe indica*—an edible summer mushroom. *Taiwan Mushrooms* 3: 14–18.

Purkayastha, R.P. and Chandra, A. 1974. New species of edible mushroom from India. *T Br Mycol Soc* 62: 415–418.

Purkayastha, R.P. and Chandra, A. 1976. Amino acid composition of protein of some edible mushroom growth in synthetic medium. *J Food Sci Technol* 3: 13–17.

Rai, R.D. 2003. Production of edible fungi. In: *Fungal Biotechnology in Agricultural, Food and Environmental Applications*. Dilip, K.A., Paul, D.B., and Deepak, B., (Eds.). CRC Press, Boca Raton, FL.

Rai, R.D. and Arumuganathan, T. 2005. Nutritive value of mushrooms. In: *Frontiers in Mushroom Biotechnology*. National Research Centre for Mushroom, Solan, pp. 27–36.

Rai, R.D. and Arumuganathan, T. 2008a. *Postharvest Technology of Mushrooms*. Technical Bulletin of National Research Centre for Mushroom (ICAR), Chambaghat, Solan, India, p. 84.

Rai, R.D. and Arumuganathan, T. 2008b. *Post Harvest Management of Mushrooms*. Technical Bulletinof National Research Centre for Mushroom, pp. 1–92.

Rai, R.D. and Saxena, S. 1989. Suitability of methods of estimation for critical assessment of the vitamin C content in mushrooms. *Mushroom J Trop* 9: 43–46.

Rai, R.D., Saxena, S., Upadhyay, R.C., and Sohi, H.S. 1988. Comparative nutritional value of various *Pleurotus* species grown under identical conditions. *Mushroom J Trop* 8: 93–98.

Romanens, P. 2001. Shiitake, the European reality and cultivation on wood-chips logs in Switzerland. In: *15th North American Mushroom Conference*, Las Vegas.

Rotaoll, N., Dunkel, A., and Hofmann, T. 2005. Activity guided identification of (S)-malic acid 1-O-ᴅ-glucopyranoside (morelid) and gamma-aminobutyric acid as contributors to umami taste and mouth-drying oral sensation of morel mushrooms (*Morchella deliciosa* Fr.). *J Agric Food Chem* 53: 4149–4156.

Roy, M.N. and Bahl, N. 1984a. Gamma radiation for preservation of *Agaricus bisporus*. *Mushroom J* 136: 124–125.

Roy, M.N. and Bahl, N. 1984b. Studies on gamma radiation for preservation of *Agaricus bisporus*. *Mushroom J* 144: 411–414.

Sabra, A. and Walter, S. 2001. *Non-Wood Forest Products in the Near East: A Regional and National Overview*. Working Paper FOPW/01/2. FAO, Rome, p. 120.

Saxena, S. and Rai, R.D. 1988. Storage of button mushrooms (*Agaricus bisporus*). The effect of temperature, perforation of packs and pre-treatment with potassium metabisulphite. *Mushroom J Trop* 8: 15–22.

Saxena, S. and Rai, R.D. 1990. *Postharvest Technology of Mushrooms*. Technical Bulletin No. 2: NRCM, Solan.

Schmitt, H.W. and Sticher, H. 1991. Heavy metal compounds in soil. In: *Metals and Their Compounds in the Environment*. Merian, E. (Ed.). VCH Verlagsgessellschaft, Weinheim. pp. 311–326.

Seth, P.K. 1973. Evaluation and selection of *Agaricus bisporus* (Lange) Sing. Strains for commercial cultivation of mushroom. *Ind J Mushroom* 1(1): 5–10.

Sethi, V. and Anand, J.C. 1976. Processing of mushrooms. *Indian Mushroom Sci* 1: 233–238.

Sethi, V. and Anand, J.C. 1984–85. Postharvest care of mushrooms. *Indian J Mushrooms* 10–11: 1–6.

Shah, H., Iqtidar, A.K., and Jabeen, S. 1997. Nutritonal composition and protein quality of Pleurotus mushroom. *Sarhad J Agric* 13: 621–626.

Shandilya, T.R. 1976. Prepare mushroom compost on wheat straw plus chicken manure. *Indian J Mushroom* 2: 43–48.

Shandilya, T.R. 1989. Paddy straw compost formulations for growing button mushroom and its comparison with traditionally made compost based on wheat straw and chicken manure. *Mushroom Sci* 12: 333–344.

Shandilya, T.R., Munjal, R.L., and Agarwal, R.K. 1987. Use of different compost quantities in relation to yield of *Agaricus bisporus*. *Indian Mushroom Sci* 2: 24–34.

Shandilya, T.R., Seth, P.K., Kumar, S., and Munjal, R.L. 1974. Effect of different spawning methods on the productivity of *Agaricus bisporus*. *Indian J Mycol Plant Pathol* 4: 129–131.

Sharma, S. and Vaidya, D. 2011. White button mushrooms (*Agaricus bisporus*) Composition, nutritive value, shelf life extension and value addition. *International Journal of Food Fermentation Technology*, 1(2): 185–199.

Sharma, M. 2012. *Development of hybrid(s) of Agaricus bisporus(Lange)Imbach and their evaluation for higher yield*. PhD thesis. Dr. Y.S. Parmar University of Horticulture and Forestry, Nauni, p. 181.

Shi, Y.L., James, A.E., Benzie, I.F.F., and Buswell, J.A. 2002. Mushroom derivedpreparation in the prevention of H_2O_2-induced oxidative damage to cellular DNA. *Teratoeg Carcinog Mutag* 22: 103–111.

Shimamura, T., Zhao, W.H., and Hu, Z.Q. 2007. Mechanism of action and potential for use of tea catechin in an antiinfective agent. *Anti-infect Agents Mech Chem* 6: 57–62.

Shukla, S. and Jaitly, A.K. 2011. Morphological and biochemical characterization of different oyster mushrooms (*Pleurotus* spp.). *J Phytol* 3(8): 18–20.

Shukla, A.N. 1994. Cultivation of Japanese mushroom shiitake in India. *Indian For* 120: 714–719.

Sinden, J.W. and Hauser, E. 1950. The short method of composting. *Mushroom Sci* 1: 52–59.

Sinden, J.W. and Hauser, E. 1953. The nature of the composting process and its relation to short composting. *Mushroom Sci* 2: 123–131.

Singh, M. Vijay, B., Shwet Kamal, S., and Wakchaure, G.C. 2011. *Mushrooms: Cultivation, Marketing and Consumption*. Directorate of Mushroom Research, Indian Council of Agricultural Research, Chambaghat, Solan, p. 246.

Singh, M. and Kamal, S. 2011. Conventional and molecular approaches for breeding button mushroom. In: *Proceedings of the 7th International Conference on: Mushroom Biology and Mushroom Products*. Arrcachon, France, 4–7 October 2011, Vol. I, pp. 35–42.

Singh, N.B. and Singh, P. 2002. Biochemical composition of *Agaricus bisporus*. *J Indian Bot Soc* 81: 235–237.

Singh, R. and Singh, S. 2006. Marketing pattern of mushroom in Haryana. *Mushroom Res* 15(1): 63–68.

Singh, R.P., Dang, R.L., Bhatia, A.K., and Gupta, A.K. 1982. Water binding additives and canned mushroom yield. *Indian Food Packer* 36(1): 32–36.

Singh, S.K., Maharaj, N., and Kumbhar, B.K. 2001. Effect of drying air temperatures and standard pretreatments on the quality of fluidized bed dried button mushroom (*Agaricus bisporus*). *Indian Food Packer* 55(5): 82–86.

Sodhi, H.S. and Kapoor, S. 2007. Recent advances in the cultivation technology of paddy straw mushroom. In: *Mushroom Biology and Technology*. Rai, R.D., Singh, S.K., Yadav, M.C. and Tiwari, R.D. (eds.). Mushroom Society of India, pp. 193–207.

Staden, O.L. 1967. Radiation preservation of fresh mushroom. *Mushroom Sci* 6: 457–461.

Stamets, P. 2005. *Mycelium Running: How Mushrooms Can Help Save the World*. Ten Speed Press, Berkeley, CA.

Steinbuch, E. 1978. Factors affecting quality and shrinkage losses of processed mushrooms. *Mushroom Sci* 10(2): 759–766.

Suman, B.C. 2001. Evaluation of single spore isolates of button mushroom *Agaricus bisporus* on long method of compost. *Mushroom Res* 10(1): 9–12.

Sun, D.W. 1999a. *Comparison of Rapid Vacuum Cooling of Leafy and Non-Leafy Vegetables*. ASAE Paper No. 996117, ASAE, St. Joseph, MI.

Sun, D.W. 1999b. *Effect of Pre-wetting on Weight Loss and Cooling Times of Vegetables During Vacuum Cooling*. ASAE Paper No. 996119, ASAE, St. Joseph, MI.

Svoboda, L., Zimmermannova, K., and Kallac, P. 2001. Concentrations of mercury, cadmium, lead, and copper in the fruiting bodies of the edible mushrooms in an emission area of a copper smelter and a mercury smelter. *Sci Total Environ* 246: 61–67.

Thakur, K. and Sharma, S.R. 1992. Substrate and supplementation for the cultivation of shiitake, *Lentinus edodes* (Berk) Sing. *Mushroom Inf* 9: 7–10.

Thompson, A.K. 2007. *Storage of Fruit and Vegetables: Harvesting, Handling and Storage.* Blackwell Publishing Ltd., Oxford, UK.

Tseng, Y.H. and Mau, J.L. 1999. Contents of sugars free amino acids and free 5-nucleotides in mushroom, *Agaricus bisporus*, during the post harvest storage. *J Sci Food Agric* 79: 1519–1523.

Tuley, L. 1996. Swell time for dehydrated vegetables. *Intern Food Ingredients* 4(1): 23–27.

Tunney, J. 1980. *Guidelines and Procedure for Making Bulk Pasteurized Compost.* Report Submitted to Compost Mother Unit of HPMC. Chambaghat, Solan, HP, India.

Upadhyay, R.C. 2011. Oyster mushroom cultivation. In: *Mushrooms: Cultivation, Marketing and Consumption.* Singh, M., Vijay, B., Kamal, S., and Warchaure, G.C. (Eds.) DMR, Chambaghat, Solan, pp. 1–10.

Upadhyay R. C. and Singh, M. 2011. Production of edible mushrooms. *The Mycota*, 10: 79–97.

Varo, P., Lahelman, O., Nuurtamo, M., Saari, E., and Koivistoinen, P. 1980. Mineral element composition of Finish Food. VII Postal, Vegetables, fruits, berries, nuts and mushrooms. *Acta Agric Scand Supp* 22: 107–113.

Vijay, B. and Gupta, Y. 1995. Production technology of *Agaricus bisporus*. In: *Advances in Horticulture.* Chadha, K.L. and Sharma, S.R. (Eds.). Malhotra Publishing House, New Delhi, pp. 63–98.

Wahid, M. and Kovacs, E. 1980. Shelf life extension of mushrooms (*Agaricus bisporus*) by gamma radiation. *Acta Aliment* 9: 357–359.

Wang, L.J. and Sun, D.W. 2001. Rapid cooling of porous and moisture foods by using vacuum cooling technology. *Trends Food Sci Technol* 12(5–6): 174–184.

Wannet, W.J.B., Hermans, J.H.M., Vander Drift, C., and Op den Camp, H.J.M. 2000. HPCL detection of soluble carbohydrates involved in mannitol and trehalose metabolism in the edible mushroom, *Agaricus bisporus. J Agric Food Chem* 48: 287–291.

Wasser, S.P. 2002. Medicinal mushrooms as a source of antitumor and immunomodulating polysaccharides. *Appl Microbiol Biotechnol* (60): 258–274.

Wasser, S.P. and Weis, A.L. 1999. Therapeutic effects of substances occurring in higher Basidiomycetes mushrooms: A modern perspective. *Crit Rev Immunol* 19: 65–96.

Weaver, K.C., Kroger, M., and Kneebone, L.R. 1977. Comparative protein studies on nine strains of *Agaricus bisporus* (Lange). *J Food Sci* 42: 364–366.

Weber, N.S. 1997. *A Morel Hunter's Companion: a Guide to True and False Morels.* Thunder Bay Press, Lansing, MI.

Wondratschek, I. and Roder, U. 1993. Monitoring of heavy metals in soils by higher fungi. In: *Plants as Biomonitors.* Market, B. (Ed) VCH, Weinheim, pp. 365–378.

Wuest, P.J. and Fahy, H.K. 1991. Spent mushroom compost: Traints and uses. *Mushroom News,* 39(12): 9–15.

Yaltirak, T., Belma, A., Sahlan, O., and Hakan, A. 2009. Antimicrobial and antioxidant properties of *Russula delica* Fr. *Food Chem Toxicol* 47: 2052–2056.

Yang, D.C. and Maguer, M. Le. 1992. Mars transfer kinetics of osmotic dehydration of mushrooms. *J. Food Process and Preservation* 16(3): 215–231.

Yapar, S., Helvaci, S.S., and Peker, S. 1990. Drying behaviour of mushroom slices. *Dry Technol* 8(1): 77–99.

Yoshioka, Y., Ikekawa, T., Nida, M., and Fukuoka, F. 1975. Studies on antitumor activity of some fractions from *basidiomycetes* I. An antitumor acidic polysaccharide fraction of *Pleurotus ostreatus* (Fr.) Quel. Isolation and structure of a B-glucan. *Carbohydr Res* 140: 923–100.

Zadrazil, F. 1999. *Flammulina velutipes*. A "dangerous" parasite and a good, cultivatable edible fungus. *Champignon* 407: 34–35.

Zardrazil, F., Ostermann, D., and Compare, G.D. 1992. In: *Solid Substrate Cultvaton*, Doelle, H.W., Mitchell, D.A., and Rolz, C.E. (Eds.). Elsevier Applied Science, London, pp. 283–319.

Zheng, L. and Sun, D.-W. 2005. Vacuum cooling of foods. *Emerging Technologies of Food Processing*. Academic Press, pp. 579–602.

Zrodlowski, Z. 1995. The influence of washing and peeling of mushrooms *Agaricus bisporus* on the level of heavy metal contaminations. *Pol J Food Nutr Sci* 4: 23–33.

13

BIOTECHNOLOGY AND TRADITIONAL FERMENTED FOODS

A.K. NATH, ANUPAMA GUPTA, BHANU NEOPANY, GITANJALI VYAS, JARUWAN MANEESRI, NISHA THAKUR, NIVEDITA SHARMA, OME KALU ACHI, POOJA LAKHANPAL, AND ULRICH SCHILLINGER

Contents

13.1 Introduction

13.1.1 Art of Food Fermentation

Food is the basic survival necessity for all human beings, and the traditional methods of food processing were aimed at food preservation and economy of fuel (Sarkar and Nout 2014; Steinkraus, 1996), so food fermentation methods arose historically from the need for processing and preservation food (Law et al., 2011; Ravyts et al., 2012). More than anything else, man has been employing microbes for the preparation of fermented food products for thousands of years. All over the world, a wide range of fermented foods and beverages are produced, which contribute significantly to the diets of many people (Achi, 2005a,b; Campbell-Platt, 1994; Steinkraus, 1994; 2004). Fermentation and drying are the oldest methods of food preservation and processing. It gives food a variety of flavors, tastes, textures, sensory attributes, and nutritional and therapeutic values (Metha and Kamal-Eldin, 2012). The availability of storable and hygienically safe food was a decisive prerequisite for the development of mankind and society (Prajapati and Nair, 2003). The skills of food fermentation are embedded in traditional knowledge systems among the native peoples of many areas of the world, and the knowledge is maintained and propagated orally. The art of fermentation practiced by the common man has continued, in spite of the scientific and technological revolution, but has largely remained confined to the rural and tribal areas due to (i) high cost or inaccessibility of the industry-made products in remote areas, (ii) the tastes of the people for the traditional fermented products, and (iii) their socio-cultural linkages with such products (Chelule et al., 2010; Thakur et al., 2004). However, with the advent of microbiology, biochemistry, molecular biology, and biochemical engineering, the art of fermentation practiced by the common man has been improved and upgraded, which has led to the rise of fermentation industries, adding quality and expanding the range of products (Thakur et al., 2004).

13.1.2 Significance of Traditional Fermented Foods and Healthy Life Style

A wide range of literature testifies to the multifaceted existing importance of traditional fermented foods and beverages. Whether of plant or animal origin, they remained important components of diets in many parts of the world, including the South Asian countries (Kabak and Dobson, 2011). Almost in contrast, some local indigenous fermented foods lack appeal and have very restricted popularity, while modern consumers prefer imported and exotic food items due to their attractive form, long shelf-life, ease of transportation, and other forms of utility associated with such foods (Achi, 2005b; Marshall and Danilo, 2012). Thus, the enhanced awareness of consumers of health, and the consequent interest in functional foods to achieve a healthy lifestyle, has established a need for food products with versatile health-benefiting properties, which is being felt all-over the world. Research on spontaneously fermented indigenous foods, their probiotic potential, and predominant microorganisms, has led to the exploration as well as exploitation of this traditional technology, with the major aim of producing consistent, safe products with longer shelf-life, which, in addition to the normal functionality of foods, also impart health benefits to the consumers. The indigenous fermented foods produced in South Asia have major functions like bio-preservation of perishables, bioenrichment of nutritional components, protective properties, increased bioavailability of minerals, production of antioxidants and omega-3-polyunsaturated fatty acids, therapeutic values, and immunological effects (Tamang, 2012). However, little exploration and exploitation of technologies used in their production has been made.

13.1.3 Traditional Technologies

Technologies are called traditional if they remain unaffected by modernization, have been commonly applied over a long period of time by the native inhabitants of a region, and constitute an important part of their cultural inheritance (Dirar, 1993). Many traditional fermented foods usually meet this criterion (Hicks, 2002). At the same time, it is essential that traditional knowledge should go together with developments in science and technology to find mutually beneficial outcomes, and many biotechnological innovations have greatly assisted in upgrading certain indigenous fermented foods to a commercial level. Several traditional fermentations, for example, Indonesian *tempe* and *soy* sauce, have been upgraded to high technological production systems and have been expanded on a global scale (becoming household products around the world and a multi-million dollar industry) due to a strong research tradition in fermented food technology (Achi, 2005a,b; Hicks, 1983). In a bid to enhance food availability and quality, and alleviate malnutrition in many developing countries, research in indigenous fermented foods has seen a marked improvement during the last few decades (Dirar, 1993; Parkouda et al., 2009; Steinkraus, 2004). For more information, excellent reviews on these aspects are available (Gadaga et al., 2013; Holzapfel, 2002; Olasupo et al., 2010; Tamang, 2010).

13.1.4 Improvement of Traditional Fermented Foods

Traditional fermented foods can be improved in a number of ways, as has also been outlined earlier (Nout, 1985). Developments made in genetics, enzymology, recombinant technology, and fermentation technology have led to our understanding of several processes and methods that can be employed to improve traditional food fermentation technology. Research priorities should include four broad categories, such as improvement in the understanding of fermentation process, refinement of the process, increasing the utility of the process, and, finally, developing local capabilities (Anonymous, 2012). Since the microorganisms are the most important component of food fermentation, understanding their physiology, genetic capability, and the conditions for their optimum growth are of great significance. The genetic engineering of microorganisms for maximum output and the development of new products and processes is an essential priority in the improvement of traditional fermented foods (Harlander, 1992).

13.1.5 Application of Biotechnology in Food Production and Processing

Biotechnology has a long history of application in food production and processing (Joshi and Pandey, 1999). It represents both traditional processing techniques and the latest techniques based on molecular biology. These techniques open up a large number of possibilities for rapidly improving the quantity and quality of fermented foods, especially the traditional fermentation processes, and can assist in upgrading the technology and changing the traditional processes into controllable, predictable, and efficient ones (Moreno et al., 2010). Not only this, it is essential in designing a biotechnology process to know the nature of the biological system to be used, the substrate to be transformed, and the operating conditions to be applied, so as to maximize the metabolic activity of the microorganisms, increase the process yield, and ensure the quality of the final product (Achi, 2005a). Significant biotechnological innovations have been made in research to understand various facets of the production of indigenous fermented foods, the microbiology and biochemistry of the fermentation, and the protocols of production, to enhance their nutritional and overall food value (Achi, 1990, 1992, 2005a,b; Eka, 1980; Moreno et al., 2010; Odunfa, 1985b; Okafor, 1983). At the same time, the safety and health of consumers must be protected, and efforts to control the levels of salt and toxins (including mycotoxins), and the occurrence of parasites and pathogenic microorganism need to be addressed (Nout et al., 2014). In this chapter, a focus is made on some economically important and well-known traditional fermented food products based on biotechnological-driven fermentations (Ravyts et al., 2012), some of which present interesting models for safe and functional traditional foods that can be upgraded. To this end, current knowledge about the production of fermented foods and applications of biotechnological knowledge, along with constraints and related issues, have been highlighted.

13.2 Industrial Development of Processes for Fermented Foods

The art of indigenous fermented food processing needs to be transformed into a very precise technology, and the steps of processing must also have provision for quality control and optimization, without losing their originality (Anonymous, 2012). Before any product is advocated for production at industrial scale, several aspects are considered. Besides the economics of the process and modeling of the product, major emphasis is laid on the scientific angles of the process. The industrial development of indigenous fermented foods and beverages can be divided into four areas: raw material development, starter development, process development, and finished product development. Research and development in these areas can certainly improve the product in terms of better quality and efficiency of the process.

13.2.1 Raw Material Development

Raw materials can be tested to find out the most suitable variety of food item and its availability for use as a substrate. For example, it is now generally accepted that sorghum cultivars with low tannin content give a better quality European-type beer (Achi, 2005a,b). Further, the raw material should be free from antinutritional factors like trypsin inhibitors, saponins, etc., when cereals, legumes, fruits, milk, etc., are the starting materials for the fermented foods. In brief, the availability of a proper quality and quantity of raw material for industrial production of any product is a prerequisite.

13.2.2 Starter Development

All the indigenous fermented foods were originally made by fermentation by natural microorganisms, and the knowledge has been transferred from generation-to-generation. Isolation, selection, preservation, or collection, and starter-making of a highly-efficient microbial strain for use as inoculum has to be made before industrialized large-scale production of any product can take place (Daengsubha and Suwana-adth, 1985; Okafor, 1983, 1990). In general, the microorganisms selected should not be able to synthesize toxins or toxic metabolites, be thermotolerant and osmo-tolerant (Anonymous, 2012). Lactic acid bacteria (LAB) and yeasts are probably the most important groups of organisms in making fermented foods (see Chapter 3). For example, pure wine yeast in grape fermentation or "tempeh rhizopus" in tempeh will make the fermentation process more precise, controllable, and promising. The development of strains with better and more stable genetic properties is a major task before the microbiologists, as they may offer nutritional benefits and compatibility to multi-strain fermentations carried out at present under non-sterile conditions (Achi, 2005a,b; Nout, 1985).

13.2.3 Process Development

This aspect has been illustrated by citing an example of how a traditional fermented food has been modified for industrial production. It is the work on the South African sorghum beer, which is an alcoholic, effervescent, pinkish-brown beverage with a sour flavor, an opaque appearance and the consistency of a thin gruel. Its preparation follows a pattern similar to that of *burukutu* or *pito* fermentation. The main steps in the brewing are mashing, souring, boiling, conversion, straining, and alcoholic fermentation (Achi, 2005a,b; Hesseltine and Wang, 1979). In the traditional process, the malt is made by soaking sorghum grains in water for 8–24 h, draining, and then, allowing the grains to sprout for 5–7 days. The malt is sun-dried and ground into a fine powder. The ground malt is made into thin slurry and boiled. A small amount of uncooked malt is added and left for one day, during which it undergoes natural lactic fermentation, where *Lactobacillus* is chiefly responsible. The mash is boiled and left for alcoholic fermentation to take place. More ground malt is added and after the fermentation on the 5th day, it is strained and is then, ready to drink.

The factory process is less complicated, though it still incorporates the lactic acid fermentation with selected LAB (lactic acid bacteria) and alcoholic fermentation steps with the top fermenting yeast *Saccharomyces cerevisiae* (Achi, 2005a,b), and leads to better fermentation and better quality of beer. So, in selecting a modification and or improvement to the process, adequate attention must be paid to its usefulness.

13.2.4 Equipment

This is the most vital component of the technology package for industrialization of the process, if the home scale methods, as employed in traditional fermented foods, are to be scaled up. To make such equipment however, is a challenge, but can be achieved, as a number of advances in the engineering sciences have been made. At the same time, the fact that traditional fermentations are carried out using vessels with unusual characteristics, such as barrels of wood, semi-porous clay (Anonymous, 2012), etc., means that the equipment should be designed accordingly for the maximum benefit of quantity and quality.

13.2.5 Finishing and Packaging of Product

After the product is finished, it needs to be bottled and packaged for proper marketing. Packaging is an essential process in the industrial production of fermented products such as South African sorghum beer. In South Africa, it is currently packaged in milk cartons, which are filled and sealed in just the same way as milk (Achi, 2005a,b; Hesseltine and Wang, 1979), and sold. Because the product is consumed in an active state of fermentation, each carton is left with an opening, large enough to allow CO_2 to escape, but small enough so that the corn fragments will seal the hole if the carton is turned on its side (Achi, 2005a,b; Hesseltine, 1983a,b). Similar

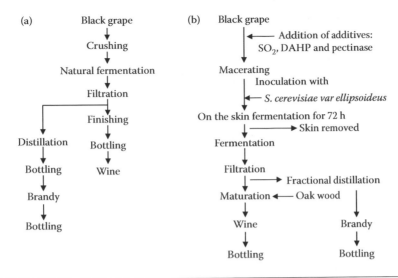

Figure 13.1 Comparison of traditional and industrial methods of wine and brandy production: (a) traditional method, (b) industrial method.

improvements in the process for wine and brandy making from grapes is illustrated by traditional technology and the modified industrial process, as shown in Figure 13.1 (Jackson, 1999; Joshi, 1997). Improvement in the production of wine and indigenous brandy from wild apricots in the Kinnaur district of Himachal Pradesh (India) has been made (Joshi et al., 1990). The improvements introduced include addition of sulphur dioxide, diammonium hydrogen phosphate, pectinase, a specific pure culture of yeast for wine making, and fractional distillation to make the brandy. More examples of improvement will be cited in the subsequent sections of this text.

13.3 Molecular Characterization of Microbial Diversity of Fermented Foods

To improve the quality of fermented foods, microbial culture screening is an important activity in fermented food research. Better species with a higher enzyme activity are still desired. Advanced molecular techniques have been explored to screen and obtain a microorganism that possesses multifunctional biotechnological characteristics—like enzyme activities, bile tolerance, and antimicrobial activity—and to improve the aroma properties of a product.

13.3.1 Application of Molecular Methods in Microbial Diversity Studies

13.3.1.1 Cultivation-Independent Methods Traditionally, the microbial composition and ecology of fermented foods are explored by using microbiological methods based on plate culturing and subsequent biochemical identification of isolated strains (Rantsiou and Cocolin, 2008). But all microorganisms are not amenable to cultivation in the laboratory, and the artificial media support the growth of only a small fraction of the microorganisms present in an ecosystem (Carraro et al., 2011). Stressed, weakened, and sub-lethally injured cells often need specific culture conditions to recover and produce colonies. Species occurring in low numbers may be efficiently out-competed by numerically more abundant and faster growing microorganisms (Hugenholtz et al., 1998). Consequently, the isolated organisms may not be representative of the community and culture-based methods, and therefore cannot accurately capture the *in situ* diversity of complex food ecosystems. Thus, for biodiversity studies of complex microbial communities, it has become mandatory to use culture-independent techniques based on the direct analysis of DNA (or RNA) from food.

Molecular techniques are the major tools for the analysis of microorganisms from food and other biological substances. The techniques provide ways to screen for a broad range of agents in a single test (Field and Wills, 1998). Its use in studies has allowed for both a more definitive analysis of the structure of the microbial community and the determination of the metabolic status of different components of microbial populations (Egert et al., 2005). Molecular techniques not relying on culturing are cost-effective valuable tools to compare microbial diversities and to monitor the dynamic changes in composition of a microbial community, leading to better understanding and improved management of the microbial processes involved in food fermentation (Van Hoorde et al., 2008). An improvement of the microbial safety due to the fast detection of potential pathogens in the food product can also be a direct outcome of this application (Justé et al., 2008). Most of these molecular methods include amplification by PCR and separation of the DNA fragments by gel or capillary electrophoresis or hybridization to specific probes (Sensabaugh, 2009). At present, the molecular marker predominantly used as a target is the bacterial gene coding for 16S ribosomal RNA. Finger printing assays allow the simultaneous analysis of multiple samples without prior knowledge of their diversity.

During the last decades various DNA fingerprinting techniques, such as denaturing gradient gel electrophoresis (DGGE), ribotyping, amplified ribosomal DNA restriction analysis (ARDRA), and repetitive element sequence-based PCR (rep PCR) have become available, and in the meantime polyphasic approaches combining classical biochemical analyses and molecular techniques, such as 16S rRNA sequencing, are frequently used for identification of bacterial isolates from food fermentations.

The molecular methods have proved extremely useful to establish biodiversity profiles of food-associated microbial communities. Our knowledge of the biodiversity of microorganisms involved in food fermentations is being extended considerably by the use of advanced tools like DNA microarray technology and quantitative PCR, next-generation sequencing and bioinformatics analyses that have become available during recent years, especially the introduction of second generation sequencing techniques, allowing parallelizing of the sequence process towards thousands or millions of sequences at once, which has revolutionized gene sequencing methodology.

Pyrosequencing provides the benefit of reduced labor time, lower reaction volumes, and extended number of sequence reads, as well as high through-put sampling (Quigley et al., 2011).

Current techniques including genetic finger printing, gene sequencing, oligonucleotide probes, and specific primer selection, discriminate closely related bacteria with varying degree of success. In addition to 16S rRNA primers and RAPD-derived PCR primers, there is a growing interest in exploiting intergenic sequences (ITS; notably the 16S–23S rRNA spacer region) as well as functional genes, such as heat-shock protein (hsp) genes, the *recA* gene, and the *Idh* gene (McCartny, 2002). A comparison of the advantages and disadvantages of PCR methods is given in Table 13.1.

Table 13.1   Advantages and Distavantages of PCR Based Molecular Methods

ADVANTAGES	DISADVANTAGES	SOURCE
• Provide ways to screen for a broad range of agents in a single test.	• Selective extraction of nucleic acids as a result of differences in the levels of cell lysis efficiency and selective amplification of target genes. • Co-migration of different fragments may occur as different sequences may have identical electrophoretic mobility.	Sekiguchi et al. (2001)
• Allows both a more definitive analysis of the structure of the microbial community and the determination of the metabolic status of different components of microbial populations • Is a valuable tool to compare microbial diversities and to monitor the dynamic changes in composition of a microbial community, leading to better understanding and improved management of the microbial processes involved in food fermentation.	• Different species may yield PCR-products which co-migrate in the DGGE/TGGE gel. • Two species *Leuconostoc mesenteroides* and *Weissella paramesenteroides* had the same melting position, and thus can not have this application. • The microheterogeneity in rRNA encoding genes present in some species may result in multiple bands for a single species and subsequently, to an overestimation of community diversity.	Ampe et al. (1999) Anonymous (2007)
• Provides microbial safety due to the fast detection of potential pathogens in the food product.	• The low sensitivity due to traditional gel staining, resulting in the loss of bands, representing less abundant community members and gels of complex communities may look smeared due to the large number of bands that could hamper the interpretation of the fingerprint. • An incomplete extension of the GC clamp during PCR amplification may result in artifactual double bands in DGGE analysis that may complicate the interpretation of the profiles.	Justé et al. (2008) Anonymous (2007) Rantsiou and Cocolin (2006, 2008)

Wherever possible, throughout the text, specific examples have been cited to illustrate the specific method. Two methods of identification which can be used are discussed here: (i) PCR-dependent methods, and (ii) PCR-independent methods.

13.3.1.1.1 PCR Dependent Methods The polymerase chain reaction (PCR) technique has gained acceptance as a powerful microbial tool. Protocols for bacterial typing using PCR techniques are becoming increasingly valuable (De Urraza et al., 2000). This can be discussed by taking an example of LAB, which have traditionally been classified on the basis of phenotypic properties, which involves physiological parameters and sugar fermentation patterns (Gevers et al., 2001; Vandamme et al., 1996). However, these tests are difficult to interpret and the techniques are also time-consuming and laborious (Mohammed et al., 2009). Moreover the results of molecular-based approaches support many of the findings derived from culture-based methods of the presence of high proportions of lactic acid bacteria (LAB) (Cabral, 2010). Among these methods, Random amplified polymorphic DNA (RAPD-PCR) have proven to be very effective to enumerate species and strains differentiation in food fermentations (Mohammed et al., 2009; Quiberoni et al., 1998). On the other hand, identification of LAB in traditional fermented foods by the use of molecular tools also offers the possibility of enhancing identification of bacterial composition in complex food fermentations. At the same time, molecular methods can complement biochemical species identification of isolated colonies (Aymerich et al., 2003; Sklarz et al., 2009). According to Arihara et al. (1993), based on the use of molecular tools strategies it may be possible to identify and then, combine the most desirable properties to construct LAB strains for probiotic application and starter culture development (Quiberoni et al., 1998). Hence, the development of PCR-based methods has opened new possibilities for rapid and specific identification of LAB.

13.3.1.1.1.1 Multiplex PCR Multiplex PCR (MPCR) methodology is based on the combination of several number of primer sets with different specificities in a single PCR reaction. It is the fastest culture-independent approach for simultaneous strain-specific detection of multiple microorganisms in complex matrices (Settanni and Corsetti, 2007). A pentaplex PCR assay was developed which allowed the simultaneous detection of 5 LAB species by direct DNA extraction from whey cultures for Grana Padano cheese (Cremonesi et al., 2011; Fornasari et al., 2006). Multiplex PCR approaches were also successfully applied to monitor the role of *S. cerevisiae* in wine fermentations (Hurtado et al., 2010; López et al., 2003) and, in combination with PCR-DGGE, to differentiate *Lactobacillus* species from sourdough (Settanni et al., 2006). Additionally, multiplex RAPD has been employed successfully for the differentiation of LAB isolated from the GI tract and the identification of probiotic strains (McCartney, 2002). As with other DNA based methods, MPCR does not provide any information on the viability of the present organisms. But MPCR assays targeting the mRNA (multiplex RT-PCR) could be an instrument to detect viable organisms.

Such techniques, if applied to indigenous fermented foods, can give interesting and actionable information.

13.3.1.1.1.2 Quantitative PCR (qPCR)   Real-time quantitative PCR (qPCR) is a method of choice for the culture-independent detection and quantification of microorganisms. The exponential increase of amplicons can be monitored at every cycle (in real time) using a fluorescent reporter, and the increase in fluorescence is plotted against the cycle number to generate the amplification curve (Postollec et al., 2011). Combined with reverse transcription (RT), qPCR can also estimate transcript amounts, and this reversed transcription-qPCR can be used to study population dynamics and activity through quantification of gene expression during food fermentation (Postollec et al., 2011). In food microbiology, the technique is predominantly being used to detect pathogens and for the detection and quantification of microorganisms participating in food fermentations, such as LAB species in fermented milk products (Falentin et al., 2010; Furet et al., 2004; Masco et al., 2007). Real-time PCR assays have also been developed for the quantitative detection and quantification of *Lactobacillus sakei* and *Leuconostoc mesenteroides* in meat products (Elizaquivel et al., 2007; Martin et al., 2006) and for the enumeration of yeasts in fermented olives (Giraffa and Domenico, 2012; Tofalo et al., 2012). Application of this test can provide better insight into the microbiology of indigenous fermented foods.

13.3.1.1.1.3 Denaturing Gradient Gel Electrophoresis and Temperature Gradient Gel Electrophoresis (DGGE/TGGE)   Denaturing gradient gel electrophoresis (DGGE) and temperature gradient gel electrophoresis (TGGE) are fingerprinting techniques based on the electrophoretic separation of DNA fragments according to their denaturing profiles. Currently, they are the most frequently used culture-independent methods to evaluate the microbial diversity of natural environments and to monitor community changes according to the environmental variations (Giraffa, 2004). DGGE uses denaturing chemicals such as urea and formamide in a polyacrylamide gel, whereas TGGE is based on a linear temperature gradient (Lynch et al., 2004). The most commonly employed target for PCR prior to DGGE is phylogenetically informative ribosomal DNA, characterized by conserved and variable regions within the gene. The 16S rDNA V3 region is the most frequently used for amplification. The *rpoB* gene, coding for the ß-subunit of the RNA polymerase, was also used as a target for PCR-DGGE (Rantsiou et al., 2004; Rantisou and Cocolin, 2006). PCR amplification of DNA extracted from mixed microbial communities with primers specific for 16S rRNA or *rpoB* gene fragments results in a mixture of PCR products which have about the same length but are different in base compositions (Nisiotou et al., 2014). These DNA molecules are subjected to increasing concentrations of the denaturing agents, and sequence differences bring about differences in their melting behaviors. Partial melting creates branched molecules with a decreased migration through the gel (Ercolini, 2004). Based on this principle, a sample containing a mixture of many different microorganisms, all with different melting domains, will result in many bands on the gel. The complexity of the profiles reflects the bacterial diversity of the sample. To prevent the complete dissociation of the double-stranded DNA, a 30–40 bp GC-rich sequence is usually attached to the 5′-end of one of the primers (Boutte, 2006; Sheffield et al., 1989). This clamp is very stable and holds the strands partially together.

One of the major advantages of the DGGE/TGGE methods is that they allow simultaneous analysis of multiple samples and the use of universal primers, which permits the analysis of microbial communities without prior knowledge of the species present in the sample (Anonymous, 2007). The individual bands can be excised from the gel and identified by sequencing, although the small size of the PCR products may not always provide sufficient information for an unequivocal classification (Manzano et al., 2002; Ovreas, 2000). On the other hand, the DGGE/TGGE techniques suffer from the inherent bias of PCR-based molecular methods.

DGGE and TGGE analyses have been used as finger printing techniques for the identification of bacteria isolated from foods and, subsequently, more frequently for analysis of the bacterial communities in foods and in monitoring differences in the populations during fermentation or between different samples (Cocolin et al., 2000; Dawen and Tao, 2011; Jiang et al., 2010; Tsuchiya et al., 1994). Applications of DGGE and TGGE analysis for finger printing in the identification of bacteria are given in Table 13.2.

Most of these studies highlight the fact that a combination of both culture dependent methods and DGGE or TGGE approaches is essential for revealing microbial diversity and dynamics during fermentation (Madoroba et al., 2011).

13.3.1.1.1.4 Single-Strand-Conformation Polymorphism (SSCP): SSCP is a molecular technique similar to DGGE/TGGE, as it is also based on the electrophoretic separation of PCR products of similar length (Ndoye et al., 2011; Thakur et al., 2004). It was originally developed in mutation research, mainly to detect novel polymorphisms and mutations in human genes (Spiegelman et al., 2005). Following denaturation, single-stranded DNA fragments are separated on a non-denaturing polyacrylamide gel or by capillary electrophoresis. Under denaturing conditions, single-stranded DNA fragments will adopt stable secondary structures according to their nucleotide sequence and their physico-chemical environment (Schwieger and Tebbe, 1998). Based on the migration of these secondary structures in the gel, PCR products of similar size can be separated and visualized (Justé et al., 2008). In contrast to DGGE/TGGE, no GC-rich clamps primers are required (Peters et al., 2000). Caution should, however, be exercised when identifying bacterial populations using SSCP peak analyses, due to the possible co-migration of different sequences (Delbès et al., 2007). This limitation can partly be overcome by the use of different sets of primers targeting various variable regions (V2, V3 of 16S rRNA) or specific groups

Table 13.2 Application of DGGE and TGGE Analysis for Finger Printing in Identification of Bacteria

PARTICULAR	SOURCE
• Developed PCR-TGGE methods for identification of lactic acid bacteria in beer and *Lactobacillus* species isolated from Italian sausages, respectively used.	Ogier et al. (2002) Fontana et al. (2005)
• TGGE analysis was used to identify different genera present in dairy products and set up a data base for lactic acid bacteria and other dairy microorganisms, with low G+C content and showed that it can be applied in complex liquid and solid dairy ecosystems.	
• DGGE and TGGE approaches have been used—mostly in combination with other molecular and culture-dependent methods—to evaluate the microbial diversity of different fermented foods including pozol, a fermented maize dough. Various cheeses from Italy, Spain, Belgium and Denmark kefir grains from different Asiatic countries nukadoko, fermented rice bran from Japan, coffee and West African cereal foods.	Ampe et al. (1999) Guyot (2012) Björkroth and Holzapfel (2006) Coppola et al. (2001), Ercolini et al. (2001), Randazzo et al. (2002), Flórez and Mayo (2006), Van Hoorde et al. (2008), Masoud et al. (2011), Chen et al. (2008), Jianzhong et al. (2009), Nakayama et al. (2007), Vilela et al. (2010), Oguntoyinbo et al. (2011)
• Studies focused to monitor the population dynamics during the fermentation process included Italian sausages artisanal cheeses from different countries, cassava, various types of cereals, berries, rice vinegar, kimchi, cocoa beans and table olives.	Cocolin et al. (2000), Dolci et al. (2010), Fuka et al. (2010), Randazzo et al. (2010), Miambi et al. (2003), Meroth et al. (2003), Weckx et al. (2010), Madoroba et al. (2011), Pulido et al. (2005), Haruta et al. (2006), Chang et al. (2008), Lefeber et al. (2011), Abriouel et al. (2011)
• The bacterial community changes during the malolactic fermentation of Spanish wine during the production of a fermented crucian carp with rice, called *funazushi* and a Philippine fermented mustard, called *burong mustasa* were also investigated using DGGE profiles.	Ruiz et al. (2010), Fujii et al. (2011), Larcia et al. (2011)

of bacteria (Delbès et al., 2007). Another major limitation of SSCP analysis is the high rate of re-annealing of DNA strands after an initial denaturation during electrophoresis. However, this problem can be minimized by the use of a 5′-phosphorylated primer allowing selective removal of the corresponding phosphorylated strand through digestion with lambda exonuclease (Schwieger and Tebbe, 1998). It has also been reported that several stable conformations out of one single DNA fragment may co-exist and result in multiple bands on the gel (Justé et al., 2008).

SSCP analysis was applied in combination with clone library sequencing to investigate the dynamics of the complex microbial community of traditional French cheeses (Chamkha et al., 2008; Delbès et al., 2007; Duthoit et al., 2003), to study the microbial populations present in brines of Tunisian olives (Chamkha et al., 2008), and in Japanese traditional fermented foods made from fish and vegetables (An et al., 2011). The diversity of soft red-smear cheese populations was also investigated using SSCP analysis (Feurer et al., 2004).

13.3.1.1.1.5 Terminal Restriction Fragment Length Polymorphism (T-RFLP) The technique called terminal restriction fragment length polymorphism (T-RFLP) analysis combines selective PCR amplification of target genes with restriction enzyme digestion, high resolution electrophoresis, and fluorescent detection (McEniry et al., 2008; Rademaker et al., 2006; Seishi et al., 2006). Small subunit rRNA genes from total community DNA are amplified using primers designed to be non-discriminating, amplifying nearly all 16 SrDNAs, or selective, targeting specific domains or groups (Lord et al., 2002; Marsh et al., 2000). One of the two primers is fluorescently labeled at the 5′-end and the PCR is followed by digestion of the amplification products with one or more restriction endonucleases that usually have four base-pair recognition sites (Schütte et al., 2008). Only the fluorescently labeled terminal restriction fragments are detected on the sequencing gel, and their size can precisely be determined by using an automated DNA sequencer (Liu et al., 1997). The T-RFLP pattern is a distinct fingerprint of the microbial community, and the obtained TRFs can be compared to the sequence database of the Ribosomal Database Project (Cole et al., 2005), allowing tentative identification of primer-restriction enzyme combinations. Multiple restriction enzymes are typically used to increase the resolution, specificity, and reliability of the T-RFLP analysis (Osborne et al., 2006).

The major advantages of the T-RFLP analysis are the high resolution of the capillary electrophoresis technology, its rapidity, and the possibility to screen and compare a large number of communities to investigate the response of the population structure to intrinsic and extrinsic factors (Nieminen et al., 2011). It is a valuable method for comparison of complex microbial communities when high through-put and high sensitivity are required, without a need for direct sequence information (Kopecký et al., 2009; Nocker et al., 2007). One of the major drawbacks of this technique is its dependence on restriction efficiency. Incomplete or non-specific restriction however may occur, leading to an increased number of fragments and, therefore, to an overestimation of the microbial diversity of the sample (Justé et al., 2008; Zhang, 2010). To avoid this problem, the amplified product from a well-characterized isolate may be used as an internal standard, which allows checking of the restriction efficiency. Additional secondary T-RFs were observed to occur in T-RFLP analysis at high frequency and were called "pseudo-T-RFs" (Blaž, 2006; Egert and Friedrich, 2003). As a result, the microbial diversity may be overestimated because of the larger number of peaks in T-RFLP profiles. However, complete elimination of the pseudo-T-RFs can be achieved by the digestion of amplicons with a single-strand-specific nuclease prior to T-RFLP analysis (Egert and Friedrich, 2003). The number of PCR cycles should also be minimized, because the formation of pseudoterminal restriction fragments increases linearly with the cycle number (Egert and Friedrich, 2003). To improve the resolution and sensitivity of T-RFLP, another approach, called fluorophore ribosomal DNA restriction typing (f-DRT) has been developed (Wang et al., 2011), based on the end-labeling of all restriction fragments from a single enzyme digestion with a fluorescent dye, using high through-put capillary electrophoresis for detection of these fragments (Wang et al., 2011). The technique has been used to study the surface microflora of smear ripened Tilsit cheeses (Rademaker et al., 2005) and to monitor population dynamics during yoghurt and hard cheese fermentation (Rademaker et al., 2006).

13.3.1.1.1.6 Amplifed rDNA Restriction Analysis (ARDRA) Amplified ribosomal DNA restriction analysis (ARDRA), also known as restriction fragment length polymorphism (RFLP) analysis of 16S rRNA genes is a fingerprinting technique that has been successfully applied to bacterial identification at species level. However, it is also valuable and reliable for ecological studies and for comparison of microbial communities (Giraffa and Neviani, 2001). ARDRA is based on the digestion of PCR-amplified ribosomal community DNA using appropriate restriction endonucleases, followed by gel electrophoretic separation of the generated restriction fragments. Multiple restriction endonucleases usually have to be used either separately or in combination. In contrast to T-RFLP, all fragments are visualized on the gel, increasing the resolution of this technique (Justé et al., 2008). On the other hand, the more complex patterns make the comparison and interpretation more difficult. A single species can produce 4–6 restriction fragments using a 4 bp cutting enzyme (Nocker et al., 2007). This, together with the limited staining sensitivity of DNA binding dyes, results in a suppression of bands from less abundant members of the microbial community (Nocker et al., 2007). In general, ARDRA is considered to be a useful technique for detecting structural changes in relatively simple microbial populations, but is not a method of choice to measure diversity of complex communities (Nocker et al., 2007; Sklarz et al., 2011).

13.3.1.1.1.7 Automated Ribosomal Intergenic Spacer Analysis (ARISA) Automated ribosomal intergenic spacer analysis (ARISA) is based on the amplification of the intergenic region between the 16S and 23S ribosomal genes. In the case of molds and yeasts, the internal region (ITS) between the small subunit and the large subunit, including the 5.8S rRNA gene, is amplified (Filteau et al., 2011). The intergenic spacer region is more heterogeneous both in length and nucleotide sequence than the flanking ribosomal genes (Justé et al., 2008). Both types of variations make the intergenic spacer regions suitable for subtyping bacterial strains and closely related species in cases where the fingerprinting of ribosomal gene sequences does not provide sufficient resolution (Nocker et al., 2007). A fluorescent primer is used in the amplification of the ribosomal intergenic spacer region, and the PCR products are analysed by an automated capillary electrophoretic system that produces an electropherogram, the peaks of which correspond to discrete DNA fragments (Cardinale et al., 2004). Potential problems associated with ARISA are the preferential amplification of shorter templates (Fisher and Triplett, 1999) and the fact that often more than one signal is generated by a single organism because of the IGS length variation within a single genome (Justé et al., 2008). As for DGGE/TGGE and T-RFLP analysis, ARISA has mainly been applied in ecological studies, and only a few applications in food matrices have been reported (Arteau et al., 2010). A multiplex automated ribosomal intergenic spacer analysis (MARISA) method has been developed and used to simultaneously analyse the bacterial and fungal microbiota composition of maple sap (Filteau et al., 2011).

13.3.1.1.1.8 Pyrosequence-Based rRNA Profiling   Pyrosequence-based rRNA profiling is a new strategy allowing high through-put and in-depth monitoring of microbial communities. Up till now, it has mainly been applied in ecological studies of soil, the deep sea, and the human intestinal tract. The method is based on the detection of pyrophosphate released during nucleotide incorporation (Margulies et al., 2005). Pyrosequence-based rRNA profiling involves the amplification of variable regions of the 16S rRNA gene V1–V9 using primers that target adjacent conserved regions, followed by direct sequencing of individual PCR products (Liu et al., 2007). It monitors the luminescence of sequencing reactions in pico-liter plates (Nakayama, 2010). A barcode-tag sequence strategy can be used to analyse amplicons from different samples in one batch. Massive parallel sequencing means that more than 300,000 sequences can be determined simultaneously (Humblot and Guyot, 2009), and thus the costs associated with sequencing can be dramatically reduced. The second advantage is that cloning of the samples is not required, thus avoiding any problems associated with this step.

Pearl millet slurries were the first fermented foods analysed by this technique (Humblot and Guyot, 2009). Then, barcoded pyrosequencing was used to evaluate the archeal and bacterial diversities of seven types of fermented seafoods (Roh et al., 2010) and to study the microbial communities of various fermented foods, including Danish raw cheeses (Masoud et al., 2011), a fermented rice bran mash, called *nukadoko* (Sakamoto et al., 2011), *narezushi*, a traditional fermented food from Japan (Koyanagi et al., 2011), *kimchi* (Park et al., 2012), and traditional Korean alcoholic beverages such as *makgeolli* (Jung et al., 2012).

13.3.1.1.1.9 Cloning and Sequencing   Individual members of the bacterial community may be identified at species level by cloning and subsequent sequencing of PCR-amplified sequences. 16S rRNA gene based libraries are used very frequently. A major drawback of the use of clone libraries is the high number of clones to be analysed to detect rare organisms against the background of a few dominant species (Nocker et al., 2007). Therefore, it is common to construct clone libraries in parallel to fingerprinting techniques such as DGGE, or clone libraries initially are screened by restriction digestion yielding different restriction types which then, can be sequenced (Justé et al., 2008). For example, Kim and Chun (2005) used clone libraries in combination with ARDRA to investigate the microbial structure of kimchi samples, and a combination of 16S rRNA gene clone libraries and real-time quantitative PCR was applied to study the structure and dynamics of raw milk microbiota and changes in the microbial community during Montasio cheese ripening (Carraro et al., 2011; Rasolfo et al., 2010).

13.3.1.1.2 PCR Independent methods

13.3.1.1.2.1 DNA Array Technology The DNA microarray (microchip) technology allows parallel analysis of hundreds or thousands of genes in a single assay. Diagnostic nucleic acid sequences, referred to as probes, are immobilized on a miniaturized support (usually a glass slide) and are hybridized with the homologous labeled target amplicons, which can then be detected (Justé et al., 2008). Since hybridization signals are proportional to the quantity of target DNA, microarrays can also quantify genes in DNA samples (Cho and Tiedje, 2002). The DNA microarray technology enables rapid and simultaneous identification of hundreds of different organisms in one assay. So far, DNA arrays primarily have been developed to detect pathogens in food and clinical samples and have been less used to identify microorganisms participating in food fermentations.

There is one report on the application of an oligonucleotide array in the analysis of raw milk bacterial communities (Giannino et al., 2009). A genome-probing microarray (GPM) was developed by Bae et al. (2005) which showed increased specificity as compared to oligonucleotide arrays. Using 149 microbial genomes as probes deposited on a glass slide, they were able to quantitatively analyse about 100 diverse LAB species involved in *kimchi* fermentation (Bae et al., 2005).

Most of the currently applied methods, however, do not deliver information on the level of physiological activity and possible metabolic capabilities of the microorganisms, so the techniques targeting the mRNA will gain more importance in the future. In order to understand the role of different organisms in fermentation, for instance, their contribution to the aroma profile, it will be necessary to study gene expression by the means of transcriptomics and proteomics. A good example is the recent study of Nam et al. (2009) showing that metatranscriptome analysis with genome-probing microarrays can be used to determine the role and contribution of the different LAB species to *kimchi* fermentation. To monitor microbial dynamics during *kimchi* fermentation, another approach using environmental mRNAs (metascriptome) in addition to the metagenome analysis has also been applied (Nam et al., 2009).

13.3.1.1.2.2 Fluorescent *in situ* Hybridization (FISH) Fluorescent *in situ* hybridization (FISH) with 16S rRNA gene probes is another method of cultivation-independent detection and identification of microorganisms in a food matrix. It combines the simplicity of microscopic observation and the specificity of DNA hybridization (Justé et al., 2008; Machado et al., 2013). Following fixation and permeabilization, microbial cells are identified *in situ* within the original food sample by visualization of intercellular probe rRNA target hybrids (Ludwig, 2007). Cell fluorescence conferred by the probe attached fluorochromes is detected by epifluorescent microscopy or flow cytometry. It is a rapid method, allowing completion of the whole procedure within a few hours (Justé et al., 2008). Besides identification, the FISH technique provides information about the distribution of the microorganisms in the food sample and enables their enumeration and visualization of cell morphology. But the technique suffers from some system inherent pitfalls, such as the difficulty of achieving permeabilization of all cells in diverse communities, and the need for an extensive knowledge of the composition of the bacterial community to be able to construct suitable probes. Moreover, fluorescent *in situ* hybridization is less sensitive than PCR-based methods (Justé et al., 2008). Nevertheless, the FISH technique has been used for the rapid detection of LAB in wine (Sohier and Lonvaud-Funel, 1998), to analyse the surface microflora of Gruyère Swiss cheese (Kollöffel et al., 1999) and to study the spatial distribution of different microbial species in the Stilton cheese matrix (Ercolini et al., 2003). It has been demonstrated that *Lactobacillus acetotolerans* is a dominant bacterium in ripening of *nukadoko* (Nakayama et al., 2007), and it has been used in combination with flow cytometry to quantify the RNA content of yeast cells during alcoholic fermentation (Andorra et al., 2011).

13.3.1.2 Culture-Dependent Methods Molecular approaches may also be applied following common culture-dependent isolation and preliminary phenotypic characterization of strains. Several DNA-based techniques are available for genotyping of isolates, and the most commonly employed targets are the 16S and 23S rRNA encoding genes (Quigley et al., 2011). They involve 16S rRNA target oligonucleotide probes and the direct sequencing of the 16S rRNA gene, as well as many fingerprinting techniques, such as RAPD-PCR, rep PCR, PCR RFLP, ribo-typing, ARDRA, PFGE, and species-specific PCR. Basically, the pattern methods rely on the detection of DNA polymorphisms between species and strains, and differ in their discriminatory power, reproducibility, ease interpretation, and standardization.

Several commonly applied techniques, including RFLP, ARDRA, ribotyping, and PFGE, make use of restriction fragment polymorphisms. Restriction fragment length polymorphism (RFLP) means that genomic DNA is digested using appropriate restriction endonucleases, and the generated fragments are electrophoretically separated (Sánchez and Sanz, 2011). Ribotyping is a variation of the conventional RFLP analysis using Southern hybridization of the DNA fragments with probes targeting rDNA and resulting in a less complex pattern to evaluate. Moreover, an automated riboprinter can be used. Amongst others, LAB from a Spanish blood sausage (Santos et al., 2005) and a Portuguese cheese (Kongo et al., 2007) were identified using ribotyping. ARDRA combines restriction fragment polymorphism with the application of site-specific PCR amplification, and has been used as a molecular tool in numerous studies, mostly in combination with 16S rDNA sequencing (Liu et al., 2007). It was shown to be useful in the identification of LAB from grape must and wine (Rodas et al., 2003), for the molecular characterization of enterococci from an African fermented sorghum product (Yousif et al., 2005), and for the classification of *Bacillus* isolates from cocoa fermentation (Quattara et al., 2011). Although 16S rRNA gene-based PCR-RFLP assays were preferentially used, *rpo* B gene sequences are increasingly being applied (Marty et al., 2012). Pulse field gel electrophoresis (PFGE) allows the separation of large fragments resulting from digestion of genomic DNA with rare-cutting enzymes. It is more discriminatory than most other techniques and, therefore, is suitable for strain typing (Sánchez and Sanz, 2011). PFGE was used for the differentiation of *Staphylococcus* strains isolated from fermented sausages (Corbière Morot-Bizot et al., 2006) and in combination with RAPD for characterization of strains involved in the malolactic fermentation of wine (Ruiz et al., 2008).

Frequently used, easy-to-perform, PCR based techniques are RAPD-PCR, rep PCR, and species-specific PCR. RAPD (randomly amplified polymorphic DNA)-PCR is a popular fingerprint method for intra- and inter-specific differentiation of LAB and other microorganisms involved in food fermentations. It uses short arbitrary primers targeting multiple sites on the genome and low-stringency conditions to randomly amplify DNA fragments which are then separated on a gel to give a fingerprint pattern specific for each strain (Quigley et al., 2011). It is frequently used for screening and a first grouping of isolates, and generally is followed by a more detailed molecular characterization (such as 16S rRNA gene sequencing) of typical representatives of the groups (Rebecchi et al., 1998). Another powerful typing method is rep (repetitive extragenic palindromic) PCR based on amplification of repetitive bacterial DNA elements. In contrast to RAPD-PCR, it uses primers such as $(GTG)_5$ against known universal sequences in the bacterial genome (Singh et al., 2009). 16S rRNA gene analysis has revealed the bacterial community structure in Kimchi (Kim and Chin, 2005).

RAPD and rep PCR have both been used for identification of LAB from fermented vegetables and bamboo shoots produced in North East India (Tamang et al., 2005, 2008) and from fermented cassava (Kostinek et al., 2005). Rep PCR-fingerprinting using the $(GTG)_5$ primer has also been successfully applied for the identification of acetic acid bacteria from fermented cocoa beans (De Vuyst et al., 2008) and LAB from raw milk Gouda-type cheese (Van Hoorde et al., 2008). RAPD-PCR analysis has been shown to be useful for the differentiation of the lactobacilli most commonly found in meat products (Andrighetto et al., 2001) and in meat and fermented sausages (Martin et al., 2006), and for the strain-specific identification of *Leuconostoc mesenteroides* used as starter culture in sauerkraut fermentation (Plengvidhya et al., 2004). The tool has been employed successfully in genotyping of starter culture of *Bacillus subtilis* and *Bacillus pumilis* for fermentation of African locusbeans (Ouoba et al., 2004).

Finally, species-specific primers are suitable tools in mono- or multiplex PCR assays, allowing a rapid and easy identification of species that are expected to be detectable in the food sample. For example, a total of 230 LAB isolates from French wheat sourdough were identified by PCR amplification using different species-specific primer sets (Robert et al., 2009).

13.4 Biofortification through Fermentation and Concentration

Biofortification of food is an integrated approach to reduce malnutrition. The concept of *in situ* fortification by bacterial fermentation provides the basis to enhance the nutritional value of food products and their commercial value. In recent years, a number of biotechnological processes have been explored to perform a more economical and sustainable enhancement of the nutritional quality of bland traditional food (Capozzi et al., 2012). By this approach, food-grade biotechnological strategy for the production of protein-enriched yam flour has been attempted (Achi, 1991, 1999; Achi and Akubor, 2000; Capozzi et al., 2012). Yam flour (*Elubo*) is an important processed yam product which is reconstituted by stirring in boiling water to form a paste (*amala*) and eaten with flavored sauces (Ige and Akintunde, 1981). Fermentation of yams to produce flour has been found to improve both product quality as well as to remove inherent coloration problems associated with the acceptability of the processed product (Achi, 1999; Ray and Sivakumar, 2009). Yam slices have been successfully fermented with *Lactobacillus* species isolated from the fermenting medium. Fermentation was found to give a product with a lighter color and a better quality product with enhanced consumer acceptability (Achi, 1992). The moisture, protein, and fat contents of the fermented flour are in the range of 7.0%–7.6%, 2.0%–3.5%, and 0.3%–0.4%, respectively, depending upon the varieties (Achi, 1999; Akingbala et al., 1995). Similarly, pretreated soy flour was used to replace 10, 20, 30, and 40% of fermented yam flour as a protein supplement. Higher concentrations of soy flour above 40% level, however, reduced the acceptability of the cooked paste (Achi, 1999; Achi and Akubor, 2000; Ray and Sivakumar, 2009). Such types of approaches can be applied very conveniently to similar products in the South Asian countries.

13.5 Fermented Food Packaging Systems and Presentation

Packaging is an essential process in the production of many food products in industrialized countries. The principal roles of food packaging are to protect food products from outside influences and damage, to contain the food, prevent physico-chemical degradation, retain the beneficial effects of processing, extend shelf-life, maintain or increase the quality and safety of food, prevent microbial spoilage, and provide consumers with ingredient and nutritional information (Anonymous, 2011; Coles, 2003; Joshi and Pandey, 1999; Marsh and Bugushu, 2007; Talwalkar and Kailasapathy, 2004; Verma and Joshi, 2000). As traditional fermented food production increases, new issues regarding problems of safety, spoilage, and sensory properties occur. The emergence of new package requirements for fermented foods is as a result of the continuous migration of people from their place of origin, population growth, and the desire to export the traditional fermented food to other cultures (Marsh and Bugushu, 2007). Fermented foods need special packaging. The required packaging protection depends on the stability and fragility of the food, the desired shelf-life of the food package, and the distribution chain. Good package integrity is also required to protect against loss of hermetic condition and microbial penetration (Kailasapathy, 2010; Lee et al., 2008). Several of the traditional packaging/presentation materials are now gradually being replaced by modern synthetic materials, such as plastics, paper, and

Figure 13.2 Example of some traditional packaging and presentation materials and modern packaging materials: (a) *Agidi* made ready for sale, (b) cooked ready-to-eat *fufu* on sale, (c) plastic package or paper cartons for processed dry *fufu*, *gari*, or *elubo* flour for sale, and (d) paper package/carton for *fufu*, *gari*, and *elubo* for sale. (Adapted from Teniola, O.D. 2009. *Role of Packaging and Product Presentation in the Acceptability, Quality Improvement and Safety of the Traditional African Fermented Foods*. Federal Institute of Industrial Research, Oshodi (FIIRO), Lagos, Nigeria.)

cans (Teniola, 2009). The traditional presentation and packaging of some traditional fermented foods is shown in Figure 13.2.

Fabricated improved fermentation trays constructed with maximum emphasis on maintenance, not only enhance sanitation and reliability for the process, but also involve minimum capital investment and operating cost (Ouoba, 2010; Sanni, 1993). Apart from the obvious effects of improving the shelf-life of the product, packaging helps to increase product popularity and general acceptability.

13.6 Production of Probiotic Functional Food

The primary role of diet is to provide sufficient nutrients to meet metabolic requirements, while giving the consumer a feeling of satisfaction and well-being (see Chapter 6). In addition, some food components exert beneficial effects beyond basic nutrition, leading to the concept of functional foods and nutraceuticals (Anonymous, 2010; Farnworth, 2008; Ribiero, 2000). A growing public awareness of diet-related health issues, and mounting evidence regarding the health benefits of probiotics including those which are dairy based (Tamime, 2005), have increased consumer demand for probiotic foods. "*Let food be thy medicine and medicine be thy food,*" the age-old quote by Hippocrates, is certainly a widely-held opinion today (Vasiljevic and Shah, 2001). Some foods and their food components modulate various physiological functions and may play detrimental or beneficial roles in some diseases (Anonymous, 2010; Koletzko et al., 1998). Fermented foods are also functional probiotic foods, and are of great significance. Probiotics are defined as live microorganisms that exert health beneficial effects to the host when administered in sufficient quantities (Quigley et al., 2011). A food can be made functional by applying any technological or biotechnological means to increase the concentration, add, remove, or modify a particular component, as well as to improve its bioavailability, provided that the component has been demonstrated to have a functional effect, such as a probiotic effect (Das et al., 2012; Roberfroid, 1998). Many traditional fermented cereal food products have been found to contain components with potential health benefits (Anonymous, 2010). Some of the major categories of cereal based functional foods contain live microorganisms and may produce or release potential health promoting compounds in the substrate medium (Nwachukwu et al., 2010). In addition to these foods, new foods are being developed to enhance or incorporate these beneficial components for their health benefits or desirable physiological effects (Anonymous, 2010). There is a considerable amount of work being done on cereal containing foods fermented with complex indigenous cultures (Blandino et al., 2003).

Studies using fermented cereals as delivery vehicles for potentially probiotic LAB for the production of probiotic foods have been reported (Angelov et al., 2006; Charalampopoulos et al., 2003; Kedia et al., 2008 Rathore et al., 2012). Cereal grains such as barley and oats are good substrates for LAB growth, which has led to the commercialization of cereal-based probiotic products (Salovaara, 2004). Strains of the *Lactobacillus* genus, such as *Lactobacillus acidophilus*, *L. reuteri*, and *L. plantarum* constitute significant proportion of cultures used in probiotic products (Charalampopoulos et al., 2003; Ravyts et al., 2012). Functional foods produced through traditional fermentation processes thus, have potential to provide food sources with increased nutritional value to South Asian countries. For more details, see Chapter 6 on the health aspects of indigenous fermented foods.

13.7 Alkaline Fermented Vegetable Proteins

A very important class of fermented products is the indigenous alkaline-fermented food condiments (Sarkar and Nout, 2014). Alkaline fermentation is the process in which the pH of the substrate increases to alkaline values, may be as high as pH 9 (Amadi et al., 1999; Omafuvbe et al., 2000; Sarkar and Nout, 1885; Sarkar and Tamang, 1995). Fermented vegetable proteins made up of legumes and oilseeds are generally used as soup flavoring condiments (Olasupo et al., 2010). The cooked forms of dietery proteins are eaten as meals, and are commonly used in fermented form as condiments to enhance the flavors of foods (Achi, 1991; Aidoo, 1986; Oniofiok et al., 1996). Apart from increasing the shelf-life and a reducing the anti-nutritional factors (Achi and Okereka, 1999; Barimalaa et al., 1989; Reddy and Pierson, 1999), fermentation markedly improves the digestibility, nutritive value, and flavors of the raw seeds. Excellent reviews on traditional fermented food condiments have been published earlier (Achi 2005b; Odunfa, 1985b; Olasupo, 2006; Parkouda et al., 2009). The methods employed in the manufacture of fermented condiments differ from one region-to-another because these processes are based on traditional systems. Various studies have been carried out to upgrade the traditional process and make it less labor intensive (Achi, 2005b; Odunfa et al., 2006).

13.8 Upgrading Fermented Cereal Products

Ogi is a fermented cereal gruel processed from maize, although sorghum or millet is also employed as the substrate for fermentation (Blandino et al., 2003). It is an example of traditional fermented food, which has been upgraded to a semi-industrial scale (Achi, 2005a).

13.9 Role of Killer Yeast in Indigenous Fermented Food

This is another application of biotechnology in indigenous fermented foods, where killer yeast is used (Cocolin and Comi, 2011). Killer yeast is the yeast which is capable of killing susceptible cells by producing toxic proteins or glycoproteins (so-called killer toxins or killer proteins). The phenomenon in yeast cells was first reported by Bevan and Makower (1963). *S. cerevisiae* can kill sensitive strains of the same species, though it has been found in the other strains also, such as *Hansenula* sp., *Pichia* sp., *Candida* sp., *Willopsis* sp., *Torulopsis* sp., *Tricosporon* sp., *Kloeckers* sp., and *Kluyveromyces phaffi* (Comitini et al., 2004a,b; Izgu et al., 2006; Vadkertiova and Slavikova, 2007). Killer yeasts have been isolated from various sources like fruits, mushrooms, decaying plants, soil, skin of poultry, and fermented products (Cocolin and Comi, 2011). Killer toxins have been classified into different types, as K1, K2 and K3. They are usually active and stable at pH 4–5 and temperatures of 20–25°C (Bendova, 1986; Heard and Fleet, 1987; Michalcakova and Repova, 1992), and cause membrane permeability changes in sensitive cells (Kagan, 1983). In some cases, they can inhibit DNA synthesis, such as in *S. cerevisiae* (Izgu and Altinbay, 1997; Schmitt et al., 1989) or stop cell division (Stark et al., 1990). Clearly visible inhibition zones surrounding the killer toxin-producing yeast can be observed, and the killer phenomenon can be confirmed with purified killer toxins, which implies a unique form of bioaction. The inhibition zone produced by the activity of killer yeast is shown in Figure 13.3.

The killer yeast toxin can be useful as a tool for the development of yeast-based biocontrol in several biotechnological applications (Marquina et al., 2002; Selvakumar et al., 2006; Silva da et al., 2008; Walker et al., 1995; Wang et al., 2007). In the food industry, killer characteristics can be conferred to yeast strains. The use of killer yeasts as starter cultures can protect the product against spoilage yeasts. They have also been considered useful in biological control of undesirable yeasts in the preservation of food (Izgu et al., 2004; Lowes et al., 2000; Petering et al., 1991). Therefore,

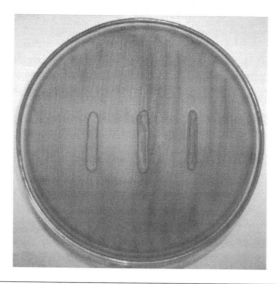

Figure 13.3 Inhibition zones of sensitive yeast strains surrounding the growth of killer yeast strains in YPD agar medium with methylene blue.

the killer yeast has the ability to control spoilage in the preservation of food. For example, *C. tropicalis* shows inhibitory effects against *S. cerevisiae* in contaminated starter cultures of the Turkish bakery industry (Izgu et al., 1997; Santos et al., 2004) and *Kluyveromyces phaffii* DBVPG 6076 killer toxin has been found effective against apiculate wine yeast from grape samples (Ciani and Fatichenti, 2001). Knowledge and understanding of each killer yeast/toxin is of importance in determining its possible use, such as in solving fermentation problems during commercial biomass production of baker's yeast (Silva et al., 2008). It has been found that K3 killer toxin gene can be transferred into *S. cerevisiae* BSP 1 to construct a baking strain resistant to killer toxin producing *C. tropicalis* contamination by the protoplast fusion technique. The new *S. cerevisiae* BSP 1 (K3) can inhibit the growth of *C. tropicalis* in a pH of 3.5–5.0 and at a temperature of 20–30°C (Izgu et al., 2004).

The phenomenon of killer yeast appears to be widely distributed within many yeast genera and such yeasts can be isolated from a great variety of food fermentation processes. Isolation of killer yeast from fermented vegetables (*pak-sien*, bamboo shoot, *sator*, and cabbage) has also been made and identified as *Candida krusei* (Hernández et al., 2008). It shows the killer activity against sensitive reference strains on YPD agar plate supplemented with 0.003% methylene blue (Figure 13.3). The killer yeast showed activity against food pathogenic bacteria (*Escherichia coli* TISTR 887, *Salmonella typhimurium* TISTR 292, *Staphylococcus aureus* TISTR 118, and *Bacillus cereus* TISTR 868) (Waema et al., 2009). *Debaryomyces hansenii*, *Klyveromyce marxianus*, *Pichia guilliermondii*, and *S. cerevisiae* have been isolated from seasoned green table olives, which had the broadest spectra of action against yeasts that cause spoilage (Hernandez et al., 2008). The effect of pH and salt on killer activity has also been studied by testing the killer strains against a killer-sensitive strain. These killer

Table 13.3 Optimum Temperature and pH for the Activity of Killer Toxins of Yeast Strains

KILLER STRAIN	TEMPERATURE (°C)	pH	REFERENCE
Saccharomyces cerevisiae	15.0–25.0	3.5–4.5	Heard and Fleet (1987)
Kluyvermyces phaffii	24.0	4.0	Ciani and Fatichenti (2001)
Saccharomyces cerevisiae	20.0	4.3–4.4	Lebionka et al. (2002)
Candida tropicalis	20.0–30.0	4.0	Izgu et al. (2004)
Pichia membranifaciens	25.0	3.0–4.8	Santos and Marquina (2004)
Pichia anomala	4.0–37.0	4.5	Izgu et al. (2006)

yeasts show their killer activity at pH 8.5 in the presence of 5% and 8% NaCl, but a high concentration (10% NaCl) decreased it (Hernandez et al., 2008). Toxins produced at any salt concentration exhibited the highest activity against sensitive strains in the detection medium. Salt has been found to have induced the formation of ion-permeable channels, killing cells by disruption of the plasma membrane function and cellular ionic balance (Santos and Marquina, 2004; Silva et al., 2008). Toxins at high salt levels can also have biotechnological applications, such as preservation of high-salt food products. Moreover, the killer activity is different at different temperature, pH, and salt level of the brine, depending upon the strains (Table 13.3) (Silva et al., 2008).

Biopreservation is considered to improve food safety without changing the sensory quality of a product. In dairy products, there is a risk of contamination by *Listeria monocytogenes*, so killer yeast strains were isolated, screened, and tested against bacteria from different sources, but mainly from dairy products. Some strains showed an inhibitory potential against the growth of pathogenic bacteria (Gorges, 2006). *Candida intermedia* was able to reduce the *L. monocytogenes* cell count by 4 log cycles, whereas *Kluyveromyces marxianus* suppressed the growth of *L. monocytogenes* by 3 log unit cycles (Gorges et al., 2006). *Williopsis saturnus* could inhibit spoilage by yeasts in cheese. It is ideal as a biopreservative and may replace chemical or antibiotics-based preservatives (Gorges et al., 2006; Liu and Tso, 2009). In brief, killer yeast and its toxins have potential for application in indigenous foods, mostly as biopreservatives.

13.10 Role of Metagenomics and Metabolomics in Indigeneous Fermented Foods

Fermented foods, a part of an important food ecosystem, harness microbial diversity and functional microbiota in the environment (Tamang, 2001). Filamentous molds, yeasts, and bacteria constitute the microbiota in indigenous fermented foods and beverages, which are present in the ingredients and are selected through adaptation to the substrate. For more details, see Chapter 3. New biotechnological methods are now applied to increase the capacity of microorganisms in terms of yield, longer time of preservation, detection of bioactive substances, etc. Metabolomics and metagenomics are new emerging fields of biotechnology that play an important role in indigenous fermented foods. These technologies interact with the metabolic pathways of microorganisms and, thus, lead to beneficial products. Metabolomics and metagenomics are derived constitutively from the suffix "OMICS" (*OM*)—the progenitor of all terminology including proteomics, genomics, transcriptomics, etc., with the ability to simultaneously measure the level of the several gene products and metabolites. So metabolic engineering has become a tool for the directed evolution of microorganisms (Noronha et al., 2000). Fermentative capacity of soybean starter has also been assessed, using metabolomics (Ko et al., 2010).

13.10.1 Metabolomics

Metabolomics is the study of metabolome in a cell at different developmental stages under different conditions. It refers to the complete set of metabolites, including hormones, secondary metabolites, etc., found within a biological system. When this technology involves biotechnology for the improvement of the cellular activities of microorganisms, *via* the use of recombinant DNA technology, then it is called metabolic engineering, as illustrated in Figure 13.4. It refers to the systematic modification of metabolic pathways for the purpose of obtaining a desired product or trait (Norouha et al., 2010).

13.10.1.1 General Methodology for Metabolic Engineering Metabolic engineering involves different steps such as

- Identify the target phenotype or trait.
- Increase the frequency of occurrence of genes that may confer the phenotype.

1. Increase the mutation frequency in producing cells by mutagen treatment (UV, x-ray, chemical mutagen) (Classical method).
2. Introduce additional gene (that may already exist or be absent in the host cell).
3. Introduce a genetic element to inactivate the gene by random insertion of extra sequences.
4. Introduce the gene that causes insertional inactivation of harmful alleles
5. Suppress the harmful genes by inserting homologous gene sequences.

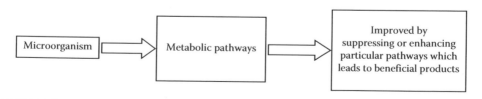

Figure 13.4 Steps in metabolic engineering.

13.10.1.2 Tool Used in Metabolic Engineering To design microorganisms for hyperproduction of a metabolite, a detailed analysis of the metabolic process is needed, which in turn requires in-depth knowledge of the various metabolities, gene products, and the regulatory system involved (Noronha et al., 2030). These include functional genomics, metabolic profiling, metabolic analysis, and metabolic control analysis (Kopka et al., 2004; Trethewey, 2004).

13.10.1.3 Separation of Metabolites To acquire the technology of various metabolites, different analytical tools are employed such as:

- Gas chromatography (GC) can separate various components, but only volatile biomolecules can be analysed by this technique.
- HPLC (high performance liquid chromatography): Compared to GC, this gives low resolution, but a wide range of analytes can potentially be measured.
- Capillary electrophoresis (CE): For charged analytes, it is the most widely used electrophoretic technique and gives higher theoretical efficiency than HPLC.

13.10.1.4 Detection of Metabolites The most common methods for identification of the abundance of metabolites present in an organism include nuclear magnetic resonance (NMR) and mass spectrometry. NMR based metabolomics were developed in the laboratory of Jeremy Nicholson at Birkbeck, University of London. NMR provides a great deal of information for every biomolecule, thus allowing identification and quantification of the differences in the metabolites. Gas chromatography mass spectroscopy (GC–MS) and liquid chromatography mass spectroscopy (LC–MS) are able to detect several hundred chemicals, including sugars, sugar alcohols, organic acids, amino acids, fatty acids, etc.

13.10.1.5 Bioinformatic Tools for Metabolomic Analysis The tools employed for the analysis of metabolomics include correlation optimized warping (COW) and chrompare (Kim et al., 2012). After detecting the metabolites present in a particular organism, whether in plants or microorganisms, metabolic engineering can be undertaken to yield the desired product.

13.10.2 Metagenomics

Metagenomics can be defined as the application of modern genomics techniques to the study of communities of microbial organisms directly in their natural environments, by-passing the need for isolation and laboratory cultivation of individual species (Anonymous, 2005). The term was first used by Jo Handelsman, Jon Clardy, Robert M. Goodman, and others, and first appeared in publication in 1998.

13.10.2.1 Metagenomic Techniques   Some of the techniques used in metagenomics include

- Sampling from habitat (microorganisms can be isolated from any environmental areas, such as the sea, mud, marine deposits, etc.)
- Filtering particles typically by size with filter membrane
- Lysis of cell wall followed by DNA extraction. Cell wall is ruptured with the help of calcium chloride or other extraction buffer to extract DNA
- Cloning of DNA into vectors and construction of libraries in bacterial artificial chromosome (BAC) vectors (BACs are used because they have cloning capacity of 100–300 kb)

Sequencing of clones by the following methods:

1. Shotgun sequencing methods: These provide information on which environment the microorganism will better adapt to, and what metabolic processes are possible in that community.
2. High through-put sequencing: This uses massive parallel pyrosequencing methods to generate shorter DNA fragments, approx 400 bp, than Shotgun sequencing methods. This becomes advantageous as it does not require cloning of the DNA before sequencing, removing one of the main bases in environment sampling.

- Assembly of sequence into contigs, which is a series of clones that contain overlapping pieces of DNA covering a specific region in a chromosome.
- Application of bioinformatics software to detect the sequences.

13.10.3 Role of Metabolomics and Metagenomics in Indigeneous Fermented Foods of South Asia

This aspect will be illustrated by taking a few examples of foods, including those of South Asia.

Cheese: Cheese is a generic term for a diverse group of milk-based products, and is produced in wide-ranging flavors, textures, and forms, and has proteins and fats derived from milk. It is produced by the milk protein casein. Typically, the milk is acidified and the addition of the enzyme rennet causes coagulation. The solids are separated and pressed into the final form. Most of the time, the enzymes are produced by fermentation using the fungus *Mucor miehei*.

Metabolic profiling technologies with GC and time-of-flight mass spectrometry (TOF-MS) have analysed the low molecular weight components, including amino acids, fatty acids, amines, organic acids, and saccharides, in hard and semi-hard cheese. The compounds that play an important role in constructing each sensory prediction model include 12 amino acids and lactose for "Rich flavor" and 4-aminobutyric acid, ornithine, succinic acid, lactic acid, proline, and lactose for "Sour flavor"

(Ochi et al., 2012). The metabolomics-based component profiling not only focused on hydrophilic low molecular weight components, but was also able to predict the sensory characteristics related to ripening (Ochi et al., 2012). Metabolic profiling of the microorganism *Lactococcus lactis* (used in cheese fermentation) under different conditions can be obtained by GC-MS and HS-MS. Such techniques are used to study different metabolites produced by the bacteria and to manipulate the metabolic pathways for better yield (Azizan et al., 2012).

Metagenomic analysis can be done using massively parallel DNA sequencing to characterize the microbial communities living in cheese. In addition to its economical importance, cheese production is also a concern for public health, with thousands of people becoming sick every year after eating improperly manufactured cheese. Description of the microbial populations and their temporal variations in cheese is thus, essential. At the Genomic Medicine Institute, Cleveland Clinic, Cleveland, DNA was extracted from cheese at four time points: immediately after insemination (day 1), at the day of sale (day 30), at the expiration date (day 90) and at day 180. Then, short fragments of 16S and 18S ribosomal RNA genes were amplified using universal primers, and obtained 100,000 sequence reads from each sample. The analysis showed that, starting from an environment relatively free of microorganisms before being inseminated with a few bacterial species, the microbial diversity increased dramatically to include multiple species belonging to several orders of bacteria and fungi. The microbial composition at the different points of time reflected the changes in the environment (e.g., pH, nutrient content, and presence of oxygen) and the interactions between the different microorganisms. The potential of this high through-put sequencing (metagenomical technique) to study the complex microbial interactions in a changing model ecosystem is illustrated by this example.

Yoghurt: This is a dairy product produced by bacterial fermentation of milk. The bacteria employed to make yoghurt are known as "yoghurt cultures." Fermentation of lactose by these bacteria produces lactic acid, which acts on milk protein to give yoghurt its texture. It is made by the addition of a culture of *Lactobacillus delbrueckii* ssp. *bulgaricus* and *Streptococcus thermophilus* bacteria, besides other lactobacilli and bifidobacteria that are sometimes added during or after yoghurt culture (Batish et al., 1997).

Biochemical impacts in yoghurt produced by the addition of probiotic bacteria (*Lactobacillus rhamnosus* GG, *Lactobacillus plantarum* WCFS1, or *Bifidobacterium animalis* Bb12), especially in metabolite production *via* NMR spectroscopy and he GC–MS or LC–MS techniques, can be used to illustrate the simultaneous growth, viability, and metabolite production of starter cultures and probiotic bacteria in yoghurt fermentation.

Fruit Based Products—Wines: Wine is an alcoholic beverage made from fermented grapes or other fruits (Jackson, 1999). Wines made from fruits besides grapes are usually named after the fruit from which they are produced (e.g., pomegranate wine, apple wine, and elderberry wine), and are generically called fruit wines (Joshi, 1997). NMR is successfully being used to characterize wine and find an association of wine metabolites with environmental and fermentative factors in vineyard and wine-making.

The metabolomics and matagenomics thus, help in the detection of various bioactive substances present in the fermented foods. Metagenomic approaches present a fascinating opportunity to identify uncultured microorganisms and to understand their biodiversity, function, interactions, and evolution in different environments (Park et al., 2011). Similarly, GC-MS and HS analysis of the metabolites produced by *L. lactis* in response to temperature and agitation contribute to the understanding of metabolic changes during environmental stresses (Azizan et al., 2012). In a nutshell, the application of new "omics" technologies to unravel the interactions in mixed starter cultures will enable the rational development of new and more effective starter culture systems and increase understanding of the microbiota present in the food (Ivey et al., 2012).

13.11 Control of Toxins in Food and Animal Feed

13.11.1 Mycotoxins

Food is liable to be contaminated with toxins like mycotoxins, putulin, etc., if proper measures are not taken. A number of fungi are known to produce different mycotoxins (Table 13.4). Formation of mycotoxins in food depends on a number of physical and biological factors, as reviewed earlier (Sharma, 1999; Sharma et al., 1980; Westby et al., 1997). The physical factors include moisture in the product, relative humidity, mechanical damage to seeds, temperature, and time of storage. The factors that determine formation of mycotoxins are the micro-environment of the food or its atmosphere, the nature of substrate, the mineral nutrition, the presence of inhibitory factors, and the treatment, if any. The biological factors that predispose the commodity to the attack of mycotoxin-producing fungi include plant stress, invertebrate vectors, fungal infection, load and strain of the fungus, varietal differences, and microbial competition. Naturally, if the raw material is contaminated with mycotoxin producing fungi, the product made out of such a crop is most likely to contain mycotoxins. Indigenous fermented foods also involve fermentation, thus, microorganisms and, hence, any contamination with undesirable microflora, can also lead to the formation of mycotoxins, even if the crop does not contain such toxins itself.

The preferred method of controlling the occurence of mycotoxins in these foods is to prevent their formation, either in the field or during storage (Coker, 1995). However, this is not always feasible, particularly when mycotoxins are produced by field fungi or during storage in uncontrolled hot and humid environments. Efforts are being made to develop modern biotechnological approaches to prevent pre-harvest crop infection for the control of mycotoxin producing fungi, including the development of transgenic

Table 13.4 Different Mycotoxins and Examples of the Reported Effects of Fermentation on Mycotoxins in Raw Materials

TOXIN	RAW MATERIAL/ PRODUCT	TYPE OF FERMENTATION	NATURALLY CONTAMINATED/ SPIKED	EXTENT OF REDUCTION	REFERENCE
Aflatoxin	Maize/kenkey	Lactic acid		None	Jesperson et al. (1994)
Aflatoxin	Sorghum/ogi	Lactic acid	Natural with B_1	12%–16%	Dada and Muller (1983) (http://www.karlsruher-ernaehrungstage.de/papers/Part1.pdf)
Aflatoxin	Wheat/bread	Yeast (dough)	Spiked with B_1	19%	El Banna and Scott (1983), Westby et al. (1997)
Aflatoxin	Milk/yogurt	Lactic acid	Natural with M1	None	Wiseman and Marth (1983)
Aflatoxin	Milk/kefir Milk/yoghurt	Lactic acid Lactic acid	As above spiked with B_1, B_2, G1, G2 and M1	Decreased None	Blanco et al. (1993)
Aflatoxin	Melon seed/ogiri	Bacillus sp.	Natural with B_1 and G1	Complete removal after 4 days	Ogunsanwo et al. (1989)
Aflatoxin	Peanut press cake/ pure mold cultures	*Neurospora sitophila* and *Rhizopus oligosporus*	Not Stated	50% and 70% respectively	Steinkraus (1983), Westby et al. (1997)
Aflatoxin	Maize/ogi Sorghum/ogi	Lactic fermentation	Natural with B_1	Greater with 70%	Adegoke et al. (1994)
Alternariol and alternaniol mono-methyl ether	Pure culture isolates from kenkey	Lactic acid bacteria	Spiked laboratory media	Reduction greater than 50% by all tested strains	Holzapfel (1995)

plants by cloning for genetic resistance to the offending fungus and the cloning of genes coding for known antifungal proteins and their tissue specific expression. The use of competitive microbes and natural inhibitors of fungi can also be developed as biological control agents for the prevention of pre-harvest crop infections. Of course, development of these strategies needs thorough understanding of the genes involved in plant disease resistance and those responsible for controlling mycotoxin production in the fungi. The cloning of mycotoxin genes could also obviate the need to isolate and study the enzymes of the pathway, while allowing understanding the effect of regulation on gene expression. Sequencing of mycotoxin genes can throw light on the evolutionary trends in mycotoxins production. Fungi with deleted mycotoxin genes could also be developed as ideal biocontrol agents. Cloning and characterization of mycotoxin genes could, therefore, help in the development of mycotoxin-resistant transgenic plants. The understanding of plant disease resistance genes can help in the control of plant infections and in the replacement of harmful chemicals through interspecific transfer.

Postharvest infection of agricultural produce could be prevented through the elimination of mold spores using antifungal agents, both physical and chemical, and the development of storage with controlled conditions of atmosphere, humidity, and temperature (Sharma et al., 1980; Zaika and Buchanan, 1987). A number of compounds have been found to affect the biosynthesis or bioregulation of aflatoxin. Many of these studies could help in hazard analysis and in identifying the critical points for preventing mycotoxins formation in foods during processing and preservation. It is also possible to develop seeds resistant to fungal infection during storage through tissue-specific expression of fungal inhibitor genes in seeds using recombinant DNA technology.

Unlike bacterial protein toxins, which can be destroyed at high temperatures, mycotoxins require the most drastic conditions for their destruction. But mycotoxins such as aflatoxin have been shown to be destructible by physical methods such as autoclaving and exposure to ultraviolet radiation, or by chemical methods. But extreme basic conditions, such as treatment with ammonia under pressure, as well as by biological methods such as enzymes and microbes, are also effective, but are known to affect the nutritional and sensory quality attributes of food as well, so these methods have limited application. Other simpler methods for the elimination of mycotoxins are the use of hand and machine sorting for the removal of infected seeds. It is essential to know that a decontamination process should essentially leave no toxic mutagenic or carcinogenic residue, while retaining the nutritive value and the acceptability of the product. It should also not significantly alter the technological properties of food, and must be technically and economically feasible (Bhatnagar et al., 1995).

Strategies to eliminate reformed toxins in a commodity are rather cumbersome and less reassuring (Park et al., 1988). These include sorting of the infected material and general detoxification of a commodity using ammonia. The conventional techniques for the prevention of pre-harvest infection include breeding for resistance to evolve germ plasm or elite varieties of crop which can resist the infecting fungus, besides the use of fungicides and insecticides. There have been a number of reports of the effects of fermentation on the aflatoxin content of contaminated raw materials (Table 13.4) (Westby et al., 1997). Nout (1994) summarized the effects of fermentation on Aflatoxin B_1. Fungi involved in food fermentations, such as *Rhizopus oryzae* (*R. arrhizus*) and *Rhizopus oligosporus*, are able to reduce the cyclopentanon moiety which results in aflatoxicol A, which is a reversible reaction. However, under some conditions (e.g., in the presence of organic acids) aflatoxicol A is irreversibly converted into the stereoisomer aflatoxicol B, which is about 18 times less toxic than aflatoxin B_1. In the conditions of a lactic fermentation (pH 4.0), aflatoxin B_1 is readily converted to aflatoxin B_2, which is also less toxic. These reactions do reduce toxicity, but cannot provide complete detoxification (Nout, 1994). Complete detoxification is only achieved when the lactone ring is broken, which corresponds to a loss in fluorescence at 366 nm, which has been used as a screening tool for this toxin. Bol and Smith (1989) used the technique to identify certain *Rhizopus* spp. that were able to degrade greater than 85% of aflatoxin B_1 present into non-fluorescent substances. The toxicity of these substances, however, is unknown. *Rhizopus* strains with the ability to degrade aflatoxin B_1 have also been reported (Kanittha, 1990; Westby et al., 1997).

Along with four major aflatoxins, several of its metabolites (M1#M2, M4, B2a, G^, etc.) have been identified and characterized. Aflatoxin B_1 (AFB) present in animal feed has been found to be excreted as aflatoxin M (AFM) in milk, which withstands processing steps like chilling, separation, pasteurization, and boiling. It gets concentrated in concentrated/dried dairy products, like *khoa*, *chhana*, condensed milks, dried milks, infant formula, etc. Unfortunately, there is no feasible method that can completely degrade the aflatoxin in milk and milk products. Some chemical methods, like treatment with hydrogen peroxide, sulphates, bisulphates, etc., and physical methods, like adsorption on particulate materials, treatment with UV light, etc., have been tried. The ability of certain strains to degrade aflatoxin offers the potential to develop defined starter cultures for this purpose. Notwithstanding this, fermentation cannot be relied upon as a means of detoxifying raw materials contaminated with aflatoxins, although, at the same time, the potential contribution of fermentation to the safety of some products should also not be ignored.

Aflatoxin content was found to be decreased during the fermentation of milk (Blanco et al., 1993; Van Egmond et al., 1977; Wiseman and Marth, 1983). *Dahi* was prepared using *Lactobacillus acidophilus* (CJ), *Streptococcus thermophilus* (C2), and *Streptococcus lactis* plus *Streptococcus cremoris* (C3) as starter culture. The aflatoxin M1 was determined in milk and experimental *dahi* samples (Table 13.5).

It can be seen that *dahi* samples contained lower aflatoxin M than the milk from which it was prepared. The maximum reduction of aflatoxin M1 was obtained using a combination of *Streptococcus lactis* plus *Streptococcus cremoris* as starter culture (CJ). When the *dahi* samples were stored at $5 \pm 1°C$ for 144 h, the aflaxtoxin M (AFM) content increased up to 72 h in the case of the *dahi* prepared by cultures C2 and C3,

Table 13.5 Effect of Fermentation on Aflatoxin M1 Content during Preparation and Storage of *Dahi*

STORAGE (H)	AF M1	STARTER REDUCTION	AF M1	REDUCTION	AFM$_4$	REDUCTION
Milk	3.00	0.00	3.00	0.00	3.40	0%
0 h (d*N)	2.10	30.00	2.12	29.33	1.27	S746
24	2.10	20.87	1.96	34.67	1.61	6347
48	2.12	30.00	1.72	42.67	1.46	464
72	2.15	29.86	1.79	40.33	1.96	–
96	2.18	27.33	1.84	44.33	1.92	0%
120	1.96	37.10	1.67	81.33	1.70	40.33
144	1.89	34.66	1.64	45.33	1.56	49.33

Source: From Wiseman, D.W. and Marth, E.E. 1983. *J Food Prot* 46: 115–118. With permission.

and upto 96 h in the case of culture C. The decrease in AFM, content in *dahi* during later stages may be due to degradation of the toxin by microbial enzymes (Bhatnagar et al., 1995; Zaika and Buchanan, 1987). Recently, Govaris et al. (2001) and Deveci (2007) studied the distribution and stability of AFM during the production and storage of yoghurt, and showed that, following fermentation, AFM was significantly lowered ($P < 0.01$) in yoghurt with pH 4.0 compared to that in yoghurt with pH 4.8 (Govaris et al., 2002).

The biological detoxification of several other mycotoxins has been also demonstrated, in particular, sterigmatocystin, ochratoxin A, patulin, rubratoxin, zearlenone, T-2 toxin, deoxynivalenol, and diacetoxyscirpenol (Smith et al., 1994). However, detailed studies have only been carried out with aflatoxins. Patulin and ochratoxin A have been reported to be reduced during the fermentation of cider and beer. A number of mycotoxins have been shown to be degraded significantly by rumen organisms. There are clearly some fermentations where the action of microorganisms can contribute to the reduction of mycotoxins from some raw materials. However, the prevention of the contamination of the raw materials should be viewed as the best long term approach (Smith et al., 1994).

In brief, there is a need for an integrated approach for mycotoxin control to be adopted.

13.11.2 Patulin

Patulin (4-hydroxy-4H-furo [3,2c] pyran-2[6H]-one), a mycotoxin, is a secondary metabolite (Figure 13.5) produced by a number of fungi (Table 13.6). Primarily, *Penicillium* and *Aspergillus* species are common to fruits and vegetables and their products (Shao et al., 2012).

Distribution of patulin production in different foods is given in Table 13.7. Most notably, apple is of major concern from food safety considerations. Patulin was first discovered as an antibiotic, but later on the scientific community realised its negative health effects, and it is now classified as a group-3 carcinogen (Joshi et al., 2013).

Patulin is found in the rotten fruits (Figure 13.6).

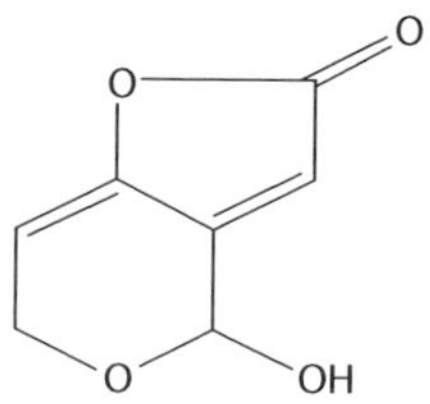

Figure 13.5 Chemical structure of patulin as a secondary metabolite (Figure 13.2) produced by a number of fungi, (Table 13.5). Primarily *Penicillium* and *Aspergillus* species are common to fruits and vegetables and their products. (Adapted from Shao, S., Zhou, T., and McGarvey, B.D. 2012. *Appl Microbiol Biotechnol* 94(3): 789–797.)

Table 13.6   Fungi Responsible for Patulin Production

FUNGAL SPECIES	SYNONYM	REFERENCE
Penicillium urticae Bainier	*Penicillium griseo-fulvum* Dierckx	Kent and Heatley (1945)
	Penicillium patulum Bainier	Chain et al. (1942)
		Birkinshaw et al. (1943)
Penicillium expansum Link	*Penicillium leucopus* (Pers.) Biourge	Anslow et al. (1943)
Penicillium cyclopium Westling		Efimenko and Yakimov (1960)
Penicillium granulatum Bainier	Penicillium divergens Bainier and Sartory	Barta and Mecir (1948)
Penicillium claviforme Bainier		Bergel et al. (1943)
Penicillium melinii Thom		Karow and Fosler (1944)
Aspergillus clavatus Desm.		Umezawa et al. (1947)
Aspergillus giganteus Wehmer		Florey et al. (1944)
Aspergillus terreus Thorn		Kent and Heatley (1945)
Byssochlamys nivea Westling	*Gymnoascus* sp.	Karow and Foster (1944)

Table 13.7   Distribution of Patulin and Patulin-Producing Fungi in Various Fruits and Fruit Products

FRUITS CONTAMINATED WITH PATULIN-PRODUCING FUNGAL STRAINS	FRUIT PRODUCTS CONTAMINATED WITH PATULIN	COUNTRIES WITH APPLE AND JUICE CONTAMINATION
Apples	Apple juice	England
Grapes	Apple-acerola juice	New Zealand
Cherries	Pear juice	United States
Crabapples	Grape juice	South Africa
Apricots	Sour cherry juice	Sweden
Persimmons	Orange juice	Turkey
Strawberries	Pineapple juice	Brazil
Nectarines	Passion fruit juice	Austria
Raspberries	Apple cider	Belgium
Black mulberries	Apple puree	Australia
White mulberries	Strawberry jam	France
Peaches	Blackcurrant jam	Canada
Pear	Blueberry jam	Italy
Plums	Baby food	
Tomatoes	Corn	
Bananas	Cheddar cheese	
Blueberries	Barley malt	
Black currants	Wheat malt	
Almonds	Bread	
Pecans		
Peanuts		
Hazelnuts		

Source: Adapted from Moake et al. (2005); Leggott and Shephard (2001); Demirci et al. (2003); Ritieni (2003); Harwig, J. et al. (1973); Sommer et al. (1974); Ware et al. (1974); Scott et al. (1972); Josefson and Anderson (1976); Brown and Shephard (1999); Gokmen and Acar (1998); De Sylos and Rodriquez-Amaya (1999); Steiner et al. (1999a); Tangi et al. (2003).

Figure 13.6   Rotten apple fruits (Courtesy of Dr. Pooja Lakhanpal).

The various risks posed by patulin to human health include acute, chronic, and cellular level health effects (Table 13.8). There is no evidence of its carcinogenicity in humans, but based on a long-term investigation in rats, the World Health Organization (WHO) has set a tolerable weekly intake of 7 ppb (µg/kg) body weight. The maximum limit of patulin in foods is restricted to 50 ppb in many countries of the world. Conventional analytical detection methods involve chromatographic analyses, such as HPLC, GC, and, more recently, techniques such as LC–MS and GC–MS (Napoli et al., 2007). The risk associated with the patulin necessitates its control and ultimate removal from food products. It has been detected in several apple products, namely apple juice, apple puree, apple wine, apple cider, and baby foods. The quantities range from 0.5 to 732.8 µg/L. Efforts to understand the basic chemical and biological nature of patulin, as well as its interaction with other food components, are being made, as it may occur in apples during both harvest and postharvest stages.

Table 13.8   Effects of Patulin on the Health of Human Beings

ACUTE SYMPTOM	CHRONIC SYMPTOM	CELLULAR LEVEL EFFECT
Agitation, convulsions, dysponea, pulmonary, congestion, edema, hyperemia, GI tract distension, nausea, epithelial cell degeneration/ disruption, intestinal hemorrhage, intestinal inflammation, ulceration	Genotoxic Neurotoxic Immunotoxic Teratogeneic Carcinogenic	Plasma membrane disruption, protein synthesis inhibition, transcription disruption, translation disruption DNA synthesis inhibition, Na-coupled amino acid transport inhibition, immunosuppressive inhibition, Interferon-α production inhibition, RNA Aminoacyl-tRNA synthetases inhibition, Na-K ATPase inhibition, muscle aldolase inhibition, urease inhibition, loss of free glutathione, protein crosslink formation, protein prenylation inhibition

Source: Adapted from Moake et al. (2005); Speijers (2004); Wouters and Speijers (1995); Riley and Showker (1991); Mahfoud et al. (2002); Hatez and Gaye (1978); Miura et al. (1993); Arafat and Musa (1995); Ueno et al. (1976), Moule and Hatey (1977); Arafat et al. (1985); Lee and Roschenthaler (1987), Cooray et al. (1982); Wichmann et al. (2002).

13.11.2.1 Methods to Reduce Patulin The principal risk arises when unfit/rotten fruits are used for the production of juices and other processed products. The process stages and conditions of the process may also effect the concentration of patulin. The nature of the processing, such as fermentation, heat treatment, clarification or additional steps of the production line, such as the use of binding materials, can help in removing patulin from products or at least reduce its concentration. Different methods used for the removal of patulin result in reduction of patulin to variable concentration, such as clarification with diatomaceous earth 0.5%, pectinase 0.02%, and bentonite 0.03%, filtration through activated charcoal 0.05%–0.3%, concentration, apple juice fermentation (with yeast), addition of sulphur dioxide 25–2000 ppm, ascorbic acid 0.05% and ascorbate, and pasteurization (60–90°C for 10 s–20 min).

The most preferred methods, however, are biological, where they can be applied (Joshi et al., 2013). Biological methods of patulin control include the degradation of patulin during yeast fermentation, and are much better understood than other decontamination methods. Stinson et al. (1978) used 8 yeast strains in three typical American processes to ferment apple juice containing 15 mg of added patulin per liter, and found that patulin was reduced to less than the minimum detectable level of 50 µg/liter in all but 2 cases; in all the cases, the level of patulin was reduced by over 99% during alcoholic fermentation. Approximately, 90% of patulin can be removed during yeast fermentation (Burroughs, 1977). In a study, 6 of 8 yeast strains reduced patulin levels to below detectable levels, while all 8 strains resulted in a 99% or better decrease in total patulin content. Control juice, on the other hand, stored for an equal amount of time (2 weeks), had only a 10% reduction (Stinson et al., 1978). In a second study, yeast fermentation reduced patulin levels completely after 2 weeks. The same study also showed that patulin levels failed to decrease significantly in juices that had been yeast fermented and then filter sterilized to remove the yeast, suggesting that active yeast, and not the byproducts, were required for the reduction (Harwig et al., 1973), as was revealed by the treatment of patulin with cyclohexamide (a protein synthesis blocker of yeast), which completely blocked protein synthesis and prevented the detoxification of patulin. Addition of cyclohexamide 3 h after patulin addition, however, reduced, patulin degradation, suggesting that the proteins synthesized within the 3-h were catalytically active against patulin and did not just bind it up in adduct formation (Sumbu et al., 1983). Moss and Long (2002) determined the fate of patulin in the presence of yeast *Saccharomyces*. It was shown that 3 strains of *S. cerevisiae* reduced patulin levels during fermentative growth, but not aerobic growth. Nevertheless, effective, biological control with yeast is limited to the products that can be fermented. Yeast are themselves sensitive to patulin—a concentration greater than 200 µg/mL, completely inhibited the yeast, thus, preventing fermentive detoxification (Sumbu et al., 1983). Still, in another study, considerable reduction in patulin content (Figure 13.7) was demonstrated by alcoholic fermentation of apple juice spiked with patulin (Lakhanpal, 2014).

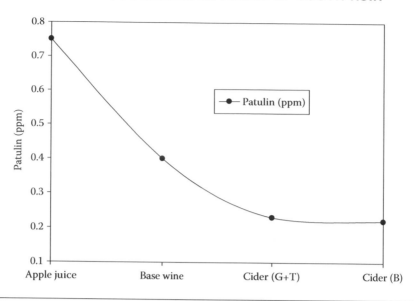

Figure 13.7 Effect of alcoholic fermentation (cider) of apple juice spiked with 0.75 ppm patulin on patulin content.

13.12 Summary and Future Prospectives

The traditional fermented foods are produced and consumed for their beneficial role in health. Indigenous fermented foods with functional properties have a large role to play in the future. All the identification, description, and characterization of different traditional foods, their production methods, and the (environmental) factors that may affect food quality and food safety, need to be considered. The microorganisms present naturally or inoculated into such foods constitute the main aspect of the production method. Though for several products the dominant microflora have been determined, this is not true for many food products, and this information needs to be ascertained by molecular biology-based methods, which are currently the basic tools in microbiological research.

The application of biotechnology in the development of starter cultures, and the use of controlled fermentation in the future could enhance nutritional properties, hygienic quality stability, and safety of traditional fermented foods, with more of their commercialization. The current use of molecular methods in microbiological research has allowed unambiguous and more reliable identification of the microorganisms involved in these fermentations, generating sufficient knowledge for the selection of potential starter cultures for controlled and better production procedures. PCR based methods have become very useful in the detection of food borne pathogens, while research to develop techniques, keeping in view the composition of indigenous fermented foods, is needed. Molecular biology-based methods, especially those culture-independent approaches, have potential to provide in-depth information on microbial diversity during fermentation. Enrichment steps are needed in many cases to recover the microorganisms from food. The future should see the development of the techniques that will ensure the food is safer, tasty, and nutritious.

As the production increases, issues related to the problems of safety, spoilage, and reduced sensory properties are expected to emerge. It is especially true with toxins naturally inherent in the food or produced by microorganisms during cultivation, processing, storage, or distribution and marketing. Biotechnology in one form or another, can meet such challenges successfully.

The development of suitable methods of packaging for traditional fermented foods has become essential, and research and development measures should be undertaken on these aspects. This will also meet the need to export traditional fermented foods from the specific regions and countries where they are produced.

It is expected that in future metabolic engineering, along with metagenomics, will play a very significant role in the processing of indigenous fermented foods. The application of biotechnological innovations will definitely lead to safe indigenous fermented foods of desired quality characteristics. These developments will further be employed to accelerate the pace of commercialization of indigenous fermented foods and their consumption, especially those with therapeutic values.

References

Abriouel, H., Benomar, N., Lucas, R., and Gálvez, A. 2011. Culture-independent study of the diversity of microbial populations in brines during fermentation of naturally-fermented Aloreña green table olives. *Int J Food Microbiol* 144: 487–496.

Achi, O.K. 1990. Microbiology of "Obiolor" a Nigerian fermented non-alcohol beverages. *J Appl Bacterial* 69: 321–325.

Achi, O.K. 1991. Effect of natural fermentation of Yams (*Discorea rotundata*) on characteristic ofprocessed flour. *J Food Sci* 56: 272–275.

Achi, O.K. 1992. Microorganisms associated with natural fermentation of prosopis African seeds for production of "okpiye". *J Plant Foods Hum Nutr* 42: 297–304.

Achi, O.K. 1999. Quality attributes of fermented yam flour supplemented with processed soy flour. *Plant Foods Hum Nutr* 54: 151–158.

Achi, O.K. 2005a. The potential for upgrading traditional fermented foods through biotechnology. *Afr J Biotechnol* 4(5): 375–380.

Achi, O.K. 2005b. Review: Traditional fermented protein condiments in Nigeria. *Afr J Biotechnol* 4: 1612–1621.

Achi, O.K. and Akubor, P.I. 2000. Microbiological characterization of yam fermentation for "Elubo" (yam flour) production. *World J Microbiol Biotechnol* 16: 3–8.

Achi, O.K. and Okereka, E.G. 1999. Proximate composition and functional properties of *Prosopis africana* seed flour. *J Mana Technol* 1: 7–13.

Adegoke, G.O., Otumu, E.J., and Akanni, A.O. 1994. Influence of grain quality, heat, and processing time on the reduction of aflatoxin B_1 levels in "tuwo" and "ogi": Two cereal-based products. *Harzard Foods Hwnan Nutr* 45: 113–117.

Aidoo, K.E. 1986. Lesser-known fermented plant foods. *Trop Sci* 26: 249–258.

Akingbala, J.O., Oguntimein, G.B., and Sobande, A.O. 1995. Physicochemical properties and acceptability of yam flour substituted with soy flour. *Plant Foods Hum Nutr* 48: 73–80.

Amadi, E.N., Barimalaa, I.S., and Omosigho, J. 1999. Influence of temperature on the fermentation of bambara groundnut *Vigna subterranea*, to produce a *dawadawa*-type product. *Plant Foods Hum Nutr* 54: 13–20.

Ampe, F., Omar, N.B., Moizan, C., Wacher, C., and Guyot, J.P. 1999. Polyphasic study of the spatial distribution of microorganisms in *Mexican pozol*, a fermented maize dough, demonstrates the need for cultivation-independent methods to investigate traditional fermentations. *Appl Environ Microbiol* 65: 5464–5473.

An, C., Takahashi, H., Kimura, B., and Kuda, T. 2011. Comparison of PCR-DGGE and PCR-SSCP analysis for microflora of Kaburazushi and Daikonzushi, traditional fermented foods made from fish and vegetables. *J Food Technol* 9: 1–8.

Andorra, I., Monteiro, M., Esteve-Zarzoso, B., Albergaria, H., and Mas, A. 2011. Analysis and direct quantification of *Sacharomyces cerevisiae* and *Hanseniaspora guillermondii* populations during alcoholic fermentation by fluorescence hybridization, flow cytometry and quantitative PCR. *Food Microbiol* 28: 1483–1491.

Andrighetto, C., Zampanese, L., and Lombardi, L. 2001. RAPD-PCR characterization of lactobacilli isolated from artisanal meat plants and traditional fermented sausages of Veneto region (Italy). *Lett Appl Microbiol* 33: 26–30.

Angelov, A., Gotcheva, V., Kuncheva, R., and Hristozova, T. 2006. Development of a new oat-based probiotic drink. *Int J Food Microbiol* 112: 75–80.

Anonymous, 2007. http://wiki.biomine.skelleftea.se/wiki/index.php/DGGE

Anonymous. http://en.wikipedia.org/wiki/Yogurt

Anonymous, 2005. http://everest.bic.nus.edu.sg/lsm2104/2005b/groupc01/public_html/intro.htm

Anonymous, 2010. Functional food, http://spearhead.bz/emedi/index.php?option=com_content&view=article&id=286&Itemid=364

Anonymous, 2011. A review of the evidence to address targeted questions to inform the revision of the Australian Dietary Guidelines.

Anonymous, 2012. National Research Council (US). *Panel on the Application of Biotechnology to Traditional Fermeneted Foods*. National Acedemy Press, Washinton, DC, pp. 199.

Anonymous, 2014. http://wheretobuychiaseeds.org/

Anslow, W.K., Raistrick, H., and Smith, G. 1943. Antifungal substances from molds; patulin (anhydro-3-hydroxymethylene-tetrahydro-1:4-pyrone-2-carboxylic acid), a metabolic product of *Penicillium patulum* Bainier and *Penicillium expansum* Thorn. *J Soc Chem Ind* (London), 62: 236.

Arafat, W., Kern, D., and Dirheimer, G. 1985. Inhibition of aminoacyl-tRNA synthetases by the mycotoxin patulin. *Chem Biol Interact* 56: 333–349.

Arihara, K., Cassens, R.G., and Luchansky, J.B. 1993. Characterization of bacteriocins from *Enterococcus faecium* with activity against *Listeria monocytogenes*. *Int J Food Microbiol* 19: 123–134.

Arteau, M., Labrie, S., and Roy, D. 2010. Terminal-restriction fragment length polymorphism and automated ribosomal intergenic spacer analysis profiling of fungal communities in Camembert cheese. *Int Dairy J* 20: 545–554.

Aymerich, T.B., Martín, M.G., and Hugas, M. 2003. Microbial quality and direct PCR identification of lactic acid bacteria and nonpathogenic Staphylococci from artisanal low-acid sausages. *Appl Environ Microbiol* 69(8): 4583–4594.

Azizan, K., Azlan, S., Nataqain, B., and Normah, M.N. 2012. Metabolic profiling of *Lactococcus lactis* under different culture conditions. *Molecules* 17(7): 8022–8036.

Bae, J.W., Rhee, S.K., Park, J.R. et al. 2005. Development and evaluation of genome-probing microarrays for monitoring lactic acid bacteria. *Appl Environ Microbiol* 12: 8825–8835.

Barimalaa, I.S., Achinewhu, S.C., Yibatima, I., and Amadi, E.N. 1989. Studies on the solid substrate fermentation of bambara groundnut (*Vigna subterranea* (L) Verdc). *J Sci Food Agric* 66: 443–446.

Barta, J. and Mecir, R. 1948. Antibacterial activity of *Penicillium divergens* Bainier. *Experientia*, 4: 277.

Batish, V.K., Grover, S. Pattnaik, P. and Ahmed, N. 1999. Fermented milk products. In: V.K. Joshi and Ashok Pandey (Eds.), *Biotechnology: Food Fermentation*. Educational Publishers and Distributors. Vol. 2, New Delhi and Calcutta. pp. 781–864.

Bendova, O. 1986. The killer phenomenon in yeasts. *Folia Microbial* 31: 422–433.

Bergel, F., Morrison, A.L., Klein, R., Mioss, A.R., Rinderknecht, H., and Ward, J.L. 1943. An antibiotic substance from *Aspergillus clavatus* and *Penicillium claviforme* and its probable identity with patulin. *Nature* (London), 152: 750.

Bevan, E. and Makower, M. 1963. The physiological basis of the killer character in yeast. In: *Proc Xlth Int Congr Genetics*, Vol. I. Pergamon Press, The Netherlands.

Bhatnagar, D., Cleveland, T., Linz, J., and Payne, G. 1995. Molecular biology to eliminate aflatoxins. *INFORM* 6: 262.

Birkinshaw, H., Bracken, A., Michael, S.E., and Raistrick, H. 1943. Patulin in the common cold. II. Biochemistry and chemistry. *Lancet*, 625.

Björkroth, J. and Holzapfel, W. 2006. Genera *Leuconostoc, Oenococcus* and *Weissella*. In: *The Prokaryotes*. Martin, D., Stanley, F., Eugene, R., Karl-Heinz, S., and Erko, S. (Eds.) Springer, New York, NY, pp. 267–319.

Blanco, J.L., Carrion. B.A., Liria, N., Diaz, S., Garcia, M.E., Domineuez, L., and Suarez, C.l. 1993. Behaviour of aflatoxins during manufacture and storage of yoghurt. *Milchtvissenschafi* 48: 385–387.

Blandino, A.M., Al-Aseeri, E., Pandiella, S.S., Cantero, D., and Webb, C. 2003. Cereal-based fermented foods and beverages. *Food Res Int* 36(6): 527–543.

Blaž, S.T. 2006. First decade of terminal restriction fragment length polymorphism (T-RFLP) in microbial ecology. *Acta Agric Slov* 88(2): 65–73.

Bol, J. and Smith J.E. 1989. Biotransformation of aflatoxin. *Food Biotechnol* 3: 127–144.

Boutte, C. 2006. Testing of primers for the study of cyanobacterial molecular diversity by DGGE. *J Microbiol Methods* 65(3): 542–550.

Brown, N.L., and Shephard, G.S. 1999. The incidence of patulin in commercial apple products manufactured and packaged in South Africa. In *The 15th Biennial International Congress and Exhibition of the South African Association for Food Science and Technology in association with the Institute of Packaging (SA)*. September 27–29. Bellville Civic Centre, Cape Town, South Africa.

Burroughs, L.F. 1977. Stability of patulin to sulfur dioxide and to yeast fermentation. *J AOAC* 60: 100.

Cabral, J.S. 2010. Water microbiology. Bacterial pathogens and water. *J Environ Res Public Health* 7(10): 3657–3703.

Campbell-Platt, G. 1994. Fermented foods: A world perspective. *Food Res Int* 27: 253.

Capozzi, V., Russo, P., Fragasso, M., De Vita, P., Fiocco, D., and Spano, G. 2012. Biotechnology and pasta-making: Lactic acid bacteria as a new driver of innovation. *Front Microbiol* 3: 94.

Cardinale, M., Brusetti, L., Quatrini, P. et al. 2004. Comparison of different primer sets for use in automated ribosomal intergenic spacer analysis of complex bacterial communities. *Appl Environ Microbiol* 70: 6147–6156.

Carraro, L., Maifreni, M., Bartolomeoli, I., Martino, M.E., Novelli, E., Frigo, F., Marino, M., and Cardazzo, B.L. 2011. Comparison of culture-dependent and -independent methods for bacterial community monitoring during Montasio cheese manufacturing. *Res Microbiol* 162(3): 231–239.

Chain, E.H., Florey, W., and Jennings, M.A. 1942. An antibacterial substance produced by *Penicillium claviforme*. *Br J Exp Pathol* 23: 202.

Chamkha, M., Sayadi, S., Bru, V., and Godon, J.J. 2008. Microbial diversity in Tunisian olive fermentation brine as evaluated by small subunit rRNA-single strand conformation polymorphism analysis. *Int J Food Microbiol* 122(1–2): 211–215.

Chang, H.W., Kim, K.H., Nam, Y.D. et al. 2008. Analysis of yeast and archeal population dynamics in kimchi using denaturing gradient electrophoresis. *Int J Food Microbiol* 126: 159–166.

Charalampopoulos, D., Pandiella, S.S., and Webb, C. 2003. Evaluation of the effect of malt, wheat and barley extracts on the viability of potentially probiotic lactic acid bacteria under acidic conditions. *Int J Food Microbiol* 82(2): 133–141.

Chelule, P.K., Mokoena, M.P. and Gqaleni, N, 2010. Advantages of traditional lactic acid bacteria fermentation of food in Africa. In: *Current Research, Technology and Education Topics in Applied Microbiology and Microbial Biotechnology*. Méndez-Vilas, A. (Ed.) Formatex Research Center, Spain, pp. 1160–1167.

Chen, H.C., Wang, S.Y., and Chen, M.J. 2008. Microbiological study of lactic acid bacteria in kefir grains by culture-dependent and culture-independent methods. *Food Microbiol* 25: 492–501.

Cho, J.C. and Tiedje, J.M. 2002. Quantitative detection of microbial genes by using DNA microarrays. *Appl Environ Microbiol* 68: 1425–1430.

Ciani, M. and Fatichenti, F. 2001. Killer toxin of *Kluyveromyces phaffii* DBVPG 6076 as a biopresevative agent to control apiculate wine yeast. *Appl Environ Microbiol* 67: 3058–3063.

Cocolin, L. and Comi, G. 2011. Killer yeasts in wine making. In: *Handbook of Enology*. Vol. II. Joshi, V.K. (Ed.), Asia Tech Publishers, INC, New Delhi. pp. 564–565.

Cocolin, L., Manzano, M., Cantoni, C., and Comi, G. 2000. Development of a rapid method for the identification of *Lactobacillus* spp. isolated from naturally fermented Italian sausages using a polymerase chain reaction-temperature gradient gel electrophoresis. *Lett Appl Microbiol* 30: 126–129.

Coker, R.D. 1995. Controlling mycotoxin in oilseeds and oilseed cakes. *Chem Ind* 260–264.

Cole, J.R., Chai, B., Farris, R.J. et al. 2005. The Ribosomal Database Project (RPD-II): Sequences and tools for high-throughput rRNA analysis. *Nucleic Acids Res* 33: D294–D296.

Coles, R. 2003. Introduction. In: *Food Packaging Technology*. Coles, R., McDowell, D., and Kirwan, M.J. (Eds.) Blackwell Publishing, CRC Press, London, U.K., pp. 1–31.

Comitini, F., Ingeniis de, J., Pepe, L., Mannazzu, I., and Ciani, M. 2004a. *Pichia anomala* and *Kluyromyces wickerhamii* killer toxins as new tools against *Dekkera/Brettanomyces* spoilage yeasts. *FEMS Microbial Lett* 218: 235–240.

Comitini, F., Pietro, N., Zacchi, L., Mannazzu, I., and Ciani, M. 2004b. *Kluyromyces phaffii* killer toxin active against wine spoilage yests: purification and characterization. *Microbiology* 150: 2535–2541.

Cooray, R., Kiessling, K.H., and Lindahl-Kiessling, K. 1982. The effects of patulin and patulin–cysteine mixtures on DNA synthesis and the frequency of sister-chromatid exchanges in human lymphocytes. *Food Chem Toxicol* 20: 893–898.

Coppola, S., Blaiotta, G., Ercolini, D., and Moschetti G. 2001. Molecular evaluation of microbial diversity occurring in different types of mozzarella cheese. *J Appl Microbiol* 90: 414–420.

Corbière Morot-Bizot, S., Leroy, S., and Talon, R. 2006. Staphylococcal community of a small unit manufacturing traditional dry fermented sausages. *Int J Food Microbiol* 108: 210–221.

Cremonesi, P., Vanoni, L., Morandi, S., Silvetti, T., Castiglioni, B., and Brasca, M. 2011. Development of a pentaplex PCR assay for the simultaneous detection of *Streptococcus thermophilus*, *Lactobacillus delbrueckii* subsp. *bulgaricus*, *L. delbrueckii* subsp. *lactis*, *L. helveticus*, *L. fermentum* in whey starter for Grana Padano cheese. *Int J Food Microbiol* 146: 207–211.

Dada, L.O. and Muller H.G. 1983. The fate of aflatoxin B_1 in the production of *ogi*, a Nigerian fermented sorghum porridge. *J Cereal Sci* 1: 63–70.

Daengsubha, W. and Suwana-adth, M. 1985. *Development of Indigenous Fermented Foods in Thailand and Other South East Asian Developing Countries: IFS/UNU. Foods in Thailand and Other South East Asian Developing Countries*. Doula Cameroun.

Das, A., Raychaudhuri, U., and Chakraborty, R. 2012. Cereal based functional food of Indian subcontinent: A review. *J Food Sci Technol* 49(6): 665–672.

Dawen, G. and Tao, Y. 2012. Current molecular biologic techniques for characterizing environmental microbial community. *Front Environ Sci Eng* 6(1): 82–97.

Delbès, C., Leila, A.M., and Montel, M.C. 2007. Monitoring bacterial communities in raw milk and cheese 1 by culture-dependent and -independent 16S rRNA gene-based analyzes. *Appl Environ Microbiol* 73: 1881–1891.

Demirci, M., Arici, M., and Gumus, T. 2003. Presence of patulin in fruit and fruit juices produced in Turkey. *Ernaehrungs-Umschau* 50(7): 262–263.

De Sylos, C.M., and Rodriquez-Amaya, D.B. 1999. Incidence of patulin in fruits and fruit juices marketed in Campinas, Brazil. *Food Addit Contam* 16: 71–74.

Deveci, O. 2007. Changes in the concentration of aflatoxin M1 during manufacture and storage of White Pickled cheese. *Food Control* 18: 1103–1107.

De Urraza, P.J., Gómez-Zavaglia, A., Lozano, M.E., Romanowski, V., and De Antoni, G.L. 2000. DNA fingerprinting of thermophilic lactic acid bacteria using repetitive sequence-based polymerase chain reaction. *J Dairy Res* 67: 381–392.

De Vuyst, L., Camu, N., De Winter, T. et al. 2008. Validation of the (GTG)$_5$-rep PCR fingerprinting technique for rapid classification and identification of acetic acid bacteria, with a focus on isolates from Ghanaian fermented cocoa beans. *Int J Food Microbiol* 125: 79–90.

Dirar, H.A. 1993. *The Indigenous Fermented Foods of the Sudan: A Study in African Food and Nutrition*. CAB International, Wallingford, OXon, U.K.

Dolci, P., Alessandria, V., Rantsiou, K., Bertolino, M., and Cocolin, L. 2010. Microbial diversity, dynamics and activity throughout manufacturing and ripening of Castelmagno PDO cheese. *Int J Food Microbiol* 143: 71–75.

Duthoit, F., Godon, J.J., and Montel, M.C. 2003. Bacterial community dynamics during production of registered designation of origin salers cheese as evaluated by 16S rRNA gene single-strand conformation polymorphism analysis. *Appl Environ Microbiol* 69: 3840–3848.

Efimenko, O.M. and Yakimov, A.P. 1960. Antibiotic from *Penicillium cydopium*. *Tr. Leningr. Khirn.-Farm. Inst.* 88; Chem Abstr 55: 21470.

Egert, M. and Friedrich, M.W. 2003. Formation of pseudo-terminal restriction fragments, a PCR-related bias affecting terminal restriction fragment length polymorphism analysis of microbial community structure. *Appl Environ Microbio* 69: 2555–2562.

Egert, M., Ulrich, S., Dyhrberg, L.B., Pommerenke, B., Brune, A., and Friedrich, M.W. 2005. Structure and topology of microbial communities in the major gut compartments of *Melolontha melolontha* larvae (Coleoptera: Scarabaeidae). *Appl Environ Microbiol* 71(8): 4556–4566.

Eka, O.U. 1980. Effect of fermentation on the nutrient status locust beans. *Food Chem* 5: 305–308.

El Banna, A.A. and Scott, P.M. 1983. Fate of mycotoxins during processing of foodstuffs. I. Aflatoxin B$_1$ during making of Egyptian bread. *J Food Prot* 46(4): 301–304.

Elizaquivel, P., Chenoll, E., and Aznar, R. 2007. A TaqMan-based real-time PCR assay for the specific detection and quantification of *Leuconostoc mesenteroides* in meat products. *FEMS Microbiol Lett* 278: 62–71.

Ercolini, D. 2004. PCR-DGGE fingerprinting: Novel strategies for detection of microbes on food. *J Microbiol Methods* 56: 297–314.

Ercolini, D., Hill, P.J., and Dodd, C.E.R. 2003. Development of a fluorescence *in situ* hybridization method for cheese using a 16S rRNA probe. *J Microbiol Methods* 52: 267–271.

Ercolini, D., Moschetti, G., Blaiotta, G., and Coppola, S. 2001. The potential of a polyphasic PCR-DGGE approach in evaluation microbial diversity of natural whey cultures in water-buffalo Mozzarella-cheese production: Bias of culture-dependent and culture independent analyzes. *Syst Appl Microbiol* 24: 610–617.

Falentin, H., Postollec, F., Parayre, S., Henaff, N. LeBivic, P., Richoux, R., Thierry, A., and Sohier, D. 2010. Specific metabolic activity of ripening bacteria quantified by real-time reverse transcription PCR throughout Emmental cheese manufacture. *Int J Food Microbiol* 144(1): 10–19.

Farnworth, E.R. 2008. *Handbook of Fermented Functional Foods*. 2nd edn. CRC Press, Taylor & Francis Group, New York, NY

Feurer, C., Irlinger, F., Spinnler, H.E., Glaser, P., and Vallaeys, T. 2004. Assessment of the rind microbial diversity in a farmhouse-produced vs a pasteurized industrially produced soft red-smear cheese using both cultivation and rDNA based methods. *J Appl Microbiol* 97: 546–556.

Field, D. and Wills, C. 1998. Abundant microsatellite polymorphisms in *Saccharomyces cerevisiae*, and the different distributions of microsatellites in eight prokariotes and *S. cerevisiae*, result from strong mutation pressures and a variety of selective forces. *Proc Natl Acad Sci U S A* 95: 1647–1652.

Filteau, M., Lagacé, L., Lapointe, G., and Roy, D. 2011. Correlation of maple sap composition with bacterial and fungal communities determined by multiplex automated ribosomal intergenic spacer analysis (MARISA). *Food Microbiol* 28: 980–989.

Fisher, M.M. and Triplett, E.W. 1999. Automated approach for ribosomal intergenic spacer analysis of microbial diversity and its application to freshwater bacterial communities. *Appl Environ Microbiol* 65: 4630–4636.

Flórez, A.B. and Mayo, B. 2006. PCR-DGGE as a tool for characterizing dominant microbial populations in the Spanish blue-veined Cabrales cheese. *Int Dairy J* 16: 1205–1210.

Florey, H.W., Jennings, A.M., and Philpot, J.F. 1944. Claviformin from *Aspergillus giganteus* Wehm. *Nature* (London), 153: 139.

Fontana, C., Vignolo, G., and Cocconcelli, P.S. 2005. PCR-DGGE analysis for the identification of microbial populations from Argentinean dry fermented sausages. *J Microbiol Methods* 63(3): 254–263.

Fornasari, M.E., Rossetti, L., Carminati, D., and Giraffa, G. 2006. Cultivability of *Streptococcus thermophilus* in Grana Padano cheese whey starters. *FEMS Microbiol Lett* 257(1): 139–144.

Fujii, T., Watanabe, S., Horikoshi, M., Takahashi, H., and Kimura, B. 2011. PCR-DGGE analysis of bacterial communities in *funazushi*, fermented crucian carp with rice, during fermentation. *Fish Sci* 77: 151–157.

Fuka, M.M., Engel, M., Skelin, A., Redzepovic, S., and Schloter, M. 2010. Bacterial communities associated with the production of artisanal Istrian cheese. *Int J Food Microbiol* 142: 19–24.

Furet, J.P., Quénée, P., and Tailliez, P. 2004. Molecular quantification of lactic acid bacteria in fermented milk products using real-time quantitative PCR. *Int J Food Microbiol* 97: 197–207.

Gadaga, T.H., Molupe, L., and Victor, N. 2013. Traditional fermented foods of Lesotho. *J Microbiol Biotechnol Food Sci* 2(6): 2387–2391.

Gevers, D., Huys, G., and Swings, J. 2001. Applicability of rep-PCR fingerprinting for identification of *Lactobacillus* species. *FEMS Microbiol Lett* 205(1): 31–36.

Giannino, M.L., Aliprandi, M., Feligini, M., Vanoni, L., Brasca, M., and Fracchetti, F. 2009. A DNA array based assay for the characterization of microbial community in raw milk. *J Microbiol Methods* 78: 181–188.

Giraffa, G. 2004. Studying the dynamics of microbial populations during food fermentation. *FEMS Microbiol Rev* 28(2): 251–260.

Giraffa, G. and Domenico, C. 2012. Microorganisms and food fermentation. In: *Handbook of Animal-Based Fermented Food and Beverage Technology*. 2nd edn. Hui, Y.H., and ÖzgülEvranuz, E. (Eds.) CRC Press/Taylor & Francis Group, New York, NY, p. 814.

Giraffa, G. and Neviani, E. 2001. DNA-based, culture-independent strategies for evaluating microbial communities in food-associated ecosystems. *Int J Food Microbiol* 67(1–2): 19–34.

Gokmen, V. and Akar, J. 1998. Incidence of patulin in apple juice concentrates produced in Turkey. *J Chromatogr* 815: 99–102.

Gorges S, Aigner, U., Silakowski, B., and Scherer, S. 2006. Inhibition of *Listeria monocytogenes* by Food-borne yeasts. *Appl Environ Microbiol* 72(1): 313–318.

Govaris, A., Roussi, V., Koidis, P.A., and Botsoglou, N.A. 2001. Distribution and stability of aflatoxin M1 during processing, ripening and storage of Telemes cheese. *Food Additives Contam* 18: 37–44.

Guyot, J.P. 2012. Cereal-based fermented foods in developing countries: Ancient foods for modern research. *Int J Food Sci Technol* 47(6): 1109–1114.

Harlander, S.K. 1992. Genetic improvement of microbial starter culture. In *Application of Biotechnology to Traditional Fermented Foods*. Report of an Ad-Hoc Panel of the Board on Science and Technology for Intl Dev. National Academic Press, Washington, DC, pp. 20–26.

Haruta, S., Ueno, S., Egawa, I. et al. 2006. Succession of bacterial and fungal communities during a traditional pot fermentation of rice vinegar assessed by PCR-mediated denaturing gradient gel electrophoresis. *Int J Food Microbiol* 109: 79–87.

Harwig, J., Scott, P. M., Kennedy, B.P.C., and Chen, Y.K. 1973. Disappearance of patulin from apple juice fermented by *Saccharomyces* spp. *Can Inst Food Sci Technol J* 6(1): 45–46.

Hatez, F. and Gaye, F. 1978. Inhibition of translation in reticulocyte by the mycotoxin patulin. *FEBS Lett* 95: 252–256.

Heard, G.M. and Fleet, G.H. 1987. Occurrence and growth of killer yeasts during wine fermentation. *Appl Environ Microbiol* 53(9): 2171–2174.

Hernandez, A., Martín, A., Córdoba, M.G., Benito, M.J., Aranda, E., and Pérez-Nevado, F. 2008. Determination of killer activity in yeasts isolated from the elaboration of seasoned green table olives. *Int J Food Microbiol* 121(2): 178–188.

Hesseltine, C.W. 1983a. The future of fermented foods. *Nutr Rev* 14: 293–301.

Hesseltine, C.W. 1983b. Microbiology of oriental fermented foods. *Annu Rev Microbiol* 37: 575–601.

Hesseltine, C.W. and Wang, H.L. 1979. Fermented foods. *Chem Ind Lond* 12: 393–399.

Hicks, P.A. 1983. The principles of food packaging, food packaging for Asia and the Pacific. *Proc. CHOGRM Working Group on Industry*, Sydney, Australia.

Hicks, A. 2002. Minimum packaging technology for processed foods: Environmental considerations. *AU J Technol* 6(2): 89–94.

Holzapfel, W.H. 1995. Use of starter cultures in fermentation on a household scale. Background paper prepared for WHO/FAO Workshop: "Assessment of fermentation", Pretoria, South Africa.

Holzapfel, W.H. 2002. Appropriate starter culture technologies for small-scale fermentation in developing countries. *Int J Food Microbiol* 75: 197–212.

http://www.karlsruher-ernaehrungstage.de/papers/Part1.pdf

Hugenholtz, P., Goebel, B.M., and Pace, N.R. 1998. Impact of culture-independent studies on the emerging phylogenetic view of bacterial diversity. *J Bacteriol* 80: 4765–4774.

Humblot, C. and Guyot, J.P. 2009. Pyrosequencing of tagged 16S rRNA gene amplicons for rapid deciphering of the microbiomes of fermented foods such as pearl millet slurries. *Appl Environ Microbiol* 75: 4354–4361.

Hurtado, A., Reguant, C., Bordons, A., and Rozès, N. 2010. Evaluation of a single and combined inoculation of a *Lactobacillus pentosus* starter for processing *cv. Arbequina* natural green olives. *Food Microbiol* 27(6): 731–740.

Ige, M.T. and Akintunde, F.O. 1981. Studies on the local techniques of yam flour production. *Int J Food Sci Technol* 16: 303–311.

Ivey, M., Massel, M., and Trevor, G.P. 2012. Microbial interactions in food fermentations. *Annu Rev Food Sci Technol* 4: 141–162.

Izgu, F.D. and Altinbay, A.Y. 1997. Identification and killer activity of a yeast contaminating starter cultures of *Saccharomyces cerevisiae* strains used in the Turkish baking industry. *Food Microbiol* 14(2): 125–131.

Izgu, F., Altinbay, D., and Acun, T. 2006. Killer toxin of *Pichia anomala* NCYC 432: Purification, characterization and its exo-β-1,3-glucanase activity. *Enzymes Microbial Technol* 39: 669–676.

Izgu, F., Demet, A., and Yasemin, D. 2004. Immunization of the industrial fermentation starter culture strain of *Saccharomyces cerevisiae* to a contaminating killer toxin-producing *Candida tropicalis*. *Food Microbiol* 21(6): 635–640.

Jackson, R.S. 1999. Grape based alcoholic beverages. In: V.K. Joshi and Ashok Pandey (Eds.), *Biotechnology: Food fermentation*. Vol. II. Educational Publishers & Distributors, New Delhi, Ernakulam. pp. 583–646.

Jesperson, A.L., Halm, M., Kpodo, K., and Jakobsen, M. 1994. Significance of yeasts and molds occurring maize dough fermentation for *"kenkey"*. *Int J Food Microbiol* 24: 239–248.

Jiang, Y., Ga, F., Xu, X.L., Su, Y., Ye, K.P., and Zhou, G.H. 2010. Changes in the bacterial communities of vacuum-packaged pork during chilled storage analyzed by PCR-DGGE. *Meat Sci* 86(4): 889–895.

Jianzhong, Z., Xialoli, L., Hanhu, J., and Mingsheng, D. 2009. Analysis of the microflora in Tibetan kefir grains using denaturing gradient gel electrophoresis. *Food Microbiol* 26: 770–775.

Josefson, E., and Andersson, A. 1976. Analysis of patulin in apple beverages sold in Sweden. *Vår Foda*, 28: 189–196.

Joshi, V.K. 1997. *Fruit Wines*. 2nd edn. Directorate of Extension Education. Dr. YS Parmar University of Horticulture and Forestry, Nauni, Solan, HP, p. 255.

Joshi, V.K., Bhutani, V.P., and Sharma, R.C. 1990. Effect of dilution and addition of nitrogen source on chemical, mineral and sensory qualities of wild apricot wine. *Am J Enol Vitic* 41(3): 229–231.

Joshi, V.K., Lakhanpal, P., and Vikas, K. 2013. Occurrence of patulin its dietary intake through consumption of apple and apple products and methods of its removal. *Int. J Food Ferment Technol* 3(1): 15–32.

Joshi, V.K. and Pandey, A. (Eds.) 1999. *Biotechnology: Food Fermentation. Microbiology, Biochemistry and Technology*. Vols. I and II. Educational Publisher and Distributors, Ernakulum/New Delhi, p. 1450.

Jung, M.L., Nam, Y.D., Roh, S.W., and Bae, J.W. 2012. Unexpected convergence of fungal and bacterial communities during fermentation of traditional Korean alcoholic beverages inoculated with various natural starters. *Food Microbiol* 30: 112–123.

Justé, A., Thomma, B.P., and Lievens, B. 2008. Recent advances in molecular techniques to study microbial communities in food-associated matrices and processes. *Food Microbiol* 25(6): 745–761.

Kabak, B. and Dobson, A.D.W. 2011. An introduction to the traditional fermented foods and beverages of Turkey. *Crit Rev Food Sci Nutr* 51(3): 248–260.

Kagan, B.L. 1983. Mode of action of yeast killer toxins: Channel formation in lipid bilayers. *Nature* 302: 709–711.

Kailasapathy, K. 2010. Packaging concepts for enhancing preservation of fermented foods. In: *Fermented Foods and Beverages of the World*. Tamang, J.P. and Kailasapathy, K. (Eds.) CRC Press, Boca Raton, FL. pp. 415–434.

Karow, E.O. and Foster, J.W. 1944. An antibiotic substance from species of *Gymnoascus* and *Penicillium*. *Science* 99: 265.

Kanittha S. 1990. *Kan khat luak chuara ti salai san allatoxin (Degradation of aflatoxin by selected molds)*. Report, Kasetsart University, Bangkok, Thailand.

Kedia, G., Wang, R., Patel, H., and Pandiella, S.S. 2008. Use of mixed cultures for the fermentation of cereal-based substrates with potential probiotic properties. *Process Biochem* 42: 65–70.

Kim, M., Ah, J., Choi I, J.N., Kim, J., Soo, H.Y., Choi, J., and Choong, H.L. 2012. Metabolomics-based optimal koji fermentation for tyrosinase inhibition supplemented with *Astragalus radix*. *Biosci Biotechnol Biochem* 76(5): 863–869.

Kim, M. and Chun, J. 2005. Bacterial community structure in kimchi, a Korean fermented vegetable food, as revealed by 16S rRNA gene analysis. *Int J Food Microbiol* 103: 91–96.

Ko, B.-K., Ki Myong, K., Young-Shick, H., and Cherl-Ho, L. 2010. Metabolomic assessment of fermentative capability of soybean starter treated with high pressure. *J Agric Food Chem* 58(15): 8738–8747.

Koletzko, B., Aggett, P.J., and Bindels, J.G. 1998. Growth, development and differentiation: A functional food science approach. *Braz J Nutr* 80: 35–45.

Kollöffel, B., Meile, L., and Teuber, M. 1999. Analysis of brevibacteria on the surface of Gruyère cheese detected by *in situ* hybridisation and by colony hybridisation. *Lett Appl Microbiol* 29: 317–322.

Kongo, J.M., Ho, A.J., Malcata, F.X., and Wiedmann, M. 2007. Characterization of dominant lactic acid bacteria isolated from São Jorge cheese using biochemical ad ribotyping methods. *J Appl Microbiol* 103: 1838–1844.

Kopecký, J.G., Novotná, M., and Ságová, M. 2009. Modification of the terminal restriction fragment length polymorphism analysis for assessment of a specific taxonomic group within a soil microbial community. *Plant Soil Environ* 55(9): 397–403.

Kopka, J., Fernie, A.F., Weckwerth, W., Gibon, Y., and Stitt, M. 2004. Metabolite profiling in plant biology: Platforms and destinations. *Genome Biol* 5: 109–117.

Kostinek, M., Specht, I., Edward, V.A., Schillinger, U., Hertel, C., Holzapfel, W.H., and Franz, C.M.A.P. 2005. Diversity and technological properties of predominant lactic acid bacteria from fermented cassava used for the preparation of gari, a traditional African food. *Syst Appl Microbiol* 28: 527–540.

Koyanagi, T., Kiyohara, M., Yamamoto, K., Kondo, T., Katayama, T., and Kumagai, H. 2011. Pyrosequencing survey of the microbial diversity of 'narezushi': An archetype of modern Japanese sushi. *Lett Appl Microbiol* 53: 635–640.

Lakhanpal, P. 2014. *Investigations on patulin in apple and apple products*. Submitted in partial fulfilment of the requirements for the degree of Doctor of Philosophy Food Technology, College of Horticulture Dr. YS Parmar University of Horticulture and Forestry, Nauni, India.

Larcia, L.L., Estacio, R.C., and Dalmacio, L.M. 2011. Bacterial diversity on Philippine fermented mustard (*Burong mustasa*) as revealed by 16S rRNA gene analysis. *Benef Microbes* 2: 263–271.

Law, S.V., Abu, B.F., Mat, H.D., and Abdul, H. 2011. A MiniReview Popular fermented foods and beverages in Southeast Asia. *Int Food Res J* 18: 475–484.

Lebionka, A., Servienë, E., and Melvydas, V. 2002. Isolation and purification of yeast *Saccharomyces cerevisiae* K2 killer toxin. *Biologija* 4: 7–9. ISSN 1392-0146.

Lee, K.S., and Roschenthaler, R. 1987. Strand scissions of DNA by patulin in the presence of reducing agents and cupric ions. *J Antibiotics* 40: 692–696.

Leggott, N.L. and Shephard, G.S. 2001. Patulin in South African commercial apple products. *Food Control* 12: 73–76.

Lefeber, T., Gobert, W., Vrancken, G., Camu, N., and De Vuyst, L. 2011. Dynamics and species diversity of communities of lactic acid bacteria and acetic acid bacteria during spontaneous cocoa bean fermentation in vessels. *Food Microbiol* 28: 457–464.

Liu, Z., Lozupone, C., Hamady, M., Bushman, F.D., and Knight, R. 2007. Short pyrosequencing reads suffice for accurate microbial community analysis. *Nucleic Acids Res* 35: 120.

Liu, W.T., Marsh, T.L., Cheng, H., and Forney, L.J. 1997. Characterization of microbial diversity by determining terminal restriction fragment length polymorphisms of genes encoding 16S rRNA. *Appl Environ Microbiol* 63: 4516–4522.

Liu, S.Q. and Tsao, M. 2009. Inhibition of spoilage yeasts in cheese by killer yeast *Williopsis saturnus*. *Int J Food Microbiol* 131: 280–282.

López, V., Fernández-Espinar, M.T., Barrio, E., Ramón, D., and Querol, A. 2003. A new PCR-based method for monitoring inoculated wine fermentations. *Int J Food Microbiol* 81: 63–71.

Lord, N.S., Kaplan, C.W., Shank, P., Kitts, C.L., and Elrod, S.L. 2002. Assessment of fungal diversity using terminal restriction fragment (TRF) pattern analysis: Comparison of 18S and ITS ribosomal regions. *FEMS Microbiol Ecol* 42(3): 327–337.

Lowes, K.F., Shearman, C.A., Payne, J., Mackenzie, D., Archer, D.B., Merry, R.J., and Gasson, M.J. 2000. Prevention of yeast spoilage in feed and food by the yeast mycocin HMK. *Appl Environ Microbiol* 66: 1066–1076.

Ludwig, W. 2007. Nucleic acid techniques in bacterial systematics and identification. *Int J Food Microbiol* 120(3): 225–236.

Lynch, J.M., Benedetti, A., Insam, H.M., Nuti, P., Smalla, K., Torsvik, V., and Nannipieri, P. 2004. Microbial diversity in soil: Ecological theories, the contribution of molecular techniques and the impact of transgenic plants and transgenic microorganisms. *Biol Fertil Soils* 40(6): 363–385.

Machado, A., Carina, A., Carvalho, A., Filip, B., Freddy, H., Rodrigues, L., Nuno, C., and Nuno, A.F. 2013. Fluorescence *in situ* hybridization method using a peptide nucleic acid probe for identification of *Lactobacillus* spp. in milk samples. *Int J Food Microbiol* 162(1): 64–70.

Madoroba, E., Steenkamp, E.T., Theron, J., Scheirlinck, I., Cloete, T.E., and Huys, G. 2011. Diversity and dynamics of bacterial populations during spontaneous sorghum fermentations used to produce ting, a South African food. *Syst Appl Microbiol* 34(3): 227–234.

Mahfoud, R., Maresca, M., Garmy, N., and Fantini, J. 2002. The mycotoxin patulin alters the barrier function of the intestinal epithelium: Mechanism of action of the toxin and protective effects of glutathione. *Toxicol Appl Pharmacol* 18: 1209–1218.

Manzano, M., Lacumin, L., Giusto, C. et al. 2012. Utilization of denaturing gradient gel electrophoresis (DGGE) to evaluate the intestinal microbiota of brown trout Salmotruttafario. *J Vet Sci Med Diagn* 1: 2. doi:10.4172/2325-9590.1000105

Margulies, M., Egholm, M., Altman, W.E. et al. 2005. Genome sequencing in microfabricated high-density picolitre reactors. *Nature* 437: 376–380.

Marquina, D., Santos, A., and Peinado, J.M. 2002. Biology of killer yeasts. *Int Microbiol* 5: 65–71.

Marsh, K. and Bugushu, B. 2007. Food packaging—roles, materials and environmental issues. *J Food Sci* 72(3): R39–R54.

Marsh, T.L., Saxman, P., Cole, J., and Tiedje, J. 2000. Terminal restriction fragment length polymorphism analysis program, a web-based research tool for microbial community analysis. *Appl Environ Microbiol* 66: 3616–3620.

Marshall, E. and Danilo, M. 2012. *Traditional Fermented Food and Beverage for Improved Livelihoods*. FAO.

Martin, B., Jofré, A., Garriga, M., Pla, M., and Aymerich, T. 2006. Rapid detection of *Lactobacillus sakei* in meat and fermented sausages by real-time PCR. *Appl Environ Microbiol* 72: 6040–6048.

Marty, E., Buchs, J., Eugster-Meier, E., Lacroix, C., and Meile, L. 2012. Identification of Staphylococci and dominant lactic acid bacteria in spontaneously fermented Swiss meat products using PCR-RFLP. *Food Microbiol* 29: 157–166.

Masco, L., Vanhoutte, T., Temmerman, R., Swings, J., and Huys, G. 2007. Evaluation of real-time PCR targeting the 16S rRNA and recA genes for the enumeration of bifidobacteria in probiotic products. *Int J Food Microbiol* 113(3): 351–357.

Masoud, W., Takamiya, M., Vogensen, F.K. et al. 2011. Characterization of bacterial populations in Danish raw milk cheeses made with different starter cultures by denaturing gradient gel electrophoresis and pyrosequencing. *Int Dairy J* 21: 142–148.

McCartney, A.L. 2002. Application of molecular biological methods for studying probiotics and the gut flora. *Br J Nutr* 88(Suppl. 1): S29–S37.

McEniry, J., Kiely, P.O., Clipson, N.J.W., Forrista, I.P.D., and Doyle, E.M. 2008. Bacterial community dynamics during the ensilage of wilted grass. *J Appl Microbiol* 105(2): 359–371.

Mehta, B. and Kamal-Eldin, A. 2012. Introduction. In: *Chemical & Functional Properties of Food Components: Fermentation Effects on Food Properties*. Mehta, B., Kamal-Eldin, A., and Iwanski, R.Z. (Eds.) CRC Press/Taylor & Francis Group, New York.

Meroth, C.B., Walter, J., Hertel, C., Brandt, M.J., and Hammes, W.P. 2003. Monitoring the bacterial population dynamics in sourdough fermentation process by using PCR-denaturing gradient gel electrophoresis. *Appl Environ Microbiol* 69: 475–482.

Miambi, E., Guyot, J.P., and Ampe, F. 2003. Identification, isolation and quantification of representative bacteria from fermented cassava dough using an integrated approach of culture-dependent and culture-independent methods. *Int J Food Microbiol* 82: 111–120.

Michalcakova, S. and Repova, L. 1992. Effect of ethanol, temperature and pH on the stability of killer yeast strains. *Acta Biotechnol* 3: 163–168.

Miura, S., Hasumi, K., and Endo, A. 1993. Inhibition of protein prenylation by patulin. *Fed Eur Biochem Soc* 318(1): 88–90.

Moake, M.M., Padilla-Zakour, O.I., and Worobo, R.W. 2005. Comprehensive review of patulin control methods in foods. *Comp Rev Food Sci Food Safety*, 4(1): 8–21.

Mohammed, M., Abd El-Aziz, H., Omran, N., Anwar, S., Awad, S., and El-Soda, M. 2009. Rep-PCR characterization and biochemical selection of lactic acid bacteria isolated from the Delta area of Egypt. *Int J Food Microbiol* 128(3): 417–423.

Moreno, G.K., Rodríguez-Lerma, G.K., Cárdenas-Marríquez, M., Botello-Álvarez, E., Jiménez-Islas, H., Rico-Martínez, R., and Navarrete-Bolaños, J.L. 2010. A strategy for biotechnological processes design: Prickly pear (*Opuntiaficus indica*) wine production. *Chem Eng Trans* 20: 315–320.

Moss, M.O. and Long, M.T. 2002. Fate of patulin in the presence of the yeast *Saccharomyces cerevisiae*. *Food Addit Contam* 19(4): 387–399.

Moule, Y. and Hatey, F. 1977. Mechanism of the in vitro inhibition of transcription by patulin, a mycotoxin from *Byssochlamys nivea*. *FEBS Lett* 74: 121–125.

Nakayama, J. 2010. Pyrosequence-based 16S rRNA profiling of gastrointestinal microbiota. *Biosci Microflora* 29: 83–96.

Nakayama, J., Hoshiko, H., Fukuda, M. et al. 2007. Molecular monitoring of bacterial community structure in long-aged nukadoko: Pickling bed of fermented rice bran dominated by slow-growing lactobacilli. *J Biosci Bioeng* 104; 481–489.

Nam, Y.D., Chang, H.W., Kim, K.H., Roh, S.W., and Bae, J.W. 2009. Metatranscriptome analysis of lactic acid bacteria during kimchi fermentation with genome-probing microarrays. *Int J Food Microbiol* 130: 140–146.

Napoli, D.L., Champdor, Ž.M., Bazzicalupo, P., Montesarchio, D., Di Fabio, G., Cocozza, I., Parracino A., Rossi, M., and D'Auria, S. 2007. A new competitive fluorescence assay for the detection of patulin toxin. *Anal Chem* 79(2): 751–757.

Ndoye, B., Rasolofo, E., Andria, M., Gisele, L., and Roy, D. 2011. A review of the molecular approaches to investigate the diversity and activity of cheese microbiota. *Dairy Sci Technol* 91(5): 495–524.

Nieminen, T.T., Vihavainen, E., Paloranta, A., Lehto, J., Paulin, L., Auvinen, P., Solismaa, M., and Björkroth, K.J. 2011. Characterization of psychrotrophic bacterial communities in modified atmosphere-packed meat with terminal restriction fragment length polymorphism. *Int J Food Microbiol* 144(3): 360–366.

Nisiotou, A., Foteini, P., and Konstantinos, K. 2014. Old targets, new weapons: Food microbial communities revealed with molecular tools. In: *Novel Food Preservation and Microbial Assessment Techniques*. Ioannis, S.B (Ed.) CRC Press/Taylor & Francis Group, New York, NY, pp. 277–312.

Nocker, A., Burr, M., and Camper, A.K. 2007. Genotypic microbial community profiling: A critical technical review. *Microbial Ecol* 54(2): 276–289.

Noronha, S.B., Yeh, H.J., Spande, T.F., and Shiloach, J. 2000. Investigation of the TCA cycle and the glyoxylate shunt in *Escherichia coli* BL21 and JM109 using (13)C-NMR/MS. *Biotechnol Bioeng* 68: 316–327.

Nout, M.J.R. 1985. Upgrading traditional biotechnological processes. In: *IFS/UNU Workshop on Development of Indigenous Fermented Foods and Food Technology in Africa*. Prage, L. (Ed.) Douala, Cameroon, pp. 90–99.

Nout. M.J.R. 1994. Fermented foods and food safety. *Food Res Int* 27: 291.

Nwachukwu, E., Achi, O.K., and Ijeoma, I.O. 2010. Lactic acid bacteria in fermentation of cereals for the production of indigenous Nigerian foods. *Afr J Food Sci Technol* 1(2): 21–26.

Ochi, H., Naito, H., Iwatsuki, K., Bamba, T., and Fukusaki, E. 2012. Metabolomics-based component profiling of hard and semi-hard natural cheeses with gas chromatography/time-of-flight-mass spectrometry, and its application to sensory predictive modeling. *J Biosci Bioeng* 113(6): 751–758.

Odunfa, S.A. 1985. African fermented foods. In: *Microbiology of Fermented Foods*. Vol. 2. Wood, B.J. (Ed.) Elsevier Appl. Sci. Publ., London, pp. 155–191.

Odunfa, S.A. 1985a. Microbiological and toxicological aspect of fermentation of castor oil seeds for *ogiri* production. *J. Food Sci* 50: 1758–1759.

Odunfa, S.A. 1985b. African fermented foods. In: *Microbiology of Fermented Foods*. Vol. 2. Wood, B.J. (Ed.) Elsevier, London. pp. 155–191.

Ogier, J.C., Son, O., Gruss, A., Tailliez, P., and Delacroix-Buchet, A. 2002. Identification of the bacterial microflora in dairy products by temporal temperature gradient gel electrophoresis. *Appl Environ Microbiol* 68: 3691–3701.

Ogunsanwo B.M., Faboya O.O., Idowu, O.R., Ikotun, T., and Akano, D.A. 1989. The fate of aflatoxins during the production of *"Oeiri"*. A West African fermented melon seed condiment from artificially contaminated seeds. *Nahrung* 33: 983–988.

Oguntoyinbo, F.A., Tourlomousis, P., Gasson, M.J. and Narbad, A. 2011. Analysis of bacterial communities of traditional West African cereal foods using culture independent methods. *Int J Food Microbiol* 145: 205–210.

Okafor, N. 1983. Processing of Nigerian indigneous fermented foods—A chance for innovation. *Nig Food J* 1: 32–37.

Okafor, N. 1990. Traditional alcohol beverages of tropical Africa, strategies for scale-up. *Process Biochem Int* 25: 213–220.

Olasupo, N.A. 2006. Fermentation biotechnology of traditional foods of Africa. In: *Food Biotechnology*, 2nd edn. Revised and Expanded. Shetty, K., Pometto, A., Paliyath, G., and Levi, R.E. (Eds.) CRC Press/Taylor & Francis, Boca Raton, FL, pp. 1705–1739.

Olasupo, N., Odunfa, S., and Obayori, O. 2010. Ethnic African fermented foods In: *Fermented Foods and Beverages of the World*. Tamang, J.P. and Kailasapathy, K. (Eds.) CRC Press, Boa Raton, FL, pp. 323–352.

Omafuvbe, B.O., Shonukan, O.O., and Abiose, S.H. 2000. Microbiological and biochemical changes in the traditional fermentation of soybean for soy-daddawa—A Nigerian food condiment. *Food Microbiol* 17: 469–474.

Oniofiok, N., Nnayelugo, D.O., and Ukwondi B.E. 1996. Usage patterns and contributions of fermented foods to the nutrient intake of low-income households in Emene Nigeria. *Plant Foods Hum Nutr* 49: 199–211.

Osborne, C.A., Rees, G. N., Bernstein, Y., and Janssen, P.H. 2006. New threshold and confidence estimates for terminal restriction fragment length polymorphism analysis of complex bacterial communities. *Appl Environ Microbiol* 72: 1270–1278.

Ouoba, L.I. 2010. Research on African alkaline fermentations for improvement of food security in Africa: Example of interactions between African and European scientists and institutions. www.foodmicro.dk/.../7._Ouoba_ICFMH_Symposium_30-08-10.pdf.

Ouoba, L.I., Diawara, B., Moa-Awua, W.K., Traore, A.S., and Moller, P.L. 2004. Genotyping of starter cultures of *Bacillus subtilis* and *Bacillus pumilus* for fermentation of African locust bean *Parkia biglobosa* to produce *soumbala*. *Int J Food Microbiol* 90: 197–205.

Ovreas, L. 2000. Population and community level approaches for analyzing microbial diversity in natural environments. *Ecol Lett* 3: 236–251.

Park, E.J., Chun, J., Cha, C.J., Park, W.S., Jeon, C.O., and Bae, J.W. 2012. Bacterial community analysis during fermentation of ten representative kinds of kimchi with barcoded pyrosequencing. *Food Microbiol* 30: 197–204.

Park, E.-J., Kyoung-Ho, K., Guy, C.J., Abell, M., Kim, S., Woon, R., and Jin-Woo, B. 2011. Metagenomic analysis of the viral communities in fermented foods. *Appl Environ Microbiol* 77(4): 1284–1291.

Park, D.L., Lee, L.S., Price, R.L., and Pohland, A.L. 1988. Review of the decontamination of aflatoxins by ammoniation: Current status and regulation. *J Assoc Off Anal Chem* 71: 685.

Parkouda, C., Nielsen, D.S., Azokpota, P. et al. 2009. The microbiology of alkaline fermentation of indigenous seeds used as food condiments in Africa and Asia. *Crit Rev Microbiol* 35: 139–156.

Petering, J.E., Symons, M.R., Langridge, P., and Henschke, P.A. 1991. Determination of killer yeast activity in fermenting grape juice by marked *Saccharomyces* wine yeast strain. *Appl Environ Microbiol* 57: 3232–3236.

Peters, S., Stefanie, K., Schwieger, F., and Christoph, C.T. 2000. Succession of microbial communities during hot composting as detected by PCR—Single-strand-conformation polymorphism-based genetic profiles of small-subunit rRNA genes. *Appl Environ Microbiol* 66(3): 930–936.

Plengvidhya, V., Breidt, F.J., and Fleming, H.P. 2004. Use of RAPD-PCR as a method to follow the progress of starter cultures in sauerkraut fermentation. *Int J Food Microbiol* 93: 287–296.

Postollec, F., Falentin, H., Pavan, Sonia, Combrisson, J., and Danièle, S. 2011. Recent advances in quantitative PCR (qPCR) applications in food microbiology. *Food Microbiol* 28(5): 848–861.

Prajapati, J.B. and Nair, B.M. 2003. The history of fermented foods. In: *Fermented Functional Foods*. Farnworth, E.R. (Ed.) CRC Press, Boca Raton, New York, London, Washington, DC, pp 1–25.

Pulido, R.P., Ben Omar, N., Abriouel, H., Lopez, R.L., Canamero, M.M., and Galvez, A. 2005. Microbiological study of lactic acid fermentation of caper berries by molecular and culture-dependent methods. *Appl Environ Microbiol* 71: 7872–7879.

Riley, R.T. and Showker, J.L. 1991. The mechanism of patulin's cytotoxicity and the antioxidant activity of indole tetramic acids. *Toxicol Appl Pharmacol* 109: 108–126.

Ritieni, A. 2003. Patulin in Italian commercial apple products. *J Agric Food Chem* 51: 6086–6090.

Quattara, H.G., Reverchon, S., Niake, S. L., and Nasser, W. 2011. Molecular identification and pectate lyase production by *Bacillus* strains involved in cocoa fermentation. *Food Microbiol* 28: 1–8.

Quiberoni, U.S.A., Tailliez, A., Quénée, P., Suarez, V., and Reinheimer, J. 1998. Genetic (RAPD-PCR) and technological diversities among wild *Lactobacillus helveticus* strains *J Appl Microbiol* 85: 591–596.

Quigley, L., O'Sullivan, O., Beresford, T.P., Ross, R.P., Fitzgerald, G.F., and Cotter, P.D. 2011. Molecular approaches to analysing the microbial composition of raw milk and raw milk cheese. *Int J Food Micrrbiol* 150(2–3): 81–94.

Rademaker, J.L.W., Hoolwerf, J.D., Wagendorp, A.A., and Te Giffel, M.C. 2006. Assessment of microbial population dynamics during yoghurt and hard cheese fermentation and ripening by DNA population fingerprinting. *Int Dairy J* 16: 457–466.

Rademaker, J.L.W., Peinhopf, M., Rijinen, L., Bockelmann, W., and Noordman, W.H. 2005. The surface microflora dynamics of bacterial smear ripened Tilsit cheese determined by T-RFLP DNA population fingerprint analysis. *Int Dairy J* 15: 785–794.

Randazzo, C.L., Pitino, L., Ribbera, A., and Caggia, C. 2010. Pecorino Crotone cheese: Study of bacterial population and flavour compounds. *Food Microbiol* 27: 363–374.

Randazzo, C.L., Torriani, S., Akkermans, A.D.L., De Vos, W.M., and Vaughan, E.E. 2002. Diversity, dynamics, and activity of bacterial communities during production of an artisanal Sicilian cheese as evaluated by 16S rRNA analysis. *Appl Environ Microbiol* 68: 1882–1892.

Rantsiou, K. and Cocolin, L, 2006. New developments in the study of the microbiota of naturally fermented sausages as determined by molecular methods: A review. *Int J Food Microbiol* 108(2): 255–267.

Rantsiou, K. and Cocolin, L. 2008. Fermented meat products, molecular techniques in the microbial ecology of fermented foods. In: *Food Microbiology and Food Safety.* Luca, C. and Danilo, E. (Eds.) Springer, New York, NY, pp. 91–118.

Rantsiou, K., Comi, G., and Cocolin, L. 2004. The *rpoB* gene as a target for PCR-DGGE analysis to follow lactic acid bacterial population dynamics during food fermentations. *Food Microbiol* 21: 481–487.

Rasolfo, E.A., St. Gelais, D., Lapointe, G., and Roy, D. 2010. Molecular analysis of bacterial population structure and dynamics during cold storage of untreated and treated milk. *Int J Food Microbiol* 138: 108–118.

Rathore, S., Salmerón, I., Severino, S., and Pandiella, S.S. 2012. Production of potentially probiotic beverages using single and mixed cereal substrates fermented with lactic acid bacteria cultures. *Food Microbiol* 30(1): 239–244.

Ravyts, F., Vuyst, L., and Frédéric, L. 2012. Bacterial diversity and functionalities in food fermentations. *Eng Life Sci* 12(4): 356–367.

Ray, R.C. and Sivakumar, P. 2009. Traditional and novel fermented foods and beverages from tropical root and tuber crops: review. *Int J Food Sci Technol* 44(6): 1073–1087.

Rebecchi, A., Crivori, S., Sara, P.G., and Cocconelli, P.S. 1998. Physiological and molecular techniques for the study of bacterial community development in sausage fermentation. *J Appl Microbiol* 84: 1043–1049.

Reddy, N.R. and Pierson, M.D. 1999. Reduction in antinutritional and toxic components in plant foods by fermentation. *Food Res Int* 27(3): 281–290.

Ribiero, H. 2000. Diarrheal diseases in a developing nation. *Am J Gastroenterol* 95: S14–S15.

Roberfroid, M.B. 1998. Probiotics and synbiotics: Concepts and nutritional properties. *Br J Nutr* 80(S2): S197–S202.

Robert, H., Gabriel, V., and Fontagne-Faucher, C. 2009. Biodiversity of lactic acid bacteria in French wheat sourdough as determined by molecular characterization using species-specific PCR. *Int J Food Microbiol* 135: 53–59.

Rodas, A.M., Ferrer, S., and Pardo, I. 2003. 16S-ARDRA, a tool for identification of lactic acid bacteria isolated from grape must and wine. *Syst Appl Microbiol* 26: 412–422.

Roh, S.W., Kim, K.H., Nam, Y.D., Chang, H.W., Park, E.J., and Bae, J.W. 2010. Investigation of archeal and bacterial diversity in fermented seafood using barcoded pyrosequencing. *ISME J* 4: 1–18.

Ruiz, P., Izquierdo, P.M., Sesena, S., and Palop, M.L. 2008. Intraspecific diversity of lactic acid bacteria from malolactic fermentation of Cencibel wines as derived from combined analysis of RAPD-PCR and PFGE patterns. *Food Microbiol* 25: 942–948.

Ruiz, P., Seseña, S., Izquierdo, P.M., and Palop, M.L. 2010. Bacterial biodiversity and dynamics during malolactic fermentation of Tempranillo wines as determined by a culture-independent method (PCR-DGGE). *Appl Microbiol Biotechnol* 86: 1555–1562.

Sakamoto, N., Shigemitsu, T., Sonomoto, K., and Nakayama, J. 2011. 16S rRNA pyrosequencing-based investigation of the bacterial community in nukadoko, a pickling bed of fermented rice bran. *Int J Food Microbiol* 144: 352–359.

Salovaara, H. 2004. Lactic acid bacteria in cereal based products. In: *Lactic acid Bacteria— Technology and Health Effects*, 2nd edn. Salminen, S. and von Wright, A. (Eds.) Marcel Dekker, New York, NY, pp. 431–452.

Sánchez, E. and Sanz, Y. 2011. Lactobacillus. In: *Molecular Detection of Human Bacterial Pathogens.* Dongyou, L. (Ed.) CRC Press, Boca Raton, FL, p. 1278.

Sanni, A.I. 1993. The need for process optimization of African fermented foods and beverages. *Int J Food Microbio* 18(2): 85–95.

Santos, E.M., Jaime, I., Rovira, J., Lyhs, U., Korkeala, H., and Bjorkroth, J. 2005. Characterization and identification of lactic acid bacteria in Morcilla de Burgos. *Int J Food Microbiol* 97(3): 285–296.

Santos, A. and Marquina, D. 2004. Killer toxin of *Pichia membranifaciens* and its possible use as a biocontrol agent against grey mold disease of grapevine. *Microbiology* 150: 2527–2534.

Santos, A. Sánchez, A., and Marquina, D. 2004. Yeasts as biological agents to control *Botrytis cinerea*. *Microbiol Res* 159(4): 331–338.

Sarkar, P.K. and Nout, M.J.R. (Eds.) 2014. *Handbook of Indigenous Foods Involving Alkaline Fermentation*. CRC Press/Taylor & Francis Group, Boca Raton, FL, p. 629.

Sarkar, P.K. and Tamang, J.P. 1995. Changes in the microbial profile and proximate composition during natural and controlled fermentations of soybeans to produce *kinema*. *Food Microbiol* 12: 317–325.

Schmitt, M., Brendel, M., Schwaiz, R., and Radler, F. 1989. Inhibition of DNA synthesis in *Saccharomyces cerevisiae* by killer yeast toxin KT28. *J Gen Microbiol* 135: 1529–1535.

Schütte, U.M.E., Abdo, Z., Bent, S.J. et al. 2008. Advances in the use of terminal restriction fragment length polymorphism (T-RFLP) analysis of 16S rRNA genes to characterize microbial communities. *Appl Microbiol Biotechnol* 80: 365–380.

Schwieger, F. and Tebbe, C.C. 1998. A new approach to utilize PCR-single-strand-conformation polymorphism for 16S rRNA gene-based microbial community analysis. *Appl Environ Microbiol* 64: 4870–4876.

Scott, P.M., Miles, W.F., Toft, P., and Dube, J.G. 1972. Occurrence of patulin in apple juice. *J Agric Food Chem* 20: 450.

Seishi, I., Nozomi, Y., Hiroshi, E., Kiwamu, M., and Fujimura, T. 2006. Soil microbial community analysis in the environmental risk assessment of transgenic plants. *Plant Biotechnol* 23(1): 137–151.

Sekiguchi, H., Tomioka, N., Nakahara, T., and Uchiyama, H. 2001. A single band does not always represent single bacterial strains in denaturing gradient gel electrophoresis analysis. *Biotechnol Lett* 23: 1205–1208.

Selvakumar, D., Miyamoto, M., Furuichi, Y., and Komiyama, T. 2006. Inhibition of fungal β-1,3-glucan synthase cell growth by HM-1 killer toxin single-chain anti-idiotypic antibodies. *Antimicrob Agents Chemother* 50: 3090–3097.

Sensabaugh, G.F. 2009. Microbial community profiling for the characterisation of soil evidence: Forensic considerations. In: *Criminal and Environmental Soil Forensics*. Karl, R., Lorna, D., and David, M. (Eds.) Springer, New York, NY, pp. 49–60.

Settanni, L. and Corsetti, A. 2007. The use of multiplex PCR to detect and differentiate food- and beverage-associated microorganisms: A review. *J Microbiol Methods* 69: 1–22.

Settanni, L., Valmorri, S., Van Sinderen, D., Suzzi, G., Paparella, A., and Corsetti, A. 2006. Combination of multiplex PCR and PCR-denaturing gradient gel electrophoresis for monitoring common sourdough-associated *Lactobacillus* species. *Appl Environ Microbiol* 72: 3739–3796.

Shao, S., Zhou, T., and McGarvey, B.D. 2012. Comparative metabolomic analysis of *Saccharomyces cerevisiae* during the degradation of patulin using gas chromatography–mass spectrometry, *Appl Microbiol Biotechnol* 94(3): 789–797.

Sharma, A. 1999. Microbial toxins. In: *Biotechnology: Food Fermentation*. Joshi, V.K. and Pandey, A. (Eds.) Educational Publisher and Distributors, New Delhi, pp. 321–345.

Sharma, A., Behere, A.G., Padwal-Desai, S.R. and Nadkami, G.B. 1980. Influence of inoculum size of *Aspergillus parasticus* spores on aflatoxin production. *Appl Environ Microbial* 57: 2487.

Sheffield, V.C., Cox, D.R., Lerman, L.S., and Myers, R.M. 1989. Attachment of a 40-base-pair G+C-rich sequence (GC-clamp) to genomic DNA fragments by the polymerase chain reaction results in improved detection of single base changes. *Proc Natl Acad Sci U S A* 86: 232–236.

Silva da, S., Sílvia, C., Lucas, C., and Aguiar, C. 2008. Unusual properties of the halotolerant yeast *Candida nodaensis* killer toxin, CnKT. *Microbiol Res* 163(2): 243–251.

Singh, S., Goswami, P., Singh, R., and Heller, K. 2009. Application of molecular identification tools for *Lactobacillus*, with a focus on discrimination between closely related species: A review. *LWT – Food Sci Technol* 42: 448–457.

Sklarz, M.Y., Angel, R., Gillor, O., Ines, M., and Soares, M. 2009. Evaluating amplified rDNA restriction analysis assay for identification of bacterial communities. *Anton Leeuwen* 96(4): 659–664.

Sklarz, M.Y., Angel, R., Gillor, O., and Soares, I.M. 2011. Amplified rDNA restriction analysis (ARDRA) for identification and phylogenetic placement of 16S-rDNA clones. In: *Handbook of Molecular Microbial Ecology. I: Metagenomics and Complementary Approaches.* de Bruijn, F.J. (Ed.) John Wiley & Sons, Inc., Hoboken, NJ.

Smith, J.E., Lewis, C.W., Anderson, J.G., and Solomons, G.L. 1994. *Mycotoxins in Human Nutrition and Health.* European Union Directorate-General XII Report 16048EN.

Sohier, D. and Lonvaud-Funel, A. 1998. Rapid and sensitive *in situ* hybridization method for detecting and identifying lactic acid bacteria in wine. *Food Microbiol* 15: 391–397.

Sommer, N.F., Buchanan, J.R., and Fortlage, R.J. 1974. Production of patulin by *Penicillium expansum. Appl Microbiol* 28: 589–593.

Speijers, G.J.A. 2004. Patulin. In Magan, N. and Olsen, M., eds. *Mycotoxins in Food—Detection and Control.* Woodhead Publishing, Cambridge, England, pp. 339–352.

Spiegelman, D., Whissell, G., and Charles, G.W. 2005. A survey of the methods for the characterization of microbial consortia and communities. *Can J Microbiol* 51(5): 355–386.

Stark, M.J.R., Boyd, A., Mileham, A.J., and Romanos, M.A. 1990. The plasmid-encoded killer system of *Kluyveromyces lactis. Yeast* 6: 1–29.

Steiner, I., Werner, D., and Washüttl, J. 1999a. Patulin in fruit juices. Part 1. Analysis and control in Austrian apple and pear juices. *Ernährung Nutr* 23: 202–208.

Steinkraus, K.H. 1983. *Handbook of Indigenous Fermented Foods.* Marcel Dekker, Inc., New Delhi.

Steinkraus, K.H. 1994. *Nutritional Significance of Fermented Foods. Food Res Inter,* 27, 259–267.

Steinkraus, K.H. 1996. *Handbook of Indigenous Fermented Food.* 2nd edn. Marcel Dekker, Inc., New York, NY.

Steinkraus, K.H. 2004. *Industrialization of Indigenous Fermented Foods,* Second Edition, Revised and Expanded. Worth Publishers, New York.

Stinson, E.E., Osman, S.F., Huhtanen, C.N., and Bills, D.D. 1978. Disappearance of patulin during alcoholic fermentation of apple juice. *Appl Environ Microbiol* 36(4): 620–622.

Sumbu, Z.L., Thonart, P., and Bechet, J. 1983. Action of patulin on a yeast. *Appl Environ Microbiol* 45(1): 110–115.

Talwalkar, A. and Kailasapathy, K., 2004. Oxidative stress adaptation of probiotic bacteria. *Milchwissenschaft* 59 (3/4): 140–143.

Tamang, B., Tamang, J.P., Schillinger, U., Franz, C.M.A.P., Gores, M., and Holzapfel, W.H. 2008. Phenotypic and genotypic identification of lactic acid bacteria isolated from ethnic fermented bamboo tender shoots of North East India. *Int J Food Microbiol* 121: 35–40.

Tamang, J.P. 2001. Kinema. *Food Culture,* 3: 11–14.

Tamang, J.P. 2010. Diversity of fermented foods. In: *Fermented Foods and Beverages of the World.* Tamang J.P. and Kailasapathy, K. (Eds.) CRC Press, Boca Raton, FL, pp. 41–84.

Tamang, J.P. 2012. Plant-based fermented foods and beverages of Asia. In: *Handbook of Plant-Based Fermented Food and Beverage Technology,* 2nd edn. Hui, Y.H. and Özgül, E.E. (Eds.) CRC Press/Taylor & Francis Group, New York, NY, pp. 49–90.

Tamang, J.P., Kailasapathy, K., and Farhad, M. 2010. Health aspects of fermented foods. In: *Fermented Foods and Beverages of the World.* Tamang, J.P. and Kailasapathy, K. (Eds.) CRC Press/Taylor & Francis Group, New York, NY, pp. 391–414.

Tamang, J.P., Tamang, B., Schillinger, U., Franz, C.M.A.P., Gores, M., and Holzapfel, W.H. 2005. Identification of predominant lactic acid bacteria isolated from traditionally fermented vegetable products of the Eastern Himalayas. *Int J Food Microbiol* 105: 347–356.

Tamime, A. 2005. In: *Probiotic Dairy Products*. Tamime, A. (Ed.) Blackwell Publishing Ltd., Oxford, U.K.

Tang, H., Jeremy, K., Nicholson, P. Hylands, J., Sampson, J., and Elaine, H. 2005. A metabonomic strategy for the detection of the metabolic effects of chamomile (*Matricaria recutita* L.) ingestion. *J Agric Food Chem* 53(2): 191–196.

Teniola, O.D. 2009. *Role of Packaging and Product Presentation in the Acceptability, Quality Improvement and Safety of the Traditional African Fermented Foods*. Federal Institute of Industrial Research, Oshodi (FIIRO), Lagos, Nigeria.

Thakur, N., Savithri, and Bhalla, N.C. 2004. Characterization of some traditional fermented food and beverages of Himachal Pradesh. *Indian J Trad knowl* 3(3): 325–335.

Tofalo, R., Schirone, M., Perpetuini, G., Suzzi, G., and Corsetti, A. 2012. Development and application of a real-time PCR-based assay to enumerate total yeasts and *Pichia anomala*, *Pichia guillermondii* and *Pichia kluyveri* in fermented table olives. *Food Control* 23: 356–362.

Trethewey, R.N. 2004. Metabolite profiling as an aid to metabolic engineering in plants. *Curr Opin Plant Biol* 7: 196–201.

Tsuchiya, Y., Kano, Y., and Koshino, S. 1994. Identification of lactic acid bacteria using temperature gradient gel electrophoresis for DNA fragments amplified by polymerase chain reaction. *J Am Soc Brew Chem* 52: 95–99.

Ueno, T., Matsumoto, H., Ishii, K., and Kukita, K.I. 1976. Inhibitory effects of mycotoxins on Na+-dependent transport of glycine in rabbit reticulocytes. *Biochem Pharmacol* 25: 2091–2095.

Umezawa, H., Mizuhara, Y., Uekane, K., and Hagihara, M. 1947. A crystalline antibacterial substance from *Penicillium leucopus* and four other strains of *Penicillium* sp. and *Aspergillus clavatus* and its probable identity with patulin. *J Penicillin* 1: 12.

Vadkertiova, R. and Slavikova, E. 2007. Killer activity of yeasts isolated from natural environments against some medically important *Candida* species. *Pol J Microbiol* 56: 39–43.

Van Egmond, H.P., Pautech, W.E., Veranga, H.A., and Schufter, P.L. 1977. *Arch Inst Pasteur 31 A Dairy Abstr* 41: 5371.

Van Hoorde, K., Verstraete, T., Vandamme, P., and Huys, G. 2008. Diversity of lactic acid bacteria in two Flemish artisan raw milk Gouda-type cheeses. *Food Microbiol* 25(7): 929–935.

Vandamme, P.P.B., Gillis, M., de Vos, P., Kersters, K., and Swings, J. 1996. Polyphasic taxonomy, a consensus approach to bacterial systematics. *Microbiol Rev* 60: 407–438.

Vasiljevic, T. and Shah, N.P. 2001. Probiotics. From Metchnikoff to bioactives. *Int Dairy J* 18: 714–728.

Verma, L.R. and Joshi, V.K. (Eds.). 2000. *Postharvest Technology of Fruits and Vegetables*. Vols. 1 and 2. The Indus Publ, New Delhi, p. 1242.

Vilela, D.M., Pereira, G.V., Silva, C.F., Batista, L.R., and Schwan, R.F. 2010. Molecular ecology and polyphasic characterization of the microbiota associated with semi-dry processed coffee *(Coffea arabica* L.). *Food Microbiol* 27: 1128–1135.

Waema, S., Maneesri, J., and Masniyom, P. 2009. Isolation and identification of killer yeast from fermented vegetables. *Asia J Food Agro-Ind* 2(4): 126–134.

Walker, G., Anne, M., Mcleod, H., and Hodgson, V.J. 1995. Interactions between killer yeasts and pathogenic fungi. *FEMS Microbiol Lett* 127(3): 213–222.

Wang, X., Chi, Z., Yue, L., Li, J., Li, M., and Wu, L. 2007. A marine killer yeast against the pathogenic yeast strain in crab (*Portunus trituberculatus*) and an optimization of toxin production. *Microbiol Res* 162: 77–85.

Wang, T., Zhang, X., Zhang, M., Wang, L., and Zhao, L. 2011. Development of a fluorophore-ribosomal restriction typing method for monitoring structural shifts of microbial communities. *Arch Microbiol* 193: 341–350.

Ware, G.M., Thorpe, C.W., and Pohland, A.E. 1974. Liquid chromatographic method for determination of patulin in apple juice. *J AOAC* 57: 111–113.

Weckx, S., Van der Meulen, R., Maes, D. et al. 2010. Lactic acid bacteria community dynamics and metabolite production of rye sourdough fermentations share characteristics of wheat and spelt sourdough fermentations. *Food Microbiol* 27: 1000–1008.

Westby, A. Reilly, A. and Bainbridge, Z. 1997. Review of the effect of fermentation on naturally occurring toxins. In: *Fermented Food Safety*. Nout, M.J.R. and Motarjemi, Y. (Eds.). *Food Control*, Vol. 8(5–6), pp. 329–339.

Wichmann, G., Herbarth, O., and Lehmann, I. 2002. The mycotoxins citrinin, gliotoxin, and patulin affect interferon—Rather than interleukin-4 production in human blood cells. *Environ Toxicol* 17(3): 211–218.

Wiseman D.W. and Marth E.E. 1983. Behaviour of toxin, in yogurt, buttermilk and kefir. *J Food Prot*, 46: 115–118.

Wouters, M.F.A. and Speijers, G.J.A. 1995. Patulin. *Toxicological Evaluation of Certain Food Additives and Contaminants*. Food Additives Series 35. World Health Organization, Geneva, pp. 337–402. Available from http://www.inchem.org/documents/jecfa/jecmono/v35je16.htm. Accessed on March 3, 2009.

Yousif, N.M.K., Dawyndt, P., Abriouel, H. et al. 2005. Molecular characterization, technological properties and safety aspects of Enterococci from "Hussuwa," an African fermented sorghum product. *J Appl Microbiol* 98: 216–228.

Zaika, L.L. and Buchanan, R.L. 1987. Review of compounds affecting the biosynthesis or bioregulation of aflatoxin. *J Food Protect* 50: 691.

Zhang, T. 2010. Application of molecular methods for anaerobic technology. In: *Environmental Applications and New Developments*. Fang, H. (Ed.) Imperial College Press, London, U.K., pp. 207–240.

14

Industrialization, Socioeconomic Conditions and Sustainability of Indigenous Fermented Foods

ANITA PANDEY, ARUN SHARMA,
CHAMGONGLIU PANMEI,
GURU ARIBAM SHANTIBALA DEVI,
LOK MAN S. PALNI, MD. SHAHEED REZA,
NINGTHOUJAM SANJOY SINGH,
OLUWATOSIN ADEMOLA IJABADENIYI,
S.S. THORAT, SATYENDRA GAUTAM,
AND SUSHMA KHOMDRAM

Contents

14.1 Introduction

Throughout the world, a wide variety of agricultural produce is fermented to produce novel foods. The major impetus for food fermentation is the need to preserve perishable food materials in a relatively stable form. This is particularly critical in tropical conditions, especially where widespread refrigeration is unavailable. Secondarily, fermentation can introduce a variety into foods in terms of taste, texture, flavor and mouth-feel (Demirci et al., 2014), improvement in nutrition (Chavan and Kadam, 1989; Chavan et al., 1988; Chompreeda, and Fields, 1981 and 1984; El-Tinay et al., 1979; Hamad and Fields, 1979), reduction in anti nutritional factors (Reddy and Pierson, 1994) as well as several other advantages (see Chapter 1 of this text) Indigenous fermented foods contribute greatly to food security, especially in the rural areas of many of the developing countries of the world (FAO, 1999a,b,c). These products not only have a wide variety of flavors and wide consumer acceptance, but are also rich in nutrients and have great economic value as dietary staples for local people.

Indigenous fermented foods have been a part of the diet of South Asia for a long time, and are consumed either as main dishes or as condiments (Steinkraus, 1996; Valyasevi and Rolle, 2002; Farnworth, 2008; Joshi et al., 2012). There is great diversity in the types of indigenous fermented foods throughout the countries of South Asia (see Chapter 2). Such foods contribute to about one-third of diets worldwide (Campbell-Platt, 1994), and with the increase in population in different countries of South Asia, indigenous fermented foods are likely to remain important to the vast majority of people in these countries, (Ijabadeniyi and Omoya, 2006) and India is no exception (see Chapter 2 for more details).

A great many traditional practices are associated with the preparation of indigenous fermented foods in certain communities in high altitude regions, and these practices need to be documented and preserved before they are lost. For example, the *Bhotiyas* of Uttarakhand in India, who once lived as nomadic pastoralists, have gradually, over the time, settled to a "permanent" lifestyle, giving up the nomadic way of life. However, as nomads they had identified that, at high altitudes, where the snow remained for up to six months, the specific conditions favored a slow fermentation process, which resulted in a high quality beverage, known as *jann*. Thus, the community used to initiate the preparation of *jann* at higher altitudes before migrating to the lower valleys during the winter, collecting the fermented *jann* on their return in early summer. Similar practices are still prevalent today in the listed tribal areas, such as Kinnaur, of Himachal Pradesh in India.

Sun-drying is a notable step in the traditional processing of foods such as *gundruk* and *sinki*, as these foods can be preserved for several months without refrigeration and are consumed during the monsoon when fresh vegetables are scarce. The dried *gundruk* and *sinki* are lighter and hence, easier to carry than the fresh produce, and the people of Sikkim still use this practice when traveling for long distances. Alcoholic beverages also have a strong ritual importance among the ethnic people of Sikkim, where social activities require the provision and consumption of appreciable quantities of such beverages. Traditional alcoholic beverages are also used in religious practices, being offered in prayer to the family God (Tamang, 2003).

Again in the Indian context, but applicable to other South Asian countries also, the attractiveness of fermented foods is underscored by the "appropriate technology" involved in their preparation. The machinery and equipment needed in the preparation of indigenous fermented foods, its maintenance, repair, and servicing, are not very complex. Furthermore, the preparation of the raw materials and the underlying principles of fermentative change are also simple. The relative stability, prolonged shelf-life, and safety of fermented foods make them ideal for areas lacking refrigeration, and for transportation to both local and urban markets (Olasupo, 2005). Such foods were developed through traditional village artisan methodologies, preserved over the years, so as to maintain the unique identity of these foods (Valyasevi and Rolle, 2002). Feasibility to improve the amino acid content of corn meal by household fermentation has also been explored (Wang and Fields, 1978). The preparation processes involve microorganisms, which play an active role in the physical, nutritional and organo-leptic modifications of the starting animal and vegetable materials (Aidoo, 1994). With a few exceptions, production of indigenous fermented foods has largely remained a household occupation, or at the most a cottage or small-scale commercial enterprise. Nevertheless, it plays a major role in popularizing and developing fermented foods and making them an engine for economic growth and social well-being.

Due to increased urbanization and the greater employment of women in the service sector, the popularity of indigenous foods has begun declining, day-to-day. The techniques used are labor intensive, time consuming, have low productivity and success, depending on the manufacturing practices employed. Product quality primarily depends on the experience of the processors, and such knowledge is not well documented. For these reasons, the technology fails to be disseminated, and risks going extinct. Since most of the processes are conducted on a trial and error basis, with little quality control, the foods are not of a consistently desirable standard and may be of doubtful safety. The simple technology used in the production of indigenous foods thus, fails to stand up in the face of such challenges.

The application of developments in science and technology, especially in microbiology, biochemistry, biotechnology and engineering principles and methods of food processing (Haverkort and Himestra, 1999) have considerable potential for improving indigenous fermented foods with respect to both shelf-life and marketability (Entsua-Mensah, 2012; Beachat 1995; Nout, 1992). Several innovative research findings like discovery oryza cystatin, fermenting with lysin extereting *Lactobacillus* (Izquerdo-Pulido et al., 1994; Newman ans Sands, 1984) can be applied to indigenous fermented foods. Similarly, biotechnological innovations have greatly assisted in industrializing the production of certain indigenous fermented foods, such as Indonesian *tempe* and oriental soy sauce, which are marketed globally (Wood, 1994; Beachat, 1995). Some of these aspects are discussed in subsequent sections of this chapter.

14.2 Socio-economic Conditions of Indigenous Fermented Food Producers

Poverty is entrenched in many countries of South Asia as there is a low literacy and high unemployment, although in these respects the countries in Asia, are better than those in Africa. The unemployment rates in China and India in 2009 were 9.6% and 10.8%, respectively (Alasoplascpas, 2010). Agriculture has made positive impact on the socio-economic conditions of people of developing countries, including those in South Asia. It still remains the highest employer of labor in most developing countries even today. In India, for example, agriculture impacts the livelihood and income security of many people, especially women. An average rural woman spends about 40% of her time and energy on agriculture and allied activities, thereby contributing greatly to the economic status of her family (Advani, 2005). The production of indigenous fermented foods is certainly an avenue for the creation of jobs for the people in rural and urban areas. As stated earlier, women have been intimately involved in the development of indigenous fermented foods in various communities, and today these are still produced at household and village levels (Steinkraus, 1996; Rolle and Satin, 2002). For some inhabitants of rural or poor communities, especially women, locally fermented foods processing and marketing is a small scale business and a means of obtaining income and creating wealth (Joshi and Kaushal, 2014). Indigenous fermentation technologies can stimulate development of the food industry because of their low cost, scalability, and minimal energy and infrastructural requirements (Rolle and Satin, 2002). However, the level of poverty and other socioeconomic constraints in the communities where indigenous fermented foods are produced are significant constraints. In most of these communities, a potable water supply, a drainage system, and sanitation are not available, leading to a considerable public health risk (Motarjemi, 2002). Low income households are, therefore, not able to implement food safety principles. Nevertheless, these Indigenous fermented foods are likely to remain a major component of the diet of people living in such countries mainly because they are socially and culturally accepted (Rolle and Satin, 2002; Joshi et al., 2012). Furthermore, the present trend towards consuming natural products with therapeutic values, especially as medicines, even in the developed, can provide a great impetus to the production of indigenous fermented foods.

14.3 Critical Considerations and Approaches for Sustainable Transfer of Technologies

14.3.1 Critical Considerations

To make the production of indigenous fermented foods sustainable, and the products of consistent quality and safe for consumption, the latest break through is made in research and development need to be immbibed (Nout, 1994).

It should however, be borne in mind that a considerable amount of hard work and commitment is needed for the successful transfer of technologies. It is the experience of Food and Agriculture Organisation (FAO) that individual preferences, rather than technical criteria, dictate the adoption of technologies and their products. It is, therefore, natural that an individual or organization will adopt any technology if it is useful to it and can generate sufficient profitability. Thus, it is basic that any technology selected for transfer must grow out of the needs and practices of its beneficiary society (FAO/WHO, 1996). The adoptability of any technology transferred, and its products, is highly dependent on the recognition by potential users of the technology of a clear advantage from its use and from its products. Once adopted, it may be necessary to provide help to entrepreneurs to acquire the technology and to develop both business and product marketing plans.

The latest research or improvements being made in the field need to be made available to sustain the transfer of technologies. To improve any technology it is essential that current technology for production of indigenous fermentation is documented. At present, it is important to assist the developing countries, who are building and improving traditional fermentation processes, in the documentation of such technologies, which are rapidly being lost (FAO, 1998, 1999a,b,c, 2000).

Domestic marketing systems for food products, including indigenous fermented foods, in most of the countries of South Asia, are poorly developed. How to utilize and access the credit facilities available both within the country and outside, it is one of the most relevant points in setting up of any food processing facility, including for indigenous fermented foods.

14.3.2 Approaches

Technology transfer to South Asian countries, as with all the developing countries, can constitute a four-phase approach: assessment of the situation in the country; selection of an appropriate scale and level of technology; implementation, demonstration, and dissemination of the technology; and, finally, the facilitation of serious competition to the imported food products, which are often of inferior nutritive value (Jones et al., 1996; Battcock and Azam-Ali, 1998; Haard, 1999). Properly-coordinated access to the technology, as advocated for developing countries, including those of South Asia, is shown in Figure 14.1. However, the approach must be suitable both economically and socially.

It is well established that the countries in South Asia are a non-homogenous group, and vary widely in their technical capabilities. Thus, prior to the transfer of any technology to these countries, an assessment of the technical, socio-economic, and institutional conditions within the recipient country, needs to be undertaken. Additionally, the potential of the indigenous food fermentation technology and the food products so made, should be undertaken by the these countries. An appraisal of the consumption of fermented foods in South Asia reveals that there are numerous indigenous fermented foods that are made and consumed (see Chapter 2 of this text for more details). The greatest advantage with indigenous fermented products is that

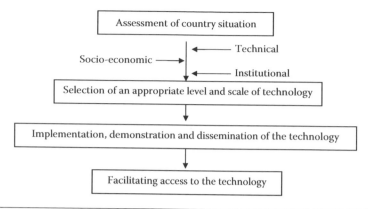

Figure 14.1   Four-phase approach to sustainable technology transfer for developing countries. (Adapted from Kumar, U., Movahedi, B., and Lavassani, K., 2008. *Towards Developing a Comprehensive Model of Technology Transfer to Developing Countries: A Comparative Analysis Approach.* ASAC Halifax, Nova Scotia, pp. 64–80; Rolle, R. and Satin, M., 2002. *International Journal of Food Microbiology,* 75(3), 181–187.)

the consumers are already familiar with these products, so there should be no problems of their acceptance.

For the transfer of the technology for indigenous fermented food preparation, an assessment of the current technology of manufacture and logistical support available in a particular country is absolutely essential. Not only this, the current processing practices within the country, and the supporting infrastructure available (energy supplies, consistent supplies of clean water, etc.) should be taken into consideration. The availability of transportation and road networks in relation to sources of raw materials storage and markets for the finished products should also be ascertained. In addition, a socio-economic assessment need to be carried out. The level of organization within the recipient country, the cost and availability of labor, consumer purchasing power, and the choice and acceptability of fermented foods all need to be determined. Facilities for the hygienic preparation of fermented foods are of utmost significance. Cultural factors, such as the food preparation and storage practices within the country, have to be determined to find out the most suitable type of fermented product for the environment into which it will be introduced (Tamang, 2009).

Institutional factors also play a major role in the successful establishment and functioning of small-scale food fermentation. The level of education within the country, locally available skills, and technical support services are also important considerations, besides the capacity for research and development within the country (Kumar et al., 2008). Besides, government policies are relevant to the transfer of technology, and the provision of incentives for the development of such technologies must be assessed. Policies for private sector investment in small-scale food fermentation plants, and price policies that allow the small-scale processor to operate profitably have to be framed (Rolle and Satin, 2002), besides the role played by social enterprises and leadership (Orozco-Quintero et al., 2010). Only technologies that meet these conditions should be considered for transfer.

14.4 Indigenous Fermented Food Production as a Small Scale Industry

14.4.1 Production of Indigenous Fermented Foods on Commercial Scale

A wide diversity of fermented foods, which vary according to staple, geographical area, and cultural preference, are produced across the globe (Campbell-Platt, 1987; Jones, 1994; Steinkraus, 1996). The diversity of fermented products, which include porridges, beverages (alcoholic and non-alcoholic), breads and pancakes, fermented meat, fish, vegetables, dairy products, condiments and many more (Campbell-Platt, 1987; Steinkraus, 1996) are produced from both edible and inedible raw materials in developing countries (FAO, 1998, 1999a, b, c, 2000). Many food fermentation processes have been developed over the years by women in specific communities in various countries of South Asia. At present, food fermentation is still undertaken as a household or village-level activity in developing countries, with a very few operations carried out at an industrial level. Small-scale food processing as a subsector can contribute significantly to the development of the rural economy (Dietz, 1999). It is especially important, with respect to indigenous fermented foods, for stimulating sustainable development in rural and semi-urban areas of various developing countries of South Asia. As an industry it can be a source of income, a means of poverty alleviation, variety in the diet, and food security for millions, providing linkages to the local suppliers of agricultural raw materials, and to income generating activities such as the manufacture of machinery, packaging, and ingredients (FAO, 1997; Glover et al., 2010).

The agricultural production sector of developing countries has traditionally received greater attention from planners and policymakers, with relatively little or no focus on the development of agro-industry horticultural, which has resulted in considerable post-harvest losses of hard-won crops of agricultural produce, especially the perishable commodities like milk, fruit, and vegetables (Verma and Joshi, 2000). Consequently, not only does the farmer not get the price for the produce, but there is also no manufacturing activity to generate employment, resulting in a poor economy and limited industrial growth. Utilizing agricultural produce to prepare indigenous fermented products on an industrial scale, and the proper marketing of the same, holds great promise for the growth and well being of the people of South Asia.

14.4.2 Technical Issues and Constraints

Due to the lack of scientific and technological know-how, fermented foods are generally evaluated on the basis of sensory attributes only (Valyasevi and Rolle, 2002). The level and scale of fermentation technologies applied in the production of these foods vary widely from country-to-country, and even place-to-place. The technologies used vary widely, from the very simple in areas that lack basic infrastructure, to the relatively sophisticated where conditions so permit. However, a majority of countries do not have a supportive technical and economic infrastructure for the development of small-scale food processing. Small-scale food processors are consequently faced with

Table 14.1　Technical Issues and Constraints Faced by the Indigenous Food Fermentation Industry

- Although the procedures and equipment used by these indigenous fermented food processes are relatively simple, the microbiological and biochemical aspects of a number of such processes are complicated and have not been fully understood.
- The physical aspects of the processes (temperature, relative humidity and level of agitation and aeration) are often poorly controlled, and production techniques are not standardized.
- In general, these processes are of low efficiency, resulting in low yields of products of variable quality, compared to the processes practiced at an industrial scale.
- Poor hygienic practices and improper handling during post-fermentation processing (e.g., drying), and at the point of sale, render fermented products susceptible to contamination.
- The shelf-life of a number of fermented products is also limited by the unavailability of appropriate technologies for post-fermentation processing treatments, such as drying, pasteurization, and refrigeration, that terminate the bioprocesses and, consequently, extend the shelf-life of such products.
- Inadequate packaging of fermented products in materials such as leaves, vegetable fibers, earthenware pots, and newsprint limits both the shelf-life and competitiveness of these products in local markets.
- Technical constraints faced by small-scale processors are compounded by a number of institutional constraints.
- Inadequacy of governmental policies which promote and support small-scale food processing.
- Insufficient raw materials and infrastructure, with limited access to external inputs and limited marketing infrastructure.
- Raw materials in most developing countries are seasonal and variable.
- Procurement of consistent supplies of raw agricultural materials for processing consequently poses difficulties.
- The bulkiness and highly perishable nature of raw agricultural produce, coupled with high transportation costs, underdeveloped road infrastructure, and inadequate and inappropriate storage facilities packaging systems discourages industrial development of the processes.
- Basic infrastructural facilities, like clean water and reliable energy supplies, required for the development of any food-processing industry, are not always accessible/available in these countries.

Source:　Adapted from Fellows, P.A., 1992. *Small-scale Food Processing.* Tech. Publications, London, pp. vii–x; FAO, 1997. *The Agro-Processing Industry and Economic Development. The State of Food and Agriculture.* FAO, Rome, pp. 221–265; Dietz, M.H., 1999. *Courier,* 174, 89–92; Rolle, R. and Satin, M., 2002. *International Journal of Food Microbiology,* 75(3), 181–187; Kailaspathi, K., 2010. *Fermented Foods and Beverages of the world,* pp. 415–434; Ogunmoyela, O.A. and Oyewolc, L.O., 1992. *Food Science and Technology,* 6, 10–13.

a number of complex technical and institutional constraints (Fellows, 1992; Rolle and Satin, 2002), and some of these are described in Table 14.1.

The high cost of utilities and materials and labour often account for well over 50% of the cost of production; see Figure 14.2, which is based on some selected food industry indicators (1991–1993).

In general, the food-processing sector in many countries of South Asia including India is given low priority by planners and policymakers, and, as a consequence, is allotted little funding (Verma and Joshi, 2000). Besides, the problems of limited access to technical books and periodicals due to high cost and a lack of foreign exchange, of limited operational, business, and marketing skills, of a lack of basic knowledge on nutrition and food safety principles, and a lack of any training programs or follow-up for upgrading these skills, contribute to the poor growth of the food industry (Rolle and Satin, 2002).

The problems for the small scale sector are even more serious. They have little economic power, are usually not able to support research, do not have the resources

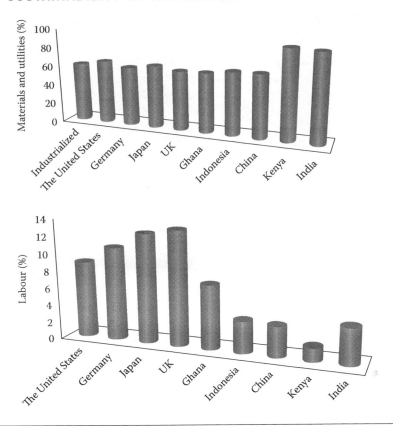

Figure 14.2   Comparison of materials and utilities (%) input for the fermentation industry in some countries. (From Rolle, R. and Satin, M., 2002. *International Journal of Food Microbiology*, 75(3), 181–187. With permission.)

required to seek technical assistance, and do not generally benefit from the technical assistance and support provided to the food-processing sector.

14.5 Need for Industralization

Traditional fermented foods and beverages offer tremendous potential for promoting health, improving nutrition, and reducing the risk of various diseases worldwide. Apart from fermented foods playing an important socio-economic role in the producing countries, they also contribute to the protein and calories requirements and sensory satisfaction of the people. However, it is essential to improve the quality and safety of indigenous fermented foods (Achi, 2005), so that it is properly controlled and upgraded and can thus, generate more income for the producers. Some examples of such technologies are cited in Chapter 13 of this text. Indigenous fermented foods offer the opportunity for scale-up to make household-confined products available to a much wider population. But there is a great need to fully understand the entire indigenous knowledge system associated with the native societies of the region, as these represent a historical continuity of "living in harmony with nature" (Singh, 2006). To upgrade and industrialize locally fermented foods requires several stages, as summarized in Table 14.2.

Table 14.2   Stages in Up-gradation of Industrialized Technologies to Produce Fermented Foods

- Isolation, identification, and determining the role of the microorganisms associated with the fermentations
- Selection and genetic improvement of microorganisms involved in indigenous fermented foods
- Improvement of the process controls employed in the manufacture of fermented foods
- Improvement in the quality of raw materials used in the production of fermented foods
- Laboratory simulation production of indigenous fermented foods
- Pilot scale production of fermented foods for proper commercialization
- Production at industrial scale

Source: Adapted from Olasupo, N.A., 2005. *Fermentation Biotechnology of Traditional Foods of Africa, Food Biotechnology,* 2nd edition. K. Shetty, et al. (eds.), CRC Press, Boca Raton, FL.

Table 14.3   Basics Criteria for Production of Indigenous Fermented Foods

- Should involve simple technology requiring only simple and easily maintainable equipment
- Should result in improved stability, shelf-life, and safety
- Should provide an inexpensive means of improving the nutrition of the community, especially children and vulnerable sections of population
- Should be an engine for economic empowerment of the rural population

Source: From Jespersen, L., 2003. *FEMS Yeast Research,* 3(2), 191–200. With permission.

All these stages are essential, but market testing and promotion of the improved fermented foods should also be included as an important stage after the production at industrial plant stage, where producers carry out the production with trained manpower. The improvement of indigenous fermented foods can also be achieved through biotechnological innovations (see Chapter 13) to industrialize locally fermented foods into commercial products, as has been advocated earlier (Valyasevi and Rolle, 2002). To have a real impact and mass appeal in a country like India, where the great majority of the people live in rural areas, producers are encouraged to meet certain basic criteria in production, as detailed in Table 14.3 for such countries.

Indigenous fermented foods meet all these criteria, hence these criteria can be applied easily, and production taken up. Nevertheless, industry and science have to work closely, sharing knowledge and common problems, thereby giving rise to research that will benefit the industry. Other improvements include the optimization of processes and the application of pasteurization to the products (Valyasevi and Rolle, 2002), as well as improvement in nutrition, where the indigenous fermented food is deficient in certain essential nutrients (Hamad and Fields, 1979; Farnworth, 2008). The improvement and industrialization of indigenous fermented food productions has the potential to develop many new fermented foods.

14.6 Sustainability of Indigenous Fermented Foods

For indigenous fermented foods to be sustainable, especially with changing consumer tastes and preference, they must be upgraded from small-scale production to a commercial or industrial level. This is important because the sustainability of indigenous fermented food is an integral aspect of food security. It can be achieved when productivity along the entire food chain—that is, from production to consumption—is enhanced (Entsua-Mensah, 2012). There is a need for appropriate research for optimization of the safety and quality of foods, besides ensuring that consumers get good value for money. But in-depth research and development are capital intensive and lacking in small-scale fermented food businesses (Valyasevi and Rolle, 2002). Other challenges to the sustainability of indigenous fermented foods are the same as for other small scale food industries, and the prioritization of financial support for research and development for indigenous fermented foods in South Asian countries has to be emphasized (Nair, 2010). However, indigenous fermented foods should be developed and improved upon not only for the local market but also for the international market (Nair, 2010), as there is growing demand for such products in view of the perceived health-promoting and therapeutic values of indigenous fermented foods. However, the safety of such goods must be improved for them to be sustainably marketed or exported to other countries worldwide. Accordingly, the introduction of quality assurance systems needs to be monitored by accredited laboratories (Sawadogo-Lingani, 2009). Hazard Analysis Critical Control Point (HACCP), an effective food safety tool, should be implemented and, if that is not possible, at least Good Manufacturing Practices (GMP) and HACCP prerequisites can be made mandatory (Alli, 2004). Active partnership between indigenous fermented foods producers, researchers, and industry is also necessary for sustainability of the products. Neither researchers nor the industry can just develop indigenous fermented foods without the cooperation of the producers in the production of such foods in a sustainable manner. There is extensive research on the development of novel fermented functional foods and functional bio-ingredients from these microbial resources through biotechnological intervention (Jeyaram, 2011). The ethnic people of Nepal, the North-Eastern states of India, the states of Himachal Pradesh and Uttaranchal, etc., consume more than a hundred varieties of traditional fermented foods produced by the age-old indigenous process of spontaneous fermentation. For research to be beneficial to such people, they and the producers must be carried along in what is known as community-based research. Such research will allow the community and the indigenous fermented foods producers to treat the project as their own and this will contribute not only to the growth and development of such foods, but will also encourage long-term viability and sustainability (Anderson et al., 2006; Cristiana and Fikret, 2010). Such research will also enable the community to keep control of the creation of employment and of the land and resources, to protect the culture, and to control the creation of wealth (Anderson et al., 2006; Berkes and Adhikari, 2006).

While promoting the concept of large-scale industrial units for the manufacture indigenous fermented foods, the significance of small-scale fermented food production and the Indigenous Knowledge System (IKS) (see Chapter 1) should not be forgotten. Indigenous knowledge system (IKS) contributes to the employment of a lot of people in the informal sector. Small-scale fermentation technologies contribute substantially to food safety, food security, and nutrition, particularly in the regions that are vulnerable to food shortages (Onofiok et al., 1996; Battcock and Azam-Ali, 1998). In addition, it reduces the dependency of urban populations on food imports, and allows farmers to sell their raw materials locally. According to Falola (2003), the indigenous knowledge system functions at the level of economic sustainability, self-reliance, and cost effectiveness. Thus, all the stakeholders should ensure that small-scale fermented food production is made sustainable and improved upon where appropriate, although large scale industrialization of indigenous fermented foods would make the products sustainable and made available to a large number of consumers. Thus, there is great need to produce traditional fermented foods beverages for improved livelihood, develop rural infrastructure and agro-industries (Elain and Danilo, 2011) in various countries in South Asia.

14.7 Consumption and Marketing System for Traditional Fermented Products

14.7.1 Consumption of Indigenous Fermented Foods

It has been stated earlier that fermented foods contribute to about one-third of the diet worldwide (Campbell-Platt, 1994). Many of indigenous fermented foods are prepared at the household level, so the data of exact consumption is not available. Cereals are particularly important substrates for fermented foods in Asia and the Indian subcontinent. Originally, fermented foods like fermented milks were developed as a means of preserving nutrients (Jayakumar et al., 2012). In almost every house in South Asian countries, curd or *dahi* is made and consumed, but precise quantities are not known. Certain closely-related products are manufactured from fermented milks by de-wheying; examples include cheese, *paneer* and *shrikhand* (Aneja et al., 2002). The manufacture of cultured dairy products represents the second most important fermentation industry (after the production of alcoholic drinks; (www.bethamsuince.com-2012). Fermented dairy drinks are reported to have grown at six times the rate of total dairy products between 1998 and 2003 in terms of value. The increasing demand from consumers for dairy products with "functional" properties is a key factor driving value sales growth in developed markets (see Chapter 6). This has led to the promotion of value-added products such as probiotic and other functional yoghurts, reduced-fat and enriched milk products, fermented dairy drinks, and organic cheese (Rudrello, 2004). Probiotic drinking yoghurt was the fastest growing dairy product sector between 1998 and 2003, followed by soy milk, (spoonable) probiotic yoghurt, milk drinks flavored with juice, and fermented dairy drinks (Adwan, 2003). Another important global trend is the increasing demand for consumer convenience. Consumers prefer foods that promote good health and prevent disease, but these foods must fit into current lifestyles, providing convenience of use, good flavor, and an acceptable price–value ratio, which defines current and future waves in the food development cycle (Chandan, 1999). Scientific and clinical evidence is also mounting to corroborate consumer perception of health from fermented milks (Adwan, 2003).

The consumption of fermented milks has generally increased around the globe over the period from 2001 to 2004 (IDF, 2005; www.Bethanorg.com, 2012). In the Indian subcontinent, fermented milk products such as *dahi*, *lassi* (a sweetened yoghurt drink similar to buttermilk), and *shrikhand* (drained curd with added sugar and flavoring) figure prominently in the diet. The demand for fermented milk products is increasing, and it has been estimated that about 10% of total milk production in India is used for the preparation of traditional fermented milk products. About 6.9% of total milk production in India is used for making *dahi* intended for direct consumption while the volume of curd and curd products was reported to be 6.0 million tones with a market value of 120 billion rupees (Aneja et al., 2002).

14.7.2 Marketing System of Traditional Fermented Products

At present, the marketing system for indigenous fermented foods in South Asia is not well organized, depending mostly on local requirements. A brief review of relevant information on this aspect is given here. Typical examples of products from different South Asian countries are also cited for illustrative purposes.

14.7.2.1 Chepashutki There is no specific marketing chain for traditional *chepashutki* in Bangladesh. The length of the marketing chain, however, varies depending on the place and season of the year. In general, the processors collect raw materials they need to produce *chepashutki* (mainly dried small indigenous fish species (SIS)) from producers, wholesalers, retailers, or via a number of intermediate traders (see Chapter 11 for more details of the product and process). Usually, no processors are involved with retailing of the processed product, which allows the entry of different commission-taking agents into the marketing chain, making the final product more expensive for the consumer (Reza et al., 2010). In most cases, these traditional fishery products are marketed without any packaging, and the consequent uptake of moisture from the environment thus increases the chance of bacterial and fungal spoilage during retailing (Wood, 1981; Reza et al., 2010).

14.7.2.2 nga-pi Products The marketing system of *nga-pi* is comparatively simple. Producers collect raw shrimp from fishermen who catch the shrimp at sea. The processors carry the shrimps by van or rickshaw to the processing area, where they sell their *nga-pi* to the wholesalers. The wholesalers then, sell the product to the retailers, whence it ultimately reaches its final destination— the consumers. In most of the cases, the producers sell their products to the specific wholesalers, as they might have previously borrowed money from them.

14.7.2.3 Fermented Soybean The selling of fermented soybean is a means of livelihood in many regions of Northeast India. The products are made at household level, so sale is also limited to homes and small shops only. It is not produced commercially due to poor marketability and improper safety measures. India has experienced a rapid increase in the production of soybean in the states of Madhya Pradesh, Uttar Pradesh, and Maharashtra. The production of soybean is increasing, while utilization remains low. The perceived flavor of fermented soybean differs from person-to-person due to strong ammonia smell. Some manufacturers however, produce an odorless *natto* or *natto*-like products to overcome the problem of flavor.

14.7.2.4 Idli, Dosa, and Dahi   The products like *Idli, dosa,* and *dahi* are made in small shops and sold locally in India and other South Asian countries such as Pakistan and Bangladesh. *Jalebi* or *pretzel* are made and sold in sweet shops (Figure 14.3c). *Dhokla* is also produced and packaged, and is available in many shops in these countries. *Dahi* is now produced by several large scale processing units and sold through normal marketing channels (Figure 14.3a). Milk products such as *dahi* and *lassi* are sold extensively in general shops throughout India. Traditionally, during the festival season, several fermented foods, such as *jalebi, bhatura,* and *kulche* (Figure 14.3d) are made and sold fresh in the countries of South Asia especially India and Pakistan (Figure 14.3).

14.8 Documentation and Popularization of Indigenous Fermented Foods

Rapid urbanization and modernization have affected the time-tested traditional technologies for the preparation of fermented foods. If we continue to ignore this important issue, the time is not far off when indigenous traditional knowledge (ITK) will disappear before it can be validated and improved upon, via science and technology, and engineering. Hence, there is a need to characterize traditional fermented products.

Figure 14.3   Different products: (a) mother dairy packed *dahi*; (b) *Maya* yoghurt; (c) preparation of *jalebi* at festivals and (d) Selling 'chat' a cereal fermented product with fermented milk 'dahi, *Bhalle, Bhatura* and *Kulche* in a movable stall in a town in India.

The first and foremost requirement in the industrialization of indigenous fermented foods is the documentation of the indigenous-cum-traditional practices involved in the preparation of them. At the same time, the indigenous practices also require "standardization" in order to obtain batch-to-batch consistency, alongside the value added to these products through scientific inputs, mainly of a microbiological and biotechnological nature. The major components of the documentation include the diversity and role of microorganisms, their succession, biochemical changes brought by them, and their probiotic value. The standardization of the physical conditions, such as temperature and moisture, is also essential. The composition and quantity of raw materials used during the process are equally important for maintaining the batch-to-batch uniformity and quality of the end product. The isolation of microbiol culture from the indigenous fermented foods such as soy sauce *koji* (Bhumiratana et al., 1980), if found effective can be utilized for preparation of such products, even at the industrial scale. Establishment of a culture-collection facility for the preservation of such important microbes, isolated from a plethora of indigenous fermented foods, will have the potential for use as the starter for other industrial applications (Benkerroum and Tamime, 2004).

Documentation of the fermented products and their value addition should lead to the validation of the processes and practices involved, so as to make them economically attractive. Initial preparation of a suitable database on the nature and the level of consumption, storage, and marketing will help in the development and management of fermented foods in the South Asian region, viable at a commercial scale. Increased awareness and the development of appropriate entrepreneurship should also help the indigenous communities. In all such endeavors, the local people, who are the real custodians of such knowledge, should get proper recognition and reward.

The challenge confronting the administrators, scientists, sociologists, extension workers, and block development workers is to bring out the advantages of fermented foods to the doorsteps of the masses throughout the countries of South Asia. To meet this broad vision, the cooperation of various agencies is needed. To be successful, the plan should be categorized under three main stages: I—Long-term goals; II—Medium-term goals; and III—Short-term goals. The long-term goals mainly involve research and development efforts on some fermented foods, as very few scientific studies have been undertaken to fully understand the course of fermentations, the flora involved, their sequence, role, and the technological steps involved as revealed by the critical appraisal of technologies discussed in various chapters of this text. Efforts to standardize the process and to establish quality and nutritional attributes, and methods for quality control and for the development of new and improved fermented foods to increase nutrient content need to be undertaken. The domestic production of food yeasts in mass quantities for upgrading the nutrient content of fermented and non-fermented foods could also be investigated. Production of selected starter cultures and their preservation for distribution in various South Asian countries could also be initiated.

Medium-term goals could involve the development of a suitable database for the known fermented foods made in various countries of South Asia. It could include details of raw materials used, the preparative steps, the fermentation conditions, end-point determination, quality attributes, shelf-life, and other details on how they are consumed. From the database, research projects on the improvement of the existing technology could be instituted and transferred to the long-term goals category.

There could be a movement involving school children wherein probiotic products like *dahi* could be prepared and marketed in the countries of South Asia. Various self-help groups (SHG) that have contributed to the economic betterment and empowerment of rural and poor women, could also be involved. Similar approaches could be made to popularize other indigenous fermented foods.

14.9 Summary and Future Thrust

Indigenous fermented foods are popular throughout South Asia, and make a significant contribution to the diet of millions of individuals. There is tremendous scope and potential for such foods to meet the growing world demand for food. For this to happen, the processes need to be refined, with a view to produce them on a larger scale, but a detailed scientific understanding of the fermentation processes is a pre-requisite. There is a need for the industrialization of indigenous fermented food production to improve the socio-economic conditions of the producers of such products. The constraints which get in the way of industrialization need to be overcome, as at present, it is progressing at a slow pace. Lessons must, however, be learned from how Indonesian *tempe* and Oriental soy sauce became industrialized as global products.

Traditional fermented food products native to a country or culture, like the fishery products of Bangladesh (*chepashutki* and *nga-pi*), are widely accepted and have high market demand. There are ample prospects for exporting *chepashutki* and *nga-pi* to other countries to earn foreign exchange. By using good quality raw materials and training the processors to maintain proper sanitary conditions, improved quality *chepashutki* and *nga-pi* can be produced. The same can be said about *angoori* from Himachal Pradesh, *chhang* from Ladakh, and *opong* from the Northeast states of India, *sinki* from Nepal, *dosa, bhatura, kulche, seera* etc. from North India and other similar products from South Asian countries.

In view of the rising global demand for alcohol for consumption, large scale production of potable alcohol is required. However, traditional methods of alcoholic fermentation in states of India like Manipur are doing the fermentation in non-sterile conditions, usually on a small-scale. The processes of production of indigenous fermented foods, being natural fermentation, are difficult to be controlled if carried out at a larger scale, and the chances of spoilage are very high in traditional methods of production. Techniques to stabilize natural fermentations would be the best option for large scale production. There is a need for research into the qualitative and quantitative production of traditional alcoholic beverages. Traditional beverages, because of their medicinal utility, need to be focused on as an alternative healing system, for example, of Yu. At present, much knowledge associated with traditional beverages is being eroded with urbanization, and many rural people have stopped alcoholic fermentation, resulting in the loss of diversity in culture (such as "Hamei"). So there is a need to document the knowledge and, collect and preserve the different microorganisms associated with alcoholic fermentation. In short, a multidisciplinary approach to the study of alcoholic fermentation in these regions of South Asia could provide significant knowledge to science and society.

In many developed countries, their important indigenous fermented foods have been well investigated, and statistical data on the production, consumption, socioeconomic role, microbiology (including selected starter cultures/protective cultures), biochemistry, nutritional profile, pilot-plant production methods, etc., are available. However, in many South Asian countries, like India, several indigenous fermented foods are yet to be investigated, except for few common fermented foods such as *idli*, *dosa*, *dahi*, etc., information on which is available. Due to the presence of various ethnic groups in South Asia including India, there are diverse food habits. However, the people who invented and preserved the age-old traditional food fermentation technology need to be reassured about the worth of their indigenous knowledge, and contributions to upgrading this technology must be made without damaging the dietary culture.

South Asian countries need to build their own resources of trained, knowledgeable individuals, who are able to apply basic microbiological principles to the production of fermented foods. Fermented foods should be recognized as part of each country's heritage and culture, and efforts should be made to preserve the methods of production. Some recognized authority (government or non-government) should take the responsibility for the collection of details and the promotion of fermented food products.

References

Achi, O.K., 2005. The potential for upgrading traditional fermented foods through biotechnology. *African Journal of Biotechnology*, 4, 375–380.

Advani, P., 2005. Agriculture sector in India. A report on the *Impact of WTO on Women in Agriculture*, new.nic.in/pdfreports/impactofWTOwomeninagriculture.pdf

Adwan, L., 2003. Fermented dairy drinks under pressure (online). Euromonitor International Archive; cited July 25, 2003. http://www.euromonitor.com/article.asp?id=1371

Aidoo, K.E., 1994. Application of biotechnology to indigenous fermented foods. *Proceedings of Technology Development Countries*, 12(2/3), 83–93.

Alasoplascpas, 2010. Unemployment rate in Asia, 2010. http://alasoplascpas.wordpress.com/2011/02/04/unemployment-rate-in-asia-2010/. Accessed November 17, 2011.

Alli, I., 2004. *Food Quality Assurance; Principles and Practices*. CRC Press LLC, Boca Raton, FL, p. 141.

Anderson, R.B., Dana, L.P., and Dana, T.E., 2006. Indigenous land rights, entrepreneurship, and economic development in Canada: "Opting-in" to the global economy. *Journal of World Business*, 41, 45–55.

Aneja, R.P., Mathur, B.N., Chandan, R.C., and Banerjee, A.K., 2002. *Technology of Indian Milk Products*. A Dairy India Publication, Delhi.

Battcock, M. and Azam-Ali, S., 1998. *Fermented Fruits and Vegetables—A Global Perspective*. FAO Agricultural Services Bulletin, vol. 134. FAO, Rome, pp. 7–12.

Benkerroum, N., and Tamime, A.Y., 2004. Technology transfer of some Moroccan traditional dairy products (lben, jben and smen) to small industrial scale. *Food Microbiology*, 21(4), 399–413.

Berkes, F. and Adhikari, T., 2006. Development and conservation: Indigenous businesses and the UNDP Equator Initiative. *International Journal of Entrepreneurship and Small Business*, 3, 671–690.

Beuchat, L.R., 1995. Application of biotechnology to indigenous fermented foods. *Food Technology*, 49(1), 97–99.

Bhumiratana, A., Flegel, T.W., Gleinsukon, T., and Somporn, W., 1980. Isolation and analysis of moulds from soy sauce *koji* in Thailand. *Applied and Environmental Microbiology*, 39, 430–435.

Campbell-Platt, G., 1987. *Fermented Foods of the World—A Dictionary and Guide*. Butterworth, London, p. 291.

Campbell-Platt, G., 1994. Fermented foods: A world perspective. *Food Research International*, 27, 253–257.

Chandan, R.C., 1999. Enhancing market value of milk by adding cultures. *Journal of Dairy Science*, 82, 2245–2256.

Chavan, J.K. and Kadam, S.S. 1989. Nutritional improvement of cereals by fermentation. *CRC Critical Reviews in Food Science Technology*, 28, 349.

Chavan, U.D., Chavan, J.K., and Kadam, S.S., 1988. Effect of fermentation on soluble proteins and *in vitro* protein digestibility of sorghum, green gram and sorghum-green gram blends. *Journal of Food Science*, 53, 1574.

Chompreeda, P.T. and Fields, M.L., 1981. Effects of heat and fermentation on the extractability of minerals from soybean meal and corn meal blends. *Journal of Food Science*, 49, 566.

Chompreeda, P.T. and Fields, M.L., 1984. Effect of heat and fermentation on amino acids, flatus producing compounds, lipid oxidation and trypsin inhibitor in blends of soybean and corn meal. *Journal of Food Science*, 49, 563.

Cristiana, S. and Fikret, B., 2010. Community-based enterprises: The significance of partnerships and institutional linkages. *International Journal of the Commons*, 4(1), 183–212.

Demirci, A., Izmirlioglu, G., and Ercan, D., 2014. Fermentation and enzyme technologies in food processing. In: *Food Processing: Principles and Applications*, 2nd edition. S. Clark, S. Jung, and B. Lamsal (eds.), John Wiley & Sons, Ltd., Chichester. 592 p.

Dietz, M.H., 1999. The potential of small-scale food processing for rural economies. *Courier*, 174, 89–92.

Elaine, M.L. and Danilo, M., 2011. *Traditional Fermented Food and Beverages for Improved Livelihoods Rural Infrastructure and Agro-industries*. Division Food and Agriculture Organization of the United Nations, Rome.

El-Tinay, A.H., Abdel-Gadir, A.M., and El-Hidai, M., 1979. Sorghum fermented kisra bread. Nutritive value of kisra. *Journal of the Science Food and Agriculture*, 30, 859.

Entsua-Mensah, R.E., 2012. Science and technology promotion and research in Ghana with regards to food preservation and safety. CSIR Ghana Annual Report 2012–2013.

Falola, T. 2003. *Ghana in Africa and the World: Essays in Honor of Adu Boahen*. African World Press: Eritrea, New Jersey and Asmara. pp. 800.

FAO, 1997. *The Agro-Processing Industry and Economic Development. The State of Food and Agriculture*. FAO, Rome, pp. 221–265.

FAO, 1998. *Fermented Fruits and Vegetables—A Global Perspective*. FAO Agricultural Services Bulletin, vol. 134. FAO, Rome, p. 98.

FAO, 1999a. *Scope for Improvement of Fermented Foods.* http://www.fao.org/docrep/x2184e/x2184e08.htm#sco. Accessed November 15, 2011.

FAO, 1999b. *Fermented Cereals—A Global Perspective.* FAO Agricultural Services Bulletin, vol. 138. FAO, Rome, p. 115.

FAO, 1999c. *Fermented Cereals.* http://www.fao.org/docrep/x2184e/x2184e06.htm. Accessed November 15, 2011.

FAO, 2000. *Fermented Grain Legumes, Seeds and Nuts: A Global Perspective.* FAO Agricultural Services Bulletin, vol. 142. FAO, Rome, p. 109.

FAO/WHO, 1996. *Fermentation: Assessment and research. Report of a Joint FAO/WHO Workshop on Fermentation as a Household Technology to Improve Food Safety,* December 11–15, 1995, WHO/FNU/FOS/96.1, Pretoria, 79p.

Farnworth, E.R., 2008. *Handbook of Fermented Functional Foods.* 2nd edition. CRC Press Taylor & Francis Group, New York, NY, p. 600.

Fellows, P.A., 1992. Food Processing Technology: Principles and Practices (3rd edition) In: *Small-scale Food Processing.* P. Fellows and A. Hampton (eds.), Tech. Publications, London, pp. vii–x.

Glover, R.L.K., Owusu-Kwarteng, J., Tano-Debrah, K., and Akabanda, F., 2010. Process characteristics and microbiology of Fura produced in Ghana. *Nature and Science,* 8(8), 41–51.

Haard, N.F., 1999. Cereals: Rationale for fermentation. In: *Fermented Cereals—A Global Perspective, 24.* FAO Agricultural Service Bulletin, vol. 138. FAO, Rome, p. 24.

Hamad, A.M. and Fields, M.L., 1979. Evaluation of the protein quality and available lysine of germinated and fermented cereal. *Journal of Food Science,* 44, 456.

Haverkort, B. and Himestra, W. (eds.), 1999. *Food Processing.* Macmillan, London, pp. 80–102.

IDF, 2005. *The World Dairy Situation.* Bulletin of IDF, vol. 399, p. 81.

Ijabadeniyi, A.O. and Omoya, F.O., 2006. *Safety of Small-Scale Food Fermentations in Developing Countries.* IUFoST doi:10.1051/IUFoST:20060993.

Izquerdo-Pulido, M.I., Haard, T.A., Hung, J., and Haard, N.F., 1994. Oryzacystatin and other protease inhibitors in rice grain: Potential use in preventing proteolysis in surimi and other fish products. *Journal of Agricultural and Food Chemistry,* 42, 616.

Jayakumar, B.D., Kontham, K.V., Kesavan, M., Nampoothiri, B.I., and Pandey, A., 2012. Probiotic fermented foods for health benefits. *Engineering in Life Sciences,* 12(4), 377–390.

Jespersen, L., 2003. Occurrence and taxonomic characteristics of strains of *Saccharomyces cerevisiae* predominant in African indigenous fermented foods and beverages. *FEMS Yeast Research,* 3(2), 191–200.

Jeyaram, K., 2011. Microbial resources associated with traditional fermented foods of North East India. http://ibsd-imphal.nic.in/IBSD%20Scientist/K%20Jeyaram/K%20Jeyaram.html. Accessed December 2, 2011.

Jones, A., 1994. The ancient art of biotechnology. *Food Chain,* 13, 3–5.

Jones, A., Hidellage, V., Wedgewood, H., Appleton, H., and Battcock, M., 1996. Food processing. In: *Biotechnology, Building on Farmers' Knowledge.* J. Bunders, B. Haverkort, and W. Hiemstra (eds.), Macmillan, London and Basingstoke, pp. 240.

Joshi, V.K., Garg, V., and Abrol, G.S., 2012. Indigenous fermented foods. In: *Food Biotechnology: Principles and Practices.* Joshi, V.K. and Singh, R.S. (eds.). I.K. Publishers, New Delhi, pp. 337–374.

Joshi, V.K. and Kaushal, M. 2014. Indigenous food products for livelihood security. In: Technologies for livelihood enhancement. Ahuja, P.S. and Chopra, V.L. (ed.). New India Publishing Agency, New Delhi, pp. 507–536.

Kailasapathy, K., 2010. In: *Packaging Concepts for Enhancing Preservation of Fermented Foods Fermented Foods and Beverages of the World.* J.P. Tamang and K. Kailasapathy (eds.), CRC Press, Boca Raton, FL, pp. 415–434.

Kumar, U., Movahedi, B., and Lavassani, K., 2008. *Towards Developing a Comprehensive Model of Technology Transfer to Developing Countries: A Comparative Analysis Approach*. ASAC Halifax, Nova Scotia, pp. 64–80.

Motarjemi, Y., 2002. Impact of small scale fermentation technology on food safety in developing countries. *International Journal of Food Microbiology*, 75, 213–229.

Nair, B.M., 2010. A Novel Brand in Food Science and Biotechnology: An attempt to develop highly value added premium products from Indian raw materials for export to the global market. http://www.nair.se/A%20Company.htm. Accessed December 2, 2011.

Newman, R.K. and Sands, D.S., 1984. Nutritive value of corn fermented with lysine excreting *Lactobacillus*. *Nutrition Reports International*, 30, 1287.

Nout, M.J.R., 1992. Upgrading traditional biotechnological processes. In: *Applications of Biotechnology in Traditional Fermented Foods*. A report of an adhoc panel of Board on Science and Technology for Intl. Development by Gaden, E.M. et al., National Academy of Science, Washington, DC, pp. 11–19.

Nout, M.J.R., 1994. Fermented foods and food safety. *Food Research International*, 27, 291.

Ogunmoyela, O.A. and Oyewole, L.O., 1992. Current status of improvised equipment for cottage processing in developing countries. *Food Science and Technology*, 6, 10–13.

Olasupo, N.A., 2005. Fermentation Biotechnology of Traditional Foods of Africa, In: *Food Biotechnology*, 2nd edition. K. Shetty, G. Paliyath, A. Pometto, and R.E. Levin (eds.), CRC Press, Boca Raton, FL.

Onofiok, N., Nnanyelugo, D.O., and Ukwindi, B.E., 1996. Usage patterns and contributions offermented foods to the nutrient intakes of low income households in Emene, Nigeria. *Plant Foods for Human Nutrition*, 49, 199–211.

Orozco-Quintero, A., Davidson, H., and Lain, J. 2010. Role of social enterprises and leadership in the management of communal resources. *Paper Presented at the 12th Biennial Conference of the International Association for the Study of Commons* (IASC Conference 2008), vol. 4, No. 1 (2010) Natural Resources Institute, University of Manitoba.

Reddy, N.R. and Pierson, M.D., 1994. Reduction in antinutritional and toxic components in plant foods by fermentation. *Food Research International*, 27, 281.

Reza, M.S., Nayeem, M.A., Pervin, K., Khan, M.N.A., Islam, M.N., and Kamal, M., 2010. Marketing system of traditional dried and semi-fermented fish product (*ChepaShutki*) and socio-economic condition of the retailers in local markets of Mymensingh region, Bangladesh. *Bangladesh Research Publication Journal*, 4(1), 69–75.

Rolle, R. and Satin, M., 2002. Basic requirements for the transfer of fermentation technologies to developing countries. *International Journal of Food Microbiology*, 75(3), 181–187.

Rudrello, F., 2004. Health trends shape innovation for dairy products (online). *Euromonitor International Archive*, October 5.

Sawadogo-Lingani, H., 2009. Value-added processing of African traditional fermented foods for improved quality and food safety. In: International Seminar 16–19 Feb, 2009. Ovagadovgou, Burkino Faso.

Singh, J.S., 2006. Sustainable development of the Indian Himalayan region: Linking ecological and economic concerns. *Current Science*, 90(6), 784–788.

Steinkraus, K.H., 1996. *Handbook of Indigenous Fermented Foods*. Marcel Dekker, New York, NY, 352pp.

Tamang, J.P., 2003. Indigenous fermented foods of the Himalayas: Microbiology and food safety. In: *Proceedings of the 1st International Symposium and Workshop on Insight into the World of Indigenous Fermented Foods for Technology Development and Food Safety*. Kasetsart University, Bangkok, Thailand, pp. 191–213.

Tamang, J.P., 2009. The Himalayas and food culture. In: *Himalayan Fermented Foods Microbiology, Nutrition, and Ethnic Values*. J.P. Tamang (ed.), CRC Press, Boca Raton, FL, pp. 1–24.

Valyasevi, R. and Rolle, R.S., 2002. An overview of small-scale food fermentation technologies in developing countries with special reference to Thailand: Scope for their improvement. *International Journal of Food Microbiology*, 75(3), 231–239.

Verma, L.R. and Joshi, V.K. (eds.), 2000. *Postharvest Technology of Fruits and Vegetables*, vols 1 and 2. The Indus Publication, New Delhi, p. 1242.

Wang, Y.D. and Fields, M.L., 1978. Feasibility of home fermentation to improve the amino acid balance of corn meal. *Journal of Food Science*, 43, 1104.

Wood, B.J., 1994. Technology transfer and indigenous fermented foods. *Food Research International*, 27, 269–280.

Wood, C.D., 1981. The prevention of losses in cured fish. *FAO Fisheries Technical Papers*, 219, 87.

Index